VEGETATION ECOLOGY OF CENTRAL EUROPE

HEINZ ELLENBERG

VEGETATION ECOLOGY *OF* CENTRAL EUROPE

FOURTH EDITION

Translated by Gordon K. Strutt

CAMBRIDGE UNIVERSITY PRESS

Cambridge

New York New Rochelle

Melbourne Sydney

CAMBRIDGE UNIVERSITY PRESS
Cambridge, New York, Melbourne, Madrid, Cape Town, Singapore, São Paulo, Delhi

Cambridge University Press
The Edinburgh Building, Cambridge CB2 8RU, UK

Published in the United States of America by Cambridge University Press, New York

www.cambridge.org
Information on this title: www.cambridge.org/9780521115124

This digitally printed version 2009

A catalogue record for this publication is available from the British Library

Library of Congress Catalogue Card Number: 83–29393

ISBN 978-0-521-23642-3 hardback
ISBN 978-0-521-11512-4 paperback

To my lifelong helper
CHARLOTTE

Contents

Foreword

To be asked to introduce Professor Ellenberg's *Vegetation Ecology of Central Europe* to English speaking readers is a privilege indeed. This book epitomizes the very best in one of the most important traditions within plant ecology. It represents an integration of basic phytosociology with the results of both historical studies and experimental work. The comprehensive description of vegetation is blended skilfully with a causal analysis, which is remarkable in two respects. First, it is based on a strong sense of the all-pervading impact of *Homo sapiens* on the landscape. Secondly, it shows a sensitive appreciation of the interacting roles of physico-chemical factors and competition between plants in determining the limits to the distribution and abundance of particular species. However, the account is concerned not only with individual species but also with the behaviour of whole communities, as seen in studies on regeneration, productivity and nutrient cycling. The rich and varied literature of Central and Eastern Europe is unlocked to the reader.

Professor Ellenberg's approach to nature was heavily influenced in the beginning by the phytosociologists Braun-Blanquet and Tüxen, and by the pollen-analyst and historical ecologist Firbas. From the start his research was clearly aimed at understanding vegetation and not simply describing it. In his doctoral research on herbs in forests in northern Germany he undertook very extensive measurements of soil factors and the light climate, and appreciated the importance for the plant of their interaction, a point sadly ignored by many ecologists even now. Later, while working with the ecophysiologist Walter, he carried out important experimental studies, particularly on the part played by competition in limiting the distributions of species. At this time he also expanded an early interest in the possibility that nitrogen supply may often have a role equal in importance to that of water supply in controlling the nature of the vegetation at a site. This theme was followed up by his students working in a wide range of forests and grasslands. Such studies at the community level led naturally to a heavy involvement in the German contribution to the International Biological Programme on ecosystem function. At the same time he initiated important studies on the nature of succession in old fields near Göttingen. This brief recapitulation of major research activities helps to explain Professor Ellenberg's concern and ability to integrate so many approaches, but two further factors must be taken into account. Firstly, he has travelled and worked extensively in the Balkans and South America, and his experiences in these 'far places' have influenced heavily his interpretation of Central European vegetation. Secondly, he has always concerned himself with both vegetation that is (or appears to be) 'nearly natural' and plant communities that are heavily managed such as meadows, pastures and weed communities. The present book not only reflects the benefit of this wide range of experience, but also demonstrates the innovative nature of the author's approach with its careful distinction between physiological and ecological tolerance, the recognition of 'ecological groups' of species, and the attribution of indicator values to many species. Despite his phytosociological training and proper concern for communities, his account is based squarely on individual species and small groups of species.

Much modern research in plant ecology is aimed at solving problems that might at first sight seem far removed from the subject matter of this book. For example, much attention is paid to the control of population size in plants of different types, to the allocation of resources within the plant, to the nature of plant–invertebrate interactions, to the recognition of 'strategies' and the study of various forms of community stability. Above all, the evolutionary significance of various traits is examined. Such studies can be immensely rewarding, but only if they are placed in a proper context, and the natural array of ecosystems, communities or species is broken up into functional groups about which worthwhile generalizations can be made. Professor Ellenberg's book gives one a sense of the immense variety and complexity of ecological phenomena in Central Europe; it provides a framework for choosing species for further critical investigations, and provides a testing ground for would-be generalizers. Because the author has drawn upon an immensely wide array of material, and yet produced a succinct and balanced account of the essential points concerning each community type, he has provided not only an invaluable text for all European ecologists, but also an inspiring example of what is needed for every continent.

Cambridge June 1985 — P.J. Grubb

Preface to the first edition

It was with some hesitation that I took on the task of writing a book about the vegetation of Central Europe, in the context of temperate vegetation ecology. The area is so varied, and the number of works available on it so great, that no one person could carry out a uniform and really adequate survey.

When nevertheless I ventured to take on the work it was in the hope that I could fill a gap in the present range of textbooks. It was clear from the outset that causal questions and results of experimental studies which are widely distributed in the literature, but only in part comprehensively evaluated, should be brought to the fore. On the other hand one can recommend many good surveys on the systematics and characteristics of the plant communities of Central European and adjacent regions. As examples may be quoted the publications (updated!) of Braun-Blanquet (1948/50), Tüxen (1937, 1950a, 1955, 1967a and c, 1974b, 1975), F. Runge (1969-, 1980), Oberdorfer (1957, 1977-83), Bohn (1981, and other volumes of the same series), Passarge (1964a), Passarge and Hofmann (1968), Szafer (1966), Szafer and Zarzycki (1972), Holub, Heiny, Moravek and Heuhusel (1967), Mayer (1974), Niklfeld (1973), Borza (1963), Soc (1960-74), Horvat, Glavač and Ellenberg (1974), and also the legends of vegetation maps compiled by Küchler (1966). It is true that these contain brief information about the site conditions of the vegetation units, which are defined by their characteristic combination of species. They cannot however go more closely into those causal, geographical, successional and historical relationships which I shall emphasise here.

To be sure, a study of this kind is only possible if one starts with a sound classification of the plant communities. Many ecological statements lose value if they cannot be related to a particular type of vegetation. Unfortunately, however, the phytosociological surveys so far published differ from each other in many details and even in more important concepts because the definition and classification of plant communities are still fluid and often changing. So I had to be content to quote more exact syn-systematic surveys only as examples. For the rest I shall stop at the alliance or association level for the bulk of the material even though the lower vegetation units are much more closely correlated with particular habitat factors than the more embracing ones.

I attached great value to clear illustrations and tried to explain the scientific problems and results to the layman also. Nevertheless a sufficient knowledge of plants must be assumed, for to describe even briefly all the species mentioned would have burst the bounds of this volume. Shortage of space also restricts the use of common names mainly when plants are mentioned in the text or in lists. Readers unfamiliar with scientific names may refer to the alphabetical index at the end of the book or to a good flora.

The nomenclature of the vascular plants follows Flora Europaea (see also Ehrendorfer 1973). The names of the remaining cryptogams are according to Smith (1978) for mosses and Wirth (1980) for lichens. In any case the author's name is added in the species index (Section E III 2b). The terms used for soil types and horizons correspond to the rules laid down by the Study Group on Site Mapping (Kohl 1971).

In order to economise on space points which apply to several vegetation types or landscapes have not been repeated. A comprehensive index of ecological concepts and vegetation types makes reference easy and may help readers to discover relationships that are not referred to expressly in the text. It has also not been possible to produce a complete literature survey. The index of references lists only those publications which have been quoted in the text, or consulted, and is not grouped under subjects but alphabetically.

Some sections have intentionally been given more weight than others whose significance may have been just as great. Woodlands, flood plains, mires, salt marshes and dunes, as well as heaths and some grassland communities are treated more thoroughly than the remaining units. However I maintain that it is better to deal with some communities in depth rather than put up with a uniform superficiality throughout. The topographical connection as well as the geographic variability of the vegetation units is only hinted at briefly, for a sufficiently clear account of the plant cover of all the regions of Central Europe would fill a separate volume. Besides it would transgress from the field of 'geobotany' (or ecology, see Ellenberg 1968) into that of 'vegetation geography' (in the sense of Schmithüsen 1959, 1968).

I hope that this book, in spite of its shortcomings, gives a picture of the many-sided nature of the Central European plant cover and an introduction to its habitat conditions, developmental causes and origin. I hope the book will also win new friends for vegetation ecology! Still more workers are needed to fill the gaps in our knowledge which we come up against with almost every new question. I am always grateful for any supplementary information or corrections.

Zurich August 1963 Heinz Ellenberg

Preface to the second edition

Fourteen years have gone by since the first appearance of this book and for seven years it has been out of print. In the meantime there has been a welcome increase in the study of vegetation ecology, and many gaps in our knowledge have been closed. Environmental problems are today discussed in the mass media and taken seriously in government policies. Ecological principles and methods are employed not only in country planning and design but also in education at all levels. Multidisciplinary team work as in the 'International Biological Programme' (see Ellenberg 1966, 1971, 1973b) and the UNESCO-programme 'Man and the Biosphere' have brought and are bringing new understanding. This is especially true for ecosystem research. Moreover, in the meantime, there has been a great flood of phytosociological literature and new proposals for classification.

The second edition tried to do justice to this development without departing from the character of the book and the main features of its layout. In order to make way for new material many descriptions and references were shortened or omitted. Many bibliographies edited by R. Tüxen have appeared and continue to appear (current volumes of *Excerpta Botanica, Sekt. B, Sociologica*, see also Dierschke 1967). On questions of phytosociological systematics I am deliberately conservative. In this context I agree with Erich Oberdorfer, who kindly allowed me to see part of the manuscript of the new edition of his *Plant Communities of South West Germany* (1977-83), and with many other colleagues. We should oppose with all our might the 'inflation of higher vegetation units' (S. Pignatti personal communication). The number of carefully written monographs on the vegetation of larger and smaller areas of Central Europe has grown in a very welcome manner.

The vegetation tables are explained in more detail than in the first edition because experience has shown that those who are not familiar with them find them difficult to 'read'. Starting with simple releves the examples progress step by step to 'synthetic' tables in which several vegetation units are brought together. In order to illustrate the geographic variation within related units these have, in some cases, been arranged in a series throughout Central Europe.

Many vegetation tables are moreover interpreted with respect to the site conditions, using 'indicator values' of the individual species as well as the average 'indicator values' of all species growing together (see section B I 4). The use of such calculations is not meant to replace exact measurements, but helps to avoid repetition and gives a quick indication of habitat relationships when more exact data are not available. As will be clear from a few examples the average indicator values are in satisfactory to close correlation with the measured values insofar as these relate to significant site conditions. In order to show what these indices are capable of telling us they have also been used to interpret some of the vegetation tables which have already appeared in the first edition, i.e. examples already known for a long time, although it would have been possible to have replaced these wholly or in part with more up to date ones.

In order to relieve the tables and text and to facilitate the 'determination' of plant communities I have extended the survey of vegetation units (classes, orders alliances etc.) in Section E III with lists of character species. Another new feature is the autecological labelling of almost all the species included in the index. Groups of 'factor indices' give information about the ecological behaviour of vascular plants in relation to light, temperature, and continentality of the climate, as well as soil moisture, acidity and nitrogen supply. The figures apply principally to western Central Europe where the ecological relationships of the species have up to now been best investigated.

Increased emphasis has been put on both autecology and synecology. Even more than in the first edition causal aspects of vegetation science have determined the character of this book, although the dynamic and historical aspects have been retained unabridged. A third edition unexpectedly became necessary after a few years, I profited by this opportunity to correct mistakes and to add new findings. Professor Gerhard Follmann kindly enabled me to update the chapters and tables on lichen communities.

As in the first edition I enjoyed the help of my wife as well as that of numerous colleagues and fellow workers with critical comments, advice and active cooperation. I should like to thank them all sincerely. I also thank my students for their stimulation and some enthusiastic readers, who incidentally asked me not to alter too much'. In any case I have taken care, in spite of the increased amount of information, to keep the book readable for the amateur.

I am particularly grateful to Roland Ulmer and his staff – especially Dieter Kleinschrot – for understanding and patience while waiting for the new manuscripts, and for the care with which these have been produced.

Göttingen, Summer 1978 Heinz Ellenberg

Preface to the third and fourth editions

New editions became necessary sooner than expected. The third edition gave me, in 1981 and 1982, the opportunity of including additional material and of making a great number of improvements. After publication of the second edition I was irked by certain inconsistencies which could harly be blamed on the difficulties of proof-reading during a research trip through Peru in 1978. Apart from that so many new publications had appeared that I would have liked to have completely rewritten certain sections. However, in order to save time and expense we decided to stick to the original layout as far as possible. Only within paragraphs or in the empty spaces next to illustrations could I draw attention to newer works which brought out corrections, refinements or new aspects. At least some descriptions of regions and overviews could be included in the bibliography.

For the fourth edition it was only possible to make a few improvements. The effects of gaseous emissions upon species composition are still not well enough known to justify a special section. Table 55 is based on Oberdorfer's *Systematik der Moor-Gesellschaften* (1983).

In order to make space for the additional material as well as for the expanded Section E III.1 (based on Oberdorfer 1977–1979) the section 'Hinweise auf Vegetationsdarstellungen interessanter Gebiete' has been omitted. This list (compiled from geographically arranged references) was, in the view of many readers, too full of gaps and attracted a good deal of criticism. Important regional publications do, however, remain in the bibliography and those that are not cited in the text are indicated with a plus sign.

The nomenclature of lichens (following Wirth 1980) and mosses (partly following Düll 1980) has been brought up to date, even though this has meant many corrections. Section E had to be entirely reset.

In the case of communities predominantly made up of lichens profound changes appeared unavoidable. Not only had the names become obsolete in the second edition, the tables and descriptions in no way corresponded with the (happily) large strides made in the plant sociology of cryptogams. Dr Gerhard Follmann, who had, with justification, criticised this defect, helped me by making important suggestions for improvement. I wish to thank him most sincerely for this painstaking help.

Otherwise I should like to thank those colleagues, fellow workers and students who have pointed out desirable changes or mistakes, especially Hartmut Dierschke, Winfried Hofmann and Barbara Ruthsatz. I thank Guntram Krauss for painstaking correction of the species index and Frau Ute Reinholt, for her careful help with the manuscript. These new editions would certainly not have been possible without the critical and understanding help of my wife Charlotte.

Roland Ulmer and Dieter Kleinschrot have done their best, in spite of vacillations, to publish speedily and economically. I hope this book will win new friends for vegetation ecology and help the specialists to 'keep the whole field in view'.

Autumn 1986 Heinz Ellenberg

Introductory survey

I The vegetation of Central Europe in general

1 The climatic situation and the vegetation of Central Europe

By Central Europe is understood here the area represented in fig. 1 together with the Alps. It consists mainly of the Federal German Republic and the German Democratic Republic but also Poland, Czechoslovakia, Austria, Switzerland, Luxemburg and Denmark as well as parts of adjoining countries. This central area of Europe is in the north temperate zone and is roughly bisected by the 50th parallel. Its climate is characterised by a change from a fairly warm, frost-free summer to a more or less cold winter, which means a resting period of several months for most types of growth. The differences between the two seasons are more or less reduced by the position of Central Europe between the oceanic western and the increasingly continental eastern side (fig. 2). Therefore the air temperature in summer seldom exceeds 30 °C and in winter only exceptionally falls below -20°C. In addition this situation causes the transition periods of spring and autumn to be significantly extended and the growing seasons for many plants to be correspondingly longer. Furthermore the fact that cyclonic rain may fall at any time of year is favourable for the vegetation of Central Europe (see Walter 1971, 1979). Only in rare catastrophic years (e.g. 1947 and 1949, also 1971 and 1973) is it exposed to longer periods of drought.

Such a climate encourages the universal **growth of trees**. Central Europe would therefore be a monotonous woodland had not man produced a colourful mosaic of cultivated land and heath, meadow and pasture and over centuries continually cut back the forest. Without man's impact only the salt marshes and the windswept dunes on the coasts, mires which are too wet and poor in nutrients, some steep rocks or screes and avalanche tracks in the high mountains as well as the heights above the climatic timber-line would be free from trees (figs. 1 and 3). Because of their taller growth trees would in the end have succeeded against all competition. Even the poor sandy heaths of north-west Germany, which many scientists at the turn of the century still regarded as inimical to trees, have since been afforested everywhere. This has removed the last doubts about the woodland character of the natural countryside. The same applies to the steppe-like grasslands of the southern-central European limestone mountains, almost all of which revert more or less quickly to scrub and woodland if grazing and burning are discontinued.

Amongst those tree species naturally present in Europe the climate of Central Europe favours summer-green deciduous trees with mesomorphic foliage and well protected winter buds, e.g. **Beech** (*Fagus sylvatica*) and Oak (*Quercus robur*). It has been designated 'Beech climate' by Köppen and covers the whole of western Europe including southern England and Scandinavia.

While the west and the centre of Europe, with their gradual change in climate and vegetation, are closely linked, the continental north-east is more sharply and obviously distinct. Here colder winters damage the buds of the deciduous trees, and late frosts not infrequently destroy the new leaves and flowers. Also in the much drier air in summer their tender leaves do not grow so well as the xeromorphic needles of Spruce (*Picea abies*) or Scots Pine (*Pinus sylvestris*). So here the evergreen **conifers** are able to suppress the deciduous trees through natural competition, whereas in most parts of Central Europe their present predominance is thanks to the care of the forester.

A very important character of the continental climate, which favours conifers over broadleaved deciduous trees is the shortness of the growing period.

Fig. 1. General distribution of the natural vegetation in Central Europe (excluding the Alps) at about the time of Christ, i.e. before it had been significantly changed by man. Based on pollen analyses; drawn by Firbas (1949) from Walter and Straka (1970), slightly modified. In this volume the adjoining Alps will also be dealt with (see figs. 3-6).

1 = Dry areas with a precipitation of less than 500 mm; mixed Oak woods with few Beech. 2 = low-lying areas with mixed Beech woods, in parts with a strong admixture of Oaks; on the North Sea coast many Alder; black circles = Pines locally dominant.
3 = low mountains with Beech, generally lacking conifers. 4 = moraine area with Beech, few Pines. 5 = mountain Beech woods with Fir and (or) Spruce (white triangles); black triangles = subalpine Beech wood.
6 = sandy soil areas where Pines dominate, in places with Oaks and other broadleaved trees. 7 = area of mixed broadleaved woodland with many Hornbeam. 8 = like 7 with the addition of Spruce.

Flood plains, mires and other special habitats are not distinguished.

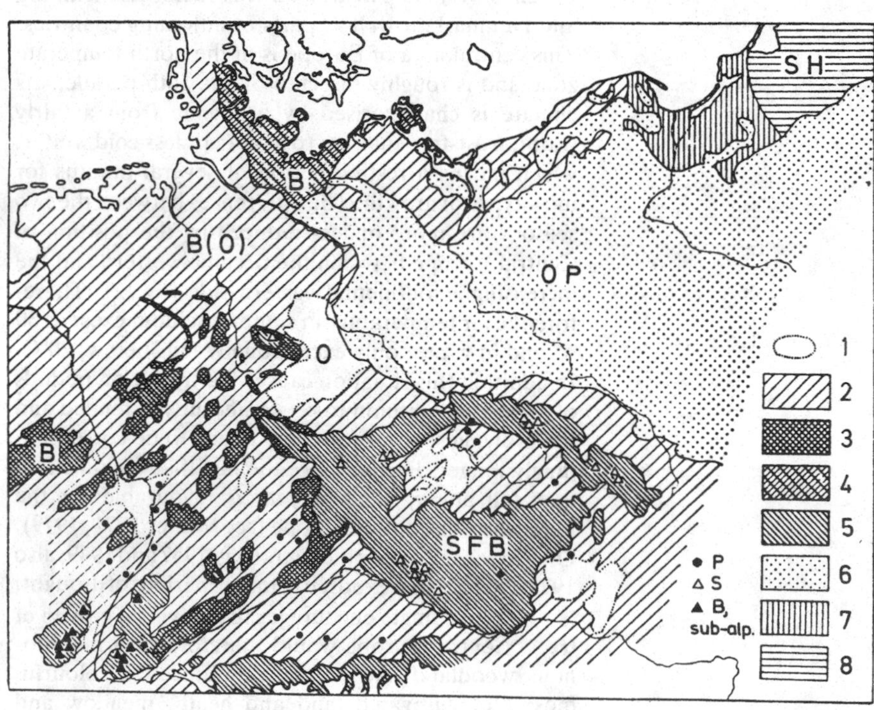

In the North also this determines the natural limit of deciduous woodland area especially in southern Sweden and Norway, where the last outposts of beech woods penetrate into the coniferous region on soils rich in nutrients. A similar interaction of climate and vegetation can be seen when climbing high mountains (figs. 3 - 5).

The wide high walls of the Alps rise up approximately in the area where the transition from the central European climate to the submediterranean climate would begin. This transition takes place gradually only to the west of the Alps, i.e. in the atlantic-submediterranean region, and to the east of them, i.e. in a really continental situation. The submediterranean climate is characterised by milder winters and warmer summers in which dry periods frequently last for many weeks. Here the beech and its associates withdraw into the cooler and moister mountains and give way to more drought-resistant trees such as *Quercus pubescens, Q. cerris, Fraxinus ornus* and *Carpinus orientalis*. Close to the Mediterranean coast, where the proximity of the warm water prevents sharp winter frosts, the deciduous woodland is finally replaced by an evergreen one. The mediterranean sclerophyllous trees native to this area, particularly *Quercus ilex*, even though rare today, are in fact more susceptible to frost than many summer-green broadleaved trees, but they can withstand much better the drought of the warmest months so characteristic of the mediterranean climate (see Horvat, Glavač and Ellenberg 1974).

Almost everywhere the woodland of Central Europe borders on areas which also have a climate allowing woodland growth. Only in the south-east is precipitation so reduced that a zone of steppe woodland forms a link leading to the steppes of southern Ukraine, Floristic elements of these naturally occurring woodland-free areas can be traced far into Central Europe and are especially predominant in the basin of the Neusiedler See near Vienna, in Bohemia and Moravia as well as in the rain shadow of the Harz mountains. As Passarge (1953a), Wendelberger (1954) and others show, under present day climatic conditions trees were nevertheless able to thrive nearly everywhere here. This can also be inferred from the map by E. Jäger (1968 fig.2) marking the gradient of continentality in Europe by means of plant geographical criteria.

Figure 4 allows us to take a closer look at the **Alpine region** in which climatic types with oceanic and continental, boreal and submediterranean characteristics meet one another and interpenetrate in many ways. In addition they are modified by increasing altitude above sea level, from the relatively warm

Fig. 2. Climatic gradient of oceanic influence in Europe as measured by the vascular plant species. Based on a coloured map by E. Jäger (1968), from Horvat, Glavač and Ellenberg 1974.

1 = oceanic and suboceanic species very numerous. 2 = less numerous. 3 = many suboceanic species right to the eastern boundary, subcontinental species strongly represented. 4 = many subcontinental and an occasional continental, but still some suboceanic species. 5 = the last of the suboceanic species die out, still many broadleaved woody plants. 6 = many widely distributed continental species; steppe woodland species dominant. 7 - 9 = continental species increasing in number, more and more desert plants. 10 = dominated by continental desert plants.

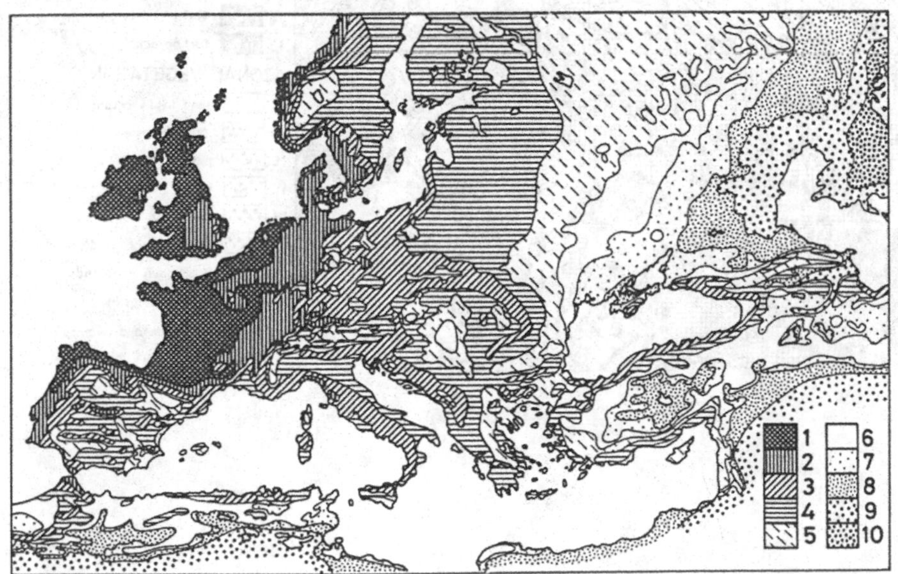

Fig. 3 The present-day potential natural vegetation of Central Europe including the Alps in an example from Lower Austria. From a colour map by Wagner (1972), slightly modified. Almost everywhere in Central Europe woodland would be dominant if left to nature.

ZONAL VEGETATION*

Colline belt

1. Warmth-loving mixed Oak woods; today almost without woods; many vineyards.
2. as 1 but with Oak-Hornbeam woods

Submontane belt

3. Oak-Hornbeam woods; today largely cultivated
4. Oak-Hornbeam woods with many Beech; largely cultivated

Montane belt

5. Beech woods, in part converted into coniferous forest; much grassland
6. Mixed Spruce-Fir-Beech woods on limestone; to a large extent maintained as such
7. Fir-Spruce woods of the central Alps, on richer soils than 8, partly converted into farmland
8. Spruce woods of the central Alps on acid rocks

Subalpine belt

9. Larch-Arolla Pine woods, today partly pastureland
10. Mountain Pine scrub, mostly on limestone, little pastureland
11. Green Alder scrub, mostly on acid rocks, much pastureland

Alpine belt

12. Alpine grassland on limestone
13. as 12, on acid rocks

AZONAL VEGETATION*

Dry habitats

14. Sand dunes with mixed Oak woods inter alia
15. Austrian Pine woods on limestone
16. Spring Heath-Scots Pine woods on dolomite, serpentine and similar
17. Montane Pine woods on acid rocks
18. Buckthorn-Pine woods and Oak-Pine woods on dry gravels, today mostly Pine forests
19. Colline acid-soil Oak-Pine woods

Wet habitats

20. Flood plains with various woodlands, parts under cultivation
21. Reed belts of the Neusiedler See and neighbouring lakes, also halophyte vegetation in the surrounding area
22. Fenland, to a large extent converted into cultivated meadows
23. Raised bogs, today drained in parts

The map which has an approximate scale of 1 : 2.2 million, covers the area from About Passau to Bratislava and from Ceske Budějovice to Graz.

* These terms are explained in section B I I a

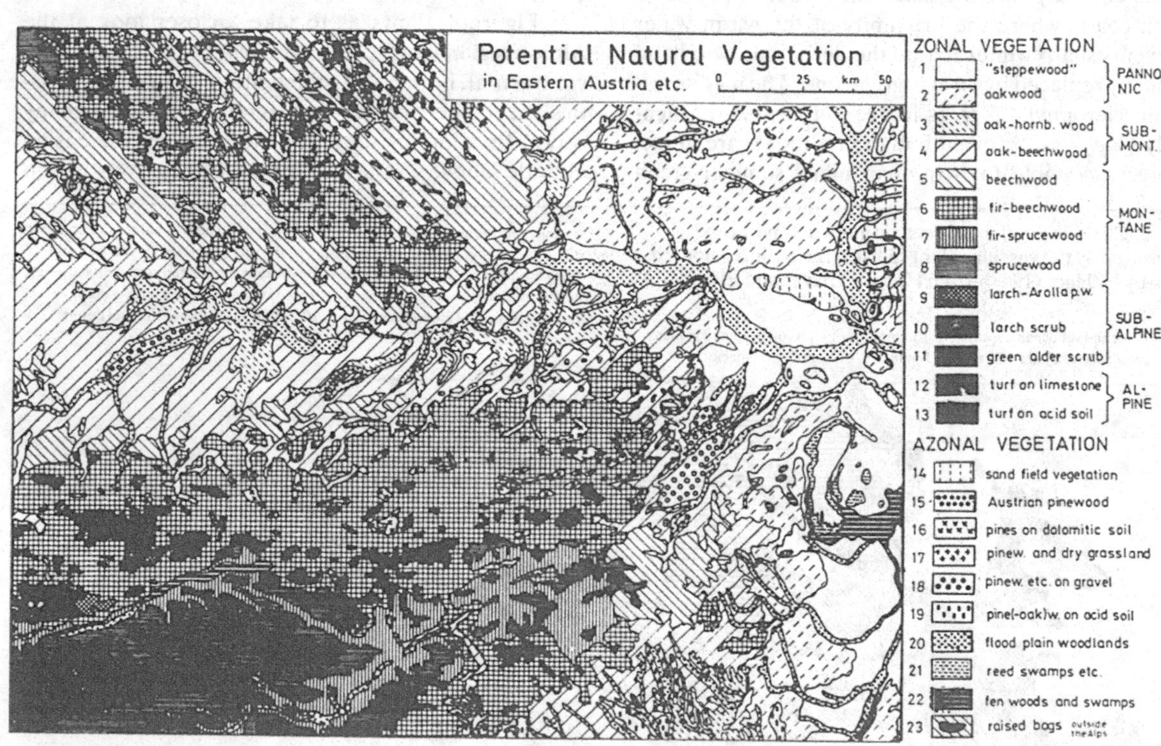

foothills right up to the eternally snow-covered peaks. Diagrams of the climate show the changes in average monthly temperature as well as in monthly totals of precipitation at selected stations. The relationships between the climatic types, represented in fig. 4a, and the naturally occurring vegetation are indicated in figs. 4b and 5, the coordinates of which correspond to the mean annual temperature and precipitation. The way in which the altitude zonation of the vegetation varies from west to east and from north to south in Central Europe is shown in fig. 6.

This overall picture of the climate and vegetation shows us that Central Europe is not a sharply defined area. On all sides, especially towards the West and the East, it merges into neighbouring territories, a fact which is also characteristic of the cultural and economic aspects. In spite of this Central Europe possesses an unmistakable individuality which cannot be inferred from a knowledge of the border areas alone.

Fig 4a. Climatic map of the Alps as an example for Central Europe. From Rehder (1965) slightly modified. The Roman numerals indicate climatic types corresponding to the 'Climatic Diagram World Atlas' by Walter and Lieth (1967). Further explanation in fig. 4b. Pure mediterranean climatic types (IV) lie outside Central Europe.

2 *The significance of plant history for the vegetation of Central Europe*

North America, Europe and Asia comprise a single plant geographic entity, the Holarctic, because in their flora they differ only slightly from one another. Europe stands out only insofar as its flora of woody plants, when compared with North America and east Asia, is poorer in genera and species. Repeated glaciations resulted in the extermination of many plants sensitive to cold, especially many tree species which were widespread here in the warmer Tertiary period. In America these were able to spread southwards unhindered by seas, and later to recolonise, whereas in Europe almost everywhere they came up against the Mediterranean Sea.

Contrary to the earlier prevailing view (still repeated in fig. 4 of the first edition), living conditions during the glaciations were so hard even in the Balkans, Italy and southern Spain, that none of the woodland communities now present in Central Europe was able to survive (Beug 1967, 1977, Šercelj 1970, Frenzel 1964, 1968). According to Fritz (1970) Beech had actually been able to migrate back into Central Europe during the interglacial periods. During the glacial periods however nearly all trees were exterminated. The climate was significantly more continental

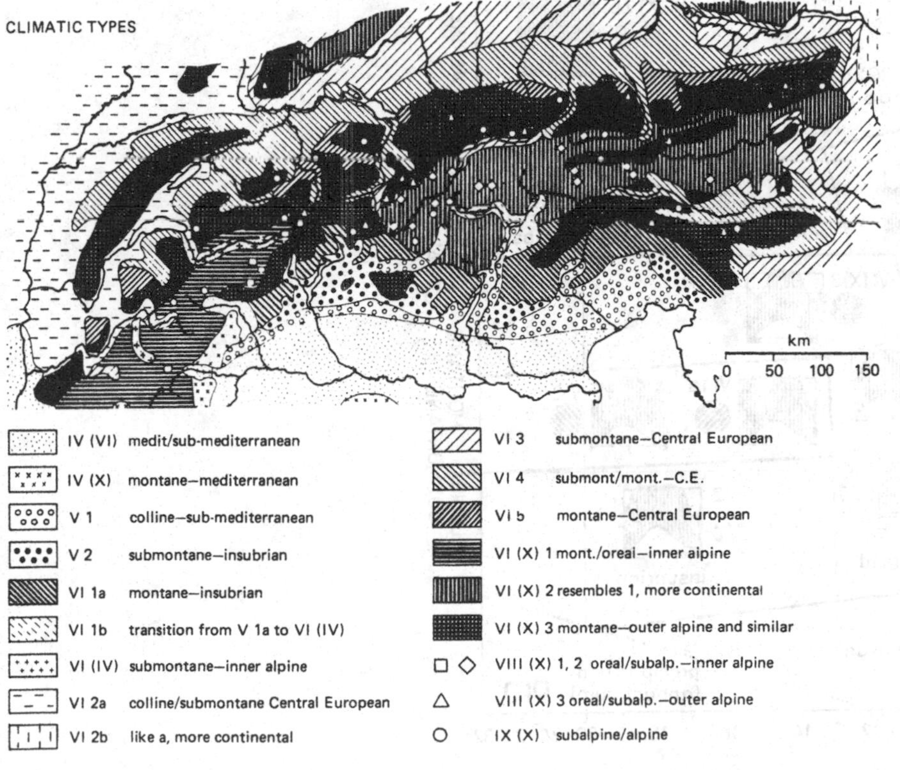

than today and favoured cold steppes. Herbs and low growing shrubs apparently suffered less heavily than the trees, or else they were able to spread out from their refuges more easily. In any case the total number of vascular species in comparable areas of eastern North America is no greater than in Central Europe including the Alps and their foothills (i.e. about 4000 species), while the comparative figures for tree species are about 120:50 (see tab. 1).

This disproportion should not be interpreted merely historically. According to Schroeder (1974) the present-day climate also plays a part insofar as the summers in comparable vegetation zones of North America are warmer than in Central Europe which on average lies some ten degrees further north. Probably this is the main reason for the absence of just those

Fig. 4b. Approximate precipitation and temperature ranges of the most important climatic types shown in the map fig. 4a. According to a survey of meteorological stations by Rehder (1965) and diagrams by Walter and Lieth (1967), with some additions in the montane to alpine areas.
In each diagram the lower boundary of the hatched area represents the average monthly air temperature from January to December (each graduation mark = 10°C), the upper boundary the average total monthly precipitation (each graduation mark = 20 mm, above 100 mm reduced by 1/10th and shown in black). The larger the hatched area, and especially the black part, the more humid is the climate. The simplified climatic diagrams refer to the following stations (see fig. 5):

V1 **colline-submediterranean** with Downy Oak and other drought-resisting deciduous trees, many Sweet Chestnuts (Bolzano 292 m).
V2 **colline to submontane-insubrian**, high precipitation, with warmth-loving Birch-Oak woods and Sweet Chestnut (Lugano 276 m).
VI **colline to submontane-central european**, a hint of mediterranean, with Beech and Oak-Hornbeam woods (Dijon 315 m); VI 3 submontane-central european, with a more or less continental flavour, with Beech and mixed Oak woods (Leoben 540 m); VI 4 submontane to montane-central european with species-rich Beech woods (Zürich 569 m). VI 1a **montane to submontane-central european**, tending to insubrian, mainly Beech (Faido, 759 m); VI 5 montane-central european, with Beech and a few conifers (Schwyz, 567 m).
VI(X) 1 **montane to oreal-inner Alps** (subcontinental), submediteranean flavour, with Scots Pine.:Larch and especially Spruce (Montana 1453m); VI(X) 2 **montane-inner Alps** (continental), mainly Spruce (Heiligenblut, 1378 m);
VI(X) 3 **montane intermediate to outer Alps**, with mixed Spruce-Fir-Beech woods (Mittenwald, 910 m).
VIII(X) 1+2 **oreal to subalpine-inner Alps** (continental), with Larches, Arolla Pines and Spruce (Davos, 1561 m).
VIII(X) 3 **oreal to subalpine-outer Alps**, with mainly Spruce (Splügen, 1500 m).
IX(X) 1 **subalpine-near the tree limit, inner Alps** with Larch-Arolla Pine (1a), intermediate and outer Alps with Spruce (1b, Rigi, 1775 m); IX(X) 2 **alpine-inner Alps** (2a) or **outer Alps** (2b, Säntis, 2500 m).

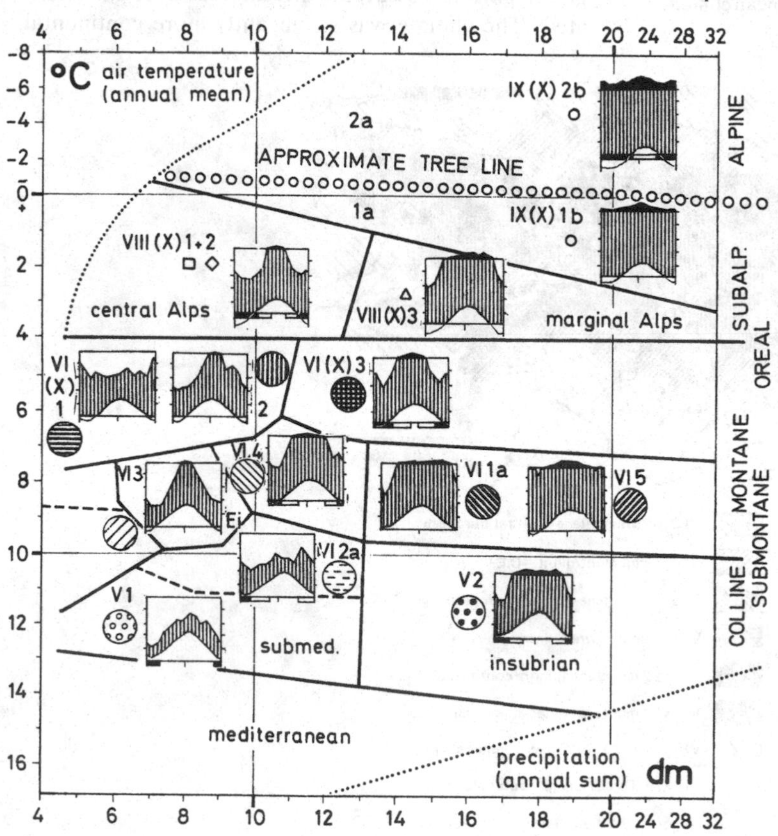

Fig. 5. Dominant tree species in the potential natural vegetation of the Alps, also showing the position of the meteorological stations named in fig. 4b. The graph has as its coordinates yearly precipitation (abscissa) and average air temperature (ordinate). After Rehder (1965), with modifications. The Roman numerals correspond to the climatic types given in fig. 4.

Ap = Arolla Pine (*Pinus cembra*). S = Spruce (*Picea abies*). P = Scots Pine (*Pinus sylvestris*). F = Silver Fir (*Abies Alba*). B = Beech (*Fagus sylvatica*). *O = the Oaks Quercus robus* and *petrea*. Q.pub = Downy Oak (*Quercus pubescens* submediterranean). Q.ilex = Hop Hornbeam (*Ostrya carpinifolia*, submediterranean) cast. = Sweet Chestnut (*Castanea sativa*).

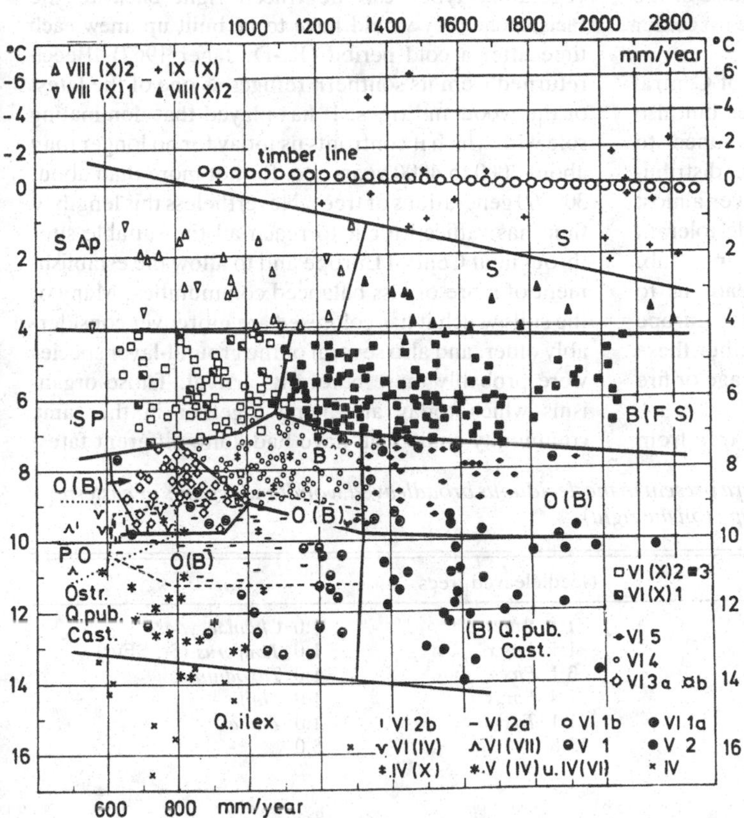

Fig. 6. Schematic cross-sections through the natural vegetation belts of Central Europe running approximately west to east and north to south.

Beech only appear in the more oceanic climatic regions, conifers in the more continental ones and the central part of the Alps. The limits of the vegetation belts rise in most cases relative to decreasing latitude and to increasing mass of the mountains. Fir is also represented in the natural vegetation of the Vosges whereas Spruce is not.

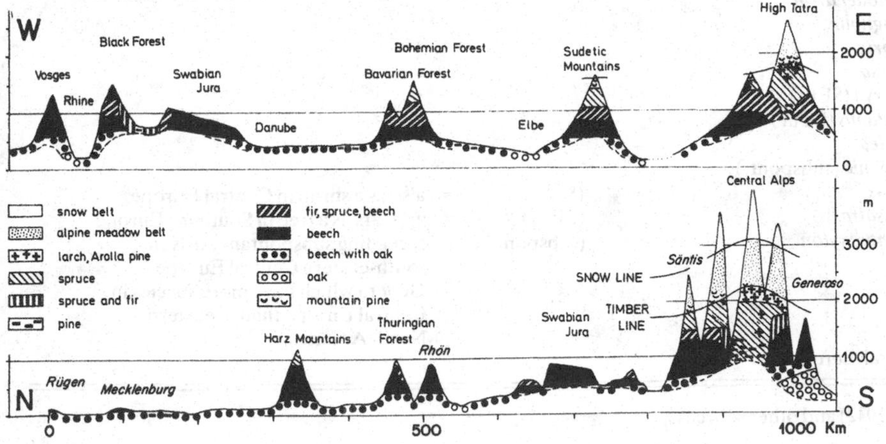

subtropical genera and species (tab. 1) which can only grow quickly enough to compete with the tree species of temperate climate where the summer temperature is sufficiently high. In botanic gardens where they are protected from such competition almost all these exotics do quite well since the winters, because of the tempering influence of the Gulf Stream, are no colder than in their North American homeland.

Apart from the trees the original flora of Central Europe is not only relatively rich in species but also quite independent. Many groups are confined to Central Europe or have their most important distribution here (tab. 2). Typically these are however almost exclusively woodland plants or at least shade-tolerant species. Moreover species of woodland clearings (tab. 2b) play a surprisingly large part. This leads us to envisage that in the natural forests of Central Europe there have always been small clearings whether these have arisen from senility of trees, wind damage or fire caused by lightning.

Since Central Europe was completely free from woodland during the last Ice Age and the climate has only gradually approached that of the present day all its woodland communities are **comparatively young** (see also Lang 1967, Walter and Straka 1970, and Sukopp 1972). Even in those places where similar vegetation types can be traced right back to the Pleistocene they would have to be built up anew each time after a cold period (K.-D. Jäger 1967). Beech returned from its southern refuges as one of the latest of the woodland trees. It has played that dominating role with which it confronts us today for no longer than about 3000 to 4000 years, i.e. for not more than about 30 – 60 generations of trees. Nevertheless this length of time has sufficed for it to reach all the suitable sites throughout Central Europe and to allow the establishment of more or less balanced communities. Many of the soils which it has colonised are moreover considerably older, and also several of the ground-layer species were probably there before its arrival. Those organisms which today are living together in the same community could well have had quite different fates;

Tab. 1. The number of species in the tree genera present in the deciduous broadleaved woodland regions of eastern North America (on the left) and Central Europe (on the right).

Broadleaved trees		Needleleaved trees	
7:5 *Acer*	2:0 *Amelanchier* (St)	1:1 *Abies*	1:0 *Chamaecyparis*
3:2 *Alnus*	1:0 *Aralia*	1:1 *Larix*	1:0 *Juniperus* (St, SEur)
5:2 *Betula*	1:0 *Asinima*	3:1 *Picea*	1:0 *Taxodium*
1:1 *Carpinus*	6:0 *Carya*	7:4 *Pinus*	1:0 *Thuja*
2:2 *Crataegus*	2:0 *Castanea* (SEur)	1:1 *Taxus*	1:0 *Tsuga*
1:1 *Fagus*	1:0 *Catalpa*	13:8	5:0
4:2 *Fraxinus*	2:0 *Celtis* (SEur)		
2:1 *Ilex*	1:0 *Cercis* (SEur)		
1:1 *Malus*	1:0 *Diospyros*	18:8	
5:3 *Populus*	1:0 *Gleditschia*	Eastern North America : Central Europe	
5:3 *Prunus*	1:0 *Gymnocladus*		
0:1 *Pyrus**	2:0 *Illicium*		
20:4 *Quercus*	2:0 *Juglans* (SEur)		
4:8 *Salix**	1:0 *Liquidambar*		
1:4 *Sorbus**	1:0 *Liriodendron*		
2:2 *Tilia*	2:0 *Magnolia*		
3:3 *Ulmus*	1:0 *Morus*		
66:45	2:0 *Nyssa*		
	1:0 *Ostrya* (SEur)		
	1:0 *Platanus* (SEur)		
	1:0 *Ptelea*		
	1:0 *Robinia* (subspont.)		
	4:0 *Rhus*		
	1:0 *Sassafras*		
	1:0 *Zanthoxylon*		
	40:0		
106:45			
Eastern North America : Central Europe			

(St) = also as a shrub in Central Europe
(SEur) = appears as a tree in Southern Europe
(subspont.) = spreading subspontaneously in south-eastern Central Europe
* = Genera which have more species in Central Europe than in eastern North America

Note: From maps by Schmucker (1942) and other sources.

Tab. 2. Vascular plants having their main centre of distribution in Central Europe.

True Central European species	Those with somewhat wider distribution
a Broadleaved woodland	
1. Trees	
Acer pseudoplatanus	*Acer campestre*
Carpinus betulus	*A. platanoides*
Fagus sylvatica	*Alnus glutinosa (fen woodland)*
Prunus avium	*Fraxinus excelsior*
Quercus petraea	*Quercus robur*
Tilia platyphyllos	*Salix fragilis (flood plain)*
Taxus baccata	*Tilia cordata*
	Ulmus glabra
2. Shrubs and climbers	
Hedera helix	* *Clematis vitalba*
* *Rosa coriifolia*	* *Cornus sanguinea*
* *R. obtusifolia*	*Corylus avellana*
* *R. rubiginosa*	*Crataegus laevigata*
* *R. villosa*	*Euonymus europaea*
Rubus hirtus	* *Rosa canina*
	R. corymbifera
3. Herbs and dwarf shrubs	
Allium ursinum	*Actaea spicata*
Arum maculatum	*Ajuga reptans*
Cephalanthera damasonium	*Alliaria petiolata*
Circaea intermedia	* *Aquilegia vulgaris*
Corydalis cava	*Anemone nemorosa* ssp. *nemorosa*
C. intermedia	*A. ranunculoides*
Dentaria bulbifera	*Asarum europaeum*
Festuca heterophylla	* *Betonica officinalis*
Galium sylvaticum	*Campanula latifolia*
Hordelymus europaeus	* *C. patula*
Hypericum montanum	*Carex digitata* ssp. *digitata*
Lathyrus linifolius	*C. montana* s. str.
Luzula luzuloides	*C. remota* ssp. *remota*
L. sylvatica	*Cephalanthera longifolia*
Lysimachia nemorum	*C. rubra*
Melica uniflora	* *Genista germanica*
* *Melittis melissophyllum*	* *G. tinctoria*
Mycelis muralis	*Hepatica nobilis*
Petasites albus	*Lamiastrum galeobdolon*
Phyteuma spicatum	*Lathraea squamaria*
Veronica montana	*Lathyrus vernus*
Vinca minor	*Lunaria rediviva*
Viola reichenbachiana	*Mercurialis perennis*
	Neottia nidus-avis
	* *Orchis mascula*
	Plantanthera chlorantha
	Pulmonaria officinalis
	Ranunculus ficaria
	R. lanuginosus
	* *Serratula tinctoria*
	Stachys sylvatica
	Stellaria holostea
	St. nemorum
	Thalictrum aquilegifolium (flood plain)
	Viola riviniana

* = *requiring more light*

Tab. 2. (contined)

True Central European	Species with a somewhat wider distribution
b Woodland clearings	
1. Shrubs	
Rubus sulcatus	*Sambucus nigra*
2. Herbaceous plants	
Atropa belladonna	*Arctium minus*
Dipsacus pilosus	*A. nemorosum*
Epilobium lamyi	*Cardamine amara*
Hydrocotyle vulgaris (swamp wood)	*Carex divulsa*
	C. hirta
	Centaurium erythraea
	Cirsium oleraceum (swamp, flood plain)
	Dipsacus fullonum
	Eupatorium cannabinum (swamp wood)
	Lysimachia nummularia (swamp wood)
	Petasites hybridus (flood plain)
	Rorippa sylvestris
	Rumex obtusifolius
	R. sangiuneus
	Sedum telephium
	Senecio sylvaticus
	Silene dioica
	Symphytum officinale
c Woodland edges ('fringes')	
Agrimonia procera	*Aethusa cynapium*
Allium scorodoprasum	*Allium oleraceum*
A. vineale	*Anthericum ramosum*
Saxifraga granulata	*Chaerophyllum temulum*
Trifolium ochroleucon	*Galium mollugo*
	Knautia arvensis
d Rocky ground and dunes	
Corynephorus canescens (dune)	*Cardaminopsis arenosa*
Sedum sexangulare	*Hiercium piloselloides* (gravel)
e Poor grassland	
Carlina vulgaris (c)	*Anthyllis vulneraria* s. str. (c) (f)
Dianthus armeria (c)	*Briza media* (f)
Hypochoeris radicata (f)	*Centaurea jacea* (f)
Ononis repens (f)	*Cerastium semidecandrum*
Orchis morio (f)	*Euphorbia cyparissias*
	Helianthemum nummularium
	Hieracium pilosella
	Orchis ustulata
	Orobanche caryophyllacea (Paras.)
	Plantago lanceolata (f)
	Prunella grandiflora
	Ranunculus bulbosus (f)
	Salvia pratensis (f)
	Selinum carvifolia (c)
	Thymus pulegioides
	Veronica teucrium
f Fertilised grassland	
	Arrhenatherum elatius[1])
[1]) originating from sub-mediterranean-montane screes	*Cynosurus cristatus* (e)
(c), (e), (f) leading to the groups indicated	*Pimpinella major* (c)

Note: From data supplied by Walter and Straka (1970) and others.

we cannot be sure that they will be inseparably bound together in the future.

3 Life forms and structural types of the Central European plant species

(a) Life forms

Summer-green deciduous trees are certainly the most competitive life forms in the flora of Central Europe and would therefore dominate the natural plant cover. However in the comprehensive species list for Central Europe with more than 3500 phanerogams, and an even greater number of cryptogams, the trees (**phanerophytes** in the definition of Raunkiaer) (see Ellenberg and Mueller-Dombois 1967a and tab. 3) make up only a modest fraction. Even shrub species (**nanophanerophytes**) constitute less than 5% although they are plentifully represented in the open woodlands.

By far the greatest number of species taking part in the building of the Central European plant communities are of low growth and belong either to the hemicryptophytes, geophytes, chamaephytes or therophytes, or to the adnate (rootless) thallophytes (mosses, lichens etc.). From this we may deduce that these more-lowly growth forms have found favourable living conditions in Central Europe for a very long time and that they have endured the inclement glacial periods better than the taller woody plants. Most of the annuals (therophytes) and probably also many grassland species have certainly been introduced by man and could not have stood up to natural conditions in Pleistocene Central Europe.

The life forms of all vascular plants mentioned in this book are given in the index (section E III).

The **hemicryptophytes** are particularly well suited to the climatic rhythm of Central Europe in that their regenerating buds overwinter in direct contact with the earth and as a rule are protected by snow. In many species, for example *Hemicryptophyta cespitosa*, the buds are so thickly wrapped around by the leaves of the previous year that they scarcely need snow protection. In mild winters indeed their leaves remain wholly or partially green. As Raunkiaer has already pointed out the hemicryptophytes make up more than half the vascular flora of Central Europe (tab. 3). He speaks of the 'hemicryptophyte climate' of most parts of Europe although this idea, derived from floral statistics, can lead to misconceptions about the natural vegetation. For looked at ecologically only some steppe and high mountain sites have a hemicryptophyte climate. In spite of this one could look upon this group as characteristic of the Central European vegetation, especially since it predominates over wide areas today in pastures and meadows as well as in many heathlands which have also been created by man.

The **geophytes** surviving with bulbs or rhizomes in the soil have the best of all life forms for protection against winter cold and summer drought. Because of this they are numerous on the continental steppes and in the Mediterranean area. In the humid-temperate parts of Europe they can spread only in those places where the hemicryptophytes, whose leaves as a rule last much longer, are not dominant. This is above all the case in deciduous woodland on fertile soil where a dense canopy in summer hinders all undergrowth because of the shade it produces. In the early part of the year, on the other hand, light can reach the ground for sufficient time and in sufficient quantity to enable the quickly growing geophytes with their low optimum temperature requirement to reach full development (fig. 113). Under natural conditions such woodland geophytes would have been common in Central Europe. Today they are retreating because the fertile soils have to a large extent been deforested and the woods on poorer soils are more open and therefore richer in hemicryptophytes.

A relatively minor rôle in the Central European flora is played by the **chamaephytes** i.e. dwarf bushes (2.7%) or herbs (6.1%) whose buds seldom lie more than 25-50 cm above the ground, but nevertheless are exposed to the winter cold and drought unless they enjoy protection by snow. It follows that they reach a

Tab. 3. Spectrum of the life forms of 1760 vascular plants of Central Europe.

%		Life form	Group %	Group
2.9	**P**	Phanerophytes (trees)	9.6	Woody plants
4.0	**N**	Nanophanerophytes (shrubs)		
2.7	**Z**	Woody chamaephytes (dwarf shrubs)		
6.1	**C**	Herbaceous chamaephytes (buds above ground)	56.8	obviously perennial herbs
50.7	**H**	Hemicryptophytes (buds at ground level)		
12.0	**G**	Geophytes (perennating below the surface)	29.4	disappearing herbs
17.4	**T**	Therophytes (short lived)		
4.2	**A**	Hydrophytes (perennating under water)	4.2	aquatic plants

Note: Calculated from data supplied by Ellenberg (1974), as percentages of the total.

greater development, on the one hand in snow-covered mountain situations, and on the other in oceanic areas with their milder winters, than they do in the more continental parts of Europe proper. This is particularly true of some of the Ericaceae such as *Calluna vulgaris* or *Erica tetralix* where the buds are barely covered and which moreover have evergreen leaves and are thus sensitive to dessication by frost.

In the cultivated countryside of Central Europe today the short-lived plants (**therophytes**, 17.4% of the total flora) cover the greatest areas although under natural conditions they would appear scarcely anywhere, and in many plant formations would be completely absent. Both the majority of our field crops and also many of their undesirable companions derive from steppes or semidesert areas where the climate is inimical to woodland.

We will look at the remaining life forms, especially those of the cryptogams when we discuss particular plant communities. There is just one feature characteristic of the Central European flora which should be emphasised at this point: amongst the epiphytes clinging to the branches and trunks of trees there are never any vascular plants, but only lower ones, namely crustose, foliose and other lichens, mosses and some algae. Higher plants can only tolerate the epiphytic way of life in a climate which is frost free and sufficiently humid, at least for more than half the year. Such conditions exist on most tropical and subtropical mountain slopes as well as many tropical lowlands.

To summarise it can be said that the climate of Central Europe favours the development of deciduous tree stands, bearing no vascular epiphytes, and where the undergrowth consists principally of spring-growing geophytes or deciduous to partly evergreen hemicryptophytes, while the chamaephytes and therophytes are unimportant.

(b) Inner rhythms

As Diels (1918) has already demonstrated by cultivating wild plants over the winter in greenhouses, the rest period of the majority of plants in our flora corresponds to an 'inner' rhythm. Even when they are not forced to do so by cold or drought these plants lose their leaves each year, and over a period of some weeks or months can be stirred into renewed growth only with difficulty. It is particularly hard to break the winter dormancy of the Beech; thus this is a true representative of the Central European climatic area. Most other woody species which shed their leaves, as well as many herbs, have an autonomous rest period which is only modified, not initiated, by the climatic rhythm (V.u.I. Kárpáti 1961).

Diels distinguishes between three significantly different rhythmic conditions in the herbaceous perennials of deciduous woodland:

1 Periodic species with a harmonious **endogenous dormancy** (*Polygonatum-Type*);
2 Periodic species but where the dormancy **can be influenced** to some extent (*Leucojum*-type);
3 Aperiodic species in which dormancy is completely brought about by **external** factors (*Galium odoratum*-type).

To the last group belong species with tropical or subtropical relations which do not form storage organs and continue growing green throughout the year in the greenhouse.

The representatives of the *Leucojum*-type are for the most part geophytes with bulbs or tubers which have originated in the mild-wintered Mediterranean area. These include *Gagea, Scilla* and *Arum* but above all many Liliaceae, Amaryllidaceae and Orchidaceae. In the greenhouse they take only a short rest of a few weeks and then renew their growth.

The majority of plants with rhizomes or other storage organs however belong to the *Polygonatum*-type whose resting period in the greenhouse can only be shortened by an insignificant amount. These are species which are widespread in Central Europe and Eurasia and have holarctic relations, e.g. *Anemone, Convallaria, Corydalis* and *Dentaria.*

On fertile soils representatives of all three groups are present in our deciduous woodland and it cannot be said that any one of them is particularly competitive. They complement each other in most of the plant communities.

Probably it is the changing day length which is one

Tab. 4. Spectrum of the anatomical–morphological structure of 1760 vascular plants of Central Europe

%	Type	Group %	Description
1.6	**su** succulent leaved	21.1	adapted to withstand periods of **drought**
19.5	**sk** scleromorphs		
53.5	**m** mesomorphs	53.5	'**normal**' structure (intermediate between sk and hg)
8.5	**hg** hygromorphs	8.5	**delicate** structure, dry out easily
13.1	**he** helomorphs	16.9	adapted to living in **wet places** or in water
3.8	**hd** hydromorphs		

Note: Calculated from data supplied by Ellenberg (1974) as percentages of the total.

of the external factors by which the annual development of many species is influenced or guided. This also applies to the very marked annual cycle of frost resistance to which we shall return in sections B III 3 and B IV 4.

(c) Morphological-anatomical structure

Along with the growth rhythms of the plant species their morphology and anatomy tell us something in general about living conditions in Central Europe (Ellenberg 1974, 1979). As tab. 4 shows by far the most dominant are the **mesomorphic** species, i.e. those whose structure shows no particular adaptations.

Soft-leaved **hygromorphic** species (e.g. *Oxalis* or many ferns) which in low air humidity quickly dry out and so are restricted to half or full shade make up only 8.5% of the total flora.

Even fewer are the more or less **succulent** species which store water in their leaves and in this way are able to survive a relatively long dry time. Since they grow only slowly they are easily overshadowed by most of the other plants and can only manage to survive in extreme situations; *Sedum* and *Sempervivum* species are examples.

Soils drying out temporarily are occupied in the first place by **scleromorphic** (more or less stiffened) plants which, when water is scarce, can reduce their transpiration rate to a greater extent than others since they possess a thick cuticle and a strong epidermis. In times of sufficient water supply however their numerous stomata permit rapid gas exchange by which the unavoidable water deficit can be made good through an efficient leaf vein network, strong vascular bundles and a relatively extensive root system. Scleromorphic vascular, plants (such as *Bromus erectus*, *Carlina* spp. and the majority of evergreens) comprise almost 20% of the total flora. Many scleromorphic plants of the raised bogs and other wet oligotrophic sites in reality are peinomorphs, that is scleromorphs caused by and adapted to lack of nutrients (see section C III 4a). In structure many species lie between mesomorph and scleromorph as in general the boundaries between the groups discussed are fluid and can be altered by slight differences in the habitat.

Gradations exist too between scleromorphs and **helomorphs**, i.e. species adapted to very wet soils with practically no oxygen. The latter have in common more or less extensive intercellular spaces in the subterranean organs (so-called lacunae or aerenchyma, see fig. 390). More than 13% of the vascular plant species of Central Europe, especially many Cyperaceae and grasses always exhibit this feature; others show it only in a substrate poor in oxygen, but many never possess it.

Aerenchyma reaches its greatest development in the **hydromorphs** (water plants) parts of which live for long periods or constantly under water (e.g. *Nymphaea, Nuphar, Potamogeton* and *Myriophyllum*).

Fig. 7. Example of the present day cultivated landscape of Central Europe: view from the Swabian Forest towards the plain of Hohenlohe near Schwäb. Hall. In the foreground Fir-Beech woods, clear felling and Spruce plantation. In the middle distance small villages and isolated farms with orchards, arable and grassland. To the right along a stream are Alders the park-like distribution of which is reminiscent of the grazing landscapes of the Middle Ages.

II The development of the plant cover under the influence of man

1 The present-day vegetation as the result of millennia of history

The picture of the natural vegetation has up to now intentionally been explained only to the point where all authors are today in agreement. Their opinions begin to differ widely when we question them about the role of woodland in particular dry areas, in heaths and in swamps or about the tree species composition of such woodland. In general ecologists would not be in agreement on the detailed features of the species mosaic of most plant communities. This is not least because Central Europe is an **ancient cultivated landscape**. As in western Europe literally no spot has been able to retain its original natural character unaltered.

The impact of man has not only resulted in the sharp boundaries between planted forests, pastures, meadows and arable land which are so characteristic of the countryside today (fig.7). Even where one might believe that purely natural forces have been at work man has often played a part. Many so-called virgin forests (e.g. the 'Neuenburger Urwald' and the 'Hasbruch' near Bremen) show traces of former use. Just those gnarled spreading trees, which strike us as being so primaeval (fig. 8), show the results of earlier pasturing with cattle, horses and swine which removed the undergrowth and opened out the woodland (figs. 8 and 9). since these stands have been protected close growing shade trees have shot up, overtopping the broad-crowned oaks and old beeches (figs. 24, 25, 77). With regard to tree shape the modern mature forest with slender trunks drawn up by competition (figs. 55, 24 Nos. 7 and 8) probably gives a more accurate picture of the natural woods of Central Europe than such supposed virgin forests and other woodlands formerly pastured, but now conserved.

Under the influence of man the microclimate and the soils have also changed along with the vegetation so that we no longer have any of the original sites today. Therefore authentic reconstruction of the original landscape, as the first settlers found it, is scarcely possible. The difficulty is aggravated because the impacts of man on different areas of Europe were not contemporaneous, and they cannot now exactly be ascertained anywhere.

Immediately after the retreat of the inland ice hunters appeared, although the extent to which they intervened in the interplay of forces could have had no lasting effect. For a greater influence we must look first

Fig. 8. Native-looking Oaks in an open grazed woodland near Sababurg to the north of Kassel some 50 years ago. In the meantime, between them bushes and trees have grown up in the open spaces where they have not been cut sufficiently frequently. Phot. Hueck.

to the end of the Mesolithic when settlement became denser and the economy more pastoral (Jankuhn 1969). During the Neolithic (about 4500 to 1800 BC, see fig. 10), many farmers had already settled in north-west Germany and in some areas of loess deposits in central and southern Germany. They carried out arable farming and reared cattle, pigs, sheep, goats and, since about 2000 BC, horses too. Very probably they allowed their animals to graze freely in the surrounding woods, in the same way as is still practised in those parts of the world where farming is run on extensive lines, such as in many mountain areas, and which was almost universally the case in Europe during the Middle Ages (see section 2 and fig. 9). The intentional or accidental spread of fire when grazing areas were burnt off, or deliberate clearance of trees by means of fire, may have contributed to the destruction of woodland. This is especially true in those valleys of the Alps where at times the dry föhn wind is blowing (Grabherr 1934, 1936, Winkler 1943, see fig. 103) and in the coniferous forests of north-east Europe with their low precipitation (Kujala 1926). Forest fires due to natural causes are very infrequent in Central Europe when compared with eastern Scandinavia or western North America. According to Baumgartner, Klemmer, Raschke and Waldmann (1967), for example, fires started by lightning in Bavaria at the present time comprise only 1% of the total which have had to be dealt with, whereas the corresponding figures for Finland and the USA are 69% and 12%.

The settled areas maintained free from woodland did not increase very greatly during the Bronze and Iron ages. Here and there however they spread onto heavier or wetter soils. Ploughshares, strengthened with iron, opened up more and more arable land, although according to Herz (1962) the wooden plough was still being used in many districts as recently as 200 years ago. In wet areas man already understood how to drain the water by a network of ditches. With sickles, not only could the harvesting of grain be carried out more efficiently, but grass too could be cut, and the first meadows must have been already in existence in the Iron Age.

Fig. 9. Mixed herds grazing in a Spruce-Pine woodland to the west of Narva (Estonia). Almost all the woodland of Central Europe, including the broadleaved woods were previously utilised in such a manner and to some extent cleared even more rapidly.

In the thickly populated long-settled countryside of Central Europe at about the time of Christ there could not have been any areas of woodland through which man and his animals had not roamed at some time. However between the more or less open areas there still stretched huge closed forests. The parts of the country occupied by the Romans were particularly intensively settled. Traces of their farms can be seen today, e.g. to the west of Stuttgart, not infrequently in the middle of what appear to be quite 'natural' woods. Here and there in those areas outside the Roman influence also, particularly on sandy soil, arable farming was more widespread than it is at the present day. This is indicated by the narrow, sharp ridges of the Iron Age, and germanic raised beds which we often come across quite unexpectedly in out of the way woods or heaths, e.g. in the Knyphauser Wald (north of Oldenburg, in East Friesland), in some Beech woods near Göttingen, and on the gravel plain north of Munich. The settlements always developed closely depending on the ecological conditions, especially on the given mosaic of vegetation types, as Burrichter (1976) pointed out for Westphalia.

Woodland was once more allowed to invade many of the ancient cultivated areas during the Hunnish invasions. However, in general, these troubled centuries meant only a short respite for the near-natural vegetation. For now the systematic developments of the Middle Ages began. The new settlers pushed resolutely into the large forests, which up to then had been influenced only marginally by grazing, and began to destroy them. Soon herds of cattle appeared on the highest peaks of the mountain areas, for example on the Feldberg in the Black Forest, and opened up the woodland particularly rapidly in those areas where it also had to contend with wind and cold (see figs. 318 and 319).

Destruction of the forests reached its greatest level in recent times when early industries demanded charcoal, preferring that made from Beech wood, and when mines, salt works and glassmaking burned up such quantities of wood so vast that we can hardly imagine them today (see e.g. Winkler 1933, Grossmann 1934, and Aust 1937). Furthermore the potash requirements of the glassworks were met by 'ash burning' from the woods. At about the same time however systematic forest management developed; areas previously destroyed were planted with trees and the primitive methods of exploiting the woodland were gradually superseded. Since quick-growing conifers were often preferred where deciduous trees would have dominated naturally, artificial stands were created which developed their own type of vegetation (see section D III 1).

This short review has already shown us that it is impossible to draw any conclusions from the present day plant communities without going further into the natural state of the vegetation. Above all it shows that it is not sufficient to know only the modern management and working methods in the woods and fields if one wants to understand the processes leading to the present vegetation mosaic. At least with regard to woodland and heathland, but also to many grassland areas and mires, we must always keep before us the earlier methods of cultivation and stock-farming and their effects. Since, in general, there are far too few facts known they will be dealt with here in detail. Then we will return to the natural vegetation and try to sketch a more exact picture of it.

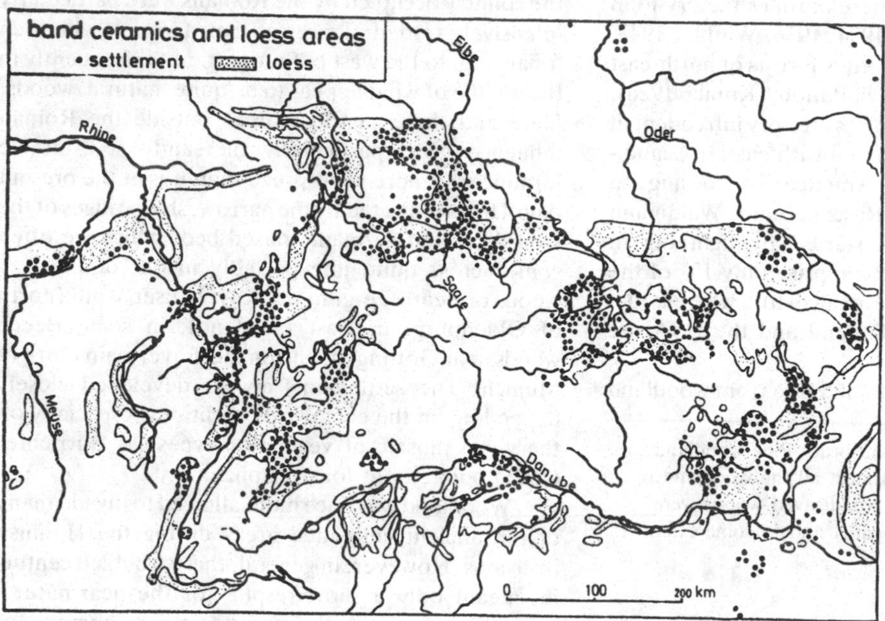

Fig. 10. In neolithic times the farming settlements were concentrated in low-lying areas with loess or, in the north, sandy soil.

Above: areas with deep loess soil and settlements of the neolithic people in Central Europe. After Clark and Jankuhn (1969), somewhat modified.

Below: A reconstruction of a village (of around 4000 years ago) near Geleen in southern Netherlands. After Waterbolk from Jankuhn (1969). The surrounding woods have been opened up by grazing (compare also Willerding 1980a).

2 Effects of extensive grazing and woodcutting on the plant cover

(a) Opening up and destruction of the woodlands

No other activity of man can be compared with extensive pasturing, including that of woodland, in the widespread and lasting effect it has produced. Although it was almost universally practised in Europe up to 200 years ago, and at one time was the main way in which the woodland was utilised, it can be studied in only a few places today (Grossmann 1927, Mathey 1900, Steen 1957, Volger 1958, Mager 1961, Hesmer and Schroeder 1963, Cate 1972).

Opportunities for study, albeit rather biased, present themselves above all in some nature reserves on the Lüneberg heath, in the Netherlands, as well as in some protected areas in central and southern Germany. In these areas sheep pasturing is used to preserve the old state of the countryside. The keeping of large animals on more or less wooded common land or private pastures can still be studied in many places in the Alps (fig. 11). Unfortunately nearly all these areas lie in the coniferous forest region and cannot be compared with the typical deciduous woodland of Central Europe. We have to go to the Baltic (fig.9), to the eastern part of Poland, to some parts of Jugoslavia, or to the southern Alps, i.e. to the boundaries of the central European region, in order to get a true picture of extensive pasturing methods in mixed deciduous woodlands. This can be supplemented by looking at the many small areas of woodland which from time to time are included in a fenced grass pasture in the north German plain, for instance in the central precincts of the 'Sachsenwald' east of Hamburg (fig. 12) or in the 'Borkener Paradies' near Meppen (Burrichter verbal communication).

In all areas the succession brought about by the farmers and their grazing animals has led from dense closed woodland, through park-like stages to open pastures altering the soil to an increasing degree (figs. 15 and 24).

Fig. 11. Remains of woods in a naturally woodland-dominated mountain region. Groups of Spruce have only managed to survive on the steepest and stoniest parts of the grazed slopes, e.g. on the Widderstein near Schröcken in the Smaller Walsertal.

Fig. 12a. Goats (and cattle) are very fond of leafy shoots and will even climb the trees. The drawing is of an engraved mussel shell exhibited in the Louvre. It dates from about 2500 BC, Tello, Near East; natural size.

b. Grazed woodland in the Sachsenwald reserve near Hamburg. Cattle have been grazing to the left of the fence; since they have been excluded from the area to the right of it saplings and shrubs have been able to grow.

When cattle, horses, goats, or even sheep break into a dense forest, consisting principally of shade-producing trees, which has not previously been grazed, they find very little to eat. They are unable to reach the crowns of the trees, the woody undergrowth is generally sparse and the ground flora offers scarcely any nutrition especially as the spring geophytes, the majority of ferns and many other typical woodland herbs are poisonous or unpalatable to domestic animals (Gradmann 1950). They wander about searching for fodder until they come to a more open place, such as could arise even in natural woodland when a large old tree falls over. Here they attack small trees and shrubs and wipe out almost all the saplings. Woods in which the tree layer consists of Oaks or other open woodland types are generally richer in undergrowth and thus offer the animals more fodder from the outset. Even here though the browsing animal eats the woody plants first.

Those who are familiar only with modern grassland farming find this difficult to accept. Nevertheless in many mountainous areas of Europe, in western Russia and in the Birch woods of northern Scandinavia the preference of animals for young tree shoots and twigs with buds is very well known. Brockmann-Jerosch (1936) has devoted a whole paper to the **edible deciduous trees** ('Futter- und Speiselaubbäumen') of Switzerland and has described how at one time a significant part of the winter fodder for the larger animals was obtained by 'lopping', that is the repeated cutting of leafy shoots (fig. 13). We are reminded of this custom by the German name 'Laube' (bower, arbor) given to the porch in front of a house, which is protected from the rain, but where air could circulate freely. Here the cut foliage was hung up to dry (Brockmann-Jerosch 1936, Guyan 1955). According to Troels-Smith (1955) about 1000 bundles of 'leaf hay' would be required for each cow for the half year because of its low nutritional value. The Ash is the tree best suited to lopping and feeding in this way (fig. 13); according to Lüdi (1955) this species gave most of the leaf hay in the Bronze Age. Its scientific name *Fraxinus* is from the Latin 'frangere' which means to break (Glässer 1969). Even today it is used in the old way in some alpine valleys where a dietetically beneficial effect is ascribed to its leaves. Species of Elm were likewise favoured but in addition Birches, Limes, Maples, Hazel and many other deciduous trees were used. The Latin name of Hornbeam also reminds us of former times: *Carpinus* can be plucked (Latin, 'carpere'). Conifers e.g. Spruce and Larch were also lopped in the Alps, but used mainly for bedding of livestock.

Fig. 13. Ash trees photographed about 1930 which have been frequently lopped (cut for 'leaf hay') near Winklern in Kärnten. The crippled trunks are overgrown with mosses. Photo. Hueck.

Beech and Oak were very seldom made into leaf hay, but animals would eat the fresh twigs at any time of the year. Once the most usual practice was to turn the animals out onto the common pasture to browse even throughout the winter when they were mainly dependent on the buds and young twigs of the deciduous trees. This gave rise to bitten-down forms of many trees, notably the Elfin Beech which we can still see today in our German highlands.

Unwittingly man and his herds were carrying out a selection of plants in favour of the species which were unpalatable or completely untouched, or, as was the case with Spruces and Pines, only suffered for a few weeks when the young shoots were free from resin. Among the conifers Silver Fir (*Abies alba*, fig 143) was eradicated most rapidly by being bitten off, as it only produces new shoots at the apex. The tree most avoided on account of its poisonous and unpleasant-smelling leaves is the Grey Alder (*Alnus incana* fig. 347) which is to be found in high mountains and in boreal-continental parts of Europe. Even today in White Russia, Lithuania, Latvia and Estonia, and also in some places in the Alps, this forms pure stands which are rarely touched by cattle and which obviously have arisen in place of former Spruce forests. Of the gymnosperms the Juniper (*Juniperus communis*) is spared by all animals, even goats and sheep since they fear its sharp needles. This slender shrub has been able to spread far and wide as a '**pasture weed**' on poor acid soils as well as on calcareous ones as it requires a lot of light but is not particular as to soil quality (figs. 18 and 27).

As the trees suitable for fodder gradually disappeared, they had to be planted and protected in the neighbourhood of settlements to supply clippings; those trees unfit for food became relatively more numerous. In Central Europe Oak held a place somewhere between the two. Because of the tannin content its leaves and twigs were not eaten so readily as those of Beech and other broadleaved trees. However pigs, which in earlier times were also driven into the woods, were very partial to the acorns and seedlings which they sought out avidly. For this reason in the later Middle Ages Oak trees had to be planted near the villages to ensure that there would be sufficient mast for the pigs. At that time of course there was no cultivation of potatoes, roots or other fodder plants.

According to Grossmann (1927) as early as Roman times a distinction was drawn between 'silvae glandiferae', that is woods in which pigs could be fed on acorns and beechmast, and 'silvae vulgaris pascuae', the usual pasture.

Closed woods of tall Beech or mixed Oak served very well as mast woods. However the rather open stands were more suited for pig feeding, producing a lot of seed almost every year. Open Oak groves were considered to be the most valuable woodland because in addition they supplied the best building timber.

Fig. 14. Parklike grazing country to the north of the Vierwaldstättersee (lake of four cantons) near Küssnacht about 150 years ago. A coloured etching by Peter Birmann in the 'Kupferstichkabinett' in Basel. On the level ground trees and bushes could survive only where they were protected against grazing by rocks or in wet places.

The browsing of woodland in its extensive form certainly damages the young growth. As time goes on this leads to opening up of the woodland since gaps in the canopy no longer close up again. All the open spaces mean better fodder prospects for the leaf-eating animals. For here those grasses and herbs which require more light can become established, and many of these have a greater feeding value than the true woodland species. This change over to shrubby grassland was very welcome to the shepherds, and they speeded it up whenever they could. Near the settlements this happened in any case by the cutting of wood for building and fires which was the unrestricted right of every commoner. So long as the woodland still dominated and there was no shortage of timber this opening was supplemented over ever widening areas by ringing the trees (i.e. stopping the transport of assimilation products to the roots which leads to death of the tree after a few years). By combined ringing, burning and pasturing the countryside in many places soon took on the appearance of parkland. The remaining trees assumed wide crowns and carried branches often right down to ground level (fig. 8). All the trees which were readily eaten by cattle looked as though their branches had been cut off parallel to the ground at a height corresponding to the reach of the animals (fig. 27a). The original woodland plants had to retreat into the shade of such trees as remained, and leave the more strongly illuminated areas to the hemicryptophytes and chamaephytes of pastures and heaths. Gradually the plant communities of the open country spread out more and more until they became completely dominant over wide areas (figs. 14-18, 26 and 27).

The speed with which the woodland destruction went on was in the first instance dependent on the intensity of grazing, burning and felling. As a rule the further away from settlements the more slowly it took place, but there is another factor: accessibility of the piece of woodland in question. Steep or very stony slopes often remained wooded for considerably longer than less-sloping or flat land (fig. 11). Here and there on such slopes remains of the original forest can be found even today, for example the Rotwald near Lunz or the Deborance in Wallis. Finally the resistance of the tree species must play a part, though even with the same species this also depends on the site conditions. A wood making slow growth on dry or poor soil turns into grass- or heathland much more quickly than one on fertile soil with favourable water content.

In regions where extensive grazing, cutting and burning have taken place over the centuries the landscape is very diversified. Pastures alternate with parklike stretches, open stands of trees and true woodland (figs. 7, 11 and 14). In addition we can imagine a village or group of farmsteads here and there, with ploughed fields and gardens protected from the freely ranging cattle by fences, walls or hedges. Then we have before us a universal picture of Central Europe as it was over a period of more than two thousand years.

Many paintings and drawings from earlier centuries show us what the old countryside looked like even better than the few actual remains of it which happened to be preserved. Lonely heaths and the compact tree forms, shown singly or arranged in groups in open pastureland, appear over and over again in the works of the Romantic artists as the expression of what was believed to be unspoilt Nature. In the form of 'English parks' this idealised view of the natural landscape became the vogue in garden design in the whole of Europe for many decades. As features of interest, animals with their herdsmen are rarely absent from the paintings and drawings of those periods, without it being realised that these were actually the creators of such attractive scenery (fig. 14). Many hunting scenes too with their gay riders and hounds charging along give us a good idea of the open nature of the 'woods' at that time and of the wide open spaces of this former extensively utilised countryside (see also fig. 15).

Fig. 15. Extensive grazing, browsing and cutting not only damage the woodland but also the soil which becomes increasingly impoverished by the removal of nutrients and by erosion. The separation of woodland and pasture, which in Europe today is complete almost everywhere, has led to an increased yield of timber and at the same time made more intensive and more productive husbandry of the grassland possible.

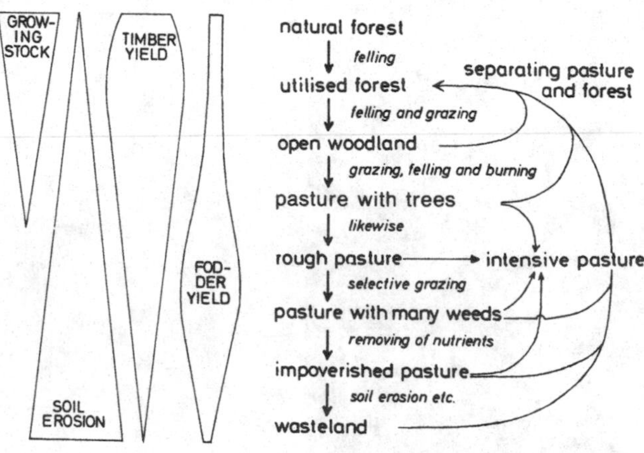

(b) Spreading of pasture weeds

The development of the landscape and its vegetation produced by earlier pasturing and timber requirements in no way stopped at the conversion of previous woodland into grass- or heathland. Already in parklike areas the animals were favouring the introduction and spread of 'pasture weeds' of which up to now only a few trees and the Juniper have been named. On soils sufficiently rich in bases Blackthorn (*Prunus spinosa*, fig. 16) is even more troublesome than Juniper, as it can spread out on all sides by means of suckers, and is sufficiently well protected by thorns from the browsing of sheep and goats to enable it gradually to succeed. Roses, Whitethorn and other prickly or thorny shrubs were less feared by shepherds. Amongst the dwarf shrubs of the Central European flora *Ononis spinosa*, *Genista germanica*, and *G. anglica* may be mentioned as pasture weeds, and amongst the herbaceous plants most species of Thistle (*Cirsium, Carduus* and *Carlina*) and the genus *Eryngium*.

Many herbs without thorns can also become weeds of pasture because they contain essential oils or other substances which animals find unpleasant. To the first group belong, for example, labiates (such as species of *Thymus, Teucrium* and *Mentha*) while most Ranunculaceae, Euphorbiaceae, Gentianaceae and Liliaceae come into the second category. Many grasses and grasslike plants are disdained by animals only because they are, at least for a part of the year, too hard or just not palatable. This group, which is numerically the strongest, includes many species of rushes and sedges (*Juncus, Carex*), also the Mat-grass (*Nardus stricta*, fig. 412), a particularly invasive weed of alpine meadows. All the ferns and mosses can be looked upon as weeds, but the only troublesome ones are those capable of high growth, particularly Bracken (*Pteridium aquilinum*).

Klapp (1938, 1971) was the first to recognise clearly that undergrazing, by selection, favoured weeds. This is true at least at the beginning of the grazing season when more food than the animals require is available. When however, as in modern rotational grazing, the animals are forced by a shortage to eat the somewhat less-palatable plants, these are kept in check by defoliation or at least by treading (fig. 28).

Extensive pasturing always results in undergrazing, especially at times when settlement is less dense. The spread of weeds is most rapid when the number of grazing animals falls temporarily following wars, epidemics, economic crises or a movement of peoples, such as occurred not infrequently in earlier times. The remaining animals then have a much wider choice and the herdsmen are not so concerned to keep the weeds down as they normally would be.

(c) Degradation of the soil as a result of extensive husbandry

The progressive increase in weeds through the centuries gradually reduced the productivity of the common pastureland. However another contributo y factor was impoverishment of the soils in areas where woodland had been destroyed. Because this process has had a great influence on the present composition of the Central European plant cover we would like to consider it in more detail. To do this we must look at sloping and level ground separately.

On slopes the opening up of woods and the trampling of animals has greatly increased soil erosion and in some places has speeded it up disastrously. Even from a distance one can recognise the characteristic appearance of a slope which has been repeatedly crossed by grazing animals. These tend to walk behind one another roughly along the contours, making paths

Fig. 16. The result of intensive grazing by sheep of the slightly arid grassland on calcareous soil near Mönsheim, in the district of Leonberg. In the foreground are Blackthorn shrubs extending as 'pasture weeds' and in the background the Oak trees have been preserved for building timber.

which they rarely leave (fig. 17). At the same time they are moving either slightly upwards or downwards so that an acute-angled network of such tracks is formed over the course of a few decades. From the appearance of the vertical slope profile these tracks are referred to as 'cattle (or sheep) steps' (fig. 378).

Today in the Alps and in many other mountainous areas extensive slopes have been patterned in the manner described. They can also be found on hillside pastures in the rest of Europe and in other parts of the world. The tracks themselves are often free from plants, and the earth has become very consolidated by treading. When it rains most of the water runs off the surface, taking with it the fine soil and in places deepening the steps into channels. Between the steps, three distinct habitats can be found on which more or less distinct plant communities have become established; the sides of the path are relatively moist, but heavily trodden; the outer edges dry out quickest, but are not grazed so closely as the 'normal' part of the slope which is scarcely ever trodden on.

On level ground the changes brought about by grazing are not immediately obvious on most soils though they may be no less damaging. As soon as the woodland is replaced by grass or heath the rain percolates the soil more quickly, washing out soluble salts and lime more thoroughly than before. On soils which were already lime free or where lime has been heavily leached there is an invasion of plant communities such as the *Calluna* heath (fig. 18), from the dead remains of which an acid raw humus develops. This effects a podsolisation, that is a progressive reduction of the clay fraction and an extreme impoverishment of minerals in the upper layer of the soil (see section D II 3a). Very poor podsol soils have come about particularly on the sandy heathlands which once were widespread from Denmark through north-west Germany as far as Belgium. As Tüxen (1933, 1957) has pointed out, under the white sand and the brown hardpan horizons of such heath podsols, some remains of the original woodland soil can still be found. The history of the vegetation and its soil can be 'read' directly from such soil profiles (see fig. 404).

The extreme impoverishment of heathland podsols and many grassland soils is not only due to the indirect influence of man initiating further soil development. Above all it is the result of his direct interference by removing material for his own use.

Fig. 17. The network of grazing paths on the dry grassy slopes of the Kyffhäuser was even more clearly recognisable fifty years ago. As a result of damage to the turf the Zechstein clay became gradually eroded by heavy rain. In spite of the periodic droughts trees and shrubs are able to flourish on both the sunny and shaded slopes (foreground). They are capable of forming woodland - even though in some places it would be very open - almost anywhere here. Photo. Hueck.

Fig. 18. Flocks are once more browsing and grazing on the sandy heathland in the Nature Reserve near Wilsede. In the background are Junipers which, unlike most other woody plants, are scarcely touched by the sheep.

Fig. 19. This is the way turves were being cut some 50 years ago on the Lüneburg heath. Photo. Backhaus.

The dwarf shrubs as well as their litter and raw humus layers, together with a little of the pale sand, were cut away with a special chopper and a small rake. The turves were stacked at the farm and used in the first place as stable litter throughout the winter. Turfcutting was repeated every 15-20 years preventing the growth of shrubs and trees and making sure that the heather did not become too old.

Demonstrably since the Iron Age, and probably before that, the farmers removed litter from the woods and heaths to improve the fertility of their ploughed land. By this 'litter raking' or 'turf cutting' (fig. 19) they removed far more mineral nutrients from the natural cycling than they would have done by just taking timber, and thus contributed more to the acidification of the soil than by felling. As a rule they did not put the litter or turves which they had taken from the commons directly onto the fields. They used them first of all as bedding for the animals which, since the Iron Age, they had kept in stalls under their own roofs during the winter.

The removal of litter deprived the wood and heathland above all of nitrogen. This was already clearly recognised by Hornsberger (1905) and also shown to be the case by Onno (1969) who studied the use of leaf litter in the Vienna Woods (tab. 5). Ehwald (1957) calculated that the removal of nitrogen by the use of litter amounted to as much as the harvesting of between a half and a full crop of Rye. In addition, according to Wittich (1951, 1954) the microorganisms, important in maintaining fertility, are reduced in numbers since they rely on the litter, and to a less extent on the older humus, for their energy. It follows that the productivity of woodland and heathland was considerably reduced (fig. 20).

Fig. 20. The removal of leaf litter reduces the growth of trees in the wood, e.g. the Pine stands growing on two areas under investigation in the Upper Rhine plain, because plant nutrients are removed from them. This is true both for relatively fertile soil (Philippsburg) and for poor soil, also for any age of tree. After Mitscherlich (1955), modified.

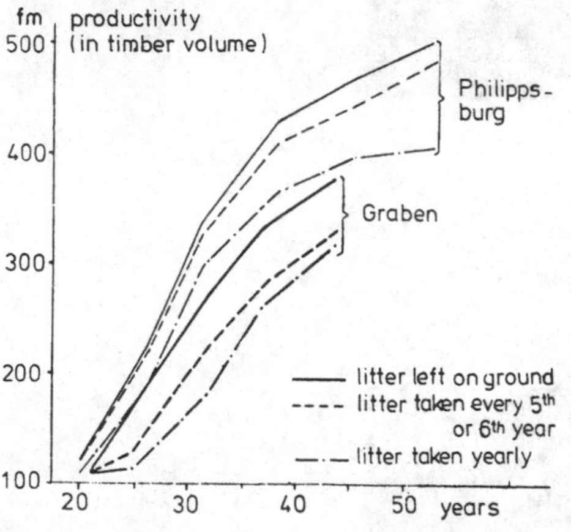

Extensively grazed woods, pastures and heaths however also suffer from a ceaseless nutrient drain which is not connected with arable farming, even though this may not be so obvious and has a slower effect. Pastures in the higher altitudes of the Alps offer the best example of this today. When they are chewing the cud, animals have the habit of seeking out certain level places or the flat tops of knolls, or they are driven to the milking places and fences in there. In such places, because of the accumulation of dung the grass grows lush and dark green. Eventually these resting places, with their large-leaved, strongly growing 'nitrate' plants (fig. 346), stand out from a distance. The fertility which is wasted here has come from the sloping pastures where the animals rarely linger after they have eaten their fill. In earlier times such resting places were also to be found in lower belts such as in 'Irrendorfer Hardt', on the Swabian Jura or in 'Wietzebruch' near Celle. It may be imagined that the luxuriant growth on such naturally manured areas was the reason for the development of grassland management including the regular use of dung.

Tab. 5. The removal of plant nutrients by taking one lot of leaf litter from Vienna Forest.

Woodland community and tree species	Nutrients removed (in kg/ha) N	P_2O_5	K_2O	CaO	MgO	Dry litter removed (kg/ha)
Querco-Carpinetum[1])						
Durmast and Turkey Oaks, Hornbeam	42	5	13	100	13	3570
Beech and Spruce	71	15	34	174	36	6830
Querco-Carpinetum luzuletosum[2])						
Oaks, Hornbeam, Chestnut, Fir	30	4	9	98	14	3800
Abieti-Fagetum						
Beech	60	4	18	151	29	5830
Beech, Hornbeam, Austrian Pine	60	7	15	143	21	7350

Note: From the researches of Onno (1969); the average of 5 samples, expressed as kg/ha.
[1]) '*Primula-Galium odoratum* type' on relatively fertile site.
[2]) on relatively poor site.

It is certain that by far the greater part of the soil in Central Europe had been more or less impoverished by removal of litter and fodder, and by leaching when the development of modern forestry and farming began at the end of the Middle Ages. While more intensive use of labour and manure has raised the productivity of arable fields over a long time, our grassland farming and especially our forestry are still burdened by the inheritance of past millennia.

With few exceptions then we must reckon that throughout Central Europe the present-day woodland soils are poorer than they were originally. They are less fertile because the removal of materials has also left them poorer in bases and therefore more acid. Above all the easily permeable soils of sandy plains or sandstone hills would have been significantly less podsolised than we find them today. These circumstances cannot be emphasised strongly enough because they have determined the vegetation and its productivity all over Europe much more than is realised by most people, even by foresters of our time.

Fig. 21. Farming activities in the Middle Ages and in more recent times have relentlessly destroyed the woodlands on the diluvial sandy plains. From a semi-schematic representation by Hesmer and Schroeder (1963), modified.

Above: the natural landscape with two small raised bogs (left) surrounded by Pine-Birch- and Alder swamp. The latter merges (centre) into a moist-soil Oak-Hornbeam wood which was preferred as a site for the village. On moraine (right) there is Beech-Oak wood and on sand Birch-Oak wood.

Below: an extensively deforested countryside with meadows, heaths and dunes, which have arisen because of the over-utilisation of the heath. The old arable land (Esch) is on the best soil. The raised bog remained untouched and is spreading. Because of this and above all through man's influence (arrows!) the relict Pine swamp has been reduced.

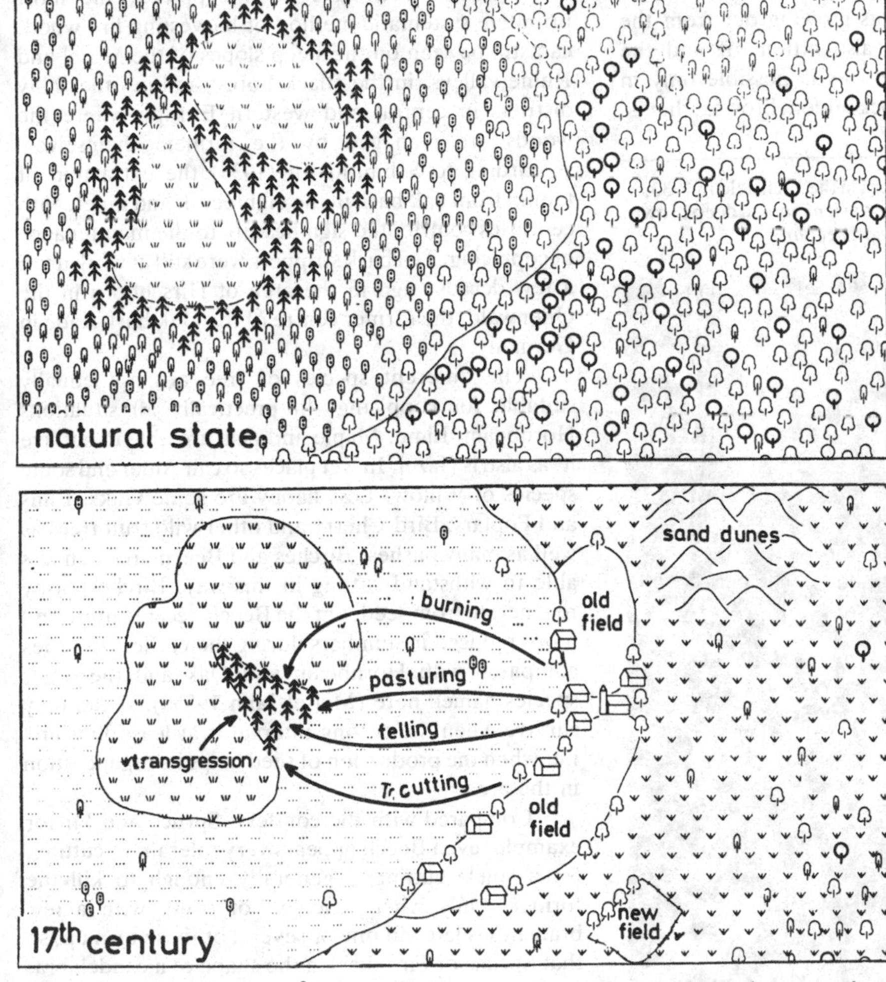

3 From coppiced woods to modern forestry

(a) Coppice and coppice with standards

At the beginning of the development as described in the previous section, the forest was an enemy of man since it confined his cultivated land and offered little food for his stock. At the end of this period not only had the might of the forest been broken over large areas, but its bounty had been squandered. 'Haphazard use and unrestricted grazing have ruined the forest. In many cases what remained no longer deserved this name' (Backmund 1941; fig. 21). Man was suffering a shortage of wood for building and burning, and depletion of the soil had led to an alarming shortage of fodder. Husbanding woodland and grassland was the only way in which the standard of life could be maintained and increased.

As long as livestock can wander over the whole of the common land, being excluded only from ploughed field and gardens, regeneration of most woody plants is prevented. The young shoots growing out from the stumps of felled trees are also bitten off and the repeated use of these regrowths is possible only in those areas where the livestock seldom browse, that is, in the more remote parts, or in those which are less accessible because of steep slope. In such places presumably the first **coppiced woods** (German 'Niederwälder') arose in which the regrowth was cut for fuel at intervals of 15-25 years (fig. 22). The yield of fire-wood was assured provided the herdsmen kept the stock out of the wood during the first years after cutting, or if more permanent protection was provided in the form of earth banks and ditches, thorn branches or hedges.

Fig. 22. A coppice near Duderstadt in the Eichsfeld in 1933. Anthropogenic *Stellario-Carpinetum typicum* on deep brown earth in place of the natural *Melico-Fagetum*.

According to Hausrath (1907) there were already a few coppiced woods in the thirteenth century. Most of them however are younger, and arose only in the sixteenth and seventeenth centuries from more mature woodland because of the excessive demand for wood as fuel. Yet other coppiced woods owed their existence to the method of farming in which fields were allowed to recuperate by returning to the wild after a few years of cropping (Schmithüsen 1934).

Woods which have been coppiced, the most primitive systematic method of harvesting fire-wood, can still be seen on the steep slopes of the Mosel and Rhine valleys, in the Black Forest and in other hilly districts in central and western Europe where the woods are managed by the farmers themselves. According to Schmithüsen (1934) the extraction of tannin from oak bark in the eighteenth and nineteenth centuries contributed significantly to the maintenance of coppicing. Tan bark strippers were still at work up to a few decades ago in the area of Hirschhorn in the Odenwald, the former centre of the Oak bark trade in Germany.

The different species of tree are not equally suitable for coppicing. In practically all situations Hornbeam, Lime, Maple and Ash are able to survive it, as also is Hazel. In wet places so can Alder and some species of Willow. Less happy are some Oaks, Elms and Poplars, Bird-Cherry and other wild fruit trees as well as many bushes. Birches and Beech are even less able to withstand cutting in this way. Birches easily regenerate from seed, but the Beech became more and more reduced in numbers during the earlier centuries compared with Hornbeam and Oak and the other species named here (Meisel-Jahn 1955a). It can only survive when the cutting interval is at least 30 years, i.e. when the production of seed helps its regeneration in the area.

Compared with the conifers, Spruce and Fir for example, even Beech appears very tolerant to cutting, for a single cutting is generally enough to kill the former. Only when a length of trunk with a few branches is left will one or several of these turn up, so that Spruce or Fir takes on the shape of a candelabra. In the marginal areas of coniferous woodland regions, for example in Poland and western Russia, but especially in the Alps and other mountains in Central Europe, coppicing has favoured the deciduous trees to an unnatural extent (Winkler 1933 and others).

Coppiced woodland was much more widespread in Europe right up to recent times than one could imagine today. Only the presence of old stumps and the occasional twisted trunk growing near the ground are reminders that many formerly coppiced woods have been allowed to grow out into mature stands.

In order to ensure a sufficient supply of building timber, as well as firewood, many communities in the Middle Ages decided to leave some of the trees to grow on instead of including them in the short cycle. For this purpose they generally selected Oaks since these provided not only food for the swine but also the chief material for the half-timbered buildings which were customary in those days. In such a 'coppice with standards' (fig. 23) Beech was again deliberately discriminated against as it was a less valuable timber for the carpenter, furniture maker or wheelwright. Hornbeam and Hazel on the other hand succeeded even better than they did under full coppicing, for they could withstand not only repeated cutting, but also the half shade cast by the large Oaks (fig. 23). To protect the young stool shoots after cutting in these open woods livestock had to be kept out for a few years as in the case of the coppice. It was however possible, as Hausrath (1907) has stressed, by rotating the yearly cut, to fence off permanently a relatively large and continuous part of the woodland.

Fig. 23. An Oak-Hornbeam coppice-with-standards near Duderstadt in 1933. The 'coppice layer' has been recently cut. The majority of trees left standing are Durmast Oak (*Stellario-Carpinetum*, as in fig. 22).

Carefully regulated management of the coppice with standards was practised above all on the heavily populated fertile loess soils where timber had become scarce at an early date. The general picture of the landscape during the Middle Ages which was described earlier holds good only for those regions which continued to be managed extensively either because of their distance from the settlements or their less-productive soils. Where there were large villages in the countryside the management of the coppice with standards as a communal activity of well proved value was kept up for a long time, and was often still in full swing right into the twentieth century.

Whoever compares such woodlands managed by farmers with close-growing tall forests on similar ground will notice, in addition to the obvious contrast in the tree layer, many differences in the species composition of the herbaceous layer. In the coppice with standards this is much richer in species and includes many light-loving plants favoured by the regular cutting of fuel timber. These light-indicator species gradually die back when the coppice is allowed to grow up into a mature forest. Many of them however, notably some grasses, persist for some time as a reminder of the earlier state of affairs (tab. 6).

(b) High forest management

Modern forestry prefers closed woods since these are more productive and supply more valuable timber than the old open grazing woods. Since its beginnings in the sixteenth century, but especially since the great impetus it received in the last century by the separation of woodland from pasture, rational forestry has tended towards monoculture and technical management. In spite of this, as was stressed in section 2a, the influence of forestry need not be seen as simply adverse to nature. Only where conifers have been introduced into the former deciduous woodland areas have quite new and unbalanced combinations of plants been created. Where the deciduous trees have been allowed to remain and develop into closed stands, on the other hand, something very near to the natural woodland has grown up (figs. 24 and 25).

Tab. 6. The influence of different methods of husbandry on the species composition of a limestone Beech wood.

Woody plants No.		1	2	3	4	5	6	7	8
Broadleaved trees									
Fagus sylvatica		●	⊙			○	●		
Fraxinus excelsior		⊙	◉	⊙		◉	◉	○	
Acer pseudoplatanus		⊙	⊙			⊙	⊙	○	
Quercus robur		○	◯	○		◯	⊙		
Carpinus betulus			◉	◉	○	◯	●		
Tilia cordata			⊙	○		⊙	◉		
Ulmus glabra		○	○			⊙	⊙		
Betual pendula		○	○		○	⊙	⊙		
Prunus avium						○			
Needleaved trees									
Picea abies									●
Pinus sylvestris									○
Shrubs									
Lonicera xylosteum		⊙	○			○		○	
Sambucus racemosa	L	⊙				⊙		⊙	◉
Salix caprea	L	⊙	○			◉	⊙	⊙	⊙
Corylus avellana		○	⊙	◉		◯	◯		○
Cornus sanguinea			○	⊙	○	○	⊙		
Crataegus spec.			◯	◉	◉	⊙	⊙	○	○
Prunus spinosa			◉	◉	●	○	⊙		
other broad-leaved shrubs			◉	◉	⊙	⊙	⊙		⊙
Needleaved shrub									
Juniperus communis				○	◉				

Herbaceous plants No.		1	2	3	4	5	6	7	8
Woodland plants									
Anemone nemorosa		◉	◉	⊙		◉	◉	◯	
Galium odoratum		◯	⊙	○		⊙	⊙	◯	○
Lamiastrum galeobodolon		◉	⊙	○		◉	⊙	⊙	⊙
Mercurialis perennis		◯	⊙			⊙	○	◉	○
Carex sylvatica		⊙	○			⊙	⊙	⊙	⊙
Hordelymus europaeus		⊙	⊙			⊙	○	⊙	⊙
Dryopteris filix-mas		○						○	⊙
acidity indicators									⊙
shade-tolerant masses									◉
Semi-shade plants									
Vinca minor			○	⊙	○	⊙	○	○	
Stellaria holostea			⊙	◉	○	◉	◉		
Aegopodium podograria			⊙	◉	○	⊙	⊙		
Dactylis glomerata		○	⊙	◉	◯	⊙	◉	○	○
Brachypodium pinnatum			○	◉	◯	○	⊙		○
Plants of open ground									
Festuca ovina and other grasses				◉	◯	○	⊙		
Cirsium acaule				⊙	◯		○		
Euphorbia cyparissias				⊙	◯		⊙		
Hieracium pilosella u.a.				⊙	◉		○		
Plants in woodland clearings									
Atropa belladonna	L	⊙				⊙		⊙	○
others	L	◉	○			⊙	⊙	◉	◉

Note: Semi-schematic and simplified (see figs. 16-18 and 22-25). The managed Beech high forest (no. 7) corresponds closely with the natural woodland (no. 1). Farmers open up the woodland by grazing (nos. 2-4) or by frequent cutting (nos. 5 and 6). ● very common. ○ common. ⊙ fairly common. ◉ occasional. ● rare. L in clearings.

Fig. 24. Transformation of the primaeval woodland through grazing, farming and forestry on a calcareous loam (brown rendzina) in the submontane belt of Central Europe (limestone Beech wood). The more or less anthropo-zoogenic plant communities in the forms 1-8 are described in more detail in tab. 6.

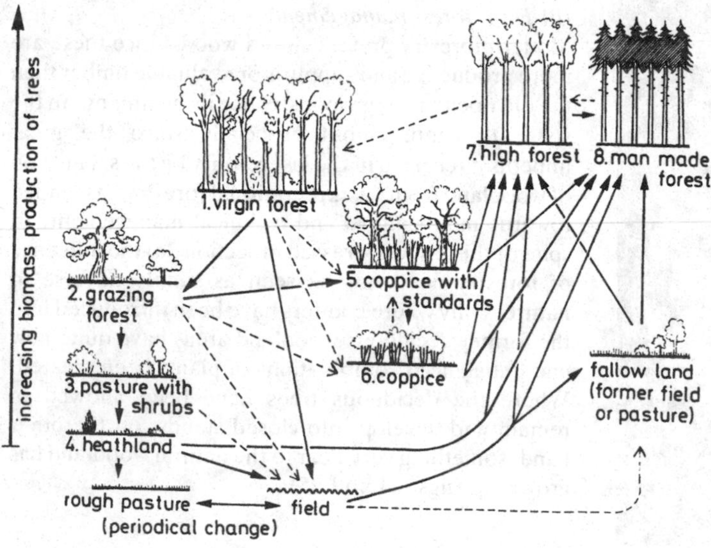

In comparison with the fundamental changes that have been wrought on the species composition of the plant communities by the replacement of coppice, with or without standards, by high forest, the replacement of deciduous by coniferous woods, and of the old rural extensive economy by intensive forestry, the different techniques of modern forest management have had little effect. The extraction method (German 'Plenterwald') where the removal of single trees aims at maintaining the high forest without interruption, and the different forms of small-area clearing (German 'Femelschlag') by removal of small groups of trees at the same time, which is more economical in labour, come nearest to retaining the character of the natural forest.

At the other extreme is the method of clear felling (German 'Kahlschlag') where, over much wider areas, a succession of ecologically very different conditions is rhythmically repeated. Immediately after complete removal of all trees there is flourishing growth of light-loving herbs, which are helped by the disturbance of the soil, and even more by the release of inorganic nutrients from the humus layer. Grasses, perennial herbs or small shrubs such as Raspberry and Bramble become dominant in the following years, but in turn these are soon supplanted by the larger bushes and trees whether these have grown from seed or have been deliberately planted. The first to appear spontaneously are the quick-growing, though generally short-lived, pioneers such as Elder, Birch, Aspen, Willows or Alder. These are finally overcome by the long-lived trees, which in most cases today means planted Spruce. In the high mountains, especially in Switzerland, clear felling is not allowed today because of the danger of avalanches and other harmful consequences. In other parts of Central Europe however it continues to be the custom although there are a number of variations on it.

Fig. 25. The Oak-Hornbeam woods which were once widespread in the Central European lowland are still managed as coppice with standards in only a few places (centre background). The coppiced layer may be allowed to grow up and become a high forest in which Hornbeam is common but where the more shade-tolerant Beech later will become dominant (right background). Often the broadleaved trees are replaced by conifers as in the Fir forest (left background) or the Spruce plantation (on the clear felled area in the foreground). 1950 near Leonberg (Württemberg).

4 The development of arable farming and the weed communities

(a) Pre-industrial agriculture

In Central Europe arable cultivation is more than 5000 years old and as we have seen already has developed in close association with stock rearing. Both forms of farming came to Europe from Asia Minor via the Balkan peninsula (Jankuhn 1969). For climatic reasons cultivation of the soil could only have begun towards the end of the Boreal period (fig. 259, No.4) at the earliest, that is at a time when woodland had already reestablished itself in Europe. It is not impossible that certain isolated areas had been kept free from trees from the start by the beginnings of a primitive agriculture carried out by man and his livestock; however the majority of fields were made out of previous woodland (Krzymowski 1939, Nietsch 1939, Firbas 1949 and others).

One could not however envisage that in the Stone Age and Bronze Age the high forest would have been cleared systematically, in the same way as has been practised by woodland dwellers since the Middle Ages in both the Old and New Worlds. It is much more likely that areas which had already been opened and become grassland or heath would have been chosen. The position of many of the raised beds previously mentioned, either isolated or distributed in the arable mark, definitely points to this. At the start the burning off of a certain area prior to cropping also played an important role. This resembled the shifting cultivation of the tropical forests and had the advantage that plant nutrients were released by the burning (Romell 1967).

Although it is certain that most of the arable land of prehistoric times is still under the plough today it has been cropped every year only for the last two centuries. Up to recent times the chief method of farming was the **three-field system** which was widespread in the Middle Ages over practically the whole of Central Europe. It was unsuitable only in the higher mountains and on many sandy soils or those fertilised by turves in the north-western plains. On the latter, a system of growing only Rye, alternating with fallow.

was developed. As is well known, in the three-field system winter cereal, spring cereal and fallow were rotated in a three-year cycle (see fig. 26).

It was important for this system that the whole of one-third of the arable land should be treated in the same way and this was enforced by law, even though it belonged to different occupiers. The basis of this constraint was not so much that an adequate network of field tracks was lacking, but rested much more on the custom of keeping the arable land available for as long as possible as common pasturage. Only during the periods from the cultivation of the field to the first tillering of the cereal crop and again from the heading stage to harvest was the arable field fenced off. This fence generally consisted of thorn branches which had been cut from the commonage, where they were weeds in the pastureland; but in some regions they were replaced by walls, earth banks or living hedges. The latter were preferred in north-western areas of Europe (see Jessen 1937). These are called 'Knicks' (i.e. laid hedge) in north Germany because formerly they were made stock-proof by bending over young trunks of broadleaved trees. This type of hedge was often planted on top of a bank of grass turves and managed in the same way as the coppices, with or without standard trees being left. Such hedges also had to be put round the arable fields of individual farmers which were situated in the commonage to protect them from the livestock. Likewise they were laid round woodland and other areas which needed to be separated from the common grazing (fig. 428). Laid hedges served this purpose as long ago as the pregermanic time, for example in Westphalia.

The livestock loved to graze on the young cereal crop which was not harmed very much provided the animals were taken off before the plants started to shoot up. It increased the amount of tillering and helped to combat weeds. This weed community on the arable land must have been much richer in perennials, especially grasses, than is the case today, since it was only disturbed by ploughing twice in three years and rarely hoed; the shallow top soil was roughly broken

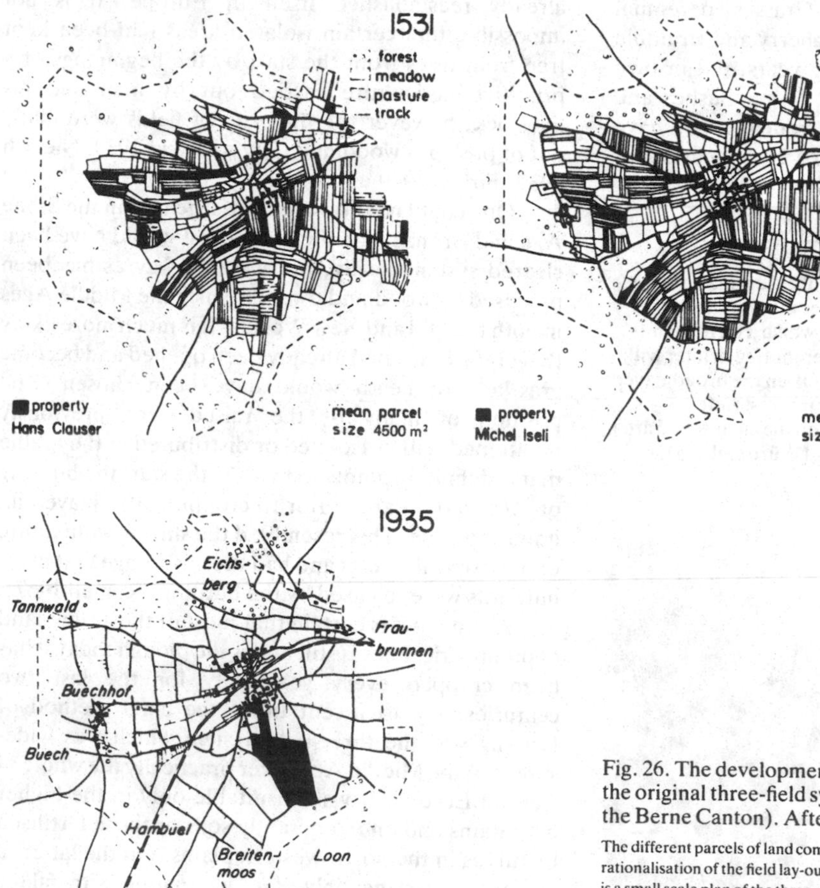

Fig. 26. The development of a Central European village from the original three-field system of husbandry (Grafenried in the Berne Canton). After coloured maps by Zryd, simplified.

The different parcels of land comprising the same holding before and after the rationalisation of the field lay-out are shown in black. On the 1749 map (top right) is a small scale plan of the three rotation blocks which were in operation from the early Middle Ages right up to 1935. The woodland supplied fuel and building material and was cut back more and more from the edge of the village.

down using more primitive tools. The cereals were not as thick or high as they are today and so were not as effective in smothering the weeds. What is more, the persistent plants capable of efficient regeneration recovered during the fallow year. In spite of its short duration grazing of the fallow land was able to determine the species composition of the weed community, making it similar to that of a grassland. Since the 'improvement' of the old three-field system in the eighteenth and nineteenth centuries by the growing of potatoes, turnips and other intertilled crops on the former fallow land, and since the introduction of more complicated crop rotations the annual weeds have become more dominant. If a field is left fallow now it takes more than a year for it to become grassed over naturally. The use of modern herbicides to combat weeds in arable land has reduced their numbers very considerably; it has also altered the weed communities qualitatively again (section D IX 2).

Earlier conditions on the arable land corresponded much more closely than they do today to those found on the grass steppes of south-east Europe where many of our weed species are native; *Falcaria vulgaris, Conringia orientalis, Caucalis platycarpos*, and *Adonis aestivalis* are examples. These must at one time have come into Central Europe as true representatives of the steppeland, but at present are only surviving on badly cultivated fields. On the other hand, the majority of the weeds of our gardens, vineyards and fertile, well hoed arable fields such as *Stellaria media, Poa annua*, or *Chenopodium album*, have originated in the nitrate-rich tide lines (section C VIb) of the sea coast or river banks as well as from other sites in river valleys or fertile marshlands.

Mediaeval farmers using the three-field system covered a wide area and were successful for so long not least because they knew now to combine extensive grazing of common land with arable farming to the advantage of both branches of husbandry. The fallow year meant not only a rest for the soil, but also that it was manured by the livestock which grazed it and was penned on it overnight. The animals in their turn found particularly protein-rich food there.

Such a **connection between livestock and arable farming** may be much older in principle than the three-field system. It probably existed right at the start of arable cultivation in Central Europe. If this were so it would explain why the arable weeds so quickly invaded ploughed land on the very scattered fields belonging to isolated settlements and why so many of the weeds are represented in practically every part of Central and western Europe. It is well known that grazing animals disseminate numerous seeds, and not only those which have some mechanism for adhering to the animal's coat, or those which remain viable after passing through the animal's gut. As Darwin has already established, many seeds and fruits are contained in the earth or dung sticking to the animal's hooves or skin (see also section D VI 2a). In this connection it can be pointed out, in particular, that in every year up to about 1950 there have been extensive sheep movements in the southern part of Central Europe. Wandering shepherds from the Swabian Jura or even from the gravel heaths of Upper Bavaria used to spend the winter with their flocks on the warm Upper Rhine plains. Increasingly however they encountered difficulties as they were no longer free to travel over the stubbles or the winter cereals.

(b) Effects of technical advances on extensive grazing and arable land

In modern times, sheep keeping is no longer necessary to maintain the fertility of arable land and has turned into an independent branch of farming. However with the import of cheap wool from overseas it became unprofitable and collapsed. Even the Danish, north German and Dutch heathlands, which had earlier supported thousands of flocks, guaranteeing the fertility of the arable land by the spreading of turves (section 2c), now suddenly had become worthless. Soon hardly any farmer could understand why his forefathers had fought for the right to occupy and use such heaths. The same thing has happened to the 'Hardts' of the Swabian Jura, the 'Grinden' of the Black Forest and other formerly grazed commons. They were allowed to lie empty, were colonised naturally here and there by shrubs, or became afforested and eventually developed into a landscape rich in trees. To be sure their character today is quite different to what it would have been naturally. Where the soil fertility and the distance from the farmsteads made it possible, part of the former heathland and poor grassland have been taken into arable cultivation which has spread quickly in many areas. The badly impoverished soil was made productive once more by the use of manure and mineral fertilisers.

In many mountainous areas, for example in the Swabian Jura and in valleys of the Alps, there has been, on the other hand, a shrinking of the area under the plough. This is particularly true of those areas where at earlier times, because of the custom of land partition and the growing density of the population, people were forced to bring into cultivation even very steep, stony and dry ground in the common mark. Since the turn of the century the yield from the better land has been rapidly increased by heavier fertilising and improved methods, making it possible to give up the poorer areas again. These were grassed out if sheep rearing was reintroduced, or they were afforested (figs. 24 and 414). However in many places the furrows, or the terraces formed on the slopes over the centuries by ploughing and soil erosion are indications that these were at one time cultivated strips. In the Alpine valleys and foothills and in parts of north-western Europe the retreat has finally gone so far that all the ploughed land has disappeared in favour of permanent grassland. This is due on the one hand to the humid climate which gives a good growth of grass for the keeping of milking cows, and on the other to modern transport which means that the farmers no longer have to be self sufficient.

In recent times we are experiencing an unexpected and rapid contraction of both arable and grassland as the smaller holdings have become unprofitable, and in more difficult circumstances, such as a mountain climate or on drained fenland, are no longer viable as farms. In the neighbourhood of large towns even some of the best arable land is no longer cultivated and has deteriorated into 'social fallow' (section D X). The maintenance of these areas becomes a problem for landscape management and vegetation ecology, and provides opportunities for studying the natural return to woodland (the secondary progressive succession). Slowly or quickly the woody plants set about reconquering the areas which had been lost to them.

At the same time the cultivation of the remaining fields becomes more and more intensive and further from nature. In spite of this, modern methods have not made it possible entirely to eradicate weeds or significantly to alter their associations (section D IX). Weeds are only being reduced in quantity, particularly on the more backward smallholdings.

5 The origin of meadows and intensive pastures

(a) Litter meadows and fodder meadows

While grazing of livestock is a very old practice in Europe, meadows requiring tools for cutting the crop are relatively recent. In the Bronze Age sickles were known, but their use was restricted to the harvesting of corn, not for cutting meadows. According to Lüdi (1955), for example, surprisingly little remains of grasses, but a lot of dried shoot of Ash and other tree twigs (see section 2a) were found in excavated settlements in northern Switzerland. Schimper, Graebner and other earlier researchers have maintained that meadow like communities could also have arisen naturally, especially in the flood plains where the passage of ice would destroy the trees. However Gradmann (1932) has already disposed of their argument very convincingly and pointed our that natural woodlands would have grown right up close to the large rivers, since these are to be found today in the Russian Taiga where the climate is much less favourable. 'Without scythe and hay harvest no meadow flora!' This saying of Schlatter made more than a hundred years ago applies to the whole of Europe with the exception of reed and sedge swamps, salt marshes by the sea and alpine meadows.

Cultivation of meadows has arisen in many different localities from extensively grazed pastureland. By about 1800 there were still such 'wood meadows' in the foothills of the Alps, according to Backmund (1941). To this day examples are preserved in many nature conservancy areas such as the 'Irrendorfer Hardt' on the Swabian Jura (fig. 27a). In Sweden and Germany they are called 'Laubwiesen' (meadows with deciduous trees), and Sjörs (1954) devoted a basic ecological study to them (fig. 27b,c). In Switzerland a distinction is drawn between 'Studmatten' which are the mown areas between trees, and 'Wytweiden' which are the proper woodland pastures. Probably before scythes were known part of the winter fodder was secured by pulling the grass (Stamm 1938). The direct effect of this method of haymaking on the plant communities was nearly the same as that of animal grazing although grazing implies a selection of species, as we have seen in section 2b. Mowing, that is cutting off practically all the assimilatory organs from all species at the same time, has quite a different effect from the gradual selection by animals. After mowing, all partners have the same start, so to speak; therefore it favours quick-growing and tall plants provided they are capable of regeneration.

To be sure the first meadows were cut only once and corresponded to the '**straw meadows**', which are still to be found here and there. These are first cut late in the year when they have already ripened to straw (section D V 1a). For the most part they occupy very wet sites, which are avoided by stock and do not need to be fenced off.

Meadows on drier soils are comparatively young; this is particularly true in the case of the '**fodder meadows**' which require a dressing of manure to replace the plant nutrients removed with the biomass. Without fertilising these can only arise in flood plains where sediment rich in nutrients is deposited on them. Of course the natural vegetation of such alluvial soils, a well grown deciduous woodland, must previously have been removed by man and his grazing animals. A further prerequisite is for livestock then to have been fenced out of the area, since only by regular mowing would the true meadow plant community have developed. It is scarcely to be wondered then that the typical fodder meadow, cut twice or more in the season, which became so widespread and popular right up to the middle of the twentieth century, is not more than about a thousand years old (Sukopp 1969). This length of time has however been sufficient to allow a plant combination to develop, which is distinguished by many good character species, though bearing no resemblance to the natural vegetation. As a result of modern farming methods, especially 'rotational grazing', it has recently been forced to retreat and is threatened with complete disappearance.

(b) Permanent pastures and those alternating with mowing

The universal intensification of farming during the last 100-150 years has finally brought about a fundamental change in the management of pastures too. The old grazing rights which had been handed down were done away with, and the commonage was either split up into privately owned plots or managed as a district forest. In areas of low rainfall, for example east and south of the Harz mountains, on the Upper Rhine plain and in the lowlands of Baden-Würtenberg, there was a changeover to keeping cattle, horses and pigs constantly in buildings for feeding. Apart from the remains of

Fig. 27a. Grazed 'browsing pasture' (*Mesobromion*) with Hornbeam and isolated Beech on the Swabian Jura to the north of Zwiefalten. As pasture weeds Juniper and Blackthorn are spreading.

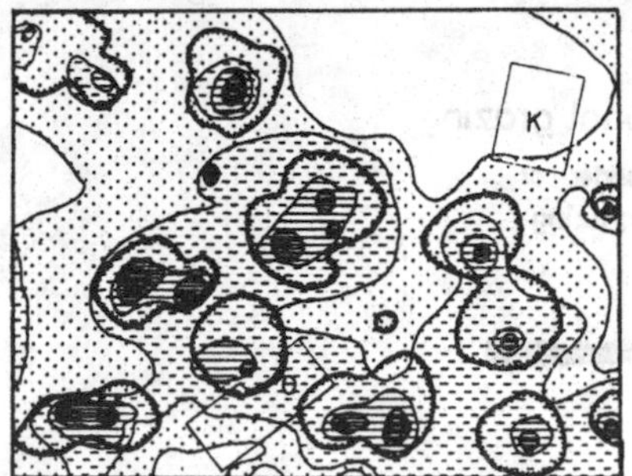

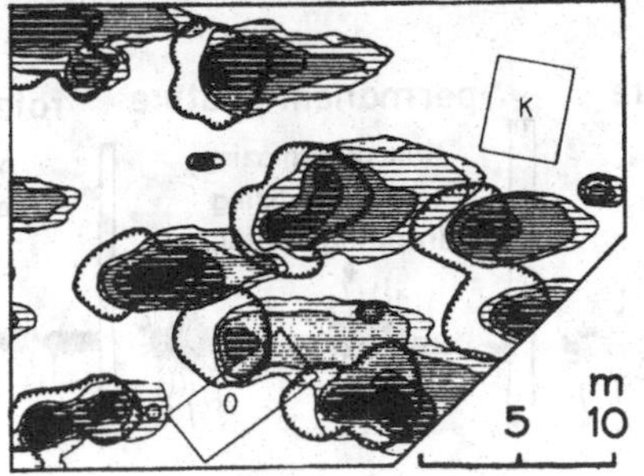

Figs. 27b and c. The amount of light reaching the ground in a Birch browsing pasture in southern Sweden varies over short distances and changes quite quickly with the passage of time (darker areas indicate less light). After Sjörs (1954), extracts.

The toothed lines delimit the vertical projection of the tree crowns. South is to the left as in fig. 27a.

b) Diffuse distribution of light when the sky is overcast (19/20.6.1949, midday).

c) Sharper shadows thrown by the clear noonday sun (21.7.1949).

The rectangles shown are experimental areas.

earlier sheep pastures there were scarcely any pastures left here around 1950. That is to say, plant communities which were principally used as pastures, and in which the combination of species differed essentially from the hay meadows, had practically disappeared from these regions. By way of contrast in the more humid and rainy areas near the coast and round the Alps the tendency was to graze all grassland with the exception of swampy ground, and to drain this if at all possible. As a consequence there was an extension of well manured pasture communities all over north-west Europe.

The first method of pasture management in general use was that of **permanent grazing**, which prevailed until about 1960. Cattle were kept on the same area throughout the whole growing period, i.e. from about April through to October, and were put into stalls only for the unfavourable months. In contrast to the earlier common grazing the livestock was kept in the pasture fields by means of hedges, ditches or fences so that gardens, arable land, forests and other parts of the farming area did not have to be protected from it. As long as the animals stay in the grazing field their dung is continually being returned to it. Therefore the only plant nutrients which have to be replaced are those removed in the form of milk or animal bodies, together with any losses caused through denitrification or leaching from the soil. As a result the general fertility of such pastures is quite good; in any case it is significantly better than that of the extensive pastures already described.

Along with the higher fertility two further characteristic features are revealed in the flora of pastures permanently grazed by cattle. They are certainly less weedy than the commons which preceded them (figs. 28, 435), but they still contain thistles and other typical pasture weeds because in early summer there is much more fodder than the cattle can eat so they are able to select the most palatable. Particularly noticeable are the rank patches (fig. 469) where the sward has been tainted by excrement and consequently avoided by cattle. These overfertilised patches of grass grow tall and dark green, while the sward in between these patches is overgrazed and towards the end of the season, often destroyed by treading of the stock. Since horses will eat grass which has been tainted by cattle and vice versa this undesirable patchiness can be avoided to some extent by mixed grazing.

In recent years grassland management has attempted to avoid the disadvantages of permanent grazing by repeated trimming of the pastures, that is, a system of grazing after raking with mowing. The ideal form of this is **rotational grazing**, in which the total pasture area is divided into 8 to 12 parts. Each paddock is grazed for a few days only but with a large number of livestock at the same time so that it is eaten down closely and evenly. As the animals are moved on to the other paddocks in succession, the herbage has many weeks in which to recover, and takes on the appearance of a meadow. In addition the paddocks are mown from time to time for hay, and this almost entirely eliminates weeds. Only some grasses and clovers can withstand this kind of treatment resulting in the most boring plant community that any botanist can imagine!

6 *Effects of technology on the vegetation*

(a) Interference with the water economy of the countryside

The many kinds of grassland that have arisen throughout the course of history, and are still to be found alongside one another, have been still further increased in number by man's interference with the water levels of the natural mires, marshes and flood plains.

No part of the large mire and swamp areas, which are so characteristic of the northern part of Central Europe and of the Alpine foothills, has retained its original water table. To study the types of wetland vegetation of the great plains we therefore have to turn to the few small relicts or to the areas from which peat has been removed and which are now regenerating. We

Fig. 28. Extensive herding favours the pasture weeds much more than more intensive set stocking or even the modern rotational grazing alternating with mowing. After Ellenberg (1952a).

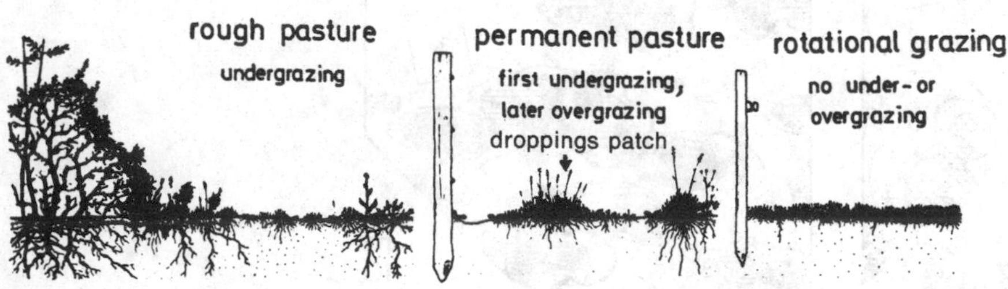

shall come back to these in detail and also discuss the fate of the mires more closely in sections C II and III.

In general it can be said that swamps and mires were certainly part of the commons but there is no question that their grazing value was low. Because of their wetness they were scarcely passable and even where they could be trodden on they offered little in the way of food. The first wetland areas to be made use of would have been the Alder fens, mainly in the neighbourhood of villages. Alder trees, which recover strongly after cutting, supplied welcome firewood. Where the trees were eliminated a sedge-rich grassland formed, the significance of which for the development of meadow communities has already been shown (section 5a).

As wood became scarcer man turned in some areas to peat cutting with, however, little success as a rule. This is because the peat from fens, which was easily accessible, had a high mineral content and burned badly. The valuable older peat from the raised bogs on the other hand is covered with younger, less valuable peat and can only be extracted after substantial drainage work. The **disturbance of mires** is thus the work of more recent times and especially of the last 250 years in which the techniques of working the moorland have been developed. The pioneers of wetland drainage and the use of peat for fuel were the Netherlanders. On the plains of Lower Saxony, one of the areas richest in mires in Europe, especially in raised bogs, practically all of these had been drained up to about a decade ago (fig. 29). Since then many of the cultivated mires have been allowed to fall fallow again because the management of them as grassland today scarcely pays.

Fig. 29. Reduction of the area of 'barren mires' from the end of the eighteenth to the middle of the twentieth centuries, mainly because of the cultivation of raised bogs, in the northern part of Lower Saxony. After Baden (1961) from F. Overbeck (1975), modified.

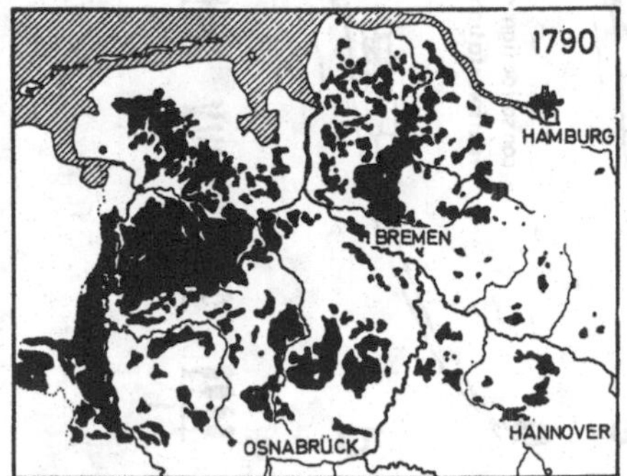

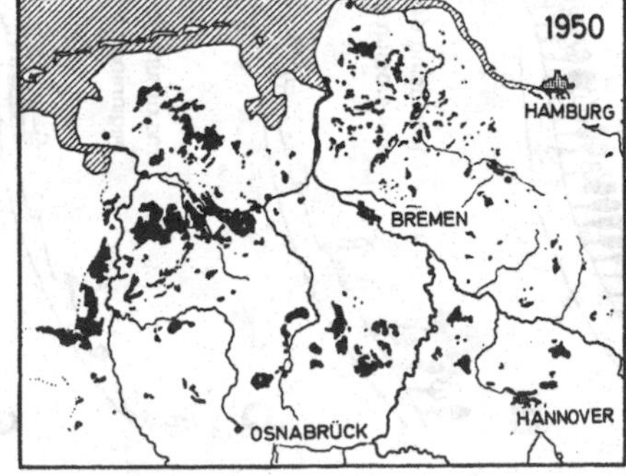

The **building of dykes** to enclose the coastal marshland and the wide estuaries began in the eleventh century spreading from Holland and West Friesland to the coast of the North Sea (Beekmann 1932). In spite of their far reaching consequences for the character of the vegetation they will not be treated until section C IV I because they apply only to a special part of Central Europe and also because they can be understood only in the context of a more exact knowledge of the habitat conditions and the plant communities in the salt marshes.

Of more general significance for almost every part of Central Europe has been man's effect on the **water economy of the large river systems**. The first effect was to be felt quite early, but this operated in an indirect way through the opening up and grazing of the woodland (fig. 30). This altered the run-off from the precipitation and increased the rate of soil erosion. As will be covered in section B V I the sedimentation of alluvial silt loam in the valley of the Weser only reached substantial proportions from Roman times onwards. Because of an increase in arable farming, large areas of the loess-covered hills in the river catchment area became denuded of trees, and exposed to a largely accelerated soil erosion. Previously gravel soils had predominated in the Weser valley giving it a character completely different from the one it has today (figs. 30 and 197). Almost every large river valley in Central Europe underwent a similar development. Besides this the flow of water in the rivers became less regular, because melting snow from the open areas thawed more rapidly, and heavy rain would run off more quickly from ground consolidated by grazing, than it did from woodland. Because of this, and also through the straightening of the rivers, the speed of the water increased, deepening the bed and allowing the groundwater table in the valley to fall as time went on. The riverside soils, although flooded occasionally, are generally no longer affected by groundwater (fig. 30). Hügin (1962) has described such

Fig. 30. Changes which have taken place in the upper (or middle) and the lower reaches of a Central European river valley because of increasing deforestation, draining, erosion and flood plain silting. Schematic, after Ellenberg (1954).

1 = Beech wood, 2 = Oak and mixed broadleaved woodland, 3 = Alder swamp, 4 = conifer afforestation, 5 = Willow scrub, 6 = other scrub, 7 = wet meadows, 8 = damp meadows, 9 = dry meadows, 10 = arable, 11 = loam, 12 = silt, 13 = peat, 14 = gravel, 15 = other soil types, 16 = average groundwater level, 17 = average height of flood water. The symbols 1 - 9 are not to scale, and the soil profiles 11 - 14 are much exaggerated.

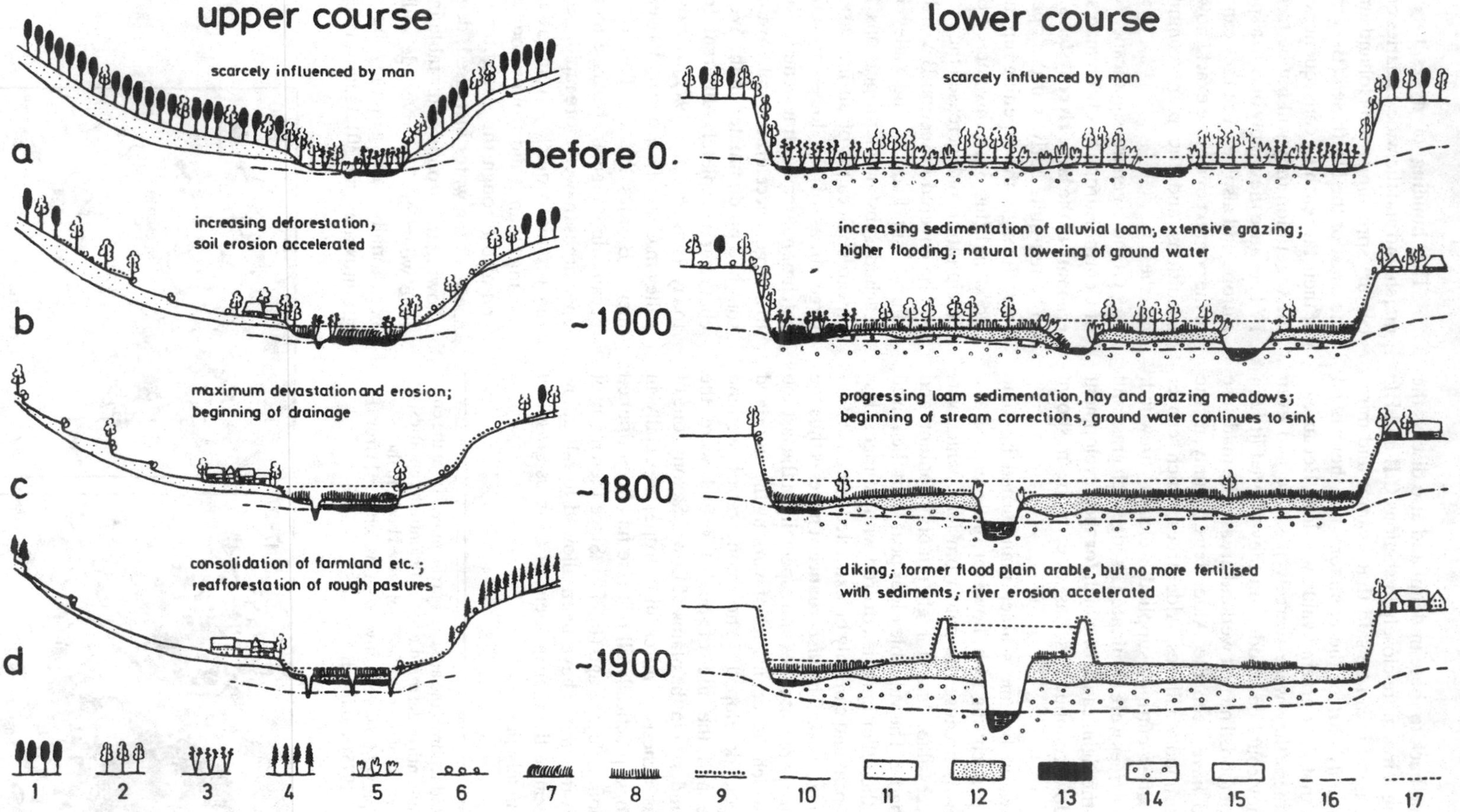

upper course

lower course

a — before 0.

scarcely influenced by man

scarcely influenced by man

b — ~1000

increasing deforestation, soil erosion accelerated

increasing sedimentation of alluvial loam; extensive grazing; higher flooding; natural lowering of ground water

c — ~1800

maximum devastation and erosion; beginning of drainage

progressing loam sedimentation, hay and grazing meadows; beginning of stream corrections, ground water continues to sink

d — ~1900

consolidation of farmland etc.; reafforestation of rough pastures

diking; former flood plain arable, but no more fertilised with sediments; river erosion accelerated

1 2 3 4 5 6 7 8 9 10 11 12 13 14 15 16 17

a change in the example of the southern part of the Upper Rhine plain which flows through gravel beds (fig. 31). Sukopp, Kunick, Runge and Zacharias (1973, fig. 32) have demonstrated convincingly what finally becomes of a river valley and its surroundings when it runs through an area of industrial aggolmerations.

The system of drainage ditches, and the straightening out of meandering watercourses of streams and rivers in the catchment area, speed up the run off after heavy and sudden downpours so much that often there is flooding in lower reaches to a greater extent than previously. Along the North Sea coast this increased so much that the dykes gave way. As a consequence of bad floods at the edge of East Friesland many marshland areas along the sandy coastal heaths were drowned.

Drainage and river management, building of dykes and cultivation of mire, reduction of forests and increasing arable farming have led to a position now in which the cultivated land has become **more and more susceptible to drought**. As Schmithüsen (1950) emphasises, there have been unusually dry summers, similar to the one in 1947, in previous centuries. These however have been less damaging because at that time the water reserves in the ground were greater, and the water requirements of the extensively managed woods and pastures, as well as of the poorly manured arable land, were less than today. In addition to this, arable farmland has become even more prone to drought since higher-yielding crop varieties, which are at the same time more susceptible, are being grown. Also modern forests suffer more in a dry period than the former natural or pastured woodlands, at least in those places where the shallow-rooted Spruce has replaced the deep-rooted Oak and other deciduous trees.

(b) Environmental stress caused by industry

Since the middle of this century and increasingly in the past two decades, the relationships between plants, animals and men and their environment are being disturbed by burdens which stem directly or indirectly from modern industry. Woods, mires, rivers and lakes, villages with their farmland, towns with their varied surroundings and spheres of action, even the whole biosphere of the Earth, with all its living communities, are being affected. Three tendencies appear in this

Fig. 31. Upper Rhine Plain with steep gradient and gravel subsoil in the flood plain before and after the straightening of the watercourse in the nineteenth century. After Reichelt (1966).

The simplified maps show the potential natural vegetation, according to Hügin. After the fall in groundwater level the gravelly river valley became drier than the loam covered terraces so that the flood plain woodlands died out and only the drought-tolerant mixed Oak-Lime woods and Downy Oak scrub were able to flourish. The extent to which the water table has fallen and the actual vegetation are shown in the schematic cross-sections.

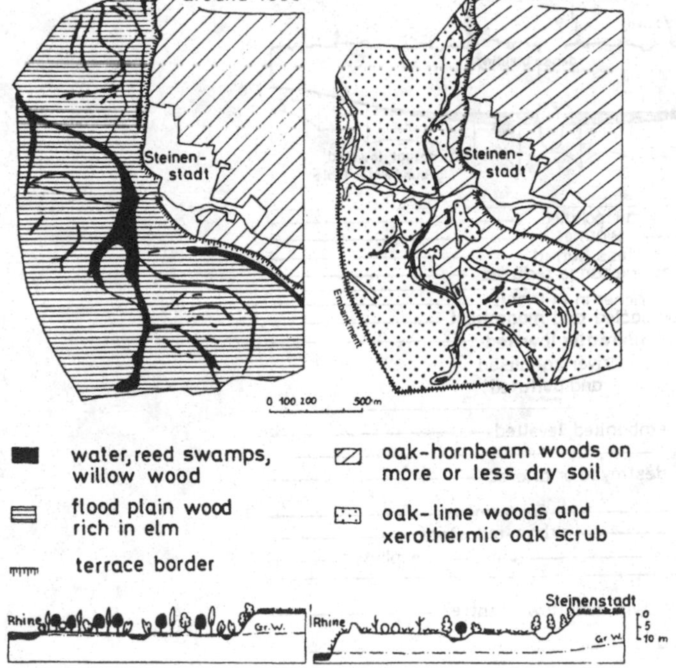

development which present new and urgent tasks for ecology:

1 Factors which previously had little effect and could have been overlooked have now grown to such **large proportions** that they are endangering many forms of life, and even man himself. The concentrations of phosphate and ammonia in the water, and carbon monoxide and sulphur dioxide in the air may suffice as examples.
2 **New** factors created by man have appeared; e.g. biocides and artificial materials which are difficult to degrade.
3 The effects of a few factors, which up to now have been limited to relatively narrowly defined areas, have recently become **widespread or even world-wide**. For example the emissions of SO_2 from the heating systems of western and Central Europe are leading to the death of trout by the introduction of sulphuric acid into the soft-water streams and other ecosystems as far away as the south of Sweden and Norway (figs. 33a and b, 34). They are also reducing the yield of woodland, which could amount in value to millions of dollars by the time the trees are harvested (see also Overrein, Scip and Tollan 1980).

Many of these factors do not appear at first glance to be causing any damage, but during the course of years or decades of accumulation the damage becomes noticeable, for example the raising of the lead content of grassland near motorways. The correct evaluation of the consequences of such damage is made more difficult because the different factors do not work in isolation, but combine together so that their synergetic effect can be much greater. Above all the symptoms are complex because man and other organisms are not affected just as individuals, but the whole ecosystem to which they belong is influenced (section 7). For example the circulation of carbon compounds, of water, of bio-elements and other substances can be altered, or the transmission of stored energy which

Fig. 32. Some changes in the environmental factors, vegetation and animal life following the establishment of a large town in a valley with a small river (e.g. Berlin). After Sukopp (1969), modified.

The following should be noted as further effects on the vegetation: damage of many evergreen trees (especially conifers) by dust and sulphur dioxide; salt damage to roadside trees following winter salting; killing of trees by damage to the roots following roadworks or by the asphalting of the surface (shortage of oxygen); but also increase in the variety of trees caused by planting or even natural regeneration; protection of frost-sensitive species from a sudden late frost; a longer growing season. Not all the effects are unfavourable for the plant world!

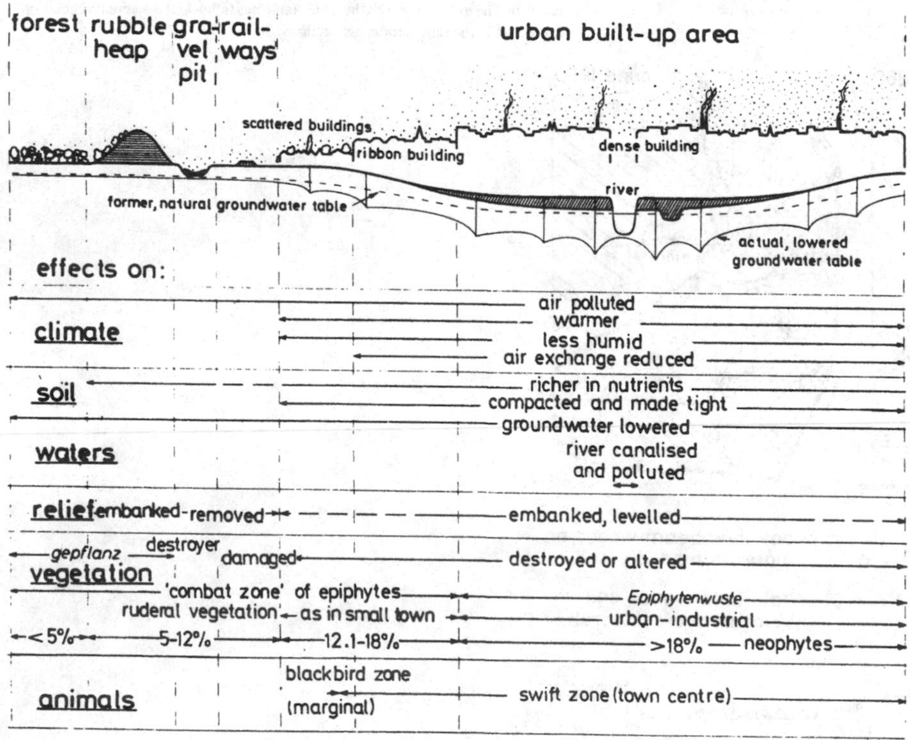

unites the green plants with the most diverse animals and microorganisms can be impaired.

It is possible to assess the results of environmental stress in many cases only through interdisciplinary research into the ecosystem as a whole (Ellenberg 1973a and other IBP publications, fig. 35). Vegetation scientists specialising in environmental interactions and causal relationships should take part in this, precisely because they are aware that man has already been altering the environment for thousands of years.

Man acting as a 'super organic factor' (as defined by Thienemann, see also Ellenberg 1976) is constantly upsetting the natural balance, or a balance which he himself has already brought about. In this way he has increased the diversity of the landscape in many areas and in many ways. This is reflected for instance in the numbers of species which have been found in the vegetation in 5 × 5 km quadrats mapping the flora of southern Lower Saxony (tab. 7, also Ellenberg, Haeupler and Hamann 1968). In the quadrats which could be said to be closest to nature, i.e. almost entirely covered with high forest, far fewer plant species are growing than in the quadrats where there is a mosaic of woods, heaths, rough grazing and other types of grassland, as well as fields with weeds and areas with ruderal vegetation which, almost entirely, have man to thank for their existence. We feel rightly that this multiplicity is a good thing and shall seek to

Fig. 33a. The pollution of the biosphere with sulphur dioxide has effects - depending on weather conditions - up to a distance of several hundreds of kilometres on agricultural and forestry ecosystems and also on stretches of water and rocks. The main effects are shown in italics; beneficial alternatives are enclosed with a broken line.

Acidification of the woodland soil can lead to a lowering of production; however this is not always just a result of the sulphur dioxide. For example in southern Scandinavia the separation of grazing from the woodland which took place several decades ago has resulted in a denser canopy of Spruce and Pine woods and an increased accumulation of raw humus. As a result the woodland soil and the water draining from it have become more acid (Malmer 1974).

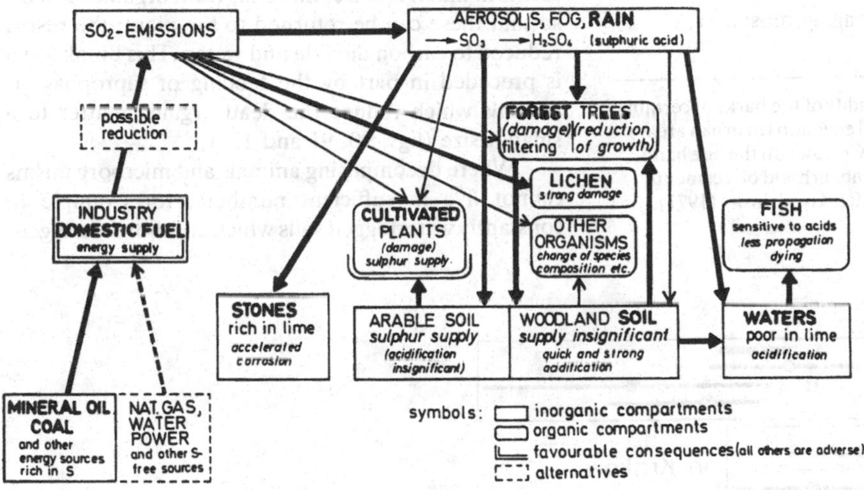

Fig. 33b. The emission of sulphur dioxide into the air has had the effect of increasing the sulphur content of rainwater and making it more acid, even on the Schauinsland in the Black Forest. The lowest average pH values occur in winter and the early spring when heating systems are being used to the greatest extent. After Rönicke and Klokkow (1974), somewhat modified.

The annual average pH value of rainwater fell in 1975 to less than 4.5 compared with about 5.5 in 1950, and is still falling. Rainwater dripping from the crowns of the trees dissolves yet more acid which has fallen on the leaves and twigs and the pH value can fall to below 3.0; where the water runs down the trunk it may be lower than 2.0 (according to Züst, lecture, 4.10.1977).

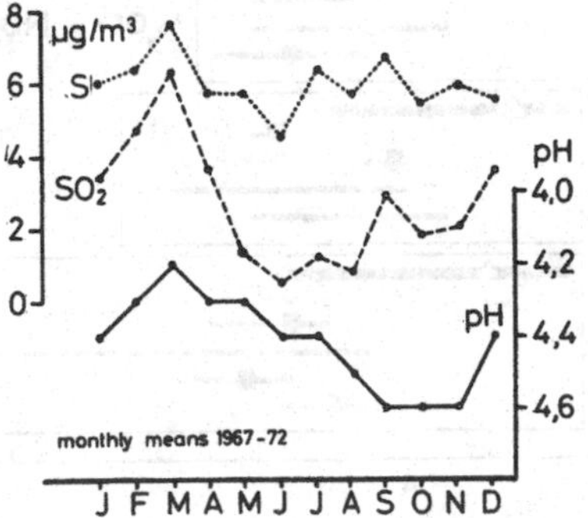

maintain, or even to increase it wherever this is possible against the tendency to uniformity which is a feature of our time. It should already have become clear from the introductory sections that maintaining in this connection does not mean rigorous conservation, but keeping up the human influence as the deciding environmental factor.

Those ecologically orientated students of vegetation can give important help also in the fight against the increasing pollution of water, air and soil – which in the end is an economic and political problem. Many plants and plant communities are sensitive indicators reacting long before the degree of pollution becomes troublesome to man (e.g. sections C 12c and VIII, also fig. 32). In the last few decades the flora and fauna of Central Europe have changed radically (Sukopp and Trautmann 1976).

We shall be returning constantly to such ecological and physiological relationships as well as to questions of the origins of the present species associations. On the other hand we shall limit the systematic reviews of plant communities to the minimum necessary for an understanding of the vegetation and the factors bringing about its changing mosaic.

Fig. 34. Changes in the natural acidity of the barks of certain tree species brought about by acid emission (in urban areas, from a steel works and in a town (Cracow) on the one hand, and by limestone dust (in the neighbourhood of a cement works) on the other, in Poland. After Grodzinska (1971), somewhat modified.

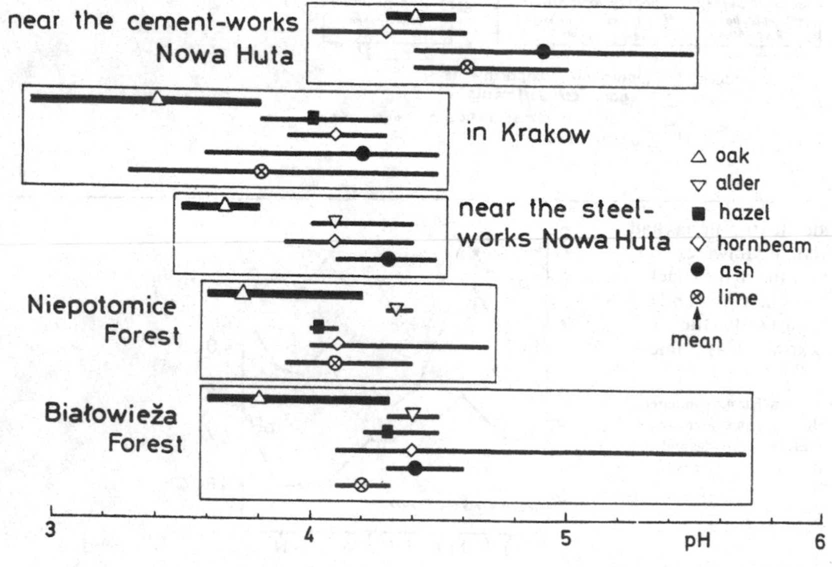

7 *Plants, animals and men as partners in ecosystems*

What we know already about the vegetation of Central Europe as well as the many consequences of environmental stresses leads to a general conclusion: plants and their communities stand in a close functional relationship with animals and man and with their common environment. Wherever plants may grow they are part of a larger operational structure, an **ecosystem** (Tansley's definition) or Geobiozönose (after Sukachev). Figure 35 shows us this structure as a system which is indeed open, but still to some extent self regulating and maintaining. This applies to all ecosystems dominated by green plants and therefore to all plant communities which are dealt with in this book.

Autotrophic plants play a central role in such a system since they capture radiant energy from the sun and convert it into chemical energy. In the form of living or dead organic matter this energy is passed on to microorganisms and animals, as well as to heterotrophic phanerogams. The activity of bacteria and fungi present is of decisive significance for the higher plants in that it releases minerals from organic remains so that these can be returned to the plant; the rest is reduced to carbon dioxide and water. This breakdown is preceded in part by the feeding of saprophagous animals which reduce the dead organic matter to a smaller size (figs. 90, 91 and 129)

Where decomposing animals and microorganisms cannot live in sufficient numbers, for example in constantly waterlogged soils which are short of oxygen,

Tab. 7. The number of plant species found as the effect of man's activities on the land increases.

Type of countryside in southern Lower Saxony	Sandy plains							Mills poor in bases							Hills rich in bases						
Number of species per 5 × 5 km (upper limit / lower limit)	151–250	251–350	351–450	451–550	551–650	651–750	751–850	151–250	251–350	351–450	451–550	551–650	651–750	751–850	151–250	251–350	351–450	451–550	551–650	651–750	751–850
I Virtually pure woodland	○		○					**O**	⊙								⊙				
II Open woodland with small settlements		⊙	●						●	○								⊙			
III —Diverse rural areas (with poor grassland, hedges, woodland, etc.)				⊙	⊙					●	○						●	●	**O**		
IV As III, but with small towns				○	○					⊙	◉							**O**	●		
V Edges of large towns (including remains of I-III)				⊙	○	⊙														⊙	◉
VI Almost entirely arable land		◉							○	○					○	◉	◉				

Note: From figures given by Haeupler (1974); the minimum numbers of vascular plants found in 5×5 km squares in southern Lower Saxony. Number of 5 × 5 km squares studied: ●1, ○2, ⊙3, ◉4, **O**5 and over.

Fig. 35. A simplified model of a complete ecosystem (i.e. a largely self-regulating interaction between organisms and their environment) for example a woodland, a meadow or a lake.

All forms of life are dependent on the abiotic factors of the environment which are operating at any particular time, such as temperature, water availability, chemicals and mechanical factors. As 'primary producers' green plants transform light energy into chemical energy on which all the 'secondary producers', i.e. microorganisms, animals and man, depend.

In maintaining a balance in the ecosystem the 'decomposers' which break down the dead remains of plants, animals and microorganisms play an important part, whether they are soil animals, fungi or bacteria. By reducing these organic remains to minerals they prevent their unrestricted accumulation. Only a small fraction of the annual production of organic matter is stored permanently as soil humus and so becomes lost to the general cycle of nutrients.

Animals which feed on living plants (phytophages = herbivores) and those feeding on other animals (zoophages = carnivores) play a less-significant role in the regulation of the ecosystem than was previously thought to be the case. Their populations are controlled in the main by environmental factors, by parasites or by internal physical or psychical limitations, or even by man.

Man fits into the food chains as a phytophage, zoophage or pantophage (eating everything); or he moulds it for his own benefit such as when he cuts wood or removes litter. However he can also consciously or subconsciously act as a 'superorganic factor' in influencing any part of the ecosystem.

The preceding sections have shown the extent to which this was and still is the case. He carries a great responsibility too for the dynamic equilibrium in almost all ecosystems in the future.

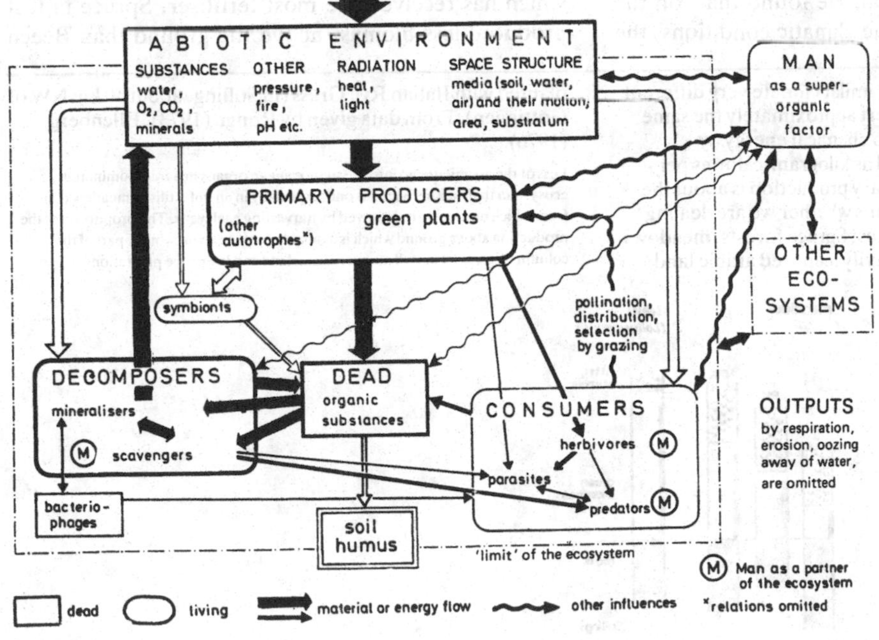

the organic substances accumulate and form mud, peat or other types of permanent humus (fig. 134). This layer covers the mineral soil and ultimately may make it inaccessible to plant roots, thus changing the whole system. In those ecosystems which do not lead to peat formation the organisms can change the environmental conditions by other means. One has only to think of the microclimate within a dense plant stand restricting the growth of its own offspring, but at the same time offering a protected place where shade-tolerating plants and animals can live without competition from those requiring more light. Plants and their associations are far better known than the other partners in the ecosystem, at least in Central Europe. Factual and practical considerations therefore allow us to put the vegetation in the foreground, and only bring in the whole ecosystem in those cases where it too is understood. This applies for example to most ecosystems in water and on the sea coast and also, more recently, to some woods and other terrestrial ecosystems (also included in the tentative classification given by Ellenberg 1973b).

During one of the most intensive ecological interdisciplinary works so far undertaken in Central Europe, the 'Solling Project' of the German Research Association (Ellenberg 1967a, 1971), a result showed up that probably has universal significance. M. Runge (1973a, b) has collected data from many co-workers in this project together with some of his own, concerning the primary production from different ecosystems over the course of the year, and has analysed the energy content of this plant material. He found that, on the same soil and under the same climatic conditions, the effectiveness of the net photosynthesis approached the same value for very different plant communities (fig. 36). If one takes the total plant production, including that of the underground parts for which it is difficult to get an accurate figure, the amount of energy incorporated by Beech woods or Spruce forests of different ages, as well as by meadows of different fertility levels, or by culture of Annual Ryegrass, in each case was round about 80×10^6 kcal per ha per annum. That is approximately 1% of the total radiant energy received, or 1.4% of that falling on bare ground during the growing period (= 3.1% of the photosynthetically active radiation).

Provided this is not just a chance result it means that the amount of radiant energy assimilated by an area of green plants is dependent on the soil and the climate, that is the site conditions, and not on the type of vegetation. A similar result was demonstrated in 1951 by Filzer who had compared the yield of different crops and forests considering the whole plants by means of conversion factors. Whether the plant stand is short or tall, whether it develops primarily above the soil or below it, whether the woodlands are nearly natural or consist of plant species introduced by man, they all have about the same capacity for converting solar energy into chemical energy provided they are completely closed, that is 'saturated' with plants. As far as production above the ground is concerned, however, in the Solling the Beech wood, which is the most natural system, is the most productive. The only other system approaching this is the Ryegrass field which has received the most fertiliser. Spruce in fact produced less biomass above the ground than Beech

Fig. 36. On similar soils and in a similar climate very different plant communities may all convert approximately the same proportion of radiant energy into chemical energy by photosynthesis. Thus, expressed as kilogram-calories per hectare per annum, the net primary production is about the same under comparable conditions whether we are dealing with Beech woods of different ages, Spruce forests, meadows manured in different ways or heavily manured arable land planted with Italian Rye Grass (in Solling, about 50 km NW of Göttingen). From data given by Runge (1973), Ellenberg (1976).

Less of the assimilates go into the subterranean organs in a tree-dominated ecosystem than in a herbaceous one, especially in an unfertilised meadow (O) from which nutrients are removed by harvesting each year. The proportion of the production above ground which is available for human use (black part of the column) is greatest in well manured grassland and in Spruce plantations.

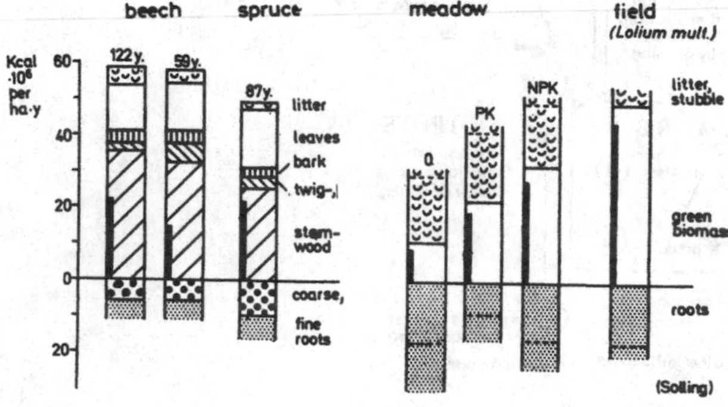

even though the cubic metres harvested by the forester were more. This is because the volume per unit weight is greater for Spruce than for Beech, and that the trunk of Spruce is straight right up to the top. Least of all in its yield of biomass usable by man is the meadow which received no fertiliser. Its ecosystem had been impoverished by the removal of hay, i.e. of organic material which can no longer be recycled, as is the case with woodland litter for example. The impoverished meadow to some extent reduces the amount of energy that can be removed from the ecosystem by subterranean storage organs as well as by large and extensive roots where it is unavailable to man. Incidently, a higher percentage of the herbaceous plants than was previously thought to be the case is changed into litter, even in meadows, and takes part in the biological cycle.

Quantitative and experimental research into ecosystems, begun by limnologists such as Thienemann and Woltereck (see Ellenberg 1973b), is still in the early stages of its development. Perhaps it will lead to the point where all vegetation units are observed and well understood as ecosystems. Then the present day hierarchy of plant communities may be replaced by a superior hierarchy of ecosystems. We do not know whether such a goal is really attainable or not, but it should always be kept in mind, especially when 'only' the vegetation is being considered.

B

Near-natural woods and thickets

I General view of the Central European woodlands

1 The main distribution pattern of the vegetation

(a) Zonal, extrazonal and azonal vegetation

As we learned at the outset Central Europe is by nature an almost unrelieved forest. The present day plant cover consists of 'replacement communities' which have arisen to a greater or lesser degree with the help of man. Although these now take up the largest area we should like to look at the individual vegetation units beginning with the woodlands. Indeed those communities which most probably are nearest to nature will be considered as giving the clearest expression of the living conditions operating in each site. We shall therefore try to present a picture of the **'potential natural'** state of the vegetation as it exists today. By this we understand the combination of species that would eventually come into existence under the prevailing environmental conditions of today if man no longer exerted any influence and if the plant succession had time to reach its final stage (Tüxen 1956a). Starting with a picture made in this way will best enable us to understand the much more diverse mosaic of the **'real'** vegetation. Of course in regions poor in woodland remains it is often not easy to define the units of the potential natural vegetation (Dierschke 1974a, Bohn 1981).

The final stage of natural succession will be different in each site according to the climatic and soil conditions. On soils which are neither waterlogged nor flooded and which do not show any other extreme properties such as a primary shortage of nutrients, the plant community present will be an expression of the climate dominating that particular area or zone. We can speak therefore of a '**zonal**' community or a 'climatic climax community' (climax meaning ladder, or the last rung of it).

The general climate may be altered by local factors, especially by the relief. Southern to western slopes are occupied by communities which require more heat, but are able to withstand drier conditions than the zonal community. As a rule they consist of combinations of species which are zonal in neighbouring areas to the south or south-east. Their occurrence in Central Europe is designated as '**extrazonal**'. Similarly those sites with a local climate which is cooler than normal have combinations of plants more or less resembling the zonal communities of more northerly districts. Units of vegetation like these which cover only a small area cannot be depicted on maps showing the overall distribution. This is even more true of special sites where the soil exerts a decisive influence.

In river valleys and on wet soils the zonal vegetation is unable to establish itself because its constituents for the most part are unable to tolerate the conditions. In such places special end stages develop which Braun-Blanquet (1928, 1964) calls 'permanent communities' (Dauergesellschaften). This term is not very appropriate because the zonal vegetation is just as, or even more, persistent as the end stages in the plant succession in the flood plains for example (figs. 186 and 3). It is better then to speak of '**azonal**' vegetation, i.e. plant communities which appear in approximately the same form in several different climatic zones because they are determined by the same extreme soil factors. This is not to say that they are completely independent of the overall climate, but that the latter exerts a less strong and obvious effect than on the zonal units. The natural vegetation of water, dunes, rocks and other special sites can be looked upon in the main as azonal.

Zonal, azonal and extrazonal plant communities form a mosaic in the landscape which can be best characterised by the zonal vegetation dominant over the greater part of the level ground. Earlier it was thought, indeed generally accepted, that the azonal and extrazonal communities were developing towards the zonal climax. This 'monoclimax theory' may now be considered disproved, although it is still being defended by some ecologists. In contrast to the earlier view, which was also the one most favoured by soil scientists, the succession and peat formation of lakeside vegetation for instance (section B V 2a), does not lead to an Oak or a Beech forest such as those dominant on the mineral soils in the surrounding area, but remains as a wet Alder wood on a peat soil (fig. 38a). It is not possible for the Alder swamp to build up the peat ground any higher above the water table. So the plants sensitive to waterlogging, which make up the zonal vegetation, cannot grow there, but only in sites which are basically drier.

In the last few decades the view has come to be accepted that zonal vegetation does not represent a single homogeneous plant association but a number of communities which differ considerably floristically and physiognomically. Right from the start the plant succession and soil formation on deep morainic loam or loess is different from that on an almost pure limestone or dolomite rock, and also different on sandstone or on a diluvian sand very poor in plant nutrients. Silicate rocks (e.g. granite, gneiss and crystalline schist) weather down and then behave in a similar way to loamy sand. The soil formation on silicate rocks normally leads to brown earth, on carbonate rocks to rendzina, on loess to parabrown earth, while sands and sandstone form acid brown earths or podsolised brown earths. As we shall see later, each of these soil types corresponds to a special type of vegetation. This is also true of marls which are rich in clay and develop in a particular way, namely to pelosoles (Mückenhausen 1977, Kohl 1971).

Where the different bedrocks and soil series occur within the same climatic zone we can expect to find at least **three** zonal plant communities (or climax communities). The zonal **limestone** vegetation, the zonal **loess or loam** vegetation and the zonal **sand** vegetation are all different, but together they make up the zonal vegetation. In accordance with the expression coined by Tüxen and Diemont (1937), 'climax group', we may speak of the '**zonal vegetation group**'. These authors were the first to recognise clearly that, for example in north-west Germany, there was no monoclimax, but a number of different climax communities on the main soil types. In north-west Germany there are three: Beech wood on limestone, Oak-Beech wood on loamy soils, and birch-oak wood with few Beeches on pure sand.

Corresponding zonal vegetation groups occur in the other parts of Central Europe, not only in the lowland, but also in every climatic belt in the mountains (section b). With this knowledge a survey of the relationships in the vegetation of Central Europe can be made clearer, if not perhaps so quickly. Figure 37 may help to give a first general idea of the situation by showing the terms already discussed alongside each other in a graphic summary.

The main divisions of this diagram give a concise explanation of the most important properties of the different sites without going into great detail. Notice in particular the 'type of humus in deciduous woodland' as being the briefest way of indicating the fertility level of the particular soil (compare Kohl 1971 and figs. 88-94):

a) **Mull** consists of clay-humus complexes in crumb form and denotes fertile soils with an adequate base content, good aeration, and a high biological activity. (The crumbs are nothing more than earthworm casts of different ages.)
b) **Moder** is a highly decomposed and active surface humus. It contains many animals, but these are much smaller than earthworms and do not work the humus layer into the mineral soil. Because of this moder is low in mineral content, but rich in plant nutrients especially nitrogen. In general it has a strongly acid reaction, its pH not being significantly higher than that of (c).
c) Mor (Danish, really Mør) or raw humus is a surface humus which is scarcely decomposed at all. Very few animals live in it and only specialised plants can live on it.
d) **Swamp-wood peat** is saturated with water from time to time and is badly aerated. However in many respects it is comparable with mull or moder in fertility.
e) **Fen peat** is even wetter, but nevertheless more or less rich in plant nutrients.
f) **Raised-bog peat** is extremely acid and poor. When saturated with water, it is the most unfavourable humus soil for root growth.

In the case of swamp-wood soils, mires and water the fertility is generally expressed by three trophic levels: **oligotrophic** (poor in plant nutrients), **mesotrophic** (intermediate), and **eutrophic** (rich). It would not be advisable to use these expressions, coined for aquatic or semiterrestrial ecosystems, for terrestrial habitats. Acid brown earth, for example, is never of low fertility under natural conditions. So the lowest level of fertility in the zonal soil sequence cannot be labelled oligotrophic, but must be mesotrophic.

The remaining pedologic and ecologic expressions which appear in fig. 37 are not absolutely necessary for understanding the next section. They will be explained later.

(b) Altitude belts of woodland vegetation

Figure 37 shows how the mosaic of the zonal, extrazonal and azonal vegetation changes according to the height above sea level. For the general view it is

Fig. 37. The dominant tree species of the zonal, extrazonal and azonal vegetation from the lowlands to the mountains in the western (suboceanic) and in the eastern (more or less continental) parts of Central Europe. (Explanation in the text).

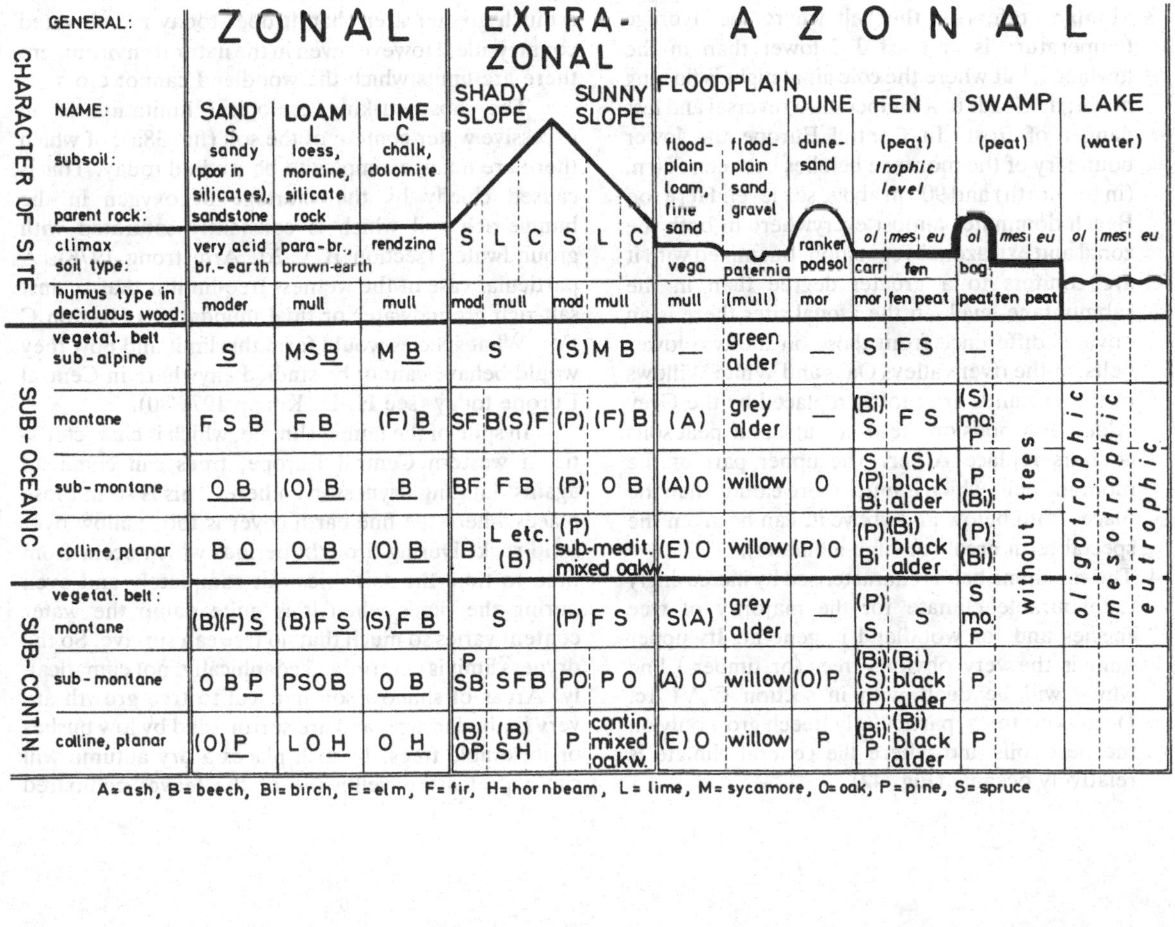

CHARACTER OF SITE													
GENERAL:	ZONAL			EXTRA-ZONAL		AZONAL							
NAME:	SAND S	LOAM L	LIME C	SHADY SLOPE	SUNNY SLOPE	FLOODPLAIN		DUNE	FEN		SWAMP		LAKE
subsoil: parent rock:	sandy (poor in silicates), sandstone	loess, moraine, silicate rock	chalk, dolomite	S ¦ L ¦ C	S ¦ L ¦ C	flood-plain loam, fine sand	flood-plain sand, gravel	dune-sand	(peat) trophic level ol	mes ¦ eu	(peat) ol	mes ¦ eu	(water) ol ¦ mes ¦ eu
climax soil type:	very acid br.-earth	para-br., brown-earth	rendzina			vega	paternia	ranker podsol	carr	fen	bog		
humus type in deciduous wood:	moder	mull	mull	mod ¦ mull	mod ¦ mull	mull	(mull)	mor	mor	fen peat	peat	fen peat	
SUB-OCEANIC vegetat. belt sub-alpine	S	MSB	M B	S	(S) M B	—	green alder	—	S	F S	—		
montane	F S B	F B	(F) B	SF ¦ B(S)F	(P) ¦ (F) B	(A)	grey alder	—	(Bi) S	F S	(S) mo. P	without trees	oligotrophic ¦ mesotrophic ¦ eutrophic
sub-montane	O B	(O) B	B	BF ¦ F B	(P) ¦ O B	(A) O	willow	O	S (P) Bi	(S) black alder	P (Bi)		
colline, planar	B O	O B	(O) B	B ¦ L etc. (B)	(P) sub-medit. mixed oakw.	(E) O	willow	(P) O	(P) Bi	(Bi) black alder	(P) (Bi)		
SUB-CONTIN. vegetat. belt: montane	(B)(F) S	(B) F S	(S) F B	S	(P) F S	S(A)	grey alder	—	(P) S	S	S mo. P		
sub-montane	O B P	PSOB	O B	P SB ¦ SFB	PO ¦ P O	(A) O	willow	(O) P	(Bi) (S) (P)	(Bi) black alder	P		
colline, planar	(O) P	L O H	O H	(B) OP ¦ (B) S H	P ¦ contin. mixed oakw.	(E) O	willow	P	(Bi) P	(Bi) black alder	P		

A = ash, B = beech, Bi = birch, E = elm, F = fir, H = hornbeam, L = lime, M = sycamore, O = oak, P = pine, S = spruce

sufficient to name the tree species which (probably) would be dominant in nature. Since the zonal vegetation of all the altitudinal belts varies with the degree of continentality of the climate, the western (or suboceanic) and eastern (or continental) areas are shown separately in fig. 37. In both areas at least four altitude belts of woodland vegetation can be distinguished (see also sections 1, 3 and 5):

1 Lowland conditions with a relatively high average temperature, but a lower precipitation favour in general the Oaks and, in continental climates on sandy soils and acid marshy ground, the Pines. This applies in particular to the **planar** level, i.e. the great plains where extrazonal vegetation has scarcely any chance of succeeding. Low lying hilly ground – known as the **colline** belt – however offers many opportunities for the establishment of extrazonal communities. In such places can be found either mixed Oak woods rich in species including submediterranean ones, or communities bearing a strongly continental imprint.

2 The **submontane** belt starts at an altitude of 200 to 300 m in the northern part of Central Europe, but not lower than 500 to 600 m in the south. The woods are generally rich in Beech except for some eastern parts beyond the area of *Fagus sylvatica.*

3 **Montane** refers to the belt where the average temperature is at least 3°C lower than in the lowlands, but where the cold air at night is flowing down; this leads to a temperature reversal and less danger of frost. In Central Europe the lower boundary of the montane belt lies between 500 m (in the north) and 900 m above sea level. Here too Beech dominates almost everywhere in both the zonal and extrazonal vegetation, but mixed with it are conifers to a greater degree than in the submontane level. In the azonal sites there is an obvious difference from those on the two lower belts: in the river valleys Oaks and White Willows of the lowlands are mostly replaced by the Grey Alder. In woods on meso- to eutrophic peat soils conifers replace Alder. The upper part of the montane belt, which is much more cloudy than the places both below and above it, can be given the special term **oreal** (or high-montane).

4 The **subalpine** belt is characterised by increasingly unfavourable climate for the majority of tree species and for woodland in general. Its upper limit is the very obvious tree- (or timber-) line which will be dealt with in section C VI 1c. Deciduous trees, particularly Beech grow only on the best soils and where the general climate is relatively oceanic (Fig. 42).

At the montane and subalpine belts there do exist level or slightly-sloping areas where the zonal vegetation can be found, but these are relatively uncommon. Extrazonal communities or variations of the zonal vegetation due to very localised climatic conditions are much more frequent. This induced Tüxen and Diemont (1937) to speak of a 'swarm' of zonal communities (or a 'climax swarm'). Examples of such a mosaic of potential natural vegetation which has been caused by local climates are given in fig. 84 which shows the communities on idealised mountain cones of calcareous and non-calcareous rocks.

Despite these local manifold variations each vegetation belt has unmistakable characters which are expressed more than anything in the combination of tree species. Since this is actually the key to understanding the woodland vegetation of Central Europe we should like to have a closer look into how it has come about. So we shall first concentrate on the deciduous forests in the submontane belt of western Central Europe, not least because there are numerous results of ecological research from this area.

(c) Wet and dry limits of woodland

In the potential natural vegetation below the climatic tree line in the high mountains, woodland would cover a much greater area than it does today in cultivated countryside. However even in the natural environment there are limits which the woodland cannot cross.

The most striking ecological limitation is an excessive water content of the soil (fig. 38a), of which there are many examples to be studied today. This is caused chiefly by the shortage of oxygen in the humus-rich soil which is constantly saturated with groundwater (section B V 2d, Armstrong 1978). A particular case of the wetness tree-limit is that against salt-rich groundwater or tidal inundations (section C IV). What species would form this limit and how they would behave cannot be studied anywhere in Central Europe today (see H.-D. Knapp 1979/80).

In spite of the humid climate, which is characteristic of western Central Europe, trees can come up against limiting dryness even here. This is so in those places where the fine earth cover is too shallow over solid rock. During drought periods which occur from time to time the soil dries out completely and even during the times when it is quite damp the water content varies so much that no tree can survive. So the dryness limit is determined edaphically, not climatically. Areas of shallow soil inimical to tree growth are very limited in size, and are surrounded by low bushes or miserable trees. In such places a dry autumn will cause premature yellowing of the leaves compared

Fig. 38a. Wetness limiting the growth of woodland along the edge of a nutrient-rich lake.

The littoral plants contribute to the build-up of mud and peat which pushes back the water frontier and enables the swamp Alder wood to extend towards the lake. Thus the Alder wood is the final stage in the terrestrialisation and no further peat is formed beyond the level to which the water rises during its fluctuations. On the adjacent mineral soils the woodland communities arrange themselves in accordance with the distance of the water table from the soil surface. The Alder-Ash wood merges into the Oak-Hornbeam wood and the mixed Beech wood. (The part of the cross-section in the water is shown with an exaggerated vertical scale.)

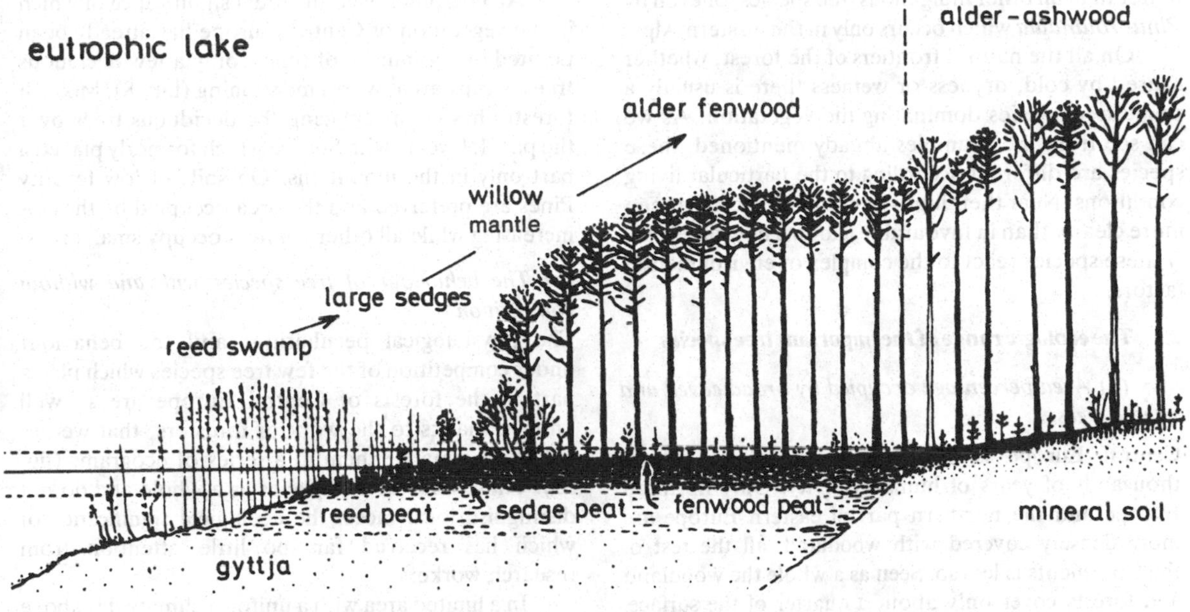

Fig. 38b. Dryness can limit the growth of woodland even in the continuously humid climate of Central Europe. This is when the amount of fine earth over rock is not sufficient to hold a reserve of water large enough to enable trees to survive the occasional drought periods.

After Ellenberg (1966) somewhat modified. Beech woods require more than about 20 cm fine soil; mixed Oak woods somewhat less. In the transition from wood to virtually bare rock shrubs form an edging 'mantle' to the woodland and outside this again a 'skirt' of herbaceous plants. (In the section the depth of fine soil is shown with a vertical scale of x10.)

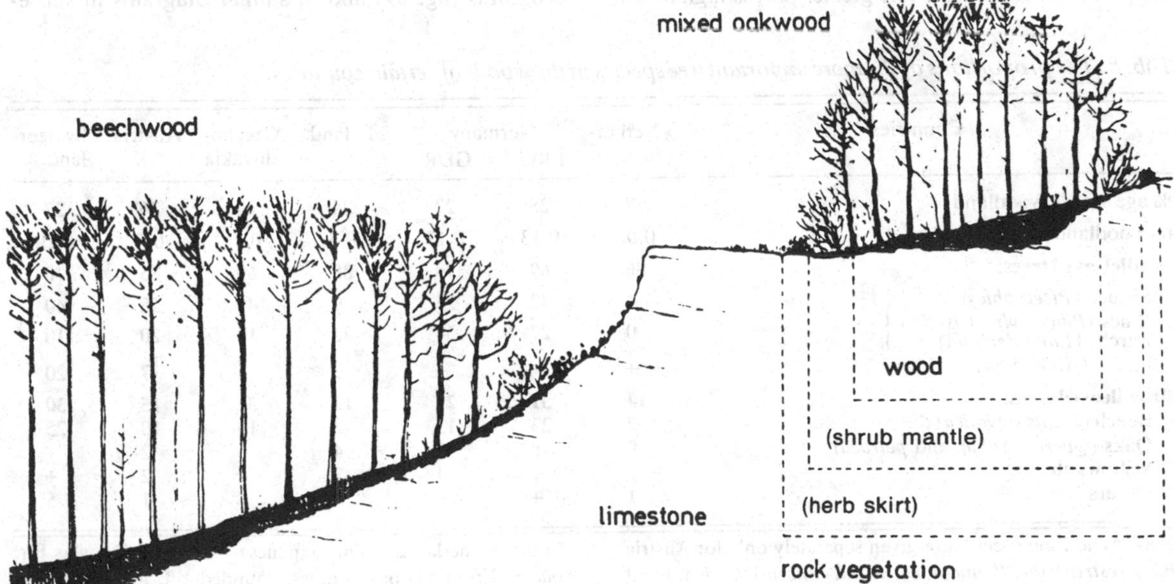

with neighbouring woodland on deeper soil. In the southern and eastern part of Central Europe the dry frontier is usually occupied by Pines (*Pinus sylvestris*, or *Pinus nigra*) which according to Künstle and Mitscherlich (1975) can withstand dry conditions better than all other indigenous tree species, or even by *Pinus rotundata* which occurs only in the western Alps.

On all the natural frontiers of the forest, whether caused by cold, dryness or wetness there is usually a single tree species dominating the vegetation. As we can see from the examples already mentioned, these species are different according to the particular living conditions. Near their limits of existence it can be seen more clearly than in favourable habitats how differently these species react to the complex of environmental factors.

2 The ecologic range of the important tree species

(a) Area percentage occupied by broadleaved and needle trees

Central Europe is still rich in woodland despite thousands of years of human activity. Only northern Europe and the northern part of eastern Europe are more densely covered with woodland; all the rest of the Continents is less so. Seen as a whole the woodland and forests cover only about a quarter of the surface, but in many mountainous areas this proportion can rise to more than half. Germany, Poland, Czechoslovakia, Austria and Switzerland together have more than 250 000 km^2 of useful forest. Of this less than 2% is coppiced and under 1% managed as coppice with standards. The greater part is high forest, but scarcely a third of it is anything like the natural forest of the area. In districts such as the northern plains, the Harz and parts of other highlands, and in the Alpine foothills the forests are dominated by alien species.

Apart from Beech, the great significance of which in the vegetation of Central Europe has already been pointed out a number of times, only a few deciduous trees occupy areas worth mentioning (tab. 8). Modern forestry has been replacing the deciduous trees over the past 150 years with Spruce which formerly played a part only in the mountains. On soils of low fertility Pines are preferred and the area occupied by these is increasing while all other conifers occupy small areas.

(b) The behaviour of tree species with and without competition

The physiological peculiarities and the behaviour under competition of the few tree species which play a part in the forests of Central Europe are so well known, thanks to the work of foresters, that we are able to summarise them in a series of ecograms (fig. 39). These refer to the species as a whole and do not distinguish local races, the ecological significance of which has received far too little attention from research workers.

In a limited area with a uniform climate it is above all the soil conditions which determine the growth and yield of the tree species. In particular the moisture and fertility of the soil are important, the latter being more or less correlated with the acidity. These two significant factors are used as the coordinates in the ecograms (fig. 39) and in similar diagrams in subse-

Tab. 8. The proportions of the more important tree species in the woods of certain countries.

Countries:	Netherlands	Germany FRG	Germany GDR	Poland	Czechoslovakia	Austria	Switzerland
% age area in woodland	7	28	27	24	34	36	27
ha woodland per inhabitant	0.02	0.13	0.17	0.3	0.4	0.5	0.19
Needleleaved trees:	**86**	**69**	**79**	**88**	**69**	**85**	**70**
Spruce (*Picea abies*)	+	42	25	9	49	58	40
Pines (*Pinus sylvestris* etc.) } Larch (*Larix decidua*) }	60	27	54	76	15	20	10
Fir (*Abies alba*)	+	+	+	3	5	7	20
Broadleaved trees:	**14**	**31**	**21**	**12**	**31**	**15**	**30**
Beech (*Fagus sylvatica*)	2	23	12	3	16	10	25
Oaks (*Quercus robur* and *petraea*)	9	8	5	4	6	2	+
Willows etc.	2	+	4	3	2	2	+
Others	1	+	+	2	7	1	5

Note: [a]The Pine species are given separately only for Austria: *P. sylvestris* 10%, *P. mugo*, *nigra* and *cembra* 1% each. Larch here amounts to 7%.

[b]In the Netherlands mainly Japanese Larch and Douglas Fir.
Source: From various sources; rounded off, all figures %, except for the second line.

quent sections. Starting at the bottom of each graph we have open water and pass upwards through very wet, wet, moist, slightly moist, and so on up to the sites on shallow soil which at times can become very dry even in the humid climate of Central Europe. From left to right are arranged the sites corresponding to an increase in the base content of the soil, that is from the most acid through to the most calcareous.

The ecograms in fig. 39 hold good for a largely suboceanic climate which corresponds to the greater part of Central Europe, and within this area to the submontane belt. Under these conditions the growth of trees is only limited absolutely by waterlogged soils on one hand, and on the other by the extremely shallow soils over rocks which dry out so easily (see fig. 38). Both limits are shown by dotted lines in the diagrams. The 'potential limits' for each tree species shown in fig. 39 are indicated by a thin line. Their physiological **optimum ranges**, i.e. the conditions of maximum growth which they could achieve where the

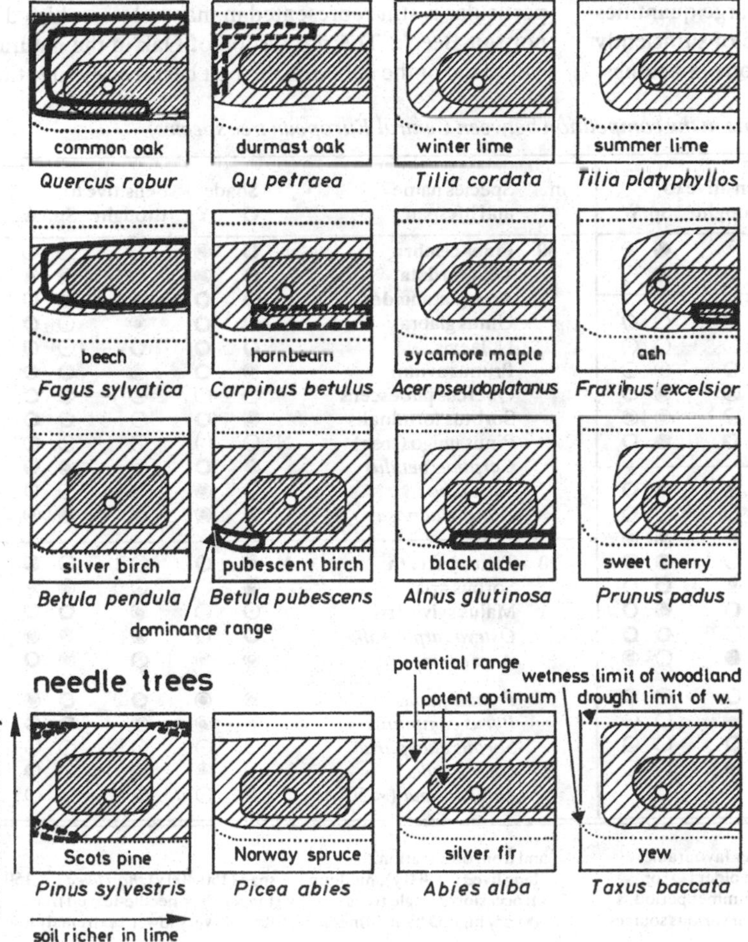

Fig. 39. The range of moisture and acidity affecting the more important tree species of Central Europe in the submontane belt in a temperate suboceanic climate.

Broad-hatched area = 'physiological amplitude' or potential range of tolerance; narrow-hatched = 'physiological optimum' range or potential optimum; area with thick black border = range where the species achieves a natural dominance to some extent under natural competition ('existence optimum' or 'ecological optimum'); a broken border = the species is co-dominant with others or (in the case of *Pinus*) this applies only in the south and east of Central Europe. For *Taxus* Leuthold (1980) gives interesting additional details.

In each of the ecograms the ordinate represents the degree of wetness of the habitat (from open water through soils of decreasing wetness to shallow soils above rock exposed to the sun which lose all moisture in drought periods – see fig. 38b). The abscissa covers the range from extremely acid to lime-rich soils. Above the upper dotted line it is too dry for any tree growth; below the lower one too wet. The small circle in the centre of the ecogram indicates the average conditions. All trees would flourish very well here but it is only the Beech which prevails under natural conditions of competition.

forester excludes competing species, is hatched in more closely than their total physiological tolerance range, their **potential amplitude**.

A fleeting glance at fig. 39 immediately shows that the conditions round about the centre of each ecogram, that is average moisture and acidity of the soil, suit all the species, and that their physiological optimum ranges more or less coincide. Under **natural competition** however the only ones to succeed are those with a relatively high growth rate and a long life span. (tab. 9). Of these in their turn the ones which can both tolerate and produce deep shade, the shade-bearing trees, are more competitive than the light-demanding trees (tab. 9). Of the shade trees Beech and Silver Fir are superior provided that the soil and climate suits them. Between the two the deciding factor is the speed of their sapling growth, and in this respect Beech has a definite advantage in the relatively mild climate of the submontane belt.

Beech can exclude all or almost all other trees from its optimal range. Only Ash, Sycamore and Norway Maple, because they produce large quantities of seedlings every year and are capable of sufficiently rapid growth, have any prospect of growing amongst them. These species are found scattered in Beech forests which are very like natural woodland. Where Beech is restricted by occasional waterlogging of the soil these other species can become dominant if at the same time the soil is sufficiently fertile to satisfy their rather large demands. In turn they drive out the Alder, which would make its maximum growth under just such conditions, into the wettest peat or partly peaty sites. Here, although close to its own physiological water frontier, this quick-growing but short-lived Alder species can achieve complete dominance (figs. 38-40).

Birches, especially *Betula pubescens*, can dominate on wet soils only when these are very acid, because here all other species flourish even worse. However since their seeds are easily dispersed by the wind and also can germinate on the poorest soil, both Birches are able to gain a footing in drier situations as pioneer trees.

Other light-demanding tree species live longer than Birches (tab. 9). Because of this property Oak in particular is still represented in many places which do not suit Beech. The proportion of Oak in the natural woodland of the submontane belt becomes greater the

Tab. 9. Some of the more important factors in the competition between Central European tree species.

m	Species name and lifespan [1]		Shade O	Shade Y	Sensitive to drought	Sp.	W:
>60	Abies alba	N	●	●	◉	●	◉
	Picea abies	N	**O**	◉	●	◉	⊙
>40	**Pinus nigra**	N	◉	○	○	○	**O**
	P. sylvestris	N	○	○	○	○	⊙
	Larix decidua	N	○	( )	**O**	◉	○
	Quercus petraea		◉	⊙	◉	◉	**O**
	Q. robur		⊙	○	○	◉	◉
	Fagus sylvatica		●	●	**O**	●	**O**
<40	**Tilia platyphyllos**		**O**	**O**	◉	◉	**O**
	Ulmus minor		◉	◉	⊙	⊙	**O**
	Acer pseudoplatanus		**O**	**O**	◉	◉	◉
	Castanea sativa		◉	◉	⊙	●	●
	Sorbus domestica		**O**	**O**	◉	**O**	**O**
	Fraxinus excelsior		◉	**O**	**O**	●	**O**
	Quercus cerris		⊙	⊙	⊙	**O**	**O**
	Populus nigra		⊙	◉	●	⊙	◉
	P. alba		⊙	◉	**O**	◉	**O**
	P. tremula (in East)		⊙	⊙	◉	○	⊙
	Betula pendula (″)		○	○	**O**	○	⊙
	B. pubescens (″)		○	○	●	○	⊙
<30	**Pinus cembra**	N	**O**	◉	○	○	○
	Tilia cordata		**O**	◉	⊙	◉	◉
	Acer platanoides		**O**	**O**	◉	◉	**O**
	Ulnus glabra		**O**	**O**	◉	⊙	**O**
	U. laevis		**O**	**O**	**O**	⊙	**O**
	Prunus avium		◉	**O**	◉	**O**	**O**
	Quercus pubescens		⊙	○	⊙	**O**	**O**
	Sorbtus torminalis		◉	**O**	⊙	**O**	**O**
	Pinus mugo (tree)	N	○	( )	○	○	⊙
	Carpinus betulus		●	**O**	◉	◉	**O**
	Salix alba		⊙	◉	●	⊙	**O**
	Alnus glutinosa		◉	◉	●	◉	◉
<20	**Taxus baccata**	N	●	**O**	⊙	⊙	●
	Sorbus aria		◉	⊙	⊙	◉	◉
	Malus sylvestris		⊙	○	◉	**O**	**O**
	Ostrya carpinifolia		**O**	**O**	◉	●	◉
	Acer campestre		◉	◉	⊙	◉	**O**
	Prunus padus		◉	◉	**O**	**O**	◉
	Pyrus communis		◉	◉	◉	**O**	**O**
	Sorbus aucuparia		⊙	⊙	⊙	○	⊙
	Salix fragilis		⊙	◉	●	⊙	**O**
	Alnus incana (< 50y.)		◉	⊙	●	⊙	⊙

Note: Maximum height and maximum life span under favourable conditions; the ability to produce shade when grown older (O) or tolerate shade when young (Y); sensitivity to a dry summer period, a late spring frost (Sp) or a keen winter frost (W). From various sources and from observation.

[a] **long lived** (> 400 y), medium life span (150-400 y) *short lived* (< 150 y); occasional single trees may live longer. N = needle-leaved trees

[b] ● very high **O** high, ◉ medium, ⊙ low, ○ very low, ( ) extremely low

poorer and drier or the wetter the soil, given that Oak itself can still flourish there. Correspondingly there are mixed Oak woods which can put up with dry conditions but still require a high base content in the soil (the so-called xerothermic mixed Oak woods), other Oak woods which can stand both dry conditions and an acid soil (the dry Birch-Oak woods) and also mixed Oak woods on wet soils. The latter may be found both on soil poor in bases (wet Birch-Oak woods) or on base-rich soil (wet Oak-Hornbeam woods).

Where the climate is humid and at the same time warm enough – as is always the case in the submontane belt – the deciduous trees flourish on almost all soils and remain more than a match for the conifers. Only towards the limits of existence for woodland, generally speaking, does the frugality of Pine give it a chance to become dominant either in very dry sites, whether calcareous or acid, or at the edge of raised bogs which are extremely oligotrophic (fig. 40).

Compared with water content, fertility and temperature other environmental factors under natural conditions play a minor rôle in the competition between tree species. At times the struggle for existence may not be mainly that between the adult trees, but is related to their powers of regeneration. Beech, in most ways so successful, may serve as an example. Like many other tree species it does not bear fruit every year (fig. 41). According to Gäumann (1935, and see also Burschel 1966a) it requires about two-thirds of the total carbohydrate assimilated in one year in order to produce a full crop of seed. In unfavourable habitat conditions it may build up this stock of chemical energy only at very infrequent intervals. However only a 'full mast' year leads to an appreciable number of seedlings. When there is only a half, quarter or a 'scatter' crop (the fruiting of single trees), practically all the seeds are eaten by mice or other rodents, Wild Boar or birds before the time for germination in the spring (Watt 1923). Even after successful germination the seedlings are threatened by many dangers. According to Burschel, Huss, and Kalbhenn (1964) for example in 1961 between one- and two-thirds of the young plants were destroyed by disease, had been eaten by deer, mice or beetles. Added to this the fact that in some years the flowers could be killed by a late frost or the seedlings destroyed by summer drought, a number of decades might go by before there was a successful natural regeneration. In this connection it is possible to see the main reason why Beech, in spite of strong growth, suddenly dies out at the limit of its distribution in eastern Europe (Steffen 1931, Wachter 1964 and Vilmos 1965). In the submon-

Fig. 40. An ecogram showing the tree species which form the woodlands on soils of varying moisture and acidity in the submontane belt of Central Europe in a temperate sub-oceanic climate. The larger the print the greater the degree to which the species takes part in the tree layer of the potential natural vegetation, as would be expected as a result of competition undisturbed by man (compare fig. 39). In brackets = only in some parts of Central Europe.

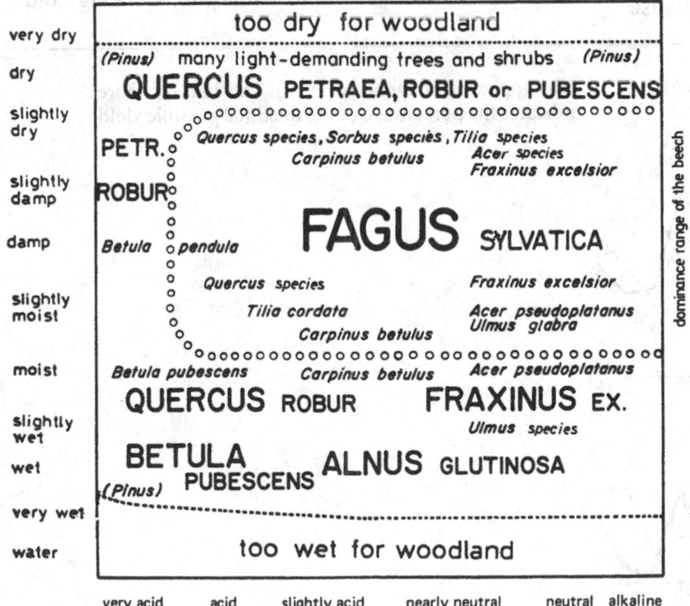

tane belt however neither the Beech nor any other native tree is threatened by such extreme conditions. Here, therefore, competition is almost always vegetative, especially amongst the young trees.

It is an advantage to all species that even within a small population they vary phenologically, genetically and ecologically. Especially at the limits of their distribution and at higher altitudes there are local races. Even in Belgium for example a subpopulation of Beech can be distinguished (Galoux 1966). Such variability increases the ecological range of the whole population without however obscuring the differences between the species (Kalela 1937).

The result of competition between the trees is shown in broad outline for the submontane belt in fig. 40. In this ecogram only those species are named which are dominant in the uppermost layers of the forest, along with a few of their more important companions. When this diagram is compared with the ecographs of individual species in fig. 39 it shows quite well that only the Beech and perhaps also Sycamore and Ash are able to assert themselves within their physiological optimum range, whereas all other species have to put up with soil conditions which are second best. The make up of our forests then under natural conditions is in no way a simple function of the inorganic environment, but much more the result of competition between those species which can live in that climate, for light, water, nutrients and other living conditions associated with the given site.

When the forester interferes with this competition by reducing or thinning out unwanted species, then such species as Pines and Oaks, which would otherwise have been suppressed, are enabled to occupy their physiologically optimal habitats. Here they attain a significantly higher yield than in those places where they achieve dominance under natural competition.

The outline of the most important combinations of tree species given in fig. 40 applies only to those habitats away from river valleys. In areas adjoining the larger watercourses which are liable to flooding there are so many diverse degrees of dampness and fertility that these must be dealt with separately (section B V).

(c) Changes in the balance of tree species brought about by climate

When one climbs from the submontane to the montane belt it can be seen that the competitiveness of Beech becomes even stronger while that of Oak and other more thermophilous trees decreases. The most obvious difference however in many mountainous areas is the advance of the Fir which to some extent takes over the role of the Oak in that it occupies those habitats which are not optimal for Beech (fig. 42). Such 'conifer-supporting' conditions, to use an expression of Kuoch (1954), are mainly on bedrocks which are poor in lime and tend to hold water. The driest places in the montane belt are occupied by Pines, but these areas are only very small. On wet soils of the montane belt Alder and Birch lose their significance and are replaced by Spruce, Fir or Pine (fig. 37).

In the upper montane and subalpine belts of the majority of mountains in Central Europe (the exceptions being those in the south-west) Beech has to surrender its dominant rôle to Spruce or other conifers which are better shaped to withstand the long cold

Fig. 41. The yield of seed from the Beech (*Fagus sylvatica*) varies from year to year more than that from other tree species and in some years can be practically nil. After Schwappach from Dengler (1930). The 'harvest figure' indicates the percentage of the maximal possible yield.

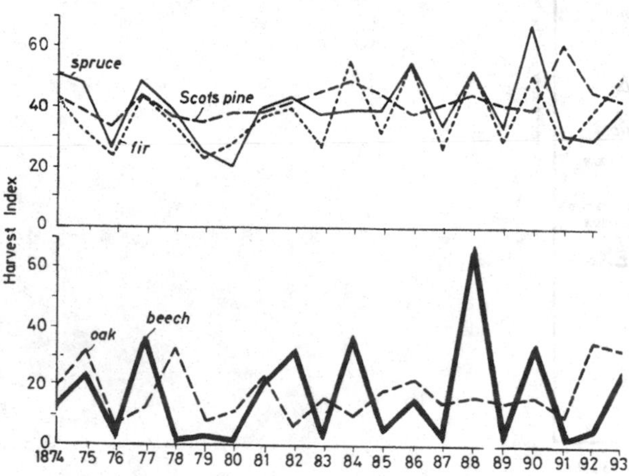

Fig. 42a. A simplified schematic cross-section of the central Alps (from about the western Swiss Jura as far as the Tessin), showing the altitudinal belts and the natural dominant tree species on average sites. Taken in part from Kuoch (1954) and Ellenberg (1966).

The snow and tree lines and the boundaries between the submontane, montane and alpine belts are higher towards the centre of the Alps as a consequence of the higher land mass and the more continental climate with its increased radiation. Here the keener winter and late frosts as well as the dry conditions exclude the Fir and, even more so, the Beech and other broadleaved trees. The tree limit in the central Alps is occupied by Larch and Arolla Pine or, in the south, by Larch alone. In contrast to the conifer-dominated inner Alps both the southern and northern outer Alps are rich in broadleaved trees. In the north the tree line is occupied by Spruce, but in the south by Beech which plays a large part in all the oceanic mountains of Europe (see fig. 6).

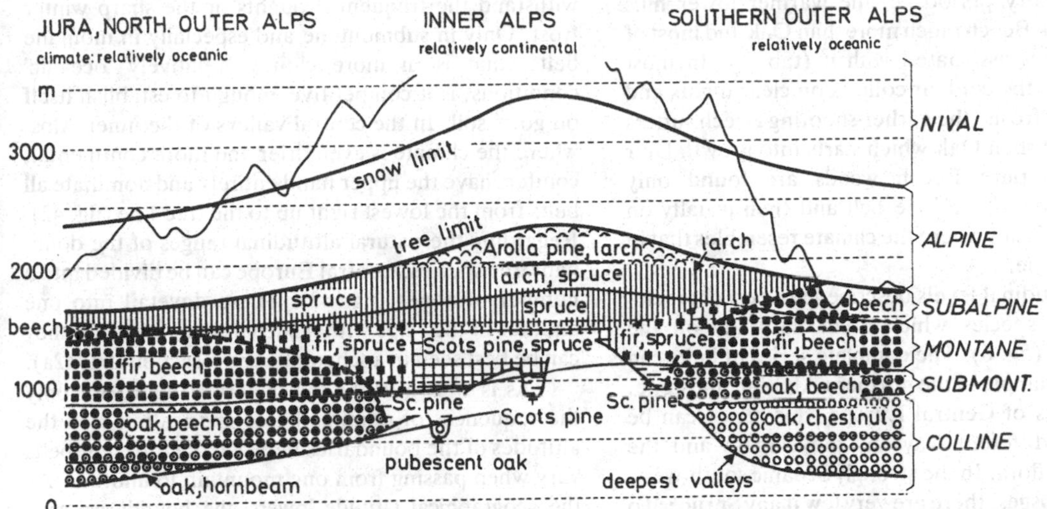

Fig. 42b. A very much simplified schematic section along the northern part of the Alps from tne submediterranean French Alps to the eastern Alps in Austria. In part taken from Wagner (1966). In the central section the altitudinal sequence of the vegetation is 'Central European' and corresponds to that shown in fig. 42a.

Because of the increasingly higher temperature of the general climate the upper limits of the different belts rise somewhat towards the south-west. However the greatest deviations occur lower down in the colline belt where the mesophilous mixed Beech-Oak-Hornbeam woods are replaced by more drought-tolerant submediterranean mixed Downy Oak woods and finally by the mediterranean evergreen hardwoods with Holm Oak (*Q. ilex*), which are better adapted to the increasing summer drought (see also fig. 4). Beech here can only achieve dominance in the cloud level while higher up Fir, Spruce and Mountain Pine (*Pinus rotundata*) take over as the climate in summer is often too dry for Beech.

Towards the east also the needle-leaved trees take on a greater significance in the outer Alpine region, not only at higher levels but also lower down. Spruce especially becomes more competitive since the general climate, even in the outer Alps with their higher precipitation, is more continental. In some places in the south-east corner of the Alps the Austrian Pine (*Pinus nigra*) plays a part (see section B IV 6 b) while here and there the southern-continental Turkey Oak (*Q. cerris*) occurs.

A section through the length of the central Alps would show almost entirely coniferous woodland with the exception of the western end where Downy Oak and evergreen broadleaved trees penetrate into the central valleys (outposts of the Downy Oak are shown in the transverse section in Fig. 42a). In the western central Alps above the mediterranean colline level there is a montane Scots Pine (*P. sylvestris*) belt, and above this are subalpine Larch or Larch-Arolla Pine woods. In the centre of the inner Alps Spruce comes in between the Pine and the Larch, and in the eastern Alps as well as in the central massif it dominates right up to the tree line.

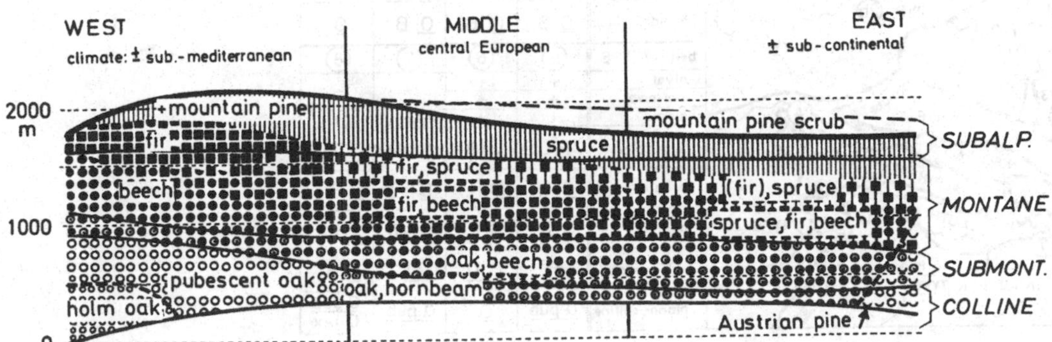

winters than the tree which is so all-powerful at lower altitudes (tab. 9).

At the lowest levels too the competitive ability of Beech tends to decline. In the colline belt, known as the vineyard level in southern Central Europe, it is noticeably weaker than in the montane and submontane. This is due more than anything to the increasing incidence of dry periods in the warmer lower hills which set back Beech much more than Oak and most of the other trees associated with it (tab. 9). In most basins, where the cold air collects on clear nights and produces late frosts, the earlier-shooting Beech suffers more damage than Oak which starts into growth later (tab. 9). So pure Beech woods are found only exceptionally in the colline belt and then usually on north-facing slopes where the climate resembles that in the submontane.

The altitudinal levels described are best characterised by the species which occur naturally on the 'average' soil (fig. 6). The complete series of belts can be seen for example in the northern Alps on limestone. In other parts of Central Europe, this series can be modified in different ways by the climate and the history of the flora. In the strongly oceanic south-west, such as the Vosges, there are very few if any Spruce. To the north of Central Europe the Fir is absent as a competitor of Beech and is partly replaced by Spruce, for example in the Harz and in the eastern part of Thuringia Forest. The choice of tree species in the north-west is even poorer where, because of the strongly oceanic influence, Pines do not occur naturally or are restricted to bog margins. In the continental climate of the east, such as in southern Poland, almost all tree species are present, but Beech is suppressed in both lowland and hilly areas because it cannot withstand the frequent droughts or the sharp winter frost. Only in submontane and especially in montane belts, that is in more cloudy, relatively 'oceanic' conditions, is it competitive enough to establish itself on good soil. In the central valleys of the inner Alps, where the climate is even drier and more continental, conifers have the upper hand entirely and dominate all belts from the lowest right up to the tree-line (fig. 42). Regarding the natural altitudinal ranges of the dominant tree species, Central Europe can be divided into a number of zones (fig. 43). These dovetail into one another in the Alps within such a small area that they can be best seen in a schematic cross-section (fig. 42a).

As is shown in the lower part of fig. 6, not only the sequences of dominant tree species but also the altitudes of the boundaries between the different belts vary when passing from one mountain to another. As the geographical latitude lowers, the boundaries as a rule increase in altitude because the average temperature is greater and the growing period lasts longer at the same altitude (section Ib). Many boundaries,

Fig. 43. In almost every group of mountains the natural vegetation belts are somewhat different, but nevertheless typical sequences may be identified. In Central Europe 7 types and several subtypes can be distinguished. After Haeupler, modified. For definitions of abbreviations, see fig. 37.

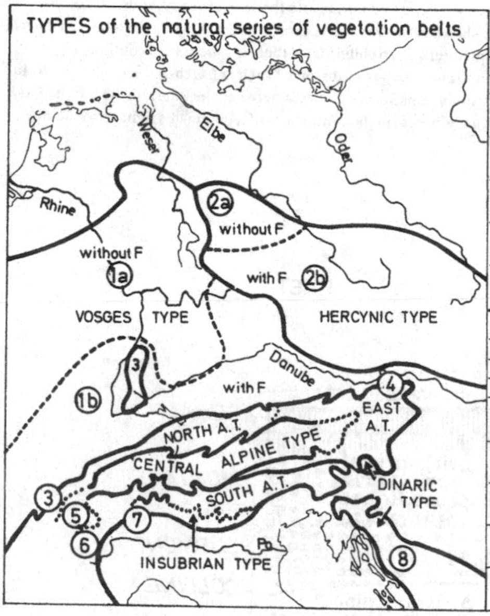

TYPES of the natural series of vegetation belts

belt / TYPE:	①	②	③	④
nival	–	–	(+)	(+)
sub-nival			+	+
alpine			+	+
sub-alpine	B(M)	S	S	S(L)
oreale	B F	S	S(L)	S F L
montane	B F	B (F)	B F (S)	B F (S L)
sub-montane	B(O)	B O	B	B (O)
planar, colline	O B	O (B)	O B	O

belt / TYPE:	⑤	⑥	⑦	⑧
nival	+	(+)	–	–
sub-nival	+	+		
alpine	+	+	(+)	(+)
sub-alpine	AP L	L (B)	B (M)	B (M)
oreale	S L	B (L)	B (F)	F B
montane	S P	B F	B (O)	B
sub-montane	P	B O	O	O B
planar, colline	Q.pub.	O	Q.pub.	Q.pub Q.ilex

particularly the upper ones, are higher where the climate is more continental without the mean annual temperature being higher. The higher boundaries in this case are due more to fewer clouds leading to greater radiation from the sun and higher summer temperatures. The inland climate with its greater amount of sunshine (even though here temperature and moisture are more changeable), and also the climate of the central alpine region, are distinguished in this way from the much more overcast regions round the borders of the Alps and along the coasts. We shall come back to this feature and go into the remaining factors when we discuss the subalpine Arolla Pine-Larch woods and the alpine tree- and snow-lines.

This survey of the distribution of the most important tree species and the interplay of edaphic and climatic factors which have brought it about has purposely been kept short in order to bring the main features of the picture out more clearly. In the remaining sections we shall become more familiar with the individual characteristics of the species composition of woodland communities.

3 The site conditions and ecological behaviour of the undergrowth

(a) Living conditions inside deciduous and coniferous forests

So far in our survey of the woodland zones and altitudinal levels in Central Europe we have only considered the trees because these are the real constructors of the woodland community. The undergrowth is largely shielded from the outside climate and lives in a microclimate which is not suitable for most of the Central European plants. In contrast with open country the **forest interior shows the following features**:

1 The light intensity, at least in the warm season, is more or less substantially reduced. Plants which require full or nearly full light in order to carry on their photosynthesis are thus starved in the shade of the forest. In some places, or from time to time, the light intensity is not even high enough for many 'woodland plants' (fig. 117). On the other hand the shade-tolerant species are protected in such places from competition by the plants which require light and so, indirectly, are favoured.

2 The canopy restricts heat exchange between the forest atmosphere and the outside air so there is much less temperature variation in the soil and the air throughout the day or throughout the year than there is in the open (fig. 44).

3 Along with the moderate temperatures there is also less fluctuation in humidity as far as the maxima and minima over a longer period of time are concerned (fig. 433). The forest climate is much more 'oceanic' in both temperature and humidity than that of the surrounding area. This means that the undergrowth communities do not indicate the degree of continentality of the general climate as is the case with the tree layer.

4 The water economy of the undergrowth is less expensive since the wind speed in the wood is always less than outside (fig. 45). In the same soil and climate therefore the plants of the woodland floor are more strongly hygromorphic (or less xeromorphic) than those of the open country.

5 On sunny days the short-term fluctuations in radiation intensity, and with it the relative humidity, can be considerably greater for the woodland plants than for those in the open, a situation which is not usually made clear. A large patch of sunlight moving across an area could raise the amount of light it received to perhaps 80% of that in the open, in contrast to the 1 – 10 % which it would normally receive (fig. 58). The effect of this strong irradiation would be to raise the temperature and the saturation deficit of the air very rapidly which for sensitive plants, for example many ferns, could be the limiting factor (fig. 75).

6 Part of the precipitation is retained by the canopy, the proportion depending partly on the nature, age and density of the stand, but also on the strength, form and duration of the precipitation (tab. 23). This interception is generally more effective in the oceanic lowland than in the mountains or in the more continental areas where the precipitation is more often in the form of a downpour.

7 The tree layer competes with the undergrowth for water and minerals since both produce a dense root growth in the upper layers of the soil. Very often it is not shortage of light but the removal of water which leads to the death of tree seedlings or herbaceous plants.

8 On the other hand the deep-rooting trees can extract minerals from the subsoil, benefitting the shallow-rooting plants through the leaf litter which is decomposed by soil animals, fungi and bacteria (fig. 115 and tab. 22).

9 Leaf litter as such can damage many plants on the forest floor by covering over their assimilating organs (for example many mosses) or by producing toxic substance near their roots.

The intensity of these different influences on the undergrowth, however, varies widely, depending on the species of trees and the density of the canopy. The

main differences between the broadleaved and coniferous forests are the following:

a) Deciduous broadleaved trees – and only these play a role in Central Europe – come into leaf a few weeks later than the undergrowth starts into growth (fig. 46). In spring therefore the latter enjoys a time of more light and warmth than later when the leaves are fully out. In autumn too, many plants can make ue of the light which now once again is reduced in intensity by only some 50% (fig. 118). In evergreen coniferous woods on the other hand the shade cast by the crown varies only according to the position of the sun. In any case radiation is less in the cooler seasons than in summer. It follows then that deciduous woodland is generally richer in spring-flowering plants whereas in coniferous stands the undergrowth is at its best in summer.

b) One effect on the undergrowth which may be attributable to the type of canopy is the way in which the light spectrum is changed as it passes through the leaves. The light quality through broadleaved trees differs from that through coniferous trees (Knuchel 1914). It has however not yet been demonstrated that this factor has sufficiently strong physiological influence to affect competition on the forest floor.

c) Just as the shading effect of evergreen conifers is more even throughout the year, so is the interception of precipitation more evenly

Fig. 44. Daily fluctuations ot soil temperatures in the open and under various woodlands during the time of warming up in the spring (5-13 April 1967). After Mitscherlich and Künstle (1970).

The daily variation is greatest where the trees have been clear felled (in the open) especially at the surface but also at depths of 10 and 30 cm. Under a mixed broadleaved (deciduous) woodland (area IV, Pine-Beech-Oak) the temperatures fluctuate more and the soil warms up more quickly before the trees come into leaf than under an evergreen needle-leaved woodland, especially where this is very dense (area III, Douglas Fir lightly thinned).

It should be noted in addition that the sequence can change during the course of the summer in the different test areas. However the soil in the mixed broadleaved woodland always remains warmer than in any of the other test areas including the clear felled one. According to Mitscherlich and Künstle this is mainly due to the insulating effect of the leaf litter on the soil.

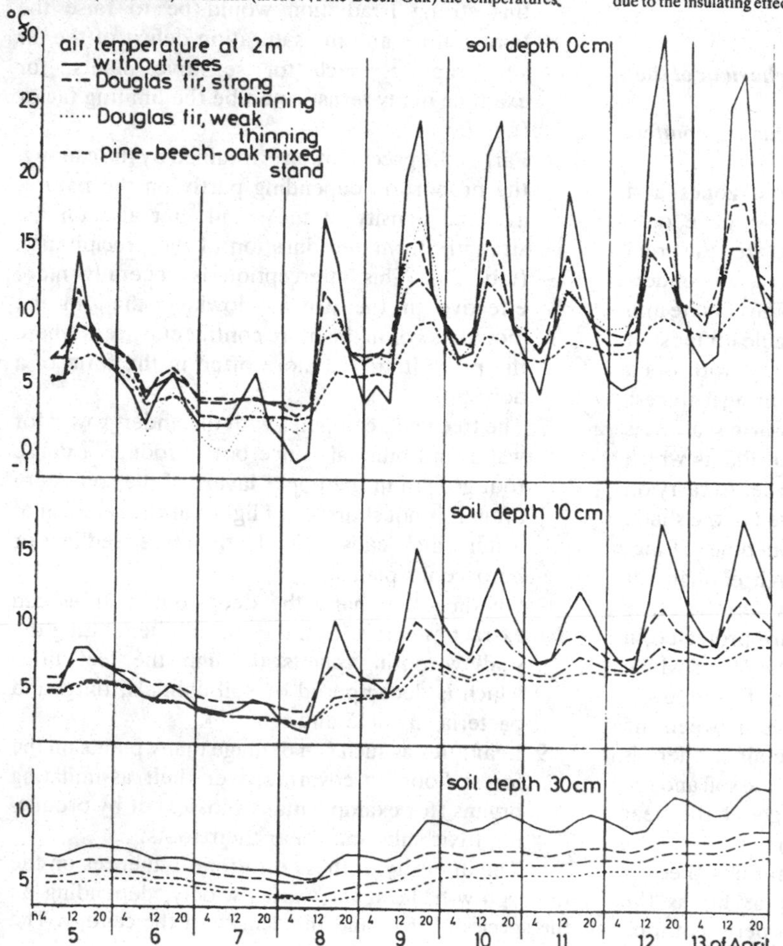

distributed over the ground. In the case of broadleaved trees, particularly Beech with its smooth bark, part of the rainfall runs down the trunk causing local wetting, and incidentally a higher acidity, where it reaches the soil (fig. 79). With conifers, especially Spruce, on the other hand practically no water runs down the trunks (tab. 23).

d) The litter from broadleaved trees differs from that from conifers including the deciduous Larch in that it consists of relatively large leaves which fall within a few autumn weeks, that is, all at about the same time (fig. 47). In doing so they cover over mosses and other low, slow-growing plants every year just at a time when the amount of light available would be greatest. Such plants are thus effectively smothered whereas the much narrower needles from confiers, falling over a much longer period of time, can be overgrown by mosses and lichens. As a rule then, broadleaved forest is poor in mosses, coniferous forest much richer. There are exceptions: firstly, open deciduous woodland which may be slow growing, where the litter does not completely cover the ground; and secondly, a highly productive deciduous woodland over very fertile soil where the layer of litter is quickly decomposed, although just those vegetation types yeld the largest quantities of litter (Bray and Gorham 1964). On the open edges of woodland or on freely exposed west or south-west facing slopes the wind may blow the leaves away creating small areas where mosses can grow abundantly (tab. 18).

e) Chemically the needle litter from conifers, particularly Spruce and Pine species, is less desirable than broadleaf litter. It contains more lignin and resin and therefore is decomposed more slowly and forms a cover of moder or raw humus on the soil surface. Fir is closer to the broadleaved trees in this respect while smong the latter Beech behaves most like the conifers (fig. 48). The formation of a layer of humus in those places

Fig. 45. Reduction of wind speed brought about by an extensive Beech wood on the plateau of Solling. After Kiese (1972), modified.

The gradients change according to the wind speed above the canopy but still maintain their characteristic form. It is always least windy close to the ground. Compared with a measuring point 5 m above the highest twigs the leafy canopy reduces the wind speeds to about one sixth, but without leaves only to about a half or a third.

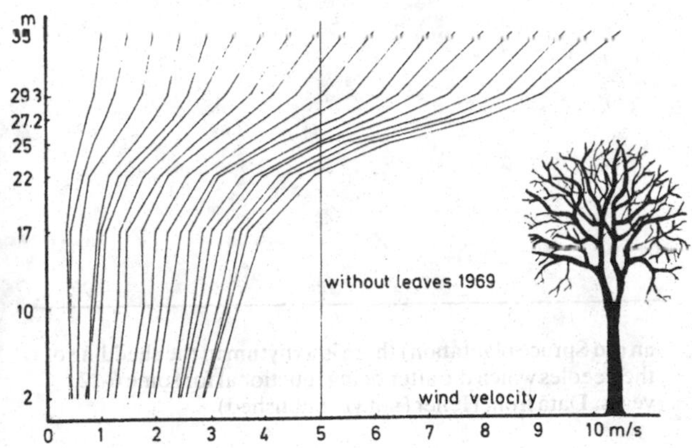

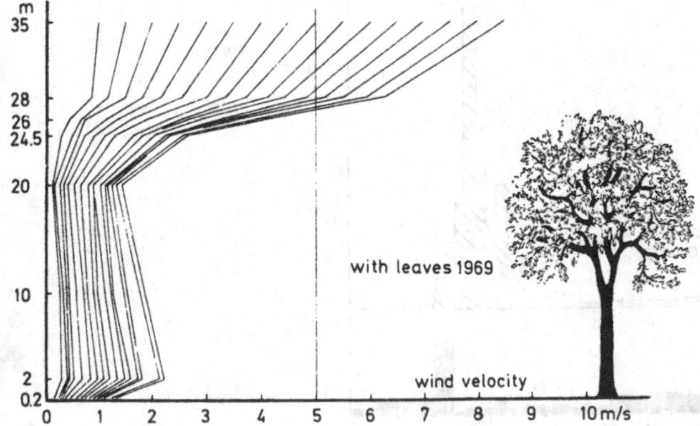

Fig. 46. The phenological development of different species in the tree, shrub and herbaceous layers of a moist-soil Oak-Hornbeam wood. After Ellenberg (1939), somewhat modified.

Black = leaves of the present year, horizontal lines = overwintering leaves, vertical lines = flowers. Average of three years of observations.

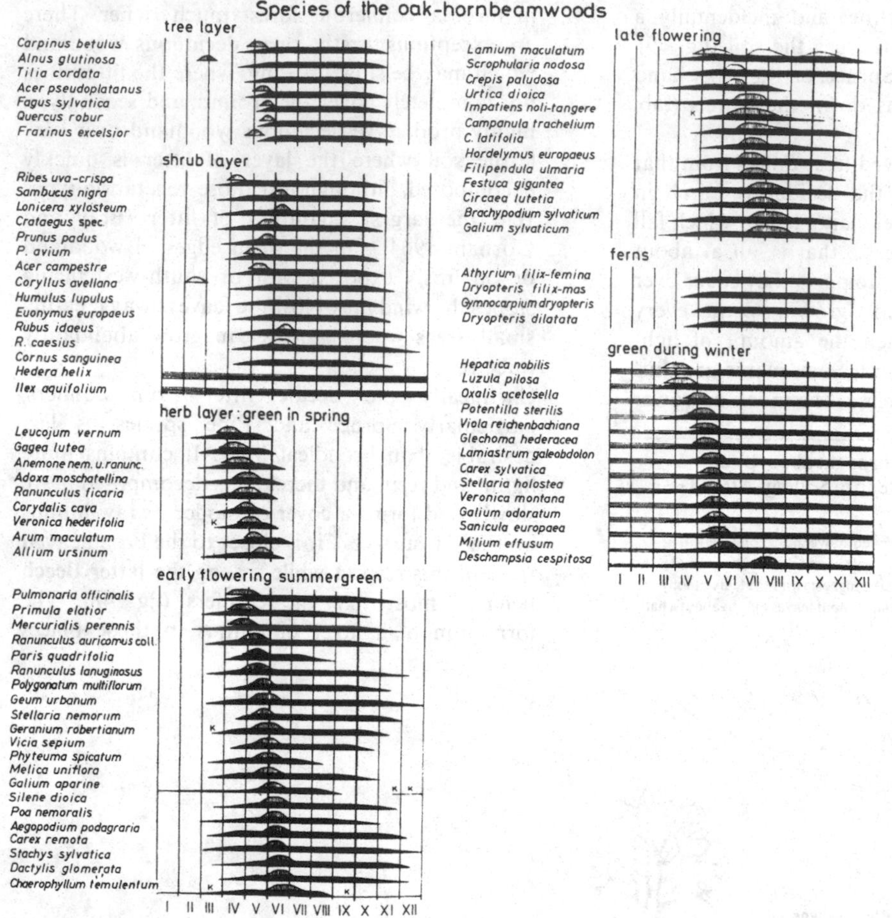

Fig. 47. In a broadleaved woodland (such as the 120 year old Woodrush-Beech wood in the Solling) the dead leaves fall within a period of a few weeks at the end of the growing season. In an evergreen needle-leaved woodland also (as in an old Spruce plantation) there is a rhythm in the shedding of the needles which die after being functional for some 6-12 years. Data from Heller (not yet published).

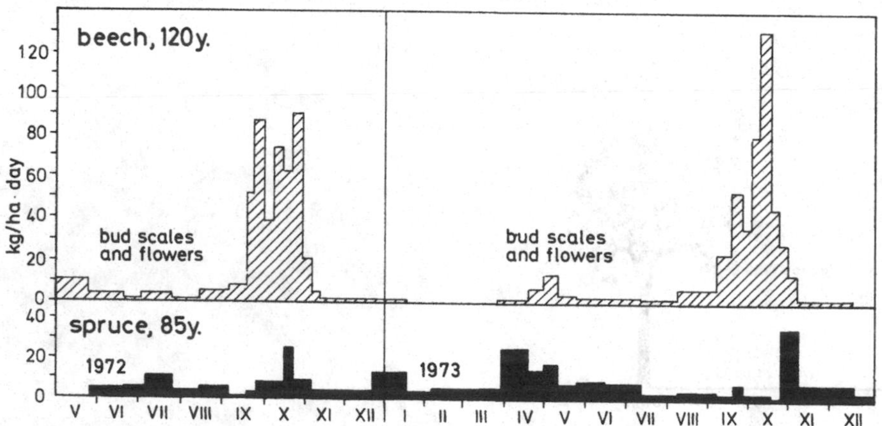

where Spruce or Pine have been planted in what was formerly deciduous woodland is the main reason why the ground flora quickly changes and those species are favoured which have their widest distribution in natural coniferous forests (section IV).

f) As a consequence of the different forest climates and the creation of their characteristic type of humus layers, the deciduous and coniferous forests can be distinguished by their fungus population, both qualitatively and quantitatively.

g) The same is true for other organisms in the woodland community, especially for the animals which live on or in the soils.

There are some species of conifers and of deciduous trees which as saplings can withstand a lot of shade, and when they are fully grown allow very little light to filter down to ground level. These are the extreme shade trees such as *Fagus* and *Abies*. At the other extreme are the sun trees for example *Betula* and *Pinus sylvestris*. The majority of other tree species can be ranked between these. However there are some species like Yew (*Taxus*) which let very little light through, but as young plants require more light than for example Abies (Hansen, verbal communication). Strictly one should make a clearer distinction between them than is usually done (but see tab. 9). The **different behaviour of the various species** modifies the effects listed under 1 to 9 and a to g above in roughly the following ways:

α) Sun trees like Oak and Pine create a woodland climate which resembles that of the open country. Only where the soil is very fertile is growth of bushes and shrubs under the trees favoured, so these have the same effect on the ground flora as a stand of shade trees (fig. 50).

β) The difference between broadleaved trees and conifers is less strongly marked amongst the light-demanding trees than it is with shade trees. In habitats where there is a low production potential the undergrowth in, say, a pure Oak wood, or even a stand of Birches, lives under very similar conditions as under Pines. This applies above all to mosses (point d), which flourish in deciduous woodland in direct relationship to the amount of light as a rule. Because of this there are practically no ground species in Central Europe which

Fig. 48. The average time taken for a natural breakdown of the fallen leaf litter of different species of broad- and needle-leaved trees under comparable conditions, i.e. on a medium brown earth. The figures on the left show the C/N ratio of the freshly fallen leaves, those on the right their pH values. After Scheffer and Ulrich (1960), somewhat modified.

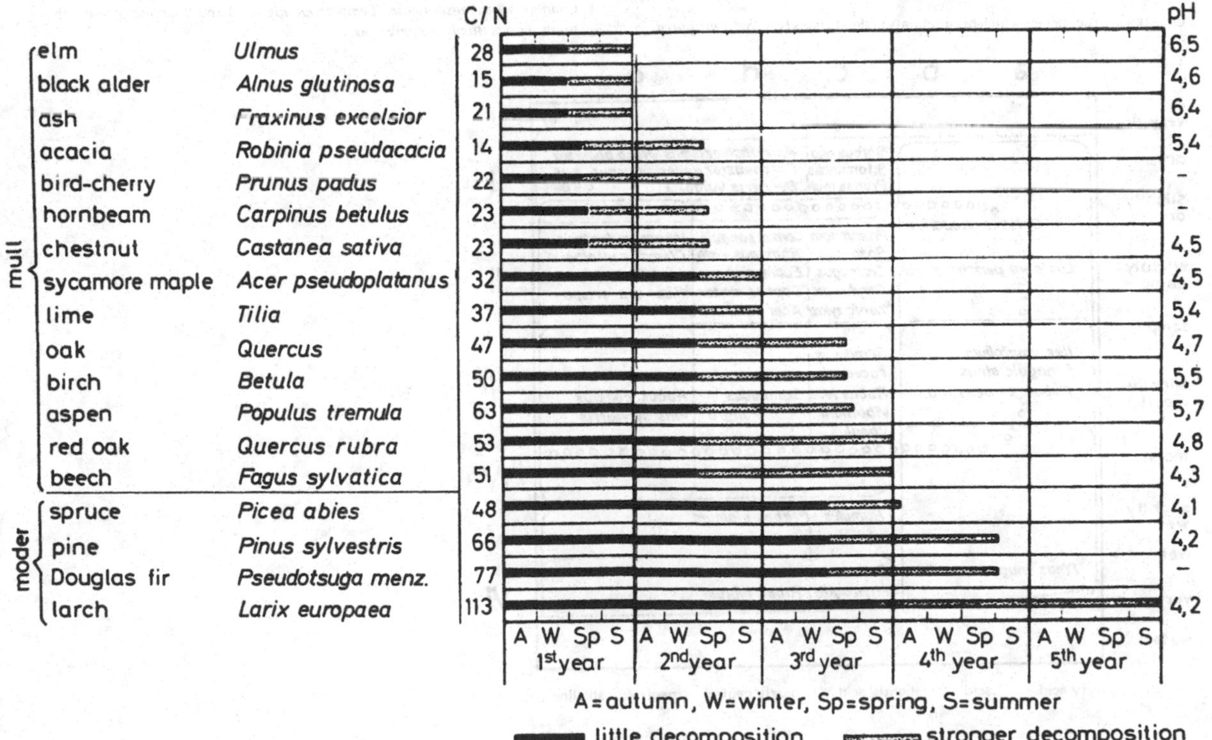

are to be found exclusively in coniferous woods, a fact which makes systematic classification much more difficult. On the other hand species which are associated with strongly shaded deciduous woodland, especially geophytes, are much more numerous. The many characteristic species of the class *Querco-Fagetea* and its subunits (section E III 1) may be taken as an example.

It scarcely needs to be pointed out that **mixed stands** of broadleaved and coniferous trees show intermediate characteristics between the extremes indicated in points a to g. However naturally occurring forests which have about the same numbers of broadleaved trees and conifers are remarkably rare. Much more often broadleaved trees will be dominant with a few scattered conifers, or it may be the other way round. This is particularly true with shade trees; the light-demanding trees show a much more complete sequence of intermediate mixtures.

Finally it must be stressed that amongst the undergrowth flora in Central European woodlands there is **not a single case where a higher plant is exclusively associated with a particular tree species**. For instance there is no such thing as an obligate 'Beech commensal' or 'Pine commensal' and so on. This fact, which has long been known to plant sociologists, has been confirmed statistically by Knapp (1958a) from a great deal of material. Where such connections appeared to exist they always turned out to be indirect associations which only held good for a particular locality. The local climate, for instance, could have been such that only one tree species could become established, or it could have been connected with particular soil conditions which locally were only met with in a single woodland type. Direct dependence, having a chemical basis, has obviously been built up between some tree genera and particular mycorrhizal fungi, for example between *Larix* and *Boletus elegans* or between *Betula* and *Boletus scaber*.

(b) The behaviour of shrubs in deciduous woodland

Figure 49, the type of diagram with which we are already familiar, serves to show the habitat relations of some 50 species of shrubs found in the woods of Central Europe. In each case the name has been placed to show the least amount of water in the soil where

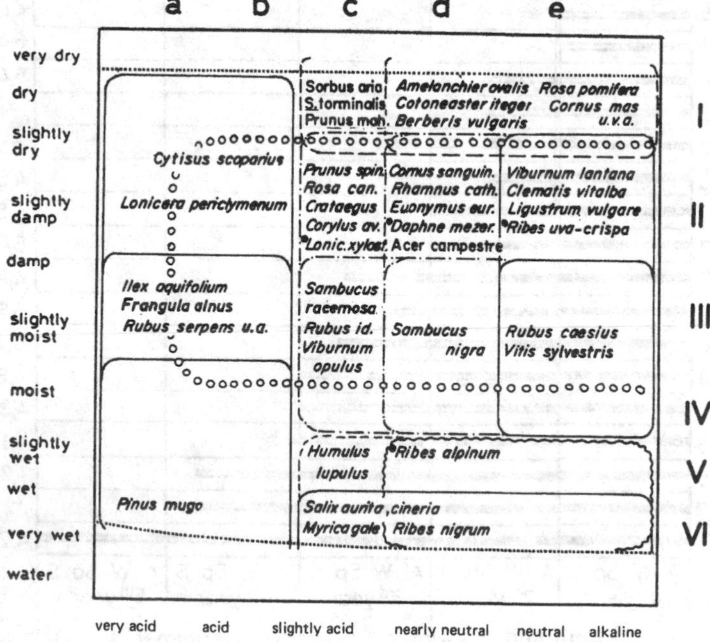

Fig. 49. Approximate areas of moisture and acidity occupied by the important shrubs, a few smaller trees and the lianes in submontane broadleaved woodland and scrub in Central Europe (cf. fig. 52). The ones indicated by a black spot prefer the semi-shade and so can be considered as true woodland plants.

Omitted are the species which occur mainly in flood plains (e.g. *Salix purpurea, viminalis* and *elaeagnos. Myricaria germanica* or *ZHippophae rhamnoides*) and also species with a very wide amplitude (e.g. *Juniperus communis* and *Sorbus aucuparia*). Almost all species of *Rubus* and *Rosa* have also not been considered. There is insufficient space in the diagram to show all the dryness indicators amongst which could also be named *Rubus canescens, R. ulmifolius, Coronilla emerus, Colutea arborescens, Clematis recta* and others. Accompanying Pine are *Chamaecytisus ratisbonensis, Lembotropis nigricans* and *Rhamnus saxatilis*; in the south-west is *Buxus sempervirens*.

that particular species is most often found in deciduous woodland. Almost all the species will also be found on damper and more fertile soils.

It can be seen at a glance that the acid soils are particularly poor in shrub species. As far as soil acidity is concerned the majority show a wide range from fairly acid through to alkaline. Juniper (*Juniperus communis*) behaves in a similar way to Scots Pine in that it is not particular as to soil so long as there is plenty of light, and is met with on dry limestone soils as well as on both wet and dry acid humus soils. Rowan (*Sorbus aucuparia*) also covers such a large range of habitats that it cannot be accommodated in the diagram.

The majority of shrubs not only make their maximum growth but also show their maximum distribution outside the shade of woods, that is round the sunny edges of woodland, in hedges or scrubland. They have recently been allocated to particular associations and orders of shrub communities to which we shall return in section D IV 1. When they do occur in real woodland these can often be traced back to the coppiced or grazed woods of earlier centuries which over a long period of time had been opened up to a greater or lesser extent.

There are no real woodland shrubs, that is species which flourish better in the shade than in the open. Some species however can tolerate shade quite well and are found more often in woods than in full sunlight. These are indicated in fig. 49 with a black spot and include the small shrubs *Lonicera xylosteum, Ribes uva-crispa, Viburnum opulus, Rubus caesius*, and *Ribes alpinum*.

(c) *The influence of soil factors on the species composition on the woodland floor*

While the general climate is only moderated in its effect on the woodland undergrowth, the latter is directly dependent on the nature of the soil. Before we go into details, the causal relationships should first of all be indicated in general terms. If we accept that all the species growing in the area could reach the place under observation and have enough time so that they could compete effectively with one another, then the outcome, that is the **species composition of the undergrowth**, would be decided above all by the following factors:

1 The nature of the soil is of primary importance, mainly in its provision of plant nutrients and water, or, where water is in excess, in the aeration and with it the provision of oxygen for plant roots. Compared with these factors the soil acidity, which has been so much researched, does not, in itself, play a determining role (Ellenberg 1958). It can be looked upon as a symptom of the general level of fertility, especially of the available nitrogen, the measurement of which is much more complicated. The pH also influences the solubility of aluminium and manganese which can become toxic for some species (fig. 117).
2 The climate within the stand, particularly the amount of light available, determines the small-scale distribution of the species.
3 The degree to which the plants are eaten by wild animals, especially Deer, may be important in some cases. Particular species or strains are preferred and can be badly damaged, while others are seldom if ever touched (Klötzli 1965).

All these factors vary during the course of the year and, within limits, from year to year too. This makes the analysis of the influence of site factors more difficult. Either they must be compared at one particular moment, in order to reduce the chance element at least in respect of the seasonal variability; or data must be assembled over a number of years, which is possible only for a few selected places.

In addition each factor does not act alone, as plants are subject to the interplay of all the factors affecting their lives. So, for example, it is not possible to speak of the 'minimum light requirement' of a particular plant species as Wiesner (1907) and other earlier authors have endeavoured to do. For the minima are dependent on the nature of the tree stand and the overall conditions of radiation, which in small-scale investigations have to be kept within narrow and comparable boundaries. Moreover they are also influenced by the water content and especially by the fertility of the soil. It can be seen in fig. 50 that the minimal light requirement of the same herbaceous woodland plant increases as the average acidity of the soil rises, other things being equal. *Stellaria holostea*, for example, requires more than ten times as much light to enable it to grow on very acid soil as it does on neutral soil. Its metabolism is much more efficient on 'good' soil as can be seen from the more hygromorphic structure of its leaves and the magnitude of its net photosynthesis. Hesselman (1917) has already recognised and stressed the fact that, as far as trees are concerned, whether one is able to tolerate more shade or less than another can only be determined for the same habitat.

The causal analysis of woodland communities is complicated most of all however because plants in their natural habitats are not living in isolation, as they are in physiological experiments. They have to compete all the time with others of their own kind as well as with numerous other species, all of which change the environment significantly. On the woodland floor too, only a few of the most aggressive species are able to occupy habitats which offer their physiologically optimal conditions. Most of the others must, like Pine be content with only a part of that environment which best suits them, or indeed with living conditions which are entirely outside their optimal range, but where no other plants are able to rival them. The relationships of some Beech wood species to the easily measured degree of soil acidity may serve as examples. *Carex alba* is a good 'lime indicator' and shows itself to be an alkaliphile when grown as a pure culture at high pH in a medium which also has other favourable properties (Sebald 1956; fig. 51). *Mercurialis perennis* is able to live in soils of lower pH than those which it actually occupies naturally. However in its optimal area it has no competitor, unlike *Carex pilosa* and many so-called 'acid indicators', e.g. *Avenella flexuosa* (figs. 51 and 94). In a pure culture *Luzula luzuloides* has an exceptionally wide range with regard to soil pH and so is potentially tolerant of soil type. Nevertheless in woodland it is almost always found on very acid soil and only very exceptionally where there is a calcareous subsoil. It may certainly be used as an acid-soil indicator, although it cannot be said to like acid soils; it is acidotolerant, not acidophile.

The same sort of thing goes for many other species and in respect of other factors such as soil water

Fig. 51. The response of some plant species to soil acidity in pure cultures and under natural competition in the herbaceous layer of south-western Central European Beech woods (compare fig. 39). The 'physiological amplitudes' (dotted area under the thick line) after Sebald (1956). 'Ecological amplitudes' in Beech woods of South-west Germany and Switzerland (cross-hatched area) from our own measurements.

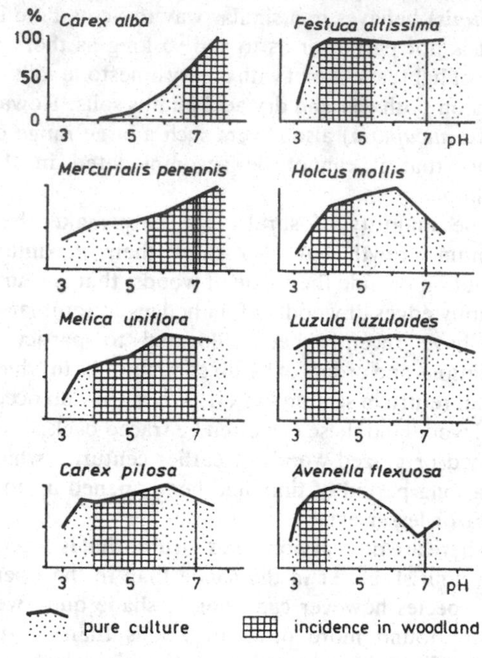

Fig. 50. In general the more acid the soil the poorer it is in available nitrogen and the greater the light requirement of any one species. After Ellenberg (1939), modified.

Average figures for the relative amount of light available are taken from numerous measurements in moist-soil mixed Oak wood communities in North-west Germany. The abscissa gives the annual mean pH values of the topsoil. Peace and Grubb (1982) confirmed the interdependence of light requirement and nitrogen supply by researches in England.

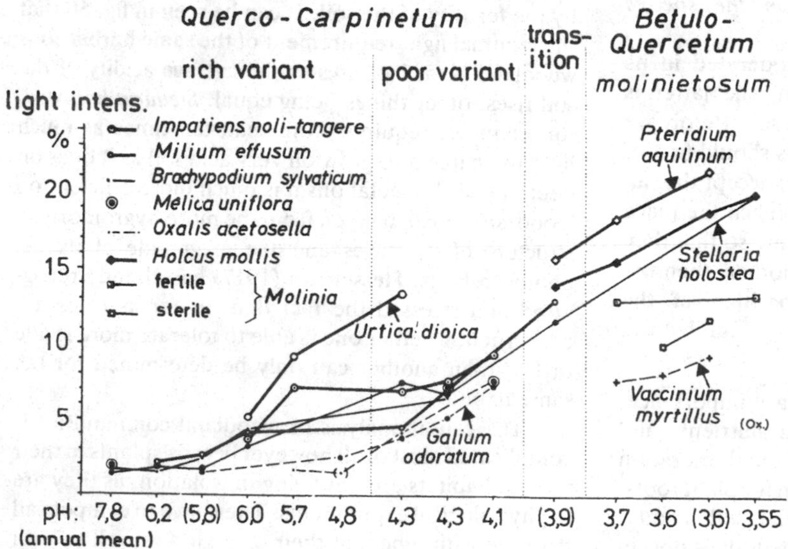

content. This means that some species need less moisture than others, but they are not 'drought-loving' species in the physiological sense. Almost all the plants which are indicators of dry conditions have been driven there, or they are plants which require warmth and have to put up with dry conditions in a habitat which has an acceptable temperature range.

All these circumstances increase the difficulty of making a causal analysis of particular plant associations, so much so that up to now we have only been able to assess it in a fragmentary manner. In most cases we have to be content with statistical information correlating the vigour of particular species, groups of species or communities, with certain factors, soil types or habitats. But we must always bear in mind that such statistical 'coincidences' are only the beginning and end of the chain of causality which decides the species composition, and do not reveal the causal links. By observation we know a lot about the **distribution** of many species, but without experimental investigation nothing about their **physiological requirements.** Unfortunately this aspect is not yet receiving sufficient attention from ecologists and plant sociologists.

4 The ecological behaviour and the ecological grouping of species

(a) Indicator values with regard to particular habitat factors

Today the pattern of distribution, the so-called 'ecological behaviour', is quite well known for the plants of the woodland floor, as well as for most of the other species in the Central European flora. Ellenberg (1974, 1979) was able to give the 'indicator value' for nearly 2000 vascular plants with respect to soil moisture, acidity, available nitrogen and also the salt and (less definitely) the heavy metal content of the soil. In addition he evaluated their behaviour towards climatic factors (i.e. temperature, degree of continentality of the climate) and to the relative illumination, which has such a decisive influence on the mosaic of species on the woodland floor. We shall often return to this information when we are considering not only the woodland communities, but also the other plant formations. It would take too much space to go into details of the indicator values here. We shall have to be satisfied with the index of all the plants mentioned in this book, where for each there is a short formula indicating its ecological behaviour. This takes the form of two groups of three figures, the meaning of which is fully explained in section E III 3. Here the principle of the system is outlined as far as seems necessary to understand the information contained in the following sections, mainly in the tables.

The ecological behaviour in respect of any of the habitat factors is expressed as a number in a nine-point scale where '1' indicates a low value for the particular factor and '9' a very high value. An x instead of a number indicates indifference or tolerance, that is, a wide amplitude for that factor or ecological dissimilar behaviour in different areas. (This system differs from previous ones which used only a five point scale, especially by Ellenberg 1950, 1952a and 1963.)

The first group of three refers to climatic factors:

L = **Light value** i.e. position in the gradient from very low illumination (1) to full light of the open country (9).

T = **Temperature value** i.e. position in the temperature gradient from arctic or alpine (1) to Mediterranean climate (9).

K = **Continentality value** i.e. predominant distribution range, from the Atlantic coast (1) to the inland parts of Eurasia (9).

The second group of three refers to important soil factors:

F = **Moisture value** i.e. position on the range from shallow soil on dry rocky slopes (1) to wet marshy ground (9); three additional numbers (10–12) indicate that the main distribution of the plant is in shallow to deep water.

R = **Reaction value** i.e. position in the range from a very acid substrate (1) to one rich in lime (9).

N = **Nitrogen value** i.e. concentrations of the plant on soils very low in available mineral nitrogen (NH_4 and NO_3) (1) to those where it is in excess (9).

Four very different plants of the herb layer will serve as examples. Soft-leaved Sedge (*Carex montana*, 554–453) is most plentiful in the half shade of open woodland (L5), where there are average heat conditions, that is, in the submontane belt in Central Europe (T5), and where the climate is suboceanic (K4). It is most competitive on relatively dry soil (F4) which is not very acid (R5) but generally rather poor in nitrogen (N3).

Yellow Archangel (*Lamiastrum galeobdolon*, 3x5–575) has a very wide distribution in respect of temperature (Tx), occupies an intermediate position in the east-west gradient (K5) but tolerates more shade (L3). It avoids both very dry and very wet soil (F5) but does not venture so far into acid and poor habitats (R7, N5) as the Soft-leaved Sedge.

As a spring flowering geophyte *Corydalis cava*, 364–688, is to be found in places which are very heavily shaded in summer and then only sparsely occupied by other plants; however it also flourishes outside woodland in places such as orchards (L3). Besides this it is suboceanic like *Carex montana*, but is widespread in relatively warm lowland areas (K4, T6). It prefers a nutrient-rich, calcareous and fairly moist soil (N8, R8, F6) and so can be looked upon as a strong indicator of high soil fertility.

Eriophorum vaginatum in contrast is not limited by the general climate (Tx, Kx) but requires good light (L7). It is an indicator of wet ground preferring an acid and very poor soil (F8, R2, N1).

For the remaining plant species of Central Europe the corresponding short formulae are given in the species index (E III 2) in order to lighten the text. Glancing over this list it may be seen that hardly any species has exactly the same ecological behaviour as another. However some may be found which are very similar, and these can be put into 'ecological groups'. Two examples can be quoted:

1. Marsh Fern group (*Thelypteris*-group)

	L	T	K		F	R	N
Calamagrostis canescens	6	4	5	–	9	5	5
Carex elongata	4	4	3	–	9	7	6
Carex laevigata	4	5	1	–	9	5	?
Osmunda regalis	5	6	2	–	8	5	5
Thelypteris palustris	5	×	×	–	8	5	6

This group consists of wetness indicators which are absent from very acid poor soil, but also avoid very rich soils (R5-7, N5-6). They prefer half shade (L4-6) and are found in places with average temperature (T4-6). The greatest variation amongst them is their degree of continentality. *Carex laevigata* is one of the few extreme oceanic species (K1) which merely has outposts reaching into Central Europe; *Osmunda* too can be looked upon as oceanic (K2) whereas the remaining species have their main distribution further to the east (K3-5) or must be considered as indifferent (Kx). The *Thelypteris*- group is mostly found in association with Alder and characterises the Alder swamp woodland (*Alnetum glutinosae*).

2. Dwarf Sedge group (*Carex humilis*-group)

	L	T	K		F	R	N
Anthericum liliago	7	5	4	–	3	5	2
Carex humilis	7	5	5	–	3	8	3
Coronilla emerus	7	6	4	–	3	9	2
Dictamnus albus	7	8	4	–	2	8	2
Geranium sanguineum	7	5	4	–	3	8	3
Peucedanum cervaria	7	6	4	–	3	7	3

These are very much indicators of dry sites and can put up with a low level of available nitrogen. They are more or less closely restricted to calcareous soil. They like a fair amount of light and so prefer the edges of woodland, but are also found within open woodland. Their distribution is suboceanic and they are concentrated in the southern lowlands and hillsides of Central Europe.

These and other ecological groups, which play a more or less important part in the plant communities of deciduous woodland in Central Europe, were already included in the first edition (1963) before the indicator values, now collected in section E III 2, were all available. They are shown in table 10 without their ecological formulae, in order to economise on space. An approximate outline of their amplitudes for water content and acidity is given in fig. 52.

It can be seen from this ecogram that the amplitudes of the herbaceous species are more restricted than those of the trees. This is perhaps not least a consequence of the fact that in the herb layer there are more than ten times as many species competing as there are in the tree layer. Without competition practically all the species would be able to live, albeit with different degrees of vitality, in most of the woodland types in Central Europe. Not only all the plants indicative of dry sites, but also nearly all the other groups, have their physiological optima where the water content, acidity and fertility of the soil are intermediate, in a similar way to the trees shown in fig. 39. The sites where they are really found in greatest concentration are those where other competitors are less vigorous. Therefore their ecological optima often deviate noticeably from their physiological optima; or in other words: **their 'ecological existence' differs from their 'ecological potential'**.

Competition among the species on the woodland floor is increased by the fact that the trees and their saplings also take part in it. The struggle between species and individuals leads to an unstable balance which can be upset by slight changes in some of the factors. Thus the species composition of the herb layer reflects every change in the conditions of the habitat. For the most part therefore it offers the best criterion for the classification and delimitation of woodland communities. The phytosociologists of Central Europe have made so much more use of this fact because the direct impact of man had a greater effect on the tree layer than on the soil cover.

(b) The evaluation of habitats using factor values

Before we go into phytosociological systematics and the problems it poses in Central Europe, and before we come to discuss individual plant communities, we should explain briefly how the indicator values just described should be used in the ecological assessment of plant communities. Here again the woodlands can serve as good examples whereby general questions, applicable also to mires, heaths, meadows and other formations, can be discussed.

Apart from extreme cases no plant community consists of just a single group of species; a number of ecological groups have always taken part in its development. This combination of species groups can give us information about the environmental conditions affecting the whole community. Thus the assessment of its habitat factors will be more objective and exact if we take into account the ecological behaviour of all species present in this community (compare the following with tab. 11, especially No. 1):

Tab. 10. Ecological groups of plants growing under broadleaved woodland in Central Europe.

In each case the groups of species are centered on the following habitats:

I Soils which dry out from time to time

a-b acid and poor
Cladonia group

L *Cladonia-species*
Carex arenaria (on sand)
M *Dicranum spurium*
Hieracium umbellatum
M *Polytrichum juniperinum*

c fairly acid
Carex montana group

Astragalus glycyphyllos
Betonica officinalis
Campanula persicifolia
Carex digitata (IIc)
C. montana
Convallaria majalis (IIc)
Hypericum montanum
Melica nutans (IIc)
Melittis melissophyllum
Polygonatum odoratum
Silene nutans
M *Thuidium abietinum*

d-e ± lime rich
Carex humilis group

Anthericum spec. (Ic)
Brachypodium pinnatum (Ic)
Buglossoides purpurocaerulea (IId)
M *Camptothecium lutescens*
Carex humilis
Coronilla emerus
Dictamnus albus
Geranium sanguineum
Peucedanum cervaria
M *Rhytidium rugosum*
Viola hirta
Vincetoxicum hirundinaria

II Soils ranging from fairly dry to slightly damp

a very acid
Vaccinium myrtillus group

Calluna vulgaris (Ia)
M *Camplyopus flexuosus*
M *Dicranum scoparium*
M *D. undulatum*
M *Hypnum cupressiforme*
M *Leucobryum glaucum* (Ia)
Melampyrum pratense
Nardus stricta
M *Pleuozium schreberi*
Vaccinium myrtillus
V. vitis-idaea

b acid
Avenella group

Anthoxanthum odoratum
Avenella flexuosa (IIc)
Carex pilulifera
Cytisus scoparius (atl.)
M *Dicranella heteromalla*
M *Ditrichum pallidum*
Festuca ovina (Ia)
Galium harcynicum (atl.)
Holus mollis
Lathyrus linifolius (IIc)
Luzula luzuloides
M *Plagiothecium denticulatum*
M *Polytrichum formosum*
Solidago virgaurea (IIc)
Teucrium scorodonia (atl.)
Trientalis europaea
Veronica officinalis (IIc)

c less acid
Anemone nemorosa group

Anemone nemorosa
M *Atrichum undulatum*
M *Brachythecium velutinum*
Carex pilosa
C. umbrosa
Dactylis polygama
Epilobium montanum
Euphorbia amygdaloides
M *Eurhynchium striatum*
Festuca altissima (IIb)
M *Fissidens bryoides*
Galium odoratum
G. sylvaticum
Hedera helix
Hieracium sylvaticum (Ic)
M *Isothecium viviparum*
Luzula pilosa (IIb)
Melica uniflora
Milium effusum
M *Mnium affine*
M *M. rostratum*
Moehringia trinervia
Neottia nidus-avis
Phyteuma spicatum
Poa nemoralis
Potentilla sterilis
Prenanthes purpurea (IIb)
Stellaria holostea
Vicia sepium
Viola reichenbachiana

d richer in bases
Lamiastrum group

M *Anomodon attenuatus*
Asarum europaeum (IIId)
Brachypodium sylvaticum
Bromus ramosus
Campanula trachelium (IIId)
Carex sylvatica
Epipactis helleborine
M *Eurhynchium swartzii*
M *Fissidens taxifolius* (IIId)
Geum urbanum
Hepatica nobilis (Ic)
Hordelymus europaeus
Impatiens parviflora
Lamiastrum galeobdolon
Lathyrus vernus (Ic)
Lilium martagon (Ic)
Melampyrum nemorosum
Mercurialis perennis (Ie)
Paris quadrifolia (IIId)
Polygonatum multiflorum
Primula elatior (IIId)
Pulmonaria officinalis
Sanicula europaea
Vinca minor

e lime rich
Cephalanthera group

Carex alba
Cephalanthera rubra
C. damasonium
M *Ctenidium molluscum*
Cypripedium calceolus (Id)
M *Encalypta streptocarpa*
Helleborus foetidus (Id)
Primula veris (Id)
M *Tortella tortuosa*

M = moss, L = lichen figures in brackets = tending towards.

Tab. 10 (*Contd.*)

III Slightly damp to damp soils

a very acid	**c less acid**	**d richer in bases**
Blechnum group	Ajuga group	Flcaria group
M *Bazzania trilobata*	*Agropyron caninum* (IIId)	*Adoxa moschatellina*
Blechnum spicant	*Ajuga reptans*	*Arum maculatum* (IIIe)
Huperzia selago	*Athyrium filix-femina*	*Ciraea lutetiana*
Lycopodium annotinum	M *Brachythecium rutabulum*	*Gagea spathacea* (IIIe)
M *Ptilidium ciliare*	M *Bryum erythrocarpum*	*Listera ovata*
M *Ptilium crista-castrensis*	M *Cirriphyllum piliferum*	M *Mnium undulatum*
	Deschampsia cespitosa	*Ranunculus auricomus*
b acid	*Festuca gigantea* (IVc)	*R. ficaria*
Pteridum group	*Geranium robertianum*	*Scilla bifolia* (IIIe)
Dryopteris carthusiana	*Glechoma hederacea* (IIId)	*Stachys sylvatica*
Galeopsis tetrahit (IIc)	*Lysimachia nemorum*	
M *Hylocomium splendens* (IIc)	*Oxalis acetosella* (IIIb)	**e lime-rich**
Luzula sylvatica	*Scrophularia nodosa*	Corydalis group
Mycelis muralis (IIc)		*Aegopodium podagraria* (IVd)
M *Plagiothecium undulatum*		*Allium ursinum* (IVd)
Pteridium aquilinum		*Anemone ranunculoides*
M *Rhytidiadelphus loreus*		*Corydalis cava*
M *Rh. triquetrus* (IIc)		*C. solida*
		Gagea lutea
		Leucojum vernum
		Rubus caesius (IVd)

IV Damp to fairly wet soils

a b ±acid	**c less acid**	**d – e± base-rich**
Molinia caerulea group	Carex remota group	Carex pendula group
Erica tetralix (atlant.)	*Carex remota*	*Astrantia major*
Molinia caerulea	*Equisetum sylvaticum*	*Carex pendula*
Potentilla erecta	*Galium aparine* (IVd)	*C. strigosa* (atl.)
	Impatiens noli-tangere (IVd)	*Chaerophyllum hirsutum*
	Lamium maculatum (IVd)	*Chrysosplenium alternifolium*
	Silene dioica (IVd)	*Ch. oppositifolium*
	Stellaria nemorum (IVd)	*Circaea alpina*
	Urtica dioica (IIId)	*C. intermedia*
	Veronica montana	*Equisetum maximum* (Vd)
		Petasites albus

V Fairly wet to wet soils

a–b ± acid	**c less acid**	**d–e ± base-rich**
Vaccinium uliginosum group	Filipendula group	Carex acutiformis group
Ledum palustre (contin.)	*Angelica sylvestris*	*Carex acutiformis*
M *Polytrichum commune*	*Cardamine pratensis*	M *Campylium stellatum*
M *Sphagnum acutifolium*	M *Climacium dendroides*	*Cirsium oleraceum*
M *S. cymbifolium* (Vc)	*Cirsium palustre*	*Crepis paludosa*
Vaccinium uliginosum	*Filipendula ulmaria*	*Geum rivale*
	Juncus effusus (Vb)	*Phalaris arundinacea*
	Lysimachia vulgaris (Vb)	*Scirpus sylvaticus*
	Lythrum salicaria	*Symphytum officinale*
	Poa trivialis	M *Trichocolea tomentella*
	Ranunculus repens	*Valeriana officinalis*
	Valeriana dioica (Vb)	

Tab. 10 (*Contd.*).

VI Wet soils

a very acid
Eriophorum vaginatum gr.

Andromeda polifolia
M *Aulacomnium palustre* (VIb)
Eriophorum vaginatum
M *Polytrichum strictum*
M *Sphagnum fuscum*
M *S. magellanicum*
Scirpus cespitosus
Vaccinium oxycoccus (VIb)

b acid
Potentilla palustris gr.

Carex nigra
C. rostrata
Eriophorum angustifolium
Hydrocotyle vulgaris
Potentilla palustris
M *Sphagnum cuspidatum*
M *S. recurvum*
M *S. squarrosum*
Viola palustris

c less acid
Thelypteris palustris gr.

Calamagrostis canescens
Carex elongata
C. laevigata
Osmunda regalis (Vb)
Thelypteris palustris

d–e richer in bases
Caltha group

M *Acrocladium cuspidatum*
Caltha palustris
Equisetum fluviatile
Galium palustre
Iris pseudacorus (Vd)
Lycopus europaeus
Peucedanum palustre
Scutellaria galericulata
Solanum dulcamara

A – F groups appearing in special habitats

A humid air, ± acid
Gymnocarpium group

Aruncus sylvester (IIIc)
Dryopteris dilatata
D. filix-mas (IIc)
Gymnocarpium dryopteris
Thelypteris limbosperma
Th. phegopteris

B humir air, ± base-rich
Lunaria group

Actaea spicata (IIIc)
Gymnocarpium robertianum
Lunaria rediviva
Phyllitis scolopendrium
Polystichum aculeatum

C high montane, much snow
Adenostyles group

Adenostyles alliariae
Athyrium distentifolium
Cicerbita alpina
Rumex arifolius
Streptopus amplexifolius

D fluctuating dry, clay soil
Carex flacca group

Calamagrostis varia (G)
Carex flacca
Molinia arundinacea (IV)

E fluctuating damp
Carex brizoides group

Carex brizoides
C. leporina

F nitrate-rich
Alliaria group

Alliaria petiolata
Anthriscus sylvestris
Chaerophyllum temulentum
Chelidonium majus
Veronica hederifolia
Viola odorata

G lime-rich, de-alpine
Sesleria group

Aster bellidiastrum
Carduus defloratus
Carex ornithopoda
Sesleriararia (Ie)
Thesium alpinum

Source: A simplified synopsis after Ellenberg (1963).

The simplest way to use the indicator values of all the partners in a plant community is to calculate average factor values, e.g. a 'mean moisture value.' This calculation will include the F value for each species as given in the index (E III 2), omitting the ones marked 'x'. For woodlands only plants of the undergrowth are included in this calculation since the trees live in an environment different from that of the understorey. In our example the average moisture value is mF 5.1, showing that the Beech stand is growing on an intermediately moist soil. The majority of species in this stand occur mostly on soil of average dampness (F 5). Many species however tend to favour damper sites (F 6) though a few are mostly found on drier ones (F 4).

In the same way the average light value, temperature value and the rest can be calculated for the same list of plants. Short formulae of this kind have been produced for many plant communities so that comparisons can be made (see e.g. Böcker. Kowarik and Bornkamm 1984). In order to demonstrate the mean values and their amplitudes for two important factors at the same time a system of coordinates may be used, similar to the ecograms already seen in section B.1 3a and b (figs. 40 and 49). These have the advantage that the indicator values of several plant communities can be compared (see e.g. Rogister 1980, 1981, Persson 1981 and Sloboda 1982).

It must be clearly understood that such diagrams and their underlying calculations only enable us to come somewhere near to the natural diversity of plant behaviour. Average factor values can only give a rough interpretation of the site conditions. In spite of this they provide us with a useful tool for estimating the relative significance of the different habitat factors. It is possible for any amateur to calculate the mean factor values from a vegetation relevé (section E III 2, or in Ellenberg 1979). This may be looked upon as an aid to the ecological evaluation of plant communities. Spatz, Pletl and Mangstl (1979), Durwen (1981). Böcker, Korarik and Bornkamm (1984) and others have worked out computer programmes for the calculation of average ecological indicator values.

5 Comments on the classification of plant communities

(a) The main units in the phytosociological system

In contrast to North America, the Soviet Union, and many other regions of the world where there are comparatively few descriptions and classifications of the vegetation available, knowledge of the plant cover in Central Europe suffers rather from too much in the way of detailed information. Even in the first edition of this book it was scarcely possible to review the flood of relevés, tables and local descriptions. Since then there have been more than a hundred thousand relevés and a few thousand publications added to the number. Under these circumstances the concern of many authors for a narrower classification of the vegetation so that local habitats can be more exactly fitted into it is only natural. However there is always the threat that the overall pattern, which should be the main aim of any classification, will be lost.

It is not possible here to elaborate on the pros and cons of the various efforts to indicate a solution or on the extensive literature, but rather to find a practicable way through the variety of Central European vegetation and its habitats. As far as the main units are concerned – the classes and orders and also to some extent the alliances and suballiances in the sense of Braun-Blanquet (1928, 1964) – we prefer to remain conservative, because we see no general advantage in

Fig. 52. Approximate natural areas of moisture and acidity occupied by ecological groups of herbaceous plants, mosses and lichens in the submontane broadleaved woodlands of Central Europe (compare with the species groups in Tab. 10).

For each of the groups I to VI the upper limit of the moisture amplitude is shown. For group I and for groups II-IV together the lower limit is also given. Regarding the soil acidity the right-hand limits for groups a and b have been drawn, but the left-hand limits for groups c,d and e (i.e. set against the greater acidity). Thus the *Anemone* and *Ajuga* groups have the widest amplitudes, whereas the *Eriophorum vaginatum* group and other wetness indicators have the narrowest.

As in fig. 40 the area within which Beech is dominant is indicated with a line of small circles.

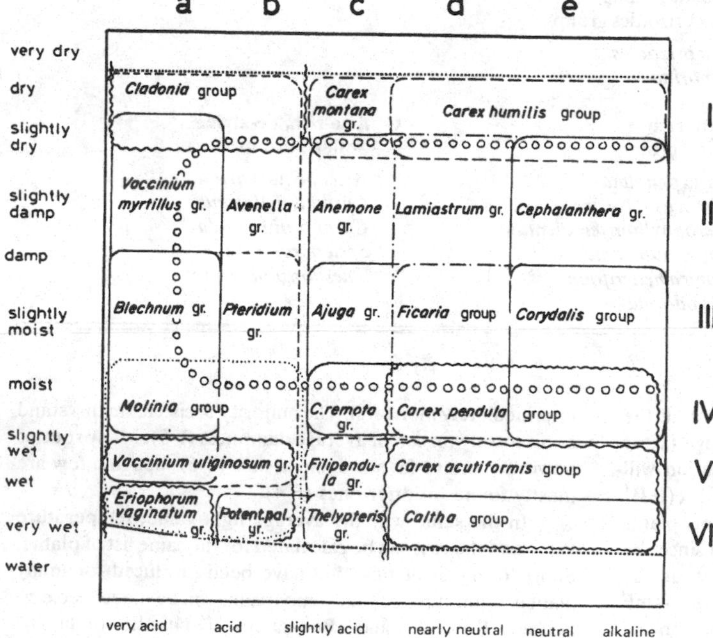

fundamental changes of these classification levels which are being advocated on many sides. As far as associations and smaller subunits are concerned we shall conform to the respective authors. There are also various devices which we shall use in some cases (these will be briefly described in section b).

The only relatively new complete survey of all the vegetation units covering a substantial part of Central Europe and based on extensive data is still that of Oberdorfer (1957). With additions and amendments it was extended to cover a wider area by Oberdorfer and his co-workers (1967) while the same author published a list of smaller, but important, alterations in 1979. These correspond to the new edition of his *Plant Communities of Southern Germany* which is now being published in conjunction with other authors (1977-1983). I have referred as much as possible to the parts already completed, especially in sections E II, E III 1 and 2. We shall follow this system to a large extent, since it covers the main areas of Central Europe including the Alps and other mountain ranges as well as the lowland areas and xerothermic slopes. Only the plant communities of the sea coast and the north-eastern plain will have to be added using information from other sources, e.g. from F. Runge (1980) and Tüxen (1937, 1956b etc).

Because of its extent our classification of the higher vegetation units has been included as an appendix (section E III 1). It differs from the usual system in that a number of classes have been combined into habitat groups such as 'Salt water and sea coast vegetation', 'Stony sites and alpine grassland' and 'Coniferous woodland and related communities'. The motive for this grouping is purely practical in that it enables the amateur to survey more easily the 45 classes and at the same time the specialist to use a decimal classification suitable for computer analysis (Ellenberg 1979, Durwen 1981).

Down as far as the alliances in section E III 1 the character species have been included. These are the species which appear almost exclusively, or at least preferentially, in a particular unit. In many of the alliances and some orders, and even in a few classes, the number of such species is in each case restricted. Thus a greater subdivision of the higher units is hardly possible if one is not to dispense entirely with Braun-Blanquet's 'character species principle' and work only with 'differential species' or other devices. This principle prevents the confusing splitting up of units and so should be accepted unconditionally, at least down to the level of alliance.

In the light of such considerations almost all the deciduous woodland of Central Europe must be included in a single class – the mixed deciduous broadleaved woodland (*Querco-Fagetea*) for which Tüxen, Wilmanns and Schwabe-Braun (1981) produced a comprehensive bibliography. The separation of this from the acid-soil mixed Oak woods (*Quercetea robori-petraeae*) which are concentrated in oceanic western Europe is difficult and there are many intermediates (see also section E II). Such intermediate types occur too between it and the Alder swamp woods (*Alnetea glutinosae*) although they occupy separate places both floristically and ecologically. The same goes for the wood and shrubland of the flood plains – the class *Salicetea purpureae*.

There is some uncertainty about the classification of the Birch woods which replace Alder in wet sites where the soil is very acid (section B V 2c). Their species composition has so many similarities with the acid-soil coniferous woods (*Vaccinio-Piceetea*) that they are usually included in this class although, at least in the west of Central Europe, they rarely contain any conifers. One could include them in the unit *Vaccinio-Piceion* as a special suballiance (*Betulion pubescentis*), although their systematic classification is of little significance for the understanding of Central European vegetation since they occupy a decreasing number of small areas. We will deal with the systematic arrangement of the coniferous woodland after we have discussed the classes *Querco-Fagetea* and *Quercetea robori-petraeae*.

Many authors, especially Jakucs and other southern Europeans have raised the order of the drought-tolerant and thermophilous mixed Oak woods (*Quercetalia pubescenti-petraeae*) up to the rank of a class, because a large number of related associations are known in the warmer parts of Europe. In this way a division into more alliances and a few orders is possible (Horvat, Glavač and Ellenberg 1974). However only a few of these radiate into Central Europe and are connected here to the native *Querco-Fagetea* woods by a number of intermediate types. From our point of view then it does not matter how the mixed Oak woods of warm dry habitats are classified. In any case they represent a special group of associations which contain many species rare in Central Europe.

Wherever possible the classes and orders have been arranged in section E III 1 according to the 'sociological progression' (i.e. with increasingly complicated life-form composition) which has been the convention since Braun-Blanquet (1928, 1964). The woods, particularly for Central Europe the characteristic deciduous woods, are therefore put at the end of the system and the coniferous woods and all other communities before them. In the text however we have chosen an opposite order following mainly the review given in section A 1, which fits in much better with the historical and ecological connections.

The ecogram in fig. 53 is a summary of the alliances and suballiances of woodland communities in the sub-montane belt of western Central Europe.

(b) Methods of subdividing the vegetation according to habitat

Classes and orders, even alliances and suballiances are relatively comprehensive units when one considers the floristic and ecological variability of the plant communities involved. The practical forester or planner as well as the ecologist or the vegetation specialist often prefers narrower units which are more closely correlated with particular habitats, without having to lose the overall picture. Work on these can be assisted in three ways which are more similar than they appear at first sight:

sociological groups of species (character and differential species,

socioecological groups (ecologically interpreted differential species),

ecological groups (see section 1 4a).

Sociological groups can be obtained by comparing relevés of plant stands as has been extensively described by Braun-Blanquet (1928, 1964), Ellenberg (1956), Mueller-Dombois and Ellenberg (1974), and others. In this way groups of differential species (separating species) are recognised which, within the total amount of recorded material, are absent from larger or smaller sections of it. If they never or very rarely are to be found outside their usual vegetation unit, then they may be described as character species, that is they can be looked upon as definite recognition features.

However the 'character species principle' which was introduced as being such an advantage in the previous section can hardly be used for the basic units in the system of plant sociology – the associations – since these have become so numerous. The probability of finding even one good character species, applicable to the whole of Central Europe, for woodland associations on 'average' habitats is very slight. If one makes it a condition (personal communication by Oberdorfer) that each association must have at least one character species, then one would only find good associations in very dry, wet, salty, or other extreme habitats and large areas of woodland would all have to be 'thrown into one pot'. In order to avoid this we must be satisfied with the floristic recognition of an association (within an alliance recognised by character species) by the presence of a definite combination of some differential species. The same method has already been in use for a long time to separate subassociations and variants (several tables in this book give examples of this).

The 'differential species principle' however can easily

Fig. 53. Approximate natural areas of moisture and acidity occupied by the alliances and suballiances of the Central European broadleaved woodland communities (cf. figs 40, 49 and 52).

The Beech wood alliance (*Fagion*) can be divided into the mesophilic suballiance of the true Beech woods (*Eu-Fagion*), the more acid-tolerant Woodrush-Beech woods (*Luzulo-Fagion*) and the Sedge-Beech woods (also called Orchid-Beech woods. *Cephalanthero-Fagion*) which are confined to relatively dry calcareous slopes. Communities of the alliance of warmth-loving mixed Oak woods (*Quercion pubescenti-petraeae*) occupy even drier habitats and go right to the dryness boundary for woodland on shallow stony soils. In very acid regions this boundary is taken over by representatives of the Birch-Oak woods (*Quercion robori-petraeae*) which are poor in species. Other types (with *Molinia*) also occupy damp to wet raw humus soils. Damp soils which are rich in bases however and also relatively dry habitats are left to the Oak-Hornbeam woods (alliance *Carpinion*). An alliance which can tolerate even wetter conditions (*Alno-Ulmion*, mixed grey Alder, Elm and Ash woods) merges into the Alder swamp woods (*Alnion glutinosa*). These, along with the Birch swamp woods (*Betulion pubescentis*, a suballiance of the acid-soil needle-leaved woodland, *Vaccinio-Piceion*) occupy the wetness boundary for woodland. Recently it has become a common practice to indicate the rank of suballiance by an ending of its own (-enion). Accordingly, e.g., the subunits of the Fagion alliance may be called: Cephalanthero-Fagenion, Eu-Fagenion, Luzulo-Fagenion.

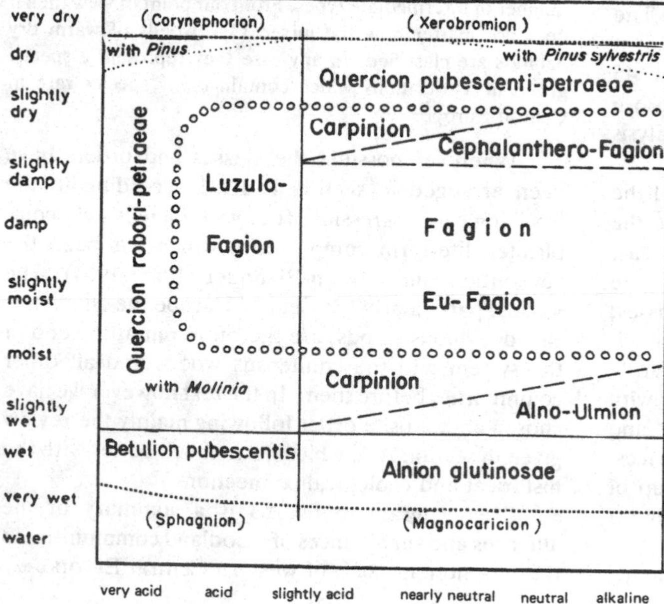

lead to the creation of numerous associations of only local significance if it is not used discretely. To avoid subjective decisions one could bring in objective criteria which must of course be fixed at random. For example one could stipulate a certain 'coefficient of presence' in the community of between 33 and 50% (Ellenberg 1956, Mueller-Dombois and Ellenberg 1974), or demand that less than half the 'combination of constant species' also appear in similar communities (Ellenberg and Klötzli 1972). Of course such aids in reaching a decision cannot replace the experience of a practised phytosociologist.

Differential species can be obtained simply by a comparison of lists or nowadays by means of a computer, that is in any case, without reference to habitat factors (Mueller-Dombois and Ellenberg 1974). Nevertheless they can be interpreted as indicators of a particular environment. Thus the first step has been taken towards the formation of **'socioecological'** groups, i.e. of species which often appear together and whose behaviour under various conditions of soil, relief, altitude etc. has been revealed by comparable observations in the field. This method appeals not least to the expert and has been used for the past 30 years in the mapping of forest habitats especially in Baden-Würtemberg by Schlenker and his fellow workers. However they speak of 'ecological' groups and justify this name even further by the measurement of important habitat factors and ecophysiological culture investigations. (Schönpar 1952, 1953, 1954, Schönnamsgruber 1959, Bogner 1966, Evers, Schöpfer and Mikloss 1968 and others).

Socioecological groups have been used as the main instruments in the classification of woods and other plant communities, especially by Scamoni and Passarge (1959), Scamoni (1960, etc.), Passarge (1964a, 1968), Passarge and Hofmann (1964), Mayer (1974), and others. Since this method is very flexible and is unaffected by criteria of phytosociological rank it has often led to the differentiation of many fairly local small associations which are difficult to survey. If one avoids a splitting of this kind one can arrive at a hierarchy of vegetation units which correspond largely to those obtained by purely sociological methods and so confirm them. This may be shown by the classification of the Beech woods, given in fig. 54, which Mayer (1974) undertook in the eastern Alps. From his total of 50 socioecological groups he has selected no less than 32 for this table but some of them which appeared only very rarely have been omitted from fig. 54 to save space.

With socioecological grouping it is possible to do justice to the edaphic, oreographic and climatic variability of an association at the same time over its total range of distribution, that is to express its change with soil conditions, altitude and the general climate.

'Ecological' species groups in the narrow sense are formed from the step by step analysis of the ecological behaviour of the species and so quite without the help of plant sociology. The method already described in section 1 4a was developed by Ellenberg (1950, 1952a) for arable weeds and grassland plants, and by P. Duvigneaud (1946) for others. By combining ecological groups of varying abundance one can characterise the vegetation units in a way similar to that in which they appear in fig. 54 using socioecological groups. In each case one arrives at units applicable within wide limits, since the basic groupings do not represent unchangeable elements but vary geographically. Each of the species remaining as a result of competition can alter the ecological behaviour of one or more of the other species in the group significantly. Regarding the area of validity of the classification neither the ecological nor the socioecological groups offer any significant advantage over the purely sociological ones. As Duvigneaud (1946) has already stressed, local variability of plant cover can be described more easily and with fewer complications with the help of ecologically defined groups of species (see also Tanghe 1971).

Ecological groups received surprising confirmation recently when Haeupler (1974) evaluated the data for the floral mapping of the southern part of lower Saxony. With the help of a computer he arranged the plant species objectively, according to the similarity of their geographical distribution, as points on the 5 x 5 km grid. In this way he obtained groups of species which correspond to a large extent with the ecological grouping of woodland plants included by Ellenberg (1963) in the first edition (see tab. 10). Plants which resembled one another in their behaviour towards important habitat factors also had a very similar geographic distribution which was sufficiently exactly expressed already in the large-scale point grid utilised by Haeupler.

If ecological, plant sociological and chorological methods all lead to similar results then the **grouping of species must be a basic feature** in the plant cover. That is why we have discussed it before describing the individual units of vegetation, and we shall constantly be coming back to it.

In dealing with the Beech woods of Central Europe in the next section we shall base it on a classification which, in its main units at least, is similar to the one given in fig. 54. This corresponds to the alliances and suballiances of the system in use based on character species (see section E III 1), and moreover appears to make sense ecologically. It also follows from the combination of constant species arrived at by Ellenberg and Klötzli (1972) by comparing numerous Swiss woodland communities with the help of punched cards. The classification of natural objects is really, and will remain, a matter of judgement based on experience. However, we may look upon a grouping arrived at along so many different ways as relatively trustworthy, and we should not hesitate to use it as a starting point for further ecological research.

II Beech and mixed Beech woods

1 *The deciduous woodland of Central Europe and the main groups of Beech woods*

(a) The order of the 'noble' deciduous woods

The largest part of the deciduous woodland of Central Europe belongs to the order of the so-called 'noble'

Fig. 54. Combinations of tree species and socio-ecological groups of undergrowth species in the eastern Alpine Beech wood communities. After Mayer (1974), somewhat altered.

The moder or Woodrush-Beech woods (*Luzulo-Fagetum*) occupy acid soils. Bilberry (subassociation *myrtilletosum*) dominates the poorest soils with Seagrass Sedge (= Quaking Grass Sedge, *caricetosum brizoides*) on those which are also waterlogged at times; the remaining three subassociations are arranged, by and large, according to the increasing fertility of the soil.
This is also the case for the sequence of brown mull or Woodruff-Beech woods (*Galio odorati-Fagetum*); the subassociations *circaeetosum* and *stachyetosum* are on damp soils; the Ramsons Beech wood (*allietosum*) leads into the slightly-damp calcareous Beech woods (*Lathyro-Fagetum*) of which three subassociations can be distinguished.
The Blue Moorgrass-Beech wood (*seslerietosum*) is one of the dry-slope Beech woods (*Carici-Fagetum*). The dry-slope Beech woods named *calamagrostietosum variae* or *veratretosum* are upper montane. The subalpine Maple-Beech wood (*Aceri-Fagetum*) is found on lime-rich but damper sites.
Some further details of the species groups can be found in tab. 10.

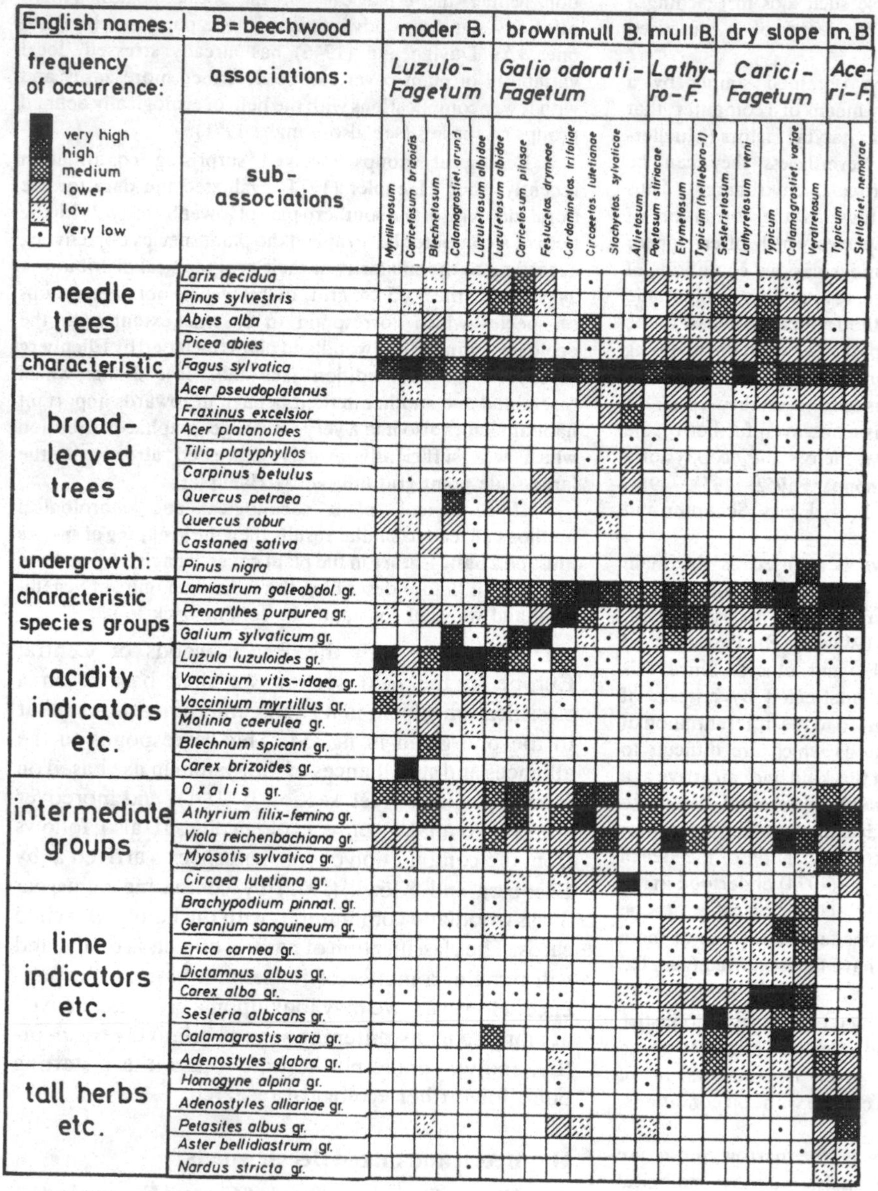

deciduous forest (*Fagetalia*) i.e. woodland dominated by trees belonging to the genera *Acer*, *Carpinus*, *Fagus*, *Fraxinus*, *Tilia*, *Ulmus* and other species demanding a relatively good supply of nutrients and water. The numerous character species of this comprehensive unit are set out in section E III 1. They are well adapted to the rhythm of life of the summer green broadleaved wood and most of them avoid poor soils as well as soils which dry out to any great extent or remain under water for any length of time (fig. 38). There are however differences of detail in its ecological amplitude.

The *Fagetalia* as a whole can occupy a wide range of habitats. Apart from real podsols and extremely wet soils practically all woodland soil types may be occupied by communities of this order. Climatic amplitude is also wide. The average annual temperature may lie between about 12° and 4°C provided that the winters are not too cold. The amount of precipitation may vary between about 400 and 2000 mm. The habitats occupied by each of the alliances are more restricted (fig. 53, see also section E III 1, 8.4):

a) **Beech and mixed Beech woods** (*Fagion*) avoid more extreme dry and wet soils than other *Fagetalia* communities and are relatively sensitive to frost damage.
b) **Sycamore and Lime mixed woods** (*Tilio-Acerion*) occupy shady slopes, screes and other special habitats within the general area of Beech woods and in some places are dominated by Ash.
c) **Broadleaved mixed woods rich in Hornbeam** (*Carpinion*) go further than the *Fagion* on to drier, wetter and more continental places, but do not venture so far up the mountains.
d) **Mixed woods rich in Ash and Alder** (*Alno-Ulmion*) require additional water either as groundwater or from flooding.

As the most extensive alliance and one which is characteristic of Central Europe the *Fagion* with its suballiances, associations and some lower divisions will be discussed in some detail.

Since in the long run Beech will suppress all other tree species on soils which are neither too wet nor too dry, not too poor nor overfertile, nor too cold, it can become dominant under very different conditions. Its optimum growth is attained in the submontane belt; nevertheless in the colline belt, and in most places on planar lowland also, it forms its own communities in which the native trees, which require a higher temperature, even here have to take a subordinate place. On the other hand it can climb further up the mountains and can even occupy the tree-line in south-west Central Europe.

In the middle and upper montane belt it must give way to the Silver Fir (*Abies alba*) whose requirements are very similar to those of the Beech in spite of a number of contrasts. Most authors regard it as a character species of the Beech wood alliance. The proportion of the two trees in various habitats can be quite different depending on altitude, continentality of the climate, type of soil and aspect of the slope. The pure or almost pure stands of Fir are separated from the Beech and the Beech-Fir woods on physiognomical, ecological and floristic grounds, as well as on grounds of forestry practice, and are dealt with alongside the rest of the coniferous woodland communities in section B IV.

Firs, like Beech, thrive on all types of bedrock although the centre of distribution of Beech tends more towards a calcareous soil. Even on fen peat *Fagus* as well as *Abies* can reach 25 m if the water table is lowered. Within quite a wide range, neither the degree of acidity nor the humus quality of the soil is a deciding factor for the presence (or absence) of either tree species. Both are to be found on slopes of all aspects as well as on plains or in valleys, provided always that the roots of the Beech have oxygen in the absence of waterlogging or flooding, whereas the Fir is more resistant to this. Almost all soil types from rendzina and pararendzina, through different brown earths and parabrown earths to podsols can be found supporting corresponding Beech wood types. These are even found quite frequently on parabrown earths and podsols which have been modified by gleying. Only true gley soils and similar wet ground, including the majority of flood-plain soils, are not occupied by Beech as long as the natural water table is maintained.

Beech would therefore be almost universally present under natural conditions in Central Europe apart from certain special habitats, most of the subalpine belt and the inner Alpine valleys which have low precipitation and very cold winters. Along with Fir in some places, it would dominate something like two-thirds of the whole area. Although most of the Beech and the mixed Beech woodlands have had to give ground to farming, and although the Beech is the 'most persecuted of all forest trees' (Hesmer 1936, Malek 1979) there are very many stands of Beech still surviving to the present day.

(b) Classification of Beech woods

Nearly all natural Beech-dominated communities, as well as some of the woods rich in Fir and Sycamore, are today included in the alliance *Fagion* by most authors. Apart from *Fagus sylvatica* itself there is not a single good, constant character species known which is applicable to the whole of Central Europe (see section E III 1). Fairly constant are some Toothwort (*Dentaria*) species, but these appear rather infrequently and are confined to the 'better' habitats, the majority in the southern part of Central Europe. Silver Fir is not found in any significant numbers outside the distribution of this alliance; within it, it is concentrated at the higher montane level.

An extraordinary large number of associations, subassociations, variants, races and similar subunits of Beech woodland communities have already been described and a review of them has been made all the more difficult because units occupying quite different habitats have been given similar-sounding names, whereas at different times the same type has often been given a variety of names. In this book we have had to restrict ourselves to the associations and a few subassociations, concentrating on those which have a wide distribution or are particularly noteworthy regarding their habitat relationships.

Because it is so extensive the Beech woodland alliance is now divided into a number of suballiances corresponding to groups of associations which are more or less closely confined to particular habitats (W. and A. Matuszkiewicz 1973, Oberdorfer 1979, Passarge 1965, Petermann 1970, Roisin 1961, So 1964b, Tanghe 1970, etc.):

1. Those Beech woods on very acid soils are particularly easily distinguished from the rest. They are usually put together in the suballiance of Woodrush-Beech woods (*Luzulo-Fagion*). They may also be called acid humus Beech woods (as in the first edition of this book) or **moder Beech woods** since the ground is always covered with a layer of acid moder humus, whether the underlying soil is an acid brown earth, a parabrown earth, ranker, or any other type (section 4 a).
2. The majority of Beech woods are found on soils where the humus is in the form of mull (section B I 1 a). A suitable name therefore is **mull Beech woods** for the extensive suballiance *Eu-Fagion*. Many authors refer to this as the Woodruff-Beech wood (*Galio odorati-Fagion*, previously *Aperulo-Fagion*). Within this unit a group of communities on rendzina (the so-called slightly-moist limestone Beech woods, sections 2 a and b) can be distinguished from another group on brown earth (the brown-mull Beech wood, section 3).
3. While the first two sections occupy habitats which have no local climatic peculiarities and enjoy a 'normal' water supply, the Orchid-Beech woods (*Cephalanthero-Fagion*) can put up with much drier conditions. Since as a rule these are associated with steep slopes, and since sedges are conspicuous in the undergrowth, the term slope-Sedge-Beech wood or steep-slope Beech wood are also used. Ecologically specific however would be the name dry-slope limestone Beech wood or, shorter, **dry Beech wood** (sections 2c-e).
4. The subalpine Beech woods in the mountains of south-west Central Europe have many species in common with the woods on the shaded slopes which are rich in Sycamore. Therefore they are placed in the suballiance Sycamore-Beech wood (*Aceri-Fagion*) or **high altitude Beech wood** (section 2 h).

In the communities of all four suballiances, Silver Fir plays a certain, but mostly subordinate, part, especially in the montane belt (section 2 f). Stands in which this conifer is naturally dominant are dealt with in section B IV 2, whether they are considered to be a suballiance of *Fagion* or rather a separate alliance.

Differences in the habitats of the four suballiances of *Fagion* correspond to the presence or absence of groups of indicator plants which can be looked upon as differential species for these suballiances. *Luzulo-Fagion* is recognised by acid indicators (groups II a and b in table 10) while a number of 'more demanding' species are absent, so that communities belonging to this group are poorer in species than the other Beech woods. The *Cephalanthero-Fagion* is characterised positively by some drought-tolerant plants (group I c-e in table 10), negatively by the absence of drought-sensitive ones, that is the more hygromorphic species. The plants of deciduous woodland which require both a high level of fertility and also sufficient water are numerous in the communities of the *Eu-Fagion* (= *Asperulo-Fagion* = *Galio odorati-Fagion*, groups IIc-e in table 10). The subalpine *Aceri-Fagion* differs from the three colline or montane suballiances in having tall broadleaved perennial herbs which thrive in the nitrogen-rich semishaded sites especially in the higher mountains (group C in table 10).

From the aspect of geographical distribution, according to the flora of Rothmaler, Meusel and Schubert (1972 etc.), the Beech woods in Central Europe consist primarily of species widespread in the southern and central parts of Europe which, like Beech itself, have a more or less montane character and a distribution in subatlantic to Central European climatic zones. In addition some boreomeridional woodland plants, i.e. those that are met with in the entire north temperate zone, play a part, although again these tend to have their main concentration near the sea. A few species extend as far as the tropical mountains (e.g. *Sanicula europea* in Usambara), or are related to

Tab. 11. Two examples of the slightly-moist limestone Beech woods in the Swiss Jura.

Serial no.:		1	2	F (moisture value)	R (reaction value)
Month of relevé		6	9		
Height above sea level (10m)		80	77		
Aspect		S	ENE		
Slope (%)		20	75		
Age of trees (years)		40	120		
Height of trees		15	28		
Canopy cover (%)		95	70		
Trees:					
V *Fagus sylvatica*		5	4	–	–
V *Abies alba*		1	1	–	–
O *Acer pseudoplatanus*		1	1	–	–
K *Fraxinus excelsior*		+		–	–
Tree saplings:					
V *Fagus sylvatica*		1	3	5	×
V *Abies alba*		1	1	×	×
K *Fraxinus excelsior*		1	1	×	7
O *Ulmus glabra*		+	+	7	×
K *Acer platanoides*		+	+	×	×
O *A. pseudoplatanus*		+		6	×
Sorbus aria			+	4	7
Shrubs (± early flowering):					
K *Lonicera xylosteum*		+	+	5	7
Daphne laureola		+	+	4	8
K *D. mezereum*		+	·	5	7
K *Corylus avellana*		·	+	×	×
Viburnum opulus		·	+	7	7
V. lantana		·	+	4	8
Rubus spec.			1	–	–
Geophytes (± early flowering):					
O *Mercurialis perennis*	m,a	3	+	×	7
K *Anemone nemorosa*	m	2	·	×	×
O *Polygonatum multiflorum*	a	1	+	5	6
O *Paris quadrifolia*	a	+	+	6	7
O *Arum maculatum*		+	1	7	7
V *Dentaria heptaphylla*	m?	2	1	5	8
V *D. pentaphyllos*	m?	1		5	7
O *Neottia nidus-avis*		+		5	7
K *Epipactis helleborine*		+		5	7
O *Lilium martagon*	m	+		4	7
K *Melica nutans*	a	+		4	7
Carex flacca	a		+	6	8
V *Cephalanthera damasonium*			+	4	7
O *Allium ursinum*	m		+	6	7
Other herbaceous plants:					
O *Lamiastrum galeobdolon*	m, a	1	2	5	7
K *Hedera helix* (steril)	a	1	2	5	×
Geranium robertianum	m		+	×	×
Total number of other species		3	5	–	–

Serial no.:		1	2	F	R
Early flowering hemicrypt.:					
O *Asarum europaeum*	m, a	2	1	6	8
O *Phyteuma spicatum*	m	1	2	5	×
O *Lathyrus vernus*	a	1	1	4	7
O *Viola reichenbachiana*	m	1	+	5	7
K *Carex sylvatica*	m	1	+	5	7
Ajuga reptans	m, a	2	+	6	×
K *Carex digitata*	m	+		4	×
O *Galium odoratum*	a	·		5	×
O *Pulmonaria officinalis*	m, a	·	+	5	8
Oxalis acetosella	a		1	6	4
Later flowering hemicrypt.:					
V *Festuca altissima*		2	2	5	3
V *Prenanthes purpurea*		+	2	5	×
Bromus ramosus benekenii		+	1	5	8
Vicia sepium	m	+	+	5	7
O *Potentilla sterilis*	a	+		5	6
O *Dryopteris filix-mas*			2	5	5
O *Euphorbia amygdaloides*	m?		1	5	7
V *Polystichum aculeatum*			1	6	6
V *Hordelymus europaeus*			+	5	7
O *Epilobium montanum*			+	5	6
Fragaria vesca	a		+	5	×
Hieracium sylvaticum			+	5	5
Solidago virgaurea			+	5	×
Athyrium filix-femina			+	7	×
Senecio nemorensis fuchsii			1	5	×

Calculation of mF and mR:

		F	R
Number of species in each moisture category	4	5	5
	5	18	22
	6	4	7
	7	2	4
Product of moisture value and species number		20	20
		90	110
		24	42
		14	28
Sum of products		148	200
Number of evaluated species		29	38
Average moisture indic. value		5.1	5.3
Number of species in each acidity category	3	1	1
	4	–	1
	5	–	2
	6	2	3
	7	15	16
	8	4	7
Average acidity indic. value		6.9	6.8

Note: The letters in front of the plant names: these show the characteristic species for the alliance V = (*Fagion*), the order O = (*Fagetalia*) and the class K = (*Querco-Fagetea*); after the plant names: m= myrmecochorous (spread by ants), a = the species spreads vegetatively.

The figures in the vertical columns 1 and 2 indicate the cover degree according to Braun-Blanquet: 5= more than 3/4 of the ground covered, 4= 3/4 – 1/2, 3= 1/2 – 1/4, 2= 1/4 – 1/20, 1= less than 1/20, + = occasional. A dot indicates that the plant occurs in a similar habitat but is absent from this particular example. The indicator values are explained in section B 1 4. The average values for moisture and acidity are calculated without reference to the cover degree.

Source: From Moor (1951); indicator values for soil moisture (F) and acidity (R) after Ellenberg (1974)

tropical species (as is *Ilex aquifolium*). Beech woods on very acid soils are not significantly different from those on less acid soils in that many so-called 'acid indicators' such as *Luzula luzuloides* and *Avenella flexuosa* have a similar geographical distribution. This is another reason why the moder Beech woods should not be separated from the other Beech woods by very deep systematic divisions as is done by Doing-Kraft and Westhoff (1959), Scamoni and Passarge (1969) and Passarge and Hoffmann (1968) for instance, in joining them with the Birch-Oak woods (*Quercetalia robori-petraeae*, section B III 5).

In order to get a clear picture of the different Beech woods we would like to start by describing some examples. We shall look first of all at the region in which each particular community is best developed or where it occurs in greatest variety. Moving out from this central area we shall, at the end of each section, try to survey the total area throughout which the particular Beech wood type is distributed. In doing this we shall point out floristic differences in connection with the habitats as well as with the geographical distribution of the association concerned, that is, its 'geographical races' (Kral, Mayer and Zukrigl 1975, etc.).

As a central part of the description we shall present the results of ecological investigations as far as these are available for the community under discussion, or for comparable units. Thus we shall endeavour to determine the factors on which the species composition is primarily dependent and to elucidate the way in which these factors operate.

2 Beech woods on rendzina and pararendzina

(a) Slightly-moist limestone Beech wood

Beautiful seminatural Beech woods still exist in Central Europe, above all on the slopes of limestone mountains. They are found extensively in the Swiss, Swabian and French Jura, on the limestone mountains in central Germany, southern Poland and Czechoslovakia, as well as in the montane belt of the limestone Alps and Carpathians, because all of these areas are difficult to cultivate. However Beeches can flourish here very well and can rejuvenate themselves without the help of man. So in describing the Beech woods of Central Europe we shall begin with the 'limestone Beech wood', in particular those of the Swiss Jura. These are in the centre of the area of distribution of Beech in Europe and rise up practically to the climatic tree line. In doing this we are able to refer to the exemplary work of Moor (1952) and Bach (1950) and also to that of J.L. Richard (1961) and others who have distinguished many associations in which Beech is dominant.

Moor describes a 'typical' Beech wood ('*Fagetum sylvaticae*') simply as a community which ecologically lies about in the centre of the other Beech wood types and shows no floristic peculiarities. In his classical vegetation monograph of the Swabian Jura as long ago as 1898 Gradmann (see 1950) called it a 'normal high Beech wood'. Such names however only apply when one is simply looking at the Beech woods on lime-rich soils and does not consider the ones on sour soils, in particular on silicaceous rocks. The latter take up a larger area of Central Europe today, and even under natural conditions they would have been more widespread than the limestone Beech woods. We shall use then the expression 'typical limestone Beech wood'. Since the majority of dry-slope Beech woods are also on limestone (tab. 12), for our example on average-damp soils, only the label 'slightly moist limestone Beech wood' is sufficiently definite.

In order to get a clear picture of the structure and species composition of this woodland type we shall take two examples from the tables published by Moor (1952) of which the first relevé was made in the spring and the second in late summer. Other differences are shown in the two lists in tab. 11. These refer to the age height and density of the two woodland stands as well as the aspect and degree of the slope. Taken together these two relevés already contain almost all the species which are to be commonly met with in the typical limestone Beech woods of the Swiss Jura in sites of average altitude. The species absent have been shown marked with a point (.), so that tab. 11 can be used at the same time as a list of the 'normal characteristric species combination' of the abstract association type.

To become familiar with the ecological evaluation of vegetation relevés we will discuss this first example more thoroughly than will be possible with the remaining communities. Like the majority of woodland dominated by Beech our two stands are real 'cathedral forests'. Like the columns of a gothic cathedral the slender smooth-barked silver-grey trunks rise up to where the branches arch out to form the canopy of leaves of regular height (fig. 55). In the twilight of the younger dense stand (No. 1 in table 11) there are neither shrubs nor young trees, so one is able to peer into the depths of the wood without difficulty. Also the mosaic of herbs, grasses and sedges on the ground of this uniform stand is rarely strong enough to completely cover with green the golden-brown carpet of dead leaves. Even in the 120-year-old stand, which has become somewhat more open by so-called regeneration cutting, the impression of a cathedral remains, although the Beech saplings form groups of various sizes under the holes in the canopy (figs. 55 and 56).

When the old trees are felled the young ones soon grow up to form real thickets which, being poor in undergrowth species, often make the mapping of the Beech wood associations more difficult and naturally do not appear in plant sociological tables. However they represent a characteristic stage in the life of Beech woods by which many species of the more open deciduous woodlands are eradicated. Whenever the trees are thinned out a carpet of herbs and grasses soon sprouts up on the floor of the forest, whether this is through the spreading of the almost starved survivors from their rootstock, or by runners, or by seeds of myrmecochorous species, which are constantly being dispersed by busy ants and which now can develop beyond the seedling stage. It is not by chance that in Beech woods we find myrmecochores (m) or else species adapted to strong vegetative spreading (a), and that many plants which are particularly plentiful in Beech woods, e.g. *Mercurialis perennis* and *Melica uniflora* combine both peculiarities (fig. 57, tab. 11).

One of the reasons why the young Beeches prevent more light from reaching the herbaceous plants is that the patches of sunlight moving across the ground are much smaller and fewer than is the case with older trees (cf. fig. 58). In addition the leaves of the younger trees expand earlier in the spring than those of the taller trees, and in the autumn they often retain a proportion of the dead leaves whereas trees in the higher layers quickly lose their leaves giving the plants on the ground, which are still green, the opportunity for increased photosynthesis.

Today the forester can alter the nature of the

Fig. 55. A tall Beech wood, poor in shrubs, on calcareous soil in the Swiss Jura (Lebern). Beech saplings are growing in the lighter place. In contrast to the majority of woods in the central and northern parts of Europe where the concentration of sulphur dioxide in the air is intolerably high, the epiphytic lichens have not disappeared from this wood.

woodland at will and the result is a varying sequence of thickets, young trees which can be used for poles, a closed wood of more mature trees or an old woodland which has clearings enabling regeneration to proceed. However a completely undisturbed primaeval Beech forest would also have a similar rhythm. This has been well demonstrated by records and transects taken by Mayer (1971) in a Beech forest near Dobra in Lower Austria which had been very difficult to reach and had remained practically undisturbed (fig. 59). As in other remains of natural forest in temperate zones it is possible to distinguish stages of development which are sequential in any particular place but over a whole area form an irregular mosaic of large patches. They are:

1. The **optimal phase** of strongest tree growth where there is a dense growth of young trees. In some places this may already be taking on the look of the 'cathedral'.
2. The **terminal phase** is a real cathedral high forest of dominating large trees. This stage normally covers the greatest area of the woodland.
3. In the **decay phase** the old trees die and break up. This progresses in three steps:
   a) The start of the decay period when odd trees die, but remain standing, already allowing stronger growth on the ground and the beginning of regeneration.
   b) As the decay phase continues many trees fall at the same time.
   c) The closing phase of the decay period is characterised by areas where the old trees are completely broken down and there is a flourishing growth of nitrogen-loving tall herbs in the clearing (section A I 2 and D III 2) and also groups of young trees.
4. The **regeneration phase** begins in 3 c although in some places it has already started in 3 a.
   a) To start with Sycamore, Elms and Ash play a large part although there are some Beech also.

Fig. 56. A Beech wood in winter, photographed at the end of February. Fresh snow shows up the twigs of the occasional young Beech and the saplings. The herbaceous layer is completely covered in snow and has to survive temperatures of only a little below 0°C.

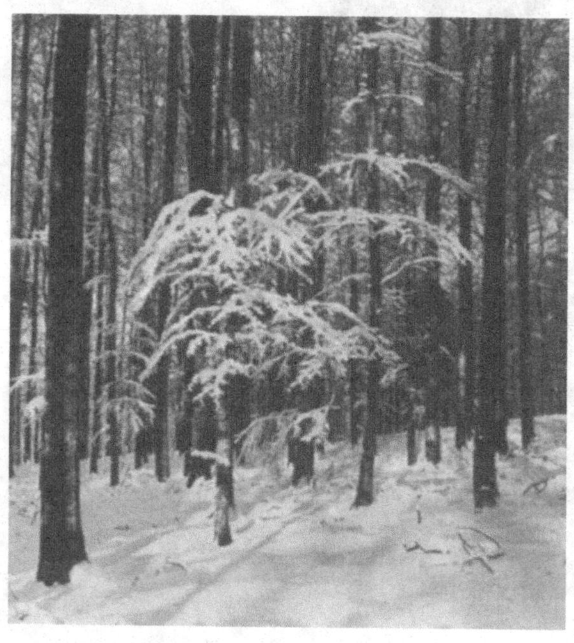

Fig. 57. The herbaceous layer of a limestone Beech wood in the Swiss Jura, photographed at the end of May.
Mercurialis perennis (right foreground) is already in full bloom, *Galium odoratum* (covering the ground) has buds, *Dentaria pentaphyllos*, in full flower, and the occasional leaves of *Allium ursinum* (left) show that there is ample moisture, at least in the spring.

b) As regeneration progresses the Beech gradually win through and achieve the optimal phase (1) once more.

The great similarity between primaeval woods and cultivated forest makes it plausible that the original vegetation was almost completely dominated by beech just as the potential vegetation would be today (Schroeder 1973). According to measurements taken by Roller (1965) the temperature variations on the ground are greater in natural forest than in one which is managed. This is due to the longer duration of the clearing phase which favours the heliophytes mentioned in table 2 b. The humidity too is less equable although wind can pass more freely through large areas of high forest.

Descriptions of the primeval Beech woods of the Lower Carpathians (Zlatnik 1935) and on the Balkan peninsula (Markgraf 1931, K.M. Müller 1929, Wraber 1952, Leibundgut 1959) had already suggested that the phases 1-4 also apply to Central Europe. It is safe to assume that the Beech tends to form 'cathedral' high forests which are almost lacking in undergrowth and where all the trees, from a mere 60 to well over 200 years old carry a canopy of practically uniform height. In those places where an odd senile tree dies the Beech saplings which germinate in large numbers after every 'mast year' (that is every three to eight years, fig. 41) appear in groups or over the whole surface. On the other hand in the dense closed stands they would perish through lack of light, through competition with the tree roots and through pest attack. Nature works in a way similar to the 'regeneration cutting' of the forester by which clean stands of even growth are produced. Single trees may succumb to parasitic fungi or other diseases, in virgin as well as in managed woodland. In this way the light-loving species are also favoured from time to time (F. Runge 1969b). Larger openings in natural woodland may be produced as a result of storm damage. These are occupied by a strong growth of competitive clearing plants similar to those following the clear felling of our coniferous forests (section D III 2). An example is described by H. Mayer (1971). In such large glades the Beech cannot regain a footing for a much longer time, so in areas where the forester values the natural regeneration of the Beech he avoids extensive clear felling.

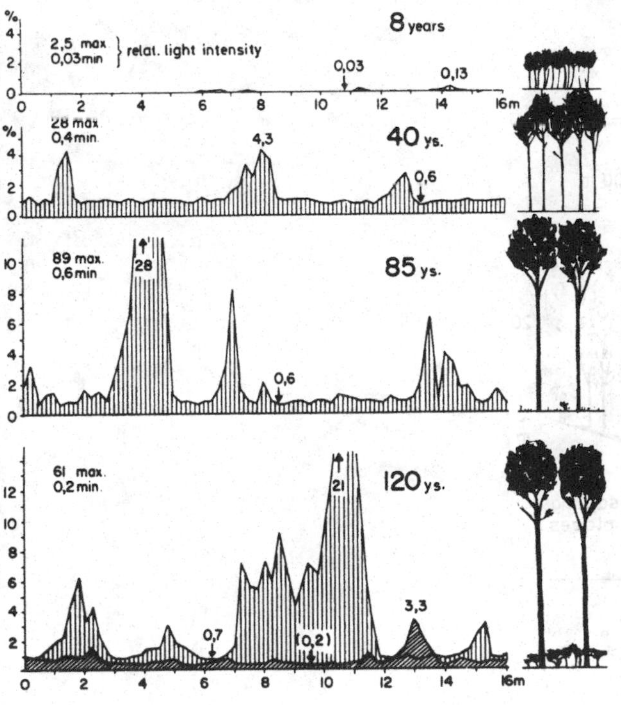

Fig. 58. The distribution of relative light intensity on the ground under a young Beech thicket, a stand with many slender stems, a mature forest and a gappy old wood.

Measurements taken at midday without cloud at 20 cm intervals along a west-east line 16 m long, at the beginning of September 1960. Diagonal hatching = under a group of Beech saplings.

Natural and managed Beech high forests thus have much more in common than may at first be imagined. They also resemble each other, and differ from other types of woodland, in having very poor moss growth on the soil. This is associated with the sudden heavy leaf fall each autumn which covers the ground with 3 to 8 layers of broad, rot-resistant leaves. Even on the most active soils this coherent litter cover, so inimical to the growth of mosses, is not broken down sufficiently to expose bare soil (fig. 48). In a Beech wood therefore mosses can only occupy stones, stumps or dead branches which stick out over the carpet of leaves. Here they form small communities which have little to do with the rest of the woodland flora and they are generally ignored in plant-sociological records. The mosses round the boles of the trees and the lichens and algae on their smooth trunks (fig. 55) also occupy special habitats and are dealt with separately (section C VIII), although they may serve as a characteristic feature of Beech woods, especially in the montane belt.

All herbaceous inhabitants of the Beech woods as well as tree seedlings must be able to penetrate or to overgrow the leaf litter cover which is renewed every autumn and is more or less persistent depending on the soil fertility. The growth of some potentially productive grasses can be prevented or hindered by this factor whereas many spring geophytes are better adapted to

Fig. 59. Transects in the 'primaeval' Beech wood at Dobra in Lower Austria (Dentario-Fagetum on brown earth) showing the decay of the old Beech stand and its regeneration. After Mayer (1971), somewhat modified.

In the 'terminal phase there are already groups of young Beech growing up. As the old trees succumb communities including tall herbaceous plants appear in the clearings but soon the growing Beech saplings take over possession of these areas.

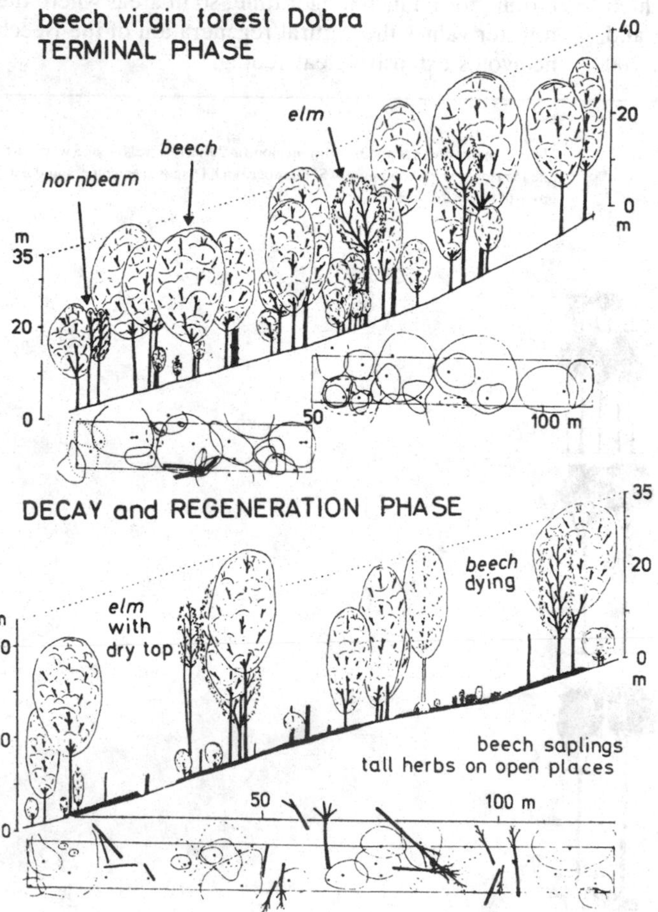

it and relatively favoured in competition (Syders and Grime 1981). This may be the reason why, for instance, *Poa nemoralis* dominates at wind-exposed woodland edges while it is rare in the interior.

If, after this essentially physiognomical survey, we return to tab. 11 and look at the species composition of limestone beech woods in more detail we see that although Beech certainly dominates it does tolerate some presence of representatives of other tree species. In the Swiss Jura Silver Fir is rarely absent although nowhere in slightly-moist limestone Beech woods is it able to establish itself in larger groups without the timely help of the forester. Ash and Sycamore too appear as single specimens. Norway Maple, Elm and other trees on the other hand are not able to keep pace with Beech and at best are confined to the undergrowth provided there are seed-producing specimens in the neighbourhood.

True shrubs are very few in older stands even when there are clearings because the young Beeches smother everything. The shrub species make such poor growth in typical limestone Beech woods that at first they may be quite overlooked. Nevertheless species like *Lonicera xylosteum* and *Daphne mezereum* can be found regularly. The Mezereon produces its pink flowers as early as the end of February or in March, but is so difficult to see that it is more often betrayed by its sweet scent than by its colour.

The ground flora of our limestone Beech woods is dominated in the spring by geophytes and other early flowering species. Many of these soon turn yellow after the Beech canopy shuts out the light in summer, e.g. *Anemone nemorosa* and *Arum maculatum*. The majority however retain their green colour and some only come into full leaf at this time, for example *Mercurialis perennis* and *Melica nutans*. As far as rhythm of development is concerned therefore they behave in a similar way to the 'early flowering hemicryptophytes' such as *Carex sylvatica* and *Lathyrus vernus*. For many hemicryptophytes which are only fully expanded after Beech has unfurled its leaves the stand No. 1 would be too dark. They are therefore much more strongly represented in the second example where the canopy is only 70% closed. Similarly it will be noticed that the only annual plant (*Geranium robertianum*) occurs here. It occasionally turns up in typical limestone Beech woods. Only a few species remain green throughout the winter. Apart from facultative winter-green plants such as *Galium odoratum* and *Polystichum* there are just *Lamiastrum galeobdolon*, the only chamaephyte, and *Hedera helix*, the Ivy creeping over the ground. Both these appear in some quantity, in stand No. 2.

None of the plants listed in tab. 11 can serve as a character species of slightly-moist limestone Beech woods. They all turn up in other associations, at least in other types of Beech wood. In spite of the absence of unique species and the small number of good alliance character species (V) the slightly-moist limestone Beech wood is sufficiently recognisable floristically as a separate association. Apart from the vigorous Beech one must take note above all of the large number of fastidious 'mull soil plants'. Almost all the herbaceous plants named in tab. 11 belong to the ecological group which avoids sour soil (R 6-8) with the exception of *Oxalis acetosella* and *Festuca altissima* which are more frequent on moder soils (R4 and 3). These establish themselves on decaying wood which forms small acid islands on the calcareous soil.

In respect of their water relationships almost all the species can be described as mesophytes (F 5 and 6). *Melica nutans. Carex digitata* and a few others are able to withstand drought more often (F 4). They only occur in stand No. 1 as this is growing on a gentle southern slope and in its habitat tends towards the Sedge-Beech wood of the dry limestone slopes (see section c). Stand No. 2 on the other hand, which occupies a steep ENE position contains some hygromorphs such as *Athyrium filix-femina* (F 7). These are more sensitive to drought and so are found mainly in shady places. This difference is also expressed in the ratio of ecological groups and in the average moisture values (5.1 and 5.3).

Strictly speaking then the typical limestone Beech wood is not an entirely uniform community. The two relevés reprinted in tab. 11 are not even extreme cases. Only 'average' examples were deliberately chosen from the 35 lists published by Moor (1952), especially with regard to altitude.

All such minor floristic deviations are very probably determined by differences in the habitat. In other words are not fortuitous. They are met with in each association and do not prevent us from establishing certain types, and tracing these throughout the whole of Central Europe. Yet it is obvious, and should always be kept in mind, how variable the vegetation types are and how many of the individual features of each plant stand are lost through abstraction. Nevertheless floristically-defined vegetation types are much more homogeneous than those determined only by the dominance of one or two species.

If we wish to specify the decisive conditions of the habitats under which the typical limestone Beech woods develop then, in accordance with Moor (1952) and Bach (1950), we can point out the following factors in the terrain:

Requirements for **all** more or less pure **Beech woods**:

1 Situated in Central Europe, i.e.
within the natural distribution of the Beech and most of the Central European woodland plants; climate of average temperature; suboceanic to weakly subcontinental.

2 Submontane to montane belt,
or a similar climate on the slopes at a lower level, i.e. the absence of keen frost in winter or early spring; relatively high precipitation, no long period of drought.

3 Soil not strongly influenced by groundwater or
flooding, i.e. no temporary or permanent saturation which would lead to shortage of air to the roots.

4 Near natural forest management, i.e.
no strong favouring of other tree species, no extensive clearing.

Requirements for **limestone** Beech woods:

5 A previous calcareous bedrock, i.e.
average to well drained subsoil, well supplied with bases, at least for the deep-rooting plants; neutral reaction (weakly acid to weakly basic)

Requirements for the **slightly moist** limestone Beech woods:

6 A stable soil, i.e.
no noticeable removal or deposition in spite of slope; leaching of lime in the uppermost soil layer.

7 Stony soil with average clay content, i.e.
containing pieces of limestone right up to the surface; average power of water retention.

Such soils are mostly of rendzina type as described by Kubiena (1953) and Kohl (1971), but not by Bach (1950) and Moor (1952) who together with Pallmann and Haffter (1933) speak of a 'humus-carbonate soil'. The rendzina is frequently brown, i.e. the fine earth fraction is poorer in lime and resembles brown earth to a certain extent, often due to a thin loess or loam cover. There are however typical limestone Beech woods also to be found on mull rendzina which is not brown. The limestone rock may come to within a few centimetres of the surface or it may be covered with a more or less thick layer of stony scree, which has been undisturbed for a long time, or with a thin layer of loess loam. In each case the root run of the Beech and its satellites extends on average only to about 20-30 cm, although individual roots will penetrate to 2 m or more into the subsoil or into rock fissures (fig. 60).

Fig. 60. Mullrendzina on a Triassic limestone formed under a limestone Beech wood (Ohmgebirge near Duderstadt). The topsoil which is only about 25 cm deep is thick with roots; occasionally the tree roots penetrate more than 2 m deep into crevices in the rock.

These kinds of rendzina react neutrally or at most weakly acidic right up to near the surface and are of average fertility. Their nitrification is constant, but not very strong (Ellenberg 1964, 1977, Grimme 1975). Waterlogging of the soil profile has never been observed even after the snow melts or a heavy downpour since the spongy texture of the base-rich mull earth with its fine pores remains stable and is very absorptive. In dry years there can be a short-term water shortage as was observed for example near Göttingen in the late summer of 1973 (fig. 63). Over large areas of the slightly-moist limestone Beech wood the *Mercurialis perennis*, *Galium odoratum*, *Asarum europaeum* and other herbs wilted although the osmotic value of their cell sap had been raised considerably. For Beeches however the water supply remained guaranteed because they tap reserves from deeper layers.

The floristic and ecological picture of the slightly-moist limestone Beech woodland that we have drawn so far holds good in its basic outline for large parts of Central and western Europe and even for many marginal areas to the north and south. An overall view with a few random samples should suffice to show this and at the same time serve as an example for many other plant associations where we shall be even less able to go into detail. Anyway completeness cannot be the aim of a book which tries to supply an understanding of all the types of vegetation over a large intensively researched area.

According to Kuoch (1954) the slightly-moist limestone Beech wood is one of the most plentiful woodland communities, not only in the Swiss Jura, but also along the northern side of the Alpine chain in the lower montane belt (600-900 m). Here it is found on limestone rocks of all kinds and also on marles and lime-rich moraines differing scarcely at all in species composition from those described in the Jura. In the northern valleys of the Swiss Alps, it is certainly still to be found on the shady slopes, but as the climate becomes drier and takes on a more continental character in the inner Alpine valleys, it becomes rapidly scarcer. Along the northern edge of the German Alps too, this type of limestone Beech woodland becomes less frequent as the overall climate becomes less oceanic. In the Chiemgau and Kitzbühel region for example conifer communities take its place towards the inner Alps much more quickly than they do in Switzerland (Mayer 1974). The tendency of slightly-moist limestone Beech woods to a 'subatlantic' distribution in the Alps is thus unmistakable (fig. 42). This is also true in the Carpathians where Beech woods, according to Zarzycki (1964) occur on very different soils.

The Beech woods on the south-eastern limestone Alps lead into those of Illyria (NW Balkans) of which I. Horvat (verbal communication) once said that in their wealth of species they could best represent the primaeval Beech forests of Central Europe. In addition to the species we find in our limestone Beech woods, the shrub layer here contains evergreens such as *Rhamnus fallax* and *Ruscus hypoglossum*, while the herbaceous layer has Illyric perennials, for example *Omphalodes verna* and *Lamium orvala*, which are cultivated in gardens further north for their lovely flowers (Horvat, Glavač and Ellenberg 1974).

It is typical of the regions to the north of the Alps that after the Ice Age new Beech woods arose which were relatively poor in species and rather monotonous. So when we describe these we shall not start out from the rich variety of the Illyric centre, but from the north-western Alpine foothill region which admittedly has fewer species and is more representative as a secondary centre for Central Europe.

If one travels northwards from the Swiss Jura or from the western part of the Swabian Jura it will be seen that floristically the limestone Beech woods become steadily poorer. Above all Silver Fir, a subsidiary but very important character species, disappears. The subatlantic-mediterranean *Daphne laureola* is already absent in the Swabian Jura, and many of the montane species, plentiful in the Alpine region, e.g. *Lonicera alpigena*, *L. nigra* and *Veronica urticifolia* soon become rare. Only one new species appears, *Melica uniflora*, more strongly competitive here, but avoiding the Beech woods nearer the Alps. By and large however the species composition of slightly-moist limestone Beech woods in similar habitats remains surprisingly constant. Even in the Eifel and the hills surrounding the north-west German plain there are stands which could be compared in every detail with the examples in tab. 11. This is especially true of the 'typical' variant – the 'herb-rich limestone Beech woodland' – in which Wood Melick is infrequent. Rühl (1960) has recorded this in the limestone Eifel, the plateaux of Brilon and Paderborn, the Osning, the Weser and Leine hills, and even the Elm east of Braunschweig (= Brunswick).

Like the stands investigated by Moor and Bach those in north-west Germany remain on the fairly level rendzinas which are mostly brown and do not contain more than average amounts of clay. In soil profiles Diemont (1938) was able to demonstrate thin layers of loess loam in many places giving rise to the formation of brown earth soils. These can also be recognised by their lower pH value and smaller pore volume and are indicated by the absence of calciphilous species. But even the most shallow rendzina soils in north-west Germany show a more acid reaction near the surface (pH 5.7-7.4, average 6.3) than those in the Swiss Jura. This fact, already recognised by Diemont, may well be associated with the more strongly oceanic character of the climate.

In the relatively humid climate of the Weser hills Diemont found the typical variant of his Wood Barley-Beech wood (which corresponds to our slightly-moist limestone Beech wood) exclusively on the south, south-west or western slopes where the local climate was drier and soil erosion more likely. On the other hand the 'eastern variant', widespread in the north-eastern Harz foothills preferred gently north-facing slopes. In the Thuringian limestone area around Meiningen with a still lower precipitation a similar community – called 'Wood Barley-Sycamore-Beech wood' by G. Hofmann (1959) – occupies only the most shady foothills and only then where the limestone is covered with a moisture-holding layer of loam.

So our slightly-moist limestone Beech wood, which is widespread in temperate oceanic climates as the climax vegetation on montane and sumbontane calcareous soils on both slopes and level ground, is restricted to fairly special sites near the boundaries of its normal range. Where the climate is wetter it occurs only on drier sites and conversely it occupies wetter places where there is less rain. Thus it serves as a good example of H. and E. Walter's (1953) 'relative constant habitat rule'.

Some stands described by Watt (1934), Tansley (1939) and others in south-east England and by Linquist (1931) and Passarge (1965) in southern Sweden can be looked upon as the last extensions of slightly-moist limestone Beech wood towards the colder north. Also on the chalk at Rügen there are similar Beech wood communities. Beech woods resembling those described here can develop in a suitable climate on lime-rich young moraine soils, as well as on base-rich basalt or other rocks which have weathered to brown earths with a mull horizon similar to rendzina. In the Alpine foothills also the '*Fagetum typicum*' of Moor (1952) is not confined to limestone in its narrow sense, according to Kuoch (1954). One should not then relate the name 'limestone Beech woodland' just to the geological substrate but to the base content of the root run of the herbaceous layer. This comes out too in the classification of the '*Cardamino bulbiferae-Fagetum*' described by Lohmeyer (1962) from the northern Rheinish slate hills. We shall see that even definite acid-soil Beech woods can be found on limestone in those places where a completely leached-out covering of loam isolates the plants from the limestone substrate (Section B II 4c).

In western Europe, especially in the centre of France, limestone Beech woods or closely related communities cover large areas, for example in the southern Cevennes, southern Dauphiné and in the Auvergne. Compared with the species composition in the French Jura, and even more so in the Swiss, they are joined by more atlantic and submediterranean floral elements as they spread further west and south, especially in the Pyrenees and in Spain (Rivas Martinez 1964). As in southern France and in the Balkans so also in Italy the Beech woods come into contact with submediterranean mixed *Quercus pubescens* woodland and possess a very rich flora.

In Poland A. Matuszkiewicz (1958) has tried to differentiate geographically between all the Beech wood communities. It transpires among other things that, of the limestone Beech woods, a Carpathian group can be separated from all the rest. This contains 'eastern' differential species, just as other communities in the same region do.

The slightly-moist limestone Beech wood may be taken as an example of how an association limited to specific soil conditions can be separated into geographic races. If one looks at Europe as a whole and in so doing tries to understant basic relationships between the vegetation and its habitats, then it is obvious that one should distinguish between as few ecologically understandable associations as possible and examine these throughout their entire distribution. This seems to be more advisable than the opposite principle, i.e. to distinguish many geographic associations (*Fagetum carpathicum, Fagetum illyricum* etc.) and divide them into edaphic subunits. Many plant sociologists share this view today, e.g. Tüxen (1960), Oberdorfer et al. (1967), Kral, Mayer and Zukrigl (1975) and Braun-Blanquet (1964).

Fig. 61. A Wild Garlic-Beech wood on a shady slope in the Swiss Jura (Lebern). In spring *Allium ursinum* forms dense carpets on the deep loamy calcareous soil. *Galium odoratum* is able to spread during the rest of the season.

(b) Wild Garlic-rich Beech woods

Limestone Beech woods have been described from different parts of central Europe which show an unusual richness in spring-flowering geophytes, particularly Wild Garlic (*Allium ursinum*) and other species of the *Corydalis* group (table 10, III e). These quick-growing herbs require a very high soil fertility and are sensitive to drying out. However they are not confined to Beech woods but form dense carpets also in Sycamore, Ash, Oak and Elm mixed woodlands provided the soil can supply their high requirements. Thus they cannot be looked upon as species characteristic of associations, but rather as conspicuous differential species for particular subassociations or variants. Here we shall restrict ourselves to the lower-order communities of the limestone Beech woodland (fig. 61) and also to a corresponding subunit of the rich brown-mull Beech wood (tab. 14, no. 1, 2).

The centre of distribution of the geophyte-rich Beech woods is without doubt those parts of northwest Germany with humid climate, where Diemont (1938) has carried out intensive studies, and generally in the western part of Central Europe. The Wild Garlic community thus shows an even more marked oceanic tendency than the typical limestone Beech woods, and likes a really moist, but not wet, soil. In contrast to the typical association the herb layer in the Wild Garlic-Beech wood is covered with a luxuriant carpet of leaves, almost without any gaps, early in the spring. In the stands studied by Diemont this was dominated either by *Corydalis cava* with its pale carmine or white bunches of flowers and its deeply divided tender leaves, or by *Allium ursinum* with broad lanceolate shining leaves and umbels of white stars. Only by closer observation does one discover the Yellow Anemone, which is rarely absent from the Wild Garlic-Beech wood in north-west Germany, the Snowflake which has been almost exterminated by plant lovers, Yellow Star of Bethlehem, equally rare, or Moschatel. Celandine and Wild Arum, which also occur occasionally in the typical limestone Beech wood, together with the indifferent Wood Anemone complete the mosaic of the early flowering plants. Scattered amongst these almost any of the herbs of the typical limestone Beech wood may turn up although they are practically swamped by the luxuriance of the geophytes, especially at the beginning of the season. Shrubs are even less in evidence than in the typical community since the Beeches appear to form a denser canopy.

Soon after this has completely closed over, *Allium, Corydalis, Leucojum, Gagea* and *Ranunculus ficaria* turn yellow and disappear almost as soon as they spring up after the melting of the snow, and the dying leaves of the garlic fill the air with the smell of onions. It would be wrong to assume though that this rapid dying down is merely a consequence of the shortage of light, for in less-shady habitats, such as a neglected meadow, in which in particular *Ranunculus ficaria* and *Corydalis cava* have been able to grow for decades, one sees that their leaves turn yellow as soon as, if not sooner than, in the wood. The death of the leaves appears to be triggered off by the rise in temperature of the layers close to the soil. Only *Arum* keeps its leaves longer than the other geophytes and in doing so can be recognised as a representative of a primarily tropical family.

In the two or three months of activity the extreme spring geophytes have not only to form leaves, flowers, fruits and bulbils, but also to collect sufficient nutrients in their bulbs, tubers or rhizomes to ensure a successful start the following spring. Such intensive production is only possible on very fertile soils and appears to depend more on the availability of nitrogen than on the lime content of the soil. In all places where representatives of the *Corydalis* group flourish one finds porous and very active mull soils which, at least in early spring, never dry out but at the same time do not hold an excessive amount of water. Both extremes would hinder not only uptake by the roots, but also nitrification, and it is quite possible that the latter is the determining factor.

Gradmann (1898, 1950) has already established that the majority of the geophytes in the Wild Garlic-Beech wood are myrmecochores. Although *Allium ursinum*, like other Liliaceae possesses an elaeosome (P. Müller 1955), according to Schmucker and Drude (1934), and also to Ernst (1979) it is not dispersed by ants. It depends on large animals (and on man) for the transport of its seeds over any distance in damp soil particles which cling to their feet. Since this is not very likely to happen it is not surprising that *Allium* has such a peculiarly patchy distribution which is verified by very exact mapping carried out by Schmucker and Drude on the limestone plateau to the east of Göttingen and recently by Ernst in Westphalia. In this connection there is the very interesting observation made by Klötzli (1965) in Switzerland that woodland communities on moist productive mull soils, to which the Wild Garlic-Beech wood belongs, are so-called 'browsing centres' for the Roe Deer which visit them regularly in large numbers. Here they eat amongst other things young Ash shoots, many protein-rich herbaceous plants and also the buds and flowers of *Lilium martagon* which are believed to act as an aphrodisiac for the bucks. Where the woodland is less rich in such plants the Deer pass through, scarcely pausing, to the next browsing centre, a habit which doubtless increases the probability of seed dispersal. Then there is the question whether the seeds of Wild Garlic could be transported by running water. However this is less likely in Beech woods than the Sycamore-Ash mixed woods at the foot of slopes and in valleys (section III 1c) and in many flood plain woods in which *Allium ursinum* is often common (section V I h).

Wild Garlic makes up for the difficulties of dispersal by mass production of seed. Schmucker and Drude counted more than 9000 on 1 m^2 which had been produced by 63 plants in 156 umbels. Some of these seeds may disappear by autolysis, as was observed by Ernst (1979). However Schmucker and Drude found in the next spring in-between the 63 large plants, 331 middle-sized and 1903 small individuals which had grown from germinating seed. Due to this high fecundity, supplemented by vegetative reproduction through offsets (Füllekrug 1971), *Allium ursinum* can cover the ground in large masses and gradually extend outwards from any suitable site where it has managed to obtain a foothold. It finds favourable conditions only where the soil is soft enough for the contractile roots to pull the bulbs deeper and deeper down each year (Ellenberg 1939). According to Ernst these contractile roots begin to act when the plant is 3 years old. In the course of 10 years the bulb can be moved down by as much as 27 cm, depending on soil resistance.

Lindquist (1931) is of the opinion (confirmed by Ernst) that luxuriant masses of *Allium ursinum* interfere with the development of other plants partly because of the dense cover of their leaves and partly because the stems and leaves collapse half way through the period of growth. Consequently the area occupied by the *Allium* in the Wild Garlic-Beech wood is always particularly poor in herbaceous partners and in tree saplings. The inhibiting effect of dying *Allium* roots, which has been noted by Grimme (verbal communication) in culture experiments, may be jointly responsible. The leaves however seem to produce no allelochemicals inhibitory to herbs like *Mercurialis perennis* or *Glechoma*, at least under field conditions (Ernst 1979).

We have looked into the biology of *Allium ursinum* in some detail not only because it is a striking member of the Wild Garlic-Beech wood, but also because it can serve as an example of the manifold interactions which operate within living woodland. Unfortunately we still know far too little about them with regard to most of the Central European species and ecosystems.

After the spring geophytes have died down the herbaceous layer is fairly full of gaps, but in plant species present the Wild Garlic-Beech wood resembles the slightly-moist limestone Beech wood so closely that in late summer it is scarcely possible to differentiate between them. In contrast to the latter however the dryness indicators never appear in the Wild Garlic-Beech wood. Instead of these, certain species may be found here and there whose optima lie on relatively damper soil and which may also serve as nitrate indicators, e.g. *Stachys sylvatica, Impatiens noli-tangere, Circaea lutetiana* or *Urtica dioica*. Definite wet indicators however are rarely met with. In spite of the absence of intensive research into the water content of the soil the Wild Garlic-Beech wood can be distinguished from the typical limestone Beech woods by the following characteristics:

damper soil, but without waterlogging
higher fertility of the upper soil
a decided loose texture of the mull soil, at least in the top 10-20 cm.

While the herbaceous layers of the different examples of the Garlic-Beech wood in Central Europe resemble each other very closely, there can be a considerable contrast in the vigour and growth of the trees. This fact, which may at first be rather surprising, results from the ability of the spring geophytes to thrive equally well on rendzina, deep-brown earths and young colluvial loams where there is a suitable local climate. Diemont (1938) therefore distinguishes between shallow soil and deep soil forms of the Wild Garlic-Beech woods.

Even in the relatively humid climate of north-west Germany *Allium*-Beech woods are usually best developed on shaded slopes. Only in the Teutoburg Forest where the Central European mountains run out to the north-west are there some stands thriving on west- or south-facing slopes as described by Diemont (and Ernst 1979). Similar observations can be made in Belgium whose climate is still more oceanic. So one can only speak of Wild Garlic-Beech wood as the zonal vegetation on calcareous soils in the region of the boundary between Central and western Europe. Usually it appears as a rather azonal community on slopes which have a humid local climate and as such it has a limited distribution even on the limestone mountains of north-western Central Europe. In the east and south it is restricted more and more to habitats with humid air and is entirely absent where the general climate is not reminiscent of western Central Europe.

With regard to soil acidity *Allium ursinum* has a much wider amplitude than *Corydalis, Leucojum* and *Gagea*. Lindquist (1931), Diemont (1938), Ellenberg (1939) and Moor (1952) have found well developed stands on soils with pH values between 5.0 and 5.5 whereas for example *Corydalis cava* flourishes only where the reaction is neutral to alkaline (above pH 6.5). However with regard to its water requirement *Allium* does appear to be more fastidious than *Corydalis* species, and Wild Garlic retreats from increasingly dry sites more quickly than *Corydalis cava* or even *Corydalis solida*. On the whole then *Allium ursinum* is the best indicator of the special habitat relationships of the wild garlic-beech woods.

(c) Dry-slope Sedge-Beech woods

While the typical limestone Beech woods and in particular the Wild Garlic-Beech wood avoid dry slopes and areas of low precipitation, there are other Beech wood communities on limestone which show their optimum development in just such places. Along with Moor (1952, 1972) we can call these 'Sedge-Beech woods' or 'dry-slope Beech woods' (*Carici-Fagetum*) since they only develop characteristically on slopes (fig. 62). Other authors, e.g. Oberdorfer (1957) prefer the name 'Orchid-Beech woods' (*Cephalanthero-Fagetum, Cephalanthero-Fagion*). All these names are equally suitable, but as we shall see, could give rise to false assumptions, if taken too literally.

Slope Sedge-Beech woods have been able to develop particularly well on Kaiserstuhl in the Rhine plain of south-west Germany. This hill complex, covered with lime-rich loess, has an exceptionally warm, dry climate for western Europe. According to Rochow (1951) the *Carici-Fagetum* here takes up no less than a third of the total woodland area and can be found on all slopes, including the shadier ones facing north and east. In the view of this authoress even on the southern slopes, where today a xerothermic Oak scrub is growing, a Sedge-Beech wood could establish itself if allowed to develop undisturbed, at least on those parts where the slope is covered with loess. Rochow gives her '*Fagetum caricetosum digitatae*' the name 'shrub-Beech wood', since the thing which distinguishes this type of Beech wood from all others at first sight is its richness in shrubs and young trees which can find sufficient light under the relatively open canopy (tab. 12). Species which require lime and tolerate the extensive dry periods are remarkably well represented, especially *Sorbus aria, Ligustrum vulgare* and *Viburnum lantana* whose leaves in part remain green overwinter. The same is true of *Rosa arvensis* which can be identified as a differential species against the other Beech woods (Schlüter 1959). The slope Sedge-Beech wood then is rarely a cathedral forest and resembles the thermophilous mixed oak wood physiognomically and floristically. However Beech naturally dominates it even though it shows rather modest growth and trunk formation. Scattered individual Durmast Oaks, Sycamores, Limes and other light-demanding or half-shade trees are present, and in places Oaks can form small groups, though mostly subordinate to Beech (tab. 12).

Fig. 62. A dry-slope Sedge-Beech wood, rich in bushes, on a calcareous sandstone facing south (Sihlwald near Zürich). Dominating are *Carex montana* and *Brachypodium sylvaticum* (broader leaves!). At the left-hand side some *Mercurialis perennis*.

Since the humidity of the habitats of the slope Sedge-Beech wood is so low there is almost a complete absence of epiphytic mosses on the trunks. Lichen communities such as *Graphidetum* and *Candelarietum* are poorly developed according to Wilmanns (1958).

In addition to shrubs the undergrowth contains a number of sedge and grass species which are generally well able to withstand the occasional drying out of the soil and the air. The most abundant are *Carex digitata, C. flacca* and *C. montana* so that the designation Sedge-Beech wood which is usually employed in Switzerland (though there it really refers to *C. alba*) appears also to be justified for Kaiserstuhl. The name can be misleading in so far as there is a lowland Beech wood in north-eastern Switzerland, dominated by *C. pilosa*, which indicates less-dry conditions and is confined to level sites.

It is worth noting, by the way, that the constancy of sedge species and of orchid species in the slope Beech woods is almost parallel, or, expressed in other words, that both groups of plants are influenced in the same sort of way by the provisions of the particular habitats. More obvious than the orchids, which are usually found singly in the undergrowth, is the gregarious Lily of the Valley (*Convallaria majalis*) which is hardly ever seen blooming as early and abundantly as here. It is usually dominant on wind-exposed rather lean sites. The communities of the suballiance *Cephalanthero-Fagion* however can only be negatively characterised, that is by the absence of broadleaved meso- to hygromorphic herbs. So the dry-slope Sedge-Beech woods make up a unified group which appears to be ecologically justified.

Moor (1972) looked upon the *Carici-Fagetum*, which he divided into not less than 15 subassociations, as the 'climax' association of the submontane belt in the Swiss Jura. In fact it is present there on all types of slope (see also Ellenberg and Klötzli 1972), but only on slopes. Thus the majority of authors see it as a 'permanent community' as defined by Braun-Blanquet. It never represents the last stage of succession on more or less level mature soils which are not disturbed

Tab. 12. The dry-slope Sedge-Beech wood and the ecological evaluation of its undergrowth.

	C	F	R	N	Cn	CN · N
Tree layer:						
Fagus sylvatica	5					
Acer campestre	5					
Quercus petraea	3					
Sorbus aria	2					
Fraxinus excelsior	2					
Acer pseudoplatanus	1					
Prunus avium	1					
Tilia cordata	1					
L *Hedera helix*	3					
Shrub layer:						
Ligustrum vulgare	5	×	8	×		
Viburnum lantana	5	4	8	5	5	25
Fagus sylvatica	5	5	×	×		
Cornus sanguinea	5	×	8	×		
Lonicera xylosteum	5	5	7	×		
Rosa arvensis	5	5	7	5	5	25
Daphne mezereum	5	5	7	5	5	25
Sorbus aria	4	4	7	3	4	12
Corylus avellana	4	×	×	×		
Crataegus monogyna	4	4	8	3	4	12
Acer pseudoplatanus	3	6	×	7	3	21
Prunus avium	3	5	7	5	3	15
Berberis vulgaris	3	4	8	3	3	9
Viburnum opulus	3	7	7	6	3	18
Quercus petraea	2	5	×	×		
Fraxinus excelsior	2	×	7	7	2	14
Tilia cordata	2	×	×	5	2	10
Sorbus torminalis	2	4	7	4	2	8
Coronilla emerus	2	3	9	2	2	4
Rubus spec.	2	–	–	–		
Acer campestre	1	5	7	6	1	6
L *Hedera helix*	5	5	×	×		
L *Clematis vitalba*	4	5	7	7	4	28

	C	F	R	N	Cn	· N
Herb layer:						
Orchids						
Neottia nidus-avis	5	5	7	5	5	25
Cephalanthera rubra	4	4	8	3	4	12
Epipactis helleborine	3	5	7	5	3	15
Cephalanthera damasonium	2	4	7	4	2	8
Sedges and grasses						
Carex digitata	5	4	×	3	5	15
Melica nutans	4	4	7	3	4	12
Carex flacca	4	6	8	2	4	8
C. montana	3	4	5	3	3	9
Brachypodium sylvaticum	3	5	6	6	3	18
Bromus benekenii	2	5	8	5	2	10
Carex alba	1	×	8	2	1	2
Other herbs						
Convallaria majalis	5	4	×	4	5	20
Euphorbia anygdaloides	5	4	7	5	5	25
Viola reichenbachiana	5	×	8	×		
Galium sylvaticum	3	4	7	5	3	15
Vicia sepium	3	5	7	5	3	15
Solidago virgaurea	3	5	×	5	3	15
Vincetoxicum hirundinaria	2	3	7	3	2	6
Anemone nemorosa	2	×	×	×		
Pulmonaria obscura	2	6	8	7	2	14
Sanicula europaea	2	5	8	6	2	12
Fragaria vesca	2	5	×	6	2	12
Angelica sylvestris	2	8	×	×		
L *Tamus communis*	2	5	8	6	2	12
Example of calculation (mN):						
Column totals	–	–	–	–	113	512
ΣCn · N : ΣCn = mN	–	–	–	–	4.5	
Av. moisture value (mF)	–	4.8			–	–
" acidity ind. value (mR)	–		7.4		–	–
" nitrogen value (mN)	–			4.5	–	–

Note: The figures in italics in this table (and also in the following tables) indicate the constancy of the species according to five levels: *5* = in more than 80% of the relevés (i.e. in this example in at least 15 of the 17 Beech stands), *4* = in 60-80%, *3* = in 40–60%, *2* = in 20–40% and *1* = in less than 20% of the relevés (many of the species in this last class have been omitted in order to save space). L = lianas.

The figures in the columns F, R and N are the indicator values for all the species in the shrub and herb layers showing their ecological behaviour as to soil moisture, acidity and nitrogen supply on a nine grade scale (see sections B I 4 and E III): 1 = very low to 9 = very high; x = indifferent. The last two columns explain the method for calculating the mean indicator value using nitrogen as an example. For each of the species which is not indifferent to the amount of available nitrogen the constancy figure (column (n) is multiplied by the N-value. The sum of these products (Σ Cn. N) is then divided by the sum of the constancy figures of the evaluated species (Σ Cn).

Judging by the mean indicator values, calculated in this way from the species present, the dry-slope Sedge-Beech wood occupies sites which are slightly drier than average for Central European soils (mF = 4.8). The lime content as well as the pH are generally high (mR = 7.4), but the available nitrogen is below medium (mN = 4.5). For communities on sunny slopes mN is often even less than 4.0 (see tab. 19, Ia).

Source: Compiled from 17 lists from the Kaiserstuhl taken by M. von Rochow (1951) on 5-20° slopes of all aspects in the submontane belt (330-460 m above sea-level); ecological evaluation after Ellenberg (1974)

by erosion. In level to slightly sloping situations loess, limestone, rubble and marl soils, the most common substrates for Sedge-Beech woods, would be fairly quickly leached of lime. Besides that, none of the precipitation would be lost through surface run off so the tree roots would be better supplied with water than on a sloping site. With a better water supply the Beech canopy would become more dense, thus increasing the competitiveness of many of the herbaceous plants of the typical limestone Beech wood. As a result, the light-loving shrubs as well as the slow-growing sedges and orchids could no longer maintain their ascendancy. So the dry-slope Sedge-Beech wood remains restricted to sloping sites which at the same time have a dry local climate.

According to J. L. Richard (1961) the most decisive factor in any habitat is the **amount of water available** during the growing season. This is often restricted on sloping, porous and lime-rich ground (Bach 1950). Substantial amounts of lime fluff which have become deposited throughout the soil profile from the drainage water indicate the frequent drying out even of the subsoil. So the habitat of *Carici-Fagetum* is moist only periodically. This means that the litter, in spite of its high base content and its diverse origin from many tree, shrub and herb species, cannot be broken down within one year. It accumulates as a dense cover over the soil restricting the amount of erosion. In places it even forms pockets of so-called dry moder which probably favour the establishment of orchids. Mückenhausen (1970) speaks of a 'lime moder-rendzina' as being a soil type frequently associated with dry slope Sedge-Beech woods. However Moll (1959) maintains that there are also normal mull-rendzina, brown rendzina and limestone brown loam to be found. On loess, lime-rich sandstone, and marls the Sedge-Beech wood occupies pararendzina in different stages of development. The deciding factor then is not the type of soil, nor its stage of development, but its water regime (fig. 63).

Even where the lime content is only average, or lime is altogether absent, the fine earth always shows a neutral reaction, i.e. the pH value is more than 6 or even over 7. Many authors agree on this for instance Winterhoff (1965), Gadow (1975) and Grimme (1975). Lötschert (1952) carried out a closer investigation into the pH values and the root layering in the soil of a *Cephalanthero-Fagetum* at the western foot of Odenwald which was only 185 m above sea level. In the smallest area (4 m^2) no less than five orchids (*Cephalanthera damasonium, C. longifolia, C. rubra, Epipactis helleborine,* and *Neottia nidus-avis*) were represented alongside the dominating Lily of the Valley. The pH values here were all over 7 (7.09 – 7.39). The decaying tree stumps and the moder pockets formed interesting microhabitats, incidentally. According to Knorre (1974) these are often damper than the surrounding area and serve as a hiding place for isopods and molluscs. In contrast the mineral soil, which often dries out, has a poorer fauna than those of the other Beech wood communities; for example the Red Worm (*Lumbricus rubellus*, see Rabeler 1962) is absent.

The C/N ratios of the soil profiles which have been investigated up to now indicate a good supply of nitrogen, at least during the time that the mineralisation is not restricted by drought. This varies from year to year, according to Grimme (see also fig. 63). The higher plants, especially the relatively shallow-rooting herbs, do not take up any more water when they are permanently wilted, but microorganisms remain active in the soil so that unused nitrate, and even ammonia, accumulate (Wetschaar 1968). The same occurs with rendzina soil of the slightly-moist limestone Beech wood, at least in dry years (fig. 63).

Long drought periods may be catastrophic for the phanerogams, as was seen for example in the years 1945 – 50 in the Swiss Jura (J. L. Richard 1961) and in the late summer 1971 on 'Lengdener Burg' near Göttingen. All the herbaceous plants of the underground wilted in the slope Beech woods even including dryness indicators such as *Vincetoxicum hirundinaria*. Trees and bushes yellowed prematurely while those in neighbouring typical limestone Beech woods and in the Wild Garlic-Beech wood remained green right into the autumn. During a drought the trees are not damaged so much because they are deeper rooting. This may be the reason why the Beech, which is quite sensitive to dry conditions, can survive in such habitats, at any rate in the montane and submontane belts. At lower levels it has to yield similar sites to the mixed Oak wood which is better able to withstand the drought (section III 4).

Slope Sedge-Beech woods are found mostly in the south-west of Central Europe, above all on the Swiss Jura and the Kaiserstuhl. Towards the north they are less frequent and poorer in species. *Cephalanthera damasonium*, which gives its name to the suballiance, is completely absent in the north-west.

(d) Herb-free Beech woods, mainly on gypsum soils

A short mention should be made of the Beech woods referred to by many authors as 'naked' (*Fagetum nudum*, survey of the literature by Slavíková 1958). The majority consists of well grown pure stands of Beech on soils rich in lime or gypsum in areas with a relatively dry climate, e.g. east of the Harz or in inner Bohemia. These herb-free Beech woods are best considered along with the Sedge-Beech woods.

All authors agree with the observation that the upper layers of soil are exceptionally thickly matted with the roots of the Beech trees which remove so much water that herbs and young trees are no longer able to develop (Slavíková 1958). Areas of 1 m^2 that had been isolated by deep cuts remained significantly damper, since the Beech roots were no longer able to remove water from them. They were then soon occupied by herbs (*Galeopsis pubescens, Galium odoratum, Mycelis muralis*) which during the course of three years could cover as much as 80% of the surface. Obviously these herbs had not been suffering from shortage of light. Most probably they had been kept away through lack of water. However Watt and Fraser (1933) have already argued that such a conclusion is not justified. They were of the opinion that decay of the roots which had been cut through had provided an increased supply of nitrogen and that these better-nourished plants would have a lower minimum light requirement (which according to fig. 50 seems to be justified). In the cases researched by Slavíková water shortage nevertheless still played a part since these referred to very dry situations. Because of occasional thorough drying out of the upper layer of the soil the leaf litter in *Fagetum nudum* is not completely worked down by animals; thus thick layers of so-called Beech 'tanglemoder' (Kubiena 1948) can accumulate.

These limestone Beech woods without undergrowth are a striking demonstration of the competitiveness of Beech on base-rich soils. They should not be confused with the herb-free stands on acid soils which are occasionally met with where Beech has been artificially brought into dominance in place of a mixed

Fig. 63. The nitrogen content and its net mineralisation relative to the water content, pH value and temperature of a rendzina under a rather dry Melick-Beech wood (*Melico-Fagetum elymetosum*). In the *Carici-Fagetum* the annual variation is similar but the absolute amounts of nitrogen net mineralisation are smaller. After Grimme (1975), somewhat modified.

In the case of the shallow rendzina on sloping ground the water content is the deciding factor even for nitrogen net mineralisation. However since the ability of the roots to absorb nitrogen is also restricted by drought there is an accumulation of mineral nitrogen in a dry period. In the dry year 1973 much of this accumulation was in the form of ammonium (NH_4) since the nitrification was restricted even more than ammonification. Where there is a better water supply the NH_4 content is reduced (and in the rendzina of a slightly moist limestone Beech wood – not shown here – is often practically nil).

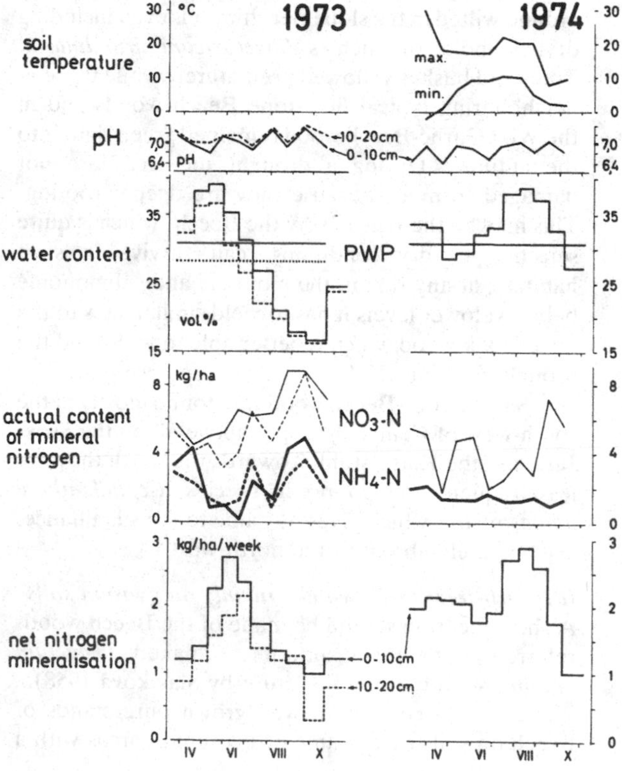

Beech-Oak wood. Thickets and closed stands of young trees in all Beech wood communities may be temporarily 'naked', even when the ground layer becomes greener as the trees mature. In genuine *Fagetum nudum* the mature stands still remain free from undergrowth if they are not severely thinned out, and the Beech roots are not damaged by surface cultivation.

(e) Steep-slope Yew- and Blue Moorgrass-Beech woods
Etter (1947), Bach (1950), and Moor (1952) described a **steep-slope Yew-Beech wood** (*'Taxo-Fagetum'*) as the definite specialist of the cool steep marl slopes in the Swiss Jura and the foothills of the Alps. Floristically and ecologically it stands near to the dry-slope Sedge-Beech woods of which we have already spoken, and to the Blue Moorgrass-Beech woods (*Seslerio-Fagetum*), but it gives quite a different impression physiognomically. Under the more or less complete umbrella of the Beeches the Yew (*Taxus baccata*) forms a second gloomy tree layer. In Switzerland which is the richest area of Central Europe for Yew, even the largest of these gnarled forms attains a height of only 15 m, but at breast height has a girth of 3.6 m (Vogler 1904). Such veterans are not much more than 300 years old, i.e. younger than one would think judging from the thickness of their trunks.

Where Yews are numerous in the undergrowth they do not permit growth of any shrubs, and their deep shadow is too dark almost everywhere for herbs and mosses also. A herbaceous layer can only develop where the tree layer is disturbed and artificially thinned out (fig. 64). Lists of species from such places are therefore heterogenic since those from undisturbed parts of the woodland contain very few species. So it is difficult to draw a clear picture of the floristic make up of the steep-slope Yew-Beech wood. Nearly all the species turning up in it can also be seen in the *Seslerio-Fagetum*. Very probably it is not a uniform association in itself, but the *Taxus* fazies of several different associations.

An important prerequisite for a sound growth of Yew is that the upper layer of trees should allow sufficient light to filter through, for *Taxus baccata* will not tolerate as much shade as is generally supposed (Leuthold 1980). In fact it is in no way shy of the light and flourishes quite well outside the woods on steep sunny rock slopes. Because of its slow growth and low competitiveness in those hilly areas where human influence is great, it is confined to impassable rocky hillsides, e.g. on the Hohenstein in Süntel. There, quite against its natural tendency, it gives the impression of being a xerophilous species.

Beech woods and coniferous mixed woods with Yew are to be found here and there throughout the whole of Central Europe apart from the regions with a strongly continental climate. It has emerged from the careful research of G. Hofmann (1958a) into woodland communities rich in Yew that *Taxus baccata* flourishes widely also in slope Sedge-Beech wood. It occurs only sporadically however in Beech woods and other woodland types on acid soil. Nevertheless according to Mayer (1974) there are also silicate steep-slope Yew-Beech woods.

In earlier times Yew was very much in demand for long-bows and crossbows as well as for carving and was exported, for example from Switzerland to England. The fact that *Taxus* is especially plentiful in the neighbourhood of old towns and castles may be connected with the protection it was afforded because of its value. Yew may have been much more common in Europe than it is today and would have occurred in many more different woodland stands if it had not remained 'outlawed' in the Middle Ages. Every year tens of thousands of trunks went to Nürnberg, the main market (Scharfetter 1938). In 1689 there were no longer any Yews fit to fell in the eastern Alps. They reappeared here as ornamental trees in the baroque gardens. A further reason for extermination of Yews from many woodlands was that of their leaves which are poisonous to horses. *Taxus* has always been cut out

Fig. 64. Steep-slope Yew-Beech wood on Tertiary marl in the Sihlwald near Zürich. Ground vegetation can only develop away from the shadow of the Yews (*Taxus baccata*).

by the carters wherever these animals have been used to draw out the timber. Its restriction to steep slopes could be connected with this practice. However Klötzli (1965) put forward yet another reason; namely that Yew is not poisonous to Roe Deer but on the contrary, is the most sought-after food tree. So where the Deer can get at it they continually eat off the young growth; thus it can only develop normally on steep slopes.

The main reason why the Yew has lost ground however must be seen as the competition for light with Hornbeam, Beech and other quick-growing shade trees which have driven it out of all the 'better' habitats (Averdieck 1971). After the Ice Ages the Yew spread before these broadleaved trees, and at one time was much more plentiful in Europe (Willerding 1968). Since the Atlantic period it has become less common, especially since the Middle Ages, and now is found only in relatively extreme habitats where it is unable to achieve full development. According to Rottenburg and Koeppner (1972) the reason it prefers shade is the sensitivity of its leaves to dry air. These close their stomata when the saturation deficit increases, which happens quickly in sunshine, much earlier and more often than do those of the Silver Fir.

While the slopes of soft marl, loess, or limestone rubble are occupied mostly by Sedge- or Yew-Beech woods, on similar dry slopes of firm limestone or dolomite the **Beech wood** is **rich in Blue Moorgrass**. This *Seslerio-Fagetum* is quite similar to the *Carici-Fagetum* but must be considered a separate association.

Its appearance is dominated by *Sesleria caerulea* (= *albicans*), already in flower by the end of February or in March. This low tussock grass holds the fine earth with its dense root system in the limestone cracks or between large stones and gradually builds up the soil (fig. 65). Compared with the *Carici-Fagetum* shrubs are less common, even though the canopy of these slow-growing Beeches is less dense than that of the Sedge-Beech wood. Apparently the water reserves are more restricted than is the case in deeper soils where the roots can penetrate the humus-free subsoil unhindered. The growth of Beech in the *Carici-Fagetum* then is considerably better than in the *Seslerio-Fagetum*, appropriately named 'Blue Moorgrass-gnarled Beech wood' by Meusel (1939, 1942, compare Rühl 1960).

Sesleria albicans has its distribution centre in the Alps, especially above the tree line where it can dominate on base-rich soils, and plays an important part as a soil stabiliser on the slopes. Although many of these Alpine Blue Moorgrass slopes are exposed to the sun and wind and loose their snow cover quite early, *Sesleria* can withstand the cold of the mountain winter without damage. It is quite wrong then to describe this grass species as thermophilous as is often done in the plant sociological literature. It is merely one which requires a lot of light and can withstand occasional drought.

In this respect there are naturally limits. On Kyffhäuser and the southern edge of the Harz mountains for example it retires to the shady steep slopes because the sunny slopes in this low rainfall region are too dry (Meusel 1939). On the other hand there are Blue Moorgrass-Beech woods, and even treeless Blue Moorgrass communities to be found also on sun-facing slopes on limestone mountains lying further to the west which receive a higher precipitation (Rühl 1960). Apparently it is not the amount of rainfall as such, but rather the water balance, which accounts for presence or absence of *Sesleria*. Otherwise it would not retreat again to shady slopes in the northern Alpine foothills, which also have a higher rainfall, but are exposed to the very drying föhn wind. This blows from a southerly direction and prevents *Sesleria* from settling also on sunny slopes. On the Swiss Jura *Sesleria* is restricted to the northern slopes in those parts which are still in reach of the föhn, whereas on the other side of the Jura (Moor 1952), just as in Swabia and Franconia (K. Kuhn 1927, Oberdorfer 1957), it can be found on steep limestone slopes whatever direction they are facing.

The Blue Moorgrass-Beech woods on sunny and shady slopes appear to be quite similar at first sight, but they can be distinguished by the other species present. Many mosses are able to grow on the damper shady side. Because of this Moor (1952) distinguishes a *Seslerio-Fagetum hylocomietosum* from a *Seslerio-Fagetum anthericetosum*. The Blue Moorgrass-Beech

Fig. 65. Blue Moorgrass (*Sesleria albicans*) is able to bind the limestone rubble and hold up the fine earth by means of its tough runners and roots. After W. Schubert (1963).

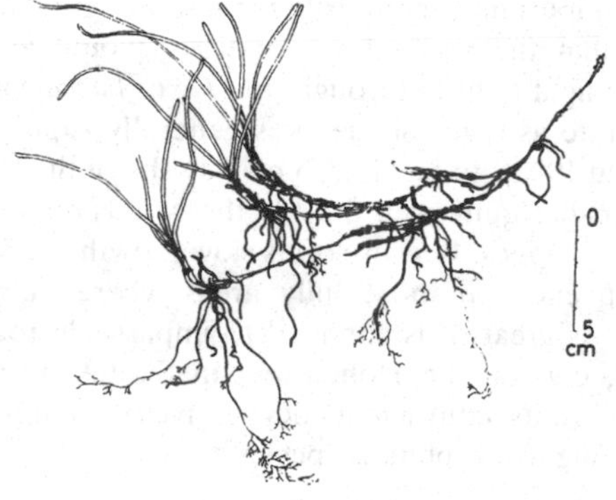

wood rich in mosses contains differential species which normally live in Spruce woodlands, namely:

Orthilia secunda	M *Hylocomium splendens*
Festuca altissima	M *Rhytidiadelphus triquetrus*
Luzula sylvatica	M *Dicranum scoparium*
Vaccinium myrtilus	

The four phanerogams as well as the three mosses enjoy the thick humus layer which collects between the grass tussocks on the damper limestone slopes. Measurements by Moor and myself show that such a surface humus has a neutral reaction (pH is often above 7 and never below 6). Following the new terminology of the soil specialist this is not a 'raw humus' but a 'rendzina moder' (Franz 1960) or even a 'tangle humus', which can also develop under Beech (Scheffer and Ulrich 1960). Underneath, this organic soil cover gradually changes into mull and is much more fertile than real raw humus with its strongly acid reaction (see Section IIa).

On the sunny side of the hill there is no accumulation of humus under Beech and there is only a weak to average development of the rendzina. Obviously, because of its high lime content, the soil here too has a neutral reaction (according to Moor pH 6.5–7.3). The Blue Moorgrass-Beech wood on the sunny slopes which is rich in *Anthericum* (St Bernard's Lily) can be recognised by representatives of the dry Pine or mixed deciduous woodland, e.g.

Anthericum ramosum	*Amelanchier ovalis*
Helleborus foetidus	*Origanum vulgare*
Polygonatum odoratum	

In those parts of Central Europe which have a lower rainfall and at the same time are less affected by the föhn wind, the differences in the species composition of the *Seslerio-Fagetum* on slopes of different aspect are not expressed so strongly as they are in the Swiss Jura.

Blue Moorgrass-Beech wood is present on steep limestone slopes throughout the whole of the distribution of *Sesleria albicans* in Central Europe, for instance also on Rügen, in Poland, Czechoslovakia and Hungary. These habitats can be looked upon as offshoots of the main region centred on the Alps or indeed as relicts of a much wider area of distribution in times when the climate was colder and fewer trees were able to exist.

Following the suggestion of Meusel, Thorn (1958) uses the expression 'de-alpine' to describe this phenomenon which applies not only to *Sesleria* itself but to the grassland and the Beech woodland which it dominates. It is worth noting that other de-alpine species also survive in Blue Moorgrass-Beech woods and on Blue Moorgrass slopes, e.g. *Aster bellidiastrum, Carduus defloratus*, and *Thesium alpinum*, which are as little thermophilous as *Sesleria*. Nor can *Seslerio-Fagetum* as a whole be looked upon as warmth loving. The two subassociations distinguished by Moor (1952) are found up to 1200 m in the Swiss Jura, that is, right up into the cool montane belt.

(f) Montane limestone Fir-Beech woods

Almost all Beech woods change in composition with increasing altitude and may be said to form 'typical' associations within certain levels. Beech can only achieve complete dominance in the lower and middle montane belt since the vigour of its growth is in close negative correlation with altitude (Manil 1963). A change also occurs in the mosaic of the ground flora. Glavač and Bohn (1971), for example, have traced this on Vogelsberg in woods where the Beech shows the same degree of dominance right up to the top of this mountain.

Within the area of the Silver Fir, provided the soil is not too poor, the upper montane belt is largely occupied by Fir-Beech woods in which the Spruce plays no great part. The change from almost pure Beech wood to one richer in the conifer is shown particularly clearly in the northern limestone Alps, much more sharply in any case than say in the Jura. First of all then we shall look at the examples described by Kuoch (1954). Comparable communities have been described by Mayer (1974) from the Austrian limestone Alps (see also Ellenberg and Klötzli 1972 and figs. 54 and 67). In the upper montane belt of the northern Alps (900–1250 m) a limestone substrate always favours the deciduous trees, so Beech is able to maintain a natural dominance. On the other hand, as the soils become less rich in lime, so the Fir takes on a stronger role. Because of this when Kuoch describes the subassociations of the Fir-Beech wood he distinguishes between a '*Fagus* variant' and an *Abies* variant'. We shall restrict ourselves to the former, that is, to the 'limestone Fir-Beech wood' (fig. 66).

This association occupies the same kind of rendzina or brown rendzina as the typical limestone Beech wood and prefers also level ground or general slopes without restriction to a particular aspect. Apart from increased representation of Fir and Spruce it hardly differs physiognomically and floristically from the pure, or almost pure, Beech wood. Shrubs are even less plentiful and Beech saplings are less commonly seen, for not only are the mast years less frequent, but conditions for the growth of Beech are no longer optimal in the upper montane belt. With increasing altitude more and more montane and subalpine plants appear in the herbaceous layer, e.g. *Petasites albus, Adenostyles alliariae* and *Ranunculus aconitifolius*. However none of these species are so constant in the limestone Fir-Beech wood that they can be used as differential species, and none are entirely absent from Beech woods at lower levels.

The presence of Fir and especially of Spruce means that the mosses on the ground are not being so continually suppressed by leaf litter as under the pure Beech canopy. *Fissidens taxifolius* is generally present and even acid-tolerant species (e.g. *Hylocomium splendens, Rhytidiadelphus triquetrus* and *Dicranum scoparium*) can be found here and there on the surface humus, occupying small pockets of moder, the pH of which rarely falls below 6 as I have found from sample tests.

Where Spruce is favoured by forestry the ground mosses increase in number and cover. The grazing of woodland in earlier times, which is still customary in some places even today, also helped Spruce as its saplings are eaten less readily than those of Fir or Beech. Because of this one can quite often find pure Spruce stands in the upper montane belt, even on shallow limestone soils where Beech and Fir would naturally have predominated. The influence of forestry often make determination of the natural tree composition still harder. Kuoch and Moor deserve particular praise for working out the rôle of Beech, Fir and Spruce in the Alps and in the Jura.

Fir-Beech woods have been described by many authors from very different parts of Central Europe, above all from the Alps (Mayer 1974, etc.). In the Sudeten mountains they occur at an altitude between 500 and 1000 m, in the Bavarian Forest from 400 to 900 m, in the eastern Black Forest from 500 to 900 m and in the north west Swabian Jura from 600 to over 1000 m. Yet only a small number of these communities is identical with the idea of *Abieti-Fagetum* as expressed by Moor. He applied this originally only to the Beech woods on limestone which had been invaded by Fir. However communities on more acid soils also warrant such a name. The classification worked out for the eastern border of the Alps by Zukrigl (1973) and by

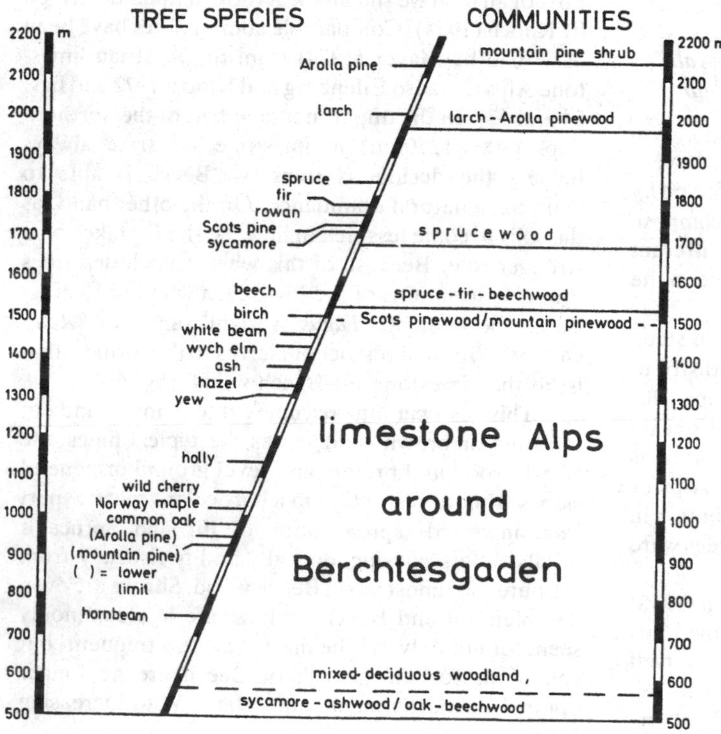

Fig. 66. Altitude limits for tree species and woodland communities on limestone in the north-eastern Alps; schematic. The broadest belt here is taken up by the montane mixed Spruce-Fir-Beech wood (600-1500 m above sea level). After Köstler and Mayer (1970), slightly altered.

The majority of the tree species ascend higher than the communities in which they are dominant; only the Beech is still dominating stands close to its absolute limit. Light-requiring trees such as *Pinus sylvestris* and *P. rotundata* can predominate only on poor and rather dry soils. Of the broadleaved trees in the montane mixed woods the Sycamore reaches the greatest altitudes.

Mayer (1974) gives a good impression of the great diversity of the Fir-Beech mixed woodland both in habitat and floristic composition (Fig. 68, see also tab. 13). In addition to changes determined by the soil must be added those brought about by climate, especially by the degree of continentality.

Zukrigl, Eckhardt and Nather (1963) have studied two primaeval Beech-Fir-Spruce woods in the limestone Alps of lower Austria, the famous 'Rothwald' on the south-eastern slope of Dürrenstein (940–1480 m) and the 'Neuwald' on a less steep site (ca. 1000 m). In the Rothwald the soil largely determines which tree is dominant: Beech on rendzina, Silver Fir on deeper soils, or Spruce on rocky slopes. The Neuwald, where the soil for the most part is a deep loamy parabrown earth-gley, is dominated by huge Firs. So even under original-forest conditions, the distribution of the main tree species in the montane belt follows the rule laid down by Kuoch. The life rhythm of these old forests resembles in part that of a Beech wood (section 2a) and in part that of natural coniferous woodland (section IV 2 d). In Dürrenstein region, the main cause of death among Spruce and Fir is being blown down by the wind. Beeches stand somewhat shorter and are not so exposed but occasionally they are torn out along with the other trees. Most old Beeches die standing and are broken down by fungal decay.

Fig. 67. Limestone Fir-Beech wood (*Abieti-Fagetum*) on a northern slope of the Lebern in the Swiss Jura. Although Fir can withstand more shade than Beech it occurs only as isolated specimens in the tree layer. In the foreground *Festuca altissima*.

Limestone Fir-Beech woods are found in all mountainous areas with a temperate suboceanic montane climate which lie within the general area of distribution of *Abies alba*, e.g. also in the mountains of Croatia and Slovenia which are outside the scope of this book (see Horvat, Glavač and Ellenberg 1974). In these Illyrian limestone mountains, as well as in the southern Alps, the Jura and the Vosges, the Fir-Beech woods do not merge into pure stands of Fir or Fir with Spruce at higher altitudes as one would expect, but the Beech and other broadleaved trees again become dominant; this is very surprising. These subalpine mixed limestone woods should be given a special section of their own, even though they are well developed in Central Europe only in the mountains to the south-west, and have a much greater significance on the limestone mountains of the mediterranean and submediterranean area.

(g) Subalpine Sycamore-Beech woods

In mountains which have a relatively mild winter with a lot of snow it is not conifers but broadleaved trees which form the tree line. Round the borders of Scandinavia warmed by the Gulf Stream these are Birches (*Betula tortuosa*). On the other hand in the submediterranean and atlantic areas it is Beech (fig. 42), while in the Vosges, western Jura, as well as some parts of the western Alps and a few other Central European mountains, it is Sycamore and Beech together.

The reasons for the deciduous trees in the subalpine belt of these mountains being free from the competition of conifers which supersede them so impressively in a more continental climate with its severe winters, are difficult to grasp and are certainly very complex. Since no clear results have emerged since my search for an explanation in the first edition (1963) I am refraining from renewed discussion of this question. In my opinion the susceptibility of all conifers to fungal attack plays a considerable role. These are often ecto- or endo-parasites on the needles (see also Mayer 1976). Snow moulds (e.g. *Herpotrichia nigra*) which have low optimum temperatures, attack young conifers under the snow covering particularly severely in mild winters, i.e. practically every winter in an oceanic climate. They kill off many needles and even whole twigs and so weaken the evergreen conifers in their competition with bare deciduous trees. Also in a warmer climate many deciduous trees, because of

Tab. 13. 'Primaeval' woods of Beech, Fir and Spruce in montane to subalpine belts in the Rothwald near Lunz in the limestone Alps of Lower Austria.

Serial no.:		1	2	3	4	5	L	T	R
Altitude of relevés (m above sea level)		960 – 1080	1100 – 1290	1120 – 1250	1320 – 1430	1270 – 1500	Ecological evaluation		
Tree layer:									
F *Fagus sylvatica*		5	5	5	5	5			
P *Albies alba*		5	5	5	5	1			
Picea abies		5	5	5	5	5			
Acer pseudoplatanus		4	4	5	5	4			
Ulmus glabra		1	1	2	1				
Shrub layer:									
h *Rosa pendulina*		3	3	3	3	2	6	4	7
F *Daphne mezereum*		3	3	2	5	5	4	×	7
F *Lonicera alpigena*		2	1	1	2	4	3	4	8
F *Rubus hirtus*		1	1	1			5	6	5
Sorbus aucuparia		1		1	4	5	6	×	4
Lonicera nigra		1	1			4	3	3	5
Herb layer:									
F *Gymnocarp. dryopteris*	C	4	1	1			3	3	4
P *Blechnum spicant*	C	2	2	1			3	3	2
F *Cardamine trifolia*		5	3	1	1		3	4	8
F *Sanicula europaea*		4	3	2	2		4	5	8
F *Viola reichenbachiana*		4	4	3	2		4	5	7
F *Athyrium filix-femina*	C	5	2	4	3		4	×	×
F *Carex sylvatica*		4	3	4	3		2	5	7
F *Galium odoratum*		4	2	5	5		2	5	×
F Dryopteris filix-mas	C	3	2	2	2		3	×	5
F *Polystichum lobatum*	C	1	2	3	2		3	6	6
F *Epilobium montanum*		2		4	1		4	×	6
Lysimachia nemorum		2		2	3		2	5	7
Veronica officinalis		1	1	2	2		5	×	2
Polypodium vulgare	C	2		1	2		5	×	2
Dryopteris carthusiana	C	4		1	1		5	×	4
P *Lycopodium annotinum*	C	4		1	1		3	4	3
Oxalis acetosella		5	4	5	5	4	1	×	4
Vaccinium myrtillus		5	4	4	4	5	5	×	2
F *Lamiastrum galeobdolon*		5	2	5	5	3	3	×	7
F *Prenanthes purpurea*		4	2	4	5	2	4	4	×
F *Dentaria enneaphyllos*		4	4	3	5	5	4	4	7
F *Polygonatum verticillatum*		4	4	2	4	5	4	4	4
F *Paris quadrifolia*		4	2	4	5	2	3	×	7
Solidago virgaurea		3	5	4	4	4	5	×	×
F *Mercurialis perennis*		3	3	3	4	4	2	5	7
I *Senecio fuchsii*		3	2	4	4	3	7	×	×
I *Fragaria vesca*		1	4	2	2	3	7	×	×
P *Huperzia selago*	C	3	1	1	1	2	4	3	3
a *Asplenium viride*		2	1	1	2	3	–	–	–
Carex digitata		1	3	2	2	1	3	5	×
a *Gymnocarp robertianum.*	C	1	1	1	2	1	5	4	8

Serial no.:	1	2	3	4	5	L	T	R
Herb layer, continued								
Adenostyles glabra	5	5	5	5	3	6	2	8
F *Helleborus niger*	2	3	4	5	5	3	5	8
Deschampsia cespitosa	2	3	5	5	5	6	×	×
E *Calamagrostis varia*	1	5	1	5	5	7	3	8
a *Valeriana tripteris*	1	4	2	5	5	7	×	8
Mycelis muralis	1	5	2	4	4	4	5	×
Luzula sylvatica	1	3	2	5	5	4	4	2
F *Euphorbia amygdaloides*	2	1	3	4	4	4	5	7
h *Adenostyles alliariae*	2		2	5	4	6	3	×
Melica nutans	1		3	4	5	4	×	7
F *Phyteuma spicatum*	1	2	2	5	2	×	×	×
Veratrum album	2	3		4	3	7	×	×
Ranunculus nemorosus	1	3	1	3	4	6	×	6
a *Moehringia muscosa*	1	2	2	1	4	3	5	9
h *Saxifraga rotundifolia*	1		1	3	3	5	×	8
Polystichum lonchitis	1	1	1	2	4	6	2	8
a *Aster Bellidiastrum*	1	1	1	1	3	7	3	8
Knautia dipsacifolia		3	1	4	5	×	3	×
Hieracium sylvaticum		2		4	4	4	×	5
F *Galium sylvaticum*		2	1	1	2	5	5	7
Poa nemoralis		1	1	1	5	5	×	5
Galium austriacum		1	1	1	4	–	–	–
Veronica chamaedrys		1	1	2	2	6	×	×
E *Erica herbacea*		2		2	1	7	×	×
Geranium robertianum			3	1	1	4	×	×
P *Homogyne alpina*			1	2	2	6	4	4
a *Valeriana montana*		1		2	2	8	×	9
Carex ornithopoda		1		2	2	6	×	9
h *Cicerbita alpina*	1			3		6	3	×
F *Lilium martagon*		2		4		5	×	7
F *Euphorbia dulcis*		1		3		4	5	8
P *Luzula luzulina*				3		3	3	×
a *Carex ferruginea*		1		4	2	8	2	8
Aconitum neomontanum				4	1	–	–	–
h *Viola biflora*				4	2	4	3	7
Ranunculus montanus				3	1	6	3	8
a *Carduus defloratus*		1		3	4	7	×	8
Primula elatior			1	4	4	6	×	7
a *Arabis ciliata*				3	2	9	2	9
Astrantia major				3	2	6	4	8
E *Cirsium erisithales*				2	3	6	?	8
Senecio abrotanifolius				2	5	7	3	7
a *Campanula scheuchzeri*				3	4	8	2	×
a *Leucanthem. maximum*					5	9	×	×
F *Melampyrum sylvaticum*		1		1	3	4	×	2
Rubus saxatilis		1			3	4	×	×
Vaccinium vitis-idaea		1			3	5	×	2
a *Phleum hirsutum*					3	8	3	7
a *Poa alpina*					3	7	×	×

Note: After tables by Zukrigl, Eckhart and Nather (1963); ecological evaluation by Ellenberg (1974, see also section B 14 and tab. 12). F = character species of the Beech woods (*Fagion*), F = of the noble broadleaved woods (*Fagetalia*), P = of the acid-soil needle-leaved woods (*Vaccinio-Piceion*, etc.), E = of the calcareous Pine woods (*Erico-Pinion*), 1 = plants of the clearings, h = species of the tall herb communities, a = species of the scree and alpine turf communities. C after the name = cryptogams (ferns and clubmosses).

Note how poor in shrubs but rich in ferns these natural woods are! Mosses and lichens are found only on stones, dead wood or trees; they have not been included in the lists.

Tab. 13. (Continued).

Communities and soils ecological values:	mL	mT	mR
No. 1-3. **Montane limestone Fir-Beech woods** in which *Abies* attains large size, but the shorter *Fagus* is mostly dominant. *Picea* and *Acer* are always present but in smaller numbers.			
1. In depressions on heavy limestone-brown loam ('*Cardamine trifolia-Galium odoratum -Oxalis-Myrtillus* mosaic complex')	3.9	4.0	5.7
2. On 25-40° slopes with ± deep rendzina ('*Calamagrostis varia-Helleborus-Adenostyles glabra*-type')	4.6	4.1	6.3
3. The same on slopes mostly less than 30° ('*Adenostyles glabra-Galium odoratum* -type'): especially rich in Beech	4.2	4.2	6.1
No. 4. **High montane limestone Fir-Beech woods** on 15-40° slopes with ± deep rendzina ('*Helleborus-Adenostyles-Luzula* sylvatica-type'): often still many Beech	4.7	3.9	6.9
No. 5. **Subalpine limestone Spruce wood** on 25-35° slopes with rendzina ('Spruce type at the tree limit)	5.1	3.6	6.5

The remains of the untouched woodland in the Rothwald are found on more or less steep boulder-strewn limestone slopes. Lime-requiring species are mixed with more or less acid-tolerating ones. These are favoured by accumulations of slowly decomposing needles, rotting wood or pockets of leached clay-rich fine soil. In hollows and on gentle slopes fine earth may cover larger areas ('Terra fusca') and favour the acid indicators. Because of this the communities 1 and 3 have the lowest mean indicator values for soil acidity (mR). However the better water-holding capacity of the loamy soil allows trees to grow well here, their denser canopy increasing the proportion of shade-tolerant species (mean light values only 3.9 and 4.2). However light-requiring species, especially tall herbs (h) and woodland clearing dwellers (1), do also appear in nos. 1 and 3 where holes are caused in the canopy when old trees collapse. Even plants of the open country (a) can find sufficient light on steep slopes especially close to the tree line (no. 5) where there are naturally many openings amongst the Spruce.

As the altitude increases the numbers of warmth-requiring plants (T 6 and 5) falls off in comparison with those which tolerate a shorter vegetative period (T 4, 3 and 2). However the mean indicator values for temperature do not fall regularly with altitude, and example no. 3 (1120-1250 m) has a higher value than no. 1 (960-1080 m). This is due to the effect of the 'warm slope' produced by the flowing down of cold air during the night. According to measurements made by Zukrigl and co-workers (using the cane-sugar inversion method) this effect is apparent between 1100 and 1250 m. Naturally the value mT is lower in the high montane and subalpine sites (nos. 4 and 5).

Looking at the vegetation systematically all the Rothwald communities belong to the order *Fagetalia* and clearly the Fir-Beech woods are in the *Fagion* alliance. In spite of the strong representation of needle-leaved trees the species typical of *Vaccinio-Piceetalia* or *Vaccinio-Piceion* play scarcely any part. In order not to anticipate the final allocation of the examples described into particular Fir-Beech or Spruce associations the authors have contented themselves in naming them provisionally according to the type of undergrowth.

Fig. 68. An ecogram of the montane Spruce-Fir-Beech woods and neighbouring communities at the eastern to south-eastern edge of the Alps. After Zukrigl (1973), somewhat altered.

Acid rocks are almost entirely absent from the eastern outer Alps, but are more common further in (in the eastern 'intermediate Alps') as well as in the more southern parts. The area occupied by pure needleleaved communities is dotted. The '*Asperulo- (Abieti-) Fagetum*' should now be called the *Galio odorati- (A.-). F.* This brown-mull Fir-Beech wood, with respect to soil qualities, occupies the central position amongst the woodland communities of the montane level, similar to that taken by the brown-mull Beech wood in the submontane and planar belts.

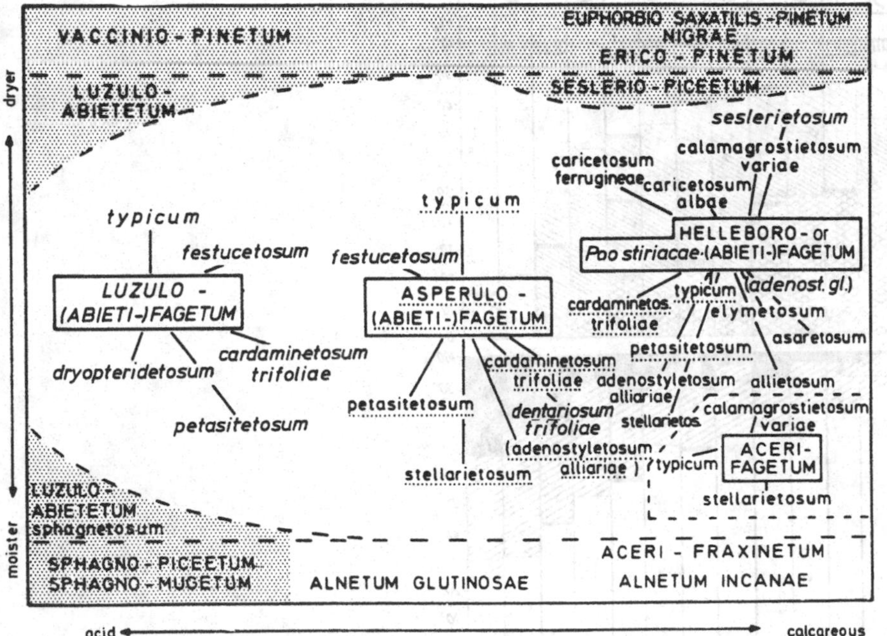

their rapid growth, are more than a match for conifers. At high altitudes with an oceanic climate the weight of snow and hoar frost is so great and widespread that evergreen trees like Silver Fir are broken down and must give way in competition with the stronger regenerating power of Beech and other deciduous trees (Pfadenhauer 1973).

The subalpine deciduous woodland is already recognisable from its physiognomy as a special formation. Whether Beech alone is dominant or whether Sycamore and other deciduous trees share in the tree layer, the trunks in *Aceri-Fagetum* always take on a 'sabre' form which goes back to their fate as young saplings. The snow cover, which is generally more than 1 m deep and may reach 3 m, slides slowly downhill and presses the young stems down to the ground as long as they are weak. Only after they have become strong enough to withstand the snow pressure do they start to grow straight up, but they always retain the curved base. A similar shape can also be seen in coppices. However in subalpine woods the bend in the trunk is always inclined downhill and is much more pronounced than that produced by sprouting after cutting off the stems. Even the oldest trees are rarely more than 15 – 20 m high and cannot achieve their optimal growth under the harsh climate in spite of the fertile soil. So natural dominance and high production in the forester's sense do not always go together even with such a competitive species as Beech.

High humidity and frequent mists, together with the slow growth of the trees, favour epiphytic lichens and mosses which often form a thick covering over the trunks and branches, but are less common near and below the winter snow covering. The significance of mist precipitation, by the way, in determining the species composition of montane and subalpine woodlands is less than one might think (Kerfoot 1968 and Kämmer 1974).

Grazing favours Spruce in competition with Beech, especially on windy hilltops where there is little snow; here Spruce would probably dominate without any help from man. Fir appears very seldom in subalpine Beech woods and only becomes more plentiful in the transition zone to the montane limestone Fir-Beech wood.

The thick snow cover usually protects the ground

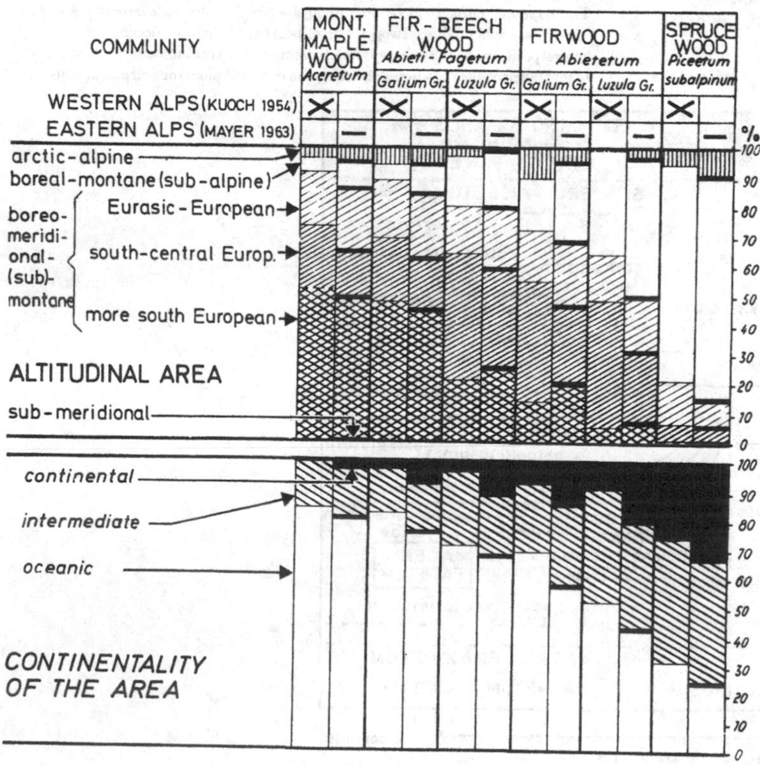

Fig. 69. Comparable woodland communities of the western and eastern Alps have different areal type spectra. The proportion of boreal and continental species is higher in the east than in the west. This is true in broadleaved as well as in mixed and needle-leaved woodlands. After Mayer (1968), slightly altered.

from frost sufficiently long so that when it melts the air temperature no longer falls below 0° C. Because of this, animal and bacterial life in the soil is unusually well developed for a habitat more than 1200 to 1300 m above sea level, and after disappearance of the snow it is immediately very active. Also even in dry weather melting of the remains of the snow maintains a moist soil and unrestricted nitrification in the upper layers of the soil so that largeleaved, more or less hygromorphic nitrate plants are able to make strong growth (fig. 70, see section C VI 7 a). These subalpine late-flowering tall perennials have their optimum in free-draining depressions just above the tree line where they are no longer shaded by trees, but where the climate is still not as bleak as on the true alpine level. They penetrate into the Sycamore-Beech wood because this forms a much less dense canopy than the montane and colline Beech or mixed Beech wood. The relative amount of light available to the perennials in the subalpine deciduous woodland is generally more than 15% while 'real' Beech wood plants have to put up with less than 5% in summer.

The following tall perennials can be found regularly in subalpine Sycamore-Beech woods (*Aceri-Fagetum*) of Swiss Jura (Moor 1952) and can serve as local character species (C) or as differential species compared with other Beech wood communities:

		additional strongly-growing nitrophiles:
C	*Rumex arifolius*	
C	*Cicerbita alpina*	*Ranunculus aconitifolius*
C	*Anthriscus sylvestris* var. *alpestris*	*Ranunculus lanuginosus*
	Veratrum album	*Petasites albus*
	Thalictrum aquilegiifolium	*Adenostyles alliariae*
	Campanula rhomboidalis (rare)	*Adenostyles glabra*

Fig. 70. Tall herbs in a subalpine Sycamore-Beech wood on the Weissenstein (Swiss Jura) in spring. In addition to *Acer pseudoplatanus* and *Fagus sylvatica* can be seen, in the shrub layer *Lonicera nigra* and *alpigena* as well as *Rosa pendulina*, and, in the herbaceous layer *Adenostyles alliariae*, *Athyrium filix-femina*, *Cicerbita alpina* and others. Epiphytic mosses grow on the tree trunks as a result of the high air humidity.

At the same time hardly any of the species of the slightly moist limestone Beech wood listed in tab. 11 is absent. The more restricted in its growth the tree layer becomes the more luxuriant and rich in species is the growth of the herb layer in these subalpine Beech woods.

Aceri-Fagetum is widespread not only in Swiss Jura (1150-1600 m) and the northern Swiss Alps (1200-1750 m), but also in the Bavarian Alps (1200-1450 m), in the Bohemian Forest (960-1190 m) and in the western Black Forest (1050-1300 m). It also occurs on the Basalt hilltops of Rhön and on the highest parts of Sauerland (Haber 1966). According to Oberdorfer (1957) it is replaced by a subalpine Spruce Wood (*Galio-Piceetum*) in the less mild climate of the eastern Black Forest. It is by no means restricted to limestone and is fairly indifferent to subsoil provided this is not too poor in bases. Nor does it seem to be confined to a particular gradient or aspect. The deciding factor for its formation would appear to be much more the oceanic subalpine climate which produces a heavy snowfall. While it appears in the more easterly parts only locally in places with a lot of snow or in moist depressions, in the western limestone Alps, in the western Black Forest and especially in the Vosges it is the dominant climax community of the subalpine belt.

The species composition of *Aceri-Fagetum* communities varies considerably under local conditions. Oberdorfer (1957), from observations in the Vosges, the Black Forest and the northern Alps, distinguishes between relatively dry, medium and wet subassociations and within each of these a number of variants. There also appears to be a subalpine Wild Garlic-Beech wood which, according to Moor, always contains fewer of the tall perennials. On scree slopes under rocky limestone edges in the Swiss Jura there is an altitudinal vicariad of the Lime-Beech wood (*Adenostylo-Fagetum*, Moor 1970, see also section III 2 b).

In many ways subalpine Beech woods are related to ravine woodlands which are rich in Sycamore and to other deciduous woodland communities occupying damp habitats. We shall return to these units, but first we shall go on dealing with other Beech wood communities.

Fig. 71. Brown-mull Beech wood on moraine loam in the Sihlwald near Zurich. *Athyrium filix-femina, Milium effusum, Galium odoratum, Lamiastrum galeobdolon* and Beech saplings may be recognised.

3 Beech and mixed Beech woods on rich to average brown soils

(a) Brown-mull Beech woods and related communities

Not all Beech wood communities on limestone mountains can be described as limestone Beech woods. Often there is a more or less deep layer of brown loess-loam or other material which is relatively poor in carbonates on top of the white cleft limestone. This brings about an obvious change in the ground flora and improves growth of the tree layer. The root zone in such brown soils or brown rendzina over limestone is not only more acid but is capable of holding much more water than the shallow stony top soil of a typical mull-rendzina. Consequently lime and dryness indicator species disappear and are replaced by those which can tolerate acid conditions but require more water (tab. 14). However these combinations of species are still included in the mull-Beech woods, i.e. the suballiance *Eu-Fagion*.

The most productive Beech woods of Central Europe grow on the deep brown or parabrown earths which are neither rich in lime nor podsolised (fig. 71). Since such soils are eminently suitable for arable farming we seldom have the chance to study their natural plant cover. This is why we started the description of Beech woods with the communities on limestone soils which still carry large areas of woodland today.

Brown earths and similar soils develop from loams of various geological origin or from crystalline rocks (section B I 1 a), and are consequently rich in silicate. Tüxen, Oberdorfer and others call the Beech woods growing on them simply 'silicate Beech woods'. Since however the acid-loving Woodrush-Beech woods (section 4a) occupy sandstones, quartz-rich slates and sands which also contain silicates (i.e. crystalline components apart from quartz) this name could easily lead to a misunderstanding. The term 'brown earth-Beech wood' used by me in the first edition is equally far from ideal. It clearly implies that these communities lie between those limestone Beech woods which have developed on rendzina or pararendzina on the one hand, and the extreme acid-soil Beech woods on the other. However acid humus, or moder, Beech woods are also found sometimes on brown or parabrown earths if these are poor in bases or podsolised. The longer name 'brown earth-mull Beech wood' would not be misunderstood and this will be used generally in its shortened form of '**brown-mull Beech wood**'. From the ecological and plant-sociological viewpoints these communities lie between the extremes (see fig. 37 and section B II 4 d). Thus one could also call them 'average Beech woods'.

Unfortunately the plant-sociological labelling of this central group of Beech woods also causes difficulties. They

Tab. 14. Brown-mull Beech woods in NW, SW and NE Central Europe on soils ranging in base content from rich to poor.

Serial no.:	1	2	3	4	5	6	7	8	9
Region[1)]	W	S	W	S	W	S	W	S	E
Tree layer:									
Fagus sylvatica	**5**	**5**	**5**	**5**	**5**	**5**	**5**	**5**	**5**
Acer campestre	1	1		3					
Quercus robur	1	4				2			
Acer platanoides	3	2	3	2		1			
Ulmus glabra	2	4	1	5		1			
Fraxinus excelsior	4	5	4	5	3	1			
Acer pseudoplatanus	2	5	4	4	1	2			
Prunus avium	1	4	1	4		2		1	
Carpinus betulus	1	5		4	1	4		4	3
Tilia cordata		2		2		2		1	1
Betula pendula		1			3	1		1	
Quercus petraea		2		5		4	4	4	
Larix decidua N		1		3		2		1	
Pinus sylvestris N		2		3		3		5	
Abies alba N		4		3		5		3	
Picea abies N		4		4		5		5	1
Shrub layer:									
Prunus spinosa		2		3					
Tamus communis		2		3					
Clematis vitalba		2		4					
Viburnum lantana		4		5					
Cornus sanguinea		4		5	1				
Ligustrum vulgare		5		5		1			
Rosa arvensis		5		5		1			
Euonymus europaeus		3		4		2			
Daphne mezereum	2	5	2	4	2				3
Viburnum opulus		5	1	5	1	3		1	
Sambucus racemosa		1				3		1	
Rubus idaeus		2		2		5		2	
R. tereticaulis		4		4		5		5	
Ilex aquifolium		3		3		3		3	
Crataegus spec.	4	5	2	5	1	1	2		
Lonicera xylosteum	1	5		5		3		1	
Corylus avellana		3		4	1	3		3	1
Sorbus aucuparia	1		1		1	4	1	2	4
Lonicera periclymenum			2		2	1		2	
Sambucus nigra				1		4		1	
Herb layer:									
demanding species									
Allium ursinum	**5**	**5**	1						
Ranunculus auricomus	5								
Glechoma hederacea		4							
Carex pendula		4							
Hepatica nobilis	4		1						
Arum maculatum	5	5	5	3					
Mercurialis perennis	4	5	5	5	1				
Euphorb amygdaloides	2	4	3	3	1				
Bromus rac. + benekenii	1	2	1	4					
Anemone ranunc.	5		5						
Euphorb. dulcis		5		4					
Melittis melissophyllum		1		5					

Serial no.:	1	2	3	4	5	6	7	8	9
Region[1)]	W	S	W	S	W	S	W	S	E
Herb layer, cont.:									
± demanding									
Primula elatior	3	5	1	3	2	1			
Stachys sylvatica		5	2	1	2	4			
Geranium robertianum		5	1	2		5			
Circaea lutetiana		5		1	1	5			
Sanicula europaea	1	2		3	1	1			
Ranunculus ficaria	1	3		1	1	1			
Lilium martagon	1	2	2	2	1				
Asarum europaeum	5		2		4				
Hordelymum europaeus	5		2		3				
Polygonat verticillatum	3		2		2				
Dentaria bulbifera	1		1		1				
Lysimachia nemorum		1				5			
Paris quadrifolia	1	5		5		4			3
Pulmon. off. + obscura	1	5		5	3				2
Aegopodium podagraria	1	3		2		1			2
Geum urbanum		2		2		3			1
Stellaria holostea	2				4				5
Species common to all units (with wider amplitude)									
Anemone nemorum	5	4	3	5	5	4	5	5	5
Galium odoratum	3	5	4	5	4	5	2	4	5
Viola reich. + riviniana	4	5	2	5	4	5	1	5	5
Lamiastrum galeobdolon	5	5	3	5	5	3	2		5
Phyteuma spicatum	3	5		5	3	3	4	3	2
Oxalis acetosella	2	3	2	2	5	5	2	4	5
Dryopteris filix-mas	3	4	2	3	3	5	2	4	5
Polygonat multiflorum	4	5	1	5		5		3	3
Ajuga reptans		5		4	1	5		1	2
Vicia sepium	4	2	1	5	2	1		2	2
Carex digitata		3		4	1	3	1	2	5
Epilobium montanum		2		2	2	4		2	1
Poa nemoralis	1			2	3	1	3	2	3
Festuca gigantea		2		1		5		1	1
Melica nutans		1		4			1		3
Hedera helix	3	5	3	5	3	5	5	5	
Carex sylvatica	3	5	1	5	5	5	2	3	
Deschampsia cespitosa	3	5		1	5	2	5	1	
Fragaria vesca	1	4		4	1	5	1	2	
Brachypod sylvaticum	2	4	1	5	1	2	1		
Lathyrus vernus	4	4		3	2		1		
Campanula trachelium	2		1	3		1		1	
Galium sylvaticum	2			2	1	1	2	2	
Epipact. helleborine	1		1	1	1	2		2	
Neottia nidus-avis		2		3		1	1		
Melica uniflora	3		1		**4**		3		
Dactylis aschersomiana	3		1		2		5		
Prenanthes purpurea		2		2		3		3	
Potentilla sterilis		3		2		2		1	
Carex flacca		2		4				1	

Tab. 14 (Continued).

Serial no.:	1	2	3	4	5	6	7	8	9
	W	S	W	S	W	S	W	S	E
Herb layer cont.:									
± acid-tolerating									
Hieracium sylvaticum		1		5	1	5	5	5	
Solidaga virgaurea	1	2		3	1	5	4	5	
Athyrium filix-femina		3	1		5	5	2	5	4
Milium effusum	1	2	1	2	5	5	2	4	5
Mycelis muralis	1	1	1	1	1	5	3	3	4
Luzula pilosa	1	2		1	5	5	4	5	4
Scrophularia nodosa			1		2	4	1	5	2
Galeopsis tetrahit			1	1	1	4		2	2
Majanthemum bifolium				2		2		3	
Luzula sylvatica				2		2		3	
Cephalanth. damason.				3				1	
Luzula luzuloides				1	1	5	5	5	
Vaccinium myrtillus						3	1	5	
Veronica officinalis						3	4	3	
Carex pilulifera						2	2	2	
Pteridium alquilinum				1		3		5	
Carex brizoides						4		3	
Galium rotundifolium						3		2	
Moehringia trinervia						4	1	2	1
Dryopteris carthusiana						5		2	4
Carex pilosa						2		2	2

Serial no.:	1	2	3	4	5	6	7	8	9
	W	S	W	S	W	S	W	S	E
Moss layer									
Ctenidium molluscum		2		2					
Eurhynchium swartzii		1	1		3	1			
Mnium undulatum				1	1	3			
Fissidens taxifolius		5	1	5	2	1		1	
Eurhynchium striatum	1	5		5	1	5	1	5	
Brachythecium velutinum	1	1		1	1	1	3	2	
Atrichum undulatum		3		1	5	5	5	5	2
Polytrichum formosum		1			3	5	5	5	2
Thuidium tamariscinum		2		1		5	3	4	
Plagiochila asplenioides		1				1	3	2	
Hylocomium splendens				1		3		2	
Hypnum cupressiforme				2		2	5	5	
Dicranella heteromalla					4		5	4	
Mnium punctatum					2		3		
Dicranum scoparium						1	3	2	
Isopterigium elegans							5		
Cephalozia bicuspidata							5		
Mnium hornum							4		
Pohlia nutans							4		
Calypogeia fissa							4		
Diphyscium foliosum							3	1	
Lepidozia reptans							3		

Source: After tables in Winterhoff (1962), Frehner (1963) and Sokolowski (1966a).

Note: [a] W = in north-western Central Europe (Göttinger Wald); after Winterhoff (1963), S = in south-western C.E. (western part of the Aargau Canton); after Frehner (1963), O = in north-eastern C.E. (the Debowo Reserve close to the eastern limit of *Fagus*); after Sokołowski (1966a). A few non-constant species have been omitted. The list of mosses in column 9 is incomplete.

Nos. 1 & 2 **Wild Gaslic-Beech woods on brown loam over limestone**, relatively base-rich (leading to the limestone Ramsons Beech wood):

1 W Ramsons-Melick-Beechwood ('*Melico-Fagetum hordelymetosum Mercurialis* variant, *Allium ursinum* fazies', Winterhoff tab. 12a)

2 S Ramsons-Lungwort-Beech wood ('*Pulmonario-Fagetum allietosum*' Frehner tab. 5b).

Nos. 3 & 4 **Rich brown-mull Beech woods**; similar to 1 and 2 but less moist in spring (leading to the slightly-moist limestone Beech wood).

3 W Dog's Mercury-Melick-Beech wood ('*Melico-Fagetum typicum, Mercurialis* variant, typical sub-variant', Winterhoff tab. II 3b).

4 S Typical Lungwort-Beech wood ('*Pulmonario-Fagetum typicum*' Frehner tab. 5a).

Nos. 5 & 6 **Average brown-mull Beech woods** on deep neutral loam (the commonest type of true brown-mull Beech woods):

5 W Wood Millet-Melick-Beech wood ('*Melico-Fagetum typicum, Milium* variant, *Dicranella*-sub-variant' Winterhoff tab. II la).

6 S Typical Quaking Sedge-Beech wood ('*Melico-Fagetum* or *Carici brizoidi Fagetum asperuletosum*, typical variant' Frehner tab. 1a)

Nos. 7 & 8 **Poor brown-mull Beech woods** on deep acid loam (leading to moder Beechwood):

7 W Woodrush-Melick-Beech wood (Winterhoff differs from the modern view in calling it 'Woodrush-Beech wood': '*Luzulo-Fagetum typicum*', moss-rich variant tab. VIII, I 1)

8 S Woodrush-Quaking Sedge-Beech wood ('*Melico-Fagetum* or *Carici brizoidi-Fagetum luzuletosum*'; Frehner tab. 1d)

No. 9 E **Average brown-mull-Beech wood** close to the Beech's eastern limit on recent moraine loam (from a table by Sokotowski with 21 records).

Throughout the whole distribution of the submontane to montane loam Beech woods the species composition depends in the first place on soil properties, especially the abundance of bases and the available nitrogen more or less bound up with it. There are correspondingly numerous intermediates between the typical limestone Beech woods and the typical moder Beech woods. The type for the brown-mull Beech woods is best shown in columns nos. 5, 6 and 9.

Regional differences stand out clearly, e.g. in the extent to which the needleleaved trees take part (N). In the north-west *Asarum, Hordelymus* and *Mélica uniflora* play a significant part.

belong to the suballiance of Woodruff-Beech woods (*Eu-Fagion*; *Galio odorati-Fagion*) as do the limestone Beech woods, but they differ from the latter, as already mentioned in the Introduction, by having a smaller number of lime indicators and a stronger representation of acid-tolerant and more or less drought-sensitive species. Thus they can be clearly recognised by the combination of species present (tab. 14). However there is not a single species which is not present either in some limestone Beech woods or in moder Beech woods, and which could be used to provide a distinguishing name. The plant sociologist finds himself in this kind of dilemma when trying to name almost all the communities which occupy the middle ground ecologically. This then is not the least of the reasons why we prefer in this book names based on habitat rather than on species.

Oberdorfer (1957) endorses Knapp (1942) in describing the brown-mull Beech woods of southern Germany as Melick-Beech woods (*Melico-Fagetum*). This name is well chosen in so far as Wood Melick is very common in Beech woods on brown earth of average quality and often forms large green patches. However it is also quite frequent in limestone Beech woods and has prompted other authors (e.g. Rühl 1960) to give the name 'Melick Beech woods' to those growing on rendzina which are poor in herbs, show relatively poor growth, and in many other respects have little in common with the silicate Beech woods. In the Beech woods of the Alpine foothills with a very similar habitat *Melica uniflora* is absent, especially in northern Switzerland where brown-mull Beech woods have been described under various names by Etter (1943, 1947) and later designated *Carici brizoidis-Fagetum* by Frehner (1963). More recently authors in the German Alpine foothills, e.g. Oberdorfer (1971) prefer the name *Galio odorati-Fagetum* (previously *Asperulo-Fagetum*). From these three names it appears that there are climatic races of brown-mull Beech woods: at least one to the north-west rich in *Melica* commonly found on suitable soils from Ardennes (Dethioux 1969) through north-west Germany (F. Runge 1969b, Winterhoff 1963 and many others) right up to Denmark (Sissingh 1970) and to Eberswalde in the GDR (Scamoni 1967b), one to the south-west (with *Carex brizoides*) and a 'typical' one in the centre as well as towards the east of Central Europe. *Galium odoratum*, which is preferred in the name of the last race, is present in all three and also in limestone Beech woods and other communities. Probably still more climatic races could be distinguished; for example *Carici pilosae-Fagetum*, whose centre of distribution is in the southern part of Central Europe, radiating into north-east Switzerland and towards Upper Hesse (Streitz 1968). Frehner (1963, see also Ellenberg and Klötzli 1972) described a montane brown-mull Beech wood in Switzerland, in which there were obviously many Spruce present, as a Wood Millet-Beech wood (*Milio-Fagetum*). This name too says very little and is used equally well for the north-west German Melick-Beech wood, which is relatively poor in species (Burrichter 1973). To avoid misunderstanding an additional word relating to the habitat should be used, for example 'brown-mull Melick-Beech wood' and 'limestone Melick-Beech wood'.

The preceding examples may have stressed sufficiently the difficulties which always arise when plant sociologists go into regional details, and at the same time try to maintain a generally acceptable nomenclature. Wherever we meet such difficulties we shall try to find a pragmatic solution.

The fertility, water content and aeration of the average brown earths from the lowland up to the montane belt and from subatlantic to temperate subcontinental climate are such that many tree species flourish on them under suitable methods of management (section B I 1). Under natural competition however the Beech is always the winner since it can reproduce so very well in such habitats. In spite of this it was relatively late before the brown-mull Beech woodlands were recognised as true *Fagetum* types, mainly because they are strongly affected by man and mostly converted into arable fields.

Like the majority of other *Fageta* brown-mull Beech woods are 'cathedral' forests, poor in shrubs and herbs (fig. 71). Table 14 quotes examples from different parts of Central Europe. Since the Beech grows strongly and creates a darker canopy than in the moder Beech wood (*Luzulo-Fagion*) Oak can rarely attain its full growth. Instead Ash, Sycamore and Norway Maple are plentiful. In the undergrowth of a mature stand most of the following mull ground flora can be found:

Anemone nemorosa	*Lamiastrum galeobdolon*
Athyrium filix-femina	*Luzula pilosa*
Brachypodium sylvaticum	*Lysimachia nemorum*
Carex sylvatica	*Milium effusum*
Dryopteris filix-mas	*Phyteuma spicatum*
Galium odoratum	*Polygonatum multiflorum*
Hedera helix	*Viola reichenbachiana*

More discriminating mull indicators which can only survive in soils with a higher lime content, are almost completely absent. Some of these can be used directly as 'negative' differentiating species for dividing the poorer brown-earth-mull Beech woods from the limestone Beech woods, e.g.:

Allium ursinum	*Mercurialis perennis*
Arum Maculatum	*Pulmonaria officinalis*
Corydalis species	*Sanicula europaea*
Dentaria species	*Stachys sylvatica*

These species are to be found only in a relatively rich

subassociation of the brown-mull Beech wood which can become established on the more fertile brown earths. They were described as *Melico-Fagetum pulmonarietosum* by Scamoni (1960) from the young moraine area of the G.D.R. and raised to the rank of association (*Pulmonario-Fagetum*) by Frehner (1963), especially because they play a big rôle in the region between the Swiss Jura and the Alps.

Table 14 also contains examples of brown-mull Beech woods on very deep and relatively acid soils. These lead into the moder Beech woods (B II 4), e.g. the Wood Millet-Beech wood of Switzerland which has already been mentioned, or the Woodrush-Melick-Beech wood (*Melico-Fagetum luzuletosum*) of north-west Germany. They contain a small number of acid indicators, e.g.:

Luzula luzuloides M *Polytrichum formosum*
Carex pilulifera

The best grown Melick-Beech woods are to be found in western Central Europe and also on the Baltic young moraines from Denmark to Pomerania. Even right on the eastern boundary of the Beech area, in what was East Prussia, there are splendid stands which are very similar to the brown-mull Beech woods lying further to the west (see the last column in tab. 14). Markgraf (1932), Hartmann (1933) and others have already recognised that the Baltic Melick-Beech woods are naturally Beech-dominated. Scamoni (1960) appreciated its significance for the Baltic area.

Thorough investigations have been carried out also in the most northerly part of its large distribution area in Denmark and southern Sweden (Lindquist 1931, and Passarge 1965).

In the Alpine approaches and to the east of Central Europe there are brown-mull Beech woods in which large areas of almost pure Hairy Sedge (*Carex pilosa*) give them a parklike appearance. This sedge spreads quickly and in all directions by means of rhizomes. It flourishes particularly well in Oak-Hornbeam woods (fig. 113) and its distribution has a clear subcontinental tendency. Even at the western tip of its area of distribution, e.g. the Kreuzlingen and Neuwilen Forest (south of Konstanz), it can form huge square-kilometre stands within which scarcely any other herbaceous plants can grow (fig. 113).

The typical brown-mull Beech wood, like the limestone beech wood in southwest Central Europe gradually admits more Firs into the community with increasing altitude. It resembles the limestone Fir-Beech wood and so need not be looked at any more closely here.

(b) Moist-soil mixed Beech woods

Just as the Wild Garlic-Beech wood corresponds to a subassociation of the typical limestone Beech wood which needs more moisture, there is also a similar 'damper' community subordinate to the brown-mull Beech wood. It has been described under different names, e.g. as *Melico-Fagetum circaeetosum* or *athyrietosum* (Oberdorfer 1957), as *Fraxino-Fagetum* (Passarge 1959a), and also as *Querco-Carpinetum elymetosum* or *asperuletosum* (Tüxen 1937, Ellenberg 1939). In these names are expressed differences not only of habitat but also of interpretation: many authors see these communities on fairly wet soil as real Beech woods, others look upon them as Oak-Hornbeam woods which are rich in Beeches. In north-west Germany where Beech can clearly be seen to play a more important part they are called 'moist-soil mixed Beech woods'. As far as the water content of the soil is concerned these woodland communities lie between the typical brown-mull Beech woods growing on soils which show no noticeable gley horizon, and the wet Oak-Hornbeam wood where Beech can no longer tolerate the occasionally very wet gley soils (figs. 38 and 72).

The soil profile of moist-soil mixed Beech woods still shows pure brown earth characteristics in the upper 30-50 cm layer, which is densely packed with Beech roots, but in the deeper layers there are more or less clear rust coloured spots and lighter reduced zones

Fig. 72 The tender-leaved Touch-me-not (*Impatiens noli-tangere*) and the equally quickly wilting Hedge Woundwort (*Stachys sylvatica*) in the 'damp' mixed Beech wood (= Touch-me-not mixed Beech wood, see fig. 75).

produced by gleization (fig. 110). One can speak of gley-like brown earth, of brown earth with stagnant gley, or brown earth with groundwater gley, according to the degree of excess water in the subsoil.

The greater amount of moisture provides growing conditions for some of the dryness-sensitive plants such as would not occur in any other Beech wood apart from the Wild Garlic-Beech wood and a few montane and subalpine communities. The following can be listed as moisture indicators (see fig. 72):

Circaea lutetiana	*Impatiens noli-tangere*
Stachys sylvatica	*Stellaria nemorum*
Urtica dioica	*Veronica montana*
Festuca gigantea	*Carex remota*

Lysimachia nemorum, *Ajuga reptans* and *Geranium robertianum* are also more numerous and luxuriant than they are in other Beech woods where the soil is not so damp. These species are not only associated with a particular degree of wetness in the soil but they also serve as indicators of high nitrate content. At the same time they point to a better general fertility and it cannot be by chance that they turn up also in the Wild Garlic-Beech woods in a similar way. The majority of these species have tender leaves and benefit from the increased relative humidity of the air which is to be found in woods with a damp soil (fig. 75).

One of the few therophytes of our woods, the Touch-me-not (*Impatiens noli-tangere*) shows how, in the moist mixed Beech wood, dampness varies from year to year. When the spring is rainy and the summer cloudy the plant germinates and develops in huge quantities and spreads into neighbouring areas which are otherwise drier. Schmucker and Drude (1934) counted near Göttingen on the 7.5.33 no less than 995 seedlings on 1 m^2 On the 14.8, in the same place, there were still 135 grown plants crowding closely together. Too high a humidity however can damage the Touch-me-not as it is attacked by mildew (*Sphaerotheca humili* var. *fuliginosa*) and does not set any seed. On dry soil it produces little or no ripe fruit. Even where the soil contains enough water it begins to wilt if the sun shines on it for more than a few minutes; in this respect it behaves like a hygromorphic fern (section c). The increased wetness can be seen in the tree layer by the increased numbers of Ash which play an important part especially in the regeneration stage. Because of this Passarge (1959a) and Scamoni (1960) use the name *Fraxino-Fagetum (balticum)*.

Such moist-soil mixed Beech woods are very widespread on the moraines of the last glaciation period and especially on the boulder clay in north-eastern Germany and in Denmark. In the older diluvial plains of north-west Germany, Holland and Belgium these serve as the best habitats for Beech. All the drier ground here has been more severely leached out and thus is poorer in bases and general fertility, while the wetter places are quite unsuitable for beech. In the moist mixed Beech wood of the Brussels urban forest there are specimens of *Fagus* over 45 m high and it flourishes here much better than in all the 'real' Beech wood communities on less-damp soils and at higher altitudes.

In the montane and submontane belts of most of the Central European mountains one rarely sees Beech woods on this kind of gleyic brown earth simply because such habitats are absent or only cover small areas. However in the Hils Mountains which have been mapped by S. Jahn (1952) they cover extensive areas of marl.

On some rather wet poor loamy plateaux of the Alpine foothills and the mountains in southern Central Europe there is a related Beech wood community to which *Carex brizoides* sets its particular stamp (fig. 73). This sedge can form pure crowded stands, looking like combed curly hair, especially in shallow depressions under Beech or conifers. Hauff (1937) observed in the Swabian Jura that these sedge stands often come about because of man's activity, in that, up to the middle of the twentieth century, *Carex brizoides* was mown from time to time and used as a substitute for the real Seagrass (*Zostera marina*) in the filling of mattresses. Like the Beech woods rich in *Carex pilosa* mentioned in section (a) the 'Seagrass'-Beech woods, *Carici brizoides-Fagetum*, have a more southerly distribution and are absent from north-west Europe especially near the coast. The *Carex pilosa*-Beech woods show this tendency more strongly than the *C. brizoides*-Beech woods.

(c) Beech and mixed Beech woods rich in ferns

One occasionally comes across dense luxurious masses of ferns in brown-mull Beech woods and related communities on steep shade slopes with a moist climate. Groups of large fronds of Buckler Fern, Male Fern and Lady Fern (*Dryopteris dilatata, D. filix-mas, Athyrium filix-femina*, fig. 74) and more rarely ferns such as *Polystichum aculeatum* and *Thelypteris limbosperma* may be dominant. In other places smaller kinds such as the Oak Fern may be widespread; this is a

Fig. 73. A Quaking Sedge-Beech wood on a somewhat waterlogged brown earth of the moraine loam in north-eastern Switzerland. The smaller shining flecks are the flowering stalks of *Carex brizoides*, the larger ones Sycamore saplings. In the background are plantations of Spruce.

Fig. 74. The herbaceous layer of a Fern mixed Beech wood in north-eastern Switzerland. *Athyrium filix-femina* (right of centre), *Dryopteris dilatata* (left and right), *Gymnocarpium dryopteris* (left and central) with *Prenanthes purpurea* (left) and, as acid indicator, *Maianthemum bifolium* (centre). The ferns are not troubled by the light so much as by longer periods of dry air conditions.

geophyte which forms expansive carpets and whose scientific name has been changed so confusingly in recent years (*Aspidium dryopteris* = *Dryopteris linnaeana* = now *Gymnocarpium dryopteris*). Why are such ferns found under some high forests and yet are absent from others which at first sight appear to be very similar sloping habitats? Obviously a number of factors must be operating.

One essential basic requirement appears to be a certain shortage of bases and with it the tendency to form a surface layer of moder humus which favours development of prothalli. In any case the ferns are found much less frequently on active mull soils on which prothalli are more likely to dry out or be eaten by animals than they are on acid, but porous, moder. Among the many fern-rich woodland communities so far investigated the reaction of the upper soil has not exceeded pH 5.5 to 6 (tab. 15). Often it approaches the acidity of the podsols, although the soil profile always shows the characteristics of slightly saturated brown earth or of ranker (humus-silicate) soil. Ellenberg (1939) found that the top soil of a fern-Beech wood in southern Lower Saxony remained below pH 4 for some months (cf. tab. 15). During the course of the year though it was always higher than that of raw humus soils (fig. 114).

A covering of more or less acid moder humus may well be a necessary condition for the establishment of ferns. It is not however a sufficient one, for there are many woods with moder soils in which few or no ferns flourish. A favourable microclimate would appear to be more important for the hygromorphous, or hygro- to mesomorphous, ferns. They prefer NE-, N- and NW- slopes facing away from the sun, but in no way do they 'love' the shade, as Lundegårdh (1954) has already stressed. They are found much more in relatively open mature forests where they can enjoy from 5 to 25% of the light that is available in the open. Nevertheless they avoid direct sunlight since this reduces the relative humidity of the air and raises the rate of transpiration from the more or less tender leaves, which have insufficient protection against cuticular water loss. Larger patches of sunlight soon cause *Gymnocarpium dryopteris*, and also *Athyrium* and other species, to start wilting. The more frequently these waves of dryness wander across the ground flora, and the more the relative humidity falls, so the competitiveness of the ferns becomes weaker compared with herbs and grasses with tougher leaves. Measurements taken by Ellenberg (1939) show how evened out are the water relationships of the ferns in a fern-Beech mixed woodland. He found that at no time of year was the soil very wet, but it always remained damp enough that the water tension never exceeded 2 bar (fig. 123).

The microclimates in the Fern-Beech wood and in a few other closely related Beech woodland communities were very thoroughly investigated by Hartmann, van Eimern and Jahn (1959) in the montane belt of the Harz mountains. Well grown ferns flourished only in those places where the sun's rays never or hardly ever shone directly on to the ground. Typical diurnal fluctuations of air temperature, relative humidity and vapour pressure are given in fig. 75 for three sample plots. In a Melick-Beech wood on a south-east slope the air temperature rises higher while the vapour

Tab. 15. Degrees of soil acidity in Fern-Beech woods of northern Central Europe.

		1 Southern Sweden	2 Teutoburg Forest	3 Leine-Weser Mountains	4 Example in 3, fluctuations throughout year
Leaf litter	O_L		4.8–**5.3**–5.9	4.7–**5.2**–6.0	
Upper soil	A_h	4.5–**4.8**–5.2	3.6–**4.4**–5.5	4.1–**4.4**–5.6	3.8–**4.2**–4.6
	A_b		3.9–**4.4**–5.8	4.0–**4.5**–5.7	3.6–**4.0**–4.4
Subsoil	B_v	4.5–**4.8**–5.2	4.3–**4.7**–6.3	4.4–**4.9**–6.5	3.7–**4.0**–4.3
	B/C_v		4.7–**4.9**–6.8	4.7–**5.1**–6.9	3.8–**4.1**–4.4
Bedrock	D		5.3–**5.8**–7.7	5.8–**6.3**–7.5	

1 Each 7-8 measurements in aqueous solution. After Lindquist (1931).
2 & 3 Each 20-30 measurements in aqueous solution. After Diemont (1938).
4 Minimum, mean and maximum values during the year 1937 in an experimental area on the north facing slope of the Süntel. After Ellemberg (1939). These values were measured in a homogeneous water suspension and as a result appear lower. They probably correspond more closely to the conditions in the soil (see Ellenberg 1958).
2 – 4 refer to deep-leached loess on shady slopes, mostly over limestone rocks.
Source: From various sources. Bolder type indicates the mean values

pressure and relative humidity sink lower than in the moist Touch-me-not-Beech wood at the bottom of a valley or the Fern-Beech wood on a northern slope. Evaporation and transpiration are correspondingly less in the latter situation.

As summer plants whose fronds first begin to unroll after the Beech leaves are almost fully out the ferns appear to be fairly indifferent to whether the trees are deciduous or evergreen or whether Beech and Fir are present in significant numbers or not. The tree layer must only allow enough light to penetrate so that the ferns can assimilate plenty of CO_2, but at the same time so little that quick-growing herbs, shrubs and young trees are restricted in their growth.

Fern-Beech woods are Fern-Beech mixed woods are to be found in all areas and at all levels in which Beech woods are distributed, but especially in the relatively oceanic climate of north-west Central Europe. In the subalpine belt fern-Beech mixed woods are met with which can be looked upon as variants of the Sycamore-Beech woods. Along with the ferns from lower belts, *Athyrium distentifolium* and *Thelypteris limbosperma* thrive well in the subalpine Fern-Sycamore-Beech wood.

(d) Beech woods rich in Wood Fescue

As far as the type of humus and the base content of the soil are concerned there is every intermediate type between the brown-mull Beech woods and the acid-loving moder Beech woods, which will be discussed in the last section (4). Some of the intermediate types are obviously dominated by Wood Fescue (*Festuca altissima*). So that these cannot be confused with limestone Beech woods or limestone Fir-Beech woods in which *Festuca altissima* also occurs, one could call the types on acid-soil brown earth Wood Fescue-Beech woods. They are to be found in nearly all Central European mountains, especially in the Harz and Fichtelgebirge (Celiński and Kraska 1969, see also fig. 99) and also in the young moraine area of the Baltic, but seem to prefer in every case a more oceanic and windy climate. Samek and Javurek (1964) described a Wood Fescue-Beech wood, rich in species, as *Festuco-Fagetum* from the montane belt even of the Czechoslovakian mountains where the local climate clearly shows oceanic tendencies.

Since Fescue-Beech woods generally occupy only small areas we shall not look at them any closer. On moraine hills on the eastern side of Schleswig-Holstein, as well as in the Elm, the Harz and other Central European mountains one gets the impression that *Festuca altissima* (fig. 99) feels at its best where the wind has piled up the Beech leaves which it has swirled away from the west and south slopes. The Wood Fescue spins a fine network of roots through the pile of moder produced in this way without penetrating more deeply into the mineral soil. The large clumps of knee-high leaves and hip-high flowering stalks can be pulled up surprisingly easily. It grows so to speak in a natural leaf compost and develops so luxuriantly that the clumps form pure stands as if they had been planted by a gardener. Such dense Fescue meadows cause great difficulties for the natural regeneration of Beech. According to Lötschert (1963) the biological

Fig. 75. Average air temperature (thick line) and relative humidity (thin line) throughout the day during good weather, also the partial pressure of the water vapour, at a height of 40 cm above the ground in three Beech wood communities near Wieda in the southern Harz (14-16.9.1953). From records supplied by Hartmann, van Eimern and Jahn (1959).

The air is coolest and dampest in the Fern-Beech wood on the north facing slope. The most-soil Touch-me-not mixed Beech wood on the valley floor has a similar microclimate. Here the absolute water content of the air is even greater. In the zonal woodland community, that of the Melick-Beech wood on a gentle south-eastern slope, the hygromorphous herbaceous plants suffer from time to time from dryness of the air.

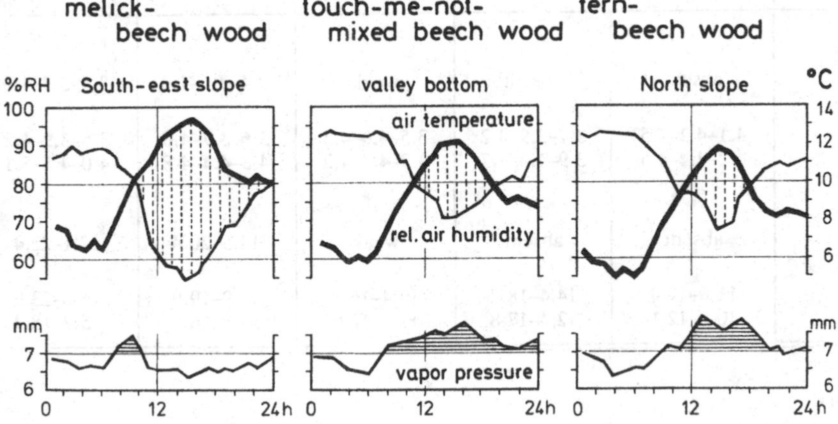

activity and the production of CO_2 by the soil are particularly high where the Wood Fescue dominates. There are bacteria in the rhizosphere of *Festuca altissima* which can fix nitrogen from the air and thus assist the nutrient turnover (Remacle 1975).

On brown earths of the montane belt within the area of distribution of *Abies alba* there are Fescue-Fir-Beech woods quite similar to the ones just described. They must be separated from the limestone Fescue-Fir-Beech woods (section B II 2 f) with which they have little in common apart from the *Festuca altissima* growing here and there and the inclusion of Firs in the tree layer (fig. 67).

In spite of the amazingly high activity in the upper soil the brown earth Fescue-Beech woods are close, both chemically and floristically, to the Woodrush-Beech woods which we shall deal with in the next section. This emerges for example from the analyses of Noirfalise (1956); he distinguishes a number of variants which are listed in tab. 16 and for each determined the pH value and the C/N ratio of the top soil. The rich variant grows in the Ardennes on soils whose pH value is over 4 and whose C/N ratio is comparable with that of mull soils. The typical variant showed a somewhat lower pH value in the top soil which together with the wider C/N ratio indicates a rather less favourable level of fertility. These results overlap those of the 'better' variant of the Woodrush-Beech wood, the typical variant of this community occupying a very acid surface humus with a poor C/N ratio. Bilberry and Wavy Hairgrass-Beech woods are even poorer regarding the C/N ratio whereas the pH value of their soils differs scarcely at all from that of other acid-soil Beech woods.

4 Beech woods and Oak-Beech woods on strongly acid soils

(a) Moder Beech woods as compared with other Beech woodlands

Communities of the suballiance *Luzulo-Fagion*, i.e. acidophilous Beech woodland poor in species, thrive on silicate rocks which are poor in bases, e.g. on granites and gneiss, sandstones or slates as well as on sandy diluvial deposits, provided that they contain a certain minimum silt or clay fraction (i.e. fine particles which will absorb plant nutrients), and that the climate is suitable for Beech (fig. 76). In the district of Germany which is richest in Beech woods today, Uslar in the Weser Mountains, where beech occupied 28.2% of the total area in 1936 (Hesmer), and also in the Harz, the Thüringia Forest and similar mountains, such acid-soil Beech woods are by far the most numerous. Oberdorfer (1957) rightly described the acid-soil Beech wood, containing scattered Oaks as 'the most important and often dominating zonal woodland type in the lowlands of Central Europe'. It has been put therefore at the centre of the most comprehensive research programme into land ecosystems which has so far been undertaken – the 'Solling project' of the German Research Association (Ellenberg 1971, see also section d).

Most places where such woodlands would have

Tab. 16. Soil acidity and C/N ratio under Wood Fescue-, Woodrush- and Bilberry-Beech woods in the Ardennes.

Communities		Brown-mull Beech woods Wood Fescue-Beech wood		Moder Beech woods Woodrush-Beech wood		Bilberry-Beech wood
		Milium variant (relat. rich)	Typical variant	*Oxalis* variant (relat. rich)	Typical variant	
pH(H_2O)[a])						
	Moder covering (O_F)	absent	absent	weak	3.4–**3.6**–4.0	3.4–**3.7**–4.0
	Topsoil (A_h), surface	4.1–**4.2**–4.5	3.7–**3.9**–4.2	3.5–**4.0**–4.3	3.6–**3.7**–4.0	3.5–**3.9**–4.2
	ca. 15 cm	4.2–**4.4**–4.5	3.9–**4.4**–4.7	3.9–**4.4**–4.2	4.3–**4.4**–4.8	4.0–**4.5**–5.1
C/N						
	Moder covering (O_F)	absent	absent	weak	14.5–20.4	17.1–22.4
	Topsoil (A_h), surface	11.0–11.4	14.4–18.7	10.4–15.7	13.9–19.0	15.0–23.2
	ca. 15 cm	10.9–12.1	12.4–17.8	12.8–12.9	14.8–16.6	15.7–18.4

Source: After records by Noirfalise (1956).
Note: a semi-bold type indicates mean values.

been the natural climax have however been taken into cultivation for a long time or planted up with Spruce and other conifers during the past 150 years. This turning to Spruce has increased more and more, especially in the highlands lying towards the east, e.g. the Erz and Fichtel Mountains and in the Sudeten highland, because Beech, once the most important fuel wood for household and industrial use has lost its value and because conifers, in demand for cellulose production, give a quicker return under forestry methods than any hardwood, especially Beech.

Our limestone mountains, especially the high Jura from Franconia right through into Switzerland, have been protected to a large extent from this changeover to conifers because the shallow-rooted Spruce does not thrive too well here, especially on steep slopes. The soil on top of the free-draining limestone gets too dry from time to time, while the deeper-rooting Beech can obtain sufficient water by penetrating into cracks in the rock. In addition the extensive use of woodland which had taken place in former times, especially the removal of litter, damaged the lime-rich soils and mull brown earths less than the silicate soils covered with a layer of moder. Spruce and Pine are better adapted to the afforestation and intensive management of impoverished or superficially acid soils than is Beech. More than 150 years have passed therefore since the latter was encouraged to grow in such places, because

Fig. 76. Woodrush-Oak-Beech wood in the Olsberger Wald south-east of Basel with *Luzula luzuloides* (centre) and *Oxalis acetosella* (in front).

Beech wood was required in larger quantities by the primitive industry of past centuries (Backmund 1941).

The development of forestry gave even more credence to the opinion, often expressed, that Beech is a lime-loving tree. This view was strongly supported however by some plant sociologists who published work in the thirties and who looked first of all for character species when classifying the different units of vegetation. In the rich limestone Beech woods they found numerous species which had their centres of distribution here and could serve as characteristic for the *Fagion* alliance. In the acid-soil Beech woods, with their fewer species, these recognition species were sought in vain. All the acid indicators which are able to survive under dominating Beech trees are found in much greater numbers in mixed Oak woods, stands of Pines and other communities on acid soil; in fact they grow better in these because they enjoy more light. This shortage of characteristic species prompted Tüxen and many other authors up to about 1945 to look upon acid-soil Beech woods, especially those in the lowlands, as artificial. That is, they thought that Beech had been introduced by foresters into the natural Birch-Oak woods or Oak-Hornbeam woods and through favoured treatment had become dominant. Only Beech woods which contained some of the *Fagion* or *Fagetalia* species named in section E III 1 were considered to be 'real' *Fagetea*. Against these Hesmer (1932, 1936) claimed on good historical, vegetational and ecological grounds that there are natural Beech communities on strongly acid soils too. Today no one doubts that this is true and in recent years plant-sociological nomenclature has taken it into account more and more.

What the acid-soil Beech woods have in common with the limestone and other *Fagetea* are their physiognomy and microclimate brought about by the dominance of *Fagus* (figs. 55, 61, 71 and 77). They form cathedral forests in which Beech saplings are to be found only in the openings. Usually there is no moss layer, or it is present only in small areas where the ground is swept by the wind, e.g. on west- or south-facing slopes or at the edges of the woodland. Apart from such unfavourable exhausted places the growth form and production of the Beech is no less than in comparable units of limestone Beech woods. For the success of the Beech it appears that the water content of the soil, which is dependent on its depth and texture, is more important than its chemical qualities.

Just a fleeting glance however is sufficient to show that soil chemistry plays a much bigger role in determining the composition of ground flora, for hardly any of the species present in limestone Beech woods also flourishes in those with acid soil. Only in the 'better' variants does one come across, here and there, the rather fastidious species, e.g.:

Anemone nemorosa	*Milium effusum*
Galium odoratum	*Luzula pilosa*
Moehringia trinervia	M *Atrichum undulatum*

Species which are fairly indifferent to soil acidity will go somewhat further:

Oxalis acetosella	*Dryopteris carthusiana*
Solidago virgaurea	*Hieracium sylvaticum*

However the dominant ones are those universally considered to be acid indicators – although of course, as with Beech itself, they cannot be considered to be acid-loving (see section B I 3 c). These species are rarely absent from old Beech stands on very acid soils:

Avenella flexuosa	*Pteridium aquilinum*
Vaccinium myrtillus	M *Dicranella heteromalla*
Carex pilulifera	M *Polytrichum formosum*
Veronica officinalis	

The vascular plants are often weakly developed through lack of light, and rarely set any seed. Besides these in the middle and south of Central Europe a few species are regularly to be found, which tend to be more frequent in the montane belt, in particular:

Luzula luzuloides(= *nemorosa* = *albida*)
Calamagrostis arundinacea

The trees and many of the herbs in acid-humus Beech woods live in symbiosis with mycorrhizal fungi which supply them with phosphorus and other nutrients. The moder cover is richer in plant nutrients and less acid than the mineral soil beneath (fig. 78). So it is certainly not by chance that the 'demanding' species which invade the moder soils are shallow rooting, or form more shallow root systems where the quality of humus is worse and its pH value is lower (fig. 78). Any additional acidification, e.g. after air pollution, by acid rain down the beech trunks, excludes these plants altogether (fig. 79). Thus acid indicators, with a wider pH amplitude (e.g. Woodrush, *Luzula luzuloides*, figs. 76 and 51) can become competitive in such places even in brown-mull Beech woods. On the other hand, according to Heinrichfreise (1981) this *Luzula* is favoured relatively on soils with a pH lower than 4 since it is more tolerant to the then easily soluble aluminium than, for instance, *Milium effusum*.

Fig. 77. Wavy Hairgrass-Beech wood on moraine loam with a covering of sand near Schwerin (Mecklenburg). Left foreground *Milium effusum*. *Avenella flexuosa* dominates but remains mostly sterile, due to lack of light. There are only few Beech saplings and no shrubs growing on the moder which already resembles raw humus.

Fig. 78. The root formation of the Spiked Rampion (*Phyteuma spicatum*) depends on the acidity of the top soil expressed as a mean annual pH value. After Ellenberg (1958).

a – c *Betulo-Quercetum molinietosum*, d – i various sub-communities of the *Stellario-Carpinetum*, mostly rich in Beech. The different positions of the rootstock come about because the main root, before it swells up, lies more or less close to the nutrient-rich soil horizons. Where the soil is very acid these are only the top centimetres of the profile which receive bases from the breakdown of the surface litter.
A further reason why the roots scarcely enter into the more strongly acid mineral soil may be the toxic effect of aluminium becoming soluble at pH below 4.5 (see M. Runge 1981).

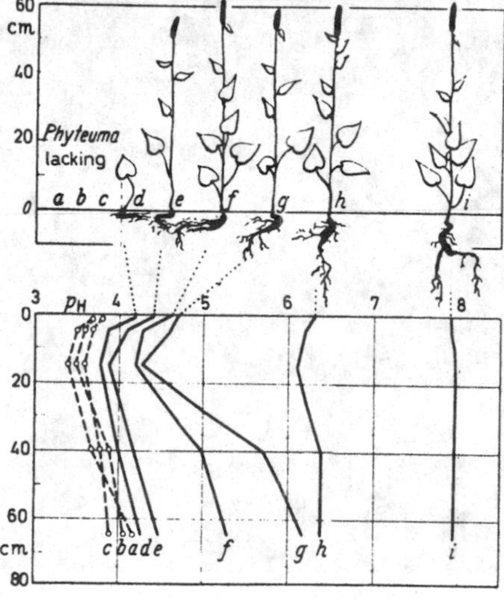

The more favourable nature of the uppermost layers of the soil which are 'manured' every year by the falling leaf litter already emerges from the pH profiles examples of which are to be found in fig. 78 and tab. 16. In nearly every case the mineral soil below the humus layer, that is at a depth of 10-20 cm, is the most acid and the pH value rises towards the surface as well as down into the subsoil although the latter is still more or less strongly acid. Although the Beech leaves have a beneficial effect on the quality of the upper soil their base content in the acid-soil Beech wood is insufficient to bring about their decomposition during the course of one year. Because of this, surface humus collects which is broken down only by arthropods and not mixed into the mineral soil by earthworms and so converted into mull. Since, according to Zuck (1952), the large worms, which are so important for the formation of clay-humus complexes, thrive neither on sandy nor on very acid soils, it is here especially where surface layers of humus are formed.

In itself however a humus cover is of no disadvantage for the growth of trees. Root fungi make it possible for all trees including Beech to make use of the acid humus. They concentrate their fine roots in this layer and in doing so come into keen competition with the shallow, rooted herbaceous plants as well as with their own seedlings. Such a shallow-rooting tree population however is more susceptible to occasional drying out of the top soil; thus its growth in places which have a dry local climate, other conditions being comparable, deteriorates more quickly than that of a population on base-rich, deep soils. Since the plant nutrient cycle, especially the release of nitrates is concentrated in the moder and in the uppermost layer of the soil which is very rich in humus (fig. 80), removal of the humus cover together with the leaf litter, which was customary in former times, has impoverished the soil much more than in mull Beech woods (see also fig. 20).

Where the chemical cycle of the acid-soil Beech wood remains closed, i.e. not robbed by removal of litter or by heavy grazing, the small amount of bases

Fig. 79. Acidification of the topsoil by water running down the tree trunks in a moderately acid brown-mull Beech wood (*Melico-Fagetum luzuletosum*). After Glavac, Krause and Straub (1971), modified (see also Koenies 1982). According to Züst (1977 verbal communication) the pH value of the water running down a Beech trunk can fall to as low as below 2 (see Fig. 33).

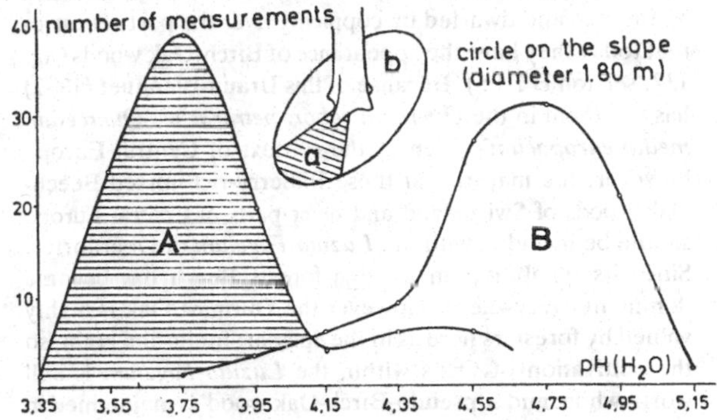

Fig. 80. The inorganic nitrogen content and the nitrogen net mineralisation in the soil profile of a moder Beech wood in Solling. After M. Runge (1974), somewhat altered.

Measured in terms of the weight of the soil sample the amount of nitrogen available to the plant (NH_4 and NO_3) is greatest in the fermentation layer (OF), but in terms of soil volume it is highest in the humus layer (OH) which contains a dense network of roots of both Beech and herbaceous plants. Here too the mineralisation, i.e. the continuous release of available nitrogen, is most rapid.

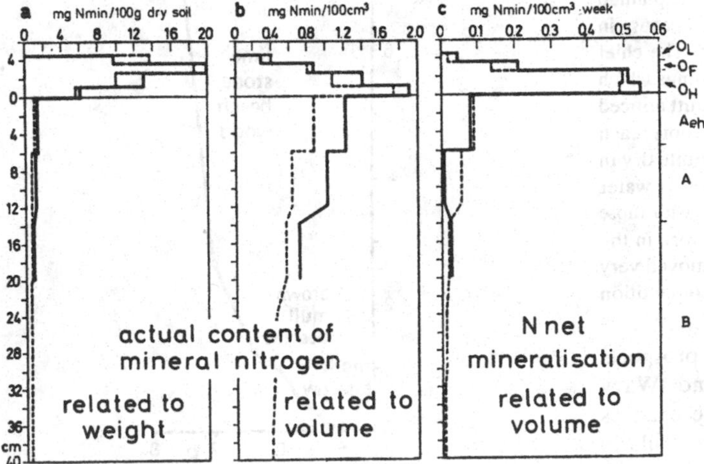

produced by the sandstones or acid slates is sufficient to prevent the formation of real raw humus (mor, see section B I 1 a). It remains in the form of a moder humus and does not lead to the podsolisation which is brought about by raw humus. Near-natural acid-soil Beech woods are thus found as a rule on brown earths or parabrown earths which though poor in bases are little if at all podsolised. Under many stands however one finds that leaching of the top soil with humic acid has been carried further because they have been previously grazed and in places litter and moder have been removed (fig. 20). Only through man's interference have Hairgrass, Bilberry and even Heather (*Calluna*), i.e. light-loving plants suited to a raw humus soil, been able to become so widespread as they are today in many woods close to settled areas (fig. 77).

On the old diluvial gravels, poor in lime, of the Irchel, the Stadler Berg and other plateau hills of north-east Switzerland, the acid-soil Beech woods have been so much opened up and dwarfed by coppicing over the past centuries that today they have the appearance of Birch-Oak woods (fig. 132, section B III 5). Because of this Braun-Blanquet (1932) has put them in the *Quercion robori-petraeae* as '*Quercetum medio-europaeum*'. Seen in the context of Central Europe however, the majority of these Bilberry-rich mixed Beech-Oak woods of Switzerland and other parts of Central Europe should be included with the *Luzulo-Fagetum* of Oberdorfer. Since its rehabilitation as high forests Beech has become dominant everywhere. However the Durmast Oak is highly valued by foresters here as in the Spessart for its fine grain, so the cultivation of Oaks within the *Luzulo-Fagetum* is still worth while, and a 'pseudo-Birch-Oak wood' is maintained in some forest districts.

According to Schmitt (1936) Beech does not regenerate very well in the Bilberry and Hairgrass stands of the High Spessart. On average he found only 14 young Beech trees per 100 m^2 in such habitats whereas there were 382 in the *Luzulo-Fagetum milietosum*, the best developed form of the Woodrush-Beech wood. This difference cannot be explained by the degree of acidity or the texture of the soils because in these respects the two habitats are quite similar. The chief cause is to be looked for in the covering of raw humus which under the *Vaccinium* can be as deep as 10 cm. Schmitt noticed that most of the tree saplings dry out before their roots reach the mineral soil. In Spessart the weather can be quite dry in spring and the raw humus retains a large part of its water reserves from the roots of trees. The latter, together with those of the dwarf shrubs and herbs, form a dense network in the humus layer, so that when water is scarce it is removed very quickly, and this becomes a dangerous form of competition against the tree saplings.

The drier the climate the worse are the prospects for Beech to regenerate in Bilberry and Wavy Hairgrass communities, and the greater the chances for the more easily satisfied tree species, particularly Oak and Pine, to succeed in the long run over this otherwise intolerant shade tree. Given the same heat and humidity, regeneration of Beech is dependent above all on humus quality and the availability of nutrients which is reflected quite closely by the pH value (fig. 81). Also F.H. Meyer (1961) found that the number of Beech seedlings per unit area increased with the pH and fertility of the soil.

(b) Climatic modifications of moder Beech and Oak-Beech woods

The species make up of the acid-soil Beech woods does not vary significantly throughout the distribution area of Beech in central and northern Europe. The factor 'acid humus' operates so effectively that the overall climatic influence is less noticeable. However certain differences do show up which are partly florogeographic and partly ecological. Regarding climatic differences the moder Beech woods of Central Europe can be divided into 5 large groups, namely:

1 A **montane-submontane** group of pure Beech woods which may be looked upon as typical of the *Luzulo-Fagion* (Woodrush-Beech wood, *Luzulo-Fagetum*, fig. 82).
2 A **submontane-colline** group of mixed Oak-Beech woods, in which *Quercus petrea* especially flourishes well and attains high quality without

Fig. 81. Regeneration of Beech on soils of differing pH in the Beech woods of southern Sweden. Ordinate: the number of 2-15 year old saplings per square metre. After Lindquist (1931), slightly modified.

The decisive reason for the low degree of germination and sapling growth on extremly acid soils may be the aluminium toxicity already mentioned under fig. 78.

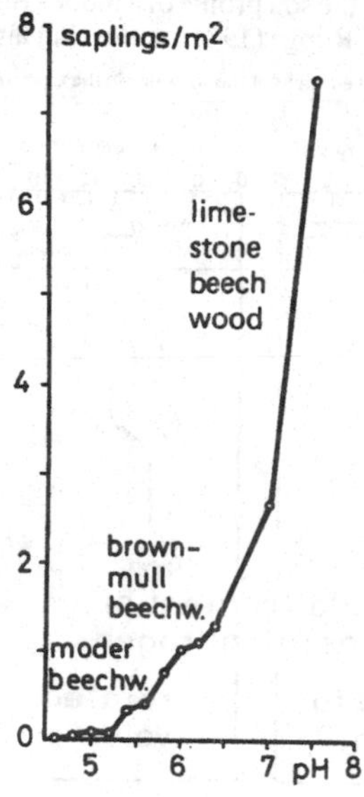

being able to achieve natural dominance. (Durmast Oak-Beech woods, previously called *Melampyro-Fagetum* by Oberdorfer, but in 1970 also included in *Luzulo-Fagetum*, fig. 132).

3 A **north-Central European-planar** group widespread in the diluvial lowlands, which is similar to the second, but appears to be poorer floristically and in part more strongly inclined towards the acid-soil Birch-Oak woods (placed in the alliance *Quercion robori-petraeae* as *Periclymeno-Fagetum* by Passarge and as *Fago-Quercetum* by Lohmeyer and Tüxen, Fig. 77).

4 An **insubrian-montane to submontane** group in the warm southern Alps with a high precipitation, e.g. in Tessin, where it appears in place of the first group (distinguished as *Luzulo niveae-Fagetum* by Ellenberg and Klötzli (1972, fig. 83).

5 A **south-Central European montane to high montane** group, with a natural admixture of firs or Spruce, leading to the montane coniferous woods (acid-soil Fir-Beech woods, insufficiently investigated but in part belonging to the *Luzulo sylvaticae-Fagetum* of Ellenberg and Klötzli 1972).

A number of species which have their main distribution in coniferous woodland can serve as floristic specifics of the last group, e.g.:

M *Sphagnum quinquefarium* and other *Sphagnum* spp.	M *Bazzania trilobata*
	M *Plagiotheciumundulatum*
	M *P. denticulatum*

These are obviously mosses which are able to develop well under conifers. The cultivation of conifers at lower levels has unfortunately obliterated the boundaries between group 5 on the one hand and groups 1-4 on the other since most of the mosses also occupy the artificial coniferous stands.

Groups 1, 2, and 5 can be distinguished from 3 and 4 by having more or less large quantities of *Luzula luzuloides. Calamagrostis arundinacea, Poa chaixii* and other montane species also are concentrated in the woods of groups 1 and 5, whereas they are almost absent from 2 and 3. In the latter two groups are found species which spread to acid-soil Oak woods, but not into coniferous montane woodland, e.g.:

	more towards the Atlantic:
Melampyrum pratense	
Lathyrus linifolius	*Lonicera periclymenum*
Frangula alnus	*Hypericum pulchrum*

In the north-west and western Central Europe group 3 is connected to Oak woods, poor in Beech, by a series of

Fig. 82. Phytomass production of a montane moder Beech wood (*Luzulo-Fagetum*) in the Belgian Ardennes divided into different compartments. All the figures are in tonnes or kg per hectare per year. After Duvigneaud and Kestemont (1977), modified.

The production above ground is about six times that below ground. For this the tree layer is mainly responsible while only insignificant contributions are made by the shrub, herb and moss layers.

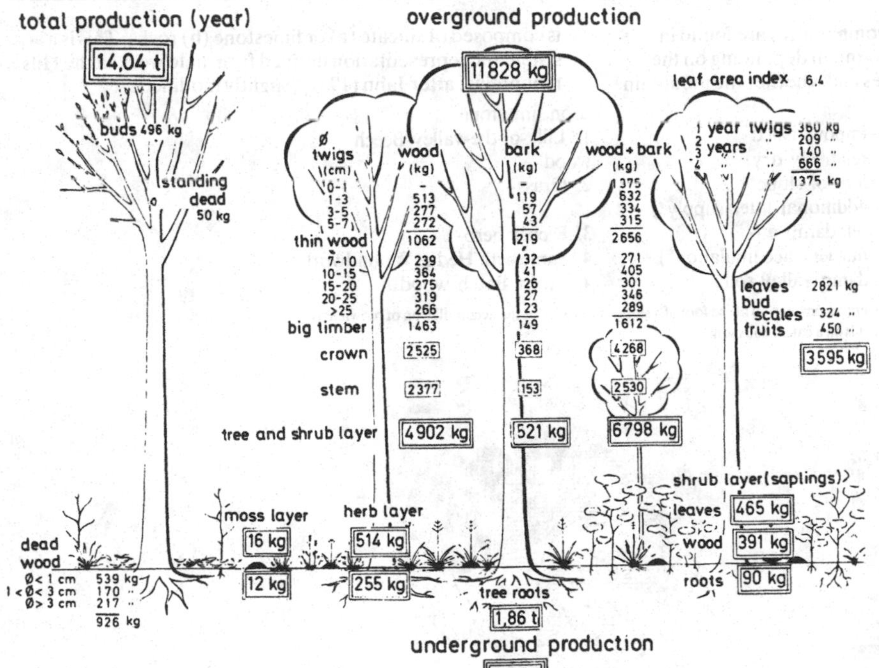

intermediate types, and for centuries has been so much influenced by man that it is scarcely possible to draw a natural line between *Quercion robori-petraeae* and *Luzulo-Fagion* (see also G. Jahn 1979).

Acid-soil Beech woods without *Luzula luzuloides, L. sylvatica* and other montane species are also present, according to Klix and Krausch (1958) in the Lower Lausitz, which previously had been considered as almost without Beech. The authors were able to show from place and field names and other historical help that Beech must formerly have been very widespread in this heathland.

Compared with geographic distribution and altitude the degree of oceanic influence in the climate on the species composition is not so clear. Oceanic elements such as *Galium harcynicum* and *Cytisus scoparius* are certainly completely absent in the east. Neither are they to be met with regularly in the western Beech woods as they are unable to withstand the shade. The behaviour of *Luzula sylvatica* (= *maxima*) is worth noting. This appears to prefer habitats with a humid atmosphere and can crowd out *Luzula luzuloides* completely from such places. Thus it is much more plentiful in mountains with a more oceanic climate such as the Vosges, the Black Forest and the outer Alps where there is a higher precipitation than in the more eastern highlands or the inner Alps. Also in Schleswig-Holstein, Denmark and west Norway there are Beech woods with *Luzula sylvatica* whereas they are nowhere to be found in the more continental climate of southern Sweden.

From Switzerland to southern Sweden and from the Ardennes to the Carpathians one can come across moder Beech woods with practically the same collection of species, but a few hundred steps up a single mountain are sufficient to show up the changes which can be brought about by the astounding variety of local climates. Although the Woodrush-Beech woods are very poor in species a whole 'swarm' of subassociations and variants can be distinguished depending on the aspect and steepness of the slope (fig. 84). On almost

Fig. 83. The distribution of the Central European White Woodrush (*Luzula luzuloides*) and the similar Snow-white Woodrush (*L. nivea*). The latter is confined more to the south-western mountains. After Gensac (1970).

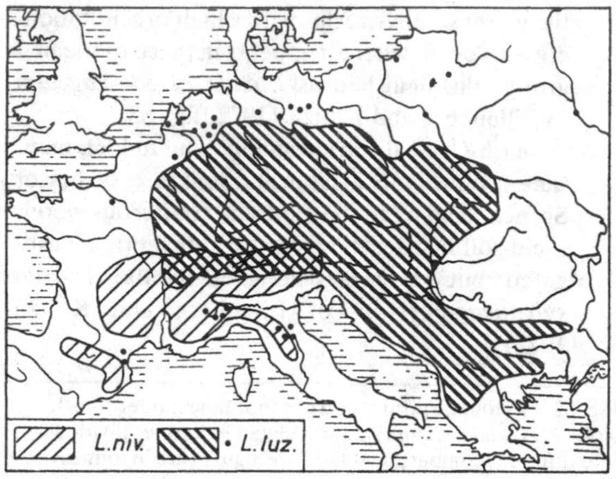

Fig. 84. Different Beech wood communities are found in various positions on a conical mountain depending on the steepness and aspect of the slopes and whether the mountain is composed of silicate (a) or limestone (b) rocks. This is a schematic representation derived from field work in the Hils mountains; after Jahn (1952), slightly modified.

a on silicate rock
1. *Leucobryum*-Beech wood –impoverished
2. Woodruch –relatively dry
3. Poor herb –foot of slope*)
4. *Calamagrostis* –additional litter supply*)
5. Fern –rel. damp air-
6. Wood Fescue –nutrient accumulation*)
7. Oak-Hornbeam wood –damp valley soil

b on limestone
1. Lilly-of-the-valley-Beech wood
2. Grass-

3. Richer herb-
4. Same with Hedge Woundwort
4. Moist Beech wood

*) The effect of even a slight amount of nutrient accumulation at the foot of a slope or because of the accumulation of blown leaves is more noticeable on soil poor in bases than it is on a calcareous soil.

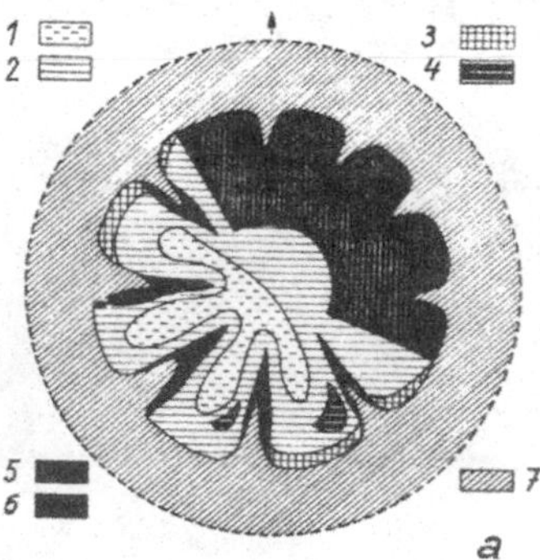

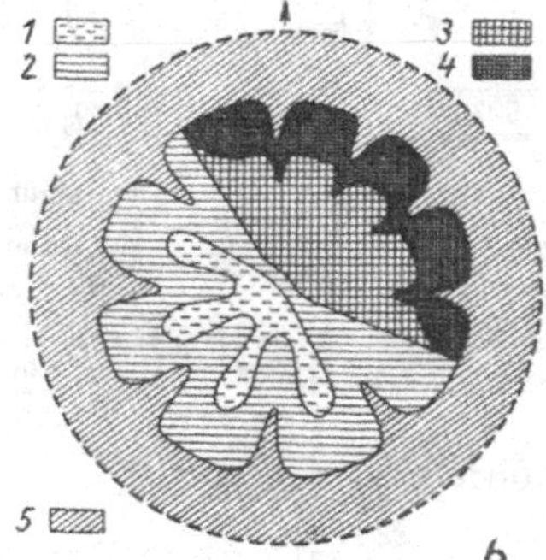

every sandstone or slate mountain with Beech stands some of the following types can be found:

a) The **typical** Woodrush-Beech wood usually occupies flat hilltops or gently shady slopes.
b) A **richer** subassociation (generally called *Luzulo-Fagetum miletosum*) prefers shady slopes and leads into the brown-mull Beech woods which dominate there.
c) A **poorer** subunit emerges on the south- and west-facing slopes; the more so as the local climate becomes drier and the more the trees have been degraded by coppicing.
d) On south-west slopes and knolls exposed to the wind Oaks are mixed with Beech as these suffer from **dryness** and low fertility. In these Durmast Oak-Beech woods the acid indicators, including many mosses, completely dominate the area.
e) Fruticose lichens and Heather (*Calluna vulgaris*) indicate totally **exhausted sites** where Beech is even less competitive. In the colline belt, under these severe conditions *Fago-Quercetum* can be replaced by a Birch-Oak wood (*Betulo-Quercetum*, see section B III 5).

This series is found to varying degrees in most of the lime-free mountains of Central Europe, e.g. in the Ardennes (Noirfalise 1956, Tanghe 1970, in Sauerland (F. Runge 1950), in the northern Swiss lowlands (Frehner 1963, Ellenberg and Klötzli 1972), in the Harz and its foothills (Tüxen 1954) and in the Thüringian Forest (Schlüter 1959). For the Hils Mountains G. Jahn (1952) has produced a schematic diagram of the different Beech wood communities on a rounded hill top (fig. 84a). In similar ways the limestone Beech woods too are dependent on the local climates of the slopes (fig. 84b). The wind-exposed ridges and crests facing south-west are also the driest and in the heavy rainfall area of the Hils region they carry a 'Lily-of-the-valley-Beech wood' reminiscent of the slope Sedge-Beech woods (section B II 2 c).

The various woodland communities which are shown in the diagrams in fig. 84 are determined by the relief, but this is not purely a function of the local climate. In addition the availability of plant nutrients always plays a part. All the ridges and overhangs are particularly impoverished, since both rainwater, flowing over the surface, and the wind, blowing away the dead litter, remove minerals (see Krause 1957, Frehner 1963). These are the 'loss areas' as opposed to the 'gain areas' the latter include bottoms of slopes, hollows and other habitats, where this material collects.

Table 17 gives the species in the communities on lime-free substrate which Schlüter recorded in the north-western part of the Thüringian Forest at a height of between 520 and 650 m. By and large the communities 1 – 8 form an ecological sequence according to the available water and with it the acidity and humus quality of the soils. No. 1 comes from relatively driest habitat, a 40° slope on porphyr, facing S. No. 8 has the most favourable water content as it is a loamy soil, with a NW aspect. The remaining stands can be ranked between these two, at any rate according to their combination of species. They are all on mica schist which forms soils deficient in colloids.

Tüxen (1954) studied a similar swarm of Beech and Oak wood communities on Burgberg near Bad Harzburg. He was able to show by measurements of the evaporation rate at the herbaceous level together with the maximum and minimum temperatures under the litter layer that the units he was able to distinguish formed an ecological sequence (cf. tab. 18). The evaporation under the high dense umbrella of the Beech trees in his Wood Melick-Beech wood was only about half that in the lower more open Birch-Oak wood where the areas of soil receiving full sunlight could reach a temperature of 50° C. Because there is less material produced, but more importantly because much of it is blown away, the layer of leaf litter is much thinner in poorer woodland types than in richer ones, and in many is absent altogether (last column in tab. 18).

Shortage of litter and the consequent poorer nutrient cycle may be considered the deciding factor in the south-west slopes of lime-deficient mountains so often carrying such stunted woodland. Where the wind can blow through freely it acts like a rake on the woodland floor. Even during leaf fall it whirls away some of the dead leaves and as long as the ground is not covered by snow it removes most of the others during autumn and spring. On steep slopes heavy rainfall can wash away some of the leaves. In each instance the topsoil is deprived of bases. In addition the soil can be invaded by mosses which are prevented from growing in normal Beech stands. Many of these mosses produce raw humus, especially the kinds which form thick cushions like *Leucobryum glaucum* and *Dicranum scoparium*, or continuous carpets like *Pleurozium schreberi* and *Hylocomium splendens*. The shortage of bases together with the drying action of the wind however restrict breakdown of the humus. In particular they favour *Leucobryum* the dead hyaline cells of which store large amounts of rainwater but rot very slowly. This chain of events explains the apparent paradox that in just those Beech woods where most of the leaf litter is lost throughout the year, the humus cover accumulates quickest. So beech leaves on lime-deficient soils should not be looked upon simply as an unfavourable influence on the soil.

Wherever surface humus is continually being

Tab. 17. Examples of acid-soil Beech-Oak woods and Beech woods on slopes in the north-western Thüringia Forest.

	Serial no.[1)]	1	2	3	4	5	6	7	8	Ecological behaviour of species
Altitude (m above sea level)		540	520	580	590	650	530	600	550	
Aspect of slope		S	SW	NE	SE	S	E	NE	NW	
Inclination in degrees		40	25	10	30	15	10	5	15	
Bedrock { P = porphyry, S = mica schist, R = red sandstone		P	S	S	S	S	S	S	R	
% of total cover tree layer		50	70	80	80	60	80	80	70	
" " " " shrub layer		15	+	2	+	+	15	+	+	
" " " " herb layer		70	80	80	70	40	90	40	60	
" " " " moss layer		40	5	2	2	2	2	–	–	F R N
Trees	*Quercus petraea*	**4**	**3**							– – –
	Fagus sylvatica		2	**5**	**5**	**4**	**5**	**5**		– – –
Shrubs	*Quercus petraea*	2	+							5 × ×
	Picea abies	1			+		+			× × ×
	Sorbus aucuparia	+	+	+		+	+	+		× 4 ×
	Fagus sylvatica		+	+	+		2	1	2	5 × ×
	Rubus idaeus			+			2		+	5 × 8
	Acer pseudoplatanus							+	+	6 × 7
Ground flora a	*Calluna vulgaris*	1								× 1 1
	M *Polytrichum juniperinum*	1								3 2 1
	M *Pleurozium schreberi*	1								× 1 1
	M *Cladonia* spec.	+								× 1 1
b	*Melampyrum pratense*	2	+							× 3 3
	M *Leucobryum glaucum*	3	+							4 1 2
c	*Vaccinium myrtillus*	**3**	**4**	+	2		+			× 2 3
	Avenella flexuosa	**3**	1	**4**	**3**	2	1		+	× 2 3
	M *Polytrichum formosum*	2		+	+		+			× 2 3
	M *Dicranum scoparium*	1			+					× 1 2
	M *Dicranella heteromalla*		+		+	+				4 2 2
	M *Pohlia nutans*		+		+	+				× × 3
d	*Luzula luzuloides*		+	+	+	**3**	1	+	1	× 3 4
	Calamagrostis arundinacea		+	+		+	**4**		+	5 4 5
e	*Oxalis acetosella*			1	+	+	2	2	2	6 4 7
	Dryopteris carthusiana			+			+	+		× 4 3
	Epilobium angustifolium			+	+°		+°			5 3 8
	Carex pilulifera				+	+				5 3 5
	Veronica officinalis					+				4 2 4
	Polygonatum verticillatum					+				5 × 5
f	*Anemone nemorosa*						+	1		× × ×
	Athyrium filix-femina						+	+		7 × 6
	Gymnocarpium dryopteris						1			6 4 5
g	*Milium effusum*							+	+	5 5 5
	Carex sylvatica							+	(+)	5 7 5
	Lamiastrum galeobdolon							2		5 7 5
	Moehringia trinervia							+		5 6 7
h	*Luzula sylvatica*								3	6 2 7
	Poa chaixii								1	5 3 4
	Mycelis muralis								+	5 × 6

Source: From records by Schlüter (1959).

Notes to Table 17

Communities named according to Schlüter; ecological evaluation according to Ellenberg (1974).

	Type of slope	mF	mR	mN
1 *Melampyro-Fagetum*, typical variant *Calluna* subvariant	S 40°	4.0	1.8	2.0
2 " " typical subvariant	SW 25°	4.8	2.6	3.1
3 *Luzulo-Fagetum myrtilletosum*, *Avenella* variant	NE 10°	5.2	3.1	4.5
4 " typical variant	ESE 30°	5.0	2.3	4.5
5 " " *typicum* typical variant	S 15°	4.8	3.0	4.2
6 " " " *Calamagrostis* variant	E 10°	5.6	3.2	4.7
7 " " *milietosum* typical variant	NE 5°	5.5	5.0	5.3
8 " " " " Poa chaixii variant,	NW 15°	5.3	3.8	5.1

The steep porphyry slope facing south (No. 1) is most impoverished; as a result the mean indicator values for moisture (mF), acidity (mR) and nitrogen (mN) are lowest here. The north-east slope near the mica schist summit (No. 7) offers favourable conditions. The remaining habitats form an irregular series between these two. Slopes facing south (nos. 1,2, and 5) always contain some dryness indicators while those facing north and east carry some dampness indicators.

removed by mechanical means, e.g. by soil erosion or on woodland paths, a discriminating type of moss flora occupies the acid loamy soil which Philippi (1963a) classifies in a special alliance (*Dicranellion heteromallae*). In addition moss communities rich in species are formed on rotten wood (Philippi 1965a, see also fig. 85).

The temperature and water measurements of the swarms of acid-soil Beech and Oak woods described above correspond roughly to those of the dry-slope Sedge-Beech woods, the Blue Moorgrass-Beech woods or even the 'thermophilous mixed oak woods' (section B 4 b) which occupy limestone hills on similar slopes. Despite this no ecologist would describe the acid-soil woods of the sunny slopes as 'thermophilous', for it is known that almost all their species are boreal and so extend into northern Europe as well as beyond the tree line in the high mountains. The parallel communities on calcareous soils and their partners, e.g. *Sesleria caerulea*, which can stand more cold than *Vaccinium myrtillus*, on the other hand are always said to like warmth and dryness. The truth is that the woodland communities of both the lime-deficient and lime-rich sunny slopes act as asyla for **light**-loving plants which are not able to live in the dark shade of the Beech canopy. In addition all sunny slope Beech woods contain a relatively large number of **dryness** indicators (tabs. 12 and 17). The mosses living on these impoverished slopes are adapted to frequent changes in their water content e.g. *Leucobryum glaucum* (Chalon, Devillez and Dumont 1977).

In the dense closed high Beech forests in other

Tab. 18. Woodland communities and habitat conditions on the Burgberg near Bad Harzburg.

Woodland communities No[a]	Measurements from 11.–23.5.1953:	Piche-Evapor. (ccm)	Temp. Max. (°C)	Tree height (m)	Leaf litter cover[b]
Birch-Oak wood (former coppice)					
1 *Betulo-Quercetum petraeae*, *Cladonia* variant		51	50	9–10	1
Moder Beech woods					
2 *Luzulo-Fagetum cladonietosum*, *Vaccinium*-Fazies		40.6	–	–	–
3 " " " typical variant		35	–	19–20	1.8
4 *Luzulo-Fagetum typicum*, *Luzula sylvatica* variant		31.7	32	23–25	3.3
Brown-mull Beech woods					
5 *Melico-Fagetum*, *Luzula*, *luzuloides* subassociation		32.3	21	28–29	4.5
6 " " typical subassociation		25.4	21	26–28	4.2

Note: [a] No. 1 corresponds to no. 1 in tab. 17; no. 2 to no. 3; no. 4 to no. 5; no. 5 approximately to no. 7. Nos. 1-4 are more or less strongly windswept and impoverished by loss of litter.

[b]) Average values based on estimations of the litter covering the soil: 1 = all leaves blown away; 2 = only a few leaves remaining; 3 = a fair number of leaves about; 4 = a good many leaves present; 5 = a large number of leaves.

Source: From records supplied by Tüxen (1954).

places in the Harz, which have been investigated by Hartmann, van Eimern and Jahn (1959), the differences in the microclimates between the stands on different slopes were significantly less than those in the more open stands investigated by Tüxen (1954). Well grown Woodrush-Beech woods and other Beech wood communities, which have not been degraded by extensive exploitation even on steep sunny slopes have a much closer and more even canopy under which the herbaceous layer can only feel a 'subdued' climate. In spite of this there are clear relationships between the type of vegetation and the meteorological measurements taken at the height of the herbaceous plants (fig. 75).

The conditions of the habitat under which the light-hungry *Vaccinium myrtillus* can settle and become dominant are realised only on very small areas in natural environments of Central Europe. However even quite isolated places within richer habitats are quickly occupied by the Bilberry as it is spread by birds. According to K.W. Schmidt (1957) its seeds will germinate only after they have passed through a bird's gut. Hares and Deer also play a part in its distrubution (P. Müller 1955) in so far as the latter always supplements its richer food with woody material. Communities with a lot of *Vaccinium* therefore constitute 'browsing centres' for the Roe Deer (Klötzli 1965) whose preference for communities in nitrogen-rich habitats we have already described in section B II 2 b).

Fig. 85. As with the majority of flowering plants the mosses too can thrive within a wide range of pH values. After Philippi (1966) slightly modified.

Both the mosses which occur naturally on acid substrates and those which grow on limestone have a physiological optimum range covering a number of units of pH value. With only a few exceptions this range is at least from pH 4.5 to 6.5. The area in which the growth rate is still more than half the maximum often extends over four pH units. This only applies however to the 'physiological behaviour' of the mosses in pure culture. The 'ecological' pH range (under the influence of numerous competitors in nature) is significantly narrower.
In the Woodrush-Beech wood mosses present on rotten wood, raw humus or bare acid soil free from leaf litter behave, with or without competition, in a similar way to the Woodrush itself (*Luzula luzuloides*, see fig. 51).

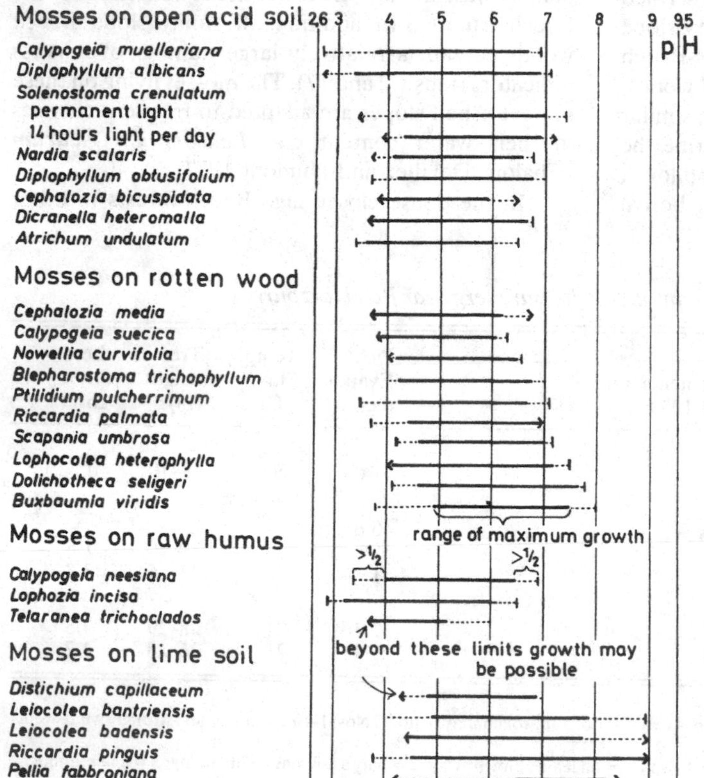

(c) Acid-soil Beech woods over limestone

It may seem surprising that even on some limestone mountains of Central Europe there are extensive stands of acid-soil Beech woods. They are only present however in those places where a layer of leached loam at least 40 – 80 cm thick covers the limestone. On the plateaux of the Franconian and Swabian Jura which have never been covered by the sea since Cretaceous times, considerable layers of calcareous rock have been leached out. According to Hauff (1937) these contain fractions up to 8% of clay or sand, and in places also flints which have originated from the petrified skeletons of sponges. Other limestone mountains e.g. around the Weser- and Leine valleys are covered here and there with more than 50 cm of loess from which the original lime content has long ago been washed out by drainage water. In the Swiss Jura, especially towards its western part, Ice Age glaciers have left behind sandy moraine soils on which Moor (1952), J.L. Richard (1961) and other authors have found acid-soil Beech woods. There are similar ones too in south-east England on acid loam over calcareous bedrock (Tansley 1939).

Hauff (1937) was the first to recognise the special position of the acid-soil Beech woods on limestone mountains and carried out a classic ecological and historical survey of them. He pointed out above all that they were true Beech woods to be compared directly with the limestone Beech woods of the Swabian Jura and not to be dismissed as artificial products. By arranging his records in an ecological sequence according to the pH value of the soil he also showed that the species composition was an expression of this factor or of peculiarities of the habitat which are related to it (see also Ellenberg 1963).

The nitrogen content in particular, according to W. Schmidt (1970) can be closely correlated with the degree of acidity, other things being equal (fig. 86). A significant though somewhat less good correlation exists with the C/N ratio of the soil. This is shown in tables 16, 19 and 25. The C/N ratio in humus increases, i.e. its N-content falls, with falling pH value. The average reaction figure (tab. 17 and section BI 4 b), which corresponds quite well with the average pH value of the soil (fig. 88), changes in the same way as the average nitrogen figure for the plant population.

(d) The nutrient supply in different Beech wood ecosystems

Slightly-moist limestone Beech woods, brown-mull Beech woods, and moder Beech woods form an ecological sequence with regard to the chemical and many of the physical properties of the soil. This has been seen already from the measurements of Noirfalise (section c) as well as from the studies by Hauff of the Beech woods on more or less sour soils in the

Fig. 86a. There is a close correlation between the total nitrogen content and the pH value of the humus topsoil in the different Beech wood communities around Göttingen.

b. In the same soils there is a weaker correlation between the organic carbon content and the pH values.

a and b after W. Schmidt (1970).

The C/N ratio too (not considered here) shows only a slight correlation with the pH. On extremely acid soils however, according to J. Scholz (1980) there is a very strong correlation with the nitrogen net mineralization. This rises from 5 kg. ha^{-1} $year^{-1}$ at pH 3.4 to 65 at pH 4.1.

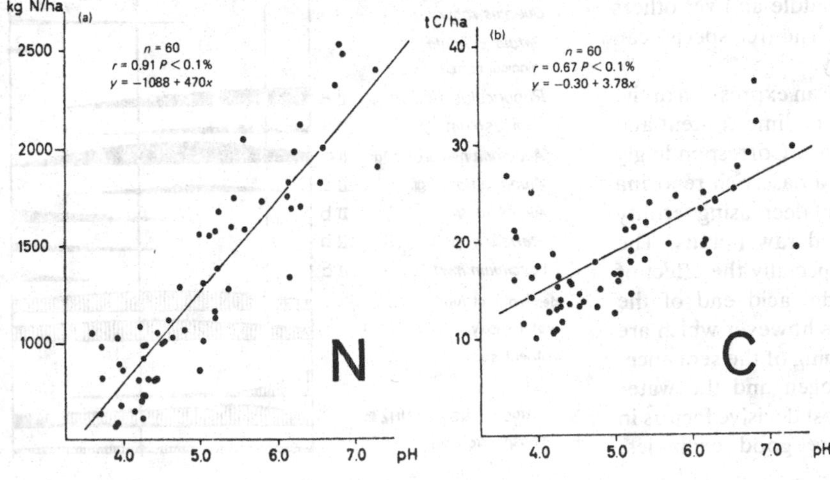

Swabian Jura. We would now like to look back at these sequences in more detail. We shall omit all the communities which occupy unusual habitats and the ones which do not represent the zonal vegetation. For submontane (and montane) sites in the highlands to the north of the Alps the sequence consists essentially of 7 stages:

	Woodland communities	**Soil types**
1	Limestone Beech wood *Lathyro-Fagetum*	Rendzina or browned rendzina
2	Rich brown-mull Beech wood *Galio odorati-Fagetum pulmonarietosum*	Brown earth of higher fertility
3	Typical brown-mull Beech wood *Galio odorati-Fagetum (typicum)*	Parabrown (or brown) earth of average fertility
4	Poor brown-mull Beech wood *Galio odorati-Fagetum polytrichotosum*	Parabrown (or brown) earth of low fertility
5	Rich moder Beech wood *Luzulo-Fagetum milietosum*	As 4 but more acid
6	Typical moder Beech wood *Luzulo-Fagetum (typicum)*	Weakly podsolised brown earth or very acid parabrown earth
7	Poor moder Beech wood *Luzulo-Fagetum vaccinie-tosum*	Podsolised brown earth or parabrown earth, also, exceptionally podsol

The proportions of the different ecological groups in the undergrowth vary from 1 to 7 as shown in the diagram (fig. 87). Some have their main concentration at the beginning, others in the middle and yet others towards the end of the sequence. The tree species can also be arranged in a similar way.

These floristic differences are an expression of the habitat conditions; by and large the lime content and the pH value fall through the series. Correspondingly the type of humus changes from a base-rich rendzina mull, through brown earth mulls of decreasing fertility to acid moder, even to very acid raw humus. The biological activity of the soils, especially the effect of earthworms, falls off towards the acid end of the sequence. There are some factors however which are not most favourable at the beginning of the sequence. The production of mineral nitrogen and the water supply to the plants, two of the most decisive factors in the suitability of a habitat, are good examples. Rendzinas as a rule are more free draining and roots do not penetrate them so deeply as the deep loams of the brown earth soils. Moder and brown-mull Beech woods do not differ so much in water balance as they do in fertility.

We will now try to identify more closely, as far as the available literature allows, the gradations in the habitat factors indicated in fig. 87. In this we shall concentrate particularly on the chemical and biological correlations.

According to Schönhar (1952), W. Schmidt (1970) and other authors there is a very close relationship between the species composition of Beech woods and Oak-Beech woods and the pH value of the topsoil (fig. 88). The 22 examples investigated near Stuttgart give a correlation curve which could be used to read off the approximate pH value from the type of vegetation.

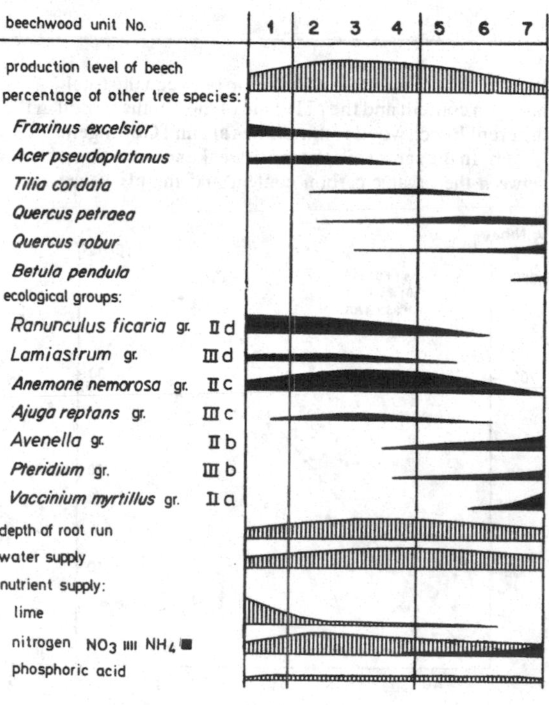

Fig. 87. Floristic and ecological gradients in the series of Beech wood communities, schematic.
1 Slightly moist limestone Beech wood
2 Rich brown-mull Beech wood
3 Typical brown-mull Beech wood
4 Poor brown-mull Beech wood
5 Rich moder Beech wood
6 Typical moder Beech wood
7 Poor moder Beech wood

The numbers of the ecological groups correspond to tab. 10 and fig. 52.

The ordinate in fig. 88 shows the 'average reaction value' (see section B I 4). Schönhar obtained this by filling in the R-value for each species on a list recording the herbaceous plants (see index of species E III 2) and calculating the average. Each pH value on the abscissa represents the yearly average obtained from a number of readings from the same soil. Single measurements gave a very wide scatter because throughout the year the pH value varies quite considerably and irregularly (more on this in section B III 2f).

From fig. 88 it can be seen that the moder Beech woods (or 'acid-soil Oak-Beech woods'), the poor brown-mull Beech woods (or 'typical Oak-Beech woods'), the rich brown-mull Beech woods and the limestone Beech woods of the Württemberg lowland actually form a continuous sequence with every possible intermediate. Such a clear correlation between the vegetation and the soil reaction can only be obtained however under constant climatic conditions; for the competition between the species changes with the heat and water balance. Within the pH span of the Beech woods the hydrogen ion concentration acts only indirectly on the plants. It can be considered simply as a 'symptom' of the level of fertility of the soil (see also fig. 50). In the first place this is dependent on the soil organisms and the type of humus they make by their activity (fig. 89).

Below pH 4.5 aluminium, which at higher pH can scarcely be absorbed by plants, becomes easily soluble in the form of $Al(OH)_3$. This is damaging to the roots of many species known not to be tolerant of high acidity (Ellenberg 1958, Ulrich, Mayer and Khanna 1979). The roots remain relatively short, but form a cluster of secondary roots. Thus the water and nutrient uptake is reduced, the more so since the phosphate and calcium balance too are disturbed, and the microbial activity is altered considerably.

The numbers of bacteria and fungi germinating per unit of organic matter are very similar in both moder and mull. Only in very acid raw humus (mor) do the bacteria find conditions less favourable. Figure 90 gives rather a false impression of the density of fungi since the fungal spore and its hyphae represent a much greater biomass than single bacteria. The breakdown of leaf litter as well as that of dead wood is brought about in the first place by fungi in all woodland communities, in moder Beech woods as well as in brown-mull and limestone Beech woods. Consequently the fungal hyphae attain a very large biomass although this remains hidden beneath the covering of leaves and in the soil. The visible fruiting bodies of fungi are only a very small part of them. Contrary to the generally held view the fungi are in no way less active or fewer in species in weakly acid or basic woodland soils than they are in strongly acid ones. According to Bohus and Babos (1967) the fungal species are much more spread out over the whole pH scale and can be put into 'reaction groups' in the same way as the higher plants (section B I 4 a). Carbiener, Ourisson and Bernard (1975) stress that communities of the *Luzulo-Fagion* are the richest in fungi, especially at moderate altitude. Nevertheless the authors name a whole list of characteristic species for the woodland communities on base-rich soils.

Fig. 88. The relationship between the mean indicator values for soil acidity (mR, as calculated from the species make up of the undergrowth, see section B14) and the average annual pH value of the soil in south-west German Beech wood communities near Stuttgart. After Schönhar (1952), modified (9 point R scale).

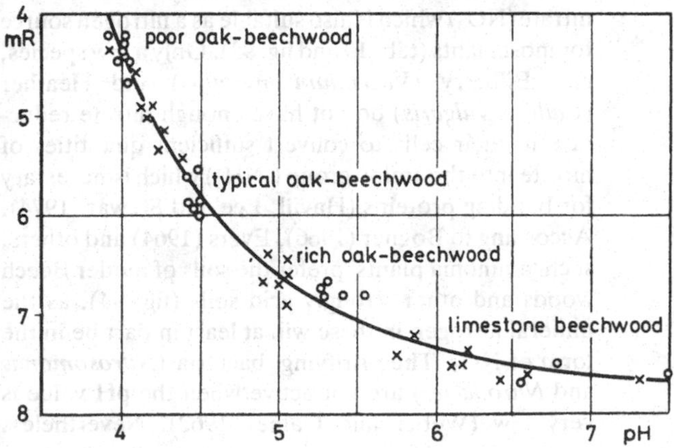

For the growth of root fungi, especially ectomycorrhiza, neutral or weakly acid mull is significantly better than the strongly acid and poor raw humus, which occurs for example on heath podsols (F. H. Meyer 1974, see section D II 3 a). Moder occupies an intermediate position in this respect. Göttsche (1972) found in the moder Beech wood of Solling many quite different mycorrhizal fungi on Beech which he put into 5 main groups according to anatomical characteristics; at first sight they can only be distinguished as brown, white and black. Other fungi play a significant part in the breakdown of the leaf litter which accrues each

Fig. 89. Mull, moder and raw humus (mor, originally Danish mør) differ significantly in their biological activity, as expressed, along with other criteria, in the mineralisation of nitrogen under comparable conditions. This difference becomes clear when the amount of inorganic nitrogen formed over a six-week period is compared with the dry weight of the organic matter. After Lemée (1966, Beech woods in the Forêt de Fontainebleau), somewhat modified.

In general the amount of mineralisation is correlated with temperature and is inhibited by dryness. In base-rich mull it is mainly nitrate which is produced whereas in mor almost all is ammonium. Both are formed in moder (see fig. 80).

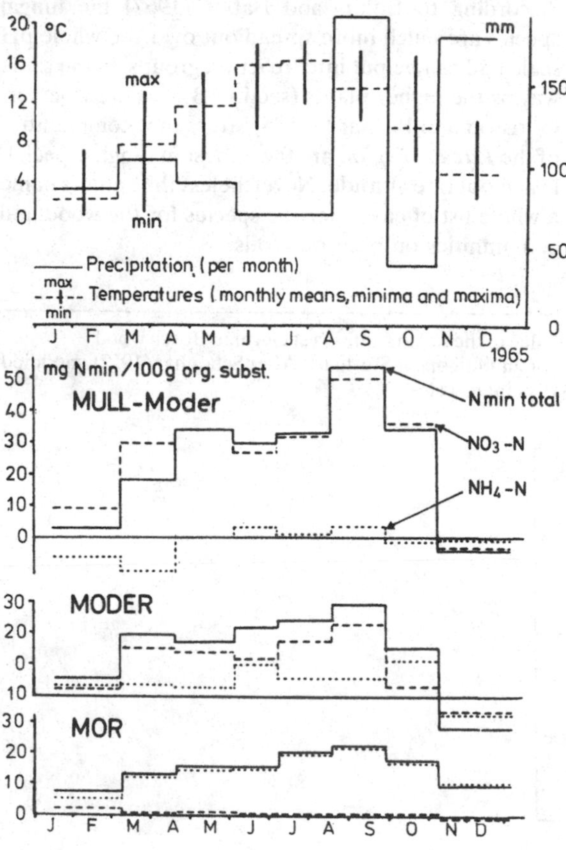

year (fig. 47). On mull soils this litter is largely worked into the soil by worms and other animals (fig. 91). In this way the mineralisation of the organically bound plant nutrients is spread out through the whole top soil, i.e. throughout a layer at least 20-30 cm thick. In strongly acid soils the animals are smaller and less strong, so the remains of the litter accumulate on the surface (fig. 80, see also Lemée and Bichaut 1973). Consequently the roots of the grasses (Lemée 1975) and herbs as well as of the trees also concentrate here. F. H. Meyer (1974) counted no less than 45 600 root tips of Beech per 100 ml in the fermentation layer (O_F) of a raw humus over podsol against only 500 in a similar volume of a brown mull. Again moder occupies an intermediate position, according to researches by Göttsche (1972). The fine roots of Beech are exceptionally active in the soil of the Woodrush-Beech wood, the intensity however varying throughout the year (fig. 92). Their increase in biomass amounts to at least 1500 kg/ha annually, i.e. about half as much as the mass of the leaves formed. The high production of roots is not least a consequence of the continuous release of minerals by microorganisms. This process is hampered to an extent with the lowering of pH. However the small droppings of arthropods which have a relatively high pH and are richer in nutrients than the leaf litter stimulate bacterial metabolism. In this way the phytophagous insects living in the tree crowns may contribute indirectly to the nutrition of these trees (Herlitzius 1975).

The provision of nitrogen for woodland plants is dependent above all on the activity of the microorganisms in the soil. Bacteria and fungi break down the proteins in leaf litter, dead roots and other organic matter, turning the nitrogen into ammonia (NH_4) which can be taken up directly by higher plants. In the soils of limestone Beech woods and most brown-mull Beech woods the NH_4 is continually oxidised into nitrate (NO_3) which is also suitable as a nitrogen source for most plants (tab. 19 and fig. 93). Only a few species, e.g. Bilberry (*Vaccinium myrtillus*) and Heather (*Calluna vulgaris*) do not have enough nitrate reductase in their cells to convert sufficient quantities of nitrate into the amino group ($-NH_2$) which is necessary for building proteins (Havill, Lee and Stewart 1974). According to Bogner (1966), Evers (1964) and others, such 'ammonia plants' prefer the soils of moder Beech woods and other strongly acid soils (fig. 94), as the mineral nitrogen in these will at least in part be in the form of NH_4. The nitrifying bacteria (*Nitrosomonas* and *Nitrobacter*) are not active when the pH value is very low (Weber and Gainey 1962). Nevertheless relatively large amounts of nitrates can occur even in most-acid woodland soils (Kriebitzsch 1978), in Central Europe at least during the summer (M. Runge 1974) and in tropical lowlands throughout the year (De Rham 1970). One must therefore assume that in addition to the usual nitrifiers other organisms might be capable of producing nitrates. These are probably not bacteria but fungi, which although acting more slowly amount to so much larger biomass in the soil that their activity is quite significant (Doxtander and Alexander 1966, and others).

The majority of microorganisms require NH_4 and NO_3 just as the higher plants do. Jannson (1958) was able to show by radioactive labelling of the N that microorganisms are more than a match for plant roots in the appropriation of available mineral nitrogen. Thus it is only the surplus NH_4 or NO_3, the so-called 'net mineralisation nitrogen', which is available in the soil for the higher plants. By means of an incubation method introduced by Zöttl (1960) and developed by Ellenberg (1964), M. Runge (1970) and Gerlach (1973) the net production of NH_4 and NO_3 can be measured in its natural form under outdoor conditions (see also fig. 63). Usually the yearly total net mineralisation of a soil is calculated by summing the results of the incubation method for all horizons per hectare per year. For the moder Beech wood in Solling for example (no. 4 in tab. 19) the N production ascertained in this way is only a little more than the amount of N taken up by the plant stand, which had been determined by Ulrich and Mayer (1973, see tab. 20) by chemical analyses of the annual biomass production.

So today we are in a position to determine reasonably accurately not only the quality but also the quantity of nitrogen available to plant communities although the production of NH_4 and NO_3 in the soil is a complicated microbiological process. For the ecological sequence of Beech wood communities looked at here, tab. 19 brings together some of the available

Fig. 90. Levels of pH value, C/N ratio and the numbers of bacteria and fungi in the upper horizons of a moderately acid, an acid and a very acid brown earth under the corresponding Beech wood communities. From data supplied by F.H. Meyer (1959).

The horizontal line indicates the boundary between the organic covoring and the mineral soil. There are relatively large numbers of bacteria living in the mull and the moder, but only a few in the mor (raw humus) which is very acid and poor in nitrogen. The numbers of viable spores and also the biomass of the fungi are about the same in the three soils. The benefit to the fungi in the poorer soils is thus only relative, not absolute.

The number of cellular myxomycetes, which is easily measurable, according to Frischknecht-Tobler et al. (1979) is closely correlated with the number of bacteria.

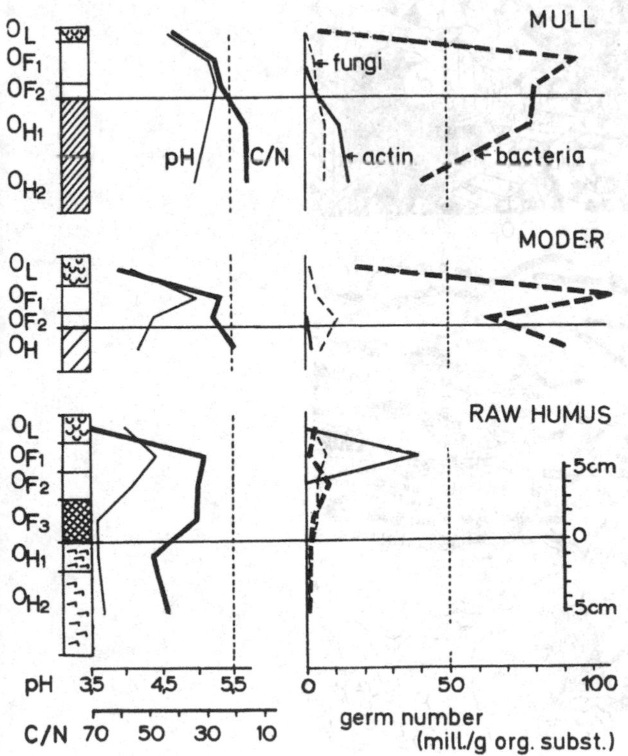

results. The highest annual totals were found in the brown-mull Beech wood growing on deep loam soil in a favourable climate; these were 238 kg N/ha in a relatively damp year and 157 kg/ha in a year with average rainfall. About 200 kg N/ha/annum corresponds to a good application of dung or fertiliser, i.e. the amount of N the farmer removes from the soil through harvesting the crop and which he must return to it. In the woodland the natural 'recycling' provides a similar amount to enable the trees to maintain a constantly high production. In an unfavourable climate, e.g. in the montane and subalpine belts this natural provision is less because mineralisation is restricted by the lower temperature and reduced by the brevity of the growing period. Thus in the Harz the production of mineral N in the brown-mull Beech

Fig. 91. The breakdown of Beech leaf litter in a biologically very active mull soil (brown earth from loess and basalt on the Vogelsberg). In order to show the full complement of animals which bring this about thin sections, some of which are to different scales, have been brought together in profile (below). In the drawing of the surface view (above) the upper layer of leaves has been removed, on the left as far as the boundary between the leaf layer (O_L) and the fermentation layer (O_F) and on the right exposing the middle of the O_F layer. After Zachariae (1965), somewhat modified.

a = leaves with traces of frass and dung from large collembola;
b = food and dung of small dipteran larvae;
c = droppings and tube (in section) of *Dendrobaena*;
d (centre of section) = droppings of large diplopods which feed on the F-leaves,
d(left) = dung balls of tipulid larvae from F-remains which are broken down in the mineral soil,
d (to the right in the surface view) = food and droppings of bibionid larvae;
e (right in surface view and left in section) = *Dendrobaena* droppings from arthropod dung and F-remains; e (section right) = enchytraeid droppings from arthropod dung and F-remains - concentrated here because they are protected from earthworms by a stone;
f = a bundle of only slightly decomposed leaves with dung from mining phtiracarids;
h (in section) = holes of *Lumbricus terrestris*;
h (surface view) = *Lumbricus* worm casts protruding from the A_h layer;
j: *Allolobophora* dung casts:
k = tubes and droppings of enchytraeids, on the left in old *Lumbricus* casts.

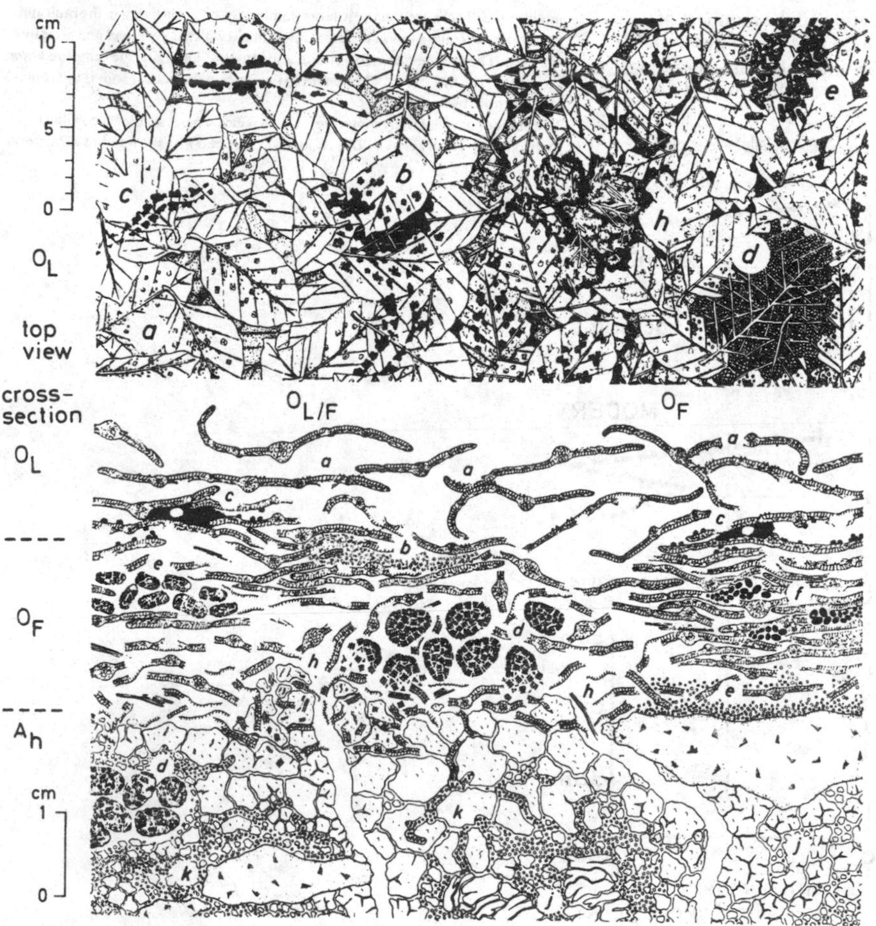

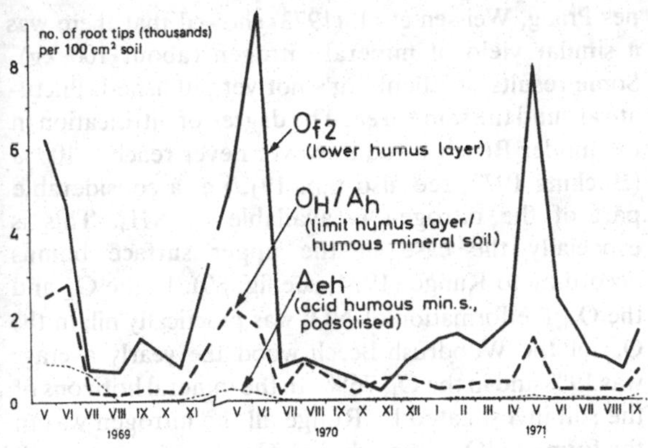

Fig. 92. Changes in the amount of root growth measured over a period of three years in the topsoil of a moder Beech wood (*Luzulo-Fagetum*, see fig. 76) in the Solling. After Göttsche (1972), slightly altered.

The number of living root tips per 100 cm² is universally greatest in the spring. In the extremely acid moder brown earth they are concentrated at all times of the year in the surface humus, especially in the Of2 layer where the mineralisation of the sold leaf litter is taking place most rapid. In the weakly bleached upper eluvial horizon (A_eh), i.e. in the top few centimetres of the mineral soil, there are only very few root tips, because the amount of nutrients available is much less (see figs. 78 and 80).

Tab. 19. The annual supply of mineral nitrogen in limestone, brown-mull and moder Beech woods.

Woodland communities and soil types	Year: Dampness	Net mineralisation (kgN/ha/y) 1972 normal	1973 dry	1974 damp	Nitrification level[a]
1. Dry-slope Sedge-Beech wood					
a sunny slope:					
Carici-Fagetum primuletosum, rendzina		–	10–45	8–**67**	V
" " *typicum* shallow rendzina		28	–	**32**	V
" " *typicum* normal rendzina		–	–	**89**	V
b shade slope					
Carici-Fagetum actaeetosum, rendzina		–	–	**94**	V
2. Slightly-moist limestone Beech woods					
a Wood Barley-Beech wood:					
Lathyro-Fagetum[c] *hordelymetosum* rendzina		80– 45	–	148	V
" " " medium deep rendzina		–	115–128	113–**187**	V
" " " brown earth over limestone		–	82	142	V
b Ramson Beech wood:					
Lathyro-Fagetum allietosum, brown rendzina		–	124	**195**	IV
3. Brown-mull Beech woods					
a wood Melick-Beech woods on the Metssner mountain					
Melico-Fagetum typicum, deep brown earths		144-57	–	188–**238**	V
" " " shallow brown earth		59- 78	–	66–70	V
b the same in the Harz (upper montane):					
Melico-Fagetum typicum, shallow brown earth		19- 77	–	24– ?	V
4. Moder Beech woods:					
Woodrush-Beech wood, very acid brown earth			(1968)	(1967)	
Luzulo-Fagetum in the Solling		–	ca. 80	ca. **130**	III
" " in the Belgian Ardennes		107			?

Note: [a]Where unspecified these are in the Göttingen area. From data in Grimme (1975), v. Gadow (1975) M. Runge (1974) and Praag et al. (1973), from Ellenberg 1977. The maxima for each community are shown in semi-bold type. Ranges of figures refer to several test areas.
Note: [b]V = more than 90% of the nigrogen is supplied as NO_3, IV = 75-90%, III = 25-50%, I = 5-25%, O = less than 5%.

[c]Described by Winterhoff (1963), who is also responsible for the other names, as *Melico-Fagetum hordelymetosum*.
[d]Calculated from the figures published by Runge (1974). He gives only the average figures of the mixing effect in preparing the soil samples.
[e]April 1972 to April 1973.

wood amounts to only 19-77 kg/ha/a. However even at this altitude there is 100% nitrification in not too acid soils because this does not stop at temperatures around 0° C (Seifert 1962).

For comparable general climates the N production in the slightly-moist limestone Beech woods is lower than that in the brown-mull Beech wood is 19 and fig. 63), particularly in those places where the fine earth is only 10-20 cm deep. Probably this is not just because of the size of the biologically active layer, but also because of the extremely small primary ammonium content of the limestone. According to Wlotzka (1962) there are only 70 g NH_3-N in 1 tonne of limestone compared with 125 g/t in sandstone and more than 500 g/t in clay rocks.

On sloping ground the ability of the microorganisms to carry out mineralisation is often restricted by drought. Because of this, slope Sedge-Beech woods are the poorest of all the Beech wood communities in the amount of nitrogen available (see also section B II 2c). In dry years the N-net mineralisation, even under dense closed stands of Beech can fall as low as 45 kg or less. In thin stands, or those impoverished by the removal of dead leaves by the wind, it can be even less.

The N availability of moder Beech woods appears to be of the same order of magnitude as that of slightly-moist limestone Beech woods. Runge (1974) found that in Solling the average for 1967 and 1968 was 112 kg/ha/year. For moder Beech woods in the Ardennes Praag, Weissen et al. (1973) showed that there was a similar yield of mineral nitrogen (about 100 kg). Some results of Ellenberg's not yet published, fluctuate around the same area. The degree of nitrification in the moder Beech wood however never reaches 100% (Bücking 1972, see also tab. 19), i.e. a considerable part of the nitrogen is available as NH_4. This is especially the case in the upper surface humus according to Runge (1974, see fig. 80). In the O_L and the O_{F1} the formation of NO_3 was practically nil; in the O_{F2} of the Woodrush-Beech wood the yearly average was 19% and in the O_H 76%. In the mineral horizons of the soil investigated by Runge all the nitrogen was in the form of NO_3 although the pH value sometimes fell to 3.0. Under these conditions Scholz (1980) found a surprisingly close correlation between the pH and the mineral nitrogen supply in Beech woods around Göttingen. The corresponding values varied from 5 kg N/ha/year at pH 3.4 to 70 kg at pH 4.1.

As already indicated the high NH_4 content of the surface humus favours the 'ammonium plants' which thrive less well when only NO_3 is offered them (fig. 94). Culture experiments by Bogner (1966) show that these are *Vaccinium myrtillus, Calluna vulgaris* and *Avenella flexuosa*. By transplanting *Avenella* and other species in different Beech wood communities Zarzycki (1968) was able to show that the form of nitrogen available appeared to have great ecological significance (see also Evers 1964). The solubility of ammonium might have

Fig. 93. The amount and form of inorganic nitrogen available at different depths in various broadleaved woodland soils near Salem (Holstein). After M. Runge (1965), modified (see fig. 80).

Mull soils and those with a proportion of fen peat, if they are not too wet, provide mostly nitrate, and the lower layers play a more or less significant part in the supply of nitrogen. It is only during times when the ground is completely waterlogged (i.e. short of oxygen) that the ammonium is not oxidised. Wet soils which are covered with water from time to time (*Alnetum*) provide little mineral nitrogen since most of it is lost through denitrification. In very acid soils with moder (*Querco-Fagetum*) or raw humus (*Betulo-Quercetum*) the proportion of ammonium is large to very large, but the total supply of nitrogen is not significantly less than in mull soils unless they are extremely acid.

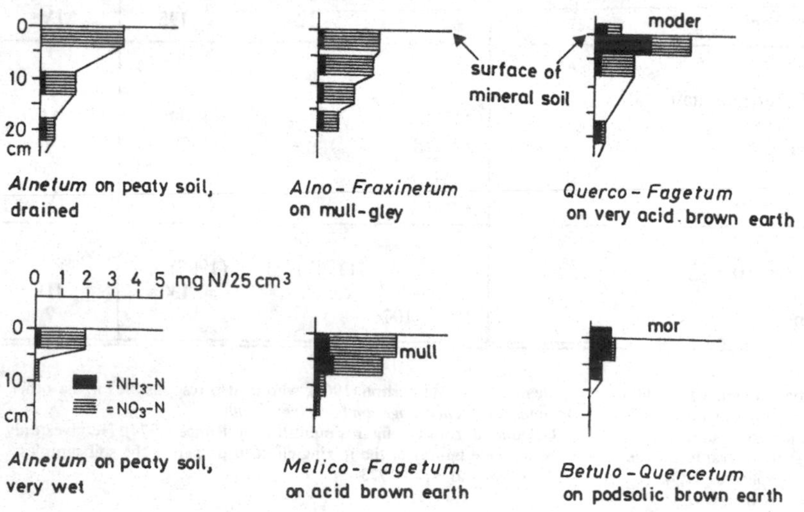

played a part in this context. Probably many acid indicators are such ammonium plants and would be indifferent to pH as such. In contrast many of the calcicoles may be 'nitrate plants', and when more or less pure NO_3 is available, that is when there is a high degree of nitrification, they are more competitive. *Mercurialis perennis*, *Campanula trachelium* and *Carex alba* are shown as examples in fig. 94. The fact that the natural proportions of NO_3 and NH_4 are so different in the soils of limestone, brown-mull and moder Beech woods corresponds strongly to the requirements of the woodland plants.

The optimum conditions of fertility for the Beech are not to be found on rendzina or pararendzina but on brown or parabrown earths with a moderately acid reaction in which a large amount of mineral nitrogen is

Fig. 94. The behaviour of some plant species in pot culture experiments when they have been supplied with equal amounts of mineral nitrogen, some with ammonium (NH_4), some with nitrate (NO_3) and some with the two forms mixed. In each case the plants were grown in soils with pH 3.3, 6.0 and 8.0, and their growth in cultivated pure stands is compared with that of plants growing in woodland soils under normal competition. In part from information supplied by Bogner (1966); from Ellenberg (1968), modified.

In the pure cultures the amount of dry matter production was taken as a measure of their growth and in each case this is given as a percentage of the maximum production for the species concerned (a second line refers to a duplicate series of experiments). In the wood the constancy of each species (in %) is taken as indicating its relative production.

In the pure cultures all the species grew best when supplied with a mixture of nitrate and ammonium. Dog's Mercury, Nettle-leaved Bellflower and White Sedge are only able to make use of ammonium in a neutral or alkaline soil whereas they thrive with nitrate whatever the pH value. In nature they avoid very acid soils where the supply of nitrogen is principally in the form NH_4.

Woodruff, Tall Brome and Wood Millet make use of NO_3 or a mixture equally well, but also make some use of NH_4 even in acid soils. In nature they are mostly found on slightly acid soils since they cannot withstand the competition on either strongly acid or alkaline soils.

The behaviour of Heather and Bilberry is quite the opposite to that of the first group. They are unable to consume large amounts of NO_3, and therefore starve on alkaline soils.

Woodrush and Yorkshire Fog, as well as Matgrass and Wavy Hairgrass, form a series of intermediates between the Woodruff and Heather groups. Under natural competition they are restricted to acid soils.

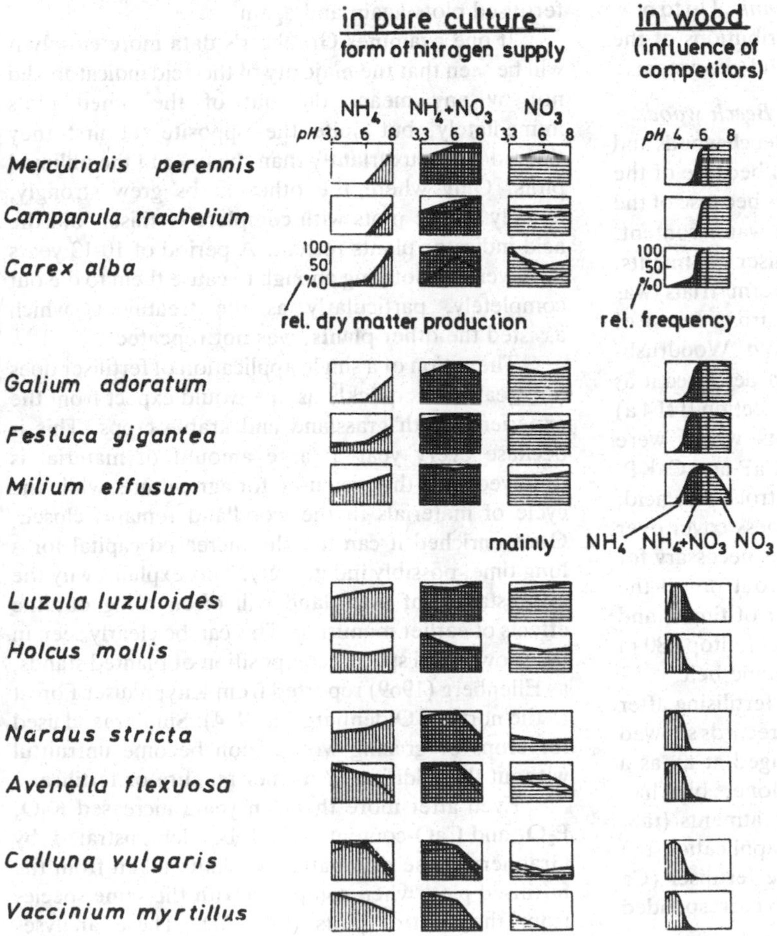

produced. Podsolised brown earths and podsols on the other hand are less favourable, especially as regards availability of nitrogen. In other words this means that the Beech is able to achieve its greatest size in the brown-mull Beech woods, which is thoroughly confirmed by forestry experience.

The 'higher production status' of silicate Beech woods compared with limestone Beech woods is further enhanced by the fact that the water capacity of the deep loamy soil on average is higher than that of the shallow stony ones, especially in a dry period. In this connection the majority of sandy podsols even are more favourable than rendzina and pararendzina since roots can penetrate deeply into them too.

This is true for the tree layer, but does not apply to the undergrowth. The herb layer, as a rule, is most luxuriant in limestone Beech woods, and becomes less and less dense with decreasing soil pH. On rendzina the herbs participate considerably in the turnover of bioelements and water. Even their interception of rain water is worth mentioning, as Schnock (1972) has measured in stands of *Mercurialis perennis*. Up to now we know fairly little about such contributions of the herb layer to the economy of woodland ecosystems.

(e) The effect of fertiliser on acid-soil Beech woods

The experimental evidence that mull Beech woods and acid-soil Beech woods differ above all because of the chemical properties of the soils and less because of the physical characteristics, especially the water content, has been obtained by a number of fertiliser treatments. One of the most informative long-term trials was analysed by Grabherr (1942a) in the urban forest of Hannoverian Münden. A 130-year-old Woodrush-Beech wood with a few Oaks (Grabherr described it as '*Querco-Carpinetum luzuletosum*', see section B II 4 a) was divided into a number of plots which were fertilised respectively with KP, CaK, CaP and CaKP. Other plots were left untreated as controls. The acid, brown earth soil has developed in a loess cover over Triassic sandstone. Ground cultivations necessary for the fertiliser treatments were carried out on all the plots. These were replicated a number of times, and the whole area covered 14.5 ha on a flat hilltop 280 m above sea level, that is in the submontane belt.

After 13 years (for the potassium fertilising after 10 years) Grabherr's plant sociological records showed that the ground flora had hardly changed at all as a result of the ground cultivations alone, but had responded strongly to the fertiliser treatments (tab. 20). As a consequence of this single application the plots which had received the complete fertiliser (Ca PK) had developed a ground flora which corresponded largely to that of the typical brown-mull Beech wood (*Melico-Fagetum*). There were many 'clearing' plants (see section D III 2 b) present which were absent from the undisturbed Beech woods, but here had apparently been helped by the sudden increase in level of nitrification. The ground flora had become adjusted to the new conditions astonishingly quickly. It is true that there were sites nearby where the invading species were already growing. However they would have had to get over longer or shorter distances with the help of ants and other animals or of the wind. Probably many of the plants had already been present as seed for a long time but because of the unfavourable soil, root competition or other reasons they had been unable to colonise before the start of the experiment.

Roe Deer could have played a special rôle in dispersal of the seed. Grabherr noticed that they browsed on the herbs and tree saplings on the fertilised plots in preference to the controls. In particular they tended to chose the species known to be rich in minerals; they also had the tendency to seek out the fertilised plots again and again.

If one examines Grabherr's data more closely it will be seen that the majority of the acid indicators did not by any means die out of the limed plots immediately, but quite the opposite; at first they thrived more luxuriantly than the ones in the unlimed plots. Only where the other herbs grew strongly, namely on the plots with complete fertiliser, did the acid indicator plants retreat. A period of 10-13 years however was not long enough to cause them to die out completely, particularly as the treatment, which assisted the other plants, was not repeated.

The action of a single application of fertiliser does not wear off as quickly as one would expect from the experience with grassland and arable crops. This is because every year a large amount of material is removed from the area used for agriculture, while the cycle of materials in the woodland remains closed. Once enriched it can use the increased capital for a long time, possibly indefinitely. This explains why the afforestation of farm land will often bring out the effects of earlier manuring. This can be clearly seen in the growth and species composition of planted stands, as Ellenberg (1969) reported from Knyphauser Forest to the north of Oldenburg (fig. 414). Similar land used for crops or grazing would soon become unfruitful without the addition of further manure or fertiliser.

Even after more than ten years increased K_2O, P_2O_5 and CaO-content could be demonstrated by Grabherr in the dry matter of plants taken from the fertilised plots when compared with the same species from the control plots (tab. 21). These analyses

Tab. 20. Effects of different fertiliser treatments on a rather moist soil moder Beech wood (Luzulo-Fagetum) on weathered sandstone near Hann, Münden.

Serial no.:	1	2	3	4	5	6
Manuring/cultivations[b]	uncult.	cult.	**KP**	**CaK**	**CaP**	**CaPK**
Tree layer						
Fagus sylvatica, 120 j.	5	5	5	5	**5**	**5**
Quercus petraea	*1*	*1*	*1*	*3*	*3*	*2*
Tree saplings						
Fagus sylvatica	*5*	*5*	*5*	*5*	*5*	*5*
Sorbus aucuparia	*3*	*3*	*5*	*4*	*2*	*4*
Fraxinus excelsior	*1*	*2*	*5*	*5*	*5*	*5*
Acer pseudoplatanus	*1*	*2*	*4*	*5*	*1*	*4*
Quercus petraea		*3*	*3*	*4*	*2*	*4*
Picea abies	*1*		*3*	*3*	*5*	*4*
Pinus sylvestris	*1*					
Larix decidua		*2*				
Shrubs						
Rubus idaeus	*2*	*2*	**5**	**4**	**5**	**5**
Sambucus racemosa		*3*	*5*	*5*	*5*	*4*
Frangula alnus		.	*3*	*2*	*1*	*2*
Ribes uva-crispa	*1*			*2*	*1*	*1*
Salix caprea					*2*	*2*
Rubus spec.					*1*	*1*
Herb layer						
Vaccinium myrtillus	*3*					
Teucrium scorodonia	*1*	*1*				
↓ [c]*Veronica montana*	*1*	*1*				
Avenella flexuosa	**5**	5	*4*	*2*	*3*	*4*
Majanthemum bifolium	5			2	2	2
Luzula luzuloides	*5*	*5*	**5**	**5**	**5**	**5**
Carex leporina	*1*	*5*	*4*	*4*	*1*	*3*
Carex pallescens	*2*	*4*	*3*	*3*	*5*	*3*
Agrostis tenuis	*3*	*5*	*4*	*4*	*3*	*5*
↓ *Poa nemoralis*	*2*	*3*	*2*	*2*	*1*	*2*
↓ *Carex remota*	*1*	*2*		*2*	*2*	*2*
Oxalis acetosella	*2*		**5**	*4*	*3*	**5**
Dryopteris carthusiana	*1*	*3*	*3*	*5*	*4*	*4*
Digitalis purpurea[d]	*2°*	*4°*	*5*	*5*	*5*	*5*
Stellaria media	*2°*	*4°*	*4*	*5*	*5*	*5*
Epilobium angustifolium	*2°*	*2*	*4°*	*5*	*5*	*5*
Galeopsis tetrahit	*1°*	*2*	*4*	*5*	*5*	*5*
Moehringia trinervia	*1°*	*5*	*5*	*4*	*4*	*4*
Taraxacum officinale	*1*		*1*	*1*	*3*	*2*
↑ *Juncus effusus*	*1*		*2*	*4*	*3*	*3*
Deschampsia cespitosa	*1*		*4*	*3*	*2*	*4*
Urtica dioica	*1°*			*4*	*5*	*3*
↑ *Hieracium laevigatum*						*2*
Scrophularia nodosa		*5*	*5*	*5*	*5*	*5*
Carex sylvatica		*4*	*4*	*4*	*4*	*4*
Mycelis muralis		*5*	*4*	*5*	*5*	*5*
↑ *Gallium harcynicum*			*2*		*1*	*1*
Epilobium montanum			*2*	*4*	*5*	*5*
Milium effusum			*2*	*3*	*3*	*3*
Calamagrostis epigeios				*5*		*3*
Festuca gigantea				*3*		*2*
↑ *Festuca altissima*				*5*	*3*	*4*
Galium odoratum				*5*	*5*	*4*
↓ *Stachys sylvatica*				*3*	*2*	*1*
↑ *Holcus mollis*				*3*	*4*	*4*
Circaea lutetiana				*2*	*4*	*3*
Stellaria nemorum				*2*	*3*	*2*
Dactylis glomerata				*2*	*3*	*2*
Senecio fuchsii				*2*	*3*	*2*
Athyrium filix-femina				*3*	*3*	*4*
Geum urbanum					*2*	
↑ *Rumex acetosella*					*4*	*3*
Veronica chamaedrys					*4*	*4*
Rumex sanguineus					*2*	*2*
Galium aparine					*2*	*2*
Ranunculus acris					*1*	*1*
Hieracium pilosella					*1*	*1*
Senecio sylvaticus					*2°*	*4°*
Lapsana communis					*2*	*3*
Poa annua					*1*	*2*
Vicia sepium						*2*
Stellaria holostea						*2*
Phyteuma spicatum						*2*
Viola reichenbachiana						*2*
Melica uniflora						*2*
Moss layer						
Dicranella heteromallu	*5*	*3*	*3*	*3*	*2*	*2*
Cladonia squamosa	*5*	*5*	*5*	*2*	*3*	*4*
Leucobryum glaucum	*2*	*3*	*2*		*2*	*1*
Dicranum scoparium	*5*	*4*	*4*	*4*	*4*	*4*
Polytrichum formosum	*5*	**5**	*5*	*4*	*5*	*5*
Atrichum undulatum	*1*		*3*	*3*	*4*	*2*

Source: From vegetation records by Grabherr (1942)[1])

Note: [a]Figures in *italics* indicate the constancy in several relevés from plots which have been treated alike. **Semi-bold type** = often occurring in large numbers.

[b] 2-6 cultivated with Hermann's woodland plough (March 1927)
Fertilisers applied per ha (single application in April 1927): Ca = 3000 kg burnt lime (52% CaO), P = 606 kg ground basic slag (= 97 kg P_2O_5), K = 30 kg potash salts (42% K_2O, not until April 1933). KP (3) had a small amount of lime since the basic slag contains about 40% CaO.
Plot size 0.69 ha., 3 replications.

[c] because of their ecological behaviour these species would be expected to be in a higher (or lower) group.

[d] ° indicates reduced vitality.

support the idea that plants which indicate low fertility do not have to avoid it. Even *Luzula luzuloides* and *Avenella flexuosa* have taken up more K_2O, CaO and P_2O_5 on the fertilised plots than on the unfertilised ones. However the 'fastidious' species, e.g. *Stellaria media* and *Digitalis purpurea* generally have a higher percentage of these minerals in their dry matter than the low-fertility indicators, when both are growing in the same soil. These facts were confirmed from a large amount of material by Schönnamsgruber (1959) and by Duvigneaud and Denaeyer-de Smet (1962).

As Grabherr's plant analyses have further demonstrated the majority of trees and herbs have been able to take up more nitrogen from the fertilised soil than from the unfertilised, even though only lime, potash and phosphate were added and they had no additional

Tab. 21. Some ecological statistics arising from the fertiliser trial described in Tab. 20 which show that a moder Beech wood has largely been changed into a brown-mull Beech wood.

Treatments		Control No. 1	Cultivated (once with wood and plough) 2 no fertiliser	3 KP	4 CaK	5 CaP	6 CaPK
Calculated indicator values[a]							
average moisture value (mF)		5.4	5.3	5.4	5.4	5.4	5.4
average acidity value (mR)		3.1	3.7	4.1	4.3	**4.4**	4.3
average nitrogen value (mN)		3.9	4.6	5.1	**5.6**	**5.6**	5.4
Soil analysis (No. 1 is missing)							
pH (H_2O)-pH(KCl)							
humus covering (O_H)		–	4.1-3.5	4.2-3.7	4.1-3.7	4.2-3.6	4.1-3.6
mineral soil (A_h)		–	3.9-3.9	4.1-4.2	4.1-4.2	4.0-4.0	4.0-4.1
Nutrients							
mg per 100 g dried soil	total N		130	130	140	150	**160**
g/l extracted in 10% HCl	P_2O_5		74	81	76	71	79
	K_2O		35	35	40	41	39
	CaO		30	33	68	72	69
Plant analysis (examples, without no. 1)							
Young Beech							
g in 100 g dry matter	Ash		2.8	3.0	3.2	3.3	3.7
	total N		0.90	1.23	1.12	1.12	1.12
	P_2O_5		0.40	0.35	0.32	0.38	0.40
	K_2O		0.49	0.46	0.46	0.47	0.47
	CaO		0.63	0.77	0.85	0.85	**1.07**
Woodrush (*Luzula luzuloides*)							
g in 100 g dry matter	Ash		7.4	7.1	7.8	7.0	6.8
	total N		1.49	1.53	1.57	1.61	1.61
	P_2O_5		0.25	0.42	0.36	0.46	0.45
	K_2O		1.76	2.10	1.98	2.15	2.15
	CaO		0.63	0.63	0.80	0.70	**0.84**
Foxglove (*Digitalis purpurea*)							
g in 100 g dry matter	Ash		10.1	14.7	14.1	13.6	16.2
	total N		1.90	2.24	2.17	2.02	1.94
	P_2O_5		0.47	0.43	0.61	0.49	0.69
	K_2O		2.20	4.05	4.10	4.41	3.58
	CaO		1.45	1.87	1.87	**2.18**	2.13

Note: [a] By the method explained in tab. 12. The evaluation of the species can be found in section E III.

Source: For the most part based on records by Grabherr (1942). Some maximum values semi-bold type.

nitrogenous fertiliser whatever. Probably the addition of lime had improved the conditions for nitrification. This is supported by the fact that nitrogen indicators such as *Urtica dioica, Senecio sylvaticus, Stellaria media* and *nemorum,* and *Epilobium angustifolium* appeared with a higher constancy in the plots which had received Ca. Thus it is not possible to draw any conclusions about the soil conditions just from the N,P,K and Ca contents of the herbaceous plants. Other parameters are required because the different species growing closely together vary to a large extent in the amplitude of their mineral content (see also Duvigneaud and Denayer-de Smet 1970a, b).

Finally if the pH values of the soils on the different plots are compared the surprising thing is that they are almost identical. A single application of lime then is insufficient to reduce the acidity of the moder brown earth and apparently it was really not necessary for the success of the experiment. We can see here a confirmation of the finding that neither the acidity of the soil **as such**, nor its Ca content within wide limits, plays a decisive role in the life of the higher plants, given that fertility is good (Ellenberg 1958). If the latter is increased in a soil which is both acid and poor in nutrients, for instance by adding N,P and K, or even just N, the trees and the undergrowth react quickly, even without the addition of lime (Burschel 1966b, Holstener-Jørgensen 1971).

The relationships described here between the soil quality and the ground flora apply, not only to Beech woods, but also to many other deciduous woodlands which we shall be looking at in the following sections. Since Beech has a very wide amplitude with respect to the chemical nature of the habitat, provided moisture and warmth are sufficient, the woodland communities where it is dominant can serve as model examples for the life conditions in most of the deciduous woodlands of Central Europe.

(f) Bioelement cycling in Woodrush-Beech woods and related ecosystems

We already know about the significance of soil acidity for the different woodland communities. More recently we have also been able to estimate their provision with nitrogen, but we still know far too little about the rôle of other elements of biological importance. Of the main plant nutrients, phosphorus appears to be available in sufficient quantities in the majority of Central European woodland communities, especially in the Beech woods. In any case W. Schmidt (1970) found no correlation between the species composition and the phosphorus content of the soil in many very different woodlands in the neighbourhood of Göttingen. Equally under limestone Beech woods as under brown-mull and moder Beech woods the amount of 'available' phosphorus varied from relatively low to exceptionally high values (the latter caused by earlier human settlement or by refuges in the woods). W. Schmidt got the same result whichever method of analysis he used; in spite of the uncertainty in the assessment of phosphorus availability which always exists for woodland soils, it may be considered then to be secured.

The cycles of phosphorus and nitrogen as well as many other elements have been worked out since 1966 especially within the framework of the 'International Biological Programme'. In this respect the Woodrush-Beech wood in Solling is one of the best researched land ecosystems and can serve as an example which may be applied to other woodland communities in this and in many other matters. Ulrich and Mayer (1973) traced the turnover of all the important elements during the course of the year (tab. 22). At least regarding additions from the air (through precipitation, line a) their data may be used for the other Beech wood communities in southern Niedersachsen (see also Steinhardt 1973 and O. Muhle 1974).

Nitrogen is brought in by the precipitation in considerable amounts (i.e. about 24 kg/ha/year, see also Lemée 1974), in the form of NH_3 (or NH_4) and NO_3. A large part of it is from the burning of fossil organic matter which contains combined nitrogen. From analysis of peat profiles it is known that the input of nitrogen before the advent of industry was only about 6-8 kg/ha/year (section C III 1 a). Large quantities of sulphur are also poured into the air in the form of SO_2 from domestic and industrial fires. This contributes to the acidification of the soil as sulphuric acid via aerosols (figs. 33 and 79). The common salt in the rainwater is however a natural product (NaCl). It is carried far inland by storms from the sea spray; as an example in Meathop Wood near the west coast of northern England, measured as sodium it amounted to 175 kg/ha/year of which 125 kg was caught by the leaf canopy as aerosols (White and Turner 1970). Magnesium too was deposited here in larger quantities – 23 kg compared with about 4 kg in Solling. Addition of calcium on the other hand was less at Meathop Wood than at Solling (13 kg against about 32 kg), since this comes mainly from dust, i.e. it is an 'inland' product. In the Woodrush-Beech wood at Solling, and indeed in other woodland communities of Central Europe, S, Cl and Na are in large measure transferred by being washed from the canopy (as much as 91, 97 and 87% respectively; see line 1 in tab. 22, and fig. 95). In contrast N and P take practically no part in this form of turnover.

The woodland uses the main plant nutrients N, P and K in an exceedingly economical way. The plants take up practically all that is available in the soil (line m, tab. 22). Only a very small part of these plant nutrients is removed therefore by drainage water from the region of the roots, namely only 6.2 kg N, no P, and 1.6 kg K (line h). In addition Bücking (1975) found that the nutrient content of water draining from the woodlands on Triassic red sandstone was relatively slight. On the other hand the water in the streams in wooded red marl and Jurassic areas can show an NO_3 content climbing up to over 20 mg/1. With potassium, which is very soluble, it is particularly surprising how slight the loss through drainage is. K is used principally in the leaves and amongst other things plays a part in the rapid regulation of stomatal opening; thus it helps to minimise unavoidable water loss during gas exchange. It is given off from the leaves in relatively large amounts and reaches the soil by means of rain falling through the canopy and running down the trunks (lines b and c, see also fig. 96).

In this connection we may refer to the calculations by Prenzel (1976, unpublished) who compared the amounts of nutrients in the plants with those in the soil

Fig. 95. The amounts of potassium, calcium and sodium washed down by precipitation each month above and below the canopy in a Belgian mixed broadleaved woodland on limestone (Virelles-Chimay, April 1964 to May 1965; no figures are available for May). From Denayer-de Smet (1966) modified.

Part of the precipitation is stopped by the canopy (left histogram, hatched). In spite of this more potassium and calcium reaches the ground in the water dropping through than is supplied to the surface of the canopy by the rain, especially in the cooler months. K and Ca are in fact given off mainly from the older leaves and the twigs. In any case it cannot be just a question of accumulated dust which is washed off by the rain, otherwise similar results would be shown with the easily dissolved sodium. (Under Oak canopies the amounts of K and Ca washed down were always more than the amounts which reached the canopy from above in every month during the year 1964-65.)

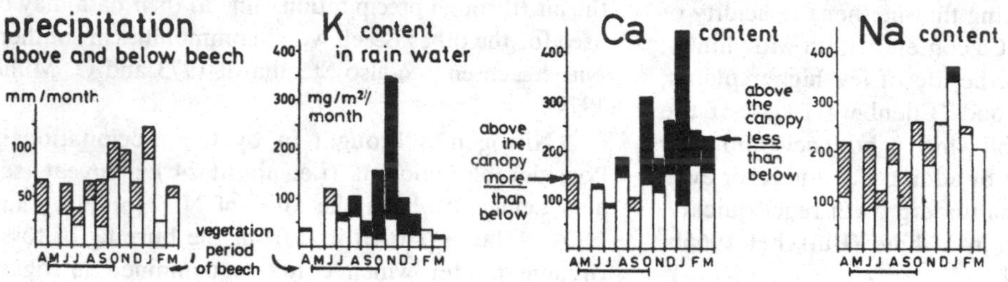

Tab. 22. The annual balance in the bioelement flow for a moder Beech wood in the Solling.

Element:	Na	K	Ca	Mg	Fe	Mn	N	P	Cl	S
a in the general precipitation	7.3	2.0	12.4	1.8	1.2	0.2	23.9	0.5	17.8	24.8
b in the canopy drip-off	11.3	18.1	26.6	3.5	1.5	2.8	22.5	0.6	**38.0**	**40.8**
c in the water running down the trunks	2.3	7.5	5.8	0.7	0.3	0.9	2.6	0.0	6.5	16.5
d in the leaf litter	0.9	21.9	15.0	1.5	2.0	6.6	53.0	4.3	0.8	3.2
e b+c+d = soil input	14.5	**47.5**	**47.4**	5.6	3.8	10.3	**78.1**	**4.9**	45.3	60.5
f drainage from the humus layer	12.9	40.4	39.8	4.7	1.3	6.5	76.8	4.8	38.3	43.9
g drainage water at 50 cm depth	6.7	2.3	14.3	2.0	0.1	4.1	5.8	0.0	36.5	12.5
h " " 100 cm depth	8.8	1.6	14.1	2.4	0.1	4.3	6.2	*0.0*	28.6	19.8
i e − h, taken up by plants	5.7	**45.9**	**33.3**	3.2	3.7	6.0	**71.9**	**4.9**	27.5	35.8
j a + c − a, turnover washed out of the canopy	6.3	23.6	20.0	2.4	0.6	3.5	1.2	0.1	26.7	32.5
k j + d, total turnover	7.2	45.5	35.0	3.8	2.6	10.1	54.2	4.4	27.5	35.7
l j as per cent k	**87**	52	57	60	30	30	*2*	*0*	**97**	**91**
m i as per cent e	39	**97**	70	57	**98**	58	**92**	**100**	61	59

Note: After Ulrich and R. Mayer (1973); all figures in kg ha^{-1} $year^{-1}$, some have been rounded off. Important and high values in semi-bold type, low values in italics.

to j: i.e. material excreted by the trees or taken up from the air.
to k: this does not include the turnover from dead rootlets, which may be quite considerable but which cannot be measured (on these grounds the figures in row i are too low).

water, and also with the amount of water transpired. N,P,K and Ca were taken up in larger quantities than would be accounted for by a simple mass transport of soil solution in the transpiration stream. In order to obtain these important nutrients the Beech roots must use a considerable amount of energy, and have a correspondingly high rate of respiration. On the other hand the plant actively protects itself against entry of Na, Cl and Al (and probably also against other poisonous substances which may be present in higher concentration). This also requires energy in order to absorb them in smaller quantities than they are available in the soil water. Only in the cases of S, Fe and Mg do the amounts taken up correspond with a mass flow of the soil water solution in the transpiration stream, but even these elements are not passively absorbed by the roots.

(g) Transpiration and primary production in Woodrush-Beech wood and other ecosystems

The transpiration stream which transports plant nutrients in such varied amounts to the leaves carries on average the equivalent of 280 mm of rainfall per annum (Benecke 1976, see tab. 23). This is only 26.5% of the total precipitation in Solling but nevertheless amounts to 2.8 million litres per hectare per year. During the same time according to Heller (tab. 24) the Beech stand produces at least 11 000 kg of dry matter. For every 1 kg which is synthesised by net photosynthesis about 180 litres of water are lost from the Beech

Fig. 96. The potassium cycle in the ecosystem of a moder Beech wood in the Solling. From data by Ulrich and R. Mayer (1973). The width of the arrows indicates the magnitude of potassium flow.

1. Only a small amount of potassium is washed out of the air (2 kg/ha/year) 2. In contrast to this the water dripping from the canopy is surprisingly rich in potassium (26 kg/ha/year) since the tree actually give off potassium. 3. About the same quantity of K is also added to the water in the soil by mineralisationof the humus (7). 4. The trees and other vascular plants remove the K so avidly (46kg/ha/year) from the zone of intense root activity (here reckoned as down to50 cm although most of the fine roots are in the top 10-20 cm) that, in spite of its high solubility, very little K reaches the lower layers (5) or the drainage water (6). The exchange of K with the solid part of the soil (8) is slight; here it has been shown as almost nil (see also R. Mayer 1971).

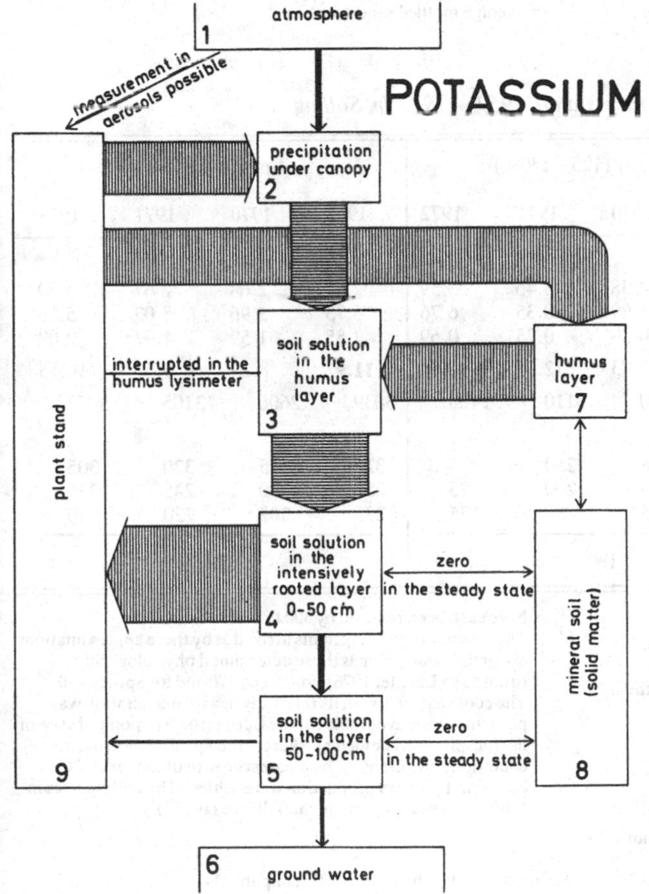

Tab. 23. The water balance of a Woodrush-Beech wood and a Spruce plantation in Solling during the years 1968 to 1972.

Year	PO[a] mm		PB mm	ET mm	D mm	ΔR Soil mm	ΔSn Snow mm	I mm	I %	IET mm	IET %
1968 (May-Dec.)	746.1	B	639.1	254.4	382.5	+ 16.8	+19	107.0	14.3	361.4	48.4
		S	*529.9*	*362.8*	*145.8*	+ *2.3*	*+19*	*216.2*	*30.0*	*579*	*77.6*
1969	1064.0	B	912.0	307.0	582.9	− 9.8	+32.0	152.0	14.3	459.0	43.1
		S	*743.7*	*383.9*	*368.9*	− *10.1*	+ *1.0*	*320.3*	*30.1*	*704.2*	*66.2*
1970	1479.1	B	1206.3	261.0	972.6	+ 23.7	−51	272.8	18.4	533.8	36.1
		S	*1152.6*	*260.6*	*890.5*	+ *21.6*	*−21*	*326.5*	*22.1*	*586.9*	*39.7*
1971	809.7	B	622.7	311.0	303.9	+ 7.8	0.0	187.0	23.1	498.0	61.5
		S	*555.3*	*310.5*	*232.2*	− *12.6*	*0.0*	*254.4*	*31.4*	*564.9*	*69.8*
1972	910.4	B	716.0	245.0	343.0	+128.2	0.0	194.4	21.3	439.4	48.3
		S	*605.0*	*307.2*	*308.4*	− *10.5*	*0.0*	*305.5*	*33.5*	*612.7*	*67.3*
Mean (excl. 1968)	1065.8	B	864.3	281.0	550.6			201.6	18.9	482.6	45.3
		S	*764.2*	*315.6*	*450.0*			*301.7*	*28.3*	*617.2*	*57.9*

Note: [a] PO = precipitation in the open, PB = precipitation below the tree canopy, ET = evaportranspiration, D = drainage water, R = reserve carried forward to following year, Sn = addition through melting snow or loss through evaporation of snow, I = interception, IET = I + ET = water given off.

Source: From data by Benecke (lecture 1976). Beech trees about 125-130 years old; Spruces planted about 90 years ago in the same area, i.e. 500 m above sea level on very acid moder brown earth in loess over average mottled sandstone)

Tab. 24. Transpiration coefficients of Beech and Spruce stands 1969 – 1972 in Solling.

a-e in t/ha/year — Stand:	Beech (125 – 130 y)				Spruce (ca. 90 y)			
year:	1969	1970	1971	1972	1969	1970	1971	1972
Net primary production								
a leaves	3.46	4.38	3.45	3.59	2.67	2.68	2.67	2.70
b trunk, branches, twigs	9.00	3.09	8.35	6.76	6.75	3.96	5.03	5.36
c roots >5 mm diam.	0.82	0.26	0.75	0.69	2.55	1.59	1.97	2.09
d total (excl. fine roots)	**13.28**	7.73	12.55	10.94	**11.97**	8.23	9.67	10.15
e **evaportranspiration** (ET)	3070	2610	3110	2450	3839	2606	3105	3072
Transpiration coefficients								
f e : d	230	340	250	255	320	315	320	305
g e : (d + 3 t fine roots)	190	245	200	175	255	230	245	235
h (e − 300 t) : (d + 3t)	170	215	180	**155**	235	**205**	220	210
h′ average 1969 – 1972	180				220			

Note: The transpiration coefficient is the number of litres of water a plant transpires during the time it is producing 1 kg of dry matter.

a - d Fairly exact figures but without the fine roots the yearly production of which may be of the same magnitude as that of the leaves.

e Fairly exact figures (see tab. 23 ET) but the transpiration and evaporation are not given separately. The latter is relatively slight from the woodland soil and can scarcely amount to more than 10% of the total ET.

f-h Rounded off to the nearest 5; f is certainly too high because no account has been taken of the fine roots and evaporation has not been deducted. In g 3 t/ha has been added in every case to take into account the growth of fine roots. Moreover in h the figures have each been reduced by 300 t.

h′ The transpiration coefficients arrived at by these approximations are of the same order as those determined physiologically (quoted by Larcher 1976): for Beech 170 and for Spruce 230. The cool year 1970 with its relatively heavy precipitation was particularly unfavourable for the Beech from the point of view of its transpiration coefficient whereas the Spruce found these conditions favourable. The greatest growth of timber and the largest net primary production were achieved by both species in 1969, i.e. a year of average rainfall (see tab. 23).

Source: From data by Heller (1977, unpubl.)1)

wood through transpiration. This transpiration coefficient is certainly considerable, but it is surprisingly small when compared with that of our cultivated plants. For example wheat requires on average some 540 litres to make 1 kg of dry matter. Even maize, which, as a C_4 plant, is able to use carbon dioxide more efficiently and thus does not need to open its stomata so widely or for such a long time, shows a transpiration coefficient of about 370. Beech works about twice as efficiently even though its photosynthetic apparatus can only use the C_3 pathway.

This also applies in principle to other tree species such as Pine and Spruce for which transpiration coefficients of 300 and 230 respectively have been determined (Larcher 1980a). In the Solling Spruce requires rather more water than the Beech and used about 220 litres per kg of dry matter assimilated. Gas exchange measurements by Schulze (1970) give us an idea as to how Beech achieves its remarkable efficiency. The leaves quickly regulate the size of the stomatal openings according to the amount of light, the relative humidity and the CO_2 concentration so that they respond immediately to every change in these factors (fig. 97). In addition the net photosynthesis of the

Fig. 97. Photosynthesis and transpiration through the day by the sun and shade leaves of a Beech in the *Luzulo-Fagetum* of Solling during a fine summer's day compared with the light intensity, temperature and relative humidity as measured in the gas exchange chamber. After Schulze (1970).

Based on the dry weight of the leaves the photosynthetic efficiency of the shade canopy is as good as that of the sun canopy even though it receives only about a tenth of the amount of light. Whenever a shade leaf was illuminated temporarily by a sunbeam it actually assimilated more CO_2 than those which were continually in full sunshine. Sun leaves reach their light saturation point at 35-40 kilolux; shade leaves are already saturated by less than 10 kilolux. Transpiration from shade leaves, again on a dry matter basis, is only about half that from sun leaves; so they are able to work more efficiently than the latter. Expressed another way: the larger the proportion of shade leaves in the canopy the greater is the net production for a given amount of water uptake. Transpiration increases as the temperature rises and as the relative humidity of the air falls; consequently it is greater in the afternoon than in the morning. Temperature and (to a certain extent also) humidity were automatically controlled in the experimental chamber to correspond with conditions outside. Thus they were continually reflecting the natural conditions. Light intensity was slightly reduced by the plexiglas chamber.

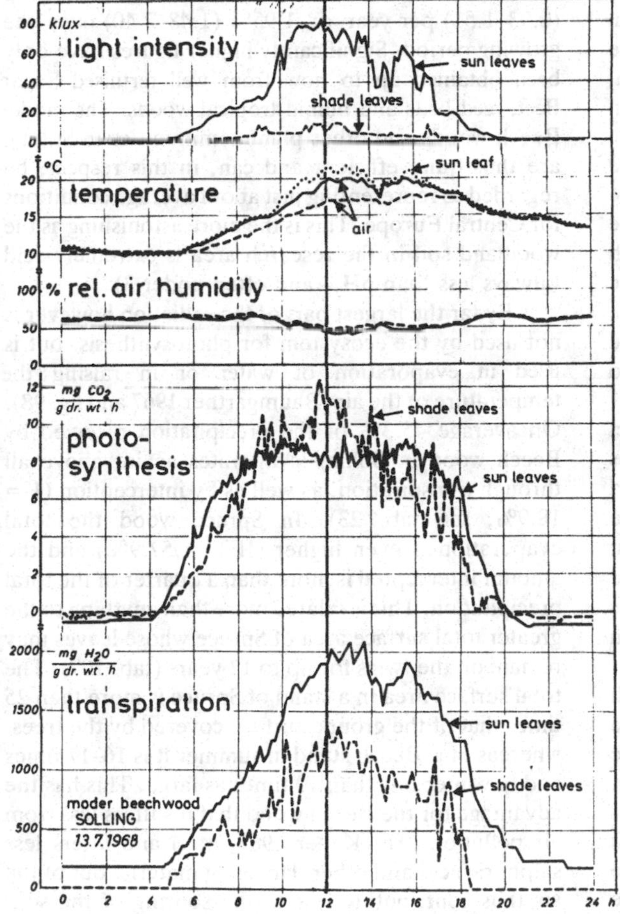

relatively thin shade leaves of the Beech is just as efficient, calculated on a dry matter basis, as is that of the sun leaves. Therefore the Beech is able to make better use of sunlight than, for example, the Common Oak which forms only about three layers of leaves, whereas the Beech in addition to three layers of sun leaves forms at least 3-4 layers of shade leaves. The amount of material produced by the undergrowth in a moder Beech wood is correspondingly insignificant (Eber 1972) and can be ignored, while that of the mixed Oak wood is quite considerable (figs. 82 and 131).

In the relatively humid climate of Solling, which at round about 500 m altitude already shows montane features, photosynthesis is rarely restricted by shortage of water. The temperature too is generally sufficient for photosynthesis which shows an optimal net production between about 8° and 24° C if other factors remain favourable. On the other hand the amount of light is often too low for the calculated miximum net photosynthesis to be achieved (Schulze 1970). This is especially true in a cloudy district such as Solling but may also apply to large parts of Central Europe. Cartellieri (1940) found that even in the central Alps, which have high levels of radiation, insufficient light was often the limiting factor in photosynthetic production.

In addition to light there are many factors in the habitat which can reduce photosynthesis but which are not subject to such short-term variation as are the climatic factors. In particular nitrogen supply, which would increase the amount of chlorophyll, and also the leaf area index (LAI) of the plant stand may determine the effectiveness of photosynthesis. In contrast, the species composition of the stand appears to be of no significance provided the leaves can unfold maximally under the actual site conditions. The Solling project showed in each case that Beech and Spruce stands, as well as meadows and a field of Ryegrass, produced approximately the same number of calories per hectare per year. Thus they used the light energy with just about the same effectiveness (M. Runge 1973a, b, see also fig. 36). This only applied however if one includes the subterranean production in the comparison – an item which is a difficult and lengthy determination.

The fraction of the net primary production which is used to build up the root system is lowest in the Beech (see fig. 82). This tree invests a particularly large proportion of its assimilate in the growth of branches, twigs and leaves, i.e. in increasing its competitiveness by overshading other plants. Spruce resembles Beech in this respect; its annual growth is concentrated in the straight trunk, which the forester can make better use of. Its timber yield is thus greater than that of Beech, under otherwise comparable conditions, although the latter shows a greater increase in biomass above the ground (fig. 36). In the ratio of above-ground to below-ground production only a crop of Annual Ryegrass (*Lolium multiflorum*), which had been well cultivated and fertilised to give maximal production approached that of the Beech wood.

According to M. Runge (1973b) each of the plant communities investigated in Solling used about 1% of the global radiation per year for their net photosynthesis. This is around 1.4% of the radiation falling in the growing period. Since approximately half the total radiation is effective for photosynthesis it could be said that on average the efficiency of light utilisation for net primary production is about 3%. Over a short period of time under favourable conditions this can rise to as much as 12%, as Ruetz (1973) showed for the Red Fescue (*Festuca rubra*) in the meadows of the Solling. Nanson (1962) found a similar figure using other methods for the moder Beech woods in Belgian Ardennes; these woods used an average of 1.23% (0.93–1.51) per year or 1.95% (1.48–2.40) over the growing period. Significantly higher figures have only been obtained up to now from well fertilised Sugar Beet, reed beds and humid tropical woods. The moder Beech woods and other plant communities in Solling are thus quite efficient and can, in this respect, be regarded as representing just about average conditions for Central Europe. This is the more astonishing as the woodland soil in the research area is extremely acid (always less than pH 4 and often under 3).

By far the largest part of the radiation however is not used by the ecosystem for photosynthesis, but is used in evaporation of water or in raising the temperature of the air (Baumgartner 1967 and fig. 98). On average 45.3% of the precipitation received by Beech wood in Solling evaporates again, above all through transpiration, as well as by interception (1 = 18.9%, see tab. 23). In Spruce wood the total evaporation is even higher (IET = 57.9%) and the amount intercepted is more than a quarter of the total precipitation. This is related more than anything to the greater total surface area of Spruce whose leaves may remain on the twigs for up to 12 years (tab. 107). The total surface area in a stand of Spruce is more than 25 times that of the ground surface covered by the trees, whereas in a Beech stand in summer it is 16–17 times and in winter less than 10 times as large. This has the advantage for the Beech wood that it suffers less from air pollution (Th. Keller 1968, etc.) and filters less sulphuric acid and other damaging material out of the air thus contributing less to the souring of the soil.

Only near the trunks is this appreciable (fig. 79).

Already from these few indications it can be seen how closely the energy, water and material flows are bound together in a woodland ecosystem and how Beech enjoys competitive advantages in so many respects. Unfortunately there are still too few results of researches into other ecosystems available in order to confirm such statements quantitatively and to be able to compare the different woodland communities. The world wide survey by Rodin and Bazilevič (1966) on production and cycling of materials in terrestrial ecosystems offers insufficient data about the vegetation for our purpose. However the recent summary of the researches in moder Beech woods and Spruce forests in the Belgian Ardennes by Duvigneaud and Kestemont (1977) confirms to a large extent the Solling observations. In addition it shows that other woodland communities differ from Beech woods not only in species composition but also in the turnover of material and energy. We shall return to this in section B III 3 h.

Fig. 98. The balance of radiation in an old Beech wood in Solling during the growing period (June 1970) and towards the end of the season (October 1970). After Kiese (1972), somewhat modified.

Of the total incoming radiation (Q) the largest part is used in the vaporisation of water, especially in summer when stomatal transpiration is high. Even during the night evaporation takes a certain amount of heat from the surfaces of the leaves and ground. The noticeable warming of the air (L) and ground (B) during the day and the opposite process during the night are further more or less important items in the radiation balance. At most only a small percentage of radiation is used in photosynthesis (see tab. 30).

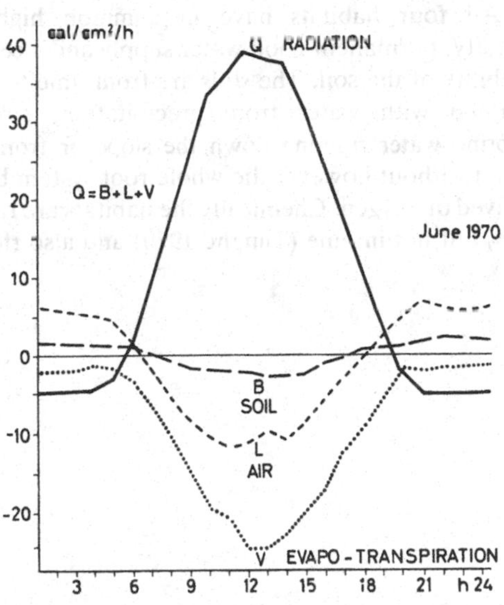

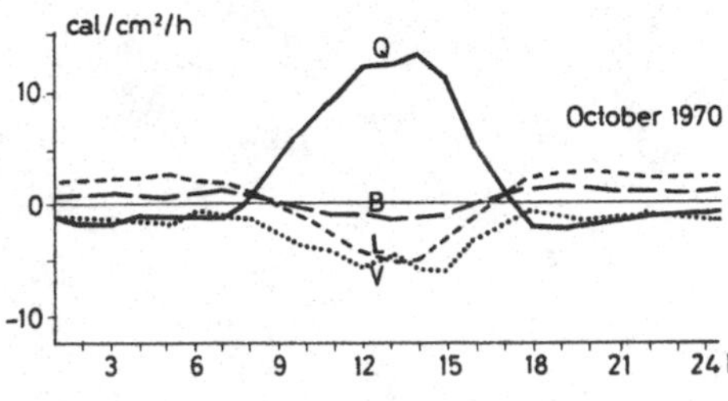

III Other deciduous woodland excluding flood plains and mires

1 *Mixed woodland rich in Sycamore and Ash*

(a) Classification of Sycamore and Ash woods according to habitat

The woodlands which are closest to the Beech woods in site and flora are the broadleaved mixed woods dominated by species of Maple (*Acer pseudoplatanus*, *A. platanoides*) and other 'noble' forest trees (*Tilia platyphyllos*, *Ulmus glabra*, etc.) or by Ash (*Fraxinus excelsior*). The Ash wood communities especially are amongst the most productive in Europe; the others too are quick growing and are an indication of a fertile habitat. Although Beech would thrive well here it is overgrown when young by Ash, Sycamore and occasionally by Wych Elm, Norway Maple and Lime. (Schlüter 1967, Moor 1975 a – c).

In spite of the reduction in numbers of Beech and, of course, Oak and other light-demanding trees the undergrowth of mixed Sycamore and Ash woods show many similarities with that of the Beech woods and also with the Oak-Hornbeam woods which will be discussed in section III 4. There are also many floristic connections with the hardwood communities of flood plains (section V I f-h). The systematic boundaries of these three groups of deciduous woodland communities are difficult to define because during the course of time they have been included in different alliances. However there is no question about including them in the order *Fagetalia* within the class *Querco-Fagetea*.

Because of their peculiar ecology we shall deal with the flood plain vegetation all together in section V and not consider it here. Apart from this the mixed deciduous woods which are rich in Sycamore and Ash are able to develop in quite different habitats:

1 **Steep slopes or ravines** with a north-west to eastern aspect on rocky or stony screes, poor in fine soil but rich in humus and more or less rich in bases in the submontane or montane belt favour the shade-slope Ash-Sycamore wood (*Aceri-Fraxinetum* according to Tüxen 1937; but better as *Phyllitido-Aceretum* following Moor 1952). The name 'ravine woodland' which is often used for this vegetation type may lead to misunderstanding since it can develop typically not only in ravines,

but on open slopes too, provided these are shady (Mayer 1969, Gadow 1975 and fig. 99).

2 At the **bottom of shady steep-sided valleys** where fertile alluvial loam has collected is a habitat which lies ecologically between 1 and 3. Here the valley-bottom Fumitory-Ash-Sycamore wood flourishes even in lime-deficient mountains such as the Harz (*Corydali-Aceretum*, see Hofmann 1965, Moor 1973 and Gadow 1975). Floristically this is reminiscent of the Fumitory-Oak-Hornbeam wood and the Wild Garlic-Beech wood. As it is rather rare we shall not consider it in detail.

3 On colluvial **deep soils at the foot of slopes**, which are very fertile and, at least in spring, moist or wet, slope-foot Sycamore-Ash wood finds conditions suited to luxuriant growth (*Aceri-Fraxinetum* according to Etter 1947). It is present in the colline to submontane belts, especially in the high precipitation areas surrounding the Alps (see also Pfadenhauer 1969).

4 Along narrow **brook channels** which have been cut into the soil, the sides of which are not flooded, but from time to time are undercut and renewed by land slips, also in similar conditions along the spring line of a slope, the brook-channel Ash wood flourishes in the submontane or planar belts (the brook Ash wood, *Carici remotae-Fraxinetum* of various authors, see fig. 102).

Fraxinus excelsior usually dominates the last named although it may only occupy a strip a few metres wide. In the slope-foot Sycamore-Ash wood too it is Ash which generally sets the tone, but other tree species are mixed in with it. On steep shade slopes *Acer pseudoplatanus* quite often gains the upper hand without being able to exclude entirely Ash and other trees, especially *Ulmus glabra*. The lists of species in the undergrowth of the shade-slope and slope-foot woodlands also have much in common. For this reason they were not separated in older writings (for instance in the thorough survey of the iterature by J. and M. Bartsch 1952). Today the two associations are united in the same suballiance (*Acerion* according to Oberdorfer 1957, 1979, or *Tilio-Acerion*). The brook Ash wood on the other hand is better included with flood plain woodlands (section V I g) although it has much in common with other communities of 'noble' deciduous trees.

All four habitats have in common high air humidity, permanent good water supply and a certain instability of the soil. The soils are from time to time saturated with water from precipitation, stream, or spring water running down the slope or from the ground without however the whole root system being deprived of oxygen. Chemically the habitats are rich in bases, but not in lime (Tanghe 1970) and also rich in

Fig. 99. Zonation of the woodland communities on a steep NW slope of the Thüringia Forest at a height of 470-520 m, i.e. in the lower Beech level at the Gömingenstein on mica slate. After Schlüter (1959), modified.

The shade-slope Ash-Sycamore wood here prefers depressions in the slope where supplies of water and nutrients collect locally. Although the rock is poor in lime the rather demanding *Corydalis cava* is able to thrive in such places in a special fazies of the fragmentary Honesty-Ash-Sycamore wood. The rocky spur is a definite 'loss site', so that only a very acid-soil Oak-Beech wood is able to become established on it. The rest of the slope receives the occasional supply of nutrients from dead leaves which are blown here from more wind-exposed places (see fig. 84), but this is sufficient to enable *Festuca altissima* to spread.

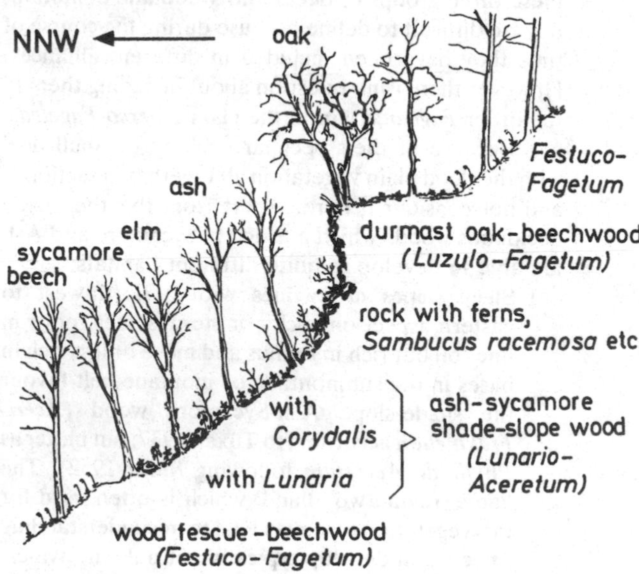

nutrients, especially nitrate. These are derived partly from the water and partly from fine earth trickling down the slope or from landslides. Above all they are guaranteed very favourable conditions for soil animals and bacteria; that means high biological activity, at least in the upper layers of the soil.

Correspondingly in nearly all mixed Maple and Ash woodlands one finds many large-leaved, quick-growing, more or less hygromorphic and nitrophilous herbs, e.g. *Urtica dioica, Aegopodium podograria, Silene dioica* and *Impatiens noli-tangere*. These only thrive with the same luxuriance in riverside woods which are frequently fertilised by mud left behind after flooding, or in subalpine tall herb communities (section B VI and C VI 7 a). The nitrophilous Elderberry (*Sambucus nigra*) too is rarely absent. Since the leaf litter is broken down within a few months the soil-dwelling mosses are able to thrive. These are indicators of biologically active mull soils, especially *Mnium undulatum* and many *Eurhynchium* and *Fissidens* species. In addition the high humidity favours growth of mosses on the tree bark and on stones. Their swelling cushions are one of the characteristics of all shade-slope woodland and many mixed Ash woods.

(b) Shade-slope Ash-Sycamore woods

The shade-slope Ash-Sycamore woods are the ones which have been best investigated floristically and ecologically and so will be discussed first. They stand close to the limestone or brown-mull Beech woods and resemble them floristically. Physiognomically however they can be distinguished at first glance by their rich mixture of tree species and mosaic of changing undergrowth from the monotonous cathedral woods. Generally Sycamore is dominant but indeed its area of distribution is greater that that of the shade-slope woodland communities. In warm situations *Tilia platyphyllos* which, according to Schmithüsen (1948) like Sycamore is very sensitive to drought can also be dominant. Other deciduous trees are present in varying numbers whereas conifers can seldom take root on the crumbling ground, or they are stunted by the quick-growing deciduous trees. In the herbaceous layer strong growths of luxuriant ferns are the first to be noticed (fig. 99), side by side with the characteristic or differential herb species (see also section E III 1, No. 8.434):

Lunaria rediviva	*Polystichum aculeatum*
Actaea spicata	*Phyllitis scolopendrium*
	Gymnocarpium robertianum

Phyllitis and *Gymnocarpium* are really inhabitants of open, but not sunny, rocky or stony slopes and may be considered as relics from pioneer communities (in this respect consult especially Moor 1975a, who has described many ravine woodland communities from the Swiss Jura, also Oberdorfer 1957).

Gadow (1975) has measured the local climates in different shade-slope Ash-Sycamore woods and compared them with those in Beech wood communities. In all cases the day extremes of air and soil temperature, and of humidity were less in the first. The evaporation and accordingly the transpiration in the shade-slope woods is noticeably small. The main reason for this is the wind speed which is generally very slight on the north- and east-facing slopes, at least during the growing period as the prevailing winds are from the south and west. In ravine-like narrow valleys the local climate is even damper but not fundamentally different from that on the open slopes facing away from the sun. Thus on the latter species which are characteristic of the 'ravine woods', namely Mountain Honesty (*Lunaria*) and ferns can find very similar living conditions.

This applies even more strongly to the soil conditions which Gadow suggests must be looked upon as the most decisive factors. In no woodland community of Central Europe is the supply of nitrate so high or so evenly distributed throughout the growing period as it is in the shade-slope Ash-Sycamore wood. This is due not least to the continuing high, but not excessive, water content of the soil in which the roots never need a high tension even in a dry year when in topsoils of most woods permanent wilting point is exceeded. Depending on the habitat and the weather the mineral nitrogen available in test areas on the Meissner were between 147 and 204, and in the Harz between 168 and 377 kg/ha/year. More than 300 kg N/ha/year represents to the farmer a more than average application of this important nutrient. Kovács (1968a) also found a very high rate of mineralisation of nitrogen in the *Phyllitido-Aceretum* in the Matra mountains of Hungary. The natural cycle of materials then in many shade-slope woodlands is extraordinarily active. This is probably true also for phosphorus and other nutrients though Gadow did not investigate these in detail. Bach (1950) has already ascertained that there are no extreme pH values in the soils of these communities, although in many cases they are lower than in Swiss Jura. The high nutrient content of the soils on the shade slopes may serve to permit a high plant production in spite of the slight amount of fine earth present on the stony slopes.

As a subalpine parallel to the *Phyllitido-Aceretum* may be mentioned the *Sorbo-Aceretum* which in places reaches right up to the tree line in the Swiss Jura (Richard 1968a, Ellenberg and Klötzli 1972). Likewise a high-montane type is the *Asperulo taurinae-Aceretum* which Winteler (1927, Ellenberg and Klötzli 1972) has described from the northern edge of the Alps (see also section 2). By the way *Acer pseudoplatanus* mixed woods are widespread also in boreal Europe, but occur only as small islands on fertile soil in the prevailing coniferous vegetation (Klötzli 1975a).

(c) Slope-foot Sycamore-Ash woods

The mixed Sycamore-Ash woods which occupy the fertile fine earth collected at the foot of slopes, are also distinguished by their high productivity. According to Etter (1947) they attain their optimum development in the wider surroundings of Zürich and occupy large areas of Sihl Forest which have never been managed as coppice or coppice with standards. In particular the wet-soil Sycamore-Ash wood (*Aceri-Fraxinetum caricetosum pendulae*) may be compared with the moist-soil Oak-Hornbeam wood (section III 3 e) and replaces it in regions with a high precipitation. 'With a lushly luxuriant carpet of herbs and an impetuous growth of trees' this type of woodland 'is one of the most productive in the whole of Switzerland' (1947, see fig. 100). Its trees need only about two-thirds of the time required by the neighbouring brown-mull Beech wood to grow to the same height, and after 100 years they have already reached an average height of 35 m (fig. 101, see Etter 1949). However the quality of the wood from trees which grow so quickly is only average, and this is as true for the natural tree species as it is for Spruce and conifers which are brought into cultivation.

The undergrowth is dominated in many places by Wild Garlic (*Allium ursinum*) or a mixture of other herbs with a high nutrient requirement. According to Ellenberg (1964) production of mineral nitrogen is very good, even in deep layers of the soil which are waterlogged from time to time. However because of a shortage of oxygen caused by the water present the ammonium is not completely oxidised to nitrate.

The so-called 'sticky-soil wood' ('Kleebwald') of Swabian Jura which has already been investigated by Gradmann (1898, 1950) is in many ways very similar to Etter's *Aceri-Fraxinetum*, but it corresponds more to the subassociation *corydaletosum* which Moor (1973) has described from Swiss Jura. Typical Sycamore-Ash woods are obviously associated with mild oceanic climates which have a high rainfall. Such a one dominates in the so-called insubrian parts of the southern Alps, e.g. in Tessin. Many of the Ash-rich woods found on the slopes described by Zoller (1960), and also Ellenberg and Klötzli (1972) stand close to this community. According to Seibert (1969) the Ash-rich slope-foot wood replaces the mixed Oak-Hornbeam wood (*Galio-Carpinetum*) which prefers a warmer climate, in the cool rainy Alpine foothills of southern Bavaria.

In the middle and to the north of Central Europe slope-foot Sycamore-Ash woods can still be found although they are less distinct (Schlüter 1967). In many dry valleys in the central hill region of Germany, for example to the east of

Fig. 100. A slope-foot Sycamore-Ash wood (*Aceri-Fraxinetum typicum*) with dominating Dog's Mercury (*Mercurialis perennis*) on the northern slope of the Blauen, NNW of Zürich.

Göttingen, they were still not yet recognised and described as anything special. Most people put them with the shade-slope Ash-Sycamore woods (section b) or the wet Oak-Hornbeam woods which can also be rich in Ash (section III 3 e). In Belgium too there are *Aceri-Fraxineta* which are closely allied to the brown-mull Beech woods (Tanghe 1970).

In the Baltic moraine hills, especially in Mecklenburg, Scamoni (1960) has met slope Ash woods (*Adoxo-Aceretum*) which he took to be a species-deficient representative of the Sycamore-Ash woods. In them nitrophilous perennials are combined with spring geophytes similar to those in the *Aceri-Fraxinetum corydaletosum* already mentioned.

(d) Brook-channel Ash woods

The Ash woods accompanying small streams were clearly recognised by W. Koch (1926) as separate associations (*Carici remotae-Fraxinetum*) and since then have been described from many areas in northern and western Central Europe as well as from west Europe. Often they are very fragmentary and intermingled with the Beech wood communities, through which they run as narrow and in places interrupted strips (fig. 102). In broader valleys the rivers are normally accompanied by Willow stands or other flood-plain woodlands.

Kästner (1941) completely questioned their existence and conceived the wet spring lines and stream banks as treeless wet areas which were only overshadowed by trees growing on the drier ground immediately to the side. These 'swamps in the woodland' (*Caricetum remotae*) of Kästner are nevertheless influenced by the forest microclimate and would have developed differently in full sunshine. Moreover Kästner's review only applies to the wettest form of the brook Ash wood, while the normal type is found on mull gley soils.

Fig. 101. In the height of its trees the slope-foot Sycamore-Ash wood (*Aceri-Fraxinetum*) far exceeds the zonal Beech wood. Each point represents the average height of a 20-tree group. After Etter (1949), somewhat modified.

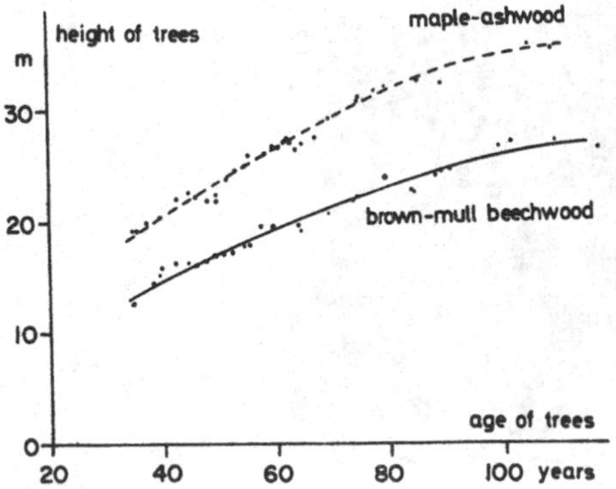

Ash is always the most plentiful tree in this community, the other trees and bushes being rarely conspicuous. In wet places it is accompanied by Alder which indicates a transition to swamp woodlands. *Carex remota* is rarely absent and flourishes particularly well here. On base-rich soils *Carex pendula* can form splendid wide clumps (fig. 102). In many years the tender Touch-me-not germinates to form dense groups. On wet patches there are carpets of Golden Saxifrage (*Chrysosplenium alternifolium*), and our largest Horsetail (*Equisetum telmateia*) seems to point to Mesozoic times. *Carex strigosa* although less common, is seen by most authors as a true character species, but Rühl (1959) maintains that it is only in the Siebengebirge near Bonn that it can be considered as a natural component of the brook-channel Ash wood. In most other districts it prefers to grow on car tracks, the edges of paths and other man-made habitats within moist-soil woodlands.

Noirfalise (1952) separated the Sedge-Ash woods into a number of geographic races. Their optimal development is in the northern part of Atlantic Europe, that is in northern France and southern England. The distribution of the east Baltic brook-channel Ash woods stretches as far as the Memel valley. In the region of Upper Rhine, in central Switzerland and in the Pyrenees the atlantic sedge-Ash wood is represented only in isolated places. Like *Aceri-Fraxinetum*, the *Carici-Fraxinetum* is also associated with an oceanic climate, however it appears to be less fastidious in its climatic and edaphic requirements. Although this community has been described so extensively there are still no quantitative measurements of the deciding habitat factors available.

2 Mixed Lime woods and their problems

(a) Mixed Lime woods in and near the Alps

As an attractive equivalent to the shade-slope Ash-Sycamore woods of the continuously humid steep slopes are the mixed woodlands rich in *Tilia* which are found growing on equally stony but drier slopes. For the most part Winter Lime (*Tilia cordata*) is dominant, but all deciduous forest trees named in the previous section also play a part. The best developed are such dry-slope Lime woods in the 'northern Alps föhn and lake district', especially on the limestone and 'Nagelfluh' conglomerate rocks above the Walen and the Vierwaldstätter Lake in Switzerland. Even here they only occupy small areas, but are so remarkable floristically, ecologically and historically that they cannot be overlooked. Trepp (1947) dedicated an exemplary monograph to these variable communities which are connected to neighbouring tree and shrub associations by intermediate types.

Most stands of his *Asperulo-Tilietum* (he called it '*Tilieto-Asperuletum taurinae*') contain species of the xerothermic mixed Oak woods (*Quercetalia pubescenti-petraeae*), but also representatives of the mixed Sycamore-Ash woods and other *Fagetalia* communities. Systematically then they lie between the two orders. Trepp hesitated in attaching them to either; today one is inclined to prefer the connection with the Sycamore-rich units in the alliance *Tilio-Acerion* of the order *Fagetalia*. As character species of the dry-slope Lime woodland applicable to northern Switzerland and the neighbouring parts of the Alps can be named (see also E III 1, No. 8.434):

Tilia cordata (plentiful)	*Euonymus latifolia*
Tilia platyphyllos (rarer)	*Staphylea pinnata*
	Asperula taurina

Along with Winter Lime, Summer Lime and Norway Maple *(Acer plantanoides)*, Ash and Elm are found regularly, but other deciduous trees only rarely.

Fig. 102. A brook-channel Ash wood (*Carici-Fraxinetum*) in the Sihlwald near Zürich alongside a small stream on recent moraine loam. A large amount of Pendulous Sedge (*Carex pendula*, summer appearance). In the background is brown-mull Beech wood.

For its apparent 'southern' species composition the warm-dry *Asperulo-Tilietum* can thank the combination of the following natural habitat factors: a high precipitation and radiation in summer, because of frequent föhn winds warm months in winter and spring (figs. 103 and 104) and occasional very dry days, increased light by reflections from the lakes, sufficient moisture and nutrients at the roots and a high mineral content of the subsoil. Alongside all these must be placed the centuries-old management as coppice woodland which protected the light-hungry new growth of the Lime trees from competition with Beech in places with a deeper soil. Since such a combination of factors is met with only in a few northern Alpine and insubrian föhn valleys (fig. 103) there are no thermophilous mixed Lime woods in the rest of Central Europe except perhaps in a form poor in species.

Ecologically the mixed Lime woods have not been investigated closely in their main area of distribution. In arranging them according to the amount of nitrogen available they may correspond to the slightly-thermophilous mixed deciduous woodland in the Tessin, in the tree layer of which *Tilia cordata* would naturally have taken part on acid rock (tab. 25, no.2). In comparison with the Beech woods (tab. 19) these are fairly well provided with nutrients, particularly nitrate, although the soils, according to Antonietti (1968) show very low pH values. The microclimate in such mixed woodland varies from a uniformly high humidity on dull days to a dry midday heat, but little cooling down at night when the sky is clear.

(b) Mixed Lime woods away from the Alps

Woods rich in *Tilia* are to be found away from the Alps, but they never occupy large areas. The Winter Lime succeeds in all those places where Beech is weakened, whether due to natural causes or to man's activity. In this respect it behaves like Hornbeam (section B I 2 b) and makes way for Beech in both dry and damp habitats. However *Tilia cordata* requires more light than *Carpinus* and thus can establish itself more easily on soils rather poor in bases than on richer ones where all trees tend to form a closer canopy. The best example for this is again under the influence of the föhn wind, e.g. to the east of the Harz (fig. 105). In relatively continental parts of the Central European lowland Winter Lime plays a significant part in mixed Cinquefoil-Oak wood (*Botentillo-Quercetum*) and in Lime-Hornbeam wood (*Tilio-Carpinetum*, section 4 c and 3 b), as well as in communities which lie between the classes *Querco-Fagetea* and *Querceta robori-petraeae*. There are also mixed Lime woods on limestone rocks which are reminiscent of the *Asperulo taurinae-Tilietum*.

W. Keller (1974), for example, has described a scree-slope Lime wood near Schaffhausen and has expressed the analogy with the name '*Asperulo odorati-Tilietum*' (today it must be called *Galio odorati-T.)*. In similar habitats in Czechoslovakia, Neuhäusl and Neuhäuslova (1968b) disting-

Fig. 103. The föhn areas of Central Europe when the wind is blowing from the SW (1, most frequent) or from the NW (2). 3 = important föhn valleys. After Flohn (1958), somewhat modified.

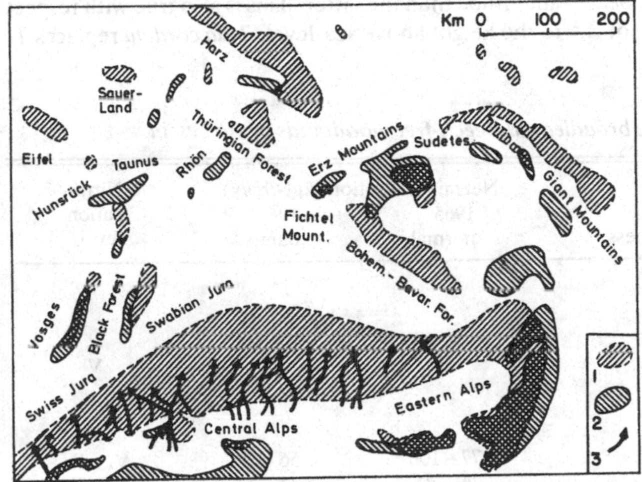

Fig. 104. Frequency of typical föhn days in a pronounced föhn valley (Altdorf) and in the pre-Alps (Zürich). After Schüepp from Furrer (1958), modified.

The curve for Altdorf shows the number of föhn days as a percentage of the total averaged over a period of 30 years. During this whole period there were only 22 typical föhn days in Zürich although frequently a föhn-like atmosphere.

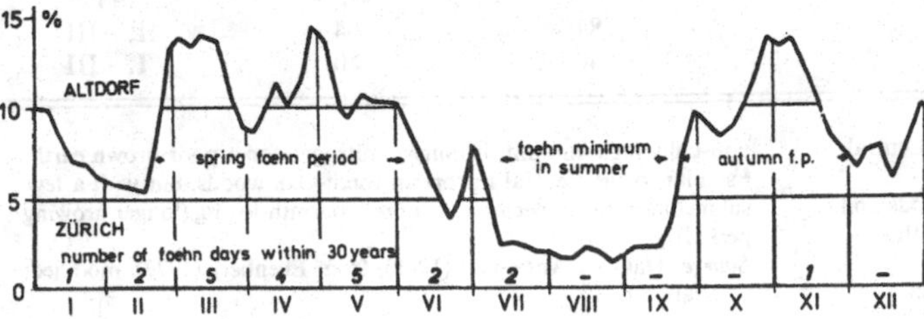

uished between two associations within the *Tilio-Acerion: Cynancho-Tilietum* on dry sunny scree slopes and *Aceri-Tilietum* on shaded ones. From the surroundings of Göttingen, Winterhoff (1965) got to know fairly similar communities.

Grimme (1975) included a stand resembling the *Aceri-Tilietum* in the Werra mountains by way of comparison in his experimental investigations into the dry-slope Sedge-Beech woods (section II 2 c). This *Tilia cordata* wood occupies a very steep east-facing slope on a limestone soil which is certainly poor in fine mineral earth, but is covered with a layer of moder more than 10 cm deep. The production of mineral nitrogen here is surprisingly good (137 kg/ha/year) and the water supply is not so bad as one would expect. The absence of Beech then can scarcely be traced to shortage of water and nutrients. Perhaps human activity has played a part similar to that in many Lime woods in Switzerland. Even in the relatively very dry climate east of the Harz *Fagus* has played an important part in the natural woodland before man's impacts became noticeable in the Late Stone Age (E. Lange 1971).

However there are mixed Lime woods on the northern boundary of Central Europe which must be regarded as natural. In the old moraine district of south-west Denmark Iversen (1958) has studied populations of *Tilia cordata* which, from pollen analysis, must go back to the New Stone Age. R.H.W. Bradshaw (1981) found nearly the same to be true for a small basin in Norfolk. In both cases we are dealing, not with a more or less basiphilous vegetation of dry and steep slopes, but with communities of acidity indicators occupying flat places where the groundwater is near the soil surface. The only similarity with the rendzinas investigated by Grimme is the thick and continuous layer of moder. It cannot be ruled out that in both cases it is this which has prevented incursion of Beech.

In contrast to Winter Lime the Summer Lime (*Tilia platyphyllos*) prefers rich soil and avoids dry habitats; thus in company with Sycamore, Wych Elm and Ash it appears in different units of the alliance *Tilio-Acerion* which have just been discussed in section 1. In intermediate habitats it can compete with Beech quite well, e.g. in the Lime-Beech wood (*Tilio-Fagetum*) described by Moor (1952, 1968) from the Swiss Jura. It also meets *Tilia cordata* in all these communities; it simply has a narrower ecological amplitude than the latter. This is also true with respect to the height above sea-level. *Tilia cordata* replaces *T.*

Tab. 25. The annual supply of mineral nitrogen in the broadleaved Beech-free woodlands of the Tessin.

Woodland units and soil types	Year: Dampness:	Net mineralisation (kgN/ha/y) 1965 normal	1966 damp	Nitrification level
Warmth-loving mixed broadleaved woods on limestone[a]				
Submediterranean Manna-Ash mixed wood,				
"*Helleboro-Fraxinetum orni typicum*", average soil		66 – 90	81 – 108	V
" " *asteretosum*", shallow soil		58	62	V
2. Temperate mixed broadleaved woods on acid rock[b]				
a Ash-Elm mixed wood, rich brown earth				1965
"*Erisithalo-Ulmetum fraxinetosum orni*"		77 – 106	56 – 86	V, IV
" " *aegopodietosum*" (cooler)		29 – 44	31 – 42	V, IV
b Ash-Oak mixed wood, poorer brown earth				
"*Querco-Fraxinetum cytisetosum scoparii*"[c]		115	98	III
" " *typicum*" (rich in Oak)		26 – 58	24 – 37	II – III
3. Temperate Birch-Oak woods[d]				
Insubrian Birch-Oak woods				1965
"*Betulo-Quercetum typicum*", poor brown earth		81	64	II, III
" " *vaccinietosum*", very acid		40	40	I, III

Note:[a] belong to the alliance *Orno-Ostryon*, but close to the Central European *Quercion pubescenti-pentraeae*.
[b] comparable with the Central European Oak-Hornbeam woods and, like them, included in the alliance *Carpinion*, but more warmth-loving.
[c] rich in leguminosae (Broom) and because of this showing an unusually high amount of mineral nitrogen for a poor brown earth.
[d] similar to the Central European Birch-Oak woods, but with a few submediterranean species and more woarmth-loving (longer growing period).
Source: Data by Antonietti (1968), from Ellenberg (1977), modified; compare tab. 19.

platyphyllos in the *Adenostylo-Fagetum* which can be considered as the montane oreal vicariant of the submontane *Tilio-Fagetum* (Moor 1973).

3 Oak-Hornbeam woodland

(a) A survey of the mixed Oak woods of Central Europe

After Beech, Oaks are the most important woodland deciduous trees in Central Europe and this is true for the potential natural as well as for the real vegetation of today. Modern forestry continually reduces their numbers as the return from them is too slow. However the farming economy of past centuries encouraged them so much – this has already been explained in section A II – that they still play a large part in the densely populated lowlands (tab. 8).

From early times the west, above all the north-west, of Central Europe has been the real home of Oak and is still so today even though the coppicing and bark-stripping businesses which were once all important here are now a thing of the past. On poor sandy as well as on wet soils which are both frequent in the north-west, according to Firbas (1949), Beech had never been able to gain the upper hand, even in the Post Glacial Period before there was any significant human influence. In the rest of Central Europe Oak is found principally in the warmer valleys and lowlands and never climbs as high into the mountains as does Beech, nor does it form woods on its own (figs. 39 and 40).

The two commonest Oak species are not generally distinguished in the statistics because they resemble each other in so many respects. However Durmast (or Sessile) Oak (*Quercus petraea* = *sessiliflora*) prefers a mild montane climate and cannot withstand severe winters and late frosts as well as Common Oak (*Q. robur* = *pedunculata*). It also withstands poor wet habitats less well but can tolerate lower illumination. In its ecological behaviour then it is more like Beech than is the Common Oak; however this tendency is only seen clearly at the edge of its distribution, e.g. at the continental eastern limit, where *Quercus robur* pushes forward much further than *Q. petraea*, and in the flood plains of the river valleys which are almost completely avoided by Durmast Oak as they are by Beech. Over the greater part of Central Europe however the two Oaks live side by side and at times even hybridise.

Downy Oak (*Quercus pubescens*) comes close to Durmast Oak systematically and there are a number of intermediate forms. Ecologically however it is clearly distinct from the other two species in that it is found only in exceptionally dry warm habitats which are frontier posts of the submediterranean vegetation. In the northern lowland it has an outpost at Bjelinek on steep slopes bordering the Oder valley. However, according to Celiński and Filipek (1958) this was probably established in the last century by the introduction of seed.

The thing that all three Oak species have in common is their sociological role in the woodland. They need a lot of light when young and when they are

Fig. 105. A semi-schematic profile of a mixed Oak-Lime wood near Halle (Saale). From left to right Beech, Lime, Durmast Oak, Hazel, Oak and Lime saplings, Hornbeam, Sessile Oak, Hazel. The vertical scale for the underground parts has been doubled. After Meusel (1952c), slightly modified.

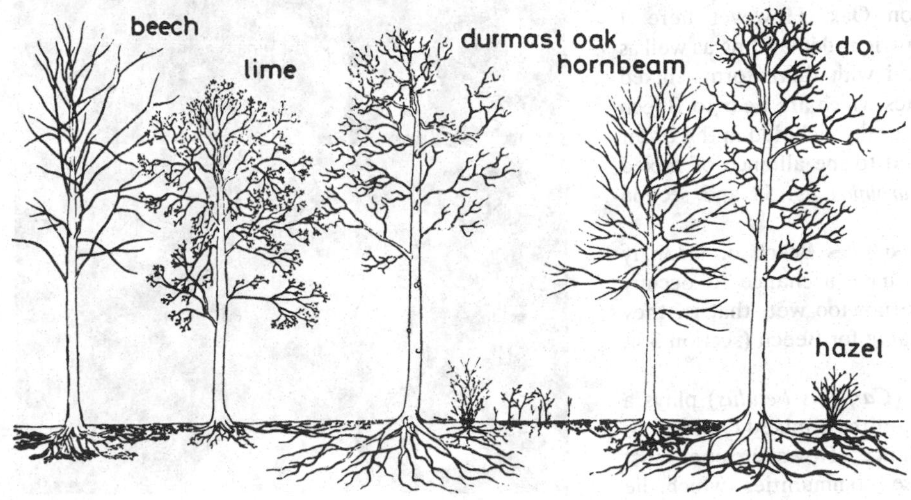

older they permit a good deal of light to pass through to the undergrowth (tab. 9). Thus Oak woods are almost always rich in a number of woody species which can develop under the broken canopy. Under natural conditions pure Oak stands occur only on the poorest soils. Since the accompanying tree species are generally much shorter-lived than the oaks which may stand for many centuries, they carry out much quicker regeneration under the Oak canopy. This natural cycle is reproduced in the type of management known as coppice with standards, but in this case there is a sudden opening up of larger areas at shorter time intervals (section A II 3 a). Because of these other light-requiring woody species and especially shrubs are very much favoured whereas in natural Oak woods the latter would be almost entirely absent.

Under natural competition Oak is eventually overcome by Beech and other tall shade or half-shade trees as it is unable, or only rarely able, to rejuvenate under their shade. As we have already pointed out in section B I 2 b, Oak can only play a significant part in a woodland when its competitors, especially Beech, are at a disadvantage. In Central Europe this is especially the case in four types of habitat:

1 In **warm dry** places the might of Oak is undisputed. In a continental type or föhn-rich climate it may be accompanied by Scots Pine. Along with a number of warmth-loving or drought-tolerant half-shade plants they form communities of open woodland belonging to the order *Quercetalia pubescenti-petraeae* (section B III 4).

2 On **very poor acid** soils Common and Durmast Oaks together form communities of the order *Quercetalia robori-petraeae* (section 5) which are close to those of the Atlantic heathlands.

3 Richer, but **damp to wet** or occasionally flooded soils favour Common Oak. However here it comes across many nutrient-demanding as well as wet-tolerant trees and with them forms mixed woodland communities which are near to Beech woods. These are generally included in the order *Fagetalia*, but assigned to the alliance of mixed Hornbeam woods (*Carpinion*, fig. 53 and section 3 e)

4 Where the **climate excludes Beech** or strongly discourages it, Oaks have a chance to occupy fertile soils which are not too wet, that is, they would have been suitable for Beech (section 3 b, fig. 106).

Here too Hornbeam (*Carpinus betulus*) plays a large part so one can also put these mixed Oak woods largely in *Carpinion* alliance. However one must distinguish between those communities which lie outside the general area of distribution of Beech and those within this area. The latter correspond floristically to the communities named under 3.; the former include elements of north-eastern, south-eastern or submediterranean floras depending on their position in Europe.

Since the *Carpinion* communities are near to Beech woods as well as to mixed Sycamore-Ash woods and lime woods we shall discuss them first. They offer a model for the geographic subdivision of a woodland alliance which is widespread over a large part of the Central European lowlands. Arranged from east to west they are:

a) **Mixed Lime-Hornbeam woods** (*Tilio-Carpinetum*) occur only outside the range of the Beech, mostly with *Quercus robur* and *petraea, Tilia cordata* and *Acer plantanoides,* in the north-east also with spruce (*Picea abies*) and other boreal species (section b).

Fig. 106. A Lime-Hornbeam wood (*Tilio-Carpinetum*) without Beech but with Spruce near Białowieża in north-eastern Poland. *Quercus robur, Carpinus betulus* and *Picea abies* can be recognised. There is no shrub layer in this dense, nearly natural woodland. Its ground vegetation is reminiscent of that of the brown-mull Beech woods.

b) **Wood Bedstraw-Oak-Hornbeam woods** (*Galio-Carpinetum*) are the next ones to the west, that is in the subcontinental part of the Beech distribution which is marked by relatively dry and warm summers (section c).

c) **Stitchwort-Oak-Hornbeam woods** (*Stellario-Carpinetum*) live in a cooler and more strongly oceanic climate, as is typical for great parts of Central Europe. These are the classical models of *Carpinion* communities formerly known as Oak-Hornbeam woods (*Querco-Carpinetum*, section c etc.).

d) **Bluebell-Oak-Hornbeam woods** (*Endymio-Carpinetum*) dominate in western Europe and are distinguished by oceanic species (Noirfalise 1968, 1969). Their area lies totally outside Central Europe.

There are many intermediates between these four associations. With increasing altitude they also dovetail into the many Beech wood communities and throughout there are differences brought about by soil conditions similar to the ones we have met with in the beech woods. In spite of this the east-west trend is clearly to be seen in the species composition (tab. 26). The changes from north to south are not so continuous as the highlands impose climatic barriers.

In **eastern** central Europe the following communities, along with others, form a north-south sequence, which becomes increasingly rich in species:

a) As already mentioned above, Lime-Hornbeam mixed woods with boreal species, especially *Picea abies*;
e) The same without the boreal species, e.g. in valley sites in the mountains of southern Poland and Czechoslovakia up to about 500 – 600 m above sea level;
f) Primrose-Oak-Hornbeam wood (*Primulo veris-Carpinetum*) in the pannonic lowland of southern Moravia and northern Hungary (Neuhäusl and Neuhäuslova 1968b), also without *Fagus*;
g) Waldsteinia-Oak-Hornbeam wood (*Querco-Carpinetum waldsteinietosum*, Jakucs and Jurko 1967) in Slovakian-Hungarian Karst, and so already inside the range of Beech;
h) Illyrian Oak-Hornbeam woods (*Querco-Carpinetum illyricum*, Horvat, Glavač and Ellenberg 1974) which could also be called Dogstooth-O-H woods (*Erythronio-Carpinetum*) in low-lying parts of the Croatian interior;
i) Moesian Oak-Hornbeam woods (*Querco-Carpinetum moesiacum*, Horvat etc.) in similar lowland sites further down to the southern limit of *Carpinus*; like (g) inside the *Fagus* range;

In **western** Central Europe, there is a strongly oceanic flavour also in the north-south sequence of Carpinion Communities:

c) Stitchwort-Oak-Hornbeam wood of the northern plain, also in some places in the southern mountains;
b) Bedstraw-Oak-Hornbeam wood in the warmer and drier lowlands, e.g. in Mainfranken and Upper Rhine plain as far as south of Basel, and around Geneva;
j) Hornbeam-hop Hornbeam wood (*Carpino betuli-Ostryetum*, Ellenberg and Klötzli 1972) in the insubrian Switzerland and adjoining Italy;
k) More strongly submediterranean *Carpinion* communities which occupy a slot between the montane Beech belt and the mediterranean lowland near the southern edge of the *Carpinus* distribution (Oberdorfer 1968), like (j) without Beech.

More mixed Hornbeam woods are met with towards Spain and Portugal, towards England and towards southern and central Sweden. But we cannot go into these here any more than into the units (d) and (f – k), which were merely mentioned as examples.

(b) Lime-Hornbeam woods outside the Beech area

A unique opportunity of getting to know seminatural Oak-Hornbeam woodlands is offered by the forest of Białowieża, an old forest of more than 4000 ha, in the part of the Polish plain which is free of Beech to the north-east of Białystok. It is one of the most thoroughly investigated woodland reserves of Central Europe. About half of it consists of deciduous mixed woodland occupying a loam soil with a more or less high water table. In a changing mosaic these are dominated by magnificent Oaks, Maples, Limes and Hornbeam. Higher and mostly sandy areas are taken over by Pine, Pine-Oak or Spruce mixed woods. Flood plain woodland and moorland are to be found in the wetter places.

Here we shall follow the description of A. and W. Matuszkiewicz (1954) who have mapped the plant associations of the whole reserve and have summarised the older literature (see also Faliński 1968, 1977). In the mixed deciduous woodland they distinguish above all three subassociations of Lime-Oak-Hornbeam woods (*Tilio-Carpinetum*): one on relatively dry sandy soil 9 *caricetosum pilosae*), one intermediate unit named 'typical', and one on soil where the water table is near the surface (*stachyetosum sylvaticae*). Summarised tables of the last two are reproduced in columns 1a and 1b of tab. 26. What is not apparent from these presence lists is the relative abundance of the species. Common Oak is rarely absent, but takes only a very small and diminishing part in the 20 – 30 m high densely closed canopy. The most plentiful are Hornbeam and Norway Maple along with other shade and half-shade trees. Scattered Spruce remind us that this wood is already within the range of *Picea*. Whoever walks through this mixed Hornbeam forest, which has been undisturbed for at least 60 years is surprised how dark it is. Apart from those places where senile trees are breaking down and allowing more light to penetrate there are practically no shrubs (fig. 106). The herb layer is so reminiscent of a mull Beech wood that it is not surprising to find character species of the *Fagion* alliance, e.g. *Dentaria bulbifera*.

It has been said that the Oak-Lime-Maple-Hornbeam mixed woodland is a 'Beech wood without Beeches'. The participation of Spruce, which here and there towers more than 10 m above the canopy is not unusual for Central European mixed deciduous woodland. *Picea* is also generally found in Sudeten highlands, in the eastern pre-Alps and limestone Alps, but also in the western part of the former East Prussia. In all these places it has approximately the same effect on the ground flora of the broadleaved forests. The relative abundance of moder indicators (such as *Gymnocarpium dryopteris*, *Dryopteris carthusiana* and many mosses) among so many well known mull soil plants (tab. 26) can be traced to the influence of the litter of Spruce needles (see also section B I 3 a).

Following the research by Pigott (1975) into the natural regeneration of tree species it can be said that the ancient Forest of Białowieża is correctly named *Tilio-Carpinetum*. Since 1923 *Tilia cordata* has regenerated so well in groups that it is going to play a greater part in the future. *Carpinus* will certainly recede although it still regenerates in sufficient numbers. Spruce and Pine can also be found at all ages. However there have been practically no young Oaks (*Quercus robur*) for over 50 years so that in the forseeable future one will no longer be able to speak of a mixed Oak-Hornbeam woodland.

Oak-Lime-Maple mixed woodland similar to that described here also occurs in Masovia (Passarge 1964b) as well as a long way into central Russia and Ukraine. As in Białowieża the climate in these eastern natural Oak-Hornbeam regions, which are completely free from Beech, is fairly dry and warm in summer, but too cold for Beech in winter. The average annual precipitation does not exceed 550 mm nor fall below 400 mm and the mean July temperature is generally over 18°C. The dryness is ameliorated during the warmer part of the year by storm downpours which give the precipitation a summer maximum characteristic of the continental part of Europe. Here the winter is quite severe with more snow than in the more western regions, where Oak-Hornbeam – or Beech – woods are predominant (Sokołowski 1962).

Further south in Poland, especially in the foothills of the Tatra, *Picea abies*, *Rubus saxatilis*, *Trientalis europaea* and other boreal species disappear from the mixed Hornbeam woods. They can be used as differentiating species of the north-eastern Spruce-Lime-Hornbeam woods against the east-Central European typical Lime-Hornbeam woods (Neuhäuslova-Novotna and Neuhäusl 1971, Szafer and Zarzycki 1972). The latter have been subdivided into a number of units which e.g. Kornaś (1968) has described from the Polish Carpathians. Some of them also contain *Fagus sylvatica* and merge into the area of the Beech (see also section a). The *Fagion* influence is even stronger in the Hairy Sedge-Oak-Hornbeam wood (*Carici pilosae-Carpinetum*, see Neuhäuslova-Novotna and Neuhäusl 1971) to the south of the Tatra.

(c) Oak-Hornbeam woods of relatively dry habitats inside the Beech area

There are also some isolated regions with a 'dry Oak-Hornbeam climate', similar to that just discussed, in middle and western Central Europe where Beech occurs. The best examples are in the rain shadow of the Harz (Magdeburg in tab. 27, group 15), of the Karkonosze mountains (Liegnitz in the same group), of Rhön and Spessart, in inner Bohemia, in South Moravia, in the Vienna basin and in the Upper Rhine plain (mainly near Mainz and Colmar, group 14). There are also traces of this relatively dry climate in many small valleys, e.g. around Würzburg. Its largest area of dominance in Central Europe however lies in the central and eastern plains of Poland and thence radiates into the valley of the Vistula as far as Gdansk, across Poznan as far as the Oder (tab. 27), and also right over as far as western Thuringia. As long as *Carpinion* woods are present in the above-mentioned and similar areas they belong for the most part to the association of Bedstraw-Oak-Hornbeam woods (*Galio-Carpinetum*, see Schlüter 1968, W. Hofmann 1966, 1968, Th. Müller 1967, 1968, Klötzli 1968a, Oberdorfer 1979, and many others). According to Müller, in south-west Germany, which is a 'pronounced optimal Beech area', they can be distinguished from Beech woods, especially by the character species of the *Carpinion* alliance (see also section E III 1, No. 8.432):

Carpinus betulus	*Carex umbrosa*
Prunus avium	*Dactylispolygama*
Rosa arvensis	*Potentilla sterilis*
Stellaria holostea	*Ranunculus auricomus* et al.

The *Galio-Carpinetum* may be recognised by the species which are relatively thermophilous and which are never present in the *Stellario-Carpinetum* e.g.:

Sorbus torminalis	*Convallaria majalis*
S. domestica	*Carex montana*
Ligustrum vulgare	*Festuca heterophylla* et al.

On the other hand *Stellario-Carpinetum*, which has also already been characterised briefly in section a, can only be distinguished by the absence of these differential species (see also Lohmeyer 1967). It just has the character species of the alliance and was named after one of them, the Wood Stitchwort, although this also occurs in *Galio-Carpinetum*.

Tab. 26. Lime- and Oak-Hornbeam woods which are influenced to varying extents by groundwater in different parts of Central Europe.

Community no.:	1		2		3		4		5
Soil dampness:	m	w	m	w	m	w	m	w	w
Management:	near natural state				± opened up				
Tree layer									
Carpinus betulus	5	5	5	5	5	5	5	5	4
Fraxinus excelsior	4	5	2	3	2	4	3	5	
Acer platanoides	5	5	2	3	2		1	2	1
Tilia cordata	5	5		1	4	1	4	4	5
and *Quercus robur*	5	4	5	5	2	5		·	4
Ulmus glabra	3	3					1	1	1
Prunus padus		1	3	4		2		·	
e *Picea abies*	5	5							1
Fagus sylvatica			5		3		3	5	
Acer pseudoplatanus			1	2	4	2		2	
Acer campestre				1	1	1	5	4	
Quercus petraea					4		5	3	2
L *Prunus avium*							4	4	1
Abies (planted)				·			4	3	
L *Robinia* (planted)							2	2	
Ls *Sorbus torminalis*							2		
Ls *Sorbus aria*							2		
Pinus (planted)									4
Shrub layer									
Corylus avellana	4	5		2	3	4	2	4	5
Euonymus europaea	1	2	2	2	2	2	3	3	3
Rubus idaeus	3	3	1	4	2	3	·	·	3
Sorbus aucuparia	2	3	1	2	1	1	·	·	3
Viburnum opulus		1	2	2	1			2	·
Daphne mezereum	2	2	·		1		1	1	1
Crataegus spec.			1	4	3	2	3	1	2
L *Cornus sanguinea*			1	2	1	2	4	4	2
L *Sambucus nigra*				2	4	2		2	
Lonicera xylosteum			·	·	2		4	3	
a *Lonicera pericylmenum*				3			2	1	
a *Ilex aquifolium*			·				3	2	
L *Rosa* spec.					2	1	5	4	
Ls *Ligustrum vulgare*	·						4	4	
Ls *Viburnum lantana*							4	2	
L *Berberis vulgaris*							2		1
e *Euonymus verrucosa*									4
Herb layer									
Moisture indicators:									
a *Athyrium filix-femina*	3	4	1	1	1	2		1	
Stachys sylvatica	2	4	1	2		5		2	
Circaea lutetiana	1	3	1	1	1	2		2	1
Impatiens noli-tangere	2	4	1	·		2		·	1
Festuca gigantea	2	3	1	2	1	4		·	3
Glechoma hederacea	1	3		1		2		·	3
Urtica dioica	4	5	3	1	3	4		·	4
Stellaria nemorum	3	4							
Carex remota		2	1	2		1			
L *Deschampsia cespitosa*			2	3	2	4		3	2
Carex brizoides						3			
Widely distributed:									
b *Viola reichenb.* and *riv.*	5	5	5	2	4	5	4	5	3
Polygonatum multiflorum	5	4	4	5	5	2	3	4	3
Anemone nemorosa	3	3	5	5	5	4	5	5	5

Community no.:	1		2		3		4		5
Soil dampness:	m	w	m	w	m	w	m	w	w
Management:	near natural state				± opened up				
widely distr., cont:									
b *Milium effusum*	4	3	5	4	3	4	3	1	2
Lamiastrum galeobdolon	5	5	5	4	5	2	1	3	5
Stellaria holostea	5	5	5	5	4	5		1	
Oxalis acetosella	5	4	3	3	2	2	1	2	4
Geum urbanum	2	4	2	3	2	2	1	2	5
Pulmonaria off. obscura	2	3	3	1	4	2	4	3	5
Carex sylvatica	1	2	3	1	2	1	3	4	1
Geranium robertianum	4	4	1	1	2	1		3	4
Ranunculus ficaria	1		5	1	3	1		4	3
Galium odoratum	5	5	5				3	3	1
Hepatica nobilis	5	5	2		2		·	1	4
Ajuga reptans	4	4		·	2	2	2	1	3
Maianthemum bifolium	4	5	1	1	3	2			3
Adoxa moschatellina	1	1	2	4	3	3		·	4
Moehringia trinervia	1	1	1		3	5		·	5
Paris quadrifolia	4	3		·	1			4	1
Sanicula europaea	3	2	·		1		1	3	1
Phyteuma spicatum	1		·		2		4	1	1
Mercurialis perennis		1	·		2		3	1	1
Scrophularia nodosa		1		·	4	4		·	4
Festuca altissima	1	1	2						
Aegopodium podagraria	5	5		·	3	2		·	5
Asarum europaeum	4	5			2		·	·	3
Lathyrus vernus	3	4			1		·	·	1
Dryopteris carthusiana	4	5	·		1		·		2
Dryopteris filix-mas	3	3	·						
Carex digitata	3	2			·		3	2	3
Differential species:									
Ranunculus lanuginosus	4	5						·	
e *Ranunculus cassubicus*	4	5							
e *Equisetum pratense*	4	5							2
Dentaria bulbifera	4	2							
Gymnocarpium dryopt.	4	3							
(e) *Carex pilosa*	4	3							1
d L *Brachypodium sylvaticum*	1			2	2	4	5	5	2
Hedera helix			5	4	1	1	5	5	
Melica uniflora			4		1		4	3	
L *Poa nemoralis*	·		2	4	5	5	2	2	3
L *Dactylis glomerata*	·	·	3	2	4	4		1	1
L *Rubus* spec.	·	·	5	2	2	2	2	2	2
e L *Fragaria vesca*	·		·		2	3	5	3	3
L *Vicia sepium*	·		·			3	4	1	
L *Melica nutans*	·				2	3	2	3	4
L *Convallaria majalis*	·		·		3		2	4	2
L *Galium sylv.* and *schultesii*					2	2	5	4	4
L *Veronica chamaedrys*	·	·	·		1		2	1	4
L *Vinca minor*					1		1	2	
L *Campanula trachelium*	·				1		3	2	
L *Lilium martagon*					3			2	
Luzula pilosa			·		2	1	3		2
Hieracium sylvaticum					1		3		
L *Galium aparine*					2	5			2
L *Galeopsis tetrahit*					2	3			2

Tab. 26 (Continued).

Community no.:	1		2		3		4		5
Soil dampness:	m	w	m	w	m	w	m	w	w
f L *Potentilla sterilis*							*4*	*3*	
L *Euphorbia amygdaloides*							*4*	*2*	
L *Solidago virgaurea*	.						*3*	*1*	.
L *Carex montana*	.						*3*		.
L *Carex flacca*							*3*	*1*	
L *Valeriana officinalis* coll.							*3*	*1*	
L *Carex ornithopoda*							*3*		
Primula elatior				.			*2*	*3*	
Listera ovata								*3*	

Community no.:	1		2		3		4		5
Soil dampness:	m	w	m	w	m	w	m	w	w
g *Stellaria media*									*4*
Bilderdykia dumetorum									*4*
Lysimachia nummularia									*3*
Galeopsis pubescens									*3*
Chelidonium majus									*3*
Lapsana communis									*3*
Viola mirabilis									*3*

L = require more light
a = mainly distributed in the west (= atlantic)
e• = mainly distributed in the east
s = mainly distributed in the south
. = occur in similar communities in the area

Explanation

The columns 1–4 each contain a community growing on slightly moist soil (m) and one on fairly wet soil (w) in the same locality.

Nearly-natural mature woods

No.1: **Lime-Hornbeam woods** in the ancient forest of Białowieża. After A. and W. Matuszkiewicz (1954, pp 42/43)
1w: wet mixed L-H-wood (*T.-C. stachyetosum*) in which the soil is often waterlogged.
1f: damp mixed L-H-wood (*T.-C. stachyetosum*) in which the soil is often waterlogged.

No. 2: **Stitchwort-Oak-Hornbeam woods** in the Eilenriede near Hannover. After Lohmeyer (1951, pp 52 and 58)
2w: Wet St-O-H-wood (*St. -C. stachyetosum*) on soils like 1. *sum*) on soils like 1m,
2f: Damp St-O-H-wood (*St. -C. stachyetosum*) on soils like 1f.

Woods which have been considerably **opened up**; coppiced, with or without standards

No. 3: **Stitchwort-Oak-Hornbeam woods** in the dry area to the east of the Harz. After Passarge (1953, pp. 11 and 45),
3m: Lungwort-Durmast Oak-Hornbeam wood (*'Pulmonaria* - Durmast Oak-Hornbeam wood') on soils similar to 1w,
3w: Woundwort-Common Oak-Hornbeam wood (*'Stachys*-Common Oak-Hornbeam wood') on soils to 1w

No. 4: **Bedstraw-Oak-Hornbeam woods** of the Upper Rhine plain. After Oberdorfer (1957, p. 424, a and b).
4m: Typical Bedstraw-O-H wood (*Galio-carpinetum*, pure form) on soils which occasionally dry out.
4w: Wet Bedstraw-O-H wood (*G. -C. circaeetosum*) on very damp soils.

No. 5: A **Lime-Hornbeam wood** in Central Poland (*Tilio-Carpinetum*, '*Galeobdolon*-variant').
After W. and A. Matuczkiewicz(1956, Ruda Forest, near Puławy, Tab. &) on predominantly damp soils.

A few less-constant species have been omitted. All the tables were constructed in the fifties when the effects of centuries of different management were still clearly apparent.

a-g = species with different types of behaviour within the Lime- or Oak-Hornbeam woods.

Tab. 26 includes examples of both associations together and allows a comparison to be made with *Tilio-Carpinetum piceetosum* from the Forest of Białowieża (no. 1 m and w) which has already been discussed. The *Stellario-Carpinetum* of the urban forest in Hannover, which has been kept in a semi-natural state for many centuries, is surprisingly similar to this and therefore has been put immediately next to it in tab. 26 (no. 2m and w). Nos. 3 m, w, 4 m, w give examples of *Galio-Carpinetum* from the dry area of central Germany and from the Upper Rhine. As in the previous cases, m is not influenced by ground water whereas w is. In the Federal Republic of Germany there are species with a more 'westerly' distribution which are not present in eastern Poland, e.g. *Lonicera xylosterum, L. periclymenum* (subatlantic), *Hedera helix, Melica uniflora* and *Galium sylvaticum*. The warm climate of the Rhineland plain also favours more or less thermiphilous elements such as *Ligustrum vulgare* and *Viburnum lantana*.

A large number of the species newly added to tab. 26 in the lists 4a and b would, according to their total distribution, be definitely expected in Białowieża Forest. They are absent there only because it is too dark. Support for this view is to be found in a list published by W. and A. Matuskiewiecz (1956a) from 'Ruda' Forest in eastern central Poland which has been added in column 5w. This list is from Oak-Hornbeam woodlands which are hardly ever affected by ground water. The stands in Ruda Forest have been severely thinned by removal of timber and by grazing in the past. Obviously this has favoured many species which are also found in the far away woods of the Harz foothills and Upper Rhine plain, but which are sought in vain in the nearby Białowieża Forest, e.g. *Crataegus*, *Cornus sanguinea* and other shrubs, grasses such as *Poa nemoralis*, *Dactylis* and *Deschampsia cespitosa* as well as other half-shade to half-light plants, e.g. *Rubus, Fragaria vesca*, *Convallaria majalis*, *Veronica chamaedrys* and *Galium aparine* (see also fig. 116).

By a comparison of the list of species in table 26 then we learn that **Oak-Hornbeam woods used by man** contain many *more plants which require light* and more species in all than those which have not been so exploited, and that floristically the former do not resemble the Beech woods so much as the latter. What we have come to regard as *Querco-Carpinetum* in

Central Europe today are almost exclusively those stands which have been changed by man's impact. Oak-Hornbeam woods managed in this way differ more from those in their natural state than the managed Beech woods do from the natural high forest. This is especially true of the numerous Oak-Hornbeam woods which have been described from areas in which, because of higher precipitation, the Beech plays a bigger part than in the drier areas.

However before we go into further details we should ask why Beech, which is otherwise so competitive in Central Europe is so ineffective in 'real' Oak-Hornbeam woods. Apparently climatic factors play a major part here since the soil conditions in these Oak-Hornbeam woods are by and large the same as those we have already come across in the Beech woods (section B II 4 d)

As long as the precipitation around the dry regions mentioned in the beginning of this section is more than about 550 mm the Beech will come in (see tab. 27, nos. 9-12, and fig. 107). However it only reaches noticeable proportions where at the same time the summer temperatures are lower, that is, where the evaporation rate may be lower and the water reserves in the soil less quickly exhausted.

In order to express the interplay of precipitation and summer temperature in an easily determined single number one can divide the average July temperature by the total annual precipitation (as an expedient this quotient is then multiplied by 1000 in order to obtain a one or two figure number). The boundary of the natural Beech area in Central Europe has a quotient of about 30 (column IV in tab. 27, and fig. 107). If it falls below 15 then the more thermophilous and drought-resistant deciduous trees drop out entirely whereas Beech becomes the most vigorous tree in the natural woodland. Even in Spessart where Durmast Oaks are famous for their quality, these require the help of the forester because when young they grow more slowly than Beech. On lower sites, which are thus warmer and drier, Oaks, on the other hand, grow as well as, or better than, Beech up to their fortieth year.

Increasing dampness then favours Beech, while

Tab. 27. Climatic data from Central European weather stations and their relationship to the natural role of the Beech.

	I Altitude (m above sea level)	II Temp. in July (°C)	III Annual precipitation (mm)	IV Ratio 1000.II : III
Subalpine Beech wood				**< 10**
1. Alps close to the tree limit	1670-1780	9.9-13.6	1670-1780	6 - 7.5
2. Vosges, Sudeten Mountains, Harz	1150-1395	10.0-13.3	1140-1930	6 -10
Beech-Fir wood				**~ 10**
3. Swiss Alps and Jura	1090-1100	14.6-14.9	1340-1530	9.5-11
4. Bavarian Alps, Black Forest	910-1025	14.0-15.5	1030-1660	8.5-14
Beech wood (with ± conifers)				**10-20**
5. Swiss outer Alps	725- 955	15.6-16.5	1165-1585	10 -13.5
6. Between the Alps and Swiss Jura	450- 570	17.4-18.4	920-1150	16 -19.5
7. Black Forest, Swabian Jura etc.	695- 910	13.7-15.5	930-1510	10.5-15.5
8. Erz and Sudeten Mountains				
Beech wood with Oak et al.				**20-30**
9. Lower parts of central Switzerland	275- 420	17.4-19.5	780- 890	21.5-23
10. South-west and West Germany	120- 400	17.0-19.5	510- 840	23 -28
11. North-west Germany	10- 60	13.3-16.8	720- 790	21 -22.5
12. Northern GDR. Northwest Poland	5- 75	16.7-17.7	610- 750	20 -28
Mixed Oak wood (± Beech-free)				**> 30**
13. Valleys of the central Alps (with Pine)	540- 550	19.3-19.5	540- 640	30.5-36
14. Upper Rhine plain, Main valley	95- 190	18.3-20.1	480- 550	33 -42
15. German and Polish dry regions	60- 130	18.0-18.6	500- 520	34.5-36.5

Weather stations: 1: Mte. Generoso, Rigi; 2: Gr. Belchen, Glatzer Schneeberg, Brocken; 3: Göschenen, Wildhaus, Sainte-Croix; 4: Mittenwald, Höchenschwandt, Todtnauberg; 5: Heiden, Affoltern (Emm.), Marsens, Seewies; 6: Zürich, Bruus, Bern; 7: Freudenstadt, Böttingen, Klausthal, Schneifel; 8: Altenburg, Schreiberhau; 9: Diessenhofen, Kreuzlingen, Basel, Genf, Aigle; 10: Freiburg i.Br., Heidelberg, Stuttgart-Hohenheim, Göttingen; 11: Münster i.W., Elsfleth, Flensburg; 12: Görlitz, Köslin, Swinemünde, Neustrelitz; 13: Sitten, Siders; 14: Colmar, Mainz, Kitzingen; 15: Magdeburg, Posen, Liegnitz. (The place names correspond to the sources in the literature.)

Source: Records taken from tables by Ellenberg (1963).

dryness restricts it more than it does Hornbeam, Lime, Oak and their accompanying trees and shrubs. The same thing appears if we follow these species which appreciate more light into the area south of the Alps. Here they go down under the boundary of the Beech-Fir belt, touching the region of submediterranean climate which is decidedly dry, while Beech remains tied to the montane climate. In the mountains of the Mediterranean, the lower boundary of Beech is certainly determined by the dryness during the main growing season, and not by the temperature during the winter and spring, which is usually looked upon as being the deciding cause of the eastern limit of the Beech's range.

The meteorological stations in the regions of the true Oak-Hornbeam woodlands (tabs. 27, 14 and 15) are not significantly colder in winter than those in the real Beech wood areas (1-12). This also applies if one compares, not the average temperatures of the coldest months, but the absolute minima. These are almost as low in the west as they are in the east of Central Europe, and do not appear to be clearly dependent on distance from the coast or on altitude in the mountains since the local climatic positions of the measuring stations are not comparable. Therefore they have been omitted from tab. 27 in order to give a clearer survey. There is really not much difference between Beech and Hornbeam in their resistance to winter temperature. This is seen above all in some older observations on the north-eastern and northern boundaries of Beech distribution.

Timm (1930), Ziobrowski (1933) and Vaarama (1941) for example were able to observe that *Carpinus* suffered more from unusually sharp frosts than *Fagus* and many other deciduous tree species. Gross (1935) went so far as to give the opinion that, in the Beech forest at its eastern limit in former East Prussia Hornbeam had been supplanted by Beech some 200 years ago because it had not been able to grow so well during a sequence of severe winters. Till (1956) showed by physiological experiment that twigs of *Fagus* and *Carpinus* died off at about the same low temperature and that their frost hardiness showed very similar yearly rhythms (fig. 108). Only late frosts in

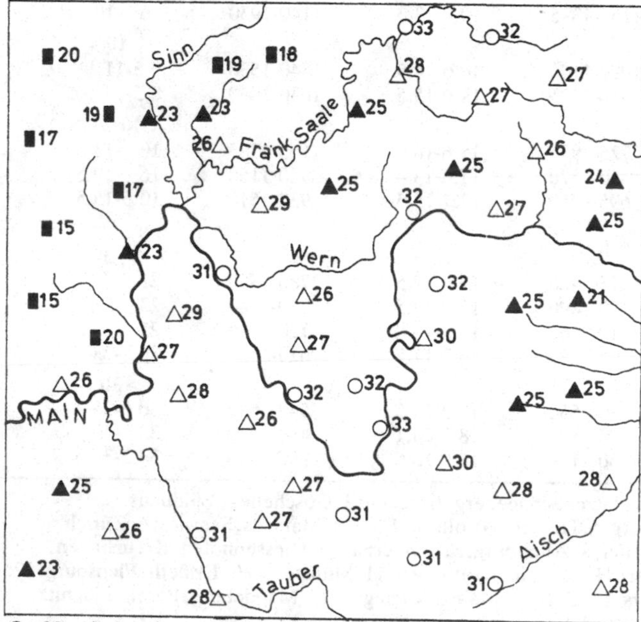

Fig. 107. The quotient obtained from the average July temperature, multiplied by 1000 and divided by the annual precipitation (see tab. 27,R) corresponds very closely to the proportions of tree species in the natural woodlands in Mainfranken. After Hofmann (1968).

True Oak-Hornbeam woods, poor in Beech, are found only in the warmest, driest areas (R > 30 in average sites). Today the area with R 21-30 is an 'Oak-Hornbeam area since it was easy to remove the naturally dominant Beech, e.g. on the Frankian plateau, in the eastern 'Bauland' and in the southern Steigerwald. In the Spessart, South Rhön and the highest parts of the northern Steigerwald Beech is completely dominant although it still tolerates other trees such as Oak (R 15-20). The district between these two extremes (R 21-25) is shared between the Oak and Hornbeam with the Beech still playing a very vital role.

May are withstood better by Hornbeam than by Beech, because it opens its leaves earlier and more quickly than the latter and so possesses tougher leaves when the dreaded night frosts appear after a long period of mild weather. If nevertheless Hornbeam does become damaged by late frost, it can recover better than Beech whose dormant buds do not break into leaf so easily as those of Hornbeam and many other deciduous trees. If Beech plantations are made in eastern Central Europe they are frequently a sacrifice to the very common late frosts which occur here, e.g. in the park of the hunting lodge at Białowieża where a few isolated *Fagus* trees have been badly mutilated and dwarfed by massive damage to the young shoots. According to Gross (1935) late frost in East Prussia killed off a lot of flowers and young fruit of *Fagus* so that a full mast year was only achieved about every 10-15 years. In southern Scandinavia also the existence of Beech is threatened by frost damage to its fruit.

Yet there are areas in Central Europe which are in no way subject to increased frost risk, but in spite of this are naturally Beech-free or poor in this species. A good example of this is the south-western Upper Rhine plain in the area round Colmar. According to Issler (1942) it is too dry here for *Fagus*. The deciding cause of the competitive weakness of Beech in the true Oak-Hornbeam landscape then is probably the **dry conditions in spring and summer** in most cases. It is well known to every forester that regeneration of Beech is much easier in a high rainfall area than in one of low precipitation and dry atmosphere. Watt (1923) reports that the relatively large Beech nuts will only germinate when they have been covered for a longer time with a film of water. The seeds of Hornbeam are significantly smaller and may require less water. The growing seedlings of Beech also require a lot of water. According to Slavíková (1958) they soon perish in dry habitats, mostly through competition with the roots of the older trees, and it is not by chance that the '*Fagetum nudum*', which was discussed briefly in section B II 2 d is found only around the boundary of dry regions. In the northern Alpine foothills, which have an unusually high precipitation for Central Europe, on the other hand, the Beech can reproduce itself under a relatively dense canopy without difficulty. Even on strongly acid soils, with a thick humus covering on which natural regeneration of *Fagus* seldom succeeds in dry areas, the seeds germinate like weeds, e.g. in the mountains round Zürich (tab. 26,6).

(d) Oak-Hornbeam woods more or less rich in Beech

Meusel (1955) has already pointed out the gradual change taking place in our deciduous mixed woodland which is experienced in passing step by step from the montane climate, friendly to Beech, over a dry climate hated by it. As an example he describes the acid-soil woodland communities of the Harz and its eastern foothills:

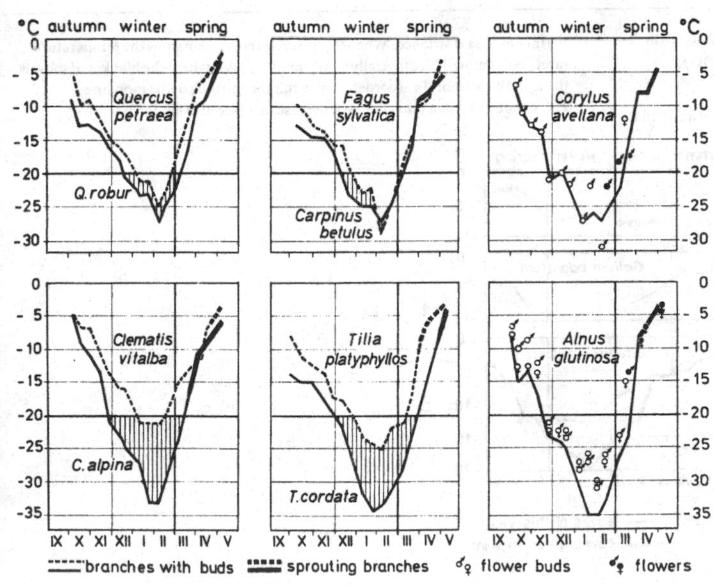

Fig. 108. The frost hardiness of trees and shrubs increases from October (1953) up to the end of the year and then starts to get less again. The twigs in spring are particularly susceptible to late frost. After Till (1956), modified (compare fig. 109).

The Woodrush-Beech wood already known to us (*Luzulo-Fagetum*) is well grown at about 500 m in lower Harz. On similar soil on the outlying lower hills Durmast Oak assumes some importance and reduces Hornbeam to the underwood or copse. According to the proportions of the main trees in the mixture, Meusel calls this community '*Luzula-Querco-Fagetum*', thus stressing that Beech is still the dominant tree. In the centre of the dry region to the north of Halle, on the other hand Beech is absent and leaves Hornbeam and Lime to form the tree layer (fig. 105). Even Hornbeam appears to be weakened by the warm dry climate and is absent in some places. Meusel speaks of '*Calamagrostis-Querco-Tilietum*' because *Calamagrostis arundinacea* obviously survives better in an extreme dry climate than *Luzula luzuloides*. The strong invasion of the drought-tolerant light-loving plants may be traced back in part to the low precipitation of their habitat and in part to the widespread use of the woodland by man. The eastern Harz foreland belongs to the most heavily settled landscapes of Central Europe since the Mesolithic Age.

In a similar way to the sequence used by Meusel it can be seen that on base-rich soils too Beech pushes right up to the boundary of the real dry area, and that its participation in the natural woodland is not gradually reduced but suddenly dies out. In the area of the transition climate the Beech still manages to play a relatively large part if it is not prevented from doing so by man's activities. This fact, familiar to every forester, was earlier as little observed by many plant sociologists as the climatic difference between pure Oak-Hornbeam woods and those which are more or less rich in Beech.

The fluidity of the proportions of different tree species in the wide climatic transition zone between the pure Beech woods and the pure Oak-Hornbeam woods makes it possible that the form of forestry known as 'coppice with standards', in which the Oaks were left as standards and the rest, including Hornbeam, were cut at regular intervals, was widespread over a great part of Central Europe (see section A II 3 a). This is true also for areas where, on the evidence of documentary records or pollen analysis, Beech was certainly dominant at one time. According to Etter (1943) the terms 'Oak-Hornbeam wood' and former 'coppice with standard' are practically identical in Switzerland, if one omits the few stands which have arisen from thermophilous or acid-soil mixed Oak woods. In the canton Bern and in the western part of the canton Aargau, where the woods have always been managed as 'high forests' there are hardly any examples of Oak-Hornbeam woods (Stamm 1938), while in those parts of the pre-Alps lying to the east and to the west of these cantons, e.g. in the canton Zürich, Oak-Hornbeam woods and even mixed Oak woods have come into existence all over as a result of the practice of coppicing with standards (N. Kuhn 1967). To the north-west of Zürich such woods can be found up to 780 m above sea level on the Heitersberg near Dietikon (see also figs. 23 and 113).

In areas where both Beech and other deciduous trees find the climate favourable it has been, and still is, man's influence which has given one or other of the species the advantage. Even the length of time between cutting in coppice management can operate to the advantage or disadvantage of particular species. According to Brouillard (1911, see also Seibert 1966) an interval of 10-12 years favours the true shrubs. With a cycle of 18-20 years Oaks become dominant as their

Fig. 109. The variation in frost resistance shown by various woodland herbs throughout the year. After Till (1956), modified.
The underground organs are much more susceptible but they too show an annual rhythm of frost resistance. When the ground is covered in snow the temperature rarely falls below 0°C, especially at the montane level where the blanket of snow is thick and persistent. In accordance with this the rhizome of the montane *Polygonatum verticillatum* is least frost resistant (see also fig. 165).

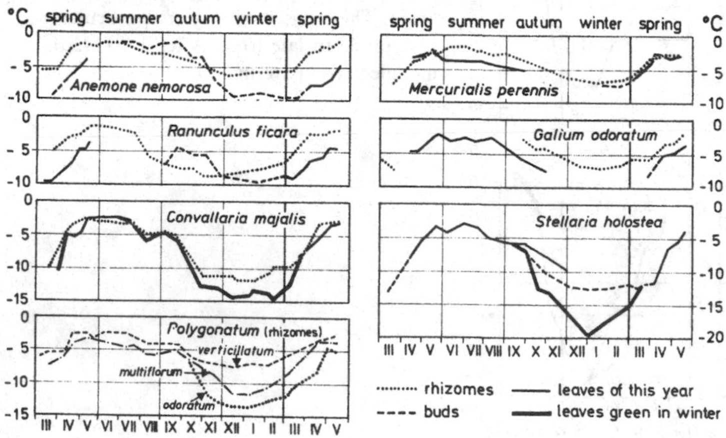

young trunks and branches shoot up freely again after cutting. The more usual interval of 20-30 years gives Hornbeam time to shade out Oak and other light-loving trees. Hornbeam was particularly valued as burning wood. Beech requires an interval of more than 30 years between cuts before it is able to assume its natural dominance. It sprouts reluctantly from the cut stumps but when it is allowed to grow old enough it produces seed before being cut down and can send its shade-tolerating seedlings into the field. Silver Fir and other conifers which cannot regrow at all after cutting are only able to survive as protected standards. Consciously or unconsciously therefore the forester's axe 'tips the scales' in the competition between the trees.

Only near the climatic limit where it is getting too dry for Beech can the situation arise that in densely populated areas this species is completely wiped out. It is not able to withstand coppicing as well here as in damper districts since the small areas of woodland which still remain between the open fields have a much drier and more continental microclimate than would have been the case in a closed natural woodland. For example the dry area to the east of the Harz mountains is shown by pollen analysis not to have been so devoid of Beech formerly as we know it is today (Passarge 1953a). Under present conditions however this area must count as a real Oak-Hornbeam countryside especially in the area where the drying föhn wind blows from time to time down from the Harz (fig. 103).

The arguments which have been used here for the Harz foothills and the Alpine foothills in Switzerland apply equally well to other areas of Europe which have a transition climate and Oak-Hornbeam woodland in which Beech is able to survive to a greater or lesser extent, e.g. in Slovakia (Magic 1968) and in Croatia (Horvat. Glavač and Ellenberg 1974). Only under such conditions can the questions: 'Oak-Hornbeam wood or Beech wood?' actually become a problem.

The '**Oak-Hornbeam wood question**' was thoroughly discussed in the first edition. According to the preceding explanations it can be answered today as follows: On brown earths which are free from groundwater every kind of intermediate is possible between pure Beech woods of the montane or upper submontane belts and mixed deciduous woods, quite free from Beech, of the dry regions of Central Europe. These intermediates do not however form a smooth sequence; on the contrary Beech generally becomes dominant where it is able to grow. Its dark canopy fundamentally changes the microclimate and the species composition of the undergrowth so that unique communities are formed. It seems advisable therefore to put these Beech-rich Oak-Hornbeam woods, or Oak-permitting Beech woods into a group of their own and to separate from them the two extremes. This is not always possible by floristic means alone since the boundaries have been displaced and aggravated by man's influence, especially by coppicing.

Modern forestry has unwittingly introduced a large-scale test of this point of view since it has changed over from coppicing to high forest management throughout almost the whole of Central Europe. Unfortunately however in practically all woodlands alien conifers have been brought in and planted on the fertile brown earths thus displacing the conditions for natural competition in a different direction. Nowhere is it more urgent to secure woodland reserves for scientific study than in Oak-Hornbeam woods within the distribution area of the Beech.

(e) Oak-Hornbeam woods on wet soils

In the less-dry lowlands of Central Europe, especially where the climatic quotient is between 20 and 30 (tab. 27) the Beech is always a possible competitor. But even here there are natural Oak-Hornbeam woods free from Beech. As already explained in section B I 2 *Fagus* cannot tolerate ground which is temporarily or permanently waterlogged, presumably because its root breathing is impaired. The seed too suffer, as Watt (1923) has stressed, from shortage of oxygen which is associated with wet conditions, and easily rot on wet ground. On the other hand Hornbeam, Common Oak, Ash, and Maple species make out much better in wet soil, given sufficient nutrients so that on wet mineral soils, especially on mull-gley (fig. 110) quite similar combinations of tree species to those on brown earths in dry areas can arise. Such wet-soil Oak-Hornbeam woods are particularly well developed in the north-west German plain and have been thoroughly researched here by Tüxen and his students. In addition their

Fig. 110. Mull-gley under a Beech-free moist-soil Oak-Hornbeam wood. After Ellenberg (1937), somewhat altered.

There is a fairly sharp distinction between the humus-rich mull on the top and the gley horizon (G) which has formed in the region of a varying water table. Its basic colour is pale grey with a bluish or greenish tinge caused by the presence of iron-II hydroxide. Shining rust-coloured spots (hatched) arise through oxidation in which bacteria probably play a part.

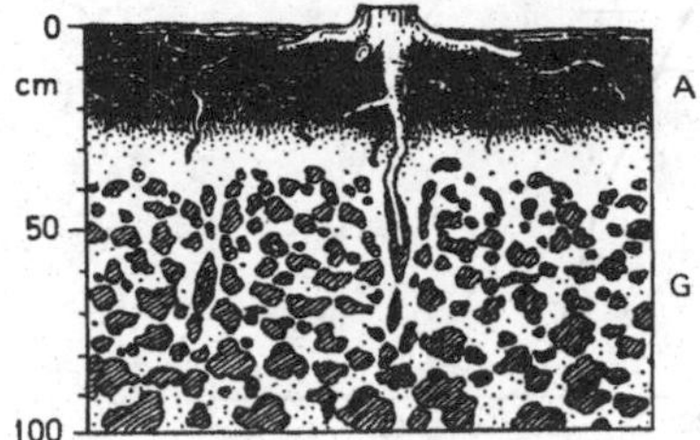

presence has been confirmed by many other authors from different parts of Central Europe. *Galio-Carpinetum circaeetosum* described by Oberdorfer (1957) from the Upper Rhine area occupies ground which does not appear to be so wet. Here Beech finds even better conditions than in neighbouring drier habitats provided the soil is mainly sandy (tab. 26, 4b). In soils rich in colloids the Beech is handicapped if the water table is nearer than 50 cm from the surface for more than 3 months of the year (fig. 111).

The wet-soil Oak-Hornbeam woods are difficult to distinguish from the slope-foot Sycamore-Ash woods (section B III 1c) and from mixed hardwoods of the flood plains (section B V 1 g). The species combination of these three woodland communities is so similar that many authors put them together in the alliance of wet-soil deciduous woods (*Alno-Ulmion*). Others combine the Oak-Hornbeam woods of damp soils with those on drier habitats in the alliance *Carpinion*, which has happened here. The main reasons for this are the general dominance of Hornbeam and the absence of an extra supply of nutrients, be they colluvial or alluvial, which is made apparent by a higher proportion of nitrogen indicators. There are no good character species for all the Oak-Hornbeam woods apart from *Carpinus* and even this is not entirely confined to the *Carpinion*.

Yet it is amazing how similar the species list of the Stitchwort-Oak-Hornbeam wood of the urban forest near Hannover (tab. 26, 2b) is to the one already discussed from the Forest of Białowieża (no. 1b) although the general climates of north-west Germany and eastern Poland differ so significantly. In both communities the Hornbeam and other half-shade trees form a dense stand through which single old Oaks tower up. Only the Spruce is absent from the natural lowland woods of north-west Germany and the Lime is less common; Sycamore and Maple come in for these and the Bird Cherry is more flourishing. Both communities have the following in common: shrubs and young trees suffer from a shortage of light, spring flowering herbs are abundant (fig. 112) and mosses are generally smothered out. For the most part the same moist-soil indicators are represented (fig. 123, e.g. *Carex*

Fig. 111. 'Lines of groundwater duration' in soils where Beech are able to grow and those which are free from Beech in central and northern Switzerland. After Klötzli (1968), somewhat modified.

Each point on the line indicates the number of weeks in the year (abscissa) during which the groundwater is at least up to a particular level below the surface (ordinate). For example under a moist-soil mixed Beech wood the water table reaches a height of 50 cm or less below the surface for only about 5 weeks, and for fewer than 5 months (20 weeks) is it closer than 100 cm to the surface. If the water level remains within 50 cm of the surface for 3 months (12 weeks) or more then the Beech can no longer flourish. This is true only for loam or loess soils in which there is a capillary zone above the water level (i.e. a layer saturated with water and poorly aerated) up to several decimetres above the measured groundwater level.

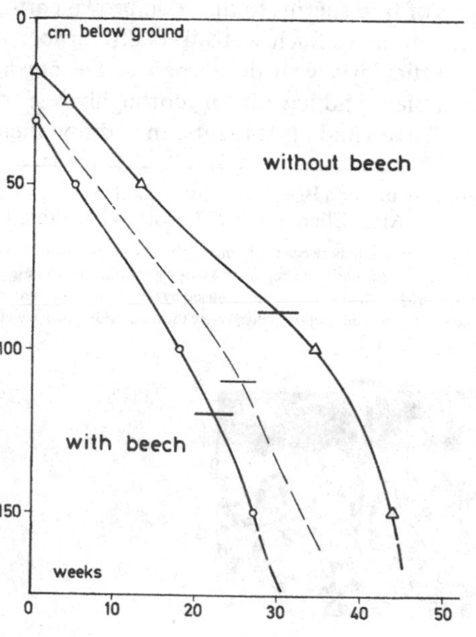

Fig. 112. A carpet of Anemones on the floor of a moist-soil Oak-Hornbeam wood (*Stellario-Carpinetum stachyetosum*) in the Eilenriede near Hannover. Spring-flowering geophytes such as *Anemone nemorosa* flourish best in a wood such as this where the ground is not covered with the dead or living foliage of hemicryptophytes. These are not able to grow in the deep shade once the trees have come into leaf.

remota, Festuca gigantea and *Stachys sylvatica.* What is more the two lists resemble one another in that definite 'light-loving' plants play no great part, even though individuals are found more regularly near Hannover than in Białowieża.

That they obviously have so much in common most probably rests with the fact that the soils in the two places are so similar; in both it is old diluvial moraines or sands strongly influenced by groundwater which converted the soils into real gleys. Another significant reason is that the urban forest of Hannover has been influenced very little by man – the least of all Oak-Hornbeam woods in the lowland to the west of the Oder. This old forest of about 650 ha has not been grazed since the Middle Ages and only used for the extraction of timber by careful management and the felling of single large trees. For this reason plants requiring a lot of light were rarely given the chance to develop, just as in the Forest of Białowieża, with the exception of oak which enjoyed special protection everywhere in the Middle Ages until 150 years ago (see section A II 3 a).

In spite of the favourable treatment received by Oak from time to time, wherever the topsoil offered more than 20-30 cm of constantly well aerated root run then Beech could invade and develop magnificent trunks. As soon as Beeches gain the upper hand they form the spacious 'cathedral' forest with which we are already familiar, and in which shade-tolerant species, such as *Anemone nemorosa*, *Galium odoratum* and *Melica uniflora*, can spread in groups (tab. 26, 2a, fig. 117). Common Oak and Hornbeam were certainly present in almost every stand of this moist-soil Beech wood (Lohmeyer 1951), but always in only small numbers.

Where the soil surface rises higher above the water table the composition of the Herb layer changes but Beech maintains its dominance. As the influence of the ground-water recedes Lohmeyer distinguishes three types of mixed Beech woods: rich in herbs, rich in Wood Fescue, and rich in Lily of the Valley (section B II 3 b) Only on the sandiest and driest sites does the Beech allow the Oak sufficient space to form Birch-Oak woods (section B III 5 b). The latter may correspond to some extent to the Pine wood communities of the sandy ridges in the Białowieża Forest which was mentioned briefly at the beginning of section a.

The described ecotone of woodland communities confirms the general rule that the tree species in the mixed Oak-Hornbeam woodland in any climate corresponding to types 9-12 in tab. 27 would be overpowered by Beech on all soils which were not too wet or too poor. However it is difficult to decide how big a role the Beech would play under natural conditions. The influence of man in this landscape in most cases began very early and must be reckoned with today even in those forests which appear to be very nearly natural.

(f) Oak-Hornbeam woods in the gradient of soil acidity

Just as there is a series of Beech woods according to soil acidity, so too there is one of Oak-Hornbeam woods. Since we have discussed in detail the sequence of Beech woods and the soil conditions which have determined them we can just state here that the conditions for the parallel Oak-Hornbeam sequence are in every way comparable. Of course many stands exist which in their species composition lie between the main types which will be described here, just as there are intermediate soil types between rendzina, brown earth or parabrown earth and podsolised brown earth.

A survey of the Oak-Hornbeam woods from the copious literature dealing with them is made more difficult by the fact that their species composition responds more strongly to differences in climate than does that of the Beech woods. This may be connected not least with the fact that the heavily utilised Oak-Hornbeam woods are relatively open and that their understorey is not so well protected by the canopy from the general climate as in the shady Beech stands.

There is a similar series of woodland types from the true Oak-Hornbeam woods of the dry continental lowlands to the real Beech woods of the submontane and montane uplands. The possible edaphic and climatic interactions between the deciduous woodland communities on soils which are not poor in bases nor influenced by groundwater can be expressed in the following scheme:

	Increasing base content →		
Falling humidity ↓	moder-,	brown-mull-,	limestone Beech woods
	moder-,	brown-mull-,	limestone Oak-Hornbeam woods rich in Beech
	moder-,	brown-mull-,	limestone Oak-Hornbeam woods

Moder Oak-Hornbeam woods have been given the name *Querco-Carpinetum luzuletosum*, e.g. by Tüxen (1937) and Etter (1943). Now they should be called either *Stellario-* or *Galio-Carpinetum luzuletosum* (section a). Brown-mull Oak-Hornbeam woods have been called all kinds of things as they occur in practically every part of western and Central Europe (fig. 113). Since they occupy a middle ground in their species make up and contain neither acid- nor lime-indicators Tüxen (1937) and many other authors speak of them as *Qu.-C. typicum* (tab. 26, 2a), or *Stellario-Carpinetum typicum*.

Limestone Oak-Hornbeam woods also bear different

names but they are less common than the other communities since the limestone rock in low-lying areas of Central Europe is largely covered with loess, moraines and other loamy fine soils (G.Hofmann 1963, Passarge and Hofmann 1968). Generally one finds limestone Oak-Hornbeam woods at the margins of Beech woods on stony slopes where coppicing has been practised. Their ground flora correspond in many ways to those of the slightly-moist limestone Beech wood but include more light- and warmth-loving species. Since the replacement of coppicing with a high forest method of management these are disappearing and the Beech is taking over the dominance in the tree layer in most cases. Probably almost all limestone Oak-Hornbeam woods have been derived from Beech woods by former coppicing or grazing.

Most habitat factors in our European woodland are subject to the rhythm of the seasons influencing not only the amount of light, the temperature, and the water content, but also many chemical factors, even the pH value of the soil. For these factors depend on the activity of soil organisms and roots as well as being directly influenced by the rhythm of temperature and humidity. The Oak-Hornbeam wood has been thoroughly investigated as an example. Therefore these processes will be discussed here although they apply just as well to other woodland communities.

Fig. 113. The appearance of an old Oak-Hornbeam wood in late spring. This is a coppice with standards on mull-gley to the south of Kreuzlingen near Lake Constance. *Quercus robur* and *Fagus* remain as standards while *Carpinus* is growing from its coppiced stools. *Carex pilosa* forms an extensive carpet on the ground.

As fig. 114 shows, the pH values of all soil horizons in all the woodland communities studied around Hannover varied throughout the year by at least a few points and in some cases by more than a whole unit (Ellenberg 1939). These fluctuations were not because the samples were taken from different places in the soil; such errors due to method were carefully eliminated. Furthermore many other investigators have obtained similar results (Ellenberg 1958). Also in the soils of other plant communities there were found to be pH fluctuations of the same magnitude. Lötschert and Ullrich (1961) pointed out that these were running parallel not only in deciduous woodland, but also in heathland and other plant formations. The main cause of these synchronised fluctuations, at least within the restricted area investigated, has been shown to be sulphuric acid brought down by precipitation (Rönicke and Klockow 1974, see also section A II 6 b). Because of the emission of SO_2 by dwelling houses and factories the pH value of the rain can fall as low as 4, even to 3.3 (Ottar 1972) or lower. This could very well explain the observations of most authors that the pH drops after rainfall, that is when the water in the soil increases. Along with this the anaerobic metabolism of microorganisms may also play a part. Previously it had been thought that this must have been the main cause (see the collection of references by Ellenberg 1958, also figs. 33, 34 and 79).

In considering such fluctuations one can take the annual curves of the pH values, or yearly averages, or possibly single measurements made at the same time, and just compare them with each other. If this is done, then one discovers that even quite similar habitats can be neatly separated, whereas pH measurements, made over a long period of time from numerous single tests, give a scatter of values which overlap to a considerable extent. For example the soils of the moder Oak-Hornbeam woods were more acid throughout the whole of 1937 than those of the brown-mull ones (fig. 114), and the pH values of the raw humus soils from Birch-Oak woods were always lower still.

One can take it from these results that there is a close connection between the species composition of the ground flora of a woodland and the acidity of the soil. This is expressed, amongst other things, by the fact that the average yearly pH value is in good correlation with the proportion of fastidious deep-rooting plants in the herb layer of the Oak-Hornbeam woods (Ellenberg 1939), or to the 'mean reaction value' which is calculated according to section B I 4 b (fig. 88)

Such correlations then are only fairly close if they are seen within related plant communities. If one tries to compare communities which differ strongly climatically, edaphically and thus floristically, it becomes obvious that it is not the pH value as such which is acting directly on plants. On the contrary it is interfering with many other habitat factors; therefore, it can best be regarded as one of the symptoms of the soil fertility. Probably the biological activity of the soil (fig. 90) and the humus content operate less indirectly. They show a clear relationship to the pH value under otherwise comparable conditions and have already been discussed in more detail with reference to the Beech woods (section B II 4 d-f).

In the sandy diluvial soils in the Netherlands and northern Germany the chemical factors just discussed are very closely connected with the nature and level of the groundwater. Where the average distance of the water table from the surface is the same then the carbonate content of the water becomes the decisive factor (tab. 28). When this is high in the Oak-Hornbeam woodland the fastidious deep-rooting herbs become dominant. When the hardness of the water is low then the strongly acid-tolerant Birch-Oak wood takes the place of the Oak-Hornbeam wood (section 5). To some extent the ground-water which is rich in lime 'fertilises' the sandy soil from below. When the water level is lowered then the upper soil becomes more acid. Hartmann (1941) was able to demonstrate this effect after just a few years through the increase in acid-indicating plants.

This quick reaction can be understood if one remembers that the lime which is taken up by the tree roots from the groundwater in the subsoil is, for the most part, returned to the soil surface in the leaf litter and worked into the root horizon of the herb layer by soil organisms (figs. 78 and 115). It can be seen from tab. 29 that the leaves of most trees living in an Oak-Hornbeam wood are relatively rich in bases and

Fig. 114. Oscillations in the acidity (measured as mg hydrogen ions per litre and as pH values) throughout the year (1937) in the top soil of several Birch-Oak woods (above the dotted line) and Oak-Hornbeam woods ('poor variant'). Curves for the 'rich variants' of the Oak-Hornbeam wood cannot be shown on the non-logarithmic soil acidity scale. They are in the region which is cross hatched, occasionally moving up into the diagonal hatching. After Ellenberg (1939), modified.

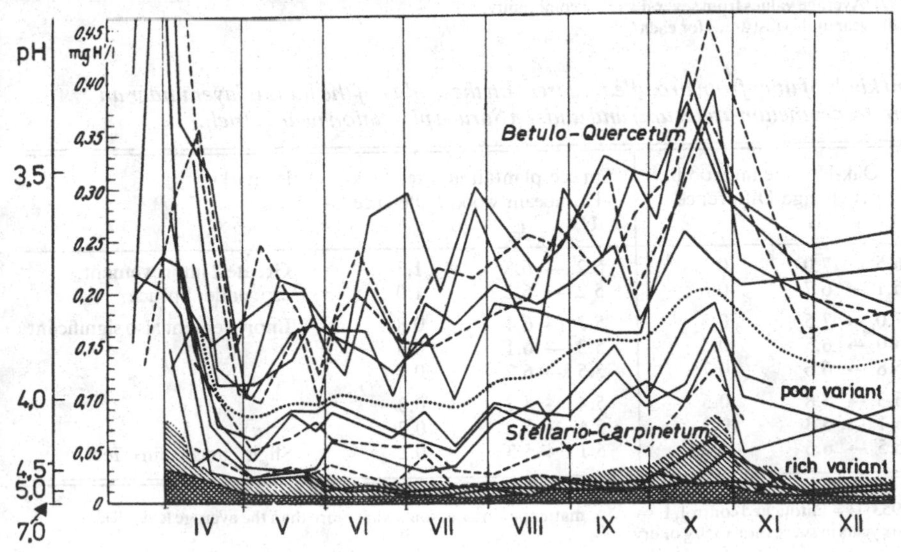

easily broken down. If for example they are spread out over the needle litter in a stand of planted Spruce the pH value of the humus layer above the mineral soil is raised considerably in six years. In this respect *Tilia* and *Acer* leaves are particularly effective and can even raise the pH value of the soil in an Oak-Hornbeam wood, whereas Oak comes right at the end of the list of the trees investigated by Leibundgut (1953, see also fig. 48). This is mainly because of the high tannin content of the Oak leaves which inhibits the rapid invasion of mineralising fungi (Harrison 1971).

The final arbiter of the fertility of a woodland is the amount of primary production. Since this can only be ascertained for trees by taking a number of measurements over many years, Ellenberg (1939) has used the leaf area index (LAI) produced as a relative yardstick. The latter can be approximately determined by collecting all the freshly fallen leaves from a number of 1 m^2 test areas immediately after leaf fall in a stand sheltered from wind, and measuring their surface (for methods see Geyger 1964). Tests made in different communities by this method showed, according to Ellenberg (1963), that wet Oak-Hornbeam woods, depending on the fertility of the soil had LAI of 6 to 9. In other words that there are, one above the other, 6 to 9 leaves from the highest canopy down to the shrub layer. This index is surprisingly high if one bears in mind that the canopy in an Oak-Hornbeam woodland is never 100% closed. It appears that the LAI depends on the habitat, but is relatively independent of the species of trees and bushes present at the time. In pure Beech woods *Fagus* spreads out the maximum possible leaf area more or less entirely on its own, allowing the undergrowth a diminishingly small part. On the other hand Oak, even in pure stands, produces leaves the total area of which is not more than 3-6 times that of the ground under the canopy. On the relatively fertile soils of the Oak-Hornbeam woods then there is always enough light remaining for the leaves of other woody plants growing between and beneath the Oaks. The contribution of the herbs to the leaf area can be neglected in the naturally dense closed stands of Oak-Hornbeam and Beech woods since they only partially cover the ground.

Tab. 28. The lime content of ground-water under Oak-Hornbeam and Birch-Oak woods in the environs of Hannover.

Communities	Lime content (in 10 mg CaO/Litre)							
Rich Oak-Hornbeam woods								
Stellario-Carpinetum corydaletosum	12.8							
" " *stachyetosum*		10.9	6.7					
Poor Oak-Hornbeam wood								
(species-poor variant of *St.-C. stachyetosum*)				5.0	2.0			
Birch-Oak woods								
(*Betulo-Quercetum molinietosum*)						1.5	0.9	0.7

Source: From data by Ellenberg (1939). Average values from several determinations at different times of the year on 1-3 test areas for each community.

Tab. 29. Influence of different kinds of litter from broadleaved trees on the acidity of the humus layer under an Oak-Hornbeam wood (Querco-Carpinetum aretosum) and under a Spruce plantation near Zürich.

Species supplying leaf litter	Oak-Hornbeam wood pH change U → L	Difference	Spruce plantation after Oak–Hornbeam wood U → L	Difference	Remarks
Winter Lime	6.5 → 7.0	0.5	5.2 → 6.5	1.3	Greatest improvement,
Sycamore	6.1 → 6.7	0.6	5.2 → 6.2	1.0	esp. under Spruce
Wych Elm	7.0 → 7.5	0.5	5.3 → 6.1	0.8	Improvement also significant
Ash	6.0 → 6.5	0.5	5.3 → 6.1	0.8	
Beech	5.8 → 6.6	0.7	5.5 → 6.2	0.7	
Black Alder	6.3 → 6.8	0.5	5.4 → 5.7	0.3	
Hornbeam	6.1 → 6.3	0.2	5.6 → 5.8	0.2	
Common Oak	6.5 → 6.6	0.1	5.1 → 5.3	0.2	Slight, uncertain effect

Source: From data by Leibundgut (1953) U = untouched control, L = covered with a layer of leaf litter for six years in succession (350 g of dry matter per m^2, i.e. somewhat more than the average leaf fall).

(g) The effects of light and temperature on the species mosaic

Oak-Hornbeam woods are rightly considered to be particularly perfect and many sided 'highly organised' biocoenoses and therefore have attracted the attention of ecologists over and over again. Most authors have occupied themselves with the living conditions of the undergrowth without touching the problem of how the combination of tree species has come into being. This applies not only with regard to the pH value and nutrient content of the soil which have already been discussed, but also to the soil dampness (section h) and to the distribution of the relative light intensity over the ground flora (fig. 116).

Fig. 115. Annual turnover of nutrients in a limestone Oak-Hornbeam wood in south-eastern Belgium. This is coppiced Hazel with standards. From data by Duvigneaud and Denayer-de Smet (1967, see also 1971) and Denayer-de Smet and Duvigneaud 1972)

The annual net increase of the phytomass contains 103 kg N, 66 kg K, 44 kg Ca and smaller amounts of Mg and P. Only a proportion of the elements incorporated during the growing season is retained, and that mainly in the woody plants including their underground organs (roots). To a lesser degree the herbaceous layer also plays a part. Even more nutrients are returned to the ground in the litter (i.e. dead leaves, decaying wood and other organic material).

The total amount of nutrients taken up by the plants can only be calculated approximately by adding together the amount retained and the amount returned to the soil. In the case of nitrogen this comes to 99 kg $ha^{-1}year^{-1}$, but since only 55 kg N reaches the soil again 44 kg must come from other sources, e.g. from the mineralisation of organic matter already present in the soil, from the ammonium and nitrate brought down in the rain (about 20 kg N) and by nitrogen-fixing microorganisms (see tab. 22).

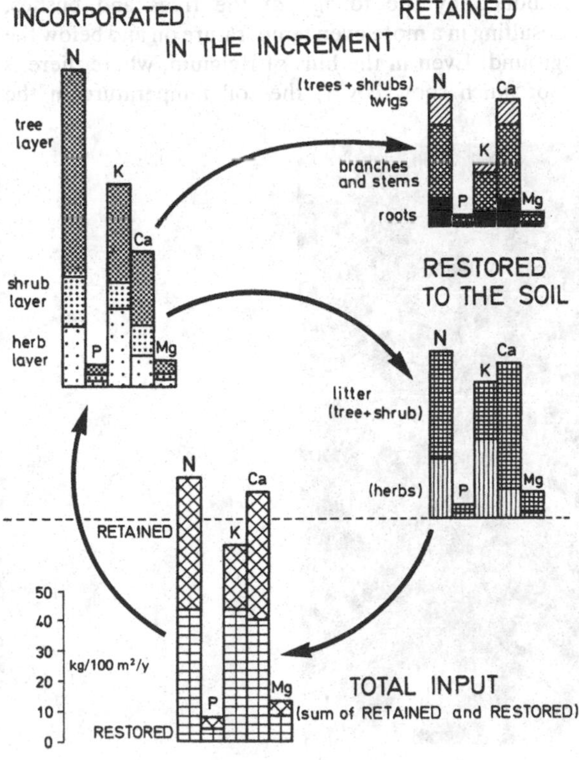

How the structure of the tree layer affects the competition of the herbaceous plants and their production requirements can best be appreciated from a 'light map' produced in summer. All the examples collected together in fig. 117 give the relative light intensity when the sky is evenly covered with cloud, so that the map is not a momentary picture of the distribution of patches of sunlight, but something near to an average illumination record. As Turner (1958) and Eber (1972) also showed the relative light intensity varies very little during the course of the day when the sky is overcast. If one compares the maps of relative light intensity measured under such conditions with the mosaic of plants present on the woodland floor, then one sees that the light intensity must be a limiting factor for them. Light-demanding, mostly tall-growing herbs, including the ferns, are found only in the clearest places; less-assuming ones like the Sweet Woodruff are content with the deeper shade. The low growing and not very competitive Wood Sorrel (*Oxalis acetosella*) is even satisfied with about 1% of the light falling on open country and is only absent in the shadow of taller herbs and dense bushes or tree saplings. Under a closed canopy even these species in relatively dark places often do not have a positive metabolic balance. Without the bonus of extra light which they enjoy in the spring before the leaves appear on the trees and in the autumn when the canopy becomes clearer they would not be able to survive (Gorshina 1975, see also figs. 46 and 118). Without competitors even *Oxalis acetosella* is favoured by a higher supply of light (Becher 1964). In the shading experiments of Packham and Willis (1977) it grew best at 27% of the full daylight and significantly less well at 6% and at 70%. In the latter case or without shade at all it is not really handicapped by too much light but by drought to which its leaves are not resistant.

If the three pairs of maps in fig. 117 are compared it can be seen how a coppice with Oak standards, a Beech wood with Oaks and a pure Beech wood differ in the amount and distribution of light they provide for the undergrowth. On the floor of the Beech 'cathedral' illumination varies by only a few percent. The distribution of light would have been even more monotonous if some individual old trees had not been removed to launch regeneration. In places Wood Melick and Wood Barley are already making use of the increase in light. The mosaic of the herb layer is somewhat more varied under the uneven canopy of the Oak-Beech wood which allows more light to reach the woodland floor. Because of this grasses and Lily-of-the-valley can become dominant so long as they are not crowded out by Beech saplings. In the Oak-Hornbeam wood with a hazel understorey the herbs can only enjoy the light coming through the tree canopy where it has not been stolen by shorter trees and bushes (fig. 116). Here the light intensity can vary by large amounts over small distances depending on the distribution of the shrub layer so that species demanding much light and those content with a little are found flourishing not far from each other. This is also true for the Hungarian Oak woods studied by Fekate (1974). One can easily imagine that the removal of firewood assists the light-requiring plants whereas if the trees and shrubs are allowed to grow undisturbed the same plants would be restricted and finally starved out.

Whereas the cutting of coppices results in a 20 to 30 year rhythm of light and shade, the grazing of woodland, which was so widespread in earlier times, created instead stable 'oases of light' which eventually progressed to become grass plots. To what degree our mixed deciduous woodlands must have been opened up in those days can be imagined by looking at the light map made by Sjörs (1954) from a Swedish wooded pasture (fig. 27). His diagrams illustrate at the same time the phenomenon well known to all photographers that the light and shade is 'harder' when the sun is shining than when the radiation is mainly diffuse. However since the shadows are moving in the course of days and seasons the amounts of light which the ground plants enjoy over a longer time are not so strikingly different. As Eber (1972) pointed out diagrams made when the sky is overcast give a quite satisfactory indication of the average light available to the ground flora.

Some very exact measurements were made in a limestone Stitchwort-Oak-Hornbeam wood, near Vireilles-Bleimont in Belgium, of the radiation during the course of the year and also throughout a number of days. Also a balance sheet was drawn up to show how the light was absorbed or reflected (Schnock 1967a b, Grulois 1968a, b, fig. 118). A large part of the energy reaching the woodland canopy is used up in transpiration. Also the purely physical evaporation of water intercepted by the woody plants from the rain and dew along with that evaporated from the leaf litter or naked soil surface accounts for a considerable amount. By comparison, photosynthesis uses very little of the radiation energy, namely about 1% of the total annual amount or less than 4% of that which could be used for photosynthesis during the growing period (see also tab. 30 and section A II 7).

Not only the light but also the temperature is subdued by the foliage of the trees and bushes, resulting in a more even temperature on and below the ground. Even in the hills of Belgium, where there is not much snow cover, the soil temperature in the

Fig. 116. Sharp differences between dark and light places on the ground in an Oak-Hornbeam wood (*Stellario-Carpinetum corydaletosum*) with a strong growth of *Corylus* in the flood plain of the Leine near Garbsen (north-west of Hannover). Nettles dominate the tall herbaceous growth to the left (summer appearance).

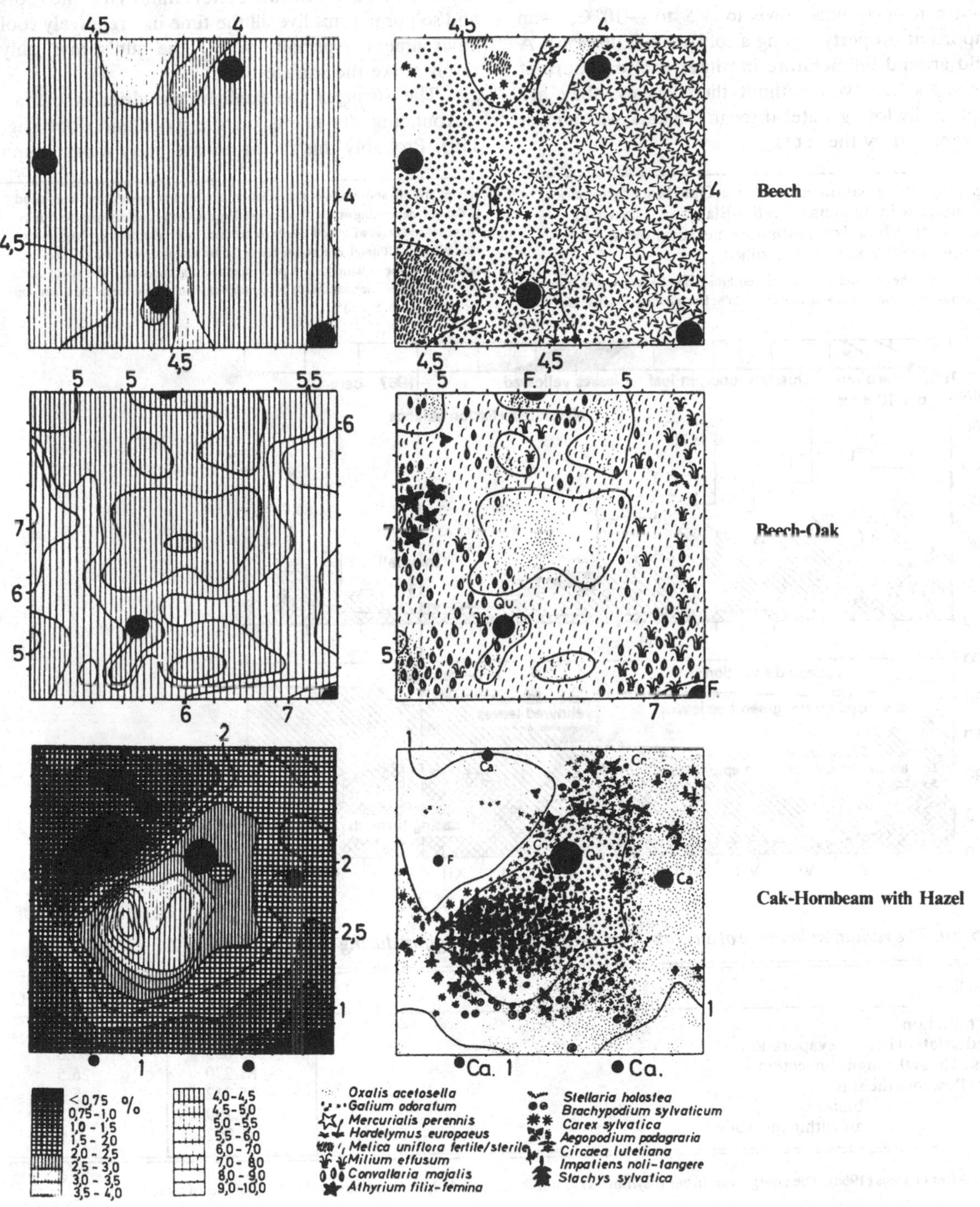

Fig. 117. The distribution of plant species compared with the light intensity (left) on 10 x 10 m test areas in a limestone Beech wood (above), a slightly-moist mixed Beech wood (centre) and a moist Oak-Hornbeam wood (below) near Hannover. After Ellenberg (1939).

The large black spots are the positions of trees these are *Fagus* (F and all the ones in the top pair of diagrams), *Quercus robur* (Qu), *Carpinus betulus* (Ca) and *Corylus* (Cr). Light intensities are shown as percentages of that in the open and are indicated round the edges of the maps. The measurements were taken around noon on overcast days at the crossing points of a metre grid and interpolated.

woods at a depth of 5 cm never falls below 0° C. even when it is much colder in the open. From this it is not surprising that the bulbs, tubers and rhizomes cannot stand many degrees of frost (fig. 109). On the other hand the tender parts of the spring plants above ground are very much more frost resistant and can survive temperatures down to −5 to −10° C, −an important property during a cold snap (Till 1956). A mild ground temperature in winter is also important for trees as, even without their leaves, they are continually losing water through the bark which must be replaced by the roots.

Most trees, in particular Hornbeam, begin to break out in spring when the soil temperature at 5 – 20 cm depth has warmed up to 6 – 7° C allowing the roots to increase their rate of water uptake (Schnock 1967a, Ellenberg 1939). In summer the soil temperature exceeds 10-15° C for only a short time. Thus the roots and soil organisms live all the time in a relatively cool environment especially where the sun's rays only rarely strike the surface directly.

The stronger the sunlight the drier is the air surrounding the plants on the woodland floor (fig. 119). Probably this factor always plays a part when

Fig. 118. The radiation balance of an Oak-Hornbeam wood on limestone in Belgium (Virelles-Blaimont), shown in ten-day totals based on continuous measurements. After Grulois (1968), somewhat modified.

A small part of the total radiation is reflected throughout the year, but only after a fall of snow is the socalled albedo more than 20%. During the growing season the canopy absorbs about 30% of the radiation, the amount varying from one period to the next depending on the cloud cover. Both in winter and summer the trunks, branches and twigs with their epiphytic lichens absorb a surprisingly large part of the radiation, although in absolute terms the amount is small in winter. The herb layer lives on the radiation which penetrates to the ground and this is about the same throughout the year, but in percentage much higher in winter than in summer (see also figs. 46, 50 and 98).

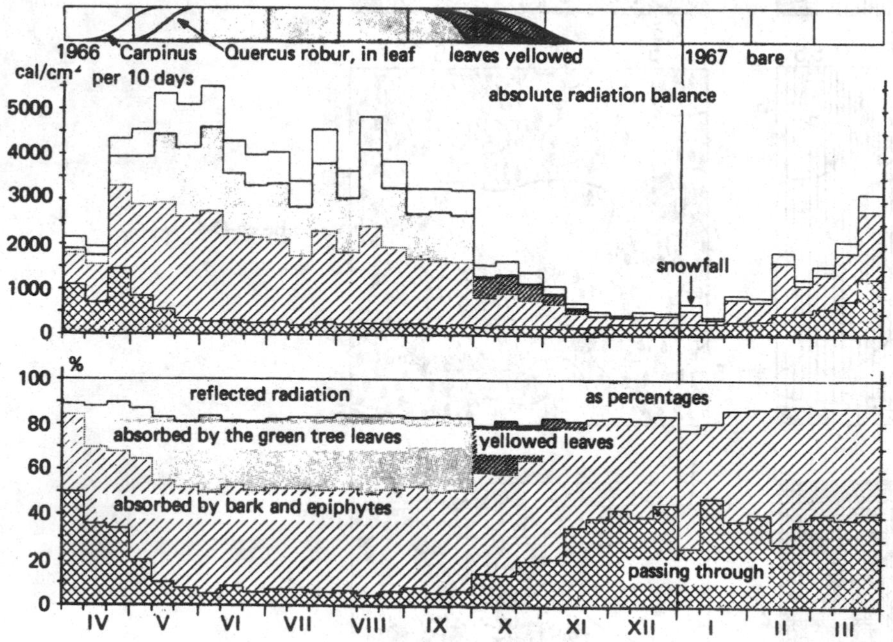

Tab. 30. The radiation balance of an Oak-Hornbeam wood in Belgium during a summer's day (12.7.1967).

Heat flow	$cal \cdot cm^{-2} \cdot d^{-1}$	as a % age of Q
Net radiation	308.543	100
Used as latent heat of evaporation	−205.053	66.6
Loss of heat through convection	− 81.270	26.5
Heat flow into the soil	− 15.312	5.1
" " " " biomass	− 3.299	1.2
" " " " air within the wood	− 1.392	0.6

Note: After Grulois (1968). The energy used in photosynthesis has not been included; this amounts to about 2% of the total radiation.

species are competing for places where the light intensity is different. The increased evaporation is at any rate partly made good by the fact that in the clearings the ground is not so densely packed with tree roots so that in dry times not so much water is removed from these areas. Small clearings are often the best habitats for susceptible plants and it cannot be by chance that here the humidity indicators are most likely to be met with. These have already been used by Tüxen (1937) to distinguish between the Oak-Hornbeam woods on damp or wet soil and those on dry soil.

(h) The behaviour of the moisture indicators

In practically all *Carpinion* communities there are subunits which indicate a high water table or waterlogging, that is a moist or wet soil. Serving as differential species indicating a better water supply are the predominantly tender-leaved plants such as Lady Fern (*Athyrium filix-femina*), Touch-me-not (*Impatiens noli-tangere*), Stinging Nettle (*Urtica dioica*), Enchanter's Nightshade (*Circaea lutetiana*) or Hedge Woundwort (*Stachys sylvatica*).

In the unusually dry late summer of 1937 Ellenberg (1939) had the opportunity to study the behaviour of such indicator plants in relation to the soil suction tension (water potential, fig. 120). The leaves of *Athyrium filix-femina, Carex remota, Festuca*

Fig. 119. As the sunlit patches move across the floor of the woodland the temperature and evaporation suddenly increase along with the relative intensity of illumination (hatched) at the herb layer. The transpiration of *Galium odoratum* (larger spots) amounts to about 2 mg per g of freshweight per min with a 'basic illumination' of 1-2%. Both are increased many times during the day. From measurements made on 22 September 1959 in a slightly-moist mixed Beech wood near Zürich.

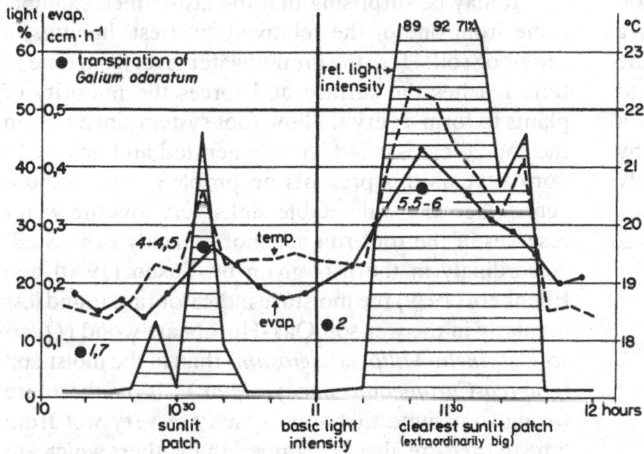

Fig. 120. The 'suction tension' (negative water potential in bar) in the soil around the roots of herbaceous plants in a moist-soil Oak-Hornbeam wood during the unusually dry September of 1937. The 'moisture indicator species' are already wilting at tensions under which species such as *Pulmonaria obscura* still appear normal. After Ellenberg (1939).

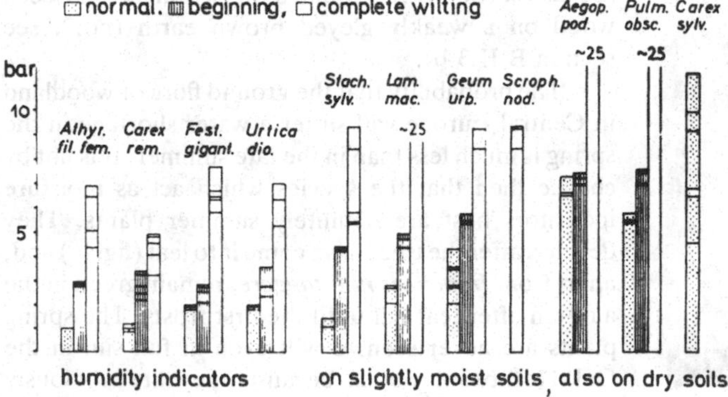

gigantea and even *Urtica dioica* started to go limp when the soil reached a water potential of –5-6 bar. whereas *Aegopodium podograria, Pulmonaria obscura* and other widely distributed woodland plants retained their normal turgidity. *Stachys sylvatica* (fig. 121), *Lamium maculatum, Geum urbanum* and other less definite moisture indicators showed an intermediate reaction between the two groups. As far as *Stachys sylvatica* is concerned these findings were confirmed by measurements of the water potential of the leaves which Rehder (1960) carried out in the dry late summer of 1959 in a Beech-rich Oak-Hornbeam wood near Zürich. The potential of the leaves on fully turgid plants was always less than –10 bar and generally amounted to only –5-6 bar. As the plants wilted the the potential was –11-12 bar and had reached –15 to more than –20 bar by the time they were completely flaccid.

In a dry period many plants on the woodland floor begin to wilt before the permanent wilting percentage (PWP) of the soil is reached. This corresponds on average to a water potential of about –15 bar. As a rule however neither the root surface area nor the cross-section of the vascular tissue are large enough to replace the water lost by transpiration during the warm hours of the day. The wilting phenomena seen by Ellenberg (1939) and Rehder (1960) then rest largely on insufficient transport capability in those plants which are not very well protected against cuticular water loss. Measurements by Pisek and Berger (1938, fig. 122) show that cuticular transpiration from cut shoots of woodland shade plants is higher in most cases than that of light-requiring ones. The highest values were obtained from *Impatiens noli-tangere* which is the most sensitive and the most strongly hygromorphic moisture indicator of all the species investigated.

Fig. 121. Individuals of *Stachys sylvatica* taken from increasingly damp sites at the time of the investigation described in Fig. 120. In the cases of the ones which are very wilted the water supply must already have been poor for some time previously.

The real moisture indicators are represented on the woodland floor in greater numbers the less often the soil dries out. In the 5 upper examples in fig. 123 which refer to stands with many of these species the soil water potential never exceeded –5 bar throughout the whole of 1937. In stands nos. 9 and 10, where the upper soil dried out for months to far beyond –3 bar they were almost or completely absent. Only *Urtica dioica* survived in no. 9 presumably because its power of regeneration is greater than that of other moisture indicators.

It may be surprising that the last-named example came from one of the relatively 'wettest' habitats, a semi-bog soil. Here the groundwater level from time to time reaches the surface and forces the majority of plants to form a very shallow root system since only in the top 20 cm is it sufficiently aerated and active. In normal years this presents no problem, but in those years when the water table sinks very low the water reserves in the root run are more quickly exhausted. Accordingly in the lists given by Tüxen (1930) and Ellenberg (1939) the moisture indicators are found less regularly in the wet-soil Oak-Hornbeam wood (*Querco-Carpinetum filipenduletosum*) than in the moist-soil (*Querco-Carpinetum stachyetosum*). It is therefore wrong to assume that soils which are very wet from time to time are always 'damper' than others which are not influenced so much by groundwater. On the contrary the safest habitats in dry years for the most susceptible plants have proved to be a mixed fern-Beech wood on a deep loess brown earth which is completely free from ground or stagnant water (no.1, see also section B II 3 c) and a moist-soil mixed Beech wood on a weakly gleyed brown earth (no.2, see section B II 3 b).

The probability that the ground flora of woodland in Central Europe will suffer a water shortage in the spring is much less than in the late summer. It is not by chance then that the species which act as moisture indicators best are definitely summer plants. They develop after the trees have come into leaf (fig.46) and, apart from *Impatiens noli-tangere*, remain green in the autumn after leaf fall until the first frosts. The spring plants are never seen to wilt even in full sun in the Oak-Hornbeam woods because they are obviously always provided with sufficient water (fig. 123). However the quickly growing geophytes are found less often on south and south-west slopes than they are under otherwise similar conditions on slopes facing away from the sun. These spring geophytes in our deciduous woodlands then can rightly be seen as indicating springtime moisture in the soil together with more or less shady sites.

(i) Animals in Oak-Hornbeam woods and other woodland ecosystems

Woodlands shelter and nourish a wealth of animal species, from large Deer down to Protozoa. Many of them prefer certain plant communities, either because they find their food plants there, or the particular climate suits them, or because they benefit from those soil properties and other habitat features which determine the species combination of the corresponding plant community. Often though most animals are not dependent on the floristic make-up as such, but on particular spatial structures, such as taller and shorter trees grouped side by side, dense undergrowth and such like. In many cases it is all the same to them whether the structures are deciduous or coniferous trees to say nothing of particular species. Very often their territory includes a number of different vegetation formations, e.g. wood and field or lake. Not infrequently the animals react to soil or climatic factors which have less significance for the plants and are therefore not expressed in the phytosociological classification with any priority.

In spite of this the phytosociological units can serve the animal sociologists as ecologically sufficiently definable areas for investigation (e.g. Rabeler 1962). The recording of a species list is certainly a good deal more difficult for them than for the botanist, because there are generally more animal species and, being motile or hidden in the soil, are not so easy to identify and count. In tab. 31 are some lists of important species which at least show that the woodland associations we have been looking at are really ecosystems and often can be quite well characterised by animal species. They have been worked out by Lohmeyer and Rabeler (1965), that is by a plant sociologist and an animal sociologist in the Weser mountains and represent only a section of much more comprehensive material. By and large the authors arranged, in order of increasing soil dampness, a

Fig. 122. Cuticular transpiration from cut shoots of shade plants (above) and light plants (below). The average of the separate measurements along the horizontal line is shown by a thicker line beneath. After Pisek and Berger (1938), somewhat modified.

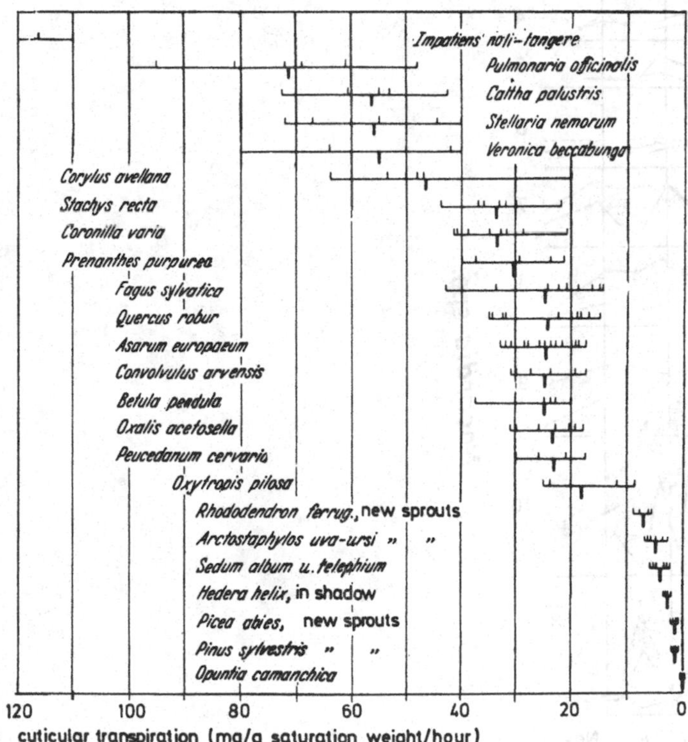

number of Beech woods, two Oak-Hornbeam woods and one thermophilous mixed Oak wood (which we would like to deal with in the next section).

Common to all these woodland communities is one group of animal species which behave like the character plants of the class *Querco-Fagetea* e.g. *Limonia modesta, Abax ater* and *Phyllobius argentatus* which certainly prefers Beech. There is no group corresponding to the order *Fagetalia*, but one which is confined to the Beech wood alliance (*Fagion*), at least as far as the animal groups investigated by Rabeler are concerned. Rather there emerges a group of differentiating species, like *Rhynchaenus fagi*, which avoids damper habitats. This group combines the Beech woods with the mixed Oak woods, while another, less drought-resistant group is common to the remaining stands. The earthworms *Allolobophora* and *Lumbricus* are well known members of this group. A few dryness- and lime-indicating species (like *Leptothorax nylanderi*) spread from the xerothermic mixed woodland to the dry-slope Sedge-Beech woods. On the other hand *Nebria brevicollis* heads a number of species which appear to be confined to the moist-soil mixed woods. The moder Beech wood (*Luzulo-Fagetum*) is relatively poor in characteristic animal species as far as one may conclude from the single example referred to. According to Funke (1973) the

Fig. 123. Fluctuations in the 'suction tension' (negative water potential) of the soil throughout the year in 11 mixed broadleaved woodlands in north-west Germany compared with the presence of moisture indicators. After Ellenberg (1939).

Thicker continuous line = upper A-horizon, thinner line with dots = lower A-horizon, dashed line = gley or B-horizon, longer dashes = values below 1 bar, dotted areas = over 3 bar (compare fig. 120).

U = Urtica dioica, I = Impatiens noli-tangere, F = Festuca gigantea, C = Carex remota, A = Athyrium felix-femina, V = Veronica montana. Large dots = present, with circle = very flourishing. Example 11 is a recently drained area. The rest are arranged in order of site "dampness".

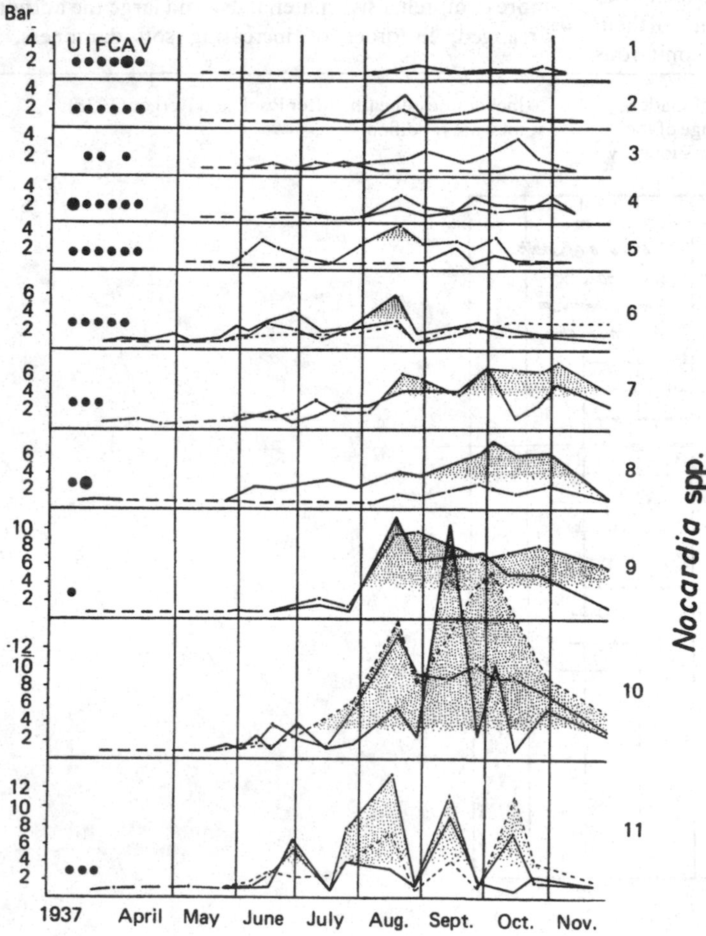

Tab. 31. Characteristic combinations of animal species in Beech and Oak woods of north-west Germany.

Animal families

Ca *Carabidae*
Ch *Chrysomelidae*
Cu *Curulionidae*
D *Dolichopodidae*
E *Elateridae*
L *Lumbricidae*
Mi *Miridae*
My *Myrmicinae*
N *Nabidae*
O *Orthoptera*
St *Staphylinidae*
T *Tipulidae*

Woodland communities		dry-slope Beech woods			mull Beech woods						Oak-Hornbeam woods				
	warmth-loving mixed Oak wood	dry-slope Blue Sesleria – Beech wood	dry-slope Sedge-Beech wood		slightly-moist limestone Beech wood		richer brown-mull Beech wood		wild Garlic-Beech wood	brown-mull Fern-Beech wood	fairly moist mixed Beech wood		moist-soil Oak-Hornbeam wood		moder Beech wood
Animal species Stand no.:	1	2	3	4	5	6	7	8	9	10	11	12	13	14	15
T *Limonia modesta* Wied.	4.4	2.2	5.9	2.3	1.4	3.16	3.36	3.5	2.4	2.4	5.18	6.17	3.11	2.7	1
Mi *Stenodema laevigatum* L.	3.13	3.18	3.9	1.2	2.2	3.16	3.30	2.13	2.5	3.4	3.7	3.6	1.3	2.5	1
Mi *Lygus pratensis* L.	2.4	4.6	2.2	2.6	1	3.3	2.3	2.6	3.8	2.7	2.4	1		2.2	1
N *Nabis pseudoferus* Rem.	2.2	3.3	4.4		1	3.4	1	1	1	1	2.3	1.3		1	1.4
Ca *Abax ater* Vill.	2.2		2.2	1		3.4	1	2.2	2.2	1		1	1.2		2.2
Cu *Phyllobius argentatus* Mg.		1	3.8	1.2	1	4.19	3.5	5.14	3.16	3.12	5.9	3.3	2.5	1.4	2.3
Cu *Rhynchaenus fagi* L.	4.6	5.11	6.25	2.6	4.11	2.5	1.5	6.59	7.89	3.42	2.2		1		4.55
Ca *Abax ovalis* Dft.	3.3	1.3		2.2	2.4	3.5	2.2	1	3.5	2.3					1
E *Athous haemorrhoidalis* F.	1		2.2	1.2	3.7	2.5		1	3.8			1			
Ca *Pterostichus madidus* Fbr.	3.5	2.2	2.2	3.4				1	4.6	1					1
T *Limonia maculata* Mg.		3.4	2.2		3.3	4.9				1.4					
Ca *Molops elatus* Fbr.			1	2.2	1	1.3		1							
E *Agriotes pilosus* Panz.			1	1	1		1								
E *Athous vittatus* Fabr.					1	5.11	1	1.3	2.6	2.3	4.24	3.3	1.2	2.8	1
T *Tipula scripta* Mg.					1	2.2	1.2			1	2.2	2.2	1	2.2	2.2
L *Allolobophora caliginosa* (Sav.)			1		3.4	5.10		2.14	3.6	4.16	6.17	4.4	5.13	1	
Cu *Apion flavipes* Payk.					1	3.4				2.2	1	1	1		
E *Athous subfuscus* Müll.						1	2.4	2.5	2.19	3.3	1.2	1.4	1.2		3.14
St *Philonthus decorus* Grav.						3.3	1	3.3	2.4		8.15	4.7	4.4	2.3	
L *Lumbricus rubellus* Hoffm.							2.4	3.3	1	2.2	6.13	3.6	4.11	3.9	3.4
Ca *Pterosticlins oblongopunctatus* F abr.							3.7	1	1	1		5.8	1	2.6	2.2
My *Leptothorax nylanderi* (F.)	8.88	3.49	3.9	6.80											
E *Limonius parvulus* Panz.	2.2	2.2	1	1											
Cu *Brachysomus echinatus* Bonsd.	1.2	1	1	1.3											
St *Anthobius sorbi* Gyll.	2.5		2.5												
O *Ectobius sylvestris* Poda.	2.2			1.2											
Ch *Hermaeophaga mercurialis* F abr.	1			1	4.12	3.8		2.2							
E *Agriotes acuminatus* Steph.					2.3	1		1		2.3	1				
Ca *Trichotichnus laevicollis* St.							2.2	4.7	2.14	3.6		1			1
Ca *Pterostichus metallicus* Fbr.							1		3.3	2.2		1			1
Ca *Nebria brevicollis* Flor.											1.2	2.2	2.2	1	
Mi *Phytocoris longipennis* L.											1	1	2.2	3.12	
D *Campsicnemus curvipes* Wied.											1	1	1	1	
Cu *Coeliodes erythroleucus* Gm.											2.2	1	1		
Mi *Cyllecoris histrionicus* L.										1.2	1	1	1		
Cu *Ceutorrhynchus rugulosus* Hbst.				1							2.2		1	1	

corresponding community on Solling has a very similar group of species.

So one can distinguish the woodland associations, and also some subassociations not only by their flora, but also by their fauna. The differentiating animals however do not form groups which are exactly parallel to those of the plants, but in our series of examples they react more strongly to the water factor than to the tree species present. Since we are dealing mostly with animals living in the soil for the whole or most of their lives this is perhaps not surprising. Many of them remain in the leaf litter as imagos, e.g. *Philonthus decorus*. Other adult beetles prefer the herb layer (e.g. *Apion flavipes*), or the shrub layer with its tree saplings, e.g. *Athous subfuscus* and *A. guttatus* and the rather eurytopic *Limonia modesta*.

Table 31 just affords a glimpse into a wide field of study. In the sociology of animals too the first step is the recording of species lists and their comparison in order to examine the differences in species combinations, thus helping to prepare for a deeper ecological understanding.

4 Xerothermic mixed Oak woods

(a) 'Relict' submediterranean Downy Oak woods and continental steppe woodlands

The development centres of the so-called xerothermic or thermophilous mixed Oak woods (*Quercetalia pubescenti-petraeae*) as well as their refuges during the Ice Ages lie outside Central Europe in the submediterranean zone of the Balkans (Horvat, Glavač and Ellenberg 1974) and southern France (Braun-Blanquet, Roussine and Nègre 1951) or even further to the south. After the Birch-Pine and the Pine-Hazel eras its representatives wandered back into our area. In the Post-Glacial warm period they extended their range onto almost all soils and climbed higher into the mountains than today (Walter and Straka 1970). However as soon as Spruce, Hornbeam, Fir, Beech and other shade or half-shade trees conquered the land they forced the open mixed Oak woods back onto less favourable dry places on which not even the Oak-Hornbeam wood could find a footing.

Quercus pubescens and other character species of the thermophilous mixed Oak woods are less resistant to cold than most of the woody plants in the temperate parts of Europe (Larcher and Mair 1968, and others). Moreover Downy Oak reaches its optimum photosynthesis at relatively high temperatures (fig. 141). Thus, the term 'thermophilous' seems to be truly adequate as well as the term 'xerothermic' which also infers a resistance to drought.

In Central Europe today we only find the xerothermic mixed Oak woods on isolated dry sunny slopes on soils which are generally shallow, but rich in lime or bases (fig. 124). The Upper Alsatian Rhine plain to the south of Colmar which has an exceptionally low rainfall is a possible exception insofar as Downy Oak here also flourishes on the flat lowlands (Issler 1937, 1942). However the ground it occupies is very free-draining limestone rubble and it gives place to Oak-Hornbeam woods where there is a deeper loam soil. So here too it is an extrazonal 'permanent community' and not a zonal climax as it is in the sub-mediterranean region, e.g. in the northern and central Provence (France).

The mixed Oak woods dominated by *Quercus pubescens* are often considered ideal examples of 'relicts' in the history of vegetation. Under their open canopy a colourful assembly of submediterranean and mediterranean varieties and even occasionally a south-east European element can be found. Such places have been centres of pilgrimage for botanists for many years. The word 'relict' however should not

---

Notes to Tab. 31

Note: After Lohmeyer and Rabeler (1965), modified. The number in front of the dot indicates the number of times the species was found out of a maximum of 8 samples per year. The number after the dot is the highest count of individuals. A single 1 indicates that one specimen was found on one occasion. Many additional species have been omitted.

- No. 1: Blue Gromwell-Oak wood (*Buglossoido-Quercetum* = "Querco-Lithospermetum" near Beverungen (eastern Westphalia); north-west German, impoverished form.
- Nos 2 - 4: Dry slope Beech woods (*Carici-Fagetum*); N.W. German, impoverished.
- 2: Slope Blue Moorgrass-Beech wood ("*C.-F. seslerietosum*") near Höxter.
- 3 & 4: Dry-slope Sedge-Beech wood ("*C.-F. typicum*") near Beverungen.
- Nos 5 - 10: Mull Beech woods (*Melico-Fagetum sens lat.*)
- 5 & 6: Slightly-moist limestone Beech wood ("*M.F. elymetosum*") near Beverungen
- 7 & 8: Rich brown-mull Beech wood ("*M.F. typicum*") at Rehburg (near Lake of Steinhude) and Neuenheerse (E. Westphalia)
- 9: Wild Garlic-Beech wood ("*M.F. allietosum*") near Herste (E. Westphalia)
- 10: Brown-mull Fern-Beech wood ("*M.F. dryopterideto-sum*") near Neuenheerse
- Nos 11-14: Oak - Hornbeam woods sens. lat. (*Stellario-Carpinetum*) on damp soils
- 11 & 12: slightly-moist mixed Beech wood (Beech-rich Oak-Hornbeam wood, "*St.-C. asperuletosum*" near Wiedensahl and Rehburg
- 13 & 14: Moist soil Oak-Hornbeam wood (*St.-C. athyrietosum*) in the area between Minden and Lake of Steinhude.
- 15: Moder Woodrush-Beech wood (*Luzulo-Fagetum*) near Neuenheerse

deceive us into thinking us into thinking that all the woods in the Post-Glacial warm period would have looked like these twisted patchy growths whose floristic wealth is in inverse ratio to their productivity and economic value. They occupy extreme habitats which even in Post-Glacial times could only carry a stunted growth whereas the woods on the better soils soon became so dense as to bring to a halt the massive spread of Hazel. It would be wrong to assume that Downy Oak at one time was dominant over the whole area up to the borders of its present day distribution, for even in the Post-Glacial periods the other two Oaks as well as Limes and Elms were also widespread and on good soils they would surely have been able to outgrow the slower Downy Oak whose strong card lies in its drought resistance.

In some regions it is possible that man has contributed to the maintenance and spread of less-competitive species. In any case the majority of 'steppe-heath woods', as they were once called by Gradmann (1898, 1950), lie in long-settled countryside. For many centuries in southern France, Jugoslavia and other distribution centres of the thermophilous vegetation, and also in Central Europe, they have been used as coppices. Or they were burned off from time to time in order to open up more grazing for sheep, goats and other farm animals. Rühl (1960, 1973) considers that all outposts of the thermophilous Oak scrub on limestone slopes in north-west Germany, which are relatively poor in species, are even degraded Beech woods. However there were at one time here too natural open sun-warmed scrubland on rocks, steep slopes and other inaccessible places, which have served as centres of dispersal for secondary xerothermic vegetation.

During the Post-Glacial warm period the species of the submediterranean mixed Oak woods were moving round the Alps, from southern France and the Balkan peninsula, many of them perhaps also using the lower passes to reach the northern side. In the same period steppes and steppe woodlands of south-eastern and eastern Europe were extending into Central Europe. Their communities too consist of plants which require a lot of light to flourish properly, but which are able to tolerate long dry periods. In their important ecological features then they are very like the species

Fig. 124. Xerothermic mixed Oak wood and 'Steppe-heath' (xerothermic open grassland) at the Lopper hill SE of Lucerne. According to the depth of the fine earth cover, the south-facing limestone slope is covered with lichens, mosses and isolated grassland species or with shrubs and stunted trees (mainly Oaks). Along the edge of the woody stands a 'fringe' of herbs and grasses has developed here and there.

of submediterranean origin except that they may be generally less susceptible to cold. Many of these continental elements have succeeded in reaching the Rhine valley, even into the Wallis and southern France, mixing there with the submediterranean plants. On the southern border of the Harz, on the Kyffhäuser (fig. 17) and in many areas lying to the south and east of these the flora of sunny slopes have very much the character of a steppe woodland, to which Meusel (1935, 1939) has emphatically drawn attention. This is even more strongly marked in the inner Alpine valleys where the xerothermic Pine woods, according to E. Schmid (1936), are made up to a large extent of elements of the 'Pulsatilla wood-steppe zone' (see also Braun-Blanquet 1961).

However it would be better in such a case to speak of a '**steppe-wood**' rather than a 'wood-steppe' (Horvat, Glavač and Ellenberg 1974). The term steppe woodland has somewhat the same meaning as xerothermic mixed woodland'. The Soviet geobotanists describe the transition zone between the closed woodland in northern Ukraine and the open steppeland in the south as 'wood-steppe' which in their sense is a park-like, very open landscape (Walter 1974). According to Funk (1927) and later authors this is in part brought about by the activity of man and his animals in eastern Europe also. The minimal amount of water required for transpiration by a closed woodland in a dry area is said by Zelniker (1968) to be 100 mm from a precipitation of about 400 mm. Woodland without open spaces can exist only if this amount is assured. This means by the way that the general climate of Central Europe today permits the growth of woodland throughout, even in its driest lowlands.

The 'islands of wood-steppe' which Meusel (1939), Wendelberger (1954) and others have described from the middle to the south-east of Central Europe, and the 'steppe-heaths' of the Swabian and Franconian Jura of which Gradmann (1898, 1950) has given such a matchless acount, form a mosaic of small areas of open dry grassland, scrub and open woodland (figs. 16, 17, 124, 125). In all three formations there is a more or less rich element of wood-steppe plants, i.e. species which occur in the eastern European transitional climate between cold woodland and steppe. These species live today however in quite different plant combinations in our steppe heaths. The habitat conditions are also different from those they meet there and have met in the Post-Glacial warm period.

With regard to the type of soil too the steppe heaths and the 'islands of wood-steppe' in Central Europe have nothing in common with the Ukranian

Fig. 125. Wintertime in the Pollau mountains (Bohemia) with open xerothermic woodland and more or less stony dry grassland. The snow cover is thin and not permanent; it offers no protection against the frost.

steppes. The dry part of eastern Europe is dominated by deep-humus black earths (chernozem) which are the most fertile natural soils in Europe. Degraded remains of them are certainly found in Central Europe, e.g. in the dry area to the east of the Harz, in the Vienna basin, and finally towards the west to the north-east of Stuttgart, the north-west part of the Upper Rhine plain, and in the foothills of the mountains near Hildesheim. Today dense woodland would be growing on these black earth remains if the farmers allowed enough room for it. After the Ice Ages the woods had actually returned here and were present when man began to interfere. This was confirmed by Havinga (1972) for the pannonic part of southern Austria too, i.e. for the part of Central Europe which is actually most like the steppes climatically. Today the water available in the loess and other deep soils of the Vienna area is without question sufficient for the growth of trees (Gruber 1973, see fig. 126).

The 'steppe-heaths' of our limestone hills, the fragments of scrub and grassland on the gypsum slopes of the Kyffhäuser, and other 'wood-steppes' of Central Europe on the other hand occupy the most shallow, dry soils to be met with in this area. Most of them are poorly developed rendzina and pararendzina or ones which had been 'beheaded' by erosion (fig. 124), where the thin crumb on steep sunny slopes dries out very badly from time to time. Where however fine earth has collected between the rocks or where there are deeper clefts the shrub or tree roots can find sufficient water to survive the rainless periods (fig. 17). According to Milthorpe (1975) the smaller plants are the first to suffer from a water shortage and can die while the growth and transpiration of the deep-rooting woody plants can go on almost without hindrance. Open dry grassland then is only able to survive naturally in Central Europe in those places where the soil is too shallow for the growth of trees and bushes. Here they are never, or hardly ever, determined in the first place by the climate as they are in the eastern steppes. Quite probably most of the dry grasslands in Central Europe under natural conditions would not be able to hold on to the places where we actually find them, and would occupy smaller areas than they do today. Wherever sheep or goats and their shepherds could and can gain access they extend the natural clearings, but rocks and steep slopes which are inaccessible are even today surprisingly thickly covered with bushes or even trees, e.g. many of the limestone cliffs where the Danube cuts through the Swabian Jura.

Fig. 126. In the dry region south of Vienna without Beech the water content of a loess brown earth under an Oak-Hornbeam wood sinks below the permanent wilting point (PWP) from July or August onwards, mainly in the subsoil (40-110 cm). However in spring there is enough water available for the trees, and the topsoil is sometimes moistened by heavy rain even in summer. Taken from Gruber (1973), modified.

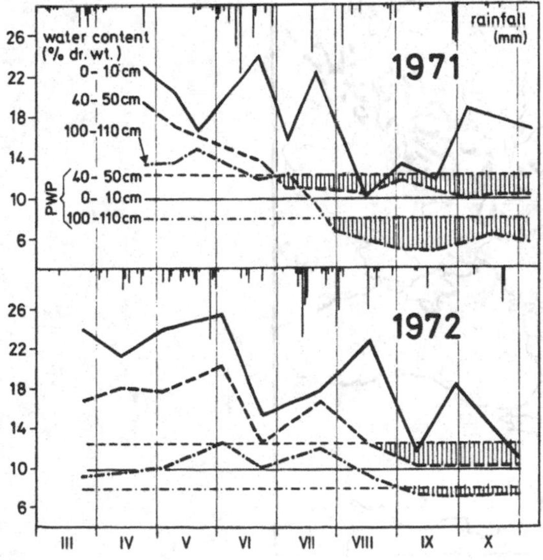

Since the grazing of woodland and coppicing have stopped many 'steppe-heaths' have become more and more closed in so that the groups of trees have become taller and darker and the shrubs have advanced onto the grassland areas. The living space of the light-hungry plants thus becomes ever smaller and nature conservation is here faced with the paradox that it is only possible to preserve floristic treasures if they are protected against the natural succession. According to Medwecka-Kornaś (1960) even in many 'steppe reserves' in Poland the shrubs spread as soon as they are placed under protection. However there are some exceptions, e.g. the 'steppe heaths' at the Hohentwiel near Lake Constance, in which the mosaic of scrub and grassland has remained stable from time immemorial. This might be due to variations in the depth of soil, or to grazing by deer and other animals, or to other reasons yet unknown (Th. Müller 1966).

(b) Xerothermic mixed Oak woods in an east-west gradient of climate and flora

Whatever one may think about the origin of the 'steppe-heaths' or 'wood-steppe islands' and especially how natural the open mixed woodlands in them might be, they represent in any case floristically the most stimulating and richest habitats in the whole of Central Europe. No group of woodland communities is so peculiar, in the truest sense of the word, as the thermophilous mixed Oak woods. They are usually put together in the order *Quercetalia pubescenti-petraeae* or in a class with the corresponding name (Jakucs 1961). These units all have numerous character species (tab. 32 and section E III 1, No. 8.42).

At present the centre of distribution of the warmth-loving mixed Oak woods lies in south-east Europe, where Downy Oak (*Quercus pubescens*, fig. 127) is also plentiful and takes part in the formation of many woodland communities. Here a more continental climate than that of Central or of south-western Europe is dominant (Horvat, Glavač and Ellenberg 1974).

The Central European thermophilous woodlands mostly belong to the alliance *Quercion pubescenti-petraeae*, whose main range is in southern France. On the calcareous soils of the southern Alpine valleys, which are warm but with a high precipitation, are to be found the Manna Ash-Hop Hornbeam woods which are outposts of the Illyrian *Orno-Ostryon*. Because this alliance occupies such a restricted area in Central Europe we shall not pursue it in any more detail (literature in Horvat, Glavač and Ellenberg 1974). The remaining alliances are all south-eastern and eastern European and must stay quite outside our discussion here. Within the *Quercion pubescenti-petraeae* five groups relevant to Central Europe can be distinguished:

1 Strongly **submediterranean – subatlantic** in character and extending into the south-western Alpine foothills (e.g. the *Buxo-Quercetum* in western Switzerland.)
2 True **Central European** xerothermic mixed Oak woods on sunny slopes more or less rich in bases, or similar places which are too dry for Beech (e.g. the so-called *Lithospermo-Quercetum* = *Buglossoido-Quercetum*),
3 **Subcontinental** mixed Oak woods on relatively base-deficient, dry, mostly level soils towards the east of Central Europe, leading into the strongly acid-soil Birch-Oak woods (e.g. the *Potentillo albae-Quercetum*),
4 **Inner Alpine** with a continental flavour, mixed woods mostly dominated by Pines (e.g. the *Pino-Cytisetum*),
5 **Southern Alpine** (e.g. Insubrian mixed Oak woods which will not be looked at any further, see Antonietti 1968).

In contrast to the other units the *Cytiso-Pinetum* has only local significance and within its range only occupies small areas (Braun-Blanquet 1961); it can be passed over here too. In order to understand the second group properly it must be compared with the

Fig. 127. European distribution of mixed Oak woods which can tolerate drought periods and are more or less warmth-loving. They are most frequent in the south and east, i.e. outside Central Europe. From Jakucs (1961), slightly modified. In the north and mainly in the northeast *Quercus pubescens* is lacking, and the entries do not belong to the order *Quercetalia pubescentis*.

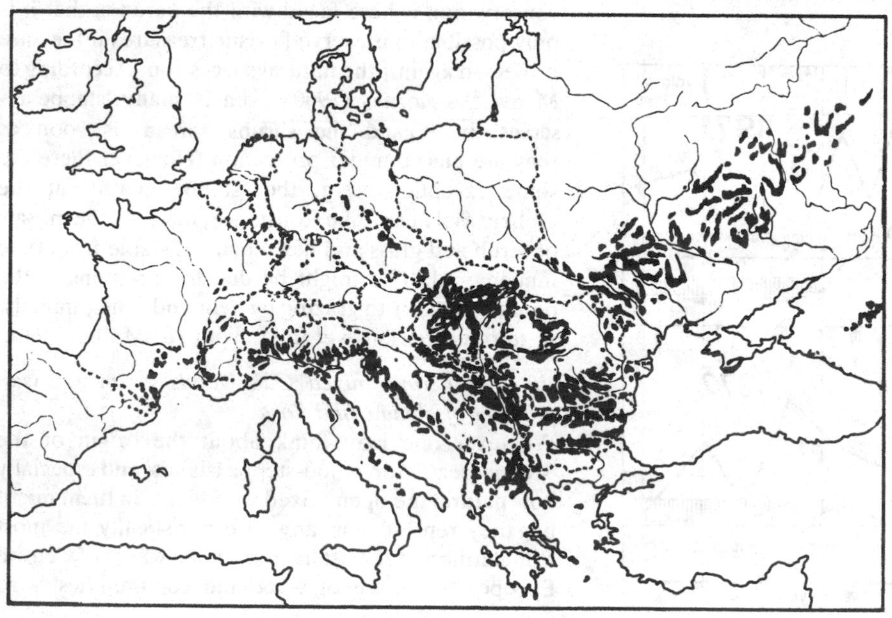

Tab. 32. West-to east gradient in the species composition of xerothermic mixed Oak woods from southern France to central Poland.

Community no.:	1	2	3	4	5	6	T	K
Trees and shrubs:								
OQ *Quercus pubescens*	5	3	5				8	3
Q. petraea		5	3	5	5	4	6	2
Q. robur		2		2	2	4	6	×
Pinus sylvestris		3			2	5	×	7
A C *Buxus sempervirens*	5						8	4
Cytisus sessilifolius	4						8	3
Lonicera etrusca	3						9	×
OP *Cotinus coggygria*	3						8	4
C *Ribes alpinum*	4						4	4
VQ *Acer opalus*	5	1					8	4
OF *Tilia platyphyllos* (t)	2	4					5	2
OP *Clematis vitalba* (Liana)	2	3					7	3
B C *Amelanchier ovalis*	5	2	2				7	4
C *Coronilla emerus*	4	3	3				6	4
OQ *Sorbus domestica* (t)	4	1	3				7	4
OP *Prunus mahaleb*	3	1	1				5	4
C *Acer campestre* (t)	2	3	4	5			7	4
OF *Fraxinus exclesior* (t)	1	3		4			5	3
C *Crataegus* spec.	1	5	5	5			×	×
C *Sorbus aria* (t)	5	5	5		1		5	2
OP *Viburnum lantana*	4	5	5		4		5	2
OP *Cornus sanguinea*	2	5	5	5	3		5	4
C *Lonicera xylosteum*	3	4	4	4	5		5	4
OF *Fagus sylvatica* (t)	3			4	2		5	2
DOQ *Sorbus torminalis* (t)	3	5	4	5	4	1	6	4
OP *Ligustrum vulgare*	2	5	5	4	4	1	6	3
OP *Berberis vulgaris*	·	3	3		1	1	6	4
C *Corylus avellana*	4	4	1	5	4	3	5	3
OP *Rhamnus catharticus*	3	3	4	2		2	5	5
OP *Prunus spinosa*	2	3	1	3	4	2	5	5
Juniperus communis	3	5	1	·	1	2	×	×
E,F *Pyrus communic* (t)		4	1		1	3	6	5
OF *Carpinus betulus* (t)		2		3	4	3	6	4
OP *Rosa canina*			4	5		2	5	3
C *Malus sylvestris* (t)				4		2	5	3
H [a]*Sorbus aucuparia* (t)						4	×	×
Betula pendula (t)						3	×	×
[a]*Frangula alnus*						3	×	5
Widespread herbaceous plants:								
d *Primula veris*	2	3	5	5	2	3	×	3
OQ *Campanula persicifolia*	3	5	3	5	3	5	5	4
T *Polygonatum odoratum*	5	4	3	4	3	4	5	5
OQ *Lathyrus niger*	2	2		3	5	4	6	4
T *Peucedanum cervaria*	2	5	1		3	2	6	4
T *Geranium sanguineum*	2	5	2	1	5	3	5	4
C *Hypericum montanum*	4	5	·		2	4	6	4
T *Trifolium alpestre*	3	3	·		4	4	5	4
OQ *Melittis melissophyllum*	5	5	3			1	6	2
T *Trifolium rubens*	3	4	·			1	6	4
T *T. medium*	2	5	·		2	3	5	4
T *Clinopodium vulgare*	3			·	3	4	5	3
Brachypodium pinnatum	1	3	5		5	2	5	5
Poa nemoralis	2	1		2	4	2	×	5
Betonica officinalis	3		1	2	3	4	6	5

Community no.:	1	2	3	4	5	6	T	K
Widespread herbaceous plants:								
OF *Hieracium sylvaticum*	4	2			4	3	×	3
Convallaria majalis	2	2	1	3	3	4	×	3
Fragaria vesca	4	3	1	2		5	×	5
OF *Viola reichenbachiana*	3	1		4	3	3	5	4
OF *Lathyrus vernus*	1	4		3		2	×	4
Differential herbaceous plants:								
a T *Lathyrus latifolius*	3						8	4
Rubia peregrina	3						9	3
OP *Coronilla coronata*		3					6	4
T *Thesium bavarum*		3					6	4
OF *Euphorbia dulcis*	4	3	·				5	2
OQ *Arabis pauciflora*	3	1	·				7	4
b *Teucrium chamaedrys*	4	4	1				6	4
AQ *Helleborus foetidus*	5	3	2				6	2
T *Inula conyza*	1	2		4			6	2
C *Hepatica nobilis*	2			4			6	4
OF *Stellaria holostea*	3			4			6	3
OF *Melica uniflora*	3			5			5	2
C *Campanula trachelium*	3		·	5			5	3
cOQ *Tanacetum corymbosum*	5	5	3	4	2		6	5
Hedera helix	3	5	5	2	3		5	2
C *Buglossoides purpuroc.*	2	2		4	2		7	4
T *Bupleurum falcatum*	1	1		3	1		6	6
OF *Galium sylvaticum*		2		4	1		5	4
e *Vincetoxicum hirundinaria*		4	3	4	1	2	5	5
T *Viola hirta*		3	5	5	4	1	5	5
T *Origanum vulgare*		3	1	4	2	2	×	3
Euphorbia cyparissias	·	3	·	3	4	3	×	4
C *Melica mutans*	·	3	·	3	1	3	×	3
C *Brachypodium sylvaticum*		3		4		2	5	3
C *Viola mirabilis*	·	3	·	3		1	5	4
C *Carex montana*		2	4	4	4	3	5	4
T *Silene nutans*		2	·	3	3	2	5	5
T *Anthericum ramosum*		3	1		4	3	5	4
Solidago virgaurea		4			3	4	×	×
[a]*Melampyrum pratense*		2			2	3	×	3
Carex humilis		3	1		2		5	5
f T *Fragaria viridis*			·	4	1	1	5	5
Arabis hirsuta				4			5	3
OF *Dactylis polygama*				4	4		5	4
Fallopia dumetorum				4		2	5	4
Dactylis glomerata				3	1	3	×	3
gAQ *Potentilla alba*		1	·		3	4	6	5
AQ *Pulmonaria angustifolia*					5	2	6	4
Ranunculus polyanthemus					2	3	×	5
Serratula tinctoria					4	3	6	5
Anthoxanthum odoratum					5	3	×	×
Festuca ovina					4	4	×	×
Luzula pilosa					3	4	×	3
OF *Festuca heterophylla*					3	2	5	4
Veronica chamaedrys					1	5	×	×
[a]*V. officinalis*					2	4	×	3
[a]*Pleurozium schreberi* (M)					1	4	×	×
OF[a] *Calamagrostis arundinacea*					1	3	5	4

Tab. 32 (Continued).

Community no.:	1	2	3	4	5	6	T	K
h s *Pteridium aquilinum*						4	5	3
Achillea millefolium						4	×	×
Pimpinella saxifraga						4	×	5
Vicia cassubica					.	3	6	4
No. of tree and shrub species in each stand	20	19	14	13	10	10		

Community no.:	1	2	3	4	5	6	
Constancy summations							
mixed Oak woods	40	33	23	17	22	17	OAQ
'noble' broadleaved woods	24	21	0	35	30	16	OF
broadleaved woods and scrub	37	37	23	42	16	18	C
scrub	29	28	31	19	16	8	OP
herbaceous skirts	25	38	13	31	35	30	T

No. 1: **Box-Downy oak scrub** in the northern part of southern France (*Buxo-Quercetum*). After Braun-Blanquet et al. (1951, p. 247, "*Querco-Buxetum cotinetosum*")

Nos. 2–4: **Blue Gromwell-mixed oak wood** in relatively oceanic regions of Central Europe (*Buglossoido-Quercetum*?)

2: in North Switzerland. After Braun-Blanquet (1932, p. 16, "Lithospermo-Quercetum").

3: in south Alsace. After Issler from Oberdorfer (1957, p. 534, "*Lithospermo-Quercetum collinum*". The name is misleading since *Buglossoides purpuro-caerulea* (=*Lithospermum p.*) is absent from the area).

4: In southern North-west Germany. After Tüxen (1937, p. 138, "*Lith.-Querc.*").

Nos. 5 - 6: **Cinquefoil-mixed oak wood** in relatively continental regions of Central Europe ("*Potentillo-Quercetum*")

5: in the northern part of the Upper Rhine plain. After Oberdorfer (1957, p. 530, "*Pot.-Querc*").

6: on the Polish plain. After W. and A. Matuszkiewicz (1957, p. 45, "*Querco-Potentilletum albae*").

C = class character species (*Querco-Fagiea*). OF = character species for the noble broadleaved woods (*Fagetalia*). OQ and AQ = character species for the order and alliance respectively of the warmth-loving mixed Oak woods (*Quercetalia* and *Quercion pubescenti-petraeae*), OP = character species of scrub and woodland edge bushes (*Prunetalia*), T = character species for the herbaceous 'skirt' to the scrub and woodland (*Trifolio-Geranietea*, see section D IV), s = acid indicator *sens lat*.

Each "**constancy total**" is obtained by adding the constancy figures for one group of character species in each column of the table. (Because of the absence of a few constant species through lack of space these summations differ slightly from the true totals, but the figures calculated for the six communities do show the real relationship between them.) No. 1 is especially rich in "mixed Oak wood plants" (AQ, OQ) but also contains species which need a good deal of light and are favoured by cutting and grazing (OP, T). In no. 4 species of the noble broadleaved woods already preponderate because the climate, close to the northern boundary of the xerothermic Oak woods, is already too damp and cool. On the other hand these are completely absent from no. 3 largely because of the low rainfall in the rain shadow of the Vosges. The Cinquefoil-Oak wood on dry level ground contains a few acid indicators.

Column **T** = temperature indicator value (see sections B I 4 and E III): 9 (very warm indicator), through 8 to 7 (warm), through 6 to 5 (temperate). Column K = continentality value: between 7 (relatively continental distribution) to 2 (quite strongly oceanic).

A - H groups of woody differentiating species, **a - h** parallel groups of herbaceous plants. The number of woody species occurring in each of the records has been calculated from the original records. It falls from 20 in the *Buxo-Quercetum* to 10 in the *Potentillo-Quercetum*.

Buxo-Quercetum and the way it has developed in the optimal and original range of our Downy Oak copses in southern France.

In tab. 32, column 1, there is a list which was put together by Braun-Blanquet, Roussine and Nègre (1951) for the southern Valentinois. The Downy Oak communities, rich in Box, are all coppiced here at 15-30 year intervals; if left undisturbed they would grow up into high forest.

The *Buglossoido-Quercetum*, like the *Buxo-Quercetum* is very rich in frost-susceptible woody species. This applies especially to the sunny slopes of the Swiss Jura which lie next to the centre in southern France (col. 2 in tab. 32). This warmth-loving Oak scrub is the only community in Central Europe which is reminiscent of the tropics at least with its wealth of woody species. But already Box (*Buxus sempervirens*), *Lonicera etrusca* and other evergreens are absent. These, in the typical *Buxo-Quercetum*, remind us physiognomically of the Mediterranean evergreen woodlands. Apart from the copses *Buxus* also appears, quite isolated, on steep limestone slopes in the Swiss Jura.

The further we advance from the Swiss Jura towards the north and north-east the poorer becomes the *Lithospermo-Quercetum* both in trees and shrubs and also in herbaceous character species of the alliance and the order of the thermophilous mixed Oak woods (cf. cols. 1-6 in tab. 32) Their loss is not made up for by the entry of the few new species with an eastern distribution. In upper Alsace even *Buglossoides* is absent; it is already too dry for it there according to Issler (1942). Certainly the name '*Querco-Lithospermetum*', coined by Braun-Blanquet (1932) was later shown not to have been a very happy one since the Blue Gromwell is also present in other woodland communities and is best developed in dry Oak-Hornbeam woods (see also Ellenberg and Klötzli 1972). For all that this widely accepted scientific name gives a correct ecological conception of these already 'temperate' warmth-loving Oak woods with their many Central European and a few eastern elements.

On the steep sunny slopes of some limestone hills to the left of the Rhine the community gradually dies out northwards, but, as Schwickerath (1958) shows in his survey, it still hangs on towards the west, dovetailed into the submediterranean atlantic *Buxo-Quercetum*. These undertones stop completely in southern Lower Saxony and in the adjoining regions (Förster 1968a and b); *Quercus pubescens* too is hardly seen here (col. 4 in tab. 32). Looking at the plant stands which are so much poorer floristically one can question whether they have anything to do with a *Lithospermo-Quercetum* in its original sense. However the connection is well established with an unbroken chain of intermediates (see also Hartman and Jahn 1967 and H.D. Knapp 1979/80). The same is true for the *Potentillo-Quercetum*.

Far away in the north-east of Central Europe there is the quite isolated subspontaneous appearance of Downy Oak at Bjelinek on the edge of the original course of the Oder 60 km south-west of Stettin. W. and A. Matuszkiewiecz (1956b) have put this shrub community, which includes *Quercus pubescens* as well as the other two Oaks, into a special association which they call '*Querco-Lithospermetum subboreale*'. It appears to be just an impoverished outlier of the submediterranean communities like the stands in north-west Germany, but at the same time it is richer in many species which appear in the steppe meadows of east Europe. Today it is clear that those Downy Oaks were introduced into the region by man (see section 4 a).

While the submediterranean accent of the mixed Oak woods falls off from south-west towards the north-east and can scarcely be detected in the northern part of Central Europe, the continental undertones increase from west to east. Already on the plains of the Upper Rhine and in Franconia one comes across the White Cinquefoil (*Potentilla alba*) and other species with a subcontinental distribution which are peculiar to mixed Oak woods in eastern Central Europe and in Ukraine.

It is remarkable and hard to understand that there are still no detailed investigations into the environmental conditions of xerothermic mixed Oak woods in Central Europe. One may mention, however, some informative measurements from the Tessin, the area along the southern border of the Alps (fig. 138) made by Antonietti (1968). The mixed Manna-Ash wood which is confined to the rare rendzinas found here has a pronounced submediterranean character, not only in its set of species, but also in its climate (tab. 25, no.1). It is surprisingly well provided with nutrients, especially nitrogen, at least in years which are not too dry. Its low productivity may be the result of water shortage in the summer which can often begin early in the year, even in the Insubrian region which has quite a high precipitation. On the other hand studies by Burnand (1976) on the ecological limitations of *Quercus pubescens* woods in the Central Alpine part of the Rhone valley are very informative in stressing their drought resistance.

(c) Subcontinental mixed Cinquefoil-Oak woods

The *Potentillo-Quercetum* is alien to the truly thermophilous *Lithospermo-Quercetum* which is associated with the southern slopes, in spite of many floristic similarities (tab. 32, nos. 5 and 6). The mixed Cinquefoil-Oak wood generally has a stronger growth and keeps to deep, more or less level, silty to loamy sands with a relatively higher water capacity (Mraz 1958). Such soils in themselves are no warmer than those of the Oak-Hornbeam woods which are so widespread in the area. They are only less fertile and consist mostly of sandy and rather acid parabrown or brown earths. Compared with the Oak-Hornbeam woods their tree growth is less dense and richer in frugal light-loving species. Consequently the herb layer enjoys relatively more light and radiant heat. It is only because of these local climatic conditions that so many xerotolerant species can settle in *Potentillo-Quercetum*, especially as the majority of stands were grazed in earlier times, which thinned them out even more. Zólyomi (verbal communication) maintains that the Cinquefoil-Oak woods present in Central Europe are certainly for the most part a result of the activity of man and his animals. Its broad but disjuncted distribution in Europe has been well documented by A.O. Horvat (1978) who gives more or less exact maps of all important occurrences. Only in the steppe-wood region, e.g. the Hungarian lowland can it be surely seen as zonal vegetation. Involuntary experimental support for this viewpoint was obtained from the previously very typical mixed Cinquefoil-Oak wood south-west of Białowieża (outside the national park) in Poland. Since this has been a nature reserve, and no longer interfered with, Hornbeams and other shade-tolerant species have spread so strongly that the xerothermal elements have almost disappeared. In order to prevent them from being eradicated entirely this destructive undergrowth must be (and nowadays is) removed from time to time, at least in part of the reserve. However it clearly would be better if sheep could also be introduced again to graze the clearings (Faliński on a conducted tour 1976).

Strictly speaking then in *Potentillo-Quercetum* of the eastern plains we are not dealing with a particularly 'warmth-loving' woodland community such as that actually portrayed by *Buglossoido-Quercetum*. The species present in both communities must in the first place be seen as light-loving and to some extent drought tolerant. Only the submediterranean elements which as we saw in lists 1-4 were gradually disappearing towards the more continental part of Central Europe can be looked upon as particularly warmth requiring or frost sensitive. The term 'thermophilous' mixed Oak woods which is often used for the order *Quercetalia pubescenti-petraeae* only applies to the Cinquefoil-Oak wood insofar as, like the other communities of the order, it does not go up into cool montane sites. In this wider sense however even the Oak-Hornbeam wood also would deserve the epithet thermophilous (figs. 3 and 43).

Since the *Potentillo-Quercetum* occupies mainly level ground which does not suffer from erosion. In spite of the low rainfall in its area of distribution the soil is superficially free from lime. It has a fairly acid reaction, whereas the soils of the *Buglossoido-Quercetum* almost always show a neutral or even a basic one Because of this there are a number of acid indicators ("a" in tab. 32), which can be used as differential species against the *Buglossoido-Quercetum*. However a few acid indicators are to be found here and there even in mixed Oak woods on lime-rich soils, e.g. *Melampyrum pratense*. These are plants which, like many others widespread on acid soils, really do not particularly need a higher acidity or a lower lime content, but are able to tolerate them. The reason they are found both on acid soils and on dry limestone soils is to be sought above all, in their dependence on well illuminated, sparsely occupied, habitats.

Mráz (1958, fig. 128) has mapped the distribution of Cinquefoil-Oak woods in Europe. His survey shows clearly that these communities prefer deep and sandy soils in their east European centre. In the plain of the Upper Rhine and other boundary areas however they occupy heavy marls with a fluctuating water balance where species of the order *Fagetalia* are not very competitive.

Here in every case water is a deciding factor. Kažmierczakova (1971) was able to observe in the Polish hill district that in the dry spring of 1967 *Stellaria holostea* and *Galium odoratum*, i.e. typical representatives of the Oak-Hornbeam and Beech woods, had been adversely affected in their production, while species of the order *Quercetalia pubescenti-petraeae* had come through it much better. Because of the greater light input the yearly production of the herbaceous layer in the *Potentillo-Quercetum* was greater than that in a comparable *Tilio-Carpinetum*, namely 3.5 against 2.1 tonnes · ha^{-1}. In many respects we may agree with the Neuhäusl and Heuhäuslova-Novtna (1969) who have investigated mixed Cinquefoil-Oak wood and other deciduous woods in the eastern part of the Czechoslovakian Elbe plain. They emphasise that the *Potentillo-Quercetum* lies between the Oak-Hornbeam wood (*Galio-Carpinetum typicum*) and the acid-soil mixed Oak woods (especially the *Luzulo Quercetum)* in pH value, lime content and other soil properties and also in species combination. However the

Fig. 128. The distribution of *Potentilla alba* and of the *Potentillo- Quercetum* in Europe. Along the edge of the eastern steppeland the Cinquefoil-Oak wood prefers sandy soils (having layers of clay to hold up the water, or a higher water table), but in the western and northern woodland areas clay soils are favoured. After Mráz (1958). modified.

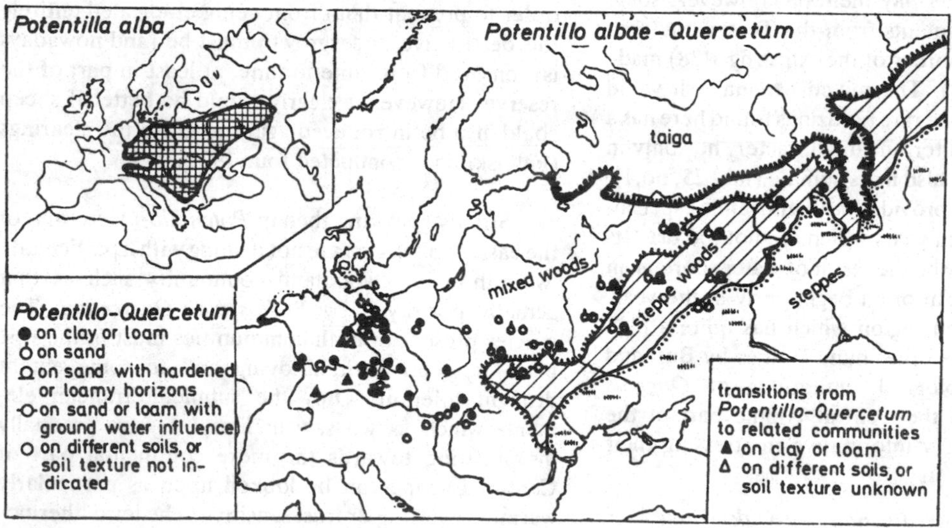

'xerothermic' elements indicate that with respect to water availability the *Potentillo-Quercetum* is worse placed than the other two communities.

(d) On the animal world in mixed Oak wood ecostems

The mixed Oak woods of Central Europe have been studied a great deal, but ecological aspects have rarely been considered. Above all we lack data which would allow us to estimate their water balance in comparison with that of other woodland communities, especially during dry years. This would probably be the deciding factor in the species composition, not only of the herbaceous ground cover but also of the tree and shrub layers. The annual fluctuations in the water content of Sedge-Beech woods and Oak-Hornbeam woods shown in figures 63 and 123 at least enable us to recognise that even in soils with a high proportion of fine particles and in a moist climate the plants can suffer from a water shortage occasionally (see also fig. 38).

Everywhere trees and bushes are able to grow, other plants and also animals, which would not be able to compete or even survive in full sunlight, are found living under their protection. This is particularly true of dipteran larvae and other soil-dwelling animals consuming the leaf litter and speeding up its breakdown (fig. 129). It is true that many of these animals and the microorganisms living with them suffer during the dry summer period; however, according to Szabó (1974, fig. 130), their numbers increase again in the autumn.

Figures 129, 130 may give an idea of the many different life forms and species which take part in the make up of a deciduous woodland community, especially in those places where the soil is not too poor in lime and fertility. Bassically this is true for all communities of the class *Querco-Fagetea*, i.e. for other mixed deciduous woodlands and for Beech woods too (figs. 89-92, tab. 31).

5 Acid-soil mixed Oak woods as influenced by oceanic climate

(a) Birch-Oak woods and related communities in Central Europe

The subcontinental mixed Cinquefoil-Oak woods which we met in section 4 c already had a number of

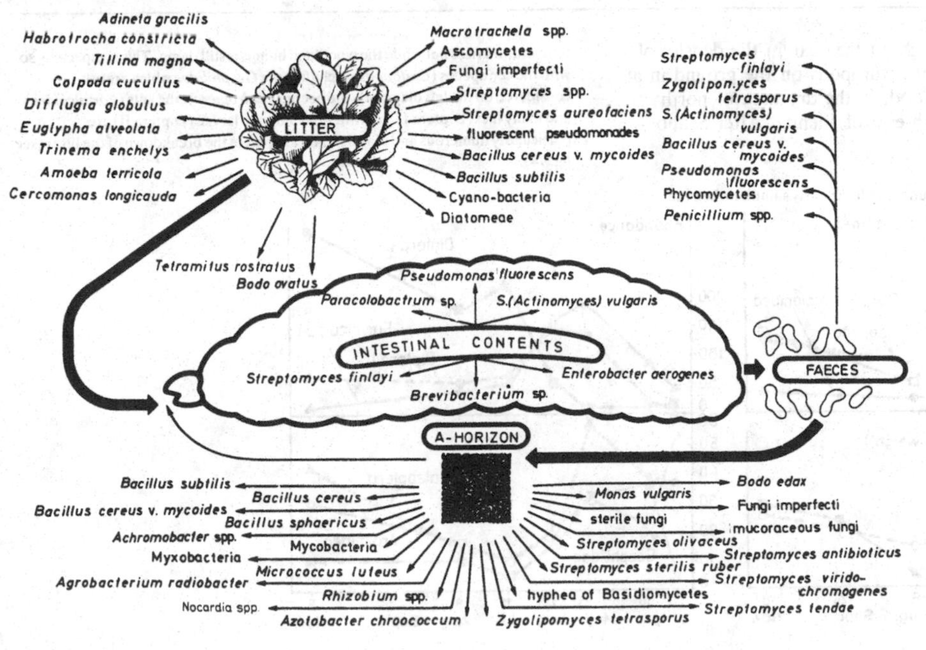

Fig. 129. Biocenoses of lower animals and plants as related to a litter-eating fly larva (*Bibio marci*), which reaches a size of 18-26 mm long and 2.5 mm wide in an open mixed Oakwood on rendzina. After Szabo (1974), somewhat modified.

The flies swarm in spring and then die after laying their eggs. The larvae which hatch after 35-40 days feed in the first place on the leaf litter (saprophage) but become omnivorous eating also living leaves and roots (phyllophage and rhizophage), dung (coprophage) and several different animals (zoophage). They contribute significantly to the formation of a fine mull which is composed largely of their droppings (see also fig. 91).

Many bacteria live even in the digestive tract of the larvae and continue to find good living conditions in the faeces, along with some lower fungi (*Phycomycetes, Penicillium*). These and other microorganisms, together with protozoa etc., take part in the breakdown of the litter (upper left).

In the lower part of the diagram is shown the community life in the mull-horizon of the rendzina. Autotrophic organisms (diatoms, blue-green algae) play only a small part in the life on the litter but are mentioned for the sake of completeness.

acid indicators and other species requiring a lot of light, but which are otherwise not too demanding. Along with these however there still live a number of the more fastidious representatives of the orders *Quercetalia pubescenti-petraeae* or *Fagetalia* which grow quite well on the loamy or marl soils with their high silicate content. The latter are less frequent in those woodlands and scrub on sunny but very lime-deficient slopes which Glavač and Krause (1969) have described from valleys in the Middle Rhine mountains. Similar acid-soil xerothermic communities may be found in other parts of Europe, e.g. in the Werra and Fulda valleys or on hill slopes north-west of Zurich. Their site conditions have not yet been studied in detail.

We would look in vain for any of these more demanding species on the glacio-fluvial sands of the early Ice Ages and on other colloid-deficient sand deposits in the northern plains of Central Europe. These quartz sands are extremely poor in silicates and had a low fertility from time immemorial. They are distinguished by a flora which is either acid loving or acid tolerant. On such decidedly acid soils in the more continental east of Central Europe Scots Pine comes into its own (section IV 6 c, tab. 33) while in the west Oaks gain the upper hand and form open mixed woodlands.

From north-west Germany and neighbouring areas such poor woodlands have been described as Birch-Oak woods (*Betulo-Quercetum*) or 'Oak-Birch woods' (Tüxen 1930). Alongside the Oaks Birches are present at least as pioneers since their seeds are easily carried by the wind and germinate in large numbers after a fire, tree felling or other disturbance and form a 'pre-woodland'. Oaks on the other hand with their heavy seeds must rely on birds which drop the acorns amongst the open Birch stands. Finally however the longer-living Oaks grow through and suppress the birches. In the older Birch-Oak woods only the modest Alder Buckthorn (*Frangula alnus*), Rowan (*Sorbus aucuparia*) and, in the western side of Central Europe, Holly (*Ilex aquifolium*) persist. According to Ovington (1963) Aspen (*Populus tremula*), noted for its high seed production, can sometimes take on the pioneer role of Birch on a somewhat better soil.

In any case the soil is not productive enough for Maple species, Hornbeam and other shade- and half-shade trees. At best only stunted specimens are to be found, which clearly demonstrate how the poverty of the soil in bases and other nutrients reduces their competitiveness. In addition the acid humus layer over the more or less strongly podsolised soil hampers their regeneration.

Since the trees and shrubs of the Birch-Oak wood thrive only moderately well and allow a lot of light to penetrate to the ground, the growth here is often quite vigorous. In some places Bracken (*Pteridium aquilinum*) produces a dense chest-high stand of fronds in the shadow of which scarcely any other species is able to survive (fig. 131). In others dwarf shrubs such as *Vaccinium vitis-idaea* and *Calluna vulgaris*, or some grasses (mainly *Avenella flexuosa*) may be dominant

Fig. 130. Fluctuations throughout the year in the density of the populations of important arthropods on the ground in an open mixed shrubby Oak wood, in the dry area of north Hungary, 50 km to the south-east of Vienna. After Szabo (1974), modified.

Many groups have their optimum abundance in the early summer when the temperature rises but while the woodland humus is still damp. This is especially so with the centipedes (*Chilopoda*, including the *Geophilidae* which are not considered on the left diagram), the beetles (*Coleoptera*), the ants (*Formicoidea*) and many flies (Diptera I). In winter the bibionid larvae (Diptera II) are particularly numerous and play a significant part in the breakdown of the litter (see fig. 74).

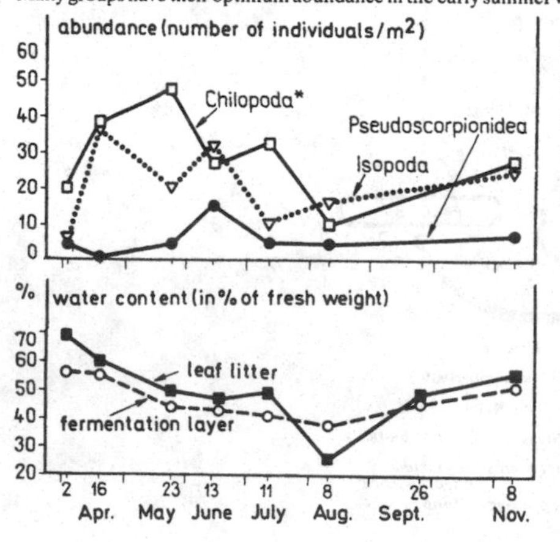

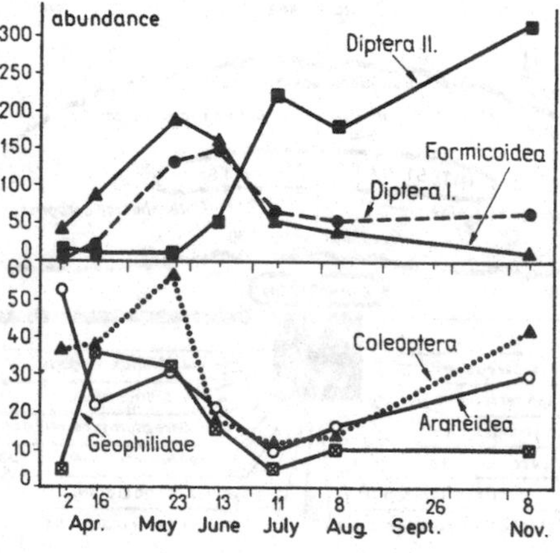

Tab. 33. West-to-east gradient in the species composition of acid-soil mixed Oak and Pine woods on diluvial sands of the northern plain in Central Europe.

Community no.:	1	2	3	4	5	6	7	K	R
Tree species combination			(B)				S		
(B = Beech, O = Oak.	(B)		O		O		O		
S = Spruce, P = Pine)	O		P		P		P		
Tree layer (also in shrub l.)									
L *Fagus sylvatica*	·	*3*	*1*	*2*				2	×
Quercus robur	**5**	**5**	**5**	*1*	*5*	*3*	*5*	×	×
Betula pendula	*5*	*5*	*5*	*3*	*3*	*1*	*5*	×	×
Populus tremula	*2*	*2*		*1*		*1*	*2*	5	×
Quercus petraea			*5*	*2*		*5*		2	×
N *Pinus sylvestris*			**5**	**5**	**5**	**5**	**5**	6	×
N *Picea abies*							*5*	7	×
Shrub layer									
L *Ilex aquifolium*	*1*	*2*						2	4
L *Lonicera periclymenum*	*3*	*4*						2	3
L *Rubus* spec. (other spec.)	*5*	*3*	*1*			*(3)*		×	×
L *Cytisus scoparius*	*2*	·		*2*				2	3
Juniperus communis	*1*	*3*		*2*	*5*	*3*	·	×	×
Sorbus aucuparia	*5*	*4*	*3*	*3*	*2*	*4*	*4*	×	4
Frangula alnus	*5*	*4*	*2*		*1*	*5*	*3*	5	2
L *Tilia cordata*						*2*	*4*	4	×
L *Acer platanoides*						*1*	*4*	4	×
Corylus avellana						*3*		3	×
Berberis vulgaris						*3*		4	8
Euonymus verrucosa						*3*		8	7
Carpinus betulus							*5*	4	×
Chamaecytisus ratisbonensis							4	6	×
Herb layer									
L *Teucrium scorodonia*	*3*	·						2	2
L *Hieracium umbellatum*	*4*	*1*						×	4
L *Galium harcynicum*	*2*	*3*						2	2
L *Corydalis claviculata*	·	*1*						1	3
L *Hypericum pulchrum*	·	*1*						2	3
L *Polypodium vulgare*	*1*	*2*	*1*					3	2
L *Holcus mollis*	·	*4*		*2*				2	2
L *Avenella flexuosa*	*5*	*5*	*5*	*3*				2	2
L *Genista pilosa*	*1*	*1*	*4*	*1*				4	2
L *Hieracium lachenalii*	*3*	·	*1*			*3*		×	4
Carex pilulifera	*2*	*3*	*3*		*3*			2	3
Anthoxanthum odoratum	*3*	·	*2*	*5*	*5*			3	5
Agrostis tenuis	*5*	*3*		*3*	*2*	*3*		3	3
Luzula multiflora	*4*			*3*	*5*	*3*		×	5
Rumex acetosella	*1*		*2*	*2*		*2*		3	2
Melampyrum pratense	*4*	*3*		*2*	*5*	*4*	*5*	3	3
Pteridium aquilinum	*3*	*3*	*1*			*5*	*3*	3	3
Calluna vulgaris	*4*	*3*	*5*	*5*	*5*	*3*	*5*	3	1
Vaccinium myrtillus	*3*	*3*	**5**	**5**	**5**	**5**	*5*	5	2
Solidago virgaurea	*3*				*5*	*3*	*5*	×	×
Maianthemum bifolium	*2*	*1*	*1*			*4*	*4*	6	3
Potentilla erecta	*2*			*1*	*2*	*1*	*4*	3	×
Festuca ovina		*2*	*3*	*5*	*5*	*5*	*5*	×	×
Luzula pilosa			*4*	·	*5*	*5*	*5*	3	5
Dryopteris carthusianna			*4*	*1*		*1*		3	4
Scorzonera humilis			*2*	*1*	*5*	·	*4*	5	5
Danthonia decumbens			*2*		*5*			2	3
N *Viscum album laxum*				*3*				3	–
Lycopodium clavatum				*3*			*2*	3	2
Hieracium pilosella				*4*	*2*	*2*		3	×
N *Vaccinium vitis-idaea*				*3*	*5*	*5*		5	2
N *Pyrola chlorantha*				*2*	*2*		*1*	5	5
N *Chimaphila umbellata*				*2*	*5*	*1*	*5*	6	5
Fragaria vesca				*2*	*2*	*2*	*4*	5	×
Genista tinctoria					*4*			3	4
W *Anthericum ramosum*					*4*			4	7
W *Peucedanum oreoselinum*					*5*	*1*	*4*	4	×
W *Polygonatum odoratum*					*4*	*2*	*2*	5	7
N *Calamagrostis arundinacea*					*3*	*1*	*5*	5	5
Convallaria majalis	*1*				*2*	*4*	*5*	3	×
N *Trientalis europaea*		*1*			*1*	*4*	*5*	7	3
Melica nutans						*3*		3	7
Carex ericetorum					*1*	*3*		7	×
Veronica officinalis		*1*			*1*	*3*	*2*	3	2
N *Rubus saxatilis*			*1*			*4*	*4*	7	×
Carex digitata						*5*		4	×
Oxalis acetosella				*1*		*1*	*4*	3	×
N *Orthilia secunda*						*1*	*4*	3	×
N *Monotropa hypopitys*						*1*	*4*	5	4
N *Goodyera repens*							*5*	7	4
Molinia caerulea							*4*	3	×
Viola canina							*3*	3	3
Moss layer									
L *Hypnum cupressiforme*	*4*	*3*						–	×
L *Aulacomnium androgynum*	*1*	*1*						–	2
Leucobryum glaucum	*2*	*2*	*1*	*1*	*1*	*1*		–	1
Polytrichum formosum	*5*	*3*		*2*		*5*	*2*	–	2
Dicranum scoparium	*5*	*4*	*3*	*4*	*1*	*1*	*3*	–	1
Pleurozium schreberi	*5*	*3*	*5*	*5*	*5*	*5*	*5*	–	1
N *Dicranum undulatum*	*1*		**5**	**5**	**5**	**5**	**5**	–	1
N *Hylocomium splendens*	*1*			*4*	*5*	*4*	*5*	–	×
Scleropodium purum					*3*			–	5
Mnium rostratum						*4*		–	5
N *Ptilium crista-castrensis*							*5*	–	3
Constancy summations L	35	34	13	8	–	6	8		
Constancy summations N	2	1	11	24	31	31	58		
Average continentality value (K)	3.2	3.0	3.5	3.6	4.0	4.3	4.7		
Average acidity value (R)	2.5	2.3	2.5	2.5	3.7	3.2	3.3		

Source: After several authors. Species which have a constancy of less than 2 and occur in only one unit have been omitted. It has not been possible to give indicator values for continentality to the mosses.

Note: Nos. 1 and 2: **Birch-Common Oak wood** *(Betulo-Quercetum typicum)*, eu-oceanic to suboceanie.

1: in central and southern Netherlands. After Meijer-Drees (1936, p. 105)

2: in north-western Lower Saxony. after Tuxen (1937, p. 128) in extreme habitats.

Nos. 3 and 4: **Pine-Oak woods with Beech** *(Pino-Quercetum)*, sub-oceanic.

3: in Brandenburg (northern Havelland). After Passarge (1957c. p. 93, "*Myrtillo-Pinetum*")

Tab. 33 (Continued).

4: in western Poland. After Preising (1943, p. 26, "*Dicrano-Pinetum eupteridetosum*").
Nos. 5 and 6: **Oak-Pine woods without Beech** ("Pino-Quercetum", better *Querco-Pinetum*).
5: between Warthe and Vistula. After Preising (1943, p. 32, "*D. -P. typicum*")
6: in the centre of eastern Poland (Putawy), after W. and A. Matuszkiewicz (1956, tab. 4, "*Pino-Quercetum berberidetosum*, var. of *Dicranum undulatum*").
No 7: **Spruce-oak-Pine wood** in the ancient forest of Białowieza, sub-continental. After A. and W. Matuszkiewicz (1954, p. 52, "*Pino-Vaccinietum myrtilli*").

L = acid-tolerating broadleaved woodland plants
N = acid-tolerating needleleaved woodland plants
W = "warmth-loving" species.

(tab.33). Dicotyledonous herbs are rare and noticeably small leaved, e.g. *Melampyrum pratense, Trientalis europaea* and *Galium harcynicum*. The Creeping Soft Grass (*Holcus mollis*) which spreads like Twitch and requires a lot of light can serve as a (local) character species of the Birch-Oak wood, which otherwise possesses few species of its own (fig. 135). Practically all acid indicators are also present in other acid-tolerant woodland communities, especially in natural coniferous woods and in acid-soil Beech woods. Only a few tall-growing and very light-demanding Hawkweeds seem to be fairly well confined to the Birch-Oak woods, e.g. *Hieracium laevigatum, H. sabaudum*, and subspecies of *H. umbellatum*. However they turn up rather infrequently and are altogether absent from most stands.

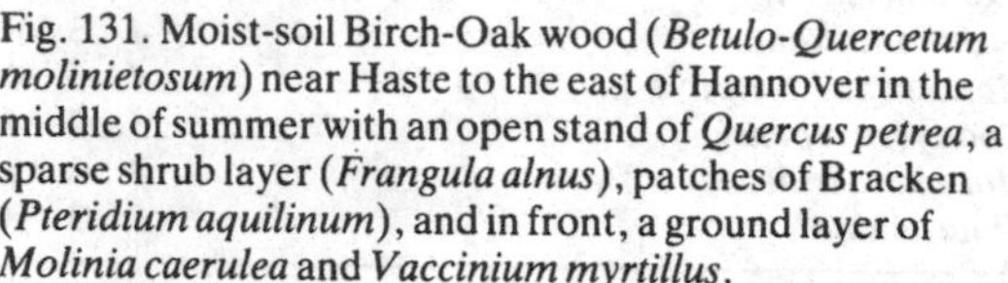

Fig. 131. Moist-soil Birch-Oak wood (*Betulo-Quercetum molinietosum*) near Haste to the east of Hannover in the middle of summer with an open stand of *Quercus petrea*, a sparse shrub layer (*Frangula alnus*), patches of Bracken (*Pteridium aquilinum*), and in front, a ground layer of *Molinia caerulea* and *Vaccinium myrtillus*.

Mosses are at a disadvantage in Birch-Oak woods, as they are in other deciduous woodland, because of the leaf litter (see section B I 3 a). However this only covers the soil with a layer of no more than four leaves deep every autumn, and as the wind blows freely through the open woodland, the litter layer is not continuous (Ellenberg 1939 and 1963). Therefore acid-tolerant mosses like *Polytrichum formosum, Hypnum cupressiforme, Pleurozium schreberi* and *Scleropodium purum* are able to develop to varying extent. None of these however is a good character species of the Birch-Oak wood.

In spite of the absence of unique species the Birch-Oak woods are easily recognisable; they can be distinguished from other woodland communities by the large number of acid-indicators, including mosses, as well as by the dominance of the Oak and the retreat of shade trees, provided this is due to natural causes.

Many of these acidiphilous Oak stands however

have become so poor in shade trees as we find them today because for centuries they have been subject to coppicing (fig. 132, see also section A II 3 a). In the north-west and west of Central Europe some of them could certainly be rich in Beech, and should be looked upon as degraded acid-soil Beech woods (*Avenello-Fagetum* section B II 4 a) or Beech-Oak woods (*Fago-Quercetum*). Meisel-Jahn (1955a) has proved that such is the case, for instance, with so-called 'Hauberge' (coppiced slopes) in the lower montane belt of the Siegerland. From pollen analyses discussed by Firbas (1949, 1952) it is evident that the role of the Beech was not so insignificant at one time as it is today on the lowlands stretching from France through Belgium, Holland, north-west Germany to Jutland (see sections B I 1 and 2).

The thing which makes the study of Birch-Oak woods in the lowlands particularly difficult is the fact that at one time they were completely opened up by grazing as well as by the removal of litter and timber which kept them almost continually as heathland (section A II 2). With the advent of modern forestry it was not profitable to allow them to grow up into high forest. The open spaces were planted up with Pines rather than Oaks since they are easy to cultivate, quick growing and more valuable economically. Consequently today there are available for study only very small and disappearing remains of what may have been a widespread vegetation type 1000-2000 years ago.

For example in north-west Germany, where Tüxen (1930, 1937) described the typical '*Querco-Betuletum*' for the first time, today it is no longer possible to find a single example on dry sandy soil corresponding to the requirements of a sample area from which a good relevé can be obtained. Only on moist soils are there still to be found seminatural acid-soil mixed Oak woods, and in some places these can be quite extensive. Our idea of a Birch-Oak wood has thus been developed in the first place from this 'moist' subassociation and is therefore biased (Ellenberg 1939).

The Birch-Oak woods of northern Belgium, the Netherlands and north-west Germany are poorer in species than the other deciduous woodland communities of Central Europe, and they are altogether rather poorly developed. The records collected together in tab. 33 give an impression of their composition. However before we look into their individual communities we must try to give a general survey of the acid-soil mixed Oak woods which are known from

Fig. 132. Dry Birch-Oak wood ("*Quercetum medioeuropaeum*" or "*Betulo-Quercetum helveticum*") which has developed from a moder Beech wood (*Luzulo-Fagetum*) as a result of coppicing; near the edge of the plateau of the Stadler mountain NNW of Zürich. *Quercus petraea* growing from the cut stumps, occasional Beech and a lot of *Vaccinium myrtillus*; appearance in spring.

Central Europe. In the lowlands one can distinguish three groups of pure deciduous types:

1 **Subatlantic** Birch-Oak woods (*Betulo-Quercetum* in a narrow sense) are particularly characteristic of and frequent in the north-western part of Central Europe.

2 In **southern Central Europe** Oak woods on strongly acid soils are rare and have all been formed through the activity of man and his animals. Left to nature they would have been rich in Beech. On warm slopes this '*Quercetum medioeuropaeum*' passes through intermediates to the next group (fig. 132).

3 **Insubrian** Birch-Downy Oak woods ('*Betulo-Quercetum insubricum*') are found in Tessin and similar places with a high rainfall and contain submediterranean species. To a large extent they have been replaced by Sweet Chestnut coppices (fig. 136)

Each of the three types, especially the last two, shows forms in the montane belt which are distinguished by Woodrushes (*Luzula luzuloides* or *L. nivea*, fig. 83) and other montane species. In addition subassociations can be distinguished within each of the three lowland types, and to some extent in the montane types also, which can be arranged according to the dampness of the soil:

a) A **typical** one (also erroneously called a 'dry' type) which is very rarely found amongst nos. 1 and 3 but is the usual one with no. 2.

b) One with **periodical moistening**, found particularly with nos. 2 and 3.

c) A **moist-soil** one (e.g. *Betulo-Quercetum molinietosum*) which is the most dominant in no. 1 and appears quite frequently in nos. 2 and 3 (fig. 131).

d) A **wet-soil** one which has so far only been established as a special unit in no. 1 (*B. -Q. alnetosum*, Burrichter 1973), but has occasionally been observed in no. 3.

Besides these purely deciduous woods one can meet woods with an increasing number of Pines as one goes from areas 1 and 2 towards the east, i.e. towards a more continental climate. Over such a gradient at least four steps can be recognised:

1 Subatlantic **pure deciduous** woods such as those already named.

2 Oak woods with **few Pines** found in the transition zone which begins somewhat to the east of the Elbe (e.g. *Betulo-Quercetum-pinetosum*, Mikyška 1963),

3 A subcontinental **Oak-Pine** wood (*Pino-Quercetum*), in which *Pinus sylvestris* can play a large part e.g. in the northern part of the German Democratic Republic (Müller-Stoll and Krausch 1968) and in north Poland.

4 Pure **Pine woods** (section B IV 6).

In each of these four groups which are determined by climate, especially a and c, a sequence of types depending on soil moisture can be distinguished. The number of types of acid-soil woodland communities then is almost as large as that of woodlands on richer soils. However because of their poverty in species it is difficult to find sufficient characteristic and differentiating species which can be used in all these climatically influenced units. On the other hand in an area of uniform climate and within limited boundaries the subunits brought about by differences in soil moisture are clearly recognisable and can be plotted easily.

The multiplicity of the mixed Oak woods is increased even further by the intermediates showing all possible stages to the better nourished, but still acid-soil, Oak-Hornbeam woods (section B III 3 f) and the moder Beech woods (B II 4). For example on the sandy soils which are somewhat richer in colloids and silicates in north-west Germany *Quercus petraea* is more plentiful than the Common Oak and often accompanied by more 'fastidious' herbs such as:

Oxalis acetosella	*Lathyrus linifolius (= montanus)*
Convallaria majalis	*Solidago virgaurea*
Teucrium scorodonia	

With the help of such differential species Tüxen (1937) separated the group of 'Durmast Oak-Birch woods' from the 'Common Oak-Birch woods'. However Common Oak can be dominant in communities where the above species are quite numerous while the Durmast Oak can still assert itself without them. In Birch-Oak woods where, in addition to these differential species, some *Fagetalia* species are also present the Beech probably takes a larger share and would indeed become dominant under natural conditions. Tüxen (1956b) therefore labelled this *Fago-Quercetum* (Beech-Oak wood). The name *Querco-Fagetum* would have expressed the natural dominance relationship more accurately (section B II 4 b).

The centre of distribution of the Birch-Oak woods lies in northern France and generally in the atlantic to subatlantic north-west Europe. Here it occupies not only quartz sand but also sandstones, granites and other crystalline rocks, slates and sandy loams with a podsolised surface layer (Tüxen and Diemont 1937). Similar communities of the *Quercion robori-petraeae* alliance turn up in north-west Portugal (Tüxen and Oberdorfer 1958) and in the British Isles as far as the centre of Scotland (Tansley 1939). In areas of France near to the coast they contain many atlantic and submediterranean-atlantic elements, e.g.:

Erica cinerea	*Luzula forsteri*
Ruscus aculeatus	*Peucedanum parisiense*
	Hyacinthoides non-scripta

The Birch-Oak woods of north-west and western Germany have only a few subatlantic species, in particular:

Ilex aquifolium	*Galium harcynicum*
Lonicera periclymenum	*Corydalis claviculata*
Cytisus scoparius	M *Aulacomnium androgynum*

Since these species penetrate far into Central Europe and the *Erica cinerea* group is absent from the Birch-Oak woods of the Netherlands, north west Germany and Jutland it is justifiable to consider the Central European Birch-Oak woods as a special association.

The largest areas of Central Europe in which Birch-Oak woods have the potential to form the natural vegetation is undoubtedly the north-west German old diluvial plain. As already mentioned there are no longer any typical Birch-Oak woods here as they have all been cleared for farm land or turned into heaths, or later into Pine forests. There are however good examples of the moist-soil Oak-Birch woods (or Moorgrass-Birch-Oak woods, *Betulo-Quercetum molinietosum*) to be found. These contain three easily recognisable moisture indicators:

Betula pubescens	*Molinia caerulea*
Erica tetralix	

These same differential species are also met with in the Pine or Spruce forests which have replaced the natural woodland. Extensive areas which would naturally have been covered with Moorgrass-Birch-Oak wood are used for the grazing of animals as they are easily put down to grass. Today more and more such land is being drained and converted into arable fields because the upper soil is rich in humus and thus contains more colloidal material than the pure sand on drier sites.

(b) The soils of Birch-Oak woods, especially in north-west Germany

Oak-rich communities of the *Quercion robori-petraeae* alliance are associated with a relatively mild, humid climate in Central Europe. They prefer subatlantic low-lying sites and only exceptionally go up into the montane belt (Noirfalise and Sougnez 1956). Typical Birch-Oak woods which have not been turned into heathland, are found in the north-western lowland on sandy parabrown earths or brown earths extremely poor in bases, but which nevertheless are podsolised only slightly or not at all. In any case the very small clay fraction is already long since washed out of the top soil by rainwater and deposited in small brownish seams between the lower layers of sand which have been intermittently deposited at one time by water from melting ice or by the wind (fig. 133). According to investigations by B. Meyer (verbal communication) these seams are brought about by the damming effect of air pockets which form in the lower layers of sand. Such depositions can be produced artificially and fairly rapidly, so K.-D. Jäger's (1970) view that they could not have occurred since late glacial times has been completely overturned. The colloid-rich seams can be recognised as ochre bands 1/2 – 3 cm wide running more or less horizontally, but in some places branching irregularly or joining up again as the result of later disturbances (fig. 133). Parabrown earths banded in this way are regularly to be met with under dry Birch-Oak woods but also under the Pine woods of eastern Europe in similar soils and under the *Periclymeno-Fagetum* of the young moraines around the Baltic. They have been thoroughly investigated and described by Tüxen (1930, 1957). According to Kubiëna (verbal communication) the material of the brown seams often derives from the remains of tertiary soil formations which were mixed into the glacial deposits, i.e. from primarily silicate-rich material which has in some places been turned into loam.

As Tüxen also recognised at an early stage the Durmast Oak-Beech wood (*Fago-Quercetum*) is found where the soil profile shows much broader brown seams than those of the typical Birch-Oak wood (fig. 133). Such 'thick-seam' parabrown earths are richer in silicates and so are more productive than the 'thin-seam' ones. They are less prone to turn to heathland, but if in the end the heath had taken over they are less strongly podsolised. Today, for the most part, they are under the plough, while the soils of the Birch-Oak woods have supported heathland for hundreds of years and now mostly carry extensive Pine forests.

The soil profiles of the moist-soil Birch-Oak woods show a completely different structure. The upper horizons are richer in humus, but generally less deep, than in the soils we have just been discussing. Whitish-grey and rust-coloured spots and stripes as well as small concretions of a dark rust or blue-black colour appear in the lower layers which are in reach of

Fig. 133. The changing soil profile found under broadleaved woodlands in North-west Germany where loess loam occurs lying over glacio-fluvial quartz sand (below and to the right of the sloping line), semi-schematic. After Tüxen and Diemont (1937), somewhat modified.

Left: parabrown earth with a compact B_t-horizon under brown-mull Beech wood (*Melico-Fagetum*) rich in Beech; centre: transition profile with coarse brown bands under Oak-Beech wood (*Fago-Quercetum*); right: narrow banded podsol-parabrown earth under Birch-Oak wood (*Betulo-Quercetum typicum*).

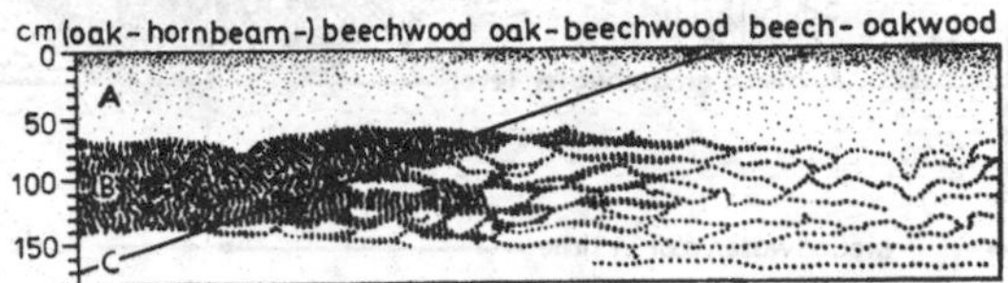

the groundwater. This gley horizon is formed by the alternation of wetting and more or less complete drying out over a long time. Such acid, sandy gley soils are given the name modergley, or, if they have been heavily podsolised under heathland at some earlier stage, they are called gleypodsols (fig. 134). They differ from the more fertile mull-gleys (section B III 3 e) only in the characteristics of the top soil, especially in the presence of a very acid surface humus and in the absence of a crumb structure. There are no obvious differences in the gley horizons even though the reasons for the different development of the upper horizons are to be found in the lime content of the ground-water (Dietrich 1958; tab. 28; fig. 134).

One thing all the mixed Oak woods of the *Quercion robori-petraeae* have in common is the extraordinarily acid reaction of their root horizons (figs. 114 and 78) which reach pH values as low as those found under *Calluna* heaths and raised bogs. As fig. 114 shows in the example of three moist-soil Birch-Oak woodlands there is also a considerable yearly fluctuation of soil acidity in the *Betulo-Quercetum*. According to Ellenberg (1939) this is due in the first place to the very slight buffering power of such soils. Each increase in bases or acids, whether produced in situ or introduced by precipitation immediately produces a shift in the pH value.

What has been said about the plant life of the acid-soil Beech woods (section B II 4) applies even more strongly to acid-soil Oak woods: the only plants which are able to flourish here are those which can put up with the higher degree of acidity and the shortage of nitrate in the soil. As the investigation into *Holcus mollis* shows (fig. 135) the species are to some extent always able to grow on richer soil. *holcus* is ousted from **the most acid soils by the less fastidious** *Avenella flexuosa* (see also fig. 51). Physiologically the latter is favoured by its ability to tolerate high cencentrations oa aluminium (which becomes soluble at low pH, see section B II 4 c and d) better than many other plants. Its growth is even stimulated by $Al(OH)_3$ down to a rather low level determined by Hackett (1962).

Almost all members of the Birch-Oak wood community live in symbiosis with mycorrhiza. The saprophytic fungal flora is also very rich in species and individuals. The therophyte *Melampyrum pratense* is a semi-parasite living at the expense of grass roots. Almost without exception then the plants of the Birch-Oak wood have a specialised nutrition; in view of the poverty of the upper soil they prosper surprisingly well. Per unit area however the total production of this community is less than that of all other woodland communities of Central Europe having a similar favourable water supply.

Between the years 1928 and 1943, because of the extreme poverty of their soils in bases and nutrients, the 'Oak-Birch woods' were dragged into the centre of an argument about the climax question in Central Europe (see first edition pp. 244-246). As this can now be considered settled we shall dispense with any discussion of it here (see section B I 1 a). The typical Birch-Oak wood is today considered to represent only the final stage in the development of the vegetation on pure sandy soils which have always been extremely poor in silicates. Loam and limestone soils can never become so extremely impoverished in Central Europe that they can compare with the 'Birch-Oak soils' and

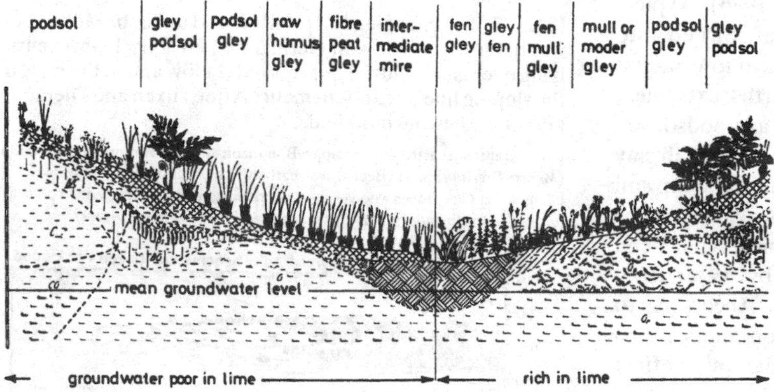

Fig. 134. Idealised sequence of the types of woodland soils on sand in the North German diluvial area where groundwater, which may be rich or poor in carbonate lies at various depths. After Dietrich (1958), somewhat modified.

Corresponding woodland communities: podsol = Oak-Pine wood, gley podsol to raw humus gley = Molinia-Oak-Pine wood or Molinia-Birch-Oak wood, fibrous peat gley and transition mire = swamp Pine-Birch wood, gley fen = swamp Alder wood, fen gley = Ash-Alder wood, fen mull gley = moist-soil Oak-Hornbeam wood, mull or moder gley = slightly-moist mixed Beech wood with Pines.

consequently they favour climax communities which are more demanding, especially those rich in Beech (fig. 37).

(c) Acid-soil Oak woods and Chestnut scrub in southern Central Europe

Right in the south of Central Europe, in the colline belt of the insubrian area on the southern edge of the Alps there are mixed Oak woods on decidedly acid soils. These 'insubrian Birch-Oak woods' resemble those in western Germany, but contain a few species which need more warmth, for example:

Castanea sativa	*Molinia arundinacea*
Quercus pubescens	*Phyteuma betonicifolium*

To a great extent the original Birch-Downy Oak woods were converted into Sweet Chestnut coppices in Roman times or during the Middle Ages (fig. 136). The Chestnut trunks were used as posts in the vineyards. The undergrowth remained largely the same as in the more natural Oak stands because the microclimate is similar. Such Chestnut communities have been described from the Tessin under various names (Ellenberg and Rehder 1962, Antonietti 1968, Ellenberg and Klötzli 1972). They are the only *Quercion robori-petraeae* woodlands in Central Europe, covering relatively large areas, which are not influenced by ground-water. Although they are found on more or less steep mountain slopes their species composition reminds one very much of the north-west European moist-soil Birch-Oak woods described above, in which *Molinia caerulea, Betula pubescens* and other moisture indicators play an important part. Here these are favoured by the high precipitation which, according to Blaser (1973), can give rise to humus-rich layers (crypto-humus-podsols) several decimetres thick even on steeply sloping sites. On neighbouring and also otherwise similar but base-rich slopes, where *Carpinion* communities prevail, the excessive soil moisture makes the Alder (*Alnus glutinosa*) competitive (fig. 139, see also Zoller 1961).

As far as Central Europe is concerned the Birch-Oak woods are a peripheral occurrence. Where mixed Oak woods are found on acid soils in the central parts, on the whole they are grouped together in a special association, the *Quercetum medioeuropaeum* following the example of Braun-Blanquet (1932).

This belongs to the group Durmast Oak woods which has already been described. However it lacks many of the subatlantic species named in section a, while at the same time it can be distinguished by the presence of:

Chamaespartium sagittale	*Potentilla sterilis*
Genista germanica	

and other species, which extend further to the east. In many stands one may look in vain for examples of the broom species so that *Quercetum medioeuropaeum* is distinguished from *Betulo-Quercetum* (*boreoatlanticum*) more on negative char-

Fig. 135. a. In pot cultures the 'physiological pH optimum' for *Holcus mollis* as measured by the dry weight of the plants, including the parts below ground, is about 6. After research by Sebald (1956).

Under natural conditions though the Creeping Softgrass is crowded out by a number of other species in soils with a pH of more than 4.5 (see b.)

b. The pH boundary determined by competition between Wavy Hairgrass and Creeping softgrass in an acid-soil Oak wood in England.

Data by Jowett and Scurfield (1952) from Ellenberg (1958).

The productivity of *Holcus mollis* (in g fresh weight per m^2 surface covered) is so low at pH 3.9 (or less) that *Avenella flexuosa* is not troubled, but above pH 4.0 this narrowleaved grass is suppressed by the broaderleaved Creeping Softgrass even though its physiological optimum is also higher than this.

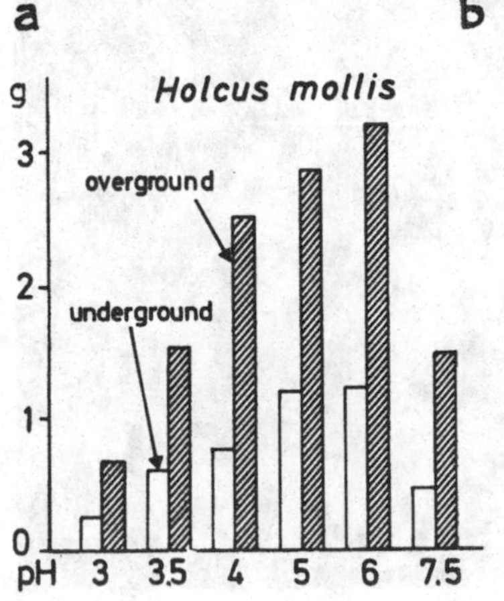

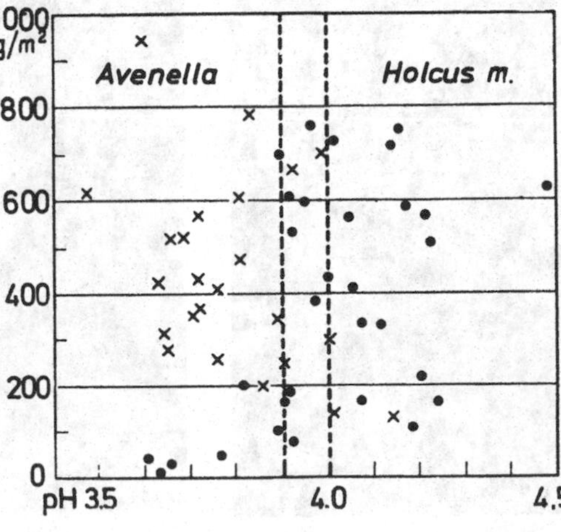

acteristics. From the impressions gained in Czechoslovakia Neuhäusl and Neuhäuslova-Novotna (1967) proposed that *Quercetum medioeuropaeum* should be renamed *Luzulo-Quercetum* because of the frequent occurrence of the Montane Woodrush. There are however floristically different montane Oak woods in the oceanic north-west which have a prior claim to this name (Noirfalise and Sougnez 1956).

The vegetation region known as Insubrian (fig. 138) to which we have already referred repeatedly lies along the southern border of Central Europe. According to Zoller and Kleiber (1971) it has taken on more and more submediterranean features ever since the warm period set in following the last Ice Age. Although the mosaic of its plant community is very exciting and has been well investigated (tab. 25; fig. 139; see also the survey by Ellenberg and Klötzli 1972) we cannot dwell on it here. As in north-west Europe there are many transitional stages between the extremely acidophilous woodlands, which clearly belong to the *Quercion robori-petraeae* and the more demanding communities which must be placed in the order *Fagetalia*, and indeed in the alliance *Carpinion*, even though here the Hornbeam does not play an important role. Zoller (1961), who first drew attention to the existence of 'better' sites in the Tessin, maintains that Ash (*Fraxinus excelsior*) and Alder (*Alnus glutinosa*) along with Oak are the naturally dominant trees. Just like the Birch-Oak woods these mixed deciduous woods too have largely been replaced by the planting of Sweet Chestnuts, mostly grafted ones (fig. 137). The finest fruit groves developed on relatively fertile sites, e.g. on the soil of the Ash-Elm woods or of the mixed Ash-Oak woods (tab. 25, nos. 2 a, b).

Cultivation of the Sweet Chestnut (*Castanea sativa*) was introduced by the Etruscans and Romans and later neglected by the Germanic tribes. During and just after the Middle Ages however it formed an important basis for the Insubrian agricultural economy. It provided not only posts for the vineyards and practically all the timber for building and burning, but also leaf litter which was raked up in the plantations and brought into the stalls to produce manure. Winter fodder was obtained for the goats and cattle by cutting and drying the young shoots (section A II 2 a). Its mealy fruits, which could be stored after they had been smoked, were used for making bread and for feeding pigs. When farming and trade became more intensive about a hundred years ago the importance of the Chestnut declined. The groves returned to Nature and the posts grew up into tall trees. The decline was speeded up even more by the arrival of the Chestnut Canker (*Endothia parasitica*), a fungus originating in Japan and reaching Italy about 1930 by way of

Fig. 136. A Sweet Chestnut coppice ("palina") with *Pteridium aquilinum*. *Vaccinium myrtilus* and *Molinia caerulea* in the site of the Insubrian Birch-Downy Oak wood near Losone in the Tessin.

Fig. 137. A Sweet Chestnut orchard ('selva') with *Fraxinus* saplings and demanding herbaceous plants in the site of the Insubrian mixed broadleaved wood near Sigirino in the Tessin before it was attacked by the Chestnut Canker. This community belongs to the *Carpinion* alliance, and apart from the Sweet Chestnuts themselves has hardly anything in common with the Chestnut-Birch-Oak wood shown in fig. 136.

America. It invades the cambium through a wound in the bark, leading to death first of the branch and then of the whole tree. The grafted selected strains all became victims of the disease while the wild stocks regenerated. The Italians are planting conifers (*Pinus strobus*) or Poplars in the places where the Chestnut has been cleared. The Swiss are trying to restore a near-natural mixed deciduous woodland more in keeping with the countryside. This is also less likely to encourage forest fires to which the Insubrian plantations are prone during dry periods (Leibundgut 1962).

IV Coniferous woodland and mixed woods dominated by conifers

1 General survey

(a) The role of coniferous trees in the woodlands of Central Europe

Central Europe is naturally a land of broadleaved woods, in which conifers only play a part in special situations and towards the north-eastern boundary of the area (fig. 140, section B I 1 and 2). It is true that wherever deciduous trees grow Firs, Spruces, Pines and Larches can also flourish (fig. 39), and indeed Scots

Fig. 138. Warm low-lying areas along the southern edge of the Alps in Switzerland and upper Italy showing the approximate boundary between the submediterranean region (relatively dry in summer and with calcareous soils) and the Insubrian one (high precipitation and mainly acid soils). After Oberdorfer (1964); see fig. 5.

In the 'insubrian' areas the naturally dominant communities are the Birch-Oak woods with Downy Oak and other warmth-loving species (alliance *Quercion robori-petraeae* = a) which have now been largely replaced by Sweet Chestnut.

The 'submediterranean' areas (b) have generally less-acid soils and because of this show more variety in their plant communities which are richer in submediterranean elements; amongst these *Fraxinus ornus* and *Ostrya carpinifolia* (*Orno-Ostryon*) play a part.

The plain which has been laid down south of the Alps by river sedimentation (c) also has a submediterranean type of climate, but because of the deep soil is wholly cultivated. Its potential natural vegetation would probably be mixed broadleaved woodland of Central European types (*Carpinion, Alno-Ulmion*). At the lower levels in the mountains (d) the landscape is dominated by Beech (*Fagion*).

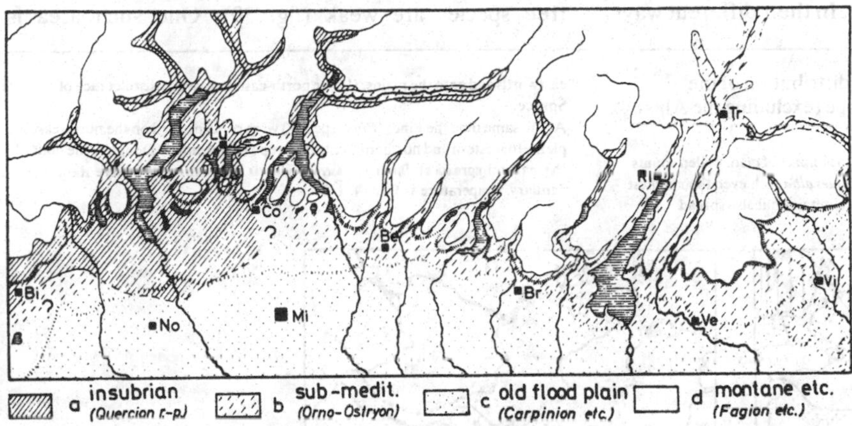

Fig. 139. An example of insubrian vegetation mosaic in the area to the north of Ascona with rocks rounded by diluvial glaciers. The relationships between the rock types, soil formations, relief and aspect are shown in a reduced diagram. After H.R. Hofer (1967) modified.

The shade slopes with more or less rich mixed broadleaved woods, containing large numbers of alder are not shown. All the mixed broadleaved woods belong to the *Carpinion* alliance, all the Birch-Oak woods to the *Quercion robori-petraeae*.

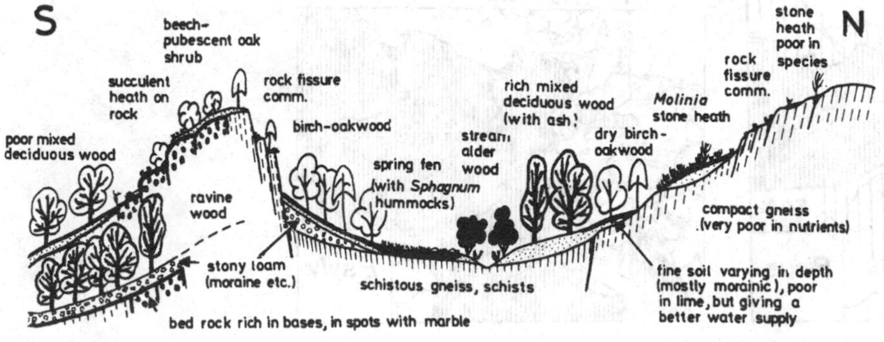

Pine has the widest physiological amplitude of all the native species. Yet over almost the whole of Central Europe conifers can achieve dominance only if they are protected from the competition of Beech and other tall deciduous trees, even including Oak. The broad-leaved trees have much stronger powers of regeneration, at least in those areas with a relatively oceanic climate. However in the hands of the forester the conifers give a much quicker return and grow relatively well. In fact as a rule conifers give a higher yield when planted in the areas where one would naturally find deciduous woodland than they do in the natural coniferous woodland communities of Central Europe (fig. 416).

Before we go into the reasons why conifers are so much less successful in open competition we would like to take a closer look at the distribution of tree species in Central Europe than we were able to do in the introductory sections AI and BI. In this connection we can use the profiles in figs. 6, 42 and 43, the ecological diagrams in figs. 4, 5, 37, 40, 68, 69 and 178, as well as the maps in figs. 1-3, 21 and 138. In their different ways these show that the ecological role of Beech is most pronounced in the suboceanic submontane belt. It recedes with both higher and lower altitudes as well as with increased continentality of the climate. The changes in the deciduous woodland communities along the gradient from moist to dry soils on level ground have already been dealt with in section B III 3. In almost all other cases it is the conifers which benefit from the reduced competitiveness of Beech. On average soils, i.e. those well supplied with water, oxygen and nutrients, Silver Fir (*Abies alba*) comes in to join Beech at the higher altitudes. Within the transition zone between the outer and inner Alps the hardier Fir becomes dominant except for soils which are very poor in bases. Spruce (*Picea abies*) which requires more light, can only become dominant on average soils where Fir has to retreat because of too heavy a frost, i.e. in the valleys of the inner Alps and also at the subalpine level.

Pines only succeed in making a significant contribution to the woodland in those places where other tree species are weak (fig. 39). One such area is

Fig. 140. Maps showing the natural distribution of the needle-leaved trees in Central Europe (excluding the Alps). After Firbas (1949), slightly modified.

After the 'post-glacial warm period' i.e. before the impact of man. pollen counts show that the Spruce (*Picea abies*) and the Fir (*Abies alba*) achieved their highest concentration in the Alps and other European mountains but also spread eastwards and onto the plains. (In the north-east there was a distinct race of Spruce.)

At the same time the Pines (*Pinus* species) were concentrated on the north-eastern plain. In western and northern Central Europe (according to Dengler) the limit of the natural spread of *Pinus sylvestris* today is approximately where the mean January temperature is 0 to + 1°C.

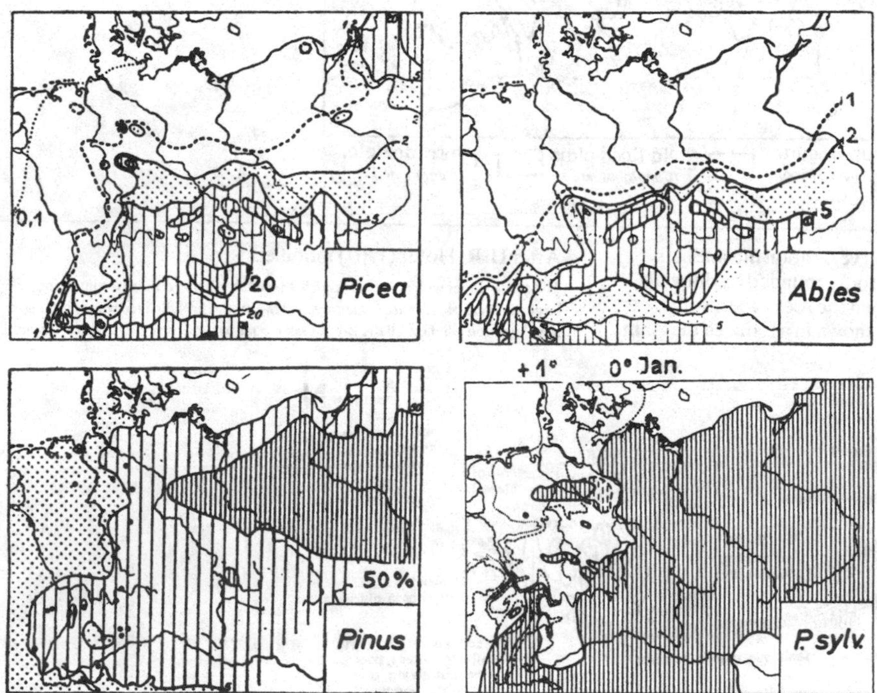

available for the Arolla Pine (*Pinus cembra*) in the high-montane and subalpine belts of the Central Alps, where the Larch (*Larix decidua*) also gets its chance as a pioneer. Scots Pine (*Pinus sylvestris*) prefers lower sites and dominates in the deepest valleys of the inner Alps which are relatively dry and warm in summer but cold in winter. The largest areas covered by natural Pine forest are on the poor sandy soils of the eastern plain. Near to the Alpine tree-line there are often extensive thickets of the Dwarf Mountain Pine (prostrate forms of *Pinus mugo*), which really cannot be looked upon as woods. The Upright Mountain Pine plays a definite role only in the montane to subalpine belts of the western Alps and this is not given in the list of tables and figures quoted above. Neither is the Yew (*Taxus baccata*) mentioned at all in them although as a participant in deciduous woodland it has already been dealt with in section B II 2 e (but see fig. 39).

The suboceanic-submontane ecogram shown in fig. 40 applies to the Central European mountains and hills north of the Alps as well as to the young moraine hills from Schleswig-Holstein to Mecklenburg. In other words it concerns more than a third of the area of Central Europe, and was used therefore as a basis for discussion of the deciduous woodland communities. Only in the lowlands of the north-east – and there only on very poor sandy or swampy ground – have the coniferous woodlands played a dominant role right from the early Post-Glacial period up to the present day (fig. 140). Thus, seen as a whole, the proportion of conifers in the natural vegetation cover of Central Europe is really quite small even when one considers that there is scarcely a deciduous woodland in the mountains or in the north-eastern lowlands where some conifers are not found either singly or in groups.

Introducing the general characteristics of Central European vegetation (section A I 1) the complex of causes which gives the conifers some competitive advantage has already been indicated. They must be looked for in the type of climate as well as in the properties of the soil and in some biotic factors, especially the influence of man. When we disregard the peculiarities of the individual species we may accentuate the following factors:

1. The **more continental** the climate, the higher the proportion of conifers under otherwise comparable conditions. This could be due to a number of factors such as keener winter frosts, and increased danger of late spring frosts – at least in the alpine valleys – stronger radiation because there are fewer clouds, a quicker transition from winter to summer, longer dry periods, etc. Without more information we cannot be sure which of these are the determining ones, and they must be evaluated later. Probably a number of factors are working together and it is possible that they are operating indirectly; for example they may be inimical to pests or diseases which attack conifers.
2. The **shorter the growing period** the better the chance of conifers succeeding within the same overall climate and on similar soils. An exception to this rule is to be found on those mountains with a strongly oceanic climate such as the south-western edge of the Alps and the Vosges. In many Central European mountains the montane and oreal belts can be referred to as the coniferous level and the lower ones as the deciduous levels. However it is wrong to think that this distinction is universally applicable as is often done. Figures 42 and 43 show clearly that the montane belts where the climate tends to be oceanic are rich in Beech and that where the climate is more continental even the lowlands may be dominated by conifers.
3. The **poorer the soil is in bases**, the greater the rôle of the conifers, especially Spruce and Scots Pine, within the same climatic range. However none of them can be considered as acidophile.
4. This tendency is accentuated the **wetter the soil** is. Water-logged soils in particular favour the Fir in competition with deciduous trees by causing a shortage of oxygen. Around the edges of raised bogs even on the oceanic north-western lowlands, there are outposts of Spruce and Pine which, although stunted, can thrive here without competition (fig. 21).
5. The **drier the soil**, other things being equal, the better the Scots Pine asserts itself. At higher levels it is replaced by the Mountain Pine. Shallow rocky ground, marls which frequently dry out, and free-draining sands poor in colloids are the places where the most nearly natural Pine stands are found in deciduous woodland areas.

With the soil factors given under 3-5 the real causes of promotion of the conifers are less apparent than in the case of the climatic conditions (1 and 2). It may be emphasised however that the advantages are almost always only relative in conferring increased competitiveness. In absolute terms the factors all lead to a reduced productivity even in the conifers, but to a lesser extent than in their accompanying deciduous trees.

Whenever we compare conifers and broadleaved trees we must bear in mind that neither constitutes a uniform group, but that the individual species each has its own physiological and ecological peculiarities. It has already been seen in tab. 9 that amongst the conifers as well as the broadleaved trees there are those which require a lot of light, those which thrive in the shade, and intermediate types. This can be seen even more clearly in tab. 34 which gives the results of experimental work. Under summer conditions (20°C) Scots Pine (*Pinus sylvestris*) only reaches a rate of photosynthesis to compensate for the respiration when the light intensity reaches 1000 lux or more (1.0-5.0 klux). Norway Spruce too (*Picea abies*) also requires more than 800-1000 lux as a rule to achieve net photosynthesis. Some evergreen subtropical broadleaved trees behave similarly e.g. *Melia azedarach*, *Ficus retusa* and *Prunus laurocerasus*, as well as many Central European deciduous trees (in tab. 34 the Dwarf Birch and the Black Poplar). However in each of the three groups there are also species with very low compensation points, such as *Taxus* and *Abies* amongst the conifers, *Myrica rubra* and *Fagus sylvatica* of the broadleaved woody plants. In respect of light saturation, that is the ability to use strong radiation for photosynthesis, the sun and shade leaves on one and the same plant often show a greater difference than exists between different species. High and low values can be found in each of the three groups.

The temperatures at which the cardinal values of the light used in photosynthesis are reached are almost the same for all the species shown in tab. 34. Very surprisingly even those plants which live in warm climates had temperature minima and optima which were scarcely any higher than those for the Arolla Pine (*Pinus cembra*), which is found high up in the severe conditions of the Alpine slopes. Similar data for a larger number of species are shown in fig. 141. These were all taken at the same light intensity of 10 klux, i.e. an average intensity which is often met with in Nature. Evergreen and deciduous trees, even Alpine herbaceous plants are so similar to one another in these respects that, for example, the values for Yew correspond closely to those of Olive; those for Beech and the Glacier Crowfoot (*Ranunculus glacialis*) are very similar, while even Spruce and Mountain Pine are not far behind. The deciduous Downy Oak (*Quercus pubescens*) requires even more warmth than the evergreen mediterranean Holly Oak (*Q. ilex*). It could scarcely be more impressively demonstrated how wide the physiological amplitudes of the majority of species are than by means of tab. 34 and fig. 141 (see also fig. 39). The separation of these species under natural conditions can only be explained by **competition** between them, the success of which is dependent on the interaction of numerous internal and external factors and has been and is still very easily influenced by man.

Since the majority of conifers are evergreen they

Tab. 34. Light and temperature values critical for photosynthesis in some woody species.

Tree and shrub species	Illumination (klux) compensation point at 20°C	saturation point	Temperature (°C) lower comp. (winter)	lower comp. (summer)	optimum (summer)	upper comp[a] (summer)	Remarks
Deciduous broadleaved trees							
Betula nana	2.3	40	–	–	17	–	'light' tree
Populus nigra	0.8	15	–	–	–	–	↓
Quercus petraea	0.5	15	–	–	–	–	↓
Q. pubescens	0.35	>50	–	–	23	45	↓
Fagus sylvatica	0.3–0.5	>10– 40	–	–5	20–22	43	'shade' tree
Evergreen broadleaved trees							
Melia azedarach	1.4	30	–	–	–	–	'light' tree
Ficus retusa	1.0	20	–	–	28	–	↓
Prunus laurocerasus	0.8	–	–6	–4	25	42	↓
Olea europaea	0.6	>50	–8	–	18	48	↓
Quercus ilex	0.6	30	–4	–	15	42	↓
Myrica rubra	0.4	40	–	–	20	–	'shade' tree
Needleleaved trees							
Pinus sylvestris	1.0–5.0	20– 60	–6, –7	–4	–	37	'light' tree
Picea abies	1.0–2.0	>30– 70	–5, –7	–4	12–18	34–37	↓
Pinus cembra	0.2–0.7	18–>30	–5	–	12–18	36	↓
Abies alba	0.3–0.6	4–>20	–7	–3	14–20	38	'shade' tree
Taxus baccata	0.3	–	–8	–5	20	41	

Note: [a] lower comp. = low temperature compensation point; upper comp. = high temperature compensation point.

Source: From a summary of the literature by Larcher (1976) and others (of. tab. 9).

can make better use of the spring and late autumn for photosynthesis than can deciduous trees (fig. 142). This advantage must be weighed against the disadvantages that their leaves are continually at the mercy of bad weather (some of them for many years), of pests and diseases, and, increasingly, of air pollution. It is just these disadvantages which may be a decisive handicap to the conifers in their competition with deciduous trees. As we shall see they are particularly vulnerable to pests when they are young, because at this time their needles have a thinner cuticle and a less scleromorphic structure than they have in the sun-bathed crowns of the taller trees.

(b) On the classification of coniferous woodland communities

Since the conifer-rich woodland communities are of less significance than the deciduous woodlands in Central Europe they will not be discussed in such detail as the latter. Also a number of questions of general interest have already been dealt with in connection with the deciduous woodland communities. Another reason for such restraint lies in the fact that with few exceptions relatively little research has been published on the ecology of confierous woodlands. Last but not least, their classification on the basis of plant sociology has not yet been generally accepted in spite of a good deal of new work having been done by, for example. Oberdorfer (1957, 1979), Hartmann and Jahn (1967). Mayer (1974). Ellenberg and Klötzli (1972). Gensac (1967, 1970). G. Hofmann (1969) and Passarge (1971).

On the other hand the difficulties of the classification are to a great extent due to the nature of the woodland communities in which conifers are dominant. As we have seen the naturally occurring conifers in Central Europe are found only near the ecological

Fig. 141. Optimal range and absolute limits of temperature for the net photosynthesis of trees and herbs from habitats of different warmth, measured under comparable conditions. After Pisek et al. (1969), somewhat modified.

The optimal range (thick horizontal line) was determined for a light intensity of 10 000 lux (=0.1 cal.cm^{-1}.min^{-1}). The material for *Quercus pubescens. Qu dex. Olea curopaea* and *Citrus limonum* came from Lake Garda, the remainder from varous altitudes in the Innsbruck area.

Evergreen (name underlined) and deciduous trees, likewise species from the high mountains (*Geum reptans, Ranunculus glacialis*) or from the warm lowland (*Qu. ilex, Olea*) showed very little difference. They were all able to assimilate very well at 15-20°C; above 25° only the warmth-loving plants.

The minimum temperature for photosynthesis in summer (dots to the left), and frost resistance (the temperature at which half the organs of assimilation were irreversibly damaged; arrow tips) was below 0°C for all species. Neither was there much difference between their heat resistance (black and white circles) while the maximum temperature for photosynthesis (crosses) was generally somewhat higher for mediterranean and submediterranean species.

Instead of *laurocerasus off.* read *Prunus lonrocerasus*.

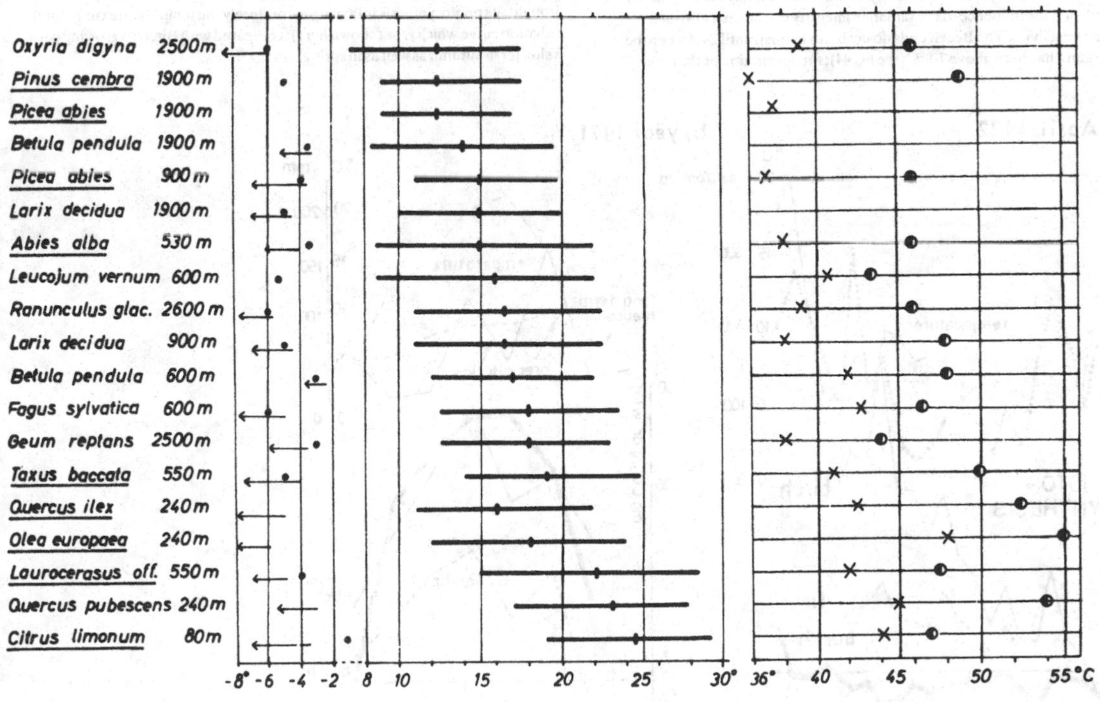

boundaries of the deciduous trees or in areas where the latter are excluded for some reason, i.e. mostly under extreme and heterogeneous conditions. But wherever conifers do manage to achieve dominance they have a homogenising effect on the character of the soil because of the needle litter which is difficult to break down and as a rule builds up an acid-humus surface layer. Acid-indicators can flourish in this even on top of limestone. However these acid-indicators are not characteristic of coniferous woodland communities, but appear almost without exception in acid-soil deciduous woodland, heaths and grassland also. It is especially difficult to separate the coniferous woodland communities from those of the Woodrush-Beech woods (*Luzulo-Fagion*) and the Birch-Oak woods (*Quercetalia robori-petraeae*).

Whenever man favours the conifers in their competition with the broadleaved trees, species combinations can arise in deciduous woodland areas which can barely be distinguished, if at all, from those in naturally occurring coniferous woods. This is especially true in districts having an 'intermediate' climate since these came under man's influence very early on. It is not possible then to establish satisfactory demarcations on a floristic basis, but only by looking into the history of the woodlands. This is very time-consuming and often cannot be localised sufficiently accurately. Nevertheless, whenever a purely floristic and sociological classification of the conifer-rich woodlands remains unsatisfactory, ecological and historical criteria should be used, even if these can be assessed only through observations and general experience. It is hardly to be expected that such methods would produce authoritative solutions, especially as the natural rôle of the coniferous species varies in different parts of Central Europe and this is reflected in the divergent opinions of the workers which are naturally based on their local experience.

Fig. 142. a. Daily totals of net photosynthesis or net respiration (negative values) for Birch, Beech and Douglas Fir (as an example of an evergreen needle-leaved tree) compared with radiation and temperature. After Künstle and Mitscherlich (1975), modified.

By April the winter dormancy of Douglas Fir is already over; the net photosynthesis of its two year old shoots varies with the radiation. The leaves of Silver Birch have already partly unfolded by 10.4 and they rapidly increase in size and productivity; the influence of the radiation intensity can be seen also on their photosynthetic activity. The Beech buds do not begin to open until 24.4 when the soil temperature has risen above 7-8°C (see fig. 44); the respiration of the buds rises and falls with the temperature.

b. Mean monthly figures for the daily net photosynthesis of the same tree species compared with the temperature and precipitation. From the same source as a (the vertical scale is half that in a).

May and June were so dry that Douglas Fir showed little or no excess photosynthesis. The reaction of Beech showed it to be the least susceptible to drought, its photosynthesis following more closely the temperature curve (or the radiation curve which is not shown here); compared with Birch its growing period is shorter in autumn as well as in spring.

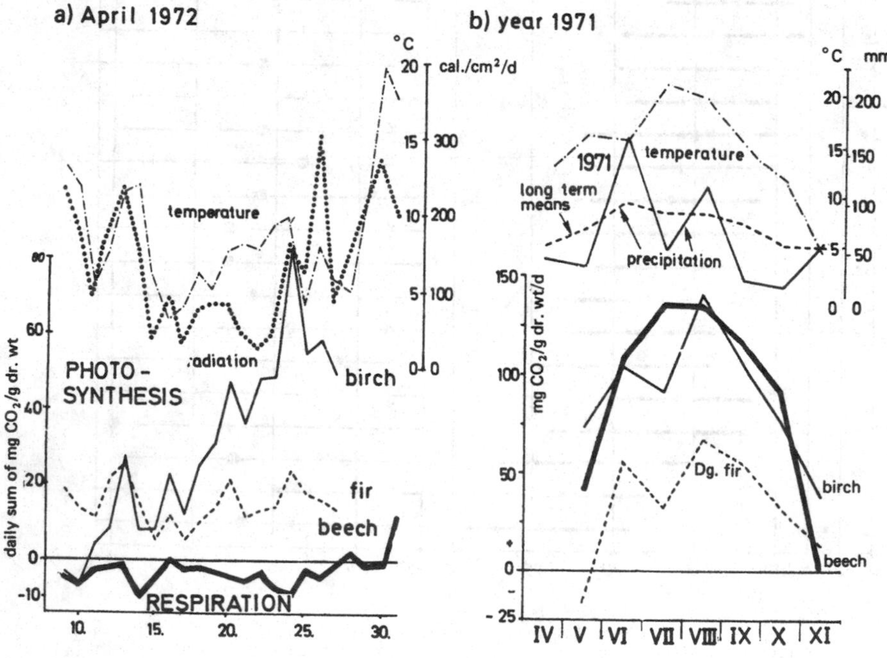

Because of these circumstances we shall strive even less than we did in the case of the deciduous woodlands to mention all the associations, subassociations and variants that have been described in the literature. On the contrary we shall deal with these conifer-rich communities from as wide an ecological point of view as possible. According to species composition and habitat conditions the coniferous woodland communities of Central Europe can be roughly divided into three groups:

1 On **strongly acid** soils with a thick covering of raw humus are the Bilberry-coniferous woods mostly dominated by Spruce or Scots Pine (class and order *Vaccinio-Piceetea*, *Vaccinio-Piceetalia*).
2 **Lime-rich** rocks in a relatively dry climate favour Spring Heath-Pine woods (*Erico-Pinetea*, *Erico-Pintalia*).
3 On soils relatively well supplied with nutrients and water are the **'Beech wood-related'** coniferous woods generally dominated by Spruce or Fir. These must be placed in the order *Fagetalia* according to the species composition of the undergrowth, but because of the climate they do not contain any *Fagus* or other tall deciduous trees.

The first-named class can be met with at all altitudes but covers much larger areas in northern and eastern than in Central Europe. The second is centred on the limestone Alps and is restricted to special sites. The third group is characteristic of the central and intermediate Alps, but is also found here and there outside this area. It causes the greatest difficulties in the systematics of plant sociology; however we shall start with a discussion of its most important representatives, the Silver Fir woods, which present relatively few problems.

2 *Silver Fir woods*

(a) The special place of Silver Fir in European woodland

Communities dominated by Silver Fir (*Abies alba*) occupy a special place which suggests that they should be considered before the remainder of the coniferous woodlands. The distribution of this species is practically confined to Central Europe and to a few other mountains with a similar climate, whereas Pines, Spruces and Larches have their main centre in the north and east of our continent or even in north Asia, with only relatively small extensions of their area penetrating into Central Europe. In addition Silver Fir has many connections with Beech with which it forms widespread mixed woodlands that it is seen by many authors as a characteristic tree of the *Fagion* alliance or the order *Fagetalia*. Of all the conifers of Central Europe Silver Fir behaves most like a broadleaved tree although – in contrast to Larch – it does not lose its dark needles in the winter, and in its growth form it resembles Spruce (fig. 143).

In spite of its ecological relationship with Beech we did not consider the woods dominated by *Abies alba* along with the Beech woods and Fir-Beech woods, but have kept them back for discussion along with the other coniferous woods. Apart from the opinion that the special characteristics of the Silver Fir woods can best be appreciated by comparing them with the Spruce woods, the following reasons are considered to be relevant:

1 Fir woods are much less widely distributed than Beech woods. It is true that, within its area of distribution, the Silver Fir is a regular companion of the Beech and in places can come down to the levels occupied by Oak. Seen on the whole however Fir and Beech are only loosely connected. Looking only at Central Europe the distributions of Fir and Spruce are much more alike in that both are widespread in the montane belt, especially in the eastern Alps and the eastern Central European mountains where they almost always appear together.
2 Fir often comes between Beech and Spruce when the woodlands are arranged according to altitude.
3 Arranged according to the continentality of the climate too, Fir takes up a position between Beech and Spruce. Thus it occupies the relatively continental Baar in the rain shadow of the Black Forest and many of the inner valleys of the Alps which are avoided by Beech because of their continental climate.
4 The Fir woods resemble the Spruce woods so much in their physiognomy and light-ecology that an unbiased observer would immediately put them together and speak of them both as 'coniferous shade woodland'.
5 Fir woods which are free from broadleaved trees are very poor in spring-flowering plants and others characteristic of Beech woods. On the other hand they are rich in species growing mainly in summer, and also in mosses which are handicapped less by the litter from conifers than by that from broadleaves tress (section B II 1a). Also in this respect Fir is similar to other confiers.
6 Where Spruce are also present in woods rich in Fir the plants associated with coniferous woods are encouraged to an even greater extent.
7 The species composition of acid-soil Fir woods is often so similar to that of Spruce woods that many

authors put them together and, in spite of the dominance of *Abies*, they describe them as *Piceeta*, i.e. they place them in the class *Vaccinio-Piceetea*.

8 From the foresters' point of view Firs are considered to be much closer to the other conifers than they are to the broadleaved trees.

In giving these reasons we have said a good deal about the nature and distribution of the Fir woods and this is also illustrated in figs. 37, 42 and 43. Pure, or almost pure, Fir woods can only arise under natural conditions when Beech is at a competitive disadvantage because the soil is too wet, or too dry, the growing period too short, the winter frost too keen, or for other reasons. However conditions must still be such that Fir is able to outgrow the remaining tree species. Both conditions can be found together in three groups of habitats:

1 in a **subcontinental climate** where the soil is more or less rich in bases (the altitude may be in the planar to montane belts);
2 in the **montane and subalpine belts** with high precipitation on clay or lime-deficient soils which encourage the growth of conifers;
3 on **waterlogged**, **acid soils** in high precipitation areas in the submontane to montane belts.

Fig. 143. A slope Spruce-Fir wood, in which selective felling is practised, on iron-humus podsol on the northern escarpment of the Pilatus near Lucerne (Schwarzenberg). On the right, an old Fir with mosses (*Hypnum cupressiforme* etc.) as epiphytes. The undergrowth is mainly Bilberry and Fir saplings. A Spruce sapling is in the foreground right.

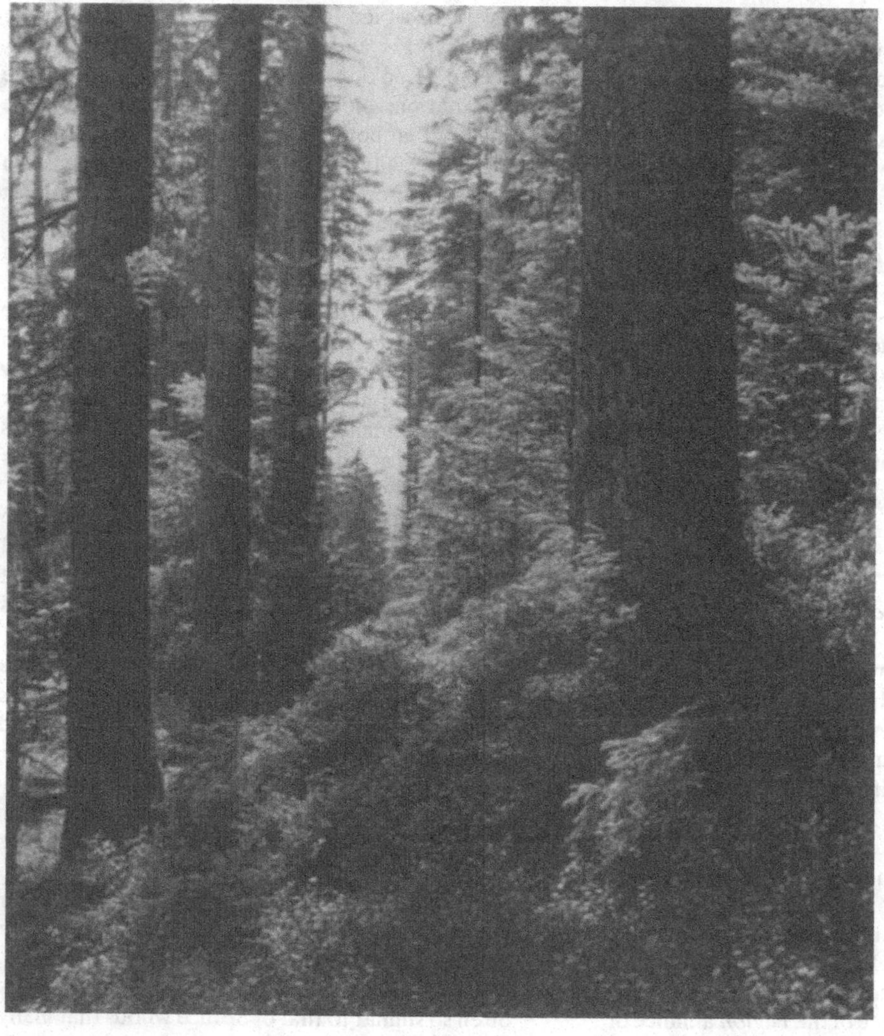

The subcontinental Fir woods (1) are closest to the Beech woods floristically although Beech itself is absent because of the rough climate. Undoubtedly they belong to the order *Fagetalia* and to the *Fagion* and are united in the suballiance *Galio-Abietion* by many authors. For the most part the montane Fir woods too can be included here. As far as the Fir woods on waterlogged and very acid soils (3) are concerned it must be questioned whether it would not be better to put these into the acidophilous order *Piceetalia*, even though *Abies alba* is accepted as a good *Fagion* character species. However this type of Fir wood is connected to the *Fagion* communities through a number of intermediates as we shall see in the next section.

Relatively slight interference by man, such as the systematic removal of individuals or small groups of trees, may allow the Fir to become dominant over the Beech in places where it would not naturally achieve this. As a result it is difficult in individual cases today to decide whether one is dealing with a natural Fir wood or one which has arisen because Fir has been favoured in a mixed woodland which was originally rich in Beech. On the other hand Fir is often hampered by man, who indirectly reduces it. Deer as well as cattle and other domestic animals have a special liking for the saplings of Fir (Klötzli 1965). The better the nutrition the more saplings overcome the browsing damage, not least because there are more seedlings per unit area (Rehfuess 1968).

(b) Fir wood communities of the Alps and their foothills

The most extensive Silver Fir woods are found today on the Balkan peninsula (Fukarek 1964) and in the Alps, particularly in Switzerland. Here they have been thoroughly studied by Kuoch (1954), Bach, Kuoch and Iberg (1954), Frehner (1963) and others. More recently they have been included in surveys by Gensac (1967, 1970), Mayer (1974), Ellenberg and Klötzli (1972) and Horvat, Glavač and Ellenberg (1974), and in part further subdivided. As in other parts of Central Europe the Fir occupies many different habitats in the Alps and on the Balkan peninsula. These range from the outer Alpine chain with a high precipitation to the dry central Alps and from the lime-rich rendzinas, through brown earths of different quality to more or less strongly podsolised or gley soils. Because of this they vary a good deal floristically just as the Beech woods do, while the trees themselves present a very constant picture with their gloomy canopy and the silver grey trunks and the group of saplings with their dark needled wide crowns casting dense shadows.

Table 35 gives an idea of this diversity. It compares the six types of Fir wood communities which are most widespread in Switzerland with each other and also with an equally widespread Spruce wood type.

The First three columns give comprehensive lists of three variants of the so-called **plateau Fir wood** (*Querco-Abietetum*, fig. 144). This is one of the waterlogged soil types (3) and is found on more or less lime-deficient loams, where the ground is level, or sloping away from the sun, in the montane belt of the 'pre-Alps' (the outer chain) at a height of 930 to 1300 m.

Columns 4-6 refer to 'real Fir woods' (*Abietetum albae*, group (2), fig. 143) which could also be called **slope-Fir woods** since they are always found on sloping ground, in contrast to *Querco-Abietetum*. Their optimal development is in the area of lower precipitation forming a transition zone between the northern marginal Alps and the dry central valleys, e.g. the Wallis. Beech is not so competitive here and may be absent altogether, but Spruce is a significant companion of the Silver Fir. The slope Fir wood is also montane, rising to about 1000-1500 m on shady slopes and as high as 1200-1700 m in the sun.

Within each of the two groups of three columns in tab. 35 the variants are so arranged that the one most resembling the Beech wood is on the left and that most resembling the Spruce wood on the right of the 'typical' association. For comparison a list from a pure Beech wood may be imagined in front of column 1 (e.g. the limestone Beech wood from tab. 11 or the rich brown earth Beech wood from tab. 14). To facilitate comparison in the other direction a type of Spruce wood, poor in Fir, has been listed in column 7. Such woodlands are widespread in the subalpine belt of the Alps.

As far as possible the individual species have been joined into groups corresponding to their ecological behaviour. The differential species have been picked out by a frame round the constancy figures. Silver Fir dominates in all six *Abieteta* and, while Spruce is also constant but with a low cover degree. The plateau Fir woods nearly always contain the occasional Beech, a tree which is very plentiful on less waterlogged soils in the pre-Alps. Sycamore is also represented here, especially on the relatively fertile and clay-rich soils of the *Lysimachia* variant (no. 1). On the other hand Larches and Scots Pines, whose true domain lies in the dry valleys of the inner Alps are able to assert themselves here and there in the dry habitats of the slope Fir woods.

The moss layer is rich in species and well developed in all Fir woods (fig. 146) while vascular plants, which can live on acid humus, such as *Vaccinium myrtillus*, *Luzula sylvatica* and *Maianthemum bifolium* also do well (figs. 143 and 145). In the 'richer' Fir wood communities (nos. 1,2,4, and 5 in tab. 35) they grow right alongside more fastidious species such as *Sanicula europaea*, *Carex sylvatica* and *Lamiastrum galeobdolon*. Such a combination of species would be quite impossible in a deciduous woodland, but here it is brought about by the surface humus formed from the needle litter and rotten wood and often forming deeper accumulations under individual

Tab. 35. Submontane and montane plateau and slope Fir woods on soils of different base content, also a subalpine Spruce wood in Switzerland. (F = Fagetalia, P = Piceealia, E = Erico-Picetalia)

Community no.:	1	2	3	4	5	6	7	T	F	R
Community groups	plateau Fir woods			slope Fir woods			Spruce	Factor indicator values		
								T	F	R
Tree layer										
F *Abies alba*	**5**	**5**	**5**	**5**	**5**	**5**	*1*			
P *Picea abies*	*5*	*5*	*5*	*5*	*5*	*5*	**5**			
F *Fagus sylvatica*	*5*	*5*	*5*	*3*	*1*	*1*				
F *Acer pseudoplatanus*	*4*	*3*	*2*	*2*						
P *Larix decidua*				*1*	*2*	*2*				
Pinus sylvestris					*2*	*1*				
Shrub layer										
Rubus fruticosus	*5*	*5*	*4*					×	×	×
Ilex aquifolium	*1*	*2*	*1*					5	5	4
F *Sambucus racemosa*	*1*	*3*	*2*	*4*	*2*	*2*		4	5	5
Lonicera nigra	*3*	*3*	*1*		*1*	*1*	*1*	3	5	5
Sorbus aria	*2*	*2*	*1*	*2*	*2*	*2*	*1*	5	4	7
Rubus idaeus	*4*	*5*	*3*	*5*	*3*	*3*	*2*	×	5	×
Sorbus aucuparia	*5*	*5*	*5*	*5*	*3*	*4*	*5*	×	×	×
Rosa pendulina	*2*			*2*	*2*	*2*	*3*	4	5	7
Corylus avellana	*1*			*3*	*2*	*2*		5	×	×
F *Lonicera alpigena*				*4*	*1*			4	6	8
Lonicera xylosteum				*1*	*3*	*1*		5	5	7
Salix appendiculata							*3*	3	6	8
Herb layer										
Moisture indicators:										
Lysimachia nemorum	*5*							5	7	7
F *Carex remota*	*3*		*1*					5	8	×
Chaerophyll. hirsutum	*3*				*1*			3	8	×
Equisetum sylvaticum	*4*						*3*	4	7	3
Ranunculus aconitifolius	*3*						*1*	3	8	5
Other demanding plants:										
F *Primula elatior*	*3*							×	6	7
Epipactis helleborine	*3*							5	5	7
Orchis maculata	*3*				*2*		*2*	×	4	8
Ajuga reptans	*3*	*1*			*2*		*1*	×	6	×
Homogyne alpina	*4*	*1*			*1*	*1*	*5*	4	6	4
F *Sanicula europaea*	*5*	*1*		*2*	*1*	*1*		5	5	8
F *Carex sylvatica*	*5*	*2*		*2*	*1*			5	5	7
F *Viola reichenbachiana*	*4*	*3*		*3*	*3*	*1*		5	5	7
F *Phyteuma spicatum*	*4*	*3*		*5*	*3*	*4*	*1*	×	5	×
F *Ranuncul. lanuginosus*	*2*			*3*	*1*			×	6	7
F *Epilobium montanum*	*3*			*3*	*1*	*2*		×	5	6
F *Galium odoratum*	*3*			*5*	*1*	*2*		5	5	×
F *Polygonatum verticill.*	*3*	*1*		*4*	*1*	*1*		4	5	4
F *Petasites albus*	*3*			*5*	*3*	*2*	*2*	×	6	×
Geranium robertianum	*2*	*1*		*3*	*1*	*1*		×	×	×
F *Lamiastrum galeobd.*	*2*	*1*		*4*	*1*	*1*		×	5	7
F *Paris quadrifolia*	*1*	*1*		*3*	*2*	*2*		×	6	7
F *Dryopteris filix-mas*	*2*	*2*		*5*	*3*	*3*	*1*	×	5	5
F *Veronica urticifolia*	*4*	*2*		*5*	*5*	*5*	*2*	4	5	7
Fragaria vesca	*4*	*3*		*5*	*4*	*4*	*1*	×	5	×
F *Gymnocarp. dryopteris*	*2*			*3*	*2*	*2*	*2*	3	6	4
± indifferent species										
Dryopteris dilatata	*5*	*5*	*5*	*4*	*1*	*2*	*5*	×	6	×
Solidago virgaurea.	*4*	*2*	*1*	*5*	*4*	*4*	*4*	×	5	×
Herb layer Continued										
F *Prenanthes purpurea*	*5*	*4*	*1*	*5*	*4*	*4*	*4*	4	5	×
Athyrium filix-femina	*5*	*5*	*2*	*5*	*3*	*1*	*3*	×	7	×
Hieracium sylvaticum	*5*	*5*	*2*	*5*	*5*	*5*	*4*	×	5	5
Oxalis acetosella	*5*	*5*	*4*	*5*	*4*	*4*	*4*	×	6	4
Maianthemum bifolium	*5*	*5*	*4*	*2*	*4*	*2*	*2*	×	×	3
in Fir woods:										
Mycelis muralis	*3*	*4*		*3*	*3*	*3*		5	5	×
Galium rotundifolium	*4*	*4*		*2*	*2*	*4*		5	5	5
Goodyera repens	*2*	*1*	*3*		*1*	*1*		×	4	×
absent from slope Fir woods:										
Carex pilulifera	*2*	*5*	*4*					4	5	3
Luzula pilosa	*2*	*5*	*2*					×	×	5
Dryopteris carthusiana	*2*	*1*	*5*					×	×	4
Thelypt. limbosperma	*4*	*1*	*2*				*4*	4	6	3
Blechnum spicant	*3*	*3*	*4*				*5*	3	6	2
P *Lycopod. annotinum*	*3*	*2*	*4*				*3*	4	6	3
Acid indicators:										
Vaccinium myrtillus	**5**	**5**	**5**	*3*	*4*	*4*	*5*	×	×	2
Luzula sylvatica	*5*	*3*	*1*	*4*	*4*	*4*	*4*	4	6	2
Veronica officinalis	*2*	*2*		*2*	*3*	*4*		×	4	2
Avenella flexuosa	*1*	*1*	*2*		*1*	*4*	*5*	×	×	2
P *Listera cordata*	*1*	*1*					*3*	×	7	2
P *Vaccinium vidis-idaea*	*1*	*2*	*2*				*5*	×	4	2
P *Melampyrum sylvaticum*	*1*	*1*		*4*	*5*	*5*		×	5	2
P *Orthilia secunda*	*1*	*1*		*2*	*5*	*3*		×	5	×
Luzula nivea				*5*	*4*	*4*		5	4	3
Montane species:										
Calamagrostis varia				*2*	*3*	*1*		3	5	8
Adenostyles glabra				*3*		*1*		2	7	8
Valeriana trypteris				*5*	*4*	*3*	*2*	×	5	8
Knautia dipsacifolia				*3*	*3*	*3*	*2*	3	6	×
further lime indic.:										
E *Carex digitata*				*3*	*5*		*1*	5	4	×
E *Carex alba*				*2*	*4*			5	×	8
F *Neottia nidus-avis*				*3*	*4*	*1*		5	5	7
Hepatica nobilis				*2*	*3*	*1*		6	4	7
Aster bellidiastrum				*2*	*3*	*1*		3	7	8
Melica nutans				*2*	*3*	*1*	*1*	×	4	7
Campanula cochleariif.				*1*	*3*			×	×	8
E *Polygala chamaebuxus*					*3*			×	3	8
E *Epipactis atrorubens*					*3*	*1*		×	3	8
Ranunculus nemorosus					*3*	*3*		×	5	6
further acid tolerators:										
Saxifraga cuneifolia				*2*	*1*	*4*		×	×	3
Polypodium vulgare				*1*	*1*	*4*		×	×	2
P *Calamagrostis villosa*					*1*	*1*	*5*	4	7	2
P *Melampyrum sylvaticum*							*4*	×	5	2
Potentilla erecta							*4*	×	×	×
Moss layer										
Mnium affine	*2*	*3*	*1*							6
Thuidium tamarisc.	*5*	*5*	*5*	*2*			*2*			×
Rhytidiadelphus loreus	*5*	*5*	*5*		*1*	*1*	*5*			2
Polytrichum formosum	*5*	*5*	*5*	*3*	*1*	*1*	*5*			2

Tab. 35. (Continued)

Community no.:	1	2	3	4	5	6	7	T	F	R
Moss layer Continued										
Eurynchium striatum	5	5	3	5	3	3				7
Plagiochila asplenoides	3	3	5	4	3	2	3			×
Rhytidiad. triquetrus	5	4	5	**5**	**5**	**5**	5			4
Hylocomium splendens	**5**	**5**	**5**	**5**	**5**	**5**	**5**			×
Hypnum cupressiforme	1	3	3	1	2	2	1			×
Ptilium crista-castrensis	2	4	5	2	1	2	5			3
Pleurozium schreberi	2	2	3	2	2	2	5			1
Mnium spinosum				3	4	4	1			×
Sphagn. acutifol.-Gr.	1	2	3				3			2
P *Bazzania trilobata*			5				1			2
P *Plagiothec undulatum*	2	1					4			2
P *Hylocom. umbratum*							2			1
Constancy summations										
F *(Fagetalia)*	73	40	16	77	48	43	12			
P *(Vacc.-Piceetalia)*	16	14	16	12	19	17	32			
E *(Erico-Pinetalia)*	–	–	–	5	14	5	1			

Note:

Nos. 1-3: **Montane plateau Fir woods** ("*Myrtillo-Abietetum*" tab. 7 of Kuoch) on the northern edge of the Alps (named *"Querrco-Abietetum"* by Frehner 1963);

1: Rich brown earth plateau Fir wood ("*Lysimachia* variant").

2: Poor brown earth plateau Fir wood (typical variant).

3: Gley podsol plateau Fir wood (*"Bazzania*-variant", previously *"Bazzanio-Piceetum"*).

Nos. 4-6: **Montane slope Fir woods** (*"Abietetum"*, tab. 10b and a) on the northern edge of the Alps:

4: slope Wood Fescue-Fir wood (*"A festucetosum"*).

5: slope Sedge-Fir wood (*"A. melampyretosum, Carex* variant") on limestone.

–: slope Saxifrage-Fir wood (*A. m., Saxifraga variant*) on rocks poor in lime.

No. 7: **Subalpine Spruce wood** (*"Picetum subalpinum sphagnetosum"*, tab. 11) on the northern edge of the alps (for comparison).

A few species which have not achieved a constancy worth noting in any of the examples have been omitted.

The **constancy totals** have been calculated as in tab. 32. They show that, systematically, the Fir woods must be included with the Beech-rich broadleaved woodlands (Order *Fagetalia*). It is only in the Whip Moss-Fir wood (no. 3), growing on very acid gley podsol, that the character species of the acid-soil needleleaved woodland (*Vaccinio-Piceetalia*) reach the same level of constancy. In contrast they are preponderant in the subalpine Spruce wood (no. 7). The Slope Sedge-Fir wood (no. 5), growing on lime-rich slopes, lies close to the dry slope Sedge-Beech wood (see section B II 2 c). It is distinguished by containing some of the species of the spring heath-pinewoods (*Erico-Pinetalia*, see section B IV 6 b).

The **indicator values** are explained in sections B I 4 and E III: T = temperature, F = dampness, R = soil acidity. For mosses only the R-value can be given. The 'ecological behaviour' of the species grouped in the table can be assessed from the factor values. Average indicator values have only been calculated for temperature (mT). They express the differences in altitude (see above).

Source: From tables by Kuoch (1954).

Fig. 144. A plateau Fir wood on acid waterlogged Riss moraine loam near Roggwil in the centre of Switzerland at about 500 m.

"Seagrass" (Quaking Sedge, *Carex brizoides*, see also fig. 73) dominates relatively clear places, as in the foreground. In the left front *Dryopteris carthusiana*, *D. filix-mas* and *Oxalis acetosella*. The old Fir trunks are thick with hanging mosses (mainly *Hypnum cupressiforme*). Spruce does not regenerate so well here as does Fir.

Spruce trees. The humus specialists have a shallow root system spreading out in the layer of moder while the roots of the mull-dwelling plants occupy the deeper soil layer.

It will be appreciated how closely related are the 'poorer' *Abieteta* (nos. 3 and 6) to the *Piceeta* from the fact that in the list of *Piceetum subalpinum* (tab. 35, no. 7) there are only four new names which do not appear in the previous columns, namely *Salix appendiculata*, a particular subspecies of *Melampyrum sylvaticum*, *Potentilla erecta*, and *Hylocomium umbratum*, the latter serving as a character species for the *Piceion* alliance. All the other acid-humus plants were already in the lists of the acid-soil Fir woods. On the other hand the *Piceeta* lack a number of species represented in the *Abieteta*, even in the poorest variants. These respond to the higher fertility and include *Sambucus racemosa*, *Mycelis muralis*, *Veronica officinalis* and *Eurhynchium striatum*. They may be used as locally differentiating species against the *Piceeta*.

Only the Silver Fir itself and possibly *Galium rotundifolium* can be considered as character species of the Fir woods. The latter shows a remarkable affinity for the Fir which has not yet been explained; it soon turns up even when Firs are planted in an otherwise deciduous woodland area, but occasionally it is also to be found in pure Spruce woods. All other 'fastidious' species have their optimum in deciduous woods which justifies the inclusion of many of the *Abieteta* in the order *Fagetalia*. However they cannot be used as character species for the association.

Already it appears from the few examples given in tab. 35 that soil and climatic conditions are as clearly reflected in the species composition of coniferous woodlands as we have already seen is the case with deciduous woodlands, especially Beech woods. Further groups of species which can serve to distinguish the different Fir wood communities shown in tab. 35 have been indicated by enclosing them in a frame. However we cannot look into all the units here, nor are we able to deal thoroughly with the rest of the Fir wood communities which have been described from Switzerland.

We shall just mention a **Horsetail-Fir wood** (*Equiseto-Abietetum*) found in the Jura and the pre-Alps, which Moor (1952) has already recognised. To some extent this contains the same moisture-indicators as are found in Kuoch's *Lysimachia* variant of the plateau Fir wood (no. 1), but otherwise it stands close to the *Aceri-Fraxinetum* both floristically and ecologically (see section B III 1 c). It occupies slopes which have a fertile, constantly moist loam in the upper montane belt and is the only Fir wood regularly to contain Ash, at least where the substrate is rich in lime.

Corresponding to the Sycamore-Beech wood of the Swiss Jura and the northern Alps (section B II 2g) in the subalpine belt there occurs a **tall-herb Fir wood** (*Adenostylo-Abietetum*) which is also distinguished by the luxuriant growth of large-leaved nitrophilous tall herbs (Kuoch 1954). This woodland community is up quite close to the tree line where the Fir suffers damage from frost and snow but regenerates rather well. Alongside the demanding tall herbs there are many acid indicators, e.g. *Vaccinium myrtillus*, *Luzula luzulina*, *Saxifraga cuneifolia*, *Melampyrum sylvaticum* and many mosses which we have already met.

At the same altitude as the tall-herb Fir wood of the northern edge of the Alps, i.e. between about 1400 and 1750 m, there is, in the Tessin mountains to the north of Lake

Fig. 145. Hard Fern (*Blechnum spicant*) and Bilberry in the herb layer of a plateau Fir wood (see figs. 144 and 146).

Fig. 146. Moss layer of a plateau Fir wood with *Bazzania trilobata* (left) and *Hylocomium splendens* (centre and right); seedlings of *Picea* (in front left) and *Abies* (centre).

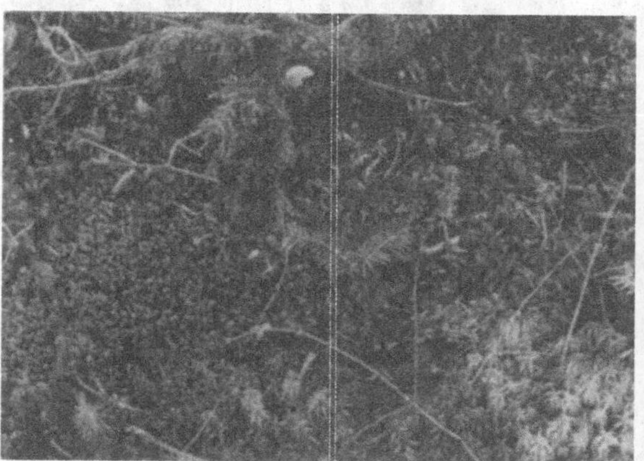

Maggiore, an 'Alpenrose-Fir wood' which leads from the typical Fir woods up to the stands of Larch at the tree line (cf. fig. 42a). Here the dwarf rhododendrons which are centred in the Larch-Mountain Pine woods of the central Alps as well as in a narrow belt above them, penetrate down as far as the submontane belt. Presumably the dovetailing of woodland and subalpine heath has been particularly favoured by the widespread grazing of the woodlands which is still being carried out in the southern Alps.

This survey of the different Fir wood communities in Switzerland, while it is by no means complete, may suffice as an introduction to the variability of their habitats. In other parts of the Alps and in other Fir-rich highlands the classification of the Fir wood communities is in many respects similar to that in Switzerland.

As the very detailed survey by Mayer (1974) shows the Fir woods in the eastern Alps are nearly always richer in Spruce. Apart from the Coral Root-Fir wood (*Dentario-Abietetum*), which is confined to moist fertile habitats, Mayer therefore speaks of '**Spruce-Fir woods**', even though he systematically classifies them as '*Abietetum*'. His three associations reflect the richness of the soils in bases, while a number of subassociations are arranged according to the soil moisture, or point to climatic peculiarities (figs. 54 and 69).

The subunits of the three associations which are shown side by side in tab. 36 cannot be compared directly with Mayer's Fir communities as they do not have the same vertical and horizontal distributions and also have many other peculiarities. To describe these in detail would take up too much space here. Our table shows that the Fir-rich woods of the eastern Alps can obviously be classified in a similar way, ecologically and floristically, to the Beech woods in the more strongly oceanic parts of Central Europe (fig. 53), or to the Oak-Hornbeam woods in the northern plains (section B III 3).

(c) Fir woods on the mountains north of the Alps and on the lowlands

Within the total area of distribution of Silver Fir there are acid-soil Fir woods and 'average' Fir woods outside the Alps and their foothills which have been described by numerous authors. We only need name the summaries of Oberdorfer (1957) for southern Ger-

Tab. 36. Spruce-Fir woods of the eastern intermediate Alps arranged according to soil acidity and dampness (in A, subassociations from fairly dry to wet) and other factors.

1 Silicate Woodrush-Spruce-Fir wood **Luzulo-Abietetum**	2 Fairly acid-soil Sorrel-Spruce-Fir wood **Oxali-Abietetum**	3 Limestone Butterbur–Spruce-Fir wood **Adenostylo glabrae-Abietetum**
A - *myrtilletosum*	–	- *caricetosum albae*
- *luzuletosum niveae*	- *festucetosum altissimae*	–
–	- *hordelymetosum*	- *pyroletosum*
- *typicum*	- *cardaminetosum trifoliae*	- *typicum*
- *blechnetosum*	- *luzuletosum sylvaticae*	- *myrtilletosum*
–	- *dryopteridetosum*	–
–	- *aruncetosum*	- *caricetosum austroalpinae*
- *sphagnetosum*	- *petasitetosum*	–
–	- *myosotietosum*	–
B On boulder scree:	- *polypodietosum*	- *asplenietosum*
In ravines:	–	- *phyllittietosum*
C - *rhododendretosum* (Alpenrose-Fir wood on lower sites in the southern Alps	- *myrtilletosum* (plateau Fir wood)	–
	- *equisetetosum* (Horsetail-Fir wood)	–
	- *adenostyletosum* (tall perennial-Fir wood)	–

Note: C = subassociations which correspond closely to those distinguished by Kuoch (1954) in Switzerland.
Source: In part taken from Mayer (1974).

many, Szafer and Zarzycki (1972) for Poland, Zlatnik (1958) for Czechoslovakia and Horvat, Glavač and Ellenberg (1974) for the northern Balkans.

Limestone Fir woods are also known from a number of different highlands, namely the Swiss Jura (Moor 1952), the south-western Swabian Jura and the Franconian Jura (Oberdorfer 1957), the mountains of Czechoslovakia (Zlatnik 1958) and the Carpathians, especially the Tatra (Zlatnik 1961). The most splendid ones are the mixed woodland communities rich in Fir growing in the limestone areas of Slovenia, Croatia and the rest of the Balkan pensinsula, but these lie outside the area we are considering.

The majority of the Fir-rich woods in the mountains of Central and south-eastern Europe are similar to those described in the previous section from Switzerland and Austria, at least with regard to their habitats. However, in the eastern foreland of the Black Forest, on the limestone plateau of Baar, there are communities of a quite unique character. Reinhold (1956) has proved historically that in this relatively dry, cool montane plain, which is afflicted by late frosts it is not Beech or Spruce but Fir which would naturally become the dominant tree. In the part of Baar nearest the Black Forest Beech is completely absent even though the altitude (750-850 m) would have been optimal for it. Either Scots Pine or Norway Spruce accompany the Fir depending on whether the soil is relatively dry or damp and cool.

This **limestone Fir wood** (designated *Piceo-Abietetum* by Oberdorfer) is closer to the slope Sedge-Beech wood, that means to the alliance *Cephalanthero-Fagion*, than is the limestone Fir wood of the Alps. Its grass-rich subassociation *brachypodietosum* in particular, which in former times may have been grazed and in part turned into semi-arid grassland, contains many of the Sedges, Orchids and shrubs with which we have become familiar in section II 2 c. The limestone Fir wood of Baar can thus be seen as a montane subcontinental parallel to the Sedge-Beech wood. On clay soils between Wutach and Eyach, somewhat further to the north, there are **moder Fir woods** of a more strongly acidophilous character (*Pyrolo-Abietetum*, Stoffler 1975).

In the most northerly and eastern part of its distribution, to the south-east of Warsaw, Silver Fir is found right down on the lowlands (fig. 140). This is probably because *Abies* is better able than *Fagus* to tolerate the continental climatic conditions. It grows here very well and it is only further south, as in the woodland reserve of Lysa Gora, that *Fagus* plays a part alongside *Abies*. On the Polish plain Fir is accompanied by Oak as it is in the centre of Switzerland, but here it is for quite different reasons. Fir can play a more or less prominent part in Beech-free Oak-Hornbeam woods, and on good base-rich soils it even forms almost pure stands which must be looked upon as natural *Abieteta* though many species of the montane Fir woods are absent.

Just as in the mountains to the south and east of Central Europe, so too on the Polish plain Silver Fir goes only an insignificant distance beyond the limit of the Beech's distribution. As Firbas (1949), Szafer and Zarzycki (1972) and others stress, after the Ice Ages the two species reached the northern limit of Fir at about the same time. So we cannot explain this distribution pattern in terms of an incomplete migration or one which came to a stop in woodlands already dominated by Beech.

It is worth noting though that wherever Fir is planted within the area of Beech, which goes so much further to the north and west than *Abies alba*, it grows very well, and at times its productivity is greater than in the natural Fir wood in montane belts. In former East Prussia, but also in the very west, near Aurich in East Friesland, there are planted Firs over 40 m high. Under natural competition Fir would have to submit to Beech, at least in the west and north of Central Europe and in the warmer sites on the highlands. Here Beech grows more quickly when young and, according to Mayer (1974) lives to a greater age (200–250 years) than Fir and other conifers (100–180 years). On cooler montane sites the opposite is the case, conifers reaching an age of 400–500 years and overtopping the Beech which hardly ever live beyond 300 years. The summer temperatures and the relative lengths of the growing period must be the reasons for this difference in the growth patterns.

On the other hand the limits of the Fir wood distribution in respect of both altitude and continentality may be related to the winter temperatures. According to Dengler (1912) *Abieteta* are rarely found in areas where the average January temperature drops below −3.5° C. However in western Carpathians there are well developed Fir woods where the figure is −6° C. In Swiss Alps there are many meteorological stations at the upper Fir wood level registering a January mean lower than −3.5° C (fig. 147). In cold winters such as 1928/29 *Abies alba* has withstood temperatures as low as −40° C without damage. However the low winter temperatures may be of significance when it comes to competition with Spruce which even survives considerably lower temperatures and pushes much further into the continental east and the cold north than Fir. In any case fig. 147 shows that Spruce woods in the inner Alpine valleys at the montane level and elsewhere in the subalpine belt have mean January temperatures of between −4 and −8°C., while Firs are dominant near those stations which record a mean January figure of only −1.7 to −4.9°C. In places with a January mean of more than about −2.5°C. it is only on water-logged soils that *Abies* can compete with *Fagus*, whose relationships with other deciduous trees we have already discussed.

While *Abies alba* has been dispersed by planting far outside its natural limits throughout the whole of Central Europe and as far as southern Sweden, in almost all parts of its natural distribution it is in more or less rapid retreat (see H. Schmid and Zeidler 1953 and the literature they quote). The reasons for this Fir dieback are complex and up to now have not been as satisfactorily elucidated as those determining the natural northern limit of *Abies alba*. Mostly it is Spruce which benefits from this retreat. It could be that Spruce has suffered less from earlier grazing and more recently from an increasing amount of browsing from Deer than Beech or Fir, or it may have benefited from changes in forest management. Spruce is favoured by clear felling. Silver Fir regenerates best under a system of single-tree extraction as the saplings can tolerate fairly deep shade and can survive for decades under conditions of low light intensity which would starve out Spruce (Mayer 1960). Vinš (1964) however is inclined to think that young Fir trees also suffer a good deal from shortage of light. In the relatively dense plateau Fir wood only 2% of viable seed developed into young plants.

The decline of Fir started by man has probably been accelerated by natural factors. Nothing certain however is known about this as the investigations into Fir woods and mixed Fir woods are still going on. The researches by Pfadenhauer (1971) into the changes in habitat factors at the boundaries of different Fir-wood communities offer no explanatory causes for the gradual disappearance of Fir which has been observed here, since they had a quite different objective.

Fig. 147. Mean January temperatures in different vegetation belts provided by Swiss meteorological stations.

Each narrow vertical line represents one station, thicker lines indicate more than one. *Abies alba* is absent from belts where the mean January temperature is less than -5°C or above zero.

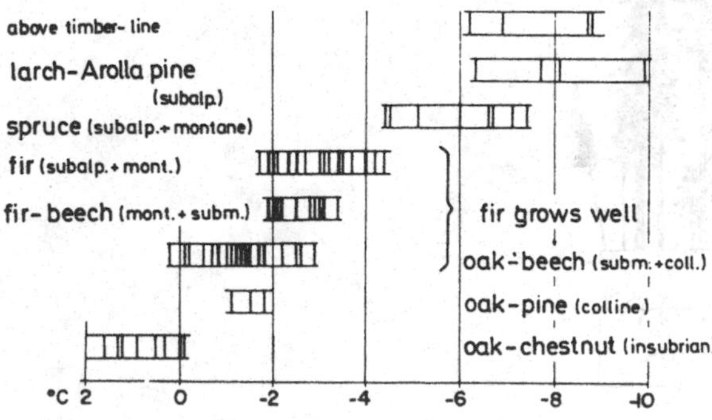

(d) The life cycles of the mixed Fir and Spruce woods
The majority of the stands rich in Fir, like the rest of the woodlands of Central Europe, have been influenced to a greater or lesser extent by the type of agriculture, or by methods of utilisation and regeneration employed in modern forestry. However there are reserves which closely resemble the original forests in south-east Europe and also odd ones in the Alps, in the Bavarian and Bhoemian Forests and in the mountains of Slovakia. They have been studied mainly from the forestry standpoint. Leibundgut (1959) used refined methods to work out the dynamics of the tree populations in the famous primaeval forest of Peručia in Jugoslavia (fig. 148).

It would be wrong, as Rubner (1926) has stressed, to look upon such original forests to some extent as though they work on the method of single tree extraction. They are much more like the forests worked by extracting groups of trees. Leibundgut distinguished a series of phases which the mixed Beech-Spruce-Fir wood lived through, and which could be observed alongside each other forming a mosaic in the forest (fig. 148, see also section B II 2 a). In the optimal phase mature trees, of different ages but all more than 100 years old, form an even, very dark canopy (fig. 149). In the decay phase the trees begin to die, at first singly and then in numbers so that it looks as though groups of trees had been removed by the forester (fig. 149). Often the stand, weakened by age, suddenly collapses as a result of a storm or the weight of snow, insect attack or fire so that regeneration takes place over large areas (fig. 149). This regeneration phase gives rise to open spaces since the shade trees have not been able to regenerate to any great extent prior to the event. In such natural clearings quick-growing light-demanding shrubs and trees such as Birch, Aspen or Willows, whose wind-blown seeds are always present, can form a pioneer woodland. These pioneers however are quite quickly suppressed by the shade trees which have germinated beneath them, or were already present as saplings. Finally they grow up again to a new optimal phase to complete the cycle. It is only in the optimal phase that Fir is able to become dominant as a rule since the deciduous trees, especially Beech, play a much stronger role in the younger stages. Where Beech can thrive it grows much more quickly than Fir or Spruce in the early years and so can gain the upper hand at least temporarily (fig. 150).

Hartl (1967) observed a similar sequence of stages in Derborence, in former times an almost inaccessible side valley of the Wallis, which carries a near natural

Fig. 148. A mosaic of the development phases of an ancient mixed Fir wood (in the Peručica Reserve) in Yugoslavia. After Leibundgut (1959), extract. 1 = regeneration phase, 2 = young stand, 3 and 4 = decay phases, 5 = phase similar to selective felling, 6 = optimum phase.

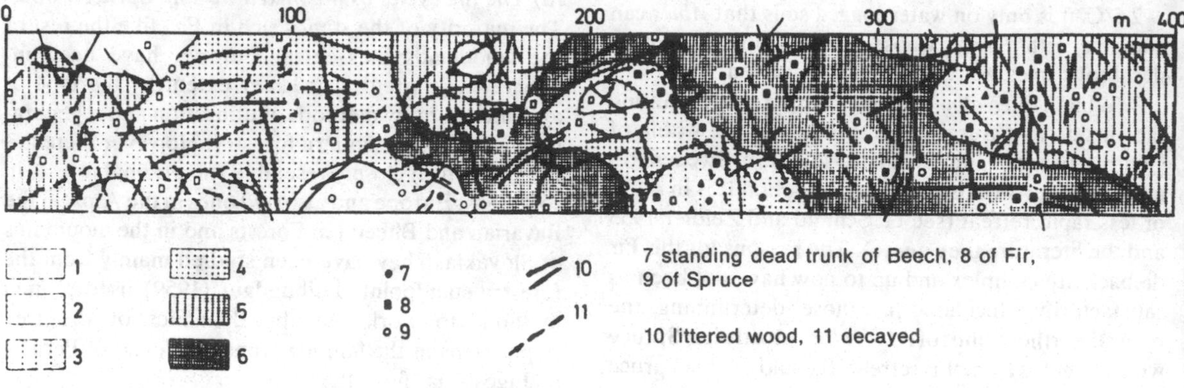

7 standing dead trunk of Beech, 8 of Fir, 9 of Spruce

10 littered wood, 11 decayed

Fig. 149. Above: the optimum phase of an ancient Beech-Spruce-Fir wood in the Klekovača massif (Dinarian Alps). The trees are of approximately the same age (160-180 years - maximum 200 y.) and probably grew up after a storm. Below: the decay and regeneration phase in the same old woodland. in the undergrowth, alongside the Firs and Spruces, there are many young Beech. After Tregubov (1941), somewhat modified.

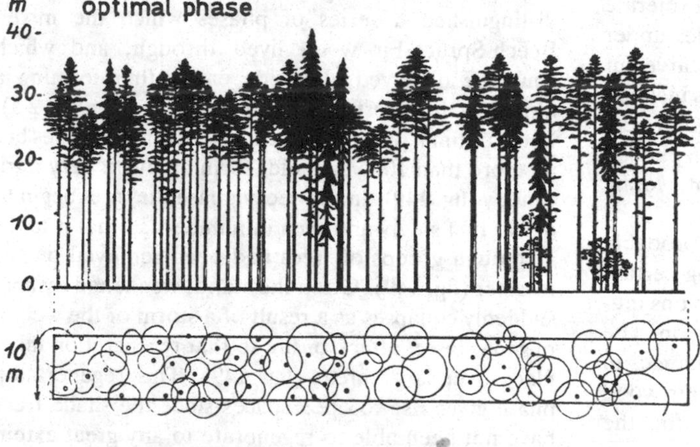

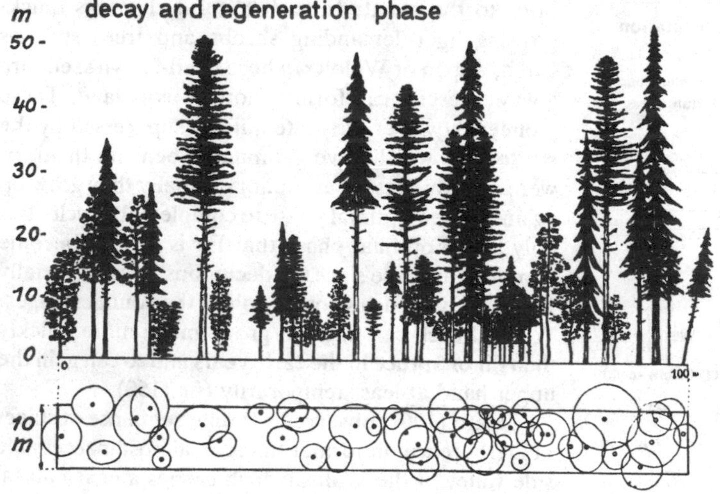

mixed Adenostyles-Fir wood (*Adenostylo-Abietetum*) on a limestone substrate. In the decay phase this is rich in species, including those of the order *Fagetalia*. The optimal phase however is so dark that there is hardly any undergrowth, but in some places an extensive carpet of moss *Mnium spinulosum*, and others). The Derborence is worthy of note since it is close to the dryness boundary of *Abies* towards the inner Alps and is very isolated. Marcet (1971) has noticed that the 'dry Firs' of Wallis are much more drought resistant than those originating from the edge of the Alps with its higher precipitation.

Spruce plays a noticeable role in the undergrowth of near-natural mixed Fir woods especially where there is a lot of dead wood lying around. Many observers agree that Spruce, in contrast to Fir and Beech, occupies rotting tree trunks in large numbers and so later appears in rows and 'on stilts' (cf. especially Eichrodt 1969). In this way it is able to escape the competition of grasses, large herbs and small shrubs, and above all, the root competition of mature trees as well as young Beech and Fir. These, although better at tolerating the shade are not able to thrive on the lignin-rich impoverished substrate. Since dead wood rots down only slowly in the cool climate of the montane belt fallen dead trees are very characteristic of the old forests here and offer a continuing chance for Spruce to come in. On the other hand in the warmer lowlands the remains break down much more quickly so that in, for example, the near-natural Forest of Białowieża (section B III 3 b), one comes across the odd dead tree still standing, but only sees pieces of rotten branches with the occasional trunk lying on the ground. In most tropical rainforests fallen dead trees are very exceptionally met with since all plant remains are broken down with extraordinary rapidity (Ellenberg 1959). Only where healthy trees have been knocked down by a storm or felled does one come across trunks lying on the ground. So any ideas we may have gained from the conifer-rich European mountain forests cannot be applied to the natural forests of other regions.

According to investigations by Rousseau (1960) in Vosges, Silver Fir regenerates best on soils with an abundant moss layer and a covering of humus. On a mull soil, even one with an acid reaction, the young plants generally die off. He was able to demonstrate experimentally that this was due to exchangeable manganese which accumulates in the upper soil by the rapid breakdown of the needle litter and reaches toxic concentrations during dry periods.

A few authors, e.g. Mayer (1960), have noticed that in mixed stands of Spruce and Fir there is a 'species exchange'. Fir regenerates particularly well under Spruce and vice versa (see also Ellenberg 1963). Although this phenomenon is not universal it is very obvious in many places but the reasons for it are still not clear.

Fig. 150. The average annual increment (as measured by the % increase in diameter at chest height 1950-1957) of *Fagus, Abies, Picea* and *Quercus* in mixed woods of the lower montane belt in the Swiss Jura near Neuenburg. Here Beech makes the best growth, especially when young. After de Coulon (1959), somewhat modified.

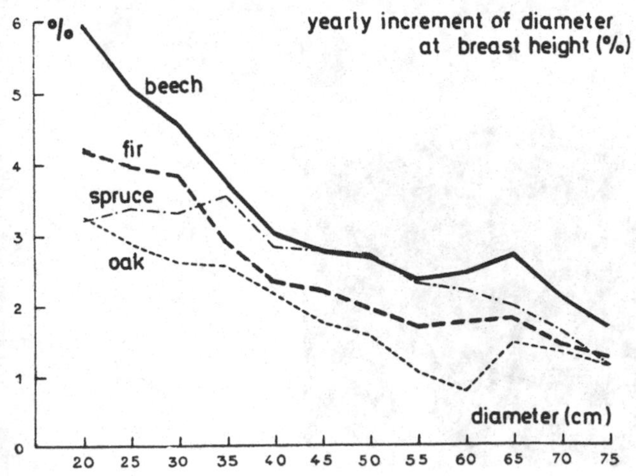

3 Spruce woods

(a) Natural distribution and habitats of the Spruce woods in Central Europe

Spruce wandered into many parts of Central Europe after the Ice Age before its greatest rivals Fir and Beech had arrived (Walter and Straka 1970). But it did not go far beyond the boundaries of its present distribution towards the strongly oceanic west, even at the time of its greatest mass development in the middle warm period. It is only modern forestry that has made it one of the most abundant timber trees in the Central European mountains, and even in many lowland areas (see sections B I 2, A II 3 a, and D III 1 c). This extensive planting of Spruce has shown that its western limit is not an absolute one. It can thrive quite well also in coastal regions. However its growth stops here at an age of about 70-80 years. In the warm-humid lowlands it is also much more susceptible to rust fungi and other diseases because its wood is softer compared with the slower growth in the higher mountain regions. In addition to this, under lowland conditions it is difficult to get Spruce to regenerate naturally in artificial plantations, primarily because of competition from the native species.

According to H.G. Koch (1958) Spruce reaches its greatest productivity neither in the lowlands nor in the natural Spruce belt, but always below the latter, e.g. in Thuringian Forest at about 600 m above sea level, i.e. in the upper submontane belt. The main reason for the productivity decrease towards lower levels is the less favourable water balance. Spruce avoids dry habitats, suffering more than other trees during drought periods because of its shallow root system. However apart from the low precipitation areas in the north-east and a few local dry islands the rest of Central Europe is humid enough for Spruce.

Fig. 151. A subalpine Spruce wood near the tree limit on the Hohe Ifen in the smaller Walsertal. Above it a belt of *Pinus mugo* scrub scattered amongst the bare limestone.

As already pointed out in sections B I 1 and 2, the natural distribution of Spruce (and Pine) woodlands show continental tendencies. In the more oceanic parts of Central Europe the milder winter and prolonged spring favour deciduous trees since conifers gain no advantage from their ability to withstand cold. In the continental regions however the growing period is relatively short and arrives suddenly. Here the possession of overwintering needles allow the Spruce to start assimilating as soon as the first few warmer days have broken their winter dormancy (Mokross unpublished), whereas the deciduous trees must first unfold their new leaves. A continental climate thus restricts the majority of deciduous trees while in an oceanic climate the stronger growth of their saplings enables them to compete successfully with Spruce. Where the climate is favourable both Beech and Fir grow up much more quickly than Spruce and are also able to tolerate more shade (fig. 150). As a result they have driven the latter into the coldest and most continental areas or into habitats where the soil is not suitable for one reason or another. So in Central Europe we find natural Spruce woods under a variety of quite different conditions (figs. 6, 42, 140 as well as 151-160):

As **zonal** vegetation on bed rocks of all kinds:

1 In the lower **subalpine** belt in the Alps (fig. 157) (except the relatively oceanic outer chains on the west and south) and in the mountains of the Hercynian type (see fig. 43).

2 In the montane belt in the relatively **continental** inner Alpine valleys where the climate is too

extreme for Beech and Fir (figs. 153 and 43, central alpine type).

B In **special habitats** restricted to certain places:

1 On stabilised screes (fig. 152), on narrow bands of rock, and occasionally in Karst boulder fields even on mountains which do not have a natural Spruce belt: (fig. 154).

2 In valleys and depressions in the montane and high-montane belts where **cold air accumulates** on clear nights, provided that the temperatures are not low enough to exclude even the Spruce (fig. 155).

3 In **'ice cellars'** at montane level, i.e. on shaded rocky screes in the hollow spaces of which cold air flows down creating conditions which favour the growth of cold-tolerant northern and alpine species (stunted Spruce communities, fig. 160).

4 Round the **edges of raised bogs** in the montane belt of the pre-Alps and on the old diluvial lowlands (fig. 218).

5 On many **moist sites** with a more or less cool continental climate, e.g. in river valleys which are no longer flooded, and in damp depressions of the montane belt, especially in the inner Alpine valleys, but also in the more easterly Central European highlands, with even a hint of them in the Black Forest (fig. 159, 211).

The Spruce woods named under A 1 and 2, and B 1 and 2 are inclined either to the *Fagetalia* or to the *Vaccinio-Piceetalia* according to the base-richness of the substrate. However the strongly acidophilous communities of the latter cover a much greater area. The communities under B 3 and 4, although seldom met with, also belong here. The Spruce-rich river valley woodlands (B5) are consistently close to their neighbouring deciduous woodland communities.

Besides this Spruce is often found, singly or in groups, **mixed** with the numerous communities of the order *Fagetalia* as long as these are in areas with a relatively continental climate, namely:

a) Oak-Hornbeam woods to the east of the Beech limit (section B III 3 b and fig. 1),

b) montane Fir-Beech woods and Fir woods (sections II 2 f and IV 2 b),

c) continental planar and montane Pine woods (section IV V 6).

We shall look more closely only at the Spruce woods which have been determined by the general climate, i.e. those named under A because the 'specialised' ones have a very limited distribution. In spite of the many-sided problems which they offer, most of these have not been sufficiently well researched.

According to Firbas (1949) the Spruce woods in Scandinavia and in the north-east of Eurasia have returned from the same glacial refuges which lay further to the east. The Hercynian-Carpathian part of the present Spruce area is genetically related to these, whereas the Alpine south-east-European Spruce woods come from more southerly refuges. Since the two populations have had no contact for a very long time they contain races of Spruce which differ quite markedly both morphologically and ecologically. The

Fig. 152. A subalpine Spruce wood with occasional Larches (*Piceetum subalpinum vaccinietosum vitis-idaeae*) in the Lower Engadin near Ramosch at the foot of a boulder fall.

Fig. 153. A montane Spruce wood on limestone on the Ifen in the smaller Walsertal, with *Cyclamen purpurascens*.

boreal Spruce woods too can only be compared in a limited way with those of the Central European mountains, although they do show some similarities. For these reasons we shall start from the Alps, which today are the centre of the many types of Central European Spruce wood communities, and not from the gigantic boreal distribution.

(b) Montane and subalpine Spruce woods

The quickest way in which we can get to appreciate the wide variation in the Spruce wood communities is to consult tab. 37 where some inner Alpine examples are compared with each other. It will be seen that they differ in a similar way to the Fir woods set out in tab. 35. Limestone Spruce woods with many fastidious species are contrasted with others poor in species and containing many acid humus indicators.

The **montane** *Piceeta* (tab. 37 nos. 1 and 2) in relatively dry valleys of the Alps occupy sites and soils which, had they been in the humid outer Alps, would have been taken by *Abieteta* or even *Abieti-Fageta*. However the continental climate of the inner Alps only allows the Spruce to grow and here it is found on soils above limestone as well as on very acid bedrocks.

Where the substrate is rich in bases many herbs grow, which are also at home in Fir and mixed deciduous woods. This is particularly true of the whole group shown in tab. 37 led by *Galium rotundifolium*. These species make it easy to distinguish a slightly-humid montane Spruce wood ('*Piceetum montanum galietosum*', fig. 153) from a dry one ('*P.m. melicetosum*'). The first is found in the less extremely continental valleys with a somewhat higher precipitation, e.g. in the Rhine valleys and in western Wallis. Here it contains the odd Fir or even Beech and for this reason Ellenberg and Klötzli (1972) call it Bedstraw-Spruce-Fir wood (*Galio-Abietetum*) and include it in the order *Fagetalia*. The majority of hygro- to mesophilous plants of mixed deciduous woodlands entirely avoid the dry montane Spruce woods of the Engadin and other dry valleys, even on relatively base-rich and loamy soils. Only a few species which can tolerate the dry conditions come in in their place, among them *Melica nutans*. The **Melick-Spruce wood** (*P.m. melicetosum* or *Melico-Piceetum*) is consequently included in the order *Vaccinio-Piceetalia*. Figure 4b gives some idea of the low precipitation of most of the inner Alpine valleys. Here the Spruce does not do so well as it does in the Bedstraw-Spruce-Fir wood. The Melick-Spruce woods in the Engadin are so open that Larch and Engadin Pine (i.e. *Pinus sylvestris* subsp. *engadinensis*) are able to grow up alongside the Spruce and only a little help is needed to bring these light-loving trees into dominance.

Large villages in the Engadin and Wallis, which still lie within the montane Spruce belt, often have in their neighbourhood grassy Larch groves. These are valued more by the farmers than are the natural Melick-Spruce woods, since Larch yields better timber and its undergrowth a good pasturage.

The Spruce valleys of the Central Alps are dominated by bed-rocks which are poor in bases. On the dry slopes it is the Melick-Spruce wood which develops, but where the water supply is somewhat better it favours the development of a more productive Spruce wood type.

This had previously not been distinguished from '*Piceetum montanum melicetosum*'. Ellenberg and Klötzli (1972) separated it as **'Speedwell-Spruce wood'** (*Veronico urticifoliae-Piceetum*), a special association. It can be looked upon as the acid-soil parallel to the montane Bedstraw-Spruce wood.

In contrast to those at the montane level the subalpine Spruce woods (*Piceetum subalpinum* tab.

Fig. 154. Natural Spruce woods and other communities on rockfall on the north-facing slopes of the Hohe Steige in the Swabian Jura. After Th. Müller (1975).

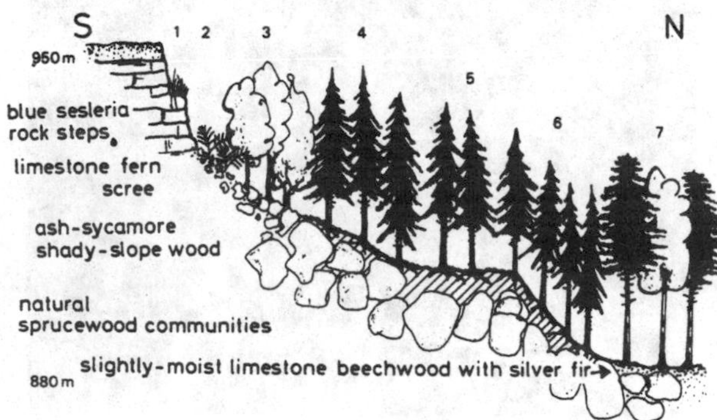

Fig. 155. Only Spruce wood can survive in the frost pocket of the Karst depression in Croatia with its relatively continental climate. After Horvat from Ellenberg (1963).

On the slopes where it is less liable to frost damage Beech dominates right up to the tree limit (3) and merges into a more or less wide belt of prostrate *Pinus mugo*. In holes where the air can be extremely cold (not shown here) these take the place of the Spruce. Fir accompanies the Beech in the upper montane belt becoming dominant on scree slopes and also in the transition (4) to the frost-pocket Spruce wood (5).

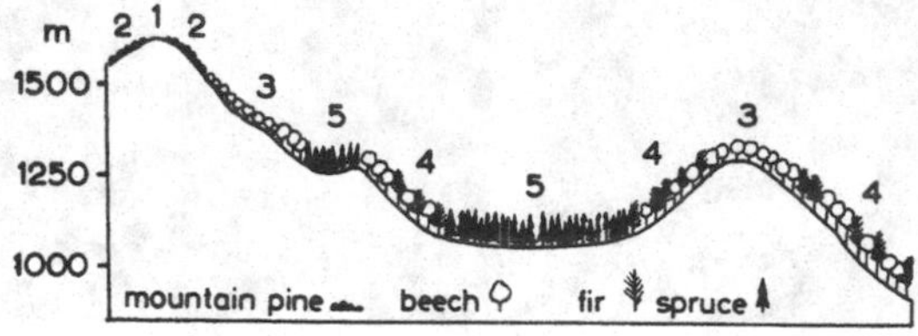

Tab. 37. Montane and submontane Spruce woods in the central Swiss Alps.

Community no.:	1	2	3	4
Vegetation belt	mont.		subalp.	
Climate h = relatively humid d = relatively dry	h	d	h	d
Tree layer				
V *Picea abies*	5	5	5	5
d 2, V *Larix decidua*		2	3	4
d 2 *Pinus sylvestris*			2	
V *Pinus cembra*			2	2
E *Pinus sylvestris engadinensis*				2
Shrub layer[a]				
F d 1 *Abies alba* (seedling)	4			
Q *Lonicera xylosteum*	2	2		
d 2 *Berberis vulgaris*			2	
Sorbus aucuparia	5	3	4	4
c 3-4 *Lonicera nigra*	1		3	2
A *Lonicera caerulea*	1		1	1
O *Rosa pendulina*	2		1	3
Herb layer				
F c 1-2 *Galium rotundifolium*	5			
F *Veronica urticifolia*	5	2	1	3
Q *Carex digitata*	5	2		
Veronica officinalis	5	2	1	1
F *Athyrium filix-femina*	4			
F *Dryopteris filix-mas*	4		1	
F d 1 *Prenanthes purpurea*	4	1	1	2
d 1 *Ajuga reptans*	3			
d 1 *Lysimachia nemorum*	3			
d 1 *Orchis maculata*	3			
F d 1 *Carex sylvatica*	3			
F d 1 *Sanicula europaea*	3			
F d 1 *Neottia nidus-avis*	3			
Q d 1 *Brachypodium sylvaticum*	3			
Deschampsia cespitosa	3			
Geranium robertianum	3			
F *Dryopteris dilatata*	3			
O *Huperzia selago*	3			
F c 1 *Circaea alpina*	1			
Q d 2 *Melica nutans*	1	5		
Q d 2 *Hepatica nobilis*	1	4		
d 2 *Knautia dipsacifolia*		5		2
d 2 *Goodyera repens*		4	2	2
F *Viola reichenbachiana*	5	4		
Mycelis muralis	5	4		
E *Carex alba*	4	5		
Campanula rotundifolia	3	4		
E *Polygala chamaebuxus*	2	3		
Luzula nivea	2	3		
E *Erica herbacea*	1	2		
Ranunculus nemorosus	3	2		

Community no.:	1	2	3	4
O *Melampyrum pratense*	1	2	1	
Carex ornithopoda	3	2		2
c 1-2, A *Hieracium lachenalii*	2	1		2
c 1-2, A *Aquilegia atrata*	3	3	1	3
Fragaria vesca	5	5	3	2
Q *Maianthemum bifolium*	3	5	3	3
Luzula pilosa	3	3	2	3
Oxalis acetosella	5	5	5	5
O *Hieracium sylvaticum*	5	5	5	5
O *Orthilia secunda*	5	5	5	5
A *Melampyrum sylvaticum*	5	5	5	5
c 3-4, A *Luzula luzulina*	3	1	3	1
A *Luzula sylvatica* var. *sieberi*	1	1	1	2
Solidago virgaurea	3	3	2	5
d 2, O *Rubus saxatilis*	2	3	3	5
O *Vaccinium vitis-idaea*	1	5	5	5
O *Vaccinium myrtillus*	3	3	5	5
O *Homogyne alpina*	1	2	5	4
F *Luzula luzuloides*	2		3	4
c 3-4, A *Lycopodium annotinum*	1	1	5	3
c 3-4, A *Moneses uniflora*	2		5	3
Gymnocarpium dryopteris	2		5	2
A *Linnaea borealis*		1	5	4
Dryopteris carthusiana			3	
c 3-4, A *Listera cordata*			4	1
V *Corallorhiza trifida*			2	1
V *Calamagrostis villosa*			5	5
Avenella flexuosa			3	4
d 2 *Clematis alpina*		2	2	5
Geranium sylvaticum				4
Moss layer				
d 1 *Eurhynchium striatum*	5			
Plagiochila asplenoides	5	2	2	
Polytrichum formosum	4		1	
c 1-2 *Cartharinaea hausknechtii*	2			
Hylocomium splendens	5	5	5	5
Rhytidiadelphus triquetrum	5	5	5	5
Cladonia pyxidata	2	2	2	2
Dicranum scoparium	4	4	4	4
Pleurozium schreberi	2	4	5	5
Peltigera canina	2		2	3
c 3-4, A *Ptilium crista-castrensis*	1	1	3	
A *Peltigera aphtosa*		1	5	4
Cladonia furcata		2	4	1
Barbilophozia lycopodioides		1	2	1
A *Mnium spinosum*		1	2	2
Cetraria islandica			2	3
Hypnum cupressiforme	1			3

Note: [a]Young specimens of the trees named in the tree layer have been omitted from the shrub-layer, and a few of the less constant species in the herb and moss layers have been left out.

Nos. 1 and 2: **Montane** Spruce woods.

1: Bedstraw-Spruce wood (*'Picetum montanum galietosum, Eurhynchium striatum-Picea* variant').

2: Melick-Spruce wood (*'P.-m. melicetosum, Carex alba variant'*).

Source: After Braun-Blanquet, Pallmann and Bach (1954), selection.

37, nos. 3 and 4), are poorer in species as a rule, but these are more constant and at times appear in large numbers, e.g. *Oxalis acetosella*, *Vaccinium myrtillus*, *V. vitis-idaea*, *Calamagrostis villosa*, and *Hylocomium splendens* (fig. 156). The Spruces are mostly stunted because they have to contend with severe winters (fig. 157). Such monotonous Spruce woods are also very widespread outside the Alps. In the Tatra and in other Central European highlands the Spruce is accompanied by the Carpathian Birch (*Betula carpatica* = *oycoviensis*) which acts as a pioneer. One of the reasons for the poor growth of these subalpine woods on acid soils is lack of nutrients. With rising altitude the supply of mineral nitrogen is reduced (Ehrhardt 1961). In the montane belt it consists only of ammonia (Stöcker 1968) The worsening of the climatical and edaphical conditions also lowers the reproduction of the trees. According to Sorg (1980) the number of Spruce saplings per unit area of otherwise comparable habitats is negatively correlated with altitude.

Sphagnum and other wetness indicators are found on waterlogged soils in all the acidophilous Spruce woods, especially in the subalpine region. We have already met the **Peatmoss-Spruce wood** ('*Piceetum subalpinum sphagnetosum*' or *Sphagno-Piceetum*) in tab. 35, col. 7. It is frequent mainly around the edge of the Alps where there is a high precipitation.

The subalpine and montane Spruce wood communities of Swiss Alps correspond to very similar ones in those parts of the Alps lying further to the east. In contrast to Beech woods and Fir woods very few species spread from their distribution in the south-east and east into the Spruce woods, and only then on base-rich soils, e.g. *Aposeris foetida*, *Helleborus niger*, *Cardamine trifoliata* and *Aremonia agrimonoides*. On the other hand some of the oceanic elements are absent from the east Alpine Spruce woods (Pignatti-Wikus 1959 and others). In the eastern Alps of Slovenia a **Woodrush-Spruce wood** is widespread on lime-deficient rocks (*Luzulo sylvaticae-Piceetum*). It has been thoroughly described by Wraber (1963) and is astonishingly similar to communities occurring further west.

Analogous to the Fir woods Mayer (1974) distinguishes a number of associations and subassociations of Spruce woods in the eastern Alps. These form communities in the montane belt in every way parallel to the units shown in tab. 36. His 'montane limestone *Adenostyles*-Spruce wood with Bed-

Fig. 156. A transect showing the species mosaic in a seminatural Shaggy Smallreed-Spruce wood with Vaccinium (*Calamagrostio-Piceetum*) on podsol in the high Harz. After Stöcker and Bergmann (1975), somewhat modified.

The micro-relief has been formed by kryoturbatic pressure shaping the sub-soil during the late Ice Age. Rain and melting snow water run off the hillocks making them less fertile; they are occupied mainly by dwarf shrubs (Vaccinium). With the water some nutrients, even though only in small quantity, were carried into the hollows so that more demanding plants, especially the Shaggy Smallreed, were able to spread here. The supply of mineral nitrogen is quite good; ammonia may even be available in excess of need (Stöcker 1979). Woodland sphagna grow in all these microhabitats in the high rainfall climate of the Harz Mountains.
O = covering of raw humus A_{he} = humus-rich bleached sand, A_e = ash grey bleached sand, B_h = humus hardpan, B_s = iron hardpan, C_v weathered rock, poor in bases.

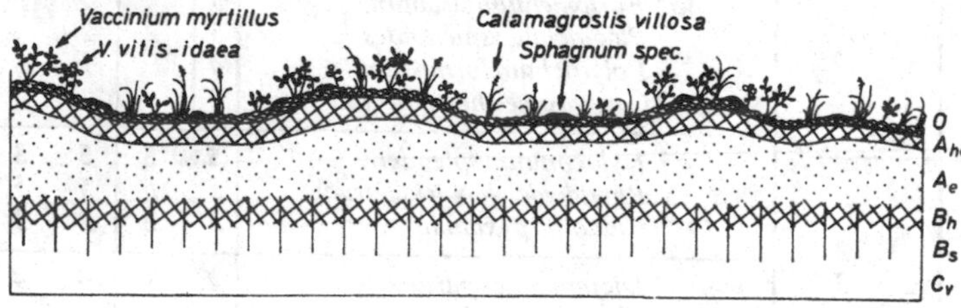

Tab. 37. (Continued).

Nos.3 and 4: **Subalpine** Spruce woods
3: Bilberry-spruce wood ('*Piceetum subalpinum myrtilletosum, Linnaea* variant')
4: Cowberry-Spruce wood ('*P. s. vaccinietosum, Peltigera-Hylocomium* variant').

Groups of species in the herb and moss layers which behave similarly in the four communities have been separated by horizontal lines.

c 1-2 = character species of the *Piceetum montanum* (cols. 1 and 2)
c 3-4 = character species of the *P. subalpinum* (cols. 3 and 4)
d 1 etc. = differential species of the subassociations
A = allinace character species (*Vaccinio-Piceion*)
O = order and class character species (*Vaccinio-Piceetalia*)
F = character species of the Spring Heath-Pine woods (*Erico-Pinetea*, etc.)
F = character species of the noble broadleaved woods (*Fagetalia* or *Fagion*)
Q = class character species of the broadleaved woodlands (*Querco-Fagetea*).

Communities 1 and 2 contain more of the character species of the broadleaved woodlands and of the Spring Heath-Pine woods than of the Spruce woods. For this reason they have had to be included in the order *Fagetalia* (see Ellenberg and Klötzli 1972).

A moist-soil subalpine spruce wood ('*Piceetum subalpinum sphagnetosum*') is included in tab. 35 (no. 7).

straw' (*Adenostylo glabrae-Piceetum montanum galietosum*) corresponds to the Bedstraw-Spruce-Fir wood (*Galio-Abietetum*) in Switzerland and his 'montane Wood Sorrel-Spruce wood with Melick' (*Oxali-Piceetum montanum melicetosum*) to *Melico-Piceetum*. As the synopsis of all the woodland communities of the eastern Alps by Mayer (1974) is readily available we shall confine ourselves to these comments. It also considers the Spruce woods of northern Italy which S. Pignatii (1970) has summarised. Hartmann and Jahn (1967) have made a survey with numerous relevés of the Central European mountainous areas to the north of the Alps, which may also be referred to. Here the **'Shaggy Smallreed-Spruce woods'** (*Calamagrostio villosae-Piceetum*) play a significant role. This is, for instance, the dominant vegetation on the Harz from 750 m above sea-level up.

Just as in the cases of the Beech woods (section B II 1 b and 4 d), the Oak-Hornbeam woods (section III 3 f) and the Fir woods (section IV 2 b), so the Spruce woods can be divided into three groups according to the acidity and the type of humus in the soil as well as to the corresponding species composition. We could call them raw-humus, moder and mull-moder Spruce woods. However these are more difficult to separate and in no case can they be given the rank of alliance as was proposed by Passarge (1971). Because the Spruce needle litter is so slow to break down the communities are very similar whether the trees are growing on lime-rich or on lime deficient soils (fig 48). Even on a mull-rendzina a surface layer of moder will build up favouring the acid-indicating plants thus accounting for the term 'mull-moder Spruce wood'. According to Stöcker (1980), however, the quality and nutrient

Fig. 157. In higher belts Spruce (*Picea abies*) certainly shows a relatively better growth than Beech. However it too can be damaged as the cold becomes more intense. Hoar frost, the weight of snow and storm winds cause the tops of the trees to break off. After Kadlus (1969), for mountains in Czechoslovakia.

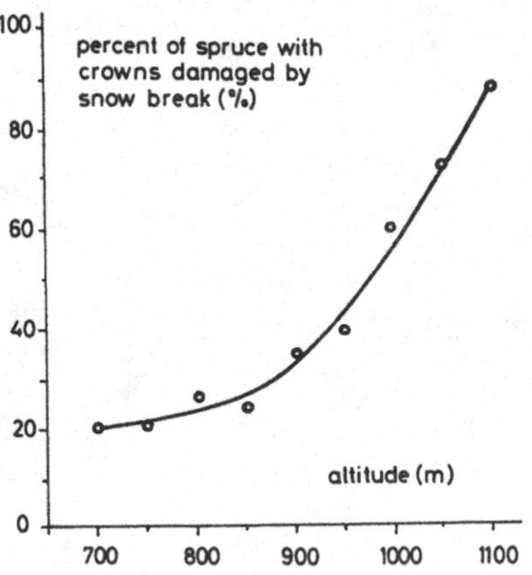

turnover of the humus layer depend more strongly on the mineral subsoil than was generally accepted before.

There are a number of old forest reserves in the subalpine Spruce woods which give us an impression of the natural state of these woodland communities. All of them are relatively poor in species. One of the finest is the ancient forest of Scatlé in the Grisons district of Breil/Brigels which lies in the Central Alps on a Verrucano scree. According to Hartl (1967) this is a *Piceetum subalpinum myrtilletosum* with raw humus. By calculations from the amount of production on this relatively poor dry soil Hillgartner (1971) put the ages of the trees at from 300 to 650 years. It was noticed that in places there were seedlings and young stocks of trees. Depending on altitude (1580 to 2015 m) the regeneration phase (see section B II 2 a and IV 2 d) of this Bilberry-Spruce wood lasted from 80 to 130 years. The optimal phase, which was mostly of one layer and very dense, remained for 200 to 260 years, while the senescent and decaying phases together lasted about as long again (100 to 160 plus 50 to 100 years). Thus between about 300 and 500 years could elapse from one crop of seedlings to another on the same piece of ground.

The subalpine original Larch-Spruce wood (*Piceetum subalpinum luzuletosum luzuloidis*) on the Lasaberg in the Austrian Lungau, which has been analysed by Mayer (1966, see also 1976) is at the start of its optimal phase and also tends towards a single layer. Since it lies at a somewhat lower level (1650 m) its vitality is greater than the majority of subalpine woodlands. Its dense stand with slim and relatively tall trunks indicates that it should be looked upon as montane rather than subalpine. Some Spruce stands in the Bavarian Forest are considered to be also nearly virgin (Petermann and Seibert 1979).

On Wimbachgries in the limestone Alps near Berchtesgaden there are screes of dolomite showing clear examples of the establishment of natural Spruce woods and their destruction by sharp erosion while still young in the succession series (Mayer, Schlesinger and Thiele 1967). Obviously Spruce is able to gain a footing on such bare stony ground quickly, and needs neither fine-earth cover nor herbaceous pioneer vegetation.

(c) On the natural rôle of Spruce on the lowlands

Spruce has spread naturally into wide areas of the lowlands as a member of mixed woodland communities. This is true not only for north-east Europe (figs. 140 and 158) but also for the whole of Central Europe with the exception of a narrow coastal belt in the north-west and a few dry areas in the centre and the south-east (see section a). In places where there is not much competition e.g. round the edges of raised bogs, in acid marshy ground and on waterlogged soils they can even become dominant at the submontane and planar levels. Dengler (1912) had already recognised such islands of Spruce going right out towards the sea on the north-western plains, and later these were confirmed by Firbas (1949, 1952). Hesmer and Schroeder (1963) and others. Grosser (1956) was able to demonstrate their presence convincingly in the Lausitz region. Recently Schroeder (1973) has succeeded in recording an isolated example at the submontane level in the western foothills of Harz near Westerhof. In the submontane belt north of the Swiss Alps Spruce is at home in well grown Fir woods on acid, waterlogged soils of the Riss-Glacial plateaux (section 2 b, fig. 144). The further east one goes into the pre-Alpine region the greater the part played by Spruce in very different habitats becomes. Already in the Białowieża Forest (see section B III 3 b, Sokołowski 1966a) and south-east of the Baltic Sea (fig. 158) *Picea abies* forms communities of its own. In Czechoslovakia *Picea* goes much further down into the warm basins than had previously been supposed (fig. 159). Here, as Malek (1961) demonstrates, it is absent only from a small part of inner Bohemia and the large Slovakian Theiss-Danube plain. However Spruce is particularly plentiful and appears regularly in deciduous and mixed woodlands of the lower slopes and valleys in the eastern Alps (fig. 3), in a similar way to Larches and Firs which have also been investigated by Tschermak (1935a and b, see also Mayer 1974).

These conifers had practically disappeared from the lower-lying woodlands by th- time the reafforestation started in the past century. The reasons for this retrogression are to be looked for, according to Stamm (1938) and others, in the practice of coppicing which has been carried out since the Middle Ages, and also in the cutting of leafy branches for fodder which has been done for more than two thousand years (section A II 2). All conifers recover less well from this continual cutting than do the majority of deciduous trees and shrubs. Because of this they were practically wiped out in the more densely populated valleys.

Fig. 158. A natural mixed woodland rich in Spruce on loamy moraine soil in southern Estonia. As a result of browsing and grazing this has become more open with grassy clearings and fewer trees. *Dryopteris filix-mas* can be seen in the foreground.

Fig. 159. A mixed Horsetail-Spruce wood on fertile gley soil in the Bohemian-Moravian mountains at 550 m. *Equisetum sylvaticum* and *Vaccinium myrtillus* dominate the ground flora. Photo. Iltis-Schulz.

The clear separation in south-west Central Europe of a pure deciduous woodland at the planar and submontane belts from the mainly coniferous woodland of the montane to subalpine belts, which was presumed to be due to natural causes, is now seen to have been brought about by man's activity and is a relatively recent phenomenon. During the past two hundred years or so a reversal of this process has been going on and now in many areas, especially on poor soils, the natural proportion of Spruce and other conifers has been exceeded (section D III 1).

(d) Habitat conditions in different Spruce wood ecosystems

Almost everywhere that Spruce is found growing in pure stands, its evergreen canopy and the needle litter, which is difficult to break down, create conditions under which mosses, dwarf shrubs and Wood Sorrel can flourish. The last named in particular does not appear so regularly and plentifully in any other community. We must not be misled by the similar appearance of most Spruce wood communities however into thinking that conditions there are uniform. The individual associations and subassociations differ, especially in different climatic areas, in the way in which we have already outlined in the discussion of table 37.

These four communities serve as very good indicators of the different climatic conditions. Within each of these climatic regions Spruce wood communities may be met with on quite different soil types. For example Braun-Blanquet, Pallmann and Bach (1954) found their '*Piceetum subalpinum myrtilletosum*' on more or less well developed iron podsols, but also on rankers (humus-silicate soils) and on rendzinas (humus-carbonate soils). So the equation '*Piceetum* = podsol', which is often repeated, in no way applies. The only thing which is common to all the soil profiles is a surface layer of humus which, as Pallmann says, makes the habitats analogous, at least to a certain degree. The average thickness of this humus layer decreases as the soil becomes drier. On many renzinas it is not real raw humus (mor) but a 'tangel-humus' as defined by Kubiëna (1953), which changes to a mull underneath and does not give rise to podsolisation. Because of this it has a less acid reaction than the mor, which forms under Spruce on most soils, above all on those rich in quartz, but poor in silicates and lime. It is only the uppermost organogenic horizons then which have a similar genetic and ecological value in all the *Piceeta*. The deeper horizons, especially the mineral ones, can differ markedly depending on their geological origin, altitude, aspect and the age of the soil. Incidentally the production of litter does not depend only on the Spruce and its partners; other conifers as well as ericaceous plants and mosses also produce a humus layer with similar properties when climatic and geological conditions are the same. The thing they have in common is the very wide C/N ratio (from 30 to over 40) of the products of decomposition, leading to a weak nitrogen cycle (fig. 48).

The mineralisation of nitrogen in such humus coverings is really very slight. According to Ehrhardt (1961) it decreases with increasing altitude and is at its lowest in the subalpine region. In spite of this Stöcker (1968) found a surprising amount of ammonia in the strongly acid raw humus of the subalpine '**Carpathian Birch-Spruce woods**' of the upper Harz. This is because the almost continuously low temperature of the soil prevents the trees from absorbing water, and with it dissolved nutrients, whereas these conditions have a relatively less-inhibiting effect on the ammonification process. Under such conditions then nitrogen does not seem to be a limiting factor. Instead even a Spruce wood community here is suffering from a lack of warmth.

However the '**stunted Spruce wood**' in the 'ice cellars' (B 3 in section a) withstands still more cold and can be looked upon as an ecological curiosity (fig. 160). J.L. Richard (1961) has studied the microclimate of one example in the Creux du Van in the Swiss Jura over the course of two years, comparing it with a Spruce wood on a rock scree and a Fir-Beech wood at about the same altitude (tab. 38). Although all three communities were growing on north-facing slopes the air temperature within the open Dwarf-Spruce stand was higher during the growing period than that in the scree Spruce wood or even in the Fir-Beech wood. But the soil temperature under the stunted Spruce wood always remained significantly lower than under the two high forest stands (tab. 38). At a depth of 160 cm it never rose above 2° C during the summer. This is certainly due to the cold air flowing constantly down between loose boulders in the subsoil of the scree. On the other hand the differences between the three stands were eliminated during the winter when their soil was protected by a covering of snow. The deciding factor then is not the lowest temperatures of the air and soil but the continuous year-round lack of warmth at the roots which reduces their capacity to take up water and nutrients. This excludes Beech and Fir and dwarfs the Spruce so that Mountain Pines and many alpine and subalpine species in the undergrowth become more competitive.

Rocky screes through which cold air flows down have also been observed in other parts of Central Europe. Long ago they aroused the interest of botanists as restricted areas where boreal and alpine plants could be found at a lower level (a survey of the literature by Furrer 1966 and further references by Wilmanns 1971). They always carry an azonal vegetation forming a solitary island in an 'ocean' of more thermophilous vegetation, and they evoke the same problems of far distance dispersal as marine islands do.

Like all higher plants Spruce eagerly takes up **nitrogen** provided it is not prevented from doing so by low temperatures. The supply may be in the form of ammonia as well as nitrate (Tamm 1975). According to Strebel (1960), Tamm (1965, 1974) and many other authors added nitrogen will increase the productivity of Spruce very considerably. Under high light intensity (40 klux) Keller (1971) found that young Spruce with a poor nitrogen supply assimilated CO_2 at only about one third of the rate of ones with a good supply. Respiration of the shoots and roots as well as the rate of transpiration were also higher. In spite of this the needles which were well supplied with nitrogen transpired more economically. As the nitrogen supply is increased the number of shade leaves formed also

Fig. 160. Rock fall and coarse scree in the Creux du Van on the Swiss Jura. On the scree the shortage of fine earth and the effect of cold air outflow result in distorted *Pinus mugo* and *Picea* (see tab. 38). On the plateau and the upper slopes are Beech and Beech-Fir woods.

Tab. 38. Air and soil temperatures in a Fir-Beech wood, a boulder scree Spruce wood and a stunted Spruce wood on the edge of the Creux du Van or on its inner screes which are subject to cold air flow (see text and fig. 160).

Community	Altitude (m)	Measurement point with height/depth (cm)	Annual mean (1958+59)	Summer mean (15.V. – 15.X.)	Winter mean (15.×. –15.X.)	Extreme temperature values: Maximum	Extreme temperature values: Minimum
Fir-Beech wood ("*Abieti-Fagetum*")	1220	air + 100	7.8	14.8	2.7	–	–
		A_1 –5	4.7	9.7	1.1	13.5	**–4.0**
		C –130	4.1	7.2	1.9	9.5	**–2.2**
Boulder Spruce wood ("*Asplenio-Piceetum*")	1190	air + 100	7.9	15.7	2.3	–	–
		OH –5	2.9	8.2	**–0.9**	12.0	**–2.5**
		C –140	0.4	2.1	**–0.8**	2.7	**–2.5**
Dwarf Spruce-Mountain Pine wood ("*Lycopodio-Mugetum*")	1190	air + 100	6.8	16.8	2.8	–	–
		OH –5	1.2	5.3	**–1.8**	8.2	**–3.0**
		C –160	**–1.5**	**–1.0**	**–1.9**	2.0	**–2.5**

Note: **Temperatures below zero are shown in semi-bold type:** (the highest values for the air temperature are framed.)

Source: From data in the years 1958 and 1959 by J. -L. Richard (1961). The temperatures are in °C.

increases and the life of the individual needle is prolonged so that the leaf surface index rises (cf. section D III 1). Pineau (1968) pointed out that the needles of well nourished Spruce are more active in the winter than those suffering from shortage of nutrients. On the other hand they are less frost-hardy (Tamm 1975). The correlations between potential productivity of *Picea abies* and the habitat factors, mainly the nitrogen supply, have been studied fundamentally by Pfadenhauer (1975).

It almost goes without saying that the ground flora of the Spruce wood also responds to the nitrogen factor. In the first place it has an effect on competition in a way similar to that in the deciduous woodlands which we have already dealt with in some detail (Karpov 1975; Treskin, Abrazho and Karpov 1975). It also influences the soil fauna and thus the breakdown of litter and the formation of humus (Wittich 1963). If there is a good supply of nitrogen and provided the soil is sufficiently base-rich even a mull-like moder can be formed through the activity of earthworms. Zachariae (1967) noticed that these never ate freshly fallen needles but only those which had lain on the ground for two or three years. Only after decay had been started by bacteria did the species of *Lumbricus* swallow the needles and break them down further. When the needles are returned as worm casts their cell structure is unaltered. Even in the podsolised soils with their low biological activity there is a certain amount of humus formation. For example Tamm and Holmen (1967) were able to show, using the C^{14} method that the B_h horizon in Swedish coniferous woodland had a maximum age of 1260 ± 60 years and that a continuous conversion of humus took place in it.

In contrast to nitrogen, phosphorus has scarcely any effect in increasing the productivity of Spruce woods as shown in trials by Holstener-Jørgensen (1970) and Tamm (1975). This nutrient, which has such a marked effect in aquatic habitats (see section C I 1), is generally present in sufficient amounts, just as it is in the Beech woods investigated by W. Schmidt (1970).

Of the common ground flora in Spruce woods Bilberry (*Vaccinium myrtillus*) is the one which has been most investigated ecologically. It demands a better nutrition than Crowberry (*V. vitis-idaea*, see Ingestad 1973). Unlike Spruce it can only absorb nitrogen in the form of NH_4 as it is unable to make nitrate-reductase (section B II 4 d). Also it is considerably more susceptible to frost and requires the protection of snow (fig. 161). In winters when there is a shortage of snow it can be killed off completely above the ground (Havas 1965), partly because of drying out through transpiration while the roots are frozen. However it is able to take up water at quite low temperatures and its shoots can even make use of the snow (Havas 1971). Chickweed Wintergreen (*Trientalis*) has a tendency towards a distinctly boreal distribution and has only a few outposts in the Spruce woods of the Alps. It requires low night temperatures which are almost a prerequisite for a sufficient biomass production (Anderson and Loucks 1973). So in the herb layer there are specialist plants which are adapted to the unfavourable conditions which the majority of Spruce wood communities have to tolerate.

Amongst the numerous fungi living in Spruce woods we find a number of mycorrhiza producers which are more or less strongly dependent on *Picea*

Fig. 161. Minimum temperatures between the shoots of Bilberry (*Vaccinium myrtillus*) and below the humus layer into which it is rooting, showing the effect of a covering of snow. After Havas (1965), modified. Dotted lines = without snow cover, unbroken lines = with snow cover; thick lines = 15 cm above ground, thin lines = 10 cm below ground.

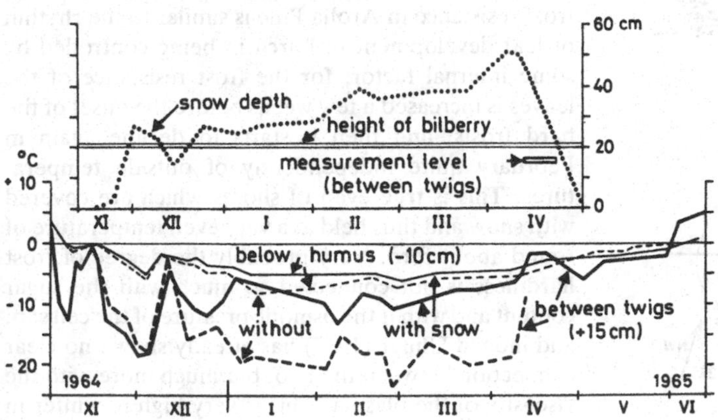

abies. Others show a close or loose association with Pines, Larch, Birch or Green Alder. E. Horak (1963) has investigated these in the subalpine woodlands of Grisons. A wider look at the relationships between fungi and higher plants in coniferous woodlands would certainly be a profitable field of joint taxonomic and ecological research.

4 Subalpine Larch-Arolla Pine woods and Larch woods

(a) Living conditions for Larch and Arolla Pine in the central alps

In the pre-Alps and on the slopes of the inner alpine valleys the subalpine Spruce woods climb to heights of between 1700 and 1900 m where the average annual temperature is only about 1.5° C (figs. 151, 66 and 162). Some Spruce may just manage to live as stunted forms protected by the snow even up above 2400 m. However where real woodland is able to survive above 1900 m it is usually domination by Larch (*Larix decidua*) or *Arolla Pine (Pinus cembra*, fig. 163). Both these derivatives of Siberian conifer type groups have raised the tree line to over 2200, even to 2450 m in some places in the continental climate of the inner Alpine 'coniferous zone' (section C VI 1 c, figs. 42, 43 and 4, Schiechtl 1970).

Resistance to very sharp frosts is a prerequisite for this remarkable ability, since during the winter months the temperature can fall below -30 and rarely fails to reach -20° C. Larch avoids the dangers of the cold time of year as it is the only one of our conifers which sheds its leaves. An endogenous rhythms triggers off the yellowing of its tender leaves quite early in the autumn and prepares its buds for the winter. If Larches from high mountains are brought down to the lowland they still keep this rhythm and come into leaf much later than the lowland races (see the accounts of this type of work collected by Kalela 1937). If, on the other hand, the latter are transplanted into the harsh climate of the subalpine belt they still follow their natural inclination to come into leaf early and to extend the growing season into the autumn, often resulting in the leaves being sacrificed to a late frost in the spring or to an early one in the autumn. Their buds too appear to be less frost-hardy than those of the subalpine races.

Pinus cembra does not avoid the rigours of winter in the same way as the Larch, but spites them with its evergreen needles. These are hardened to withstand a temperature of -40° C during the depth of winter without damage (fig. 165). Such needles are 'ice hardy', i.e. able to withstand the formation of ice in their interior (Larcher 1963b). As is the case with almost all plants in our latitudes however their frost hardiness changes during the course of the year. In July, in their natural habitats, they have to withstand a temperature which only falls a few degrees below zero and even those which have been acclimatised by holding them for some time previously at low temperature never achieve their winter resistance in the summer (Pisek 1950). Apparently this rhythm of frost resistance in Arolla Pine is similar to the rhythm of leaf development in Larch in being controlled by some internal factor, for the frost resistance of the leaves is increased a few weeks before the onset of the hard frosts, and then it starts to decline again in February quite independently of outside temperatures. This is true even of shoots which are covered with snow and thus held at a very even temperature of round about zero. Very probably the degree of frost hardiness is not connected so much with the sugar content and with it the osmotic pressure of the cell sap, and indeed Ulmer (1937) has already shown no clear connection between the two, but much more with the viscosity of the plasma. This is very high in winter in

Fig. 162. Spruce trimmed into flag shapes by icy blasts on a steep north-west facing slope in the Czechoslovakian Tatra. Woodland outposts like this can spread vegetatively amongst the surrounding mountain Pine scrub when branches, pressed to the ground by the weight of the snow, form roots. After Myczkowski (1973), somewhat modified.
The development of such growth forms and of similar ones in avalanche tracks has been studied in detail by Schönenberger (1978).

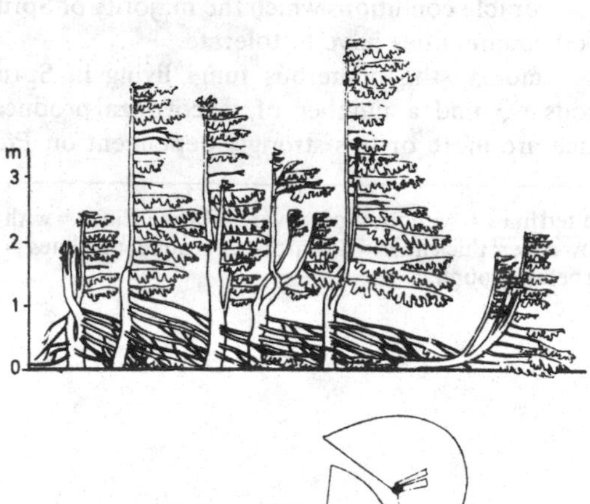

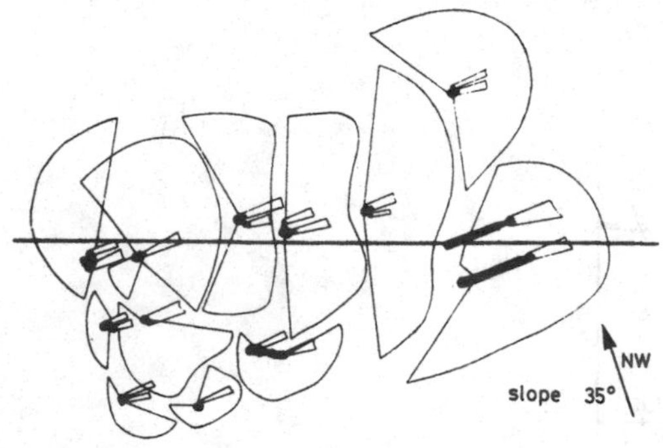

Pinus cembra, and the water content of the needles is at a minimum at this time. Therefore the cell interior never freezes through completely (Tranquillini 1958), but step by step up to about 50% of the total. The yearly rhythm of frost hardiness is controlled above all by the day length, that means largely independently of the temperature (fig. 165). This can be shown in greenhouse experiments where a high even temperature is maintained. Also under these conditions the frost hardiness is relatively much greater in winter than in summer, although the temperature may reach only – 15°C instead of less than – 40° C in the open.

Many plants of the undergrowth would succumb to conditions of temperature which present no danger for the Pine. According to Ulmer (1937) the Alpenrose, for instance, can only survive frosts down to –28° C. Nevertheless it can live at higher altitudes than the furthest outposts of Pine going right up to the lower part of the alpine belt. Here it survives in places where it is certain to be protected by snow. If insufficient

Fig. 163. A grazed stand of Arolla Pine in the Aletsch Forest above Brig. In the clearings is an Alpenrose heath community (*Rhododendron ferrugineum*). The old Pine trunk in the foreground shows fire damage.

Fig. 164. Growth forms of Arolla Pine (*Pinus cembra*) from the lower montane belt right up to the tree limit. After Jugowiz (1908), modified.
1 = free standing on a manured pasture (1080 m), 2 = free standing in a south-eastern position sheltered from storms (1650 m), 3 = in a mixed woodland with Spruce (1350 m), 4 = with Larch on an exposed north-west site (1650 m), 5 free standing in a position similar to 4, 6 = in the uppermost 'combat zone' with extreme conditions (2050 m).

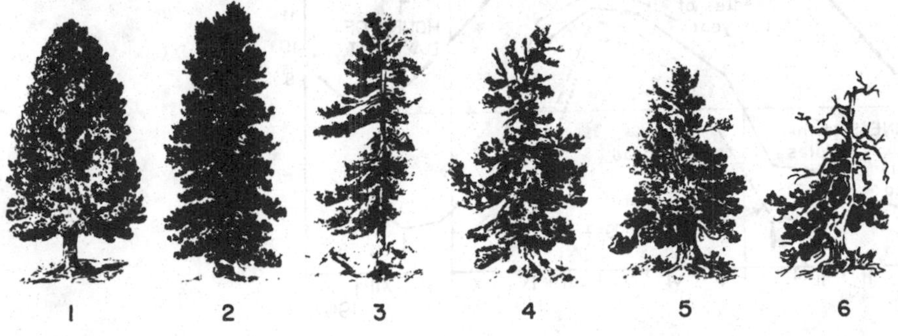

snow falls or is blown away by the wind then the plant's shoots are frozen or it succumbs entirely to drought caused by the frost (fig. 334). In the Larch-Arolla Pine woods of Upper Engadin Pallmann and Haffter (1933) found that *Rhododendron ferrugineum* was only well developed outside the area of the tree canopy since this held back a large part of the light snowfall. It requires a snow depth of more than 70 cm if it is to come through unscathed, and the majority of the other evergreen dwarf shrubs of the Larch-Arolla Pine wood are equally susceptible.

In spite of its evergreen needles Arolla Pine has almost as short a growing season as Larch. This was very nicely shown by a series of measurements carried out by Turner (1958) of the gas exchange throughout the year up near the tree-line above Innsbruck (see also Tranquillini and Machl-Ebner 1971 and fig. 165). The needles are covered by thick cuticle and in winter the stomata remain closed so that too great a water loss is prevented. What little transpiration does take place at this time can be replaced by water from the deeper roots because under snow it is only the uppermost layer of the soil which might be frozen. Even on sunny days the stomata remain closed so that any carbon dioxide produced by respiration can be reassimilated. During this time the amount of chlorophyll is often so small that the needles appear yellowish. Only when the night temperature no longer falls so low do the needles awake from their winter rest and then as soon as the upper roots too are able to come into action the gas exchange rate quickly increases. However it does not reach its maximum until July and by September the night frosts bring about a return to the economical winter regime. Whenever the growing period is extended by even a few weeks and the temperature is slightly higher there is a marked increase in production. This is reflected in the wider annual rings in these subalpine trees (fig. 166). Where the temperature is high enough then light intensity can become a limiting factor for photosynthesis as it is for the trees at lower levels (figs. 167 and 97).

Since suitable growing conditions rarely last for more than three months the annual growth is very slight and it takes at least 100 years before an Arolla

Fig. 165. Changes throughout the year in the frost hardiness of the needles of a young *Pinus cembra* on the Patscherkofel (2000 m) showing its relationship to the air temperature, depth of snow, and length of day. After Schwarz (1968 and 1970), combined.

Frost hardiness is at a minimum in early August. Young needles at this time are damaged if the temperature falls below -2°C; two and three year old needles can stand down to -4°C. At -6°C 50% of the older needles are killed and below -9°C none survive.

In November and December similar needles show a high degree of frost resistance and can withstand temperatures below -40°C. This early increase in hardiness enables them to survive a keen frost at this time even if they are not covered by snow.

When the snow melts in April or May the hardiness of the needles is only about half of what it was at the beginning of winter. Exceptionally severe late frosts then can seal their fate (the lowest air temperature at a height of 2 m was -24°C in January both in 1966 and 1967).

By and large the changes in frost resistance correspond to those of the duration of daylight so that the plants are prepared for the coming season independently of the weather conditions. In addition, cold periods strengthen the frost resistance, warm periods make the needles more tender.

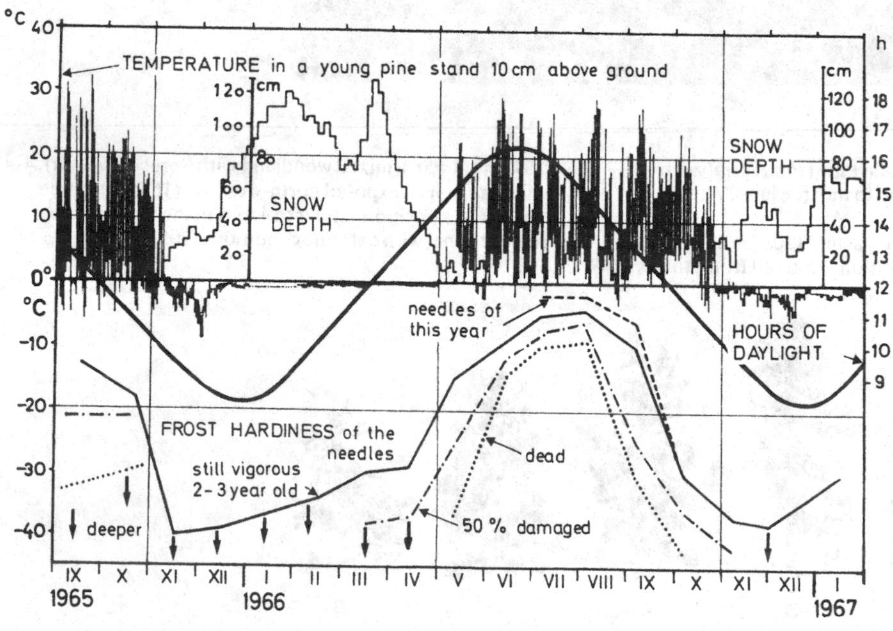

Pine in the upper subalpine belt reaches 10 m in height. The wood is very dense and resistant to pests and diseases. This means that such Pines can live to a great age – Hess (1942) has counted more than 1000 annual rings on a number of occasions in the Aletsch reserve – and finally they can reach an awe-inspiring size. Larch grows somewhat more quickly and seldom survives for more than 300 to 400 years (maximum 700). During this time it can reach a height of about 25 m. Both trees grow much better and more evenly in warmer climates, even those of the high mountain races. This is well known in the case of Larch as it is often planted at colline and planar levels. It also occurs naturally in mixed woodlands down in the valleys of the inner Alps. However Arolla Pine can also grow at lower altitudes where it makes fine trees with large crowns bearing no resemblance to the romantic forms produced by frost, storm and lightning strikes (fig. 164). Larch too grows very slowly under the climatic conditions of the subalpine belt, so slowly in fact as could hardly be imagined in view of its performance at lower levels.

Auer (1947), following a many-sided investigation, came to the conclusion that even in the Upper Engadin, one of the largest areas for Larch in Central Europe, *Larix decidua* is 'never under its optimal conditions for growth'. In common with Arolla Pine however it is able to withstand the severe climatic conditions better than all other trees which, under more favourable circumstances, would crowd it out. The Larch-Arolla Pine woods of our high mountains then are an impressive example of the fact that the dominating species of many plant communities don't 'like' their habitat but are the only ones which can

Fig. 166. Growth of subalpine trees depends very much on the warmth of the growing season. Because of this the width of the annual rings in Arolla Pines and Larches follows closely the pattern of the mean summer temperatures. After Köstler and Mayer (1970), modified.

The temperature measurements (above) refer to appreciably higher altitudes than those of the tree sites. Below are the sequences of the average width of the annual rings of *Pinus cembra* at two different sites, and of *Larix decidua* during the same period.

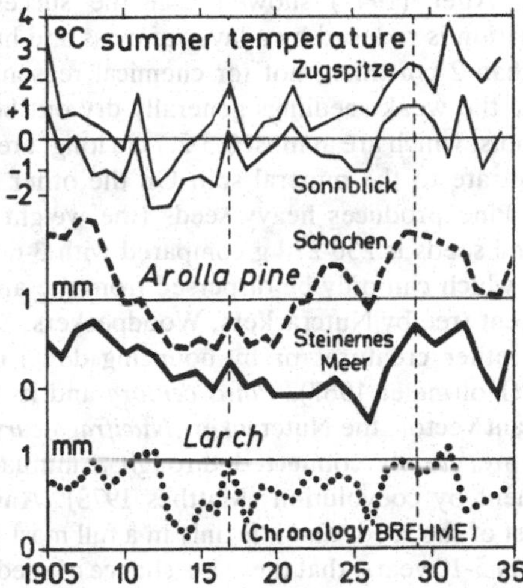

Fig. 167. The net photosynthesis of *Pinus cembra*, given a sufficiently high temperature, is governed mainly by light intensity. During the course of the day there is a close correlation between the two. Light saturation is reached only at about 25000 lux, and on 16.IX.53 this figure was exceeded only for a very short time. After Tranquillini (1955), somewhat modified.

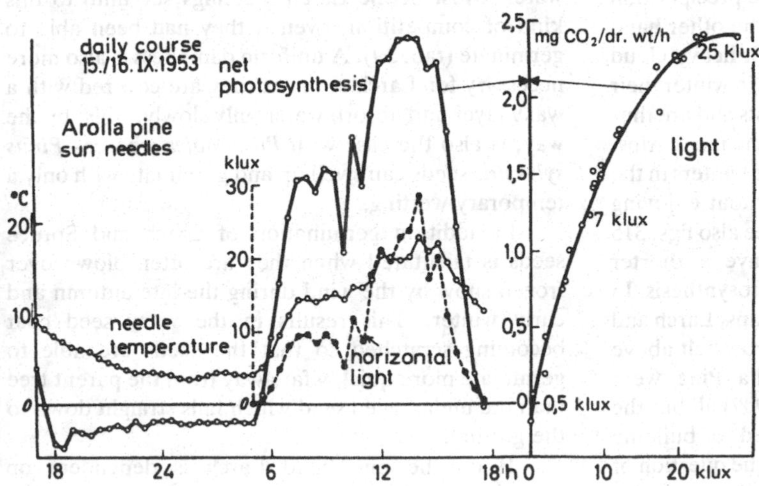

manage to exist and regenerate in the particular habitat. The attempts of many of the older authors to draw conclusions about the overall 'requirements' of a plant species from a study of its distribution and dominance must have led to very unsatisfactory results in the case of such 'displaced' species.

Regarding the soil requirements of Larch it is apparent from a thorough survey by Tschermak (1935a) that it shows no preference for either limestone or acid rocks. Provided other conditions are acceptable, it can thrive on any type of soil except for very wet ones. Arolla Pine too is not particular as to soil (Tschermak 1935b; Furrer 1955) although in the Swiss Alps it is principally found on lime-deficient rocks. The impression that it has a preference for this substrate has been brought about by the fact that its main area of distribution lies in the inner Alps which are mainly made up of older rocks. However there are quite good stands of Arolla Pine also to be found on carbonate bedrock in the Dolomites, the Swiss National Park and in other parts.

The reason why Arolla Pine and Larch have only been able to form a distinctive subalpine woodland zone in the inner Alps and not in the pre-Alps is to be looked for in the different climatic conditions, especially the amount of sunshine in summer. Tschermak (1935a) has already shown that the amount of precipitation under which the Larch can grow varies considerably. For example it thrives in the driest part of Switzerland near Grächen which has an average precipitation of only 513 mm per annum, but also in upper Tessin where more than 2000mm have been recorded (fig. 175). The thing that both places have in common though is the small amount of cloud in summer so that their soils warm up relatively quickly by irradiation (see fig. 321). May in particular is a sunny month in Tessin where most of the precipitation is in the autumn and early spring. On the other hand the northern pre-Alps are covered with heavy cloud for long periods during the summer, but in winter their peaks rise up above the layer of high mists and are thus exposed to hard radiation frosts as are the inner Alps. Thus trees have to endure equally severe winters in the outer Alps as they do in the centre without enjoying the compensation of a warm summer (see also figs. 315 and 320). As a consequence they have a shorter growing period in which to carry out photosynthesis. In the pre-Alps, in contrast to the central Alps, Larch and Arolla Pine are only able to form a narrow belt above the Spruce. Actually stands of Arolla Pine were planted here, but according to Rikli (1909) all but the few remaining have long since been used for building or furniture making. We shall return to the question of the significance of climate for the position of the tree-line in the Alps (section C VI 1 c).

Larch and Arolla Pine are similar in their ability to make use of the continental subalpine climate and to occupy habitats of varying lime content, but their roles in the formation of the Larch-Arolla Pine woods are exactly opposed. In this respect they can be compared with Birch and Oak in the *Betulo-Quercetum*: the light-loving Larch is definitely the pioneer while the longer-living Pine, which can tolerate shade much better, is the tree of the optimal and senescent phases. The seeds of Larch are easily blown by the wind and germinate best on bare ground. Experiments carried out by Auer (1947) showed that the success of germination is reduced by a layer of moss and humus more than 2 cm thick, not for chemical reasons but because the weak seedlings generally dry out before their roots, which are at most 0.5-2.7 cm long, are able to penetrate to the mineral soil. On the other hand Arolla Pine produces heavy seeds (the weight of a thousand seeds is 250-270 g compared with 3-6 g for Larch) which can only be dispersed from the area of the parent tree by Nutcrackers, Woodpeckers, Squirrels or other creatures or by bouncing down rocky slopes (Holtmeier 1967). *Pinus cembra* and its most important vector, the Nutcracker (*Nucifraga caryocatactes*) are closely connected through a mutual development by coevolution (Matthes 1978). Animals eat most of the seeds and it is only in a full mast year, i.e. every 5-10 years, that there is a chance of seedlings appearing. The seeds contain enough foodstuff to grow a radicle down to a depth of 5-10 cm. Quite often a number of seeds will germinate close together as they came from a single cone. In this way the seedlings are more competitive against other tree species, but also against the herbs and grasses which would rob them of water. Most of the Larch seedlings succumb to this kind of competition, even if they had been able to germinate (tab. 39). A uniform dampness is also more necessary for Larch seeds as they are covered with a waxy layer and absorb water only slowly. This, by the way, is also the case with *Picea abies* whereas *Pinus sylvestris* seeds can swell up and germinate with only a temporary wetting.

In addition germination of Larch and Spruce seeds is facilitated when they are often blown over frozen snow by the wind during the late autumn and early winter. This results in the waxy seed coat becoming scratched so that the seeds are able to germinate more quickly far away from the parent tree than the undamaged seed which falls straight down to the ground.

If on the one hand Larch is dependent on

continuous damp conditions if its seeds are to germinate and the first roots to gain a footing in the soil, then on the other hand these same conditions can have an adverse effect on the young plants themselves. Where it is damp they do not become woody so quickly as they do when it is drier (tab. 40) and so are more easily attacked by fungal diseases. Arolla Pines in their turn become stronger much more quickly and thus are rarely attacked. Since dry air assists in lignification (Auer 1947) this may have a good deal to do with the fact that Larches develop more plentifully on sunny slopes than on shade slopes which are mostly seen to be dominated by Arolla Pine in undisturbed stands. As they grow older Larch saplings become less resistant to drought and do not tolerate extreme conditions (Tranquillini 1968).

In this connection it is necessary to correct an often repeated error. Since *Pinus cembra* saplings are more often seen in closed communities and old woods than they are in the open many authors have come to the conclusion that the seeds require a layer of humus in which to germinate. However where there are seed-bearing stands of Arolla Pine on a slope above a recent moraine soil, e.g. in the Aletschwald or in the north-eastern part of the Morteratsch valley, young Arolla Pines can be found coming up along with Larch on the bare ground which has almost no humus in it (fig. 356). It is just that the heavy Pine seeds do not have as good an opportunity as the anemochorous Larch seeds of reaching such places. Braun-Blanquet (1964) has pointed out that on the young moraines of the Morteratsch glacier, which are still very poor in humus, there could be real Alpenrose-Arolla Pine woods within 150 years. Stands of Larch then are not absolutely indispensible as pioneers of the Arolla Pine woods even though in most cases they are prior to the latter in the course of succession.

Pure Larch woods are rare under natural conditions since within the distribution area of Arolla Pine they are sooner or later invaded, and finally dominated, by this more shade-tolerant species (see table 9). Wherever Larches dominate large areas in the high mountains they are in danger of being attacked by the larvae of *Zeiraphera diniana* which cause them to lose their needles early (fig. 168). A thorough investigation of this pest in Switzerland has led to a conclusion which has a significance beyond this special case. It appears that the abundance of caterpillars present in any one season is neither due to the predation of carnivorous insects or birds, nor to the interaction of climatic

Tab. 39. Germination and the death of Larch seedlings amongst different types of undergrowth.

Ground vegetation[a] in pots, well lit and watered	Germination rate %	Surviving plants (as a % of germinated seeds) to 105 days	170 days	(as a % of all seeds)
bare soil, i.e. without competition	64	*77*	*73*	47
covered with moss	55	*83*	*82*	45
Moss and herbaceous plants	45	*91*	*0*	0
dense herbaceous layer	43	*2*	*0*	0

Note: [a] The denser and taller the growth present the fewer the number of seeds germinating and the quicker these succumb to competition.

Source: After Auer (1947).

Tab. 40. Speed of lignification of Larch seedlings under different conditions.

Ground cover[a] (under natural conditions)	No. of lignified seedlings as % of those present *no. of days after germination* *56*	*62*	*68*	*75*	*83*	*105*
1. raw humus without vegetation	–	50	92	96	100	100
2. bare mineral soil	–	20	88	92	97	100
3. moss cover (*Polytrichum*)	–	–	28	80	78	96
4. dwarf shrubs	–	–	15	42	54	90
5. thin grass with moss	–	–	–	10	25	96

Note: [a] Once the seedlings have reached a certain age lignification goes on more quickly the drier the air. In the case of (1) and (2) higher temperatures and saturation deficits occur more frequently than in (3) to (5). Moreover in (4) and (5) the wind speed is reduced.

Source: Data taken from Auer (1947).

factors with diseases (internal factors) and parasites of *Zeiraphera*. As has been proved by Fischlin and Baltensweiler (1979) the regression of the *Zeiraphera* population is caused by lack of food for the caterpillars. They can no longer eat the needles of Larch trees which had been damaged by them in the foregoing years, since these needles are much harder than those of healthy trees. Thus the rhythm shown in fig. 168 was indirectly produced by the pest insects themselves.

We have dwelt so long with the origins and fate of the Larch and Arolla Pine woods because there has hardly been so much detailed research performed on other woodland communities, and because it is instructive to realise the difficulties with which these trees have to struggle on their subalpine site. In addition we should consider that Chamois and Deer prefer to eat young Larch and Pine shoots, that old trees are frequently split by lightning and blown over in storms, and above all that man with his axe, fire and cattle has given no protection to the trees in the areas of his alpine meadows. Summarising we cannot but appreciate with respect what vitality these plants possess.

Fig. 168. Premature yellowing of Larch in the subalpine belt of the central and southern Alps is brought about mainly by the Larch Moth (*Zeiraphera diniana* Guen.)

About every ten years the moth population (shown here on a logarithmic scale) explodes to devastating numbers. There is no obvious correlation with climatic factors such as temperature or sun-spot cycles, nor do the numbers of predators appear to exercise any significant control. The reduction in the numbers of Larch Moths is brought about much more by lack of food caused by the *Zeiraphera* caterpillars themselves. Larch trees, the needles of which have been grazed in several successive years tend to form very hard and short needles poor in proteins. These are nearly unpalatable for the caterpillars, which then starve or become easily attacked by disease and parasites. The latter are not the primary cause of the rapid regression of the *Zeiraphera* population (as was believed till a few years ago) since in some of the Alpine valleys the regression took place without any noticeable increase of parasites or diseases. See Fischlin and Baltensweiler 1979.

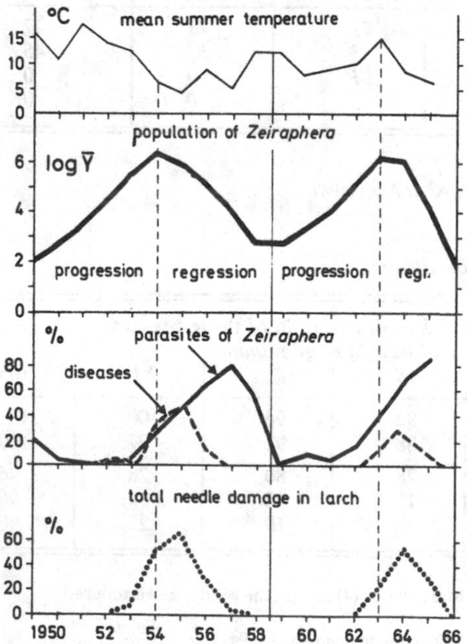

(b) Larch-Arolla Pine woods in the Alps and in Tatra

The most extensive and best investigated Larch-Arolla Pine woods are to be found in the upper valleys of the Rhone and the Inn, especially in the upper Engadin. To get an idea of the species taking part we shall use some of the relevés from the classic work of Pallmann and Haffter (1933).

Following the lead of Braun-Blanquet these authors have named the Larch-Arolla Pine wood as '*Rhodoreto-Vaccinietum cembretosum*' (tab. 41, no. 1) and have compared it with a '*Rh.-V. extrasylvaticum*' (no. 3) i.e. treeless Alpenrose stand. They look upon the undergrowth, an ericaceous heath which needs the protection of a blanket of snow, as the significant element in this community and see the trees merely as additional partners which have come in. Since Arolla Pine also occurs in the grass-rich subcommunity, ('*Rh.-V. calamagrostietosum*', no. 2) it cannot be considered as a differential species for the '*cembretosum*'. This view, which goes back to that of Schröter (1926), takes into account the open parklike nature of the Larch-Arolla Pine woodland which hardly restricts the development of the light-requiring Alpenrose. Without interference of man, however, the Larch-Arolla Pine woods are only so open right up at the tree-line. A little way below this level the trees would be able to grow more densely if they had not been prevented from doing so by man and his animals. This can be seen for example in the Aletsch Forest above Brig where access is very difficult because of the glacier and the steep rocky cliffs. Here *Rhododendron ferrugineum* is rare and scarcely flowers in the shade of the dense tree stand, but the less light-hungry *Vaccinium myrtillus* was able to spread on the ground. Within living memory the Bilberry has been so abundant in the Aletsch Forest and fruits so well that it has provided the inhabitants of Riederalp with a more valuable source of income and food than the cutting of timber and pasturage. The setting up of the Aletsch reserve was almost wrecked because the district did not want to give up its use of the Bilberry. Similar mature Arolla Pine stands, poor in *Rhododendron* but rich in *Vaccinium*, can be found in other places e.g. on the sample plot from which the relevé no. 1 of tab. 41 was taken. They represent the end stage of the natural succession in which the light-demanding Larch scarcely plays a part. Thus the Larch-Arolla Pine woods should be called *'Rhododendro-Cembretum'* (Ellenberg 1963) or better *'Larici-Pinetum cembrae'* (Oberdorfer 1979, as to the classification see also section E III 1, No. 7.214.4).

Mayer (1974) has classified the Larch-Arolla Pine woods of the eastern Alps in more detail. He distinguishes an association on lime-deficient bedrocks ('silicate Larch-Arolla Pine wood') and one on subsoils rich in carbonate ('limestone Larch-Arolla Pine wood') in which *Rhododendron hirsutum* plays an important part. Only the first forms dense shady stands, and these are either

- with Bilberry (*myrtilletosum*) or
- with Woodrush (*luzuletosum*).

The majority of stands are more or less open; according to the undergrowth the following subassociations are found on acid rocks:

- with Rust Red Alpenrose (*rhododendretosum ferruginei*),
- with Shaggy Smallreed (*calamagrostietosum villosae*, on steep slopes),
- with Mountain Pine (*'mugetosum'*),
- with Dwarf Juniper (*juniperetosum*).

On substrates rich or poor in lime but relatively moist and infertile there is also a subcommunity

- with tall herbs and Green Alder (*Alnus viridis*).

This leads on to the Green Alder shrub (section VI 7 c) When there is excessive soil moisture and a shortage of lime a

- bog-edge peatmoss-Arolla Pine woodland (*sphagnetosum*)

can be found. Acid rock screes may have been occupied by

- lichen-Larch-Arolla Pine wood (*cladonietosum*).

The distribution of the Larch-Arolla Pine woods corresponds approximately to that of the Central European race of Arolla Pine, which can be clearly distinguished from that in northern Asia. It is mainly

Tab. 41. Two different stands of the subalpine Larch-Arolla Pine wood and an Alpenrose heath in the Upper Engadine close to the tree limit.

Stand no.:	1	2	3	L
Trees:				
C *Pinus cembra*	3	1		(5)
Larix decidua	1	2		(8)
Shrubs:				
C *Lonicera caerulea*	2	1	+	5
A *L. nigra*	+			3
O *Rosa pendulina*	(+)	+		6
Dwarf shrubs:				
O *Vaccinium myrtillus*	5	3	2	5
O *V. vitis-idaea*	1	+	+	5
C *Linnaea borealis*	(+)	1		5
C *Rhodod. ferrugineum*	·		4	7
O *Vaccin. uliginosum*			2	6
Loiseleuria procumbens			+	9
A *Empetrum hermaphr.*			+	8
O *Juniper. c.* ssp. *alpina*			+	9
Herbs:				
A *Calamagr. villosa*	2	5	+	6
Avenella flexuosa	+	2	1	6
C *Luzula syli.* ssp. *sieberi*	+	1	·	3
O *Homogyne alpina*	(+)	1	+	6
A *Luzula luzulina*	+	·		3
Oxalis acetosella	1	1	+	1

Stand no.:	1	2	3	L
Herbs (cont.):				
O *Hierac. sylvaticaum* ssp.	(+)	+	·	4
Peucedan. ostruthium	(+)	·	+	6
Anthox. odoratum	·	+	+	×
A *Melampyr. sylvaticum*	·	+	+	4
Solidago v. ssp. *minuta*	·	+	+	5
Dryopt. carthusiana	·	·	+	5
O *Huperzia selago*	·		+	4
A *Lycopod. annotinum*	·		+	3
Viola biflora	·		+	4
Polystichum lonchitis			+	6
Avenochloa versicolor	·	·	+	9
Veratr. album var. *viride*	·	·	+	8
Potentilla aurea	·	·	+	8
A *Hierac alp.* ssp. *halleri*	·		+	8
Galium pumilum		·	+	7
Campan. scheuchzeri		·	+	8
Diphasium alpinum			+	8
Ligusticum mutellina			+	7
Phytemua hemisphaeric.			+	8
Arnica montana			+	9
Juncus trifidus			+	8
Cardamine resedifolia			+	8

Stand no.:	1	2	3
Mosses:			
Brachythecium starkei	+		
Pleurozium schreberi	4	3	2
Hylocomium splendens	2	1	1
Rhytidiad. triquetrus	+	3	+
Dicranum scoparium	+	+	+
D. muehlenbeckii	+	+	+
A *Barbilophozia lycop.*	+	1	1
Polytrich. formosum	+	+	
P. alpinum			+
P. juniperinum			+
Lichens:			
A *Peltigera aphthosa*	(+)		1
Cetraria islandica		1	+
Cladonia pyxidata		+	+
Cladina arbuscula	·		+
A *Cladonia macroceras*			+
mean light indic. value mL (calculated from the species present, i.e. without regard to cover degrees and those occurring on the outside)	3.6	5.0	6.4

Note: [a] No. 1: **Bilberry-Larch-Arolla Pine wood** (*'Rhododendro-Vaccinietum cembretosum'*) on the southern shore of the Campfersee, 1820 m, NW 25°.
No. 2: **Smallreed-Larch-Arolla Pine wood** (*'Rh.-Vacc. calamagrostietosum'*) above Lejmarsh, 1870 m. NNW 30°.
No. 3: **Rusty Alpenrose heath** (*'Rh.-Vacc. extrasylvaticum'*) on the eastern slope of Piz Albana, 2225 m. NNE 30°.

Systematic evaluation of the species:
C = character species of the *Rhododendro-Vaccinietum*
A = character species of the alliance Vaccinio-Piceion
O = character species of the order Vaccinio-Piceetalia
The figures in columns 1 – 3 indicate cover degrees according to the Braun-Blanquet scale (+ to 5). The cross in brackets means that the species is present in the area where the relevé was made; a dot indicates that the species is present in other records of the same subassociation.
L = light indicator value (see sections B I 4 and E III), only given for vascular plants. Those plants requiring a high light level are more and more represented going from community no. 1 to no. 3; but under their protection the shade-tolerant plants can also thrive. In calculating the average 'light figures' the trees have been omitted.
Nos. 1 and 2 differ so much from no. 3 that it would be better to consider them as belonging to a different association (Larici-Cembretum or Larici-Pinetum cembrae).

Source: Examples taken from tables in Pallmann and Haffter (1933).

found in the inner Alps and here it takes over the subalpine belt. The upper limits of this are shown in figs. 42 and 6. In the Carpathians, especially in the high Tatra, only fragments of the Larch-Arolla Pine woods are to be found (Szafer and Zarzycki 1972). Because of the thoughtless pursuance of cutting and pasturing they have been completely wiped out from the eastern Carpathians (Soó 1930). All the isolated appearances in Central Europe are connected with their Late-Glacial distribution area, which stretched from the south-eastern Alps to the lowlands of the Balkan peninsula. Macroscopic remains and high pollen counts of Larches, Arolla Pines and some of their companions have been discovered on the lowland plains of Hungary showing that at one time *Larici-Cembretum* occupied places which today have very different vegetation (Scharfetter 1938, Zólyomi 1953 and Havinga 1973). Whether *Rhododendron ferrugineum* was present at that time seems doubtful. This small shrub, which today is so characteristic of the Alpine highlands, arrived comparatively late after the end of the last Ice Age to occupy its present habitats (Zoller 1960). It is very likely that at first it formed a belt of heathland above the tree line and only penetrated the woodland as it gradually became opened up through human influence during the last 2000 years (fig. 319).

The Larch-Arolla Pine wood forms a valuable protection against avalanches. Since this has been recognised and its growth encouraged, *Pinus cembra* has quickly increased in numbers and height. It has already become so plentiful in the Upper Engadin and in some other central Alpine valleys that consideration has now to be given to protecting the Larch which is more highly valued as a forest tree. One thing can be recommended to achieve this objective and that is to stop the removal of single trees but to take them out in groups so that larger clearings are made. These certainly would favour the light-demanding Larch. The effectiveness of this measure would be considerably increased if the humus layer and competitive plants were removed during a good seed year in order to prepare artificially areas of bare soil to act as seedbeds for Larch. It is just in the most valuable sections where natural catastrophes such as storm felling of old trees, avalanches, landslips and fires from lightning strikes, which would have the same effect, play a very small part. Auer even advised the maintenance of cattle browsing and exclusion of the animals only up to the point where the growing Larch trees would no longer suffer damage. So here the forester must strive by any means to hold up the natural succession at the stage of the pioneer woodland. Only the woods expressly required for avalanche protection should be allowed to become so rich in Pines that they correspond to the optimal phase of the natural Larch-Arolla Pine wood.

(c) Larch woods in the southern Alps

At the subalpine level in some of the limestone and silicate massifs from the south-eastern Alps to the Maritime Alps there are no naturally occurring Arolla Pines (figs. 72a and 43, type 6). Even Spruce which plays such an important role in the northern outer Alps, and in most places there goes up to the tree-line, retreats here. So Larch, which is less susceptible to intermittent dry spells, becomes the dominant tree in the southern pre-Alps. The drought periods correspond to the rhythm of the mediterranean climate. This favours Larch indirectly, even below the subalpine belt, especially on rocky sites.

The subalpine Larch woods were rather late in being described as separate communities since they have much in common with the Larch-Arolla Pine woods of the Central Alps. However since they show many 'southern' characteristics, both floristically and ecologically, Mayer (1974) has stressed the independence of the 'limestone Larch woods' (*Laricetum*). In the south-eastern Alps he distinguishes the following subassociations:

- limestone Alpenrose-Larch woods (*Laricetum rhododendretosum hirsuti)*,
- limestone-rockfall Larch woods (*L. asplenietosum*),
- limestone-steep slope Larch woods (*L. rhodothamnetosum*),
- limestone Larch meadow woodland (*L. luzuletosum sylvaticae*).

Only the first of these can be considered as a zonal community; the rest are permanent communities occupying special habitats. On many limestone plateaux at 1600-1800 m above sea level *Laricetum rhododendretosum* forms real woods some 20-25 m high. They let in so much light that the Hairy Alpenrose is able to become dominant on the ground. Falls of rock with large boulders are first colonised by Mountain Pine, and the Larch stand which follows remains open. Larch can go down to 800 m on steep rocky shaded slopes, i.e. down to the region of the dark Fir or Beech woods, because it has no real competition on such stony dry soil.

Screes of small but hard limestones on shade slopes carry a woodland community also dominated by Larch. However a layer of moder forms in places making possible a meadow-like undergrowth. Along with *Luzula sylvatica* and *L. luzulina* there may be for example *Calamagrostis villosa*, *Sesleria albicans*, *Festuca rubra*, *Carex ferruginea* and *C. firma*, each of which could be locally dominant. Only at the foot of the Larch trees, which may be several hundred years old, are there suitable habitats for *Vaccinium vitis-idaea*, *V. myrtillus*, *Lycopodium annotinum* and other plants which live on acid humus. Mayer had already recorded the remains of such an ancient stand in 1957 on the northern side of the Hochkönig in the Salzburg limestone Alps. These naturally occurring Larch-meadow woodlands should not be confused with the Larch meadows which are found near the villages. The latter have often been produced from Spruce-rich woodland by mowing and grazing and a management which has favoured the Larches. Such grassy Larch groves can be found on limestone rocks as well as acid ones (Putzer 1967).

The subalpine Larch woods on silicate rocks in the south-western Alps have still not been closely investigated.

Fig. 169. Upright Mountain Pines in the Purple Moorgrass-Pine wood (*Molinio-Pinetum*) growing on Tertiary marl of the Albis Ridge near Zürich. Bare soil is constantly being exposed through slippage. This alternates with thick turf and groups of low trees. *Pinus sylvestris* can also dominate in place of the upright Mountain Pine (*P. mugo*) visible here. Photo by Furrer.

5 Mountain Pine stands other than those of mires

(a) Communities of the upright Mountain Pine

In the upper subalpine belt of the pre-Alps the Larch-Arolla Pine woods are often replaced by communities of Mountain Pine. To the west of a line from Lake Constance to the Lake Como these mostly include the upright form of *Pinus mugo* (= *P. montana* grex *arborea* Tubeuf = *P. rotundata*). In the western Alps and Pyrenees this is the only representative of the species. In the eastern Alps, the Carpathians, the Dinaric Alps and other high mountains to the east, on the other hand, a prostrate form (*Pinus mugo* s. str. = *P. montana* grex *prostrata* Tubeuf) produces a more or less extensive scrubland above the tree line which here is generally occupied by *Picea* or *Fagus*. Both the upright form (fig. 169) and the genetically distorted prostrate form (fig. 170) are indifferent as to soil. Since they require a lot of light and are thus non-competitive they have become restricted to the poor or dry places which are left them by the more demanding trees.

Stands of upright Mountain Pine can reach a height of 8-20 m so that, in contrast to the prostrate form, it is not protected by snow for the greater part of the time. In the Swiss National Park Mountain Pine forms the largest part of the woodland area. Here Braun-Blanquet, Pallmann and Bach (1954) have studied them from the aspects of plant sociology, soil science and climatology. In reference to this extensive work and to the literature it quotes we can restrict ourselves to some of the items of general interest (see also Kurth, Weidmann and Thommen 1960 as well as Mayer 1976).

The majority of the Mountain Pine woods in the Swiss National Park are secondary, many of them

Fig. 170. Mountain Pine scrub (*Pinus mugo*) with *Rhododendron hisutum* on limestone above the tree line on the Hohe Ifen at about 1850 m above sea level.

having probably arisen from the disturbance of original Larch-Arolla Pine woods. Iron smelting which has been carried on since the Middle Ages in the vicinity of the Ofen pass, and also charcoal burning required a large amount of wood. In the seventeenth century the whole of the remaining woodland on the salt-works of Hall in the Tyrol was sold and, where it was accessible, thoughtlessly cleared of timber. By damming up the small amount of water in the river Spöl and its tributaries it was possible to float the timber down to the Inn in the spring. Part of the woodland of the National Park had also fallen victim to a forest fire, traces of which can still be seen in the soil profiles. In this case the majority of the Mountain Pines are now about the same age and the area has about it the appearance of a pioneer woodland.

However not all these monotonous *Pinus mugo* woods can be looked upon as the results of man's interference. Very probably there were extensive stands of Mountain Pine before man came to exert his influence on the poor dolomite soils where the precipitation is extraordinarily low for high mountain conditions (900 mm at an altitude of some 2000 m near Buffalora!) The reason the Larch-Arolla Pine wood has not reestablished itself more quickly since the setting up of the National Park in 1914 may be in part due to the excessive increase in numbers of Chamois, Red Deer and Roe Deer, whose droppings can be seen all over the place by an observant visitor. These animals prefer to browse on the Arolla rather than the Mountain Pine thus slowing down the development of the climax woodland. In addition they help some of the unpalatable plants of the undergrowth such as *Erica herbacea*, *Sesleria albicans*, *Polygala chamaebuxus* and *Carex humilis*.

Contrary to the expectations of many visitors the Swiss National Park then in the montane and subalpine belts offers something quite different from undisturbed primaeval nature. However it does make possible, as no other place in the Alps does, the study of the development and plant sociological units of Mountain Pine woods.

J. and G. Braun-Blanquet, Trepp, Bach and Richard (1964) distinguished two main communities at the subalpine level (between about 1600 and 2200 m): the **Spring Heath-Mountain Pine wood** (*Erico-Pinetum mugi*) on warmer slopes and the **Hairy Rhododendron-Mountain Pine wood** (*Rhododendro hirsuti-Pinetum mugi*) where it is distinctly colder and where more snow collects (tab. 42, nos. 5 and 6). Each of these can be further divided into subassociations and variants depending on the finer differences in the water and temperature economies.

Closely related to the Spring Heath-Mountain Pine wood is the **Dwarf Sedge-Engadin Pine wood** (*Carici humili-Pinetum engadinensis*) which is dominated by a subspecies of the Scots Pine (tab. 42, no. 4). In the Swiss National Park and its surroundings it occupies the warmest slopes causing them to stand out clearly from the dull Mountain Pine woodland as the tall slim trunks of the Engadin Pine are smooth and yellowish red. It leads down to the Spring Heath-Pine woods at the montane level which we shall be discussing in section 6 b.

(b) Prostrate Pine scrub under different habitat conditions

In the high mountains of Central Europe the Prostrate Pine scrub, which rarely exceeds 3 m in height (fig. 170) occupies a larger area than the stands of Upright Pines. Particularly in the limestone and dolomite areas of the eastern Alps, but also in Carpathians, they cover extensive slopes and are often thought to be basophile. In no way however are they restricted to limestone and can be found in many different sites which are inimical to the growth of trees but which can still be occupied by scrub. The following are possible **reasons for the absence of trees:**

1 Too low a winter temperature from which the Prostrate Pines are protected by a covering of snow. These could be:
   a) above the tree-line (figs. 170 and 316),
   b) below the tree-line in karst sink holes and deep depressions from which the cold air is unable to flow away (a reversal of the altitude belts similar to that shown in fig. 155).
2 Too deep and late melting snow drifts below the tree-line, e.g. along the edge of avalanche tracks or on the lee of mountain ridges provided these places are not too wet for the Prostrate Pine (fig. 348).
3 Too strong a wind which does not allow the trees to grow properly; this only operates on a small scale on very exposed hill tops.
4 Too little fine earth over smooth and hard rock
   a) on humps in the ground which have been polished by ice movement or on rock faces,
   b) on large boulders, e.g. where there has been a rock fall.
5 Too wet and poor a soil around the edges of a raised bog, or on its hummocks (see section C III 1, especially fig. 249).
6 Repeated destruction of trees
   a) by avalanches and snow slips (figs. 171, 172),
   b) by rockfalls and land slides (rare),
   c) by felling, i.e. human influence.

Tab. 42. Pine woods and Mountain Pine woods at various altitudes on limestone rock.

Serial no.:	1	2	3	4	5	6	T	R
Dominant Pine species: A = Austrian, S = Scots, E = Engadine, M = Mountain	A	S		E	M			
Altitude from (100 m)	7	6	6	18	15	18		
to (100 m)	12	17	9	21	23	21		
Trees (also in shrub layer):								
E *Pinus nigra*	5						7	9
P. sylvestris	2	5	5				×	×
P *P. sylvestris* var. *engadinensis*				4	1		–	–
P *P. mugo (arborea)*				4	5	5	3	×
P *P. cembra*	3	5	2	1	2	1	2	4
Larix decidua	1	1		2	2	2	×	×
Shrub layer:								
Q *Cotoneaster tomentosus*	4						5	9
Q *Amelanchier ovalis*	4	4			1		7	×
Juniperus communis	4	5	5				×	×
Q *Sorbus aria*	4	2	5				5	7
Q *Acer pseudoplatanus*	3		5				×	×
Q *Berberis vulgaris*	1	4	2	2			6	8
Q *Corylus avellana*		3					5	×
Q *Viburnum lantana*		4	5				5	8
Q *Ligustrum vulgare*		2	4				6	8
Q *Fagus sylvatica*			3				5	×
Populus tremula			3				5	×
Q *Salix caprea*			3				×	7
Sorbus chamae-mespilus		1		1	1	4	3	8
Juniperus intermedia				4	?		–	–
P *J. communis* ssp. *alpina*					4	2	2	7
Herb layer:								
Genista pilosa	5						5	2
Carduus crass. ssp. *glaucus*	5						–	–
Scabiosa lucida	4						3	8
Galium austriacum	4						–	–
Q *Cyclamen purpurascens*	4						6	9
Valeriana tripteris	4						×	8
Campanula cespitosa	4						2	8
Globularia cordifolia	4			1			×	9
Vicia cracca ssp. *gerardii*		2					–	–
E *Thesium rostratum*		2					6	9
E *Pyrola chlorantha*		1					5	5
Brachypodium pinnatum		4	4	1			5	7
Punella grandiflora		3	5				×	8
Anthericum ramosum		3	5				5	7
P *Goodyera repens*		4	3				×	×
M *Lathyrus pratensis*		3	4				5	7
M *Molinia arundinacea*		1	5				5	×
M *Succisa pratensis*			5				5	×
M *Gymnadenia conopsea*			5				×	8
Galium verum			5				5	7
Scabiosa columbaria			5				5	8
Potentilla erecta			4				×	×
Trifolium montanum			4				×	8
Carex flacca	1		5		1		5	8
Knautia dipsacifolia			5	1	2		3	×
Pteridium aquilinum		1	4				5	3

Serial no.:	1	2	3	4	5	6	T	R
E *Calamagrostis varia*	5	4	5				3	8
Campanula rotundifolia	4	4	3				×	×
Buphthalmum salicifolium	4	2	4				5	9
Teucrium chamaedrys	3	3	3				6	8
Carex humilis	1	4	1				5	8
E *Saponaria ocymoides*				5			4	9
Leontod. hisp. var. *crispatus*				4			–	–
Hieracium bupleuroides				3			×	9
Ephrasia salisburgensis				3			×	8
Laserpitium kr. ssp. *gaudinii*				5	2		–	–
Biscutella laevigata				4	3		×	7
Leucanthemum maximum				4	3		×	×
Centaurea scabiosa var.				4	2		×	8
P *Arctostaphylos uva-ursi*				3	2		3	×
E *Crepis alpestris*				2	2		×	8
Campanula cochleariifolia			1	5	2	3	×	×
E *Daphne striata*				4	5	2	3	8
Valeriana montana				2	3	3	×	9
Dryas octopetala				2	2	4	×	8
P *Homogyne alpina*					4	5	4	4
Vaccinium myrtillus		1			3	4	×	2
P *V. uliginosum*					2	5	×	1
P *Pyrola rotundifolia*					3	4	×	5
P *Luzula sylvatica* ssp. *sieberi*					2	4	3	2
P *Calamagrostis villosa*					2	2	4	2
P *Moneses uniflora*					2	1	×	4
E *Rhododendron hirsutum*					1	5	×	7
P *Rh. intermedium*						2	×	6
P *Arcostaphylos alpina*						5	3	×
Campanula rapunculoides	3			3			6	8
Thesium alpinum	3			2	1		3	×
E *Coronilla vaginalis*	3			3	1		5	9
Teucrium montanum	4	3		3			7	9
Leontodon incanus	4	2		3	2		×	9
Hippocrepis comosa	4	3	4	4			5	7
Thymus serpyllum ssp.	4	2	3	4			–	–
Euphorbia cyparissias	3	2	4	4			×	×
Galium lucidum	3	4	1	2			×	8
E *Polygala chamaebuxus*	5	5	5	5	4		×	8
E *Epipactis atrorubens*	2	5	5	5	2		×	8
Lotus corniculatus	4	4	5	4	5		×	7
Carduus defloratus	4	2	2	5	3		×	8
Phyteuma orbiculare	4		2	1	2		3	8
Carlina acaulis	3	2		4	3		×	×
Sesleria varia	5	2	3	5	5	5	×	8
Carex ornithopoda	4	2	3	3	3	4	×	9
P *Orthilia secunda*	1	2	1	1	4	5	×	×
E *Erica herbacea*		5	5	4	5	5	×	×
Hieracium bifidum		5	2	3	5	3	×	8
H. sylvaticum ssp.							–	–
E *Gymnadenia odoratissima*		2	3	4	3	1	×	9
E *Carex alba*		3		2	5	3	5	8
Melampyrum pratense ssp.		4		1	4	2	–	–
P *M. sylvaticum*		3		1	2	2	×	2

Tab. 42. (Continued)

Serial no.:	1	2	3	4	5	6	T	R
Moss layer:								
Rhytidium rugosum	4						–	8
Ctenidium molluscum	1		4				–	8
Scleropodium purum		2	3				–	5
Tortella tortuosa	4	3		4	3	3	–	X
Hylocomium splendens		4	4		3	5	–	X
Moss layer (contd).								
Rhytidiadelphus triquetrus		4	3		3	4	–	4
Dicranum scoparium		3	2		3	4	–	1
Pleurozium schreberi		4	2		4		–	1
Cetraria islandica					5	5	–	2

E = species of the Spring Heath-Pine woods (*Erico-Pinetalia, Erico Pinion*)
P = species of the acid-soil Spruce and Pine woods (*Vaccinio-Piceetalia*)
Q = broadleaved woodland species (*Querco-Fagetea*)
M = meadow plants, especially species of the Purple Moorgrass meadows (*Molinio-Arrhenatheretea, Molinion*),
T = indicator value for temperature, R = for soil acidity (reaction), according to Ellenberg (1974; some of the lower taxa have not been evaluated. A few species which occur infrequently have been omitted.

	Ecological evaluation	mT	(mRv)	mRvm [1])
No.1:	**Black (or Austrian) Pine wood** on the eastern edge of the Alps (*Chamaebuxo-Pinetum nigrae*) after Knapp (1944):	4.8	(7.9)	7.9
No.2:	**Spring Heath-Pine wood** (*Erico-Pinetum*) in Graubünden, after Braun-Blanquet, Pallmann and Bach (1954):	5.1	(7.5)	6.9
No.3	**Purple Moorgrass-Pine wood** (*Molinio-Pinetum*) on the north-eastern edge of the Swiss Alps, after Etter (1947):	4.8	(7.6)	7.3
No.4:	**Small Sedge-Engadine Pine wood** (*Carici-Pinetum engadinensis*) in the Swiss National Park, after Braun-Blanquet and co-workers (1954):	4.2	(8.0)	8.0
No.5:	**Spring Heath-Mountain Pine wood** (*Erico-Mugetum*) in the Oberengadin, after Braun Blanquet and co-workers (1954):	3.4	(6.6)	5.8
No.6:	**Hairy Alpenrose-Mountain Pine wood** (*Rhododendron hirsuti-Mugetum*), as no.5.	3.3	(5.7)	5.2

[1]) mT = average temperature value, which, as a rule, falls with increasing altitude (see also tab. 13). It is higher in the valleys of the inner Alps than at other sites with the same altitude (compare no.2 with no.1 and no.3). Even at altitudes of over 1800 m the mT reaches values of more than 4 on dry sites in the Swiss National Park.
(mRv) = average reaction value when only the vascular plants are taken into account.
mRvm = the same when the mosses are also included in the calculation.

The species of the Spring Heath-Pine wood are, for the most part, lime indicators and predominate in communities at lower levels. The acid-tolerant plants of the needle-leaved woods only play a rôle at greater altitudes.
The Austrian Pine wood (no.1) and the Engadine Pine wood (no.4, dominated by a subspecies of the Scot's Pine), live in a particularly dry climate, in which hardly any surface humus forms over the limestone. In cooler damper climates the top soil quickly becomes poorer in lime and richer in humus so that acid-tolerant plants, especially mosses, can obtain a foothold. For this reason the values for mRvm here are smaller than those for (mRv).
The Purple Moorgrass-Mountain Pine wood (no.3) which occupies unstable marl (see fig. 169) is rich in plants of the broadleaved woods (Q). Because its tree layer is unstable some plants from the unfertilized meadows (M) and from the slightly-arid grassland (many of the remaining species) find a natural home here.

Often a number of these factors work together, e.g. 1 and 2, 3 and 6, or 2 and 4. The remaining factors may be quite dissimilar, especially the edaphic ones or, in the cases of 4 to 6, climatic ones too. So these Pine scrubs which at first sight appear to be so similar contain some very different types of undergrowth, especially as cattle are allowed to roam through some of them. On special sites they are reminiscent of the Spring Heath-Pine woods or the Hairy Alpenrose-Mountain Pine wood communities (sections B IV 6 b and 5 a). In others they are rather like the Larch-Arolla Pine woods (sections 4 b), the subalpine Spruce or Beech woods (I V 3 c and II 2 g) and in yet others they resemble the *Rhododendron* heaths (D II 4b) or stands of tall herbs (C IV 7a). Many of them contain species from submontane grassland or other neighbouring communities. This great variety is increased even more by the fact that subalpine pastures and heaths are invaded by the Prostrate Pines where they are looked upon as weeds. Nevertheless the majority of Prostrate Mountain Pine communities can be placed in the order *Vaccinio-Piceetalia* since their needle litter creates almost the same sort of conditions as one finds in coniferous woodland. This is why they have been mentioned along with the Spruce- and Larch-Arolla Pine woods.

At this point we must be satisfied with a general reference and should not try to discuss in detail any individual examples since a comprehensive work on their plant sociology has not yet been attempted. But we shall return to the Prostrate Pine wood formations when we discuss the alpine tree line and when we deal with the subalpine Green Alder scrubland.

6 *Pine woods other than those of mires and flood plains*

(a) Site-dependent groups of Pine woods in Central Europe

In spite of, or even because of, its modest demands, our common Scots Pine (*Pinus sylvestris*) dominates or participates in very different plant communities (figs. 173, 37, 39 and 140). It surpasses all other tree species in the multiplicity of habitats it occupies. From the edge of the tundra in the far north and from the tree line high up in the Alps where stunted specimens occupy at least some outposts at an altitude of 2250 m, right down to the warmest valleys of the inner Alps; from the washed out sands of north-east Europe to the limestone screes of the pre-Alps; from föhn parched rocky ridges, through the marl slopes, which are wet in spring but dry out over the summer, to the spongy wetness of the raised bog; everywhere Scots Pine can find a sunny spot or an uncontested corner. It occupies these with far-flying and quick-germinating seeds and manages to maintain its hold in spite of its open canopy.

To the different habitats correspond diverse growth forms – whether they are genetically determined varieties, races or ecotypes or merely modifications brought about by the environment. These can vary from giants with slim unbranched trunks like Spruce reaching a height of over 40 m (table 9) to wrinkled dwarfs which at first sight no one could believe were 150 to 300 years old. Where the climate is mild in winter there is a tendency for more compact forms to grow, while the tall ones tower up where the winters are colder and the climate is more continental. Kalela has surveyed the many experiments which have been carried out with trees from different sources. According to him, because of the widespread dispersal of the pollen, the whole of the European population can be pollinated from all sides in a continuous exchange of genes, so that only in exceptional cases are particular races limited to specific habitats.

Where the climate is strongly oceanic as in Holland or north west Germany, or in the insubrian area along the southern edge of the Alps, Pine cannot hold its own against Birch and Oak which are naturally dominant here, even on the poorest soils. In spite of this, Scots Pine is the most dominant feature of the landscape in Lower Saxony since it was and still is preferred for the afforestation of heathland. In other parts of Central Europe too they are, along with the Spruce, one of the favourite forest trees because of

Fig. 171. Subalpine and montane Spruce woods which have been destroyed by an avalanche above Zernez in 1952. To the left is a gully where avalanches regularly occur. The woodland near the tree line is occupied by stands of Larch and Arolla Pine. These have been increasingly so thinned out by grazing that finally they were no longer able to hold back the mass of snow. Photo, Lüdi.

Fig. 172. Spruce torn down by an avalanche in the spring of 1951 and deposited in the valley below Stuben on the Arlberg. In the background Prostrate Mountain Pines in an area which is affected by avalanches almost every year.

its rapid growth and useful timber. This forestry practice gave *Pinus sylvestris* the opportunity to grow in sites which correspond to its physiological optimum and so attain its maximum growth rate (fig. 39).

In its main area of distribution, the huge northern European-Siberian coniferous region, Pine has proved to be especially useful as a pioneer after the forest fires which are often started by lightning in drought years. For example in the Spruce zone to the west of the Urals Kortschagin (quoted by Walter 1974) has proved that all Pine trees in a particular stand are the same age and must have germinated in the different areas in the years 1715, 1765, 1790, 1825, or 1895. Similar observations have also been made in Scandinavia (e.g. in the Muddus National Park, Sweden) as well as in the near natural coniferous areas of North America. Here the strict firefighting by well organised parachutists ('smoke jumpers') during the last two decades has put nature conservation into a precarious position. Since these young idealists have succeeded in preventing all wood fires, they are endangering the regeneration of open Pine stands and with this an abundant light-demanding flora and fauna. There are no more new burnt areas to which the Pine seeds can quickly be blown and successfully germinate without the hindrance of a deep humus layer or the competition of shade trees. Trees of seed-bearing size survive most fires in sufficient numbers because *Pinus sylvestris* and related species have a thick bark which offers effective protection against fires sweeping through the floor of the forest, especially when the trees are older. The spongy resinous needle litter and the bulky dead wood on the floor of the Pine stands facilitate the starting and spreading of fires in dry periods so much that even in the relatively humid climate of north-west Germany young Pine forests are often burnt down over wide areas, e.g. in the year 1975 when the summer was very dry in the Lüneburg heathland. The more broadleaved trees there are mixed with the conifers the less is the danger from fires, but also the less the chance the Pine has of regenerating (see sections B IV 1 a and 3 a).

Fig. 173. A natural Pine wood on sand (*Dicrano-Pinetum*) in Masovia (NE Poland) showing a lot of natural regeneration and the occasional Juniper – evidence of earlier grazing (see also figs. 9 and 184).

However there is certainly no other tree species which can provide any dangerous competition for Scots Pine on soils which are very poor and at the same time either dry or very wet in the less oceanic climates, especially in the north-east of Central Europe or in the boreal coniferous region. Here it is naturally the dominant tree even though it may not grow very well, and here too it does not require forest fires to maintain its dominance. This is shown in the transects in fig. 174 of an ancient Pine forest which was carefully recorded by Huse (1965). The young, optimal, senescent, decay and regeneration phases alternate in time and space with each other in a way similar to that in the primaeval Beech and Fir woods with which we are already acquainted (figs. 59, 148 and 149). At the same time these changes indicate alternation between a period of intensive biomass production and one where production is much less, but reproduction is stepped up. According to Krauklis (1975) this is true for many boreal woodlands. It is a self-regulatory mechanism and, taken over a wide area, ensures the stability of an ecosystem which is poor in species.

The many different habitats occupied by Scots Pine, its tolerance in allowing light-loving plants of the open country into its lightly shading stands and, not least, the way in which it has been favoured by man, all combine to make the separation and classification of the Pine wood communities a particularly difficult problem. In Central Europe, but also in other parts of its large area, this has still not been solved satisfactorily. Above all the widely distributed communities have scarcely any characteristic species since they are a meeting place for members of very different woodland, moorland and grassland communities without them entering into an exclusive partnership with the

Pine. Many of the plants of our dry grassland and damp meadows *(Festuco-Brometea* and *Molinietalia,* section E III 1) in any case have their natural home in open Pine woods (fig. 169). They may be used then in characterising these woodlands floristically even though they thrive better in the open grassy communities. In the Alps under the canopy of Scots Pine we may find representatives of the alpine belt together with ones from the warmer valleys or, even from the submediterranean region. Such groups of species combine so regularly that this freakish situation may almost be looked upon as identifying certain communities (tab. 42).

In accordance with Braun-Blanquet (1961), W. Matuszkiewicz (1962), Passarge (1963a), Poldini (1969), Sokołowski (1970), Ellenberg and Klötzli (1972), Mayer (1974) and others the communities dominated by *Pinus sylvestris* in Central Europe can be divided into 7 groups on grounds of habitat and flora. These correspond to the same number of alliances:

1. **Steppe-Pine woods** (*Pulsatillo-Pinetea*) as rarities in the warmest sites of the inner Alpine dry valleys i.e. in a distinctly continental climate with a low rainfall (figs. 175, 42, 320 and 321);
2. **Xcrothermic mixed Oak-Pine woods** in the colline belt of low rainfall regions in the north-eastern lowlands (*Quercion pubescenti-petraeae* which has already been mentioned in sections III 3 a and 4 a);
3. **Marl-slope Moorgrass-Pine woods** on marl slopes with a fluctuating dryness in the submontane to montane belts near the Alps (*Molinio-Pinetum*, some stands belonging to *Erico-Pinion* or *Molinion*, fig. 177);
4. **Spring Heath-Pine woods** on more or less lime-rich dry slopes and screes in the submontane to high-montane belts (*Erico-Pinion*, fig. 176);
5. **Acid sandy-soil Pine woods** of the subcontinental north-eastern diluvial landscape (*Dicrano-Pinion,* fig. 173 and section 6c); There is also a moist-soil subunit (*molinietosum,* fig. 183);
6. **Dune Pine woods**, the natural climax in the succession of dune colonising and stabilising in the eastern part of Central Europe (*Corynephorion*, *Koelerion albescentis* or *Koelerion glaucae*, in part also *Dicrano-Pinion;* section C ▽ 1 d);
7. **Mire Pine woods** round the edges of raised bogs and on wooded raised bogs, as well as in very poor

Fig. 174. The phases in the development of an ancient pine wood, on poor sandy soil in Norway, correspond to those of other woodland communities (compare figs. 59, 148 and 149). In no way therefore does Scots Pine depend on fires for its regeneration. After Huse (1965), modified.

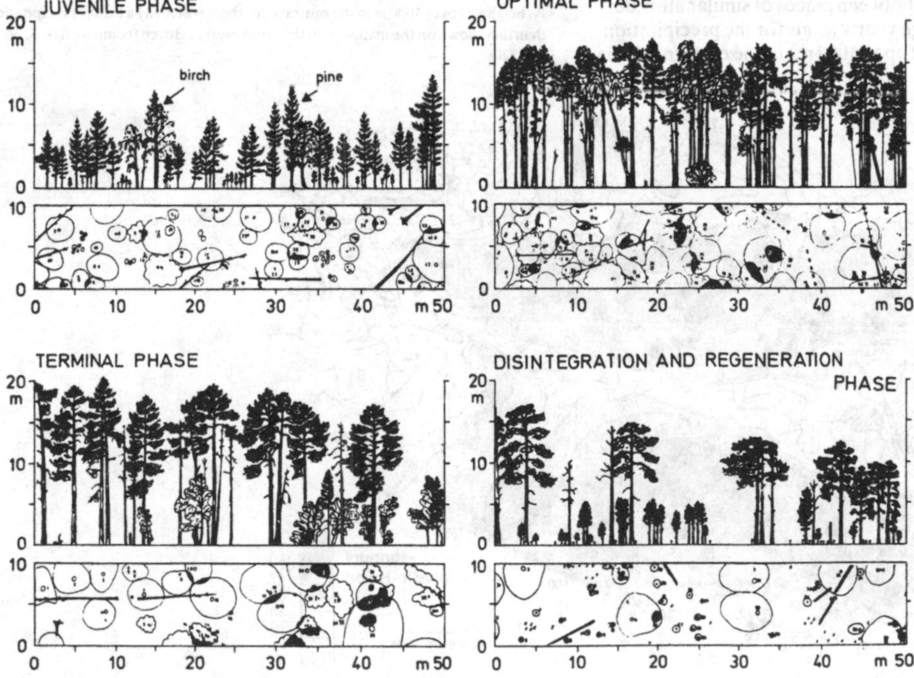

fens where the climate is more or less continental (*Sphagnion fusci* and in part *Dicrano-Pinion*; sections C III 1 and 2, B V 2 c).

The first group holds a decidedly unique position, but there are intermediates between it and 2. The Moorgrass-Pine woods carry out exchanges with the moist-soil meadows (*Molinion*) but also dovetail in with the dry-soil Beech woods (*Carici-Fagetum*). The Spring Heath-Pine woods (4) have a very characteristic flora and are regarded as a class of their own (*Erico-Pinetea*). The acid sandy soil Pine woods (5) occupy similar habitats to the Birch-Oak woods, and they gradually merge into these as they go from east to west. These are followed by the dune-Pine woods, which however contain species of open dune grassland in more or less large numbers and should be looked upon foremost as the end stage in the dune succession. The Pine woods of the more or less wet ground (7) are also best understood as the climax of an azonal succession, and we shall deal with them in this connection.

The first four groups of communities were above all the ones which Gams (1930) and many other authors (up to 1950) had in mind when talking about 'relict Pine woods'. Just like the zerothermic Oak communities these open woodlands have become the refuges of less competitive and shade-avoiding species. So they could be seen as evidence of formerly more widespread types of vegetation. Korshinskij, the creator of the relict idea (quoted by Gams 1930) thought originally of woodland from the Tertiary whose species had survived in the areas which had not

Fig. 176. Spring Heath-Pine wood (*Erico-Pinetum*) on rendzina growing on a sunny slope in the Inn valley below Innsbruck. Already in February the snow has gone except for a few small patches. *Pinus sylvestris, Sesleria albicans* and *Erica herbacea* (= *carnea*) form open stands.

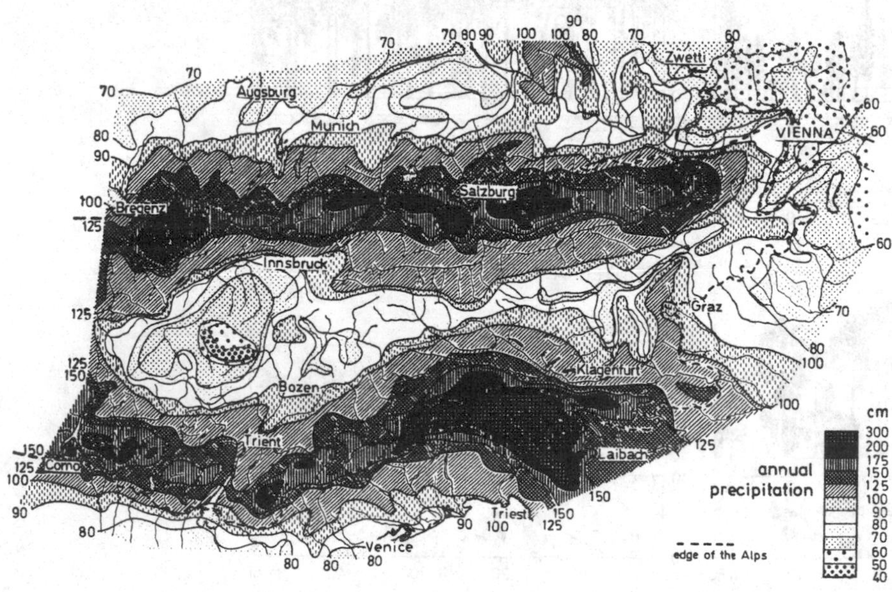

Fig. 175. Precipitation is considerably greater around the edges of the Alps than it is towards the centre, especially when comparison is made between places of similar altitude. Shown here are the average yearly totals for the precipitation in the lower montane belt (up to 1000 m) in the eastern Alpine region as well as in the submontane and planar foothills. After Mayer (1974) somewhat altered.

At heights of over 1000 m in the central Alps there is actually a higher precipitation than that shown on the map, e.g. to the north-west of Bozen (compare figs. 5, 36 and 320).

been glaciated during the Ice Ages. However over almost the whole of Central Europe the 'relict Pine woods' are occupying areas which had surely been glaciated. After the last Ice Age they had to re-occupy extreme habitats which had previously been free from woodland. In the true sense of the word then they are not relict woodland. Many of these *Pinus* communities are even quite young. They are proved to have spread out in recent times under the influence of man, e.g. the Pine woods of Wallis and Engadin (as well as the Mountain Pine stands in the Swiss National Park, section 5 a). Oberdorfer (1954) stresses that the sandy soil Pine woods have considerably extended their distribution in the Upper Rhine plain through woodland grazing and afforestation of open ground. This has largely been at the cost of Oak-Hornbeam woods and dry-soil Orchid-Beech woods. During the course of the centuries these secondary stands have also acted as refuges for the light-requiring, but otherwise undemanding, plants of the steppes and the Mediterranean area. The 'relict Pine woods' of the Bohemian Forest described by Mikyška (1964) are mostly secondary ones too, and surely did not survive the Ice Ages in situ.

Although floristically the first four types of Pine woods are quite different they stand closer to one another than they do to the sand and mire Pine woods. The former could be united by terming them 'species-diverse' or 'southern' Pine woods. The last three may be considered together as '**species-poor**' or 'northern', or even 'acid-humus Pine woods' since they are found on very acid ground and contain only acidophiles or plants which can at least tolerate acid conditions. Speaking of 'relict Pine woods' one mostly thinks of the types 1-4. Incidentally from a certain point of view the sand Pine woods (6) and the raised bog Pine woods (part of 7) could merit the name 'relict Pine woods' as well, although no one uses it in this connection. Just

Fig. 177. An open Moorgrass-Pinewood (*Molinio-Pinetum*) on marl above the entrance to the Aar gorge near Innertkirchen. Where the terrain is steep and stony it merges into a Blue Moorgrass slope; where the soil is deeper and settled it leads to a mixed Oak wood.

like types 1-4 they are confined to special habitats through the competition of Spruce and Beech, and here they provide a refuge for the weakly competitive species which were at one time more widespread in Central Europe. In this case however we are not dealing with plants from the mediterranean, sub-mediterranean or south-eastern regions, but with those from northern and Alpine floras (e.g. *Linnaea borealis* and many of the *Pyrolaceae*).

We shall leave the question of systematics in abeyance and now turn to some examples of Pine woods from the Alps and from the plains of north-east Germany and Poland. These are the two areas of Central Europe which are richest in Pines.

(b) Communities of Scots Pine and Austrian Pine in the Alps

In the dry valleys of the inner Alps Scots Pine is dominant in many different habitats in the colline to montane belts. Consequently the species composition of the different communities varies much more than appears at first sight. This is true even for limestone habitats as can be seen from tab. 43. On one hand they support **'steppe-heath Pine woods'**, i.e. open stands through which the bright sunshine can pour down and which are the home of many of the plants which flourish in dry places. Excellent examples are to be found in the wide deep valleys of the western Alps: Briançonnais, Maurienne, and Tarantaise where they have been described by Gensac (1968). Yet there are also Pine wood valleys throughout the central and eastern Alps as far as the Vintschgau. Braun-Blanquet (1961) has devoted a monograph to these in which he also deals with the rest of the vegetation and the species-rich flora of these warm continental sites. Since this is easy to follow we shall not repeat here what he says. There are famous examples above Siders (Sion) in central Wallis, in the Pfynwald, an area of old rock falls on the driest slopes of which *Euphrasia viscosa* and other species of mediterranean origin are mixed with widespread dryness indicators such as *Carex humilis* and *Teucrium chamaedrys*.

Some steppe-heath Pine woods have also been described from the relatively continental Austrian inner Alps with their equally low precipitation (fig. 175 and tab. 43). Here too the stunted tree layer on more or less poor soil is made up of *Pinus sylvestris*.

If one climbs higher up the southern slopes of the central Alpine Pine valleys one finds that the stony ground is sparsely occupied by Pines whereas the deeper moister sites carry closed Spruce or even Fir woods. Here *Pinus sylvestris* assumes a slimmer form than it does in the even drier steppe Pine woods. But it does not become very tall, grows slowly and intercepts very little light from the shrub and ground layers. Such dry places in the submontane and montane belts are the province of the **Spring Heath-Pine woods** (*Erico-Pinetum*, fig. 176). List no. 2 in tab. 42 represents a number of stands from Grisons where this community is also widespread. The Spring Heath (*Erica herbacea = carnea*) impresses every visitor to the inner Alps

Tab. 43. Scots Pine associations in the central Alps on soil which is never wet; also Spring Heath associations in the south-eastern Alps.

Soil acidity:	very acid	fairly acid	neutral to alkaline	
Altitude belt:				
Montane	acid Small Sedge- Pine wood *Antherico (liliaginis-) Pinetum*	slightly-acid Spring Heath- Pine wood *Vaccinio-Pinetum*	carbonatic Spring Heath- Pine wood *Erico-Pinetum*	carbonatic Dwarf Sedge- Pine wood *Carici humilis-Pinetum*
Submontane	acid Oak -Pine wood *Pino-Quercetum roboris*		[Hop Hornbeam-Austrian Pine wood *Ostryo-Pinetum nigrae*]	
Colline	steppe heath Milk Vetch-Pine wood *Astragalo-Pinetum*	steppe heath Restharrow Pine wood *Ononido-Pinetum*	steppe heath Downy Oak- (Austrian Pine-) Pine wood *Cotino-Quercetum* [with *Pinus nigra*]	

Note: Each of the associations is divided into several subunits; for example, along with the widely distributed typical *Vaccinio-Pinetum* there are a number of subassociations in special habitats. These include *myrtilletosum, rhododendretosum ferruginei, piceetosum, vaccinietosum (vitis-idaeae), callunetosum* and *cladonietosum*, with dominating *Pinus nigra*.
Source: Taken from Mayer (1974) 1)

because its flowers, which have been almost completely formed in the autumn, are turned red by the sun as early as January or February on the warm slopes. As in the lower alpine belt it is always accompanied by other early flowering species, mainly *Sesleria albicans*, *Carex ornithopoda* and *Polygala chamaebuxus* (section C VI 2 a). Plants of the lower lying dry grasslands as well as *Calamagrostis varia* are also well represented. These tend to be found more and more with Scots Pine as they climb up into the subalpine level (compare no. 2 with nos. 4-6 in tab. 42).

In the Visper valleys, north of Montblanc, with their extremely low rainfall, and on the sunny slopes of many other central Alpine valleys the montane and subalpine belt of Spruce can be entirely missing. On such dry slopes the Spring Heath-Pine woods come into direct contact with the subalpine Larch-Arolla Pine woods which at these points descend somewhat further. The *Erico-Pineta* are typical and cover large areas on dolomite, serpentine and other slow-weathering rocks. According to Eggler (1955) they occur also on serpentine in Steiermark. They are also common all over the montane belt of the central Alps, especially on lime-rich soils (tab. 43). *Erica herbacea* is not confined to calcareous soil; it may even be dominant on soils of moderate acidity, but here it is mixed with acid indicators such as the ones we have met in the Spruce woods. Schweingruber (1974) found numerous intermediate types between typical Spring Heath-Pine woods and various acid-soil Pine woods both near the Vierwaldstätter See and in the Alpine foothills south of Berne.

Extensions of the submontane *Erico-Pinion* can be found as far as the southern edge of the Upper Rhine plain and the Bavarian pre-Alps, that is, about as far as the effect of the föhn wind is felt according to Oberdorfer (fig. 103). Further north the group of 'dealpine' species (*Sesleria*, *Carduus defloratus* etc., section B II 2 e) is mostly associated with Beech. Only *Erica carnea*, *Polygala chamaebuxus* and a few other species remain in fairly close connection with the Pine at greater distances from the Alps, for instance on the poor serpentine soils of northern Bavaria whose Pine woods, according to Gauckler's (1954) description, doubtless belong to the *Erico-Pinion*. Where Zöttl (1952b) found the Spring Heath-Pine woods in Bavarian pre-Alps they were restricted to lime-rich very rocky steep slopes and even here only on the driest places facing south and west.

Austrian Pine (*P. nigra*) is at home in the marginal mountains of the south-eastern Alps, but it has been spread throughout Central Europe by afforestation. It prefers a lime-rich and rather dry soil under cultivation as it does in nature. According to Wendelberger (1963a) it is concentrated in the submontane and colline belts between about 300 and 700 m above sea level in its south-east Alpine area of distribution. However under natural competition it is only able to assert itself here against the xerothermic mixed woodlands on slopes exposed to the wind and sun, the more since it survives drought periods better than *Pinus sylvestris* (Künstle, Mitscherlich and Hädrich 1979). Where it appears to be dominant in more sheltered and shady places one can be sure that the forester has played a part (Wendelberger 1977 verbal communication). Perhaps pure stands of Austrian Pine do not occur naturally in Central Europe. In any case Mayer (1974, see tab. 43) speaks of a 'Downy Oak steppe-heath Austrian Pine wood' at the colline level because, at least in the shrub layer, there are representatives of the submediterranean alliances *Quercion pubescenti-petraeae* or *Orno-Ostryon*. According to Horvat, Glavač and Ellenberg (1974) in south-eastern Europe, the main area of distribution for *Pinus nigra*, there are also Spring Heath Austrian-Pine woods where the plants of broadleaved woodland are virtually absent.

Looked at systematically the steppe Heath-Pine woods are closer to the mixed broadleaved woodlands than they are to the continental Pine woods of eastern Europe. To do justice to their unique position they should be given a class of their own (*Pulsatillo-Pinetea*).

The Spring Heath-Pine woods are also floristically so unique that they are placed in a special class (*Erico-Pinetea* with a single order *Erico-Pinetalia*, and the single alliance *Erico-Pinion*). Between the *Erico-Pinetea* and the *Vaccinio-Piceetea* there are however a number of intermediates, depending on the pH-value of the soil; but *Erica herbacea* is very acid-tolerant and in a dry climate can even be found in the communities of the acidophilous alliance *Dicrano-Pinion* (section c).

The systematic arrangement of the Pine stands in heavy clay and marl soil (group 3 in section 2) causes the greatest difficulties. We should turn now to these more or less basiphilous Moorgrass-Pine woods (*Molinio-Pinetum*). As is already apparent from the example given in tab. 42, col. 3, there are at least three groups of plants mixed up in the Moorgrass-Pine woods, namely species from the Spring Heath-Pine woods (E), the mixed broadleaved woods (*Querco-Fagetea*, Q) and the poor lime-rich meadows (M).

Soft clay marls in the rainy area surrounding the Alps mainly in Switzerland are a typical habitat for the **Moorgrass-Pine woods**. *Pinus sylvestris* finds even worse conditions for growth here than it does on stony slopes. Such soils are repeatedly soaked with water, so the roots are unable to obtain sufficient oxygen and remain near the surface. However in the dry summer period the topsoil dries out putting the sparse growth under a water stress. Where land slips or the erosion of periodical watercourses prevent deep disintegration of the marl this gives rise to a very open colourful mosaic of gnarled Pine tree stands and tall grass meadows rich in herbaceous perennials (fig. 177). According to the degree to which the marl is periodically saturated with water, some of these meadow patches are dominated by the same plants found in dry Spring Heath-Pine woods. Other patches are made up of species indicative of varying soil moisture such as *Molinia arundinacea*, *Succisa pratensis*, and further representatives of the order Molinietalia (most of which are included in col. 3 in tab. 42; see also section D V 5 a). In the Karwendel mountains *Molinia* has been helped by fires which previously often broke out on such slopes (Grabherr 1936).

Etter (1947) and Rehder (1962) have described and arranged in subunits the Moorgrass-Pine woods of northern Switerland, mainly from marls in the submontane belt around Zürich. A similar grass-rich Pine wood community appears on the steep marl slopes in the Swiss Jura up as far as the montane belt. According to Zoller (1951) it contains a larger number of alpine species, e.g. *Carex sempervirens*.

Even in the 'Fallätsche' a nature reserve on the Uetliberg to the west of Zürich, at only 660-750 m there are a number of species which have their distribution centre in the subalpine and alpine belts (Fabijanowski 1950, Dafis 1962). Because of the repeated slipping of this steep marl slope they can always find open spaces and have been able to keep going here since the Ice Age even though the whole of the surrounding area is densely wooded. Undoubtedly these plants are real relicts of the last Ice Age. By comparing detailed records of small areas Fabijanowski has been able to work out the cyclical succession from the bare marl through to the closed Moorgrass-Pine wood. The individual stages, and above all the different aspects of the slopes, lead to very dissimilar microclimates, so that the proximity of species, which ecologically are opposites, together with the unusual length of the species list, can be understood quite well.

Rehder (1962) uses as an example another part of the Uetliberg, the 'Girstel', to demonstrate all the intermediate communities, between the Moorgrass-Pine wood and the dry-slope Sedge-Beech wood which also occurs in this area (*Carici-Fagetum*, see sections B II 2 c and fig. 177). These 'transitional' communities do not correspond to phases of a succession in time, but form a rather stable mosaic corresponding to the actual site conditions. Pines and their floristic companions could only establish themselves where the habitat is unfavourable for the majority of Central European trees and herbaceous plants.

(c) Sandy Pine woods of the lowlands in comparison with Birch-Oak woods

In contrast to the majority of *Erico-Pinion* communities all those of the *Dicrano-Pinion* inhabit soils ranging from acid to very acid (tab. 33). In spite of this they are able to produce a greater biomass than the former, since the soils are deeper and give a better water supply. Most subsoils are silicate-deficient sands of various origins, especially fluvio-glacial and aeolian, with a low groundwater table.

If we look at the natural woodland vegetation of such sites as we pass from Holland over north-west Germany and Brandenburg as far as eastern Poland we shall see that the proportion of Pines increases with increasing continentality of the climate (figs. 140 and 178). As tab. 33 shows the species composition of the shrub, herb and moss layers too change over this climatic gradient.

In Holland and West Germany (lists 1 and 2 in table 33) the Birch-Oak woods, which we have already discussed in section B III 5, are the dominant natural vegetation. It is only in these that the subatlantic floristic element is well represented. In the sequence we are following its members can serve as indicators for 'broadleaved woodland conditions' (L in tab. 33). Just as the precipitation falls off rapidly to the east of the Elbe river from more than 650 mm to under 550 mm so the subatlantic species disappear quite sudden-

Fig. 178. Approximate proportions of Scots Pine and other trees making up the natural woodlands on soils of different clay content in the diluvial plain between Holland and East Poland. The numbers of the west-east gradient (1-7) correspond to the regions shown in the columns of table 33.

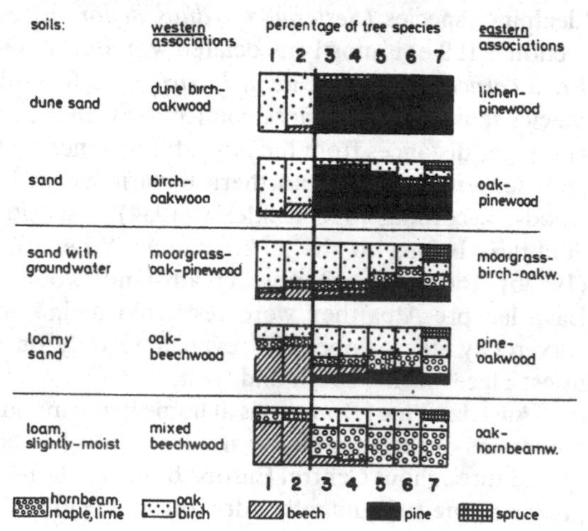

ly (fig. 2). At the same time Oak, although still always present, must hand over the dominance to Pine (figs. 179 and 180).

In northern Havel country with its low precipitation a natural **Oak-Pine wood** occurs on poor dry sand (list 3). It does not occupy a very large area since the majority of the habitats of northern Havelland are somewhat more fertile or have a better water supply, conditions which favour the Oak. What is worth noting is that here neither acid-tolerant elements of deciduous woodland nor 'coniferous woodland plants' (N) with a tendency towards a more continental distribution contribute to the mixed Pine woods in any significant numbers. However, as Passarge (1957c) has worked out in detail a rapid change in the flora takes place within this area from west to east.

To the right of the Oder river the percentage of 'eastern' species is considerably higher on similar habitats (list 4). Here we are still in the area of the Oak-Pine wood, which can be deduced from the occasional appearance of *Holcus mollis*, *Avenella flexuosa* and *Genista pilosa* as well as *Cytisus scoparius* and *Fagus*. Further eastwards beech is completely absent whereas Oak and other broadleaved trees can still compete with Pine.

Even in the middle and east of Poland *Quercus robur* takes on an accompanying role. The almost pure Pine forests which are the rule today in the sandy areas must be looked upon as artificial products from all we know about vegetation history (see especially Firbas 1949 and 1952 as well as Scamoni 1960). The typical Oak-Pine wood of Central Poland (list 5) can still be unreservedly compared with the communities already discussed in respect of its soil conditions. Like the Birch-Oak wood it occurs on quartz-rich sands with a more or less clearly defined 'narrow seam' subsoil (section III 5 b) which shows no noticeable signs of groundwater influence.

Lists 6 and 7 are not fully comparable in this respect because the soil here is somewhat richer (in the records made by Puławy from an Oak-Pine wood in east-Central Poland) and somewhat damper too (in the list from the Spruce-Oak-Pine wood in Białowieża Forest). This information is taken from the descriptions of the profiles given by the authors cited in tab. 33, but it can also be seen in the species composition. In list no. 6 there are for example *Tilia cordata*, *Corylus avellana*, *Carex digitata* and *Melica nutans* as new additions and in list no. 7 *Molinia caerulea* is shown with high frequency. Unfortunately no better comparable examples are available to show the purely subcontinental character of the species composition.

It has already been mentioned in section III 3 b that the Forest of Białowieża comes within the distribution of Spruce while lists 1-6 do not show any Spruce. List 7 represents the most continental community in our sequence, the **Spruce-Oak-Pine wood** (fig. 181). Like *Fagus* in the west, *Picea* plays only a small part in the canopy on poor sandy soils, but it can speed up erratically the percentage participation of coniferous woodland species as it builds up a more effective humus layer than *Pinus sylvestris*. The Common Oak still holds its own at Białowieża in spite of the Spruce coming in, which we take as further

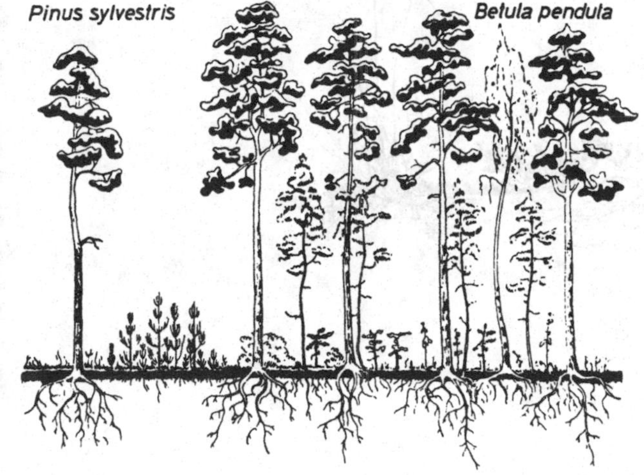

Fig. 179. The layer structure and regeneration of a natural Pine wood in Poland. After Pawłowski (1959, see also fig. 173).

evidence that there are no natural pure coniferous woodland communities in any of the lowland habitats of Central Europe that have so far been discussed.

Looking at tab. 33 (and fig. 178) as a whole it is astonishing how little the Birch-Oak woods of the west and the Oak-Pine woods of the east have in common floristically. Although both groups of communities occupy very similar soils their ground flora differs, except for a few acid indicators which are not particular as to the company they keep (e.g. *Vaccinium myrtillus*, *Calluna* and *Melampyrum pratense*). These are insufficient as a reason for putting the *Quercion robori-petraeae* and the *Dicrano-Pinion* together in one class. Such a union would be even less justified on the grounds of physiognomy and geography. It is much more logical to give the acid-soil Oak woods their own class (*Quercetea robori-petraeae*) whose main area of distribution lies in atlantic western Europe (the '*Quercus robur-Calluna* belt' of E. Schmid, 1936). The acid-soil mixed Pine woods on the other hand certainly belong to the class of eurosibirian coniferous woods (*Vaccinio-Piceetea*).

In the dry subcontinental areas between the Elbe and the Vistula the proportion of Pine in the make up of the natural woodlands depends on the properties of the soil, especially on its silicate and colloidal content and its water supply. The pH value and lime content appear to be of less significance.

Only on sand which has been poor right from its origin does Scots Pine dominate alone, and only on this does the raw humus derived from its needles cause natural podsolization. Such conditions exist for instance, in southern Slovakia (Bublinec 1974). The podsols found in north-west Germany, however were formed before the *Calluna* heathland was afforested with Pine (see fig. 404 and section D III 1 c). Pure stands of *Pinus sylvestris* are mainly to be met with on wind-blown sand. Here very meagre **lichen-Pine woods**

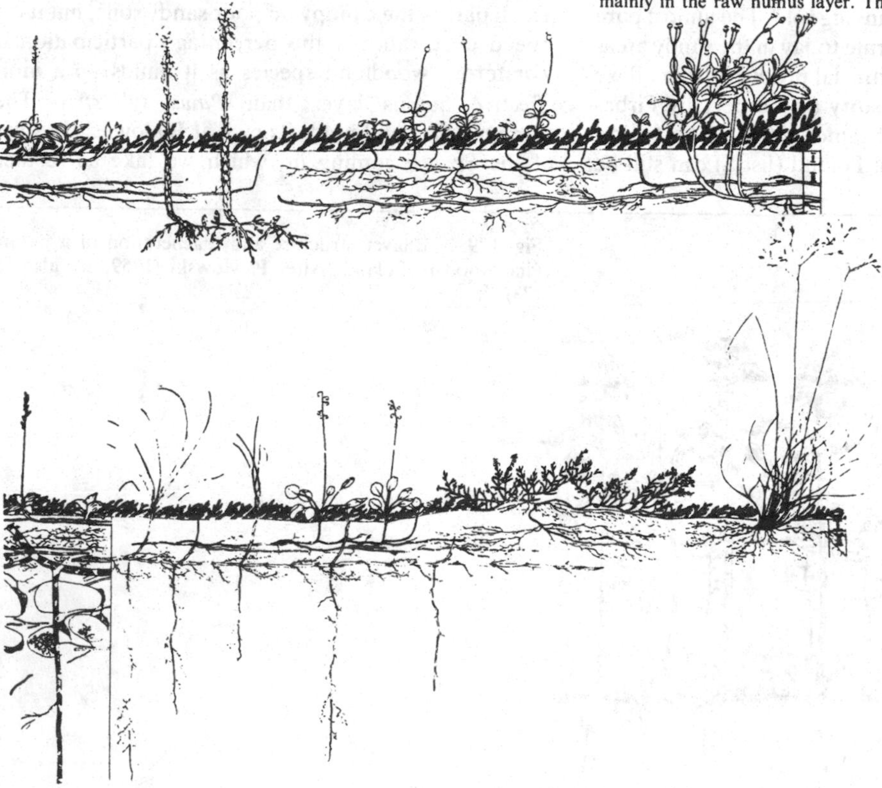

Fig. 180. Sections through the herb and moss layer of an Oak-Pine wood (*Pino-Quercetum*) on gravelly sand on the Schaabe (island of Rügen). After Meusel (1952a), slightly modified.

Below, from left to right: *Goodyera repens* with Pine roots beneath, *Carex arenaria, Pyrola chlorantha, Calluna vulgaris* and *Avenella flexuos*.
Above: *Goodyera, Monotropa hypopitys, Moneses uniflora* and *Chimaphila umbellata*. Only *Goodyera. Calluna* and *Avenella* root mainly in the raw humus layer. The moss layer is continuous.

are dominating, e.g. in the northern diluvial plains (Krieger 1937, see also section C V 3 b) and in the Slowakian Záhorie (Ružička 1961).

'Normal' sands which are free from groundwater are taken over by the Oak-Pine woods (tab. 33, lists 3-5) which have already been dealt with. Here Pine must put up with Oak as a partner. The higher the silicate content of the sand the more strongly do Oaks grow. According to Kundler (1956, see fig. 182) sands of average grain size are generally poorer in silicate than either coarse or fine sands and are consequently more favourable for the competitiveness of Pine. Above all though Kundler was able to show that the nutrient content of the sand was related to its geological age. The further one goes towards the north over the eastern-Central European lowland the younger are the glacial deposits and the less likely one is to find woods on sandy soils which are naturally dominated by Pines. The competitive strength of the broadleaved trees increases towards the Baltic coast not only because the sands become more fertile but also as the climate becomes more oceanic. The continentality gradient going from south to north obviously operates in a similar way as the east-west gradient which was discussed above (figs. 140, 178 and tab. 33, also fig. 2).

If the root run of the woodland lies within the range of the groundwater Pine has difficulties in asserting itself against Oak even on silicate-deficient sands in those regions which have a low precipitation. Thus Moorgrass-Birch-Oak woods on moist soils reach much further towards the east than typical Birch-Oak woods (compare the middle diagram in fig. 178 with that above it). Certainly they take in some Pines as partners when these are present in the neighbourhood, but the pure Pine stands with *Molinia* which are frequently met with are the product of forestry (fig. 183).

The competitive ability of broadleaved trees, other things being equal, also increases with the proportion of clay mixed with the sand. The more loamy the root run the more Beech can succeed in the west and north, or Hornbeam in the east, along with their companions (lower part of fig. 178). These shade and half-shade trees completely exclude Pine from fertile brown earths. According to Passarge (1953b) *Pinus sylvestris* is at a disadvantage even more on fertile soils with a high water table since in such sites it seems to be more susceptible to attacks from noxious fungi (e.g. *Trametes pini*).

So Pine, Oak or more demanding broadleaved trees play varying parts in the composition of the woodlands on the diluvial lowland to the north of Central Europe, depending on fertility of the soil and degree of continentality of the climate. It is difficult to

Fig. 181. Spruce-Pine wood in north-eastern Poland on diluvial silicate-rich sand. *Pinus sylvestris* produces good specimens whereas *Picea abies* achieves only second quality. *Vaccinium myrtillus* dominates the ground flora.

Fig. 182. The silicate content of plateau sands from different Ice Age periods in the lowlands of north-east Germany. After Kundler (1956) modified.

The older the diluvial sand the poorer it is in silicates. The 'silicate index (Si) is a relative value determined microscopically.

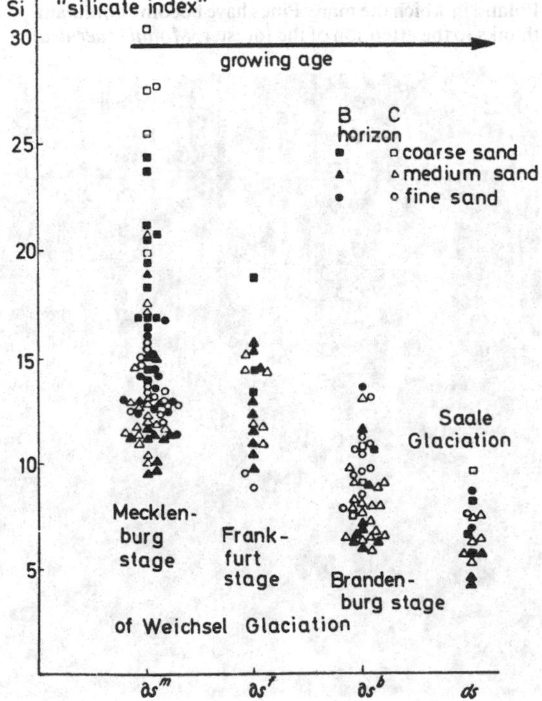

express these changing circumstances in a plant-sociological system since they have been affected so much by man, sometimes in ways which are not clearly recognisable. This interference has altered the balance of species in the undergrowth as well as in the canopy.

According to Passarge (1958) there are some remarkable differences in the ecological behaviour of many of the herbaceous plants over the climatic gradient from west to east. For example *Luzula pilosa* is very rare in the western atlantic region and here it is so fastidious that it can only survive in Oak-Hornbeam woods. In Poland on the other hand it is one of the most frequent woodland plants and prefers poorer soils, e.g. those of the Moorgrass-Birch-Oak woods with Pines. Just the opposite is Ivy (*Hedera helix*), which is fairly tolerant as to soil in Holland and north-west Germany and climbs up the trees. Further east however it is confined to the Oak-Hornbeam woods and creeps on the ground. Susceptibility to winter frosts may play a part here, whereas the Woodrush reacts primarily to humus quality and the amount of light and is probably more vulnerable to competition. The changes in the 'sociological amplitude' in *Moehringia trinervia* are similar to those in *Hedera*. Other species can be compared with *Luzula pilosa*, e.g. *Maianthemum bifolium*, *Hepatica nobilis* and *Carex digitata*. Unfortunately a closer ecological and, above all, experimental investigation of these interesting findings is still not available.

Alongside the pure and mixed Pine woods we have already looked at other types all of which cannot be mentioned or discussed in more detail here. For example there are **Beech-Pine woods** in Mecklenburg and Pomerania, a combination which is rarely found in Central Europe (Scamoni 1960). Between Lysa Gora and Warsaw, where the Fir comes down to the plain, it is mixed with Scots Pine in a **Fir-Pine wood** which Preising (1943) considers a separate association. As with many other vegetation units such dovetailing can be quite well understood given a knowledge of the more frequent communities.

In different Pine wood communities in the north-east of Central Europe, *Juniperus communis* is so common that one can straightway speak of a '**Juniper-Pine wood**' (fig. 184). Singly or in groups this evergreen coniferous shrub often occupies large areas which are scantily covered with tough grass, under the open canopy of Pines. Woodlands of this appearance have arisen because of the grazing with sheep, goats and other domestic animals which was at one time customary and may be still carried out even at the present day. The browsing, treading and dung killed

Fig. 183. A Purple Moorgrass-Birch-Oak wood in western Poland in which the many Pines have become dominant thanks to the attention of the forester. *Molinia caerulea* appears especially in the lighter places such as the clear felled area which has temporarily become heathland.

out dwarf shrubs and mosses allowing Sheep's Fescue, Bent-grass and other rough pasture plants to penetrate the woodland. Over the course of the centuries the browsing animals also eradicated the Oaks and other broadleaved trees and shrubs, leaving young Junipers untouched. Only when protected by large groups of Juniper are the original members of the community able to hold their own, and begin to reoccupy the habitat. 'Juniper-Pine woods' have grown up in the neighbourhood of large settlements in Brandenburg district, in Poland and the former East Prussia, but are also met with here and there on the dunes of the Upper Rhine plain. They may be regarded as the subcontinental equivalent of the subatlantic *Calluna* heath of north-west Germany, and like this they are becoming more and more threatened by modern forestry practice. Incidentally there are also stands of the Spring Heath-Pine woods which are rich in Juniper, e.g. in the valley of the Inn between Landeck and Innsbruck. Here too we are not dealing with a special association, but only with Juniper facies of the zonal or azonal vegetation units which have been more or less influenced by former husbandry.

Fig. 184. A 'Juniper-Pine wood' to the south-west of Białystok in eastern Poland. *Juniperus* spreads as a pasture weed; Sheep's Fescue (*Festuca ovina*) which dominates the ground flora is also encouraged by the grazing, because it is easten only in spring.

V Tree and shrub vegetation of flood plains and peat lands

1 Flood plains and their vegetation

(a) Living conditions and plant formations in flood plains

Up to now we have not considered the river valley woodlands in our presentation of the woodland communities of Central Europe even though many of them are very similar to the Oak-Hornbeam woods and the Sycamore-Ash woods which we have already discussed. Spring Heath-Pine woods and other woodland types of drier ground also have a close association with river valleys. However we shall understand these woodlands better if we look at them as the final stages of an ecological sequence as well as of a dynamic process which is common to all plant formations living under conditions of more or less frequent flooding (fig. 185 and tab. 44, also fig. 186).

Although in appearance and species composition the plant communities of flood plains may be so different, they are all – or were all – dependent on the conduct of the river water. The fluctuation in the amount of water can vary from five-fold to fifty-fold according to the section of the river involved, although in Central Europe there is no regular dry period. When the flow is at its minimum the river only occupies a small part of the valley, but at its highest the whole river valley may be flooded for a time. Only the plant communities and soils within this area of maximum inundation, are considered to be part of the riverside plain, or 'flood plain'. Wherever this overriding ecological factor no longer has an effect sooner or later this becomes noticeable in the species composition of the vegetation.

Since the melting of the snow in the high mountains sets in much later than lower down the streams fed by it usually carry most water in June and July (fig. 187). For this reason there are generally no meadows in the river valleys near these high mountains, but only pastures or woodland, the use of which is not upset so much by the high water in summer. In rivers flowing from the smaller mountains on the other hand, especially those in the west of Europe with its relatively mild winters and small amounts of snow, the normal surge of water is already past by this time, so that in the lower stretches of the Ems, Weser, Elbe and Oder extensive meadows are cultivated and the woodland has long ago been almost entirely changed into grassland. The further east one goes the longer the rivers remain frozen in winter and the longer they carry floating ice which can do so much damage to the banks in spring when the river is in spate. The rhythm of the rise and fall of the water level varies with the catchment area although at any time there may be catastrophic flooding following unusually heavy rain. In the Pupplinger Au above Munich, for example, on 13 September 1956, a time of year when the Isar normally carries very little water, the water level rose by 2.80 m in 24 h and a few days later had fallen back to its normal height.

All the plants and plant communities which live in flood plains are subject to such changing conditions, be they regular or quite unforeseen. The majority are capable of withstanding **occasional flooding** without permanent damage, or they are able to regenerate quickly. In addition many can withstand **dry periods** either by following the retreating water with their roots or by reducing the transpiration rate. Under certain circumstances however dry periods can do more damage to riverside plants than normal flooding.

The uncertainty of the situation however is compensated for to some extent by the **exceptionally fertile conditions** which occur in flood plains more than in any other set of habitats (fig. 188). Each flooding adds nutrient salts and solid deposits to the soil, increasing its fertility so that one could look upon this as a natural manuring. This is most marked at the 'drift line', that is where the remains of plants and animals are deposited along the edge of the highest floods. Such deposits are mostly rich in protein and are damp so they are quickly broken down into minerals. Hardly

Fig. 185. The turbulent Isar near Icking/Wolfratshausen ('Pupplinger Au'). The swiftly flowing river splits up into many channels when it reaches the gravelly flood plain below the Alps, changing its course with each succeeding flood water. In this way it creates habitats for some very different plant communities, from an unstable gravel to a woodland; these are constantly changing their position, often within a short time span. Photo. Heering.

any other site in nature is so rich in nitrate as is this kind of high water mark during the year following its deposition. Since the detritus is rarely completely smothered with the grasses and herbaceous plants growing there it provides an opportunity for quick-growing annuals such as is found nowhere else in Nature. The seeds of these light-loving nitrate plants are often washed up along with the other material.

The higher the soil level stands above the average water level the less often are the plant communities

Tab. 44. A summary of the morphology and vegetation of the flood plains in Central Europe from the Alps to the North Sea, schematic (compare figs. 185 and 186).

RIVER SECTION	ALPINE VALLEYS	ALPINE FOOTHILLS	LOWLAND	NORTH SEA ESTUARIES
Morphology	mostly upper or middle reaches	mostly upper, some lower reaches	lower reaches	river mouth (estuary)
in general	predominantly erosion	erosion and sedimentation	predominantly sedimentation	predominantly sedimentation
Water speed	fast flowing	average	slow	slow; changing direction with tide
Main sediment	gravel	sand	clay	silt and sand
Time of year when there is most water	summer	early summer also winter	winter rarely summer	winter
time of day	early afternoon	–	–	twice daily
Source of summer spate	melting snow and glaciers	as left, but sometimes as right	unusually heavy rain	as left and also storm tides
Form of watercourse	many small channels	several arms with some meandering	few arms but many large bends	funnel-shaped mouth, on tidal flats small channels
Areas with still water and silting up	absent	infrequent (large old courses)	numerous (old watercourses)	absent
Vegetation upper	–	hardwood trees	hardwood trees	hardwood trees
layers	Grey or Green Alder	Alder or Willow	–	–
lower	Willow bushes herbs	Reed beds annuals	Willow trees river Reed beds annuals	Willow trees tidal Reed beds –
on the river bed	*Chondrilletum*	*Polygono-Chenopodietum* and (or) *Polygono-Bidentetum*		–
lower part of river valley	*Chondrilletum*	*Agropyro-Rumicion*-communities		*Bolboschoenetum maritimi*
transition zone	*Salici-Myricarietum*	*Phalaridetum*	*Phalaridetum*, also *Phragmitetum*	*tidal phragmitetum*, also *Phalaridetum*
Willow and Alder communities	*Salicetum elaeagno-daphnoides*	*Salicetum triandro-viminalis and Salicetum albo-fragilis, Populetum*		the same but without Poplars
Willows Grey Alder	*Calamagrosti-Alnetum incanae*	*Equiseto-Alnetum incanae*		–
'Hard' wood communities				
below	–	*Ulmetum* and other mixed woodland		the same without Elms
above	–	*Ulmo-Quercetum*		
in old water courses (oxbows)	–	in places as right	land-forming succession leading to *Alnetum glutinosae*	in some places as left

Note: The German 'Hartholz' is not the same as the English 'Hardwood'. It means broadleaved trees giving good timber (e.g. *Quercus Ulmus, Fraxinus*) and does not include *Alnus. Salix, Populus* etc. which are called 'Weichholz', 'Softwood' in English, however, means conifers.

damaged by flooding and the longer they can enjoy an undisturbed growing period. On the other hand there will be a less constant supply of plant nutrients and the average level of the water table in the soil will be lower. The most favourable conditions for herbs and shallow-rooting woody plants are therefore to be found round about the centre of the habitat sequence shown in figs. 186 a and b, provided these species are able to survive an occasional inudation (I. and V. Kárpáti 1971a).

In actual fact the sequence is never as clear cut and stable as it appears to be in the schematic representations. These are shown merely to make it easier to survey the changeable habitat conditions which the river brings about by erosion, sedimentation and change of course. Figure 186 makes it plain that the different riverside formations cannot be combined in a single scheme. For a river having its source in the high mountains we must go along with Moor (1958) in distinguishing at least five sections (tab. 44):

1 The **stream near its source** is narrow and quite quickly cuts a channel into the rock, depending on the type of geological formation. It is accompanied along its way by hygrophilous perennials or other communities which can tolerate the wet

Fig. 186. Above: a schematic cross-section through the complete series of flood plain vegetation occupying the middle reaches of a river in the Alpine foothills. The Grey Alder wood can be found at the same level as the Willow wood.

Below: a schematic longitudinal section through the sequence of vegetation of a flood plain from the Alpine valleys to the coastal lowland in relation to the mean water level throughout the year (small dots) and during the summer (larger dots) as well as the range between high and low water.

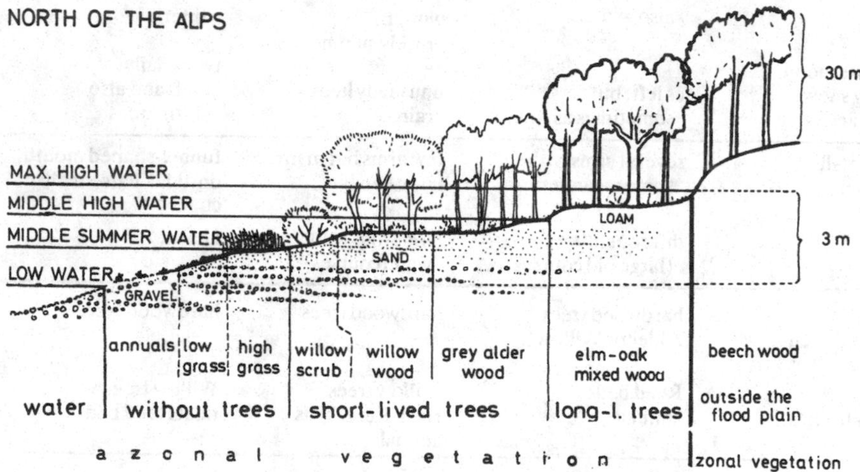

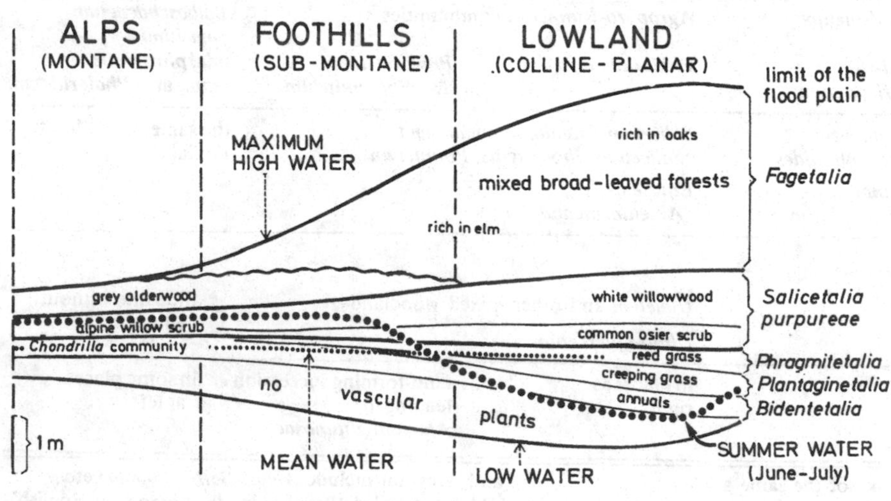

conditions but it only forms small flood plains where it is impeded in its course by a particular rock formation.

2 In the **upper reaches** the flood plain may also be absent but generally it accompanies the river as a narrow strip. The quickly flowing water deposits mainly gravel, or at the most sand, and there is only shallow flooding of the valley. Shrubs may even colonise the gravel banks thrown up in the middle of the stream, but such plants must be very active in rooting and capable of regeneration. The valleys at somewhat higher levels are occupied by soft-wooded trees and shrubs, especially the Grey Alder (*Alnus incana*).

3 In the **middle reaches**, which usually start in the larger mountain valleys or in the foothills, it is possible to distinguish in addition to the river bed a 'soft wooded' and a 'hard wooded' zone. We shall be going more thoroughly into these three vegetation levels and the conditions determining their species composition.

4 The **lower reaches** have a gradient of less than 0.3%; the river takes a much more winding course and builds up a valley with mostly fine-grained material. The vegetation levels resemble those in the middle reaches, although they mostly differ from them in the absence of the Grey Alder. There are also many more old water courses which gradually become filled up in a way similar to that seen in lakes.

5 Special conditions exist in the **estuaries**. Where they pour their water into a lake or a sea without

Fig. 188. The texture (a), pore volume and air content (b), phosphoric acid and lime content (c) and the pH values (d) of the soils at the levels of the Willow and mixed broadleaved woodlands in the flood plain of the March near Strážnice. After Mezera (1956), modified.

Broken line = *Populo-Salicetum*, solid line = *Querco-Ulmetum*. In addition the amount of nitrogen available to the plants is increased by a larger proportion of fine particles, good aeration and a neutral reaction of the soil as well as by the frequent deposit of sediment rich in nutrients.

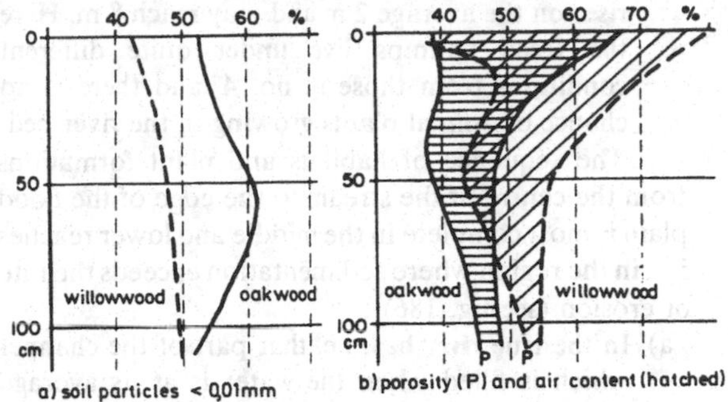

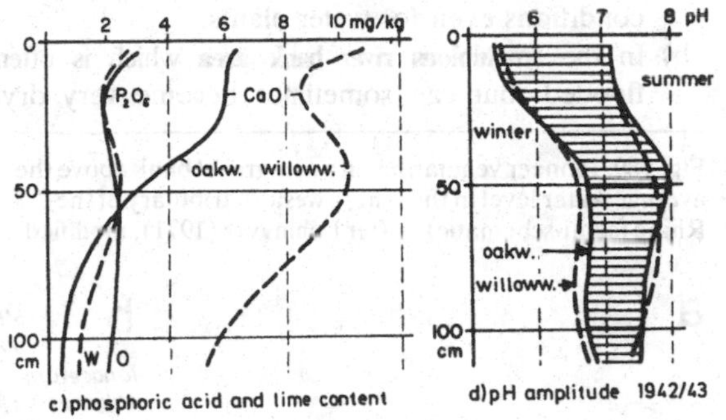

Fig. 187. Fluctuations in the rate of flow of three rivers throughout the year at points between the Swiss Alps and the lowland. After Heller (1969), modified.

The **Danube** has its source in the Black Forest and the Swabian Jura, but it gets most of its water from the Alps. As far down as below Vienna the greatest average flow is in summer, i.e. at the time when the snow and glaciers in the high mountains are melting most rapidly ('glacier dominated').

In the case of the **Rhine** the glacier domination exists as far as Basel, but below this the oceanic rainfall over the lesser mountains and the Central European plain has a greater and greater influence.

The **Rhône** too is noticeably affected by the high mountains as far as Lake Geneva but beyond this it comes increasingly under the influence of the mediterranean rainfall pattern with a pronounced minimum in late summer.

The large rivers of eastern Central Europe (not shown here) are controlled mainly by the melting snow lying on the plains and have their highest floodwater in the spring (April/May).

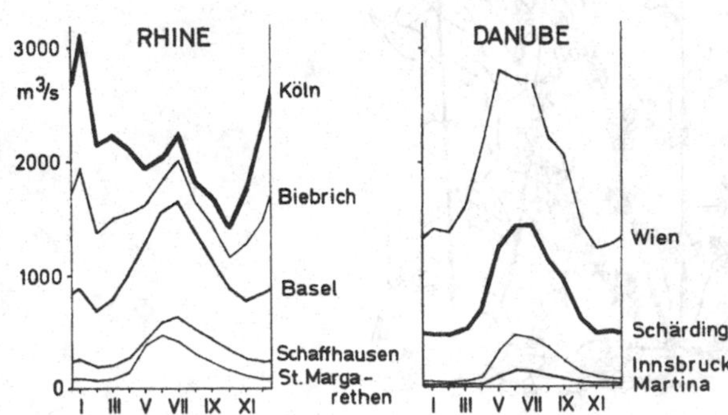

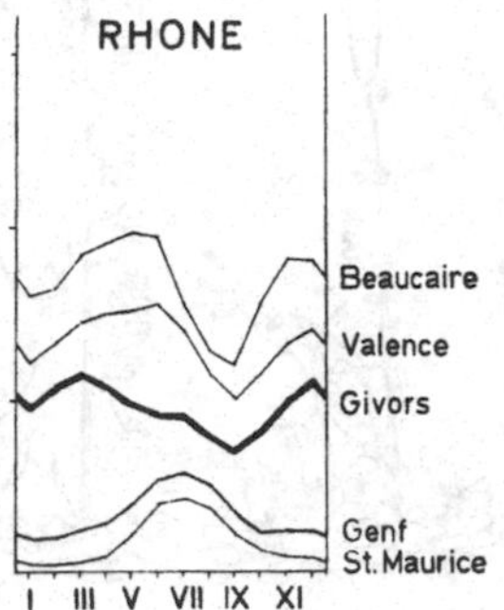

any marked fluctuations in the water level a wide delta is formed under natural conditions where Reed swamps and Willow woodlands find a fertile mud habitat. However in rivers flowing into the North Sea the twice daily ebb and flow of the tide prevents the formation of such a delta. Instead an estuary is formed, often extending more than 100 km upstream. In Hamburg for example the tide rises on the average 2 m and may reach 8 m. Here the Reed swamps live under quite different conditions from those in no. 4, and there is no chance of annual plants growing in the river bed.

The sequence of habitats and plant formations from the centre of the stream to the edge of the flood plain is most complete in the middle and lower reaches i.e. in the region where sedimentation exceeds the rate of erosion (see fig. 186):

a) In the true **river bed,** i.e. that part of the channel which is filled when the water is at its average height, there is nowhere for land plants to find a permanent home and there are unfavourable conditions even for water plants.

b) In the **amphibious river bank area** which is often flooded, but can sometimes become very dry, some herbaceous plants are to be found, at least in sections 3 and 4 referred to above. Of these only the short-lived annuals are able to set seed (fig. 189). Mainly species of *Chenopodium* and *Polygonum* find a place free from competition in the gravel or on coarse fertile sand. In the lower reaches of the river, where the soil is richer in fine particle, species of *Bidens* also play a part and generally the annuals are better developed here than in the middle reaches where the banks often become very dry in August. In the transition to the next higher level there may be here and there rapidly growing grasses or grass-like plants forming low carpets. However, like the occasional plants lower down the bank, their numbers and their vitality vary from year to year, and if the water level stays unusually high for some reason they are scarcely able to develop at all (fig. 189).

c) Above a certain degree of flood risk tall grasses and grasslike plants can gain a footing. They anchor themselves with a meshwork of rhizomes and spread out vegetatively (figs. 190, 191). In the middle reaches and sometimes also in the lower reaches such **river-bank reed strips** are found.

Fig. 189. Pioneer vegetation on a new gravel bank above the average water level in the Ahr, a western tributary of the Rhine (semi-schematic). After Lohmeyer (1971), modified.

a. The first year is dominated by therophytes – mainly species which are also weeds of arable land. Some perennials are present too but remain relatively obscure.
b. A year later large biennial and perennial herbs dominate the scene; *Agrostis stolonifera* remains in the undergrowth.

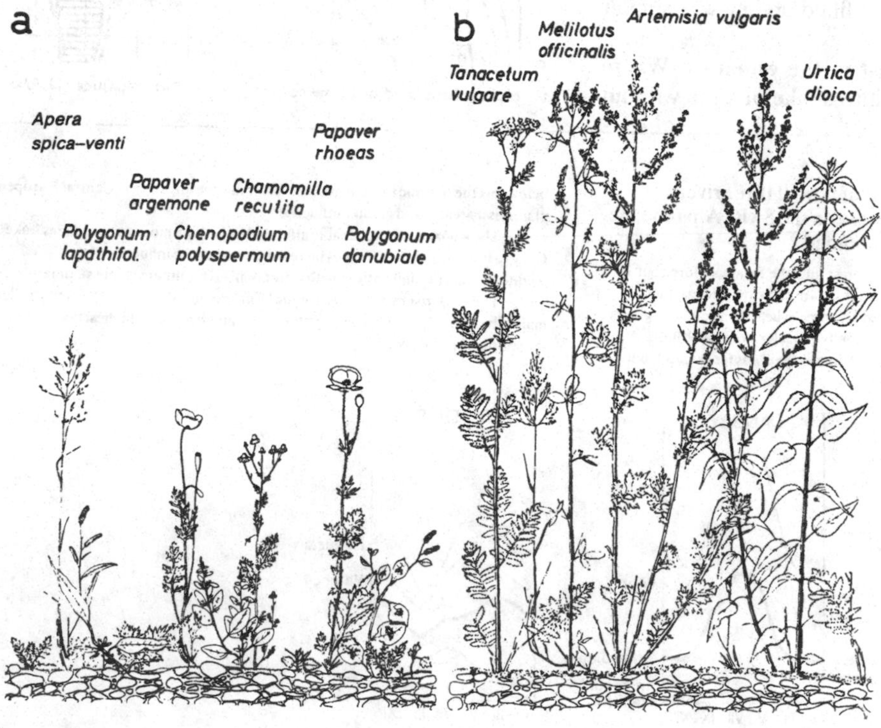

Regularly these are dominated by Reedgrass (*Phalaris arundinacea*) whose limp stalks tolerate being bent over by the stream better than does the stiff Upright Reed (*Phragmites australis*). The differences between reed communities bordering still and flowing waters have been well elaborated by Kopecký (1966). The *Phalaris* communities, in spite of being poor in species, reflect clearly the particular habitat conditions, also those caused by the altitude above sea-level (Niemann 1965, Kopecký 1969). The *Phalaridetum* may be absent altogether from banks which are trampled a lot, and only form wide continuous bands where they are not damaged too much mechanically. The establishment of isolated colonies of river reed beds is often started by a turf being washed up at high water and so is largely a matter of chance.

d) In a similar way the first pioneers of the riverside woodlands can become established. Thee are first and foremost very rapidly growing Willow shrubs such as *Salix purpurea* and *S. triandra* (fig. 192). This **riverside Willow scrub** does not succeed in settling every year even though it produces a quantity of air- and waterborne seeds. This is because the seeds only retain their germination power for a few days and even then will only germinate where the soil is damp, but not flooded. The seedlings also require a lot of light and thus would not be able to succeed in an established reed bed.

e) The Willow scrub can only cover large areas where there are extensive low banks of gravel or sand. Otherwise it just forms a narrow band which joins the reed-strip or the almost bare bank to the **often flooded riverside wood** (in German: Weichholz-Auenwald) on somewhat higher ground. This type of woodland, which is never absent from any natural river valley, consists of quick-growing trees whose wood is relatively soft and less enduring (fig. 193). But these are broadleaved, not coniferous trees, which in English, are called 'softwood'. Notable amongst them are some fine species of Willow, e.g. *Salix alba, S. fragilis* and their hybrid *S. rubens* as well as *S. triandra* which are present in all planar to submontane river valleys throughout Central Europe. In montane sites and in the foothills of the high mountains i.e. in all districts where the high water comes in summer, the riverside woodland is dominated by Grey Alder (*Alnus incana*, fig. 194). It is not yet really known what rôle the Poplars would naturally play, especially *Populus nigra* and *P. alba*, in the riparian woods of Central Europe (section B VI d).

f) The highest level of the area liable to flooding in the middle and lower reaches of the rivers is naturally also taken up by woodland. Because this is rich in strong persistent tree species, it is known as the **rarely flooded riverside wood** (in German: Hartholz-Auenwald, fig. 195). This is flooded only when there is unusually highwater and therefore the undergrowth is more 'forest like' than that under the Willow trees. In the latter many plants of the reed bed and others not usually found in woods are able to grow in the open places. Since this hardwood formation is only rarely flooded and the soil is so fertile it has been largely grubbed out and used for grassland. The majority of the remaining woods are dominated by Ash (*Fraxinus excelsior*) which flourishes quite well in the transition zone between the 'soft' and 'hard' flood plain woods. In the upper, drier parts of the latter Elm species (*Ulmus glabra, U. laevis*) and also Common Oak play a part. But Beech, which is otherwise so competitive, is found only in very special circumstances within the area liable to flooding (section B V 1 h).

g) When the river is in spate it lays down more sediment on its banks than it does further away so that relatively dry places are to be found close to

Fig. 190. A 'river Reed-grass' community (*Phalaridetum*) on the gravel bank of the river Aare (Switzerland). In front of it are fragments of the Bentgrass sward (*Agrostis stolonifera*) and, in the dried up river bed, annuals (*Polygonum lapathifolium*).

the river (fig. 195). The further away from the river one goes the less the water table varies in the soil. At the edge of the flood plain the soil water if often near the surface since it collects here from the higher bank and from the terrain outside the flood plain (fig. 30). This is especially so with those valleys which were cut out in the later Ice Age and are now much too wide for the river which is flowing through them. In some places extensive marginal areas have been turned into **floodplain mires** corresponding to those outside the river valley (section V2).

h) Banks within the river valley which are **never flooded**. The vegetation on these corresponds to the zonal one for the area and has already been discussed (see fig. 18b and 30).

Fig. 191. Growth on the river banks under conditions of varying current speed and water height. These examples are from the small river Studený (a) subjected to snow melting in the High Tatras and from the lower reaches of the Adler (b, c) in the hills of north-eastern Bohemia. After Kopecký (1969), modified.

a) Following melting of the snow in the high mountains the strong mountain stream carries most water in the early summer. Because of this the 'subriparian' area (below mean water level) is practically devoid of plants. Only in the 'lower riparian area is it possible for a sward of creeping species (with *Agrostis stolonifera, Barbaraea vulgaris* and *Juncus articulatus*) to become established. The bank Small-reed sward (*Calamagrostietum pseudophragmitis*), a community found only in south-eastern Central Europe, overgrows gravelly sandbanks in the 'upper riparian' area where Tamarisk (*Myricaria germanica*) maintains a foothold. Extensive beds of Butterbur (mostly *Petasites officinalis*) often form a 'skirt' to the *Myricaria-Salix incana* scrub.

b) The impact slope of the river bank has to withstand strong currents and variations in water height. Because of this the area of low water and below carries no vegetation. On the banks which are free from water for a time annuals such as *Polygonium lapathifolium, P. danubiale, Chenopodium rubrum* and others are able to grow. The Reedgrass beds offer a certain amount of protection for the bank. On the edge of the bank which is rarely flooded there is a Sedge community based on *Carex buekii* which is also restricted to south-eastern Central Europe. This retreats where Willow scrub lines the bank; this is overgrown by a veil of *Calystegia sepium, Cuscuta* species and similar plants.

c) Above an artificial barrier the vegetation zones are reminiscent of those in a shallow lake; a covering of Duckweed with scattered Water Lilier, a Reed bed with Water Dropwort and a border of Flowering Rush (*Butomus umbellatus*), a Reed-Sweetgrass swamp (*Glycerietum maximae*) and a Sedge swamp (*Caricetum gracilis*) with *Phalaris*.

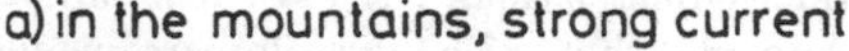

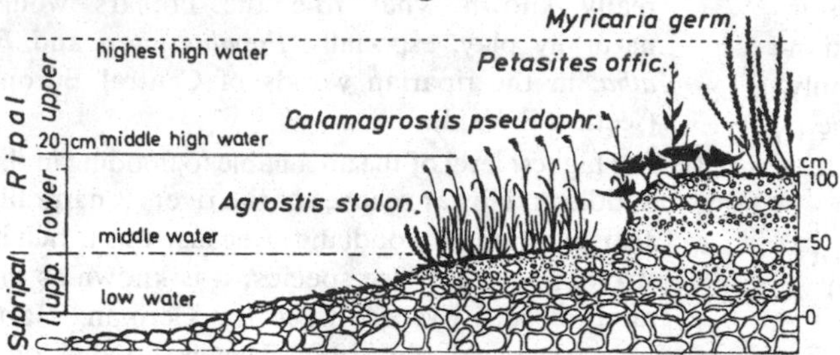

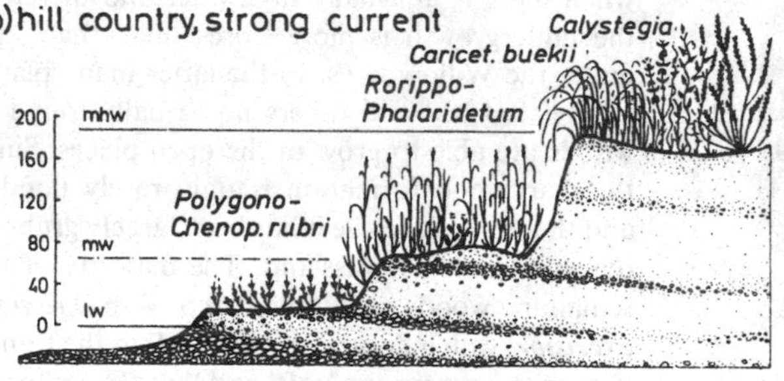

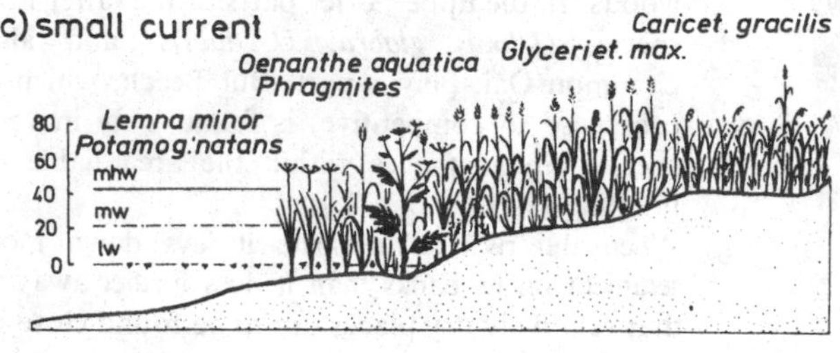

i) Many large river valleys in Central Europe are to be found in **old gravel terraces** which during or between the Ice Ages, or shortly afterwards, were themselves flood plains. Their vegetation today is clearly different from the zonal one but at the same time it is only slightly reminiscent of that of the present flood plains, and indeed can show a striking contrast to them. Their soils, which are gravelly or of coarse sand and very free draining, very often dry out and allow only a stunted open growth. Near the Alps a special type of Spring Heath-Pine woods dominates such gravel terraces (figs. 209 and 210).

This general picture of the vegetation levels in the upper, middle and lower reaches of the rivers is true, not only for Central Europe, but also basically for all humid regions of the Earth which are covered with broadleaved woodland. It is particularly true for the wide river valleys in regions of tropical rainforest. The sandbanks and old channels of the upper Amazon and its tributaries, which dry out when the water is low, are occupied anew every year with loosely distributed therophytes (b), and on the embankments of the river there is thick grassland corresponding to our river reed strips (c). Bushes similar to Willow (d) form the first woody covering along the side of the stream, but this soom merges into a woodland of large broadleaved trees with soft wood, these corresponding ecologically to our Grey Alders. A tall 'hard' broadleaved woodland dominates that part of the valley which is flooded less frequently (f). This contains a number of shade trees and is physiognomically similar to the rainforest on the 'firm ground' (terra firma), although it differs from it floristically (Ellenberg 1959, 1985).

(b) The dynamics of flood plain vegetation

In the tropics, as in Central Europe, the river valleys are undergoing constant change. On tight bends the river undercuts the ground which it had itself deposited at one time. In doing so it can even pull down tall trees. On the gently sloping inner shore on the other hand the sequence from the annual flora, throught the grassy banks and the soft woody plants back to the tall valley woodland is accomplished in broad zones. The river swings this way and that in its wide valley, now tearing down and now building up in an ever-repeated cycle. This dynamism can still be seen very well in the humid tropics and other sparsely occupied woodland regions of the Earth, standing in sharp contrast to the stability of the natural woodland outside the river valleys. For a long time now in Central Europe the water engineers have taken the rivers in hand to such an extend that the effects of erosion and sedimentation correspond only in a very few places to those which would occur in Nature. Because of the straightening and damming of the rivers the 'hard' wood formations in particular are being affected (figs. 30–32). This the basic question which arises from a study of the flood plain vegetation can scarcely yet be answered with certainty in Central Europe. It is this: can the community of annual plants

Fig. 192. A Purple Willow scrub (*Salicetum purpureae*) and fragments of a Reedgrass bed (*Phalaridetum*) on a sand-covered gravel bank in the Enns (in the 'Gesäuse' below Admont, Austria) at low water in September. In the right foreground are Willow seedlings, mainly *Salix alba*, which have grown from seed deposited in June.

in the river bed be connected to the broadleaved 'hard' wood community by means of a step by step development or is the sequence of plant formations discussed above merely an expression of a spatial juxtaposition of habitats which the river has created simultaneously? In short, is it a question of **succession** (in time) or **zonation** (in space)?

Siegrist (1913) and other pioneers of flood plain research, but also Seibert (1958) and many more recent authors, maintain the first point of view though they do concede that alongside genuine succession long-term stable zonation is often present. Moor (1958) sees it as almost entirely a matter of zoning and like him many vegetation experts have become sceptical over what at first glance appear to be such clear and convincing succession diagrams.

In any case what is certain is that bare soil, newly deposited by a river, or left behind when it has suddenly burst through to cut off a loop, very quickly becomes invaded by plants. For the Pupplinger Au, to the south of Munich, Seibert (1958) was able to show from a comparison of aerial photographs that in only 34 years a 'soft' deciduous woodland composed of Grey Alder had become established on ground which had been free of any plants (fig. 196). This astonishingly rapid development no doubt began with a therophyte flora similar to that which each year occupies parts of the river bed. However there was no step by step succession of Reed stand. Willow scrub. Willow woodland and only then the Grey Alder, but the latter achieved dominance after just a few years. This was because the surface of the new gravel bank lay so far above the average height of the groundwater table that *Alnetum incanae* was able to establish itself from the

Fig. 193. A 'virgin forest' of White Willow, i.e. the first stand of trees to colonise a new island in the Aar which was formed in 1940 and had not been disturbed up to the time the photograph was taken in 1963.

Salix rubens (S. alba x fragilis) and *S. alba* are dominating in the foreground *Festuca gigantea* and *Urtica dioica* with *Phalarisarundinacea* behind.

start. In a similar manner White Willow woodland appears on new sandy islands in lowland rivers. The *Salix* stands can reach a height of 12-15 m in 20-25 years as Moor (1958) quotes and Heller (1969, see fig. 193) confirms for northern Switzerland. The *Salix alba* (or *S. rubens*) germinates at the same time as the species of *Bidens* and *Polygonum* in such cases and is therefore one of the original pioneers. These examples only apparently support the 'succession hypothesis' and, on closer inspection, discredit it.

As observations on wild rivers in the tropics teach us, changes in the levels in a river valley happen very suddenly, due to the very strong currents when the river is in high spate. The surface contours formed in this way over the course of just a few days or weeks remain when the water falls and take on a covering of vegetation in different ways depending on height above the water table. At the next severe flooding such ground which may only recently have been formed could be completely destroyed again along with the first stages of its vegetation, or it could be changed erratically into other kinds of habitat. On the other hand it is always possible that it may remain untouched for a longer time and then the plant community which finds the most favourable conditions there will be the one to express itself undisturbed. In every case it is the **river** which determines the allocation and succession of the plant communities. At these lower levels the vegetation itself contributes very little and does not build up the soil step by step so that each year it retains a fresh layer, but invades the new habitats which have suddenly been created. A number of habitats with different initial conditions are often invaded at the same time. Colonisation of the new land which is set going in this way can certainly be looked upon as a series of natural successions, but similar ones are also found even on new land which is not subject to flooding. So they are not evidence of a genetic coherence in flood plain formations, and the advocates of the zonation hypothesis are generally right as far as the lower ground in river valleys is concerned.

Fig. 194. A Grey Alder flood plain wood on the Inn near Strada (eastern Swiss central Alps) in late autumn. Flood water has eroded the sand and gravel loosening and carrying away part of the Alder stand. View looking upstream after the water had receded.

It is only the last two members of the flood plain series, the 'soft'- and the 'hard-wood' deciduous formations, which may be connected by a succession in a number of cases, especially where the water course keeps to the same bed over a longer period of time. Normally the river is incapable of throwing up gravel or sand banks so high that they would reach the level of the hard broadleaved woodland from the outset. Instead it gradually raises the level of the ground by depositing layer after layer each time the water level is high enough to flood this area. As the level is raised the ground is flooded less often and not so deeply. The speed of flow is also less, especially where it is slowed down by a thick layer of surface plants. Thus the finer particles in suspension have time to settle. A section showing the profile through a wooded flood plain soil shows that the lower layers, which are only a little above the average height of the river, are generally of coarse sand or even gravel and show an oblique layering characteristic of those formed by rapidly flowing water (fig. 191). In the upper part of the profile the soil becomes fine grained, i.e. consisting of sand, silt and clay or loam and the layers are arranged horizontally. However under a typical 'hard' broad-leaved wood the uppermost horizon is never layered as the raising of the surface by added fine deposits goes on more and more slowly. Over the course of decades the layers are continually being disturbed by the roots and especially by earthworms. These are quite well able to live in the soil here because even during the few days or weeks when the ground is flooded there is still sufficient air for them. The soil profile of the flood plain woodland thus indicates that a gradual succession has taken place, but this has not been brought about by the vegetation alone ('autogenic') but has been helped by sedimentation from the river ('auto-allogenic').

However one must always remember when studying such river valley soils that a change from unevenly deposited gravel to a fine-grained flood plain loam has not infrequently been brought about by geological processes working alongside the historical development of the vegetation. The river gravels were mostly laid down in the Central European lowlands during the Diluvium period, whereas the valley loam is a product of historical times, largely following the destruction of woodland and the resulting increased soil erosion in the catchment areas of the rivers (figs. 30 and 197). So after all it may be that formation of the valley loam in this case too has coincided with a succession from a 'soft' to a 'hard' woodland.

Many Oak or Ash woods now occupy sites previously held by Willow or Alder communities just because the river level has been lowered and with it the groundwater table. This may have been due to artificial straightening of the river or because its bed has been deepened through natural causes, so that in relative terms the soil surface is now high enough not to be flooded so often. The water level could indeed drop so low that a one-time Willow stand, for example, is no longer flooded. In this case, because of the permeability and dryness of the soil, which would have very little fine earth in it, the site would only be able to support a drought-tolerating community such as a Pine wood or a xerothermic Oak wood (figs. 195 and 31).

Taken as a whole one can say that the series of plant formations and communities which can be observed in flood plains have been produced in different ways. In order properly to understand the vegetation of the river valleys one should distinguish the following processes:

1 Changes in the vegetation during a **long-term stable water pattern** especially a constancy in the

Fig. 195. Fairly natural woodland communities from the bank of the Danube across to the landward side of the flood plain near Vienna: below are the names of the different units as they are normally used in Austria. After Margl (1971), somewhat altered.

The high gravel banks, covered with a thin layer of fine earth are know as 'Heissland' (hot ground). Trees growing on these are mostly unable to reach the ground-water. The first colonizers are the Purple Willow (*Salix purpurea*) together with the more drought-resistant Felt Willow (*S. elaeagnos*) and Black Poplar (*Populus nigra*). These are followed by Silver Birch (*Betula pendula*) and Common Oak (*Quercus robur*), and, where it is not so dry, mixed woods with Oak and Small-leaved Lime (*Tilia cordata*).

On the more or less sandy or silty ground, which from the outset has been somewhat lower lying, the Purple Willow or Common Osier scrub (*S. viminalis*) is followed by a 'slightly-moist' or a 'damp' Willow wood with White Willow (*S. alba*) or by a 'slightly-moist' or a 'damp' Poplar wood with mainly White Poplar (*P. alba*). The broadleaved woods ('Harte Au') which finally become established likewise have different compositions depending on the amount of soil moisture. Elm (*Ulmus minor*), Ash (*Fraxinus excelsior*) or even White Poplar or Oak may be dominant in varying proportions. Bird Cherry (*Prunus padus*) is conspicuous in the shrub layer in the damper places.

The Grey Alder (*Alnus incana*), so plentiful in and near the high mountains, is rare downstream of Vienna. It is no more confined to a particular stretch of the river valley than are most other tree species. Since it is so shallow rooting, it avoids the gravelly flood plain soils in the lowland where the rainfall is slight (see fig. 201).

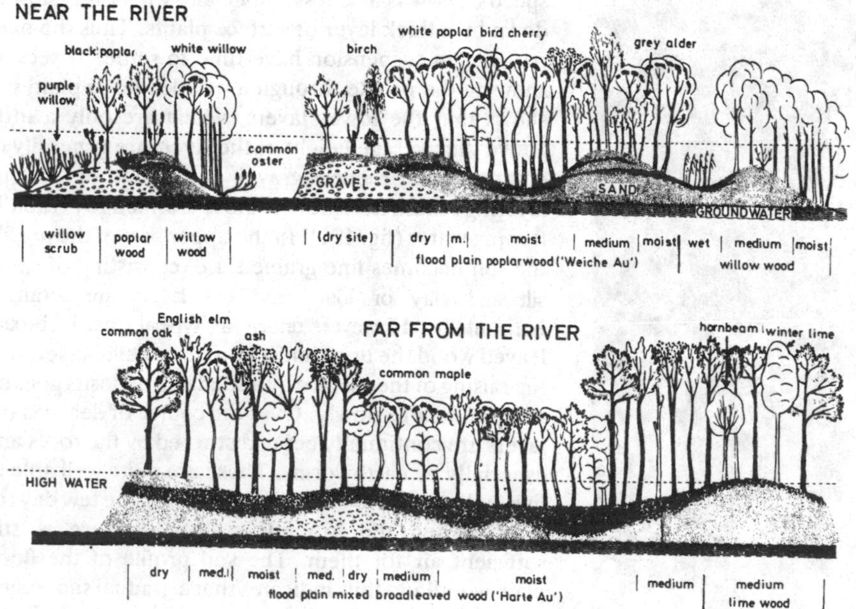

average high, mean and low water levels. These are:

a) The **colonisation of new land** which has been deposited by the river (or by man) or newly exposed through removal of the surface layer. On these areas which at first have no vegetation a real vegetation development takes place which is primarily an autogenic succession corresponding to the conditions of the particular habitat. This generally takes place quite rapidly since seeds and other regenerative parts of the species making up the final community are present right from the start.

b) A **sudden covering over** of the habitat and its vegetation by a catastrophically high flood water. This happens when the plant community at a low level is covered over so deeply with a mass of fresh sediments that only a few species are able to survive as relicts. Apart from this difference it is really another case of the colonisation of new land as in a).

c) A **gradual covering** of the soil and vegetation by sand and silt brought about during flooding by the plant stand quietening down the water so that the particles are deposited ('auto-allogenic succession').

2 Changes in the vegetation brought about by a **change in the water regime**, especially by a lowering of the mean- and low-water levels.

a) **Gradual transformation** of the habitat following continued erosion or a slow lowering in the amount of water coming down the river, and similar causes. In the case of these purely 'allogenic successions' the vegetation itself in no way acts as an instigator, but is forced to adjust gradually to the changing conditions.

b) **Sudden changes** in the habitat which are generally caused by man, such as the building of reservoirs or a series of dams. Occasionally however there are natural causes acting in a similar way. In any case the changes are so catastrophic for the vegetation that many of the species are killed off straight away. Most of the others may be crowded out through competition with new arrivals which are able to thrive under the new conditions.

Fig. 196. Changes in vegetation between 1920 and 1956 in the 'Pupplinger Au' south of Munich (see also fig. 185) at the point where the Loisach (left) flows into the Isar. To the right are shown the areas which are now flooded by high water during the summer spate. After Seibert (1958), modified.

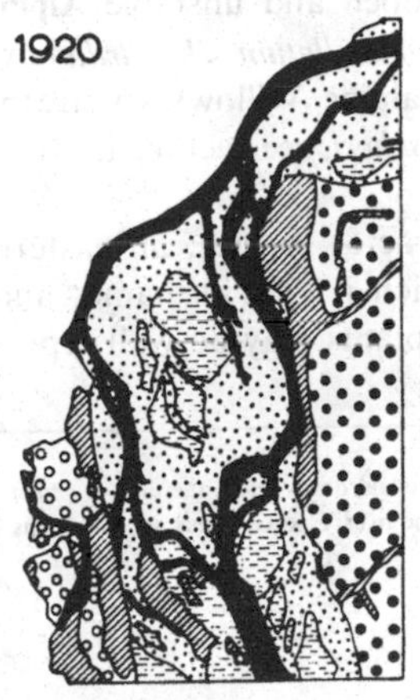

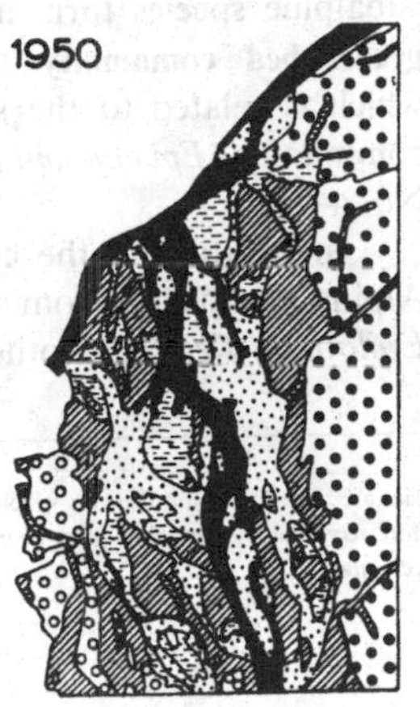

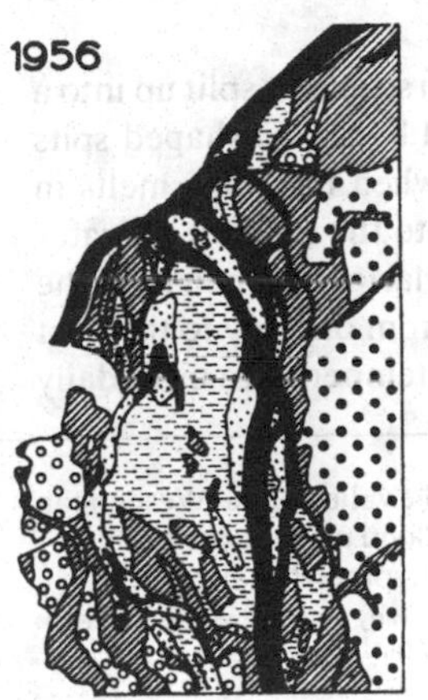

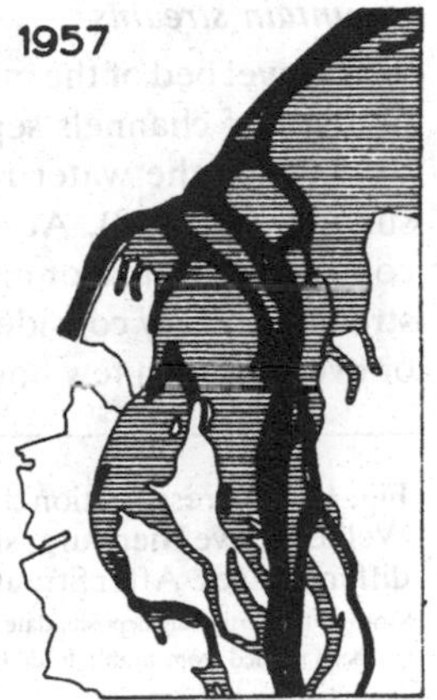

Through the processes named in 1, the final stage of the appropriate flood plain succession is reached more or less rapidly. This could be, for example, a Grey Alder wood or an Oak-Ash wood. However this would never develop any further into the climax (or zonal) vegetation (see section B I 1 a) because its roots are still within the influence of the high water, even if not actually flooded. Also there is the possibility with the constant pattern of water flow that the end stage may be destroyed again by erosion. In principle then the succession within the river valley is cyclical. Only the second group of circumstances could trigger off further development to a zonal climax vegetation. This however has nothing to do with the succession within the flood plain and should be looked upon as something quite distinct.

Long-term geological changes, the natural alteration in the course of the river and constant erosion, as well as man's interference, all work together so that even today our river valleys are landscape areas with strongly dynamic soils and vegetation (fig. 185). They remain subject to the unpredictability of natural forces and it is just this which makes the study of the last remnants of near-natural vegetation on the flood plains so stimulating.

(c) Tamarisk, Sea Buckthorn and Sallow Scrub by mountain streams

The gravel bed of the mountain stream is split up into a number of channels separated by shield-shaped spits over which the water rushes when the snow melts in summer (fig. 198). According to the amount of water coming from snow or melting glaciers the height of the stream can vary considerably from one day to the next or even within a few hours. Its temperature has a daily rhythm over a number of degrees, but on average is always less than that of the air (fig. 199). It is only in winter that the temperature of the mountain stream is higher than that of the air, whereas in the foothills and in the lowlands the monthly average temperature of the water is higher than that of the air throughout the year.

There is rarely a difference of more than 2 m between the lowest water level in the mountain stream in January and the highest level, which normally occurs in July. Thus it is considerably less of a difference than in the lowland rivers. However when the mountain streams are in spate they rush down with great force carrying with them mud and stones which are deposited according to size in various places. The amount of such burden is always more than one part per thousand parts of water and may be as high as one third of the water volume.

It is no wonder then that the stream bed remains for the most part free of plants, since the growing period available after the summer high water is short and cool. In spite of this the occasional plant can be found on a gravel island. These are mostly plants which have been washed down from higher levels, either as seeds or whole plants, and have managed to gain a footing on the downstream side of gravel banks where finer sand and silt can be deposited. Here alpine and subalpine species form an open and unstable **Alpine gravel bed community** (*Chondrilletum chondrilloidis*) which is related to the subalpine Willowherb stream community (*Epilobietum fleischeri*, see section E III 1, No. 444).

In addition to the character species, an eastern Alpine composite from which it gets its name, and *Epilobium fleischeri*, other plants may be well repre-

Fig. 197. A cross-section through the valley of the Weser near Wellie (above Nienburg) showing flood plain deposits of different ages. After Strautz (1959), somewhat altered.
Most of the fertile silt deposits date back to the early Middle Ages; their material has been eroded from arable fields (see fig. 30). Only along the edge of the flood plain are there clay-loam deposits which were laid down before the time of Christ. More recent silting has taken place nearer the river, especially since its course has been straightened.

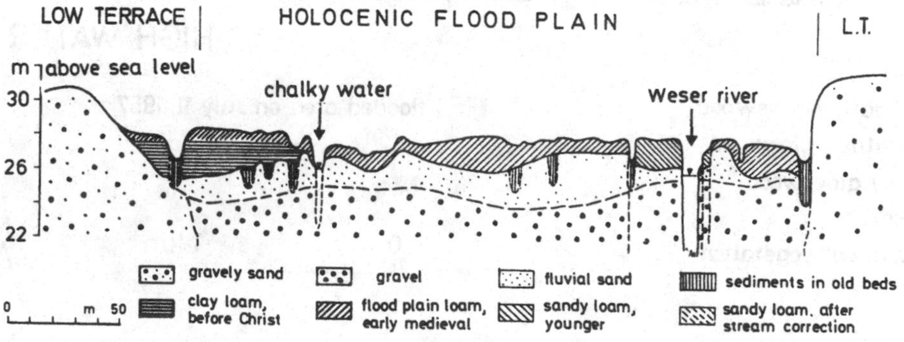

sented (e.g. *Gypsophila repens* and *Saxifraga paniculata*). Both communities belong to the class of limestone scree communities (*Thlaspietalia rotundifolii*) the associations of which can be recognised by the presence of basiphilous species able to colonise bare ground such as *Linaria alpina, Hutchinsia alpina* and *Rumex scutatus*. The inconstant alpine plants which are brought down by the water have only a slight influence on further development of the communities since they are not able to stabilise bare ground (section C VI 8 a).

Myricaria germanica germinates freely in the *Chondrilletum* and is often included in this community. However, according to Moor (1958) it cannot survive here permanently as Willows, Buckthorn and Grey Alder often gain a footing. These too would thrive better at a somewhat higher level where their seedlings would suffer less from submersion. The low growing **Willow-Tamarisk scrub** (*Salici-Myricarietum*, alliance *Salicion eleagni*, E III 1, No. 8.111) appears most often to fill the gap between the herbaceous community and the taller Willow stand. Moor stresses that this community is only found on silt and never on coarse sand or gravel (fig. 200). *Myricaria, Salix purpurea* (in the alpine variety *gracilis*) and *Salix elaeagnos* are able to anchor themselves in the gravel with strong tap roots branching out horizontally at the groundwater level in spring and autumn (Jeník 1955). *Salix daphnoides* may also take on a similar growth form in this community. Not only are the three pioneers able to withstand the violent flow of the water and being partially covered with the sand but they can also tolerate the cutting and the browsing by goats which still goes on.

Where the gravel shoals stand higher above the mean water level Willow-Tamarisk scrub is generally soon replaced by **Willow-Buckthorn scrub** (*Salicetum eleagno-daphnoidis*). This is one of the most striking communities in the alpine valleys and therefore was

Fig. 198. The gravel valley of a river in the montane belt in the high mountains (Wimbach in the Berchtesgaden district with a view of the Palvenhörner). Right foreground *Chondrilletum*; centre right *Salicetum elaeagno-daphnoidis*.

one of the first to be described. The tough stemmy silver-leaved Buckthorn, with its creeping roots continually seeking fresh ground and its orange-red water-borne berries which are so rich in vitamin, is very much a pioneer of lime-rich bare ground. As such it turns up in some quite different communities, for example in the *Salix repens-Hippophaë* scrub on the dunes of the North Sea islands. Because Buckthorn lives in symbiosis with nitrogen-fixing actinomycetes in its root nodules it is independent of humus or mud; its growth form is shown in fig. 201 (see also section E III 1, alliance *Salicion repentis* No. 8.414).

Fig. 199. Daily variations in the level and temperature of the water in the glacier-fed Visp near Stalden (700 m) in April, June and August 1919. After Lütschg (1928), from Ellenberg (1963), modified. The suspended bars indicate precipitation in mm at Saas-Fee/Zermatt.

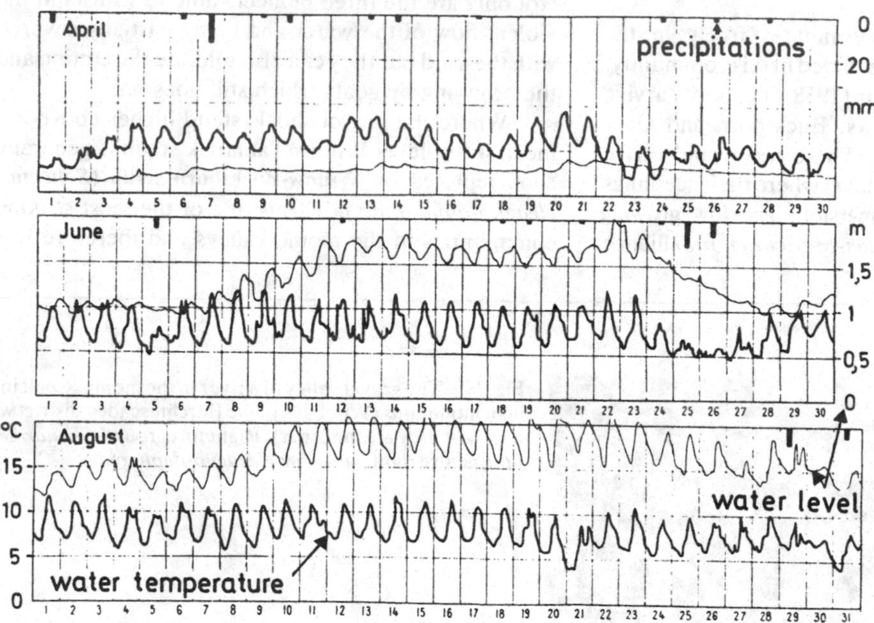

Fig. 200. Distribution of grain size and vegetation units on an island in a river at the montane level in the Alps (schematic). After Moor (1958), somewhat modified.

1 = Alpine gravel bed community *(Chondrilletum)*.
2 = Willow-Buckthorn scrub *(Salicetum elaeagno-daphnoidis)*.
3 = Grey Alder wood *(Calamagrosti-Alnetum incanae)*.
4 = Willow-Tamarisk scrub *(Salici-Myricarietum)*.

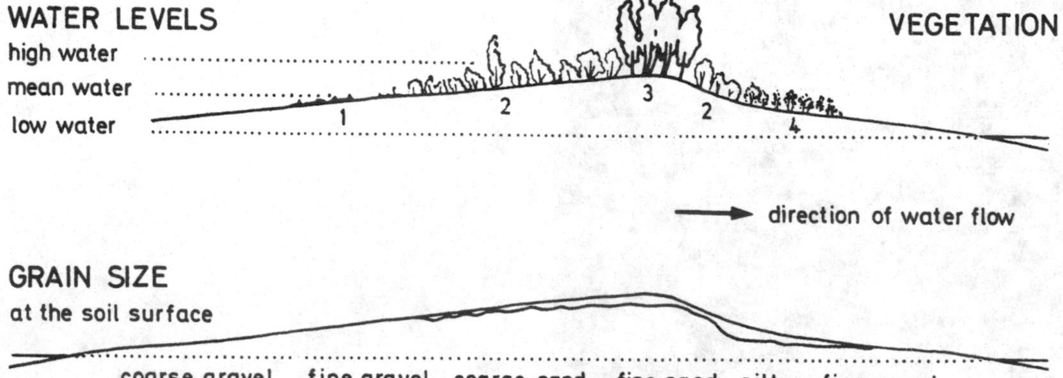

The gravel which is colonised by *Chondrilla*, Willow-Tamarisk scrub and by sea Buckthorn-Willow communities is usually very poor in humus and must be described as calcareous raw river-deposit (Rambla). Even in areas with acid bedrocks the river water is generally so rich in lime that all soils in its reach have an alkaline or at least neutral reaction.

(d) Osier scrub, White Willow wood and Poplar-Willow wood

The Sallow-Buckthorn scrub of the subalpine and montane gravel beds does not accompany the streams down into the submontane belt. In the warmer valleys *Salix eleagnos* itself is capable of growing into a 20 m high tree (Müller and Görs 1958). As soon as the water flow becomes less rapid and finer particles of sand and silt can be deposited, then the valley soils begin to have a better water-holding capacity, and other species of Willow are now able to flourish. Almond Willow (*Salix triandra*) comes in, Purple Willow (*S. purpurea*) is often dominant in its normal larger form and, together with other small-leaved Willows and their hybrids, they form a dense scrub. Finally we are left with an **Almond Willow-Osier scrub** (*Salicetum triandro-viminalis*) in which hardly any montane or subalpine species are present. This accompanies the rivers right down into the plains almost as far as the sea (fig. 186 and section E III 1, No. 8.112).

Fig. 201. The course of development of Tamarisk, Sea Buckthorn, Purple Willow and Grey Alder near the Inn at about 700 m. After Schiechtl (1958), modified.

Myricaria germanica anchors itself deep in the gravel bed of the torrential mountain stream; individual plants do not often live beyond about 10 years. At the other extreme *Alnus incana* develops more slowly when young but lives for more than 50 years; its roots are very shallow and dense in the fine-grained topsoil of the higher levels in the flood plain. *Hippophaë rhamnoides* spreads by means of root suckers.

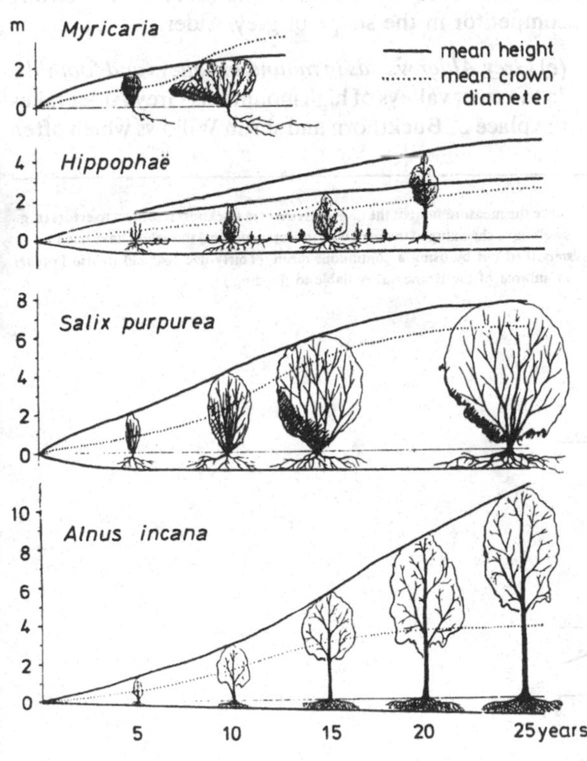

Moor looks upon the Almond Willow-Osier scrub merely as a 'mantle community' (see section D IV 1), i.e. a narrow shrub strip coming between the river bed and the really dominant community of the flood plains, the **White Willow wood** (*Salicetum albae*). Actually White Willow is often found with Sallow and other species even on the lowest river banks wherever Willows are able to gain a footing (fig. 193), but it is only on the slightly higher ground, where flooding is less frequent and more gently so that finer silts are deposited, that it grows into those fine broad trees, more than 20 m tall which give the Central European river valleys their particular charm. As the tallest and longest living Willow *Salix alba* dominates alone or with it hybrid *Salix rubens* (figs. 193, 195). The pure form of Crack Willow (*S. fragilis*), which is often put forward as a character species, is only found along those smaller rivers where the gravel is deficient in lime.

Almost all Willow species hybridise and this makes it very difficult to identify them. Since they are most often mixed in with the parents to which they are most closely related the species lists which have already been published do not give a correct picture of this genetic complexity. Mang (verbal communication) has observed from Willows planted along river banks that the hybrids are often stronger and less likely to be attacked by pests than are their parents. With Poplars too the hybrids are preferred because of their higher productivity and hardiness. These observations possibly explain why *Salix* and *Populus* species are so plentiful in flood plains, which are the most 'threatened' woodland habitats in Central Europe and very much at the mercy of chance events.

A question to which there is no really clear answer today is how far the **Black Poplar** (*Populus nigra*) is a natural participant in the White Willow woods. It is only native in the south-east of Central Europe but has been planted all over, even as far as the coasts of the North Sea and the Baltic. In many cases such plantings are of the hybrids which go under the collective name of Canadian or American Popular. While the Black Poplar is not affected by the wind in itself, and indeed is often planted as a wind break round the houses in marshland areas, nevertheless it suffers from flooding which can be widespread in the low lying areas near the coast. When their root runs are completely saturated with water they are easily blown down by a strong wind with the result that they rarely reach their full height in flood plains. Whether because of this or for some other reason there would probably be no naturally growing Black Poplars in the river valleys of the north-western lowlands. For this reason Tüxen (1955) in north-west Germany prefers the name *Salicetum albo-fragilis* rather than *Salici-Populetum* found in the older writings. In the east of Central Europe, i.e. on the Vistula and the lower Danube, *Populus nigra* is certainly a natural partner of *Salix alba*, and W. Matuszkiewicz and Borowik (1957) as well as Jurko (1958) can correctly speak of Poplar-Willow woods. Examples of these from Czechoslovakia are given in tab. 45, nos 1–5.

In contrast to Willows which can form adventitious roots in the water, Poplars and other kinds of trees are adversely affected by flooding (fig. 202). This temporary setback however is compensated for by the extra fertility which the flooding leaves behind.

White Poplar (*P. alba*) has an even weaker claim to citizenship in Central Europe than Black Poplar. Its home is in submediterranean Europe (*Populion albae*) and stretches from there into the large river valleys of the Danube plain (Jurko 1958). In Germany it has been planted or spread sub-spontaneously. Its presence indicates a somewhat drier habitat than does that of the Black Poplar (fig. 195).

Little definite can be said about the undergrowth of the White Willow woods since it is now practically impossible to find one in its natural state; they have often been heavily thinned out by coppicing or pollarding. Looked at from the point of view of distribution geography the White Willow woodlands with or without Poplars are strikingly rich in cosmopolitan and alien species which tend to be dispersed by activities on the rivers. European and Central European species which set the tone in other communities, or elements from particular climatic areas (e.g. continental or mediterranean) on the other hand are very much less apparent.

Under stands of White Willow or of mixtures of Willow and Poplar several ecological groups of plants can be found which differ quite widely depending on such things as the mean water height in the river, how far the stand is from the main stream, the position of the water table and the type of soil. So many sub-communities and variants have been described that we are unable to go into them here (tab. 45, also Wendelberger-Zelinka 1952 and Horvat, Glavač and Ellenberg 1974).

White Willow woodlands are widespread in river valleys of Central Europe and can be looked upon as the characteristic feature of their landscape in the planar, colline and submontane belts. In the most north-easterly corner of Central Europe as well as in the foothills of the larger mountains, especially in the Alps however the tree Willows meet a very strong competitor in the shape of grey Alder.

(e) Grey Alder woods in mountain valleys and foothills

In the river valleys of high mountains Grey Alder takes the place of Buckthorn and shrub Willows which often

Fig. 202. High water during the growing period results in a restriction of growth for the majority of flood plain trees such as Poplar. After Kern (1970), somewhat altered.

Since the measurement of the daily increment in thickness is subject to errors (e.g. swelling or shrinking of the bark with its water content) the values have been smoothed out by using a continuous mean. (Thirty-five year old hybrid Poplars in an area of the Rhine valley liable to flooding.)

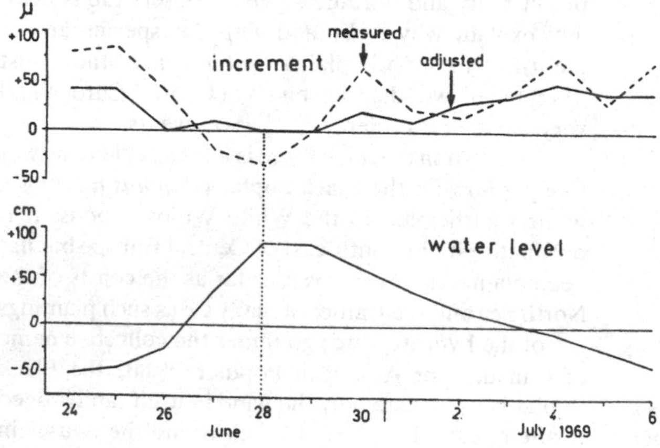

Tab. 45. *Flood plain woods dominated by Willow, Popular, Elm or Oak in the Czehoslovakian lowland.*

Community no.:	1	2	3	4	5	6	7	8	9	10
Tree layer:										
Salix alba	4	4	3	1	1	3				
Populus nigra + americana	3	5	4	4	4	3	3	3	1	1
Alnus glutinosa	3	4	5	4	3	5	3	4	2	1
Q *Ulmus minor*	3	3		5	4	5	5	5	3	3
F *Fraxinus excelsior* (kult.)				2	2	5	4	4	4	5
Q *Acer campestre*				2	1	3	5	5	4	5
Quercus robur				2	3	4	4	3	5	5
F *Tilia cordata*				2	1	3	3	5	5	5
F *Carpinus betulus*						2	1	2	2	4
Shrub layer:										
Humulus lupulus (Liana)	3	1	2	2	3					
Sambucus nigra	1	1		5	4	3	3	2		
F *Prunus padus* (often planted)	3	2	3	5	5	5	5	5	4	2
Q *Euonymus europaeus*		2	2	3	2	2	2	3	2	1
Q *Cornus sanguinea*				3	3		2	4	1	2
Q *Crataegus* spec.						1		4	3	1
Herb layer:										
Bidens tripartita	3									
P *Lycopus europaeus*	3									
M *Galium uliginosum*	3									
P *Alisma plantago-aqu.*	3									
Polygonum amphibium	3									
P *Glyceria maxima*	4	2								
Polygonum hydropiper	4	4								
Solanum dulcamara	3	3								
P *Carex gracilis*	4	2	2							
P *C. riparia*	4	2	2							
Caltha palustris	3	1	2							
P *Mentha aquatica*	3	2	2							
P *Scutellaria galericulata*	3	1	2							
M *Myosotis palustris*	4	2	4							
M *Stachys palustris*	4	3	2							
P *Iris pseudacorus*	5	5	3	1		1				
M *Lythrum salicaria*	4	3	5	1						
M *Lysimachia vulgaris*	5	4	5	1	1		1			
Ranunculus repens	5	5	4	1		1			2	
F *Stellaria nemorum*	2		2	4						
Rumex obtusifolius	4	3	2	4						
Galeopsis tetrahit	4	1	2	2	2					
F *Impatiens noli-tangere*	3	3	3	3	3	1				
P *Poa palustris*	5	3	3	3	3	2				
P *Phalaris arundinacea*	5	5	4	5	3	2			2	
M *Alopecurus pratensis*	2	1		2		3				
Impatiens parviflora	3	3	3	4	5	3	1	3		
Urtica dioica	5	5	5	5	5	5	5	4	1	
M *Poa trivialis*	2	3	3	4	3	5	1	3		
F *Rumex sanguineus*	2	1	3	2		3	2	3		

Community no.:	1	2	3	4	5	6	7	8	9	10
Alliaria petiolata	4	3	3	5	5	5	5	3	4	
M *Angelica sylvestris*	3	4	3	2	2	1	2	1	2	
F *Agropyron caninum*	2		2	3	2	2	3	3	5	
M *Taraxacum officinale*	2	1	2			1		2	3	
Rubus caesius	5	5	5	5	5	4	5	2	3	2
Symphytum officinale	5	3	5	4	1	2	2	1	5	2
Q *Glechoma hederacea*	3	4	5	5	4	5	5	5	5	4
Galium aparine	4	3	4	5	5	5	5	4	5	3
M *Lysimachia nummularia*	3	3	5	1	2	2	2	2	3	2
Deschampsia cespitosa	3	3	5	3	2	5	4	3	4	5
M *Filipendula ulmaria*	3	3	5	4	3	4	2	3	3	5
F *Geum urbanum*	1	1	5	4	4	5	5	4	3	1
F *Festuca gigantea*	2	1	3	5	3	4	5	4	3	2
Ranunculus ficaria	1	1	2	2	2	4	2	3		2
Silene dioica	1		3	3	3	3	3	2	2	1
Veronica hederifolia	1		3	1	2	4	1	4	2	1
Q *Moehringia trinervia*		2	4		1	3	1	3	2	
M *Heracleum sphondylium*			2	3	2	1		2	3	1
F *Milium effusum*			2	2		1	2		2	3
Q *Aegopodium podagraria*			2	4	5	5	5	5	5	5
M *Dactylis glomerata*			2	4	3	4	5	5	5	4
Veronica chamaedrys			4		1	3	4	3	5	5
F *Scrophularia nodosa*			2		1	1	2	1	5	4
Q *Campanula trachelium*			2		1	1	2	2	5	4
F *Circaea lutetiana*			2	1	3	2	2	1		
Lamium maculatum			2	2	1	4	4	3	2	
M *Anthriscus sylvestris*				4	4	4	3	3	4	
Omphalodes scorpioides				1		5	3	5		
F *Adoxa moschatellina*				2		2		4		
Allium scorodoprasum					1	3	1	2	3	
F *Paris quadrifolia*					1	3	2	2	2	
F *Corydalis cava*				1	1	4	2	4	1	1
F *Stachys sylvatica*					1	3	3	3	3	1
Q *Anemone nemorosa*						3	1	3	4	2
Q *A. ranunculoides*						3	1	3	4	2
Q *Brachypodium sylvaticum*						2	2	4	5	5
Poa nemoralis						2	1	1	5	5
F *Gagea lutea*						3	1	4	1	1
F *Viola reichenbachiana*						1	3	3	3	4
Ajuga reptans						2	4	3	4	3
F *Stellaria holostea*					1	1		4	2	2
F *Pulmonaria officinalis*								3	1	1
F *Carex brizoides*							1		3	1
F *Lathyrus vernus*								1	3	3
Hypericum hirsutum								1	3	2
F *Melampyrum nemorosum*									2	2
Ornithogalum umbellatum						1			3	
M *Lathyrus pratensis*									3	

have prepared the ground for it by the gradual accumulation of sand and gravel. *Alnus incana* creates so much shade that all these light-loving pioneers are eliminated. Besides this it can reach a height of over 10-15 m, while *Hippophaë* and *Salix purpurea* rarely exceed 3 m in mountain valleys (fig. 201). The **montane Grey Alder wood** (*Alnetum incanae*) soon becomes dominant, even in those places where hardly any more humus has collected than under the Buckthorn-Willow scrub (fig. 194). Like *Hippophaë* though *Alnus* posesses root nodules and so can make use of atmospheric nitrogen.

In valleys within the inner Alpine coniferous region (see fig. 42a) Grey Alder communities are the only natural broadleaved woodland, maintaining their place largely because of the summer flooding and their significance in protecting the banks. In these places they have become the meeting places of many species which are normally found in deciduous woodland communities (*Fagetalia* section B II 1 a), especially those amongst them which like moisture or can withstand wet conditions. So in the Grey Alder stands we come across a surprising number of the species which we already know from the wet-soil Oak-Hornbeam woods and the Sycamore-Ash woods, e.g. *Fraxinus excelsior, Sambucus nigra, Euonymus europaeus, Viburnum opulus, Rubus caesius, Festuca gigantea, Stachys sylvatica, Circaea lutetiana, Scrophularia nodosa, Impatiens noli-tangere* and *Aegopodium podograria*. Many of these wet indicators can be looked upon as local character species of *Alnetum incanae* as by and large they are not present in most of the other *Fagetalia* communities. Apart from *Alnus incana* itself however the Grey Alder woods have no species which is characteristic throughout (see section E III 1, No. 8.43 and 8.433).

Although scarcely any nitrogen is carried to the Grey Alder wood by the flood water its soil must be quite rich in nitrate since parts of the nodule-carrying roots are dying and being mineralized. In any case many of the species already named can be spoken of as nitrophilous. The same is true of others which are always present, e.g. *Prunus padus, Agropyron caninum, Geranium robertianum, Urtica dioica, Galeopsis tetrahit, Solanum dulcamara*, and *Carduus crispus*.

Riverside Grey Alder wood is the only community of the montane valleys which accompanies the river right down into the foothills showing a practically constant combination of species (fig. 186). However, according to Müller and Görs (1958) it is quite possible to connect the different altitudes with certain differential species. For example in the mountains of Württemberg they are:

in the high-montane form	in the submontane form
Sorbus aucuparia	*Ulmus* species
Crepis paludosa	*Populus alba*
Viola biflora	*Ranunculus ficaria*
Ranunculus aconitifolius	*Arum maculatum*
Geranium sylvaticum	*Allium ursinum*
and other sub-alpine tall herbs	and other species of the *Corydalis* – group

Moor (1958) called the montane riverside Grey Alder wood in Switzerland *Calamagrosti-Alnetum* and the submontane one *Equiseto-Alnetum* (fig. 200).

The reasons why Grey Alder woods do not occur in the lower stretches of all Central European rivers are still not clear. The timing of highest water may be a decisive factor. If this happens in summer, as it does in the higher reaches of the river, following the melting of snow and ice on the mountains, then the relatively

Tab. 45 (Continued)

Source: **After tables by Mráž and Šika (1965), modified.**

Note: P = Plants of Reed and Sedge communities (*Phragmitetea*)
Q = Plants of broadleaved woodland (*Querco-Fagetea*)
F = Plants of 'noble' deciduous woods (*Fagetalia*)
M = Plants distributed in meadowland

Nos. 1–3 **Poplar-Willow flood plain wood** (*Salici-Populetum*); relatively often flooded:

		Average moisture value
1:	with large Sedges (*'magnocaricetosum'*); wettest form, mostly dominated by White Willow	7.6
2:	typical (*typicum*); often dominated by Black Poplar or by planted hybrids (*Populus americana*)	7.3
3:	with Wood Avens ('*geosum*') often dominated by Black Poplar, leading to nos. 4 and 5.	6.8
Nos.4-5:	**Poplar flood plain wood** ('*Sambuco-Populetum*'); flooded less often than nos. 1-3;	
4:	with Wood Stitchwort ('*stellarietosum nemorae*'), still relatively wet, mostly dominated by Poplar	6.5
5:	typical (*typicum*); generally dominated by *Ulmus minor* or by (planted) Ash	6.4
Nos.6-8:	**Elm flood plain wood** ('*Omphalodo-Ulmetum*', named after the Blue-eyed Mary); rich in *Fagetalia* species	
6:	with White Willow ('*salicetosum*'), most have Alder: dominated by Elm or Ash (planted).	6.3
7:	typical (*typicum*); most are dominated by Elm and with many Bird Cherry	6.1
8:	with Lungwort ('*pulmonarietosum*'); dominated by Elm, leading to nos. 9 and 10.	5.8
Nos.9-10:	**Flood plain Elm-Oak wood** (*Ulmo-Quercetum*): seldom flooded, predominantly *Fagetalia* species	
9:	with Cow Parsley ('*anthriscetosum*'); relatively open and with meadow plants	5.8
10:	typical (*typicum*); a dense valley wood dominated by Oak, stands near the Oak-Hornbeam woods	5.7

Species which did not achieve a constancy of more than 2 (= present in up to 40% of the samples) have for the most part been left out. Where species often cover a large proportion of the ground the constancy figure has been printed in semi-bold type.

shallow roots of the Alder will be well supplied with water (fig. 201). If the high water is already past by the end of the spring, or it is very irregular, then the Willows which are more capable of adjusting to the drier periods will have advantages in competition.

The sites where Grey Alder is able to achieve dominance in the middle reaches of the river are about the same in respect of the river water level as they are for most of the Willows, according to Heller (1963, see also fig. 186). In the Danube valley near Wallsee it can even lie deeper (Wendelberger-Zelinka 1952). In many cases Alder is merely arriving first, just by chance, and therefore becomes dominant. The waterborne seeds of Alder remain viable for relatively longer than those of Willows, and they are also ripe at a time when transport is available to disperse them. Since this is also at a time when the water level is falling, the Alder seeds find just the right conditions on the emerging bare ground. Willows on the other hand, especially those species like *Salix alba, S. fragilis* and their hybrids which are the ones most likely to come into competition with *Alnus,* fruit mostly in June when the rivers are in spate and there is no bank or shoal exposed from them to gain a footing. In addition Willow seeds do not remain viable for more than a few days so they are able to withstand neither a lengthy journey downstream nor prolonged exposure to conditions unsuitable for germination. Best of all they germinate on the banks of silt or sand which have recently been exposed by the falling water, after the seeds have been blown there by the wind during a time of good weather (fig. 192). It is obvious that the chance of this happening is greater in the lower reaches of the river than in the middle and upper reaches. Either the general lack of warmth or the very low winter temperature makes it impossible for *Salix alba* or *S. fragilis* to live in the montane river valleys therefore they are only found up to an altitude of 800-900 m above sea level.

Stands of Willow and Alder slow down the flow of flood water and so act as sediment traps especially near the river banks (fig. 194). Because of this their soils are quite young in the upper layers, and usually clearly stratified. At some places buried humus layers are to be found. Typologically the soils of Grey Alder and White Willow woods are known as grey calcareous riverside soils (flood plain rendzina, calcareous paternia). Generally they have an alkaline to neutral reaction and apart from the occasional rust-coloured spot do not show any brown colouration.

(f) Stream Ash-Alder woods and Alder-Ash woods

The undergrowth of the majority of riverside Grey Alder woods has a decidedly mixed deciduous woodland character. Among the trees too there is the occasional broadleaved specimen, especially the Ash, although it is only on the edge of the submontane level that the Ash becomes so strong as to dominate. On deeper soils, with a higher proportion of fine earth, Elms (*Ulmus minor* and *U. laevis*) come in too. Müller and Görs (1958) have clearly portrayed the gradual transition to a Grey Alder-Ash wood using examples from the valleys of Iller and Argen. The high-montane perennials disappear in the middle section of these rivers and it is only at the river mouths that the growing season is long enough for the spring geophytes to develop fully and thus determine the appearance of the riverside woodlands.

Outside the Alps it is Black Alder (*Alnus glutinosa*) rather than Grey Alder which is found accompanying the Ash. It replaces both Grey Alder and Ash as a stabiliser for the river banks since it is able to tolerate the oxygen shortage better and can send its roots deep into the ground below the water table (section 2 d). In the **stream Ash-Alder wood** (figs. 203 and 205), which was first recognised as a separate community by Lohmeyer (1957), Black Alder grows better than it does in the Alder swamp although it undoubtedly dominates the latter (section 2 b).

The scientific name *Stellario-Alnetum glutinosae* is thus justified for the stream Ash-Alder wood but it is not really appropriate since it does not take Ash into sufficient account and leads to the impression that we are dealing with a member of the *Alnion* alliance. However species of mixed broad-leaved woodlands are as dominant here as they are in the Grey Alder woods of the river flood plains so that without doubt it must be classified in the *Alno-Ulmion (Fagetalia)*. According to Tüxen and Ohba (1975) only *Stellaria nemorum*, which can withstand short-term flooding quite well, can be considered a character species. It requires a fertile soil and cannot withstand drought. Recently a *Ribo-Alnetum*, characterised by *Ribes rubrum*, has been separated from *Stellario-Alnetum*. This Spring Alder wood prefers places where water is coming to the surface in hillside depressions or on peaty ground which is rarely under water. Möller (1970) has described a similar community from the young moraine area of Schleswig-Holstein as '*Cardamine amara*-Alder wood'. In mountainous country away from the Alps *Stellario-Alnetum* is replaced by montane communities (such as *Chaerophyllo-Alnetum glutinosae*) which can be considered parallel to the *Alnetum incanae* (see also Rühl 1964).

Where the banks of streams or small rivers are so low that they are often flooded and fertilised with deposits there is often a belt of **Butterbur riverbank community** with its luxuriant foliage (figs. 204 and 205). *Petasites hybridus* often penetrates into Grey or Black Alder woods. Therefore this eye-catching community, which has not been investigated ecologically to any extent, is mentioned here. It is also of significance in stabilising the river bank.

The **Black Alder-Ash wood** (*Alno-Fraxinetum*, previously called *Pruno-Fraxinetum* after the Bird Cherry) is related to the stream Ash-Alder woods. It can however take over large areas outside stream or river valleys. For instance it dominates in the Spree Forest, the most important area of Alder woodland in Central Europe where the vegetation has been thoroughly described by Krausch (1960) as well as by Scamoni (1954), Passarge (1956a and others). As may be seen in tab. 46, cols. 4-6 the Alder-Ash wood can be divided into a number of subassociations, from the even wetter Alder swamp woodland to the wet-soil Oak-Hornbeam wood, which for the most part has been brought into cultivation. *Alnus glutinosa* can reach a height of over 30 m in 90-100 years in the Alder-Ash wood, although it has to share these optimal conditions with Ash. Its trunks are as slim and solid as those of Spruce and Pine, and have been used in a similar way for the construction of log houses, some of which can still be seen on the lonely Spree Forest farms.

In many river valleys Black Alder is found in Poplar-Willow woods, especially where only a small amount of sediment is deposited by flooding. Column 3 in tab. 45 represents an example. The species combination of this 'Wood Avens-Poplar-Willow wood' where Black Alder is generally dominant, reminds us of that in the Alder-Ash wood, especially that shown in tab. 46, no. 5. As in all other mixed Black Alder woods so far mentioned these have a herb layer of tall grasses

Fig. 204. A bank community of Common Butterbur in the montane Schwarzwasser valley (Smaller Walsertal). *Petasites hybridus* and *Cirsium oleraceum* (in front).

This community is classified by Oberdorfer (1977, 1979) as belonging to the allinace of shaded woodland skirts (*Aegopodion*, see Section D IV 2b and E III 1, no. 3.523; see also fig. 205).

Fig. 203. The Ash-Alder wood (*Stellario-Alnetum*, with *Fraxinus escelsior*, *Alnus glutinosa* and *Acer pseudoplatanus*) is often preserved along the sides of a stream, e.g. in the southern Black Forest. On the other hand most of the flood plain woods at higher levels are converted into meadows (montane *Arrhenatheretum*, see fig. 456b).

Tab. 46. Swamp woods rich in Alder and Ash in the upper Spreewald arranged with a decreasing level of water saturation.

Community no.:	1	2	3	4	5	6	F		N
Tree layer:									
Alnus glutinosa	**5**	**5**	**5**	5	5	5	9	=	×
Prunus padus		4	5	5	5	5	8	=	6
Ulmus laevis		3	5	5	5	5	8	=	7
Fraxinus excelsior			5	5	**5**	**5**	×		7
Quercus robur				2	5	5	×		×
Acer pseudoplatanus						4	6		7
Tilia cordata						2	×		5
Shrub layer:									
Salix cinerea	**5**						9	~	4
Salix pentandra	5						8	~	4
Frangula alnus	2	5	5				7	~	×
Humulus lupulus	4	5	5	5	4	3	8	=	8
Rubus idaeus		2	3	5	5	3	5		8
Ribes nigrum		1	1	3	3		9	=	5
Viburnum opulus			2	2	1	2	×		6
Sambucus nigra				3	5	3	5		9
Euonymus europaeus					2	4	5		5
Ribes rubrum						3	8		6
Herb layer:									
Phragmites australis	4						10	~	5
Rumex hydrolapathum	5						10		7
Mentha aquatica	5						9	=	4
Ranunculus lingua	3						10		7
Acorus calamus	3						10		7
Bidens tripartita	3						8	=	8
Carex elata	5	4	5				10	~	4
Solanum dulcamara	5	5	2		1		8	~	8
Thelypteris palustris	**4**	5	**5**	2			8		6
Calamagrostis canescens	4	4	5	3	2		9	~	5
Lysimachia thyrsiflora		5	5				9	=	3
Carex vesicaria		2	3				9	=	5
Dryopteris carthusiana		2	4				×		3
Peucedanum palustre	5	5	1	4			9	=	4
Galium palustre	3	5	5	4			9	=	4
Scutellaria galericulata	2	2	2	3			9	=	6
Glyceria maxima	5			4			10	~	7
Lysimachia vulgaris	3	4	4	4	2		8	~	×
Lythrum salicaria	5	5	5	3	2		8	=	×
Stachys palustris	4	3	5	4	2		7	~	7
Carex riparia	3	4	3	3	2		9	=	4
Lycopus europaeus	2	5	5	5	2		9	=	7
Calystegia sepium	5	5	5	5	3		6		9
Symphytum officinale	5	5	5	5	3		8		8
Iris pseudacorus	5	5	4	5	4		10		7
Carex acutiformis	4	**5**	**5**	5	4		9	~	5
Phalaris arundinacea	3	1	**5**	5	4	4	8	=	7
Caltha palustris	1	1	3	5			8	=	×
Juncus effusus		1	3	4			7	~	3
Carex elongata		2	5	5	2		9	~	6
Cirsium palustre		3	4	5	5		8	~	3
Impatiens noli-tangere		5	5	**5**	5	4	7		6
Urtica dioica		2	4	5	5	5	6		8
Rubus fruticosus			4	3	4	4	–		–
Deschampsia cespitosa			3	5	5	5	7	~	3
Filipendula ulmaria			4	5	4		8		4
Angelica sylvestris				4	3		8		×
Valeriana officinalis				4	1		8	~	5
Myosoton aquaticum				5	4		8	=	8
Cirsium oleraceum				4	3		7		5
Rumex sanguineus				3		2	8		7
Galium aparine				5	5	5	×		8
Festuca gigantea				5	5	5	7		6
Geum urbanum				5	5	5	5		7
Ranunculus ficaria				5	5	5	7		7
Poa trivialis				5	5	5	7		7
Lamium maculatum				5	3	4	5		5
Circaea lutetiana				4	3	3	6		7
Glechoma hederacea				5	5	**5**	6		7
Geranium robertianum				4	4	5	×		7
Moehringia trinervia				3	5	5	5		7
Galeopsis tetrahit				3	4	5	5		7
Anemone nemorosa					3	4	×		×
Brachypodium sylvaticaum					2	5	5		6
Stachys sylvatica					1	3	7		7
Milium effusum						5	5		5
Anthriscus sylvestris						4	5		8
Bilderdykia dumetorum						4	5		6
Moss layer:									
Calliergon spec.		2	4						
Eurhynchium spec.				5	4	5			
Mnium undulatum					2	3			
Atrichum undulatum						3			
Groundwater depth (cm):									
highest (approx.)	.	>0	.	>0	0	50			
lowest "	.	90	.	.	150	150			
Stratum thickness (cm):									
mud (approx.)	.	.	.	.	25	+			
peat "	.	50	.	.	100	0			
Constancy totals:									
flooding indicators	36	41	38	45	21	7			
fluctuat, moisture ind.	46	35	46	40	20	5			
Mean moisture values		8.2		7.5		6.0	mF		
	8.7		8.0		7.0				

No. 1: **Alder-Sallow scrub** on coarse thin peat (*'Alnus-Salix' 8.7 cinerea-Ass., Rumex hydrolapathum* subass.)

Nos 2 and 3: **Alder Swamp wood** (*Carici elongatae-Alnetum*)

2: 'Normal' A.-s. W. (*Symphytum officinale subass.)* on 8.2 normal fine peat

Source: After tables by Passarge (1956).

and sedges, together with thin-leaved dicots, which also enjoy damp conditions with a good supply of nitrogen. According to analyses carried out by Duvigneaud and Denayer-de Smet (1970a,b) these species are almost always rich in nitrogen, but low in potash. Although the flood plain woods rich in Black Alder have no character species in the strict sense their ground flora is obviously different from that of the Alder swamp woods or the real Willow woods on the one hand, and the less frequently flooded mixed broadleaved woods, on the other.

The majority of Black Alder stands which are met with in Central Europe today correspond ecologically to the *Alno-Fraxinetum* even if originally they were Alder swamps and still do not contain any other trees today. A slight reduction in the amount of flooding is sufficient to allow Ash and other tall broadleaved trees to become competitive. Wet meadows which are no longer used, and where Alder has been allowed to grow for a number of decades, also develop towards the *Alno-Fraxinetum* although this succession takes a long time.

(g) Ash-rich stands in other flood plain woodlands

Where Ash forms large stands in the river valleys we can be sure that these are outside the areas of the 'soft' wood formations which are frequently inundated (figs. 186 and 195). In the Danube valley in Slovakia, the

Fig. 205. Zonation in the flood plain of a woodland stream with a submontane climate and on lime-deficient rock, (the middle reaches of the Bröl in the Eifel). After Lohmeyer (1970), slightly modified.

1. *Aegopodio-Petasitetum*, a pioneer community on gravelbanks or at the side of the stream;
2. *Phalaris arundinacea* reedbed;
3. *Stellario-Alnetum glutinosae*, an ash-alder wood frequently flooded, with *Stellaria nemorum* (far left), *Aegopodium podagraria, Athyrium filix-femina* etc.

4 – 6. *Stellario-Carpinetum*, Oak-Hornbeam woods with *Stellaria holostea* (far right);
4. *St-C. periclymenetosum* (with Honeysuckle) on relatively dry habitats, poor in bases, with *Luzula pilosa, Maianthemum bifolium*, etc.;
5. Q. -C. typicum, on average habitats, with *Polygonatum multiflorum*, etc.;
6. *Q. -C. stachyetosum* on soil relatively rich in bases and with a higher groundwater table, with *Stachys sylvatica, Primula elatior*, etc.;
7. *Luzulo-Fagetum* outside the flood plain and the influence of groundwater, with *Luzula luzuloides*.

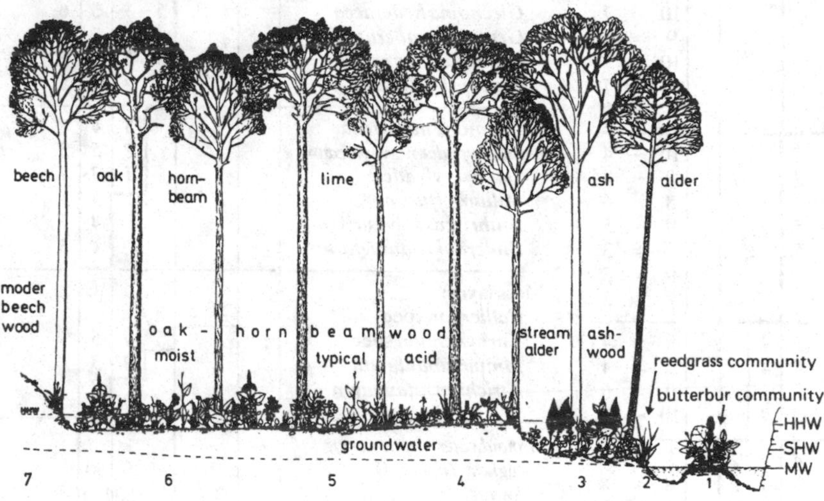

Tab. 46. (Continued).

		mF
3:	Ash rich a.-s. (*Phalaris* variant) on a deep more strongly decomposed covering of peat	7.5
Nos. 4: – 5:	**Alder-Ash wood** (*Pruno-Fraxinetum*)	
4:	Yellow flag-a.-a. (*Iris pseudacorus* sub-ass) on somewhat muddy peat	7.5
5:	Oak a.-a. (*Quercus robur* variant) on peat covered with a layer of mud	7.0
No.6:	**Oak-ash wood** ('*Pruno-Fraxinetum typicum*'), on loamy sand covered with a layer of mud on the edge of the wet plain.	6.0

(Nos. 1–3 are swamp woods, 5 and 6 flood plain woods, 4 lies between the two). Some less constant species have been left out.

F and N are moisture and nitrogen indicator values. = indicates extensive flooding and ~ indicates fluctuating wetness.

The constancy totals for = and ~ have been calculated in the same way as in tab. 12 (i.e. without reference to the tree layer). F,N and mF are explained in more detail in sections B I 4 and E III 2 (mF has been calculated without the tree layer). Judging from the species present and the average moisture value unit no.5 corresponds closely to unit no. 3 in tab. 45 which is often dominated by Black Alder, and no. 6 corresponds fairly well to the typical Elm flood plain wood with many Bird Cherry trees (no.7 in tab. 45).

only part of Central Europe where 'hard' broadleaved flood plain woods grow under fairly natural conditions, Ash plays an important part in the middle one of the following three levels:

- On the lower, frequently flooded level Populars and other 'soft' trees still dominate *Fraxino-Populetum).*
- *The middle level is dominated by Elm or Ash (Ulmo-Fraxinetum).*
- The upper one is dominated by Common Oak *(Ulmo-Quercetum convallarietosum).*

The latter merges into the xerothermic mixed Oak woods outside the flood plain (*Ulmo-Quercetum caricetosum albae* and *buglossoidetosum*, see Horvat, Glavač and Ellenberg 1974).

Recent evidence makes it doubtful whether we can transfer to the western part of Central Europe the picture of the natural flood plains which we have obtained from the relatively continental parts. Streitz (1967) proved that *Fraxinus excelsior* had been spread on the Rhine riversides in Hesse by sowing and planting during the past 150 years far beyond the area where it would occur naturally: 'Originally it would have been completely absent from the flood plain woodland (*Fraxino-Ulmetum*) whereas today it occupies the largest area (40% of the total) (p. 165). According to Streitz if these woodlands had been left to nature the main trees would undoubtedly be Common Oak and European White Elm (*Ulmus laevis*). These would be accompanied by *Populus alba, Pyrus malus, Acer pseudoplatanus, A. campestre* and even *Alnus glutinosa*, but not by *Fraxinus excelsior*. Dister (1980) however, to whom we owe a thorough ecological investigation of the Rhine flood plain reserves east of Worms, does not agree with this opinion. Among other interesting findings he attributes to *Fraxinus* a much greater competitiveness under natural conditions.

If the view of Streitz is correct then many of the flood plain Ash woods or Ash-Elm woods, at least in the area of the Upper Rhine, could be looked upon as afforested, i.e. mainly man-made communities (section D III). It would appear that further research into the history of forestry and flood plain woodlands is urgently required to clear up this matter. Such investigations must also take into account the fact that Ash has always played a relatively greater rôle in coastal regions (and in western Europe) when compared with Elms and Oaks, for example in the lower Ems valley (Behre 1970). Ecologically it is hard to see why Ash can maintain an important place in the south-east and the north-west, but could not do the same in the area between. Possibly in the Rhine valley it was often damaged by the high water table which must have been the rule here before work on the river resulted in it being lowered (figs. 30 and 31). The raising of the water table by 60-80 cm caused by the damming of the Danube near Offingen led to premature death of many Ash trees (Seibert 1975).

(h) Oak-, Elm- and Beech-rich flood plain woods

In northern Central Europe the rivers, at some distance from the mountains, are usually only in spate during winter and early spring and the water table is quite low during the summer months. Here the naturally dominant tree of the 'hard' broadleaved riverside woodlands is Common Oak. This is clearly shown by the few remaining examples. Flood plain **Elm-Ash-Oak woods** or pure Oak woods have been described by Carbiener (1970a, 1975) from the Alsatian Upper Rhine, by Schwickerath (1951) from the Erft in the neighbourhood of Aachen, by Trautmann and Lohmeyer (1960) from the middle stretch of the Ems, by Ellenberg (1939) from the Leine and Innerste near Hannover, by Passarge (1956) from the Elbe and by Meusel (1952c) from the Saale. In the flood-forests of the Alsatian Rhine plain live nearly 50 species of trees and shrubs. As J.-M. Walter (1981) has shown on 'architectural profiles' these form manifold stands in a small-scale mosaic. In both their floristic and structural diversity they resemble tropical flood plain woodlands. In the far north-west the comparable communities are less diverse. On the Ems, for instance, there are no Elms at all in the river-valley woodlands, whereas in the drier regions of eastern Central Germany they appear strongly. Seen as a whole though even here Oak is dominant. Since many of the flood plain Oak woods have been coppiced in the past they are rather like the wet-soil Oak-Hornbeam woods both in appearance and in their species composition (figs. 206 and 207).

All 'hard' broadleaved flood plain woods grow on deep, more or less loamy brown soils belonging to the type 'brown vega'. The brown material is not weathered in situ but is washed down from the whole catchment area and deposited in the wide valleys of the lower reaches (allochthonous vega). The cutting down of woodland and its conversion into arable land has greatly accelerated this process and much of the brown flood plain soil has been deposited relatively recently. For instance in the Weser valley near Hameln the greater part of the loam has been deposited on top of buildings which were occupied in Roman times (fig. 30). However Strautz (1959) was able to show that in some places there had been pure loess-like loams available in the lower Weser valley even before this.

Fig. 206. The old course of the Rhine at the Reis Island near Mannheim with a Waterlily community, Reed beds and a broadleaved flood plain wood with Oak, Alder and Ash ('*Fraxino - Ulmetum*') in June. Photo. Landesbildstelle Baden.

Fig. 207. The appearance in spring (above) and in high summer (below) of the herbaceous layer in a flood plain elm-Oak wood in the eastern Harz foreland. From Meusel (1951) extracts.

Above: *Galium aparine. Corydalis cava. Stachys sylvatica, Gagea lutea. Veronica hederifolia, Arum maculatum, Allium ursinum, Ranunculus ficaria, Milium effusum*. Below: *Galium, Stachys, Impatiens noli-tangere, Milium*.

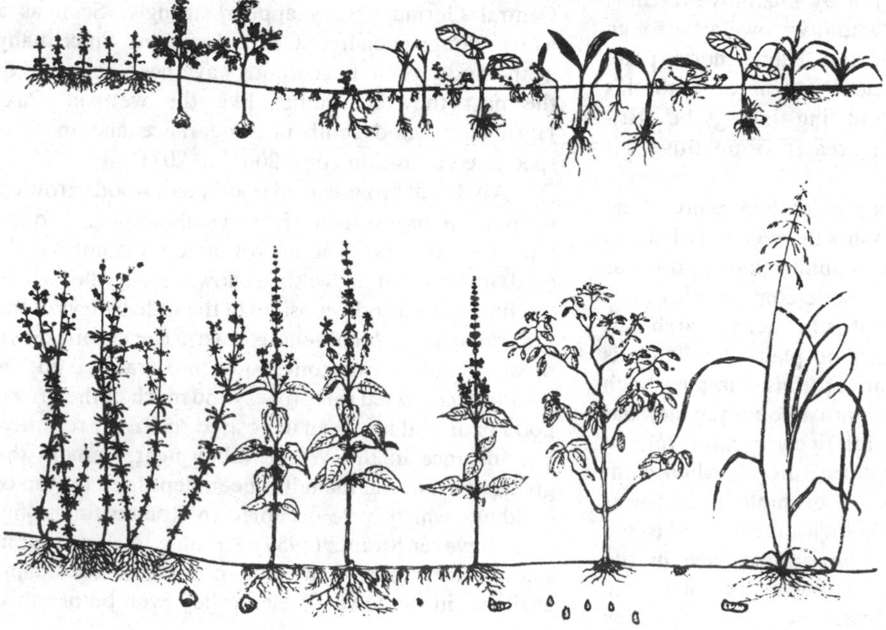

They had begun to cover the glacial or early post-glacial gravels more than 2000 years ago (fig. 197). However the deposition of loam in the river valleys was speeded up more and more with increasing density of human settlement. Deep layers of loam have only been formed in the Leine valley since the Iron Age (Firbas and Willerding 1965, see also Oelkers 1970). The accumulation of loam in the Elbe valley can be looked upon as a natural process which has merely been accentuated by man's activities, just as it has been in the Weser and Leine valleys (Strautz 1962).

Naturally this only applies where there is such loess-loam or other fine grained soil in the catchment area of the river. This is true in the cases of the Erft, Rhine, Weser, Leine, Oker and other rivers of North-West Germany, but is not so in the case of the Ems which is mainly sandy. Such permeable soil dries out very quickly after each winter high water and does not retain its moisture into the spring like the flood plain soils with a high clay content do. According to Trautmann and Lohmeyer (1960) because of this Beech can even grow within the area affected by the high water. On loamy soils Beech is rarely met with inside river valleys and can generally be considered as a tree which avoids this habitat. Figure 208 shows that there is a whole sequence of mixed Beech wood communities corresponding more or less to the brownmull mixed Beech woods on the moraine soils (section B II 3). The more frequently these **flood plain Beech woods** are flooded the richer their soils become in nutrients and colloids and the more the species requiring a higher level of fertility and moisture are able to flourish in them. In addition to these reverside Beech woods there are also Oak woods and Willow woods in the sandy Ems valley, but these are found at lower levels. In parts of old river courses which have been cut off from the main river Black Alder fen woodlands have developed. We shall look at these in section V 2.

How dangerous flood plains can be for Beech, even in those places in the middle reaches of the Ems where it has flourished for decades, was shown by the unusually persistent high water which occurred in July 1956. Soon after this catastrophe many Beeches of all ages died in the areas which had been inundated. Their leaves became discoloured and the bark broke up at the bases of the trunks. Later this split allowing the entry of secondary diseases which contributed to the destruction of many trees. Other trees managed to recover partially or completely, and the prospects are that gradually quite a strong natural regeneration will be able to take place. The death of the Beeches was probably due to the active fine roots being killed by shortage of oxygen while the soil was flooded. In any case the deciding factor was that the high water came during the growing period and lasted for more than a week. Previously the high water in summer had lasted only a few days and the trees had been able to survive this without damage, as they had the flooding in winter, even though the water had risen to a higher level than it had in 1956.

Thus this unusual natural experiment proved to be very informative about the ecology of Beech. Only a river like the Ems with a small catchment area and lying in a low rainfall district, where high water in summer is very exceptional, would not have dealt a death blow within a few years to any Beech venturing into its sandy flood plain. *Fagus* can also be found in similar circumstances in the valleys of many small rivers and streams in the Münsterland of Westphalia where there is a similar water regime to that of the Ems (Wattendorf 1964). It is found particularly associated

Fig. 208. Oak and Beech woods in the sandy flood plain of the middle Ems. A schematic section from the river bank across to the edge of the diluvial terrace (right).

The unusual highwater in the summer of 1956 which led to the death of many Beech reached a height of only halfways between the average high water and the highest level. After Trautmann and Lohmeyer (1960), somewhat modified.

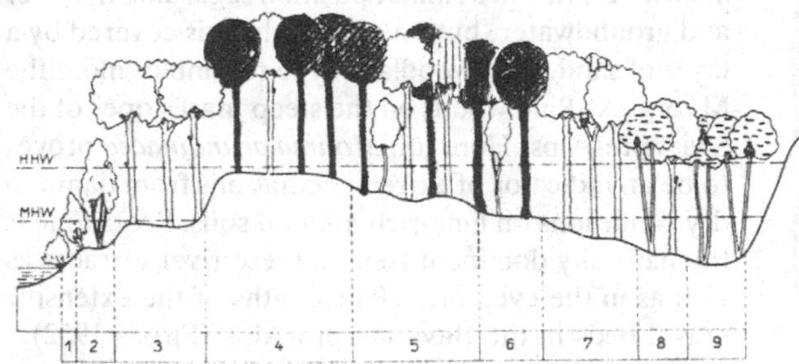

1 = Osier scrub, 2 = Willow wood, 3 = typical flood plain Oak wood, 4 = mixed Beech wood, poor in species, and brown-mull Beech wood with *Lonicera periclymenum*, 5 typical mixed *Circaea*-Beech wood and typical brown-mull Beech wood, 6 = *Glechoma* form of 5, 7 = flood plain Oak wood rich in *Impatiens*, 8 = flood plain Alder-Oak wood, 9 = swamp Alder wood. Black = *Fagus*, hatched = *Fraxinus*, broken lines = *Alnus glutinosa*.

with Ash in stands which are regularly flooded in winter, but only occasionally and never for long periods in the summer. Neither of these appear to damage the Beech.

Among the trees inhabiting the upper parts of the river valleys in Central Europe Beech occupies one extreme position, and Grey Alder (*Alnus incana*) the other, regarding the annual variation in the water levels. Beech can only tolerate high water in winter and is killed by unusually persistent flooding in summer. Grey Alder on the other hand is in danger when the summer spate does not reach its usual height and it flourishes only in those places where this regularly occurs. Some other trees behave like Alder in this respect, or, like the Common Oak, are indifferent to the amount of summer flooding. Those which, like Beech, cannot tolerate extensive flooding are conifers (which will be discussed in the next section) and also a number of shrubs and herbaceous plants. The following species are examples from the Danube near Wallsee. They have been found by Wendelberger-Zelinka (1952) on refuge hummocks for the animals, but not on the general level of the valley floor. They are:

Berberis vulgaris	*Galium odoratum*
Viburnum lantana	*Lathyrus vernus*
Daphne mezereum	*Mercurialis perennis*
Euphorbia amygdaloides	*Galium sylvaticum*

According to Trautmann and Lohmeyer (1960), except for those at right hand they are also absent from the Beech-rich reverside woodlands of the Ems. The inclusion of the Grey Alder woods and of mixed broadleaved woods of the river valleys in the order *Fagetalia* therefore should not lead to the supposition that all the character species of the latter can be found flourishing in river valleys.

Fig. 209. A Purple Moorgrass-Spring Heath-Pine wood (*Dorycnio-Pinetum molinietosum*) on a limestone gravel terrace in the Isar valley below Mittenwald. Grazing has given rise to clearings with Moorgrass meadows and slightly arid grassland.

(i) Pines and other conifers in river valleys

Like Beech, Conifer trees too can be found in flood plains only in very exceptional circumstances, since none of them seems able to withstand flooding for any length of time. This is true even of Scots Pine although it is able to colonise raised bogs and other wet sites. On the contrary Pine demonstrates its ability to withstand very dry and poor conditions. **Dry valley Pine woods** (*Dorycnio-Pinetum*) are only found on very free-draining calcareous gravel beds in some river valleys of the Alps and their northern foothills, for example in the Rhine valley above Chur and along the Isar south of Munich. Seibert (1958) has investigated them in some detail in the 'Pupplinger Au', a nature reserve on the Isar above the mouth of the Loisach (see fig. 196).

These stunted Pine heaths on old gravel beds near the river should not be designated flood plain woodland since they are only typical in those places which are no longer subject to flooding or are, at most, reached by the high water for a day or two in exceptional years. The water table in the gravelly subsoil, which is very deficient in fine particles, is generally more than 1.5 m below the surface and rarely rises to within 1 m, so it remains out of reach of the plants in the undergrowth. The drought-tolerant ground flora as well as the stocky forms of the trees are very reminiscent of the Spring Heath-Pine woods discussed in section IV 6 b (fig. 209). Both communities belong to the alliance *Erico-Pinion*.

Seibert distinguishes two communities in the Pupplinger Au where each of them occupies large areas (figs. 210 and 196). The Dwarf Sedge-Spring Heath-Pine wood (*Dorycnio-Pinetum caricetosum humilis*) is typical for the gravels poorest in fine particles and has a species composition similar to the Engadin dwarf Sedge-Pine woods of the Central Alps. The **river-valley Moorgrass-Pine woods** (*D. -P. molinietosum*) grow in a similar position regarding the river and groundwater, but on gravel which is covered by a layer of sand. This woodland type resembles more the Moorgrass-Pine woods on the steep marl slopes of the Swiss pre-Alps. Here too *Molinia arundinacea* proves to be an indicator of extreme variations from damp to dry conditions on lime-rich mineral soils. Scots Pine is the naturally dominant tree on these river terraces as well as in the even drier Pine heaths of the extensive gravel beds in the Bavarian pre-Alps (Firbas 1952).

Data given by Seibert (1958) show that the river and groundwater levels in the region of the Pupplinger Au have fallen by about 1.10 m since 1900. Today only the very highest floods occasionally come up to the old levels, but because of their spasmodic occurrence have scarcely any influence on the water supply to the plant communities. This rapid lowering of the water table is thought to be the reason why the herbaceous pioneer communities have been replaced directly by Pine woods and even why the undulating nature of the ground has been maintained (fig. 210). If the average water levels had remained constant these local variations in the height of the ground would have been levelled out by sedimentation at high water. Moreover Grey Alder would have assumed its usual dominant role in these montane river valleys. Industrial demands on the water economy of the Isar have speeded up the natural processes of erosion which can be seen in many Central European river valleys as a result of a gradual reduction in water flow since the last glaciation. In the region of the Pupplinger Au this has led to the formation of 3 or 4 alluvial terraces, each 1 to 1.5 m higher than the previous one, which are no longer subject to flooding. Wherever these old gravel terraces are not brought into cultivation they carry Spring Heath-Pine woods.

The majority of the dry Pine woods in the river valleys are very open and here and there they are interrupted by large areas of grassland dominated by *Bromus erectus* or *Molinia*. These park-like areas have arisen because of the cutting of bedding rather than through the removal of timber, according to Seibert. Also the grazing of sheep and goats was at one time very widespread on the gravel plains and old river valleys in upper Bavaria. More than fifty years ago the Pines began to reconquer the grassland areas so that now many of the rare light-demanding plants will die out, if man does not interfere.

Nowhere is **Spruce** a tree of the flood plains. When many authors speak of 'riverside spruce woods' they are not referring to woodland innundated from time to time but to communities which are formed along deep watercourses cut into wide depressions or the remains of terraces, and which contain moisture-indicating plants such as *Equisetum sylvaticum* (see fig. 159). At most *Picea abies* sometimes invades an area liable to flooding from neighbouring slopes where it is dominant (fig. 211). Mondino (1963) has rightly suggested that Spruce woods should not be looked upon as the climax in a river valley succession. Even a 'development' of Grey Alder to Pine woods in the river valley is often only hypothetical. Further development to a Spruce wood is possible only if there is a lowering of the water table.

(j) River valleys as migration routes for plants

Both in the natural woody landscape of Central Europe and in the present-day cultivated landscape, river valleys make very good routes for the migration of plants which are native to other vegetation belts or

Fig. 210. A vegetation profile in the gravel plain of the Isar near Wolfratshausen ('Pupplinger Au', see also fig. 196). After Meusel (1940).

On deep lime-rich sands to the left and right is Purple Moorgrass-Spring Heath-Pine wood with stands of *Molinia* (1), *Calamagrostis varia* (2) and *Brachypodium pinnatum* (5), also with *Gymnadenia conopsea* (3), *Laserpitium latifolium* (4), *Linum viscosum* (6), *Cypripedium calceolus* (15), *Pleurospermum austriacum* (16), *Epipactis palustris* (17) and *Astrantia major* (18).
In the centre on coarse gravel of an old river bed is a poorly grown Dwarf Sedge-Spring Heath-Pine wood with Junipers and a dwarf shrub community with *Dryas octopetala* (10), *Thesium rostratum* (11), *Dorycnium germanicum* (12) and *Leontodon incanus* (9 and 13). In the transition zone: *Festuca amethystina* (7), *Erica herbacea* (8) and *Daphne striata* (14).

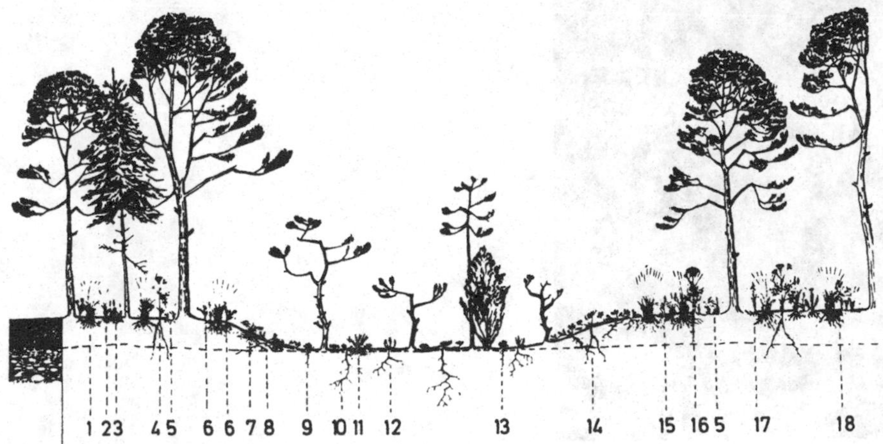

floristic areas (Tüxen 1950b). A number of special features of riverside habitats contribute to this:

1 The **transporting ability** of the flowing water which can carry not only seeds and vegetative propagules but also whole plants and turves,
2 **Reduced competition** on bare stretches of river bank or temporarily flooded places,
3 The **formation of new land** when the river is in spate,
4 A **good supply of water and nutrients** which allow even very demanding species to develop quickly in such places,
5 The ability of **animals** living on or near the river to transport seeds.

This is particularly true of birds which can transport small seed in mud clinging to their feet, beaks and feathers equally well upstream as down. It would be difficult to think of any other way in which an 'upstream migration' of plants which do not have windborne seeds could be brought about; a phenomenon which has often been observed and commented on.

All five factors are most effective near the bank of the winding river since those parts of the flood plain which lie somewhat higher, and are naturally more or less heavily wooded, slow down the flood water and reduce its carrying capacity. Even the ice which can tear off turf from the edges of terraces and so produce bare soil in relatively dry sites would not be able to maintain a continuous migration route in the woodland at some distance from the river bank.

The groups of plants which in the first place are able to spread much further and more rapidly with the help of rivers are the montane species mentioned in section c, several short-lived nitrate plants and many perennial garden escapes which are becoming increasingly at home here.

Some waterborne **alpine and subalpine plants** (German: Gebirgs-Schwemmlinge) are able repeatedly to travel far down into the foothills along the rivers

Fig. 211. Spruce and Grey Alder (some of which are dying) near Strada on the driest part of an island of the Inn. It is now hardly ever flooded by the highest of high waters because the river has cut itself a deeper channel. Although there are river channels on both sides this Spruce habitat is already outside the flood plain.

and can flourish there for a time in spite of the warmer climate and the occasional flooding. This is further evidence to support the idea that they are prevented from maintaining a foothold below the belts in which they normally occur by the competition of denser and taller plants in the montane and colline belts. *Gypsophila repens* for example has travelled down the Isar as far as Landshut and Landau, down the Iller to Ulm, the Rhine to Speyer, the Tessin to Bellinzona, and down the Isonzo to below Görz, that is, almost as far as the Adriatic Sea. Before 1918 *Campanula cochleariifolia* was occasionally found as far down the Rhine as Ottenheim (155 m) and the Isonzo as far as Lake Garda. From the High Tatra migration of plants does not reach the same proportions as in the Alps since the catchment areas of the rivers in the subalpine and alpine levels are smaller. However they have been the subject of a model piece of research and presentation by Walas (1938, see fig. 212).

Principally it is the plants of the alpine gravel beds and the subalpine tall perennials and spring flora which are carried down into the valley by floodwater in the late spring and early summer. Most of these species are wind disseminated in their normal habitats and possess no special features which would fit them for water transport. Nevertheless wind plays scarcely any part in their removal to the lowland areas. For instance the tiny seeds of *Saxifraga aizoides* can float on the surface for 17 days and can remain in water for 38 days without damage, according to Hegi, especially when the temperature is low. Other species succumb more quickly but even these have a chance of being carried a long way because the rivers flow so rapidly in the mountains. The average speed of an Alpine river amounts to about 2.25 m/s which is enough to carry a plant 200 km in 24 h. Some species actually germinate during transport and are already seedlings when they are deposited on land. More rarely others are torn out of the ground by the floodwater and are transported as fully grown plants.

The seeds or other viable portions of plants carried like this by the river are set down on sandy slopes or gravel banks as the water recedes. During the year following the great flood of 1934 for example, Walas found *Arabis alpina, Poa alpina* and other species, which are otherwise confined to the high mountains, on newly formed gravel banks. Such plants can survive for a number of decades under favourable conditions. Many of them show an amazing viability in the lowlands, also thriving happily in botanic gardens. However under natural conditions they usually disappear quite quickly as they become overgrown by the lowland plants, or as the river erodes their temporary resting place, or if they are discovered and collected by eager botanists! So they are dependent on being constantly renewed from the mountains or from favourable intermediate sites. Should these be cut off by a series of dams their localities become wiped out one after another.

The second group of plants migrating along the

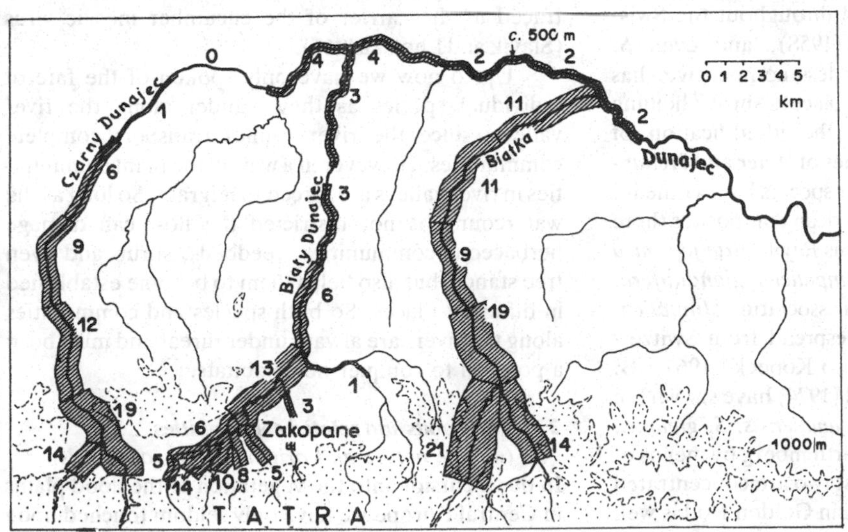

Fig. 212. The numbers of alpine and subalpine species which have been carried down by the water and have settled along the rivers flowing to the north from the High Tatras. The number in each section of the valley is also indicated visually by the width of the shaded band. After Walas (1938), somewhat modified.

river do not have so far to come as the ones which float down from the high mountains, and therefore are found in larger numbers. Several **annual nitrate plants**, widespread today on ploughed land, rubbish dumps and other ruderal places probably have their natural habitat on the lower river banks which cannot be occupied by perennial plants (figs. 189 and 190).

By far the most conspicuous and also the youngest group of river valley migrants are the growing number of **garden escapes** which are all characterised by strong growth, requiring good light and high fertility. These are mostly perennials of the genera *Aster, Solidago, Helianthus* and *Rudbeckia* (members of the Compositae) growing to 1-2 m in height. Most of them originate in north America and were introduced into European gardens as decorative plants between one and three centuries ago. First of all they became naturalised on ruderal places but soon invaded river valleys, especially where these were not persistently flooded. They managed to gain a footing along the water's edge or in clearings in Willow or Alder stands, smothering out the native perennials such as *Urtica dioica* and *Senecio fluviatilis*. Sometimes they are in strong competition with each other and are often very persistent in dominating the places where they first appeared. Obviously chance still plays a large part in their distribution, but they are becoming more and more a permanent element of our vegetation and are giving rise to the formation of some new plant communities. Most of these can be included in the alliance *Calystegio-Alliarion* (section E III 1, No. 3.52).

Solidago canadensis is often sown for pheasant food and has already become naturalised in many parts of Central Europe. It is occupying fallow land outside the river valleys to an increasing extent (see section D X 1 b). *S. gigantea* has spread throughout the Swiss valleys, according to Moor (1958), and even *S. graminifolia*, in spite of being less competitive, has become quite plentiful in some places. Since Thellung (1925) published his key to the identification of cultivated and naturalised species of *Aster* and *Helianthus* in 1913 a number of further species have come in. Alongside these and other American composites there are newcomers from Asia such as *Reynoutria japonica* and, from the East Indies, *Impatiens glandulifera*. After the latter Moor named an association (*Impatienti-Solidaginetum*) which is widespread from Switzerland to Czechoslovakia (see also Kopecký 1967). B. and K. Rüdenauer and Seybold (1974) have shown the exact distribution of *Solidago canadensis, S. gigantea* and *Helianthus tuberosus* in Württemberg by means of point maps showing that the last two are concentrated in the river valleys. The Canadian Golden Rod is well on the way to becoming a general weed of urban wasteland (or its adornment!)

Since these new arrivals are tending more and more to take possession of open spaces. As they often increase the difficulties of afforestation they are becoming a nuisance. Interesting as it may be for ecologists to follow the fate of these new plants in the indigenous communities, they have also been able to help combat them. A simple method is by the removal of light. As Moor noticed, the majority of these species are sterile in the shade, they become stunted and finally succumb altogether. So it is only necessary to help the trees in the beginning, and this is best done by mowing the weeds down and leaving the dead remains spread on the ground where they smother any new herb growth. Following promising North American experience Zwölfer (1974) examined the possibilities for the biological control of Golden Rod.

Several of the alien and some native large herbs were earlier considered to be characteristic species of the river Willow scrub, or the riverside Willow woods since they were present in large numbers in the more open parts of these communities. However, as Tüxen (1950a) has clearly recognised, they also form independent, even though rather unstable communities avoiding the shade. They are now included in the alliance *Calystegion*. The common name **'veil communities'** (German: Schleiergesellschaften) applies in particular to the *Cuscuto-Calystegietum* which climbs up other herbaceous plants and low shrubs so that in late summer they appear to have a veil drawn over them. In many river valleys in the south-east of Central Europe *Echinocystis lobata* is widely distributed in such communities. This member of the Cucurbitaceae from the north-east of North America has been definitely traced as the carrier of the cucumber mosaic virus (Slavík and Lhotská 1967).

Up to now we have only spoken of the fate of individual species as they wander along the river valleys, since the river cannot transport complete communities. However in a way all the plant communities in river valleys are forced to migrate. So long as the watercourse is not restricted the flow can damage herbaceous communities, reedbeds, shrub and even tree stands, but also helps them to become established in different places. So both species and communities along the rivers are always under threat and must be in a position to conquer new habitats.

2 Fen woods and related communities

(a) The nature and origin of fen woodlands

In our incursion into the vegetation of the flood plains in Central Europe we have only seldom touched upon woodlands rich in Alder, even though these would take up large areas as the natural vegetation of the broad river valleys of the northern diluvial plain (fig. 30). Postponing a description of them to this point seemed advisable on a number of grounds:

1 Very few of the woodlands consisting chiefly of *Alnus glutinosa* are true riverside woods which are regularly flooded and thus supplied with sediment. The exception is the Ash-Alder woods on the banks of streams (section I f).
2 Certainly many Black Alder communities have grown up, and still do develop, in river valleys, especially in old river beds and in the wet parts along the edge of broad valleys, but they are not confined to such places. They are also to be found round lakes and ponds which may be fed only by small rills or by groundwater.
3 The systematic and ecological position of the various kinds of Central European Alder woods can only be appreciated if one knows the 'real' *Alnetum glutinosae* (fig. 213).

This Alder wood however is not a flood plain wood but a swamp or **fen wood** (German: Bruchwald) and can be distinguished, like all woods on wet peaty soil, from the former by a number of significant conditions in the habitat:

a) Fen woods grow on ground where the **water table is constantly at round about ground level**. The height of the water very rarely alters by more than 1 m and is generally much less (fig. 214), the opposite being the casc in flood plains (see also Klötzli 1969a).
b) The soils of fen woods are generally **only inundated in early spring** when snow melts in their immediate vicinity. After that they remain wet for a long time in contrast to the soils of the river valleys which dry out a few days or weeks after flooding.
c) Flooding of fen woods does **not bring with it much sand and silt** and so does not cause raising of the soil level or mineral enrichment of the soil. Whenever such addition of inorganic sediment does occur then the species composition approaches that of the flood plain woods which have already been described.
d) True fen woods contain **at least 10-20 cm of peat** (fig. 38). That is they have an upper soil which consists mainly of organic matter formed in situ. In contrast all true river-valley woods are associated with mineral soils.

Of course, every conceivable intermediate type can be found between fen and flood plain woodlands. In addition there are woodland communities linking

Fig. 213. A wet Alder swamp wood (*Carici elongatae-Alnetum*) in spring at the edge of a pond near Chorin (north-east of Berlin). In the foreground there are bud scales and other organic detritus floating on the water. Photo. Hueck.

the fen woods with those on damp mineral sòils not subject to flooding such as the wet-soil Oak-Hornbeam woods, the wet-soil Birch-Oak woods (figs. 116 and 131) or the stream Ash woods.

Flood plain woodlands are always found on relatively base-rich substrates, since even rivers flowing over very old rocks bring down sediments rich in lime and hydroxyl ions. Fen woods on the other hand can occur on very acid soils. However the Alder woods themselves do require a certain lime content in the groundwater which it can get from its surroundings by effluents or surface run-off. Where such an input does not happen and the groundwater itself is also deficient in lime then only a very acid, nutrient-deficient, peat can be formed on which Alder is unable to flourish. So it gives way to Birch or the still less fastidious Scots Pine (figs. 216 and 217). Under intermediate conditions mixed woodlands with different proportions of the three trees can be found, adding to the number of different types of Alder wood.

Fig. 214. Lines indicating the persistence of the groundwater at different levels in some of the swamp communities in central Switzerland (cf. fig. 111). After Klötzli (1975b).

For example in the Sphagnum-Alder swamp wood the groundwater is at or above ground level for 13 wooks/year but never floods to a depth of more than 8 cm. The groundwater is lowest in the Clubmoss-Birch swamp wood and also shows the most variation her relative to the others, although in absolute terms it is quite small, namely between depths of 25 and 50 cm. Between the two extremes come the Sphagnum-Pine swamp wood and the Willow-Birch swamp wood. The former is rather like the woodland along the edge of a raised bog.

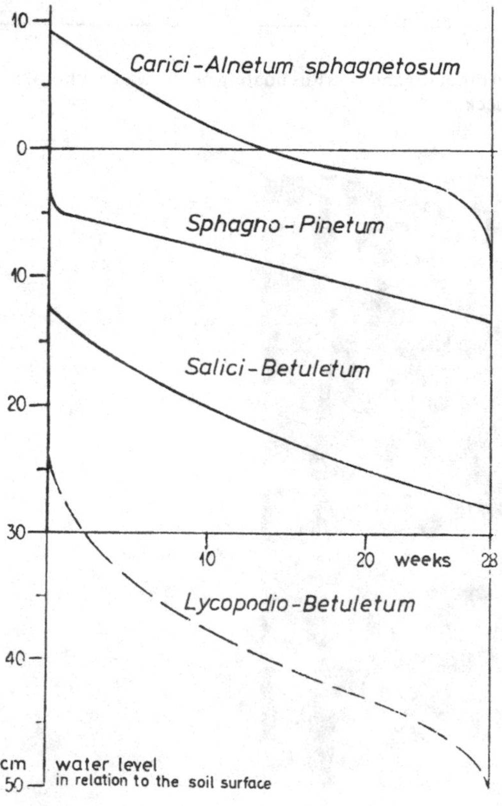

In various ways man has multiplied the variety of Alder woods, chiefly by lowering the water level. Thus he caused a succession to mixed Ash-Alder woods, wet-soil Oak-Hornbeam woods containing Alder and finally communities which are quite independent of groundwater. The previous widespread use of the woods for coppicing, which Alder endures very well, has favoured light-loving plants. Many stands of *Alnus glutinosa* are at present particularly rich in grassland species since they have been brought about by the afforestation of meadows which had become worthless, or they had taken over themselves after mowing had been discontinued.

All 'true' fen woods grow on wet soil rich in organic matter, i.e. on peat. This peat normally cannot be formed above the average water level since as soon as air can get at the organic remains they are more or less quickly decomposed. Only in those places where the soil water is very acid is the activity of decomposing microorganisms so restricted that a covering of raw humus can accumulate. Finally, with the help of certain *Sphagnum* species, even a raised bog can be formed. We shall return to this special case in section C III 1; in Central Europe it is the rare exception (see also fig. 259).

By digging or boring into the ground further layers of peat are very often found below that produced in the fen wood itself. The upper layer contains the remains of wood and cones, and the degree of decomposition is higher than that of the lower 'dead' layers. These have been formed from swamps of sedges or reeds and may still contain their roots, leaf sheaths or densely matted rhizomes (fig.38). In such cases the fen wood can be looked upon as the climax of a so-called **'hydrosere' succession** (see fig. 38a). A typical example of a complete succession of this kind would start with a layer of lake deposits (gyttja) i.e. of inorganic mud containing some remains of animals and plants. This is followed by at least several decimetres of reed and sedge peat. Only the uppermost 2-3 dm will normally be of fen wood peat if the average level remained constant during the whole succession.

Often the layer of woodland peat is thicker due to tectonic sinking of the area, or from some other natural cause which has led to a gradual raising of the water level in relation to the soil surface. In some diluvial valleys of northern Germany and Poland, such as the lower Elbe, the water table has risen because of sedimentation which raised the level of the river bed and its banks, but not the general level of the very wide valley. It is in this and similar ways that the largest areas of fen woodland and the deepest accumulations of peat have come about, and not by the silting-up of lakes.

Profiles of such peat fens in the old river valleys show that from time to time the river has changed course and that the conditions in any one place have alternated between those of a flood plain and those of a swamp or fen. So while many fen woods grow over an extended series of peat layers others have been formed in shallow waterlogged depressions or along the very shallow shores of lakes and show an incomplete series. In some places the peat layer is only about 20 cm deep over the mineral subsoil, in others it may be several metres thick, but in each case the species composition of the fen wood may be very similar. This is because the majority of the roots do not penetrate into the waterlogged subsoil, and the levels within which the ground water fluctuates are practically the same in both cases.

It is obvious that many fen woods must be considered as the climax stage in the vegetation development when the water regime remains the same, and that they would not develop any further towards zonal vegetation unless the water level falls, thus having a decisive effect on the habitat.

(b) Fen Alder woods

True fen woods of Alder occupy such extremely wet ground that the majority of plants found in the broadleaved woodlands cannot survive. Systematically then they are very isolated and form a distinct order (*Alnetalia glutinosae*) although this is made up of only a few associations, and all these must be considered as belonging to the one alliance (*Alnion glutinosae*). Like *Alnus glutinosa* itself this alliance is widespread in Central Europe (fig.215). Bodeux (1955) has named the following as character species for the order and the alliance:

Alnus glutinosa	*Lycopus europaeus*
Salix cinerea	*Calamagrostis canescens*
Salix aurita	*Solanum dulcamara*
Salix pentandra	
Ribes nigrum	M *Sphagnum squarrosum*
Ribes spicatum	M *Trichocolea tomentella*

By working through the whole of the European literature Bodeux has been able to complete successfully the difficult task of sorting out the 'true' fen Alder woods from the many woodland types containing Alder which have been described. Of the 547 relevés he considered only 187 to be sufficiently pure and these can be grouped into 4 associations (fig.215):

1 The **West European** fen Alder wood (*Carici laevigatae-Alnetum*) still to be found in the Eifel and in the western part of the Upper Rhine plain,

2 The **Central European** fen Alder wood (*Carici elongatae-Alnetum medioeuropaeum*), of which Passarge (1978) also distinguishes a montane race,

3 The **North-east European** fen Alder wood (*Carici elongatae-Alnetum boreale*) which is distributed in

Fig. 215. As azonal communities the European Alder swamp woods are so similar to each other that only two associations can be distinguished – the atlantic *Carici laevigatae-Alnetum glutinosae* and the subatlantic to subcontinental *Carici elongatae-Alnetum glutinosae*. However Central Europe is distinguished by having its own subassociation of the latter (*medioeuropaeum*). Along the eastern border are the more continental *orientale* and to the north-east the *boreale* subassociations. After Bodeux (1955) and also Matuszkiewicz and Traczyk (1958), modified.

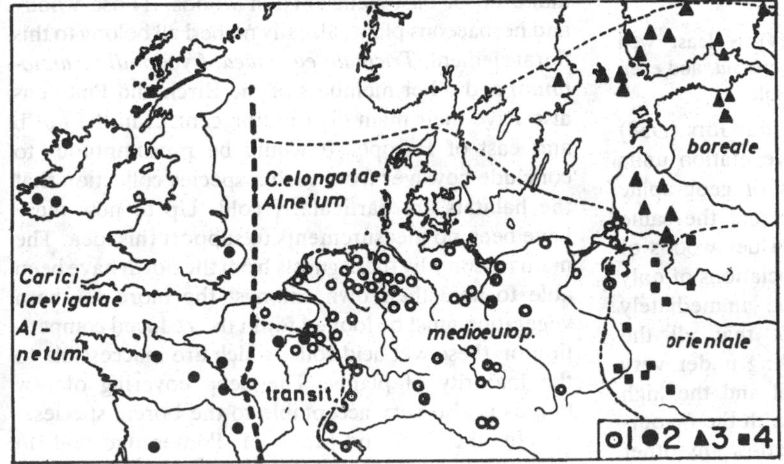

Associations of Black Alder:
1 Central
2 Western
3 North-eastern
4 South-eastern

the eastern Baltic countries as far as southern Finland,

4 The **East European** fen Alder wood of eastern Poland and western Russia which is very similar to the Central European type. In contrast to Bodeux's opinion W. and A. Matuszkiewicz and H. and T. Tracyk (1958) consider that it should not be a separate association (*Dryopteri cristatae-Alnetum*) but only warrants the rank of a subassociation (*Carici elongatae-Alnetum dryopteridetosum cristatae*). It could also be named after its geographic distribution (*C. el.-A. orientale*) (fig. 215).

Certainly none of these associations has its own character species, but the alliance character species can be seen as such locally, since the four fen Alder wood types are practically mutually exclusive. Their geographic pecularities are expressed in the differential species:

The *Carex laevigata* group is limited to the western association (2) with only occasional forays further east (fig.215,2):

Carex laevigata	*Valeriana procurrens*
Osmunda regalis	*Scutellaria minor*

The *Carex elongata* group is common to the remaining associations:

Carex elongata	*Carex vesicaria*
Thelypteris palustris	*Geum rivale*
Peucedanum palustre	M *Rhytidiadelphus triquetrus*
Rubus idaeus	

The north-east European Alder fen (3) always carries a few Spruce and its companions, especially those of the *Picea* group (fig.215, 3):

Picea abies	*Orthilia secunda*
Rubus saxatilis	(*Carex tenella*)

The eastern European *Alnetum* (4) is least well differentiated, namely only by *Dryopteris cristata*, and even this is present in only about half the examples.

Schwickerath (1942) and Müller and Görs (1958) proposed to designate the different vegetation units which are only separable by means of geographic differential species as 'races' of one and the same association. In this case all the fen Alder woods of Europe would be classified as subassociations of only one association. Then it would be immediately apparent from the systematic rank that all the European fen Alder woods were living under very similar conditions. The very wet soil and the high humidity within the stands, together with the favourable nutrient supply, so much overshadow any other factors that the species make up in the fen Alder woods remains virtually the same from the Bay of Biscay to White Russia and from Croatia to Finland. There is no other woodland community of the lowlands which responds so little to variations in the general climate as this one. The fen Alder woods therefore are a very good example of azonal vegetation units (see section B I 1 a) and in this respect are only exceeded by some communities of water plants (section C I).

(c) Fen Birch, Pine and Spruce woods

Alder fens and other woodland types which are dominated by Alders always indicate a groundwater rich in bases. When the concentration falls below a minimum of about 0.1 mg CaO/l *Alnus glutinosa* can no longer compete successfully with *Betula pubescens* or *Pinus sylvestris*. So when there is a deficiency in bases under somewhat similar conditions the Alder fen is replaced by a **fen Birch or Pine wood** (fig. 214). Which of these less nutrient-demanding trees achieves dominance depends on the climate. Birch prevails in the north-west and Pine in the more continental east. In the mountains and in the north-east of Central Europe Spruce too mixes in under natural conditions. All the base-deficient swamp wood types however resemble each other ecologically and floristically so much that we shall deal with them all together here.

Trees do not flourish very well in the acid swamps and the canopy is very open. However in spite of the good light only a few tall shrubs are found, and these are all acid- and water-tolerating species like *Frangula alnus* and *Sorbus aucuparia*. On the other hand the ground cover is relatively luxuriant and consists of dwarf shrubs like *Vaccinium uliginosum*, *V. myrtillus*, *V. vitis-idaea*, *Ledum palustre*, and hygrophilous mosses.

Boreal species play a noticeably large part in the make up of base-deficient fen woods. Those woody and herbaceous plants already named all belong to this floral element. *Trientalis europaea*, *Lycopodium annotinum* and other members of the Birch and Pine fens also have their main distribution centres in the north and east of Europe. It would be presumptuous to conclude however from such a species collection that the habitats are particularly cold. Up to now there have been no measurements to support this idea. The main reason why these guests from the north have been able to hold their own amongst the more southern vegetation must be looked for in the reduced competition on these wet acid soils, which are inaccessible to the majority of plants. The deep covering of raw humus is also very acceptable to the boreal species.

In the lakeland area of Pomerania and in

north-east Poland Birch and Pine fens are often found close together. According to W. Matuszkievicz (1963) *Betuletum pubescentis* prefers shallow, mostly small depressions with only 20-60 cm depth of peat, or even just a humus-rich mineral soil. *Vaccinio uliginosi-Pinetum* on the other hand is found where there is more than 1 m of peat in larger depressions. The species show some preference for one or the other as follows:

mainly in fen Birch woods:	mainly in fen Pine woods:
Betula pubescens	*Pinus sylvestris*
Frangula alnus	*Ledum palustre*
Dryopteris carthusiana	*Vaccinium uliginosum*
Trientalis europaea	*Calluna vulgaris*
Oxalis acetosella	*Eriophorum vaginatum*
M *Polytrichum commune*	*Andromeda polifolia*
	M *Aulacomnium palustre*

All these species however can also be found in the other type. As Wojterski (1963, fig. 217) pointed out the continental small shrub *Ledum palustre* and the subatlantic Bog Myrtle (*Myrica gale*) are found growing together in the coastal region of West Kassubia.

Fig. 216. A Birch swamp wood with Spruce leading to a poor Alder swamp wood (behind) near Riga in early spring. The peat is saturated with stagnant water almost up to the surface.
The classification of the Birch swamp woods is difficult because, in spite of being broadleaved, their floristical composition is similar to that of acidophilous needle-leaved woodlands (*Vaccinio-Piceion*, see section E III 1, no. 7.212). Perhaps they may be separated as a particular suballiance (no. 7.212.3).

There is a continuous series of intermediate types between the fen Pine woods and the woodlands around the edges of otherwise treeless raised bogs or the Pine stands on wooded raised bogs (see sections C III 1 b and 2 b). Presumably these have been included in the comparison drawn up by Matuszkiewicz which we have just discussed; this could account for the much thicker peat layers when compared with the Birch fens.

Fen Spruce woods are also frequently found in conjunction with raised bogs, lying between these and the woodland on mineral subsoils (fig. 218) or with Alder stands in the surrounding swamp (fig. 250). Sokołowski (1966b) described such **fen Spruce woods** from the ancient Białowieża Forest as *Sphagno girgensohnii-Picetum typicum*. They are also found along with Pine woods on very acid sandy soils with a high water table (*Vaccinio myrtilli-Pinetum molinietosum*). In wetter and somewhat more fertile parts of these wet-soil Pine woods a *Pinus*-rich subassociation of the fen Spruce wood can develop *(dryopteridetosum)*. That *Picea abies* becomes dominant in many fen woodlands is probably climatic. In the first place Spruce fens are a feature of the montane and subalpine levels in the mountains and in the higher parts of the pre-Alps, as in Upper Bavaria. However the example from the Białowieża Forest shows that they can also occur in the lowlands

Fig. 217. *Ledum*-Pine swamp wood (*Ledo-Pinetum*) at the edge of a raised bog near Riga. *Ledum palustre* can be recognised in the left foreground and *Chamaedaphne calyculata* in the centre; both dwarf shrubs have a more or less continental area of distribution.
The Pine swamp woods belong to the class *Vaccinio-Piceetea* as do the majority of other needle-leaved woods and may be distinguished as a particular suballiance (see section E III 1, no. 7.212.2). However they form such a unique group that at least they should be distinguished as a separate suballiance (see also fig. 216).

especially along the edge of the extensive Spruce wood region of the Boreal zone which extends into the north-eastern part of Central Europe. Spruce also plays a part in the fen Birch woods and appears spontaneously as far towards the suboceanic west as north-western Germany where it has been described by Hesmer and Schroeder (1963) occupying lowland habitats.

(d) The oxygen and base content of the water in fen wood peats

Where a stream winds its way through poor fen Birch and Pine woods, or a spring wells up through the nutrient-deficient peat, Alders are found skirting the moving water, and tall herbaceous plants take the place of the slow-growing species found on the acid humus. Emerging groundwater and quickly flowing surface water appear to be prerequisites for a good growth of Alder while Birch and Pine seem to prefer stagnant water. Often the beneficial effect of moving water has been ascribed to its oxygen content, an opinion which has not been substantiated by incontrovertible evidence. However, Hesselman (1910) had already proved the contrary to be true. With modern methods Janiesch (1981) came to the same conclusions:

The oxygen content of open water is always very high near the surface whether of quickly flowing streams and rivers or standing water where the wind agitates the surface as in lakes or pools, and whether the water is rich or poor in nutrients or humus (tab. 47). Soil water running into a hole which has been dug deep into peaty ground in either rich or poor fen woods contains much less oxygen, although oxygen is by no means absent altogether. Where the groundwater is welling up to form a spring the oxygen content falls off from the surface down into the column of water and where this is very deep or has a high humus content there may be no oxygen at all near the bottom. All water in holes or springs is in contact with the air for a few minutes or even hours and can take up oxygen, so measurements taken near ground level cannot lead to any conclusions about the oxygen state of the water in the undisturbed peat where the plant roots are growing.

Samples of water taken **from the peat** contain practically **no oxygen**, whether taken in a rich or a poor fen wood. Oxygen can only be detected very near to the surface or in the immediate neighbourhood of springs and quickly flowing streams (as well as in layers of gravel or sand which are free from humus).

The amount of oxygen available to the roots in waterlogged peat of whatever kind is thus at a critical minimum and this must be counteracted by the plants by the use of an internal aeration system. Conduction of oxygen to the roots from aerial parts of the plant by means of intercellular air spaces is only found in herbaceous plants in Central Europe. These helophytes are able to root in soils which are constantly waterlogged or at the bottom of lakes. Apart from Alder and some Willows, woody plants can grow on either peat or on mineral soils only when these are not completely saturated during the growing period, even if the non-saturated depth is only 10 or 20 cm. Since the

Fig. 218. Alder and Spruce swamp woods at the edge of a young raised bog in the Upper Bavarian foreland of the Alps.

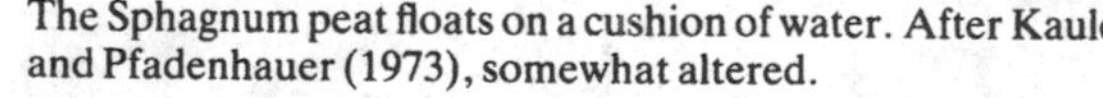
The Sphagnum peat floats on a cushion of water. After Kaule and Pfadenhauer (1973), somewhat altered.

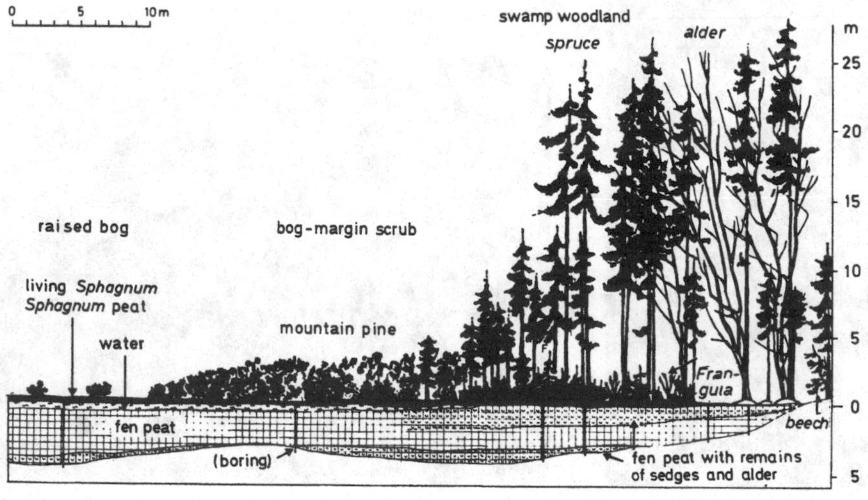

fen woods in Central Europe are normally only under water during winter and early spring, that is when the trees are dormant, little damage is done to the roots.

The roots of Alder are obviously very soft and, according to Köstler, Bruckner and Bibelriether (1968), contain a good deal of air in the xylem which is in direct communication with the outside air through large lenticels in the bark of the lower trunk. Also the intercellular spaces of the phloëm of inundation-tolerant trees are able to transport oxygen downwards (Hook and Scholtens 1978). This enables the alder to withstand prolonged flooding to a depth of a few decimetres, but should the water level rise above the uppermost lenticels during the summer the tree would succumb in a few weeks. Many species of Willow also have air-containing xylem, e.g. *Salix alba*. When they are flooded they can quickly produce adventitious roots which are able to get oxygen from the surface water – rather like the pneumatophores of tropical swamp and mangrove trees. Because of this they are able to survive high water levels for weeks and even months. Such Willows and Alders growing on river banks are able to send their roots deep into the soil thus stabilising the banks (Lohmeyer and Krause 1974).

Other flood plain trees (e.g. *Quercus robur*, *Fraxinus excelsior*, and *Populus* species) confine their roots to the layer above the average summer water level and are not very firmly anchored on the river banks. Nevertheless, according to Dister (1980) *Quercus robur* was able to survive exceptional inundations of the Rhine lasting up to three months even in summer.

The reason that peat or humus-rich soils saturated with standing or moving water contain no oxygen must not be looked for only in the roots and the microorganisms using it up (fig. 219). Hesselman (1910) had aleady demonstrated this in a very simple investigation which unfortunately has not been sufficiently heeded. He enclosed samples of soils and peats from various sources in flasks along with well aerated water and measured the amount of oxygen remaining after various lengths of time. After three days all the oxygen had been used up and this was as true for the calcareous mull soils and the peat supporting a well grown fenwood as it was for the less fertile peats of poorer sites (tab. 47). Samples of humus and peat that had previously been sterilised absorbed oxygen almost as well as unsterilised samples. So the oxygen loss seems to be primarily a chemical reaction.

Thus groundwater which is in contact with peat or a humus-rich soil for some time loses its oxygen due to various causes. This is true not only for stagnant water, but also for water flowing quite quickly; the latter will not be able to supply plant roots with oxygen shortly after its entry into such an oxygen-absorbing medium.

The beneficial effect of moving groundwater on

Tab. 47. The oxygen content of surface and groundwater under various conditions.

Source of sample	O_2 content (cm^3/1) s[a]	sat.[b]	Temp. (°C)
River water			
near the surface	**7.4**	7.7	10
near the bottom	**6.8**	7.7	10
Swamp at the edge of raised bog			
near the surface	7.2	7.7	10
Rich Spruce swamp (fed by spring)			
spring water	**5.0**	8.3	7
hole connected to stream	2.9	8.0	8
hole freshly dug in peat	1.4	7.7	10
groundwater in wooden well near the surface	2.0	8.2	7
near bottom of well	*0.7*	8.4	6

Source of sample	O_2 content (cm^3/1) s[a]	sat.[b]	Temp. (°C)
Poor Spruce swamp (with stagnant water)			
pool water	**6.1**	8.0	8
from a *hole* freshly made in peat	2.6	8.0	8
Groundwater in wooden well near the surface	*0.4-0.8*	8.0	8
near bottom of well	*0 -0.4*	8.2	6- 7
in both rich and poor Spruce swamp			
Water in the peat (sucked from a depth of 20 cm)	*0*	7.3-8.2	7-12

Note: [a]from the test sample; **bold type** = almost saturated; *italics* = poor in oxygen
[b]saturated by shaking with air
[c]numerous samples taken from various swamp soils in such a way that the water did not come into contact with air. Water in the peat is always without oxygen.

Source: From data supplied by Hesselman (1910) from Sweden.

mire and swamp vegetation must be due to other reasons. According to present day knowledge an explanation is to be found in the content of soluble minerals, especially bases. These neutralise the acids produced by anaerobic organisms and so prevent souring of the peat. Alder fen peat is always less acid than that of Birch fens through which neither surface nor soil water flows. Consequently it is richer in bacteria (fig. 219) and even contains many earthworms which are able to live in it as long as the water table falls below the soil surface (fig. 220). The activity of the soil organisms in Alder peat is so great that the remains of plant structures can hardly be recognised.

In favour of the view already put forward in section c that the **base content of the soil water** is the deciding factor for mire and swamp vegetation is the fact that fen Alder woods and other eutrophic communities are found even in places where the soil water is completely stagnant. This is so for example in some small depressions which have no outlet for their water, especially in young moraines, where the soil is rich in lime. As long as rain and melting snow water drain into these depressions with sufficient bases then Alder woods, or other lime- and water-demanding plant communities are well able to flourish. Further evidence has been obtained from manurial experiments on nutrient-deficient and very acid bogs with stagnant groundwater. McVean (1959) succeeded in growing Alder from seed on a raised bog after he had treated it with mineral phosphate. Raabe (1954) monitored a succession in Holstein which led from a raised bog and a Birch fen to an *Alnetum* even without such an introduction of seedlings. This succession had been gradually brought about by waste water running on to the raised bog and its marginal fen and swamp complex from the surrounding settled area. In both cases the groundwater had been stagnant before as well as after the treatments.

In spring fens which have very quickly flowing water Armstrong and Boatman (1967) were able to detect oxygen at a depth of 16-18 cm whereas in normal fen soil it was only present down to 6 cm (measured by

Fig. 219. Alder swamp peat contains many more living bacteria and is more quickly broken down than the organic deposits in Birch and Pine swamp woods. After Haber (1965), somewhat modified. The number of bacteria may be understood as a relative measure of their activity.

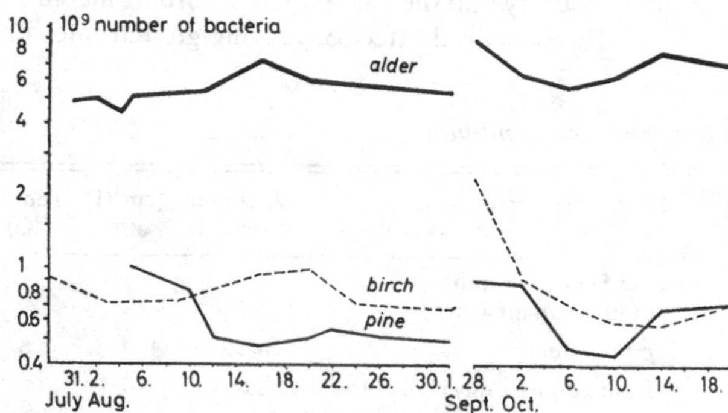

Fig. 220. The relative abundance of earthworms in soil samples from different woodland communities rich in Alder compared with others in Bavaria. After Ronde (1951), from Ellenberg (1963), modified.

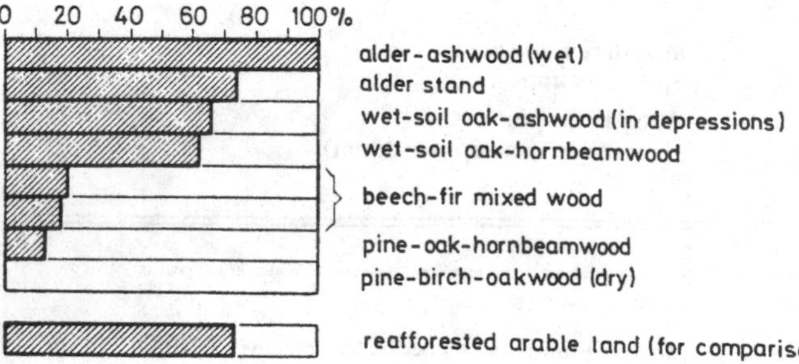

Redox potential). The growth of Purple Moorgrass (*Molinia caeruliea)* varied correspondingly. Thick stands of grass-like or herbaceous helophytes had carried oxygen down into the topsoil through the aerenchyma in their rhizomes and roots (see section A I 3 c) bringing about for example the oxidation of divalent iron, and the appearance of a red-brown colour around these organs.

The ecology of fen woods, sedge fens and other swamp communities has been given new aspects by the researches of Janiesch (1981) mainly concerning the physiological behaviour of sedges and other herbaceous plants. *Carex* species of more or less waterlogged habitats (e.g. *C. pseudocyperus*, *elata* and *acutiformis*) differ fundamentally from those on periodically moist or dry sites (e.g. *C. remota* and *sylvatica*) with respect to their demands on the nitrogen and iron supply.

In contrast to the second group, the first one can use nitrate only in very limited amounts because the nitrate-reducing enzyme is almost totally lacking in their roots and leaves. This behaviour corresponds to the well known fact that waterlogged soils are oxygen deficient and can only offer ammonia to the plant roots (Blume, Friedrick, Neumann and Schwiebert 1975, Ellenberg 1977).

Species such as *Carex elata* are highly resistant to divalent iron and prefer soils rich in this, whereas it is toxic to others. They regulate their iron uptake by oxidising FeII to FeIII in their rhizosphere by excreting O_2, which is delivered through the aerenchymatic system of the whole plant. On the other hand *Carex sylvatica* and corresponding species are almost unable to do this. Therefore they need oxygen in the close surroundings of their roots, coming from outside, that is normally through a good soil aeration. As we have already seen the site conditions cover the needs of the plants living there in this context also.

C

Other near-natural formations

I The vegetation of fresh water, its banks and springs

1 *Successions starting from open still water*

(a) Types of aquatic ecosystems based on nutrient supply

We have already touched upon the plant communities of standing and flowing water when discussing the fen woods (section B V 2 a) and during our survey of the flood plains (section B V 1 a). Now we shall look at these a little more closely although we do not want to dwell too long on them because, since the classic work of Thienemann, Naumann, Ruttner and others, the study of limnology has developed into a science of its own. The number of transactions dealing with the subject has become almost too large to survey. Important publications are listed in the text-books of Thienemann (1925, 1956), Gessner (1956), Ruttner (1962), Elster (1963, 1974), Schwerdtfeger (1975), Colterman (1975), Odum (1980, see also J. Overbeck 1972), Schwoerbel (1980) and others, so it is unnecessary for us to give further references.

Here we shall concentrate on the phanerogams and the larger green cryptogams, that is on the macrophytes which grow close to the banks. These have been dismissed with little study by most limnologists who have paid much more attention to the phyto- and zooplankton and to other heterotrophic organisms which play such important roles in the food webs of the aquatic environments.

One of the main reasons why lakes, more than any other ecosystems, have stimulated the study of their structure and function lies in the fact that they are relatively sharply delimited (Ellenberg 1973b). The food chain can be followed in an unbroken sequence from the microscopic primary producers to the large carnivorous fish. Two facts stand out from the results of such researches which are important for the understanding of the nutrient turnover: Firstly any organic excreta or remains of plankton are mineralised by bacteria for the most part before they sink to the bottom of the lake (J. Overbeck 1970). Secondly the green macrophytes, i.e. higher plants and large algae, are very seldom eaten when living (Thienemann 1956). Water animals get their nourishment from them mostly after the death of these plants. Even snails do not graze on the stems and leaves of submerged macrophytes but on the epiphytic algae covering them. The leaves of water-lilies floating on the surface or the robust haulm of the reeds seldom show any trace of having been eaten before they die down in the autumn adding a new supply of organic matter to the layers already present.

This cycle is disturbed only by man's activity, either regularly or just occasionally. For instance when he cuts the Reed in early summer as well as in winter, or allows his cattle access to the banks and the shallow water round the edges of the lake. The effect of grazing down and treading the young shoots in the spring is to confine the green belt of Reeds to the deeper water out of reach of the animals, leaving the shallower margin nearly devoid of vascular plants. This naturally slows down the rate at which the ground is built up by the accumulation of organic matter. On the other hand man can increase the amount of inorganic nutrients and also the organic matter in streams, rivers and lakes through pouring sewage into them or by fertilising the adjacent fields from which drainage water runs into the water courses. As a result of this the amount of growth in the water is far higher than it would be under natural conditions. In recent years this 'eutrophication' has accelerated very significantly, contributing to the actual environmental crisis (fig. 222).

Each new lake, pond, ditch or similar piece of water however also develops naturally more or less quickly from a nutrient-deficient habitat to one which is rich in nutrients, that is from an oligotrophic to a **eutrophic** type (tab. 48). Mineral nutrients, especially nitrogen and phosphorus compounds on which the

Fig. 221. A succession of plant communities contributes to the filling up of a eutrophic lake (an old course of the river Reuss) in central Switzerland. The dominant plant with floating leaves is *Nuphar lutea*; in some places the leaves protrude above the surface. Of the Reed-bed plants *Schoenoplectus lacustris* grows in deepest water. In the foreground are blankets of algae, *Hydrocharis morsus-ranae* and *Glyceria maxima* which indicate the inflow of additional plant nutrients (eutrophication). At the far lakeside is a belt of *Phragmites* and *Alnus glutinosa*.

Fig. 222. The most significant factor in the rapidly increasing enrichment of our inland waters (socalled eutrophication) is phosphate. The origin of this additional phosphate has been determined for Lake Constance (upper part) from 1930 to 1974. After Wagner (1976) from Elster (1977).

Primary production in aquatic ecosystems is limited in the first place by the amount of available phosphorus. Since about 1950 this scarce nutrient has been added to many lakes in rapidly increasing amounts so that formerly oligotrophic lakes have become increasingly quickly eutrophied. Phosphate-containing washing powders have played the largest part in this.

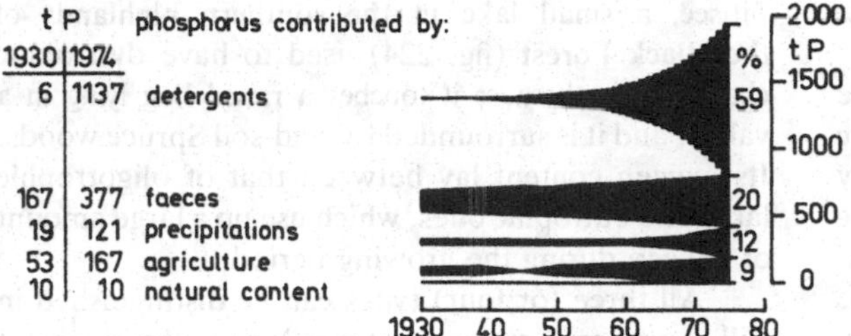

species composition and productivity of the vegetation depends, are gradually obtained by runoff water from the surrounding countryside. Plants growing in oligotrophic stretches of water do not produce much organic matter, while those in eutrophic sites produce so much each year that the margins of a lake or pond visibly become built up with it.

Waters rich in lime as well as those deficient in it can start out as oligotrophic ecosystems. Naumann (1927) and others united them into a single type, but in tab. 48 they have been separated since they show significant differences both as habitats for plants and as parts of the countryside. The **lime-rich oligotrophic** lakes always remain poor in available phosphate even though this mineral is being supplied continuously. It is locked up in the form of tricalcium phosphate and as such can only be absorbed by plants in minimum amounts (Gessner 1939). Lime-rich and **lime-deficient**

Tab. 48. The general characteristics of the main types of inland still waters.

Type	Water colour	Clarity	pH	N	mg/l P_2O_5	Soil type	Type of bank	Distribution in Central Europe
1. **oligotrophic-lime rich**[a]	blue to greenish	very clear	>7.5	tr.	0	calcareous gyttia or chalk propedon[c]	mostly steep	limestone mountains and their foothills
2. **eutrophic**	dirty grey to blue-green	more or less turbid	≥7	>1	>0.5	sapropel, gyttia near banks rapid land formation	shallow banks	moraine and loess areas also numerous in other places[b]
3. **oligotrophic-lime deficient**[a]	greenish to brownish	clear	<7 >4.5		traces	gyttia or propedon[c]	mostly steep	silicate mountains, lime-deficient sandy regions[d]
4. **dystrophic**	yellowish to deep brown	very turbid	<5		0<0.5	dygyttia or dy (peat mud) quaking bogs	shallow	raised bog and acid humus regions[e]

Note: [a] The oligotrophic type has been separated here into one with a high lime content (1) and one poor in lime (3) since these subtypes are very different in their plant ecology. All types are connected by intermediates, especially 1 with 2, 3 with 2 (mesotrophic) and 3 with 4.
[b] More and more eutrophic waters are also being formed from oligotrophic and dystrophic ones by the drainage into them of sewage and manures.
[c] i.e. bare underwater soil, poor in organic matter (Franz 1960).
[d] Large lakes with a copious water flow also in the moraine areas, e.g. in the foreland below high mountains.
[e] Rare in Central Europe and only found in pure form in shallow pools turf diggings in raised bogs. Intermediates towards the acid oligotrophic type however are frequent and even occur in the subalpine belt.

Source: After Naumann and others.

oligotrophic ecosystems differ as habitats not only for the higher plants but also for the plankton and the ground algae. This is obvious from the numbers of species which are arranged according to pH value of the medium in fig. 223. Lime-rich oligotrophic lakes certainly have a low primary production, but a large number of algal species are concerned with it.

The trophic level of a water body can also be defined in terms of the quantity and kind of organic matter present (Aberg and Rodhe 1942, quoted by Elster 1963). This is little in the case of oligotrophic waters, and often much more in eutrophic ones. Eutrophism usually arises when there are large amounts of soluble nutrients coming in, but can also result from the input of organic matter broken down in the water. The larger the primary production or the inflow of organic matter the more heterotrophic microorganisms and animals are able to live. Since these all require oxygen the trophic grade of a lake can also be deduced from its 'oxygen profile' (fig. 224).

In limnology a third main type of aquatic ecosystem is recognised; this is known as **dystrophic**. Its water is coloured brown by acid humus, which however is not produced in situ, but washed out of the raw humus layer in the surrounding woods or heaths, or from the peat of a raised bog (section C III). Such strongly acid peaty waters are particularly widespread in northern Scandinavia, but are rarely found in Central Europe, and then mostly as secondary formations, e.g. as peat cuttings in a raised bog. The Titisee, a small lake in the southern highlands of the Black Forest (fig. 224) used to have dystrophic characteristics because it touches a raised bog lying in a valley, and it is surrounded by acid-soil Spruce woods. Its oxygen content lay between that of oligotrophic lakes and eutrophic ones, which use up a large amount of oxygen during the growing period.

All three (or four) types can be distinguished in still as well as in moving water as long as these are not salty. Rather than describe them at length their different characteristics have been compared in tab. 48. Naturally there are intermediate types. For instance in the pre-Alps many watercourses and lakes are lime-rich **mesotrophic**, i.e. clear and base-rich with average nutrient content. In the old diluvial area of north-western Germany lime-deficient mesotrophic or dystrophic-oligotrophic shallow lakes, ponds, pools and streams can be found alongside the typical oligotrophic ones. In the area of the moder Beech woods, Birch-Oak woods or acid-humus coniferous

Fig. 223. The lakes round Bremen may be arranged in a series from alkaline to very acid. The number of species in most of the algal groups declines as the pH value of the water falls. This is particularly true of the diatoms in which the number of species gets steadily smaller from the Zwischenahner Meer to the Bullensee. After Behre (1956), modified.

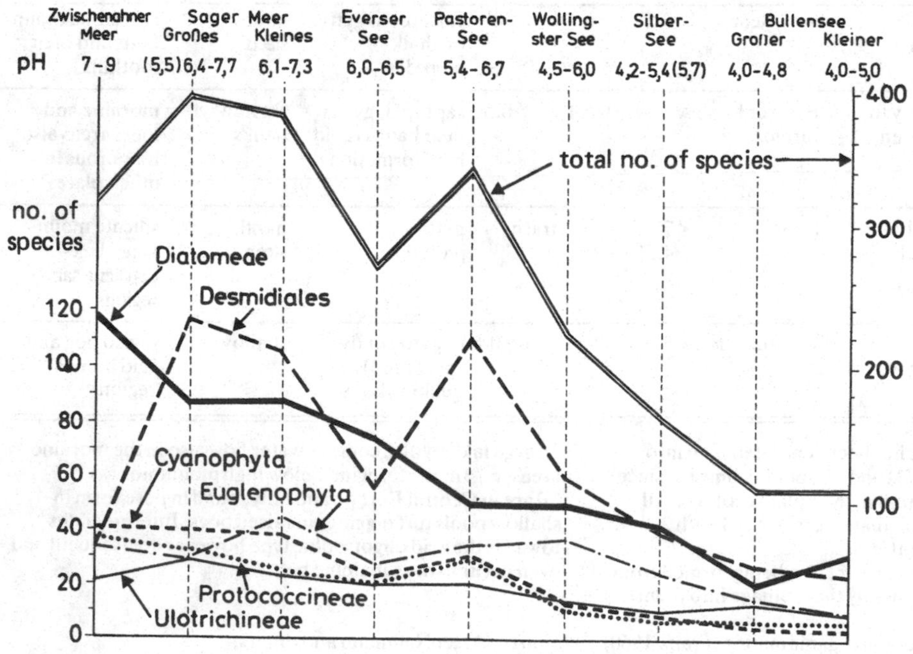

woodlands the subsoil generally has sufficient minerals to maintain the streams at least in an oligotrophic state; they then do not show dystrophic characteristics.

The types described in tab. 48 are only found in inland waters where the salt content does not exceed the concentration which can be tolerated by fresh water plants. **Brackish and salt waters** can also be oligotrophic, mesotrophic or eutrophic. However they are never markedly lime-deficient or rich in acid humus. We shall mention the brackish waters in connection with freshwater Reed swamps (section C I 2 a); but they will be discussed above all as features of the sea coast (C IV 1 and 2). One thing which distinguishes the salt water ecosystems from those of the freshwater is the presence, in the latter, of numerous blue-green algae able to fix atmospheric nitrogen. These rarely appear in seawater (Nees, R.C. Dugdale, Goering and V.A. Dugdale 1963).

In contrast to phosphorus then nitrogen is rarely deficient in our inland waters (Feth 1966 and T. Overbeck 1972). According to El-Ayouty (1966) N-fixation by Cyanophyceae is especially vigorous on submerged soils and along the water's edge.

As Lohhammar (1938), Lumiala (1945) and others have already argued Naumann's typology, which was first proposed for open water, can only be applied with reservations to the marginal regions in which we are primarily interested. Because of changing ground conditions and inflow, as well as uneven water exchange, different parts of the same stretch of water can have different trophic grades. So it cannot be expected that the distribution of particular units of vegetation will always correspond with that of the limnological types since the life of the vascular plants is dependent as much on changes in level and on lateral movement of the water as on the mineral supply. In spite of this the plant ecologist should bear in mind the trophic grade of the habitat he is studying and its relationship with the whole body of water.

Fig. 224. Profiles of the oxygen content of the water in lakes of different trophic levels at the beginning (April/May) and end (August to October) of the growing season before 1950, i.e. before the onset of the period of rapid eutrophication caused by human activities. After Elster (1963), modified.

In the relatively shallow eutrophic lakes the plentiful supply of oxygen present in the spring is largely used up during the stagnation stage in summer, especially near the muddy bottom where a lot of animals and bacteria live.
On the other hand the clear oligotrophic lake remains rich in oxygen to a considerable depth since the amount produced by green algae during photosynthesis outweighs that used by heterotrophic organisms.
The dystrophic lake contains so much acid humus which has been washed into it from the surrounding mires or swamp woods that even in spring the breakdown of this organic matter uses up a considerable amount of the oxygen present. However because of the unfavourable living conditions the numbers of animals and micro-organisms are relatively small so that, at the end of the summer the dystrophic lake is richer in oxygen than the eutrophic one.
Today all the lakes chosen here as examples are richer in nutrients than they were in 1935 or 1949.

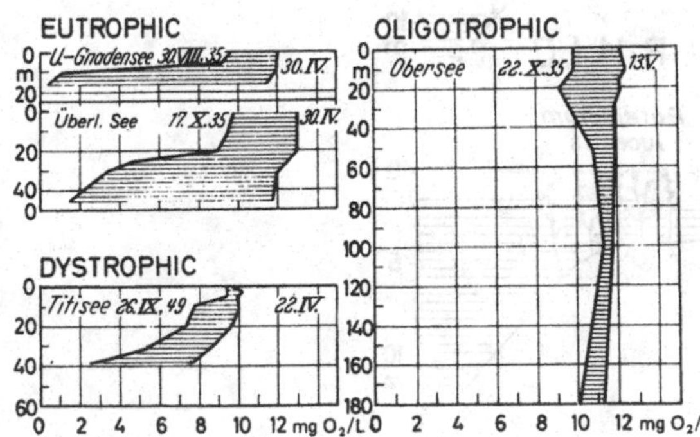

(b) Sequences of vegetation in different types of still water

Higher plants and green cryptogams are not able to live in either fresh or salt water if they do not have enough light for photosynthesis. In eutrophic and dystrophic water the light intensity often falls below the minimum at a depth of less than 2 m. In oligotrophic water too it diminishes with increasing depth and is completely lacking below a depth of about 50 m.

Of the plants in contact with the soil or those actually rooting in it the ones in deepest water are those which are able to live always completely submerged, such as Characeae and mosses in lime-rich clear lakes or brown and red algae in the sea. Some phanerogams too may be found a number of metres below the surface, for example species of the genus *Zostera* in salt water (section IV 1 c), *Zannichellia* and *Ruppia* in brackish water and *Potamogeton* in fresh water. The majority of higher plants though appear for a time on or above the surface, even if this is only during the flowering period. They become however only a feature of the landscape if their stems and leaves persist in the air for some time (fig. 221). Partly in accordance with Heijný (1960) and others we can distinguish the following life forms in relation to habitat:

I **Free floating** water plants (pleustophytes)
  1 Below the surface:
    at most sending their generative organs above the surface (e.g. species of green algae, and *Utricularia*, see figs. 228, 232).
  2 2. On the surface:
    the leaves have a direct gaseous exchange with the air at least for part of the time (e.g. *Hydrocharis*, *Salvinia* and most species of *Lemna*, fig. 232).

II **Rooted water plants** (true hydrophytes)
  1 Assimilating organs completely submerged (e.g. *Elodea*, *Zanichellia* and many species of *Potamogeton*, fig. 225).

2 Partly submerged, partly above water: Their leaves may be floating on the surface or constantly submerged, but generally can put up with either condition, alternating between the two (e.g. species of *Nuphar* and *Nymphaea*, fig. 221).

III **Lakeside plants** (littoral helophytes)

1 Can carry on photosynthesis while submerged: a small group of reed swamp plants which can grow a long way out into the water (e.g. *Schoenoplectus lacustris*, *Equisetum fluviatile*, figs. 233, 256).

2 Can only assimilate above the water surface: their leaves die if submerged for long (the majority of reed plants, e.g. *Phragmites*).

IV **Swamp plants** (true helophytes): for some weeks or months their root run is not waterlogged but from time to time they have to tolerate long periods of flooding; they are really to be considered as land plants.

1 Amphibious plants which can still go on assimilating under water: a rare form only exceptionally met with (e.g. *Polygonum amphibium*, fig. 479).

2 Can only assimilate in air: most plants of the sedge fen (e.g. *Carex elata*, *C. gracilis*, figs. 234, 237).

Glück (1934 and later) has already demonstrated by observations of disturbed habitats and by culture

Fig. 225. Vegetation zonation along the banks of Lake Constance at different trophic levels (semi-schematic). High water occurs in early summer, i.e. following the melting of the Alpine snow. This large and deep lake was previously more or less oligotrophic (see figs. 222 and 224, Über linger See). After Lang (1967b), somewhat modified.

Above: Gravelly places which have still remained **oligotrophic** have no Reed belt. The mean high-water mark carries a creeping sward of Bent or Stunted Canary-grass. The rare Shore Hairgrass community (see fig. 227) and the Shore Weed sward (*Littorello-Eleocharidetum*) are only covered with water intermittently. The communities of water plants too are poorly grown.

Middle: The greater part of the shoreline today has **mesotrophic** conditions with beds of Stonewort (*Charetum asperae*). In places these are suppressed by Nymphweed communities (*Najadetum*) or, in deeper water, Pondweed ('*Potametum*' which is a short form of '*Potamogetonetum*'). A broad belt of Reeds along the shore is characteristic, with outposts of Club Rush (*Schoenoplectus lacustris*). The *Phragmitetum* stands dry in the autumn and does not form any peat. The adjoining Tufted Sedge community (*Caricetum elatae*) is also adapted to the changing water level.

Below: Where **eutrophication** has taken place through the introduction of sewage the Reed is suppressed in places by Reed Sweet-grass (*Glyceria maxima*) or Lesser Bulrush (*Typha angustifolia*). The area of shallow water becomes overgrown by a Horned Pondweed community and the light-loving Stonewort disappears entirely.

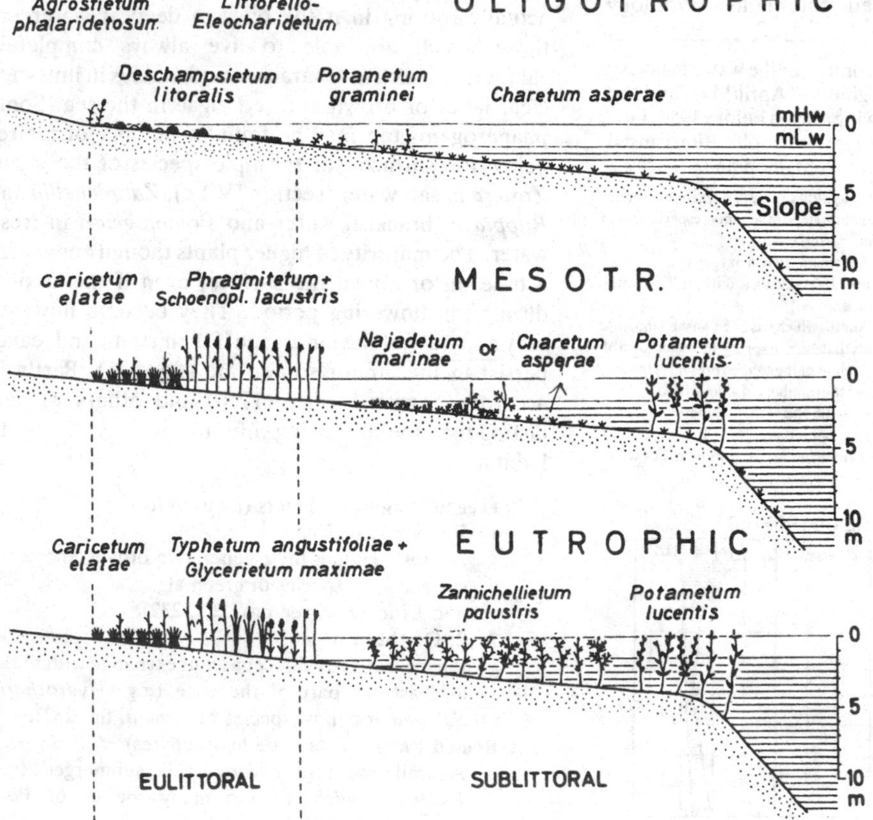

experiments that the majority of species are fairly plastic with regard to whether they can live by or in the water, and could be assigned to a number of the above groups. However under the pressure of competition they are generally confined to one group. In any case the variations of the water table play an important part in determining the species composition of macrophytic water plant communities (Hejný 1962).

Since the different life forms venture to different distances out or down into the water they usually become arranged in belts along the lake margin. For example in more or less **eutrophic freshwater** (figs. 221, 225, see also section E III 1, No. 1) there could be:

a) Subaqueous Stonewort swards (*Charion asperae*),
b) Subaqueous Pondweed meadows (*Potamogetonion*),
c) Floating-leaved communities (*Nymphaeion*),
d) Coverings of free-floating Duckweeds (*Lemnion*), only found here and there between c and e on calm water,
e) Reeds (*Phragmition*),
f) Sedge fens (*Magnocaricion*),
g) Alder fen (*Alnion glutinosae*) as the last stage in the peat formation process.

In **lime-rich oligotrophic** water there is a similar though poorer sequence. Free-floating plants are usually absent, but on the other hand the submerged ground plants, especially subaqueous Stonewort swards are well grown and develop over large areas in the clear water.

Lime-deficient oligotrophic still waters reduce the floating plant cover to practically nothing. The rooted plants with floating leaves and the reed-swamp communities too are poorly developed. On their clear humus-deficient beds there are extensive growths of grasslike phanerogams giving rise to the following type of zonation (figs. 225-227):

a) Subaqueous mosses and algae (may be absent),
b) Submerged low grass-like communities (*Isoëtion* and others),
c) Stunted floating-leaved communities (*Nymphaeion*, mostly absent),
d) Gappy Reed stands (*Phragmition*, generally with a and b mixed in with it, fig. 228),
e) Sedge fens (poor *Magnocaricion* or other sedge communities, e.g. *Caricetum lasiocarpae*)
f) Poor fen Alder, Birch, or Pine woods as final stage (no raised bog!).

Quite different plant communities appear in the **dystrophic** waters (fig. 229):

a) Submerged peat moss communities (*Sphagno-Utricularion*)
b) Peat moss hollow communities (*Rhynchosporion albae* or other)
c) Peat moss hummocks (*Sphagnion fusci* or other)
d) Raised bog.

Because of their close association with raised bogs we shall not discuss these until section C III.

Fig. 226. The texture and nutrient content of the soils where some of the communities shown in fig. 225 are growing. Data taken from Lang (1967b)

Left: spectrum of soil grain size; centre: total nitrogen content; right: total phosphorus content as a percentage of soil dry matter.
In each case the right end of the black bar shows the minimum value and the right end of the hatched bar the maximum value found in from 2 to 6 samples (except in the case of gravel, when the diagram is reversed). Most of the soils are composed mainly of sand or silt, and their clay content is so small that it could not be shown. The communities of the oligotrophic shoreline (1 and 2) root in a gravel soil which is extremely poor in N and P. All the other soils are richer in finer particles and in nutrients. The *Phragmitetum* has a broad amplitude; it appears in some places where the ground is quite poor, but also where the fertility is much higher. The best soil conditions are found in connection with the eutrophic vegetation (6 and 8). The soils of the *Najadetum marinae* (5) are noticeably rich in nitrogen.
The C content (not shown here) follows roughly the N content; the Ca content is high for all habitats.

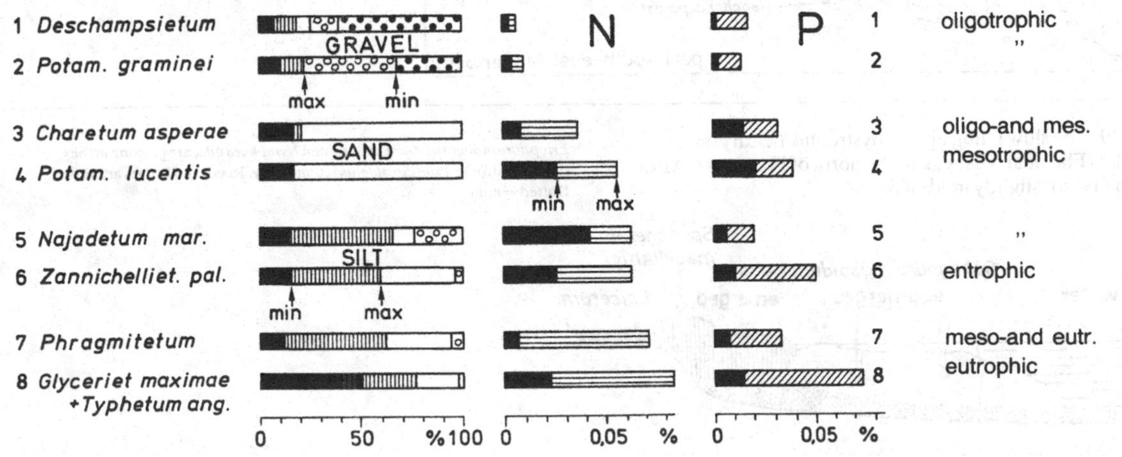

Fig. 227. Fluctuations in the water level of Lake Constance during a dry (1949) and a wet year (1936) and in an average (1937-60), along with the profile of vegetation on an oligotrophic shoreline near Konstanz-Eichorn (see also fig. 225). After Lang (1967b), slightly modified.

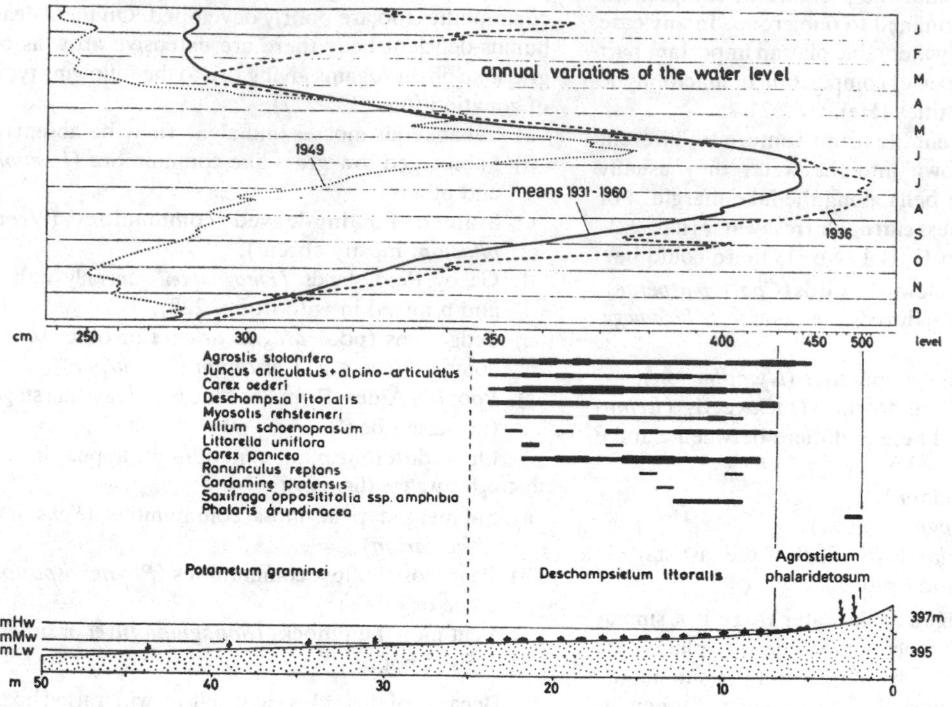

Fig. 228. Vegetation zonation of an acid pond amongst heath dunes on Sylt. The Heather humus from the surrounding *Ericetum tetralicis* makes the water somewhat dystrophic. After Jeschke (1962), modified.

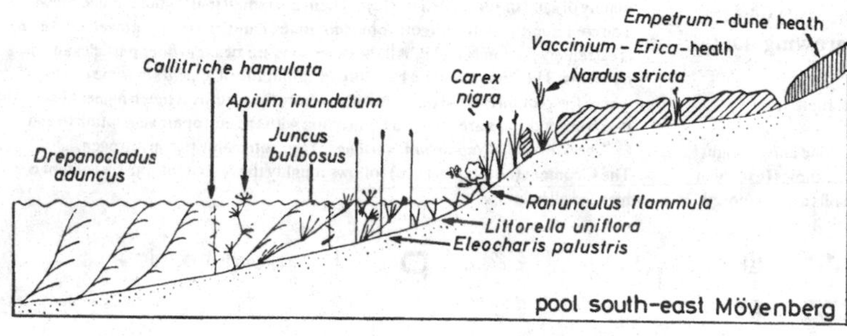

Fig. 229. The slow filling up of a dystrophic heath pond (Blankes Flat near Vesbeck to the north of Hanover). After Tüxen (1958), slightly modified.

Eriophorum angustifolium and stunted *Nymphaea alba* are growing in the submerged *Sphagnum cuspidatum* community. Reeds and Sedges are absent. Dotted = mud.

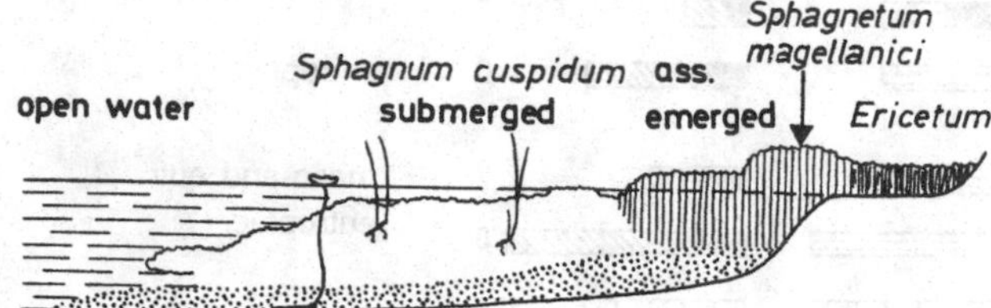

The different girdles of communities which have so far been pointed out only in general terms are partly a zonation and partly a succession. In the oligotrophic and even in the mesotrophic still waters they remain more or less stable, provided the water levels or the properties of the water do not change fundamentally (F. Runge 1969a). In dystrophic bogland depressions too they may persist for decades or even centuries with no significant alteration (fig. 265). Only in eutrophic waters are they always an expression of a more or less rapid silting up or peat formation brought about by the plant communities.

(c) Subaqueous meadows of oligotrophic and eutrophic lakes

The majority of water plant communities consist of relatively few species. They live under extreme conditions where not many macrophytes are able to grow. These few however have the opportunity to spread over large areas since they meet hardly any competition at first. It is often just a matter of chance which species of a particular life form happens to get to a place first. This species determines the dominance for a long time since it is difficult for others to establish a footing. There is thus a mosaic of facies in water plant communities for which no particular reason can be seen in the habitat. The list of accompanying species too varies from one place to another so much, and only a few of them are worth noting as appearing regularly (see e.g. tab. 49). According to Tüxen, the fact that aquatic communities are so seldom completely developed is the main reason for the absence of any classification which is universally recognised to-day.

This also applies to the subaqueous communities of shallow oligotrophic lakes which have managed to retain many weakly competitive relict plants. The main distribution centre for these is in the boreal-subatlantic region, and they have been studied in southern Sweden in particular by Lillieroth (1950), Malmer (1960-62), and others, and in Denmark e.g. by Iversen (1929). Following W. Koch (1926) they have been put together in the order of Shoreweed communities (*Littorelletalia*) for which according to the classification of Oberdorfer (1977, see also Dierssen 1975 and section E III 1, No. 1.3) the following species are characteristic:

Deschampsia setacea	*Littorella uniflora*
Elatine hexandra	*Ranunculus oblongus*
Juncus bulbosus	*R. polygonifolius*
	Veronica scutellata

They occupy clear acid water which is poor in nutrients, not likely to be deeply frozen and where the littoral zone is covered with water the whole time, or at least in summer, and has a sandy or gravelly soil. In Central Europe such habitats are found above all in the sandy region of the north-west. It is not pure chance that these oceanic communities have been studied most completely in west Europe (by Scoof-van Pelt 1973) where many of them are amphiphytic. The **Quillwort-Lobelia pond** (*Isoëto-Lobelietum*) as conceived by Tüxen (1937) is not infrequently met with in the Birch-Oak wood area of north Germany, the Netherlands and Denmark, but also in northern Poland and in the Lausitz. Its character species are *Lobelia dortmanna*, *Isoëtes lacustris* and *I. echinospora. Litorellion* communities gradually die out towards the Alps and south-east Europe.

The reason that the Quillwort communities do not grow deeper than about 1.2-1.7 m below the water table, in spite of the water being very clear, has not yet been investigated. Although *Isoëtes lacustris*, *Littorella uniflora* and *Lobelia dortmanna* naturally live in acid waters they are not damaged by water with an alkaline reaction and can be cultivated for months in lime-rich water from a eutrophic lake (Roll 1939). Their absence from the shores of nutrient-rich lakes is doubtless due to removal of light by the more strongly growing plants such as those with floating leaves and the members of the Reed swamp communities. In addition they probably cannot tolerate for long the epiphytic algae which soon cover all submerged organs under eutrophic conditions.

The bottom of unpolluted lime-rich lakes, gravel pits and the like is often covered with dense **carpets of chandelier algae** (*Chara* and *Nitella* species). The inclination of these plants to 'keep themselves to themselves' was for Krause (1969) one of the bases for his hierarchy of the Characeae communities (see section E III 1, No. 1.18). It is worth noting how closely these are linked with the ecological divisions of the upper Rhine plain. K. Behre (1966) has discussed in depth the sociological behaviour of other large algae species in inland waters, especially in the litoral zone.

(d) Communities of rooted plants with floating leaves

When the floor of the lake has been raised by the formation of gyttia, or if the water is originally shallow enough, then plants with floating leaves are able to come in. The more these expand the more they suppress the plants growing on the floor, not only because they reduce the amount of light getting to them, but also through the continual deposition of dead remains which tend to smother them. So

Tab. 49. Rooted and free-floating waterplant communities in eastern lower Saxony together with floating Fern communities from southern Central Europe.

Community class:	**Potametea**						**Lemnetea**									
Serial no.:	1	2	3	4	5	6	7	8	9	10	11	12				
Alkalinity[h'] low (up to 1.5)	0.7		1.5				1.5					–	Ecological evaluation			
average		2.4		3.3				1.8			–	–				
					3.6				(3.0)		–	–				
high (over 4.0)						4.6				4.6	–	–				
Trophic level (oligo-, meso-, eutrophic)	o	e	m	e	e	e	m	e	e	e	e	e	T	R	N	
Myriophyllum alterniflorum	*4*												×	3	3	P-poor
N *Nuphar lutea*	**5**	**5**											×	6	×	
P *Potamogeton obtusifolius*	*5*	*2*	*2*										5	6	4	P-poor
P *P. alpinus*	*2*	*1*	*1*	*1*									–	–	–	P-poor?
N *Hottonia palustris*	*4*	*1*	**5**	*3*									6	5	4	P-poor
H *Stratiotes aloides*	*3*	*1*	*1*	*3*							*1*		7	7	6	P-poor
N *Utricularia australis*	*4*	*1*	*1*	*1*			*2*						6	5	4	P-poor
L *Riccia fluitans* (M)			*1*	*3*			**5**						–	–	–	P-poor
N *Myriophyllum verticillatum*		*1*	*4*	*3*				*3*					6	6	7	
Callitriche hamulata		*1*	*2*	*1*					*4*				–	–	–	
H *Hydrocharis morus-ranae*	*4*	*2*	*5*	**5**		*2*	*4*	*5*			*3*	*3*	6	6	5	
L *Lemna trisulca*	*5*	*3*	*5*	**4**	*1*		**4**	**5**		*2*	*3*	*2*	×	7	6	
Po *Ranunculus peltatus*	*5*	*2*	*4*	*2*	*5*								–	–	–	
N *Potamogeton natans*	*4*	*5*	*4*	*2*	*2*	*2*							4	7	6	
P *P. crispus*		*4*	*2*		*5*	*5*							5	7	6	
Po *Elodea canadensis*	*5*	*4*	*5*	*4*	*4*	*2*		*5*		*2*			6	×	7	
P *Callitriche platycarpa* f. *natans*	*5*	*3*	*3*	*2*	*4*	*3*	*2*	*5*	*5*	*2*			–	–	–	
L *Spirodela polyrrhiza*	*5*	*5*	*5*	*5*	*5*	*5*	*5*	**5**		*5*	*3*	*2*	6	×	7	
L *Lemna minor*	*5*	*5*	*5*	*5*	*5*	*5*	*5*	*5*	**5**	*5*	*2*	*1*	×	×	×	
Po *Potamogeton friesii*	*2*	*2*	*1*		*5*	*5*				*4*			×	6	6	P-rich?
Certophyllum demersum	*5*	*4*		*3*		**5**			*5*				7	8	8	P-rich
L *Lemna gibba*		*3*		*3*	*5*	*5*				**5**			6	7	8	P-rich
Po *Myriophyllum spicatum*		*2*			*2*	*3*		*3*					×	8	×	
P *Potamogeton pectinatus*		*1*			*4*	*2*				*2*			×	7	7	
P *P. pusillus*		*1*			*2*	*3*				*1*			5	7	8	P-rich
H *Utricularia vulgaris*		*1*											×	6	6	
N *Nymphaea alba*		*1*											×	7	7	
N *Polygonum amphibium* f. *natans*		*1*									*1*	*1*	×	×	7	
P *Potamogeton lucens*		*1*										*1*	×	7	8	P-rich?
Glyceria fluitans					*5*	*5*							×	×	7	
Sium latifolium					*2*	*2*							×	7	8	
Oenanthe aquatica					*1*	*1*							6	7	6	
Butomus umbellatus					*2*	*3*							×	×	8	P-rich
Alisma plantago					*1*	*2*							×	×	8	
L *Ricciocarpos natans* (M)							*2*						–	–	–	P-poor
Po *Montia fontana*									*2*				4	5	4	
Nasturtium officinale									*2*				4	7	7	
Zannichellia palustris									*2*				×	7	6	
L *Salvinia natans*											**4**	*2*	8	7	7	P-rich?
L *Azolla filiculoides*												*5*	9	×	8	P-rich

Source: From tables by Weber-Oldecop (1969) and Th. Müller and Görs (1960).

nutrient-rich waters are recognisable by a particularly well developed belt of floating leaves while in lime-deficient oligotrophic waters the inconspicuous Quillwort meadows grow right up to the water's edge.

The associations of plants with floating leaves are grouped together in the *Nymphaeion*. They are among the best known of the aquatic communities, and still widespread in Central Europe even though many lakes and ponds have been drained. Therefore we shall study their species combinations in more detail.

A first glance at table 49 shows how poor in species are all the floating-leaved communities. Only those which have been considered as being characteristic are fairly well represented while most of the ones listed after them are rare.

The commonest floating-leaved association in Central Europe is the **Water Milfoil-Yellow Waterlily community** (*Myriophyllo-Nupharetum*) and this is moreover the richest in species (tab. 49, no.9; see also fig. 221). Besides the Yellow Waterlily it also contains the larger flowered White Waterlily (*Nymphaea*) which supplants the former in shallow water, but cannot grow down to 3-4 or even 5 metres as can *Nuphar* with its underwater leaves. For the same reason the latter is able to cope better with a fluctuating water level or slowly moving water provided there are

---

Tab. 49. (Continued)

Note: [a] (Compare also section E III 1, 1.1 and 1.2)
L = species of the Duckweed covers (*Lemnetea, Lemna* communities)
H = species of the Floating Frogbit covers (*Lemnetea, Hydrocharis* communities).
P = species of the Pondweed communities (*Potamogetonion*, Po = *Potamogetonetalia*)
N = species of the Water Lily communities (*Nymphaeion, Potamogetonetalia*)

[b] The ability of the water to neutralise acid determined by titration of 100 ml water with 0.1 N HCl. (Mean value for the amplitudes, which were shown graphically for each community by Weber-Oldecop 1969).

T = indicator value for temperature, R = for reaction, N = for nitrogen according to Ellenberg (1974), explained in sections B II 4 and E III. For some species the ecological behaviour is not yet sufficiently well known.

Communities (nos. 1-10 in eastern Lower Saxony) ecological evaluation:	mT	mR	mN
Nos. 1-2: **Waterlily communities** (*Myriophyllo-Nupharetum*; alliance *Nymphaeion*) mainly bottom rooting:			
1: Broadleaved Pondweed-Waterlily comm. (*Potamogeton obtusifolius-Nuphar* ass.) nutrient poor	5.9	6.0	5.5
2: Typical Milfoil-Waterlily comm. (*Myriophyllo-Nupharetum*) relatively nutrient rich	5.7	6.8	6.5
Nos. 3-6: **Floating plant communities** (*Hydrocharis* comm. etc.); some are bottom rooting:			
3: Water Violet comm. (*Hottonietum*); shallow water, fairly poor in nutrients, partly shaded	5.7	6.2	6.0
4: Floating-Frogbit cover (*Hydrocharitetum*)	6.1	6.5	6.5
5: Still Water Crowfoot comm. (*Ranunculetum peltati*)	5.5	6.9	7.0
6: Hornwort comm. (*Ceratophylletum demersi*); very polluted shallow water	5.5	7.0	7.1
Nos. 7-10: **Duckweed covers** and other communities of free-floating plants (*Lemnion*):			
7: Floating liverwort comm. (*Riccietum fluitantis*)	6.0	6.2	5.8
8: Floating Duckweed comm. (*Lemno-Spirodeletum*)	6.0	6.7	6.4
9: Duckweed cover, few species ('*Lemnetum minoris*')	5.7	7.1	6.7
10: Gibbous Duckweed cover (*Lemnetum gibbae*)	5.9	6.7	7.0
Nos. 11-12: **Floating fern covers** (in Baden-Württemberg); decidedly warmth-loving (to the *Lemnion*):			
11: Duckweed-Swimmingfern cover (*Spirodelo-Salvinietum*)	6.8	7.6	6.3
12: Duckweed-Algal Fern-cover (*Lemno-Azolletum*)	7.6	7.1	7.1

The average temperature of still water throughout the year is higher than that of the surrounding soil. Because of this more or less thermophilous plants can live in it. In any case they have a higher warmth requirement than those in the surrounding submontane and planar Beech woods. Even in eastern Lower Saxony they lift the average temperature value to considerably above 5. Relatively warmth-loving communities here are the Frogbit, the floating liverwort and the Duckweed covers (nos.4, 7 and 8). In the warmer parts of south-western Germany the Floating Fern covers thrive and already show submediterranean characteristics (nos. 11 and 12).

By and large the average reaction value of the plant communities rises along with the alkalinity of the water. However there is an even closer correlation of the water vegetation with the average nitrogen value which, with reservations, can serve as an expression of the trophic level. The latter is really decided by the available phosphate content, but this was not measured by Weber-Oldecop. Nevertheless it can be taken as confirmed that communities nos.1, 3 and 7 live in relatively clean water while nos.6 and 10 (and even 12) indicate a high level of pollution. Species which, according to Weber-Oldecop (1969), Wiegleb (verb.) and my own observations, can only tolerate water relatively poor in phosphate are indicated in the last column by 'P-poor' while the pollution indicators are shown as 'P-rich'.

sufficient nutrients available. Alongside either or both Waterlilies can usually be found the Pondweed *Potamogeton natans*, the Common Hornwort (*Ceratophyllum demersum*), and some Milfoil species.

In order to understand the ecological and sociological behaviour of the aquatic plants it is necessary, perhaps more than with any other group of plants, to be aware of the structure and function of their organs. A few examples will serve to illustrate this.

According to Gessner (1951) Waterlilies transpire rather less than most helophytes, but just as much as, or even more than, the majority of land plants. It is interesting to note that the water given off from the upper sides of the leaves is replaced either through hydathodes on the lower surface or through root pressure, not by an active tension as is the case with most plants. A severed Waterlily leaf wilts when held out of the water even when the cut end of the stalk is still under the surface, whereas a leaf still connected to the rhizome remains turgid provided the air is not too dry. Because of this the leaves of Nymphaeaceae which are gradually exposed and surrounded by air when the water level falls are not damaged (fig. 221). By way of contrast many other aquatics, especially those with submerged leaves, very quickly dry out under the same conditions.

The connection between the leaves of the Waterlilies and their rhizomes does not just ensure their water supply and that they are held in place; it also provides a better supply of nutrients. The surface water of almost all lakes, even eutrophic ones, is much poorer in nutrients, especially phosphates, than is the oxygen-deficient upper layer of the ground (Gessner 1956, Burckhardt and Burgsdorf 1962). This peculiarity in the distribution of nutrients may account for the absence of free-floating plants from oligotrophic waters and the fact that they flourish best in polluted water with a superabundance of nutrients.

A system of intercellular air spaces which gives buoyancy to the leaves and also runs through the leaf stalks in most aquatic plants, supplies the roots with oxygen so that these can develop sufficiently well to anchor the plant (Williams and Barber 1961). According to Armstrong (1975) almost all water and swamp plants have a more effective aeration system than land plants. The rootstock of the Waterlily contains so much air that if it is pulled out of the ground it will float, and so could be transported to a new place. However it would only manage to establish itself here if it was thrown up on wet soil and could root firmly before it was flooded again and carried away even further. For this reason vegetative dispersal plays very little part in the case of the Waterlily or of other plants with a similar construction.

In a comparable way the seeds of *Nuphar* and *Nymphaea* remain floating for a long time because of the air contained in their spongy gelatinous bracts; when the bracts rot away the seeds sink down to the bed of the lake (Glück 1934). The seed or fruits of the majority of aquatic plants behave in the same way, e.g. species of *Potamogeton*, *Ranunculus* and *Najas*. Some others though have seeds which are heavier than water right from the outset, e.g. *Hippuris*, *Myriophyllum* and *Trapa*. This may have something to do with the fact that these three genera have a very erratic distribution; where they do occur they are present in large numbers.

Several aquatic plants increase and are dispersed by means of winter buds (so called turions) after the main part of the plant has died, e.g. *Hydrocharis morsus-ranae*, *Myriophyllum verticillatum*, all the species of *Utricularia* and many grass-like Pondweeds (*Potamogeton obtusifolius*, *pusillus*, etc., see fig. 230).

The most effective long-distance dispersal mechanism for the seeds or vegetative parts of water

Fig. 230. The phenological development and life forms of the higher aquatic plants in Lake Constance in relation to the water temperature. After Lang (1967b), somewhat modified. *Zannichellia repens* is a subspecies of *Z. palustris*.

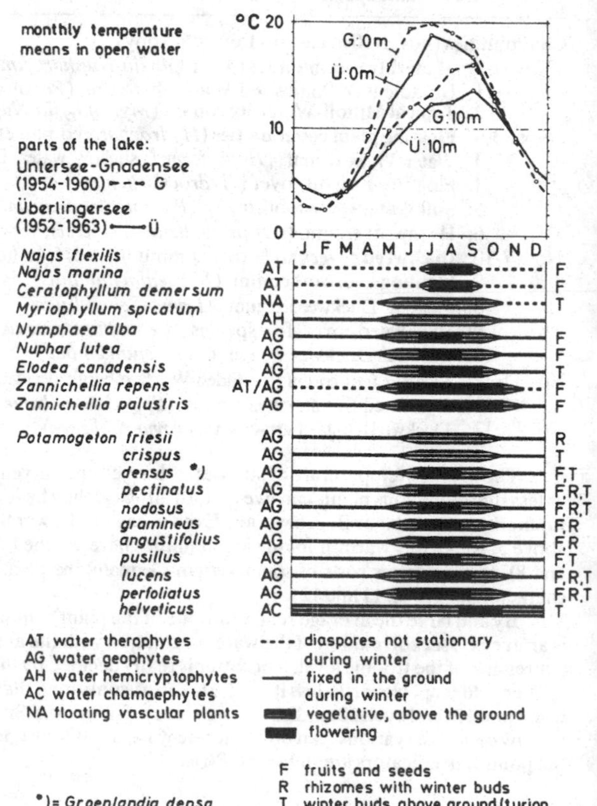

plants is that carried out by water birds such as ducks (tab. 50). These are chiefly responsible for the fact that the plant communities in wet places are much more uniform than those of any other habitat over the whole of Europe, and indeed over much of the world where conditions are similar. In addition they make sure that any new aquatic habitats are quickly invaded and occupied by a complete representative flora.

European relevés of the *Myriophyllo-Nupharetum* which have been collected together by Müller and Görs (1960, see Ellenberg 1963) come from southern France, the area of the Upper Rhine, Upper Swabia, Lower Bavaria, the hill country of western Saxony, the Elbe-Elster region, Holstein, Mecklenburg, Uckermark and Neumark, the Upper Spree Forest, Silesia and the surroundings of Stockholm. The same association has also been recorded by other authors of whom we can name but a few (and this is also true of the communities which follow). The number could easily be increased from the centre of Switzerland and West Germany for example (see tab. 49, section E III 1, No. 1.2).

Detailed descriptions of aquatic plant communities have been made by, amongst others, Hilbig (1971a), Jeschke (1963), Knapp and Stoffers (1962), Lang (1967b, 1968), Philippi (1969a, b), Schrott (1974) Weber-Oldecop (1969), and Wiegleb (1976). As far as the nutrient-rich waters are concerned they agree in general with Müller and Görs. The classification proposed by Den Hartog and Segal (1964) contains many more higher units, but usually leads to the same associations A comparative ecological characterisation has also been elaborated by Pott (1979).

Eutrophic standing waters apart from those in cool climates then provide a very uniform habitat for floating-leaved plants throughout the whole of Central Europe and beyond.

In warmer, mostly rather shallow lakes and oxbow lakes ('Altwasser') one comes across the Water Chestnut (*Trapa natans*). In the Post-Glacial warm period it was distributed much further northwards than it is to-day (Walter and Straka 1970). Its hard barbed nuts were formerly used for food and, like Hazelnuts, were ground down in special wooden mills (Apinis 1940). Experimental and ecological investigations by Apinis showed that the seeds of the common Water Chestnut, *Trapa natans* ssp. *natans* (L) Schinz, germinate at temperatures over 12° C., but only after a resting and maturation period at a lower temperature (1.5-10° C) lasting until December. At temperatures below -8 to -10° C they soon lose their viability. They germinate in the dark and in an oxygen-deficient medium – conditions found in mud – and most readily where the medium is alkaline. However, oddly enough, the young plants are sensitive to lime and cannot tolerate pH values over 7.9 any more than they can those below 4.0. The optimum pH for their development lies between 5 and 7. Conditions which suit *Trapa* then can only be found in a warm temperature climate which is subcontinental, such as on the Balkan peninsula. It is here that the Water Chestnut communities have their main distribution and achieve their greatest diversity today, especially in the lowlands of Hungary. Apinis investigated them as

Tab. 50. Seeds from various plant communities which are eaten and dispersed by ducks.

Water plants	Swamp plants	Plants of waste and cultivated land
Ceratophyllum demersum	*Alopecurus geniculatus*	*Atriplex patula*
Hippuris vulgaris	*Cirisium palustre*	*Bromus sterilis*
Myriophyllum species	*Eleocharis palustris*	*Chenopodium album*
Potamogeton, many species	*Galium palustre*	*Galium aparine*
Ranunculus aquatilis	*Juncus inflexus* and other spp.	*Hordeum distichon* (Barley)
Ruppia species	*Polygonum amphibium*	*Triticum aestivum* (Wheat)
	P. hydropiper, nodosum	and others
Reed bed plants	*Ranunculus repens*	
Bolboschoenus maritimus	*Rumex conglomeratus*	
Glyceria fluitans	*Taraxacum palustre*	Woody plants
Phalaris arundinacea		*Alnus glutinosa*
Phragmites australis	Other sward plants	*Betula* species
Schoenoplectus lacustris	*Carex hirta* and other spp.	*Crataegus monogyna*
Sparganium species	*Holcus lanatus*	*Quercus* species
	Lolium multiflorum	*Rosa* species
Saltmarsh plants	*Medicago lupulina*	*Rubus* species
Armeria maritima	*Phleum pratense*	*Sambucus nigra*
Salicornia species	*Poa trivialis* and other species	and other species
Suaeda maritima		

Source: From observations by Olney (1963-1967) in Gillham (1970), modified.

far as their northern limit in Latvia. In Central Europe some oxbow lakes in the Upper Rhine plain are the most favourable localities, especially in the northern part of this region which is also noted for other examples of continental incursions.

In the cooler montane climate the Yellow Waterlily communities are less well developed since the water is generally poorer in nutrients. The shallow lakes of the pre-Alps often contain a community of the White **Dwarf Waterlily** (*Nymphaeëtum minoris*). While the floating-leaved communities of eutrophic and relatively warm still waters are characterised by *Potamogeton crispus* and *Myriophyllum spicatum*, in the less rich ones are found *Potamogeton obtusifolius* along with *Myriophyllum alternifolium* (tab. 49, no. 1). In dystrophic peat and pools or peat cuttings which are only 50-100 cm deep with a dy-type peat mud, *Potamogeton natans* forms a species-poor community with the Overlooked Bladderwort (*Utricularia australis*). These **Bladderwort pools** have scarcely anything in common with the luxuriant Waterlily ponds with which we started.

(e) Duckweed covers and other free-floating communities

On those parts of the surface of oxbow lakes, ponds, pools and bigger lakes, which are protected from the wind, a number of free-floating plants can multiply, but where the water is disturbed they are not able to form a recognisable cover. The richer the water is in nutrients the more luxuriously do these carpets of floating plants flourish (see section d); otherwise they are fairly independent of the nature of the water (Klose 1963). The **Duckweed covers** (class *Lemnetea*) are the subject of a monograph by Tüxen (1974b) in which he also evaluates the whole of the literature. Within the order *Lemnetalia* he distinguishes two alliances and numerous associations which are only characterised by the species of Duckweed or floating liverwort which is present at the time. The following scheme shows the relationship of the different communities to the base and humus content of the water and also to its depth:

According to Tüxen there is no '*Lemnetum minoris*' as had previously been described. The Lesser Duckweed can turn up in all the communities (tab. 49) and is only dominant in those which have recently come into existence, or which have become poor. Gibbous Duckweed covers (*Lemnetum gibbae)* and Great Duckweed covers *(Spirodeletum)* often fill up the spaces between the leaves of the *Myriophyllo-Nupharetum* so that they can only be separated on the basis of their different life forms. Liverworts which are suspended in the water or float on the surface (*Ricciocarpetum* and *Riccietum)* live in the half shade between the stalks of Reed or in holes within a flooded sedge fen, i.e. in shallow and relatively humus-rich water. The Ivy-leaved Duckweed community suspended in the water (*Lemnetum trisulcae*, fig. 232) is found in situations between these and those of the more-demanding communities.

All free-floating aquatics are warmth requiring and prefer waters which remain ice free. Thus the *Lemnetea* communities are richest in the atlantic and submediterranean regions and become poorer as they extend into Central Europe. Only *Lemna minor* goes up to an altitude of more than 1000 m in the mountains. In the mild Upper Rhine plain and in other warm valleys ferns of tropical origin accompany the less pretentious Duckweed. These are the **floating ferns** *Salvinia natans* and the rarer *Azolla filiculoides*. (see e.g. Philippi 1969a), tiny ferns which are also well adapted to their floating life and have a number of features showing this. In no way should they be seen as phylogenetically primitive. Only the vegetation units which are formed by them, as well as by Duckweeds and floating liverworts, can be considered as primitive because of their simple structure. For this reason they are usually placed at the beginning of the plant sociological system (see section E III and tab. 49, nos. 11 and 12).

The small size and rapid growth of the floating plants have led to many experiments being carried out on them from which some conclusions can be drawn about the strategies of plant competition. According to Harper (1961, see fig. 231) separate cultures of *Spirodela polyrrhiza* have a significantly higher pro-

	Gibbous Duckweed comm.	Ivy Duckweed comm.
increasing pH value and base content ↑	*Lemna gibba* *Spirodela polyrrhiza*	*Lemna trisulca* *Ricciocarpon natans* *Riccia fluitans*
	.... →	increasing humus content decreasing water depth

duction per unit area of water surface than those of *Lemna gibba*. In spite of this the latter becomes dominant in mixed cultures because a large part of it projects above the water surface and shades the *Spirodela* which is mostly submerged. Clatworthy and Harper (1962) also included *Lemna minor* and *Salvinia natans* in their investigations. In pure culture *Lemna minor* has the highest production of all the species, and because of this one can understand why it is so plentiful and widespread. In mixed cultures however it was suppressed by each of the others and only plays the part of a stop gap in the Duckweed covers. When the water is warm enough *Salvinia* dominates all the others, being able to form higher and denser mats with its leafy stems which cling closely together. At the other extreme is the rootless Dwarf Duckweed (*Wolffia arrhiza*, fig. 231) which also likes the warm water and is rare in Europe. When Wołek (1974) compared it with the other Duckweeds it turned out to be the weakest competitor of them all, which perhaps accounts for the fact that it takes only a small part in Duckweed communities. Allelopathic influences, which can be so easily evaluated in the case of floating plants, play only a very minor role in the competition between them, when compared with morphological and physiological peculiarities, according to Wołek.

Fig. 231.
a. A morphological comparison of the Duckweeds: *Wolffia arrhiza* (without roots), *Lemna* species (one root) and *Spirodela polyrrhiza* (many roots). After Wołek (1974) modified.
b. In a pure culture *Lemna gibba* grows less vigorously than *Spirodela*. In spite of this when the two are growing together *L. gibba* is able to suppress *Spirodela* since it floats a little higher in the water thanks to its larger air spaces. Where the water surface is densely covered with Dugweeds *L. gibba* is pushed up above *Spirodela*. After Clatworthy (1960) from Harper (1961), somewhat modified.

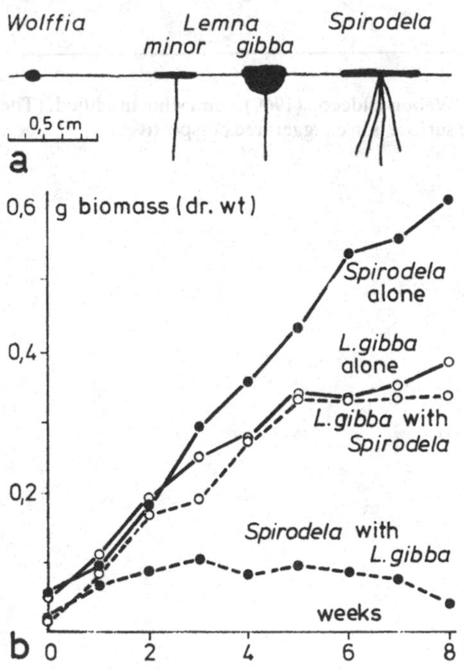

Floating macrophytes have always dominated the Duckweeds wherever they have been able to develop. This is especially true of the always free-floating Frogbit (*Hydrocharis morsus-ranae*) and the sometimes rooted Water-soldier (*Stratiotes aloides*), which grow in nutrient-rich shallow ponds of the lowland flood plains (fig. 232). The **Water-soldier-Frogbit community** (*Hydrocharitetum morsus-ranae*) is often mixed up so thoroughly with the *Myriophyllo-Nupharetum* or some other bottom-rooting floating-leaved community that it is difficult to separate them. Where there is a lot of river traffic the leaves of the firmly rooted plants are damaged or torn off whereas the Water-soldiers merely part and then float together again. In the end these are the sole survivors as could be seen at one time in the canals of Havelland.

(f) Reed swamps of non-brackish still waters

Plants of the Reed swamps which tower high out of the water and can build dense stands blocking out the light like a forest canopy will soon suppress any growth remaining near the water surface. This is the case with the Common-reed (*Phragmites australis*), which dominates at an average water depth of 1.2 to 2 m. Individual tubelike runners will grow out to another metre deep in the mud, but the shoots from these do not reach their normal size and only get stronger when silting up has reduced the depth. The furthest limit of the Reed stands then is almost always determined absolutely, i.e. by physiological factors and not by competition. Without doubt *Phragmites* is the most competitive of all the Central European water plants, to some extent comparable to Beech amongst land plants. Its ecological and sociological amplitude is equally large. According to Krausch (1965a) it extends from lime-oligotrophic and acid-oligotrophic to eutrophic waters and from deep water to ground which from time to time is above the water surface. Pietsch (1965) found *Phragmites*, *Typha latifolia* and other Reed plants even in an opencast working of the Lausitz brown coalfield in an extremely acid water (pH 2.9-3.0).

The communities formed around *Phragmites* are correspondingly numerous and varied. A rich literature is available on them in addition to the works mentioned already and in the following: Balátová-Tuláčková 1963, Hilbig 1971a, Weisser 1970. The different aspects of Common-reed swamps (order *Phragmitetalia*) and their varying habitats are however similar throughout wide areas of Europe from Spain (Bellot-Rodriguez 1964) to Finland (Eurola 1965) and from the British Islands (Wheeler 1980), down into the Danube delta (Krausch 1965b) which has the largest continuous stands. Even in the highland of south-west Japan Horikawa and co-workers have found parallels.

In calm lakes the Common Clubrush *Scirpus lacustris* subsp. *lacustris*) is a pioneer which goes out even deeper than *Phragmites* since its green stems are able to go on assimilating even when they are submerged by high water (see section b). Its black rhizomes which creep under the ground can sometimes be found at a depth of 5 m. Where there are strong waves the Clubrush cannot establish an outpost as its stems, filled with a spongy pith, are more liable to be bent over than those of the Reed which become lignified and stiffened with silica after having grown up in late spring. *Phragmites* and the *Typha* species are like helophytes in which the aerial parts have become strongly xeromorphic (Geyger 1964), and they transpire in a similar way to land plants (Krolikowska 1975).

Where *Phragmites* grows well it can reach a height of 3.5 m above the water surface and cut out more than 99% of the light from it (F. H. Meyer 1957). This allows only a small number of the character species of the order and alliances shown in tab. 51 to grow alongside. The only exception to this is when the two species of Reedmace manage to become dominant in a few places, for instance in nutrient-rich muddy bays or shallow lakes which dry out in most years. *Typha* reproduces from seed much more successfully than *Phragmites* and can cover bare mud from the outset with a thick turf.

For *Phragmites* which has such strong vegetative growth, reproduction by seed is a weak point. It can certainly produce plenty of seed according to Hürlimann (1951), and these retain their awns which enable them to be dispersed by wind as far as the long-haired seeds of *Typha*. However a fungal parasite (*Claviceps microcephala* according to Luther 1950) affects the ripening of the seed and, as Bittmann (1953) showed by many experiments its germination percentage is very small. The seeds do not ripen until the end of February and remain attached to the dead haulm of the parent plant until the end of April. Since they can float on the water for many days and since they require full light, plenty of oxygen and continuous damp to germinate, their best chance of becoming established is at the water line on bare ground. One occasionally comes across young *Phragmites* plants in such a position from which they grow deeper into the lake by means of runners. Many a broad belt of Reed may have originated from a single successful germination.

The seed germinates quickest at high temperatures (27–36°C. according to Hürlimann). Vegetative

Fig. 232. Layering in floating plant communities, especially in the Frogbit-Water-soldier cover (*Hydrochari-Stratiotetum*). After Weber-Oldecop (1969), somewhat modified. (The view of the surface is in exaggerated perspective.)

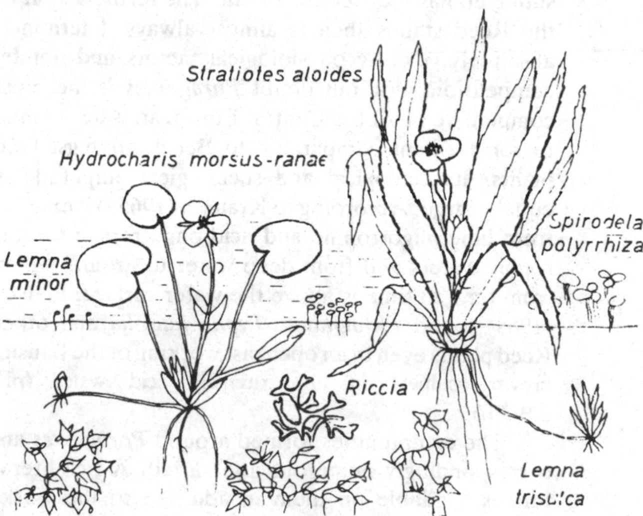

growth only begins when the temperature has risen to about 8–10°C so that the young shoots appear hesitantly after the woods and meadows have already been green for some time. In trial tanks without water they were killed in January at a temperature of −8.7° and in March could only tolerate −0.8° (Dykyjová et al. 1971). *Typha latifolia* and *Scirpus lacustris* behaved in a similar manner. *Phragmites* and other Reed swamp plants however require a certain length of rest period in winter to be able to develop normally (Dykyjová. Véber and Pribáň 1972). Its need for warmth imposes a clear limit on the distribution of this Reed in southern Scandinavia. Finland and in the montane belt of the Central European mountains, whereas it grows well in subtropical regions and its distribution is world wide.

There is plenty of information from many workers on the productivity of Reed stands. In the Neusiedler See south of Vienna, Sieghardt (1973) found that the annual amount of dry matter produced above ground was similar to that from a crop of wheat (about 17 t/ha), and that nearly 5% of the photosynthetically active radiation had been utilised (for comparison see sections B II 4 g and D V 2 b). The denser the stand the higher the productivity; it fell off both on the lakeward and landward sides (Ondok 1970). It is difficult to give reliable average yields because these vary so much

*Tab. 51. Synopsis of the Reed and Sedge beds and their character species.**)

Character species of the class and order of **Reed and Sedge beds** (Phragmitetea, Phragmitetalia):

Acorus calamus	*Poa palustris* (4)
Alisma plantago-aquatica	A *Sagittaria sagittifolia* (1)
Equisetum fluviatile	*Scirpus mucronatus*
Iris pseudacorus	*S. lacustris* ssp. *tabernaemontani* (2)

Character species of the four alliances:

1. **True Reed swamps** (Phragmition)	3. **Stream Reed swamps** (Sparganio-Glycerion)	4. **Large Sedge swamps** (Magnocaricion)
Butomus umbellatus	*Apium nodiflorum*	A *Carex appropinquata*
Eleocharis palustris (K)	A *Berula erecta*	*C. distans*
A *Glyceria maxima*	*Epilobium parviflorum*	A *C. paniculata*
Hippuris vulgaris	*E. roseum* (?)	A *C. pseudocyperus*
Oenanthe aquatica (K)	*Glyceria fluitans*	A *C. riparia*
A *Phragmites australis* (K)	*G. plicata*	A *C. vesicaria*
Ranunculus lingua	*Nasturtium officinale*	A *C. vulpina* (?)
Rorippa amphibia	*Scrophularia umbrosa*	*Circuta virosa*
Rumex aquaticus (K)	*Veronica anagallis-aquatica*	*Cyperus longus*
R. hydrolapathum	*V. beccabunga*	*Eleocharis uniglumis*
A *Scirpus lacustris*		*Galium palustre*
Sium latifolium (?)		*Oenanthe fistulosa*
Sparganium emersum (K)		*Peucedanum palustre*
Sp. erect. ssp. *polyedrum* (K)		*Scutellaria galericulata*
A *Typha angustifolia*		*Teucrium scordium*
A *T. latifolia*		
		4a in the suballiance Caricion elatae:
		A *Carex elata*
2. **Shore Rush-Reed beds** (Bolboschoenion)*)	In Reed and Sedge swamps	A *C. rostrata*
		Lysimachia thyrsiflora
		Senecio paludosus
A *Bolboschoenus maritimus*		
Scirpus pungens	A *Cladium mariscus*	4b sa Caricion gracilis:
S. triquetrer	A *Phalaris arundinacea* (K)	A *Carex gracilis*

Note: A = character species of a community (generally an association) which is named after it (usually the dominant species).
(K) = looked upon as a class character species by many authors.
(l) etc. can also serve as character species for the relevant alliance.
*)Some of the species are no longer looked upon as character species (see section E III 1, No. 1.4).
Source: After Tüxen, Preising, Oberdorfer and other authors.

from year to year depending on the weather; in England for example Mason and Bryant (1975) found that a yield of 10.8 t/ha in 1972 contrasted with one of 5.5 t/ha in 1973. Round the Plattensee which is relatively poor in nutrients *Phragmites* yielded only 0.05-0.3 t/ha (I. and V. Kárpáti 1971b). Yields similar to those from the Neusiedler See have been determined also in the Danube delta the vegetation of which is comprehensively described by Krausch (1965b). Here *Phragmites* is used extensively for obtaining cellulose. According to Rudescu (remarks made in a discussion on 8.9. 1970) the Reed has been mown here for over 100 years without any noticeable falling off in production. To some extent at least this is because before the Reed is cut in the autumn about two-thirds of its biomass and most of the important elements are translocated to the rhizomes for storage (Sieghardt 1973, Westlake 1975). These contain more N,P and K throughout the year than the stems and leaves (Dykyjová and Hradecká (1976). In addition it must be borne in mind that in the Danube delta 'all the nutrients have collected which have been washed out of Central Europe during the past 5000 years' (Rudescu).

In certain respects the Reed swamp represents a natural monoculture since *Phragmites* can dominate over several square kilometres without there being any other plant of similar height to be seen. However Straškraba (1963) states that in a fish pond in southern Bohemia the Reed contributes only 70% of the primary production. Algal growths on the submerged stems account for 21% and the phytoplankton in the shaded water for 7%. In the Neusiedler See the submerged floating Bladderwort (*Utricularia vulgaris*) also plays a certain part in the primary production (Maier 1973). In spite of these partners the Reed swamp must be seen as a community very poor in species. Usually it is surprisingly stable and it has managed to survive the complete drying out of the lake several times. In different places in the hugh *Phragmites* stands around this lake small circular areas collapse and die, especially where the rhizomes have formed a thick mat. In a way these so-called 'Lacken' (gaps) can be compared with the natural clearings which originate from the dying of tree groups in ancient forests (see figs. 59, 148, 149 and 174). Under natural conditions they close up again within a few years. It is not known whether the formation of these holes in the Reed stand is due to poisonous substances accumulating in the rhizomes. McNaughton (1968) was able to show this kind of autotoxic effect in the case of *Typha latifolia* but only during the germination and seedling stages.

Recently larger and **spreading gaps** are being produced with increasing frequency in Reed swamps near to towns or on lakes with heavy tourist traffic (Sukopp and Kunick 1969, Sukopp, Markstein and Trepl 1975). Apart from a lowering of the water level the main reasons for this are **mechanical** such as traffic by ships and boats, eddies caused by structures erected on the banks, bathing activities, as well as by sewage and blankets of algae which are produced in large amounts as a result of eutrophication and are thrown into the Reed beds during storms. These factors all cause damage because they break over the young haulm interrupting the passage of oxygen down to the rhizomes which then die and rot. Klötzli (1971) thinks that the Reeds have been increasingly more prone to mechanical damage since the middle of this century because the stems have a less well developed sclerenchyma due to the luxuriant supply of **nutrients**. In the ground where stands of Reed were breaking down Klötzli and Züst (1973) found an unusually high level of mineral nitrogen. The death of the Reed can usually be prevented by erecting barriers against the Blanket Weed and the floating debris being washed ashore. Leippert (1978) studied the history of the *Phragmites* dieback in the extremely shallow Dümmer lake (North-west Germany). She came to the conclusion that here the custom of cutting the Reed in former times had been decisive for its stability. Taking off the dead haulms regularly in winter encouraged the formation of strong fresh haulms resistant to mechanical damage (see also Grünig 1980 and other papers in the same volume).

Compared with *Phragmites* all the other plants of the Reed swamp are of little significance, at any rate in most still-water lakes. The Reed Sweetgrass (*Glyceria maxima*) is favoured by strong eutrophication although it does not grow so high as *Phragmites* (Lang 1967b). The **Reed Sweetgrass swamp** (*Glycerietum maximae*) is usually an equally intolerant pure stand. According to Westlake (1966) it also resembles the *Phragmitetum* in translocating and storing more than half the net-photosynthetic production in the underground organs, reaching its highest biomass in late summer.

Numerous animal species, according to M. Vogel (1981) mainly invertebrates, have adapted to the 'natural monocultures' of *Phragmites*. None of these animals multiplies excessively, and the 'damage' caused by them remains insignificant. Also in this respect then the Reed swamp does not conform with our usual idea of a monoculture.

The Branched Bur-reed (*Sparganium erectum*) is one of the most productive plants in the temperate zone (Dykyjová and Ondok 1973), but it does not form dense stands and seldom exceeds 1 m in height. Thus it plays only a subordinate role in Reed swamps, but it leads on to the Large-sedge swamps which we shall discuss next. The **Saw Sedge swamp** (*Cladietum marisci*) occupies a unique place in that it only grows on lime-rich ground and has a low nutrient requirement, but it cannot tolerate large variations in the water level. Görs (1975) in a discussion of the literature suggests that the typical subassociations of the *Cladietum* should be put into the *Phragmition* alliance, but the rest of the subcommunities are inclined strongly towards the *Magnocaricion*. *Cladietum* reached its maximum distribution in the early Post-Glacial period when the Central European lakes were mostly lime-rich and oligotrophic (Hafsten 1965). The *Sphagnum palustre* variant in particular can be considered a 'living fossil community' according to Görs. The Saw Sedge is sensitive to high water levels and according to V.M Conway (1937) it dies from shortage of oxygen if it is cut off under the water. In this respect it is much more sensitive than *Phragmites*, and it is such a weak competitor that on many fronts of its world-wide distribution it is disappearing more and more.

Where the helophytes of Reed and Sedge swamps grow well they are able to clean polluted water, at least to a certain extent (Seidel 1966). Either they absorb phenolic and other noxious substances, or the bacteria in their rhizosphere decompose them. Of course these 'ecochemical effects' are connected with a high water loss through the luxurious transpiration of those swamp plants (Kickuth 1970).

(g) Large-sedge swamps

As the above-ground biomass of the Reed swamps dies off every year it raises the level of the lake bottom more and more with organic remains. In still water, clay particles and organic colloids also sink amongst the haulms helping to consolidate this layer. As the root run is raised in this way it emerges more and more often and becomes more suitable for plants which demand less water, but which may be more sensitive to prolonged and deep submersion. So on the landward side the *Phragmites* stands tend to become weaker; the haulms are not so tall and are more isolated. Gradually tall sedges and their accompanying flora take over (fig. 233).

These Sedge swamps however have so many species in common with the Reed swamps that they are put into the same order and class (*Phragmitetalia*, *Phragmitetea*). Within these they form only one special alliance (tab. 51), which along with the *Carex* species is characterised above all by a sub-species of *Galium*

Fig. 233. Multiple zonation of plant communities at a nutrient-rich lake in southern Germany after the natural woodland has been destroyed. After Ellenberg (1952).
The Waterlily community, the Reed bed and the Tufted Sedge are still fairly natural but the Small Sedges and Moorgrass communities have been maintained by regular mowing, which has prevented the development of an Alder swamp wood, an Alder-Ash wood or a moist Oak-Hornbeam wood. The low yield from the unmanured litter meadows, especially the Small Sedge community, is primarily a result of shortage of air in the perpetually waterlogged soil which leads to denitrification.

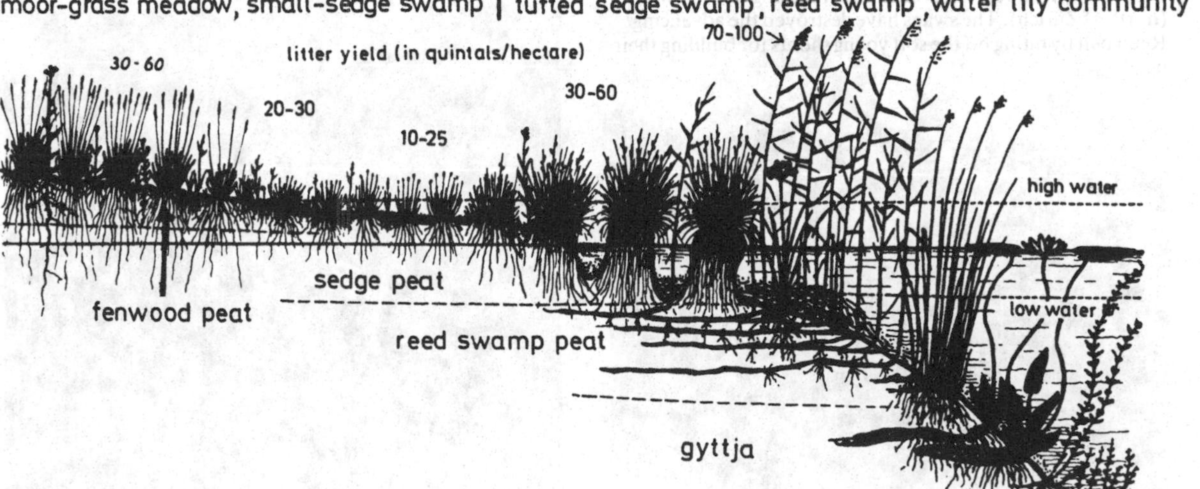

palustre adapted to flooded conditions and by *Poa palustris*.

Depending on the height and duration of flooding as well as on the water quality the Large Sedge swamps have different combinations of species. As a rule one particular Sedge species becomes dominant and this can also serve as character species. Such pure stands, as a rule, produce much biomass and correspondingly much litter (Wheeler and Giller 1982). By this they accelerate the peat formation if they are not regularly cut.

The **Tufted Sedge swamp** (*Caricetum elatae*) can tolerate the largest variations in the water level and today it still occupies wide stretches of shoreline in the lakes of the pre-Alps and in south-east Europe. In earlier times it was valued as a source of litter for housed livestock. In Hungary it also flourishes in places where the water level does not vary much but remains over long periods of time at about or slightly above the level of the leaf bases (Kovács 1968b). It is very striking even to the non-botanist with its big columnar tussocks (or 'stools') growing up to a height of as much as 1.2 m above ground (fig. 233). Many people will have experienced jumping from one stool to another at high water in early summer and discovering that even those furthest out will still carry the weight of a man (fig. 234). In the autumn it is possible to walk between them with dry feet and to notice that the stools consist for the most part of roots and leaf sheaths of the Tufted Sedge (fig. 235). On the landward side the stools grow closer and closer together until only small water-eroded depressions remain between them. It was at this well developed stage that the farmers preferred to mow them for litter and in doing so prevented them from developing further. Under natural conditions they would soon be occupied by Black Alder and Sallow, developing fairly quickly into an Alder fen (fig. 236, see Ellenberg and Klötzli 1967).

In the north-east of Central Europe the *Caricetum elatae* is rare and is generally replaced by the **Sharp Sedge swamp** (*Caricetum gracilis*). The Sedge giving its name to this community does not form tussocks, but an even sward which, from a distance, resembles a fur as the wind blows all the leaves over in one direction. Only when the Sharp Sedge swamps are trodden by cattle do they become tussocky. Usually they are flooded in the early part of the year and in summer they tend to be less wet than the Tufted Sedge swamps although it can sometimes be the other way round (compare Balátová-Tuláčková 1957 and Kovács 1968b; see fig. 237).

The productivity of the Sharp Sedge swamps does not quite come up to that of the best Reed swamps, although it can be about the same as the average. Baradziej (1974) found a net primary production of 5.5 t/ha per annum in Poland; stands with much *Iris pseudacorus* even reached 8 t. According to Gorham (1974) the above-ground biomass of Large Sedges swamps, when well supplied with water, is closely correlated with the summer temperature in central and northern regions. The extremes are almost 15 t/ha in

Fig. 234. Tufted Sedge bed (*Caricetum elatae*) with a high water level in late spring in an old peat digging near Stadel (north of Zürich). The swans have destroyed the advancing Reed belt by biting off the soft young shoots for building their nest.

Fig. 235. The columnar clumps of Tufted Sedge stand up like thick palms on the dried-out bed of the pond in early winter near Blitzenreute (Upper Swabia). Between them is a stunted growth of Reed.

the relatively warm lowland and 2 t/ha in the montane belt or in the subarctic zone. These figures only apply to relatively nutrient-rich habitats where the organic topsoil is biologically active (see Ambrož and Balátová-Tuláčková 1968).

In the area liable to flooding around lime-deficient oligotrophic or moderately dystrophic lakes only less dense and shorter Sedge swards can grow in which the Bottle Sedge (*Carex rostrata*) is most often dominant. The **Bottle Sedge swamp** is found even around the small lakes in the centre of some raised bogs as well as in their marginal swamps ('lagg', see figs. 249 and 247) where there is no Reed because of the shortage of nutrients.

Numerous other Large Sedge communities have been described from Central Europe in addition to the three already mentioned. These can be classified under one or other of the two recently separated suballiances, namely the Tufted Sedge swamps *sens. lat.* (*Caricion elatae*) and the Sharp Sedge swamps, *sens. lat.* (*Caricion gracilis*, see tab. 51). This subdivision is justified in so far as the first-named suballiance contains a number of differential species, e.g.:

Carex rostrata	*Peucedanum palustre*
C. lasiocarpa	*Potentilla palustris*
C. diandra	*Menyanthes trifoliata*
C. paniculata	

On the other hand the suballiance *Caricion gracilis* can be recognised only by the name species and the absence of the above species, together with the

Fig. 237. Sharp Sedge bed (*Caricetum gracilis*), consisting of little more than the named Sedge and a few Reed stems, covers large areas in the Bille valley to the east of Hamburg.

Fig. 236. Alder is already present in the Large Sedge bed. *Sparganium erectum*, *Eleocharis palustris*, *Festuca pratensis*, *Ranunculus flammula* and *Galium palustre* can also be recognised. Photo, Iltis-Schulz.

reduction in the proportion of *Carex elata*. The ecological causes of the bipartition and the habitat factors determining the individual communities are not immediately apparent because each of the suballiances comprises deeply inundated as well as less wet communities. The decisive habitat conditions are above all to be looked for in chemical factors, especially in the availability of nutrients and bases, and in part also in climatic features. This is set out in more detail in tab. 52 in which the average factor values (see section B I 4 b) for 'nitrogen', 'reaction' and 'moisture' are given along with those for 'temperature' and 'continentality'. These have been calculated from the tables published by Balátová-Tuláčková (1976) from southern Slovakia.

All the Sedge swamps included in tab. 52 under the suballiance *Caricion elatae* have relatively low average N values, that is they contain mostly species which are competitive only when the supply of nitrogen is low. This applies almost entirely to the differential species named above. Apparently the ones least well nourished are *Caricetum diandrae* (mN 3.5), *C. appropinquatae* (mN 3.6)

Tab. 52. Ecological evaluation of the Large Sedge communities in the March plain in southern Slovakia.

Suballiances Associations Subassociations		Mean indicator values[a] N	 R	 F	 T	 K	%[b] fw	Tab no.[c]
sa Caricion elatae (= *rostratae*)								
Caricetum rostratae		4.3	4.4	**9.3**	4.8	3.8	30	3
Caricetum diandrae		3.5	5.2	8.7	4.6	**4.7**	25	5
Caricetum appropinquatae		3.6	5.9	8.6	4.9	4.0	31	6
Caricetum elatae	Subass. 1	5.1	5.7	**9.7**	5.1	3.5	**38**	2 a
" "	" 2	4.9	5.9	**9.1**	5.1	3.5	**42**	2 b
" "	" 3	4.4	5.6	9.0	5.1	3.6	**43**	2 c
Caricetum paniculatae		4.8	6.6	8.7	5.1	3.8	21	7
sa Caricion gracilis								
Caricetum ripariae	Subass. 1	5.2	6.6	**9.7**	**5.9**	3.5	11	9 a
" "	" 2	5.4	**6.9**	**9.3**	**6.0**	3.8	13	9 b
" "	" 3	5.2	6.8	**9.3**	**5.8**	3.9	15	9 c
Caricetum vesicariae		5.2	5.8	9.0	4.8	4.0	22	10
Caricetum gracilis		5.3	6.5	8.7	5.0	**4.7**	24	11
Caricetum vulpinae		**6.1**	**7.0**	8.3	5.4	4.3	**39**	12
Caricetum distichae		5.0	6.6	8.2	5.1	4.5	28	13
Phalaridetum		**6.2**	**6.9**	8.5	5.2	4.3	28	8

Note: [a]**Mean indicator values calculated from tables by** Bálátová-Tuláčková (1976): bold type = relatively high values, in frames = relatively low values. Calculated according to Ellenberg (1974).

N = mean nitrogen value (from 1 = indicator of nitrogen shortage to 9 = high nitrogen indicator),
R = mean reaction value (from 1 = acid indicator to 9 = lime indicator),
F = mean moisture value (from 1 = dryness indicator through 9 = indicator of very wet soil to 12 = species only living in relatively deep water),
T = mean temperature value (from 1 = cold-tolerant species to 9 = one requiring very warm conditions),
K = mean continentality value (from 1 = oceanic species to 9 = continental one).

Each mean indicator value has been calculated with reference to all the evaluated species and their cover as average from all the relevés of a community table. The calculations have been made by Spatz et al. (1979) with the help of a computer programme (see also sections B I 4 and E III).

[b]The number of species indication a fluctuating water regimen (according to Ellenberg 1974) in percent of the mean total number of species in a relevé of the community.

[c]**In the book by Bálátová-Tuláčková (1976) in which the** authoress gave some measurements of soil factors from random samples taken in May 1963 and in May 1964. The picture of the habitat conditions derived from these data was only partly informative in so far as the number of micro-organisms present and the measurement of their nitrifying power was related to the weight of soil and not to its area (kg/ha).

The factor which distinguishes most clearly the different suballiances is the calcium content of the water; for the communities of the suballiance *Caricion elatae* (or *rostratae*) this always turns out to have a low value (20–60 mg Ca per litre), whereas in those of the *Caricion gracilis* it is higher (108–416 mg/l). This tendency is expressed in the pH values of the soil (S) and the water (W) only for some of the communities:

Caricion elatae	S 4.7–7.6	W 5.1–7.1
Caricion gracilis	S 6.3–7.6	W 6.6–7.1
(*Caricetum rostratae*	S 4.7	W 5.1)

and *C. rostratae* (mN 4.3). The acid indicators are concentrated in the *Caricetum rostratae* and *diandrae* (mR 4.4 and 5.2 respectively). On the other hand pronounced indicators of nitrogen deficiency and of low pH values are absent from the suballiance *Caricion gracilis* so that the N-indices for the corresponding stands have an average value of more than 5.1 and the reaction indices generally exceed 6.0. In this respect *Caricetum paniculatae* is similar to *C. gracilis* whereas *C. vesicariae* is more like *C. elatae*.

Within both suballiances there are communities which are almost continuously under water, i.e. their average moisture index exceeds 9.0. This is especially true of the 'wettest' subassociations of the Tufted-Sedge swamp (*C. elatae*- mF 9.7) and of the Greater Pond-Sedge swamp (*C. ripariae* – mF 9.7). The latter seems to differ from the Tufted Sedge swamp not only in the nutrient level, but especially in the amplitude of the fluctuations in the water level, which is expressed in the percentage of indicators of periodical wetting present in the community table. *Caricetum ripariae* has an average of only 11-15% whereas in *Caricetum elatae* this amounts to 38-43%. The relatively least-wet habitats are those of *Caricetum vulpinae* (mF 8.3) and *C. distichae* (mF 8.2) which form a transition to the damp meadow communities. Finally the Reedgrass Community *(Phalaridetum) occupies a special place. This community of the river banks, which is poor in species but best nourished, is not included with the Large Sedge swamps by most authors (see fig. 190, 191).*

Climatic response is also very interesting as far as it is reflected in the average factor indices. All three subassociations of *Caricetum ripariae* are relatively warmth loving (mT 5.8 to 6.0) and thus are less frequent in the northern than in the southern parts of Central Europe. On the other hand communities such as *Caricetum diandrae*, *rostratae* and *versicariae* (mT 4.6–4.8) are able to penetrate further into the cooler regions. *C. elatae* (mK 3.5–3.6) and *C. ripariae* (mK 3.5–3.9), rich in species but relatively frost-sensitive have a mainly oceanic distribution, contrasting with *C. diandrae* and *C. gracilis* which have a surprisingly high continentality figure (mK 4.7). However there are high and low figures for both the climatic and wetness factors in both suballiances. As has already been stressed the main difference between these suballiances is probably one of nutrient supply, but unfortunately adequate data are not yet available.

It is worth noting that all the communities listed in tab.52 can be found within a relatively small area in southern Czechoslovakia (or at least they could be found before the carrying out of an extensive drainage programme). This area then can serve as a model for the Sedge swamps of Central Europe. The differences in the supply of nitrogen, as indicated by the average factor indices, and in the way the water levels fluctuate result from inequality in flow of river water. The climatic peculiarities may be traced to local site conditions in the countryside, e.g., accumulation of cold air on clear nights.

2 Plant communities of running waters

(a) Reed swamps in tidal reaches of the North Sea estuaries

Very unusual conditions are found in the Central European rivers where they enter the North Sea in that they are dominated by the influence of the tides. In such estuaries large Reed stands form the outposts of the higher plants, which have to tolerate continuous and extreme fluctuations of the water table. Therefore these habitats and their communities will be dealt with along with the other communities of the *Phragmitetalia*. Zonneveld (1960) has devoted an exemplary monograph to the largest freshwater tidal area in Europe, the Biesbosch in the Brabant district of the Netherlands. In Germany Kötter (1961) and F.H. Meyer (1957) have dealt with the plant communities of the Lower Elbe.

The tides in the North Sea hold up the flow of the rivers twice daily in the estuaries to the extent that the current is reversed. However the salt water is only able to turn the river brackish for a short distance in the lower part of the estuary, since the ebb tide supported by the river washes it out again every six hours. For example the water of the Elbe is always 'fresh' above Glückstadt, even though at Hamburg the difference between high and low tide averages 2 m (fig.238). When the wind is blowing towards the land in a north-westerly gale combined with a spring tide (see fig.286) the amplitude can reach as much as 8 metres here. A gale blowing away from the shore combined with a neap tide brings the opposite extreme.

Phragmites can live very well under these conditions. Nowhere does it produce more vigorous growth than in the estuaries where the mud is one of the habitats with the richest supply of nitrogen in the whole of Central Europe (F.H. Meyer 1957). Usually this fertile soil can only be occupied by *Phragmites* where there is a depth of less than 0.5m of water at mean tide, that means where it is uncovered for several hours. The muddy and silty soil contains worms and other air-breathing animals since the floodwater does not cover it long enough to drive all the air out of the spaces in it (fig.239).

In the spring, before and during the time the Reed is shooting up lower growing plants make use of the fertile soil. The Marsh Marigold luxuriates here as nowhere else (up to 1 m!, fig.240), and the Celandine, with its equally bright yellow flowers, covers the ground as it does in a flood plain woodland. Eventually the Reed overshadows these early flowering plants causing them to die down or at least to stop flowering. In many respects the tidal **Marsh Marigold-Reed swamp** differs from the Reed swamps of the lakes and ponds. It even appears that in the Reed stands and neighbouring communities on the landward side in the Elbe estuary, an endemic species has been able to establish itself. This is *Deschampsia wibeliana* which, according to Weihe and Reese (1968) can be distinguished from *D. cespitosa* by its earlier flowering and by a number of morphological and anatomical features.

Only the **Shore Rush** (*Bolboschoenetum maritimi*, fig.241) ventures deeper into the river estuaries than the Reed stands. It is sometimes erroneously described as the 'brackish water Reed swamp' but Kötter has shown that it also lives in quite fresh water. It is only

Fig. 238. Zonation of vegetation on the sandy banks of the Elbe estuary below Hamburg in the tidal freshwater region, semi-schematic and in part not to scale. After Kötter (1961), modified.

So much organic material is deposited from drift at high water that nitrophilous algae (*Vaucheria*), annual ruderal plants (e.g. *Bidens* species) and moisture-loving tall perennials are able to flourish. Above this driftline however apart from isolated wetter spots there is a more or less acute shortage of nutrients.

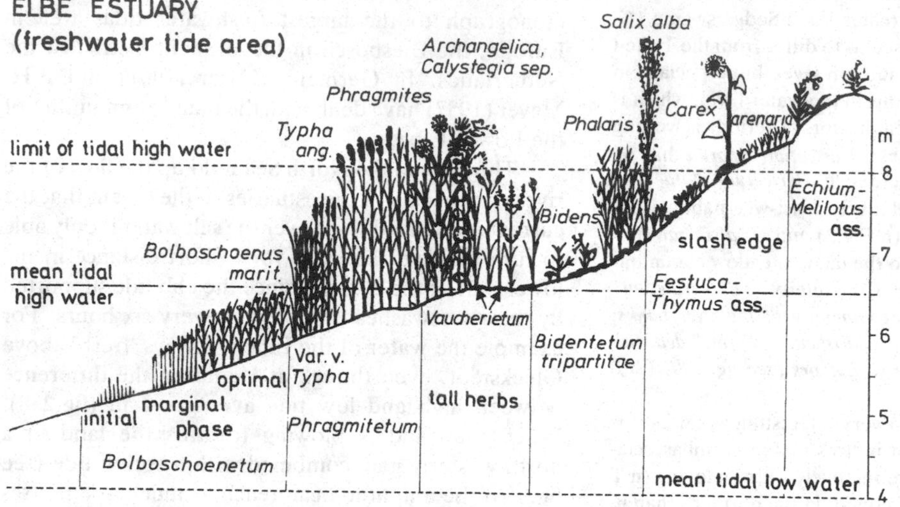

Fig. 239. High water levels, groundwater levels, air content and earthworm activity in the soil of a tidal Reed bed in the Biesbosch (Rhine delta). After Zonneveld (1960), somewhat modified.

1 = well aerated and plenty of humus, 2 = still aerated, 3 = reduced (not aerated), 4 = rusty marks (around old roots etc.), 5 = *Lumbricus*, 6 = *Dendrobaena*, 7 = *Allolobophora*, 8 = enchytrids, 9 = high water level above the ground, 10 = groundwater levels (Sept. – Oct. 1953). Since the ground is covered with water for only a short time, air so essential for the life of the worms, is not entirely lost from the soil.

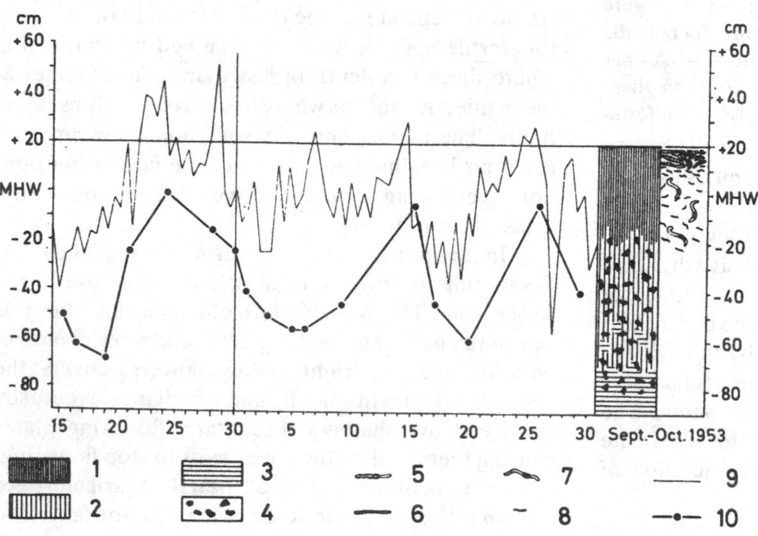

the large fluctuations in the water level caused by the tides which lead to its continued existence in the estuaries. In the absence of these conditions it would not be competitive either with *Phragmites* on the one hand or, on the other, with the floating-leaved communities which would come in, if the water levels did not vary so much.

Besides the Shore Rush other medium-sized members of the *Cyperaceae* can form special communities, above all the Club Rush *Scirpus lacustris* ssp. *tabernaemontani*. At about 1.20 m below mean high water even the *Bolboschoenetum maritimi* comes to a halt. There is little chance for Waterlilies or other plants with floating leaves to grow, nor can the submerged Bladderworts, Charas and similar plants survive the mechanical demands of the twice daily change in direction of the flow and the heaving of the tide. Free-floating plants too find it impossible to remain because of the violent storms which often afflict this area near the North Sea.

The part played by the Rush and Reed stands in the formation and establishment of new marshland can be seen in the scheme clearly set out by Zonneveld (fig. 242). It also takes into account the differences brought about by dyke building, draining, and farming practices such as mowing, pasturing and ploughing. Section C IV in which we look more closely at the formation of marshlands on the sea coasts should also be consulted.

A special form of the Shore Rush is salt-tolerant and can live in brackish water whether or not the water level fluctuates a great deal. It is more resistant to salt than the Common Reed (Borhidi 1970) although *Phragmites* can also put up with fairly brackish water without damage. In this case too we are dealing with a salt-tolerant ecotype; it can be distinguished from the normal erect *Phragmites* by its lower growth, even after several years of being cultured in the same soil by Dykyjová (1971).

On the landward side of the tidal Reeds instead of a Large-Sedge swamp there is usually a strip of Reed-grass (*Phalaris*, fig. 238) or a high-water mark flora followed by a pasture or river-side woodland. We have already partly discussed these communities in connection with the river banks and we shall also come back to them. First though we should like to turn to other habitats and plant communities which merit our attention in or near flowing waters.

Fig. 240. A lush growth of Marsh Marigold (*Caltha palustris*) in the freshwater tidal Reed bed on the island of Schweinesand in the Elbe near Blankenese (Hamburg). In front *Phalaris arundinacea*, behind *Typha angustifolia*.

(b) Communities in running water and at its banks

To the organisms living in them running waters offer more nutrients and oxygen than still waters having the same qualities. The water flow is continually replacing these bioelements and there is rarely a shortage even of

Fig. 241. A Shore Rush bed (*Bolboschoenetum maritimi*) in the freshwater tidal region of the Elbe island Nessand near Blankenese which has been built up by dumping dredged material. In the foreground centre *Scirpus lacustris* ssp. *tabernaemontani*; behind groups of *Phalaris* and on the higher bank masses of *Phragmites*.

Fig. 242. The build up of land and the vegetation succession in a freshwater tidal delta (Biesbosch in Brabant, schematic). After Zooneveld (1960), somewhat modified.

1 = medium sand, 2 = coarse sand, 3 = sand with clay, 4 = clay, 5 = alternating layers of sand and clay, 6 = peaty clay, 7 = rush peat, 8 = *Sphagnum* peat, 9 = Sedge peat, 10 = pumping mill for drainage, 11 = drain outfall with non-return flap, 12 = high and low banks (dykes).

M.H.W. = mean high water, M.L.W. = mean low water, I – V = build up of land as far as M.H.W., e = open arm of river, h = canal, l = a river channel is cut and allows the formation of fen and raised bog, VI f-g = with summer dykes, VII = with winter dykes, VIII = after the intake of the whole area (the different soil layers settle unevenly).

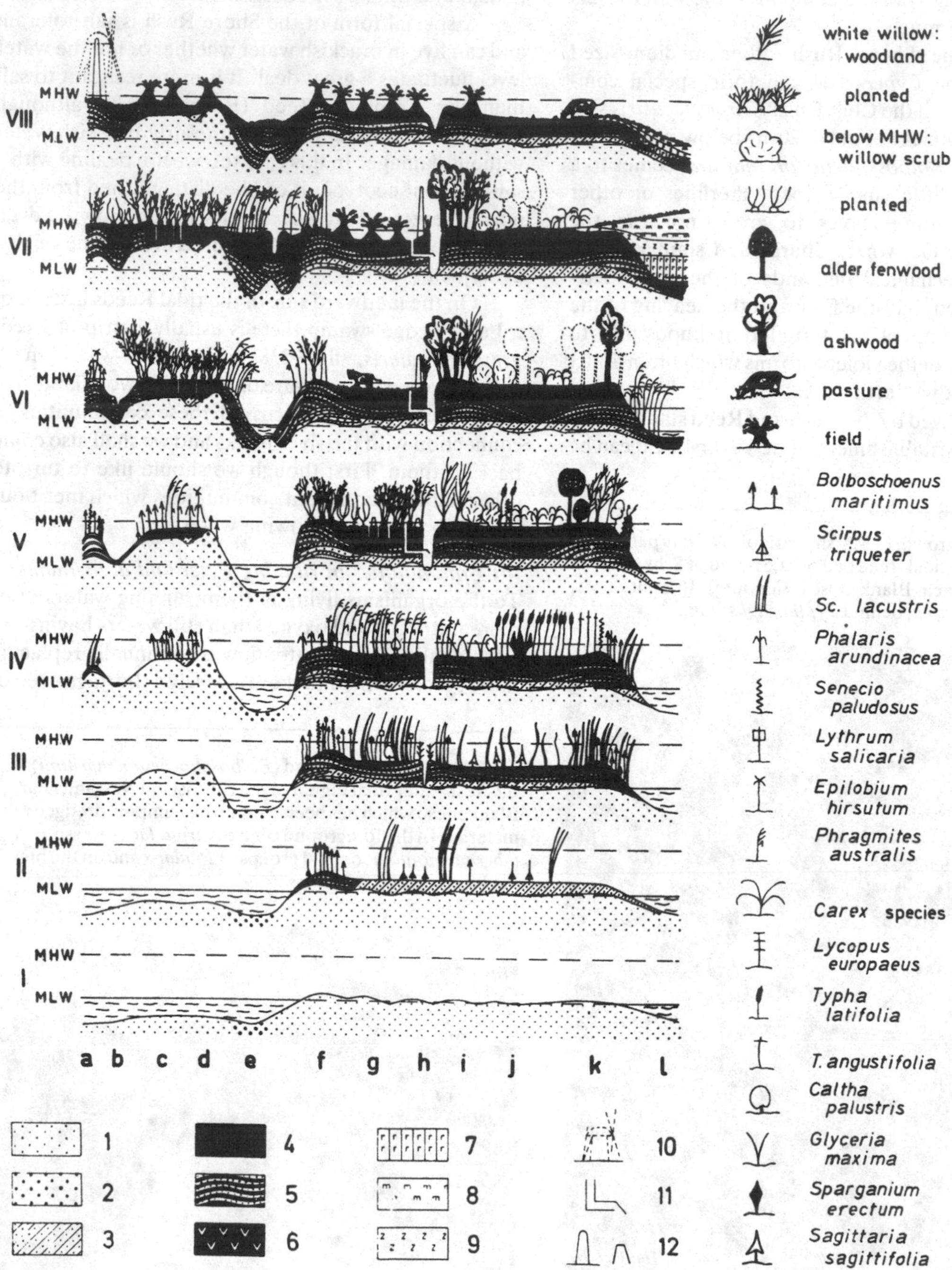

those which are in low concentration. However in itself the current is an 'extreme environment inimical to life' in the words of Thienemann (1922). Only a few plants and animals are able to survive in it. In the 'lenitic biotopes', i.e., the sheltered places where the current is weak, there is much more life than in the 'lotic' ones where flow is stronger. However species which are able to survive in the latter avoid competition from other species (Haslam 1975).

Roll (1938) showed, for example, how the different water plants distribute themselves between the two adjacent environments in one and the same stream. Still-water regions had plant communities resembling those of a eutrophic lake while the parts with more or less strong currents had either no plants at all or a quite different community (figs. 243-245).

The types of aquatic plants which were distinguished on the basis of their photosynthetic organs in section 1b also appear in or on running water. Here however they are represented by other species or by forms, varieties or subspecies more adapted to the current. Some plants which have round or relatively broad leaves in still water are able to produce long flowing grasslike 'strap-leaves', e.g., the Arrow-head (*Sagittaria/sagittifolia* f. *vallisneriifolia*, see also fig. 243) and the following species:

Fig. 243. Water Crowfoot community in the stream running from a pond to the north of Bratislava. *Ranunculus aquatilis* and *Glyceria fluitans*. Photo. Schultz.

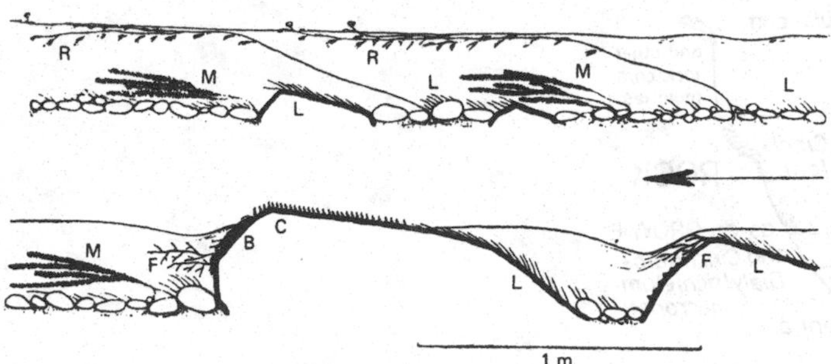

Fig. 244. Distribution of plants brought about by the currents in a mountain stream with stony bed (Ourthe in the Ardennes). After Vanden Berghen (1953). R = *Ranunculus aquatilis*, M = *Myriophyllum* (weaker current!), L = *Lemanea* (red alga), F = *Fontinalis*, B = *Brachythecium rivulare*, C = *Cinclidotus fontinaloides* (F, B and C are mosses).

living in **quickly flowing** water in Schleswig-Holstein:

higher plants:	algae (A) and mosses (M):
Butomus umbellatus f. *submersus*	A *Cladophora glomerata*
	A *Hildenbrandia rivularis*
Glyceria fluitans f. *submersa*	A *Hydrurus* sp.
Potamogeton nodosus	M *Hygramblystegium irriguum*
Ranunculus trichophyllus	M *Rhynchostegium rusciforme*
Sium erectum f. *submersum*	

flourishing only in a **weaker current** (between 13 and 70 cm/s):

Elodea canadensis	*Potamogeton crispus*
Glyceria maxima f. *submersa*	*P. lucens*
Lysimachia nummularia f. *submersa*	*P. nitens*
	P. perfoliatus
Phalaris arundinacea f. *submersa*	
Sparganium erectum f. *submersum*	

Weber-Oldecop (1969) used the example of eastern Lower Saxony for a survey comparing the plant communities of running with those of still water. In 1977 he developed a running water typology for the whole of Lower Saxony on a floristic-sociological basis. These were somewhat altered by Wiegleb (1977) by taking in ecophysiological criteria.

Liverworts and mosses form their own communities along the stony banks of streams and rivers as well as on rocks which may stick out of the water from time to time. The examples in figs. 245 and 246 must suffice very briefly to indicate these.

(c) The effect of pollution on freshwater vegetation

Running water and still waters alike are being increasingly subject to the influence of chemical and organic pollutants originating from houses, farms and factories. The eutrophication which the more or less imperfectly purified waste waters bring about has already been discussed in section C I 1 a. This has led to rapid and sometimes catastrophic changes in the species composition of all the trophic groups belonging to the ecosystem, especially the plant communities. In many cases the number of species has been reduced whilst total primary production increased remarkably. The dead remains of the plants along with organic matter brought in by the drainage water have favoured the saprophytic microorganisms as well as the saprophagous animals (see fig. 247, also fig. 35). Such changes can be used for assessment of the water properties, at least according to the relatively rough 'saprobial' system of Liebmann (1962).

Many higher plants also react very sensitively, especially the submerged ones. A sign of pure lime-rich **running water** is the presence of the Stonewort *Chara hispida* and the Fen Pondweed

Fig. 245. The vertical amplitudes of some characteristic lichens on stones in a stream in the Taunus (near Frankfurt/Main) shown in relationship to the fluctuations in water height and to the degree of shading. After Ried (1960), slightly modified.

Black = in deep woodland shade (green shade), hatched = in half shade (blue shade), plain = in sunshine.

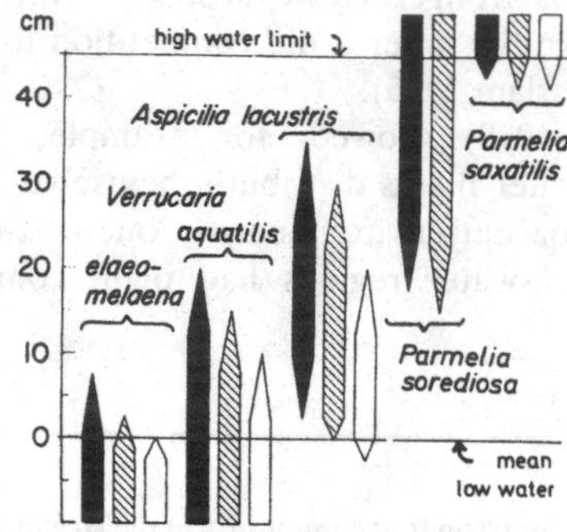

Fig. 246. Zonation of the Water Moss communities along the bank of the Mosel on natural rock (right) and on a stone-faced dam (left), semi-schematic. After Hübschmann (1967), modified.

Conditions for mosses are very unfavourable between summer highwater and the maximum highwater mark. However new substrate is rapidly occupied up as far as the height of the summer water.

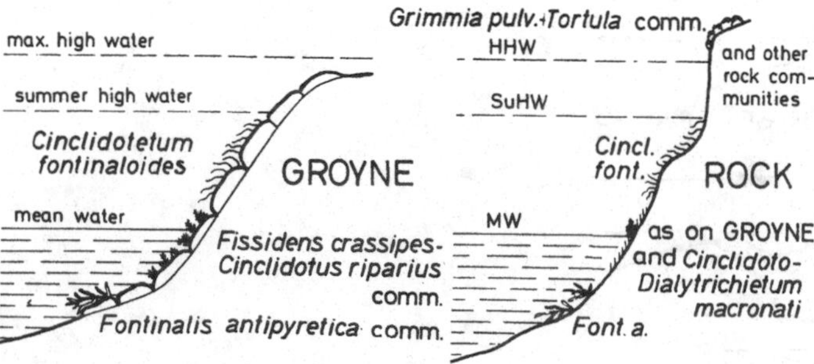

Potamogeton coloratus (see tab. 53 and fig. 248, also Kohler, Brinkmeier and Vollrath 1974). The clear brook community characterised by the latter was previously very plentiful in the northern pre-Alps and the centre of Switzerland, but during the course of the last two decades it has become a rarity. Instead a community of the river Water Crowfoot (*Ranunculetum fluitantis)* has extended its range. For a longer time this has been a common feature of the streams and small rivers of northern Germany (see Wiegleb 1976). Measurements by Kohler indicate that the deciding factor favouring *Ranunculus* and damaging *Potamogeton coloratus* is the NH_4 ion concentrations, while even the latter is not sensitive to NO_3. Ammonium actually reaches high levels in the drainage water, especially from settlements and cattle farms where natural manure is no longer used. Kohler (1976, see fig. 248) showed by transplant experiments that *Potamogeton coloratus* is damaged directly and not merely crowded out by competition. *Groenlandia densa* is not so susceptible to the effects of NH_4-ion and can exist in water with average pollution (tab. 53). However it dies during the course of a few months if it is transplanted to a more severely affected stretch of the river.

Many species of Starwort (*Callitriche*) have

Fig. 247. Relative numbers of the different feeding types in the saprobial system after Liebmann (1962). From Elster (1966), somewhat modified.

In heavily polluted (polysaprobial) water the main life forms are **saprophytes** and detritus eaters which get their nourishment from dead organic matter, as well as animals which eat bacteria and other saprophytes (the numbers of species are only relative).

In cleaner (oligosaprobial) waters it is the **autotrophs**, mainly green algae which predominate. These support a chain of consumers. In waters which are rich in nutrients but not burdened with organic material (mesosaprobial) the primary producers find their most favourable living conditions.

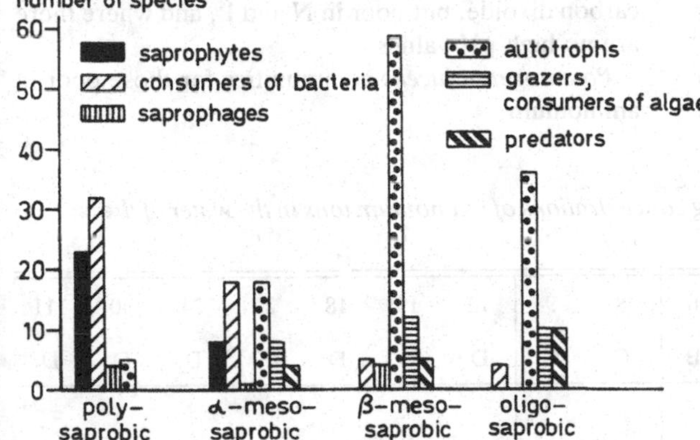

Fig. 248. Deterioration of species of Pondweed after they have been transplanted into more polluted parts of the same river (the Moosach in Upper Bavaria). After Kohler, Zeltner and Busse (1972), somewhat modified.

Ammonium and phosphate are introduced with sewage, especially in zone D which starts in the town of Freising. In experimental aquaria also the addition of extra ammonium leads to the blackening and death of the plants. Transplanting into clean parts of the Moosach (B, 18–14.5 km above Freising) or into a couple of tributaries (A) had no harmful effect on the plants.

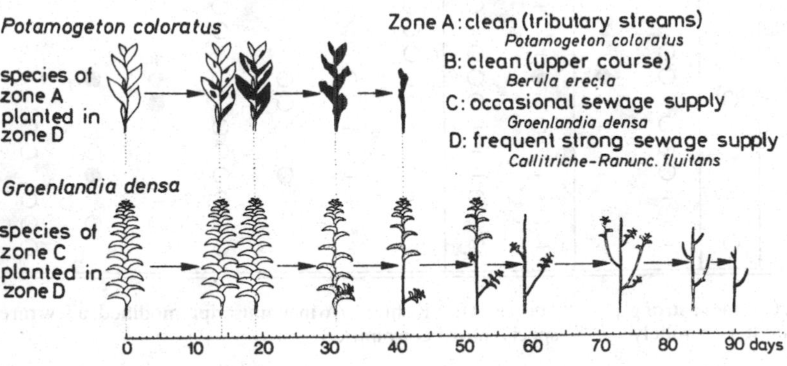

proved to be particularly resistant to the pollution of running water by sewage. However since these are difficult to identify they are not very useful as indicators. Nevertheless any luxuriant growth of the genus can be looked upon as a sign of pollution. They are often the last of the higher plants to persist in our streams (see Grube 1975). Taken as a whole the macrophytic vegetation of our flowing waters is reflecting quite clearly the degree of their pollution. Weber (1976) for instance, has given detailed information about the Hase river, and Weber-Oldecop (1977) has developed a typology in this context.

In **still waters** a dense covering of floating Duckweed species, especially *Lemna gibba* and *L. minor* is a good indication of pollution (tab. 49). We have already discussed the indicator value of both rooted and free-floating aquatics in sections I c and e. Since various natural chemical and physical factors affect the species composition in addition to any waste water it is difficult to pick out distinct species as definite pollution indicators. In spite of their widespread and thorough investigations Weber-Oldecop (1969) and Wiegleb (1976) arrived at only a very broad classification. The same is true for the tables of indicator values which Pietsch (1980, 1982) has thoroughly elaborated for the German Democratic Republic. However all authors found a relatively close correlation between the species combination of still-water communities and the degree of alkalinity, or the electrical conductivity of the water which is bound up with it. Under otherwise comparable conditions this factor follows rather closely the degree of pollution with sewage water, which not only adds organic material and plant nutrients, but also bases. Wiegleb summarised his findings in the form of a short ecological analysis of the most frequent species. He attributed high indicator values to the following communities:

– *Lemnetum gibbae* for waters rich in phosphate and mineral nitrogen, especially ammonium,
– *Ceratophylletum demersi* for those rich in nitrate,
– *Zannichellietum* for those very rich in calcium and influenced by phosphate,
– *Hottonietum* and *Stratiotetum* for waters rich in carbon dixoide, but poor in N and P, and where there are no high pH values.
– *Potamogeton lucens* communities for those poor in ammonium,

Tab. 53. An ecological sequence according to increasing concentrations of ammonium ions in the water of the Moosach.

Average ammonium content (0.01 mg NH_4/1) Ecological river zones	3 A	5 B	6 B	8 C	9 C	12 D	13 D	18 D	22 C?	24 D	40 D	111 D
Pogamogeton coloratus	●											
Chara hispida (alga)	●	–										
Mentha aquatica f. *submersa*	○	○	○	○								
Sparganium emersum + erectum	○		–	–	○							
Groenlandia densa				○	○							
Potamogeton natans var. *prolixus*				–	○							
Hippuris vulgaris				–	○							
Scirpus lacustris f. *fluitans*					○							
Callitriche obtusangula (spp. agg.)					○	○	○	○	●	○	●	●
Ranunculus fluitans + X trichophyllus				○	●	○	●	●	●	●	●	●
Zannichellia palustris ssp. *repens*				–	○	○	–		○			
Potamogeton crispus					–	○		–	○			
Potamogeton pectinatus		○	–	–	○	○	–					
Betula erecta f. *submersa*	○	○	●	●	○	○	○	○	○	●	●	
Fontinalis antipyretica (moss)	○			–	○		–	○		●		
Ranunculus trichophyllus		○	–	○	–		–		○	–		○
Agrostis stolonifera f. *submersa*		○		○	–	–	–		○		○	○
Elodea canadensis		○	–	–	–	–	–	○	○		●	
Veronica anagallis-aquatica		–		–	○	–	–		○	–		○
Nasturtium officinale		–		–	○				○	○	○	○
Phalaris arundinacea, submerged	○	–		–	○	–	–		○			○

Note: A = clean, rich in lime; B = weakly, C = more strongly, D = strongly polluted by sewage. – infrequent, ○ = widely distributed, ● = plentiful.

Source: After Kohler 1976 in manuscript, modified; a few rare species have been omitted.

– *Juncus bulbosus – Sphagnum* communities for very acid, bicarbonate-deficient peatland waters and those formed following open-cast mining.

3 *Spring areas and adjacent swamps*

The sources of our streams and rivers are distinguished by a manifold group of plant communities. They offer the most constant set of conditions which it is possible to get since they maintain practically the same temperature summer and winter. This corresponds to or slightly exceeds the average annual temperature for the particular district. Also the chemical properties of the water change little, with the exception of the karst springs which occupy a unique position.

Since spring water remains rather cool in the warmer parts of the year, and during the cold months is relatively warm and never freezes, it is a meeting place for alpine-arctic or subalpine-boreal organisms on the one hand and frost-sensitive atlantic or even submediterranean ones on the other. Hydrobiologists and plant sociologists have repeatedly occupied themselves with these fascinating biotopes. Thienemann (1922) first distinguished the following types of spring based on the way in which the water emerges:

1 **Rheocrenes** (gushing springs) in which the water spurts out of horizontal or downward sloping strata and immediately races down into the valley. Their vegetation resembles more or less that of streams.
2 **Limnocrenes** (spring basins) in which the water wells up from below. The large karst outflows in particular have similarities as plant habitats with the oligotrophic still waters with a corresponding lime content.
3 **Helocrenes** (seepages or swamp springs) where the water seeps up through the ground forming a swamp. This type is very interesting to botanists in carrying a number of plant communities arranged in belts around the wettest area. Its borders may lead into a normal swamp or mire vegetation.

The lime content of spring water is more important in determining the species composition than is the case with still waters. The two extremes are:

a) **soft** water springs which are chemically reminiscent of acid oligotrophic lakes,
b) **hard** water springs in which tuffs are formed by carbonate-assimilating plants.

In the case of the seepage springs and also where the gushing springs spread out, very characteristic vegetation units correspond to these two chemical types, namely (see also section E III 1, No. 1.5 and 1.6):

water	a) lime-deficient soft water spring vegetation	b) lime-rich spring tuff vegetation
	Montion	*Cratoneurion commutati*
saturated swamp	acid Small Sedge fen	calcareous Small Sedge fen
	Scheuchzerietalia	*Tofieldietalia*
	Caricion nigrae	*Caricion davallianae*

We shall have to be satisfied with these brief indications at this point but there is a very good comprehensive monograph by Maas (1959) on the spring vegetation of Central Europe, including the Alps, and by Warnke (1980) on that of Denmark. In particular the plant communities of the lowland springs are described in detail as these were the subject of investigations by Maas in the Netherlands and by Warnke (1980) on the Jutland peninsula. Mosses as structural elements in spring as well as in fen plant communities have been studied by Kambach and Wilmanns (1969).

Outside the Alps in Central Europe plant communities associated with springs are not very often met with, but above the alpine tree line they make up quite a large proportion of the total area, and they are much more diverse than in lower regions (Philippi 1975). For this reason we shall look at them more closely in connection with the vegetation of the high mountains where they often mingle with those of the alpine meadows and the screes (see section C VI 1 g and 6 c). In the lower mountains, e.g., in the Thuringia Forest (E. Lange and Schlüter 1972) and in the region between Stuttgart and Schwäbisch Hall (Sebald 1975) the spring vegetation is normally overshadowed by trees. Woodland spring mires have been described also from the lowlands and the foothills of the Alps, e.g., by Pfadenhauer and Kaule (1972) around the Chiemsee.

The Small-sedge communities occurring in the spring swamps are maintained largely through man's activities. If they were not mown occasionally or annually they would become overgrown with *Phragmites* or completely suppressed by Willow bushes and fen trees. At the same time their vegetation is very similar to that of the fens or the 'intermediate' mires to which they are floristically closely related. We shall return to them therefore when we discuss the mires and meadows.

II Woodless fens and intermediate mires compared with other mire types

1 A survey of the European mire vegetation

(a) Types of mires based on their physiognomy and ecology

Mires are accumulations of peat covered with vegetation. The plants are growing on almost pure humus which contains only small amounts of minerals. At least during the time of its formation the peat was saturated with water preventing the further decomposition of plant remains through shortage of oxygen. These organic debris have come either from plants which resemble those still growing there, or from earlier vegetation types forming stages in a peat formation process as described in section C I 1 b. Mires which have been drained cease to carry a peat forming vegetation.

Since most mires in Central Europe are now under cultivation or have been influenced by man in some other way their original condition can now often only be reconstructed from a study of the soil profiles. Depending on the conditions under which they were formed and the age of the mires there are so many differences in the sequence, thickness and properties of the peat layers that it is almost impossible to put them into any kind of order (F. Overbeck 1975). To use the present day vegetation as an indication of the type of peat on which it is growing can only be done with reservations since the species composition depends in the first place on the fluctuations in the water level and on the chemical composition of the water saturating the relatively shallow root run.

Conditions under which the plant cover is living for the time being change from one mire to another and can be very dissimilar even within the same mire complex. Because of this our mires are exceptionally varied both **physiognomically and floristically** (see table 54). On many of them woodland comes in as the climax of their succession. We have already discussed these fen woods in section B V 2. Most of the 'grass mires' too we have met in the form of Reed or Sedge swamps and similar communities along the lakesides (section C I 1 f and g). In section D II 2 a we shall return to the 'heath mires' dominated by dwarf shrubs. Large areas are overgrown with 'moss mires' which will be dealt with in their own main section (C III) since these have been exhaustively studied for many decades.

In a narrower sense the term mire is restricted to the grass and moss mires on which there are no trees or bushes. Fen woods for example are often not referred to as mires by laymen since, as we have seen, they do not always grow on thick peat. However like many peatland heaths and Sedge fens they lead on to 'anmoors' (half-bogs), i.e., to a mineral soil which is very rich in humus and carries an intermediate type of vegetation.

Since mires are a product of plant communities it was suggested that they should be named after them. However the only satisfactory method of characterisation of mires in their natural state is one based on **plant sociology** and even this runs into difficulties. In most mires a number of communities of dissimilar systematic rank are to be found growing alongside each other. Since they tend to fit together as a mosaic or a series of belt-like zones into higher units, a survey of them is made easier if a **genetic-ecological** mire typology takes precedence over one based merely on vegetation.

Since the cultivation of mires started a distinction has been drawn between 'Niedermoor' (= fenland) and 'Hochmoor' (= raised bog). The differences between them can be seen particularly well in the lowlands near the North Sea and in the pre-Alps. The surface of a **fen**, or low mire, follows that of the groundwater and is by and large horizontal. The raised bog forms a more or less clear dome shape above the surrounding land (see figs. 263, 264) and makes its own water table. For this reason it is more easily drained than a fen which never possesses a natural run off round its edge. Low mires are fed by groundwater for the most part and so can arise even in the driest parts of Central Europe e.g., in the rain shadow of the Harz mountains. Raised bogs on the other hand are dependent on sufficient precipitation and air humidity. They are formed by the growth of certain species of peat moss (*Sphagnum*) which can produce large spongy cushions rising above the general water level. We shall discuss the conditions under which these rise over the general ground water level in section C III 1 a. Then we shall also get to know the different types of **raised bog** which are to be found in Europe. Here only the main types will be indicated briefly following the arrangement suggested first by Osvald (1925). Advancing from the extreme oceanic to the relatively continental climate it is possible, according to F. Overbeck (1975), to distinguish five main types of raised bog which become progressively less hostile towards the establishment of trees:

1 Blanket bog
2 Flat raised bog
3 Plateau raised bog
4 Shield raised bog
5 Woodland raised bog

The plateau raised bog may also be called the 'true' raised bog. In Central Europe it is naturally the most plentiful, especially in the area where Beech dominates the woodlands. In the lowlands a big true raised bog can rise to a number of metres above the very wet zone round the edge, where the water draining from the raised bog collects and joins the water draining from the surrounding mineral soils (see fig. 249). These surrounding swamps (called 'lagg' following the Swedish example) consequently show

Tab. 54. A synopsis of the different types of fens and bogs with their plant communities.

Type of mire	Fen (and related "anmoors") eutrophic	lime-rich oligotrophic	lime-deficient oligotrophic	Intermediate mires oligotrophic to dystrophic	Raised bogs dystrophic
Peat characteristics: pH of peat	ca. 4.5–7.5	above 7.0	ca. 3.5–5.0	ca. 3.5–4.5	ca. 3.0–4.2
lime content	average to high	very high (lake chalk)	low	very low	extremely low
Content of other minerals	high	low to average	low	very low	extremely low
Nitrogen availability	good to very good	poor to fair	fair to poor	very poor	extremely poor
Average decomposition degree	very high to fair	fair (to high)	fair (to high)	low to fair	very low
Dominant growth forms: Trees and bushes	**Alder fen** *Alnetum glutinosae* Willow-Alder Buckthorn scrub *Frangulo-Salicetum*	rare	Birch-Alder fen Spruce-Alder fen and similar intermediate communities	**Birch fen** in the west *Betuletum pubescentis* **Pine fen** in the east *Ledo-Pinetum* and others	**wooded raised bog** in the east *Sphagnetum pinetosum* raised bog edge woodland, bog Pine scrub
Dwarf shrubs	–	–	–	**Cross-leaved Heath mire** *Ericetum tetralicis* (only in the north-west)	**dwarf shrub hummocks** *Erica* or *Calluna* stadia
Tall grasses	Reed beds, *Phragmitetum* and others			(Purple Moorgrass stadia after draining and burning)	
Tall grass-like plants	**Tufted Sedge swamp** *Caricetum elatae* **Sharp Sedge swamp** *Caricetum gracilis*	**Saw Sedge swamp** *Cladietum marisci*	**Bottle Sedge swamp** *Caricetum rostratae*	Slender Sedge **swamp** *Caricetum lasiocarpae*	–
Short grass-like plants	–	Bogrush beds beds *Schoenetum nigricantis* calcareous small Sedges beds *Caricetum davallianae*	**acid small Sedges beds** *Caricetum can.-nigrae*	White Beak-sedge communities *Rhynchosporetum*	**Deergrass bog** *Trichophorum-Sphagnum* communities in the west
Sward-forming mosses	–	Brown Moss-rich communities (*Drepanocladus-*, *Calliergon-* species or other Hypnaceae, also Bryaceae and other non-Sphagnum) only fragmentary in Central Europe		**Peatmoss hollows** *Scheuchzerietum* and other communities with *Sphagnum cuspidatum recurvum* and similar	**Peatmoss hummocks** *Sphagnetum papillosi* in the west *Sphagnetum magellanici Spagnetum fusci* in the east

the characteristics of a fen. Raised bogs in the mountains also mostly have surrounding swamps; they can occur in hollows and on slopes as well as on level ground (fig. 264). Most of the real raised bogs in their natural state are free of trees. Only the better-drained marginal slopes near to the lagg carry a more or less dense stand of trees (fig. 249 and 250).

In the continental part of Europe, where at times the climate can be very dry, trees are able to establish themselves even in the centre of raised bogs, to an extent that the whole mire may eventually become wooded raised bog (see section 2 b). On the other hand a very wet climate with cool summers, as in Ireland and northern England, leads to a bog on which trees cannot grow at all. Here conditions are favourable for the formation of a blanket bog growing luxuriantly over everything and covering the whole terrain. There are only suggestions of this type in Central Europe and only then at the montane and subalpine levels. The flat raised bog too, besides being difficult to identify, is also hardly ever found here. According to Osvald (1925) it can be distinguished from the 'true' raised bog by the absence of a lagg in southern Sweden; more or less swampy heaths in the surrounding country merge into the treeless raised bog without any obvious boundary. We shall talk about the 'shield' raised bog in section C III 2 a. It is rather like the plateau raised bog, but possesses deeper concentrically arranged hollows. The distribution centre for this type is in the north-eastern part of Europe.

A contemporary of Osvald, v. Post (1925), recommended a simpler mire typology. All the fens and the intermediate mires, which we have yet to discuss, he described as **'topogenous'** since they depend on where the water collects and therefore on the topography of the country. True raised bogs on the other hand are **'ombrogenous'** i.e., dependent on rainwater. Fens can also be said to be **'soligenous'** (mainly determined by the ground).

Fig. 249. View from the marginal slope of a raised bog in Upper Bavaria (with Mountain Pine, dwarf shrubs and Cotton-grass) towards the wet (lagg) zone with fen and intermediate mire and on to the surrounding swamp woods in the distance.

Ecologically, raised bogs and fens differ above all in their content of nutrients and bases in turn leading to the differences in pH value which can easily be measured (figs. 251 and 252). Within the series from fen to raised bog not only do the pH and the lime content fall but also the contents of potassium, magnesium phosphorus and nitrogen (Waughman 1980). The latter is offered mainly as ammonia (see also section 4 a).

Fens and raised bogs are not sharply divided from each other, but can blend into one another both in space and in time. Therefore when one wishes to stress the successional aspect one may also speak of a 'transitional' mire ('Übergangs-moor'). If one merely wants to indicate a mire type occupying the middle position with regard to vegetation or habitat then the name **'intermediate' mire** ('Zwischenmoor') can be used. We prefer the second since it is often difficult to prove a genetic connection, which can rarely be recognised in the countryside without further evidence. Where the ground is poor in bases an intermediate mire can develop from the outset, i.e. without a previous fen.

(b) The sociological classification of tree-deficient mires and swamps

In mires and swamps where the top layers are too wet for the growth of trees or where there are no trees for some other reason, many different plant communities can be found. According to Oberdorfer (1977) these can be arranged into 6 orders divided amongst 3 classes

Fig. 250. A semi-schematic section through a raised bog and its adjoining fen near the Alps south of the Chiemsee. After Leiningen (1907), somewhat modified.

I. General Survey: A-B = man-made meadows on mineral soil; B-C = Spruce and Pine swamp wood merging into the woods along the edge of the raised bog; C-D and E-F = marginal slopes of the raised bog with Birch; D-E = raised bog with Pines, Birch and (in the centre) Mountain Pine; F-G = transition to swamp wood; G-H = Spruce and Alder swamp wood; H-J = Sedge beds (G-J = fenland).

II. Detail of the transition from raised bog to fen: K-L = raised bog with Mountain Pine and Birch; L-M = surrounding slope with Birch; N-O = Spruce swamp wood; O-Q = transition to the Alder swamp wood (Q-R); R-S = vegetation of the filled-up lake (N-S = fenland in the broad sense, i.e. including the swamp woods).

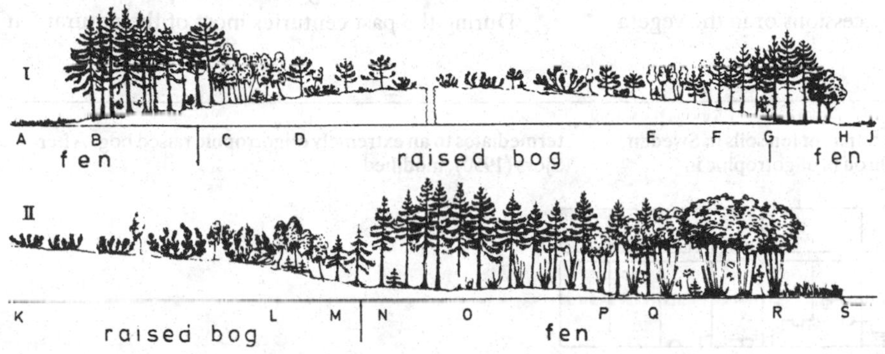

Fig. 251. The total amount of nutrients per hectare in a fen, an intermediate mire and a raised bog (the amounts of phosphorus and especially of nitrogen available to the plants are significantly less). From data in Brüne (1948).

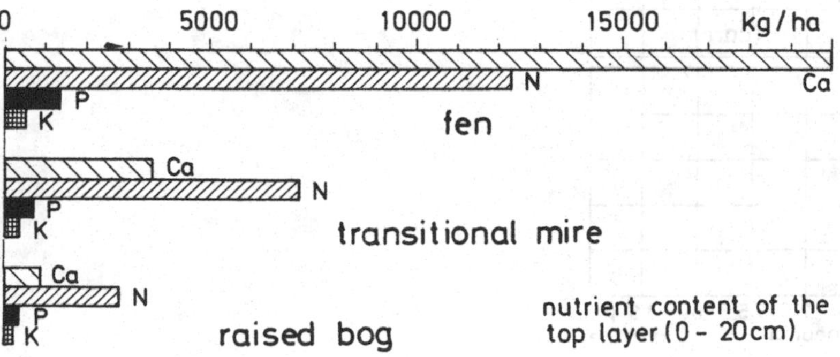

as follows:

Raised bog and heath mires *(Oxycocco-Sphagnetea)*	heath mires of western Europe (*Sphagno-Ericetalia*) raised bog communities (*Sphagnetalia magellanici*)
Intermediate mires and Small-sedge fens, including the raised bog hollows (*Scheuchzerio-caricitea*)	bog hollows and intermediate mires (*Scheuchzerietalia*) acid-soil Small-sedge fens (*Caricion nigrae*) } fens calcareous Small-sedge fens *(Tofieldietalia)* } fens
Large-sedges and Reed swamps (*Phragmitetea, Phragmitetalia*)	already dealt win in sections C I 1 f and g.

The most important alliances and associations for Central Europe within the 5 orders listed above are mentioned in tab. 5. We shall discuss them in the following sections but look more closely only at the vegetation of the raised bogs (*Sphagnetalia fusci* and *Scheuchzerietalia*, section III).

2 Natural and semi-natural Small-sedge fens

(a) Lime-rich Small-sedge fens

Most of the treeless fens have already been met with as members of peat-forming successions or in the vegetation of springs (section I 1 and 3). In this way we were able to work out their genetic position and the peculiarities of their habitats better than if we had treated the fens as a closed group. Table 54 can serve as an easily understood reminder of this and will help to summarise which systematic units belong to fen vegetation. The wooded fens which we have already met in section B V 2 are also set out in tab. 54 in order to facilitate comparison and to name all the vegetation units falling under the general concept of 'mire'.

During the past centuries most of the natural fen

Fig. 252. Acidity of the water in different fen soils in Sweden, from a eutrophic lime-rich fen through oligotrophic intermediates to an extremely oligotrophic raised bog. After Sjörs (1950), modified.

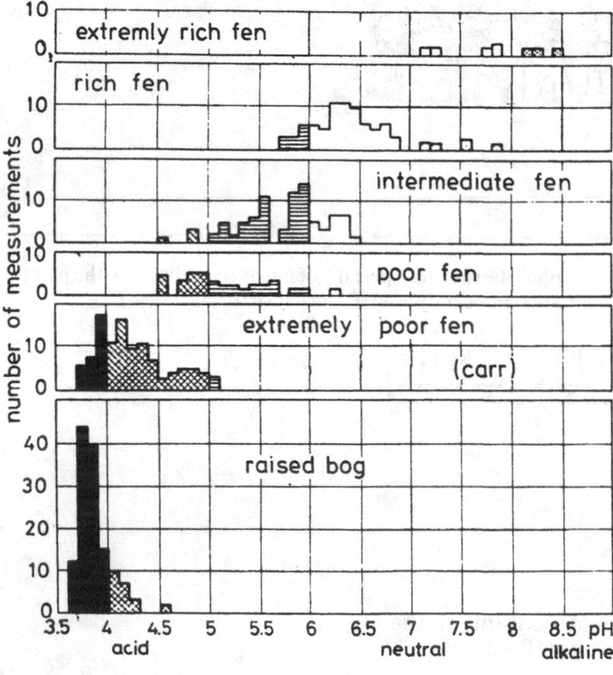

woods have been replaced with grasslike communities. By means of occasional mowing these are prevented from reverting to woodland, although apart from this they are very little influenced by man. They can be recognised by the dominance of small sedges, rushes, Deergrass or Cottongrasses and could be called Small-sedges fens (fig. 253). Because of the lack of competition from quick-growing plants they have become the refuge of numerous helophytes otherwise rare in the Central European flora. This particularly

Tab. 55. Systematic survey of raised bog, intermediate mire and fen communities after various authors. CAPITALS = *classes*, bold = *orders (see also E.III.1).*

Braun-Blanquet (1950 u.a.) Swiss Central Alps	Tüxen (1955) N.W. Germany	Oberdorfer (1983)[1] BRD and DDR	English names of the Oberdorfer list
OXYCOCCO-SPHAGNETEA **Sphagno-Ericetalia** *Ericion tetralicis*	OXYCOCCO-SPHAGNETEA **Erico-Sphagnetalia** *Ericion tetralicis* *Ericetum tetralicis* *Sphagnetum papillosi*	OXYCOCCO-SPHAGNETEA **Sphagno-Ericetalia** *Ericion tetralicis* *Ericetum tetralicis* *Sphagno compacti-Trichophoret. german.*	RAISED BOGS and HEATHS **West European heaths** Erica heaths and 'anmoors' Cross-leaved Heath swamps
Ledetalia palustris *Sphagnion fusci* *Sphagnet magellanici* *Sphagnetum fusci*	*Sphagnet magellan.* **Sphagnetalia fusci** *Sphagnion fusci* *"Trichophoro-Sphagnetum fusci"*	**Sphagnetalia magellanici** *Sphagnion magellanici* *Sphagnetum magellan.* *Eriophoro-Trichophoretum cepitosi* *(Sphagnetum fusci)* *Pino-sphagnetum* *and other Communities*	**Eurosib raised bog hummocks** Europ. raised bog hummocks red hummocks brown hummocks Pine raised bog
SCHEUCHZERIO-CARICETEA **Scheuchzerietalia** *"Stygio-Cricion limosae"* *Caricetum limosae* *Rhynchosporetum*	SCHEUCHZERIO-CARICETEA **Scheuchzerietalia** *Rhynchosporion* *Scheuchzerietum* *Rynchosporetum albae* *Caricetum lasiocarpae*	SCHEUCHZERIO-CARICETEA **Scheuchzerietalia** *Rhynchosporion* *Caricetum limosae* *Rhynchosporetum albae* *Caricion lasiocarpae* *Caricetum lasiocarpae* *and other Communities*	INTERMEDIATE MIRES and FENS **floating mats, hollow and intermediate mires** hollows Mud Sedge hollow White Beak-sedge hollow Northern Small Sedge hollow Slender-Sedge swamp
Caricetalia nigrae *Caricion nigrae* **Caricetalia davallianae** *Caricion davallianae* *Caricetum davallianae* *Schoenetum nigricantis* *Caricion bicolori-atrofuscae*	**Caricetalia nigrae** *Caricion canescenti-nigrae* *mehrere Assoziationen* *Caricion davallianae* *Schoenetum nigricantis* u.a. Assoziationen	**Caricetalia nigrae** *Caricion nigrae (=fuscae)* *Caricetum nigrae* *and other Communities* *Eriophoretum scheuchzeri* **Tofieldietalia** *Caricion davallianae* *Caricetum davallianae* *Schoenetum nigricantis* *and other Communities* *Trichophorum alpinum* *Caricion bicolori-atrofuscae*	**Acid-soil Sedge swamps** Central Eur. Common Sedge swamps Common and White Sedge swamps Cottongrass swamps (alpine and subalpine) **Calcareous Small Sedge swamps** European Davall's Sedge swamps Davall's Sedge spring swamps Black Bogrush swamps Sedge swamps of the central Alps

[1]) After Oberdorfer, E., 1983 Pflanzensoziologische Exkursionsflora, 5th ed. Sphagnetum papillosi is rarely found in Central Europe

applies to the lime-rich Small-sedge fens (*Tofieldietalia*, see tab. 55). The ecological and sociological behaviour of *Tofieldia calyculata* has been studied by Sebald (1964).

Such floristically exciting semi-natural fen communities actually have become extremely rare. They have their natural habitat at the swamp springs (helocrenes) in the subalpine belt, and it is here that they achieve their richest development (see section C VI 6). At lower altitudes alpine and subalpine species are absent from the Small-sedge fens which are brought about and maintained by mowing. Instead some meadow plants came in, such as Purple Moorgrass (*Molinia coerulea*) and Devil's-bit Scabious (*Succisa pratensis*), but only where the water level had been lowered could they achieve dominance and eventually lead to a real meadow community (*Molinio-Arrhenatheretea*).

In the pre-Alps, for example on the broad flat banks of the Bodensee, a special community of the *Tofieldietalia* occurs, i.e. the **Black Bog-rush fen** (*Schoenetum nigricantis*), to which Zobrist (1935) has already devoted a monograph from Switzerland and adjacent regions (see also Görs 1967 and Klötzli 1973). According to Kloss (1965) Bog rush fens and other lime-rich fens occur even in Mecklenburg, i.e., in the far north of Central Europe. **In the British Isles** a *Schoeno-Juncetum subnodulosi* is widespread but common, and the same is true for *Carex davalliana* spring swamps (Wheeler 1980). It should be emphasised that *Schoenus nigricans* is by no means restricted to habitats rich in lime. Near to the sea it flourishes also on blanket bogs (see section 1a) where it profits from nutrients supplied by ocean spray. According to Boatman (1962), Sparling (1967) and Wilson (1959) it mainly receives K, Mg, and possibly also N and other bio-elements.

Fig. 253. Calcareous Small Sedge bed with *Eriophorum latifolium* alongside a Sharp Sedge bed (*Caricetum gracilis*, shining leaves in the background) and the Reed belt of a pond in the Rot valley to the east of Crailsheim.

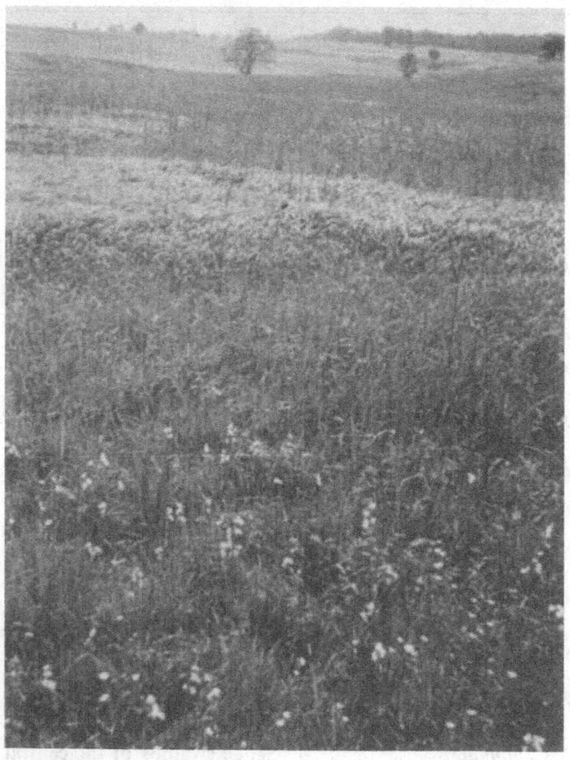

The wettest subunits of the *Schoenetum nigricantis* are certainly inimical to tree growth. The same is true for some of the **Davall-sedge fens** (*Caricetum davallianae*) at any rate for those directly on the edge of springs actively depositing tuff (Görs 1963, W. Braun 1968). By means of pollen analysis Moravec and Rybníčková (1964) have proved in the Bohemian Forest that a fen Alder wood had been converted into a *Caricetum davallianae* obviously through human impact. Such calcareous Small-sedge fens are only to be found in southern Central Europe, mainly in the Alps. **The true *Caricetum davallianae* needs a rather oceanic** climate; further east there are related communities (fig. 254, see also Kopecký 1960).

All communities of the lime-rich Small-sedge fens are fairly worthless to the farmer as they have a very low yield and at the most could only provide a small amount of litter (fig. 233). So for the most part they have been drained and turned into more productive meadows. The lime-deficient sedge fens give a somewhat better kind of fodder, but their yield also is low and they have been gradually disappearing as they have either been improved or afforested.

(b) Acid-soil Small-sedge fens

The Small-sedge fens which have a good or above average supply of lime look quite similar physiognomically to the ones whose species combination indicates a shortage. These acid Small-sedge fens (*Scheuchzerietalia*) contain fewer species than most of the *Tofieldietalia* communities. Their main centre of distribution lies in the lowlands of the north-west, but they are also found in the south of Central Europe including the Alps, often as a **Common Sedge fen** (*Caricetum nigrae*, see fig. 255). They are a step towards the intermediate mires and indeed in some cases they are included with these by many authors.

3 Intermediate mires and their problems

(a) Intermediate mires without trees

Treeless intermediate mires range from extremely oligotrophic to fairly dystrophic. Morphologically they still belong to the fens but they are often found in close association with raised bogs and sometimes they can lead to these genetically. This idea of an intermediate mire has been applied in so many ways and defined differently by various authors. However H. Paul and Lutz (1941) have tried to clarify the position from the point of view of plant sociology and peatland science. We shall follow their ideas in all important matters.

Most of the treeless intermediate mires which have been described by Paul and Lutz from the pre-Alps are joined like the acid Small-sedge fens, to the class *Scheuchzerio-Caricetea*. Within this they belong to the order *Scheuchzerietalia*, whose main distribution lies in northern Europe. It embraces not only bog hollows which are included in the *Rhynchosporion* (tab. 55), but also the intermediate mire communities par excellence which are always present in the 'lagg' surrounding a raised bog (see section C II 1). They are united in a single alliance, the *Eriophorion gracilis*, which is mainly distributed in Fennoscandia.

The most plentiful community in this intermediate mire alliance is the **Slender-Sedge swamp** (*Caricetum lasiocarpae*), an open turf of unique appearance. The Slender Sedge grows up to over knee height from a wide meshed network of rhizomes, its narrow leaves bending over gently in a characteristic manner (see tab. 51 and fig. 256). Accompanied by just a few other

Fig. 255. Fenland communities on a base-deficient slope with springs in the gneiss area of southern Black Forest (schematic). After Philippi (1963), somewhat modified.
The Parnassus-Flea Sedge community (centre) grows where there is a fair amount of groundwater movement while the Common Sedge swamp (left) can tolerate the stagnating acid groundwater of the valley floor. The Matgrass sward (right) is independent of the groundwater; its soil dries out from time to time in summer.

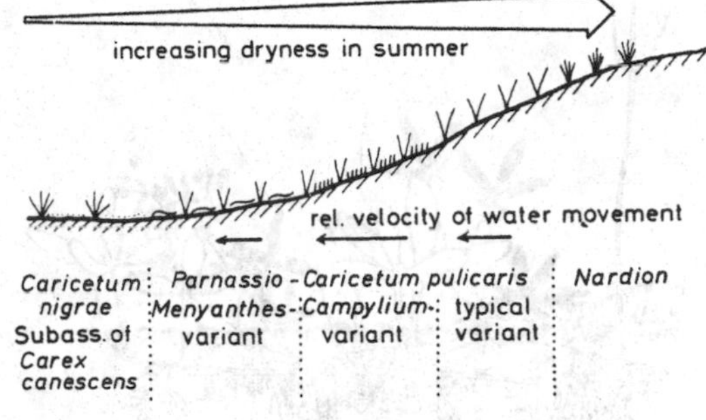

Fig. 254. The distribution of Davalls Sedge communities (calcareous Small-sedge swamps) in Central Europe as classified by Moravec (1966), modified. Many of the eastern communities penetrate towards the west. The true *Caricetum davallianae* is confined to the south-western part of Central Europe.

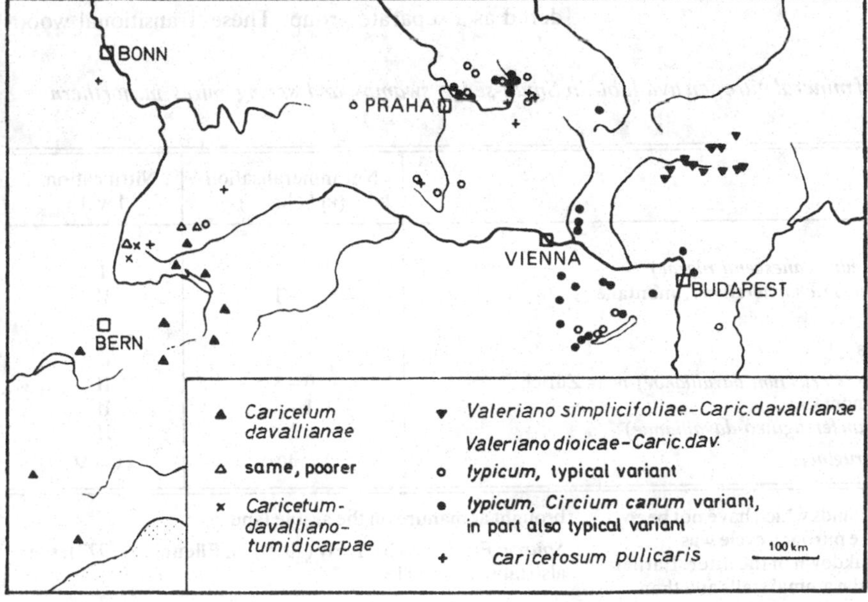

species it occupies not only great parts of the laggs but also oligotrophic shallow ponds and in suitable habitats it is distributed as far as Hungary (Kovács 1962, which also contains the European literature; see also section E III 1, No. 1.613).

Mosses often produce the most biomass, not only in the swaying Slender Sedge community, but also in those of other intermediate mires. In the Deergrass spring swamp (*Chrysohypno-Trichophoretum*) for example these are branching mosses (Hypnaceae) especially in the relatively wet subassociations (fig. 257).

The Slender Sedge and the Deergrass communities are entirely without trees. The same applies for the **Rannoch-rush hollows** (*Scheuchzerietum palustris*) and the **White Beak-sedge hollows** *Rhynchosporetum* of raised bogs, peat cuttings and wet moorland. Only in their driest forms, which can only be produced by drainage, is it possible for Willows and Birches or other trees and shrubs of oligotrophic fen woods to come in.

The same is true of the intermediate mire communities which are richer in peat mosses and are much more like raised bogs than those which have already been discussed. The lists Nos. 1-3 of *Sphagnum parvifolium* communities given in table 57 are from that part of Sonnenberg bog in the Harz which is not raised. They are somewhat richer in species than the true raised bog communities. According to a careful investigation by Jensen (1961) there is a step by step transition from the spring fen in the uppermost part of the mire slope (No. 1) to the lowest, purely ombrotrophic part (No. 5, see also fig. 258). This gradual transition corresponds to a nutrition gradient.

Fig. 256. A quaking bog with *Equisetum fluviatile, Menyanthes trifoliata, Potentilla palustris*, and *Carex diandra* (*Caricetum lasiocarpae*). After Vanden Berghen (1952), slightly modified.

The swimming *Sphagnum* mass is given an additional lift by the air moving in the rhizomes of *Menyanthes* (Coult 1964).

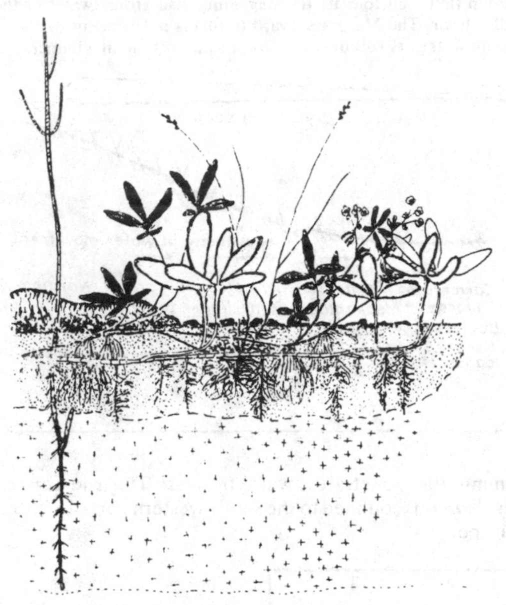

(b) The sociological status of 'transitional woodland mires'

According to Kubiëna's (1953) key for the identification of soils the treeless intermediate mire had to be included in the underwater humus formations. It was only the wooded intermediate mires that he considered as a separate group. These 'transitional wood-

Tab. 56. The annual amount of mineral nitrogen available in Small-sedge swamps and spring mires in northern Switzerland.

Plant communities and soils	Net mineralisation (kgN/ha/y.)	Nitrification level
1. Acid-soil Small-sedge swamps		
a. Common Sedge sward (*Caricetum canescenti-nigrae*)	0 – 2	I
b. Deergrass mire (*Tomenthypno-Trichophoretum*), montane	0 – 1	II
2. Calcareous Small-sedge swamps		
a. Davall's Sedge spring swamp (*Caricetum davallianae*) near Zürich	0 – 5	II
the same in W. Switzerland, montane	1 – 5	II
the same, subalpine (*Caricetum ferrugineo-davallianae*)	2	II
b. Bogrush fen (*Schoenetum ferruginei*)	0 – 40[a])	IV – V

Note: [a] The higher values refer to stands which have not been mown for a long time and where the nitrogen cycle was reactivated through microbial breakdown of the litter. Earlier the mowings were used as litter in the animal stalls and then brought as manure on the arable land.

Source: From data by Yerly (1970), in Ellenberg (1977); see also tabs. 19 and 118.

land mires' ('Übergangs-Waldmoore') he included with the raised bogs as semi-terrestrial formations since the uppermost layer of humus is scarcely affected by groundwater if at all. We have already dealt with these acid-humus Birch or Pine fen woods in section B V 2 c and compared them with the more or less eutrophic fen Alder woods. Lutz and co-workers (1957) studied the successional position of these acid-humus fens in Upper Bavaria including questions of microbiology and animal sociology.

Seen from the plant sociological point of view the transitional woodland mires cannot be considered as a uniform group. Some of the suboceanic fen Birch woods are very similar to wet-soil Birch-Oak woods. Others may be looked upon simply as Birch stages of raised bogs. On the other hand fen Pine woods with

Fig. 257. Moss-rich spring-fed intermediate bogs are generally found on soils which are permanently waterlogged (see also fig. 255). From data by Rybniček (1964). In the Bladderwort community (1) the groundwater level is always round about the soil surface. In the Deergrass bog (2) it does not fluctuate much more apart from the Beak Sedge subassociation (2c). The calcium content of the water falls throughout this sequence.

1. *Scorpidio-Utricularietum*
2. *Chrysohypno-Trichophoretum*
 a. *utricularietosum*
 b. *typicum*
 c. *rhynchosporetosum albae*

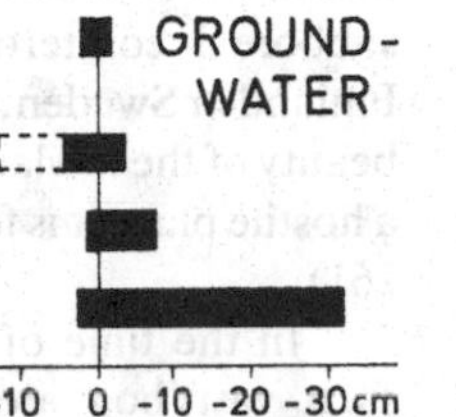

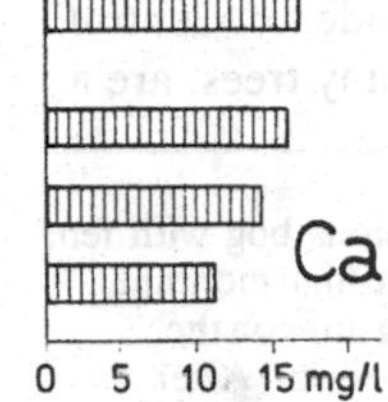

Tab. 57. Intermediate and raised bog communities on the Sonnenberg mire in the Harz at about 800 m.

	Interm.			Raised	
Serial no.:	1	2	3	4	5
Species common to all:					
Eriophorum vaginatum	5	5	5	5	5
Vaccinium oxycoccus	5	5	5	5	5
M *Sphagnum magellanicum*	5	5	5	3	4
M *S. rubellum*	3	4	4	5	5
M *Aulacomnium palustre*	2	4	3	4	1
Drosera rotundifolia	1	1		1	1
Calluna vulgaris		3	4	5	5
Andromeda polifolia	1	3	2	5	5
Trichophorum cespitosum	2	1	1	3	4

	Interm.			Raised	
Serial no.:	1	2	3	4	5
Distinguishing species:					
Molinia caerulea	5				
Trientalis europaea	4				
Eriophorum angustifolium	3	5			
M *Polytrichum commune*	2	2			
M *Sphagnum apiculatum*	1	2			
M *S. parvifolium*	5	5	5		
Vaccinium uliginosum	1	2	2		
M *Calliergon stramineum*		1	1		
Empetrum nigrum	1	4	5	5	
M *Polytrichum strictum*	5	4	5	2	

Note: Nos. 1-3: **Intermediate bog** communities, described as fen by Jensen but assigned to the same association as the raised bogs (*Sphagnetum megallanici-rubelli sphagnetosum parvifolii*):
1: *Molinia* variant; weakly mesotrophic (bordering on oligotrophic), with springs;
2: *Eriophorum*-variant; fairly oligotrophic, wet;
3: typical variant: oligotrophic, less wet than 1 and 2.

Nos. 4-5: **Raised bog** communities (*Sphagnetum magellanici-rubelli typicum*):
4: *Empetrum* variant; very oligotrophic, otherwise like 3 and 5;
5: typical variant; extremely oligotrophic, strongly peat forming.

Species with a very low constancy have been omitted.

General remarks

The formation of the raised bog began here about 1600 B.C. and went on until the peat was 3 – 3.5 m deep (maximum 5.4 m). No. 1 lies in the upper part near a spring at the foot of the Bruch mountain producing water that is certainly very acid but contains some minerals. Nos. 1–5 form a series on the gently sloping surface of the raised bog in which the mineral content falls gradually without any significant change in the acidity (about pH 4). Topographically no. 5 is lowest (see fig. 258). Like most of the other species common to all the communities *Sphagnum magellanicum* and *S. rubellum* grow best in no. 1. Here though they do not form a raised bog type peat since their dead remains are rapidly decomposed. The distinguishing species can be looked upon as indicators to different degrees of minerals in the soil water.

Source: After tables by Jensen (1961).

Ledum palustre are partly aligned with the acid-soil Pine woods and partly with the continental wooded raised bogs (*Sphagnion fusci*, see section III 2 b). This splitting of a group where the soil conditions are rather uniform does not satisfy the ecologist. Nevertheless it expresses the dynamic tendencies of individual stands much better, for not all transitional wooded mires end up as raised bogs (see fig. 259).

III Raised bogs and the communities connected with them

I *True raised bogs*

(a) Conditions necessary for the creation of raised bogs

In the woodland-dominated countryside of Central Europe the raised bogs, with scarcely any trees, are a world unto themselves, not only with regard to their unique appearance, but also the way in which they come about, their economy, and the organisms able to exist in these ecosystems.

Unfortunately it is becoming almost impossible to study 'living' raised bogs, even in flat coastal landscape which is richest in bogs. Apart from some very small remains they have been cultivated or at least drained. To get a clear idea of this vegetation type which was once so widespread one has to climb up into mountains such as the Harz, Hohe Venn, Rhön, Black Forest and Krkenose, seek out remote parts of the Alps, or of the north European countries. Only if one can wander for many kilometres, sometimes having to jump from one hummock to another, over one of the huge bogs in the subboreal coniferous woodland region of Estonia, Finland or Sweden, can one appreciate the quiet lonely beauty of the lowland raised bog, and understand what a hostile place it is for the growth of trees (figs. 260 and 261).

In the time of Grisebach (1846) there were still such vast bog areas in the north-western part of Europe. 'At the borders of Hannover and Holland', he wrote, 'striding across the trackless bog from Bourtange I came to a point where the flat horizon formed a complete circle as though I were on the high seas, with no tree, nor bush, nor hut, or anything taller than a child standing out above the apparently endless wilderness. Even the villages, hidden in the Birch-woods, which for a long time appear like blue islands in the distance, eventually sink below the clear horizon'. To day we must be glad that there are at least a few small areas left on the plains of north-west Germany and the Netherlands which remain for scientific study, such as the 'Grosse Moor' and the 'Weisse Moor' near Kirchwalsede in the district of Rotenburg, Hannover (Jahns 1962).

Each true raised bog is a microcosm and it deserves this name much more so than a lake since it depends only on local precipitation for its water supply and for any nutrients. All other types of mire, or of water, are supplied from outside by ground or drainage water from the surrounding countryside. The shape of mires depending on water from mineral soil (Du Rietz 1954) is determined by the groundwater level and so is more or less flat (see section II 1 a). In contrast the raised bog lies like a large cake over an ancient lake filled up with peat, or swampy ground (fig. 259). It has its own water table which is quite independent of that in the surrounding ground. This is also true in those cases where the raised bog does not project clearly above the level of the surroundings. In this case it is the peculiar bog vegetation which betrays its real nature.

Fig. 258. Schematic map of an ombro-soligenous bog with fen, intermediate and raised bog complexes in the high montane Spruce belt of the Harz (e.g. the Sonnenberg mire on the southern slope of the Bruch mountain in the Harz). After Jensen (1951), modified. (See table 57.)

The spring arising above impermeable quartzite was the cause of the bog formation which was thus originally soligenous, i.e. began because the soil was permanently waterlogged. Although the spring water is very acid (pH 4) and poor in nutrients it is continually supplying sufficient minerals for the upper part of the mire still to have the characteristics of a fen. As it flows down the slope the water becomes poorer and poorer in minerals and finally completely oligotrophic. Parallel to this sequence the Moorgrass fen (with *Molinia caerulea*) becomes a Cottongrass fen (with *Eriophorum angustifolium*) and finally a Bog Bilberry fen (with *Vaccinium uliginosum*) which can be looked upon as an intermediate mire. In any case the adjacent strip (called 'intermediate fen' by Jensen) certainly deserves the name.

In the lower part the bog is purely ombrogenious, i.e. supplied by rainwater. In the Crowberry complex it may still occasionally receive mineral water which favours the growth of some more soil demanding plants, but this is no longer the case in the typical raised bog. Since because of air pollution the precipitation in recent years brings with it more and more nutrients, especially nitrogen and calcium, the Crowberry complex has started to spread. In this way a raised bog can lose its oligotrophic character.

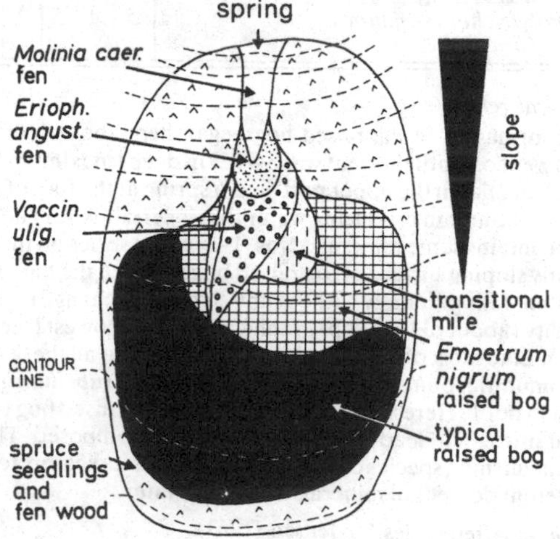

Fig. 259. Development of a raised bog in the diluvial lowland of north-western Germany, semi-schematic. From a coloured chart by Overbeck, modified.

1-3 = Late Ice Age: 1 = old tundra without trees (about up to 10 000 B.C.). 2 = Birch-Pine wood Allerӧd period (some 10 000 to 9 000 B.C.). 3 = tundra with some trees. (to about 8 000 B.C.) 4-6 = warm times following the glaciation: 4 = Boreal (about 6800 to 5500 B.C.) with a rich growth filling up still waters, and thick woodland on mineral soil; 5 = Atlantic period (5500-2500 B.C.) with higher temperatures and rainfall leading to formation of raised bogs; 6 = Sub-boreal (about 2500-600 B.C.), formation of older 'black peat' rich in Cottongrass and Heather. Beech begins to spread. 7-8 = Subatlantic following the warm period: 7 = 'white peat' formation and strong arching of the peat mass; 8 = a violent end to natural mire growth through drainage and peat cutting since the seventeenth century. The vertical scale of the peat layers is exaggerated; the trees are not to scale.

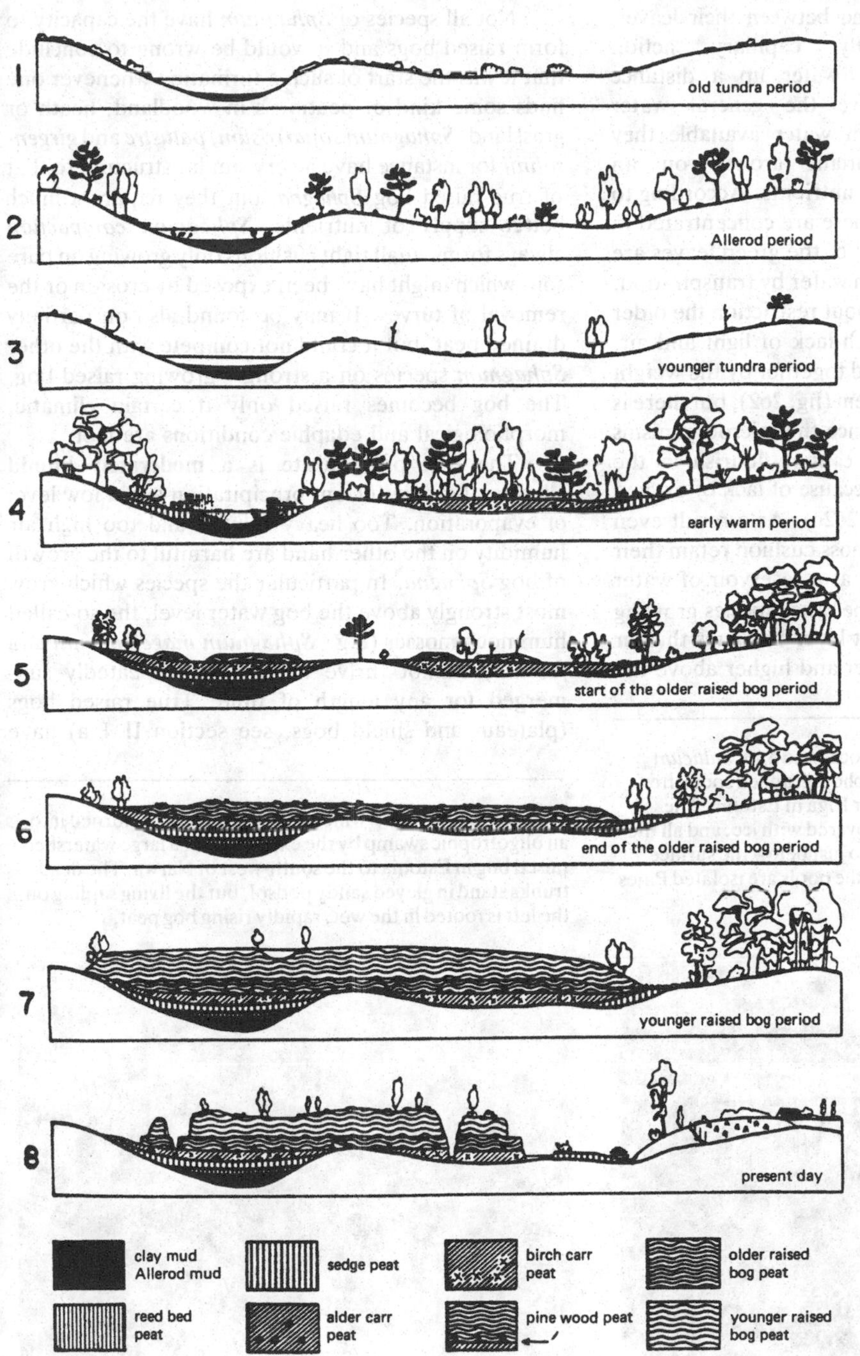

The formation of raised bogs is made possible by the sponge-like structure and the ability of some *Sphagnum* species to live on very small amounts of nutrients. Peat mosses such as *Sphagnum cuspidatum recurvum, magellanicum, rubellum* and *fuscum* can store up as much as 10–20 times their own volume of water in their hyaline cells and between their leaves, branches and stems. By capillary action they are also able to draw water up a distance of several centimetres above the general water level. When there is sufficient water available they maintain a relatively luxuriant growth on an astonishingly small amount of nutrients. According to Rudolph and Brehm (1966) these are concentrated in the growing heads where most of the green leaves are freely exposed and loose much water by transpiration. As the tips go on growing without restriction the older parts of the plants die through lack of light and air. These collapse and are pressed together by the weight of the younger stems above them (fig. 262), but there is scarcely any decomposition since the microorganisms which would bring it about cannot flourish in the waterlogged *Sphagnum* beds because of lack of oxygen and bases (Burgeff 1961, fig. 262c). As a result even the deeper layers of the peat moss cushion retain their open spongy structure, acting as a reservoir of water for the living *Sphagnum* and the other species growing with it. Continuously the peat layer becomes thicker and lifts the bog surface higher and higher above the surrounding mineral soil. Built up in this way the wet peat moss cake can eventually reach a height of several metres above the original ground surface. The maximum height of old raised bogs varies as a rule between 2 and 5 m, but thicknesses of over 10 m have been recorded (fig. 263).

Not all species of *Sphagnum* have the capacity to form raised bogs and it would be wrong to conclude that it was the start of such a formation whenever one finds some kind of peatmoss in woodland, heath or grassland. *Sphagnum squarrosum, palustre* and *girgensohnii* for instance have a very similar structure to that of true raised bog *Sphagna,* but they require a much better supply of nutrients. *Sphagnum compactum* always forms small tight cushions only growing on bare soil, which might have been exposed by erosion or the removal of turves. It may be found also on partially drained peat, but it could not compete with the other *Sphagnum* species on a strongly growing raised bog. The bog becomes raised only if certain climatic, morphological and edaphic conditions are met.

The first prerequisite is a **moderately humid climate**, that is sufficient precipitation and a low level of evaporation. Too heavy rainfall and too high air humidity on the other hand are harmful to the growth of bog *Sphagna*. In particular the species which grow most strongly above the bog water level, the so-called hummock mosses (e.g., *Sphagnum magellanicum* and *fuscum*) cannot thrive if they are repeatedly submerged for any length of time. True raised bogs (plateau- and shield bogs, see section II 1 a) have

Fig. 260. *Sphagnum fuscum* hummocks and *Sph. balticum* hollows form a small-scale mosaic photographed soon after the spring thaw on a raised bog near Riga in Estonia. The ponds in the background are still covered with ice, and all the hummocks contain cores of ice up to just below the surface (see also fig. 273). On the banks of the pools are isolated Pines up to some 6 m tall.

Fig. 261. A Pine wood which has been killed and turned into an oligotrophic swamp by the expansion of a large watershed raised bog in Estonia to the south-west of Narwa. The dead trunks stand in gleyed sandy podsol, but the living sapling on the left is rooted in the wet, rapidly rising bog peat.

arisen therefore mainly in suboceanic and montane climates, that means in southern Sweden, south-west Finland, countries further south along the Baltic and the North Sea, on most of the central mountains in Europe, the northern pre-Alps and at the montane and high-montane levels of the outer Alpine chain.

According to Gams (1962) actively growing raised bogs occur in the northern Alps only up to about 500-800 m above sea level. Because of the significantly higher radiation in summer they are found up to between 1000 and 1600 m in the central Alps. This indicates that a **sufficiently high temperature and length of the growing season** is a second important condition for the formation of raised bogs.

All areas rich in raised bogs however have a relatively cool climate. This fact seems to contradict what has just been said about the need for warmth, which has also been found to be the case experimentally, but higher temperatures increase the rate of peat decomposition. At the same time they restrict the development of *Sphagna* by lowering the relative humidity mainly during dry periods which often occur in the warmer parts of Europe. For this reason typical *Sphagnum* raised bogs are absent from warm continental valleys of the east and south-east. In mediterranean and submediterranean climates, which are decidedly dry in summer, raised bogs can only exist in the montane belt which stays cloudy even in the hottest season. A cool climate then favours the formation of raised bogs indirectly because it is accompanied by a **higher and more constant relative humidity**. However the colder it is and the shorter the growing period the

Fig. 262. a, b. The way in which Sphagnum Moss and higher plants grow and then become embedded in the peat of a raised bog, semi-schematic (a Cottongrass, b Cranberry). After Grosse-Brauckmann (1963), modified.
c. Profile of the peat horizons in a growing raised bog; schematic, smaller scale. After Grosse-Brauckmann and Puffe (1964), modified.

The growing *Sphagna* are green; dying parts turn yellow and lie down horizontally. The upper layers still have oxygen and contain large numbers of aerobic microorganisms which carry out a certain amount of humification (brown colour) and mineralisation of the more easily broken-down fractions. At a depth of only a few decimetres the young peat is permanently saturated with water and contains no oxygen (smells of hydrogen sulphide when dug out).

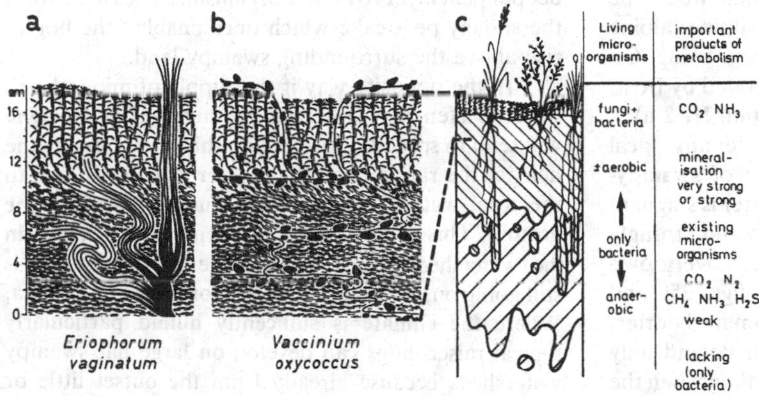

Fig. 263. The development of a true raised bog (plateau type) on the site of a swamp wood in southern Finland. After Aartolahti (1965), somewhat modified. Many raised bogs have grown up in such a way, also in the plains of north-western Germany.

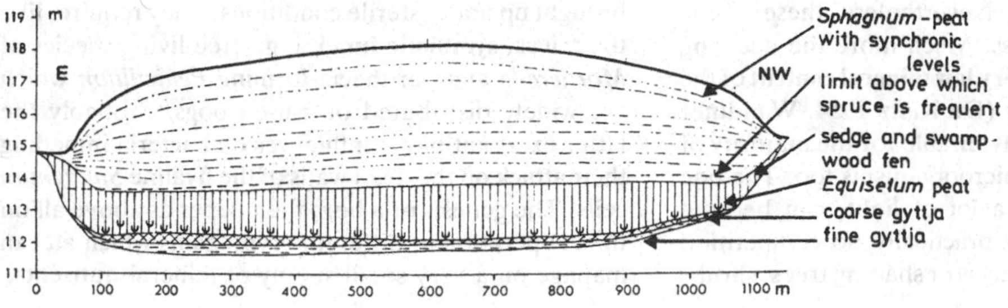

less growth is made by the mire. Thus no true raised bogs can originate above the tree line in the mountains nor in a subarctic climate as, e.g., in northern Scandinavia. In addition to the lack of warmth here other growth-restricting factors are the large amount of water produced by melting snow and the ground movements caused by frequent freezing and thawing. These lead to the formation of bog islands and ridges in the so-called 'Aapa-mires' of northern Fennoscandia.

Raised bogs in Central Europe were first able to start forming after the Ice Age as the climate became warm enough and there was sufficient rain, i.e., in the 'Atlantic' period (about 5500 to 2500 B.C.). Before that only muds (gyttjas), fens and swamp peats were being formed (fig. 259). Pop (1964) considers it to be proven that, geologically speaking, *Sphagnum* raised bogs are a very recent phenomenon and, at least in Europe, were first formed during the post glacial period. There are nowhere remains of peat which could unquestionably be assigned to raised bogs either from the period between the glaciations or from the Tertiary or older periods. At its peak the post glacial warm period was warmer than the temperatures of today (fig. 319). At that time raised bogs were growing up at higher altitudes and latitudes than would be possible at the present time. These subalpine, alpine and subpolar bogs are no longer actively growing. On the contrary they are being gradually eroded by frost, pressure of snow, water and wind (section III 2 a).

In a cool-temperate humid climate any local waterlogging can trigger off the formation of '**swampy-ground raised bogs**' provided that the water is stagnant and poor in bases. Even lime-rich fens and strongly growing trees on rich soils have finally been overgrown by raised bogs under such conditions (figs. 259 and 261). On the other hand where the climate is drier, e.g., in the Brandenburg region and in Poland only '**ancient lake raised bogs**' arise, and only then when the water becomes very lime-deficient. The majority of raised bogs no longer reveal their origins, and it is necessary to carry out a number of borings to reconstruct their history (F. Overbeck 1975).

Sufficient warmth and dampness are without doubt necessary conditions for the formation of *Sphagnum* raised bogs. Nevertheless these alone cannot initiate the process. Much more the deciding factor in each case is the **very low mineral content** of the water saturating the moss (Du Rietz 1954, Wandtner 1981), because this is unfavourable for the majority of other plants, and many microorganisms too. The bog *Sphagna*, which require a lot of light, can become dominant only if there is practically no competition from either other moss species or shading trees, shrubs or herbaceous plants, and if they are not covered by tree litter or other plant remains. The raising of the bog results from the fact that dead remains of *Sphagna* and other plants are scarcely decomposed and therefore accumulate in the course of time. The rate at which these organic remains are broken down is strongly correlated with their content of lime, phosphate and nitrogen (Coulson and Butterfield 1978, see also fig. 251). On this depends the activity of those animals and microorganisms taking part in the break-down.

Raabe (1954) was able to show that when an undrained raised bog near to an inhabited area in Schleswig Holstein became more eutrophic without any change in the water regime (see section B V 2 d), the mere increase in nutrients was sufficient to restrict peat formation by *Sphagna* though these kept growing vigorously. This check is brought about by two significant factors working together in a way which might be difficult to follow: The general eutrophication allows stronger growth in all plants, also in swamp and woodland species able to shade out the bog plants. In addition the better nutrient supply stimulates break-down of the dead *Sphagnum* cells as well as of the deeper peat layers by microorganisms. Thus it destroys the spongy peat cake which once enabled the bog to rise above the surrounding swampy land.

In the opposite way if the supply of mineral-rich water to a fen or swamp is cut off then the bog *Sphagna* will tend to spread. They will do this more quickly the heavier the rainfall and the greater its contribution to the total water available in the upper layers of the ground. This is most likely to happen after change in the watershed caused by tectonic movements as is still going on, for example, in the north-east of Estonia. Where the climate is sufficently humid particularly typical raised bogs can develop on large flat swampy watersheds because already from the outset little or no mineral-rich water flows into these area (fig. 261).

One additional condition for peat mosses to thrive has been little appreciated up to now, but, according to Burgeff (1956, 1967) appears to be necessary wherever these mosses are growing (and not just on raised bogs): *Sphagna* do not develop normally, when they are brought up under sterile conditions. They require, like the ericas, **symbiotic fungi**, e.g., free-living species of *Mortierella* and perithecia-forming *Penicillium* which are widely distributed on raised bogs. Probably the latter exert antibiotic influences on bacteria impeding their attack on the dead walls of the hyaline *Sphagnum* cells. The genesis of a bog, then depends above all on the cooperation of different cryptogams which all can manage on a very small supply of mineral nutrients.

However Rudolf (1963) was successful in cultivating *Sphagna* permanently in the laboratory under really sterile conditions. Even the species typical for raised bogs grew well and did better than in their natural habitat when they were fertilised with nitrogen. This proves that physiologically they **do not need oligotrophic conditions.** In the raised bog ecosystem, however, they survive the lack of nutrients better than other mosses and most phanerogams since they are helped by their symbionts (see also section 4a).

Sphagna capable of forming raised bogs are by no means confined to them. Otherwise, bogs could not originate from fens. Typical bog *Sphagna* are found in large quantities on intermediate mires, for instance in the lagg, which we already met in section II 1a and 3a (see also figs. 249 and 264 and table 57). Watershed raised bogs however which extend rapidly over fens and swamps may lack such intermediate mires. Raised bogs which are still growing in a wide depression or in flat countryside push this wet border zone continuously outwards (fig. 264). The water-logging kills off the trees by cutting off the oxygen supply to their roots (fig. 261). In this way even a fertile and base-rich soil can be turned into a swamp and finally into a bog.

Whenever one wants to get to a typical raised bog one usually has to wade through the more or less waterlogged lagg. On the bog itself in dry weather one could walk about in light shoes without getting one's feet wet. To do this one only has to notice the vegetation units, which clearly indicate the wet and the dry areas making up a small-scale mosaic in most of the bog types.

(b) Vegetation mosaic and succession on a true raised bog

On a typical convex raised bog there are as a rule several complexes of habitat which can be readily distinguished by their vegetation.

In large living raised bogs the **centre** or 'regeneration complex' is the wettest part as the general surface is nearly horizontal (figs. 259 and 263), and the rainwater takes a long time to flow outwards. Here numerous **'hollows'** ('Schlenken') usually develop; these are shallow depressions filled with water or at least with moss and peat completely saturated. Bright green *Sphagnum* together with a few, mostly grass-like, phanerogams remind one of the vegetation of intermediate mires. Such depressions are rarely more than a few metres across or more than a few decimetres deep (fig. 270).

Forming a mosaic with these are cushion-like **'hummocks'** ('Bulten') 0.5 to 3 m across, formed of reddish, brownish or yellowish *Sphagna*. Each of these represents a raised bog in miniature. Depending on the position relative to the bog water level they are occupied by tussocks of Cottongrass or Deergrass or by ericaceous dwarf shrubs (fig. 249). Obviously many of these phanerogams can grow here only if the root run is not continuously saturated with water, but becomes aerated from time to time.

The mosaic of hollows and hummocks with various intermediate stages suggests interpretation as

Fig. 264. Different types of montane raised bog in the Harz – shown in schematic profiles along the length and across. After Hueck (1928), somewhat modified.

I = Plateau bog (a few definitely watchglass-shaped arched bogs). II = Sloping bog (numerous examples; mainly growing down the mountain side; see also fig. 258). III = Saddle bog (watershed bog, originates from water course; several examples). IV = Ridge bog (watershed bog not receiving any mineral soil water, e.g. the large bog on the Bruch mountain). V = Basin bog (does not occur in the Harz).

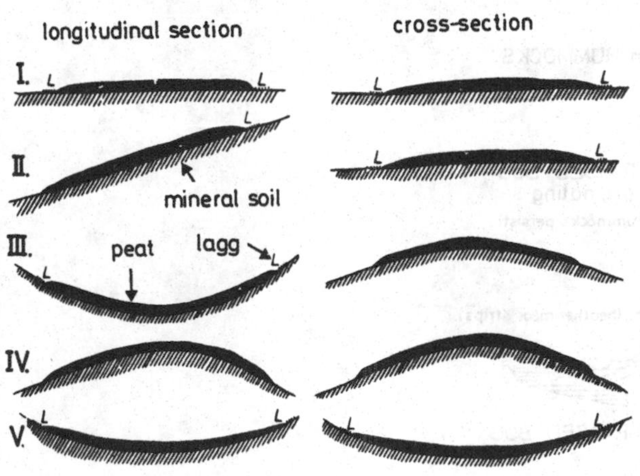

phases of a 'cyclic succession'. This previously widely held view has been shown to be wrong in most cases. Through exact observations and analyses of peat profiles a number of workers have found that such a type of succession is very much the exception (figs. 265, 266). Critical investigations were going on at about the same time and quite independently by Walker (1961), Jensen (1961), F. Overbeck (1961, 1975), Eurola (1962), Casparie (1969) and others. What started off as a hummock often remained so for hundreds of years, and what began as a depression stayed a hollow so that in many bogs the two types making up the mosaic grew up together at more or less the same rate (fig. 265, see Boatman and Tomsinson 1977, Boatman, Goode and Hulme 1977, 1981). Nevertheless Tolonen (1966) did find a cyclical succession in time and space showing a repeated change from hummock to hollow and vice versa when he studied a section from the Atlantic Period in the raised bog of Varrassuo in southern Finland.

In profiles of plateau bogs there have been observed over wide areas repeated alternations between 'regeneration stages', when there has been rapid *Sphagnum* growth and very little decomposition, and 'ripening stages' when the bog surface has remained fairly stable and the peat has been broken down and darkened to a greater extent. Casparie (fig. 266) explained this by assuming that the run off from the bog surface had changed in different places. Obviously these large scale changes in the rate of humus formation were irregular in time and not necessarily climatically determined as had often been assumed (see also F. Overbeck 1961, 1975).

The growth rate of a *Sphagnum* cushion can be roughly determined from looking at the remains of flowering plants rooting in it. They become smothered if they do not grow their rhizomes nearer the surface or form a new rosette of leaves each year (Bertsch 1925, fig. 262 a and b). Depending on local habitat conditions there could be an addition of several centimetres a year. However as the new growth of the moss carpet is itself overgrown and pressed down by young shoots, the dead remains of the older *Sphagnum* are pushed over from a vertical to a horizontal position and compressed to a small fraction of their original thickness (fig. 262). From measurements of the age and thickness of peat layers it appears that the thickness of peat added each year is at most some millimetres.

According to Burgeff (1961, see fig. 262) one can usually distinguish four layers in a profile through a growing raised bog: The first one is green consisting of

Fig. 265. The development of a ridged raised bog ('Kermi bog') with permanent waterfilled depressions. This type is widely distributed along the north-eastern border of Central Europe. Schematic. After Aario (1932) from Eurola (1962), modified. The hollows between the ridges (Finnish 'Kermi') are also known as 'Flarks' (Swedish).

The shield-shaped arching raised bog is furnished with long waterfilled depressions (see fig. 269) arranged concentrically at right angles to the slope and alternating with ridges carrying *Sphagnum fuscum*. Dwarf shrubs (*Calluna, Rubus chamaemorus* etc.) isolated stunted trees, especially Birch or Pine can grow on these ridges (see fig. 268).
Aario was already aware that most of the depressions (hollows as well as ponds) have existed in the same places for centuries and had not taken part in any cyclical succession with the hummocks or ridges. This kind of deep waterfilled depression is formed, at least in part, because of the ice pressure in the intensely cold winters. They are absent from the raised bogs of the milder climate of the more oceanic parts of Europe (see figs. 266 and 271-273).

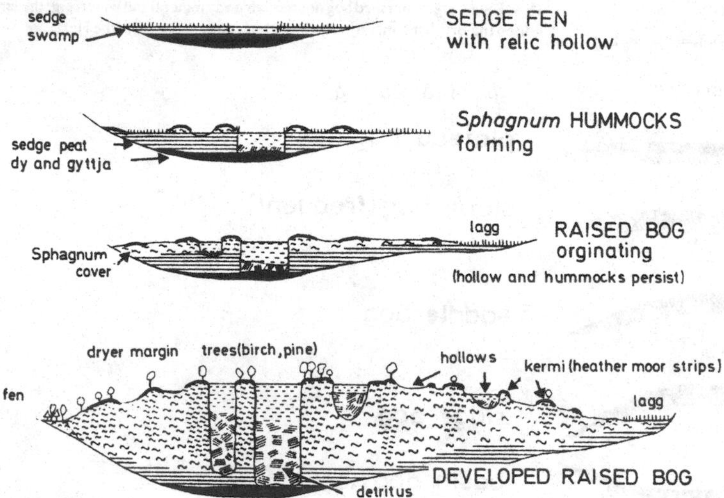

living *Sphagnum* and has many roots growing in it. These roots also penetrate into the next (grey) layer in which the *Sphagnum* is dying off and being pressed together. Below this is a brown or black layer where Burgeff has traced the colour partly to the formation of melanin and partly to oxidation of lignin and resin accompanied by the formation of humic acids. The fourth and lowest layer is yellow and rich in hydrogen sulphide which can be identified by its smell and by the fact that it turns silver black. Its upper boundary is sharper under the hollows than under the hummocks; where there are trees it is very irregular. Since it is completely devoid of oxygen it no longer contains any mycorrhizal fungi. The hydrogen sulphide comes from the breakdown of protein.

Except in the wettest hollows the upper part of the *Sphagnum* cushion where atmospheric oxygen can penetrate contains an extremely dense mat of ericaceous roots with their symbiotic fungi. These are of great significance for the breakdown of the dying *Sphagnum* since they use the pectin and some of the cellulose. In addition to mycorrhiza there are aerobic bacteria living in the aerated layers of the *Sphagnum*. Their growth is stimulated by the fungi, which appear to play part in a protein decomposition.

The number and size of the hollows get less as one

Fig. 266. Stages in the growth of a raised bog in north-western Central Europe. As an example is shown the carefully analysed section covering 50 m in length of such a bog near Emmen (Netherlands). Growth has been irregular but it shows no cyclical alternation between hummocks and depressions. After Casparie (1969), modified.

In order to clarify the sequence the same profile face has been drawn four times, and that at intervals of between 640 and 730 years (intermediate stages are not shown). Since about 1550 *B.C.* lighter-coloured *Sphagnum cuspidatum* peat has been spreading over the oldest, well decomposed darker peat. In other places on the profile face more strongly decomposed peat, rich in Heather or Cottongrass, remained dominant right from bottom to top until about 600 B.C. At that time a generally wet period set in leading to the rapid formation of large areas of weakly decomposed relatively pale-coloured *Sphagnum* peat (socalled white peat). Layers formed since 100 A.D. are not shown here because they had been disturbed. Permanent deep water-filled depressions, typical of the socalled 'kermi' bog, are absent here (compare fig. 265).

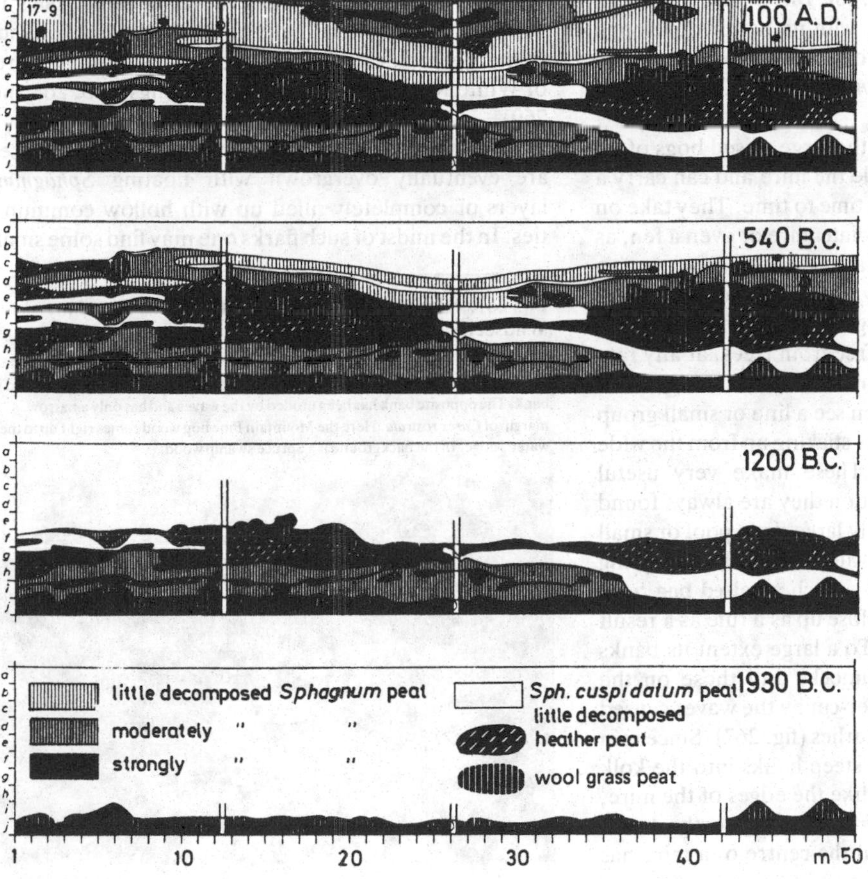

goes from the centre of the mire towards the edge until finally the hummock community rich in ericaceous species dominates over large areas. Some of the larger hummocks carry a number of lichens (fig. 270); in some places even Birch and Pine find a footing, but they are so stunted that after a hundred years they might have become no more than 1 m tall. Such a combination of heath-like plants is known as a **'standstill complex'** (F. Overbeck 1975) Here the growth of the mire has come to a halt largely because a part of the rain and melting snow runs off the steeper slope and is not stored up for use in the dry period. Where the run off cuts into the peat sometimes '**erosion complexes**' can arise with wet gulleys and their well drained banks which obviously favour the growth of trees (fig. 249). In some places where the peat has been deeply furrowed by running water and has then dried out intermittently the wind has been able to scoop out large troughs. However this only happens on those bogs which have dried out so much that their centre is no longer growing.

The driest parts of a raised bog are its '**marginal slopes**' running down fairly steeply into the 'lagg' and here and there cut into by deep run-off gulleys. Here a relatively dense woodland of Birch and Pine can become established which, with its better developed trees and undergrowth of dwarf shrubs, has little in common with the *Sphagnum*-rich communities of the remaining complexes.

The run-off gulleys of the large raised bogs often start from a place well inside the mire and can carry a large amount of water from time to time. They take on the character of an intermediate mire or even a fen, as is also the case with the 'laggs'. These grooves also have a Swedish name; they are called 'rullen'.

Apart from their marginal slopes most plateau raised bogs are practically free from trees, at any rate as long as the central level area is still in active growth. However one can quite often see a line or small group of stunted Birch or Pine trees sticking up from the wide flat expanse of the bog. These make very useful landmarks and on investigation they are always found growing close to an unusually large deep pool or small pond filled with brownish water and dark mud going deep into the peat (fig. 265). Such a **raised bog 'eye'** (German 'Kolk') does not close up as a rule as a result of the growth of the mire. To a large extent its banks have the character of hummocks and those on the north and east sides are under-cut by the waves caused by the prevailing south-westerlies (fig. 267). Since rain can drain freely from these steep banks into the kolk they are relatively dry and, like the edges of the mire, can offer a secure, albeit frugal, living to the trees. Often a single kolk found in the centre of a mire has arisen through the separation of the peat mass falling away in all directions rather as a lump of stiff porridge would split under its own weight. Where the raised bog has been formed over the site of a former lake the kolk can be seen as the remains of the lake. On a bog which started as a swamp kolks may indicate places where lakes had been overgrown by the mire. However other kolks are found just above those places where the solid ground beneath the mire is highest. It appears then that there could be a number of different reasons why kolks appear where they do.

In the eastern Central European mountains, in Scandinavia and round the Baltic, places which are subject to keen frosts, one quite often comes across large groups of small kolks arranged in parallel lines on the raised bog. Perhaps they are better seen as elongated deep permanent hollows which alternate with the regular surface level of the mire (figs. 268 and 269). These permanently water-filled furrows are 1-4 m wide with a vegetation similar to that of the hollows. They are known as **'flarks'** and are separated from each other by banks or 'ridges' (Finnish **'kermi'**) some 0.5 – 5 m in width. Such 'flark complexes' are typical of many of the rounded **shield or kermi-raised bogs** in southern Scandinavia, central and southern Finland, in the Baltic Soviet Republics and the adjoining parts of White Russia as well as in Poland (figs. 265, 268 and 269).

In older flark complexes the open water surfaces are eventually overgrown with floating *Sphagnum* layers or completely filled up with hollow communities. In the midst of such flarks one may find some small

Fig. 267. A large dystrophic pool in the raised bog of the Wildseemoor near Kaltenbronn (Black Forest).

In the foreground is a floating Cottongrass community (quaking bog) with *Eriophorum angustifolium* and *Sphagnum cuspidatum* along the wind-protected bank. The opposite bank has been eroded by the waves and has only a narrow margin of *Carex rostrata*. Here the Mountain Pine bog wood comes right up to the water's edge. In the background is a Spruce swamp wood.

hummocks which are obviously still in the process of being formed. One must take care however not to consider these young *Sphagnum* hummocks as being the first stage in the succession towards a 'ridge' with its solid peat soil on which even trees may grow. Hummocks of such a height and dryness would never arise during the growth of a plateau raised bog. The *Sphagnum*-free heathland community, rich in lichens, which forms on many of the kermi resembles much more the dwarf shrub heaths on mineral soil or woodlands of the mire's sloping edge both ecologically and in their species composition. They can be seen as natural examples of the stages produced by the drainage of raised bogs which in so many cases is now done deliberately by man.

Fig. 268. A complex of elongated depressions (Flark, see also fig. 269) on a raised bog near Riga shortly after the onset of the spring thaw. The ice begins to melt first on the bank exposed to the south.

It is probable that solifluxion has played a part in the formation of the flark complexes. At any rate they are most widespread and well developed on raised bogs where there are severe winter frosts and frequent fluctuations in temperature. They appear to have been absent from the large mires, at present under cultivation, in the region around the North Sea coast with its milder winters. In the Harz, the Black Forest and other western mountains of Central Europe they have not achieved the typical appearance of the ones in the Karkonosze Mountains (Riesengebirge) and other continental highlands.

(c) Vegetation of hollows and kolks

By the alternation of hummocks and hollows, flarks and kermi as well as of places which have been eroded and others which have been drained, the living raised bog presents a richer mosaic on such a small scale of habitats and biocenoses than almost any other plant formation on earth. This spatial diversity is produced by the permutation of only about 30 phanerogams and a similar number of mosses and lichens, of which only about a third are plentiful. This is because the extreme oligotrophic living conditions of the raised bog bring about a rigorous selection (tabs. 57 and 58).

A survey of the plant combinations found on raised bogs can best be made first of all by arranging them in order of decreasing water surplus. As was seen in the previous section however not all the steps in this ecological sequence are a result of vegetation dynamics, and without further evidence they must not be considered as a succession series however obvious it may appear.

A The **kolks** (deep pools) and the **hollows filled with water** for a long time contain interesting plankton

Fig. 269. The Flark complex of a raised bog in the Memel delta. After Hueck (1934), modified (see fig. 268).

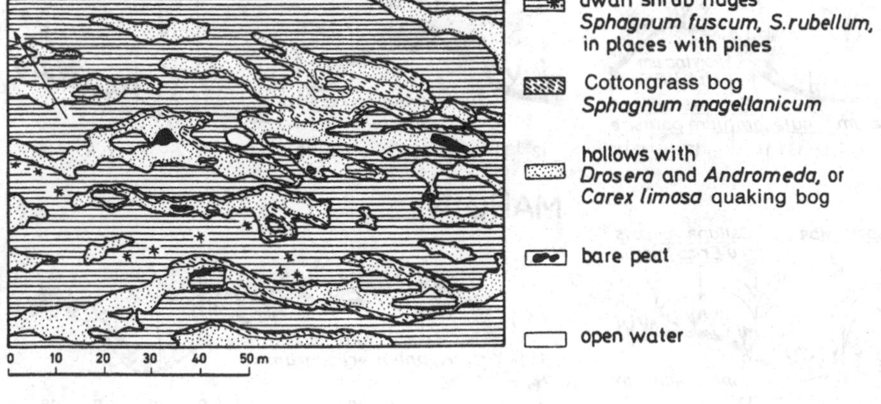

communities of desmids, diatoms and other small organisms with modest requirements which we are not able to look at more closely. They have been the subject of study by Höfler (1951), Loub, Ure, Kiermayer, Diskus and Milmbauer (1954) and Fetzmann (1956, 1961), and already investigated ecologically by Redinger (1934). Large algae and higher aquatic plants are generally absent. Only big kolks and water-filled peat cuttings are sufficiently well supplied with nutrients to allow communities of the Least Burreed-Bladderwort (*Sparganium mimimum-Utricularia intermedia* association), or even the White Waterlily to grow.

B Peat formation in such humus-brown strongly acid bog waters is not initiated by higher plants as it is in the more fertile lakes, but by *Sphagna* which form **quaking bogs** (figs. 229 and 256) and by phanerogams criss-crossing them with long rhizomes (marked R in the following list). The most abundant species in these Rannoch-rush quaking bogs (*Scheuchzerietum palustris*, called *Caricetum limosae* by many authors) are the following (see also fig. 270 and table 58):

mosses:	phanerogams:
Sphagnum cuspidatum f. *submersum*	*Eriophorum angustifolium* (R)
	Rhynchospora alba
S. *recurvum* var. *majus*	*Menyanthes trifoliata* (R)
S. *dusenii*	*Carex rostrata* (R, only in Kolks)
Drepanocladus fluitans	*Scheuchzeria palustris* (R)
Cephalozia fluitans	*Carex limosa* (R)

The last two serve as local characteristic species but are not met with so often. They have a subarctic and pre-Alpine distribution and are almost completely absent from the plains around the North Sea. Where they are found they usually form extensive open stands. They can withstand a high degree of acidity and constant submersion which enables them to compete successfully with the other small sedges and reeds.

Fig. 270. Mosaics of vegetation growing in zones of different moisture levels in the Esterweger Dose, formerly one of the largest raised bogs of NW Germany. After Jahns (1969), modified. Each of the four schematic sections is 6 m long with a greatly exaggerated vertical scale (compare fig. 272).

In the **centre** of the bog hollows with *Sphagnum pulchrum* and White Beak-sedge alternate with low hummocks carrying *Sphagnum magellanicum* and Heather. In part of the centre where the water level is not disturbed are socalled 'Blänken' or **dystrophic pools** around the edges of which, because of a slightly higher nutrient level, fen plants such as the Bottle and Mud Sedges, Moorgrass and Crowberry grow together with some more demanding mosses.

In the dry **marginal zone** of the bog from which the surface water can drain away there have been no real water-filled hollows. On the hummocks which often dry out there are drought-tolerant species such as *Leucobryum glaucum*, *Hypnum ericetorum* and *Cladina portentosa*.

A wide **transition zone** shows intermediate properties but also has its own character. For example the hollows here contain *Sphagnum tenellum* (= *molluscum*), and round their edges in some places *Sphagnum papillosum*, which once dominated in large areas of the bog, has maintained a footing. The figures refer to the scientific names of the numerous plant communities of which these are examples:

1 = *Sphagnetum cuspidato-obesi*;
2 = *Cuspidato-Scheuchzerietum, Carex* subass.;
3 = the same, typ. subass.;
4 = *Sphagnum cuspidatum-Eriophorum angustifolium* group;
5 = *Rhynchosporetum sphagnetosum cuspidati*, typ. var., *Sphagnum pulchrum*-rich form;
6 = the same, *Sphagnum tenellum*-rich form;
7 = the same, var. of *Erica tetralix*, typ. subvar. *Sphagnum pulchrum*-rich form;
8 = the same, *Sphagnum tenellum*-rich form;
9 = the same, var. *Erica tetralix, Cladonia* subvar., *Sphagnum tenellum*-rich form;
10 = *Sphagnetum papillosi*, subass. of *Rhynochospora alba*;
11 = *Sphagnetum magellanici*, subass. of *Rhynchospora alba*, typ.var.:
12 = the same, var. of *Calluna vulgaris*;
13 = the same, typ. subass., var. of *Calluna vulgaris*;
14 = the same, subass. of *Aulcomnium palustre*, var. of *Empetrum nigrum*;
15 = the same, var. of *Eriophorum vaginatum*, typ. subvar.;
16 = the same, subvar. of *Cladina portentosa*;
17 = *Sphagnetum fusci*, form of *Aulacomnium palustre*.

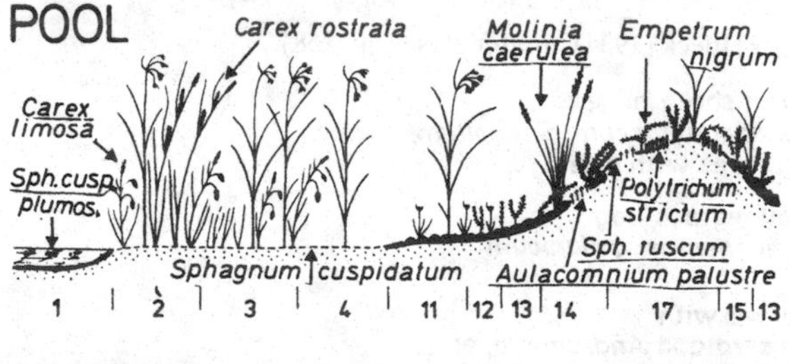

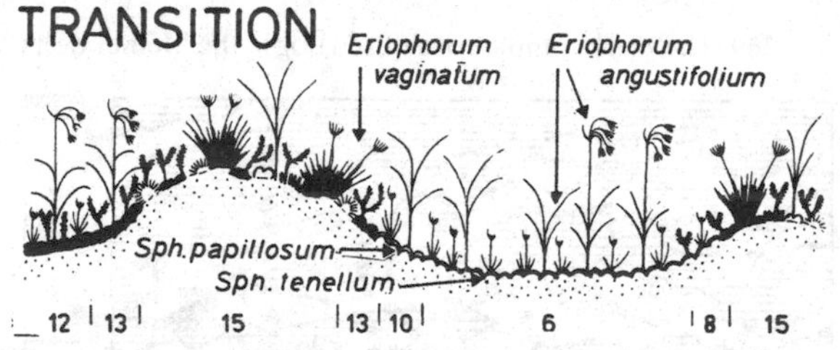

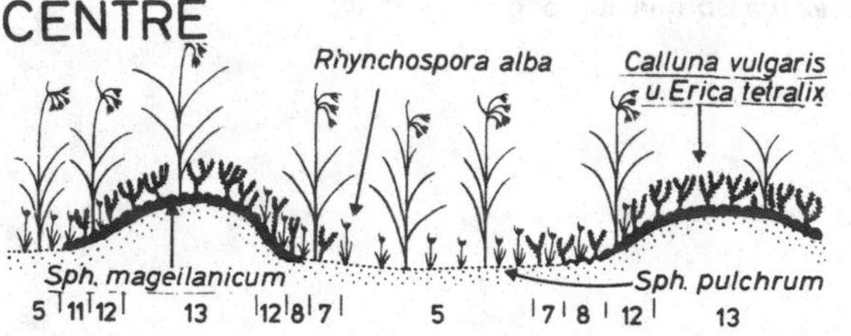

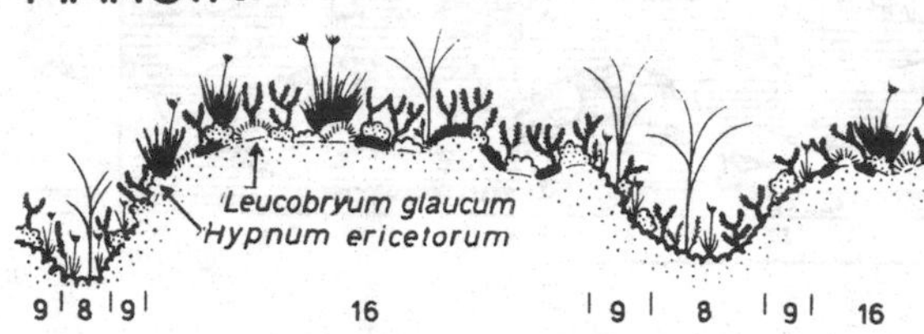

C In **shallow hollows** which are permanently full of water the *Sphagna* do not form a floating cover but fill up the water from the bottom (figs. 270, 271). Usually such hollows are poorer in species and can only be included with reservations in the *Scheuchzerietum* whose typical examples are found in flarks, in the lagg and around the edges of dystrophic lakes, i.e., where there is a more permanent body of water.

If the depressions are filled up with *Sphagnum* or if the floating mass of plants in the kolks becomes so thick that it can carry the weight of the mosses then their stems will grow up above the surface mutually supporting each other. *Sphagnum cuspidatum* changes from the *submersum* into the *elatum* form. Now even hummock-forming *Sphagna* are able to establish themselves, along with other plants which cannot tolerate constant submersion, such as *Sphagnum magellanicum*, *S. rubellum*, *Drosera rotundifolia* and *Vaccinium oxycoccus*.

D Many hollows which are only **temporarily filled with water** and alternate with the hummocks in the growth complex are shallower than the ones already discussed, and their ground is more solid. Melting snow and heavy rain wet such depressions but they dry out again during the summer and are then susceptible to erosion by the wind. They are the typical habitat of the Beak-sedge community (*Rhynchosporetum albae*) which is also found under similar conditions in places where peat turves have been removed or on peaty heathland (section D II 2 a). A reddish brown alga of the class Conjugatae *(Zygogonium ericetorum)* often covers the bare peat like a cloth. *Sphagna* are hardly ever present and only then in the hollows where water remains for a longer period (*Sph. cuspidatum* and *subsecundum* are the commonest). Higher plants present are those which can germinate when it is wet, but which can tolerate the extreme variations in dampness, especially:

Rhynchospora alba and *fusca*
Drosera intermedia
Lepidotis inundata (= *Lycopodiella i.*)

Along with these *Eriophorum angustifolium* and

Fig. 271. A detailed map of the positions of the flowering plants and mosses on an area 6 x 10 m of the western Ahlenmoor, a typical raised bog to the south of the Lower Elbe. After Klaus Müller (1965), modified (compare this with the sections through the Esterweger Dose, fig. 270).

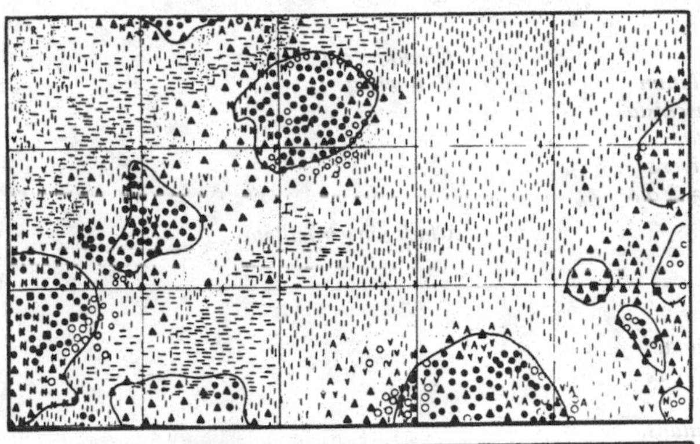

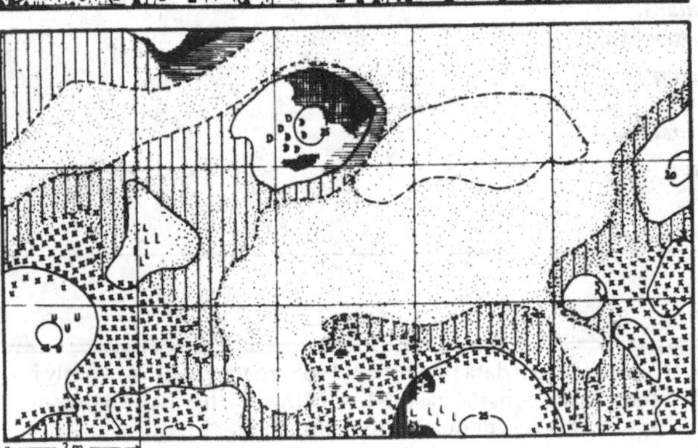

Tab. 58. Species of vascular plants, mosses and lichens on raised bogs of the north-western German plain.

Vascular plants (20)	**Mosses** (39) (L = Liverwort)	**Lichens** (22)
Hummocks:		
plentiful: *Calluna vulgaris* *Eriophorum vaginatum*	*Sphagnum rubellum* *Polytrichum strictum*	*Cladonia floerkeana* *C. chlorophaea*
frequent: *Pinus sylvestris* (dwarf) *Trichophorum cespitosum* (W and montane)	*Sphagnum parvifolium* *S. acutifolium* *S. molle* *Dicranum bonjeani* *D. undulatum* *Leucobryum glaucum* *Pleurozium schreberi* *Hypnum ericetorum* L *Compylopus piriformis* L *Lophozia ventricosa* L *Mylia anomala* L *Calypogeia muelleriana*	*Cladonia bacillaris* *C. incrassata* *C. pleurota* *C. subulata* *C. fimbriata* *C. crispata* *C. glauca* *Cladina arbuscula* *C. mitis* *Hypogymnia physodes* (Epiphyte)
rare: *Ledum palustre* (E)	*Sphagnum fuscum* *S. plumolosum* *S. imbricatum*[a] *Dicranum scoparium* *D. undulatum* L *Calypogeia neesiana*	*Cladonia macilenta* *C. gracilis* *C. pityrea* *Cladina rangiferina* *C. tenuis*
Hummocks and Hollows		
plentiful: *Andromeda polifolia* *Erica textralix* (W) *Vaccinium oxycoccus* *Eriophorum angustifolium* *Narthecium ossifragum* (W) *Drosera rotundifolia*	*Sphagnum magellanicum* *Aulacomnium palustre* L *Odontoschisma sphagni*	*Cladina portentosa* *Cladonia squamosa*
frequent: *Empetrum nigrum* *Drosera intermedia*	*Pohlia nutans* *Sphagnum apiculatum* L *Cephaloziella elaschista* L *Gymocolea inflata* L *Cephalozia connivens* L *Calypogeia sphagnicola*	*Cladonia degenerans* *C. uncialis*
rare: *Rubus chamaemorus* (NE) *Dactylorhiza maculata*		*Cladonia verticillata*
Hollows:		
plentiful: *Rhynchospora alba*	*Sphagnum cuspidatum* *S. papillosum* *S. tennellum* *Cladopodiella fluitans* L *Cephalozia macrostachya*	
frequent: *Drosera anglica*	*Sphagnum pulchrum* L *Telaranea setacea* L *Cephalozia lammersiana*	
rare: *Rhynchospora fusca* *Carex limosa* *Scheuchzeria palustris*[a]	*Sphagnum balticum* *S. compactum*	
Kolks and rullen:		
a further 11 species	a further 8 species	

Note: [a] significantly more plentiful in the past.
See also section E III 1, no. 1.611 and 1.71.

Source: From data by Müller (1965) rearranged. W = only in the west, E = in the east, NE = in the north-east. All the species also occur outside the raised bog.

Molinia caerulea (in the north-west) or *Eriophorum vaginatum* and other plants which also spread on to the hummocks are seen.

(d) Vegetation of hummocks, kermi and marginal slopes

The most important community of the raised bog is that of the **colourful Sphagnum hummocks** *(Sphagnetum magellanici)* (tab. 57). It is the really characteristic association and its common name is a reminder that during the greater part of the growing period the *Sphagna* in it are of a more or less bright red or brownish yellow colour (see fig. 283). It is mainly made up of *Sphagnum magellanicum* (= *medium*) and *S. rubellum* along with other moss species (see tabs. 57 and 58, figs. 270 and 271). Liverworts grow amongst them, e.g., *Odontoschisma sphagni, Campylopus piriformis* and *Mylia anomala*.

Depending on the height of the hummock surface above the mire water level different sub-communities or stadia of *Sphagnetum magellanici* can be distinguished mainly by the colour of their mosses and the kinds of phanerogams they contain. They continue the ecological sequence which we started in the previous section. As can be seen from fig. 270 however it is not just a single sequence. Even in the same raised bog there are a number which differ according to their distance from the centre and from a kolk. In this respect they reflect above all the differences in water nutrient supply. We shall have to content ourselves with the examples given in more detail in fig. 270.

E Around the **edge** of the hummock there is a transition zone where both the plants of the hollow and the hummock can be found. Amongst the mosses the majority are still *Sphagnum cuspidatum, apiculatum, pulchrum* and other green or yellow species. Of the phanerogams only the Sundew *Drosera rotundifolia* and grasslike plants are present and these occur also in the hollows. Dwarf shrubs find only a hesitant footing, but closer inspection reveals more individuals than would appear at first sight, especially of the Bog Rosemary (*Andromeda polifolia*).

F Wreathing the **base of taller hummocks** there is sometimes a band of *Sphagnetum magellanici sphagnetosum rubelli* in which the delicate dark red sphagnum, susceptible to lack of light or too much water, is dominant. It lies between E and G but may be absent altogether or substituted by other transitional communities.

G The **typical** *Sphagnetum magellanici* has a brownish or greenish red colour. This swollen-leaved species grows from 10 to 30 cm above the water level of the adjacent hollows. It contains the phanerogams found also in the previous subassociation and the following one. The dwarf shrubs however are not so well grown as they are on the taller hummocks and do not overshadow the *sphagna*. In the east of Central Europe as well as in many mountains *Sphagnum magellanicum* is generally replaced by the shining brown *S. fuscum* (tab. 57) although this is also to be found on many suboceanic bogs (fig. 270).

H The ocasional drying out of the upper *Sphagnum* layers allow the **dwarf shrubs** to become dominant, and in some places the dark green *Polytrichum strictum* also. Heather *(Calluna vulgaris)* is almost always present. In the north-west *Erica tetralix* also comes in and if there is a better than average supply of minerals the Crowberry (*Empetrum nigrum*) may join them. The hummock mosses suffer from shade and a shortage of water but are still able to survive under the dwarf shrubs.

I The tallest hummocks and also the **kermi and the dry edges** of the erosion gulleys are taken over by *Sphagnetum magellanici vaccinietosum* which is completely dominated by dwarf shrubs and often contains trees too. Acid-tolerant heath and woodland mosses (e.g., *Leucobryum glaucum*) and many species of lichens (see tab. 58) join and partly replace the peat mosses. This subassociation leads on to the swamp wood like communities of the marginal slopes and of the forested raised bogs. Besides the three dwarf shrubs named in H the following are met with:

Vaccinium uliginosum (more in the mountains and in the north)	*Ledum palustre* (in the east and north)
Vaccinium vitis-idaea	*Myrica gale* (in the north-west)
Vaccinium myrtillus	*Betula nana* (a rare glacial relict, more frequent in the north)

The following trees may be present (figs. 249, 250, 260):

Betula pubescens	*Pinus mugo* (in mountains and the pre-Alps)
Pinus sylvestris (var. *'turfosa'* in lowland bogs)	*Picea abies* (in mountains; very stunted)

In their shadow and because of the coniferous needles the *Sphagna* may be completely forced out so that there is doubt as to whether this community is still a *Sphagnetum* or should be looked upon rather as a swamp wood (see section B V 2 c).

The significant factor determining the species composition of the different stadia in the hollows and on the hummocks is the relative bog water level (tab. 59). Of the *Sphagnum* species mentioned under E to I only *S. cuspidatum* and, to a lesser extent, *S. tenellum* can go on growing when submerged. All the real hummock mosses must at least have their heads above the water, even *S. magellanicum* which has the widest amplitude, and *S. rubellum* which dominates in F. All those living on the hummocks can withstand drought periods better than waterlogging (fig. 278). In stage I the water level can even fall to more than 1 m below the surface for a time without the species composition being affected.

2 Other types of raised bog

(a) Oceanic, subcontinental and subalpine raised bogs without trees

In the description of true raised bogs and their vegetation we have repeatedly had to return to the differences in geographic behaviour of their species. The further we go away from Dutch and the north-west German plain or from the submontane and montane level the more obvious do these differences become, until in the end they are so large as to demand the separation of special associations and alliances. Because of the small number of species existing on raised bogs it is less a question of the presence of new character species than one of presence or absence of differential ones which justifies the establishment of new units.

The characteristic hummock community of the true raised bogs, *Sphagnetum magellanici*, is superseded by other communities in the west and the north-west and also in the high mountane to subalpine levels in the mountains. However the geographical variability of the hollow communities is much less and we can ignore them.

In **western** Central Europe, e.g., on the Hohe Venn, the atlantic *Sphagnetum imbricati* and *Sphagnetum papillosi* (fig. 274, cf. fig. 270) replace *Sphagnetum magellanici* in some raised bogs. Both communities tolerate or demand more moisture and develop only on sites with a slight covering of snow and short mild winters.

Sphagnum fuscum is dominant in the **east and north** (tab. 58). It resembles *Sphagnum rubellum*, but is of a pure brown colour, and can tolerate long cold winters. It is also apparently able to tolerate the occasional drying out much better than the mosses of the *Sphagnetum magellanici*. Its boreal companions, namely the Cloudberry (*Rubus chamaemorus*) and the Mountain Crowberry (*Empetrum hermaphroditum*) are plentiful in the Scandinavian countries, in former East Prussia and the Baltic lands, but are absent from the subcontinental mountain and lowland bogs which are also rich in *Sphagnum fuscum*. *Sphagnetum fusci* often forms steep hummocks in which lifting by frost plays a part. On the other hand the atlantic *Sphagnum* communities which are never deeply frozen seldom grow up into large hummocks (cf. figs. 272, 273, 260 and 265).

Fig. 272. A subatlantic completely treeless Deergrass raised bog, the Esterweger Dose between Emsland and Oldenburg as it was 50 years ago, i.e. before it was drained. The peat was up to 13 m thick. The pale tufts on the shallow hummocks are *Narthecium ossifragum*, the dark ones *Calluna vulgaris*. In the hollows, e.g. in the foreground to right and left, *Rhynchospora alba*. Photo, Tüxen.

Tab. 59. The range of groundwater levels in Sphagnum stands on raised bogs[a] of the north-west German plains.

Hummocks:	Groundwater depth (cm)	**Hollows:**	Groundwater depth (cm)
Sphagnum fuscum (rare)	< 19 – **33** – > 46	*S. papillosum*	0 – **15** – 28
S. acutifolium	25 – **37** – 50	*S. apiculatum*	0 – **15** – > 30
S. rubellum (plentiful)	7 – **29** – 50	*S. balticum* (rare)	0 – **13** – 26
S. imbricatum (very rare)	< 20 – **28** – > 36	*S. pulchrum*	0 – **14** – 27
S. magellanicum (very plentiful, also in hollows)	5 – **27** – 49	*S. tenellum* (frequent)	0 – **12** – 23
		S. cuspidatum (plentiful)	< + 15 – **5** – 24

Note: [a] i.e. the least, average and greatest distance of the water level from the surface of the bog. < +15 means that the surface may be covered with water to a depth of up to 15 cm.
Source: After K. Müller (1965).

The **high-montane and subalpine** raised bogs of the high mountains in the centre of Europe are not entirely without *Sphagnum* communities (figs. 258, 273 and tab. 57). They do however play only a subordinate role to the Deergrass (*Trichophorum cespitosum* subsp. *austriacum*); in the west ssp. *germanicus*) which dominates large stretches of the more or less level bog surface turning a uniform yellowish brown colour in the autumn. Such Deergrass bogs (*Trichophoretum cespitosi*) are no longer growing in height which may be due to climatic effects or because the water economy has been changed artificially. According to Issler (1942) the Deergrass, which now covers the whole of the Tanneck raised bog in the Vosges formed only small patches around the edge a hundred years ago. In a similar way a number of *Sphagnum* bogs in the Black Forest (J. and M. Bartsch 1940) and in the Hohe Venn (Schwickerath 1944) have become overgrown with *Trichophorum cespitosum* following draining.

Fig. 273. A schematic section through *Sphagnum papillosum* hummocks of an atlantic raised bog in the Ardennes. The lense-shaped blocks of ice were still present after several days of thaw on 27 Dec. 1949. They are much less persistent and smaller than the ones in subcontinental raised bogs. After Vanden Berghen (1951).

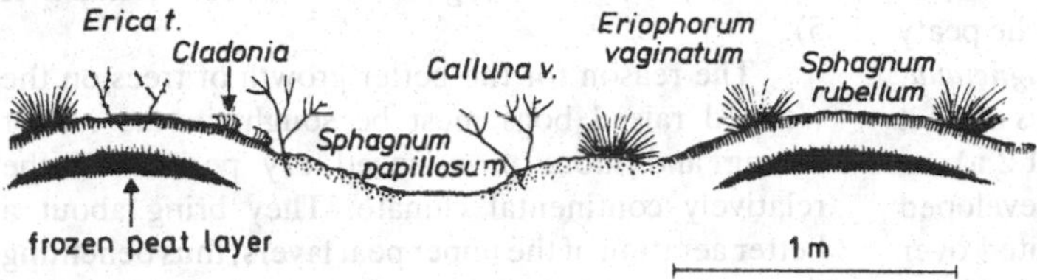

Fig. 274. The composition of the younger Sphagnum peat (so-called white peat) in the raised bogs of north-west Germany in so far as analyses are available. After Overbeck (1975), somewhat modified.

On the western and northern plains the main peat-producing mosses in former centuries were *Sphagnum imbricatum* and *Sph. papillosum*. In present-day communities on the surface of the raised bogs these species play an insignificant part but species of the Acutifolia group (*Sph. rubellum, Sph. fuscum* etc.) and also *Sph. magellanicum* are the dominant ones – as indeed they have been from earlier times in the south and east (see figs. 270 and 271).

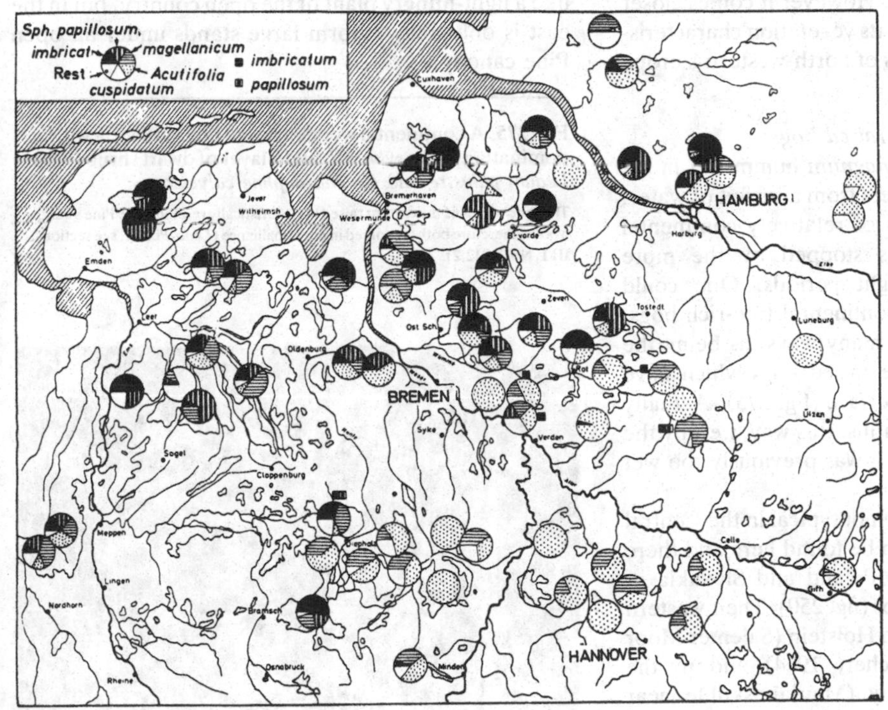

However the Deergrass community is only indirectly favoured by drainage. It still requires a lot of water and can only succeed in places with a frequent precipitation and a moist atmosphere, i.e., in oceanic climate (fig. 270). It rapidly disappears from the majority of raised bogs in the low rainfall lowlands whenever they have been drained. *Scirpus cespitosus* and *Sphagnum compactum*, its frequent satellite, thrive best on more or less bare peat which may occasionally dry out, but is constantly being saturated again, and where it can no longer be overrun by quick-growing hummock mosses.

The *Scirpetum cespitosi* of the Hohe Venn and other western raised bog areas should, according to Schwickerath, no longer be considered as members of the order *Sphagnetalia fusci*, but, along with the peaty soil atlantic *Erica* heaths, of the *Erico-Sphagnetalia*, which could well be named *Ericetalia tetralicis* since it contains so little *Sphagnum* (cf. section D II 2 a).

The Deergrass communities are best developed on blanket bogs which are naturally distributed over large areas in the atlantic northwest of Europe, e.g., western England, Scotland and Ireland. Morrison (1959) has described some examples from northern Ireland. The best investigated mire complex in north-west Germany, Sonnenberger Moor in the upper Harz which has been so carefully mapped by Jensen (1961), is certainly rich in Deergrass in its oligotrophic parts (fig. 258). However it comes closer to the plateau raised bogs in its vegetation characteristics than to the blanket bogs of north-western Europe (see tab. 57).

(b) Subcontinental wooded raised bogs

While the growth of the *Sphagnum* hummocks in the subatlantic blanket bog suffers from a surfeit of water, in the wooded raised bogs of relatively continental north-eastern Europe it is stopped by the more frequently occurring drought periods. One could mentally picture these subcontinental tree-rich bogs, similar to swamp woods in many ways, as being the marginal woodlands of true raised bogs which have spread out over the whole surface (fig. 275). Actually many of them have arisen in just this way, i.e., by the drying out of a surface which was previously too wet for the growth of trees.

Wooded raised bogs are not typical in the Central European area, but they can be found here and there especially in Brandenburg, Poland and Slovakia as well as in parts of the pre-Alps (fig. 250). Their western outposts are in south-eastern Holstein (Salemer Moor near Ratzeburg, see Lötschert 1964) and in the north-east of Lower Saxony (Maujahnskuhle near Lüchow). Often they have arisen from the filling up of a dystrophic lake (cf. section C I 1). In contrast to the treeless centres of the true raised bogs their surface is covered with a more or less open stand of Pines or Birches. These grow only slowly and rarely form a closed canopy. Nevertheless they give a definite woodland appearance to the mire. In the lowlands Pine (*Pinus sylvestris* var. *turfosa*) is the usual dominant species, accompanied by, and in some places replaced by Birch (*Betula pubescens* and *pendula*). Mountain Pine (*Pinus mugo*) gives a gloomy appearance to the bogs on the Upper Bavarian plateau. The upright form often grows round the edge, but the prostrate form in the centre (fig. 250, see Lutz 1956). The latter is also favoured by a more continental climate, just as in the high mountains (see section B II 5).

The reason for the better growth of trees on the wooded raised bogs must be sought in the longer lasting and more pronounced dry periods of the relatively continental climate. They bring about a better aeration of the upper peat layers, thus benefiting the growth of tree roots and their mycorrhiza. The sparse canopy for its part does not have such a bad effect on the growth of *Sphagnum* as it would have under oceanic conditions. In fact the light shade could be beneficial for bog *Sphagna* in preventing an extreme lowering of the relative humidity. This is also seen in Heather (*Calluna vulgaris*) which in the west behaves like a light-hungry plant of the open country, but in the east is only able to form large stands under an open Pine canopy.

Fig. 275. A continental wooded raised bog in Polesia with dominating *Pinus sylvestris* and a layer of dwarf shrubs with *Ledum palustre* and *Chamaedaphne calyculata*.

The Pine woods of the eastern raised bogs floristically resemble the Pine swamp woods; therefore both are joined in one suballiance (*Ledo-Pinion*, see section E III 1, No. 7.212.2).

The ecological problems presented by the wooded raised bogs have still not been solved by these discussions. There is so little *Sphagnum* growing on the majority of these bogs that it is difficult to imagine how they could have grown to their actual height. Certainly they are not raised in the centre, but the larger ones have several metres of *Sphagnum* peat, although this is darker and in a more advanced state of decomposition than the 'younger moss peat' of the classic raised bog (cf. fig. 266) and generally it contains more remains of Cottongrass and trees. Is the climate drier today than it was? Has the growth of most wooded raised bogs fallen victim to man's activities? Or is it more likely that Kulczynski (1949) is right in thinking that the development of wooded raised bogs consists of an alternation between a wet growing phase with few, if any, trees and a drying out phase favouring the growth of trees? The researches of Müller-Stoll and Gruhl (1959, see also Ellenberg 1963) in the Moosfenn near Potsdam support this last view. At present there are plant communities, similar to those in the hollows of typical raised bogs, to be found also on most of the wooded raised bogs, but only around pools or kolks. The small-scale mosaic of hummocks and hollows so characteristic of the true raised bog is nowhere to be found on a wooded bog. The quaking bog community on the kolk edge, similar to that found on the intermediate mire, gives place directly to a fairly wet Cottongrass stage which can last a long time and occupy large areas where tree growth is not possible. Systematically it can be seen as a wet sub-community of the *Eriophoro-Sphagnetum recurvi*. It is very obvious even from a distance because of the abundance of *Eriophorum vaginatum* whose tufts of white seed hairs shine far and wide in the summer and whose competitive tussocks remain green into the autumn.

The Cottongrass-*Sphagnum* bog can change into a Cottongrass-Pine wood or a Cottongrass-Birch wood while the ground flora remains much the same. Less-wet places are occupied by a dwarf shrub-Pine wood with a good deal of *Ledum palustre*, and difficult to distinguish from swamp Pine woods (fig. 217). Occasionally one comes across dead stands of Pine on peat which is so wet that the trees could not possibly have grown on it. This shows that in the wooded raised bogs waterlogging phases can occur even today.

According to Steffen (1931), Lötschert (1964), F. Overbeck (1975) and many others most of the wooded raised bogs in Central Europe today owe their existence to conscious or unconscious interference by man, some of which goes back a long time. Where the climate is relatively dry it only needs some drainage to be carried out in the surroundings of the mire to deal the final blow to the growth of bog *Sphagnum* already near the limits of its existence. Wooded raised bogs and true raised bogs are often found close together in the Baltic countries so they must be formed under similar climatic conditions. This is true near the coast as well as in more continental areas. Many raised bogs in western Central Europe and in the pre-Alps which only a few decades ago were completely free from trees now carry a vegetation similar to that of the eastern wooded raised bogs. Here it is doubtless drainage which has brought the growth of practically all the lowland raised bogs to a standstill.

3 The cultivation of raised bogs and its effects

(a) Past and present methods of cultivation

Economic exploitation of raised bogs near the coast first started in the Netherlands in the sixteenth century when the peat was extracted as fuel for towns and industry. A canal was dug for the peat barges and also acted as a main drain for the ditches. Later it would become the main traffic line of the 'peat colony' farming (German: 'Fehnkolonie') the land from which the peat had been cleared. Such peatland communities also became established in north-west Germany, especially near the town of Emden.

On bogs which lay at some distance from a town, making it uneconomical to cut and transport the peat, there developed first of all an extensive system of burning cultivation in which the mire was drained, but only down to a certain level, and the surface set alight. Then Buckwheat or some other undemanding crop was sown in the ashes. Many raised bogs were destroyed by this method of robbing them of what little fertility they had, made possible by the fact that they could be drained easily, thanks to their raised position relative to the surrounding land. Repeated burning also prevented the growth of trees so that the bog retained its open character. Peat cuttings were only made near to villages, covering their individual requirements. These were generally dug round the edge of the raised bog where it was easiest to lower the water level (see fig. 276).

In the nineteenth and twentieth centuries however this unproductive use gave way to a planned **'German raised bog cultivation'** which has determined the landscape character of the majority of the north-west German bogs as they are today. Farms were laid out on the drained raised bog which carried on a very productive form of agriculture. The peat was not removed, but good crops were grown by application of lime and fertiliser. Today liming is no longer practised as this led to a rapid breakdown of the loose moss peat and to a compressed structure of the soil crumb. The experiences of the peatland research station at Bremen show that wheat can be grown successfully at a pH of 3.5 provided the fields are well fertilised (Ellenberg 1958). Today the houses of old raised bog settlements are surrounded by tall Birch, Ash and Pine trees and strong trees also grow alongside the roads. Very often a characteristic undulating road surface resulting from irregular sagging of the peat is the only indication that one is passing from a solid sandy soil on to a former raised bog. The sinking of large bog areas, by the way, can be of the order of several metres, an indication of how wet the spongy peat was at the time of its active growth.

The present world economic situation means that it is no longer profitable to farm the drained raised bogs and in part their use has been discontinued (see sections D X 1 a and 2 a). One must consider then whether they could be afforested or put to some other use. Perhaps it might even be possible where a bog has not been altered too much, to fill in the drainage ditches and revive the *Sphagnum* communities.

(b) Drained raised bogs turning into heath- or woodland

Remains of the former vegetation are rarely found on many cultivated mires. Perhaps in a few peat cuttings there may be stages developing similar to those in the bog hollows with plenty of *Eriophorum vaginatum* or even also a few hummocks with *Sphagnum magellanicum*. A few years after it has been drained the old surface of the raised bog turns into heathland. That is it becomes covered with a vegetation similar to that of the hummocks, or even of pure *Calluna* heath. If left undisturbed trees would come in, mainly Birch and Pine, and finally Oak (to become a *Betulo-Quercetum molinietosum*). If the secondary bog vegetation is burnt from time to time tussocks of *Molinia caerulea* become dominant because they regenerate more quickly than any dwarf shrubs (fig. 277).

Fig. 276. Peat diggings in a raised bog in Upper Bavaria. The top layer of 'white peat' which is hardly decomposed at all is thrown into the ditch on the left while the 'black peat' which has decomposed much more and is better for burning is dug out in briquettes, spread out to dry (right) and finally stacked to complete the drying out process (background right). The *Sphagnum* peat is so impervious that it takes days for the trench to fill up with water. In the background Purple Moorgrass and Birch communities.

Since many raised bogs have been drained but not taken into intensive cultivation some very different heaths, grassland communities or woodlands have developed. Unfortunately they are only a monotonous substitute for the colourful diversity of the undisturbed raised bogs, the life of which is irredeemably destroyed by even a small lowering of the water level in the peat (figs. 277, 278). Under natural conditions it is true that in dry summers the water level can fall so that the *Sphagnum* growth is stopped (see e.g. fig. 278). However when the rain returns the *Sphagna* can recover, and in damp warm periods they can grow even more luxuriantly.

It is interesting to observe the behaviour of different *Sphagnum* species on hummocks and in hollows when water is gradually withdrawn (fig. 279). First of all the *Sphagna* in the hollows die since they form a rather open community which allows free entry for dry air thus increasing the transpiration rate (fig. 280). The ones to hold out longest are some hummock *Sphagna*, especially just beneath the top of the hummock where they can store the water running from it and live in the half shadow of the dwarf shrubs. Even when drainage is slight it can slow down the growth of most *Sphagna*. Figures 280 and 281 together show the reasons why different species replace one another following slight changes in the amount of water available, even under natural conditions.

4 The structure and living conditions of raised bog phanerogams

(a) 'Xeromorphy' and nutrition

While the *Sphagna* suffer and die when a raised bog is drained the phanerogams flourish all the more. Taller plants with deeper roots increasingly dominate the better-aerated peat until finally a woodland emerges if it is not prevented from doing so by man's activities (fig. 277).

At the same time the morphology and anatomy of the dominating species change in a striking manner. On wet sites it is only the underground parts of the plants in the hollows which show the typical structure

Fig. 277. Stages resulting from conditions brought about by the increasingly effective drainage of a north-west German raised bog, semi-schematic. After Ellenberg (1954), somewhat modified.

Even just a slight lowering of the bog water level is enough to restrict the growth of the *Sphagnum* moss and favour the dwarf shrubs and other 'hummock plants'. With deeper drainage trees can grow better and better. The peat shrinks when it is no longer saturated with water.

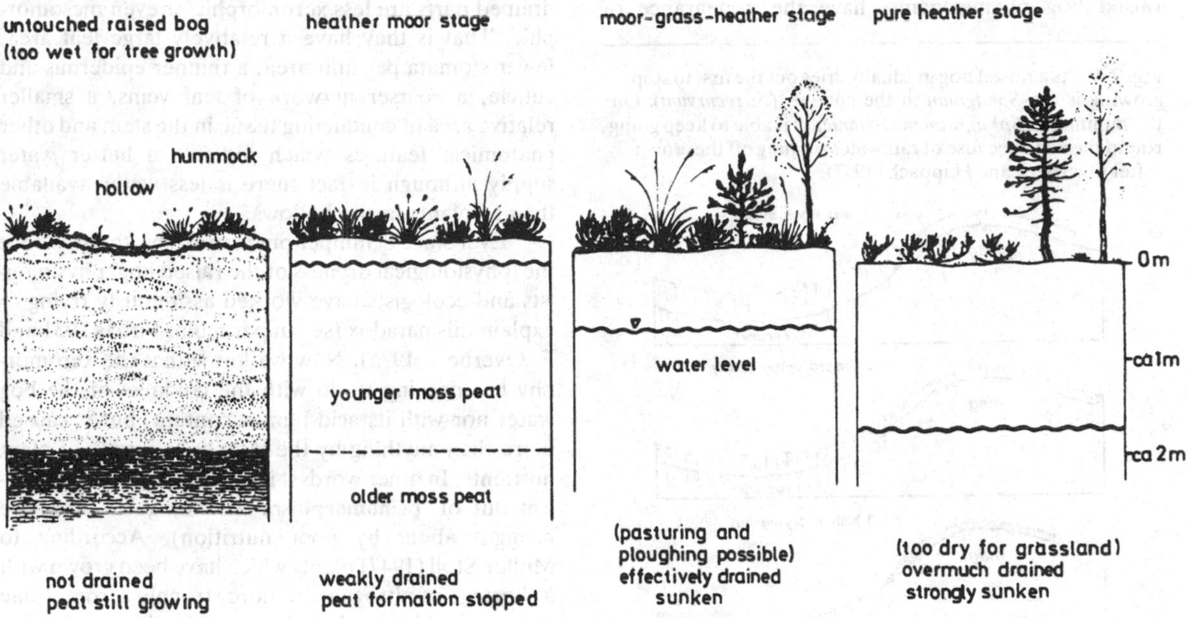

untouched raised bog (too wet for tree growth)

heather moor stage

moor-grass-heather stage

pure heather stage

hummock

hollow

younger moss peat

older moss peat

water level

0 m

ca 1 m

ca 2 m

not drained
peat still growing

weakly drained
peat formation stopped

(pasturing and ploughing possible)
effectively drained
sunken

(too dry for grassland)
overmuch drained
strongly sunken

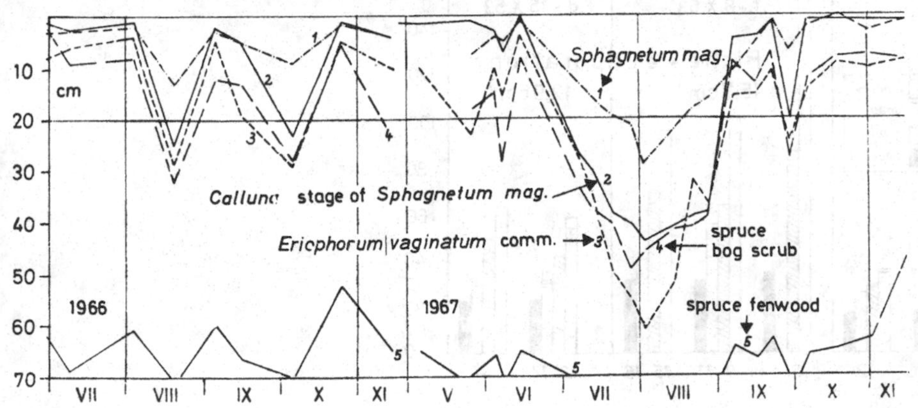

Fig. 278. The water table in the raised bog fluctuates naturally. When compared with a normal year (1966) the relatively dry years (e.g. 1967) can result in a failure of the peat mosses which do not grow any more (mountain bog in the Thuringia Forest). After Schlüter (1970), modified.

1. In the primary stage of the Peat Moss communities (*Sphagnetum magellanici*, round the edge of a pool) the water level of the bog remains mostly close to the surface. If exceptionally it drops below a depth of 20 cm the *Sphagnum* dries out and becomes pale in colour as its hyaline cells fill with air ('bleached moss'); 2. Dwarf shrubs (e.g. *Calluna vulgaris*) find a footing in the *Sphagnetum* where the water level falls deeper more frequently. 3. Cottongrass heath (*Eriophorum vaginatum* community) is an indicator of habitats where the water level fluctuates more erratically. 4. Under such conditions as these even a Spruce scrub may exist though with difficulty (*Piceo-Vaccinietum uliginosi*). 5. A Spruce swamp wood where *Picea* grows quite vigorously can only become established where water never stays near the surface for any length of time (the stand at Rennsteig is drained to a certain extent).

associated with helophytes such as large intercellular air spaces in the cortex and a relatively weak vascular system (see section A I 3 c). In contrast the parts of the plant above the ground, in the case of the majority of raised bog phanerogams, have the appearance of plants which are characteristic of very dry habitats such as dry grassland and steppes. They are 'xeromorphic' even though they are growing on wet soil or even in water. Surprisingly the species growing on the better drained parts are less xeromorphic or even mesomorphic. That is they have a relatively large leaf area, fewer stomata per unit area, a thinner epidermis and cuticle, a coarser network of leaf veins, a smaller relative area of conducting tissue in the stem and other anatomical features which indicate a better water supply although in fact there is less water available than to plants in the hollows.

Ever since Schimper put forward his theory about the 'physiological dryness of the raised bog' physiologists and ecologists have worked assiduously trying to explain this paradox (see in particular Firbas 1931 and F. Overbeck 1975). Now we know that the xeromorphy has nothing to do with the coldness of the bog water nor with its acid humus content, but is caused more than anything by the very poor supply of plant nutrients. In other words it is not a case of xeromorphism but of **'peinomorphism'** (a change in structure brought about by poor nutrition). According to Müller-Stoll (1947) plants which have been grown with a shortage of nitrogen are more strongly xeromorphic than ones which, under otherwise identical conditions, have received more nitrogen (fig. 282). In the case of *Andromeda polifolia* the plants with the most strongly developed xeromorphic characters were the ones which had been grown under wet conditions poor in nitrogen and not the ones under dry conditions poor in nitrogen or dry conditions rich in nitrogen (Simonis

Fig. 279. As a raised bog gradually dries out the first to stop growing is the *Sphagnum* in the hollows *(S. recurvum)*. On the hummocks *Sphagnum magellanicum* is able to keep going round the edge because of rainwater running off the crown. After Overbeck and Happach (1957).

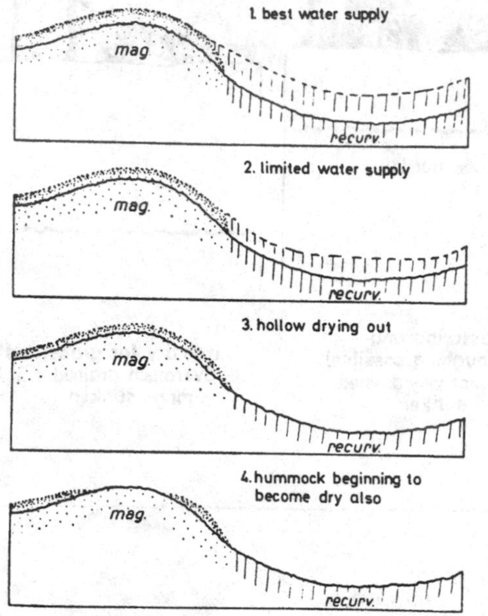

Fig. 280. Evaporation from *Sphagnum recurvum, magellanicum* and *rubellum*, also from *Leucobryum glaucum* during periods of one week in experimental jars standing in the open, in which the water was at depths varying from 2 to 16 cm below the moss surface. The results on the extreme right are from a jar with no standing water. Horizontal lines inside the columns are individual trials. After Overbeck and Happach (1957), modified.

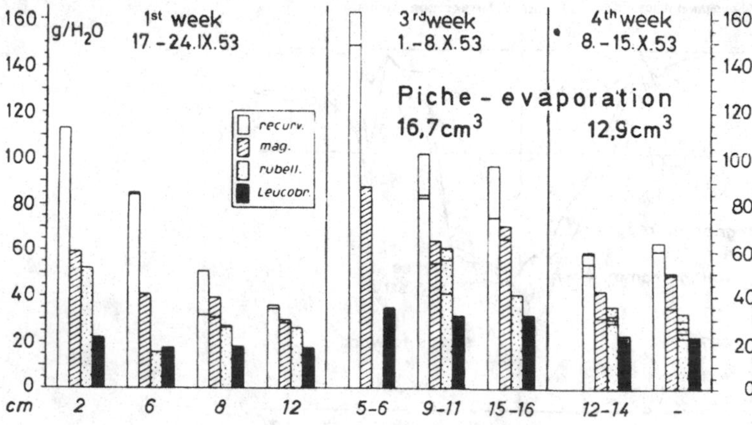

1948). These results are in accordance with the fact that most phanerogams growing in the hollows are more strongly peinomorphic than the ones on the hummocks. The leaves of many raised bog phanerogams are not only small and hard but also evergreen (e.g. *Andromeda, Vaccinium oxycoccus, Erica* and *Calluna*. Their longevity according to Small (1972) enables them to use the mineral nutrients (mainly N and P) for a number of years. The same author found that the deciduous perennials economise by transferring those nutrients into their subterranean organs more effectively than related species outside the raised bog do.

Living raised bogs are actually very poor in nutrients, especially phosphate and nitrogen. There are a number of reasons for this and they combine to increase the adverse effect on the higher plants. Firstly the bog can obtain no nutrients from the ground because of the thickness of the layers of dead peat which insulate it from the original surface and because no water can drain onto it from the surrounding lower countryside. Secondly the nutrients incorporated by the living *Sphagnum* do not become available for the other plants as there are not sufficient free-living microorganisms available to break down the dead organic matter (Burgeff 1961). Because of this the raised bog phanerogams almost without exception have mycorrhiza which can release the proteins from dead plant remains, and these are only effective in the well aerated top soil. A third reason why the raised bog is so poor in available nitrogen is its high water table which restricts the root run of most of the phanerogams, especially the trees. Probably the wetness and the acidity acting together provide a fourth reason for the poverty of the raised bog in that they prevent the breakdown of organic substances which may arrive on the bog from outside, such as pollen grains, the bodies of animals and their droppings. It is significant that the Sundews, which can make direct use of these sources of nitrogen are the only more or less mesomorphic

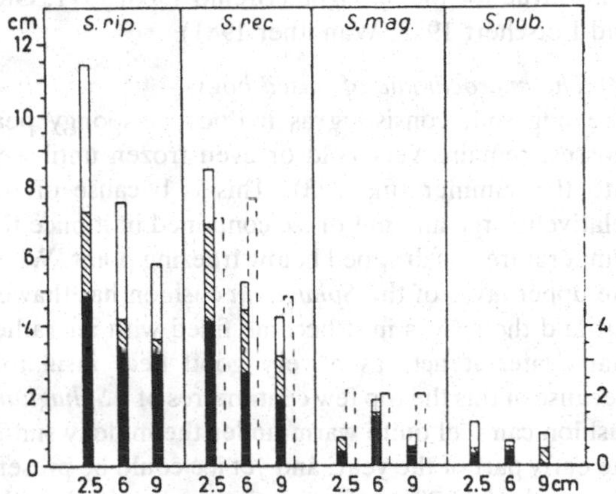

Fig. 281. Growth in length of some *Sphagnum* species in culture jars when the water level was held at depths of 2.5, 6 and 9 cm below the surface of the moss. After Overbeck and Happach (1957).

Black = growth between 22.7 and 23.8, hatched = up to 23.9, white = up to 21.10.1954. All the jars were in the open air but protected from rain by a glass roof. Broken lines = freely exposed to the rain. *Sphagnum riparium, recurvum, magellanicum* and *rubeilum*.

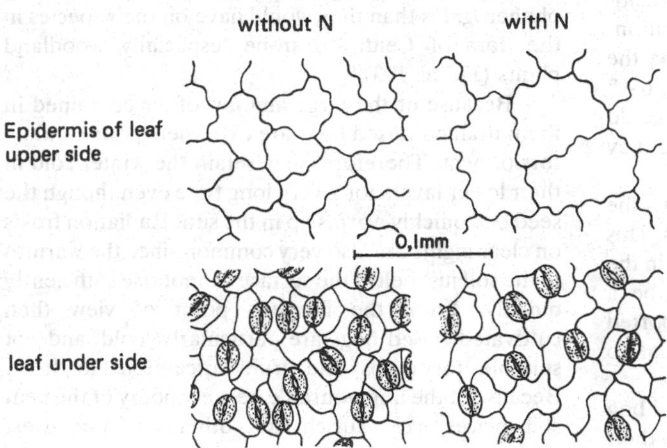

Fig. 282. The effect of fertilising with nitrogen (ammonium sulphate) on the leaf structure of Cranberry (*Vaccinium oxycoccus*). Without N the structure is 'peinomorphic', with it rather mesomophic. After Müller-Stoll (1947) from Ellenberg (1963).

phanerogams living in the wet *Sphagnum* communities. Finally it must be borne in mind that the nutrients collected and metabolised by living plants are not continuously recycled as they are in a woodland, but become locked up in the peat as 'dead capital'. So even the small amount available does not form a closed cycle and the plants are dependent on a continuous supply from outside (see Malmer 1980).

A sure, but limited, nitrogen supply is provided by some blue-green bacteria living on peat mosses. According to Alexander (1975, see also Stewart 1970) they can fix more than 10 kg/ha/year in a bog in northern Sweden. But the major supply comes from the air in the form of pollen and other organic dust, and above all from the precipitation which washes ammonia and other nitrogen compounds down on to the surface (Geister 1975). At one time the amount of 'manure' supplied in this way was about 6 kg per ha per annum in Central Europe and even less in northern Europe. However because of air pollution this figure has increased considerably since about 1950; on the Solling it was measured at 20 kg/ha in 1972 and the increase is accelerating. It is to be feared that as long as fossil fuels are being burned the concentration of NH_3, NO_2 and NO_3 compounds will continue to increase. This unavoidable eutrophication will lead to the live raised bogs of Central Europe losing their extremely oligotrophic character and ceasing growth. However, the differences in the cation content of the various raised bog communities, as described by Lötschert and Gies (1973), will probably not disappear in the future.

According to Brandt (unpublished) the mineral nitrogen content of the *Sphagnum* cover is already correspondingly high. As Wandtner (1981) measured in 30 West German raised bogs, actively growing hummock mosses (*Sphagnum magellanicum, S. rubellum, Polytrichum strictum*) accumulate heavy metals (Cd, Pb, Cu, Zn) in rapidly increasing quantities, because of the generally accelerated air pollution. The peat profile of British blanket bogs reflects the variations of the lead production during the last 6–7 centuries. As Liwett, Lee and Tallis (1979) also could show, this is not true for copper and zinc because they are washed out more or less completely.

In many raised bogs, e.g., on the Harz, the calcium content has also increased in recent years. This is no doubt due to the emissions of cement works in the area. The real limiting factor now seems to be a shortage of phosphorus which can only be transported by air in very small quantities. However I know of no research having been done on this.

The peinomorphic structure of typical raised bog phanerogams makes it possible for them to keep their stomata open even on dry days. This allows them to transpire copiously, and to some extent the large volume of water passing through them helps to compensate for the very low nutrient content (Firbas 1931). On the other hand the thick cuticle and the efficient closing of the stomata enable them to avoid water loss through transpiration whenever the soil becomes dry or frozen hard, which is certain to happen in those winters when there is little snow on the ground.

Since the raised bog phanerogams have high transpiration rates, the reflections of Armstrong (1981) are void. He supposed that their 'decreased transpiration' could be an advantage 'by reducing the velocity at which phytotoxins approach the root system.' Quite to the contrary, the strong transpiration stream carries bioelements into the growing parts. This is also true for the *Sphagna* (Brehm 1968, 1971, Gies and Lötschert 1973, Wandtner 1981).

(b) The microclimate of raised bogs

The bog soil, consisting as it does of spongy peat mosses, remains very cold or even frozen until well into the summer (fig. 273). This is because of the relatively large amount of ice contained in it once the temperature has dropped below freezing point. When the upper layer of the *Sphagnum* cushion has thawed out and the spaces in it become filled with air rather than water it acts as a very good heat insulator. Because of this the top few centimetres of a *Sphagnum* cushion can feel quite warm under the midday sun in the early part of the year, and yet ice could be present at a depth of 10-20 cm. Even as late as June or July the temperature at this depth is less than 10° C and could be approaching zero (see fig. 283 and Ichmeidl 1962, 1964). However these low temperatures and the violent changes in temperature of the top soil have a much less-inhibiting effect on the water uptake of bog phanerogams than they would have on their species in the flora of Central Europe, especially woodland plants (Firbas 1931).

Because of the large amount of air contained in them drained raised bogs are extremely poor conductors of heat. Therefore, they retain the winter cold in their lower layers for a very long time even though the seed bed quickly warms up in the sun. Radiation frosts on clear nights are also very common since the warmth in the soil just below the surface cannot rise sufficiently quickly. From the farmer's point of view then cultivated raised bogs are particularly 'cold' and not suitable for crops and fruit susceptible to frost. Because of the unfavourable heat economy of the peat soil places like Aurich and others in north-west

Germany, which are situated between drained raised bogs, have a surprisingly low night temperature in spite of their nearness to the coast (see tab. 60).

No doubt the richness of raised bogs in boreal species has something to do with the peculiar microclimate, although admittedly they are also favoured by the extremely acid humus. Important representatives of this northern element are *Scheuchzeria palustris, Rhynchospora alba, Scirpus cespitosus*, species of *Drosera* and *Vaccinium, Andromeda polifolia, Ledum palustre* and *Chamaedaphne calyculata*. Frost-susceptible plants with an Atlantic distribution, such as *Erica tetralix, Myrica gale* and *Narthecium ossifragum* are found only on bogs which are poorer in peat mosses and wetter; in other words where the top soil is rarely air filled. Bog plants like *Eriophorum vaginatum* which form dense tussocks enjoy a relatively warm microclimate (Chapin and collaborators 1979); this favours the nutrient supply indirectly by accelerating the mineralisation of litter and peat.

Fig. 283. a. The temperature within the tips of Peatmoss shoots (*Sphagnum magellanicum*) fluctuates markedly (see also fig. 267). From measurements by Rudolph in the Kaltenbrunner Moor (Schleswig) in Overbeck (1975), modified.

b. The red colour of many Peatmoss species is due to the presence of anthocyanin in the cell walls the production of which is favoured by the fluctuating temperature. In winter this pigment turns brown; in spring the new tips are bright green. After Rothe (1963), somewhat modified.

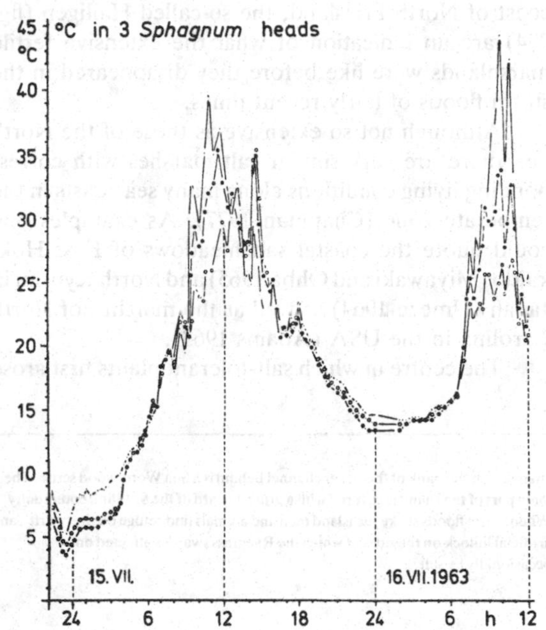

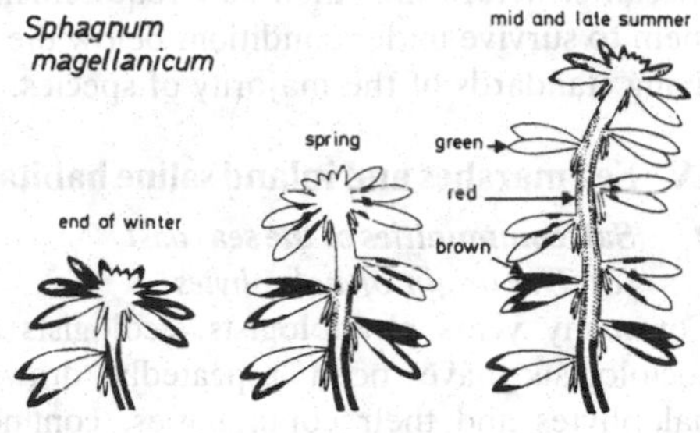

Tab. 60. The influence of drained raised bogs on the minimum air temperature at different times of the year in Lower Saxony.

Months	North Sea Coast Borkum	North Sea Coast Wilhelms-haven	Bog districts Aurich	Bog districts Bremer-vörde	Inland Bremen	Inland Hildes-heim
January	0.0	− 1.1	− 1.8	− 2.4	− 1.3	− 2.0
April	4.6	3.7	2.7	2.0	3.6	3.4
July	14.3	13.0	11.5	11.3	13.0	12.5
October	8.0	6.7	5.3	4.6	6.2	5.7

Note: Throughout the year the raised bog districts are the coldest!

Source: The monthly means over several years of the minimum daily temperature in °C, from reports of the German Met. Office (Reichsamt für Wetterdienst 1939).

The local climate of a raised bog is relatively 'continental' including the temperature maxima. On sunny days the air above it is quickly warmed up to a relatively high degree. This leads to an evaporation rate which is only 20-30% less than that on rocky slopes or similar very dry habitats (Firbas 1931).

According to Rudolph (1963) the red colouration of many *Sphagnum* species is brought about by the extreme temperature fluctuations. The colour is due to an increase in the amount of certain anthocyanins in the cell walls. This colouration is intense in the late summer and autumn when practically every night the raised bog is subject to a frost (see fig. 283) whereas the ground temperature in the neighbourhood still remains above freezing.

From many points of view then the plants of the raised bog are living in an extreme habitat. Only their special structure and their low requirements enable them to survive under conditions below the minimum living standards of the majority of species.

IV Sea marshes and inland saline habitats

1 Salt communities of the sea coast

(a) The origin of halophytes

For many years physiologists, ecologists and plant sociologists have been repeatedly drawn to the halophytes and their communities, confined to the battleground between the sea and the land. A rich literature (see Adriani 1958, Beefting 1965, 1968 and Chapman 1975) enables us to get a good idea of the extreme conditions under which they live and the causes of their ecological behaviour even though many questions remain unanswered. For this reason we shall deal with them thoroughly here. Even more so because the sea coast beyond the dykes is one of the few places where we can experience truly wild nature still untamed by man. A detailed account is also justified by the fact that the tidal flats (or 'Watten') of the North Sea German Bay are amongst the most extensive in the World (Linke 1939). Of the whole tidal zone of the North Sea which is several kilometres wide, up to a maximum of 17 km, only a small part is suitable for the growth of higher plants. This includes the salt marsh behind the chain of islands which protects the flats from the open sea, and the marshland running out as a narrow, often interupted band, in front of the dykes on the coast. A few marsh islands without dykes off the coast of North Friesland, the so-called Halligen (fig. 284) are an indication of what the extensive fertile marshlands were like before they disappeared in the storm floods of fairly recent times.

Although not so extensive as those of the North Sea there are very similar salt marshes with corresponding living conditions along many sea coasts in the temperate zone (Chapman 1977). As examples one could quote the coastal salt meadows of East Hokkaido (Miyawaki and Ohba 1965) and North Kyushu in Japan (Umezu 1964) as well as the marshes of North Carolina in the USA (Adams 1963).

The centre in which salt-tolerant plants first arose

Fig. 284. The Hallig (small island in the tidal flats) Langeness in North Friesland is fully exposed to the North Sea and to some extent shows what the natural marsh vegetation is like. Photo, Landesbildstelle Hamburg about 1950.

Foreground left is a frequently flooded Saltmarsh-grass sward with a small channel; on the bank of the larger channel behind is a Sea Wormwood scrub. The main part of the island is covered with a grazed sward of the Sea Pink community. When storm floods strike the island men and animals find refuge on the 'warft', an artificial hillock on the sides of which the Ryegrass sward is affected only occasionally by salt.

is probably not in Europe. Related communities which however are much richer in species are to be found in depressions and salt pans of the Asiatic semi-deserts and steppes and in the continental dry landscapes of other parts of the earth, also on the warm coast of the Mediterranean with its lagoons. For example of the 30 species of Sea Blite (*Suaeda*), most of which are perennial, only one grows on the coasts of the North Sea and the Baltic and this is an annual. The Marsh Samphire (*Salicornia europaea* agg.), one of our most important land-building pioneers is also a therophyte and (together with a few recently distinguished subspecies) represents the last outpost of a genus which has many perennial species already living in south-west Europe. Apparently the majority of the Mediterranean salt plants cannot stand our hard winters. These include the *Chenopodiaceae* to which *Salicornia* and *Suaeda* belong. The annuals though can overwinter in the form of seed which in the case of *Salicornia* does not germinate until April. Winter frosts, together with the water stress they cause, are probably the main reason why there are no mangrove swamps along the sea coasts in the temperate zone. Only at the uttermost margin of the salty floodings of the North sea are trees able to get a footing (Ranwell 1974). These are mainly Black Alders which would also form the natural forest limit at the Baltic coast. Frost resistance is usually high in our herbaceous halophytes and this is independent of the salt/sugar ratio (Kappen and Ullrich 1970).

The salt-resistant plants of the coasts are distributed world wide by the sea currents and by birds. How effective migrating birds are in their dispersal can be seen in the inland salt meadows which we shall discuss in a special section. Any newly emerged salty habitat which may be quite isolated is occupied in a very short time by all the different species of halophyte which would be able to grow there (see Aellen in Hegi III 2, 2nd edn). Compared with the terrestrial salt flora in Central Europe the marine one is richer in species and life forms, especially as far as cryptogams are concerned (Gessner 1957, Schwenke 1969). Nevertheless we shall dwell only briefly on the plant communities in the sea. Their significance for us is above all that they help to explain the surprisingly high productivity of the tidal flats and sea marshes which in part depend on them.

(b) Living conditions in the tidal flats

All land halophytes are extreme light-loving plants and are only able to develop in open places where they are not overshadowed by other plants. Nowhere can they find habitats so free from competition in the natural landscape of Europe as those which are available in large measure on the flat tidal coasts. The removal and deposition of sand and silt never ceases in the area of the ebb and flow of the tides, even though it is only storms blowing in the direction of the land which result in obvious land building. This applies both to the outer coasts exposed to the sea currents and the winds and to the quieter inner coastline behind the chain of dune islands.

Only specially equipped organisms can cope with the twice daily alternation of being covered with water and then exposed to the air. As is the case with all

Fig. 285. Zonation of land plants at the German North Sea coast depends on their position relative to the mean highwater mark, semi-schematic. After Meyer in Schütte (1939), modified. Samphire (*Salicornia*) penetrates down to about 30 cm below mean highwater mark, or even a bit more (see fig. 287). Sea Aster (*Aster tripolium*) and Sea Blite (*Suaeda maritima*) hardly ever go below it; Saltmarsh-grass (*Puccinellia maritima*) and Sea Arrowgrass (*Triglochin maritimum*) are confined to a higher level.

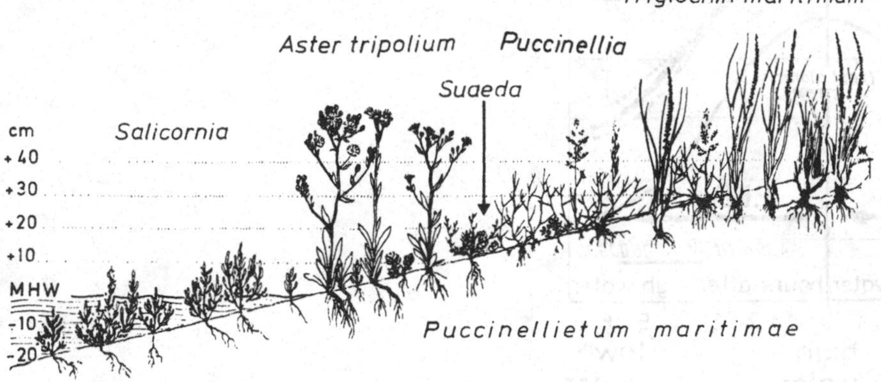

extreme habitats one finds only a few species here, but the number of individuals is immense. Going away from this critical zone both towards and away from the sea the numbers of both plant and animal species able to survive increases with distance. In order to understand this we must familiarise ourselves with the tides.

On the German North Sea coast the average tide, that is the difference between the mean low water (MLW) and the mean high water (MHW) is about 1.5-3.75 m (fig. 286). It is greatest in bays such as the Jadebusen and the Dollart and in the river estuaries, because in such places the incoming water is dammed up. At the time of the spring tides, i.e. when the gravitational pulls of the moon and sun are acting together, the spring tide high water (HWST) is some 25 cm higher and the low water (LWST) correspondingly lower. At neap tides the amplitude is correspondingly reduced. Westerly storms can raise the height of the tide by several metres while easterly storms can lower them by as much as 2 m. The maximum difference in levels then can be as much as twice normal (fig. 286). Both extremes can prove fatal for many organisms.

The most difficult place to live is at the MHW mark where the dry exposed time (DT) lasts as long as the time when it is covered with water (WT). Land organisms here are in danger because of the water while water organisms have to suffer long periods when their element is lacking. If in addition the waves here are strong and the ground unstable there will be practically no life. On the sandy outer shore therefore land phanerogams can only find a footing some 0.5-1.0 m above the MHW and over large areas they leave the sand to the mercy of the wind. However on the inner shore they go down to below MHW and prevent the sand being blown since the surface of the flats seldom dries out.

By **tidal flats** (or briefly flats) we usually mean the transition zone between sea and land which is covered at MHW and exposed at MLW. Linke (1939) defines it

Fig. 286. Normal and extreme variations in the tides at Wilhelmshaven with the terms usually employed on the North Sea coast. After Schütte (1939), modified.

The term flood does not refer to the high water level but to the period of about six hours during which the tide is rising from low water to high water; the ebb is when the tide is falling. The average tidal range at Wilhelmshaven is 3.59 m (from 2.1 m below 'normal zero' to 1.5 m above it). The actual range is greater when there is a spring tide, i.e. when the sun and moon are pulling in the same direction, and smaller when it is a neap tide. Storms from the west drive the sea into the Jade bay leading to very high storm tides, especially when combined with a spring tide; storms from the east have the opposite effect. The difference in height of the water between the two extremes can be more than 9 m.
The most significant tidal mark ecologically is the mean highwater mark (MHW, see fig. 285). It lies 1.54 m above normal zero.

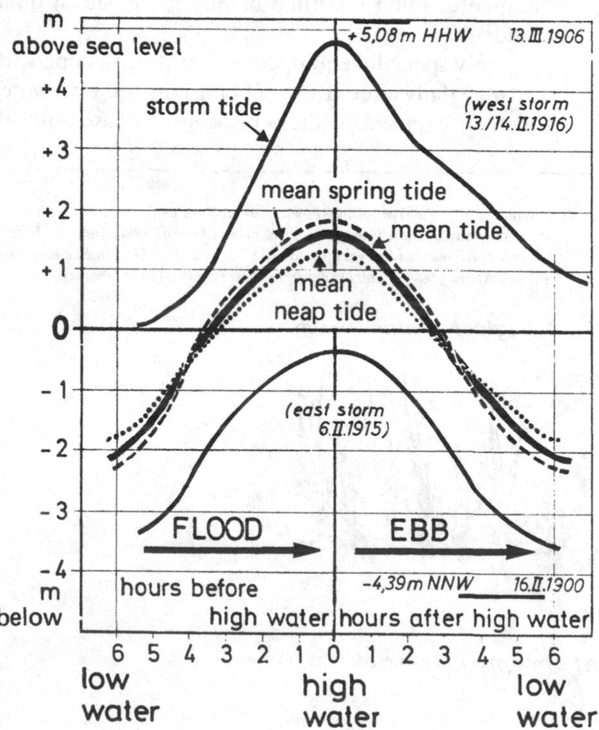

more exactly as the tidal shelf, i.e., the whole of the gently sloping surface lying protected from the direct influence of the sea and limited by the HWST and LWST. Only storms cause much wave damage on the flats and the flood tide normally creeps noiselessly over them without any surf. When the tide is on the ebb the water gathers in meandering 'channels' (or tidal inlets, German 'Priele') which start off being very shallow, but as they join together become wider and deeper until they form a deep channel which even at low tide does not lose all its water. Behind each of the larger islands of East Friesland there is a watershed between the two systems of channels along which one can walk or drive to the mainland without much difficulty when it is exposed by the tide going out.

The mud as well as the sand flats are settled by only a small number of species but for these it is an exceptionally favourable living space. The water and the sand contain a large amount of organic substances (tab. 61) including nitrogen and phosphorus compounds. Gessner (1939) found in pools of the mud flats near Husum 80 to 100 mg P per cubic metre of water, but 20-40 mg in the area of the sand flats. Most of the tidal flat soil is also rich in lime (tab. 61) which exists chiefly in the form of ground up shells of mussels and snails. Generally speaking the fertility of the flats increases towards the land since the carrying power of the flow water falls off in this direction and the finer particles richer in organic matter are deposited higher up (tab. 61). In this way sand flats and mud flats can be distinguished with silt flats coming between the two.

How then does this large amount of organic matter come to be in the shallow seas flowing over the flats? The bulk of the organic sediment stems from the sea itself. It has been able to collect in the sea over the flats because this part is protected by a chain of islands and is very shallow throughout. Most of the redistribution of the material takes place over short distances and is mostly confined to the flats themselves. According to Brockmann (1935) this is apparent also in the luxurious growth of the plankton flora, whose diatoms, peridinea and similar families are principally represented by species peculiar to the flats. They form the main food of the many bottom-living animals and are thus a source of the dead and living organic matter concentrated on the flats.

However this source alone cannot have provided the whole of the huge amount of humus and its raw material which is present in the water and mud of the tidal flats. Of greater significance is that brought down by the rivers which empty into the sea in this area. Even before the pollution resulting from man's activities dead organic material and living organisms were carried down by the rivers. Many of the freshwater algae and plankton animals are killed by the salt water.

In any case most of the organic material in the North Sea tidal flats is a result of the photosynthesis of **Phanerogams.** In the Jade Bay (Linke 1939) these are in the first place the remains of raised bogs which had once grown up outside the saltwater area and from time to time had been carried out to sea when the latter had broken through. These fossil and extra-marine sources of organic matter are however, as I see it, of at most local significance. The main producers which are constantly at work are the Eelgrass meadows (*Zostera* communities) growing in shallow water at the outer edge of the sand shelf. We shall return to these presently. In the Danish fjords the remains of *Zostera marina* can make up as much as 57% of the organic detritus (see Gessner 1957). There are also gigantic *Zostera* beds in front of the coasts of Holland and Belgium (Dieren 1934). Their dead parts are carried eastwards along the coast

Tab. 61. The distribution of particle size, lime content and the amount of organic matter in tidal flats of the North Sea coast.

Soil type on flat		Sand	Sandy mud	Mud
Particle size-fractions				
coarse sand	> 1 mm	0.5	< 0.5	–
sand	1.0 – 0.1	**71**	24	8
coarse silt	0.1 – 0.05	24	**49**	23
silt	0.05 – 0.01	2	17.5	**42**
fine silt	< 0.01 mm	2.5	> 9	27

Soil type on flat		Sand	Sandy mud	Mud
lime content	min	0	0	0
	mean	**2.6**	**6.7**	**10.6**
	max	8.5	13.3	21.5
Organ. matter	min.	0	0	3.5
	mean	**1.6**	**4.3**	**9.5**
	max.	5.5	12.6	21.5

Note: Average and extreme values from a large number of samples in each case from the area of the Jade bay. Figures rounded off; percentage of dry weight.
Source: Data from Linke (1939).

by the sea as far as the German Bay. In the southern North Sea where *Zostera* meadows cover 'only' about 100 000 ha they produce around 0.5 million tons of dry matter. The meadows of the shallow sea then yield more per ha per annum than the majority of meadows on land (on average 5 t/ha compared with 2.5-5 t/ha of dry matter, see also Jacobs 1983). Their high production may be a result of the richness of the water in minerals, nitrogen, phosphorus and other nutrients which the *Zostera* receive particularly on the ebb tide. Since a large part of the leaf of the Eelgrass is torn off by the sea and carried on the flood over the flats, the photosynthesis of *Zostera* species must be seen as the most important anabolic factor in the cycle of material in the shallow seas in this area. Because of the increase in pollution of the rivers by organic material and artificial fertilisers the productivity of the shallow seas near the coasts will probably continue to increase.

The movement of the seawater over the flats stirs up the inorganic sediment and the organic detritus to such an extent that it is sometimes not possible to see the bottom through only 20 cm of water. The surface of the water appears a dirty grey-blue to browny-grey and to anyone coming in from the sea this denotes the beginning of the shallow water. Linke (1939) has made measurements showing that a litre of this water can hold up to 0.93 g of material in suspension even when there is no wind. When there are waves, and along the edge of the rising flood tide, this amount is increased many times. In the shipping lane out of Wilhelmshaven Linke found that 20 cm of sediment had collected in a glass tube placed 50 cm below low water and he calculated from this that a sedimentation of 3 m per annum was possible.

The water flowing over the flats then is not short of material for land building but it is necessary for this to be deposited and not washed away again by the ebb tide.

(c) Communities living below the mean high-water mark

The organisms of the tidal flats assist sedimentation and binding of the mud in different ways. The many plankton feeders such as the bivalves (e.g., *Mytilus*, *Ostrea*, *Scrobicularia*, *Macoma*, *Cardium*), worms (e.g., *Arenicola marina*) and crustaceans (*Corophium*) play an important role. Along with the digestible material these animals take in a lot of useless stuff which they pass out in the form of pellets of faeces which are less liable to be washed away than the original fine particles. The tubes and passages they make in the mud or sand do not damage the surface as was once thought to be the case by Schütte (1936). This is because as the tide goes out air can get into these tunnels, oxidising and strengthening their walls. Wohlenberg (1937) compared the U-shaped tubes of the Mud Crab directly with the rods in reinforced concrete.

The **coating of blue algae and diatoms** on the surface of the mud flat are now seen to be of great significance. In some places they make the surface so slimy that one could easily slip up on it (König 1972). As soon as sediment is laid down on this coating by the flood tide the algae move through the deposit onto the surface again, attracted by the light, and there they again produce a binding slime layer during the dry period. This covering protects the new sediment against erosion by sticking it together and smoothing the surface. If the surface is kept dry for a longer time, which can happen when there is a persistent easterly wind, this layer of slime protects the mud against water loss and cracking due to drying out. Where the diatoms flourish well they can hinder the activities of the tiny mud crabs, forcing them to move lower down into deeper water where they are covered with sea water for longer periods, but where there is not enough light for optimum growth of the algae. In turn the diatomaceous covering is grazed by gasteropods (*Hydrobia*, *Littorina*) whose dung pellets add to the sedimentation.

Higher plants only play a part in the sedimentation, and thus in the formation of new land, when they are able to grow in dense stands. Between these the water covering them comes to rest. This is the case at or below MLW on the one hand or above MHW on the other. In the lower area the '**Eelgrasses**' (*Zostera*) are the most important plants in these submarine meadows.

Zostera communities flourish only when they are exposed to the air for not more than 2-3 hours. For this reason they are found in a deeper zone than the diatoms or the *Enteromorpha* species which appear for a time in the early part of the year. They are sometimes found higher up on the flats, but here they occupy shallow depressions where the water stays a little longer (fig. 287).

Going in a seaward direction one first encounters a **Dwarf Eelgrass zone**. As the exposure time is reduced *Zostera noltii* is replaced by *Z. marina* var. *stenophylla*. With its longer broader leaves this is more competitive, but at the same time more susceptible to drying out. Below MLW it is joined by individuals of the typical form of *Zostera marina*, the real eelgrass. However in many areas of the flats the latter is absent altogether. In the Jade bay for example Linke found only *Zostera noltii* when he carried out a survey of the communities.

Accompanying *Zostera noltii* a form of *Fucus vesiculosus* without bladders is occasionally found which Wohlenberg named *Fucus mytili*. It attaches itself to the shell of the Edible Mussel and in this way is able to maintain a foothold on the pebble-free flats. However it is not very secure and is easily dislodged during a storm whereas the *Zostera* species, deeply rooted in the silt, are able to survive.

According to Tüxen (1974b) extensive **Eelgrass or Grass Wrack meadows** (*Zosteretum marinae*) are sensitive to strong currents and so they preferentially occupy watersheds or quiet bays at a depth of some 0.5-3 m below mean low water, i.e., outside the real flats. Here the water is not nearly so murky as it is over the higher mud flats because when the tide comes in it brings with it some water from the open sea. A consequence of the better illumination is a higher production by the Eelgrass meadows than by the Dwarf Eelgrass or brown algae communities. The significance of this for the nutrient supply of the flats has already been assessed in section b.

Mention should be made of underwater communities which turn up here and there although they are of little importance for the general economy of the flats. The *Ruppietum maritimae* enjoys fairly clear still sea water some 20-100 cm deep which it finds in some of the small depressions in which water remains during DT. It can tolerate a wide range of salt concentrations (according to Tüxen 1974b 0.5-64‰ Cl^-) In the Baltic the *Ruppion* alliance even has three communities (Lindner 1974).

(d) Samphire flats and Saltmarsh-grass swards

If one accompanies the incoming tide across the flats then at about 40-25 cm below the mean highwater mark one meets the first land plants, isolated individuals of the succulent Glasswort, *Salicornia* (fig. 288). As Iversen (1953), Willi Christiansen (1955) and others have pointed out this is always the polyploid *S. dolichostachya* (figs. 288, 289) which can grow in wetter places than the more heavily branched diploid form (*S. europea* sens. strict.). Older authors considered *S. dolichostachya* to be young plants of the normal form, but Wohlenberg (1931) noticed that both types germinated together when the soil surface had warmed up sufficiently in the spring and that their growth habits were different right from the start. Millions of seeds of both Glassworts are washed up by the sea and carried further inland by the storm floods. Some seeds may also be transported on the underside of ice-floes which lay for some time on a Glasswort flat. In the **low-lying Samphire flat** (*Salicornietum dolichostachyae*) the name species is dominant and under natural conditions it is the only higher plant (fig. 288).

In some places **Cordgrass** (*Spartina townsendii*, see Kolumbe 1931 and fig. 290) which has been planted since 1927 on the German sea shore is dominant on the mud or sand flats. This striking tall broadleaved perennial grass probably appeared around 1870 spontaneously on the English coast as an amphidiploid hybrid (Ranwell 1972). The polyploids have proved to be unusually persistent on the seacoast, but *Spartina* has not proved so effective in speeding up land formation in the German Bight as had been hoped. This is because it forms tussocks which remain isolated and by confining the water running between them actually increase the erosion rate by the ebb tide. Also unexpectedly *Spartina* has spread onto the higher

Fig. 287. Dwarf Eelgrass (*Zostera noltii*) covers the sand flat between single Samphire plants (*Salicornia dolichostachya*) about 40 cm below mean highwater. The small pale heaps are made by Sand Worms (*Arenicola marina*). Photo, Wohlenberg.

coastal meadows of North Friesland, reducing their value for grazing. In the Leybucht of the island of Norderney on the other hand *Spartina* and *Salicornia* are about equally effective in increasing the rate of sedimentation (Michaelis, in a lecture 9.11.74). Along the English coast Cordgrass is a much more effective pioneer of the lower salt marsh than Glasswort (Tansley 1949 and 1968). But it dies out unless sedimentation occurs (Goodman 1960). The reason it behaves so differently in different areas has not yet been explained. The amplitude of the tide may play a part; this continually increases as it travels along the English Channel. In contrast to Cordgrass all the other salt-marsh plants and communities behave remarkably similar on the English and the continental North Sea coast.

In pure stands of *Salicornietum dolichostachyae* an initial stage can be distinguished in which the Glasswort plants cover only a fraction of the area (fig. 288) and an optimal stage when they grow more closely

Fig. 288. Isolated plants of Samphire (*Salicornia dolichostachya*) at about 30 to 40 cm below mean highwater on a bare flat over which frequently a **strong** ebb current flows. They do not help with silt deposition; on the contrary they lead to the formation of small pools. Photo, Wohlenberg.

Fig. 289. Numerous seedlings appear in April from the seed pods of a dead Samphire plant which have lain in the mud throughout the autumn and winter. Photo, Wohlenberg.

together to form a sward (fig. 291). This starts about 30 cm below MHW. In the degenerating stage other halophytes such as *Aster tripolium* and *Puccinellia maritima* (fig. 292) come in as well as *Salicornia europea*.

Contrary to a widespread view isolated Glasswort plants do not assist sedimentation. They are only exceptionally found growing on little mounds, as was at one time thought to be typical. Much more often a small hole can be seen round each plant from which the ebb tide has washed out the silt and mud not firmly held by the roots (fig. 288). It is only when the population of samphire plants in the optimal phase reaches several hundred per m² that the water is brought to a standstill near the mud surface (Wohlenberg 1931). In particular the dung pellets of the Mud Crab and other animals living on the bare mud flats below arc carried up on the flood tide and deposited to form a regular layer among the Glasswort plants. In order to win new land more quickly one could establish a **Samphire sward** on suitable higher places by sowing the seed. The raising of the soil level is brought about in quite a passive way in the Samphire stands since these short-lived plants do not root into the new sediment. Only permanent plants

Fig. 290. Cordgrass (*Spartina townsendii*) has rhizomes and two kinds of roots, branched surface feeding roots submerged with the tide from time to time, and anchoring deep roots (these long roots and the aerial parts of the plant have been trimmed back). After Oliver, in Ellenberg (1963).

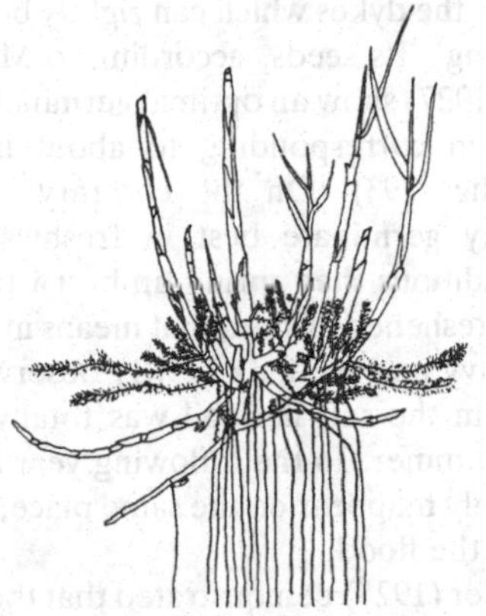

Fig. 291. Following on above the Samphire the common Saltmarsh-grass (*Puccinellia maritima*) forms a more or less dense sward from about 15 cm below to some 25 cm above the mean highwater mark in that part of the North Frisian coast where land is being reclaimed from the sea. It is seen here growing along an embankment shortly before high water.
Rows of stakes (as in the background) may accelerate the silting-up process and the plant succession considerably. (Compare the natural succession documented in table 62!)

Fig. 292. Sea Aster (*Aster tripolium*) in the Saltmarsh-grass sward about 20 cm above mean high water. Photo, Wohlenberg.

The species composition corresponds to the 'optimal stage' of the *Puccinellietum* (as in the relevé of quadrat no. 8 on the island of Tritschen in 1971, see tab. 62).

can stabilise the soil of the flats by binding the new layers with their roots and runners. Such plants also protect the ground from winter storms whereas the Glasswort plants die in the autumn and so are out of action for half the year (fig. 289).

The most efficient marsh builder is the **Saltmarsh Grass** (*Puccinellia maritima*). This grass with its striking blue-green colour can establish itself once the flats have become silted up to about 20 cm below MHW. Mostly though it first comes in at a slightly higher level (fig. 291). The following year it increases its holding at the expense of the short-lived Glasswort and spreads out to form a thick turf, growing out in all directions and catching fresh material at every tide until it forms a shallow mound. On this its leaves spend only a short time submerged and can carry on their photosynthesis almost without hindrance. The yield of the Saltmarsh Grass is correspondingly considerable; according to Stählin and Bommer (1958) it can amount to as much as that of a good pasture. However one cannot generalise about this because of the great variation within the species (Dreyling 1973).

Storm tides bring in coarser material which can cover the Saltmarsh Grass swards to a depth of as much as 10 cm. However it is rarely smothered by this as it can send out new shoots and roots to take in the additional layer. On the other hand heavily silted areas or those which have been covered by material along the tide mark may provide the opportunity for *Salicornia europea* and *Suaeda maritima* to germinate because they find open spaces. Following the raising of the ground level, conditions continue to become more favourable for those species which are less well able to withstand prolonged flooding (tab. 62). The Saltmarsh Grass sward leads into the marshland meadow and marks the upper limit of the tidal flats (figs. 291 and 292). Economically the *Puccinellietum* is invaluable not merely as a producer of new land, but as a grazing area for sheep outside the dykes. In addition, like all grass swards well provided with nutrients it produces a hay rich in protein.

The Glasswort continually loses ground as the *Puccinellietum* develops since its light-requiring seedlings find less and less open ground for their germination. It could in itself thrive on much higher ground as long as this is not too dry. It has occasionally been found up to 1 m above MHW in places where the turf has been removed and the ground laid bare. On such sites it is only represented by the branched form (*Salicornia europea* sensu stricto.). According to Schreitling 1959 it attains its best growth on the bare ridges of the newly enclosed marshes where it is no longer flooded by the sea and where the ground gradually loses its salt. Thus the boundaries of the Samphire communities are determined by competition, except for the limit towards the lower levels of the tidal flats.

Salicornia is one of the few genera living on the land outside the dykes which can rightly be referred to as 'salt-loving'. Its seeds, according to Montfort and Brandrup (1927) show an optimal germination in a salt concentration corresponding to about half that of seawater (fig. 293). On the contrary, says Ungar (1962), they germinate best in freshwater. Under natural conditions they spring up best when the flats have been freshened by rain, that means in springtime. Jeffries, Davy and Rudmik (1981) observed that the seed stock in the salt flat soil was totally exhausted already in summer. In the following year a *Salicornia* stand can only reappear on the same place, if seeds are brought by the flood.

B. Keller (1925) demonstrated that the Glasswort grew better when salt was added and was more 'drought resistant' with salt than without. *Salicornia*

Fig. 293. The germination of halophytes and glycophytes in different concentrations of sea salt. After Montfort and Brandrup in Stocker (1928), somewhat modified.

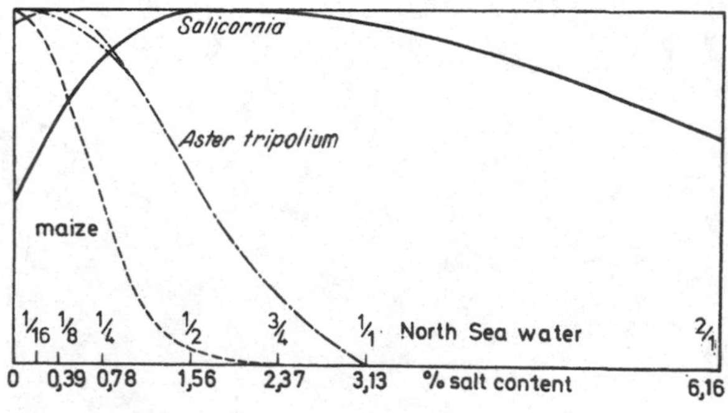

The optimum for *Salicornia europaea* lies in concentrations of salt equivalent to between half and threequarters of that in the North Sea. On the other hand *Aster tripolium* is more like maize with its optimum in freshwater.

Tab. 62. The succession of salt marsh communities on the bird sanctuary island of Tritschen where there are no protective dykes or other constructions, 30 km to the north of Cuxhaven between 1970 and 1974.

a Communities around the mean high-water mark		Salicornietum → Puccinellietum					Puccinellietum Initial → Opt.					Puccinellietum Opt. → Degener.				
Permanent quadrat no.		1					7					8				
Ground level rise (cm)						3					3					2
Year	197	0	1	2	3	4	0	1	2	3	4	0	1	2	3	4
Vegetation cover	(%)	20	60	75	80	85	80	85	90	95	95	95	100	100	100	100
Number of species		1	1	2	4	4	5	6	5	6	6	7	9	9	9	9
Salicornia dolichostachya		2	4	4	4	2	3	3	2	2	2	2	+	1	+	+
Puccinellia maritima				+	2	4	3	3	4	5	5	4	3	2	2	2
Spartina townsendii					+	+	+	+	+	1	+					
Suaeda maritima					r	r	+	+	+	1	+	1	1	1	+	+
Aster tripolium							+	+	1	+	+	1	2	2	2	+
Spergularia media								+		+	+	1	1	2	+	1
Glaux maritima												2	3	3	3	2
Festuca litoralis*												1	2	3	4	4
Plantago maritima													+	+	1	+
Triglochin maritimum													+	+	+	+
Average salt indicator value (mS)		3.0	3.0	3.0	3.0	3.0	2.9	2.8	2.9	2.8	2.8	2.5	2.2	2.1	2.0	1.9
Average moisture ind. value (mF)		9.0	9.0	8.8	8.6	8.4	8.5	8.3	8.3	8.2	8.2	7.7	7.4	7.3	7.2	7.1
mF for the years 1970-74				8.8					8.3					7.3		

b Communities *above* the mean high-water mark		Juncetum gerardii Init. → Degen.					Juncet. gerardii Opt. → Degen.					Artemisietum → Juncet. gerardii				
Permanent quadrat no.		11					12					9				
Ground level rise (cm)						5					7					3
Year	197	0	1	2	3	4	0	1	2	3	4	0	1	2	3	4
Vegetation cover	(%)	100	100	100	100	100	100	100	100	100	100	100	100	100	100	95
Number of species		13	12	7	8	8	8	7	7	10	6	9	6	5	7	6
Salicornia dolichostachya		1	r			(r°)										
Puccinellia maritima		1	1													
Suaeda maritima		r	+													
Aster tripolium		1	1	1	1	+	+°									
Spergularia media		1	1	r								+				
Glaux maritima		3	3	2	2	3	1	2	1	1		1	1	2	1	+
Festuca litoralis*		+	2	4	5	5	1	3	3	5	4	1	2	3	5	4
Plantago maritima		1	1	1	2	1	1	1	1	+	2	r			1	+
Triglochin maritimum					r	r	+	1	1	+	+					
Odontites litoralis*					r		3	3	2	2					+	
Agrostis maritima*		1	2	2	+	1	2	2	3	1	+	2	2	2	2	3
Juncus gerardii		2	2	2	+	+	3	2	2	1	+	1	2	3	2	2
Parapholis strigosa		2	2													
Atriplex hastatum		r	r													
Artemisia maritima												4	4	2	2	1
Centaurium pulchellum												+	+			
Average salt indicator value (mS)		2.1	1.9	1.5	1.5	1.5	1.5	1.4	1.4	1.3	1.4	1.3	1.3	1.4	1.3	1.3
Average moisture indicator value (mF)		7.2	7.1	6.6	6.5	6.8 (6.6)	6.4	6.2	6.2	6.2	6.4	6.1	6.0	6.2	6.1	6.2
				6.8					6.3					6.1		

Note: In addition there occur with smaller degrees of cover: in no.11, 1970, *Spergularia marina*; no. 12, 1973, *Centaurium litorale*, *Sagina nodosa*, *Plantago major*; no.12, 1974, *Sonchus maritimus*; no.9, 1970, *Bryum pallens*. Numbers are given in the scale devised by Braun-Blanquet (see tab.11); r = very rare, (r°) = outside the real test area, very rare and less vigorous.

The average indicator values have been calculated on the basis of the presence of the species without regard for the cover degrees (see section B I 4, evaluation of individual species in Ellenberg 1974 and in section E III 2. With regard to the salt factor only four indicator grades can be distinguished: 3 = always indicating a salty soil, 2 = generally indicative of salt, 1 = tolerating salt, but more frequent on soils without salt, 0 = never or very rarely on salt soils.

Source: From the relevés of permanent quadrats by A. Schwabe (1975), rearranged.

europaea can be cultivated in a 15% solution, i.e. 4-5 times more concentrated than seawater (Schratz 1936). Its ability to withstand high salt concentrations serves it in good stead especially in those places which are not often flooded, and which lose some water through evaporation in the summer so that the soil solution becomes very concentrated (fig. 294).

However the Glasswort will only grow where this solution is available in sufficient quantity since, in contrast to other succulents, it transpires as freely as many mesomorphic land plants (Stocker 1925). Its shoots have a large number of stomata per unit surface area and only a thin cuticle. Its succulence is associated with the fact that the salt which is taken up is not excreted, but stored in the cells. Thus its degree of succulence increases during the course of the year; in habitats where the salt concentration is low Glasswort is less succulent from the outset than it is in places rich in salt. As is the case with other halophytes its transpiration coefficient is reduced by higher NaCl concentrations (Önal 1971). If *Salicornia* is grown in a culture solution containing no common salt it requires about 500 cm^3 of water to produce 1 g of dry matter which is similar to the amount used by wheat and other mesophytic plants. On the other hand in a salt concentration similar to that of seawater it only requires less than half the amount, so as far as water is concerned it has a similar economy to that of Beech and other trees (see section B II 4g). The Saltwort (*Salsola kali*) is especially economical in its use of water with a transpiration coefficient amongst the lowest to be measured when growing in a 1% salt solution. It occupies the tide line on sandy beaches which are liable to dry out from time to time (see section C V 1b).

Fig. 294. Root growth of halophytes and glycophytes in different concentrations of sea salt. After Montfort and Brandrup in Stocker (1928), somewhat modified.

1/1 = 3.1% solution of sea salt. Maize is a glycophyte, *Aster tripolium* a facultative halophyte, *Salicornia* a 'true' halophyte.

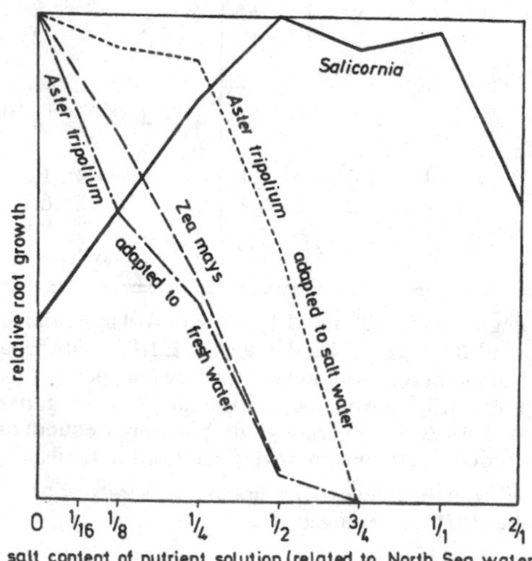

The **Sea Aster** too (*Aster tripolium*) is decidedly succulent when growing in salty places which are daily affected by seawater; but where the salt concentration is not so high its leaves are thin and have a structure almost like that of a mesomorph. Physiological researches by Montfort and Brandrup (1927) and by Schratz (1934) showed that the optimum salt concentration for its germination was very low. When the salt content in the soil water exceeded 0.5-1.0% its germination rate was reduced and a concentration above that of seawater (3.2%) was damaging to *Aster* (figs. 293 and 294). In spite of this the Sea Aster is found in largest numbers on the seaward side of the Saltmarsh Grass sward down towards the Glasswort communities (fig. 285). Its behaviour in natural conditions then is contrary to its physiological tendencies, a fact which Iversen (1936) showed emphatically. The main reason it is found growing near the MHW mark then is because it cannot compete in the denser sward, but also because the salt concentration where it is constantly being washed by the tide cannot rise so high as it may do further inland during dry weather periods. Since the Sea Aster is constructed like a swamp plant, with air spaces in its roots, stems and leaves, it can put up with repeated submersion and a saturated soil. Iversen even compared it with the mangrove, so well is it provided with air in the spongy tissue of its root and stem cortex, a structure which enables it to live on the flats even though it is at the limit of its salt tolerance. Its occasional appearance near the high tide mark or in semi-ruderal places with a low salt content correspond to its physiological optimum, but here mechanical factors must first provide a space for it.

The Saltmarsh Grass (*Puccinellia maritima*) shows a distinct adaptation to the tidal rhythm, a feature which is shared by other plants growing on the flats and outer marshes. Weihe and Dreyling (1970) varied the time during which different clones from the Saltmarsh-Grass sward were submerged in a series of pure culture investigations. In every case maximum production was achieved under the natural rhythm (see also fig. 295). In pot cultures arranged by Weihe (1980) *Puccinellia maritima* was able to support higher salt concentrations than *Festuca rubra* subsp. *litoralis*,

Lolium perenne and *Poa pratensis*. However, its productivity was also highest under freshwater conditions; it is nothing but a facultative halophyte.

The Glasswort, Saltmarsh Grass and Sea Aster are further examples which demonstrate how varied are the reasons why different plants of a community come together in the same habitat and how much their natural distribution is influenced by competition. Similar connections between salt concentration and succulence as we have seen with *Salicornia* and *Aster tripolium* are also observed in other land halophytes, e.g. *Armeria maritima* and *Plantago maritima*. Most of them are also like the Glasswort in that their seeds germinate better after they have been in seawater for some time, or at least their germination is not so adversely affected as is that of glycophytes (Ungar 1962). This behaviour which is very appropriate for plants of flats and sea marshes has been seen in for example *Puccinellia maritima*, *Aster tripolium*, *Glaux maritima*, *Plantago maritima*, *Spergula maritima*, *Suaeda maritima* and others. With the exception perhaps of *Salicornia* the optimum conditions for germination and growth of our land halophytes are in fresh or only slightly salty water (figs. 293 and 294). On the whole there are many eco-physiological parallels between the glycophytes and the (for the most part facultative) halophytes. However the salt-resistant plasma structure limits the productivity and with it the ability of the plants to compete (Neuwohner 1938). The halophytes are confined to soils containing high salt concentrations only **because many plants which cannot tolerate salt are more competitive in other habitats** (Chapman 1975, Goldsmith 1973).

Fig. 295. The height range in the distribution of individual Saltmarsh-grass plants (wide hatching) and in that of the grazed Saltmarsh-grass sward (*Puccinellietum maritimae*, – narrow hatching) compared with the time during which they are covered with water and with the frequency of flooding in the estuary of the Eider. The broken curve for the average tides from 1961 to 1965, also the continuous curves for the number of times each level was flooded during the growing period (May to October 1966) and during the whole year are drawn in for comparison. After Weihe and Dreyling (1970), modified.

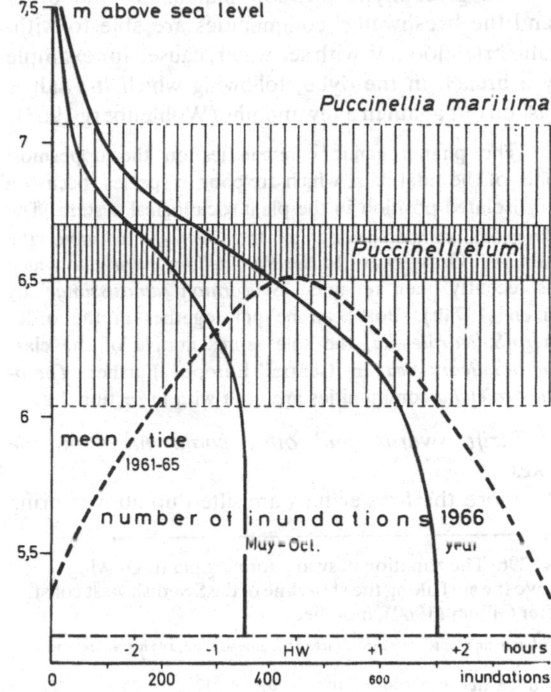

Tab. 63. The transpiration coefficients[a] of halophytes in solutions of different salt concentrations.

Halophytes (C_3)	NaCl concentration in the nutrient solution 0.0	0.5	1.0	2.0	3.0%	**Glycophytes** (for comparison)[b]			
						C_3-herbs:		trees (C_3):	
						Rice	680	*Oak*	340
						Sunflower	600	*Birch*	320
						Wheat	540	*Beech*	170
Salicornia europaea	512	–	271	239	200	C_4-herbs:			
Suaeda maritima	591	426	356	–	–	*Maize*	370	*Pine*	300
Spergularia salina	600	526	504	–	–	*Millet*	300	*Spruce*	230
Salsola kali	345	142	132	–	–	*Purslane*	280	*Douglas Fir*	170

Note: [a]Grams of water transpired for every gram of dry matter produced.
[b]During photosynthesis the C_3 plants use only ribulose 1,5-diphosphate as a CO_2 acceptor in the dark reaction. C_4 plants also make use of phosphoenolpyruvate and remove more CO_2 from the air – down to a half or a quarter of the concentration utilised by C_3 plants. Therefore they need not open their stomata as wide as the latter are forced to do, and so lose less water for the same amount of CO_2 assimilated. Nevertheless many C_3 plants are also able to manage economically with water, e.g. trees (see also table 24) and most of the salt plants. This is probably due to a thicker cuticule and a quicker regulation of the stomata.

Source: From investigations by Önal (1971). Some comparable figures for glycophytes from Larcher (1976).

Schreitling's (1959) observations on the reclaimed land in the Gotteskoog, in North Friesland, where sometimes the halophytes were left undisturbed for a few years after closing the dyke, also supports this viewpoint. Where the land had not received any fertiliser the members of the halophyte communities persisted even after all the salt had been leached out of the soil. Only where the land had been farmed and manured were they soon crowded out by the quicker growing glycophytic meadow plants. On the other hand the 'freshwater' communities are able to withstand brief flooding with sea water, caused for example by a breach in the dyke, following which the salt is washed out again in a few months (Wohlenberg 1963).

The phanerogamic communities on the uppermost zones of the tidal flats, which are poor in species, occupy a very isolated position in the plant sociological system. The Glasswort stands belong to the *Thero-Salicornion*, the Saltmarsh Grass sward to the *Puccinellion maritimae* which has recently been renamed '*Salicornion ramosissimae*' by Tüxen (1974b). Both can be put together in the order *Thero-Salicornietalia*, the sole representative of the class *Thero-Salicornietea* in Central Europe. Further *Thero-Salicornietalia* communities are dealt with in section 2 a.

(e) Thrift swards and other communities outside dykes

The more the former flats are silted up above spring tide high water the more often can the soil be leached by rain and temporarily desalinated. This washing out of salt is helped by the fact that the mineral particles deposited by storm floods are less fine than those brought in by normal tides. Consequently species that can tolerate only low salt concentrations find more and more favourable conditions as the marshland level is built up. The Saltmarsh Grass has prepared the ground for them, but becomes overshadowed and almost entirely crowded out. It is only competitive where its greater salt tolerance can be used to advantage (fig. 296). It could be that its greater water requirement, or its weaker drought resistance, also plays a part.

The so-called middle marshland meadows start at about 25 cm above MHW on the North Sea coast but are still liable to flooding during storms. On this level the most important community is the **Thrift sward** (*Armerietum maritimae*). It can be studied in the upper channel zone and on small islands (fig. 284) along the coast of North Friesland and along the marshes of the East and West Friesland dune islands on the side facing the flats. It is also referred to as a Sea Fescue sward since the weakly succulent form of Red Fescue (*Festuca rubra* subsp. *litoralis*) and Creeping Bent (*Agrostis stolonifera* var. *maritima*) are by far the most dominant plants. The leaf rosettes of Thrift, *Armeria maritima* which lie pressed close to the ground, and the

Fig. 296. The zonation of sward-forming plants growing above the surf along the shoreline of the Swedish west coast. After Gillner (1960), modified.

The dotted line represents the surface of the ground along a 45 m transect and shows the level at which the plants are growing. From this and the vertical lines indicating the daily range of the tides it can be seen that where *Galium verum* for example was growing the surface was covered by the tide only once during the summer of 1948 whereas *Puccinellia maritima* was covered 120 – 140 times.

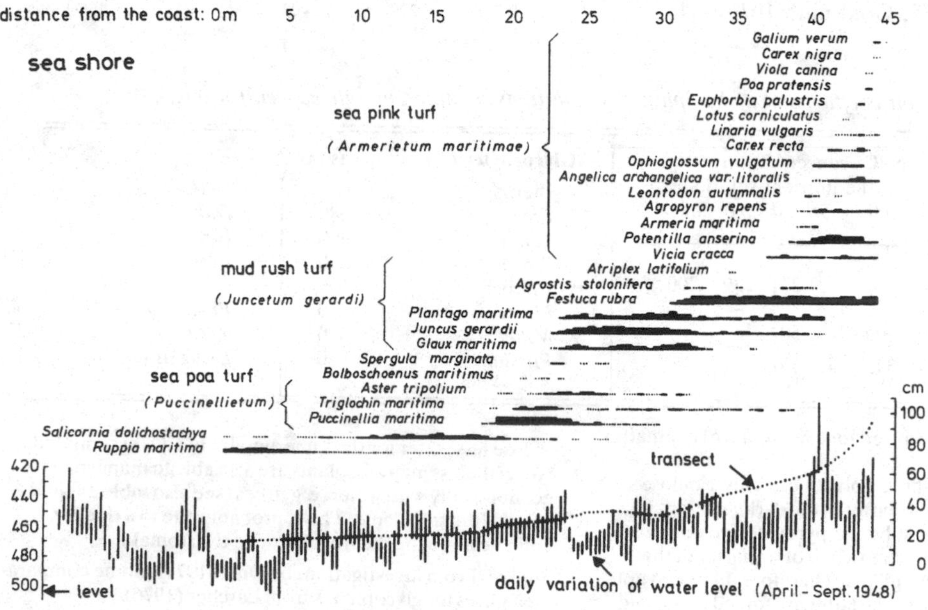

very succulent leaves of Sea Plantain (*Plantago maritima*) both have a grass-like appearance. The remaining species shown in fig. 296 and listed in tab. 62 play only a subordinate role in the community but can sometimes be looked upon as character species. The Thrift sward stands out at a distance because of its wealth of flowers. In May it is dotted with the delicate pink heads of *Armeria* and in July with the pale lilac curled spikes of the Sea Lavender (*Limonium vulgare*). The fruiting stalks of both species tower over the thick rich green sward unless they have been previously mown in July. Most of the other partners remain inconspicuous even when they are present in fairly large numbers, e.g. *Glaux maritima*. Populations of *Armeria* from habitats poor and rich in salt differ in their degree of succulence as well as in their productivity. When cultivated by Goldsmith (1978) in mixed stands on a fertile loam without saltwater, the coastal race was suppressed by the inland race.

Where Thrift swards are heavily grazed they pass over to **Mud-rush** pastures (*Juncetum gerardii*, section 2 a). According to Behre (1979) these already existed near the North Sea 2000 years ago.

When there are still functioning incipient channels (or 'tidal inlets') in the marshes these are accompanied in places by a grey-green strip of tall perennials which enjoy a good supply of nitrogen and are also salt tolerant, especially the silver-grey *Artemisia maritima* and the yellowish green *Halimione portulacoides*. *Limonium vulgare* and *Cochlearia anglica* are also found in this **Sea Wormwood scrub** (*Artemisietum maritimae* fig. 284) in much greater numbers than in the typical Thrift sward. This fragmentary drift line community grows on the somewhat higher sandy banks which are laid down by the storm tides along the sides of the channels. The remains of plants and animals are deposited here, and the good aeration of the drift line accelerates their nitrification. Similar communities are to be found in the transition from dunes to marshlands, e.g., on Wangeroog (Klement 1953), but here nitrophilous species preponderate (see section V 1 b).

Apart from the tall herbs on the drift line and good Saltmarsh-grass stands the Sea Fescue sward represents the optimum production of the land outside the dykes. With increasing height the soil fertility decreases because humus-deficient sand is deposited here revealing a characteristic pale colour in a soil profile. Sandwiched between the layers of sand there may be darker layers where mud has been deposited leading to the growth of plants and the formation of humus, but such layers are very thin. Because of this type of subsoil the grasslands outside the dykes approach those of the dry sands (section C V 1 c) in their appearance and species combination. Layers of pure sand more than 20 cm thick indicate that one is dealing with wind-borne sand which had been blown down from an adjacent dune or washed up from the sea for a short distance. Such transition formations are found in many forms at the foot of island dunes (see section C V 1 and fig. 303).

In the plant-sociological classification all variants of *Armerietum maritimae* and *Artemisietum maritimae* belong to the alliance *Armerion maritimae* (tab. 62). This has its main centre of distribution along the tidal coastlines. However many species connect it with the brackish water swards of the Baltic coasts and communities of corresponding habitats on the North Sea islands and inland sites which are included in the *Juncion maritimi*. *Juncion* and *Armerion* make up the order *Juncetalia maritimi* (cf. tab. 62).

In the boundary area with a fluctuating salt content between the salt swards and the manured marsh meadows a group of communities in which the **Sea Pearlwort** (*Sagina maritima*) plays a part are found in places where the thick growth of grass has been destroyed by being trodden or for some other reason. These associations, consisting mainly of low annuals and covering only small areas have been put together into an independent class, the *Saginetea* by Tüxen and Westhoff (1963).

Before going into the Fescue meadows outside the dykes we should remind ourselves briefly of the ways in which man has influenced the succession along the edge of the salt flats.

(f) Man's influence on the formation of marshes

Before the start of dyke building along the North Sea coast storm tides were able to penetrate deeply into the interior as they were not held up by restricting barriers. The maximum tidal range at that time was very probably smaller. In order to make the houses and farms safe from the floods they were built on a small hillock consisting of mud and cow dung called 'Wurt' or 'Warft' (Körber-Grone 1967); such otherwise unprotected settlements can still be found on some 'Halligen' off the North Friesland coast (fig. 284). 'Hallig' and 'Warf' are unknown features on English coasts.

Starting off from Holland the 'golden ring' of dykes was closed about 1000 to 1200 A.D. (Beekman 1932). It enabled fertile land to be drained by means of sluices which open during the ebb tide and are closed automatically by pressure of the sea water when the tide is in flood. However the drained mud soil sank so that its surface was below the MHW mark. Then to drain it effectively pumps had to be installed. Before the discovery of more powerful machinery lines of windmills which at one time were such a characteristic feature of Holland and Friesland were used for this purpose.

From the fourteenth century onwards repeated catastrophic breaches in the dykes took place which even now have not been entirely explained. Since the fascinating account by Schütte (1939 in which earlier authors are quoted), tectonic sinking of the coast has been thought to be the main reason. However this could not be proved by accurate levelling in a refined triangulation system for the German North Sea coast (Gronwald 1953). The increase in the tidal amplitude caused by the dykes and the sinking of the enclosed land acting together have had a greater effect. Breaks in the dykes occurred more frequently in the past than they do today because the dykes then were much weaker and had steeper sides.

The remains of dwellings, plough furrows, cattle tracks and soil horizons showing roots of salt-sensitive plants which can be found today out on the flats below MHW are not necessarily indications of a general sinking of the coast or the tilting of a tectonic plate. It can certainly be proved however that there were earlier coastal sinkings, e.g., in the Stone Ages, the evidence for which is the fact that peat deposits of raised bogs can be found on the floor of the North Sea far beyond the area of the tidal flats and more than 5 m under the present MHW.

Thorough drainage of marshland which had been enclosed centuries earlier and today lies quite a distance from the coast can unwittingly lead to salination of the groundwater, by removing the back pressure of the freshwater in the ground thus letting seawater seep back through the sandy subsoil. Eggers (1969) has made an exact study of such a case, by means of vegetation maps and soil analysis, in the largest area of reclaimed land in North Friesland, the Gotteskoog which in 1562 was still salt flats with small marsh islands. Helophytes have reacted very quickly to the recent oversalting of the groundwater.

2 Saltwater and brackish water communities of the Baltic coast

(a) Peculiarities of the salt vegetation along the Baltic

In the postglacial 'Litorina period' the Baltic was still a bight of the ocean similar to the North Sea, but today it is blocked off from the Atlantic by the Danish islands so that even in the Kattegat the tidal range is on average only 30 cm and at Rügen it hardly amounts to 10 cm. The salt content of the Baltic falls off from the sounds towards its eastern end. Here because of all the freshwater flowing into it from Oder, Vistula, Niemen and other rivers the water can only be described as brackish. In the region of the sounds the salt content varies considerably, depending on which is stronger, the inflow from the Atlantic or the outflow from the Baltic. This balance shifts according to the general circulation of wind and water. Gillner (1960), who has thoroughly investigated the vegetation and its living conditions on the west coast of Sweden, has also devoted a comprehensive survey to this question.

Even though the daily tides are scarcely noticeable in the Baltic the water level does not always remain at the same height. Yearly variations, which also occur in the North Sea, are even more obvious in the Baltic. The lowering of the sea water table which normally occurs in April and May helps the germination of Glasswort and other pioneers, while the higher levels in summer assist in their water economy. Winter storms naturally create the greatest disturbance. Easterly ones raise the water level in the fjords of Denmark and Schleswig-Holstein and drive the salt water far inland, while westerly storms drive the water away from the coast and make the freshwater influxes more effective. For all that the maximum tidal variation at Flensburg amounts to 2.5 m.

Shallow coasts with mud rich in plant nutrients are more extensive in the western part of the Baltic than in the east. They are similar to the flats in the North Sea especially in the vicinity of islands such as Fehmarn and Hiddensee, and they have a similar origin in that they are partly made up of organic material.

Since the water in the Baltic is generally clearer than that over the flats in the North Sea there is a richer flora of algae and phanerogams growing on the sea bed down to greater depths. They have been studied intensively by Kornaś, Pancer and Brzyski (1960) in Danzig Bay and by Schwenke (1964, 1969) in Kiel Bay.

All in all the living conditions and the organisms show a west-east gradient along the Baltic coastline from ones which are very similar in character to those of the North Sea to ones resembling a brackish lake. On the west coast of Sweden the plant communities and their succession resemble those of the southern North Sea to an astonishing degree (Gillner 1960). Of course since the land rises more quickly from the coast here woodlands can come down nearer to the sea. A few northern species and grassland communities form a link with the boreal type of coastal vegetation, e.g., in Norway (fig. 296). In the Schlei (Steinführer 1945), on the Graswarder near Heiligenhafen (Schmeisky 1974) and on the island of Fehmarn (Raabe 1950) too one can find practically all the species and communities we have already met in the German Bight. However as Raabe has shown, many of the species combinations have reached the eastern or south-eastern boundary of their distribution in this area. While *Salicornietum dolichostachyae* still occupies large areas in the

Kattegat and the *Puccinellietum maritimae* here goes down to the open sea with a sharper lower limit, on Fehmarn a pure Glasswort flat is no longer represented and the Saltmarsh Grass meadow occupies only small areas which are kept open by man or his cattle. Instead of these the *Bolboschoenetum maritimi*, i.e., a **brackish water Reed bed** has taken over the water front going down to below the average high water (tab. 51, see also section C I 2 a). Next to this on the landward side is a **Sea Rush sward** (*Juncetum maritimi*) which in contrast to the similar communities on the islands of East Friesland contains species barely able to tolerate the salt.

The largest areas of the Baltic shores are occupied by the **Mud-rush pasture** (*Juncetum gerardii*) on rather higher ground which is seldom flooded (fig. 297). As Schmeisky (1974) showed by fenced-off experimental areas on the Graswarder near Heiligenhafen, the species combination of this weakly halophilous community has been brought about mainly by selection through cattle grazing. Remove this factor, and Red Fescue, Saltmarsh Grass and, in damp places, Reeds soon suppresses the Mud-rush and its satellites. If the sward is just mown and not grazed these dominating pasture weeds are likewise suppressed so it is more correct to speak of *Juncetum gerardii* as a **Mud-rush pasture** rather than 'meadow' which has been customary up to now.

Fig. 297. A section through a Saltmarsh Rush meadow (*Juncetum gerardii eleocharidetosum*) on the Darss peninsula (Baltic coast). After Fukarek (1961).

From left to right: *Eleocharis quinqueflora, Glaux maritima, Juncus articulatus, E. qu., Blysmus compressus, G.m., Blysmus rufus, Triglochin palustre*.

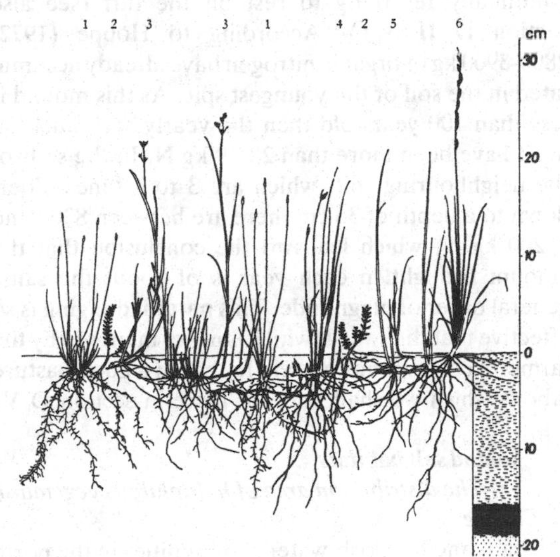

A number of subassociations and variants of *Juncetum gerardii* can be distinguished which are similar to those in the North Sea region, but not identical (Fukarek 1961). One comes across sandy drift lines on Fehmarn and the coast of Schleswig-Holstein which carry Sea Wormwood scrub (*Artemisietum maritimae*) dominated according to Raabe by a salt-tolerant variety. This community too dies out towards the east although individual plants of *Artemisia maritima* are to be found as far as the Niemen mouth.

In all the *Salicornieta* which have been described from Fehmarn, Rügen and other Baltic beaches *Salicornia europaea* s. str. (section 1 d) is dominant. As Gillner pointed out these branched forms of Samphire occupy a relatively restricted place which is seldom covered by the sea and is kept free from plants either mechanically or by the high salt concentration. According to Gillner *Salicornia* is the plant which can penetrate furthest into these local salt deserts; *Suaeda maritima*, the Annual Sea-blite, can also survive here. This rather zerophytic **Sea-blite-Samphire community** should not be confused with the *Salicornietum dolichostachyae* which always occupies damp places.

(b) Plant succession on a new silt island in the Baltic

Living conditions most resembling those of the North sea mud flats were found temporarily on the island of Bock to the south-west of Hiddensee. This island was constructed in 1945 from dredged up silt and mud (Voderberg and Fröde 1958, 1967, see also Ellenberg 1963). Already by 1946 Samphire had established itself everywhere on the bare ground (the two species were not distinguished). It remained longest on the wet shore line, but higher up it was crowded out by *Puccinellietum maritimae* or by species of the *Juncetum gerardii*. The Saltmarsh-grass on Bock grew more luxuriously than anywhere else in the Baltic region. Perhaps the absence of grazing played a decisive part in this, along with the high nutrient level of the newly deposited mud. The fertility of the Baltic coast line does not normally come up to that of the North Sea flats, but in this case it was increased, albeit for a short time, by the humus-rich dredgings.

The island of Bock is important for the understanding of coastal vegetation in other ways too. It was occupied by plants with an almost explosive rapidity presumably because of the masses of viable seeds floating in the sea and deposited on the new land shortly after its formation. The effectiveness of this method of dispersal could not have been demonstrated better. At first most of the new arrivals were to be found all over the island and only later, because of competition, were they restricted to their present position. In this respect too the island was an informative large experiment. The more competitive species asserted themselves very quickly on the fertile soils and had already **achieved a balance within a few**

years, so that only 10 years after the island building most of the salt plant communities were showing typical features. In addition examples of natural formation of new land were to be seen on the coast. The succession from *Puccinellietum* to *Juncetum gerardii* was completed stepwise in just a few years. Today it is becoming more and more clear that woody plants may find a footing on all those parts of the island which are not permanently salty. Where there is only a small tide a woodland can go right down to the coast.

There are not so many different habitats on Bock as there are on the large diluvial island of Hiddensee which has been thoroughly investigated by Fröde (1950). However the shore line vegetation of the new island is scarcely inferior to that of its old neighbour in its diversity of communities and species.

There are also well grown salt plant communities on the Darss peninsulla (a monograph by Fukarek 1961), but they get progressively poorer the further east one looks along the Baltic coast. This gradient in the flora cannot be traced to historical grounds, as was done in the past without any hesitation. The example of the young island of Bock supports much more strongly the view that the present living conditions and the means available for seed dispersal are sufficient to explain the mosaic of communities and species existing on the shores of the Baltic.

(c) Plant succession and living conditions on sandy new land

If the succession on the silt island of Bock reminds one of the conditions along the North Sea flats, then the nature reserve of 'Graswarder' off Heiligenhafen offers an opportunity to study the plant succession which is typical of the Baltic coast.

As the mixture of sand and fine pebbles drifts along the coast it is thrown up from time to time into spits, i.e., banks of sand which are laid down behind a small promontory. Each new layer is thrown up higher than the older ones and so is not flooded to the same extent when storms from the east or north raise the sea level.

At the foot of the oldest spits, in spite of continuous contact with the brackish water, up to more than 1 m of peat was laid down, probably formed largely from a Reed bed. In a part of the Graswarder which is fenced and no longer being grazed such **brackish Reed swamps** have developed, followed by lush Red Fescus stands. Where cattle can graze communities of short grasses rise up resembling those on sandy islands in the North Sea. At the slope foot which is frequently saturated with salty water, the Saltmarsh Grass *(Puccinellia maritima)* dominates. Above this one can recognise different subassociations of the **Mud-rush pasture** *(Juncetum gerardii)* in which *Festuca rubra* plays a more or less important part. The youngest spits are so high that glycophytes dominate here, especially *Lolium perenne.*

Schmeisky (1974) has mapped these different vegetation units and investigated their ecology. The degree of salination is measured by chlorine content of the soil water (fig. 298). This varied both with the time of year and from one year to another but it was obviously related to the species composition, and in fact proved to be the dominating factor.

The amount of nitrogen in the soils which are repeatedly saturated with salty water is surprisingly high. According to Zimmek (1975, see also Ellenberg 1977) the mineralisation of nitrogen starts relatively late in the year, mainly due to the low temperature of the soil. Nevertheless during the growing period under a *Festuca rubra* grassland of average dampness 177 kg/haN accumulates as nitrogen, and this was found to be so in grazed areas as well as in those left unstocked. In the damper subassociations of *Juncetum gerardii* the net nitrogen mineralisation is just over 75 kg/ha/y (also mainly in the form of NO_3). In *Puccinellietum* it amounts to only 24–31 kg N and here, because of a shortage of oxygen, NH_4 predominates. Gerlach (1978) in a critical investigation into the incubation method by which the N-mineralisation is determined found that the values obtained even from very wet soils always corresponded closely to the amounts of nitrogen taken up by the plant cover.

Very likely a large part of the nitrogen circulating in the ecosystem of Graswarder, comes from the dung of seagulls which nest here in their hundreds and are continually returning to rest on the turf (see also section D II 3 c). According to Hoppe (1972) 2800-3900 kg of organic nitrogen have already accumulated in the soil of the youngest spit. As this mound is less than 100 years old then the yearly accumulation must have been more than 28-39 kg N. In the soils of the neighbouring spit, which are 3 to 4 times older, down to a depth of 36 cm there are between 8300 and 17200 kg N, which leads to the conclusion that the amount brought in each year is of about the same general order of magnitude. This natural dunging is so effective that the sward (which is never manured by the farmers) corresponds in yield to that of a good pasture (about which we shall find out more in section D VI 1 a).

3 Inland salt habitats

(a) The distribution area of halophilous vegetation in Europe

The salt- and brackish-water communities in the north

and west of Central Europe radiate out from the North Sea and the Baltic. Where they are found inland, they are in connection with natural or artificial salty springs which are present on a particularly large scale in the Halle–Halberstadt–Merseburg area. They are absent from the geologically older Baltic-Russian continental shield since the petrographic and tectonic prerequisites are missing mesozoic salt-rich sediments and tertiary anticlines.

The further towards the south and east these inland salt habitats lie, the more their flora differ from those of the coast, and the richer they become in species. Consequently Wendelberger (1950) has recognised four main areas of salt vegetation outside the Mediterranean, namely: 1 the sea coasts, 2 inland salt habitats of Central Europe, 3 the pannonic area and 4 Roumania. In the first two those halophytes predominate which are very widespread and somewhat indifferent as to climate (cosmopolitan, eurasian, circumpolar). On the other hand in the third and fourth the commonest plants are those of the dry-warm continental area. Endemic species too have developed in these remains of the tertiary Tethys Sea. The coastal vegetation contains atlantic elements which do not get as far as Roumania. In the pannonic-balkanic regions those species which are able to tolerate the high proportion of soda in the soil assume dominance. Along the coasts of the North Sea and the Baltic only common salt plays a decisive part.

Suaeda maritima may count as a soda-plant. By the Neusiedler See it can occupy ground which in summer sparkles with efflorescent gypsum fur. *Salicornia europea* on the other hand does not relish these extreme conditions according to Wendelberger. It occurs only sporadically and then only on soils with less soda, preferring those which are kept open artificially. Under these conditions then *Salicornia* and *Suaeda* do not occur together as they do in communities on salt-rich dry habitats near the sea coast.

(b) Salt vegetation of inland Central Europe

Communities are found in the salt-rich inland habitats

Fig. 298. The chloride content of the soil water throughout the year under some grazed sward communities on a coastal spit near Heiligenhafen (Baltic) where the tide is barely noticeable. After Schmeisky (1974), somewhat modified.

Going from the Saltmarsh-grass sward where the surface is only some 20 cm above mean seawater level, through different subcommunities of the Mud-rush pasture (*Juncetum gerardii*) up to the top of the spit with its sandy gravel soil the effect of the salt falls so quickly that an almost glycophyte community (*Lolium perenne* trodden turf) can thrive (*Juncetum* with *Leontodon*)

All the soils had less chloride in 1969 than in the next two years, especially in the autumn of 1970 when seawater washed over the top of the spit. The chloride content fluctuates considerably during the course of the year, especially in the topsoil. As a rule the salinity rises during a dry period whereas abundant rainfall can more or less wash out all the salt. Nearly every year this leaching out of the salt can be seen to go on during the summer too.

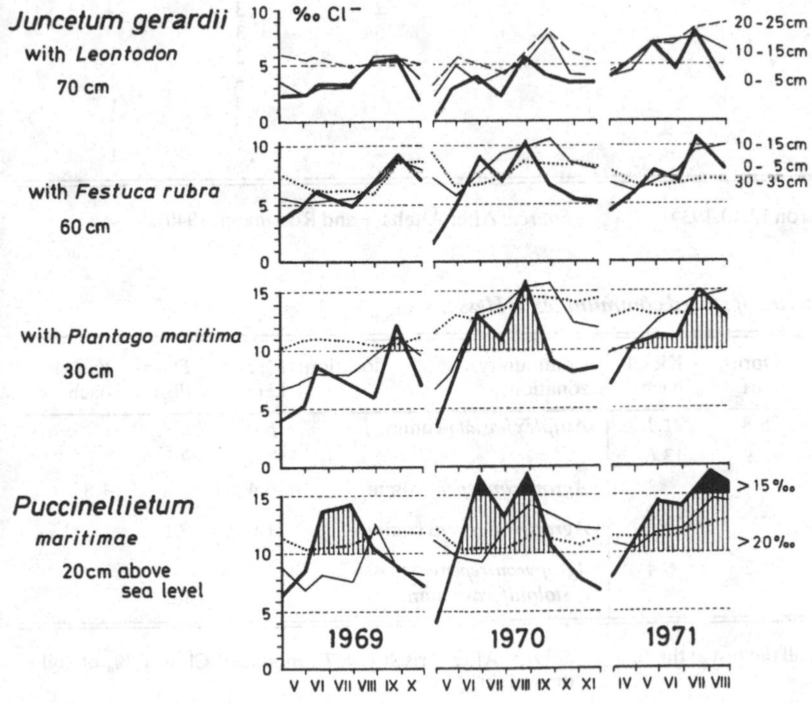

of Poland and Germany which have a similar combination of species as those of the North Sea and Baltic coasts. The examples in tab. 64 form an ecological sequence. They are mainly sward communities which are reminiscent of the Mud-rush pasture *(Juncetum gerardii)* and are made up principally of taxa which though certainly salt-tolerant are not really halophilous. Where the salty water level is high enough the brackish Reed swamps (*Bolboschoenetum maritimi*) can flourish (see section C 12 a and fig. 238). The transition to normal meadow or pasture communities takes place gradually, clearly reflecting the falling off in the salt concentration of the soils (tab. 64).

We shall have to forgo a systematic survey of the numerous associations and subunits which have been described, but instead will mention some new publications dealing with different areas of Central Europe. Raabe (1965) has described salt meadows in the Treene lowlands near Sollbrück in Schleswig-Holstein. Glahn and J. Tüxen (1963) analysed a changing small-scale mosaic of salt-plant communities and their soils near the town of Lüneburg. Succow (1967, see also fig. 480) found a number of more or less halophilous plants amongst the grassland communities in the glaciofluvial Ziese valley in East Mecklenburg.

The salt vegetation of eastern Lorraine was described in detail by J. Duvigneaud (1967). The British inland marshes are related to the coastal ones, but poorer in species. Lee (1977) distinguished four communities which correspond well to those in Central Europe.

Krisch (1968, see also tab. 65) studied the communities and the degrees of saltiness which have formed since about 1930 under the influence of the potash industry in the Werra valley near Bad Salzungen. According to Steubing and Dapper (1964) just in the top 10 cm of soil were concentrated 9.4 tonnes NaCl/ha on which a meadow of *Juncus gerardii. Triglochin maritima* and *Glaux maritima* was growing. In the roots of this stand (dry weight 38 t/ha) were 2.1 t NaCl/ha. The branches and leaves are significantly poorer in salt then than the roots.

At the margins of the pannonic plain, e.g., in the south of Slovakia (Vicherek 1964), in the Austrian

Tab. 64. The sequence of plant species in relation to a graduation of the salt concentration in sward near Artern/Unstrut.

No.	1	2	3	4	5
Vegetation cover (%)	0	70	90	100	100
NaCl concentr. in soil water (‰)	13.8	9.7	5.4	3.5	2.5
Salicornia europaea		5	2	3	+
Suaeda maritima		+	5	3	
Halimione portulacoides			+	2	1
Puccinellia distans				+	4
Spergularia media				+	1
Artemisia maritima					+
Aster tripolium					1

Note: Samples taken at a depth of 2-10 cm on 12.10.1934

Source: After Altehage and Rossmann (1940).

Tab. 65. Salt concentrations under a series of sward communities in Hesse.

Community zonation (location:)	Merkers	Dorndorf	Kieselbach	Community zonation (location:)	Merkers	Dorndorf	Kieselbach
Bare soil	41.1*	28.8	21.1	*Atriplex hastata*-comm.	16.0	6.7 5.5	
Spergularia marina comm.	18.4	18.9	13.6	*Agropyron repens* comm.	6.4		4.8
Triglochin maritimum comm.			11.0	*Agropyron repens* comm.	1.9	3.1	
Agrostis stolonifera-comm.		7.2	6.4	*Agropyron repens-Agrostis stolonifera*-comm.		1.2	

Note: *Sample taken end of August 1964, all the rest at the beginning of October 1964.

Source: After Krisch (1967), modified. Cl as a ‰ of soil dry weight.

Marchfeld (Wendelberger 1964) and by the Neusiedler See (Stocker 1960) the halophilous communities have a quite different character. These must be seen as extensions of a southern European vegetation which has been comprehensively described by Horvat, Glavac and Ellenberg (1974).

The majority of the plants found in the continental salt-plant communities on soils which dry out from time to time are no more obligatory halophytes than those of the sea coast. Culture experiments carried out by Weissenböck (1969, see tab. 66) show that salt-plants including *Aster tripolium*, *Artemisia maritima*, and *Plantago maritima* grow significantly better in garden soil poor in salt than in a salt-rich one. On the garden soil they accumulated only a little common salt so that the Na/K ratio in the cell sap remained low, as did the osmotic pressure. The leaf-succulent Pepperwort (*Lepidium crassifolium*) in contrast needed the sodium so much (and probably also the chlorine) that it extracted a large amount even from the garden soil. Here it grew badly and the leaves, instead of being succulent, were hard (Weissenböck called them 'peinomorphic' since their rather xeromorphic structure had been produced through a shortage of nutrient, cf. section C III 4 a). Its optimum growth was achieved in soil which contained salt, though not so much as the soil from which *Lepidium* originally came. It would appear that to subject the other partners in the continental salt communities to such tests could lead to informative results.

Tab. 66. Growth characteristics and cell sap properties in relation to different soil salt concentrations.

Habitat:		Salt content high	Salt content lower	low (garden)	Remarks
Soil salt content	Cl	0.03	0.01	0.005	
(g/100g dry soil)	Na	0.30	0.17	0.03	
Growth characteristics:					
Aster tripolium					
height (cm)	H	5 – 20	20 – 30	**30 – 40**	
leaf length (cm)	L	1 – 3	3 – 6	10 – 12	
leaf width (cm)	B	0.5 1	0.8 1.2	1.5 – 2	Succulent on
leaf area (cm^2)	F	1 – 4	4 – 8	15	salty soil
Artemisia maritima	H	8 – 15	10 – 20	**30 – 40**	Mesomorph on
(leaf much divided)	L	5 – 8	7 9	8 – 10	garden soil
	(B)	2 – 3	3	3 – 5	
Plantago maritima	L	9 – 12	10 – 13	10 – 13	Always fairly
(prostrate)	B	0.3 – 0.4	0.3 – 0.4	0.3 – 0.4	succulent
Lepidium crassifolium[1]	H	5 – 10	**10 – 15**	3 – 7	Succulent on
	L	2 – 4	5 6	1.2 – 1.5	salty soil,
	B	1.5 – 2	2 3	0.4 – 0.6	peinomorph on
	F	2 – 4	8 – 11	0.8 – 0.9	garden soil
Nature of cell sap:					
Aster					
Na/K ratio	Na/K	3.23	3.36	0.07	
Osmotic pressure (atm)	π^*	30	27	14.3	Low Na
Artemisia maritima	Na/K	–	3.10	0.09	requirement
	π^*	–	30	12.2	
Plantago maritima	Na/K	8.30	7.13	0.22	
	π^*	27	25.4	9.9	
Lepidium crassifolium[1)]	Na/K	**12.00**	1.80	**0.54**	High Na
	π^*	32.9	22.2	15.5	requirement
In the soil water under *L.c.*	Na/K	1.76	1.70	0.23	

Note: [a] According to Weissenböck's definition *Lepidium c.* is an obligate halophyte; the other species are facultative halophytes.

Source: From data by Weissenböck (1969).

V Dunes and their vegetation successions

1 *Coastal dunes*

(a) Conditions leading to the formation and destruction of dunes

Dunes can only be formed where wind blows over stretches of bare sand. Before the woodland returned to Central Europe after the Ice Ages there were such drift sand areas widespread inland also, but today they are found almost exclusively near the coasts.

A constant supply of sand, the second prerequisite for any dune formation, is best guaranteed along the shallow 'outer coast' of many North Sea islands and the Baltic. Here the sand is moved by the sea at right angles to the main stream in broad waves or spits along the strand. All finer fractions are washed out of the original material and come to rest on the flats or in quiet bays, i.e., on the 'inner coast'. The surf throws the average sized sand grains (0.25-1 mm diameter) right up over the mean highwater mark, mixed with ground up and whole mussel shells, bodies of plants and animals or other flotsam.

During fine weather or when the wind is strong the uppermost layer of this foreshore sand dries out quickly. One can see it being blown and flying about in long streaks moving inland whenever the **wind speed** exceeds about 6 m/s (Wiemann and Domke 1967). On stormy days in summer the sand may be carried away right down to the surface of the capillary water layer. In this way a more or less broad gradually advancing strand develops which only the surf from storm floods is able to surmount.

In dry periods and when the wind direction is parallel to the strand the foreshore sand can be thrown up into small barchanes (crescentic dunes) like those characteristic of deserts. Dieren (1934) observed such migrating barchanes from time to time along the Dutch coast as small unstable structures which have nothing to do with the gigantic moving dunes which we shall deal with later (section 2). Generally however the dry sand just glides in a thin layer over the surface, with its pattern of ripple marks.

It first comes to rest in the wind shadow of some obstruction such as objects washed up from the sea or plants which had been able to find a footing there. The 'tongue-dunes' formed in this way are rarely higher than a few cm or dm. Only some strong-growing gregarious grasses help to build higher and permanent dunes able to stand up to the storm tides. These tall grasses grow up layer upon layer with the blown sand which settles on them.

The most successful **dune builder** is the Marram Grass (fig. 299). Sand Couch and Lyme Grass and a few other species prepare the way for it. As Warming (1907) and Reinke (1903, 1912) first recognised these plants are not only able to trap the sand between their stalks and hold it fast with a dense network of roots, but also to keep on growing through, should they be covered by an unusually thick layer of sand during a

Fig. 299. Marram Grass dunes (white dunes) on the Baltic coast. In the foreground left is a young quickly-growing dune, in the background right an older one partly damaged by storms. Here can be seen harder, more-stabilised, layers.

storm (figs. 299 and 300). As we shall see they actually need a constant supply of sand in order to grow normally.

In order to illustrate the extent to which the wind speed is reduced by Marram Grass dunes we can use the relative figures determined by Lux (1964, see tab. 67) from numerous measurements on the island of Sylt. Even on the crest of a dune the wind speed between the clumps of Marram is cut down to a third.

A series of plant communities takes part in the formation and further development of the coastal dunes. Where the strand is still undisturbed these are arranged in a characteristic order in zones one behind the other. These **zones** are largely brought about by a **succession** which is similar for both the North Sea and the Baltic since the differences in salt concentration no longer have any bearing on the dunes themselves.

As is shown schematically in fig. 301 there is first of all a 'drift line' (or strand-line) with a scattered growth of luxuriant annual nitrophytes at this extreme high water mark. A bit further inland Sand Couch and Lyme Grass form low 'primary dunes' which later become steep crested main or 'white dunes' following the invasion by Marram Grass. The white dunes connect up to form a bank which can get up to over 10 m in height. Still further inland the sand has come largely to rest and forms fixed or 'grey dunes' with a plant cover consisting of short grasses and low herbaceous plants as well as of mosses and lichens. Where these more or less humus-rich grey dunes turn into dwarf-shrub heathland they are known as 'brown dunes'. Quite often bushes will be found growing on the lee side of the white dunes forming 'scrub dunes' which can also build up small hills growing out of the 'slacks', i.e., wet depressions within the dune landscape. Today most coastal dunes are not wooded, but

Fig. 300. The development of Marram Grass dunes, schematic. *Ammophila* forms fresh rhizomes at higher and higher levels as it is covered by sand. After Paul (1944).

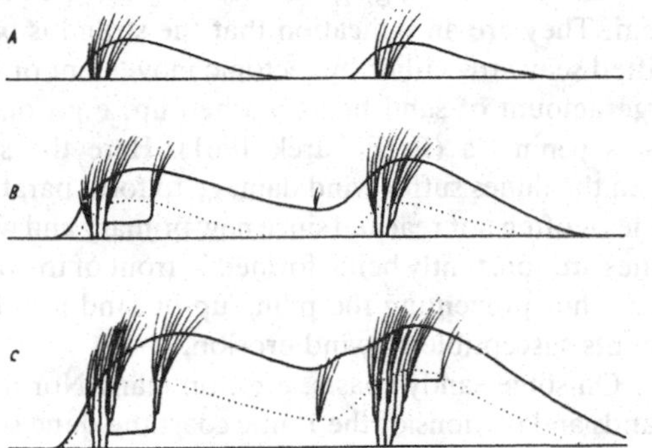

Fig. 301. The succession of vegetation on dunes along the North Sea coast, schematic. In the absence of any interference by man or domestic animals trees, or at least bushes, would be found growing in the valleys and on the brown dunes (heath dunes) and probably also on the grey dunes (see fig. 310).

Leaching of lime takes place rapidly once the dunes are stable from the grey dunes inland. Where there are large dune complexes it is possible to extract reasonably fresh groundwater. This is rainwater which has drained through the sand and collected as a 'cushion' above the heavier salt water.

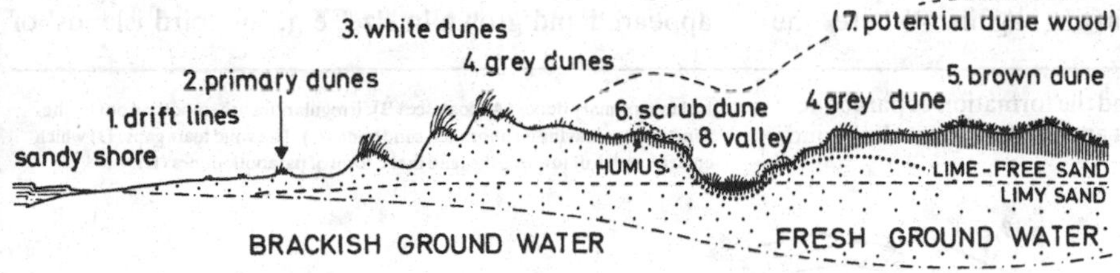

Tab. 67. Relative wind speeds on the shore and over dunes of different exposure.

Measurement site	Open strand	Marram Grass stands *(Ammophiletum)* windward slope	on the ridge	lee slope open	lee slope dense	foot of lee side
50 cm above the ground	100	**103**	81	69	75	73
10 cm above the ground	**85**	60	33	42	18	**50**

Source: After measurements by Lux (1964) on Sylt, expressed in percentages of the highest average value on the flat shore.

as we whall see a woodland could gradually develop on the brown, grey or scrub dunes.

On long stretches of the mainland coast and of the islands the strand line remains fairly stable. The constant supply of sand enables the white dunes to form huge groups of hills which are continually being bound together by the Marram Grass and accompanying plants, a process which is now helped artificially by planting. The higher these hill dunes become the more easily they are damaged by storms. Here and there one sees fresh mussel-shaped holes which have been torn out by the wind. These quickly become long tongues (fig. 302, D and E) which are usually restablished by plants growing out from the edges, but at the tips of the tongues the sand is moving too quickly and the plants cannot gain a foothold. The prevailing west to south-west wind draws these tongues out into long troughs with parabolic sides. Finally whole groups of such parabolas of different ages and sizes are on the move together, occasionally joining or interfering with each other (fig. 302 E). The hill dunes have become a field of parabolic dunes. Inland these can reach greater heights than the bank of white dunes nearer the coast, but under natural conditions the outer slopes of the parabolic dunes remain covered with vegetation, and sooner or later they will be brought to a complete standstill when their apices finally peter out. The plant communities of the parabolic dunes rarely have the same character as those of white dunes, but correspond much more closely to those of grey dunes, scrub dunes or other older stages. Under natural conditions they would eventually be taken over by woodland.

Only in those places where the plant cover has been destroyed by man or by the sea can the wind-borne sand start moving again. Where the damage assumed catastrophic proportions the mass of sand could pile up into 'wandering dunes' devoid of vegetation as is the case on the large sand spits of łeba and Kurskiy. It is probable that many areas of parabolic dunes and all wandering dunes have arisen through the action of man reaching a peak in the sixteenth and seventeenth centuries when inconsiderate overexploitation of the coastal dunes became a real nuisance, mainly in Holland and Denmark (Warming 1907).

There are many deviations from the normal dune succession. Individual stages may be absent, sometimes a number are missing. Quite often one can see several dune strips of similar proportions one behind the other of increasing age as shown by the growth on them. They are an indication that the strand is being shifted seawards either by tectonic movement or by a large amount of sand being washed up, e.g., on the Darss peninsula (F. Fukarek 1961). Here the stage when the dunes suffer wind damage to form parabolic dunes is often not reached since new primary and white dunes are constantly being formed in front of the older ones, thus preventing the piling up of sand into high mounts susceptible to wind erosion.

On some sandy coasts, e.g., on many North Sea islands and sections of the Baltic coast the dune series is incomplete. White or grey dunes and even older stages are found right by the tide line (fig. 309). Here the sea and the wind are destroying what they once helped to build, carrying the sand to less exposed places where it can be used for new dune building. In this way the islands are slowly but gradually moving eastwards (fig. 303). Some islands have disappeared completely within historical times while others have appeared and grown in size, e.g. the 'bird islands' of

Fig. 302. Dune development and the formation of parabolic dunes on coasts running at right angles to the prevailing wind, schematic. After Paul (1944).

Isolated primary dunes (A) coalesce (B); irregularities are smoothed out by the wind and sea leaving an unbroken sand bank (C). The wind tears gaps (D) which enlarge and drift towards the land in the form of parabolic dunes (E).

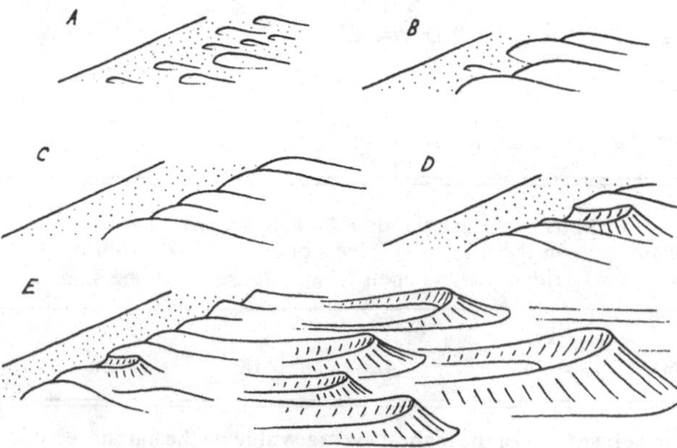

Memmert and Scharhörn (Tüxen and Böckelmann 1957). On the dune coasts, just as on the flats or in the river valleys, there is a constant building up and breaking down unless it is prevented by costly artificial protective measures.

The dune landscape is sometimes further diversified when the sea breaks through or deep cuts are made by the wind resulting in more or less wide valleys. Such channels may be blown out right down to the capillary layer above the groundwater and should this happen in summer when the water level is low these depressions may be so deep that later they fill with water to form lakes and mires (fig. 301, see also fig. 228).

Along the strand and in the dunes the habitat conditions can vary both in time and from place to place. There is every intermediate stage to be found from very rich to extremely poor in nutrients, from wet to very dry sites and from rapidly changing to stable places. The amount of salt is very variable and in addition there are several factors introduced by human activity such as grazing and planting. The geographic and climatic gradients from west to east also affect the dune vegetation. A comparison of maps of the northern part of the island of Sylt by Straka (1963, see also Jeschke 1962) shows that only a few decades are required to remodel the mosaic of dune and strand communities more or less completely. Nowhere is the dynamic nature of vegetation more apparent than in the marshes and on the dunes of the sea coasts (Ranwell 1972).

There is no room here to describe in detail the colourful diversity of all the plant communities. We must limit ourselves to the main stages in the natural formation of dunes and their further development and try to look into the factors determining them.

(b) Drift lines, primary dunes and white dunes

It is difficult for the majority of plants to establish themselves on the strand. The sea is constantly bringing seeds to the shore which have been blown or washed into this huge reservoir. By no means all, however, can survive a long submersion in seawater; they may be more suited to dispersal by wind, birds or men.

According to a number of authors cited by Dieren (1934) most plants growing on the tidal flats have seeds which germinate well after 36 or more days in salt water, e.g., *Salicornia* and *Puccinellia*. Seeds of grasses such as *Agropyron junceum* and *Atriplex litoralis* have also proved to be resistant. In the case of *Cakile maritima* salt water even increases the germination percentage from 57 to 100. The wind-dispersed seeds of *Senecio vulgaris* too remain viable after months in salt water.

These plants and other apparently equally salt-resistant species are found in large numbers growing on the drift line thrown up on the strand by an extraordinarily high tide. They germinate here only when the soil is quite warm in late spring, but then they grow very rapidly thanks to the lively nitrification going on in the decaying seaweed bed mixed up with the sand. The **therophytic drift line occupants** in particular are pronounced nitrophiles and are also salt tolerant, for example *Cakile maritima*. Tüxen (1950a) named a vegetation class after this crucifer (*Cakiletea*) and this includes a large number of the coastal drift line communities. Some of our garden weeds are probably

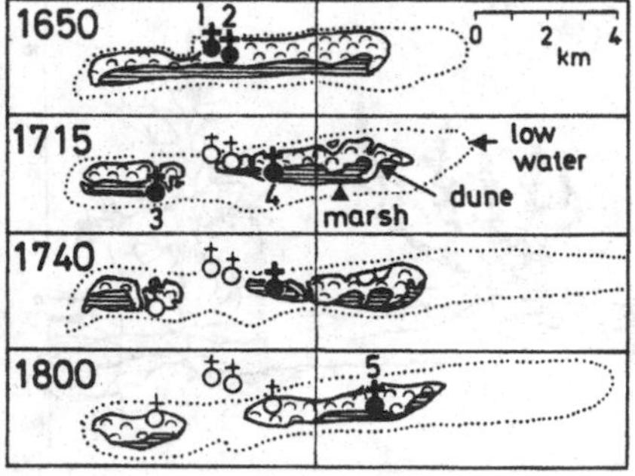

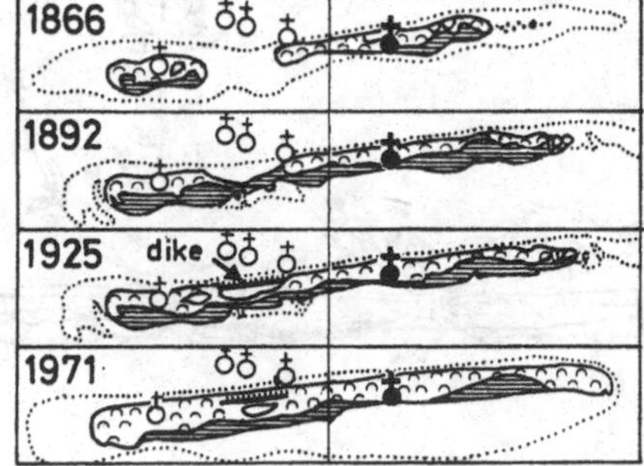

Fig. 303. Dune formation and destruction which have taken place on the North Sea island of Juist since 1650. Simplified from a publication on sale and from topographical maps.

The movement of the island towards the south and west can be followed from the positions of the submerged churches. The enlargement and strengthening of the sea defences in the twentieth century not only stabilised and extended the dunes but also the surrounding ground exposed at low water. A freshwater lake, the 'Hammer', which is now protected from the storm flood tides by a dyke, indicates the place where the sea broke through splitting the island in two at the beginning of the eighteenth century. The formation of marshland was possible only where there was protection from the direct action of the sea. Parts of the marsh have been progressively covered over by dunes.

at home here, e.g., *Senecio vulgaris* and *Matricaria maritima*, but on the coast they take on special forms or are special varieties. As character species of this class of inconstant therophyte communities, the following may serve (though never found all together):

Cakile maritima	*Matricaria maritima*
Salsola kali var. *polysarca*	f. *litoralis*
Atriples hastata subvars.	*Polygonum aviculare* var.
salina and *oppositifolia*	*litorale*

After they have ripened their fruit in the autumn these plants generally die and so avoid the winter storm tides. Perhaps their seeds will find new favourable conditions in other drift lines. The *Cakiletalia* communities then are neither spatially nor floristically stable, but are constantly renewing themselves. Very exceptionally they are involved in dune building and only then where the Sea Couch (*Agropyron junceum*) becomes dominant in them. Its seeds, like those of *Cakile*, *Salsola*, etc., are carried by wind and sea and germinate at the tide line (Pijl 1969). The germinating conditions are worse for all plants outside the open nutrient-rich drift line because pure sand dries out quickly and frequently on the surface. Only in car tracks or footprints did Dieren once find seedlings of Sea Couch in large numbers and assumed quite rightly that here they had found a favourable sheltered microclimate. However the majority of the seedlings were subsequently washed away by the surf during a high tide. It is only where the waves are spent that they would stand a chance of survival.

Wherever *Agropyron junceum* has reached the groundwater with its roots and permeates the sand with a dense network of rhizomes and roots its bent haulms begin to collect the sand and form small **low dunes**. Sometimes *Honkenya peploides* also shows the same strategy. One would scarcely believe that this small semi-succulent could root so strongly and deeply as Steubing (1949) established on Hiddensee (fig. 305). For several years the specimen she studied repeatedly put out roots which formed a new network in the freshly deposited sand, anchoring the plant and binding the soil. Here and there *Elymus arenarius* accompanies the pioneers of the primary dunes and for this reason the *Agropyretum juncei* of older authors has recently been renamed *Elymo-Agropyretum*. The Lyme Grass is a fairly broad-leaved nitrophile which thrives best in semi-ruderal places as well as on low white dunes (tab. 68 and fig. 306).

Marram Grass (*Ammophila arenaria*) comes in as a rule when the primary dunes on the North Sea coast have reached a height of more than 1 m since it can only stand a little salt near its roots. Because the groundwater under the flat strand and the forward dunes still contains salt (fig. 301) it can only grow here when the primary dune has been raised up to form a cushion of fresh water. On the coast of the Baltic which is much lower in salt the Marram can grow right down to the flat strand and displaces *Agropyron* as a dune builder right from the start.

While the amount of haulm produced by *Agropyron junceum* and also *Elymus* is relatively small per unit area (fig. 306) the Marram Grass grows into thick tall clumps (fig. 299) which soon suppress the pioneers and at the same time quickly collect sand. The **young white dunes** with their steep sides rise up held together by thick rhizomes and the network of roots. The *Elymo-Ammophiletum* is richer in species than the

Fig. 304. Species of the drift-line community (*Atriplicetum litoralis*) root only in the upper layer which is well aerated and rich in nitrogen, to a depth of about 8 cm. After Fukarek (1961).

From left to right: *Cakile maritima*, *Chenopodium glaucum*, *Atriplex hastata*, *Salsola kali*, *Atriplex litoralis*.

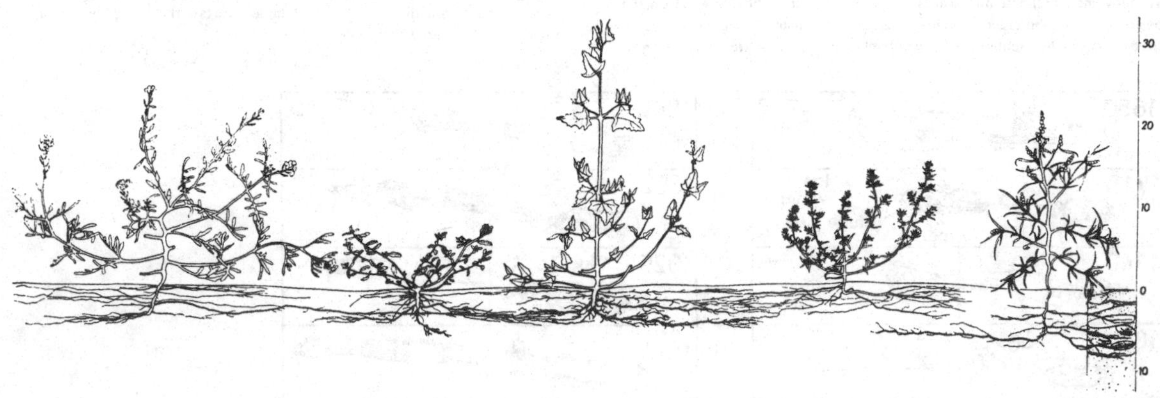

primary dunes (tab. 68) and some of the newcomers prove to be less tolerant of salt, e.g., *Ammophila baltica* and *Lathyrus maritimus*.

Sand dune plants such as *Agropyron junceum, Elymus arenarius* and *Ammophila arenaria* are so thoroughly adapted to the life of the dunes that they need a constant supply of additional sand in order to grow properly. Dieren and others have observed repeatedly that these plants suffer if they are not covered several times each year by some fresh sand. They are normally quick growing and rich in chlorophyll but if they do not get fresh sand blown up from the sea strand their leaves become narrow and yellow. They do not grow well on dunes which are some distance from the coast even if these are still rich in lime, while flourishing on white dunes near the strand even though they may be poor in lime (although this is exceptional). According to Hope-Simson and Jeffries (1966) the growth of *Ammophila* (and also of *Corynephorus*) increased whenever they formed new roots following a fresh application of sand. However the same dune grasses in the new Botanic Garden of Göttingen thrive very well on loess soil rich in both lime and nutrients and do not fall off even though there is no addition of fresh material. In sand dunes the plants suffer from a shortage of nutrients (e.g., nitrogen, potash and phosphorus) if these are not brought in by fresh sand. According to Lux (1964) a covering os sea sand 40 cm deep contains as much P_2O and K_2O as would be supplied to farmland by an average application of manure. Part of the nutrients would also be supplied in the sea spray blown over the dunes (van der Valk 1974). The nitrogen requirement of Marram Grass seems to be particularly high. Even with plants grown on sea sand the leaves were a fresher green and they produced more flowers after they had received additional mineral nitrogen (Lux).

Amongst the psammophytes the species already named and their partners are nutrient-demanding types. This is shown by their capacity for rapid growth. Even during the winter, provided it is not freezing, they grow more quickly than the species which also inhabit inland dunes. The less-demanding species are only competitive on the coastal dunes when the fresh sand supply is no longer available. One may conclude that the character of the dune vegetation rests largely on the productivity of the organisms in the sea, in the same way as the fertility of the tidal flats does.

Although the wind carries nutrients and builds up sand in which the salt content is reduced it makes life difficult on the coastal dunes. The exposed organs of the plants are tugged to and fro and subjected to the abrasive action of the sand. Those near to the surf are liable to be sprayed with salt, at times even to the extent that they become encrusted with it. Both these

Fig. 305. The root system of *Honkenya peploides* in a primary dune. After Steubing (1949), in Ellenberg (1963).
The horizontal rhizomes at a, b and c and d, each with a network of roots, have been formed close to the surface but were then covered with more sand.

Nearly all vascular plants of the primary and white dunes are able to penetrate new layers of sand with their shoots and roots stabilizing them soon after deposition. This is mainly true for grasses and grasslike species (*Elymus, Ammophila, Agropyron, Carex arenaria*) but also for herbs (e.g. *Sonchus arvensis* and *Lathyrus maritmum;* see also fig. 308).

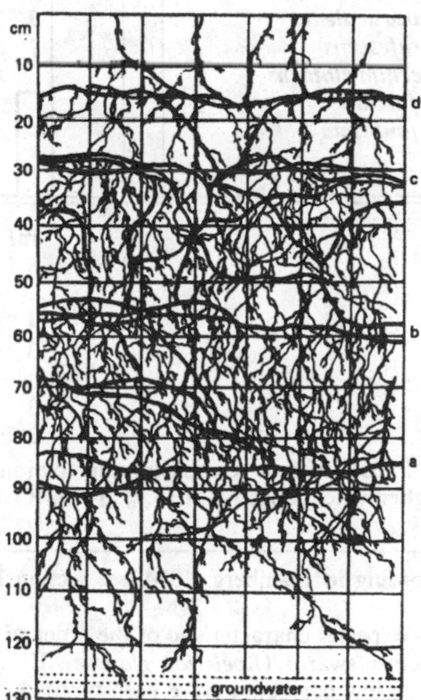

Fig. 306. Primary dunes of the Baltic coastline with *Elymus arenarius* and *Sonchus arvensis*.

The latter is a variety typical for coastal sites, which, according to Pegtel (1976), is quite different from the Sow Thistle of arable ground.

factors would damage or kill the majority of meso- or hygromorphic plants. Martin (1959) showed experimentally that with North American coastal plants the xeromorphic species were more resistant as a rule. Scarcely any plant organ is so elastic and at the same time able to keep the salt at bay as the rolled leaves of *Ammophila*. However it may not be just on these grounds that its xeromorphism has a positive competitive value. It is true that there is plenty of water available for the deep-rooting Marram along the Atlantic coast, but in this wind-exposed habitat the variations in temperature and relative humidity are very considerable (fig. 307).

In the *Elymo-Ammophiletum*, the rather nitrophilous *Elymus* is usually more plentiful on the lower dunes nearer the sea, i.e., in the earlier stages of succession. *Ammophila arenaria* dominates the higher white dunes. It is often accompanied by *Ammophila baltica*, a hybrid between *A. arenaria* and *Calamagrostis epigeios*. *Eryngium maritimum*, *Oenothera ammophila* and *Lathyrus maritimus* may also be named as characteristic species. However in spite of

Tab. 68. Primary, white and grey dunes of the East Frisian islands.

Dune formations:	P.	W.		Gr.			
serial no.:	1	2	3	4	5	F	R
Grasses and similar							
Cp *Agropyron junceum*	*5*	*1*				7	7
A *Elymus arenarius*	*3*	**3**	*3*			6	7
Cw *Ammophila baltica*		*3*	*1*			4	×
Cw *Ammophila arenaria*		**5**	**5**	*5*	*4*	4	×
Festuca rubra ssp. *arenar.*	*3*		*5*	*5*	*4*	4	7
Corynephorus canescens			*4*	*5*	*4*	3	3
Carex arenaria			*2*	*5*	*5*	4	2
F *Phleum arenarium*				*5*	*2*	2	7
G *Koeleria glauca*				*5*	*5*	3	7
Luzula campestris				*2*	*5*	4	3
Aira praecox					*4*	3	2
Other phanerogams							
Cp *Honkenya peploides*	*4*					6	7
Sonchus arvensis f.	*2*	*1*	*3*			5	7
Cw *Eryngium maritimum*	*1*	*3*	*1*			4	7
Cw *Hierac. umb.* f.*linariifol.*		*3*	*3*			–	–
A *Oenothera parviflora*		*2*	*3*			3	×
Cw *Lathyrus maritimus*		*2*	*3*			4	7
Hypochoeris radicata			*3*	*4*	*2*	5	4
F *Jasione montana*			*3*	*4*	*3*	3	3
Viola canina var.			*2*	*3*	*4*	4	3
Cg *Hierac. umb.* f.*armeriaef.*				*4*	*2*	–	–
Silene otites				*3*	*1*	2	7
F *Myosotis stricta*				*3*	*2*	3	4
F *Saxifraga tridactylites*				*2*	*1*	2	7
G *Vicia lathyroides*				*2*	*1*	2	3
Erophila verna				*4*	*2*	4	×
Galium mollugo				*5*	*3*	5	×
Cg *Viola tricolor* ssp. *curtisii*				*5*	*4*	3	5
F *Cerastium semidecandrum*				*4*	*4*	4	×
Cg *Lotus cornicul.* f. *crassifoli*				*5*	*5*	4	7
F *Sedum acre*				*3*	*3*	2	×
F *Trifolium arvense*				*2*	*2*	2	2
Rumex acetosella					*3*	5	2
4 *Galium verum* f. *litorale*					*3*	4	7
Rubus caesius					*3*	7	7
Mosses and Lichens							
Syntrichia ruralis f. *rur.*				*5*	*2*	–	–
Cornicularia aculeata				*3*	*2*	–	–
Peltigera rufescens				*2*	*3*	–	–
Brachythecium albicans				*2*	**4**	–	–
F *Cladonia furcata*					*3*	–	–
Cladonia fimbriata					*3*	–	–

Note:	mF	mR
1 **Primary** dune, Sand Couch dune	5.7	7.0
2 & 3 **White dunes**, Marram Grass dunes		
2 typical, relatively young	4.5	7.0
3 Red Fescue-Marram Grass dunes	3.8	5.4
4 & 5 **Grey dunes** Silvergrass shore dunes		
4 typical, i.e. still somewhat wind blown	3.3	5.0
5 rich in cryptogams, stable and richer in humus	3.7	4.7

No. 1 is still affected by groundwater (mF 5.7); the higher dunes are drier and colonised by more or less drought-tolerant species (mF 4.5 to 3.3). No. 5 enjoys more protection from the wind and a soil with a higher water capacity (mF 3.7). The mR values for the different communities show up very clearly the progressive removal of bases.

Source: From data by Tüxen (1937) and others. *Ecological evaluations according to Ellenberg (1974, see section B I 4).

*Cp, Cw, Cg = local character species of the primary, white and grey dunes, O = species characteristic of the order of Marram grass dunes (*Ammophiletalia*), G = character species of the alliance of Hairgrass swards (*Koelerion glaucae*), F = character species of the order of dry sand swards (*Festuco-Sedetalia*). A few species with low constancy have been omitted.

their relatively high nutrient level the typical *Ammophila* dunes are poor in species because the movement of the sand is too strong for most species. It is only during the transition to grey dunes that the number of species begins to increase markedly.

(c) Grey dunes and their further development

Drift lines, primary dunes and white dunes are all eutrophic habitats, but the amount of additional fertility falls off as we progress step by step from the first to the last. For the dunes furthest from the strand, which rarely receive additional sand it falls off practically to nil. Now the unfavourable properties of the colloid-deficient sand at last become apparent; it is easily leached by drainage water, it absorbs very little in the way of nutrients and it has a low water capacity. However in the course of time the amount of humus increases, and so do both the water- and nutrient-holding capacities. Minimum soil fertility seems to be reached in the first stage of the grey dunes.

The development of grey dunes from the white takes longer than for the white dunes to develop from the primary dunes. The building up of an *Ammophila* dune may be completed within a few years; but it needs some 10-20 years for the first step towards the grey dunes to be accomplished. In the beginning the partners of *Elymo-Ammophiletum* persist, but they become infiltrated by a type of Red Fescue (*Festuca rubra* ssp. *arenaria*) and other low-growing sward-forming plants so that the amount of ground covered and the number of species is greater in the *Elymo-Ammophiletum festucetosum arenariae* than in the typical community (tab. 68 and fig. 308).

Gradually the Marram and its associates become more and more stunted. As long as the sand contains lime numerous species which would lose their competitive strength at a lower pH-value can live in the **dune short grassland**. Because of this it may be joined with grassland communities of the class *Festuco-Brometea* (see section D I).

Today the majority of authors put the Red Fescue grey dune in the alliance of the **Grey Hairgrass communities** (*Koelerion arenariae*), within the order *Festuco-Sedetalia* (tab. 68). Sometimes they have also been put together with the acid-soil Grey Hairgrass dunes (*Corynephorion*).

Grey Hairgrass communities are able to occupy very lime-deficient dune sand and to stabilise it again if it is disturbed. They are found above all then in places where storms have cut into older dunes. On the East Friesland islands they come right up to the *Elymo-Ammophiletum* over a broad front. Tüxen (1956c) described the Grey Hairgrass communities of the sea dunes as *Violo-Corynephoretum maritimum* (in the *Corynephorion*). As can be seen from tab. 68 however they contain so many species of the *Koelerion* and of the *Festuco-Sedetalia* that it would be better not to unite them with the markedly oligotrophic *Corynephoretea* of the inland dunes which are much poorer in species (section 3 b).

Only where white dunes develop into grey dunes does the pH level play a part in the determination of the coastal vegetation. In the soils of the land outside the dykes as well as in the sand near the sea the reaction is neutral or weakly alkaline. Since seawater itself contains so many bases and is well buffered this

Fig. 307. Fluctuations in temperature and relative humidity during the course of the day over a Marram Grass dune on the island of Terschelling. Below: cloud cover, sunrise and sunset. After Heerdt and Mörzer Bruijns (1960).

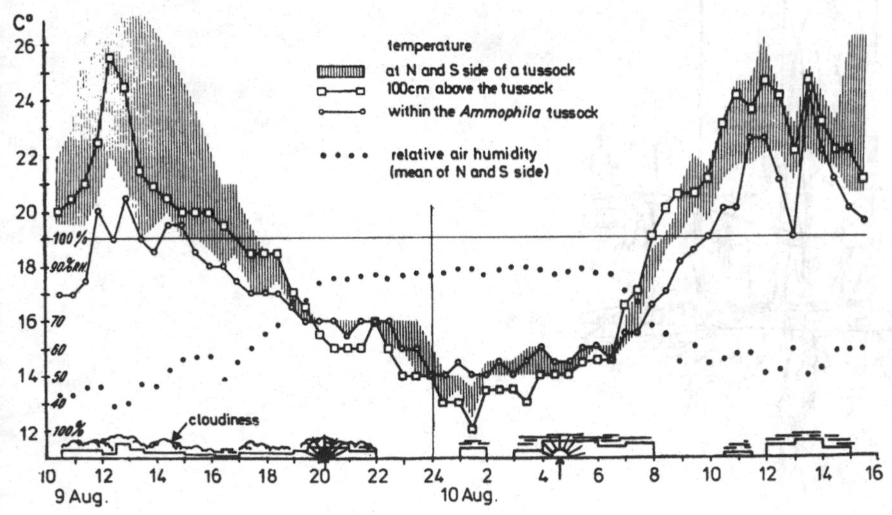

applies even where the deposits contain very little lime right from the outset. Salisbury (1922) has already shown that the topsoil became progressively more acid as the dunes got older (see tab. 69). This is an easily-determined pointer to the general lack of nutrients and to the increase in the amount of acid humus. In themselves pH values of about 6 or 5 are not insurmountable limitations for the majority of land phanerogams, but rather can be seen as favourable for their growth.

Both the *Koeleria* and the *Corynephorus* communities, instead of having strong rhizomatous geophytes, are dominated by hemicryptophytes, the roots of which spread out a long way in all directions (fig. 312). Alongside these there are also species with deep tap roots, e.g., composites like *Leontodon saxatilis*. All plants of the short grass dune swards are more or less xeromorphic, and this may be connected with the shortage of nitrogen as well as with the mechanical demands of the wind. Water is always available for the plants in the grey dunes, apart from the topmost centimetre. From time to time however the evaporation rate on the sunbaked sand rises very considerably, but not so high as on the white dunes where the wind has freer access and the sun heats up the ground even more (fig. 307).

Wind is quite a significant factor even for the fixed dunes since it can start the sand moving again whenever the surface is damaged. If the mechanical disturbance is extensive the **Grey** Hairgrass clumps tend to come in along with the Sand Sedge. On the other hand the Marram very rarely turns up, presumably because of nutrient deficiencies (see section b).

Cryptogams can only keep their place on the sand dunes in the shelter of grasses. Neither mosses nor lichens can stand being covered up with wind-borne sand more quickly than they are growing. Orthotropic cushion-forming mosses are relatively better suited to keep pace and grow up through any accumulating sand. As Warming (1907) has already shown they bind

Fig. 308. A section through a grey dune sward on the Darss (*Elymo-Ammonphiletum festucetosum*). After Fukarek (1961).

From left to right: *Calamagrostis epigejos*, *Ammophila baltica*, *Hieracium umbellatum stenophyllum*, *Festuca rubra arenaria*, *A.b.*, *Carex arenaria*, *Galium verum*, *A.b.*, *Viola tricolor curtisii*, etc. The position of the rhizomes shows that there has been a covering of additional sand.

the sand with their rhizoids. According to Koppe (1969) the Star Moss (*Syntrichia ruralis*) often forms extensive carpets on younger grey dunes (tab. 68). Where the sand has been more severely leached *Racomitrium canescens* is likely to be met with. Both of them make a significant contribution to the accumulation of acid humus in the topsoil. Less productive in this respect are the fruticose lichens (*Cladonia*, *Cladina* and *Cetraria* spp.), which are found here and there amongst the Grey Hairgrass on very dry but stable sand. Since they are more numerous on the inland dunes than near the sea we will leave discussion of these unique plants until section 3 b.

Should older dunes be opened up again by the wind then Crowberry (*Empetrum nigrum*) often takes part in the succession since it is able to withstand a certain amount of deposited sand. In this way **Crowberry heaths** may arise and we will compare these with other dwarf shrub heaths in section D II 2 b.

If the grey dune rests undisturbed for a longer time under a dense sward and a moss cover then the topsoil becomes richer in humus and rather stable. This allows plants to come in which are not able to tolerate being covered with sand. One of the first to spread on a completely stable dune is *Calluna*, mostly accompanied by other humus plants. These are favoured by sheep grazing, which was a common practice in the dune country at one time.

On the older parts of the East Friesian islands and also on some islands of North Friesland a type of dune grassland is common near to habitations which in a number of respects does not fit in with the usual types of short grass swards. Its relatively higher yield as a sheep pasture and its balanced composition is a result of earlier farming practice and the addition of humus in the form of dung. Tüxen (1956c) called this sward *Agrostieto-Poetum humilis* (earlier *Festuca capillata-Galium litorale* ass.). The commonest species are the Sand Bent Grass, Sheep's Fescue and Yellow Bedstraw. Strangely all the real dune grasses such as *Festuca rubra arenaria* and *Ammophila arenaria* which are otherwise so widely distributed are absent here. *Rumex acetosella* is plentiful as a one-time arable weed. Also *Trifolium arvense* and *Achillea millefolium* may be seen as relicts from fallow land. The *Agrostis-Poa humilis* ass. then should not be thought of as a natural stage in dune succession.

Willis (1963) described some obvious changes in the white and grey dunes in north Devonshire brought about by repeated application of mineral fertiliser. The average height and biomass increased considerably in two years and grasses such as *Festuca rubra* and *Poa pratensis* extended at the expense of other phanerogams and mosses. Natural grey dune communities then are mainly determined by the paucity of nutrients. The Marram Grass (*Ammophila arenaria*) proved to be particularly appreciative of the extra nutrition and was not crowded out. In contrast to nitrogen, phosphorus and potassium, trace elements had no obvious effect on the vegetation. Pemadasa and Lowell (1974) came to similar conclusions in respect of therophytic dwarf grasses and herbs such as *Aira caryophyllea*, *A. praecox* and *Erophila verna* which live during spring time in the sparse sward of grey dunes. They grew considerably better when given additional N or P, best of all both NP; but K had little effect, presumably because the spray from the sea provided it in sufficient quantities.

To summarise once more the vegetation units of the coastal dunes reference should be made to the

Tab. 69. The acidity and the base content of the top soils of coastal dunes.

Dune type	Community	pH (H_2O) value Spiekeroog	Netherlands	V%	Remarks
Embryonic dune	*Elymo-Agropyretum,* Initial phase	6.7 – 8.3	7.6	97.7	base-rich
" "	" " Optimal phase	6.0 – 9.0			
White dune					
primary	*Elymo-Ammophiletum typicum*	6.7 – 8.6	7.6	93.5	
secondary	" " "		7.4	84.3	
White/grey dune	" " *festucetosum*	6.7 – 7.8	6.8	76.2	gradual reduction in base content
Grey dune	*Tortulo-Phleetum jasionietosum*	5.1 – 6.7			
"	*Violo-Corynephoretum*		6.3	71.6	
"	blow-outs in older dunes	5.0 – 6.2			
Brown dune	*Polypodio-Empetretum*		4.6	38.8	poor in bases
"	*Empetro-Callunetum*		4.6	35.5	

Source: After data from Wiemann and Domke (1967, Spiekeroog) and from Steubing and Westhoff (1966, Netherlands coasts). V = degree of saturation of bases in the soil.

succession scheme shown in fig. 301. The probable further development to a natural woodland which is the subject of our final discussion has already been noted there.

(d) Scrub dunes and dune woodlands

On most West and East Friesian islands the *Elymo-Ammophiletum* can go over directly into a scrub community. This **Sea Buckthorn-Creeping Willow scrub** (*Hippophaë-Salicetum repentis*) occupies mainly the lee side of narrow white dune ridges (see the instructive vegetation maps of Baltrum, Tüxen 1956c, and of Spiekeroog, Wiemann and Domke 1967). Protected against the wind Sea Buckthorn reaches a height of up to 2 m whereas the Creeping Willow seldom comes above knee height. However the latter is better able to stand being covered with sand, and it ventures right up to the tops of the dunes which it helps to build. Here its twigs often look as though they have been shorn off, and in dry years they die back, mainly due to the high salt content of the sea wind.

The distribution centre of Sea Buckthorn (*Hippophaë rhamnoides*) is in central Asia. Its subspecies *fluviatilis* and *carpatica* prefer the continental south and east of Europe whereas subsp. *rhamnoides* is confined to the Atlantic and Baltic coasts (see also section B V 1 c). According to Pearson and Rogers (1962) both the Sea Buckthorn shrubs and their fruits can stand rather low temperatures, a property which of course is of no advantage on the coastal dunes. The reason it can settle on dunes is no doubt the lack of competition in its pioneer role on the one hand and its symbiosis with nitrogen-fixing bacteria on the other. It also relishes a sufficient lime content in the sand which is available here. Sea Buckthorn occupies any places which suit it with extraordinary rapidity as it is spread by birds. Its seeds have a 95-100% germination after they have passed through a bird's gut. *Hippophaë* prefers to grow in valleys between the dunes where it can reach the groundwater with its tap roots. When it grows on higher dunes Stocker (1970) found that in addition to an upper rootstock there was always a lower one which had spread out in the permanent dampness of the capillary water.

The *Hippophaë-Salix repens* scrub is almost always connected with dune valleys, but there it develops into a **damp-soil Creeping Willow scrub** (see fig. 301). Braun-Blanquet and De Leeuw (1936) have described such a community from the West Friesian island of Ameland as *Acrocladio-Salicetum repentis*. A number of moisture indicators such as *Acrocladium cuspidatum*, *Drepanocladus* species and *Hydrocotyle* show that fresh groundwater on the average is only about 45-50 cm below the surface and at times much closer to the top layer of this humus-rich sandy soil. *Hippophaë* is only exceptionally seen here, and in other ways too this very variable community has little in common with the scrub dune described before. In spite of this both communities dominated by *Salix repens* var. *argentea* lie close together.

Paul (1944, 1953) showed that on the Kurskiy Zaliv (Kurische Nehrung) *Salix repens* prefers to germinate on damp bare sand. In lower dunes such places are only available when a valley has been freshly excavated by a storm. In larger dunes, which build up a groundwater layer of their own *Salix repens* invades the foot zone moistened by seepage water. *Juncus balticus* too, which is mentioned by Braun-Blanquet and De Leeuw as being associated with Creeping Willow scrub, prefers to occupy strips of sand which are temporarily water-logged. Since Creeping Willow can easily grow through any sand which is blown over it, the dunes formed by this shrub may get up to 10 m in height. No one would suspect that such a Willow dune had originated near the groundwater. So one must agree with Paul when, unlike Braun-Blanquet and De Leeuw, he does not see the scrub dunes as being stages in normal dune succession.

Young Buckthorn-Willow scrubs indicate that the dune sand has not yet been leached of lime, at least in the upper layer. Their companions remind one of the short grass communities of the grey dunes. However the sand will soon be leached out and in the half shade of shrubs acid-tolerant woodland mosses come in such as *Dicranum scoparium*, *Hylocomium splendens* and above all *Hypnum cupressiforme*. Acid-humus vascular plants will also become established, e.g. *Polypodium vulgare*, *Pyrola rotundifolia* (var. *arenaria*), *Veronica officinalis* and *Hieracium umbellatum*. The **Polypody-Willow scrub** (*Polypodio-Salicetum repentis*) so formed is included in the order of the Blackthorn hedges and scrubs (*Prunetalia spinosae*) by Tüxen (1952, 1956b). On sunny slopes of island dunes of East Friesland one comes across a related, but less acid-loving community, the **Burnet Rose scrub** (*Rosa pimpinellifolia-Salix repens* ass.) which belongs to the same alliance (*Salicion repentis*).

Since the East Friesian islands are moving relatively quickly from west to east all their dunes are relatively young (Schütte 1939, Gessner 1957, Tüxen 1956c, see fig. 303).

These islands have always been heavily populated so there was never an opportunity for the dune scrub to develop into a true woodland. On the other hand on many of the relatively old dunes in Holland, according to Westhoff (1950), a number of communities can be distinguished in which both shrub and tree species take part. Here they tend to form **acid-soil Oak woods** ('*Quercetum atlanticum loniceretosum*' and '*convallarietosum*' in the alliance *Quercion robori-petraeae*) on dry dune sands.

In the dune areas further to the east, especially on the Baltic coast, Scots Pine is more and more frequent. Certainly *Pinus sylvestris* has been planted in large numbers ever since the nineteenth century. However it would probably have appeared here also under natural conditions (cf. section B IV 6 c). Wind-distorted specimens have ventured right on to the crest of old dunes close to the sea (fig. 309). East of the Oder mouth according to Wojterski (1964a, b) **Pine woods** even form part of the zonal vegetation on old duneland, which in some places is very extensive. Many of the subcommunities associated with Pine are rich in Crowberry (*Empetrum nigrum*) and this distinguishes them from Pine woods growing further inland (fig. 310 and section B IV 6 c).

Fig. 309. Wind-affected Pines, including plantations, on an old dune on the Polish Baltic coast. The sea has broken down the dunes in places and storms have set the wind-blown sand in motion again (centre) enabling the Grey Hairgrass pioneer sward to return.

Fig. 310. A transect through the dune series with Crowberry-Pine woods on the Polish Baltic coast. After Wojterski (1964), somewhat modified.

Compared with the North Sea dunes (fig. 301) the herbaceous dune communities are poorer in species and not so varied. Pines very soon initiate the formation of woodland with a subcontinental character. Nearer the coast though oceanic species such as *Erica tetralix* and *Myrica gale* play a surprisingly large part. The tall moving dunes of the south-eastern Baltic coast have originated from the destruction of dunes similar to these (see fig. 311).

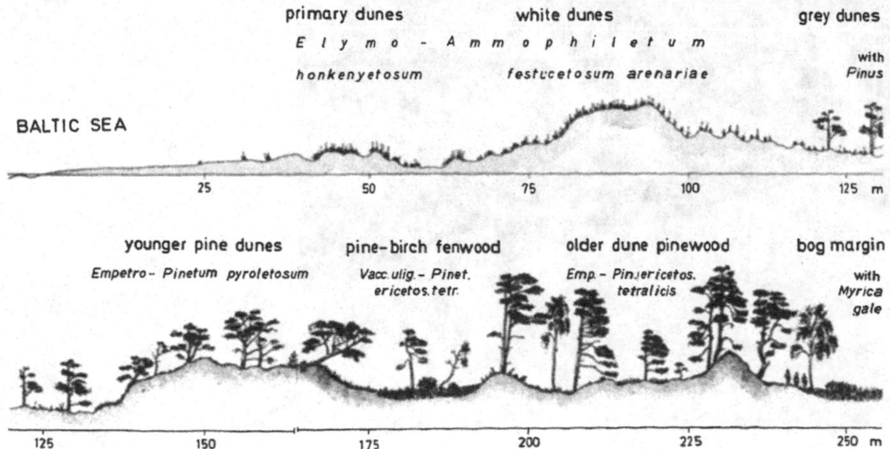

Many doubt that coastal dunes could have become wooded under natural conditions. In view of the numerous successful attempts at afforestation it can scarcely be disputed that in spite of the strong wind the growth of trees is possible. It is also known that on the exposed Norwegian islands Birches will grow quite near to the windlashed salt-sprayed edge and form small woods taking on a streamlined form (Wassén 1965). Here however they are growing on solid ground while on the dunes the blown sand must be an unfavourable additional factor. The rarity of trees in the coastal dunes on the North Sea islands makes one hesitate to believe that their potential vegetation would be a woodland. But van Dieren (1934), a Dutch authority in this context, after his thorough historical study came to the conclusion that the freedom of the dunes from trees is 'an illusion after all'.

The final solution to this problem could only be found in a large reserve protected from the influences of men, sheep and deer. At present there is no such area which has been protected for a long enough time. Experiences on the Łeba spit of land (supervised by Wojterski, 1976) indicate that this experiment will probably turn out to favour the trees although it is also very stormy there. Certainly one would not expect that trees could cover the whole area all at once. Probably the moister dune valleys and hollows would act as nuclei for a natural development to woodland both on the North Sea and the Baltic coasts. With regard to amount of available water and richness in bases, favourable conditions exist here for establishment of Alder, Birch, Oak and other broadleaved trees, and in a continental climate for Pine too. These trees would certainly grow up in the shelter of the dunes. When they got older on the other hand they would in turn shield the sandy soil from the wind. As a consequence the trees could now find it easier to become established on the drier dune sands than on the open grassland or scrub dunes. Only where this natural succession is not disturbed by man would it have any chance of being completed.

2 Moving dunes free from vegetation

(a) The origin of large wandering dunes

In the Middle Ages and more recently cutting of timber and firewood and grazing of animals on the dunes along the coasts of North Sea and Baltic went on to such a frightful extent that it led to the creation of desert-like migrating dunes. Paul (1944, 1953) carried out a thorough investigation into the largest area of moving dunes in Europe on the Kurskiy land-spit (Kurische Nehrung). Due to lack of space we cannot discuss his findings in detail (for this see also Ellenberg 1963). In principle what he found also applies to wandering dunes which are more easily accessible, e.g. on the Łeba spit in Poland (see fig. 311).

Fig. 311. The leeward slope of the large moving dunes on the Łeba spit of land is smothering the swamp Alder wood along the edge of the lagoon and is even narrowing the belt of Reeds.

The chain of sand hills which was set loose some 200-250 years ago on the Kurische Nehrung is more than 80 km long and between 400 and 800 m wide. The bare dunes average 35 m in height and go up to a maximum of 70 m whereas overgrown dunes only exceptionally go above 25 m. Those dune hills which have not already been fixed move 3-12 m in an easterly direction each year leaving behind them a more or less level area, the 'Palve', which is only a few metres above sea level. Any sand which is not saturated with water or which has not been stabilised by plants is driven by the wind up the 4-12° western slope of the huge dune ridge and tipped over onto the eastern slope at an angle of 30-35°. During the course of time individual parts of the chain which are particularly rich in sand, and tower over neighbouring weaker places, tended to become independent and form barchanes. Nevertheless it is just these which were first fixed and recolonised by plants.

Up to the seventeenth century a closed Pine woodland covered most of the Kurische Nehrung. It grew on old parabolic dunes some of which have not yet been covered by the moving dunes. Humus layers can be seen on the exposed windward side of large dunes indicating previous woodlands. On slopes which have been afforested these old woodland soils may be identified even from a distance because of the better growth of Pines. Generally the woodland soils follow an irregular sequence on the slopes with paler layers of sand several metres thick between them. These horizons tell a significant part of the changeable history of the wandering dunes.

In the first half of the nineteenth century the Kurische Nehrung was a single huge field of sand. The moving dunes had swallowed up villages and farmland, even threatening to silt up the adjacent lagoon and the harbour of Memel. First of all, to limit the addition of further sand, Marram Grass was planted along the outer coastline, resulting in the formation of an unbroken dune bank, the so-called pre-dunes. *Ammophila* would not spread on to the bare migrating dunes, and *Elymus arenarius* was even more reluctant. Both preferred to remain on the coast where they could get fresh supplies of fertile sea sand. Only very exceptionally would the Marram move inland and this was in places where there was very heavy deposition of new sand. Paul found it in some places on the ridge of the moving dunes. It also dominates here and there on the migrating dunes in the Polish Leba reserve which are not quite so high, but equally impressive (fig. 311). Planted Marram grows so badly on the impoverished sand as a rule that it acts merely as a mechanical obstruction in contributing to its stabilisation.

The short Grey Hairgrass and Fescue swards which are to be found everywhere on the Palve plain left behind by the moving dunes, grow quite well in spite of the shortage of nutrients, but will not produce anything more than a very slow build-up of sand. So in order to stop the movement of the large sand hills there was nothing else for it but to erect a grid of faggot hedges on windward slopes in the gaps of which were planted Scots or Mountain Pines. Most of the dunes have been brought to a standstill in this way and the Palve has now been almost entirely planted up with trees. This reafforestation can be seen as further proof that neither wind nor blown sand are able to prevent the growth of trees on the coasts (see fig. 309).

(b) Succession on the 'reverse banks'

Placed under Nature Conservation a few of the large dunes are still moving eastwards, e.g. in the Łeba reserve at about 6 m a year. On the Kurische Nehrung, along the western foot of the large dunes there arose a system of so called 'reverse banks' ('Gegenwälle' in German). In the 300-900 m wide zone which has been left behind by the migrating dunes since the start of their movement, there are about 20 parallel sand ridges 0.5-2 m high. The plants on them fix the sand which is blown down from the large dunes by an **east** wind, hence the name 'reverse banks'. Since Paul was able to determine the age of the reverse banks by various methods these offer an unique opportunity **to measure the rate at which grey dune succession proceeds.**

The formation of a *Festuca rubra arenaria* stage here takes more than 3 but less than 10 years. The coastal *Corynephoretum* reaches its optimum in 25-30 years. Mosses and lichens never act as pioneers. Only when the sand has come completely to rest, after a number of decades, do the lichens gradually come to occupy more than 25% of the ground area. Where, in different succession stages, the surface is damaged mechanically, *Carex arenaria* with its long rhizomes finds favourable conditions. Grazing or heavy Deer browsing lead to a sward dominated by Sheep's Fescue. However these species-rich communities are only found on reverse banks more than 75 years old.

All things considered succession on the poor sand of reverse banks is very slow. Under the less oligotrophic conditions, that is nearer to the sea, Reinke (1903), Warming (1907), Tüxen and Böckelmann (1957) and others all agree that it only takes a few years for a bare stretch of sand to develop into the optimal stage of an *Elymo-Ammphiletum* white dune.

Paul found practically no woody plants as pioneers on the reverse banks. For all that it is worth noting that pine seedlins establish themselves, above all in lichen swards. They could be looked upon as the precursors of a lichen-Pine wood which may form the natural climax on older dunes in eastern Central and northern Europe (see section B IV 6 a, c).

Similar reverse banks are also to be seen where the large dunes have moved away on the Łeba Nehrung and on the island of Sylt, although they are of smaller dimensions and not so easily recognisable as such.

3 Inland dunes

(a) Origin and distribution of inland dunes

Older coastal dunes which are cut off from any further supply of sand have in fact already become inland dunes. Parabolic dunes which have moved inland for several hundred metres are no different in their structure from the sand hill situated further inland. In some parts they can be of a similar age and carry a very similar vegetation. Nevertheless they are closely connected with the coastal dunes, and it seemed advisable to treat them in this context.

The separation of coastal and inland dunes is also justified for floristic reasons because even on the oldest, completely leached sea dunes which are well beyond the reach of any additional sand one finds some plants indicating the nearness of the coast and not to be met with further inland. All the many species, subspecies, varieties and forms with the Latin names *'maritima'*, or *'litoralis'* and with common names referring to the strand or the sea etc. could serve as examples (tab. 68). Seeds of these coastal plants are carried inland in large numbers and some odd specimens are always turning up in vegetation relevés. Only *Ammophila* is no longer a definite 'coast indicator' since it is planted so much on the inland dunes. However in most cases it has disappeared after a few years even on the lime-rich sands of the Upper Rhine plain.

The majority of inland dunes arose during the last Ice Age or in the early Post Glacial period. In any case the material of which they are made comes from those times. On the northern plain where the old rivers could not have brought much lime the dunes were deficient in both lime and nutrients from the outset. Large areas of inland dunes are generally found on the eastern or north-eastern sides of the old diluvial river valleys, e.g. the Ems, Weser, Aller, Elbe, Oder and Vistula. We can learn a lot from nature reserves where there is still airborne sand such as the one to the east of Warsaw.

The inland blown sand originated from the broad rivers carrying the melting water from the inland ice. These provide a supply every summer when the prevailing west and south-west winds blow the coarse sand out of river beds which are no longer flooded and spread it out over the adjacent terraces and moraines, while the finer material is carried even further and deposited as finest sand, silt or loess. Originally the glacial sand fields were flat and not built up into dunes with the help of binding vegetation. The most striking dune hills found today came into existence for the most part in the Middle Ages when men unwittingly set them in motion again. The cutting down of trees and sheep grazing were the main causes of this sand beginning to blow. Only a few decades ago the turning of sheepwalks into blown sand areas could be seen going on in Emsland and other heathland areas. In the heath park near Wilsede it is still happening today though to a very small degree.

In the broad Upper Rhine plain there are also widespread dune areas, especially in the central and northern parts east of the Rhine, but also south of the Main (the 'Mainzer Sand'). In contrast to the dunes of the old diluvial north European plain however the sand here is richer in lime since the material of origin was also richer in lime and the process of leaching has not reached such an advanced stage. The difference between the two is reflected so strongly in the vegetation that we must discuss the two areas separately. Compared with the vegetation of coastal dunes on the East Friesian islands, or even with that of the north-west German inland dunes, that of the Rhineland plain is more like grassland and richer in species. We shall deal with these therefore in section D I 4 c.

The third area of Central Europe rich in dunes is the Marchfeld, which, like the Upper Rhine plain, has a relatively young lime-rich sand, but floristically has a distinct south-eastern flavour. The inland dune vegetation of north-eastern Germany is rather similar to that of the north-western part, but lacks a few atlantic elements. It links up with the less oligotrophic and more strongly continental dune communities which have been described from near Warsaw (Juraszek 1928). Since it is impossible within the scope of this book to go into all the different types we shall look only at the north-west German inland dunes in greater detail.

(b) Grey Hairgrass pioneer stages and fruitcose lichen carpets

The sands of inland dunes on the northern diluvial plain are leached of lime to a considerable depth and show a medium to strongly acid reaction. According to measurements carried out by Krieger (1937) pH values lie between 3.2 and 4.6, i.e. in the same range as those of acid-soil woodland and heath. These dune soils have been very poor in plant nutrients right from the start and because of their deficiency in colloids they are extremely free draining. Whenever they are exposed to the sun their surfaces dry out quickly and are then at the mercy of the wind. Scarcely more than 25 species of phanerogams in Central Europe can live under such a combination of unfavourable conditions. Nevertheless it is the higher plants which first colonise the bare sand and not lichens and mosses as was once erroneously presumed. The slow-growing cryptogams certainly make small demands, but they do require a stable substrate such as rocks, boulders, bare raw humus or sand protected from the wind to enable them to develop. On bare sand which is blown along and carried away by the wind they are no more able to thrive than in stablilised places. Tüxen (1928) was the first to recognise clearly this fact and he has produced a clear survey of the succession on the north-west German inland dunes.

Fig. 312. A section through a Grey Hairgrass community rich in lichens (*Corynephoretum canescentis cladonietosum*) on the Darss. After Fukarek (1961).

From left to right: *Polytrichum piliferum*, *Corynephorus canescens*, *Cladonia sylvatica*, *C.c.*, *Cladonia foliacea*, *P.p.* In older stands of this community the lichens may become dominant.

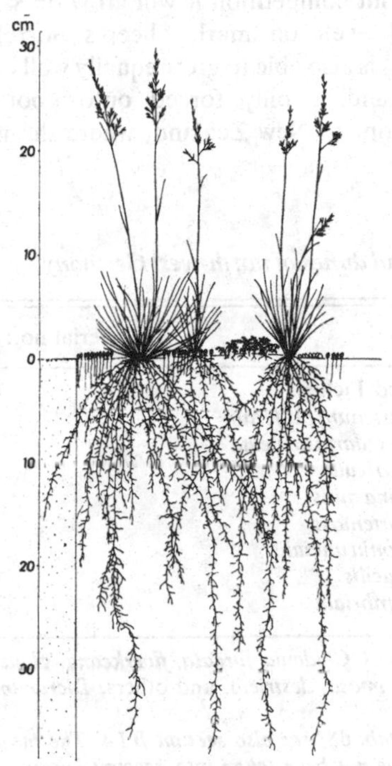

Seeds of Grey Hairgrass (*Corynephorus canescens*), brought by the wind, are able to germinate on sand which is exposed, but not too severely blown about, and is thoroughly moistened right up to the surface. This could be either in the late summer or in the spring after it has been raining for some time. The young grey-green bristle-leaved clumps collect the moving sand into small heaps into which the grass sends more roots. These dense rooting systems can penetrate in all directions down to a depth of over 50 cm (fig. 312) so that the plants are very well anchored and are able to draw moisture and nutrients from a large volume of sand. Early the following year they dust their numerous tiny seeds from the silvery white panicles and many of these germinate close to the mother plants.

After three years the Grey Hairgrass sward can become so dense that other plants which cannot tolerate the blown sand are able to come in. On the other hand *Corynephorus* starts to degenerate, apparently not because of increased competition but because, like *Ammophila*, it grows more strongly when it can get a constant supply of new sand. Here too it is a question of nutrition as Lux (1964, see tab. 70) was able to show by a series of fertiliser applications to a

Tab. 70. Effect of different fertiliser treatments on pure stands of Grey Hairgrass in dunes.

Corynephorus: Fertilisation level	No. of clumps (on 12.5 m^2)	Degree of cover as a % age of the area	Degree of cover as % of unmanured	Haulm length of the five strongest	Remarks
Unfertilised	123 – 143	5.7 – 6.2	100	6.7 – 7.0	
PK	219 – 220	6.7 – 7.4	109 – 121	5.5 – 7.8	Nitrogen is the most effective nutrient
NK	225 – 231	12.5 – 14.2	202 – 231	9.8 – 11.6	
NP	288 – 322	15.4 – 22.0	250 – 331	11.5 – 14.8	
NPK	295 – 365	**15.8 – 18.0**	256 – 292	**12.6 – 13.6**	
NPKCa	348 – 366	15.5 – 18.0	251 – 292	11.0 – 12.2	

Source: From investigations by Lux (1964) on Sylt.

pure stand of Grey Hairgrass on sand from the grey dunes. *Corynephorus* grew best when all three of the main elements N, P and K were added, whether or not lime was also given. Nitrogen and phosphorus were more effective than potash.

A few other phanerogams are able to take part in the initial stages of the **Grey Hairgrass pioneer sward** (*Spergulo -Corynephoretum*) of the subatlantic inland habitats; these are mainly spring therophytes such as *Spergula vernalis* and *Teesdalia nudicaulis*. These small shallow-rooted ephemerals produce their seeds very quickly. They are natives of the submediterranean to atlantic coastal area and they give the *Corynephorion* communities of the north German inland areas a distinctly oceanic look (tab.71). More permanent species are represented by *Carex arenaria* and *Rumex acetosella* whose seeds are also wind dispersed and are covered over by a sprinkling of sand. From the edges of new holes in the dune the Sand Sedge sends its rhizomes down to a depth of 10-20 cm into layers which seldom dry out but are still porous and well aerated. This habit enables it to survive when the uppermost layer is often completely dried out and, especially in the warm summer months, cannot support any plant growth. Rosette-forming plants, e.g. *Hypochoeris glabra* turn up only here and there.

Corynephorus is distributed throughout the whole of Central Europe but within this area it is only found on porous sandy soils with a more or less strongly acid reaction. In culture experiments too it grows best on acid sand although Ph. Paul and Richard (1968) found that without competition it will grow on sand rich in lime and even on marl. Sheep's Sorrel (*Rumex acetosella*) is also able to grow equally well on all kinds of soils and is only forced onto poor sand by competition. In New Zealand, where the number of

Tab. 71. Grey Hairgrass swards and fruticose lichen coverings on inland dunes of north-west Germany.

	Serial no.:	1	2	3	F	R	N
Grasses and Sedges:							
	Corynephorus canescens	**5**	**5**	*3*	3	3	2
	Carex arenaria	*4*	*2*	*1*	4	2	2
	Agrostis canina	*1*	*4*	*2*	×	3	2
Da	*Festuca ovina* s. str.?		*3*	*3*	–	–	–
Other phanerogams:							
Cs	*Spergula morisonii*	*5*	*4*	*3*	2	×	2
Cs	*Teesdalea nudicaulis*	*3*	*5*	*1*	3	1	1
	Hypochoeris glabra	*2*			3	2	1
	Jasione montana	*2*	*3*		3	3	2
Cs	*Filago minima*	*1*	*3*		2	4	1
	Rumex acetosella	*4*	*5*	*2*	5	2	2
Da	*Hypochoeris radicata*		*4*		5	4	3
Da	*Hieracium pilosella*		*4*		4	×	2
	Calluna vulgaris			*2*	×	1	1

	Serial no.:	1	2	3
Mosses and Lichens[a]				
	Polytrichum piliferum	*5*		**4**
	Cornicularia aculeata	*3*	*2*	**5**
Cc	*C. muricata*		*1*	*2*
	Cladina mitis	*3*		*3*
	C. portentosa		*2*	**4**
Cc	*Cladonia uncialis*			*4*
	C. gracilis			*4*
Cc	*C. firmbriata*			*4*

[a]*also in no. 3 Cladonia furcata, floerkeana, glauca, chlorophaea, squamosa, destricta,* and others, *Dicranum scoparium.*

[b]***compare tab. 68; see also section B I 4! The mosses and lichens have not been taken into account because their evaluation is not sufficiently recorded.***

Note:		mF	mR	mN
1 & 2:	**Grey Hairgrass inland dunes**			
1:	Grey Hairgrass pioneer community on moving sand	3.2	2.3	1.8
2:	Brown Bent-containing humus swards on still sand	3.5	2.7	2.0
3:	**Cover of fruticose lichens** (*Sp.-C. cladonietosum*; better as an independent association *Cladinetum mitis*) on stable, very dry poor sand.	<3.2	<2.1	<1.8

Cs = local character species of the *Spergulo-Corynephoretum* (see section E III No. 5. 221)
Da = differential species of the *Sp.-C. agrostietosum*
Cc = local character specvies of the *Cladinetum mitis;* many other lichens can serve as differential species from the Grey Hairgrass communities.

The special place of the fruticose lichen communities can be seen also by the extremely low average moisture (mF), reaction (mR) and nitrogen (mN) values. These would be even less if it were possible to include the cryptogams in the calculation along with the phanerogams.

Source: From tables by Tüxen (1937); ecological evaluation after Ellenberg (1974)b.

competitive plants is smaller, Sheep's Sorrel is a troublesome pasture weed, as it is in England.

Like many other plant species of Central Europe *Corynephorus* has a tendency towards a suboceanic or oceanic distribution. This is related to the anatomy of Grey Hairgrass (Berger-Landefeldt and Sukopp 1965). Apart from its dense and wide-spreading root system it has no adaptations which would enable it to live under water stress (see fig. 312). Since the Grey Hairgrass leaves cover only a small proportion of the total area (see tab. 71) and the quickly drying surface sand then reduces further water loss from the soil there is always a sufficient supply of water available in a climate near to the sea. Even on fine days the amount of water transpired is quite small in relation to the surface area of the soil (400-800 g/m^2 per day) and requires only an eighth to a quarter of the total radiant energy (fig. 313).

Not all Grey Hairgrass swards are pioneers on raw sand, i.e. where there is no humus. In many cases they occupy former woodland or heathland soils which are already rich in acid humus, but which have been exposed to the wind by the destruction of the plant cover. On such riper sandy soils the succession progresses more quickly, for instance around Nürnberg (Hohenester 1967).

Other grasses can spread alongside the Grey Hairgrass tufts when they have reached their optimum development and fixed the sand. Most commonly a xeromorphic form of *Agrostis canina* as well as *Festuca ovina* subsp. *vulgaris*, *Hypochoeris radicata*, *Hieracium pilosella* and *Jasione montana* participate in an open **Brown Bent-Grey Hairgrass sward** (*Corynephoretum agrostietosum caninae*, tab. 71). When grazed by sheep this can develop into a half-dry grassland (see section D I 4 c) and finally turn into heath in which *Calluna* takes over more and more. All these succession stages live under oligotrophic conditions. Phosphoric acid is particularly lacking in acid sandy soils and this shortage is not balanced out by the fact that the phosphorus is very quickly circulated in grass and heath communities of the dunes (Atkinson 1973).

Undisturbed this succession could pass into a sand Birch stage and then lead to a **Birch-Oak wood**. East of the Elbe *Pinus sylvestris* is a pioneer of dune woodland which is more or less rich in Pines. Since Pine has been abundantly planted on the heaths in north-west Germany it has become the best pioneer and dune stabiliser here too. New wind damage can delay the formation of woodland, and the development has to start all over again from the beginning especially where animals or man are continually exposing the sparsely covered blown sand. For example in the district of Meppen (Emsland) around 1780 there were, according to Hesmer and Schroeder (1963), 165 km^2 of bare blown sand. No less than 7.5% of the surface of this administrative region had become the habitat of Grey

Fig. 313. Changes in the microclimate of a Grey Hairgrass sward (*Corynephoretum*) throughout the day on poor dune sand near Berlin-Wannsee. After Berger-Landefeldt and Sukopp (1965), somewhat modified.

By and large the air temperature near the surface of the stand (18 cm above ground) follows the amount of radiation from the sun and sky, but because of turbulence there are quite rapid fluctuations. The same is true of the air humidity, measured here as water vapour pressure. Only the soil itself carrying a sparse growth heats up quickly and, in spite of its light colour, reaches 40°C on the surface.

It was not possible to plot a continuous measurement of the evaporation from a sheet of green blotting paper nor the transpiration from *Corynephorus*. The lines E and T join up points representing readings made at intervals; the high values of T early in the morning were probably because of dew on the leaves, but the other fluctuations cannot be explained.

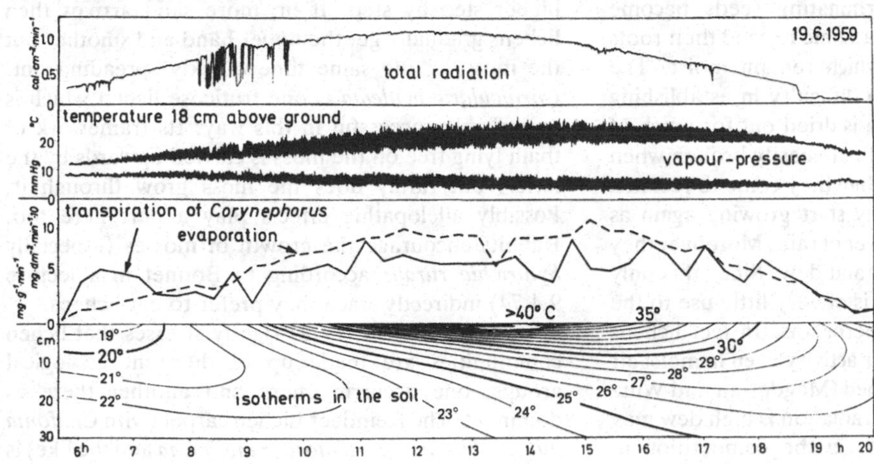

Hairgrass because of overgrazing by sheep. Today there are only a few remaining pockets near the Dutch border and these are disappearing.

However there are some inland dune areas where the plant cover has not been damaged and yet they have remained treeless for many decades. This is especially true for areas covered by **fruticose lichens** such as the brown *Cornicularia* spp. and the grey, olive-green or brownish grey *Cladina* and *Cladonia* spp. locally dominating at least some square metres. They differentiate a degeneration stage of the Grey Hairgrass sward (*Corynephoretum cladonietosum*) from other communities, (tab.71 and fig. 312). In the eastern part of north-west Germany and in Brandenburg fruticose lichens may even cover many acres of rolling duneland. To the ecologist this poses an interesting problem which has not yet been solved: what gives the tom thumb lichens the power to overcome so obstinately the attack of grasses, sedges, herbs, shrubs and trees in spite of the general climate which is so favourable to the formation of woodland?

The high acidity and the absence of buffers in the sand (Krieger 1937), its deficiency in humus and nutrients, and its more or less coarse texture cannot be considered as reasons, since these factors are of the same general order of magnitude in adjacent grassland and woodland communities. Neither can relief play a decisive role since lichen cover can be found on level ground and on gentle slopes facing in all directions. Usually they are absent only from steep shady slopes and ravines, an observation which leads us to consider the water economy of the fruticose lichen habitats. Although it can rain at any time of year in the climate of Central Europe, ranging as it does from subatlantic to subcontinental, there are also dry spells which can occur at any time from spring through summer and into autumn. In only a few hours the surface of sand dries out after rain so that germinating seeds become dessicated if they have not had time to send their roots down into deeper layers which remain moist. The lichens in their turn find no difficulty in establishing themselves on ground which is dried out for much of the time. Pieces of their thalli are easily broken when dry, and carried by wind. When they come to rest in a sheltered spot they promptly start growing again as soon as there is another shower of rain. Moreover they are able to make use of fog and dew. Since this only falls on the upper surface it is of very little use to the seeds and seedlings of phanerogams. In fact lichens can swell up and resume their activity even in moist air before the dew point is reached (Mägdefrau and Wutz 1951). On clear nights when radiation is high dew may become an important factor in the competition in favour of lichens since it bridges the time when water is scarce, whereas vascular plants cannot make use of water condensing on their surface.

The deciding weapons which the lichens may utilise in the fight against phanerogams, mosses, and tree mycorrhiza probably are the numerous specific chemical substances they produce, some of which may become toxic to phanerogams. This was recently found to be the case quite independently by Brown and Mikola (1975), Ramant and Corvisier (1975) and Fabiszewski (1975); but they were working with species different from the ones we are looking at here. Follmann and Nakagava (1963) had already observed that relatively low concentrations of such lichen toxins inhibited the germination of seeds.

Tüxen (1928) has noticed that wheel marks on lichen heathland are often the site of a dense growth of Grey Hairgrass seedlings by the following summer. From similar gaps Sand Sedge and other sward-building plants grow out into the lichen carpet from time to time. Krieger and Paul have shown that amongst fruticose lichens seeds of *Pinus sylvestris* germinate best because their roots grow down very quickly. In this way Pine can develop more easily than other trees. So in the end the lichen community may be overgrown and crowded out. Krieger (1937) also found that the species of *Cornicularia*, *Cladina* and *Cladonia* which he investigated are not able to tolerate a shortage of light.

Krieger noticed that fruticose lichens do not often invade the bare dune sand directly, but generally settle amongst cushions of *Polytrichum piliferum*. This moss can stand a small amount of blown sand settling on it and consequently it is the first conspicuous cryptogam to join the optimal *Corynephoretum*. As long as its cushions have sand trickled over them amounting to 0.5-2 cm each year it continues to flourish and grow higher step by step. If no more sand arrives then lichens gradually get the upper hand and smother out the moss, at the same time quickly spreading out. *Cornicularia aculeata* is one fruticose lichen which is remarkably successful in this way. Its framework of thalli lying free on the moss is carried upwards by the latter, but rarely does the moss grow through it. Possibly allelopathic effects play a part here too. Rabbits encourage the growth of mosses (especially *Syntrichia ruralis* according to Bonnet in a lecture 9.4.74) indirectly since they prefer to eat lichens.

Sommer (1970, 1971) rightly stresses that lichen communities are made up of different ecological groups, one occurring here and another there as dominant. The Reindeer Lichen carpet (with *Cladonia mitis*, *arbuscula*, *portentosa*, *rangiferina* and the like) is especially sensitive to being trodden on and requires full illumination. A group of lichens requiring somewhat less light includes *Cladonia furcata*, *verticillata*, *uncialis*, *gracilis*, *squamosa*. This in turn leads to lichens with brown podetia (*C. coniocraea*, *chlorophaea*, *subulata*, *ochrochlora*, *degenerans* and similar ones) which can best put up with interference by man. Cladonias with red podetia (*C.floerkeana*, *pleurota*, *macilenta*) prefer the raw humus on old rotten stumps, thus occupying only small and unstable areas.

Ecologically all fruticose lichen carpets have the common property that they can develop in habitats which are low in nutrients, often dry out and are difficult ones for higher plants to occupy. They can remain for a long time in places where they are not damaged, covered with sand, or overshadowed. Above all they are constant mainly in a temperate oceanic or subcontinental climate where the rainfall is below average. In mountainous areas with a higher rainfall they are absent from edaphically similar habitats, e.g. on heath podsols. Apparently their ability to survive dry periods without damage does not count sufficiently here, and they are unable to survive in competition with the luxuriant mosses and higher plants.

Fig. 314. Altitude zonation of vegetation in Unterengadin below Zernez. On the right above the church inner alpine dry grassland and pine woods. Montane and subalpine spruce woods on the original-rock cliffs in the middle distance, interrupted by rocks and avalanche tracks. Yellow Oatgrass meadows on the remains of the river terraces. The Inn has been diverted by the gravel delta of a tributary, now overgrown by Grey Alder. In the background left the tree-line and alpine grassland level on original rock, in the centre the steep cliffs of Piz Linard which, like Piz Buin (right) reach up into the nival level. (Compare Fig. 316).

VI The vegetation above the alpine tree line

1 *Introductory survey*

(a) Vegetation belts in the high mountains

In the lowland areas of Central Europe the only natural habitats in which trees cannot grow are salt meadows, bare sand dunes, raised bogs, Sedge fens and Reed swamps. Once above the tree line set by the climate in very high mountains, turf-forming hemicryptophytes and chamaephytes, or light-living cryptogams can develop on all kinds of soils, provided the snow and frost give them enough time to do so. With a gay variety of species, and communities which can change even over short distances, they occupy some very different habitats at this alpine level.

Most of the climatic features becomes less favourable the higher one goes (figs. 314, 315). Meadow-like communities, i.e. consisting of herbaceous plants growing densely together, dominate in the middle, or true, **alpine** belt (fig. 316). In the transitional zone leading down to the communities of crooked trees and elfin woodland is a more or less wide belt of dwarf shrubs. This may be disignated the lower alpine belt. Going upwards the closed swards become broken up into islands which are finally lost amongst stony screes and rocks. As fig. 316 shows in a schematic form this transition lies partly within the upper alpine belt and partly in the sub-nival. It is characteristic of the latter that the phanerogams rarely occur in closed stands although there are still numerous species.

Where the annual snowfall exceeds the rate at which it melts, on average, the **nival** belt begins. Its lower boundary varies with the aspect, slope and other local climatic factors and cannot be exactly set out on the terrain. The lower boundary of the alpine belt is also difficult to determine in detail. It has been obscured in many places by human activities which extended the area of alpine meadows for several hundred metres down into the valleys (see fig. 319).

However even under natural conditions grassland strips free from trees are found stretching from the alpine down to the montane belt. Besides the stream gulleys with their fan-shaped screes (fig.348) there are the narrow avalanche tracks which have cut down through the wooded slopes of many valleys for distances of from a hundred to a thousand metres (figs. 171, 172 and 317, see also Jeník 1958). They connect the herbaceous communities of the high mountains with the montane to colline meadows and pastures which man had produced after cutting down the natural woodland (see section A II). The 'avalanche-meadows' in the woodland belts serve as important homes for the cultivated and semi-cultivated grassland flora (see section D V 7). Alpine species have to give way here to more competitive warmth-requiring, but equally light-loving plants. The trees which would otherwise shade them are kept at bay by the force of avalanches thundering down these tracks nearly every year.

(b) The flora of the alpine and nival belts and its history

Typical alpine plants can get much further down than

Fig. 315. Average changes in some important climatic conditions as the elevation above sea level increases in the eastern intermediate Alps. After Harder, Lauscher and Steinhauser in Turner (1970), modified (see also fig. 320).

Favourable areas within the coordinates are accentuated by dots, unfavourable ones by hatching. For example at a height of 1800 m (i.e. near the tree line) snow falls on average 100 days in the year, the snow cover persists for more than half a year and reaches a depth of more than 1.5 m. Here the average temperatures are for January -6°C, for the whole year less than +2°C and for July about +10°C. At this altitude there are no longer any 'summer days' (with mean temperatures over 25°C) and the ground surface is dry for only about a quarter of the days in the year on average.

The mean soil temperature at 1.2 m depth corresponds to the mean air temperature only at sea level. Because of the insulating effect of the snow this comparison is increasingly in favour of the soil temperature with greater altitude. At 1800 m for example the difference between air and soil temperature is already about 2°C.

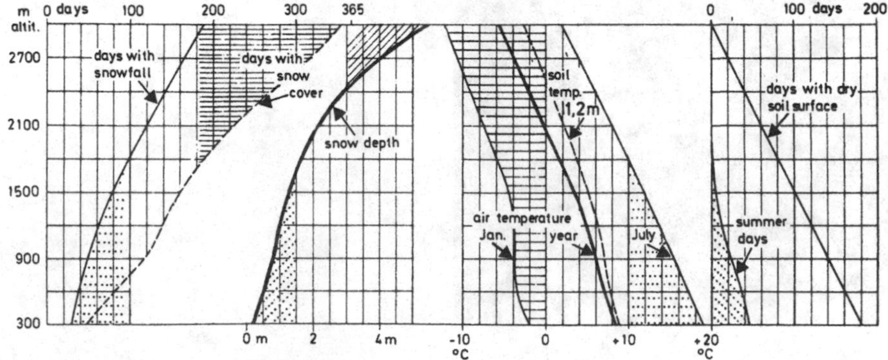

the avalanche tracks with the help of fast-flowing streams (see section B V 1 j) and on rocky outcrops interrupting the forests in the milder belts (tab. 72). *Saxifraga azoides*, *Linaria alpina* and some others flower and fruit here very well, which is an indication that they are not confined to higher levels for physiological reasons but just because they are not sufficiently competitive lower down. Against this it must be said that the majority of alpine plants lose their characteristic form or do not grow at all in the lowland. The Edelweiss (*Leontopodium alpinum*) for example in gardens grows much taller and the hairy cover on its leaves is much sparser so that it no longer lives up to its name. On the other hand plants which are at home lower down such as *Taraxacum officinale* and *Leontodon hispidus* appear more stunted and highly coloured the further they venture up the mountain.

This change in the appearance of a species which is distributed throughout a number of altitudinal belts is more than just a modification in all but a very few cases. Generally there are different genetically determined ecotypes, races or even higher taxa, and a significant proportion of their characteristics are retained when alpine plants are grown under lowland conditions. The flora at the alpine level then is much more independent than may be gathered just from looking at the list of species. A more detailed and critical list will have numerous forms, varieties or subspecies carrying the name *'alpina'* or *'nivalis'*. In some measure these narrowly defined taxa can be regarded as character plants of certain alpine associations.

Alongside some 400 lowland phanerogams which have been modified or considerably altered at the alpine level there are just about as many true high mountain plants. For the most part these are also at

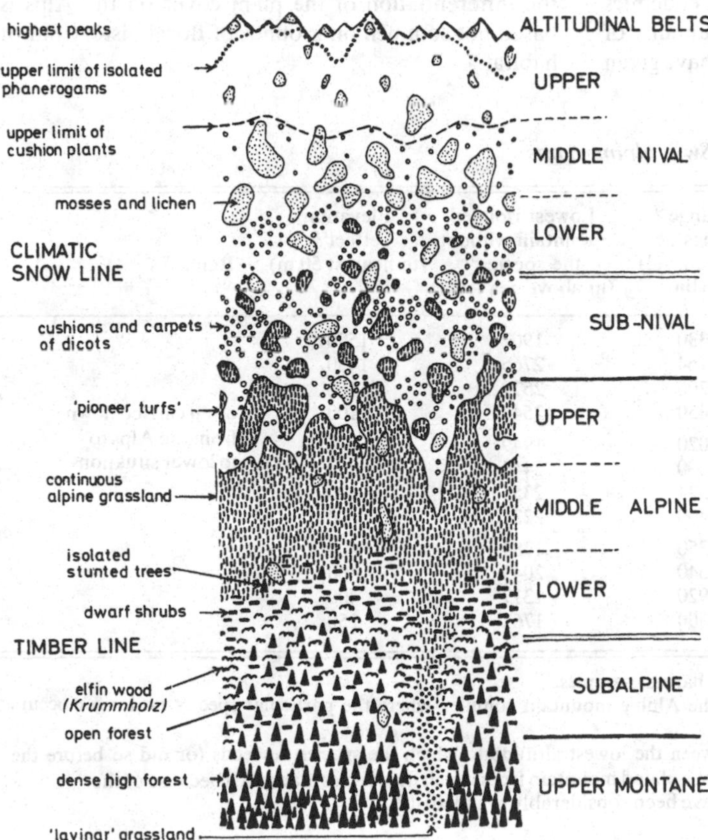

Fig. 316. Altitudinal sequence in the pattern of plant formations from the upper montane to the nival belts (schematic). To the right is a depression in the slope down which avalanches pour into the valley every year (compare fig. 317). Taken in part from Reisigl and Pitschmann (1958).

home in the high mountains of Asia, in the Pyrenees or in the Mediterranean mountains and have arisen from lowland families during the Tertiary Period.

Many of the true alpine plants are found occasionally at lower levels (tab. 72) but here they rarely reproduce themselves and only appear from time to time in those places where seed or vegetative parts have been provided from the permanent distribution area at a higher level. This one can conclude from the 'elevation effect' (as defined by Steenis), i.e. the difference in height between the highest mountain top from which the species is absent within its overall area and the lowest record of its casual appearance in a higher, larger mountain complex. As tab. 72 clearly shows the elevation effect is at least a few hundred metres and can be over 1000 m for many species.

The arctic-alpine element, which is so richly represented in our high mountains, is a reminder that during the Ice Ages tundra dominated much of Europe. Seen as a whole the flora of the Alps and of other high mountains which have been heavily glaciated is quite young and contains few endemic species. As Poelt (1963) stressed this also applies to the lichens which grow on the rocks right up as far as the snow line. Only isolated parts of the Alps are richer in endemics and in this they resemble the Balkan mountains of which Horvat, Glavač and Ellenberg (1974) have given a comprehensive account. Here and along the southern border of the Alps large blocks of mountain remained ice-free the whole time, enabling many plants and animals to survive even the Riss Ice Age in which the glaciers in the central Alps formed a practically continuous shield with tongues reaching far down into the foothills. Many relicts have only been able to spread out a little way from their places of refuge. This can be pinpointed by the rare small endemics which are still surviving, e.g. *Saxifraga tombaeensis* and *S. vandellii* in the Insubrian Alps or *Paederota lutea* and *Phyteuma sieberi* in the region of the Karnian Alps (further examples in Pitschmann, Reisigl and Schiechtl 1959, Sutter 1969, see fig. 363). Others are distributed throughout large parts of the eastern Alps, e.g. *Thlaspi alpestre*, *Rhodothamnus chamaecistus*, *Senecio carniolicus*, *Gentiana pannonica* and *Potentilla clusiana*. On the other hand a migration back from the west is shown by the present distribution of *Senecio incanus*, *Gentiana purpurea*, *Potentilla grandiflora*, *Trifolium thalii*, *T.alpinum*, *Plantago alpina* and a few other species which are also represented in the Pyrenees (in this connenction see Merxmüller and Poelt 1954, etc.). By and large though the differentiation of the plant cover on the Alps is based not so much on grounds of floral history but of habitat.

Tab. 72. Examples of the 'elevation effect' in Swiss alpine plants.

Species	Altitude range[a)] of species (m above sea level) Max. – Min.	Lowest isolated mountain without the species[b)] (m above sea level)	Elevation effect[c)] (to nearest 50 m)	Remarks
Tanacetum alpinum	3300 – 330	1902	1550	
Campanula cenisia	3300 – 1164	2765	1450	
Saxifraga biflora	3200 – 1200	2578	1350	
Achillea nana	3400 – 1450	2545	1100	often carried down from the Alps to to lower situations
Soldanella pusilla	2900 – 1020	1949	900	
Gerum reptans	3600 – 1600	2444	850	
Trisetum spicatum	3700 – 1524	2353	800	
Ranunculus glacialis	3600 – 1400	2231	800	
Loiseleuria procumbens	3300 – 1350	1784	400	
Eritrichium nanum	3500 – 1640	2045	400	
Salix reticulata	3400 – 920	1317	400	
Chamorchis alpina	3700 – 1500	1762	250	

Note: [a]Throughout Switzerland, according to herbarium samples.
[b]i.e. the lowest individual mountain away from the Alpine mountain complex where the particular species no longer occurs within its general distribution area.
[c]as defined by van Steenis, i.e. the difference between the lowest altitude to which the species descends (or did so before the construction of a dam) and the altitude of the lowest isolated mountain from which the species is absent. (because of the incompleteness of herbarium records the figures have been considerably rounded off).
Source: From data by W. Backhuys (1968).

Compared with the Alps the Tatra and other high mountains of Central Europe are much poorer in species. In the Alps one finds that the plant communities too are better stocked. They appear as impoverished parallel versions on the isolated peaks which tower above the tree line of the other mountains. Therefore we shall confine our attentions to the Alps and point out to those who are particularly interested in the Tatra the exhaustive monographs of Szafer and his co-workers (1962) and of Pawłowski and co-workers (1929). Nor are we able to go into the alpine elements in the vegetation of Jura, Vosges, Black Forest, Bavarian-Bohemian Forest, Karkonosce and other lower mountains of Central Europe.

(c) Woodland and tree lines as lower limits of the alpine region

At all times of the year the climatically determined limit for the growth of trees on our mountains is a characteristic feature of the landscape (figs. 317 and 316). In this connection what we understand by trees are all woody plants taller than the average depth of snow and freely exposed to the damaging effects of the winter weather.

It is not possible to give a particular height for what should be called a tree. However anything with a trunk 2 m tall has the right to be so called, according to Holtmeier (1967). Other authors take a minimal value of 5 or 8 m and because of this their tree limit is drawn in a different place. It is even more difficult to define the term 'wood' completely accurately. According to Jeník and Lokvenc (1962), who in turn support Vincent, a woodland would have a canopy covering at least 50% of the total area, the trees should be at least 5 m tall and the stand should occupy at east 1 ar. These standards too are set rather too high and would deprive many stands of trees even in the lowlands of the name woodland. When there is a coherent network of roots in the areas between the trees then even if the canopy is only 30-40% one can speak of a wood, albeit an open one (Ellenberg and Mueller-Dombois 1967b).

Generally one distinguishes between a 'forest line' (or '**woodland limit**') where a more or less closed

Fig. 317. An avalanche track in the subalpine Spruce wood on the north-facing slope of the Steinwasser valley below the Susten Pass in late summer showing natural meadows and prostrate trees or shrubs.

stand of many individual trees comes to a stop and a 'tree line' (or '**tree limit**') up to which isolated individual trees are able to climb. Where they are protected by snow or on crags exposed to the sun occasional Mountain Pines, Larches or Spruce venture even higher, but only as slow-growing dwarfs disappearing beneath the snow cover which protects them during winter-time (fig. 318; see also Holtmeier 1971). A line joining their highest outposts is called the '**cripple limit**'. These limits, especially those of the woodland and individual trees are set either by nature or by man and his livestock. Following the scheme proposed by Köstler and Mayer (1970) the natural limits can be distinguished by the different factors which were primarily responsible for them:

– the **general climate**

– the **local climate** (e.g. piling up of snow, strong wind, accumulation of cold air and similar factors)

– **edaphic** (e.g. smooth rock, large boulders, or swampy soil, etc.)

– **orographic** (avalanche tracks, a wall of rocks, stream erosion or similar).

Carbiener (1963, 1964) was of the opinion that in some places the woodland limit was caused by cryoturbation (i.e. alternate freezing and thawing of the soil), e.g. in the Vosges. However this cannot be taken as confirmed because it has been noticed in many mountains of the Earth that such a cryoturbatic movement of soil can also set in after the woodland has been destroyed by man or animals, and this even several 100 m below the climatic limit, e.g. in Greece (Hagedorn 1969).

It has long been a disputed question amongst plant geographers whether a woodland under natural conditions stops at a certain height, above which it is completely excluded, or whether it merges into an area of single trees so that a climatic tree limit can be determined above the climatic woodland limit. Schröter (1926) and many other authors both before his time and since have inclined to this latter view. The present day landscapes in Alps, Tatra and other high mountains in Europe appear to indicate that they are right (figs. 318 and 151). On the other hand Scharfetter (1938) and an increasing number of younger authors see this situation as a consequence of worse soil conditions or extensive grazing or selective wood cutting (see section A II 2): Where **one** tree grows then others can grow near it if man and animals permit, provided the soil is deep enough. A change of the general climate cannot be so abrupt that directly beside a living tree, several metres tall, which has been growing for centuries, conditions could be too rough for the development of another tree. Therefore the woodland, when untouched by man, gives way to small groups of trees rather than to isolated individuals. The latter would have to live under a harsher climate than a

Fig. 318. The upper woodland limit in the Andes at about 40° lat. S. (temperate zone) is still for the most part in its natural state compared with that in the Alps. In the Argentinian National Park of Nahuel Huapí the Southern Beech (*Nothofagus pumilio*) becomes more and more stunted as it approaches the limit but even here it forms dense stands. It is only where trees have been cut and grazing practised that a 'tree limit' or timber line distinct from a 'woodland limit' is to be found. After Ellenberg (1966).

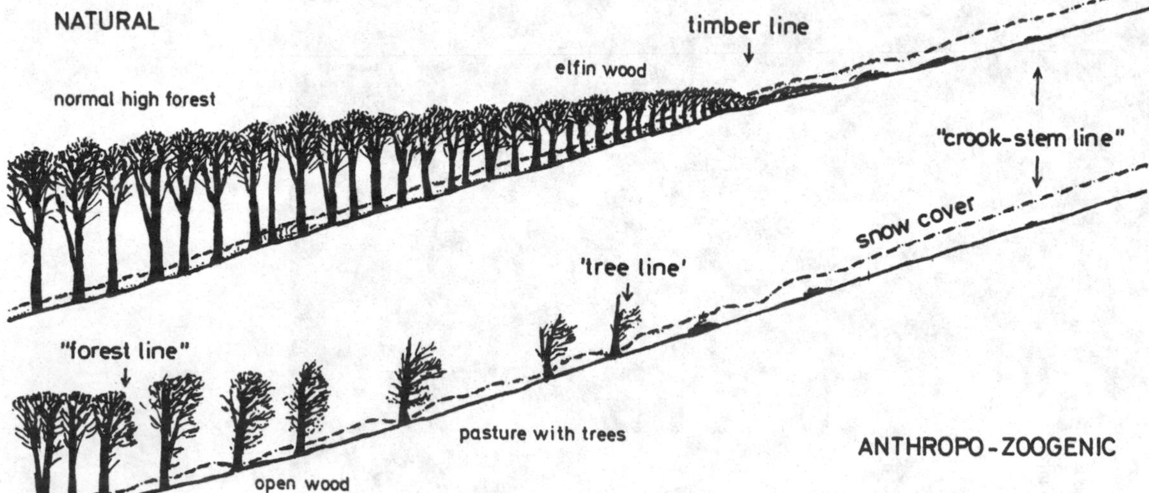

group of trees which is better protected from all sides, the stems becoming shorter towards the outer edges of the stand. Only on rocky slopes, such as a bare Karstic limestone, are trees forced to grow isolated as tall individuals even without any human interference, because here they are confined to the splits in the rocks. Only in these do they find sufficient fine earth, because the formation of humus at the altitude of the tree line goes on very slowly and trees cannot invade the stony areas between the rocks as they do at lower levels. According to Scharfetter's conception then the climatic forest line and the climatic tree line are identical.

Observations in mountains which have come much less under man's influence, e.g. in Norway and the Argentinian National Park Nahuel Huapí support Scharfetter's point of view (see fig. 318). The trees making up the woods (e.g. *Nothofagus pumilio*) become shorter as the altitude increases, and they form an almost closed front against the alpine level (see also Neuwinger 1967). Only in the neighbourhood of San Carlos de Bariloche and other settlements is the picture the same as it is in the Alps. However here, during the course of the last few decades, the woodland has been eaten into by axe, cattle and fire just as it has been in Europe for a number of centuries. So it is only through man's influence that the woodland limit and the tree limit have been separated (fig. 318). In the Alps there are still a few sharply defined closed woodland limits to be found, even on slopes of deep soil which are not too steep, e.g. here and there in the Upper Engadin (Holtmeier 1967), on the left side of the Sertig valley (Nägeli 1969), in the Tyrolean Radursch valley (Köstler and Mayer 1970) and in the Kauner valley (Tranquillini 1967). 'On a regular slope with better soil and a less than average amount of extensive grazing the woodland and tree limits coincide over long stretches in the eastern Alps too' (Mayer 1976, transl.).

The woodland boundary in the Tyrol for example has been lowered by activities of man and his livestock by an average distance of some 208 m (maximum 220-400 m, Friedel 1967). However, as Kral (1979, see fig. 319) has shown for the Dachstein massif not only man and animals, but also changes in the general climate can affect the altitude of the woodland limit. Here the actual woodland boundary is about 235 m below the potential one as determined by the climate. The Alpenrose heaths which extend between the woodland and tree limits in the outer Alps are considered to indicate areas which have been woodland by Lüdi (1921) and most of the more recent

Fig. 319. Changes in the tree and woodland limits brought about by the climate and by man's activities since about 1500 B.C. on the plateau and northern slope of the Dachstein massif, determined on the basis of pollen analysis of peat bogs. After Kral (1971), somewhat modified.
Altitude and time scales are logarithmic. In the middle warm period (Atlantic – about 3000 B.C. – not shown here) the climatic woodland limit was about 2200 m and since then has sunk slowly, with fluctuations. In the Bronze Age it was about 1975 m and several times since then has exceeded 1950 m even in the Middle Ages. In cooler periods the natural woodland limit retreated down the slopes and the Hallstatt glacier advanced (arrows along the top). Since the beginning of the Iron Age the woodland limit has been depressed to below 1600 m by cutting and grazing. In the pollen diagram the anthropo-zoogenic destruction of the woodland can be recognised by the higher fraction of pollen other than that of trees, especially that of 'pasture indicators'.
Before the influence of man the dominant trees were Spruce, Arolla Pine and Siver Fir. Larch only became numerous later as a light-loving pioneer penetrating further and further down into the valleys. The Mountain Pine scrub, which under natural conditions forms only a narrow belt above the mixed Arolla Pine wood, has also increased in area. At lower levels Fir has been almost eradicated in favour of Spruce, and Beech too has been pushed back. AP Arolla Pine, B Beech, F Fir, L Larch, MP Mountain Pine, S Spruce.

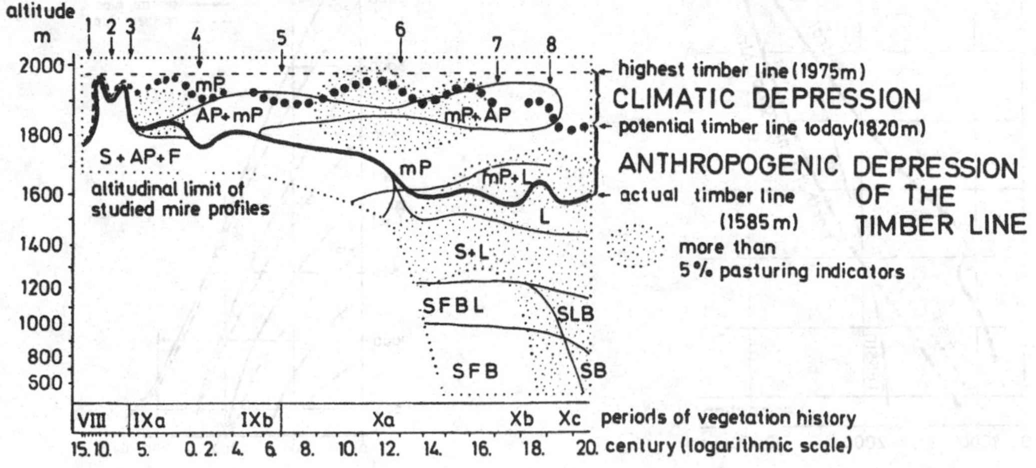

authors. Probably at one time the dense subalpine woodlands were closed right up to the upper limit, only being interrupted by rocky slopes or by streams and avalanche tracks as illustrated in figs. 316 and 317. According to Zoller (1960) pollen from the *Rhododendron* species in the Alps is only found in peat deposits which have been laid down in relatively recent times, long after man had begun to exploit the Alps.

Many investigators have worked on the problem of **natural causes of the altitude limits of woodlands** and of distinct tree species. More recently ecophysiologists in particular have taken up the challenge (e.g. Tranquillini, Pisek, Larcher, Turner, Th. Keller and others). Brockmann-Jerosch (1913) and Furrer (1923) were already well aware that no one single factor could be responsible but the whole character of the climate must be considered. In the inner Alps with their more continental climate and also in the central Tatra trees are found much higher than they are in the outer Alps (figs. 6 and 42). This principle also applies if a single tree species is considered, e.g. Spruce which is present in all parts of the Alps. In Eurasia and America too woodlands are favoured by a more continental climate, reaching further northwards and higher up the mountains in the interior of Eurasia and America than they do nearer the sea (see esp. Hermes 1955).

Very probably the summer temperature during the daylight hours plays a deciding role here but also the soil temperature from which, according to Havranek (1972), depends the root activity and the water uptake. Figure 320 shows that the number of days with an average temperature of 10°C or more, i.e. with favourable conditions for photosynthesis, is significantly greater in the central Alps than in the outer Alps at the same altitude. A similar situation is seen with the length of time the temperature is above 5°C. The extension of the growing period in more continental conditions is partly due to the so-called mass effect (i.e. the higher average altitude of the inner Alps), but above all to the increased number of hours of sunshine in a relatively continental climate with its fewer clouds (fig. 321). This promotes not only the growth of trees but especially the ripening of their new branches and needles (or, in deciduous trees, the buds). In the case of Spruce, for example, it takes three months before the walls of epidermal cells and the cuticle reach their final thickness under montane conditions (Lange and Schulze 1966). The fate of a tree in winter depends on this ripening as Michaelis (1932 and 1934a, b) was the first to point out clearly. Organs with an insufficiently thick epidermis and cuticle dry out, especially during the late winter, when they are warmed by the sun but

Fig. 320. The climate of the outer Alps is relatively oceanic, while that of the inner Alps is more continental, i.e. at the same altitude it has less precipitation and more radiation (see also figs. 4-6, 42a, 321 and 315).

a. The increase in precipitation with altitude is at first significantly greater in the northern Alps and the Swiss Jura than in the inner Alps. Above 3500 m the precipitation is about the same all over. In the insubrian Tessin it is only the lower levels which are distinguished by their unusually high precipitation. After Uttinger and Nägeli (1969), modified (see fig. 4).

b. The number of days in which the average temperature exceeds 15°, 10°, 5° and 0°C in the Swiss inner Alps and in the northern outer Alps, interpolated from monthly averages. At the woodland limit there are about 100 days with an average temperature of more than 5°C and this is true both of the outer Alps (at about 1800 m) and of the inner Alps (at about 2200 m).

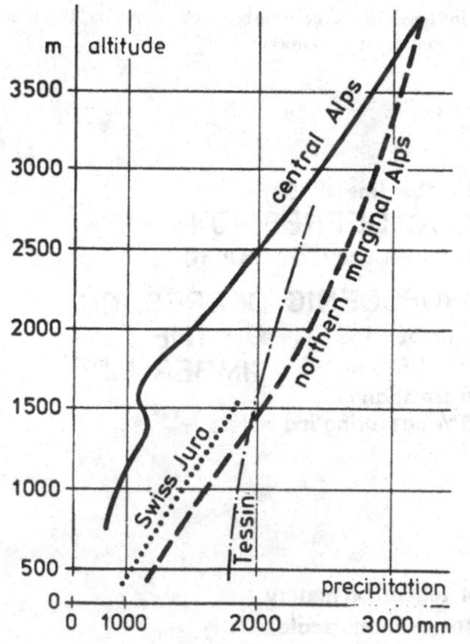

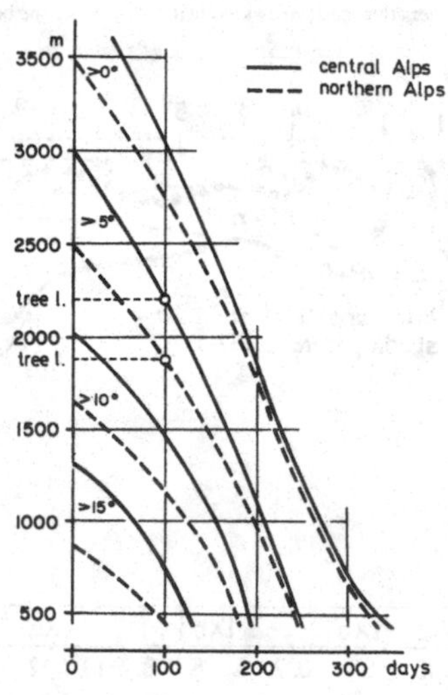

the roots are still in frozen soil from which they can get no water (see also Baig and Tranquillini 1976).

Generally speaking then the interaction between the growth conditions in summer and the water stress in late winter, caused by frost drought during fine weather, may be the real deciding factor in limiting the tree growth. Only this explains the sharp caesura which seems to contradict the gradual change in the general climate with increasing altitude (Tranquillini 1979, Larcher 1963a, see fig. 315).

The correlations were impressively shown in measurements taken by Tranquillini (1957) at the woodland limit on Patscherkofel near Innsbruck (tab. 73). The needles of young Arolla Pine which have

Fig. 321. On sunny days, which are the rule in the inner Alps, the air, plant and soil temperatures fluctuate much more than they do on cloudy days which prevail in the outer Alps. The difference between the more continental and the more oceanic climatic conditions operates particularly through the amount of warmth available to the trees during the growing period. After Turner and Nägeli (1969), somewhat modified. The measurements were made at the same place in August 1953 near the woodland limit on the Patscherkofel near Innsbruck.

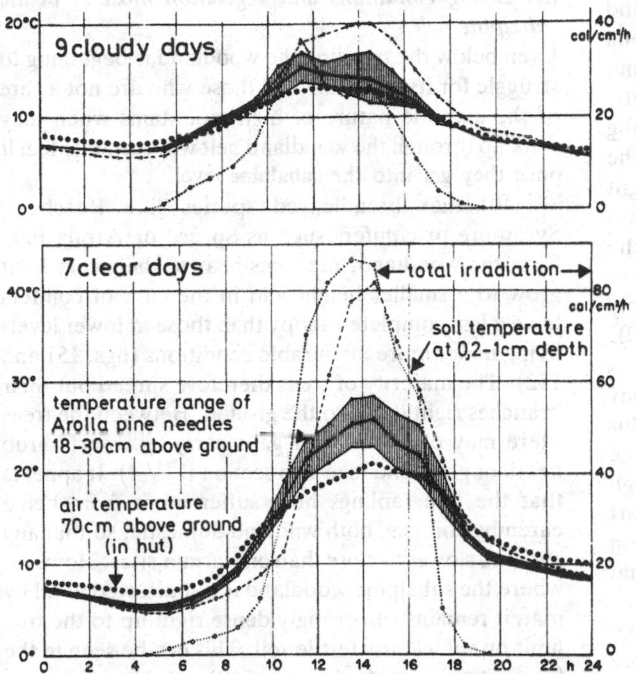

Tab. 73. The water content and osmotic values of needles on young and old Arolla Pine (Pinus cembra) at the end of February, i.e. after surviving the winter on the Patscherkofel near Innsbruck.

Altitude:	Below the woodland limit			At the tree limit
Needles:	in the snow[a]	above the snow		above the snow
Tree age:	young	old[b]	young	young
Minimal water content of needles (as % of the dry weight)	155	ca. 125	115	78 (below 70 lethal!)[c]
Maximum osmotic pressure (π^* in atm.)	23	26.5	30	42.5

Note: [a] The needles are protected against water loss by the snow cover but, because of the high humidity are in danger of fungal attack.
[b] The older trees had more water available than the younger ones because they root deeper and the ground was only frozen to a depth of 35 cm.
[c] Irreversible damage is caused when the water content reaches 70% or below.
Source: From data by Tranquillini (1957).

overwintered under the snow show a relatively high water content and a low osmotic pressure potential. Where Pine needles of a small tree protrude above the snow their osmotic value is higher and their water content less. The water balance in a taller tree growing in a similar situation is more favourable because its roots go deeper into the ground where they can find liquid water even in winter. The leaves of young trees growing near the tree line show figures for water content and osmotic pressure at the end of February, i.e. at the end of the deep winter, which approach critical values leading to death. At greater altitudes the needles would be even less resistant and in a sunny late winter would inevitably dry out. Tranquillini and Machl-Ebner (1971) found that intact twigs of Arolla Pine persisted for longer in their winter dormancy and that this had something to do with a change in the cytoplasm. Their photosynthesis is restricted during the cold months and does not compensate for the respiration during this time. If such twigs are brought into a warmer place there is a delay of some days before they become active. On the other hand at the submontane level of northern Switzerland the photosynthetic ability of young Arolla Pines is noticeably high throughout the whole winter (Th. Keller 1970).

According to Schwarz (1970) the frost hardiness of Arolla Pine, Alpenrose and very probably of many other plants depends on the day length (fig. 165). This appears to be very practical when one considers the frequent changes in the weather mainly in high mountains. A frost resistance induced at least in part by day length is safer than one which depends only on temperature changes. M. Moser (1967) noticed that the frost hardiness of young shoots was improved or reached more quickly with better nutrition. Even the ectotrophic mycorrhiza could also play a part therefore in helping to maintain the trees at their climatic limit. There are probably similar connections which apply to other trees too, but they have not yet been so thoroughly investigated as in *Pinus cembra*, nor even looked into at all. In each case they could depend on **an interaction between favourable growth conditions in summer and the drought caused by frost in winter**. This has been confirmed recently for Spruce which, according to Tranquillini (1974), requires at least three months of active growth in order to survive at the tree limit. The cuticular transpiration of its seedlings proved to be a function of the degree to which they have developed by the start of winter. Early sprouting strains then have an advantage over later ones.

Physiognomically and ecologically one must look upon the woodland or tree limit as the start of the alpine level. However floristically and sociologically it is the upper boundary of the stunted trees and shrubs which is a more significant frontier because woodland plants find enough shade or half-shade even under low bushes. Indeed many of them can be found along with the dwarf shrubs persisting as much as 300 m above the tree limit in depressions where the snow collects.

Living conditions become progressively worse with increasing altitude (figs. 315, 320) Any attempt to fix boundaries here is very arbitrary because the pieces in the mosaic of plant communities and their habitats become smaller and smaller, and the vegetation cover varies also with the firmness and stability of the soil as well as with snow depth and wind conditions.

(d) Living conditions and vegetation mosaics in the subalpine belt

Even below the tree line the woodland is beginning to struggle for its existence and those who are not aware of the extreme nature of high mountains when they walk up through the woodland belt will certainly feel it once they get into the **subalpine** level.

Whether broadleaved species like Beech or Sycamore or conifers such as Spruce or Arolla Pine form the woodland, the trees near to their very limit grow to a smaller height and in the case of conifers have a less complete canopy than those at lower levels living under more favourable conditions (figs. 151 and 152). The majority of trees therefore spread out their branches right down to the ground. Between the trees there may even be small gaps carrying dwarf shrub heath or grassland (see also section B IV 4). It appears that the tree saplings here suffer more from being eaten by animals, both wild and domestic, so that any gaps are slower to close than on warmer sites. However where the subalpine woodland is not affected at all by man it remains surprisingly dense right up to the tree limit on sufficiently fertile soil. This can be seen in the South American Andes (fig. 318) as well as in some places in the Alps (Mayer 1975).

One kind of damage which subalpine woodland suffers is from the weight of hoar frost and snow (fig. 157). Broken branches in the crown are the rule rather the exception, especially with evergreen conifers. Trees with short branches and a very slim form (fig. 317) are less liable to this type of damage and in most mountainous and subpolar areas they form a population which is obviously adapted to these conditions. When reafforestation was carried out at high levels in the Harz and other mountains where nearly all trees had been cut down for mining, long-branched lowland types of Spruce trees were planted since at that time, about a hundred years ago, the significance of different ecological races was not yet appreciated. A considerable amount of damage was suffered by these broad-crowned trees so that many more gaps appeared in their stands than would have been the case with the original woodland type (Heynert 1964, Fiedler and Nebe 1969, etc.).

Wherever a woodland in the high mountains becomes opened up, whether from natural causes or by the incursions of man and animals then the **snow becomes a deciding ecological factor** in another respect. When snow is falling the wind, which is often very strong, blows it along until it becomes trapped by lee slopes, depressions in the ground, stones, groups of shrubs or trees or other obstacles, while it can hardly settle at all on the windswept surfaces and at most forms a shallow layer here (fig. 328). On the one hand the snow is a protection against cold and transpiration losses, but on the other it has effects which are not so favourable, and these will be thoroughly discussed in section e. The result is a **mosaic of fragments of different plant communities** which are not distributed irregularly but form a very characteristic pattern, some examples of which are shown in figs. 324 and 317 (see also Wagner 1965, 1970).

In such a subalpine mosaic habitat conditions which are more like those of the montane belt can be found alongside ones which are definitely alpine in nature. Thus it is not surprising that plant species and even whole communities from the two levels take part in the mosaic. The montane floral and vegetational characteristics normally predominate at least as far as the area they cover is concerned.

The individual communities in this mosaic have already been largely discussed in connection with the other woodland communities, namely:

- subalpine Sycamore-Beech woods (section B II 2 g),
- subalpine Spruce woods (B IV 3 b),
- subalpine Larch-Arolla Pine woods (B IV 4),
- subalpine Prostrate Pine scrub (B IV 5 b).

The remainder connect the subalpine with the alpine belt and are more easily understood when looked at in connection with the latter especially:

- subalpine Green Alder scrub (C VI 7 c)
- subalpine to lower-alpine dwarf shrub heathland (C VI 4)
- subalpine to alpine grassland (C VI 2 & 3)
- subalpine tall perennial herb communities (C VI 7),
- high-montane to lower-alpine spring swamps and fens (C VI 6),
- montane to alpine scree communities (C VI 8),
- montane to alpine rock crack communities (C VI 9 a)
- subalpine to nival cryptogam communities, especially the lichen covering of stone (C VI 9 b).

A small-scale mosaic of plant communities is characteristic not only of the Alps, but is also found in less extensive mountain ranges, such as the Tatra (J. Horak 1971), Harz (Stöcker 1967) and Swiss Jura (J.L. Richard 1968a).

Wherever in the subalpine belt the snow cover is several metres deep and does not melt completely until later in the year it prevents the establishment of trees locally. In the case of conifers snow mildew and other cold-tolerant parasitic fungi contribute to prevent tree growth. In mountains where there are particularly heavy snow falls, such as the coastal ranges of north-western USA, parts of the northern Alps e.g. the Allgäu (Knapp 1960) and the eastern ranges of Japan (Ogasahara 1964), the woodland dissolves into groups of trees at its upper limit. According to Fonda and Bliss (1969) these groups are able to form and to regenerate above the normal limit only in the years which are unusually warm and with a low precipitation. However on such snow-rich mountains the isolated trees near the natural tree line are just as rare as on mountains with a 'normal' amount of snow. Here too then one should not speak of tree limits, but in every case of a '**tree-group limit**'. Such groups of trees do not always arise from seeds, but may come from runners which spruce and some other conifers are able to produce (Kuoch and Amiet 1970).

Dolines (i.e. funnel-shaped pits of various sizes which are a feature of limestone or gypsum mountains) offer very special conditions in the subalpine belt. They are not only places where snow accumulates, but where cold air tends to collect. Thus they are particularly unfriendly sites for trees to establish themselves. Because of this dolines favour alpine plant communities which can occupy them well below the normal tree line for the area (Gensac 1968a).

Special ecological problems are also presented by the attempts at reafforestation in those areas close to the tree line which have been denuded of trees through agricultural methods. A number of thorough scientific investigations have been carried out here in connection with tree planting as a protection against avalanches. These studies recently added greatly to our understanding of subalpine vegetation as well as the limits of trees and woodland. As examples we can quote the works of Aulitzki (1962, 1963a,b, 1965), Hampel (1963), Stern (1965) and other authors mentioned in these publications.

Since differences in the local and microclimatic conditions play a decisive role in the growth of planted trees near to the general climatic tree line, a number of long-term programmes have been dedicated to study how much these conditions can vary over small distances, a subject which up until then had been neglected. We refer only to the exemplary research series carried out in the Dischma valley (Engadin) by Nägeli (1969) and Turner (1970) as well as near Obergurgl in Tyrol by Friedel (1967) and Aulitzki (1962). Their measurements, mapping and other work give a lot of information about conditions at the alpine level. We shall discuss these in greater detail in the next section.

(e) The climate in the alpine belt and its local variations
In order to understand the unique nature of alpine vegetation we have to remind ourselves of the climatic conditions under which it has to live. Since Schröter (1926) they have been discussed by many authors, more recently by Flach (1963, 1967), E. Winkler (1963), Rehder (1965), Larcher (1980b) and by those just mentioned in the preceeding section. The alpine climate shows the following main characteristics (see also fig. 315):

1. The dominating factor in the environment is the **short growing period** (fig. 320b). Every additional 100 m in altitude normally means a shortening of the growing period by about 6–7 days. Average daily temperatures exceeding 5°C are required before most woody plants can add to their total dry weight. In the Swiss Alps for example this figure is attained on 195–210 days each year at a height of 1000 m above sea level, but on only 85–120 days at 2000 m (see fig. 320b).
2. Not only is the growing period shorter but it is also **colder** when measured at 2 m above the ground in the shade. The föhn wind, which has such a warming effect in many deep valleys of the Alps, is not warm when it comes down from the snow line and only causes damage. Within the alpine belt it does not raise the temperature or speed up the melting of snow. With increasing altitude the effect of the mountain shadow on the climate of a valley or of a terrain north of a rock wall is greater since the differences between air and soil temperatures grow larger (Böhm 1966).
3. The soil surface and the layer of air close to it are warmed up much more by the sun than would be expected in view of the low air temperature in general (figs. 322, 323, see Aulitzki 1961a, b, Turner, Rochat and Streule 1975). Because of the thinner air the radiation is much more intense than at lower levels. Near to the ground plants then grow in a particularly **favourable microclimate**. This applies especially to trellis-like shrubs such as *Salix retusa* and *Dryas octopetala* which cling closely to the rocks and stones. Where the ground is dark in colour the temperature can even rise to over 65°C, causing heat damage to seedlings and young plants (Kronfuss 1972).
4. In the thin air at great heights the loss of heat at night through radiation is also remarkably high and night frosts are much more common at the alpine level than lower down. This is especially true in spring and autumn when the climate alternates between frost at night and a thaw during the day. A beneficial effect of the **low night temperatures** is to reduce respiration and so conserve the products of photosynthesis. They also reduce the amount of sugar which is turned into starch. The relatively high sugar content, in its turn, increases the resistance of leaves to cold and assists in the formation of anthocyanins. The green parts of most alpine plants have a bluish or violet shade and in autumn they take on beautiful bright colours. Since the soil and the plant roots in it remain significantly warmer at night than the air (fig. 321) guttation is common, much more being seen here than anywhere else (Frey-Wyssling 1941).
5. During the time when the alternate freezing and thawing is most prevalent many habitats are covered in snow. The **depth of the snow** varies very much from one place to another because the wind is slowed down mainly by the configuration of the land and hardly at all by the plants growing on it.

Fig. 322. Total radiation in the high mountains (Stillberg near Davos, solid lines) when the weather is sunny but with white clouds can be up to 10% higher than the solar constant (i.e. the sun's radiation falling vertically outside the earth's atmosphere). When there are no clouds the total radiation is much less even in the clear mountain air (thicker curves). At lower levels (near Zürich, broken lines) under comparable conditions it is always lower than in the high mountains. After Turner (1970), modified.

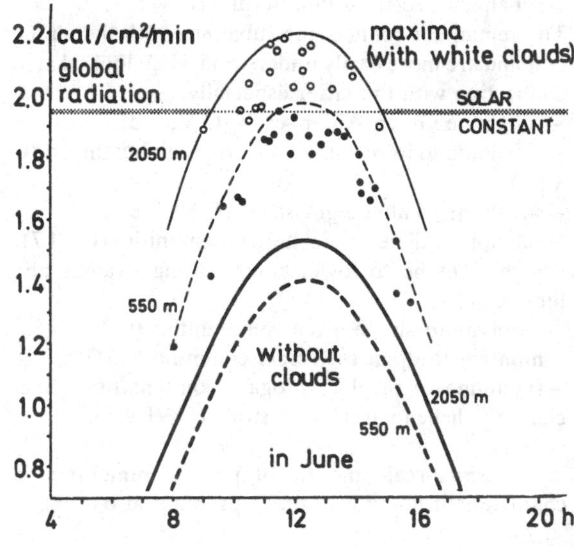

Depressions therefore remain covered with snow for much longer while rises are cleared earlier in spring or may even remain free from snow throughout the winter (fig. 324). The mosaic of plant communities faithfully reflects that of the snow persistence (fig. 325). In the woodland belts snow rarely drifts to this extent since trees and shrubs temper the force of the wind.

6. An **insulating snow cover** of more than 50 cm ensures that the temperature will rarely fall below 0°C (Rübel 1912, fig. 165). Under its protection even frost-tender plants can survive the whole

Fig. 323. Temperatures throughout the day at a depth of 1 cm in the ground at different aspects in the region of the woodland limit near Obergurgl (Ötztal, 2000 m) compared with the air temperature at a height of 2 m above the horizontal surface. After Turner (1970), somewhat modified. A bare dark ground surface on the sunny slope reached 80°C!

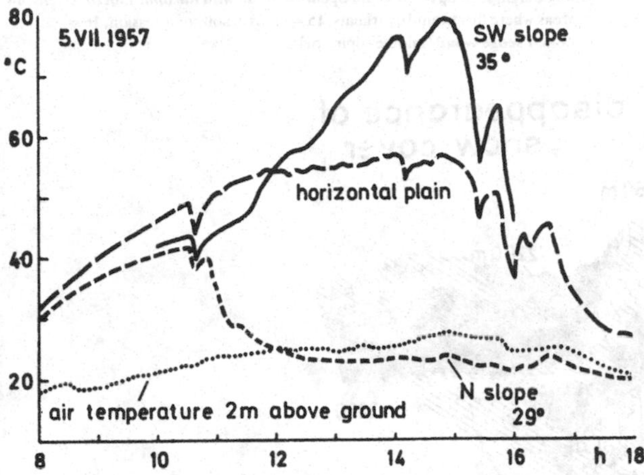

Fig. 324. Vegetation at the alpine level, affected by the relief, snow distribution and soil character, varies within a small area (see also figs. 325 and 328).

Near the Stuttgart cabin in the Lechtal Alps one can find swards growing on shallow or deep calcareous soils (e.g. right foreground) or on soils which are more or less deficient in lime. In many gulleys and at the foot of some of the slopes the snow may lie until the beginning of August while on nearby sunny slopes it may already be gone in April; the windswept ridges may even remain free from snow throughout. Homogeneous dense sward communities are able to form only on level areas and in wide valleys.

winter. If sufficient light gets through to the ground the plants remain green and can even carry on CO_2 assimilation since snow is porous and allows gas exchange. Depending on the nature of the snow a covering of 20 cm lets through about 2-20% of the total radiation (fig. 341). When the thickness is increased to 30 cm the average light getting through is often as low as 1%, but according to F.B. Salisbury (1975) some light can even penetrate to a depth of 200 cm and this is

Fig. 325. In the region of the woodland limit and in the alpine belt the distribution of the plant communities depends very much on the duration of the snow cover. A map showing the dates by which different areas become clear of snow corresponds largely with one of the vegetation. After Friedel from Aulitzky (1961b), modified.

Above: an interpolated map of the dates by which different areas are free of snow between 2050 and 2250 m above sea level near Obergurgl.

Below: A simplified vegetation map of the same area. 1 = more or less horizontal areas, 2 = windswept lichen heaths, 3 = windswept *Loiseleuria* heaths, 4 = Crowberry heaths, 5 = Alpenrose-Bilberry heaths, 6 = dense Alpenrose heaths, 7 = Prostrate Pines, 8 = gaps in the Alpenrose undergrowth of the subalpine wood, 9 = scree vegetation, 10 = old Arolla Pine stands, 11 = young Arolla Pine, 12 = young Larch, 13 = open pioneer Alpenrose scrub with *Racomitrium canescens*, 14 = areas where the ground overheats, 15 = areas of soil compression, 16 = Crooked Sedge sward (middle-alpine belt).

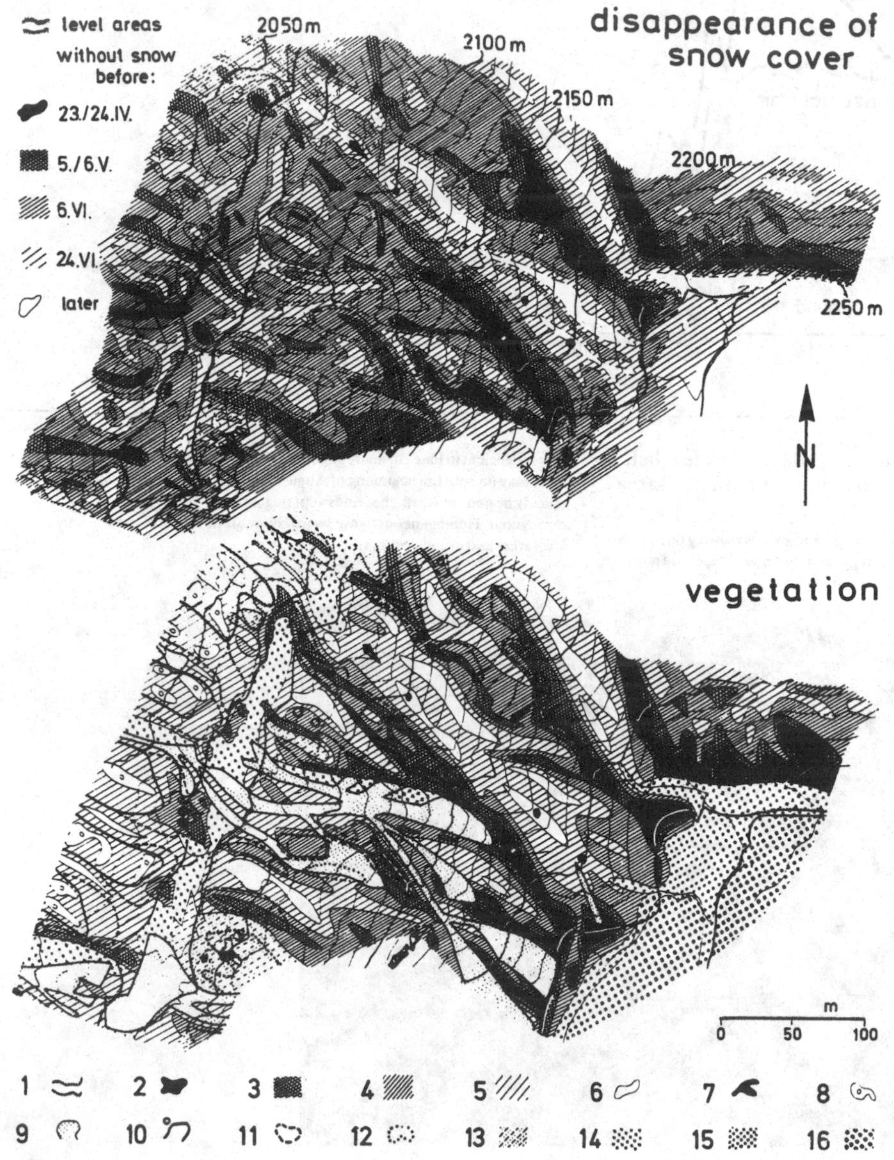

mainly the blue light which stimulates the formation of chlorophyll and, in many species, the germination of seeds. In any case the leaves are protected from drying out when they are under snow whereas those exposed to wind and frost soon turn yellow and may even be killed (section c). Because of the snow covering many more plants are able to overwinter while still retaining their green leaves, even though they may become somewhat paler, than those in the meadows of lower belts which are much warmer and from which the snow disappears from time to time in winter. Rübel recognised many years ago that the majority of alpine grasslands were 'evergreen'. It is because of these circumstances that the ibex and chamois are able to find green food throughout the winter by scraping the few decimetres of snow away with their hooves.

7. Since for the most part snow only melts away when the sun is standing high in the summer sky alpine plants enjoy a **considerable amount of warmth at the start of their growing period**. Shortly after the last snow has gone they are able to photosynthesise normally since they still have green leaves which hardly suffered under the snow (Cartellieri 1940, Brzoska 1973). In spite of this most species take at least 3–4 weeks before they flower (Rompel 1928) because they grow slowly (see points 3 and 11). Only in the species in which the flower buds are already formed, such as *Crocus albiflorus* and the soldanellas do they unfold immediately the snow has thawed (figs. 436 and 340).
8. For most plants the alpine climate offers more extremely **fluctuating conditions** than the montane climate, at any rate as far as temperature is concerned (see points 3,4,7, and fig. 365). Even in the relatively oceanic outer Alps the change from winter to summer occurs so suddenly that in lowland areas it can only be matched by an extremely continental climate, e.g. in western Siberia. In contrast to the continental lowland climate however the high mountains are always subject to cold spells brought on by cyclonal conditions and which can lead to snow and frost.
9. Except in special places (point 10) the alpine plants **seldom suffer a water shortage**, as the precipitation increases with altitude up to about 3000 m (fig. 320a). Almost all fine soils above the tree line contain such a reserve of water in spring that it lasts throughout the growing period. When there is a shortage of rain the water from melting snow is often an additional source of water for many plant communities.
10. Both in places with a little snow and in those with a lot the water balance of plants can be put under considerable strain by a **strong wind** and higher temperatures when at the same time the root run is frozen (see also Tranquillini 1970). Areas quickly cleared or perpetually free from snow present living conditions which are very 'continental' regarding their water economy. So it is easy to understand why many of the high mountain plants show features similar to the flora of Asian steppes especially in the higher taxa and in their life forms. Even in summer above the tree line there can be large differences in wind speed over quite small distances. These affect the photosynthetic achievements of the plants (fig.336). According to Caldwell (1970) *Rhododendron ferrugineum* reacts to a wind speed of less than 1 m/s by movements of its stomata. As soon as the speed exceeds 1.5m/s these close altogether even when there is plenty of water available.
11. Since the **light is relatively strong and rich in ultraviolet rays** at high altitudes the majority of plants are shorter than those at lower levels. However the main reason for their small growth in height lies in the fact that the temperature rapidly falls off above the ground surface and at night it is normally very low (fig. 321). The intensive radiation in late winter can even lead to destruction of chlorophyll and damage to those parts of the xeromorphic Arolla Pine needles which are exposed to the sun (Turner and Tranquillini 1961). Contrary to a widespread view the light intensity in the high Alps is by no means always excessively high. Research by Cartellieri (1940), W. Moser (1973), Brzoska (1973) and others showed that light is often the limiting factor for photosynthesis, especially on cloudy days (fig. 368).
12. The temperature fluctuations which have already been mentioned in points 3 and 4 speed up the weathering of the soil and lead to **occasional solifluxion**. This soil flowing is helped by the excessive amount of water supplied by melting snow (fig. 326). According to J. Schmid (1955) frost-induced movement of the soil takes place in all the climatic belts in Central Europe, but it happens much more frequently in the high mountains where the slower growing vegetation sometimes cannot hold the soil together on sloping ground. The top soil which from time to time is saturated with water flows down and is dammed up by 'festoons' of sward, the origins of which have been studied in the Alps by Thomaser (1967) and – very thoroughly in a monograph – by

Zuber (1968, see fig. 327a), but also in the Vosges by Carbiener 1970b, see fig. 327b). Together with the large relief energy the alpine and subalpine climate ensures that many habitats remain unstable. It is only on level ground, which is confined to small areas, that the soil has been able to ripen undisturbed since the Ice Ages.

13. Amongst the mechanical damage to which the vegetation is exposed must be included the **grinding action of snow**. On bare exposed ground

Fig. 326. Signs of solifluxion on moraine slopes in the upper Fimber valley (Silvretta) in the lower alpine belt. The horizontal places without humps are taken over by swamp communities, especially beds of Common Sedge. In the areas with humps Matgrass swards predominate. The humps have diameters of 0.5 to 1 m; trampling by grazing animals has had some influence in their formation.

Fig. 327. Left: Because of solifluxion on calcareous moraine slopes festoons of vegetation can be found occupying the brow of a moving cake of soil or one which has come to rest. After Zuber (1968), somewhat modified.
Movement of the soil occurs especially in the early part of the year when it is saturated with water from melting snow. Pioneer stages of the *Seslerio-Sempervireturn* at the upper subalpine level in the Swiss National Park.

Right: On acid rock too such festoons can occur through solifluxion in the subalpine level, e.g. on granite in the Vosges. After Carbiener (1970), modified.
Vaccinium myrtillus, *Luzula desvauxii* and *Pulsatilla alpina* occupy the brow position along with a dense layer of mosses. On the surface of the tongue of soil are pioneer mosses to which Matgrass and other vascular plants join in the transition towards the brow.

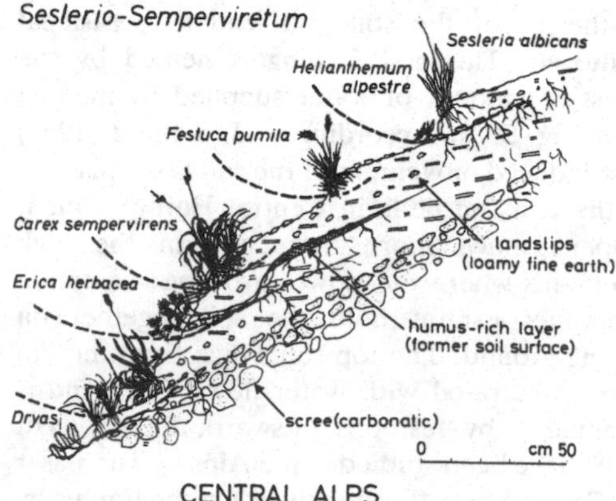

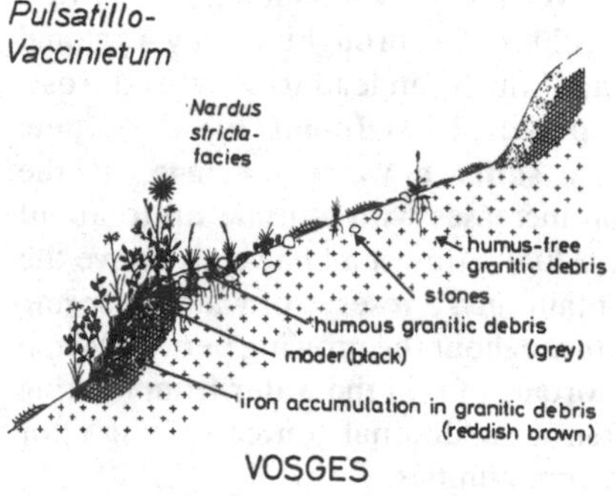

and on the surface of a snow cover hard crystals of ice act as a grinding powder when driven by strong winds. This can damage or even kill any shoots sticking out through the snow surface whereas those which the snow has protected remain alive and form a dense green table (fig. 162). The wind speed and above all the number of storms is much higher than in the lowland, even when compared with the sea coast. According to Geiger (1950) figures taken from weather stations in West Germany show an average wind speed of 4.8 m/s on the coast, 3.6 m/s inland and 7.9 m/s on mountain tops. Storms, with a windspeed greater than 17 m/s, follow the same pattern occurring at frequencies of 1.4, 0.3, and 10.6% respectively. However not all trees growing in the form of a table or which have been reduced to a height of 0.5–1.5 m must be considered as victims of climatic destruction. Often it is browsing animals which are mainly or even wholly responsible for damage of this kind, as became apparent from the discussion between Holtmeier (1967, 1968) and Turner (1968).

14. The **slipping of the snow cover** in the form of a small snow slide or a large avalanche damages tall woody growths more than anything but by doing so it favours smaller plants of the alpine belt (figs. 317, 171 and 172).

Many of the climatic peculiarities mentioned in points 1–14 also apply to the tundra north of the polar tree limit in Scandinavia, especially nos. 1, 5, 6, and 7 (Ives and Barry 1974, see also fig. 315). However there the air is not so thin and the radiation inwards and outwards is less intensive (3 and 4); and the light is less rich in ultraviolet rays (11). On the other hand the days are much longer in summer, and there is hardly any night frost at this time of the year. Nevertheless the difference in temperature between winter and summer in the far north is less than in the Alps because the sun is always lower in the sky. Consequently the ground, apart from steep southerly slopes, receives much weaker radiation, and the deeper soil layers remain constantly frozen. In spite of a much lower precipitation the plants in the tundra rarely suffer from a shortage of water (9). Just the opposite is the case because the ground is generally flat and the permafrost prevents the soil water from draining away. Coupled with the low rate of evaporation, there are many more swampy areas in the tundra than in the high mountains. It is little wonder then that the flora of the more stony and hilly tundra resembles that of the Alps rather more than does that of the open tundra plains.

The main difference between the arctic and the alpine climates is that the **permafrost**, while it is the rule in the former, is **seldom** found in the latter. At lower latitudes the sun is higher in the sky and the mountains get more radiation during the winter. Therefore in the Alps, even though there is a greater outward radiation during the night, on balance the soil is several degrees warmer than the air. Moreover most parts of the mountains are afforded a better protection by the snow cover, the thickness of which is between 2 and 8 times that lying on the tundra in winter (see also figs. 165 and 161). Only at the bottom of shaded holes (e.g. dolines) which hold the cold air or in screes of large stones, through which the cold air can flow down freely (see section B IV 3 d), is ground ice formed and sometimes persists throughout the summer.

Permanently wet and cold conditions in the soil restrict the breakdown of humus and the mineralisation of nitrogen. Consequently peaty formations with peinomorphic helophytes (see section C III 4a), low-growing dwarf shrub heaths and lichen communities are much more widespread in the Arctic than in the Alps and similar high mountains. Mosses find the conditions in the tundra quite favourable in so far as they are able to make use of warm sunny days outside the normal growing period (Oechel and Sveinbjörnsson 1975) Poor carpets of fructicose lichens (*Cladonia, Cladina* and *Cetraria* communities) are found only on windswept sites in the high mountains. In the more normal habitats of the alpine belt the mesomorphic meadow plants with their greater nitrogen requirement tend to dominate and overshadow the lichens (see esp. Rehder 1970).

Presumably it is the better supply of nutrients along with the warmer conditions which make the alpine vegetation appear so much more vigorous than that in the arctic regions; but there are no comparable figures available from tundra regions to substantiate this. Only the effect of organic and inorganic dust contained in the snow has been studied. When the snow melts this collects in the hollows and has some nutrient value (section 5).

(f) The development of soil and vegetation in the alpine belt

The removal of soil by water and wind, glacier movement, solifluxion due to alternate freezing and thawing, slipping of screes and falling rocks, flowing mud when fine soil becomes saturated with water and other erosion processes, go on continuously and extensively in high mountains like the Alps and Carpathians. The higher one climbs the more bare rocks and screes dominate the landscape, especially in those areas where the rock is more susceptible to splitting by frost, such as limestone and dolomite, (figs. 329, 331 and 359).

A consequence of this is that it is possible to study the beginnings of soil and vegetation development in many parts of the alpine belt. On the other hand ripe stable end stages are seldom met with. These would be restricted to gentle slopes and completely flat areas which are the exception here. In spite of this, or perhaps because of it, the question of the natural vegetation climax at the alpine level is constantly discussed. In order to understand the problem correctly we must make ourselves familiar with the peculiarities of soil formation above the climatic tree line.

As we have learned in the previous section the general climate of this belt is one of copious precipitation and cool, i.e. decidedly humid. All soils are leached by drainage water and therefore denuded of easily soluble salts including carbonates. Since the breakdown of organic matter is not completed within six months it tends to accumulate. Where the underlying soil is acid the lignin-rich litter forms a moder and finally a very acid raw humus (mor). On limestone on the other hand a weakly acid or neutral raw humus is produced which looks rather similar. Using a local name Kubiëna (1948) called it 'tangel-humus'.

On standing rocks or large boulders humus collects in cracks, on ledges or in small depressions, i.e wherever foliose or fruticose lichens, mosses or even phanerogams are able to gain a footing. However because of the slow plant growth at this altitude the humus layer also grows very slowly. Generally speaking the chemical aspect of soil formation, as opposed to the physical decomposition of the rock, needs a long time under high-alpine conditions. On solid rock it has scarcely passed the stage of a 'protoranker' (on silicate), or a 'protorendzina' (on limestone). Even on scree slopes it has progressed little further than this insofar as these arose during the late Ice Age or shortly afterwards and have been very short of sand and clay fractions right from the outset.

Well developed soil profiles are only found on moraines or on colluvial accumulations of fine earth at the foot of slopes, in karst depressions and similar places. Strictly speaking though these deep soils cannot be considered as autochthonic and thus as final stages which could have developed over the whole area. In the central Alps many of the fine soils can be seen as the remains of Tertiary soil formations, e.g. the terra rossa (a red soil which is rich in clay and crystallised iron hydroxide) often found over chalk. In many places the **alien cover of fine earth** forms a more or less thick sheet over the native rock. At first glance it is difficult to distinguish it from the products of rock weathering produced in situ. Only by looking at it more closely to discover small pieces of mineral which could not have come from the underlying rock can its true nature be established. The soil and vegetation development is quite different on such 'moraine sheets' or 'colluvial covers' compared with the bare rocks since from the very beginning they are able to provide better conditions of plant nutrition and water supply. Therefore they do not as a rule have a pioneer community of lichens and mosses, but are able to support phanerogams from the start.

Not only the original material but also the course of the soil formation process varies from one place to another. This face should be heeded when one tries to draw conclusions about a change in time (a succession in a narrow sense) from an investigation of change in space (a zonation or succession in a broader sense), both in respect of the soil and the vegetation. Braun-Blanquet and Jenny (1926) have pointed out how close the relationship is betweem the soil type and the plant association which grows on it. In what has become a classic exposition they describe the overlapping pH amplitudes of the '*Firmetum*', '*Elynetum*' and '*Curvuletum*', i.e. the Cushion Sedge (*Carex firma*) pioneer sward on limestone, the Naked Reed (*Elyna myosuroides*) sward on soils of medium acidity, and the Crooked Sedge (*C. curvula*) sward which can tolerate very acid soil (sections 2 b, c and 3 a). They interpret this series as a true succession, i.e. as a soil and vegetation development which could be run through from start to finish in one place.

Włodek, Strzemieński and Ralski (1931) have already cast doubts on these conclusions as far as the Tatra is concerned. They found pieces of granite and quartz in the top soil of acidophilous swards over limestone, an indication that the fine soil had not arisen from the carbonate bed rock but from morainic material which had been transported to the site. In another connection Szafer (1924) spoke of a 'floating covering of granite over the limestone massif of the Tatra'. As far as the northern limestone Alps were concerned Ellenberg (1953a) came to similar conclusions as Włodek and his co-workers. No case has been recorded so far where an acid-loving alpine community is to be met with on pure limestone unless there is a thin moraine cover or a colluvial silt layer in which the plants are rooting. A layer of raw humus is never thick enough at the alpine level to be sufficient in itself to prevent the lime-rich subsoil from having its effect on the vegetation (Wagner 1965).

Because of the extremely active physical weathering and erosion by water and wind the early stages of soil formation differ much more in high mountains than they do in the lowland. Since up there soil formation proceeds very slowly, there is an even smaller probability than in a warmer climate that on all the different substrates it would lead to one and the same climax. On this basis too we agree with Wagner (1958) and others in dealing separately with the vegetation on limestone rocks, chalk, dolomite, marl and other base-rich soils on the one hand and that of the rocks poor in lime on the other.

(g) Ecological and phytosociological classification of alpine vegetation

Frost, ice, snow, water and wind have all combined to make and are still changing large numbers of small habitats at the alpine level. Their number is increased even more by geology and relief as well as by man and his livestock. In order to simplify the general survey, in fig. 328 an attempt has been made to show the most important participating habitats in this changing mosaic in a semi-schematic transect.

Whoever strides up over the tree limit into the open world of the high mountains for the first time is bewildered by the wealth of unknown plants and plant communities which he meets step by step. As a student of vegetation he would be well advised to concentrate first of all on the few species combinations which can be clearly distinguished from their surroundings. Moving stony screes and stable grassland areas have few species in common. Also the damp cool hollows from which the snow is loth to depart even in July are well separated from the flowering carpet which became clear of snow some weeks earlier. It is no surprise that such little snow valleys were among the first to be seen as special vegetation units. As early as 1799 Hoppe had described them from the Alps near Berchtesgaden (see section 5).

Such obviously different vegetation types however seem to be exceptions in the alpine belt. A merging of various communities is more common than 'pure' associations. Therefore a survey is difficult to obtain as long as one has neither the experience nor the courage to simplify this great diversity in a sensible way. Here more than anywhere expert guidance is needed if one wishes to understand the mosaic of plant communities and not just to marvel at the beauty of individual flowers.

The main guidelines for the present day classification of the alpine vegetation are based on the work of Schröter and his students, especially Brockmann-Jerosch, Rübel, Braun-Blanquet, Lüdi, Gams and E. Schmid. Above all we shall be following here the classical survey of the plant communities of the Grisons (Graubünden) which Braun-Blanquet published in the years 1948 to 50 (and 1969).

For almost all alpine plant communities Braun-Blanquet found a **large number of character species**. This is true not only for the units of higher rank, especially the alliances, but also for individual associations so that it is often possible to recognise these by the presence of several species which are either confined to them or reach their best development here. By contrast in the lowlands one usually has to look at

Fig. 328. Characteristic plant formations and their habitats in the alpine belt showing the different kinds of soil and also the depth and duration of the snow covering (semi-schematic). Each habitat carries various plant communities depending on the base content of the soil (see tab. 74 and the following ones).

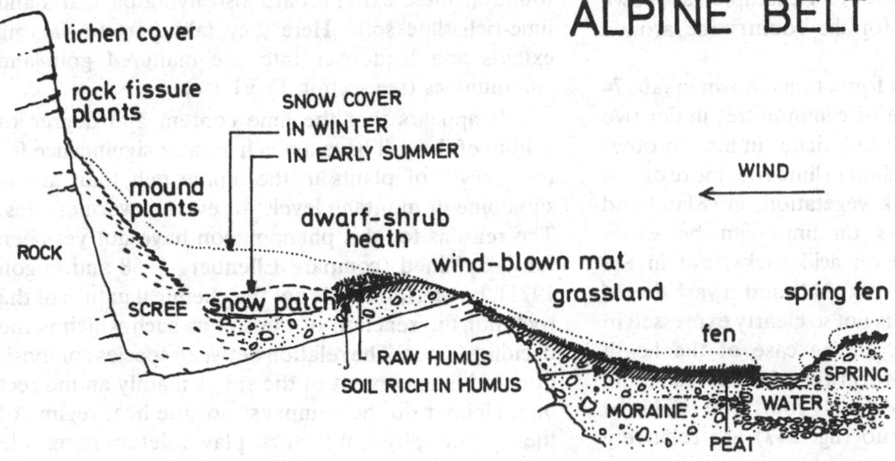

the whole combination of species in order clearly to allocate a community to a particular association. Only in very dry, very wet, or otherwise very extreme habitats are the vegetation units of the warmer belts well characterised. That which here is the exception becomes the rule in the severe climate above the tree limit. To some extent all plants are living in extreme habitats in the alpine belt, even when the soil is quite deep and without any special features.

The wealth of character species in the alpine vegetation may well have been an important factor in deciding Braun-Blanquet (1928, 1964) to base his system on the 'constancy principle', whereas the geobotanists in areas where there are fewer special habitats and a relatively poorer flora e.g. in the north, west and east of Europe have concentrated more on the dominance of certain species or on other criteria (see Müller-Dombois and Ellenberg 1974). In order to stress the floristically unique place of the alpine plant communities in most of the following tables we shall give a summary of the hierarchy of character species rather than lists showing the whole combination of species and their constancies (see esp. tables 75, 78, 70, 81-84).

Most of the classes, orders and alliances in Braun-Blanquet's classification coincide with those of other authors based on physiognomic-ecological factors. Since those who are not familiar with the alpine flora may find it easier to appreciate the main groups of habitats rather than a hierarchy of unfamiliar species we will start from the ecological relationships illustrated in fig. 328 and explained in more detail in terms of plant sociology in tab. 74. The individual plant formations and the way in which they have been altered by human exploitation can incidentally be distinguished with relative accuracy from aerial photographs such as Haefner (1963) for example has shown in a 'photographic key' for the countryside around Davos.

Each of the groups of formations shown in tab. 74 is arranged in a sequence of communities under two main substrate headings: those richer in lime or other carbonates and those deficient in lime (i.e. more or less acid). In the cases of rock vegetation, grassland and swamps the communities on lime can be easily distinguished from those on acid rocks, but in the communities of tall herbs (fig. 344) and dwarf shrubs (fig. 337) the type of rock is not so clearly expressed in the floristic composition. In the case of the heath shrubs this is because they form a continuous insulating humus layer in which they mainly root. The tall herbs and the Green Alder scrub (fig. 347) are relatively tolerant of the subsoil, and this is because running water supplies them with nutrients from time to time. With spring-fed communities and fens too the plants are much more responsive to the carbonate content of the water than of the underlying rock. Tables 75 and 77-82 are arranged in a similar manner to 74; in this way it is possible in each case to compare the two lists of plants from lime-rich and acid soils. The vegetation types influenced by water supply, duration of snow cover and other factors can also be easily compared.

In reality then the series of habitats depicted in fig. 328 is made up of two parallel series with alternative plant communities. The 'lime-sequence' is found not only on pure carbonate bed rock such as oolitic limestone and dolomite, but also on marls and young moraines as long as their top soil is still rich in lime. The second sequence is confined to rocks poor in bases such as granite, gneiss, sandstone, lime-free slates and leached moraines.

It is not surprising that the pure forms of each series are only found on the extreme types of rock. Recently communities which occupy a middle position have been described as legitimate associations by a number of authors, first by Zollitsch (1966). **Marls, lime-containing slates** and other rather soft geological formations have been weathered down to a soil which is generally deeper than over harder rocks. They carry dense meadows dominated by plants requiring a good water and nutrient supply (see fig. 329). Consequently they are the most valuable for agricultural use and have already been under the influence of farmers and their livestock for a very long time. In spite of their practical importance we shall not treat them as a third group. As long as the two extreme vegetation types are well known then it will be quite simple to recognise the intermediates, the more so since many of the species found at these extremes are also living on marls and lime-rich slate soils. Here they take part to varying extents and lead over into the manured grassland communities (see section D VI 1 c).

It appears that the lime content and degree of acidity of the soil have a much greater significance for the growth of plants in the alpine belt than at the subalpine or montane levels, or even in warmer sites. The reasons for this phenomenon have not yet been fully explained (compare Ellenberg 1958 and Gigon 1971), but certainly it is not the chemical nature of the rock nor the reaction of the soil as such which is the deciding factor. The relation between species composition and lime content of the soil is mainly an indirect one. Neither do the dampness nor the heat regime of the soil, i.e. physical factors, play a determining role since both the calcicolous and acid-indicating plants have a much wider climatic amplitude than that corresponding to the microclimatic contrast between two adjacent soil types, one with lime and the other without.

A difference in the water-holding capacity of the two soil groups may have a certain significance. On the whole limestone and dolomite mountains are drier than those made of silicate rocks since they have more vertical cracks through which water can drain away. They are also more inclined to have rock faces and scree slopes with little fine earth (fig. 349). Consequently the most common vegetation types on limestone mountains are those of screes and shallow soils, while those requiring more water are less frequent. Well developed lichens on the surface of rocks or phanerogams rooting in cracks are also less common in limestone areas because the rock surfaces are weathered so rapidly that these slow-growing plants do not have time to mature. However the differences between

Tab. 74. A synopsis of habitat types and the main plant communities of the alpine belt.

Main formations	Plant communities on **Carbonatic rocks**	**Lime-deficient** rocks[a]
1 **Rock – Lichen covering** *Rhizocarpetea geographicae* *Protoblastenietea immersae*	carbonate rock Lichen covering *Verrucarietalia parmigerae*	silicate rock *Umbilicarietalia cylindricae*
2 **Rock-fissure comm.** *Asplenietea*	carbonate r.f.c. *Potentilletalia caulescentis*	silicate r.f.c. *Androsacetalia vandellii*
3 **Scree communities** *Thlaspietea rotundifolii*	carbonate s.c. *Thlaspietalia rotundifolii* *Thlaspion rotundifolii*	silicate s.c. *Androsacetalia alpinae*
4 **Snow patch swards** *Salicetea herbaceae*	base-rich s.p.s. *Arabidetalia coeruleae*	acid-soil snow patches *Salicetalia herbaceae*
5 **Alpine swards**	calcareous alpine carpets *Elyno-Seslerietea* *Seslerietalia variae*	acid-soil alpine swards *Caricetea curvulae* *Caricetalia curvulae*
a) on shallow soil	Blue Sesleria slopes and similar *Seslerion variae*	Variegated Fescue slopes *Festucion variae* (also montane)
b) on deep soil	Rusty Sedge meadows and others *Caricion ferrugineae*	Crooked Sedge swards[b] *Caricion curvulae* (only alpine)
6 **Wind-swept places**	Naked Reed comm. *Oxytropi-Elynion*	Creeping Azalea carpets[c] *Loiseleurio-Vaccinion*
7 **Dwarf shrub heaths**[d] (subalpine – low alpine) *Vaccinio-Piceetea*	Bearberry heath comm. *Arctostaphylos alpina* Hairy Alpenrose heath comm. *Rhododendron hirsut.*	Crowberry heath *Empetro-Vaccinietum* Rust-red Alpenrose heath comm. *Rhododendro-Vaccinietum*
	Dwarf Juniper heath comm. *Juniperus alpina*	
8 **Fens** and similar *Scheuchzerio-Caricetea*	calc. Small Sedge fens *Caricetalia davallianae*	acid Small Sedge fens *Scheuchzerietalia*
9 **Spring communities** *Montio-Cardaminetea*	calc. spring comm. *Cratoneurion commutati*	soft-water spring comm. *Montio-Cardaminion*

Note: [a] There are intermediate soils and communities, especially on the screes and in the alpine swards.
[b] In the older literature the alpine Matgrass swards were also included here ('*Nardetum*')
[c] No swards and no true dwarf shrub heaths but espalier heaths
[d] Are classified along with the needle-leaved woods since they have so many raw humus plants in common; the systematics have still not been satisfactorily cleared up (see also table 75 and section 4 a).
Source: Mainly based on Braun-Blanquet (1948/50) section E III 1, no. 4.2, 4.4 – 4.7, 7.212.4 and 1.5).

Tab. 75. A synopsis of the alpine sward communities and their character species.

Alpine grasslands:	**Carbonate-soil**	**Acid-soil swards**
Classes	Elyno-Seslerietea l *Astragalus alpinus* g *Carex ornith.* var. *elongata* g *C. rupestris* *Dryas octopetala* *Gentiana nivalis* *Saxifraga adscendens* and other lime indicators	Caricetea curvulae[b] *Aniennaria dioica* ssp. *borealis* *Arnica montane* var. *alpine* *Botrychium lunaria* (?) g *Juncus trifidus* g *Luzula spadicea*
Orders represented in the Alps:	Seslerietalia albicantis *Acinos alpinus* *Alchemilla hoppeana* l *Anthyllis alpestris* *Arabis ciliata* *Astragalus frigidus* *Bupleurum ranunculoides* *Carduus defloratus* *Gentiana verna* var. *compacta* l *Hedysarum hedysaroides* *Helianthemum nummular.* ssp. *grandiflorum* p *Pedicularis verticillata* *Phyteuma orbiculare* *Polygala alpestris* *Pulsatilla alpina* *Scabiosa lucida*	Caricetalia curvulae[b] *Achillea moschata* g *Agrostis rupestris* *Armeria alpina* **D** g *Carex curvula* ssp. *eucurvula* *Gentiana acaulis* g *Koeleria hirsuta* *Laserpitium halleri* *Minuartia recurva* *M. sedoides* *Phyteuma hemisphaericum* *Poa violacea* *Primula integrifolia* *Pulsatilla apiifolia* *Ranunculus pyrenaeus* l *Trifolium alpinum* *Veronica bellidioides* and other acid indicators
Alliances on **static soils** snow-covered until relatively late:	Caricion ferrugineae (alpine-subalpine) *Astrantia major* (?) g *Carex capillaris* **D** g *C. ferruginea* p *Pedicularis foliosa* g *Phleum hirsutum*	Caricion curvulae[b] (alpine-nivale) *Androsace obtusifolia* p *Euphrasia minima* g *Juncus jacquinii* *Leontodon helveticus* (?) *Silene exscapa* (?)
Associations:	**Rusty Sedge meadow** *Caricetum ferrugineae* **Violet Fescue sward** *Festuco-Trifolietum thalii*	**Haller's Fescue sward** *Festucetum halleri* **Crooked Sedge sward** *Caricetum curvulae*[b])
Alliances of subalpine-alpine swards on **shallow-soil slopes** clear of snow early:	Seslerion albicantis[c] g *Carex atrata* *Gentiana orbicularis* *Globularia nudicaulis* *Helianthemum alpestre* *Leontopodium alpinum* *Ranunculus thora*	Festucion variae[c] (rare in northern Alps) *Bupleurum stellatum* p *Euphrasia alpina* p *E. pulchella* *Potentilla grandiflora*
Associations:	**Cushion Sedge pioneer sward** *Caricetum firmae* **Blue Sesleria-Evergreen Sedge slope** *Seslerio-Caricetum sempervirentis*	**Variegated Fescue slope** *Festucetum variae*

Tab. 75. (Continued).

Alpine grasslands:	**Carbonate-soil**	**Acid-soil alpine heaths[d])**
Alliances on **wind-swept** knolls and other places rarely snow-covered:	Oxytropi-Elynion[c]) *Antennaria carpatica* g *Carex capillaris* var. *minima* *Cerastium alpinum* *Draba siliquosa* **D g** *Elyna myosuroides* *Gentianella tenella* *Pulsatilla vernalis* *Saussurea alpina* (?)	Loiseleurio Vaccinion[d]) (of the Vaccinio-Piceetea) resp. Cetrarion nivalis **D** *Loiseleuria procumbens* f *Alectoria nigricans* f *A. ochroleuca* f *Cetraria ericetorum* f *C. cucullata* f *C. nivalis* *Thamnolia vermicularis*
Associations:	**Naked Rush windy corner** *Elynetum (alpinum)*	**Mountain Azalea wind-carpet** *Loiseleurio-Alectorietum ochroleucae*

Note: [a] In this table and in most of the following summaries only the species characteristic of the classes, orders, alliances and some associations as they are valid for the Swiss Alps have been listed and not those 'companions' (in the sense of Braun-Blanquet) which turn up in numerous different communities. Many of these accompanying species such as *Campanula scheuchzeri*, *Crepis aurea*, *Plantago alpina*, *Poa alpina*, *Polygonum viviparum*, *Soldanella alpina* and *Selaginella selaginoides* are frequent and more or less closely limited to the alpine and subalpine belts. Thus they help to mark out the vegetation of the high mountains as something special.

g = grasses or grass-like, l = leguminous, p = semiparasite, f = lichens, m = mosses. **D** = dominant species.

[b] The alpine acid-soil swards merge into the Matgrass swards (*Nardion*) in the subalpine belt; the two have many acid indicators in common.

[c] The grasses after which these alliances are named i.e. *Sesleria albicans* and *Festuca varia* also appear in other communities (see e.g. section B II 2 e). Thus they cannot act as character species either for the alliances or for higher units. Perhaps they may be seen as characteristic for the Blue Moorgrass-Sedge slopes and the Variegated Fescue slopes since in these they are usually dominant. The Evergreen Sedge (*Carex sempervirens*) can be looked upon in a similar way.

[d] Floristically the wind-swept acid dwarf shrub heath is closer to the needle-leaved woods than to the alpine swards. In many respects it forms an ecological parallel with the *Elynetum*. Recently it is classified as *Loiseleurio-Alectorietum*, i.e. as an association mainly formed by lichens and belonging to the alliance *Cetrarion nivalis*.

[e] This has recently been considered as a class of its own distributed throughout the whole of the holarctic (*'Carici rupestris Kobresietea bellardii'*, Ohba 1974).

Source: From data supplied by Braun-Blanquet (1948/50), Oberdorfer (1957) and others (the evaluation of some species has changed: see E III 1, 4.6 and 4.7 and 4.712.4).

the plant associations on deeper soils rich or poor in lime cannot be explained by the peculaier properties of more ore less bare lime or silicate rocks.

The **causes of the obvious differences between grassland communities on lime-rich and on acid soils** have been the subject of lively discussion since the days of Wahlenberg (1813), Unger (1836) and Sendtner (1860). In an attempt to throw some light on the problem Gigon (1971) has carried through a number of experiments involving transplanting and pot culture. For these he chose an alpine meadow above Davos at an altitude of 2355 m where slopes with similar aspect on calcareous and acid soils lie next to each other.

One experiment involved transplanting from one soil type to the other. For example Blue Moorgrass (*Sesleria albicans*) which is typical of subalpine and alpine grassland on calcareous soils was planted on a very acid soil covered with Matgrass (*Nardus stricta*), while the latter was planted in the calcareous area. Competition was prevented by removing all adjacent plants, and as long as this was done both species grew quite well even on the soil foreign to them (see fig. 330). Of course, at the beginning of the trial all the plants suffered a check on being transplanted because the roots had to be freed from the original soil. This was also true of controls which were replanted on their home ground. During the following years *Sesleria* grew practically as well on the acid soil as on the calcareous, but *Nardus* on the latter grew less vigorously and turned yellowish (that means it became chlorotic). The Matgrass was suffering from a shortage of iron, a phenomenon also often observed in warmer climates when acid-tolerant species are grown in calcareous soil. It can be demonstrated quite easily that it is the plant's inability to obtain enough iron from a soil rich in carbonates by giving iron in a chelated form (in a colloidal solution) when the chlorosis disappears and the plant growth obviously improves.

Fig. 329. Soft marl with good alpine pasture (right) and rugged dolomite slopes with screes and the remains of woodland (left) on the Widderstein (2536 m) in the Smaller Walsertal. Photo. Metz.

Fig. 330. In culture experiments acid indicators such as *Nardus* and *Geum* will also grow in calcareous soils, while lime indicators like *Sesleria* and *Scabiosa* will grow on strongly acid silicate soil. However when grown together there is mutual suppression even within the first year, and after four years this is complete (further details in the text). After Gigon (1971), modified. The relative yield removed from each pot is shown in the ordinates.

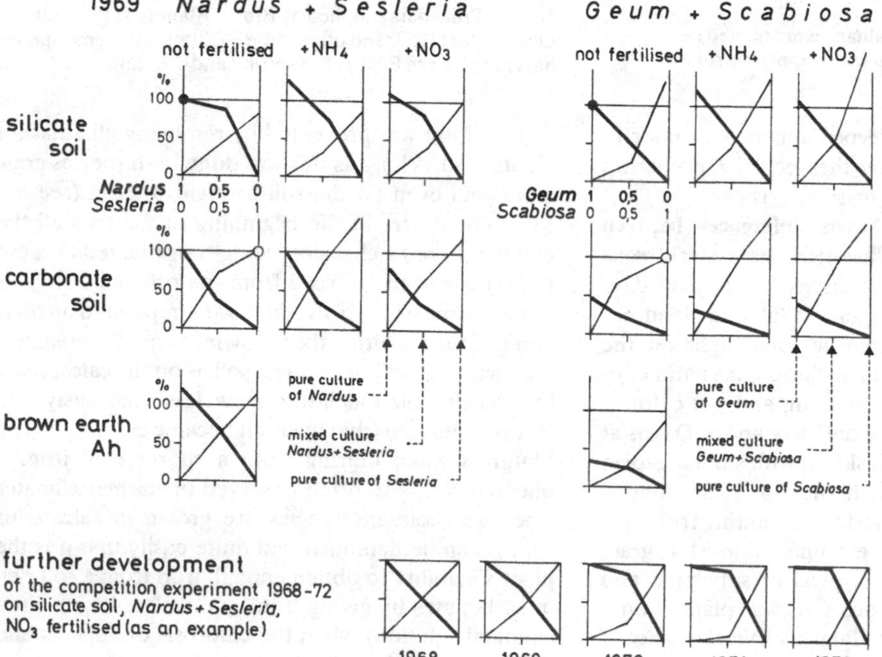

Other species behaved in a similar way to *Sesleria* and *Nardus*. In order to confirm his findings, Gigon brought the same plants and soils down to Zürich (at about 550 m) and planted them up in square pots which stood close together in order to eliminate the edge-effect. Under these conditions where he was able to exercise greater control, he planted up both types of plants in the two kinds of soil as pure cultures and in two and two mixtures. As can be seen from fig. 330 all the plants grow more or less well in pure culture even on the 'foreign' soil, but in the mixed culture the calcicoles soon asserted their dominance when grown on the calcareous soil and the calcifuges likewise on the acid soil.

After experiments carried out by Bogner (1966) with woodland plants it was assumed that the form in which nitrogen was offered to the plants could be significant for their ability to thrive in soils of different pH values (section B II 4 d). According to Rehder (1970) only ammonium is produced in acid soils of the alpine and subalpine Matgrass pastures whereas the calcareous soils with Blue Moorgrass offer nitrogen mainly in the form of nitrate. (In spite of the low temperatures nitrification goes on actively at the alpine level as Labrone and Lascombes (1971) also found.) However Gigon (1971) discovered in his pot experiments that providing *Nardus* with nitrate was not harmful in any way, and that ammonium could be given to both calcifuges and calcicoles in either soil without any disadvantage accruing (see fig. 330). The results are all the more remarkable in that Matgrass in particular is a 'nutrient avoiding' poverty-indicator (Ellenberg 1974, 1979), and also the other plants used by Gigon in his experiments disappeared from the alpine meadows when these were heavily manured.

It appears then that, in the final analysis, it is competition which decides the way in which the plants of alpine and subalpine pastures behave, just as it plays the decisive role in woodlands and meadows or on the farmed land with its weeds at lower altitudes where earlier experiments have been carried out (see sections B I 2 b, D I 6 b and V 6). Gigon has demonstrated the fact that it is **the behaviour of the 'acid-soil plants'**, such as Matgrass, which **is decisive** in determining the distribution of species in the alpine grasslands. Where the soil is sufficiently acid to suit them they crowd out the 'lime-soil plants' which could grow quite well here if competition were excluded. Where the soil is rich in lime, on the other hand, the vitality of the acid-soil plants is so weakened that they die out in time even without competition in places where they have managed to gain a footing at first. In those species tested by Gigon neither the lime content nor the acidity of the soil has any effect on germination. It is their later growth which decides their fate, as Fossati (1980) recently confirmed. Probably aluminium, becoming soluble at low pH, adds to the handicaps of chalk plants on very acid soils (M. Runge 1981).

Many of the species growing in the subalpine calcareous and acid-soil grasslands are not so sensitive as the ones just mentioned, and in fact many of them appear to the keen observer to be rather indifferent as to soil type even under natural conditions (see Ellenberg 1979). Gigon found a lot of plants which occur mainly in alpine calcareous grasslands also growing on soils where the pH value was only about 5.5; that means for purely chemical reasons, there could be no free lime.

Our general discussion on the behaviour of lime- and acid- indicators was necessary in order to avoid repetition when we turn to the following more detailed treatment of alpine plant communities on substrates rich or poor in carbonate. Since it is the grassland formations which are so characteristic of the alpine belt we shall begin with these. After that we shall turn to the snow bed vegetation, and to the dwarf-shrub heaths, both of which dovetail in so much with the grassland communities. For the same reason we shall follow with the fens, spring area communities and tall herb stands. It is only when we are familiar with all these communities that we can understand the successions which start from young moraines, from wet ground, from moving screes and from bare rock. We shall defer a consideration of the stone scree communities until last, since these lead naturally on to the nival belt, where, as we shall learn; they have their greatest distribution.

2 Subalpine and alpine grassland on carbonate soils

(a) Blue Moorgrass-Evergreen Sedge slopes

In the eastern Alps the outer chain is made up largely of limestone and dolomite while the inner mountains are older and non-calcareous. From about the Swiss frontier westwards however the difference is not so clear cut. Carbonate rocks also form large complexes in the inner Alps, e.g. the Swiss National Park, which has already been the subject of intensive study for more than 50 years (Baer 1962), while lime-deficient rocks permeate the limestone-rich outer chains.

Wherever limestone soils occur in the alpine belt one meets Blue Moorgrass-Evergreen Sedge slopes rich in species and with a bright show of flowers which always sends mountain walkers into raptures (fig. 331, Schönfelder 1970). This goes for other high mountains in Central Europe too, e.g. the High Tatra and the Carpathians (Szafer and co-workers 1972, Zlatnik

1928, Beldie 1967, Horvat, Glavač and Ellenberg 1974). Along with Curved Sedge swards, which dominate the acid rocks (section 3a), the Blue Moorgrass-Sedge slopes (*Sesleria albicans-Carex sempervirens* ass., abbreviated to *Seslerio-Semperviretum*) are thus the most widespread grassland associations at the alpine level. Both these natural pasture communities are very characteristic of the High Alps since they are absent from the lower European mountains and from the tundra. It is true that many of their species found in the Alps are also to be met with in the far north, but here they are partners of other communities. This is mainly due to the fact that the alpine belt enjoys a much higher insolation than the northern tundra (see Cernusca 1977).

Arctic limestone grasslands and those at the alpine level both belong to the class *Elyno-Seslerietea* whose characteristic species are given in tab. 75. In Central Europe this class is represented only by the order *Seslerietalia* and the species by which this is recognised are also given in tab. 75. Since the Blue Moorgrass dominates in so many of the communities it cannot be used as a character species for any one unit. Its ecological amplitude is extremely wide and it appears even in Beech woods and on rocky mountain slopes at the submontane level (section B II 2 e). Very few other alpine elements accompany it so far, and the rest of the species combinations here would hardly remind one of the *Seslerio-Semperviretum* (see Schönfelder 1972).

Fig. 331. Cushion Sedge pioneer sward (*Caricetum firmae*) on limestone crag, a stepped Blue Moorgrass slope (*Seslerio-Semperviretum*) and a dense Rusty Sedge sward (*Caricetum ferrugineae*) on a deeper and damper soil (left foreground) in the Smaller Walsertal (northern limestone Alps).

The **Blue Moorgrass-Evergreen Sedge slopes** have so much in common with the Cushion Sedge swards, which will be dealt with in section b, that they must be put together in the same alliance, the *Seslerion albicantis*. As can be seen in many nature conservation areas Edelweiss (*Leontopodium alpinum*) is well developed in both communities. However it has had to retreat to more rocky parts for it was too easily accessible on the grassland, and has been practically wiped out by visitors. It can serve as a local character species for the *Seslerio-Semperviretum*. Braun-Blanquet (1948/50) also names *Oxytropis montana* (although this is more plentiful on scree slopes than in such swards), and *Pedicularis rostrato-capitata* as further character species, along with a few rarities.

Amongst the species by which the alliance and the order can be recognised some of the most obvious are the yellow-flowering *Helianthemum grandiflorum* and *H. alpestre* and *Anthyllis alpestris*. In different shades of blue come the Gentians, the Ball Flowers (*Globularia*), the Rampion (*Phyteuma obiculare*) and some others. Red flowers are not so common, but a few character species have them, e.g. *Carduus defloratus* and *Pedicularis verticillata*. The last named is a semi-parasite, sending haustoria into the root vascular system of sedges and grasses from which it extracts mainly water. *Sesleria albicans* and *Carex sempervirens* are by no means confined to the community which is named after them but they do make up a large part of its total production. With their dense root systems the Blue Moorgrass and the Evergreen Sedge collect the fine earth on steep stony slopes and slow down the erosion rate. According to Lüdi (1948), Thomaser (1967) and Zuber (1968, see fig. 327) not only is the flowing of the snow, but also that of the soil easily seen on such slopes. The soil is exposed to frost and sun for a long time because as a rule it becomes clear of snow early. The two dominant species can withstand the occasional drying out, which can always be a danger on the steep sunny slopes, without permanent damage. *Erica herbacea* (= *carnea*), the spring Heath, *Polygala chamaebuxus* and *Carex humilis* also appear to be particularly resistant in this respect. They are more strongly represented in the Blue Moorgrass swards of the drier central parts of the Alps, e.g. in the Swiss National Park, than in the damper outer chain.

Since the dominant species are very effective in trapping fine earth the Blue Moorgrass slopes always take on a slightly stepped appearance. On these small terraces which are more or less level and hardly shaded at all the less competitive partners in the community are able to establish themselves. This is why the *Seslerio-Sempervireta* are so rich in species and why their composition can vary so much. Grazing animals tread down the steps into narrow tracks and continually damage the surface but even more damage is done by the freezing and thawing and the wetness of soil on lime-rich moraine soil after the snow has melted. This results in pieces of the sward sliding down revealing bare soil. This usually happens in May or June and occasionally in the autumn. Cases of such 'garland solifluxion' can be studied very well in the Swiss National Park, e.g. on Munt La Schera and in the Val dal Botsch (fig. 327). Where the pieces of turf pile up they form a coarse network of strands consisting only of densely rooted fine earth and not at all typical of the Blue Moorgrass sedge slopes.

The more effective the solifluxion and erosion the more the soils of alpine Blue Moorgrass slopes can be described as proto-rendzinas or proto-pararendzinas. More stable and thicker swards produce a mull rendzina (Zöttl 1966, see fig. 60) and because of their better water-holding capacity they often have representatives of the Rusty Sedge meadow (section c) coming in. In the northern limestone Alps this sort of invasion is not infrequently seen whereas in the central Alps it is more the typical association or the subassociation *caricetosum humilis* which is met with.

Many of the dolomite and limestone massifs are surrounded by huge scree slopes reaching right down into the subalpine belt (fig. 349). As long as the *Seslerio-Sempervіretum* can maintain a footing on them it follows them right down into the warmer valleys. The Blue Moorgrass slopes of the Juras at the high montane and subalpine levels are reminiscent of those of the Alps, and they lead on to the submontane and montane Blue Moorgrass slopes of the Central European limestone mountains which have already been mentioned in section B II 2 e (J.L. Richard 1972).

A special alliance of limestone grassland, the *Caricion austroalpinae*, belonging to the order *Seslerietalia* has been described by Sutter (1962, 1969) from the Insubrian area of the southern outer Alps. The four associations belonging to this, which are found between Lake Como and mt Grappa are characterised by some endemic taxa. In the Carpathians the *Seslerion albicantis* communities are represented by closely related units of the *Seslerion bielzii*, which Pawtowski (1935, 1959) named after an endemic variety of the Blue Morrgrass and has described in more detail.

The main partners in the *Seslerio-Sempervіretum* are unable to colonise large bare areas without the help of pioneers which we shall discuss in the next section. On stony carbonate soils of the higher peaks an impoverished subassociation of the *Seslerio-Sempervіretum* is found in which *Festuca pumila* or *Silene acaulis* is dominant.

(b) Cushion Sedge and Mountain Avens carpets in extreme habitats

On limestone rocks which become free from snow early in the year and get the full force of cold and wind the Blue Moorgrass slope communities cannot grow properly even though *Sesleria* itself and many of its partners are able to withstand the extreme local climate (figs. 331, 332). It is here that the dome-shaped cushions of the **Cushion Sedge** *(Carex firma)* are successful. According to Rehder (1976) they are able to root into the remains of their dead parts inside the cushions; this means that the plants are highly economical with the nutrients contained in their self-made humus. On the drier carbonate rocks of the southern Alps this species is replaced by *Carex mucronata* which is even less sensitive to drought.

These gappy slow-growing communities of the outposts of vegetation have already been described and evaluated by Kerner von Marilaun (1863). *Carex firma* and other low-growing rosette or cushion plants such as *Gentiana clusii, Saxifraga caesia, Crepis kerneri* and *Chamorchis alpina* serve as character species. The *Caricetum firmae* (*Firmetum* for short) can be found on many limestone and dolomite peaks, but is often very fragmented. Its roots are not active enough to withstand solifluxion and form strands like the *Seslerio-Sempervіretum* (section a) but are just passively torn away (Zuber 1968).

On stable stony ground however *Caricetum firmae* is usually a permanent community and has endured as such for centuries without being able to develop any further towards a closed sward. This is because of the extreme shortage of fine soil which is subject to drought and wind erosion. It would be wrong then to see it as a pioneer in a succession series of communities. However in less-exposed places where there is a young raw calcareous soil cushions of *Carex firma* do prepare for a subsequent sward formation. Incidentally the Cushion Sedge is by no means a lover of dry conditions and can grow on quite wet ground. Where the soil is deeper but still rich in bases the Cushion Sedge stands dovetail in with the Naked Rush sward (*Elynetum*), although this mixture too is not a question of a succession (see section 1 e).

Very often the **Mountain Avens** (*Dryas octopetala*) acts as a pioneer both for the *Caricetum firmae* and the *Seslerio-Sempervíretum*. This tough dwarf shrub, which is also found in subarctic regions, anchors itself deeply in rock cracks or between the stones of a scree, covering the surface with an espalier of branches. Its dead leaves contribute to the formation of a protorendzina rich in humus which may then be occupied by other phanerogams.

(c) Wind edges with the Naked Rush

Like *Caricetum firmae* the **Naked Rush sward** (*Elynetum*) is able to live in dry habitats exposed to strong winds, but it needs sufficient fine earth offering a good root run (fig. 332). Such 'windy corners' can be recognised from a distance with their shining brown colour, produced by a dense sward of the tough-stemmed Naked Rush (*Elyna myosuroides*).

Amongst the character species of the *Elynetum* can be numbered, according to Braun-Blanquet, not only *Carex atrata, Dianthus glacialis* and *Potentilla nivea*, but also *Erigeron uniflorus* and *Saussurea alpina*, which overlap into other alpine and subalpine grassland. The Naked Rush itself, locally very characteristic in the Alps, under European aspects can only serve for recognising the class as it is also very widespread in arctic and subarctic regions where it forms related communities (see also tab. 75[e]).

The *Elynetum* is often found on wind-exposed but deep soil such as a protruding ridge or the edge of a larger turf area subject laterally to storm-force winds. The icy blasts very often undercut the edges of these swards in a way similar to that of a swiftly flowing river undercutting its banks. The turf itself with its matted roots resists erosion more stubbornly than the fine earth beneath it. Such formations are caused by the wind following damage to an otherwise closed sward by such events as solifluxion, earth slip, heavy treading, skiing on snow-free spots, or being cut artificially. Part of the fine earth which is blown from the subsoil may be trapped by the *Elyna* turf and restabilised. *Elyna* however plays scarcely any part as a pioneer and soil stabiliser on the wind-whipped bare soil of the exposed hill tops. Its strength lies mainly in defence and most Elyneta must be seen as permanent communities or as a stage in the degeneration of a sward which had formed originally under less extreme conditions.

On windy corners where the Naked Rush sward is found evaporation is always higher than on normal sunny slopes. This is true even in more shaded positions or on days when the wind is only slight to moderate. Because of the cooling effect of evaporation the higher wind velocity results in more dew falling during the night or in the early morning than is the case in less exposed places. As is also the case with the grasslands on continental steppes the water economy

Fig. 332. Cushion Sedge sward (*Caricetum firmae*, left), wind-swept Naked Rush edges (*Elynetum*) and fragments of a Blue Moorgrass sward (*Seslerio-Semperviretum*, right) form a mosaic, including hollows still full of snow at the beginning of August, on the weathered limestone of the Ifen (Smaller Walsertal) at about 2100 m.

of *Elynetum* is under less stress in the mornings than at midday.

The large number of high-alpine species (such as *Oxytropis montana* and *Draba fladnizensis*) which are able to live in such Naked Rush habitats are responding to a continental microclimate according to Meusel (1952b) and E. Schmid (1961). Like *Elyna* itself, the *Elyna* communities have their main distribution in the southern Siberian mountains. Incidentally Meusel points out in this connection that *Elyna* is not confined to deep soil, but also occupies stony ridges, especially those made of stone with a medium lime content. *Elyna* mingles with elements of steppeland heaths preferably on the lower peaks of the northern limestone Alps (see section B III 4 and DI 2).

There are large areas of the Naked Rush swards in Iceland and in northern Norway, especially at the foot of slopes and sometimes only a little way above sea level. This area is another centre of their distribution, and they occur where the thin covering of snow is repeatedly blown away during the winter. Whether this absence of a snow covering is a necessary condition for the growth of *Elyna* or whether this species is just able to withstand such conditions better than other sward-forming plants, which in turn would be more competitive under the protection of snow, has still to be investigated experimentally.

(d) Rusty Sedge meadows and other mesophllous carbonate-soil grassland

On lime-rich rocks in the alpine belt of the outer Alps one sometimes meets with a more vigorous grassland, which is worth mowing, and is correctly named natural meadow, but it only occurs where a number of favourable conditions combine. The climate must have sufficient cloud and rain and the soil must be deep enough so that it never dries out. Where the climate is drier then to some extent this can be counteracted by a locally shady site or a certain amount of groundwater being available. In any case the snow must remain to protect the meadow until no more keen frosts are to be expected. On the other hand the snow must not lie so long that the summer, which in the alpine belt is short, could not be fully used for photosynthesis. These preconditions are only to be met with on gentler slopes or in depressions on the steeper ones, and even here only in the lower part of the alpine belt and in the subalpine belt where it is free from trees. Such conditions are best realised in the more oceanic marginal Alps (fig. 331), and it is here that all the communities to be discussed in this section are richest and typical. They are also met with in the Swiss Jura although here they are rather fragmentary.

It can be seen from tab. 75 that these natural hay meadows on calcareous soils are placed, along with the *Seslerio-Semperviretum*, in the order *Seslerietalia*. However they are put into their own alliance, the *Carion ferrugineae*. The typical **Rusty Sedge meadow** (*Caricetum ferrugineae*) has been investigated in depth by Lüdi (1921, 1936) in the Lauterbrunnen valley and on the 'Schynige Platte' in the Bernese Highlands *Carex ferruginea*, with its shallow rhizomes and its long leaves hanging down the slope like combed hair, forms a dense productive sward. Like the Blue Moorgrass slope the Rusty Sedge meadow also contains many species with bright flowers, amongst which a number of Leguminosae are particularly noteworthy. The alliance can be recognised by some of these, namely *Hedysarum hedysaroides, Astragalus alpinus, A. frigidus*, and *Lathyrus laevigatus*. Mesomorphs with taller growth and broader leaves such as *Astrantia major, Centaurea montana* and *Anemone narcissiflora* indicate a good water and nutrient supply. According to Rehder (1970) though the average production of mineral nitrogen in the top soil is no greater than with the Blue Moorgrass-Sedge or the Naked Rush swards (see tab. 76). The higher production of *Carex ferruginea* rests in the first place with the favourable water balance (Gökçeoğlu 1975).

Gradually the Rusty Sedge meadow merges into even more productive communities of the tall herbs (section 7 a). The presence of *Dactylis glomerata* and *Pimpinella major* indicates a connection with the Yellow Oatgrass meadows (*Trisetetum*) of the lower vegetation belts (section D V 3 b).

Meadows of the fastidious *Carex ferruginea* are seldom met with in the central Alps and then only here and there on small areas with a locally better water regime. They are almost entirely absent from the Swiss National Park, one of the driest parts of the Alps. Best developed examples on the other hand are found in the outer chain of the Alps where the rainfall is highest, but even here they are not on the very free-draining pure carbonate rocks but rather on calcareous slates and similar rocks which are softer and tend to be weathered more easily (fig. 329).

On such deep soil there are intermediate grassland types lying between the Rusty Sedge meadows on the damper soils and the Blue Moorgrass communities which are always found on relatively dry ground. These intermediate communities have been described from the northern Swiss Alps as **Violet Fescue pastures** (*Festuca violacea-Trifolium thalii* ass., tab. 75). They are similar to the 'Milkwort pastures' (*Leontodon hispidus* communities, see section D IV 1d) and with more intensive farming will change into these. Usually they have been or still are grazed and only mown in places where access is difficult – so-called 'wild hay patches'.

The soil carrying this Violet Fescue grassland has the character of a deep-browned rendzina or pararendzina. It can sometimes be almost like a brown earth and occasionally has gley-like characteristics. Very often the top soil no longer has a neutral reaction but is slightly acid, and, because of this, one often finds acidity indicators such as *Geum montanum, Potentilla aurea, Gentiana acaulis, Leontodon helveticus* and *Ligusticum mutellina* which have wandered in from the grasslands on acid rocks. The Violet Fescue pasture is rather like the brown mull Beech woods at lower levels in Central Europe (section B II 3a) in taking up an intermediate position between the communities on lime-rich soils and those on acid soils. However it is less frequently met with than the latter because in the higher mountains the unstable conditions mean that places where a deep soil of near neutral reaction can form are the exception.

Like so many of the plant communities which have been previously highly valued by farmers in high mountains, both the Violet Fescue and the Rusty Sedge meadows are now considered no longer worth cutting or even grazing. Most of them are no longer being mown and the dead leaves and stalks form a slippery surface down which the accumulated snow can easily slide. Quite often the snow brings with it stems and even young trees which have become frozen into it, tearing them out of the ground (fig. 498). Thus the abandonment of extensive husbandry in these areas can lead to increased danger of avalanches and erosion (see Laatsch and Grottenthaler 1973 and Spatz 1975).

3 Alpine and subalpine grassland on acid soils

(a) Crooked Sedge swards

Over crystalline rocks and other lime-deficient substrates the soils are generally poorer in nutrients, but have a better water-holding capacity than those in similar situations over lime-rich rocks. Apart from bare rocks on the higher peaks these acid soils in the alpine belt are usually completely covered with carpets of plants.

Apart from the transition zones between the subalpine and alpine levels which are rich in dwarf shrubs and Matgrass, it is the **Crooked Sedge** *(Carex curvula)* which is the strongest of the sward plants (fig. 333). Its olive-brown leaves never have a fresh green colour, and its twisted tips are nearly always dead because they are attacked by the ascomycete *Pleospora elynae*. They give the *Caricetum curvulae* (abbreviated to *Curvuletum*) a monotonous, impoverished appearance which can be recognised even at some distance. It is hardly ever eaten by cattle and even the less-demanding sheep and goats find it insufficient most of the time.

According to Trepp (1950) the feeding value of the Crooked Sedge pasture is only about 3.6% of that of the Violet Fescue pasture. Only the Cushion Sedge community has a lower value (1.2%), whereas the Blue Moorgrass-Evergreen Sedge pasture must be valued at considerably more (27.8%). Although these are only approximate figures varying from place to

Tab. 76. The supply of mineral nitrogen to alpine and subalpine plant communities.

* = extensively grazed areas; ** = areas frequently sought out by cattle; *italics* = yield up to 3t/ha; **semi-bold** = yield over 5t/ha

N-supply classes N available per annum (kg/ha)	I up to ca. 25	II up to ca. 50	III up to ca. 100	IV up to ca 250
Mostly shallow lime-rich soils	Seslerietum I (shallow soil)	Firmetum	Elynetum**	
↓		*Seslerietum I* (deep soil)	*Seslerietum II**	
↓	*Car. ferrug. I**	*Caricetum ferrugineae II***		*Poetum***
↓	Rhod.-Mugetum			
↓	*Nardetum**			**Rumicetum****
Mostly deep soils, partly leached	Rhod.-Vaccinietum			**Alnetum vir.**

Source: After Rehder (1970).

place and calculated from estimated data they are of the right order and comparable with each other. From them we can readily understand why common grazing can be carried out more intensively on marls and slates than on hard rocks whether these are crystalline or calcareous (see figs. 329 and 324).

The Crooked Sedge sward is very characteristic of the Alps and is absent from the north as well as from the Tatra and the mountains in the Balkan peninsula. Outside the Alps it is replaced by other communities of the order *Caricetalia curvulae*. To some extent the role of the Crooked Sedge is taken over by *Juncus trifidus* and *Oreochloa disticha* in the '*Trifido-Distichetum*' of the Tatras (Pawłowski 1935). In the Alps however these species play a very humble part alongside the all powerful *Carex curvula*.

Flowering plants appear only timidly in the *Curvuletum* carpet with its perpetual autumn colour. The most easily seen are the yellows of *Leontodon helveticus, Arnica montana, Potentilla aurea* and *Senecio incanus* ssp. *carniolicus* or of the tiny annual semiparasite *Euphrasis minima.* var. *minor* which germinates in large numbers. Light and dark blues are dotted about in between by *Veronica bellidioides, Phyteuma* species and *Campanula barbata*.

All the partners in the Crooked Sedge sward and related communities are able to tolerate very acid soil and to fill their nitrogen requirements from the raw humus. This is true not only for *Carex curvula* itself but also for all the species by which the classes, orders and alliances given in tab. 75 can be recognised. Gilomen (1938) distinguishes between a Crooked Sedge of the acid rocks (ssp. *curvula*, pH 4-6.8) and a limestone Crooked Sedge (ssp. *rosae*, pH 4.8-8.6) which is distributed mainly in the western Alps.

According to Braun-Blanquet the typical *Curvuletum* contains *Oreochloa disticha, Potentilla frigida, Hieracium glanduliferum* and other character species as well as differential species such as *Ligusticum mutellinoides* and *Gnaphalium supinum* which require a certain minimum moisture level. Where the wind can penetrate freely and frequently blows the snow away the sward becomes patchy and even shorter so that dwarf shrubs and lichens of the Azalea wind heath (section 4, see fig. 328) can gain a footing. These dominate in a lichen-rich *Curvuletum cetrarietosum*, especially *Cetraria ericetorum, C. cucullata, Thamnolia vermicularis* and *Alectoria ochroleuca*. In places where the soil is deeper or somewhat richer in lime *Elyna myosuroides* can become dominant.

Most frequently *Curvuletum* becomes mixed with *Salicetum herbaceae* (section 5 b) especially where the snow lies longer, leading to the formation of snow patches. Here *Luzula alpino-pilosa, Polygonum viviparum, Salix herbacea* and others serve to separate this moist-soil Crooked Sedge sward from the normal one.

Fig. 333. Crooked Sedge sward (*Caricetum curvulae*) on a gneiss-rich moraine soil on the Susten pass. Left foreground *Vaccinium uliginosum* and *Loiseleuria procumbens*, in the centre *Leontodon helveticus*. (The detail is 80 cm wide.)

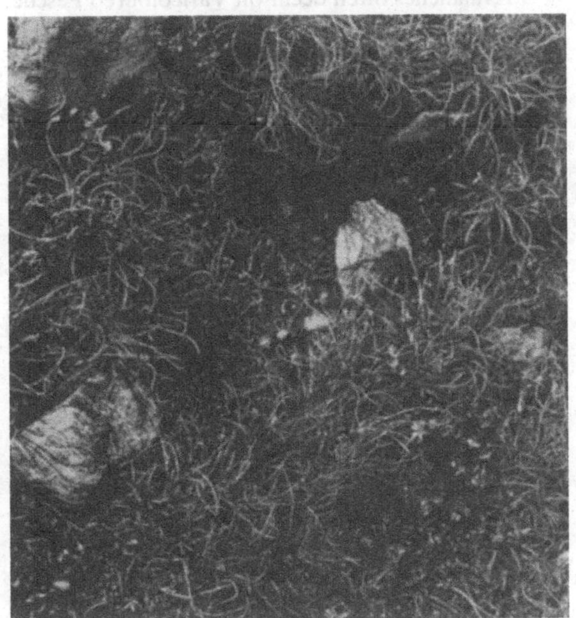

As a widely distributed grassland community in the alpine belt the Crooked Sedge pasture has many times been the subject of thorough investigation (Rübel 1912, 1922, Braun-Blanquet and Jenny 1926, Oberdorfer 1959, Rehder 1970). From these the following significant points arise:

Although *Carex curvula* itself, even the extremely acid-tolerant subspecies, occasionally spreads over onto neutral soils the *Caricetum curvulae* is always associated with an acid substrate (see also section 1 f). The pH value of its humus-rich topsoil rarely exceeds 5.5 and generally it is about 4.

In spite of this, Crooked Sedge communities are by no means always found on ripe acid-humus soils. Quite often the soil type is a younger ranker (humus-silicate soil) which has been very poor in lime from the outset. Occasionally it is also met with on a pararendzina if the surface layer has been leached of lime. Only exceptionally does the raw humus layer exceed a thickness of 3-5 cm. In the few cases where there is a greater accumulation of humus this usually means that we are dealing with a sub-fossil fen or a colluvial humus accumulation. The typical *Curvuletum* itself is unable to form a deep layer of humus, because of its low productivity. The profiles of the moist-soil *Curvuletum* as well as of the snow coombs in contact with them are surprisingly poor in humus (see section 5 a). Horizons of bleached sand and hardpans do not form in an alpine climate since the leaching of the soil by rainfall is too weak and too brief. It is only under woodland and heath in the subalpine belt and lower that we find typical humus-iron podsols with their coffee-brown hardpan.

The water supply of the Crooked Sedge swards is always sufficient, as Körner and his co-workers (1980) have proved. But the light is often insufficient for optimal photosynthesis, even in the central Alps where the sky is frequently clear (Körner 1982). In this respect the living conditions of the high alpine vegetation are not fundamentally different from those in the lowlands (see also section VII 2 f).

It has already been mentioned in section 1 f that the Crooked Sedge sward is occasionally found over carbonate rock, but only where it is insulated from it by an alien layer of fine soil poor in lime. So one should no more think of it as the climax community for the whole alpine belt than one would today consider the Birch-Oak wood to be the climax for all lowland areas (see section B 1 J a and III 5). However on acid soils at the true alpine level the *Curvuletum* certainly represents the end stage of a natural succession.

Well developed *Curvuletum* is found only at middle and upper alpine levels, that is at rather high altitudes. In the lower part of the alpine belt in the central Alps the *Caricetum curvulae* gives way on deep soils to a closely related very variable association, the *Festucetum halleri*. It will only just be mentioned since its distribution is very limited both vertically and horizontally. On acid soils it appears to form a parallel to the calcareous soil's *Festuco-Trifolietum thalii* (section 2 d). However because of their dissimilar physical soil requirements *Caricetalia curvulae-* and *Seslerietalia*-communities cannot be closely parallel. The same goes for *Festuca varia* and *Sesleria albicans* slopes.

(b) Varicoloured Fescue slopes and similar acid-soil grassland

Patchy grass steps, so characteristic of Blue Moorgrass slopes on limestone, are much rarer on slopes of acid rocks. They are mostly to be found in the southern and central Alpine chains, i.e. in regions where the steep sunny slopes are most likely to dry out at least occasionally. Brockmann-Jerosch (1907) has already investigated extensive slopes of this kind in the Puschlav.

Here the Varicoloured Fescue (*Festuca varia*) is widespread. It is a grass whose clumps trap the fine earth in a similar way to Blue Moorgrass and from a distance it could be mistaken for *Sesleria albicans*. Closer inspection however reveals that it does not have broad *Poa*-like leaves as has the Blue Moorgrass but long and hard rolled ones with needle points that may damage both men and grazing animals. Moreover the leaves are so smooth that without nailed boots one could easily slip down a Varicoloured Fescue slope whereas one can easily run up and down a Blue Moorgrass slope as though on real steps.

Within the *Caricetalia curvulae* Braun-Blanquet distinguished a special alliance, the *Festucion variae* (tab. 75). Its most important alpine to subalpine association is the *Festucetum variae* which prefers relatively dry and only moderately acid soils. Apart from the dominating Fescue species it has only a few rare character species. Varicoloured Fescue, like Blue Moorgrass, forms communities from the alpine level right down into the valleys, i.e. from about 2 800 m down to 800 m. Accordingly its partners change, but because of its very competitiveness it shuts out many species and only tolerates a rather monotonous combination. As Markgraf-Dannenberg (1979) stresses, *Festuca varia* comprises many smaller taxa whose ecological behaviour differs as regards warmth, acidity and other factors. Only *Festuca acuminata* is found exclusively on acid soils; it also belongs to the *F. varia* aggregate.

The most common companion of *Festuca varia* is *Carex sempervirens* and in this respect too there is a parallelism between the Varicoloured Fescue and the Blue Moorgrass slopes (see section a). In the northern Alps where *Festuca varia* does not occur *Carex sempervirens* can become dominant on lime-deficient slopes where it too forms stepped grassland. Within the Blue Moorgrass communities on limestone *Carex sempervirens* indicates habitats where the soil is relatively fine and here it is quite often accompanied by acidophytes (Lüdi 1948). In its ecological behaviour then it lies somewhere between *Sesleria albicans* and *Festuca varia*. Interpenetration between *Festucion variae* and *Seslerion albicans* on intermediate soils is not infrequent.

Avalanches often occur on Varicoloured Fescue, Evergreen Sedge and Blue Moorgrass slopes. The snow rushing down every year extends the habitat of these light-loving communities down into the montane level. A comparison of the three step-forming grasslands on different rocks, different gradients and in different parts of the Alps would be very informative, but it has not yet been carried out.

(c) Matgrass swards

In the acid-soil grasslands of the lower alpine and the subalpine belts the Matgrass (*Nardus stricta*) is dominant so often that one is inclined to speak of a **Matgrass belt**. *Nardus* is one of the commonest plants especially in the outer Alps with their higher rainfall and on heavily grazed common pastures (fig. 41 2, 324). Crooked Sedge gives way to Matgrass here because it grows more slowly and cannot take advantage of its superior tolerance of the cold and drought. Moreover Matgrass is less palatable to livestock than Crooked Sedge and behaves as a pasture weed.

Consequently Matgrass is one of the most successful plants in the rough grazing on lime-deficient soils, and this is so not only at altitudes up to the alpine belt, but also in the lower mountains of Europe and at all levels right down to the heathland areas on the plains (section D II 4 a). Generally speaking *Nardus* forms stands which are similar in appearance, poor in species, rather colourless and floristically difficult to characterise. As already mentioned in section 1g it avoids lime-rich soils because here it suffers from a disturbance in its nutrient balance and especially from an iron shortage (see also fig. 330).

Oberdorfer (1959) distinguished the following grassland communities in the German Alps:

true alpine	*Curvuletum s.l*	Crooked Sedge level
lower alpine	*Curvulo-Nardetum*	Dwarf shrub level
subalpine	*Aveno-Nardetum s.l.*	Larch-Arolla Pine level
high montane	*Nardetum alpigenum*	Spruce level

Apart from the first, one must put all these communities in the alliance *Eu-Nardion* and the order *Nardetalia*. Together with the dwarf shrub heaths of the lowlands they form the class *Nardo-Callunetea*.

Because *Nardus* is indirectly favoured by grazing, it is able to penetrate other communities in the subalpine and alpine belt, communities which do not belong to the *Caricetea curvulae* such as snow coombs and small sedges fens (fig. 255). Thus, given that the other species are weakened by grazing, Matgrass is fairly ubiquitous. One should not be deceived by the dominance of the Matgrass and must try to trace back the plant combination to its original natural composition. This is generally possible from the presence of some persisting character species of the former vegetation type, as Aichinger (1957) has demonstrated with some quite heterogenous examples.

Although the Matgrass pastures have arisen in many different ways they can all be improved by manuring although, as we saw in section 1 f, *Nardus* itself responds positively to nutrients. The improvement takes place more rapidly if the grassland is rested from grazing for a while and mown. In this way fodder grasses and herbs are assisted whereas under a grazing-only regime they are overshadowed by the Matgrass. At the alpine and subalpine levels these more desirable grazing plants include *Festuca nigrescens* (= *rubra fallax*), *Phleum alpinum*, *Poa alpina*, *Trifolium pratense* and *Leontodon hispidus*. When manured and better cared for, the typical high mountain *Nardetum* would pass through a subassociation *trifolietosum* into a 'milkwort pasture'. Along with other partly natural grazing lands this will be discussed in more detail in section D VI 1 d.

4 Dwarf shrub heaths of the lower alpine and subalpine belts

(a) The Creeping Azalea wind heath

A few dwarf shrubs are all that survive of the many woody plants up at the true alpine level. These include some Willows (*Salix* species), the Mountain Avens (*Dryas octopetala*) and *Loiseleuria procumbens*, a prostrate kind of Azalea. It is only in particular habitats that these low and slow-growing outposts are able to compete successfully with herbaceous plants whose leaves they are able to overshadow in summer. Stronger and taller shrubs such as Dwarf Juniper (*Juniperus nana*) or the *Rhododendron* and *Vaccinium* species may restrain the herbs in a narrow transition belt between the subalpine and alpine regions.

Physiognomically three forms of dwarf shrub communities can easily be separated: the carpet-like soil cover of Creeping Azalea or Avens, the meadow-like stand of *Salix herbacea* with subterranean stems, and the real dwarf shrubs growing to some 40 cm. The former two correspond to extremes of snow covering (see figs. 334, 335 and 338). The herbaceous willow lives in well protected snow coombs (section 5), whereas *Loiseleuria* is hardy against wind and cold and can thrive without any snow cover at all. According to Larcher (1953) it can withstand a temperature of −40°C, as indeed can Spruce, Arolla and Mountain

Fig. 334. In the region of the tree line the average wind speed is determined by local contours. On the Stillberg near Davos at 2150 m. After Turner (1970) somewhat modified.

The wind speed is particularly high over ridges where low Creeping Azalea Heath is growing (*Loiseleurietum*). Here there is no protective layer of snow in winter and the presence of the lichen *Alectoria ochroleuca* is an indication of the dry cold conditions. A 2 m high *Pinus cembra* slows down the wind almost entirely near the surface. In a 5 m deep depression a Rhododendron scrub (*Rh.ferrugineum* and *Calamagrostis villosa*) enjoys considerable protection from the wind (and therefore a deeper and more persistent snow cover).

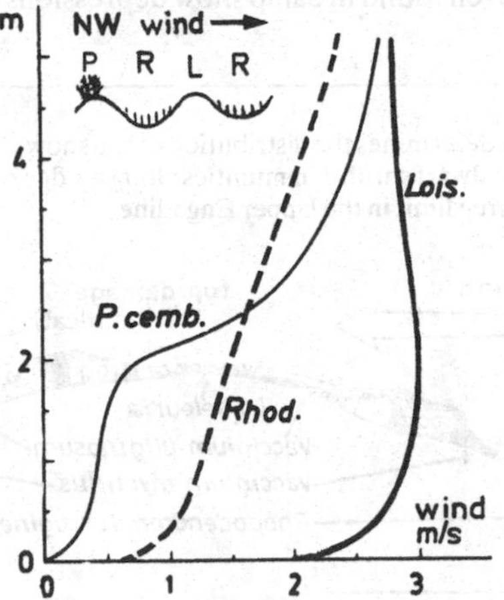

Pines. Amongst the evergreen dwarf shrubs only *Juniperus nana* and *Arctostaphylos uva-ursi* are as frost hardy. All the rest, especially the ericas, are killed by the frost unless they remain covered by a blanket of snow (fig. 161). Even *Loiseleuria* and the vacciniums often associated with it avoid the more extreme continental regions where the frost falls below −40°C and so they have a tendency towards a more oceanic distribution. Of the *Vaccinium* species *V. uliginosum* reaches the greatest altitudes and is nearest to *Loiseleuria* in its behaviour. As far as its microclimate is concerned the **Creeping Azalea wind heath** or 'chamois heath carpet' (*Loiseleurietum*) corresponds most closely to the *Elynetum* and the *Caricetum firmae* on less-acid soils which are equally hardy (see tab. 74). The latter contains *Dryas octopetala* which to a certain extent, is comparable with *Loiseleuria*.

Such wind-swept habitats are sometimes very dry in winter as well as in summer. Dew which frequently falls here is very important for the plants to survive the warmer time of year. Lüdi (1948), for example, proved this for the 'wind edges' of the Schynige Platte (section 2 c). The rolled leaves of *Loiseleuria* are able to suck up liquid water quickly because of two hairy, weakly cutinised channels on the undersides running like capillaries down from the leaf tip. Larcher (1957) observed that Creeping Azalea leaves could take in water from freshly fallen snow. Although it is better able to reduce its transpiration than other dwarf shrubs it is directly dependent on snow or ice melting in its surroundings during the times of alternate freezing and thawing, in order to withstand the frost drought. It takes water up not only with its leaves but even more through numerous adventitious roots borne on the stems. Actually the *Loiseleurietum* prefers places which, although still windswept, are not very steep and where water can collect in places. Specimens of *Loiseleuria* are even found in damp snow depressions if there is no competition from taller plants. *Loiseleuria* espaliers growing out over large stones often show drought damage whereas the ones lying over humus remain sound. This is because the humus soaks up water whether from rain, mist or dew, or from melting snow running over it from time to time. This humus store can then supply leaves which are tightly pressed on its surface. Such a water source is also made use of by foliose and fruticose lichens which appear in large numbers in the Creeping Azalea heath.

Since exposed places often catch the sun and are warmed up during winter, living plant organs above the ground lose relatively large amounts of energy through respiration. *Loiseleuria* is able to cope with this, according to Larcher, Schmidt and Tschager (1973), since it stores large amounts of fat and has a high calorific value (up to 5820 cal/g of dry matter, see also tab. 77). However, according to Tschager and co-workers (1982), the physiological rôle of fat accumulation in alpine Ericaceas is not yet quite clear. In the alpine belt, annual production of *Loiseleuria* is very small, e.g. in a plot on the Patscherkofel (1.1 t/ha) because of the dominating severe conditions in its natural habitat. Not more than 150 m lower down, in the subalpine level, the yield of a virtually pure *Loiseleuria* stand was three times this amount, and that of a Bilberry heath (*Vaccinium myrtillus-uliginosum* stand, see tab. 77) four times as great. The last-mentioned communities on the PatscherKofel above Innsbruck are made and maintained by the activity of man and animals and would revert to the Larch-Arolla Pine wood under natural conditions. In respect of the total living phytomass that of the subalpine heaths is only twice that of the alpine *Loiseleurietum* (tab. 77) because leaves and stems of Creeping Azalea are very long lived.

The **lichen species** found in *Loiseleurietum* are the same as those in *Elynetum cetrarietosum*, but the other

Fig. 335. The wind determines the distribution of the snow and the zonation of dwarf shrub communities along a ridge on the slope near the tree limit in the Upper Engadine, semi-schematic. After Holtmeier (1971), modified (cf. fig. 334).

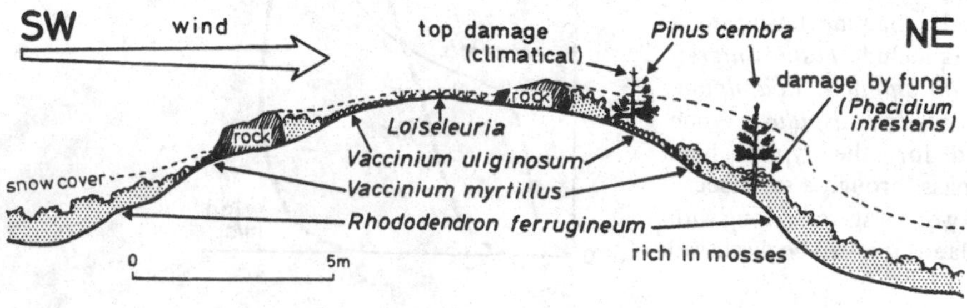

plants accompanying Azalea have little in common with those in the Naked Rush sward. There are many more raw humus plants able to tolerate half shade, such as *Vaccinium uliginosum*, *V. vitis-idea*, *V. myrtillus*, *Empetrum hermaphroditum*, and *Huperzia selago*. Even though these are stunted and tousled with the wind they reach a relative high covering in most Azalea heaths. Therefore species lists of such extremely thin carpets are rather similar to those of ordinary dwarf shrub heaths or the acid-soil coniferous woodlands, so that Braun-Blanquet has put all these associations together in the same order (see tab. 74). Thus we have a situation where, by following the floristic principle in the classification of plant communities, the consequence appears absurd from the physiognomic and ecological standpoints: wind-whipped espaliers above the climatic tree limit being united with high forest which is widely distributed in the lowland. Only at the rank of the alliances are the two distinguished floristically (*Loiseleurio-Vaccinion* and *Vaccinio-Piceion*).

Gams (1927), Klement (1955) and others take the differing view that the main emphasis should be placed on the combinations of lichens when the Creeping Azalea wind-swept heaths are being classified. They put all the wind-hardy fruticose lichen communities together in an independent association (*Thamnolietum vermicularis*) with the following character species:

Alectoria ochroleuca	*Thamnolia vermicularis*
A. nigricans	*Dufourea madreporiformis*

These live on the raw humus produced by higher plants, but otherwise do not depend on them. Therefore the same lichen community may penetrate stands of *Elynetum*, *Curvuletum* or '*Loiseleurietum*'. In many cases it seems difficult to decide whether such small-scale mosaics should be looked upon as phanerogamic or cryptogamic vegetation units. Recently phytosociologists have returned to the rather intermediate interpretation first given by Du Rietz (1925): As lichen species are often dominating in the extremely wind-exposed habitats, *Loiseleuria* and other phanerogams are simply their companions. The *Loiseleurio-Alectorietum ochroleucae* then is to be joined with the arctic-alpine alliance of fruticose

Tab. 77. Standing phytomass and its net production along with the energy content in subalpine dwarf shrub heaths on the Patscherkofel.

	Vacci-nium-heath[a]	Loise-leuria-heath[a]	Loise-leuri-etum[b]
Altitude (m above sea level)	1980	2000	2150
Phytomass (t/ha)			
above ground			
living	10.5	11	8
dead	2.5	1	0.7
below ground			
living	25	22	8
dead	15	6	0.6
litter	8.5	11	9.3
living phytomass total	35.5	33	16
Net production (above ground)			
Annual production (t/ha)	4.8	3.2	1.1
the same as a % age of the living phytomass	13.5	9.6	6.9
Net production rates per day (g/m²)	1.3	0.9	0.3

	Vacci-nium-heath[a]	*Loise-leuria-heath*[a]	*Loise-leuri-etum*[b]
Altitude	1980	2000	2150
Energy content (Lcal/g)			
Dwarf shrubs:			
Vaccinium species	5.0 – 5.2	–	5.2
Calluna	–	5.3	–
Loiseleuria	5.6	5.8	5.7
Herbs:			
Avenella flexuosa	4.7	–	–
Antennaria spec.	–	–	4.7
Primula minima	–	–	4.9
Plants parts below ground	5.2	5.3	5.3
Mosses:			
Pleurozium schreberi	–	*4.5*	–
Lichens:			
Cetaria islandica	4.2	4.3	4.3
Alectoria ochroleuca	–	4.4	4.4
Thamnolia vermicularis	–	–	4.6

Note: [a] communities which have replaced *sub*alpine woodland; dense stands.
[b] natural windy heaths free from trees; do not cover the ground completely.

Source: From data in L. Schmidt (1974; rounded off averages for the years 1970–72).

lichens on rarely snow-covered acid humus (*Cetrarion nivalis*) which is best developed in the arctic lichen tundra (see also Wirth 1980 and tab. 75).

Where a mosaic of small humps ('thufurs') has been formed by a combination of frost heaving and cattle treading, exceptionally also in the Alps, the lichen heaths and the *Loiseleuria* carpets take over the humps while the hollows carry *Rhododendron* and *Calluna* heaths which need more protection from the wind. This has been well described already by Bolleter (1920, see figs. 326 and 327).

The Creeping Azalea wind-swept heaths rarely occupy such large areas as they do for example on the Koralpe at the south-eastern edge of the Alps. Generally they develop in quite small patches, and even the wind shadow of a stone can favour the growth of sward plants or large dwarf shrubs. Thus there are all transition stages from almost pure lichen communities with very little *Loiseleuria*, through a dense *Loiseleuria* carpet to typical Crooked-Sedge or Matgrass swards and Naked Rush wind-swept corners where there may be the odd dwarf shrub but hardly any lichens.

All these combinations of species produce a covering of very acid raw humus with a mostly favourable water regime. Another thing their habitats have in common is that they are clear of snow very early in the season and only exceptionally do they have its protection for a longer time. These extreme specialised communities fade away in the subalpine level very soon where the snow is more evenly distributed as there is more protection from wind. The upper limit of their distribution is brought about by the increasing persistence and depth of the snow at greater altitudes. In addition the Creeping Azalea and fruticose lichens can only spread on stable ground where the surface is not being constantly disturbed by erosion or solifluxion. They require a long time to cover the ground and form the characteristic small-scale mosaic of the Creeping Azalea wind heath since they grow so very slowly.

(b) Crowberry-Bog Bilberry heaths

Many of the species of the Creeping Azalea carpet are also to be found in the Crowberry-Bog Bilberry heaths (*Empetro-Vaccinietum*) of the subalpine and lower alpine levels, as well as in the northern tundra although these are covered with a blanket of snow during winter and so are able to develop into full-sized dwarf shrubs. However Bog Bilberry heaths cover such small areas and are generally so heavily grazed that it is difficult to separate them from the Matgrass or the Crooked Sedge swards (fig. 335).

Those who come from the northern part of Central Europe and are used to seeing the Bog Bilberry *(Vaccinium uliginosum)* in Birch fenwoods and in the marginal woodlands of raised bogs are surprised to find it growing just as strongly in relatively dry places in the high mountains. Certainly up here it rarely grows to a height of more than 20-30 cm, but it fruits heavily and can also spread vegetatively quite quickly on acid soils. Thus one soon comes to appreciate that its real areas for development are to be looked for above the tree limit and in the northern tundra. As the tundra receded from Central Europe in Late Glacial times and was replaced by woodland the Bog Bilberry retreated to the open stands of Birch on the mires thus becoming separated from its companion the Mountain Crowberry (*Empetrum hermaphroditum*) which requires even more light. On some of the higher mountains of Central Europe relicts of this boreal and alpine dwarf shrub can still be found (Schubert 1960 and Stöcker 1965a).

Thanks to the work of Pallmann and Haffter (1933) and of Lüdi (1948) we are now quite familiar with the chemical properties of the soils on which the subalpine **Crowberry-Bog Bilberry heaths** grow in the Oberengadin the *Empetro-Vaccinietum* grows on real raw humus over a crystalline substrate. The humus is very acid with a pH value of between 3.5 and 4.8, much of it lying below pH 4. According to Lüdi (1948) it can also be found on a substrate containing carbonate, e.g. on the slaty dogger limestone of the Schynige Platte. In constrast to the sunny Blue Moorgrass-Sedge slopes it is restricted here to shady slopes where the cool damp climate favours the accumulation of surface humus. This can reach a thickness of 30-50 cm but frequently has pieces of limestone mixed in with it and may be quite rich in nutrients. However its pH value lies around 4.5 or less. A few relicts of the pioneer vegetation such as *Sesleria albicans, Dryas* and *Salix retusa* needing a more calcareous soil point to a better type of humus. Because of its shady position the humus remains damp or wet throughout the whole year although it is quite porous and free draining. Unfortunately we have no information at present about the water economy of the *Empetro-Vaccinietum* on sunny slopes of older rocks. It would be very interesting to be able to compare this with the mire habitats of *Vaccinium uliginosum* in north Germany.

In his monograph on the subalpine dwarf shrub communities of the Aare catchment area Schweingruber (1972) distinguishes 5 associations just in the Bernese Oberland. Each of these has 1-3 subassociations and can again be separated into up to ten variants. These units reflect local climatic conditions and also edaphic and anthropozoogenic effects. If in addition to the Crowberry-Bilberry heaths we also include the Prostrate Pine scrub which we have already discussed in section B IV 5 b, together with the Dwarf

Juniper and Alpenrose heaths which will be dealt with in the next section, then the whole position can be summarised for those communities occurring between 1600 and 2200 m above sea level:

Mainly in shade	**Mainly in sun**
1. Crowberry-Bilberry heaths *Empetro-Vaccinietum* *—sphagnetosum* *—hylocomietosum* *—cetrarietosum*	2. Dwarf Juniper heaths*) *Junipero-Arctostaphyletum* *—callunetosum*
3. Hairy Alpenrose-prostrate pine scrub *Rhododendro hirsuti-Mugetum* *—salicetosum retusae* *—globularietosum*	4. Spring Heath-prostrate pine scrub *Erico-Mugetum* *—rhamnetosum* *—rhytidiadelphetosum*
5. Rusty Alpenrose heath *) *Rhododendro-Vaccinietum* *—seslerietosum*	
In sun and shade: 5. *Rhododendro-Vaccinietum* *) *—typicum* *—leucobryetosum*	*) dependent on husbandry (pasture weeds!)

The subalpine dwarf shrub heaths on isolated mountains in northern Central Europe, such as those investigated by Stöcker (1965b) in the Harz, contain little or no *Vaccinium uliginosum* and are more like the grassland heaths of more continental mountains, e.g. the *Cetrario-Festucetum supinae* which was described by Jeník (1961).

(c) Alpenrose heaths

Because of their fine display of red flowers the Alpenrose heaths are noticed much more than the other dwarf shrub communities in the region around the tree limit. However they are not true representatives of the alpine vegetation for they really belong to the subalpine belt and only venture as outposts into the alpine level when they can be afforded the protection of snow during the whole winter. Because of this we have already dealt with them in section B IV 4 b. Since the *Rhododendron* heaths can only remain free from trees through man's activity we shall also return to them in section D II 4 b.

Of all the communities replacing subalpine woodlands the *Rhododendron* heath comes closest to the wood in its productivity. In many respects though *Rhododendron ferrugineum* is more sensitive than the trees. It reduces its photosynthesis and transpiration more rapidly when the wind speed increases (fig. 336), and without the protection of snow it can be killed by drought in winter, or in early spring if the snow goes and it is exposed too soon (fig. 337).

According to data so far available it appears that the dwarf shrub heath formations are more productive than the unmanured grasslands. Measurements carried out by Bliss (1962) on Mt Washington in the north-western USA for example show that they produce 283g/m^2 of dry matter above ground each year compared with 176 g/m^2 for a sedge sward and no more

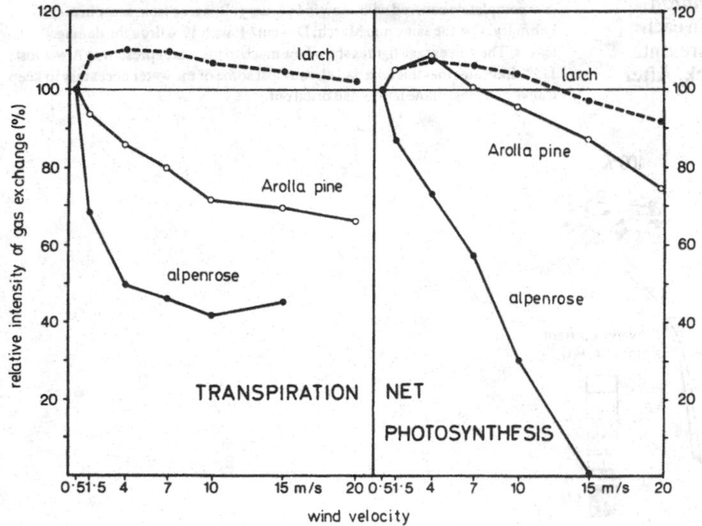

Fig. 336. Larch responds to the demands imposed by an increased wind speed by increased transpiration and (up to 10 m/s) by more net photosynthesis. On the other hand *Pinus cembra* and *Rhododendron ferrugineum* close their stomata to some extent resulting in a reduction of gas exchange. After Tranquillini (1970).

The ordinate shows the relative amount of gas exchange compared with that in practically still air. The experiments were carried out with young plants in a wind tunnel; other conditions were kept constant and optimal.

than 67 g/m^2 for a community of cushion plants. According to Wielgolaski (1975) the productivity of dwarf shrub heaths is never more than 10g/m^2 in the high alpine or the deep arctic regions whereas on good ground in the subalpine or subarctic they can go up to 1000 g. The dwarf shrub communities are also at the top of the league when it comes to their calorie and fat content (Bliss 1962). Taken as a whole the subalpine and alpine plants store more reserves of high calorific value than comparable plants at low levels in the temperate zone or even in the tropics. They are thus better prepared to come through an unfavourable period, especially a long winter, without damage caused by lack of energy.

Incidentally *Rhododendron ferrugineum* and other representatives of the dwarf shrub heaths from close to the tree limit are also able to live at lower levels. Isolated outposts occur for example on the submontane ridges of the Swiss pre-Alps under protection. In Tessin they form large stands right down as far as the neighbourhood of the Chestnuts. Mayer (1970) described *Rhododendron* pine woods from low-lying parts of the Pustertal in the southern Tyrol where they were occupying boulder-strewn shady slopes between 800 and 1600 m. Similar open woods rich in Alpenrose are also to be found in the serpentine region of Steiermark. In each case they are on poor soil with a moist climate where there is little competition.

5 Snow patches and related communities

(a) Moss-rich snow patches

Much as the sward plants and the dwarf shrubs require the protection of a snow covering during the winter their growth is restricted if the snow should linger too long (fig. 324). Depressions where the snow hangs about are always saturated with snow-melt water and have a characteristic plant covering poor in species (see section 1 f). They are called snow-patches (German 'Schneetälchen'). Where the ground is free from snow for less than two months there is scarcely sufficient time for phanerogams to exist (fig. 338), but because of the damp soil mosses are able to develop. Lichens are less in evidence because they cannot use their greater drought resistance to any advantage. Where the soil surface is exceptionally rough a fruticose lichen community may dominate, the *Stereocauletum alpini* (in the alliance *Solorinion croceae*, see Wirth 1980).

The purest moss-snow patches (*Polytrichetum sexangularis*) are to be found in the upper alpine belt of the high silicate massifs. In the Scandinavian mountains too they can cover wide depressions with plush dark green carpet. Their character species are given in tab. 78. The dwarf *Ranunculus pygmaeus* is rarer in the Alps but a very typical Nordic guest.

Kiaeria falcata (= *Dicranum falcatum*) dominates in the short-lived initial communities. The plant able to survive longest under snow is the liverwort *Anthelia juratzkana* which is covered by white felt-like wax excretions. Where the ground is clear of snow somewhat earlier *Polytrichum sexangulare* or even *Pohlia commutata* becomes more frequent. All these mosses are also represented in snow patches with phanerogams. However, they are generally absent from limestone because this, being so well drained, dries out as soon as the snow is gone.

Fig. 337. Water loss from the Alpenrose (*Rhododendron ferrugineum*) after the disappearance of the snow. In each case the determinations were carried out on 3–4 representative leafy shoots on the Patscherkofel near Innsbruck. After Larcher (1963), modified somewhat.

A = completely covered with snow in February; B = free from snow in mid February; C = the same mid March; D = cut off with 10% drought-damaged leaves. The percentage figures show how much of the water present at A was lost; 115% indicates that this twig had already lost some of the water necessary to keep it alive and most of the leaves had dried out.

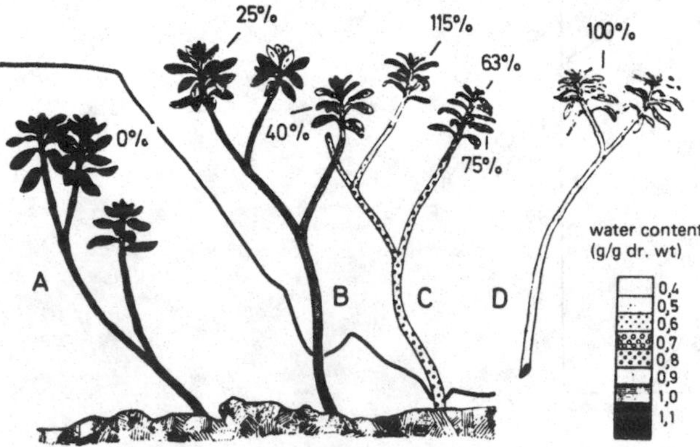

J. Braun (1913) has already noticed that the soil in the snow patches contains less humus than one would imagine at first glance. The amount of material produced by the mosses is only small although quite a large amount of organic dust is left behind by the snow (fig. 340). But this acts as a manure because in spite of the short season it is broken down by insects and numerous bacteria so that most of it is mineralized. So the snow-bed communities do not lack nutrients. Because the soil is saturated with water for such a long time a wet soil type is built up which, in spite of the shortage of lime, has only a slightly acid reaction.

(b) Dwarf Willow snow patches

Where the period without snow lasts around 8 weeks or a little more relatively quick-growing phanerogams are able to spread, especially *Salix herbacea*, the Dwarf Willow (figs. 338 and 339). Like most of the snow-patch specialists it spreads primarily vegetatively with its subterranean stems into the boundary zone of the moss community which has already been described. Only when it is clear of snow for more than 3 months is it able to flower and fruit. Its hairy seeds are dispersed by convection currents or storm winds but, like all the other willows, they will only germinate if they land on a wet but well oxygenated substrate within a short time after ripening. The water draining from the melting snow is thus an important prerequisite for the establishment of the Dwarf Willow in a snow patch.

In the Scandinavian mountains, where the surface remains moist right into the early summer because the sunshine is less strong, *Salix herbacea* is found in some very different communities and is not confined to depressions or 'hollows'. Because the snow is so persistent and the soil is always wet all snow patches are decidedly cool. According to Dahl (1951) this appears to be a necessary condition for the success of *Salix herbacea* and other nordic-alpine plants. Even on the plains near Oslo it cannot endure the summer temperatures.

Since *Salix herbacea* grows best on fairly acid soils the communities formed by it are generally found on crystalline rocks or others which are poor in lime. Like the Crooked Sedge sward, with which it is connected through a number of intermediate types depending on the length of time the site is free from snow, the Dwarf Willow snow patches are characteristic communities of the silicate central Alps. They are most numerous in the middle and upper alpine belts.

Along with the dominating Willow species *Gnaphalium supinum*, *Alchemilla pentaphylla* and *Arenaria biflora* are the principal character species of the *Salicetum herbaceae*. *Soldanella pusilla* (not *S. alpina*, which prefers warmer places with more nutrients) and *Sibbaldia procumbens* occur also in

Fig. 339. Dwarf Willow snow patch (*Salicetum herbaceae*) at the time of maximum growth. (The photograph covers a width of about 50 cm.)

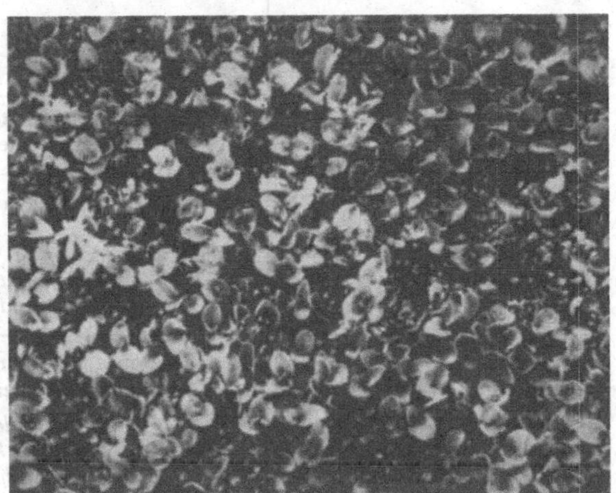

Fig. 338. Zonation of the snow patch vegetation, according to the persistence of the snow cover on the Minorjoch, Bernina, 2450 m. After Braun-Blanquet (1951).

1 = Moss snow patch (*Polytrichetum sexangularis*), 2 = typical Dwarf Willow snow patch (*Salicetum herbaceae*), 3 = a form with *Gnaphalium supinum*, 4 = the same with *Ligusticum mutellina*, 5 = Crooked Sedge sward (*Caricetum curvulae*).

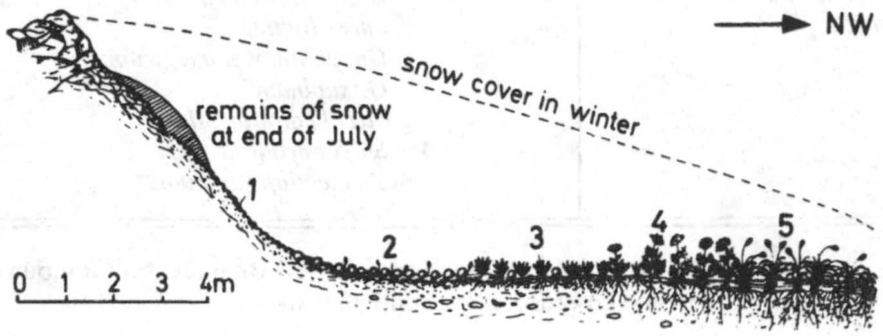

other acid-soil snow-patch communities, and so they are useful for characterising the alliance *Salicion herbaceae* (tab. 78). In addition there may be found a number of widespread phanerogams which do not grow only in snow patches, e.g. *Poa annua* ssp. *varia, Poa alpina, Polygonum viviparum* and *Potentilla aurea*.

All these species are tiny, with thin leaves susceptible to drying out. They prefer a more or less acid soil though some are indifferent. Their leaves and shoots often push through the snow before it is completely melted, e.g. the violet-fringed bells of *Soldanella* (fig. 340). The strong June sunshine warms up all the dark objects, including humus and plant parts, under the snow by several degrees above zero. As soon as the snow has melted down to a thickness of some 11-18 cm there is sufficient light for photosynthesis to take place (Curl, Hardy and Ellermeier 1972) (fig. 341). Thus the growing period of the snow-patch plants can be several weeks longer than the time during which they are actually free of snow. *Salix herbacea* reaches light saturation at 40 k lux (Cartellieri 1940) and so behaves in a similar way to the Beech in this respect. *Primula glutinosa* will even go on increasing the amount of gas exchange up to 60 k lux. Photosynthesis is so lively in most plants of the snow patches that

Tab. 78. A synopsis of the alpine snow patch communities with their character species.

Class: (only a few species in common!)	**Snow-bottom communities** Salicetea herbaceae *Alchemilla fissa* *Cerastium cerastoides* *Sagina saginoides* *Veronica alpina*	
Orders:	**Calcareous snow patches** Arabidetalia caeruleae	**Acid-soil snow patches** Salicetalia herbaceae
Single alliances: Character species of order and alliance	Arabidion caeruleae g *Carex atrata* *Ranunculus alpestris* *Saxifraga androsacea*	Salicion herbaceae *Sedum alpestre* *Sibbaldia procumbens* *Soldanella pusilla*
Pioneer communities Arabidetum snow-covered for a long time, Polytrichetum for a very long time	**Rock-cress snow patch** Arabidetum caeruleae *Arabis caerulea* *Gnaphalium hoppeanum* *Hutchinsia alpina* ssp. *brevicaulis* *Potentilla brauneana*	**Moss snow patch** Polytrichetum sexangularis **D** m *Anthelia juratzkana* m *Gymnomitrium varians* **m** *Moerckia blytti* **m** *Pleuroclada albescens* **D** m *Polytrichum sexangulare* Ranunculus pygmaeus
Permanent communities: snow-covered for shorter period	**Espalier Willow community** Salicetum retusae-reticulatae *Gentiana bavarica* **D** *Salix reticulata* **D** *S. retusa*	**Dwarf Willow snow patch** Salicetum herbaceae *Alchemilla pentaphylla* *Arenaria biflora* g *Carex foetida* *Gnaphalium norvegicum* *G. supinum* g *Luzula alpino-pilosa* **D** *Salix herbacea* *Tanacetum alpinum*

Note: **D** = more often the dominant species
m = moss, g = grasslike.
[a] Compare section E III 1, no. 4.5.

Source: From data supplied by Braun-Blanquet for Graubünden and Oberdorfer for South Germany.[1]

they need only one month free from snow in early summer in order to fix enough carbon for their survival.

Instead of the mosses typical of the *Polytrichetum sexangularis*, a moss which often becomes dominant in the Dwarf Willow communities is *Polytrichum juniperinum*. This species is easily satisfied and can withstand occasional drying out without damage. Its presence indicates that the soil surface could sometimes become very dry towards the end of the exposed period. However the herbaceous plants and Willows very rarely suffer from a water shortage because they are rooting into the more or less silty or sandy subsoil saturated with melt water.

(c) Snow patches on limestone

We have already stressed in the previous section that the snow-bed communities on silicate rocks are usually purer and richer in character species than on carbonate rocks. However in valleys and on slopes of the limestone Alps too there are hollows and depressions where the snow remains longer than average (fig. 332). This is particularly the case at the foot of extensive screes, and frequently the scree communities which will be discussed in section 8a merge into those of a snow patch (fig. 324). For this reason they were previously united systematically, but Braun-Blanquet (1969) made the limestone snow-beds a special order (*Arabidetalia coeruleae*) in the class *Salicetea herbaceae* although it has only few character species in common with the silicate snow patches (tab. 78). *Salix herbacea* and other acid-tolerating species do gradually invade the snow patches on a lime-rich substrate because the melt water collects fine earth in these hollows and washes the lime out of it. Local accumulations of crystalline foreign material could have the same effect. Szafer (1924, see also section 1 f) found no genuine limestone snow-beds in the Tatra especially as the higher parts of these mountains are mainly composed of silicate rocks.

There is no moss-rich community on limestone to correspond to the *Polytrichetum sexangularis*. This is probably due in the first instance to the fact that the snow patches here are mostly stony and dry out more quickly, coupled with the presence of cracks in the rocks through which the melting snow can drain away. There is certainly also a geomorphological reason which is operating: in the upper and middle alpine belts where snow patches are most common, bare rocks and stony slopes dominate the limestone and dolomite massifs whereas in the silicate mountains more stable and softer forms are the rule and scree slopes the exception. The best calcareous snow patches are formed on clay marls which morphologically behave in a similar way to the silicate rocks.

The **Bluish Rock-cress snow patch** *(Arabidetum caeruleae)* represents a parallel with the *Salicetum herbaceae* at least as far as the period without snow cover and the supply of organic and mineral nourishment are concerned. For a time in late summer however its water economy is rather less favourable. Even in the paces where this community is only fragmentary it can be recognised by the white starlike flowers of *Ranunculus alpestris* decorating the cal-

Fig. 340. *Soldanella alpina* (about 10 cm high) bursting through the melting snow on the Pastura di Lagalb. Dust collected by the snow adds nutrients to the soil. Photo, Ganz and Rübel.

According to Bogenrieder and Werner (1979), this snow patch plant has little resistance to frost, but can assimilate CO_2 effectively around freezing point. The deep and persistent snow cover protects it against frost during winter and cools the habitat until mid-summer.

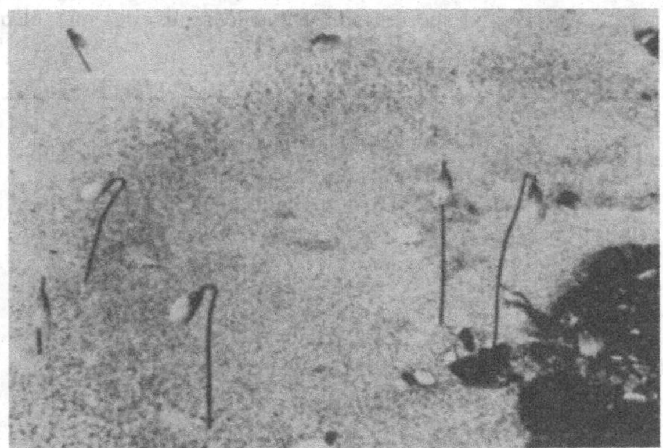

Fig. 341. The amount of radiation absorbed by different types of snow (as a percentage of the total radiation). After Sauberer and Härtel (1960).

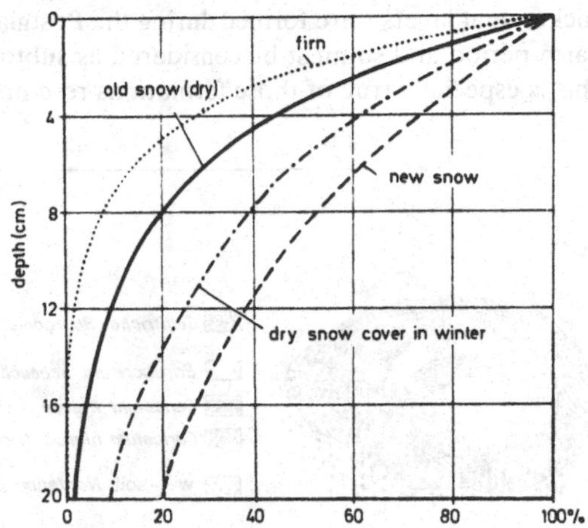

careous snow patches, often with a dense carpet. The rest of the character species (see tab. 78), like those of the acid-soil, have rather insignificant flowers.

Where there are large limestone rocks in the hollows Glacier and Net-leaved Willows form their espaliers over the stony surface. The **Espalier Willow snow patch** *(Salicetum retusae-reticulatae,* see tab. 78) is seldom found completely free from species of the *Arabidetum caeruleae*. It enjoys a somewhat longer snow-free period and more warmth, but is drier and therefore less favourable for mesomorphic herbs and mosses than is the pure Bluish Rock-cress snow patch.

6 Subalpine and alpine fens and spring-swamp communities

(a) Common Sedge fen and peat formation in acid still water

Although the soil of snow patches often becomes quite waterlogged no organic layer is met with on its surface. In the alpine belt, peat can form only in those places which are free from snow long enough for a noticeable amount of plant material to be produced, and where the ground is saturated or even under water for the whole of the growing season. Such collections of standing water are found here and there where a depression has no outlet and the soil consists of fine impervious material, such as in the basin at a terminal moraine (figs. 342 and 353). Since the maximum altitude of moraines in the central Alps is about 2700 m above sea level where woodlands are absent then the formation of fens is quite possible both biologically and geomorphologically. However for reasons given in section 1 d they are much rarer here than in the arctic and subarctic tundras.

Under present-day climatic conditions peat is formed very slowly in the alpine belt. Many of the thicker peat layers were formed during the Postglacial warm period and so must be considered as subfossil. This is especially true of those formations resembling raised bogs which are never formed above the tree limit at the present time.

Since the surface layer of fine soil in the older moraines even on limestone mountains has become leached of bases, practically all **standing water** in the alpine and subalpine belts is acid or at least deficient in minerals. Here and there it is occupied by *Sparganium angustifolium, Callitriche palustris, Eleocharis quinqueflora, E. acicularis, Ranunculus trichophyllus, R. reptans* or other species of the *Callitricho-Sparganietum* (fig. 342). This community of the clear lakes is seen as an impoverished representative of the class *Litorelletea* which is distributed mainly in north-western Europe (section C I 1 c).

Most of the small, acid, still-water lakes above the tree limit are fringed with a stand of *Eriophorum scheuchzeri*. The spherical seed hair-tufts of this **Cottongrass** are well known to all alpine walkers (fig. 342) the shining white standing out in sharp contrast to the dark waters. The dense stands of Cottongrass which begin to shoot up soon after the ice has melted contribute significantly to the peat accumulation when their remains sink to the bottom each year. In some places brown mosses, (e.g. *Drepanocladus exannulatus*) take part in peat formation. Apart from the name species *Eriophoretum scheuchzeri* contains few phanerogams. At most one might come across *Carex nigra, C. rostrata, Juncus filiformis* and *Eriophorum angustifolium*, which are also found at lower levels in fens of the class *Scheuchzerio-Caricetea*.

The Cottongrass girdle leads on to a **Common Sedge fen** (*Caricetum nigrae*) for which character species at the alpine and subalpine levels are given in tab. 79. This association is the most important current producer of peat in the alpine belt (fig. 342).

Caricetum nigrae appears in many forms; *Carex nigra, C. panicea* or another Sedge could be present in different places along with varying amounts of *Juncus filiformis* or *J. alpino-articulatus*, also *Eriophorum scheuchzeri* or *E. angustifolium* and even *Trichophorum cespitosurum*. Where the water is less acid species from the calcareous fens may also

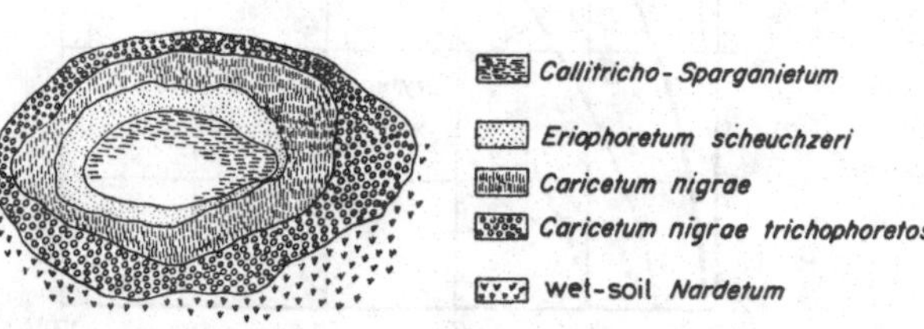

Fig. 342. Zonation of the vegetation surrounding a moraine lake poor in lime and nutrients in the lower alpine belt (semi-schematic). After Braun-Blanquet (1954), somewhat modified.

join in. The Sedge sward often becomes restricted to hummocks which are formed by cattle coming down to drink. Most of its species have creeping rhizomes and so, when undisturbed, have a tendency to form a spread-out, open sward rather than dense clumps.

Species from the Crooked Sedge and Matgrass swards penetrate the Small Sedges fen from its outer edge (fig. 328). Under natural conditions however this happens only when the mineral soil gradually passes over into the lake. On the peat soil the plants from drier habitats only find a footing when the fen has been drained (see also section C I 1 b and fig. 38a).

We can look upon the *Caricetum nigrae* in the alpine belt as a permanent community. Its own potential can no more enable it to develop into a climax community than that of the *Alnetum glutinosae* at the lowland lakes of Central Europe.

(b) Calcicole Small Sedge and Small Rush fens

Calcicolous swamp plants can grow at the alpine level only where fresh base-rich spring water is flowing constantly or where a valley soil is frequently flooded by calcareous mud brought down by a stream. Where

Tab. 79. Subalpine and alpine fens and spring swamps along with their character species.

Class:	**Fens and spring swamps** Scheuchzerio-Caricetea	
(many species extend from the lowland to above the woodland limit)	g *Carex dioica* g *C. flava* g *C. hostiana* g *C. panicea* (?) D g *Eriophorum angustifolium* g *Juncus alpino-articulatus* g *J. triglumis*	D *Menyanthes trifoliata* *Pedicularis palustris* *Swertia perennis* D g *Trichophorum cespitosum* *Triglochin palustre* etc.
Orders:	**Calcareous Small Sedge fens** Tofieldietalia (= Caricetalia davall.)	**Acid Small Sedge fens** Scheuchzerietalia
Alliances Order and alliance character species (V = alliance)	Caricion davallianae *Dartsia alpina* (?) g *Carex capillaris* var. *typica* g *C. frigida* (V) g *C. oederi* m *Campylium stellatum* *Dactylorhiza traunsteineri* m *Drepanocladus intermedius* g *Eleocharis quinqueflora* *Epipactis palustris* (V) *Eriophorum latifolium* (V) *Pinguicula vulgaris* g *Primula farinosa* (V) *Selaginella selaginoides* *Tofieldia calyculata* (V)	Caricion nigrae g *Agrostis canina* (rare) m *Calliergon sarmentosum* m *C. stramineum* *Cardamine pratensis* var. *hayneana* D g *Carex canescens* g *C. echinata* D g *C. nigra* *Ranunculus flammula* g *Trichophorum alpinum* *Viola palustris*
Swamp swards extending to the alpine level:	most form **tufa**: **Davall's Sedge fen** Caricetum davallianae and Caricion maritimae[a] D g *Carex bicolor* g *C. maritima* etc. *Tofieldia pusilla*	do not form **tufa** but peat: **Common Sedge fen** Caricetum nigrae g *Carex norvegica* (rare) g *C. paupercula* (rare)
Land-forming vegetation around still water	(for explanation of symbols see tab. 78)	**Cottongrass edge** (alpine) Eriophoretum scheuchzeri m *Drepanocladus exannulatus* D g *Eriophorum scheuchzeri*

Note: There are intermediates between the alliances and between the associations. [a]) See E III 1, no. 1.622.

Source: From data supplied by Braun-Blanquet, Oberdorfer and others. (See also section C II 2.)

there are springs they contribute to the formation of calcareous tufa. Brown mosses and blue algae are even more effective in separating out the lime and it is only exceptionally that they form pure organic layers.

Alpine and subalpine calcareous spring fens are the natural centres of distribution for the alliance *Caricion davallianae* and especially for the *Caricetum davallianae* (see section C II 2 a). They contain several otherwise rare species and are easily characterised (see tab. 79). However in the outer Alps which have a high precipitation many of the partners are also found in the swards of *Seslerietalia* or in other communities. *Pinguicula vulgaris* and *Tofieldia calyculata* for example are not restricted to the areas around springs in the northern limestone Alps, but are also to be found on slopes where rain water trickles down from time to time bringing with it a fresh supply of lime. They behave in a similar manner in the Scandinavian mountains too wherever there are rocks containing sufficient bases.

We should just mention here the communities of the *Caricion incurvae* which are close to the nordic *Caricion bicolori-atrofuscae*, but are not often found in the Alps (see tab. 74). They appear in small patches and with a varied combination of species.

There has been practically no work done on the ecology of alpine communities of the *Caricetalia davallianae*. Only Yerly (1970) has compared them with other Small Sedge and Small Rush swards in respect of their nitrogen supply and other factors of the subalpine and montane belts in the outer Alps near Freiburg (see fig. 253 and tab. 56).

(c) Spring fens rich in mosses

More striking than the Small Sedges fens fed by spring water are the communities which are rich in mosses. These too are to be found near to springs, but also form carpets along the banks of the lively ice-cold mountain streams which one crosses away from the beaten track. These communities are associated with quick flowing, well oxygenated clear water which seldom rises above 5°C.

At first sight they are reminiscent of the *Polytrichum* or *Dicranum* snow patches. However the spring fens have little in common with the *Salicetea herbaceae* either floristically or ecologically since they are not covered with snow until late in the year and enjoy a long season of exposure. Even in the coldest winters the springs and many of the stream sides are free from ice. Where the water comes from inside the mountain it remains cool even in the warmest weeks of the year so that organisms in and alongside it live under almost constant conditions. *Bryum schleicheri*, one of the most characteristic species of the moss-rich softwater spring fens, is one of the mosses most sensitive to drying out although others in the same genus are very resistant to drought (fig. 343).

The fur-like moss cushions and tapestries are made up principally of species of *Bryum*, *Cratoneuron* and *Philonotis* (tab. 80). Against this background the daintily branched red tinged *Saxifraga stellaris* stands out particularly, while *Saxifraga aizoides* is noticeable with its yellow to orange-brown flowers. It thrives abundantly here but is also found in other habitats. Dwarf forms of *Caltha palustris, Deschampsia cespitosa* and *Cardamine amara* are often well represented.

These swamps at spring and stream sides form a vegetation class which is very isolated and is made up of only the single order. Impoverished communities occur right down into the lowland. They show a much better development in the high mountains and can be easily placed in two parallel alliances: one which is attracted to lime, always rich in mosses and normally forming tuff (*Cratoneurion commutati*) and the other common on the older rocks (*Cardamino-Montion*, or preferably *Montio-Cardaminion*).

The **hardwater-spring moss carpet** (*Crataneuro-Arabidetum bellidiflorae*, tab. 80) can lead into the lime-rich Small Sedges fens which have already been discussed *(Caricetum devallianae)*, because in producing large amounts of tufa it becomes raised above the trickling water in places. The two communities often form an interwoven mosaic which may change to and fro during the course of the year, but in other places one gets the impression that the zonation from the stream water, through the moss carpet to the Small Sedges

Fig. 343. The resistance to drying out shown by different Cushion Moss species as the relative humidity of the air is gradually reduced. From research by Abel (1956), modified.

P = primary, i.e. kept moist beforehand, V = predried, RH = relative humidity. Dark: mosses living.
Bryum schleicheri which is common in alpine-subalpine spring swamps is the most sensitive to dry conditions. The same is true for the class character species *B. pseudotriquetrum* (= *ventricosum*).

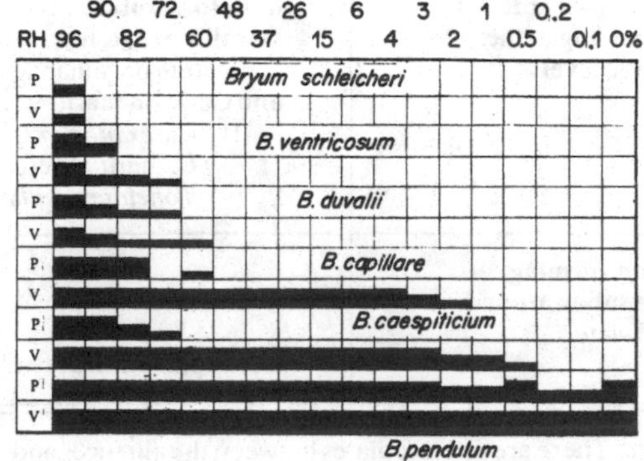

sward remains stable for a longer time. Sometimes a community dominated by the Ice Sedge (*Carex frigida*) is found lying within this series. According to Oberdorfer (1956) it maintains a remarkably constant species composition in the Alps, in the Pyrennees and also in the Black Forest.

Two associations can be distinguished on the older rocks – the **Large Bittercress spring swamp** (*Cardaminetum amarae*) and the **softwater-spring moss carpet** (*Bryetum schleicheri*, tab. 80) The presence of the latter indicates a spring which flows out gently and trickles over the surface and is restricted to the central Alps and the Pyrenees.

7 Tall-herb communities and Green Alder scrub

(a) Subalpine and alpine tall-herb communities

While spring water in the high mountains generally contains only small amounts of plant nutrients, rainwater which has collected on the soil surface is much richer, especially in nitrogen and phosphorus. In hollows and gulleys of the subalpine and lower alpine belts the nutrients brought in by the overflow water can become so concentrated that they give rise to an unusually fertile soil. In contrast to the snow patches these sites become free of snow early in the season. Where they keep their moisture during the warmer time of year they allow the most luxuriant plant growth that it is possible to have in the high mountains. This subalpine or alpine tall-herb community looks like a lowland garden (fig. 344). Since so many favourable conditions must coincide for such rich-green large-leaved herbaceous communities to come into existence they are rarely met with above the tree limit. Usually we meet them at the foot of slopes, in depressions or along stony channels, but not on hillocks or flat places where cattle rest and which therefore receive an excess of dung (see section b).

The tall-herb communities are included in the class *Betulo-Adenostyletea*, i.e. in a unit whose best development is in the subalpine and subarctic regions. Species such as Alpine Sow-thistle (*Cicerbita alpina*), the wood Crane's-bill (*Geranium sylvaticum*) and Plane-leaved Buttercup (*Ranunculus platanifolius*) with its bunches of white flowers are found both in the region of the boreal Birch and conifer woodlands and in the Central European mountains. In the Alps there is only the order *Adenostyletalia;* its character

Tab. 80. Subalpine and alpine spring swamp communities where there is rapidly flowing cold water, and their character species.

Class and order:	**Cold-water spring communities** Montio-Cardaminetea Montio-Cardaminetalia m *Bryum pseudotriquetrum* *Caltha palustris* var. *minor* **D** m *Cratoneuron decipiens* *Deschampsia cespitosa* var. *alpina*	*Epilobium alsinifolium* m *Mniobryum albicans* **D** *Saxifraga stellaris* etc.
Alliances:	**spring-tufa communities** Cratoneurion commutatae **D** m *Cratoneuron commutatum* m *C. falcatum* **D** *Saxifraga aizoides* (?) and character species of the following association	**Soft-water spring communities** Montio-Cardaminion *Alchemilla coriacea* m *Brachythecium rivulare* *Epilobium nutans* m*Mnium punctatum* (?) *Stellaria alsine*
Moss-rich spring communities	**Calcareous spring moss carpet** Cratoneuro-Arabidetum soyeri *Arabis soyeri* m *Philonotis calcarea*	**Soft-water spring moss carpet** Bryetum schleicheri **D** m *Bryum schleicheri* m *Philonotis sericea*
Herb-moss spring communities	**D** = often dominant m = moss	**Large Bittercress spring swamp** Cardaminetum amarae **D** *Cardamine amara* *Montia fontana* ssp. *amporitana*

Source: From data supplied by Braun-Blanquet, Oberdorfer and others.

species include magnificent flowering forms such as the *Aconitum* and *Adenostyles* species, *Veratrum album* and *Carduus personata*. All the tall-herb communities grow very vigorously but, in terms of primary production and height, the European ones are surpassed by those in north-east Asia (Walter 1981).

Many of these large herbs come down regularly into the subalpine and high-montane conifer and broadleaved woods. *Veratrum* and the *Aconitum* species are pasture weeds in some places since they are avoided by cattle and are able to withstand the full sun. The majority of the species listed in tab. 81 then are by no means true to the order *Adenostyletalia*, but taken together they characterise it quite well.

According to Braun-Blanquet (1948/50) some of the subspecific taxa of *Alchemilla* and the splendid *Cirsium spinosissimum* are some of the best character plants, but even these are widely distributed as pasture weeds far outside their natural range (fig. 345). At the foot of sunny rocks where the warmth reflected from the stones melts the snow away much earlier than usual this decorative thistle climbs up as high as the upper alpine level with just a few of its partners.

(b) Resting places for cattle and game

Pasturing is so widespread in the high mountains of Central Europe that there are now scarcely any tall-herb stands which are not influenced by the manurial value of cattle or sheep dung. Without this addition it is probable that these luxuriant communities would be much rarer in the alpine belt. One can observe all the intermediate stages from the near natural tall-herb communities to the resting places of domestic animals (fig. 346) where the plants depend for their existence on the excrement of ruminants. Here and there one comes across heavily dunged places under natural conditions, and these are the resting places of Chamois and other large herbivores. For this reason we are discussing the resting places here even though the majority carry a combination of plants which is brought about by the influence of man.

Alpine Dock *(Rumex alpinus)* usually dominates in the neighbourhood of cowsheds as well as the cattle resting areas and the milking places in the high mountain pastures. This giant sorrel, although avoided by cattle, is readily eaten by goats, and in the past its leaves and rootstock were an almost indispensable pig fodder (Wendelberger 1971). Apart from this only *Senecio alpinus* is worth mentioning as a characteristic plant of the *Rumicetum alpini*, but the Alpine Ragwort is absent from many parts of the Alps (see also fig. 376).

Through the enrichment by dung over decades or even centuries the ground carrying the resting place

Fig. 344. A community of tall herbs in the Alpine garden on the 'Schynige Platte' above Interlaken. In front *Gentiana lutea*. In the background are the snowy peaks of the Eiger, Mönch and Jungfrau.

Fig. 345. The Spiniest Thistle (*Cirsium spinosissimum*) is avoided by grazing animals and so is able to spread from the tall herb community (at the foot of the cliff in the background) into the Violet Fescue sward.

community has become extremely fertile. According to Rehder (1970) up to 250 kg N/ha per annum is mineralised and the aerial parts of the plants produce more than 5 t/ha of dry matter during the year. Since the farm yard manure is no longer piled up outside the stalls, but is spread out over the surrounding pasture-land and since rotational grazing has been introduced nearly everywhere, the tall-herbs are deprived of their most important site factor. In spite of this they hang on stubbornly unless they are mown too often, or, as is often the case nowadays, killed off with chemical herbicides. In the Swiss National Park for example there are former resting places which have had no more dung since 1913 but their community of tall perennials still occupies the same areas as it did before. The main reason for this is certainly the biological cycle of bio-elements which retains the plant nutrients that were once concentrated here and continues to make them available for use. Added to this is the fact that Deer, Marmots and other wild creatures have increased so much in this area where they are completely protected that they have taken over the role of the farm livestock.

Tab. 81. Subalpine and alpine moist-soil scrub and tall-herb communities and their character species.

Class, order and alliance:	**Tall-herb and damp scrub communities** Betulo-Adenostyletea, Adenostyletalia, Adenostylion alliariae	
subalpine to the lower alpine levels C = Class-character species (i.e., also in northern Europe)	*Aconitum napellus* (?) *A. vulparia* (?) **D** *Adenostyles alliariae* *Athyrium distentifolium* **D** *Cicerbita alpina* (C) *Crepis pyrenaica* *Doronicum austriacum* *Epilobium alpestre* **D** *Geranium sylvaticum* (?C) *Milium effusum* var. *violaceum* (C)	*Peucedanum ostruthium* **D** *Ranunculus platanifolius* (C) s *Rosa pendulina* (?) *Rumex alpestre* (= *arifolius*) s *Salix appendiculata* s *S. arbuscula* s *S. hastata* *Saxifraga rotundifolia* *Senecio nemorensis* *Tozzia alpina* *Viola biflora* (?) etc.
Associations on both **lime-rich and lime-deficient soils** (but better developed on the latter)	**Tall-herb community** (without shrubs) Adenostylo-Circerbitetum (also alpine)	
	D *Aconitum napellus* ssp. *vulgare* *Poa hybrida* (?)	**D** *Alchemilla vulgaris* (several subspecies)
	Green Alder scrub (with tall herbs) Alnetum viridis (mostly subalpine)	
	D s *Alnus viridis* *Achillea macrophylla*	*Hieracium jurassicum* *Stellaria nemorum* ssp. *montana*

Note: s = shrub, D = often dominant. New character species list – see section E III, no. 6.3 1.

Source: From data supplied by Braun-Blanquet, Oberdorfer and others.

(c) Green Alder scrub in the subalpine and lower alpine belt

Just as the alpine tall herbaceous perennial communities are connected with the resting place communities through a series of intermediates, so they are with the Green Alder or Prostrate Alder scrubland. This is the reason why we have not dealt with the *Alnetum viridis* along with the other communities of stunted trees (in section B IV 5 b), especially the Mountain Pine scrub, even though they have a similar altitude distribution to the latter and often replace it (see fig. 348).

Green Alder requires relatively more moisture and thrives best on rather impervious silicate rocks and schists (see L. Richard 1969). It often forms quite extensive stands on shade slopes as it suffers less from moving snow than Spruce, Arolla Pine and Larch with their upright trunks. Since it is avoided by all kinds of grazing animals it is a harmful weed of productive pasture land with a favourable water supply and has to be continually removed which is not an easy task (fig. 347).

Green Alder is largely replaced by the Prostrate Pine on limestone and dolomite rocks since the latter is better able to develop on free draining soils which are sometimes very dry. From this there came the idea that *Alnus viridis* was a calcifuge and *Pinus mugo* a calcicole. In fact both species are indifferent as to the soil reaction – a point that has already been stressed in section B IV 5 b. Where the slopes and knolls are of lime-deficient rock, but are nevertheless dry then the *Pinus mugo* dominates, e.g. at the Grimsel Pass; and where a carbonate-rich rock carries a soil with a good water-holding capacity, e.g. in the Smaller Walsertal, it is entirely overgrown with Green Alder (fig. 347). The latter appears to be the more strongly competitive of the prostrate trees. Where the habitat is favourable it smothers out the slower-growing young Pines which require more light.

Since *Alnus viridis* has nitrogen-fixing symbionts in its roots the ground on which it is growing becomes fairly rich in nitrates. According to Rehder (1970) up to 250 kg/ha of mineral nitrogen can be delivered to the soil each year. Because of this the tall perennials can come in wherever there is enough light, and a suitable amount of water reaching the Alder slope. In particular the edges of Prostrate Alder scrub often have a dense luxuriant growth of tall herbs. In fact it would seem that this is their real home and that tall perennials form independent communities only where the Alder is unable to follow them either because of the short growing period or the winter cold or for some other reason, or where man prevents it.

Not all the Green Alder stands are rich in tall herbaceous perennials so they cannot all be considered as belonging to the *Adenostyletalia*. Some of them are more

Fig. 346. Section through a *Rumex alpinus* cattle resting place community in the High Tatra. After Smarda (1963), somewhat modified.

Like all plant communities living under extreme conditions, the resting places are poor in species. The excessive nitrogenous manuring leads to the dominance of species with large, thin leaves (see also fig. 486). The roots and leaves of the Alpine Dock serve as food for pigs. Its leaves are certainly eaten by goats but avoided by cattle which are the most numerous grazing animals. Because of this and the constant addition of excrement the Dock can develop more vigorously than any other herb in the higher mountain region. The character species of the Alpine Dock communities are given in E III 1 (No. 3.512).

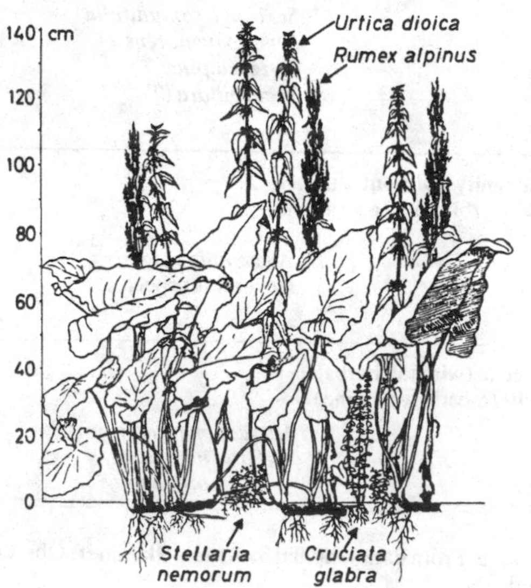

Fig. 347. Green Alder scrub (*Alnetum viridis*) on the marl slopes of the Fellhorn in the Smaller Walsertal. The Green Alder has largely taken the place of the subalpine Spruce wood of which there is a remnant on the steep stony slope (right foreground). It is only on the less steep slopes that it still pays to cut this pasture weed.

reminiscent of the grassland community where *Alnus viridis* has become dominant as a pasture weed giving rise to an unstable combination of species which could be compared with those in certain tree plantations (see section D III 1). The systematic relationships of these Green Alder communities which are poor in large perennials has still not been cleared up. Green Alder are often to be found in large numbers in the *Caricetum ferrugineae* which likes well saturated soils. However it is also frequent in combinations with *Rhododendron* heaths or other acid-soil communities.

Alnus viridis enters into certain combinations of woody species along the banks of subalpine and high montane rivers. From about 1300 to 1500 m upwards it replaces *Alnus incana* as the dominant shrub on river sides which are not flooded very frequently. It appears to withstand the cold much better than Grey Alder or most of the Willow species which accompany the rivers in the lower reaches. Only a few other shrubs are found accompanying the Green Alder in the upper-montane valleys, the most commonly met with being *Salix daphnoides* (Moor 1958).

Alnus viridis is absent from the Scandinavian mountains. At corresponding sites near the tree-line low-growing *Salix* species with mostly silvery leaves attract attention; their role is very small in the Alps and Carpathians. Oberdorfer (1979) recently distinguished the subalpine Willow scrubs (*Salicion arbusculae*) as an alliance of their own, but with rather few character species.

(d) Subalpine Small-reed meadows

The Small-reed swards of the alliance *Calamagrostion arundinaceae* are related to the tall-herb communities in some ways. Carbiener (1969) sees them, in part at least, 'as primary meadows', i.e. as climax communities. They occupy habitats similar to those of the Rusty Sedge meadows (*Caricetum ferrugineae*, see section 2d).

These 'natural meadows' as Carbiener points out in his first survey of them, are extremely heterogeneous, in the literal sense. Floral elements of diverse sources and from different ecological formations meet here, especially some from the dwarf shrub heaths, tall-herb communities, subalpine and montane woodlands, woodland edges and steppe woods as well as representatives of the half-dry grasslands and the manured meadows, e.g. *Centaurea scabiosa*, *Carlina vulgaris* and *Avenochloa pubescens* on the one hand and *Heracleum sphondylium*, *Pimpinella major*, *Leucanthemum vulgare* and *Tragopogon pratensis* on the other. In addition several 'adalpine' species as defined by Schönfelder (1968), i.e plants distributed in the pre-Alps, but not in the alpine belt, also come in. However it is usually the acid-tolerant Small-reed *Calamagrostis arundinacea* which is dominant. This tussock grass also plays a large part in the open subalpine coniferous woods. The Small-reed meadows which are very rich in species, reach their optimum development on south to east facing slopes with a gradient of 25-50°, i.e. lee slopes where huge snow cornices build up and slide down preventing the establishment of trees, even though it is below the climatic tree limit.

Carbiener is inclined to accept that grassland areas such as these are the natural original home of our manured meadow communities although he also admits that some of the species may have come in more recently under the influence of man, i.e. as secondary

Fig. 348. Green Alder (*Alnus viridis*) and Mountain Pine (*Pinus mugo*) form areas of scrub extending down well below the tree limit on the north-facing slope of the Bernina valley where there are frequent avalanches. From a sketch in perspective by Holtmeier (1969), modified.
Green Alder prefers the places where water is readily available on this acid scree soil leaving the Pine to occupy only the crests and summits. Trees (*Larix europaea* and *Pinus cembra*) are able to survive only on the ridges between the avalanche tracks, although under natural conditions they would be more numerous.

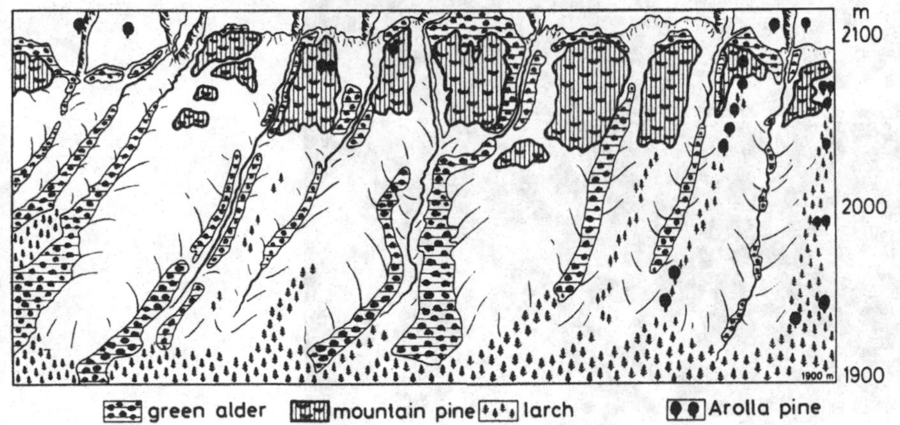

partners in the *Calamagrostietum*. Further evidence is required if this question is to be resolved.

8 The colonisation of scree slopes and glacial moraines

(a) Plant succession on carbonate- and silicate-screes

Apart from the snow patches and a few of the grassland communities all the vegetation units so far discussed have their optimum development in the lower alpine or even the subalpine belts. If they reach the nival belt at all this is only in a very fragmented condition. In contrast the vegetation of rocks and screes to which we shall now turn finds its most unrestricted developmental possibilities high above the tree limit and above the glacial moraines which consist mainly of finer particles. The communities on rocks and screes are formed mainly of very definite light-requiring plants which would soon die under the competition of a dense growth, or, as Zöttl (1951a) has shown experimentally, could not even germinate on other substrates. Schröter (1926) described the occupants of the stony ground communities as 'refugees' from the competition for existence.

Very extensive screes, some dating back to the Ice Ages, others more recent, cover large areas of the peaks towering up above the snow-line especially in the dolomite and limestone massifs (fig. 349). Their supply of stones of all sizes came, and is still coming, from the rock faces above them. During the mornings in early summer and autumn one can hear a constant clatter of the pieces falling down. These have been dislodged by water freezing in cracks during the night and then being thawed out by the sun. Where the stone fall is particularly active only a few plants venture, but nevertheless one is amazed at the diversity of life which manages to develop on these steep unstable slopes.

The first impression they give is of a habitat which is less hospitable than it really is. If one follows the stems of the plants down through stones one finds them rooting in pockets of finer soil which have accumulated in some places. These may be covered by only 10-20 cm of bare dry stones (fig. 350). Under such conditions the phanerogams thrive best where they have lain protected by the winter snow for 7 or 8 months and then when the snow melts they have a good water supply far into the short summer. According to Zöttl (1953) even at this altitude the open stony screes facing the sun can offer a remarkably warm habitat where the radiation is nearly vertical.

Fig. 349. Sliding and stabilised limestone screes with growth of varying density in the lower alpine belt of the Lechtal Alps close to the Stuttgart cabin. Sharp-edged rock debris of all sizes cover the subsoil which itself is much richer in finer particles. In the background snow is still lying in August in some depressions and on shaded slopes.

Today many of the screes have a negligible quantity of new material added to them each year and are no longer moving. The vegetation has had the time, often over a number of centuries, to overgrow them completely. However these screes which have come to rest, as well as the lateral moraines and other formations consisting of fine earth with only a shallow covering of stones are not our first consideration.

Within the large group of **scree-slopes** we are able to distinguish those which are still '**active**' (i.e. continually being changed by the addition of new material), those which are '**slipping**' (no longer active but still unstable through the the agency of animals or man stepping on them or through other small disturbances, e.g. frost lifting the surface and trickling water from the melting snow), and those '**at rest**'. These last generally have a gradient of less than 37° (for coarse material) or of 27° (for finer material) because during the course of time they have been flattened out somewhat. Depending on the average diameter of the stones one could refer to them as 'block screes' (25 cm), 'coarse screes' (2-25 cm) and 'fine screes' (0.2-2 cm). The factor which has greatest influence on how easily they can be occupied by plants however is the quantity and distribution of even finer particles. The least favourable are the '**block-screes**' which arise because of rock falls, i.e. single events (Mayer 1964). Further information about the different kinds of screes in the sub-nival, alpine and subalpine belts can be found in works by Zöttl (1951a), Zollitsch (1966) and other authors whom they quote.

The vegetation of the 'active' screes which are still in the process of formation and are still sliding down can be divided into a group of **limestone-scree communities** (*Thlaspietalia rotundifolii*) and a group of silicate scree communities (*Androsacetalia alpinae*, tab. 82). Each of these orders has several character species whereas the number of scree plants found on all the different kinds of stone (character species for the class *Thlaspietea rotundifolii*) are relatively few.

The scree plants are met with in many different combinations because their root runs are so unstable and heterogenous. Representatives of the **Pennycress scree** (*Thlaspietum rotundifolii*) are found on limestone and dolomite screes in the alpine and nival belt. The **Mountain Hawkbit slope** (*Leontidetum montani*) occupies finer, damper clay-slate screes. Limestone screes and marl slopes which have a more or less fine, moist soil and which go down into the subalpine belt are taken over by some subassociations and variants of the **Alpine Butterbur slope** (*Petasitetum paradoxi*).

Screes composed of large stones which dry out on the surface are characteristic of the dolomite and hard limestones, but are found only exceptionally in the silicate Alps. Where they appear they are only sparsely populated with fragments of the alliance *Senecion leucophyllae* which is much better developed in the Pyreenes.

As a rule conditions for growth on the alpine **silicate screes** are more favourable from the outset since they are much richer in fine sandy material holding the water better than is the case with the limestone screes. Silicate screes also soon come to rest and then they are densely covered by communities belonging to the *Androsacion alpinae*. Even the rather stony **Rock Jasmine scree** (*Androsacetum alpinae*) does not indicate that new material is still being added to the scree, and therefore it is not comparable with the *Thlaspietum rotundifolii*. All the other communities which have been described from silicate substrates up to now prefer stable slopes (see tab. 82 and section 3 b).

The **Mountain Sorrel scree** (*Oxyrietum digynae*) is found on silicate screes poor in humus and in similar habitats

Fig. 350. Root layers under a dense stand of the limestone scree community (*Thlaspeetum rotundifolii*) on the Glärnisch, 2250 m. After Jenny-Lips (1930), modified.

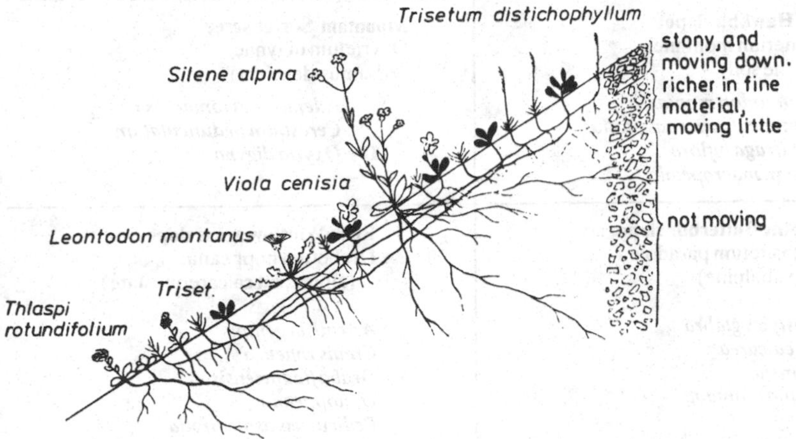

Tab. 82. Synopsis of the alpine scree communities and their character species (compare section E III 1, no. 4.4).

Class:	**Scree communities** of the high mountains Thlaspietea rotundifolii *Calamintha nepetoides* *Campanula cespitosa* D *C. cochlearifolia* *Galeopsis ladanum* *Gypsophila repens*	*Hieracium staticifolium* *Linaria alpina* D *Rumex scutatus* *Saxifraga biflora* *Scrophularia canina*
Orders	**Calcareous** scree comm. Thlaspietalia rotundifolii *Arabis alpina* D *Doronicum grandiflorum* D* *Gymnocarp. robertianum* *Leontodon hisp.* var. *hyoseroid.* D *Saxifraga oppositifolia* (?)	**Silicate** scree comm. Androsacetalia alpinae * *Cryptogramma crispa* *Epilobium anagallidifolium* g *Poa laxa* *Ranunculus glacialis*
Alliances dominating in the Alps: D = locally abundant * = fern g = grass	Pennycress scree comm. Thlaspion rotundifolii (subalpine to nival) D *Achillea atrata* *Cerastium latifolium* * *Cystopteris regia* * *Dryopteris villarsii* *Galium helveticum* *Hutchinsia alpina* D *Leucanthemum atratum* g *Poa alpina* var. *minor* * *Polystichum lonchitis* g *Trisetum distichophyllum* *Valeriana supina* *Viola calcarata*	Rock Jasmine scree comm. Androsacion alpinae[a] (alpine to nival. relatively damp) *Cardamine resedifolia* *Cerastium uniflorum* D *Geum reptans* *Saxifraga bryoides* Dg *Trisetum spicatum* [a]the alliance *Senecion leucophyllae*, which is more abundant in the drier Pyrenees, has not been considered here.
Up to the nival level	**Pennycress scree** Thlaspetum rotundifolii (alpine to nival) *Moehringia ciliata* *Papaver alp.* ssp. *sendtneri* *Saxifraga aphylia* D *Thlaspi rotundifolium* *Viola cenisia*	**Rock Jasmine scree** Androsacetum alpinae (high alpine to nival) *Androsace alpina* *Gentiana bavarica* var. *subacaulis* *Saxifraga seguieri*
Alpine associations of damper sites:	**Mountain Hawkbit slope** Leontodonetum montanei (rich in fine soil) *Leontodon montanus* *Ranunculus parnassifolius* *Saxifraga biflora* ssp. *macropetala*	**Mountain Sorrel scree** Oxyrietum digynae (humus-deficient)[b] *Adenostyles tomentosa* *Cerastium pedunculatum* D *Oxyria digyna*
Other associations (*Drabetum hoppeanae* belongs to the order *Drabetalia* with a single alliance *Drabion*)	**Alpine Butterbur slope** Petasitetum paradoxi (subalpine) D *Adenostyles glabra* var. *calcarea* g *Poa cenisia* *Valeriana montana*	**Rock Whitlow grass slope** Drabetum hoppeanae (alpine, on calcareous slate) *Artemisia genipi* *Crepis rhaetica* *Draba fladnicensis* *D. hoppeana* *Pedicularis asplenifolia*

Note: [a] The Brown Woodrush sward (*Luzuletum alpina-pilosae*) which grows on humus-rich soils has been omitted.

Source: From data supplied by Braun-Blanquet, Oberdorfer and others.

on young moraines. Slopes which are richer in humus and remain covered with snow for a longer time are often carpeted with the **Alpine Woodrush sward** (*Luzuletum alpino-pilosae*). This open hygromorphic meadow-like community generally leads into the Dwarf Willow snow patches or into moist subunits of the Crooked Sedge swards.

Communities of the **calcareous slate slopes** (*Drabetalia hoppeanae*) are seen by Zollitsch (1966) as special units lying between those on limestone and on silicate material. Merxmüller and Zollitsch (1967), with the approval of E. Pignatti (1970a) have raised these units to the rank of a new order. As can be seen from fig. 351 there is no difference between the communities of the calcareous slate slopes (*Drabion*) and the limestone slopes (Thlaspion) in respect of the soil acidity, although both contrast with those of the alliance *Androsacion alpinae*. The main ecological differences between two first named alliances are to be seen in their water balance, and probably also in their nutrient supply.

All the plants of the screes and similar bare ground must have very good means of dispersal in order to reach the isolated newly formed potential growing places. Zöttl (1951b), Söyrinki (1954) and earlier authors have made a thorough investigation of this aspect. They all agree that the majority of the species produce plenty of seed, in spite of the short growing period, and that most of these are wind dispersed. In some species the whole plants change into 'snow runners' which the wind drives over the snow surface (e.g. *Thlaspi rotundifolium*).

Zöttl came to realise through sowing experiments that, although the germination was good, only a few seedlings were able to develop because they were unable to find suitable conditions, or they died off in the tender seedling stage. Thus, on the one hand chance plays a large part in the colonisation of the scree slopes. On the other hand species foreign to the association completely failed, even when they were sown artificially, the hard conditions operating a selection even during the germination stage. Once the plant has become established it spreads vegetatively and may be carried down with the moving scree (fig. 350). There is no lack of nutrients according to Zöttl's (1952a) investigations because even the limestone scree has sufficient for the plant's requirements, and the dirt accumulated in the snow cover represents a significant addition here as it does in other sparsely populated bare soil habitats. Since the water supply is generally abundant mesomorphic phanerogams can predominate also on sunny slopes; they are often strikingly tender and thin-leaved with a relatively rapid growth rate.

The many growing strategies of scree plants which have been distinguished by Schröter (1926), Jenny-Lips (1930) and other authors can be put into three main groups: 1 Those which move **passively** along the scree, 2 Ones which **hold up** the scree material and 3 Those which **creep over** the scree. Many plants which finally are able to bring the scree to a stop behave passively at first, and many of the ones which creep over the surface also help to secure the sliding ground. Only the surface creepers accumulate a layer of humus and eventually lead to the formation of a scree-rendzina (or a scree-ranker on acid material). Finally this process may be completed by sward plants.

However, succession to a sward community is only possible after the stone fall has practically ceased and the scree has by and large stopped moving. This stabilisation can often be seen to start in the 'shadow' of a large boulder which protects a longitudinal strip of the scree from the effects of the downward movement and the fall of fresh material. The *Seslerio-Semper-viretum* in particular readily occupies such sites. Finally the whole of the scree can be taken over once it has ceased to be active (see fig. 331).

Before inferring that adjoining strips of different types and densities of vegetation are an example of succession in time one must always ascertain whether they are occupying ground of similar structure. This is certainly not the case in the majority of the 'pioneer' and 'climax stages' which have been described in the literature. J. Braun (1913) stressed long ago that generally speaking the open scree vegetation in the upper alpine and nival belts must be considered as local permanent communities. The true climax vegetation

Fig. 351. There is no difference between the soil acidity of the scree communities on limestone (*Thlaspion*) and on calcareous slate (*Drabion*) but the two together contrast with that of the silicate scree communities (*Androsacion alpinae*). After Merxmüller and Zollitsch (1967), modified.

The habitat of the silicate screes moreover is moister and much more stable than that of the limestone screes.
Calcareous slate screes are intermediate in this respect.

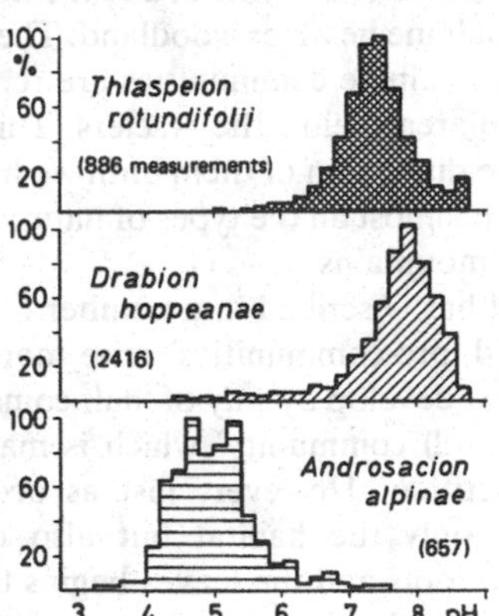

of the particular altitude zone can only develop on soils which from the outset are favourable, that means rich in finer particles. Nevertheless in the absence of accurate investigations we cannot go usefully into the question of succession.

Stone scree communities play only a minor role away from the high mountains. As relatively frequent may be mentioned the **Hemp-nettle community** (*Galiopsietum angustifoliae* see Schönfelder 1967) for which man has extended or created new habitats, e.g. in Thüringia (Hilbig 1971b). The **French Sorrel scree** (*Rumicetum scutati*) is a pioneer on dry screes in south-western Central Europe (fig. 381). The **Limestone fern scree** (*Gymnocarpietum robertiani*, see figs. 154 and 362) on the contrary needs shaded damp conditions. According to Oberdorfer (1977) all these three communities belong to the order or Rough-grass scree communities (*Achnatheretalia calamagrostis*) which is named after the **Rough-grass scree** (*Achnatheretum calamagrostis*) distributed in the lower and warmer paths of the valleys in the Alps.

(b) Successions on young moraines in the alpine and subalpine belts

Nowhere can succession be studied more profitably than in the valley below the front of a large glacier. As in other mountains of the world during the past 100 years or so the majority of glaciers in the Alps have been gradually receding, following a reduction in the supply of firn snow. According to Zingg (1952) they have suffered a reduction in area by as much as 28% in Austria and 30% in Graubünden between the years 1895 and 1940.

This melting of the glacier ice however does not mean that we are enjoying a long term improvement in the climate such as occurred in the Post-glacial warm period. It is much more likely that we are living in a time which is somewhat cooler on average, but in which the glaciers are alternately advancing and retreating. According to Heuberger (1968) the glaciers were very extensive during the years 1400-1200 B.C. and 900-300 B.C. and again in 100-750 A.D. Since about 1250 they have been even further back than they are at present. For example in the Sonnblick and Glockner areas there are fifteenth and sixteenth century mines still covered by glaciers today (Klebelsberg 1949), and old irrigation ditches in Wallis can be seen above the level of the present day outflow of several glaciers. Heuberger records a particularly large advance of the Rhône glacier about 1600 the movements of which have been accurately measured (see fig. 354). Other glaciers advanced even further down the valleys during 1820-1850, and about this time the frontier of the perpetual snow moved down by some 100 m on average. After a period of rapid melting new forward thrusts were observed from about 1890 to 1900 and 1910 to 1920 which probably had some connection with the cloud of volcanic dust from the Krakatoa explosion which persisted for a long time at a height of about 30 km (compare also fig. 319).

The last mentioned decade was the coolest we have so far experienced this century. A few smaller advances have occurred since and these have left moraine banks which can be dated (fig. 352). These have not happened since 1940 because almost all glaciers have revealed a fresh area of moraine or rock each year, the amount depending on the weather conditions. Since 1964 more and more glaciers have started to advance again (F. Müller 1980) but they still have in front of them a series of soils from very young to centuries old.

These soil series in front of the glaciers offer a large scale natural experiment to the scientist studying vegetation or soil development. By comparing the records from moraines of different ages he can obtain a good idea of the course and the velocity of succession. A more exact method would be the observation of changes taking place inside permanent quadrats such as the ones laid down by Lüdi (1945) near the large Aletsch glacier or by Jochimsen (1963, 1970) in the Tyrolean Alps.

Depending on the climatic and soil conditions, the size of the glacier and other circumstances the development of both the vegetation and the soil goes on in different ways and at different speeds (a review of older investigations by Lüdi 1958a). The succession always starts off with a **pioneer stage** resembling the scree flora, even where we are dealing with moraines which extend down into the subalpine level. After a while the initial, more or less gappy vegetation is nearly always followed by a **sward stage** which in some cases resembles the alpine grassland and in others the snow patches or spring communities. This is also the case in deeper valleys, i.e. in warmer conditions, but here **woody plants** soon come in and the flora develops towards subalpine heath or woodland. The majority of subalpine and alpine communities are represented in the moraine areas below the glaciers. This is why we have left the discussion of them until we have become familiar with almost all the types of natural vegetation in the high mountains.

Friedel has described in a number of works how the so-called 'pre-communities' arise more or less by chance. They develop by way of 'half communities' to a balanced 'full community' which is maintained by keen competition. However, just as occurs on the screes, not only the habitat but also competition between the roots and the leaves begins to affect the plants right from the early pioneer stage; and this biotic

Fig. 352. The Stein glacier with its tongue basin as seen from the large terminal moraine.

The shadow has just reached the gateway of the glacier. On the left the lateral moraine formed about 100 years ago when the glacier was at its greatest height is clearly visible. In the foreground and to the right can be seen the lateral and terminal moraines which have been formed since 1920. The most extensive moraine steps lie to the right outside the photograph. Further details can be seen in the generalised diagram in fig. 353.

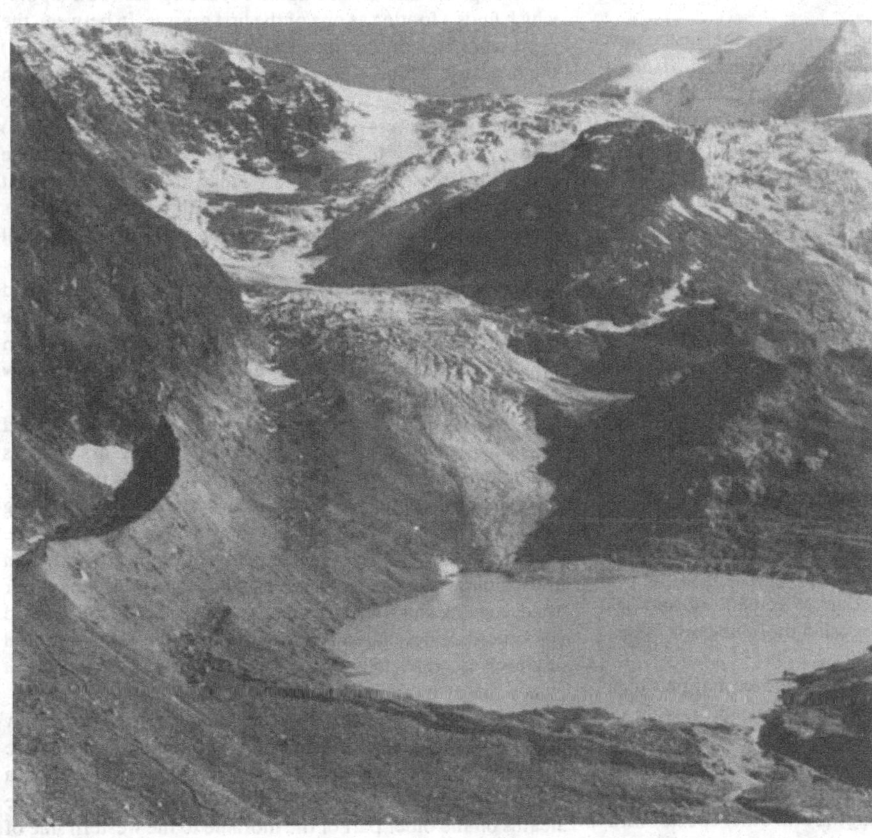

Fig. 353. A semi-schematic longitudinal section of a glacier similar to that shown in fig. 352. After Streiff-Becker (1944), from Ellenberg (1963).

When the glacier recedes because of the melting of the ice then the drumlins (RH), lateral moraines (SM), basal (GM) and terminal moraines are all available for occupation by plants. The firn line (FL), i.e. the boundary above which the snow does not melt even in summer, lies somewhat lower than the climatic snow limit.

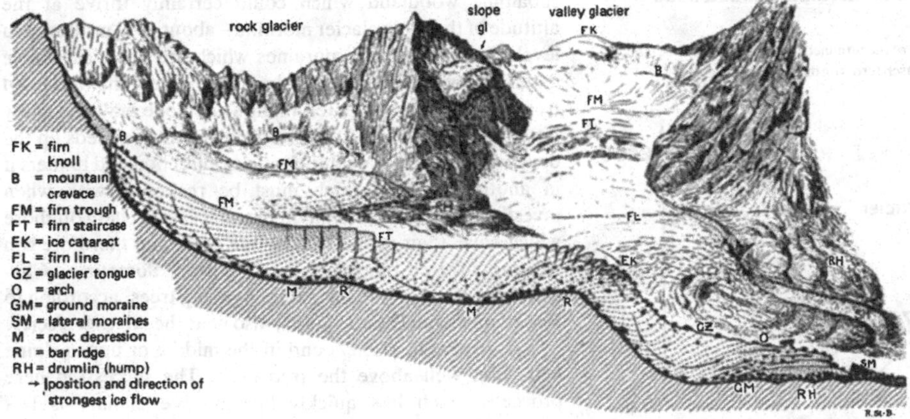

factor continues to influence selection among the newcomers which arrive from near and far. As a consequence each glacier valley has its own peculiarities and we must end with a brief introduction to certain examples.

Our first choice is the **Stein glacier** below the Susten pass road between Göschenen and Innertkirchen (fig. 352) since it is easily accessible and, being a small glacier, can be readily surveyed. Thus it is better suited to the first excursion than many of the better known glaciers. Like most of the bigger glaciers the Stein glacier is fed by a number of firm snowfields. Its melting tongue feeds a small lake through an exit hole which opens up from time to time. The broad lake basin was dug out by the glacier before 1920 and became surrounded by a high bank of moraine. From the position of this bank it can be seen that the lake was previously some 5.5 m higher than it is today. On 31st July 1956 it broke through the moraine wall and tore out a wide stony stream bed which is now bridged just above the hotel. In the twenties it was touch and go whether the hotel would have to be evacuated as the glacier was extending relentlessly down the valley. Now one has to walk for a good quarter of an hour in order to reach it from the hotel. A few years ago the glacier began to advance again, but still quite slowly.

On its right side where the lateral moraine falls steeply away there are only the remains of dead ice, but on its left side the glacier has had ample room to spread out and has formed a long echelon of small parallel moraine banks (figs. 352 and 353).

The sandy or stony material near to the ice margin which has only been exposed for a few months is still devoid of plants, but where the moraine is a year old a careful search will reveal the first seedlings of *Epilobium fleischeri*, one of the most plentiful pioneers provided there is sufficient moisture in the ground. *Oxyria digyna* and other species with wind-dispersed seeds are also found at this stage. During the following two or three years the ground, which still has no mosses or lichens, becomes dotted with the pink-flowered Willow Herb, groups of brownish-red Mountain Sorrel and the soft green clumps of Alpine Meadowgrass (*Poa alpina*) which finds a footing as one of the first grasses. In between the representatives of the silicate scree community (*Oxyrietum*) the brownish-yellow stars of *Saxifraga aizoides* shine through where the ground is somewhat damper. Quite often one also sees the blue flowers of the Alpine Toadflax (*Linaria alpina*) which, like *Arabis alpina* (fig. 355), is at home on limestone screes. Its presence here is an indication that the recent moraine, in spite of originating from acid rocks, still contains carbonates in its top soil. Where it is mainly sandy mosses will form the first carpet within a year, soon followed by lichens. The commonest species are *Racomitrium canescens* and *Stereocaulon alpinum*. Thus **at quite an early stage there are different communities** in different parts of the moraine, and in no way can it be chance alone which determines how the raw soil will be occupied once the ice has retreated from it.

As long as the lateral moraines are not too stony a sward of phanerogams begins to develop after 10-20 years and this gradually becomes more dense and richer in species. Occasionally cattle may wander over it to graze, leaving behind droppings which contain the seeds of clovers. These will germinate and the plants spread out concentrically. In those hollows between the banks of moraine where the water tends to hang about species of the Common Sedge fen can find a foothold. Where it is not so wet the Espalier Willows spread themselves, especially *Salix serpyllifolia* and *S. reticulata* which are at home in the limestone snow patches.

However more than 30 years must elapse before a few upright woody plants grow up here and there. Typically the most likely will be *Rhododendron ferrugineum* since this is a pasture weed avoided by cattle. This Alpenrose forms dense heaths on the older part of the moraine to the western side of the Stein glacier valley to which no date can be attached. The well defined lower boundary of this dwarf shrub heath marks the start of the tract of land which has been clear of ice for a long time. Further up individual upright specimens of the Mountain Pine grow between the large boulders. These may be looked upon as the remains of a former extensive subalpine woodland which could certainly thrive at the altitude of the Stein glacier moraine – about 1800 m above sea level. On the younger moraines which are more accessible and with fewer large boulders the cattle prevent the woodland trees from coming in. After a long search one may find a tiny bitten-down Spruce. Thus the grazing of cattle speeds up the development of the vegetation in one direction and hinders it in another, a fact which must be reckoned with when investigating the succession on moraines, especially on those which have a fair proportion of fine earth and therefore can become covered quickly with vegetation. Sometimes Chamois contribute to the browsing of young trees, according to Bodenmann and Eiberle (1967) also near the Aletsch glacier.

Many small glaciers end in the middle or upper alpine belts, i.e. well above the tree limit. The succession here proceeds much less quickly because the climate is less

Fig. 354. Between 400 and 200 years ago the Rhône glacier filled the wide valley but during the past 120 years has receded up into the slopes. It advanced for a short time between 1912 and 1921 leaving behind on its further retreat a terminal moraine. From data supplied by Mercanton and Renaud (1955), in Ellenberg (1963).

Abscissa = time, ordinate = distance of the terminal moraine or the front of the glacier from the healing spring at Gletsch (cross on the left).

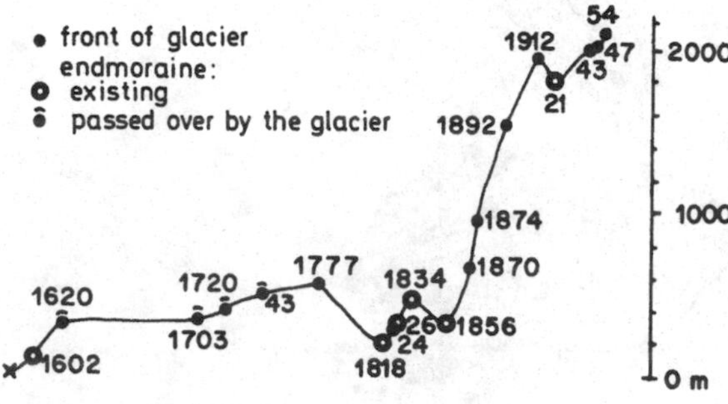

favourable; it resembles that of the screes. The terminal moraine of the **Hintereisferner** for example lies between 2400 and 2250 m above sea level, and here the number of species is added to very slowly over the years, the more so since this moraine material is relatively dry (tab. 83). On the other hand Friedel (1938a) established that there is a succession series leading to the communities of snow patches or springs. The development on the dry moraine follows a somewhat different course on the shaded side of the valley than it does on the sunny side. Thus the general and local climates, the amount of fine earth available, and the water supply of this new land exert a selection which is already apparent at a very early stage.

On the left lateral moraine of the large **Aletsch glacier** which also contains a lot of fine earth, grazing is no longer carried on and the progress of a natural succession can be followed much better (fig. 356). It has been described in detail by Lüdi (1945) and J.L. Richard (1968b). Unfortunately its first stages are difficult to distinguish because this gigantic glacier has shrunk by some 100 m during the past century, and its lateral moraines actually form very steep unstable slopes which are continually sliding away. Only in a few places do they remain as terraces of different ages. On these the first plants are herbaceous phanerogams as on the end moraines of the Stein glacier (fig. 357 and tab. 84). Mosses and lichens appear a few years later and soon after that dwarf shrubs, bushes and trees, especially Larch, become more prominent. Depending on the character of the subalpine climate the development quickly proceeds in the direction of a Larch-Arolla Pine wood so that the grasses and herbs are only temporarily in possession of a major proportion of the surface.

Many of the large boulders being carried on the long and broad central moraine of the Aletsch glacier are already covered with lichens before they are deposited on the moraine. Here they can be distinguished from the smoothly polished naked boulders which are set free from the melting ice. On the central moraine of the Unteraar glacier numerous phanerogams are able to thrive whereas they are found only occasionally on that of the Aletsch glacier. Lüdi (verbal communication) has even seen woody plants growing on such unstable habitats.

The **Morteratsch glacier** above Pontresina retreated very quickly as can be shown by means of a number of dated boulders. Its terminal moraine is covered with numerous large stones and only in the older part has it any depth of fine soil. This has given rise to a vegetation mosaic where hollow places with a rich sward and sometimes even Larch and Arolla Pine are mixed in with collections of boulders that are still almost bare and less stony humps which have a gappy plant cover (see Lüdi 1958b, Braun-Blanquet 1951). Here it is certainly necessary to keep the influence of soil conditions and of the time available for the colonisation strictly apart. Thus the large glaciers in the Swiss Alps which are so impressive and the object of many excursions offer much more difficult conditions for a comparative study of succes-

Fig. 355. Mountain Sorrel community on an approximately 10 year old moraine of the Stein glacier which although mainly siliceous does contain some limestone. *Linaria alpina* and *Epilobium fleischeri* (left), *Arabis alpina* (centre) and *Oxyria digyna* (right).

sion than do the smaller ones. Unfortunately this applies also to the **Rhône glacier** the movements of which have been measured more accurately than any other glacier in the world (see also figs. 354 and 358, Friedel 1938b).

The majority of the glacier moraines which have been closely examined by ecologists up to now are composed mainly of stones poor in lime. Only the **Hüfi glacier** (Lüdi 1934) brought with it limestone material, and here the development of the vegetation follows a course most similar to that on the limestone screes. Unfavourably however the bottom of the valley here, which in some places has been scraped clean by the glacier, is a gneiss which is deficient in lime. Thus the differences in carbonate content, pH value and the amount and quality of humus which Lüdi found when comparing the young alluvial soils with those of the older humps shaped by the glacier cannot be simply a result of a progressive soil development. In the case of the rounded humps and the fine earth which covers them we are dealing with a material which was lime deficient from the start. However the **Upper Grindelwald glacier**, which also stretches right down into the woodland belt, has left behind it a uniform material in the end moraine and here Lüdi (1945) found that ripening of the soil needs much more time than normally expected by pedologists and ecologists.

If the leaching and acidification of soils goes on so slowly in the Bernese Oberland with its high precipitation at 1250 m how much more slowly must it progress in the alpine belt, another 1000 m higher, where all the soil-building processes are restricted (see section 1 f). Only enrichment with humus makes relatively rapid progress on the terminal moraine, especially at the montane and subalpine levels where trees, dwarf shrubs and woodland mosses are able to contribute to

Tab. 83. Some stages of the vegetation development before an alpine glacier, the Hintereisferner in the Ötztal Alps, on a dry moraine scree exposed to the sun.

	Distance from glacier (10 m)		2	3	4	5	8	10	25 → about 1890	45	60
	% age cover		+	+	+	1	3	10	40	90	95
	Stone debris and scree plants:										
	Cerastium uniflorum	C	+	+	**1**	**2**	**1**	+	+		
g	*Poa laxa*	H	+	+	**1**	**2**	**1**	+	+		
	Cardamine resedifolia	H			+	+	+		+	+	+
	Linaria alpina	H			+	+	1	+	+	+	+
	Saxifraga bryoides	C			+	+	+	1		1	1
	Cerast. arv. ssp. *strict.*	C			+	+	1	+	+	+	+
	Epilob. anagallidifol.	H			+	+		+		+	+
	Snow-patch plants:										
	Tanacetum alpinum	H		+	1	1	+	1	1	1	1
	Cerastium cerastoides	C			+		+		+	+	1
	Sagina saginoides	H				+	+	+	1	+	+
	Arenaria biflora	C				+		+			+
	Salix herbacea	Z					+	+	+	1	2
m	*Polytrichum sexangul.*						+			+	+
	Sedum alpestre	C						+	1	+	+
	Gnaphalium supinum	H						+	+	+	+
	Other **mosses:**										
	Pohlia spec. (short-lived)			+	+	+	+	+	2		
	Furnaria hygromete. (")			+	+			+			
	Polytrichum juniperinum				+	+	+	2	2	3	3
	Racomitrium canescens					+	+	1	2	1	+
	Polytrichum piliferum						+	+		+	+
	Black liverworts							+	2	2	+
	Lichens:										
	Stereocaulon alpinum								+	+	+
	Cetraria islandica									+	+
	Cladonia pyxidata										+

	Distance from glacier (10 m)		2	3	4	5	8	10	25 → about 1890	45	60
	% age cover		+	+	+	1	3	10	40	90	95
	Sward plants sens. lat.										
g	*Agrostis rupestris*	H			+	+	1	2	3	1	2
	Silene accaulis	C					+	+	2	2	1
g	*Festuca varia*	H					+	+	2	1	2
g	*Poa alpina*	H					+	+	+	1	2
g	*Deschampsia cespitosa*	H					+		+	+	
l	*Trifolium pallescens*	H					+	+	2	**4**	**4**
g	*Luzula spicata*	H						+	+	+	+
g	*Festuca rubra*	H						+	1	1	2
g	*F. halleri*	H							+	2	+
	Campanula scheuchzeri	H							1	+	1
	Leontodon hispidus	H							+	+	+
	Senecio abrotanifolius	C							+	+	+
g	*Nardus stricta*	H							1	2	2
g	*Anthoxanth. odoratum*	H							+	1	1
	Thymus serpyllum	C								2	1
l	*Lotus corniculatus*	H								2	1
l	*Trifolium badium*	H								1	1
g	*Phleum alpinum*	H								+	+
l	*Trifolium thalii*	H									+
	Semiparasites:										
	Euphrasia minima	T							+	2	1
	Bartsia alpina	G								+	+

g = grasslike, l = leguminous, m = moss
Life forms: C = herbaceous chamaephyte, Z = woody chamaephyte, H = hemicryptophyte, G = geophyte, T = therophyte.

Note: Chamaephytes, hemicryptophytes and some mosses are pioneers; lichens and plants with specialised nutrition appear relatively late.

Source: From data in Friedel (1938). The figures show quantities according to the Braun-Blanquet scale; semi-bold = species which characterise that stage in the vegetation development. Sample areas 50–100 m^2.

it. Even without their help though organic material collects on the young moraines coming partly from the remains of phanerogams and cryptogams growing there and partly from organic dust trapped in the snow. The amount of combined nitrogen found in the topsoil is a good measure of the extent of this organic enrichment (fig. 358). Gradually the amount of circulating nutrient capital is increased until an optimum is reached which depends on the climate and the nature of the soil.

Much can be learned from the successions of vegetation and soils on the end moraines of the arctic glaciers which, compared with those in Central Europe, are scarcely influenced at all by man (Ives and Barry 1974). As an example we may quote some data supplied by Viereck (1966) from the pebbly moraine of the Muldrow glacier in Alaska. Here there is practically no more bare soil to be seen after 25-30 years, and after 200-300 years all the species of the climax vegetation are present. However the stages which have reached an equilibrium are at least 5000 years old and the ripening of the soil has taken an equally long time.

Fig. 356. Lateral moraines of the Great Aletsch glacier (the biggest glacier in the Alps) below the Aletsch Forest (left). Photo, Lüdi. Many Larch and some Arolla Pine had already occupied the older moraines (centre) by 1944. The younger moraines (right) are still without trees.

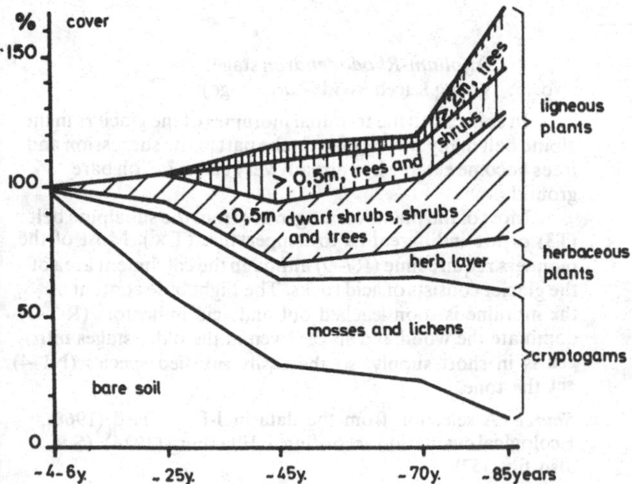

Fig. 357. The degrees of cover in the different vegetation layers in test areas on lateral moraines of increasing age by the Great Aletsch glacier (see fig. 356). After Lüdi (1945), modified. More than 95% of the surface is still bare after 4–6 years. But after 85 years only about 10% of bare ground remains; woody plants together cover almost 90%. Since the layers overlap in parts the sum of the areas covered by all the layers comes to more than 100%.

9 The vegetation of rocks and boulders

(a) Communities occupying rocks and crevices

The higher one climbs into the mountains the more the landscape is dominated by precipitous rocks. However even the highest ridges and the steepest rock faces are not without their plants provided they are free from snow for a few months in the year (fig. 359). Such bare rocks are the domain of the drought-resisting hardy lichens. Phanerogams are only able to occupy cracks and crevices in the rocks or piles of rubble, and only a few individuals are able to reach the highest peaks (see fig. 366).

Tab. 84. The succession on the lateral moraines of the Aletsch glacier in the subalpine woodland belt.

Serial no.:	1	2	3	4	5	T	R	N
Time since the retreat of the glacier (years) min.	5	10	30	60	100	Ecological evaluation		
max.	10	30	60	100	?			
% age cover: Tree layer	–	–	–	5	10			
Shrub and herb layer	10	10	60	90	40			
Moss layer	–	10	10	30	10	T	R	N
Herbaceous pioneer group:								
Oxyria digyna	1	1				1	3	×
Cerastium peduncul.	+	1				–	–	–
Linaria alpina	1	1	·			×	8	2
Arabis alpina	+	+	·			2	9	3
Achilla moschata		+				–	–	–
Willow-moss pioneer group:								
S *Salix appendiculata*	·	+	·	·	·	3	8	5
Z *S. reticulata*		+	+	·	·	3	9	3
Z *S. herbacea*		+	·	·		2	3	4
Z *S. retusa*		+	1	1		2	8	4
S *S. helvetica*		+	1	+	·	–	–	–
S *S. hastata*		+	+		·	3	7	4
S *S. foetida*		·		+	·	–	–	–
Z *S. serpyllifolia*			3	+		2	9	2
S *S. purpurea*			·	+	+	5	8	×
M *Racomitr. canescens*		2	1	2	2	–	–	–
Leguminous group:								
Trifolium badium	·	2	2	2	·	3	8	×
T. pallescens	+		2	2	1	–	–	–
Lotus cornic. var. *alp.*			1	+	1	×	7	3

Serial no.:	1	2	3	4	5	T	R	N
Trees:[b)]								
Larix decidua		H	+	**1**	**3**	×	×	3
Picea abies		H	+		**2**	3	×	×
Pinus cembra		H	+	1	+	2	4	×
Betula pendula			H	+		×	×	×
Acid-humus plants:								
Empetr. hermaphrod.			+	1	·	3	4	?
Vaccinium vitis-idaea			+	·	+	×	2	2
Pyrola minor			+	·	·	×	3	2
Orthilia secunda			+	+	2	×	×	2
Calluna vulgaris			·	+	·	×	1	1
Rhodod. ferrugineum			·	2	·	×	2	2
Vaccinium uliginosum			·	2	·	×	1	3
V. myrtillus				+	+	×	2	3
Avenella flexuosa				+	+	×	2	3
Luzula sylv. ssp. *sieberi*				+	+	3	2	3
p *Melampyrum sylvaticum*				+	1	×	2	2
Huperzia selago				+		3	3	5
Latecomers:								
S *Sorbus aucuparia*					+	×	4	×
S *Salix caprea*					+	×	7	7
Dactylorhiza maculata					+	×	×	?
M *Dicranum scoparium*					1	–	1	–

Note: [b]semi-bold = in the tree layer
H = in the herb layer
+ and I = in the shrub layer.

Note: These are lists from individual sample areas; the figures indicate cover degrees according to Braun-Blanquet's scale (see table 11). The examples selected all have approximately the same aspect (NW to W) and slope (50–70% = ca. 25–35°); they lie between 1680 and 1970 m, i.e. at the level of the subalpine Larch-Arolla Pine wood of which there is a significant stand remaining above the lateral moraine (the so called Aletsch Forest).

· = present in other lists from the same stage. S = shrub, Z = dwarf shrub, M = Moss, p = semi-parasite (specialised nutrition).

No. 1: **scree: pioneer vegetation** (*Oxyria-Cerastium* stage),
2: the same with young woody plants (*Oxyria-Cerastium-Salix* stage),
No. 3: **Willow scrub**-leguminous stage (*Salix-Trifolium* stage),
4: the same with Alpenrose and young trees (*Salix-Trifolium-Rhododendron* stage),
No. 5: **Young Larch wood** (*Larix* stage).

In contrast to the terminal moraines of the glaciers in the alpine belt many woody plants take part in the succession and trees become established already very early, i.e. on bare ground.

Most of the plants coming in are from the subalpine belt (T3) or are indifferent as to temperature (T×). Most of the pioneers require lime (R7–9) although the catchment area of the glacier consists of acid rocks. The slight base content of the moraine is soon leached out and acid indicators (R 1–3) dominate the woodland stage. Even in the older stages nitrogen is in short supply; so the easily satisfied species (N1–4) set the tone.

Source: A selection from the data in J-L Richard (1968). Ecological evaluation according to Ellenberg (1974). (See also fig. 357).

Fig. 358. In the younger moraine soils (left) less nitrogen has been collected than in the old soils of Central Europe (right), which at least since the end of the latest glacial period were covered with closed stands of vegetation. After Zöttl (1964), partly from data supplied by Jenny, somewhat modified.

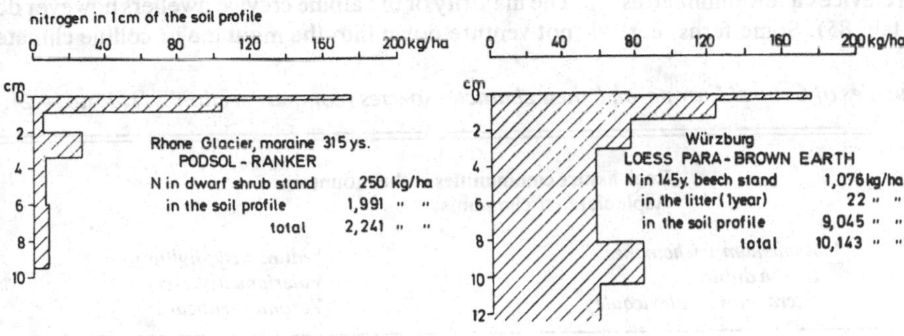

Fig. 359. Rugged crags, firn basins ground out by the ice (right), loose screes and moraines (foreground) are the elements forming the alpine level in the High Tatra also. Photo, Osveta. The dark patches on the moraine are *Pinus mugo*. The rocks carry a coating of lichens; phanerogams inhabiting the fissures cannot be seen from a distance. The Lomnitz peak (2634 m) and the Skalnatého lake (1752 m).

There are a surprising number of species to be found however at such sites in the subalpine right up to the nival belts. All of these are satisfied with a small amount of fine soil lodged in crevices a few millimetres or a few centimetres wide (tab. 85). Some ferns, e.g. *Asplenium trichomanes, A. ruta-muraria* and *A. septentrionale* are more frequent in the lowlands where they have mostly taken possession of joints in walls. The majority of the alpine crevice dwellers however do not venture down into the montane or colline climate

Tab. 85. Rock fissure communities of Central Europe and their character species (compare section E III 1, no. 4.2).

Class: (relatively few species in common!)	**Rock fissure communities** in the mountains Asplenietea trichomanis * *Asplenium trichomanes* *Draba dubia* *Hieracium amplexicaule*	 *Sedum dasyphyllum* *Valeriana tripteris* *Veronica fruticans*
Orders:	**Carbonate** rockfissure comm. (Potentilletalia caulescentis[c])	**Silicate** rockfissure comm. Androsacetalia vandellii
Alliances: Character species of the orders and alliances (A) = alliance	Potentillion caulescentis * *Asplenium ruta-muraria* g *Carex mucronata* (A) * *Cystopteris fragilis* (A) s *Daphne alpina* *Draba aizoides* *Erigeron alpinus* ssp. *glabratus* (A) g *Festuca alpina* (A) g *F. pumilla* (A) *Hieracium bupleuroides* (A) *Kernera saxatilis* (A) *Minuartia rupestris* (A) *Moehringia muscosa* (A) *Poa glauca* *Potentilla clusiana* (A) *Primula auricula* (A) s *Rhamnus pumila* *Saxifraga paniculata* *Silene saxifraga* *Valeriana saxatilis*	Androsacion vandellii[b]) * *Asplenium adiantum-nig* (A) * *A. septentrionale* *Epilobium collinum* (A) *Erigeron gaudinii* (A) *Primula hirsuta* (A) * *Woodsia ilvensis* (A) (Fissures in silicate rocks offer a less favourable habitat than those in the more easily weathered carbonate rocks) [b]This alliance also occurs in the Pyrenees, the others are Central European.
Associations in the **alpine** belt;	**Swiss Rock Jasmine fissure** Androsacetum helveticae *Androsace helvetica* *Draba ladina* *D. tomentosa*	**Many-flowered Rock Jasmine fissure** Androsacetum vandellii *Androsace vandellii* *Artemisia mutellina* *Eritrichium nanum* *Minuartia cherlerioides*
Associations mainly in **lower belts:**	**Stalked Cinquefoil fissure** Potentillo-Hieracietum humilis (montane to subalpine) g *Festuca stenantha* *Hieracium humile* *Potentilla caulescens*[c]) [c]The species giving its name to the association does not occur in the alpine belt!	**Red Alpine Primrose fissure** Asplenio-Primuletum hirsutae (insubrian to alpine) *Erysimum rhaeticum* *Phyteuma scheuchzeri* *Saxifraga aspera* var. *intermedia* *S. cotyledon*

Source: From data supplied by Braun-Blanquet, Oberdorfer and others. Away from the Alps and Tatras there are many other communities not mentioned here.

Note: s = espalier shrub, * = fern, g = grass or grasslike.

even where there are suitable rock faces. We shall come back to the reasons for this.

The chasmophytes (crevice dwellers) of the subalpine to nival belts occupy the sunny side and enjoy the favourable thermal conditions of such fully irradiated habitats. As early as May the rocks above the tree line can have a temperature up to 35°C, and it often exceeds 20°C during the next few months (Wetter 1918). Owing to their high heat capacity, the solid rocks remain at a higher temperature than that of the surrounding air throughout the night. Vertical south-facing rock walls receive more direct radiation in spring and autumn than they do in summer and no situation above the tree line has such a long growing period as here. However since there is no covering of snow in winter the alpine crevice plants must be able to survive a temperature of -20° C. or lower. The crevices on the shaded side are permanently iced up at the nival level and for this reason are unsuitable for higher plants.

Oettli (1904) and Wetter (1918) have already shown from many thoroughly investigated examples that practically every individual plant occupies a rock face because of special conditions. Nevertheless certain broad types of behaviour can be distinguished (see fig. 361a).

Some algae and many lichens (fig. 360) are able to establish themselves just on the bare surface of the rock. Lichens can penetrate tiny cracks with their rhizoids which hold them firmly. Wetter called these **'rhizolithophytes'** to distinguish them from the **'exolithophytes'**, i.e. the algae which just adhere to the surface. Several algae, particularly the Cyanophyceae, can penetrate some millimetres deep into the surface of limestone rocks dissolving the carbonate, and thus living as **'endolithophytes'**. Diels (1914) was the first to carry out an ecological investigation into these specialists. On dolomite ridges of southern Tirol, he found that different species arrange themselves according to the amount of light available, and some of them must be seen as genuine shade plants. Only a centimetre or so away the exolithophytes manage to make a bare living as light-loving xerophytes. A few lichens too are able to penetrate limestone and form so-called 'internal crusts'.

While certain algae and lichens can thrive easily on the bare rock, provided they have enough moisture, mosses and higher plants need a more or less thick humus and fine earth layer in order to establish themselves. Thus they can be called **'chomophytes'** (detritus plants) *sensu lat.* The fine material could have fallen down from soil lying above the habitat or, over a longer period of time, it can have built up from organic and inorganic dust which has settled here. Depending on whether the fine material lies on a shelf of rock or in a crevice the vascular plants are distinguished as either chomophytes, *sensu strictu*, or **'chasmophytes'** (crevice plants).

Fig. 360. Crustose lichen community (*Rhizocarpetum alpicolae*) on acid rock in the Ötztal Alps (2500 m). Photo. Follmann. Typical of silicate in the high mountains. The dominating species is *Rhizocarpon geographicum* (appr. natural size).

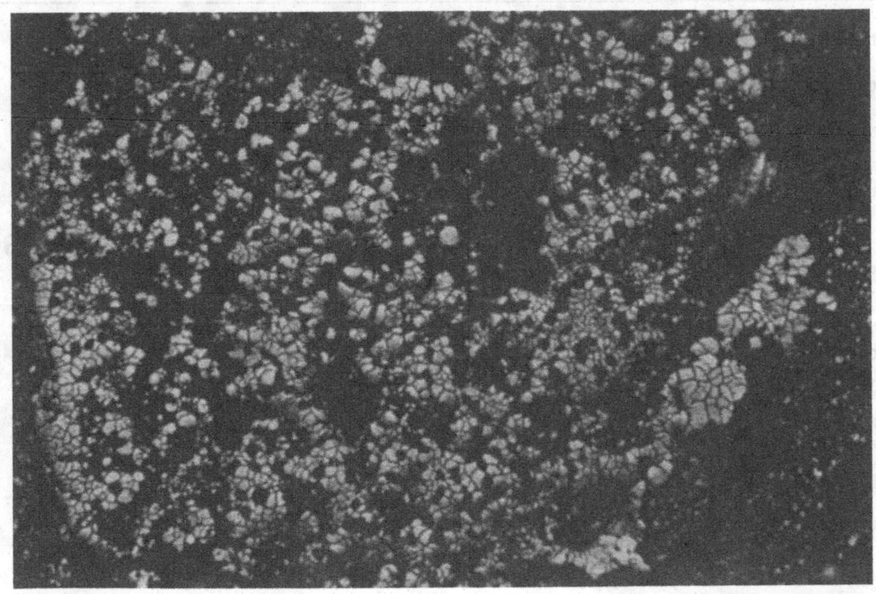

From studies of rock plants in Czechoslovakia Čeřovský (1960) produced a more detailed classification which differs in certain respects. In the case of the **cryptogams** he distinguishes:

- Exo- and endo-lithophytes,
- Chasmolithophytes (e.g. many mosses)
- Micro-lithophagophytes (i.e. lichens and mosses which actively attack the solid rock)

Amongst the **vascular plants** there are:

- Macro-lithophagophytes,
- Exo- and endo-chasmophytes,
- Exo- and endo-chomophytes.

By and large the amount of available humus together with the water and nutrient supply increases as we come down the list.

In both calcareous rocks (Oettli) and the crystalline ones (Wetter) the humus contained in the crevices is the home of numerous animals feeding on detritus, such as insect larvae, worms and woodlice, and it provides a very fertile soil. Often the roots of different chasmophytes are in strong competition here and make use of every corner available. Contrary to the evidence the rock plants too are only exceptionally so isolated that competition does not make some selection. This is true at least for the alpine climate and at lower altitudes. In the more hardy nival conditions it may be possible that competition between species has little or no significance.

In their manner of growth the inhabitants of the rocks are not particularly adapted to their exposed habitat. Almost all the species growing in the neighbourhood could appear on rocky crags. But only those which are easily satisfied and hardy can remain there for any length of time. Dense cushion plants such as *Androsace helvetica* (fig. 366) and many mosses are best able to put up with the extreme temperature variations combined with being dried out from time to time.

They climb high into the mountain peaks, but are accompanied and even surpassed by other species which can achieve the same without such structural peculiarities; by rosette plants (e.g. *Draba* species or *Kernera saxatilis*) and grass-like ones (*Poa nemoralis* and *Carex mucronata*). Some species are semi-succulent and economise with stored water, for example *Sedum dasyphyllum, Saxifraga paniculata* and *Primula auricula*. Wetter showed that the last named was able to go without water for more than a month.

Espalier shrubs are surprisingly rare on steep rock faces, even where there are lots of crevices. Such dwarf shrubs which nestle closely against the stones only grow well where the humus from their dead leaves can accumulate, i.e. on gently sloping rocks and over stones sticking out from a moraine. This is especially true for the 'classical' espalier plants *Salix reticulata, S. serpyllifolia, S. retusa, Dryas* and *Globularia cordifolia*.

Most of the things we have already noted about the occupants of scree communities also apply to the chasmophytes. Species which are dispersed by the wind dominate these habitats. According to Pikula (1963) the proportion of such species increases with

Fig. 361. a Types of rock plants (petrophytes). After Wetter (1918), modified.
b Important rock habitats, especially for lichens. After Frey, from Ellenberg 1963.
All kinds of different conditions are provided by the strength of sunshine, exposure to the wind, water trickling, hollows, fissures, snow cover etc.

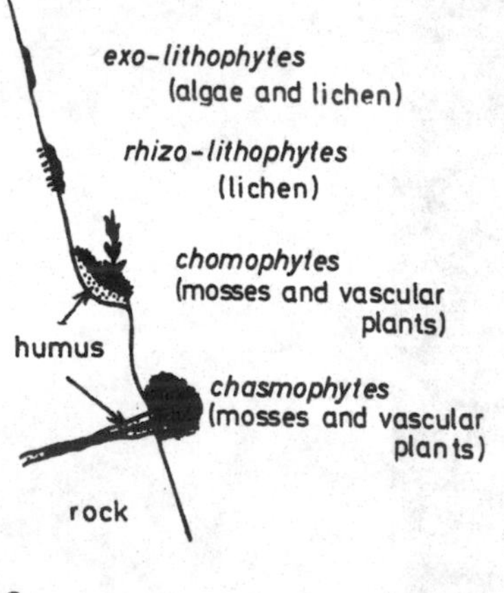

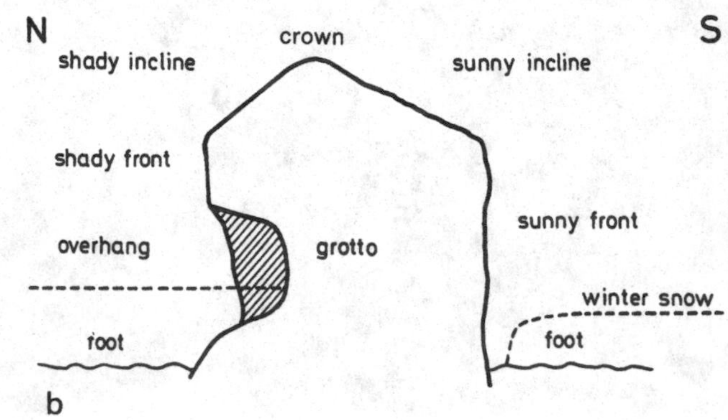

altitude. They produce a lot of seed, especially in a sunny year, as Nosova (1975) showed in the Pamir district. A few rock-crevice specialists are dispersed by insects, especially by ants, but it is only in the lower belts, from the subalpine downwards, that these play a noticeable part.

On rocks the habitat conditions can change over very short distances so that here one is rarely dealing with a single community, but rather with a whole vegetation complex, which makes classification very difficult (Stöcker 1965b). In every case the chemical nature of the rock is important. Many of the character species of the **limestone-rock communities** listed in tab. 85 (*Potentilletalia* and *Potentillion caulescentis*, fig. 362) are definite calcicoles. On the other hand only a few representatives of the **silicate-rock communities** actually prefer a lime-deficient stone (*Androsacetalia* and *Androsacion vandellii*). Walls which have been made of sandstone or crystalline rock carry a calcareous flora in their joints because they are built with mortar. According to Meier and Braun-Blanquet (1934) both orders can be split up into a number of associations. The communities showing the greatest floristic diversity are not found in the Alps but in the submediterranean area where relicts from the tertiary have survived the Ice Ages. This also applies to some extent in the southern Alps where the rock-crevice communities are especially rich in endemic taxa (Sutter 1969, see fig. 363).

J. L. Richard (1972) agrees with the proposal put forward by Oberdoofer and co-workers (1967) that the alliance *Potentillion caulescentis* should be split up and narrowed down in the following manner.

1. *Potentillion caulescentis* s. str. (Limestone crevice communities of **warmer** places which includes *Hieracium humile, Asplenium ruta-muraria, Kernera saxatilis*, etc. along with the named Cinquefoil.
2. *Cystopteridion* (limestone crevice communities of **shady cooler** places) with the bladder-ferns (*Cystopteris fragilis*

Fig. 362. Plant communities on the ridge of a peak in the Swiss Jura with vertical hard limestone layers. The schematic section shows their dependence on the relief, the amount of fine earth available and the aspect. After J.L. Richard (1972), modified.

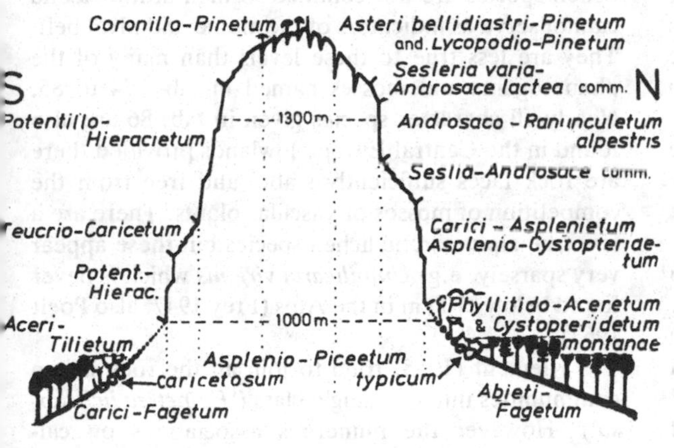

Fig. 363. Examples of rare rock-crevice communities in the insubrian Alps (Val Lori, Valsassina). After Sutter (1969), modified.

The *Potentilio-Telekietum* occupies sunny rocks and is endemic in the shaded area of the small map. One of its character species, *Moehringia bavarica*, subsp. *insubrica* has an even more restricted distribution. The locally rarer, but more widespread, Maidenhair Fern community occupies shady overhangs.

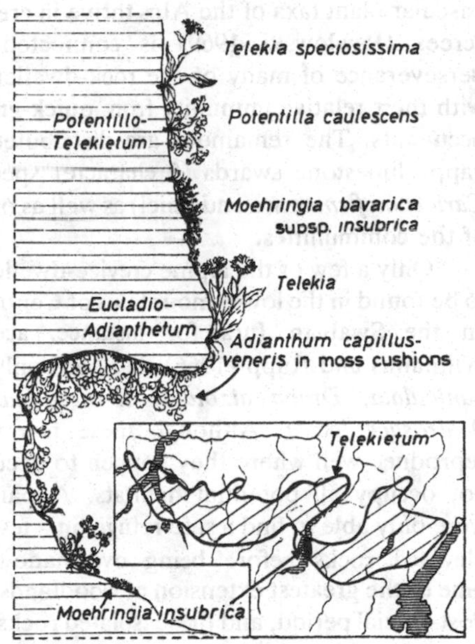

and *C. regia*) as well as *Asplenium viride, Phyllitis scolopendrium* and *Carex brachystachys*.

Richard puts these ecologically very obvious alliances into the order *Asplenietalia rutae-murariae*.

Jurko and Peciar (1963) recommend that the moss-rich plant cover of shady rocks in the western Carpathians should be separated off into a new class (*Polypodietea*). However this posesses practically no species of its own and is primarily composed of woodland plants, e.g. *Polypodium vulgare, Oxalis acetosella* and *Calamagrostis arundinacea* as well as mosses including *Hypnum cupressiforme, Dicranum scoparium* and species of *Polytrichum*. Incidentally it is not a true inhabitant of rock crevices (made up of chasmophytes) but rather a collection of exo-chomophytes, as defined by Čeřovský (see above) which come close to ground dwellers.

As long as the rock remains unchanged its crevice-dwellers form a stable community. Their roots soon form a mat running through the available fine soil and scarcely any newcomers can find a foothold. It has already been stressed that the steep rock faces cannot collect any humus except where there are small plant cushions. Under these circumstances each further development endures for thousands of years. Rock-crevice communities are thus permanent communities in the truest sense. Sometimes the mosses arrive first but it is the phanerogams which are the real pioneers since lichens and algae covering the stony surface are really occupying quite a different habitat which they also maintain for a long time.

The fact that 35–40% of the 350–400 endemic vascular plant taxa of the Alps thrive in crevices or on screes (Pawłowski 1969) is connected with the perseverance of many of the rock-dwellers and also with their relative immunity from quick-growing new occupants. The remainder are distributed over the gappy limestone swards (5 character species of the *Caricetum firmae* are endemic!) as well as over the rest of the communities.

Only a few of the alpine crevice-dwellers are also to be found in the lower mountains of Central Europe. In the Swabian Jura for instance, according to Wilmanns and Rupp (1966) there are only *Saxifraga paniculata, Draba aizoides, Kernera saxatilis* and *Androsace lactea*. Although these few thrive and reproduce well where they happen to occur, they do not occupy all potential habitats. Apparently they were only able to find a safe refuge on a few relatively elevated rocks before being overshadowed in the time of the greatest extension of woodlands during the Post-glacial period, and these isolated rocks have been the centres from which at least *Draba* and *Kernera* could spread. It seems that in spite of good germination the seeds of such crevice-dwelling plants apart from *Saxifraga*, have only a very slight chance of becoming established under suitable conditions.

(b) Lichen crusts and 'ink-stains'

It is with some justification that Gams (1927) considers the lichens, especially the navel-lichens (*Umbilicaria cylindrica* and others), which cover the surfaces of rocks and stones nearly everywhere as the most typical expression of the alpine-nival climate. Provided they can obtain sufficient moisture from time to time they prosper here in spite of cold and drought. Their main competition is from mosses but most of these cannot develop without the shade of trees. Poisonous gases and smoke which kill off the lichens near our towns and industrial areas do not reach lethal concentrations in the barren lonely world high up in the mountains. Thus the lichens are not present anywhere else in Central Europe in such profusion as they are above the tree limit.

Frey (1947, 1969) considers that these modest cryptogams are the oldest inhabitants of our mountains. Even during the Ice Ages there were rocks sticking out above the ice and snow and no doubt these were available for the growth of lichens. Since the crustose lichens grow extremely slowly in the nival and high-alpine climate Frey calculated that many of them must be thousands of years old. Perhaps they are still occupying the same places as they did during the Ice Age!

In spite of their venerable age and their wide distribution in the high mountains the majority of the lichen species are not confined to high altitudes and cannot serve as indicators of the alpine and nival belt. They are less true to these levels than many of the phanerogams and mosses named in tabs. 74 to 85. Nearly all character species given in tab. 86 are also found in the Central Europe lowlands provided there are rock faces sufficiently stable and free from the competition of mosses or vascular plants. There are a few true alpine-arctic lichen species but these appear very sparsely, e.g. *Umbilicaria virginis* which is never found below 2700 m in the Alps (Frey 1947, also Poelt 1963).

Klement (1955) tried to join all the rock lichen communities into one single class ('*Epipetrea lichenosa*'). However the numerous associations on calcareous and silicaceous rocks do not actually have any real character species in common. The uniqueness of the limestone lichen flora compared with that of other rocks is also stated for regions outside the Alps, e.g. for Britain (Bates 1978). Therefore we should follow Wirth (1972) and Roux (1978) in distinguishing at least two vicariant classes, the **carbonate-rock lichen crusts** (*Protoblastenietea immersae*) and the **silicate-rock lichen crusts** (*Rhizocarpetea geographici*, fig. 360).

As the synopsis in table 86 shows, the crustose lichen communities in the subalpine to nival belts are

Tab. 86. Abundant Rock lichen communities and their character species, in the upper montane to the nival belts.

Classes:	Lichen growth on **carbonate rocks** *Protoblastenietea immersae* I *Protoblasternia immersa* I *P. calva* I *P. incrustans* *P. rupestris* *Lecanora dispersa* and others	Lichen growth on **silicate rocks** *Rhizocarpetea geographici* *Rhizocarpon geographicum* *Acarospora fuscata* *Lecanora atra* *L. badia* *Lecidea lactea* and others
Orders:	Lichen growth on open rain-swept carbonate rocks *Verrucarietalia parmigerae* I *Verrucaria parmigera* I *V. baldensis* *Caloplaca ochracea* *Catillaria lenticularis* I *Petractis clausa* and others	Lichen growth on open rain-swept silicate rocks *Umbilicarietalia cylindricae* U *Umbilicaria cylindrica* *Haematomma ventosum* *Lecanora intricata* *Lecidea lapicida* *Schaereria tenebrosa* *Sporastatia testudinea* and others
Alliances of **crustose lichen** communities on steep hard rocks poor in nutrients:	on steep carbonate rocks ('internal crustose lichens') *Hymenelion caeruleae* *Hymenelia caerulea* I *Polyblastia deminuta* *Rhizocarpon umbilicatum* *Thelidium decipiens* I *Verrucaria dufourii* and others several associations, e.g. *Arthopyrenietum saxicolae*	on steep silicate rocks ('external crustose lichens') *Rhizocarpion alpicolae* *Rhizocarpon alpicola* *Fuscidea kochiana* *Lecidea aglaea* *L. armeniaca* *Rinodina polyspora* and others several associations, e.g. *Rhizocarpetum alpicolae* ('snow level community')
Alliances of **foliose lichen** communities on rocks poor in nutrients:	Gelatinous lichen stands with navel **leathery lichens** on calcareous rocks wet for a short period with seepage *Collemion tuniformis* G *Collema tuniforme* G *C. cristatum* G *C. multipartitum* U *Dermatocarpon miniatum* *Placynthium nigrum* and others several associations, e.g. *Verrucario-Placynthietum nigri*	**Navel lichen** stands on sunny relatively dry silica rocks *Umbilicarion cylindricae* U *Umbilicaria cinereorufescens* U *U. crustulosa* U *U. polyphylla* U *U. proboscidea* C *Cornicularia normoerica* H *Hypogymnia intestiniformis* B *Pseudephebe pubescens* several associations, e.g. *Umbilicarietum cylindricae*
Associations of **semi-crustose and foliose lichen** communities on nutrient-rich **bird resting rocks:**	in limestone mountains *Caloplacion decipientis* P *Caloplaca decipiens* P *C. cirrochroa* P *Buellia canescens* P *Xanthoria elegans* and others several associations, e.g. *Xanthorietum elegantis*	in mountains of silicate rock *Rhizoplacion chrysoleucae* S *Rhizoplaca chrysoleuca* S *R. melanophthalma* *Dimelaena oreina* B *Ramalina capitata* and others several associations, e.g. *Ramalinetum capitatae*

no less diverse and well characterised than phanerogamic communities in rock fissures, on screes or in snow-beds. Since they are very dependent on the quality of their substrate and on the local climate they can be used as reliable indicators of their habitat conditions. The *Rhizocarpetum alpicolae* for instance develops only on those parts of steep crystalline rocks which remain covered with snow for a long time; so it is a good 'snow level' community.

Since only a few readers will be familiar with the lichen species, the different growth forms have been shown by a different letter in tab. 86. Nearly always it is the crustose forms which dominate and most of these are '**external**' (exolithophytes, fig. 360 and 361a). Only on the limestone rocks are some species able to dissolve the stone forming chelates, and actually penetrate the rock itself – the '**internal crustose lichens**' (endolithophytes). Foliose lichens are less frequent on limestone than on crystalline rock. This is probably due in the first place to the more permeable, drier nature of the limestone, although the more rapid corrosion of the surface, giving a less stable substrate, could also play a part. Because of their slow growth at this altitude the foliose lichens probably have insufficient time in which to extend before the surface decays and crumbles away. In addition the majority of the *Umbilicarias-Cetrarias* and *Ramalinas* are definitely acid loving.

The most luxuriant lichen growth is to be found on rocks which are sticking up and are used by birds as resting places. Similar observations are to be made at resting places of marmots. The droppings favour nitrophilous lichens, such as *Caloplaca* and *Xanthoria* species, eyecatching with their bright orange colour. The excrements also neutralise to some degree the acids of crystalline rocks, enabling even lime-loving lichens to become established. In spite of this there are two parallel **coprophilous lichen alliances**: the lime-loving *Caloplacion decipientis* and the *Rhizoplacion chrysoleucae* which is widely distributed on acid rocks (tab. 86).

A closer study reveals that the lichen communities react, not only to chemical factors, but also to water availability, the amount of sunshine and other local climatic factors. Frey (1933) has given an illuminating classification of the rock habitats (fig. 361b). Even without the influence of bird droppings the flat top of a rock often carries growths different from those on the sloping or vertical surfaces. On the shaded sides can be found dominating species which are absent from the sunny sides, but it is the growth on the **overhangs** (re-entrants) or in dark grottos which show the greatest degree of specialisation. Here one meets, for example, the *Lecideetum lucidae* which forms its own alliance, (*Leprarion chlorinae*). The species giving its name to this association is an acid indicator which stands out because of its sulphur-yellow colour. All of the species of the *Lecideetum* are 'powder crusts' unable to resorb liquid water.

In lower mountains there are also manifold lichen communities on rocks, screes, etc., but nowhere as dominating as in the alpine and nival belt. Therefore we put up with references of literature, especially Wirth (1972, 1975, 1980), Schubert and Klement (1961).

Under a limestone overhang the lichens grow only in those places where water trickles over the surface from time to time. Communities of gelatinous lichens belonging to the *Collemnion tuniformis* (tab. 86) can develop under such favourable conditions. They are also found round the edges of the '**ink stains**' to which Jaag (1945) has devoted a comprehensive monograph.

These obvious dark strips and patches that can be seen from a long distance, on crystalline rock faces in particular, are formed by blue-green algae (which have no cell nucleus, and therefore are recently classified with the bacteria). The Cyanophyceae require constant wetness, at least for several weeks, for their development, whereas quickly flowing water which may be carrying pebbles along with it will rub them off and wash them away. Suitable habitats therefore are on those bare rock surfaces where water is seeping out of more or less horizontal cracks for some months or weeks. During active growth these autotrophic bacteria assume a reddish or blue-green colour, but when they dry out they form striking blue-black stripes. Even these specialist crytogamic communities are quite rich in species and can be separated according to the lime content of their substrate and other factors. Generally *Gloeocapsa ralfsiana* dominates on the silicate rocks and seldom appears on limestone; the

Tab. 86. (Continued).

Note: Life forms of the lichens (all species without a symbol form an external crust):

Crustose lichens: I = internal crusts (penetrating limestone), P = *Placodium* type (effigurated crust), D = disk lichens (peltate crust).

Foliose lichens: G = gelatinous lichens, H = Hypogymnia type, U = *Umbilicaria* type.
Fruticose lichens: C = *Cetraria* type; B = band and beard lichens (*Ramalina* and *Usnea* type).

Source: From data in Klement (1955), Wirth (1972), Roux (1978) et al., compiled by Follmann.

opposite is true of *G. sanguinea*. Although these rock-dwelling blue-green algae are of great interest to the ecologist, and their communities must be reckoned as well developed vegetation types, we must forego a more detailed discussion of them here.

As Jaag (1945) emphasises large stretches of rock faces in the high mountains are just bare without any algae, lichens or other autotrophs. However a certain amount of organic dust settles on them and this is sufficient for heterotrophic bacteria which can survive on this small amount of humus. Blöchliger (1931) has isolated several different taxa from the surface of karstic limestone which appeared to be completely free from organisms. With their peculiarly thick cover of mucus bacteria are better able to withstand long dry periods, the enemy of all rock-dwelling organisms.

VII Plant life in the region of eternal snow

1 The boundaries and divisions of the nival belt

(a) The climatic and orographic snow limit

'Plant life at its furthest limits' (Braun 1913, see also Pisek 1963), mainly that within the nival belt of the high mountains, already had that past master Heer (1884) under its spell, and it has had many researchers since (literature given in Reisigl and Pitschmann 1958). More than 250 species of phanerogams venture up into what at first sight appears to be just a lifeless desert of snow and rock (fig. 364) to say nothing of many lichens on rocks and boulders.

All the autotrophic plants here depend on sites where there is an unusually small amount of snow lying. They may even be free from snow throughout the year, e.g. on ridges exposed to the wind and sun, or on the surface of steep rocks. Precipitation, normally in the form of snow, is very high at these altitudes (fig. 315).

The division between the alpine and the nival belts is considered to be the **climatic** snow limit, i.e. an imaginary line above which, taking one year with another, more snow falls than melts (see figs. 42 and 316). On 'horizontal surfaces with normal exposure' (Zingg 1954), apart from exceptionally warm or dry years, there would be snow lying throughout the year. Since such places are very rare in the area of high and mostly steep peaks it is difficult to draw the snow limit exactly in many parts of the mountains.

In reality the lower limit of the permanent snow is an extremely changeable line depending on the aspect, slope and type of surface. The **'orographical'** or local snow line may be 100 m lower in one place or more than 1000 m higher in another compared with the theoretical limit (fig. 314). Even the highest alpine peak can produce snow-free places. Nevertheless it is ecologically justifiable to proceed from the climatic snow line since it is here that the region starts where one of the most extremely severe general climates in Central Europe is a dominating factor.

The lower limit of the continous snow cover which can be determined on a distinct day – it could be called

Fig. 364. Even in the nival belt there are snow-free places where plants can grow, e.g. on the Zinal-Rothorn (4221 m) in the Wallis Alps.

the 'actual snow limit' – varies considerably in our geographical area with the time of year and the prevailing weather (fig. 365). During the winter it may sink for a long time to below sea level. At the end of the winter and in the spring it climbs up the mountain slopes, always oscillating according to the weather conditions, until it reaches its highest level towards the end of the summer. This is the orographic snow line for the year in question. Similarly in the autumn as the snowfall increases it retreats in a fluctuating manner down the mountain.

Among a number of methods of determining the climatic snow line Klebelsberg (1948) prefers the 'firn method'. One asks oneself at what altitude does the snow form a snow field or glacier? For the mountain peaks in the Alps using this method the height of the snow line above sea level is 2 500 m on Säntis which is 2504 m high, but 3350 m on Gran Paradiso (height 4504 m). Where the mountains are of greater size and afford more shelter the climatic snow line is higher. It rises from the outer chain of the Alps towards the middle in a way similar to the tree limit (see figs. 4,6 and 42). Hermes (1955, 1964) has collected together information showing that these two limits are always about 800-900 m (700 – 1000 m) apart. On a profile through the eastern Alps, put together by Paschinger (1954), the two limits show an almost exact parallelism. The climatic snow limit is much less dependent on aspect than is the orographic limit. It lies some 200-400 m higher on slopes facing the sun than on those in the shade.

Zingg (1954) doubts whether the "firn method" is really justified since the accumulation of snow to some extent includes remains from former times. For example by extrapolation of observations made over a period of 17 years on the snow covering flat areas between heights of 530 and 2540 m he came to the conclusion that the snow line on the Weissfluhjoch above Davos should be about 3200 instead of 2900 m.

In general the altitudes of the snow line suggested by Zingg are above those of the majority of other authors. However for the period 1920-1950 at least his figures are in close agreement with the observations of Braun-Blanquet (1951, 1957), Reisigl and Pitschmann (1959) and other botanists who noted that the phanerogams have all wandered further up the mountain during the last few decades (tab. 87). It appears that the high-lying snowfields do not react so quickly or sensitively to changes in the general climate as do the tips of glaciers (section VI 8 b) or the plants.

Nothing shows more clearly than the fact that **flowering plants** during the last few decades **have moved further up the mountains** (see table 87), which factors impose a limit on their ability to survive: these are warmth and the length of the growing season. As

Fig. 365. Altitude variations in the 'actual snow limit' (dotted) and the frost limit (minimum air temperature 0°C) near Davos (o = snow line as seen from the Säntis). After Gasser from Ellenberg (1963), somewhat modified.

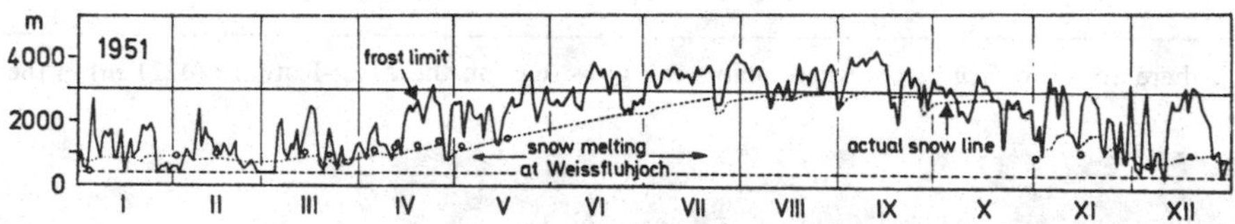

Tab. 87. Since 1911 the nivalflora has climbed higher in the Engadine and Ötztal.

Place:	L		S	
Altitude from	3350		3460	
to	3414		3472	
Observation year:	19 11	47	11	54
No. of species:	*8*	*11*	*5*	*11*
Ranunculus glacialis	v	v	v	v
Saxifraga oppositifolia	v	v	v	v
Saxifraga aspera ssp. *bryoides*	v	v	v	v
Androsace alpina	v	v	v	v
Gentiana bavarica var. *subacaulis*	v	v	v	v
Poa laxa	v	v		+
Draba fladnizensis	v	v		+
Tanacetum alpinum	v	v		
Festuca halleri		+		
Cerastium uniflorum		+		
Saxifraga exarata		+		
Luzula spicata				+
Draba dubia				+
Potentilla frigida				+
Erigeron uniflorus				+

Note: v = species already present, + = new arrivals.
Source: L = on the Piz Linard, after Braun-Blanquet (1957), S = on the Hinter Seelenkogel, after Reisigl and Pitschmann (1959).

we shall see these even nullify the effect of the type of rock from which the soil is derived, an effect which is still obvious in the alpine belt.

(b) The nature of the nival belt and its subdivisions

As we saw in the montane and alpine belts, so at the nival belt also the vegetation becomes progressively poorer in species as the climate gets less favourable (see fig. 316).

At the **subnival,** which at one time was not distinguished from the true nival level, it can already be seen that the phanerogams no longer form large closed swards. There are however many **small sward patches** each covering an area of some 1 – 2 m^2 in which the species composition is similar to that of the Crooked Sedge sward or some other alpine grassland type, and they could be considered as fragments of these communities. Reisigl and Pitschmann (1958) and others speak of them as 'pioneer swards' or even 'pioneer-stage swards' but such terms should be avoided because of their implications of a dynamic process. In the majority of cases they are stable and will not develop any further into a larger, continuous grassland since the area suitable to them is limited. According to Braun (1913) they have usually originated in the nival belt and have rarely developed from other communities; their species composition has probably been in a state of equilibrium for centuries. The mosaic of sward patches and bare ground then must be looked upon as the climax in the vegetation development just as the lichen coverings and other rock-dwelling communities are.

The sward patches are also met with here and there in the real nival belt, but with increasing altitude the number of grasses and grass-like plants becomes less, and gradually dicotyledons, in the form of hemicryptophytes and chamaephytes, become entirely dominant. The majority of these are species we have met before in the rock and scree communities at the alpine level. There is no species of higher plant which is confined to the nival area.

Figure 316 indicates where patches of cushion-forming chamaephytes are the prominent vegetation type. They develop in the nival belt only on soil which is sufficiently deep but is not slipping or being weathered away. Normally they form an open carpet amongst which other individual plants, mostly hemicryptophytes, find a home. These **'dicot-carpets'** are characteristic of the **lower nival** belt. Here they represent permanent communities both on firm rock and on screes, whereas in the alpine belt they usually appear as pioneer stages.

The cushion form tends to reduce the extreme differences in temperature and moisture which occur on the sunny slopes due to the intense radiation, inwards during sunny days and outwards on clear nights (fig. 366). Water relations however play a minor role in controlling growth and distribution of cushion plants in the high Alps (Körner and Moraes 1979). Above all a dense cushion makes possible the accumulation of plant remains which would otherwise be blown away by the strong winds. Thus, inside the cushions with their moderated microclimate, there

Fig. 366. Rock Jasmine cushion (*Androsace helvetica*) on the Padella summit; on the right *Carex firma*. Photo, Ganz and Rübel.

develop islands of humus in which numerous animals and microorganisms can exist and work. The cushion plant itself is able to make full use of the nutrient and water reserves in its 'filling' by means of a mass of lateral and adventitious roots. The storage ability of the plant varies depending on its height and spread. Rauh (1939) distinguishes a number of morphological types:

1 The **rosette cushion** (e.g. *Saxifraga paniculata* and *Sempervivum* species) is least able to accumulate organic matter and form soil. It consists of a dense group of rosettes, the dying older leaves of which contribute to the formation of humus. According to Oettli (1904) they can only occupy rocks which already have a covering of humus or fine earth.
2 The **creeping cushion** (e.g. *Saxifraga oppositifolia* and *S. bryoides*) also remains low and accumulates only a thin layer of humus. Its growth habit however helps in vegetative dispersal because its side branches may be broken off by the movement of soil through frost or on a scree slope.
3 Within the **tussock cushion** (e.g. *Carex firma* and *Saxifraga androsacea*) the main roots and old shoots die each year, and adventitious roots grow through the mass of humus which is being formed in the interior of the more or less strongly rounded tussock.
4 The **radial flat cushion** (e.g. *Silene acaulis* and *Minurtia sedoides*) behaves in a similar way except that the main root remains alive for longer and anchors the cushion deeper. It often develops into the next type.
5 The **hemispherical cushion** (which Rauh called the ball cushion) with its tightly packed shoots grows upwards almost as quickly as it extends laterally. Usually it is firmly held in a crevice by means of a long branched lignified main root (e.g. *Androsace helvetica*, fig. 366). It is the best protected of all the types against drought but like no. 1 less well adapted to spreading.

In the **middle nival** belt the flowering plants find it increasingly difficult to form groups. The cushions of which we have spoken above and also a few other specially adapted herbs often appear only singly and any fragments of sward are almost entirely absent. Mosses may cover larger areas where there is fine earth, but such patches are at constant risk through solifluxion. The moss cushions do better in wide crevices or on ledges.

In the **upper nival** belt, where dicot cushions are lacking, only a few more or less isolated phanerogams persist, and these soon arrive at their absolute limit. However not even the highest peaks, although poor in soil, are entirely without vascular plants. Where they are not permanently covered with snow they have become the realm of the undemanding crustose and foliose lichens. For this reason the upper nival belt can also be called the thallophyte level or the **lichen level**.

2 *Peculiarities of the nival habitats and their plant communities*

(a) Sward fragments more or less indifferent as to soil

The commonest grass-like plant on silicate stone in the subnival and nival belts is the **Crooked Sedge**. Many of its partners accompany it into the colder regions; indeed it is just here where they are particularly faithful. According to Braun (1913) the *Curvuletum* in the nival belt contains no less than 23–26 species on average in the Swiss inner Alps; Reisigl and Pitschmann (1958) had similar results from their investigations in the Ötztal Alps. During the cold months of the year the crooked sedge sward is often even better protected by snow at this height than lower down. On the other hand it requires a minimum of three summer months free from snow in order to develop satisfactorily and this condition is only exceptionally fulfilled above the climatic snow line. Generally the low sward patches are found on low hillocks or on sunny slopes which are protected from the strongest winds.

On ledges which are swept clear of snow by the wind the **Naked-rush sward** is still able to survive and, like the *Curvuletum* the *Elynetum* also has a rich flora up here. In addition fragments of the Blue Moor-grass-Sedge sward can also be found as far up as the lower nival belt. The only woody plant to feel at home in the region of eternal snow takes part in this community – this is *Salix serpyllifolia* which sheds its leaves in the autumn. It can accumulate humus in a similar way to the creeping cushion plants and, according to Braun, may reach an age of over 100 years.

In many of the sward patches at the subnival and nival levels we come across mixtures of species which we know from the alpine belt as being calcicole or calcifuge, but as altitude increases the majority of the phanerogams become indifferent as to soil type, and this has been confirmed by Braun (1913), Schröter (1926), Wendelberger (1953b), Friedel (1956) and others. From a general consideration of the behaviour of plants near the limits of their existence one would have expected the opposite, i.e. a closer adherence to their optimal habitat. Actually even in the alpine belt it is not the lime content nor the soil acidity as such which play a deciding rôle. Wherever competition becomes weaker or disappears altogether because the individuals are isolated and grow extremely slowly there for the majority of species any soil is right provided it offers them a firm foothold.

This reduction in the degree of adherence to a particular soil reaction is all the more curious since up here the original differences between the types of rock cannot have been reduced by soil maturation or by the introduction of soil from another type (see section C VI 1f). Because of the low average temperature, the long duration of frost and the steepness of the terrain, any soil formation at the level of permanent snow can scarcely progress beyond its initial stages, i.e. the physical breakdown of the rock and a certain amount of local humus enrichment. Even the leaching of lime is so restricted here that no one has suggested up to now that it could lead to a levelling up of the soil properties (Braun-Blanquet 1958).

(b) Nival rock and scree communities

Rock ledges, crevices and the foot of rock faces exposed to the sun, also screes, if they are not moving very much and contain a proportion of fine soil, are the only safe places at great heights where phanerogams can grow. Thus rock and scree plants (tab. 82 and 85) make up the largest part of the nival flora. The cushion-forming dicotyledons are especially well represented in the lower and middle nival belts. Plants are completely absent from the frozen north faces or steep shade slopes.

By far the most widespread areas carrying plants have a covering of lichens (tab. 86) which have already been the subject of discussion in section VI 9 b. However, even these, which are so resistant to extreme drought and cold, become poorer in species and in some places are completely absent.

(c) Other vegetation types in the nival belt

Because of the rugged terrain and the absence of deep clay soils it is almost impossible to find any swamp and spring plants or similar communities of wet places in the nival belt of our high mountains. This is in direct contrast to the position one finds in the arctic where wet-soil vegetation often prevails. Even snow-beds are mostly sought in vain since one of the most important conditions for the presence of these communities is hardly ever extant at this altitude, that is the effect of snow melt water trickling over a waterlogged soil for a sufficiently long time during the snow-free period. For all we find are only fragments of small **moss snow patches** in the lowest part of the nival belt on slates which have weathered to earth or on a similar substrate. *Polytrichum piliferum*, i.e. a species which is exceptionally drought-resistant, and also widely distributed at lower altitudes, is often dominant on such sites.

Where wild animals or grazing sheep have their resting places sufficient nutrients can accumulate, even above the snow line, for **tall herbs** and other **plants of the resting areas** to come in, e.g. *Cirsium spinosissimum* and *Poa alpina*. A high proportion of the species reported as having moved further up the mountains during the past few decades belong to these groups. For this reason Braun-Blanquet (1957), who carefully documented the gains and losses in the flora on Piz Languard and Piz Linard, considered the increased grazing of domestic and wild animals, along with the warmer climate, as the significant causes of those changes (tab. 87).

Finally a peculiar 'snow and ice flora' the **Cryoplankton** must be counted amongst the communities of the nival belt. The group has been dealt with comprehensively already by Huber-Pestalozzi (quoted in Schröder 1926). The best known member of this floating flora is the 'red snow' (*Chlamydomonas nivalis*) which turns up over the whole world in places where there is perpetual snow. This unicellular alga colours more or less large areas a raspberry red from time to time. Unlike the phanerogams and the majority of the cryptogams at the nival level it can exist only in the snow, and a temperature higher than –4°C will kill it. On the other hand in its resistant stage it can withstand –36°C without damage. As in the case of the water plankton, we are not able to look any further into the cryoplankton here.

(d) Plant species reaching the highest altitudes in Central Europe

In the Ötztal Alps which have been investigated by Reisigl and Pitschmann (1958) and where, according to Klebelsberg (1949), the climatic snow limit is between 3000 and 3100 m, the following numbers of angiosperms have managed to live above 3000 m:

Over 3500 m	3 species	Over 3200 m	60 species
Over 3400 m	34 species	Over 3000 m	102 species

The three highest records are grasses, namely *Poa laxa* (3680 m), *Festuca halleri* (3516 m) and a small form of *Festuca rubra* (3516 m). On the Finsteraahorn (4275 m) and other peaks in the western Central Alps some of the species named in tab. 87 can be found at over 3800 m. Here the snow line is in any case to be put somewhat higher on average than in the Ötztal Alps, namely at over 3200m. According to Braun (1913) the following dicots are to be considered the flowering plants which grow at highest altitudes (shown in this case relative to the snow line)

more than 640 m higher:	*Ranunculus glacialis*
about 610 m higher:	*Androsace alpina*
	Saxifraga bryoides
	Saxifraga exarata
about 600 m higher:	*Saxifraga oppositifolia*

Cushion-building plants and grasses have shown themselves to be particularly hardy phanerogams, but even they are exceeded by *Ranunculus glacialis* (fig. 367), which does not possess any obvious adaptation. Nature uses various means to attain the same end; here this statement is verified once more.

In the case of the Tatra with their lower altitude correspondingly very much lower maxima are attained. A comparison with figures collected by Pawłowski (1931) is very informative. On the 2663 m high Garlach the highest climbing plants are:

up to 2663 m:	*Senecio incanus* ssp. *carniolicus*
Minuartia sedoides	*Luzula spicata*
Cerastium uniflorum	*Poa laxa*
Saxifraga bryoides	*Festuca ovina* ssp. *supina*
S. carpathica (endemic)	*Oreochloa disticha* et al.
S. moschata	up to about 2650 m:
Primula minima	*Ranunculus glacialis*
Gentiana frigida	*Sedum alpestre*
Tanacetum alpinum	*Campanula alpina*
Doronicum clusii	*Poa alpina* ssp. *vivipara* et al.

Thus in the Tatra, as in the Alps, the cushion plants rendezvous on the highest peaks with grasses and other plants which are not so obviously adapted to live in an extreme environment. All the taxa named here, as far as they occur in the Alps, are found there at heights up to over 3000 m.

The cryptogams living at high altitudes include species of the genera *Rhizocarpon, Umbilicaria, Parmelia* and *Lecidea* (Frey 1947, 1969). In the subnival and nival belts of the Tauern massiv he found no less than 151 species of lichen, 101 of them on only 11 peaks. Mosses are not entirely absent from the peaks studied by Reisigl and Pitschmann (1958) nor from the Grosglockner which Friedel (1956) studied. On the Weisskugel in the Ötztal Alps for example the first named authors found:

Gymnomitrium coralloides	*Polytrichum sexangulare*
P. piliferum	*Racomitrium lanuginosum*
Grimmia donniana	*Bryum kunzei*

(e) Methods of dispersal amongst the nival flora

How have the flowering plants and cryptogams which are found growing on isolated spots near the mountain peaks managed to make the lonely journey across the permanent snow? Many of them obviously are able to move quite quickly, otherwise they could not have reacted so promptly to the improved weather conditions of the past few decades. Braun (1913) had already determined that most of the species living at the nival level were anemochorous. This is true of all the cryptogams and about two thirds of the phanerogams. Many of them can be looked upon as 'snow-runners'; the frequent storm-force winds blow away their propagules which roll and slide often for long distances over the snow surface. Pieces of the thallus which have become detached act for the lichens while the higher plants are dispersed by fruits, e.g. almost all grasses, many sedges (*Carex curvula, C. sempervirens* et al.) and several species of the genera *Luzula, Juncus* and *Saxifraga*. Other phanerogams, especially the plants on resting places which have already been mentioned are carried by animals.

Not a few of the plants growing in the nival belt depend on insects for pollination. These are blown off course up into these lonely heights where they find a favourable microclimate during the summer days, at least near to the ground. It is also surprising how many of the plants at this altitude manage to set and ripen their seeds, especially in more favourable years. According to Braun about 80 species produce seed with a high germination capability, amongst them:

Silene acaulis (w)	*Sibbaldia procumbens*
Minuartia verna (w)	*Geum reptans*
Arenaria biflora (w)	*G. montanum*
Thlaspi rotundifolium	*Linaria alpina*
Cardamine residifolia	*Erigeron uniflorus*
Draba fladnizensis	*Anthoxanthum odoratum*

Many of these are 'overwinterers' (w) in which the fruits remain on the plant until spring and only then do they reach their full powers of germination. About a fifth of the nival plants behave in this way.

Even therophytic phanerogams happen to set seed above the snow line, an indication of how their habitat is favoured by the local microclimate. Braun has recorded the following from the Swiss Alps:

Sagina saginoides	*Gentianella tenella*
Sedum atratum	*G. campestris*
Euphrasica minima	*Gentiana nivalis*

The *Euphrasia* named here holds the record for the highest fruiting annual dicotyledon. In the Antener

Fig. 367. Glacier Crowfoot (*Ranunculus glacialis*), the flowering plant climbing highest in Europe, on a silicate scree of the Ortler region (the photograph is about 50 cm across).

Erzhaupt it has reached a height of 3500 m (Vaccari, quoted by Reisigl and Pitschmann 1958). However perennials are definitely favoured because once they have established themselves they can retain their place in a bad seed year.

Many of the nival phanerogams rely on a constant supply of new seed from the alpine or lower belts because, although they can grow quite well above the snow line they have not been found to flower there. Examples of these are:

Juniperus nana	*Arnica montana*
Aconitum napellus	*Senecio doronicum*
Trifolium repens	*Saussurea alpina*
Astragalus frigidus	
Vaccinium myrtillus	
V. vitis idaea	

The list on the left comprises zoochorous species which are carried by birds and other wild vertebrates and also by domestic grazing animals.

By and large then the nival vegetation is capable of an independent existence. One could imagine that it has survived the Ice Age on the mountain peaks. In any case it is dependent to some extent on the nutrients which are brought to it by the air, either in the form of organic or inorganic dust or by gases such as ammonia which dissolve in the melting snow.

(f) Photosynthesis and metabolism in some nival plants

There are two ecophysiological questions concerning the nival flora which can now be answered thanks to the Austrian contribution to the International Biological Programme (Ellenberg 1973): How great is the photosynthetic production of these plants which manage to endure such extreme conditions, and how do they economise in the use of the organic material they have produced?

After a large amount of basic research work, especially his series of continuous measurements using air-conditioned cuvettes on the Hohe Nebelkogel (3184 m) W. Moser (1970, 1973) found rather surprisingly, that the nival plants have the same optimal temperature for photosynthesis as lowland plants (see fig. 368). This is true not only for the Glacier Crowfoot (*Ranunculus glacialis*) on which a particularly large number of measurements were made, but also for *Saxifraga bryoides*, *Leucanthemopsis alpina* and others.

In fact the optima are higher with stronger

Fig. 368. The Glacier Crowtoot does not differ from many lowland plants in the way temperature affects its net photosynthesis. After Moser (1970). When light is restricted (10 000 lux) the optimum lies between about 12 and 20°C; with more light it rises to some 20–24°C. Measurements in the laboratory on plants taken from 2700 to 3000 m up on the Schrankogel or cultivated on the Patscherkofel at 2000 m.

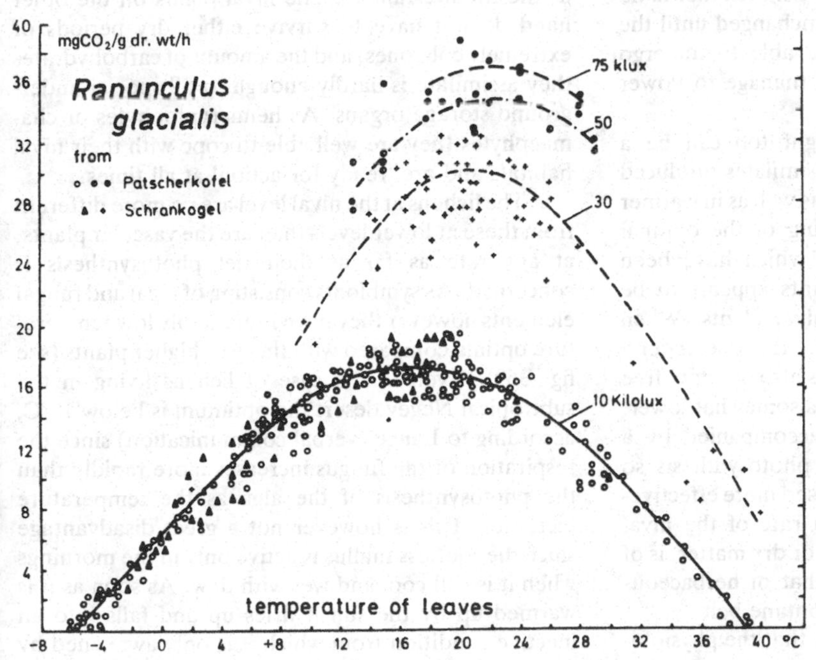

illumination, and such temperatures are reached in the midday sunshine when the nival plants are clear of snow, although this may be for only a few weeks or even days per year. Thus the leaves near to the ground are able to build up new material relatively quickly. In favourable years this enables the plants to produce more leaves and even flowers. On most days the temperature will be significantly lower but at least a positive balance can be achieved. Quite often the plants are covered in snow again after a few fine days, and the sudden transition from cold to warm conditions can be repeated several times during the summer. But it might be as much as two years before a plant gets enough sun to show a profitable photosynthesis once more.

The reason that these enforced periods of rest do not lead to starvation is that the plants, covered in snow, are maintained at a low temperature, varying around freezing point but occasionally falling below it, and there are scarcely any respiration losses. In addition they carry out a remarkable economy which Brzoska (1973) for example has demonstrated with *Ranunculus glacialis* and other species for several months. At the end of the summer when they are definitely snowed in for several months the assimilated material is transferred from the leaves and stems down into the underground organs which are less affected by frost and suffer practically no losses through respiration. These provisions are even taken to the point where buds which have not reached a certain stage in their development are dismantled. On the other hand the more advanced ones remain unchanged until the next warm period when they are able to undergo further development and perhaps manage to flower and fruit.

Not just temperature but light too can be a limiting factor in the amount of assimilates produced during photosynthesis, in the nival as well as in warmer climates (see fig. 368). The shifting of the optimal conditions for net photosynthesis which have been observed for almost all green plants appears to be particularly appropriate for the nival plants. When illumination is low (such as would be the case under a thin snow cover or on dull cool days also on snow-free places) the optimum temperature is somewhat lower. A better illumination is always accompanied by a higher temperature optimum for photosynthesis so that sunny summer days can be utilised more effectively. On such days the assimilation rate of the nival plants, whether related to leaf area or dry matter, is of the same order of magnitude as that of herbaceous plants in the lowlands or in the montane belt.

To summarise then we can say that the physiological make up of the **nival phanerogams** as far as photosynthesis is concerned is not significantly different from that of other herbaceous chamaephytes, hemicryptophytes or therophytes. In their morphological-anatomical structure and in their resistance to cold, they can also be compared with the lowland plants, especially the **spring-flowering woodland plants**. The latter are able to withstand being covered with snow again even when they have reached the flowering stage, and they use cool sunny days for their photosynthesis.

The herbs of the broadleaved woodlands differ from their relatives in the high mountains in only one respect. They have amongst them many geophytes whereas this life form is remarkably rare in the nival region. Of the plants named in section d as being the ones growing highest on the mountains only *Linaria alpina*, which normally produces overwintering buds near the surface, is also able to produce them deeper in the ground. So it is probably no mere accident that just this facultative geophyte is observed to survive for many years along the banks of our lowland rivers after being carried down by water from the mountains (see section B V 1 i). In my opinion the geophytes are adapted not so much to survive the winter as to live through unfavourable summers. They are able to rest during the months when there is very little light in our broadleaved woods, and if necessary they can also survive a period of drought. In fact the geophytes tend to be concentrated in the areas with dry summers such as the Mediterranean. The nival plants on the other hand do not have to survive either dry periods or extremely cold ones, and the amount of carbohydrates they assimilate is hardly enough to fill special underground storage organs. As hemicryptophytes or chamaephytes they are well able to cope with their nival habitats and are 'ready for action' at all times.

The lichens at the nival level are no more different from those at lower levels than are the vascular plants, at any rate as far as their net photosynthesis is concerned. As symbionts consisting of algal and fungal elements however they have remarkably low temperature optima compared with those of higher plants (see fig. 369). Even in the case of lichens living in the subtropical Negev desert the optimum is below 10°C, according to Lange (verbal communication) since the respiration of the fungus increases more rapidly than the photosynthesis of the alga as the temperature increases. This is however not a great disadvantage since the rootless thallus is active only in the mornings when it is still cool and wet with dew. As soon as it is warmed up by the sun it dries up and falls into an inactive condition from which it is only awakened by the next supply of water. In their dry state even tropical lichens are extraordinarily resistant, not only to heat but also to cold. The nival rock lichens behave in the same way. They must however be capable of withstanding temperatures down to –40°C, because, although they are soaked with water quite frequently they do not enjoy the protection of a snow cover. According to Kappen and Lange (1970) this does not present them with any difficulties since the majority can withstand low temperatures in both the damp, active state and in the dry one.

VIII On the epiphytic vegetation in Central Europe

1 *In general*

During our journey through the natural vegetation of Central Europe we have been concerned repeatedly with plant communities dominated by cryptogams, particularly in water, on raised bogs and in the area above the climatic tree line as well as on rocks and screes. These habitats have one thing in common, they cannot become occupied by trees, except perhaps by a few stunted specimens. However even in woods and forests, with a tall dense stand of trees, algae, lichens and mosses also play an important part, not to mention the fungi whose universal presence is apparent only when they send their fruiting bodies up through the humus of the woodland floor (see Tüxen 1964). Such ground-dwelling cryptogams must count as part of the phanerogam community insofar as they play a significant part in the cycles of nutrients and energy in the ecosystem (see section A II 7).

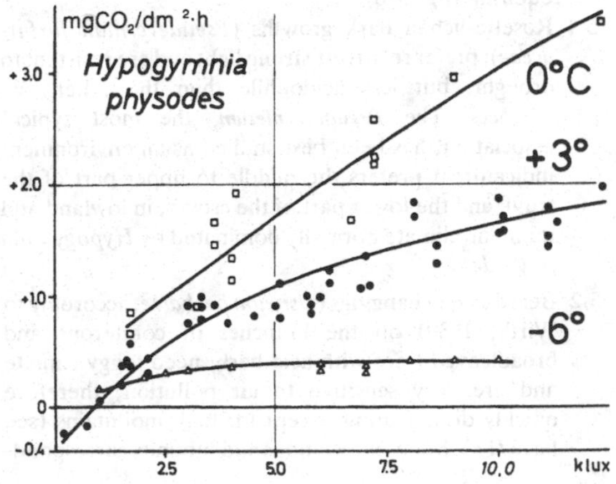

Fig. 369. Lichens are particularly productive at low temperatures.

The net photosynthesis of an epiphytic foliose lichen (on Beech in the *Luzulo-Fagetum* of the Solling) is greater at freezing point than at +3°C or higher temperatures when the respiration of the fungal component becomes progressively greater than the photosynthesis of the alga. Even at -6°C there is some assimilation but this does not increase when the light intensity is raised at this temperature. After Schulze and Lange (1968), somewhat modified.

A much greater degree of independence, from the floristic and ecological standpoints, can be ascribed to the epiphytic communities which use the trees merely as a substrate although they are to some extent dependent on the tempered climate within the wood. It has been noticed more and more during the past few decades that these bark-dwellers, which are less obvious and are generally omitted from records of the woodland flora, represent the earliest and most sensitive indicators of air pollution. There is now a rich literature concerned with the eco-physiological reactions of certain of the lichen and moss species and with the practical use which can be made of these as indicators on a statistical basis (e.g. Arzani 1974, Barkman 1958, Bauer 1973, De Slover and Le Blanc 1968, Düll 1973, Frey 1958, Jürging 1975, Kirschbaum 1972, Klement 1971, Kunze 1972, 1974, Le Blanc 1969, Lötschert and Köhm 1973, Lötschert, Wandtner and Hiller 1975, H. Muhle 1977, Nash and Nash 1974, Skye 1968, Søchting and Johnson 1974, Thiele 1974 and Wilmanns 1966). On the other hand there are also a large number of studies available about the species composition of the Central European epiphyte communities. These were started by Frey, Ochsner, Militzer and earlier researchers long before anyone forsaw the retreat of that cryptogam vegetation caused by air pollution. Barkman's monograph (1958) brings all these older works together and gives such an excellent ecological and phyto-sociological survey that a brief summary of it will be sufficient here. More recent work has only added and changed small points (e.g. Barkman 1962, 1968, Fabiszewski 1967, 1968, Seaward 1977 and Wirth 1980).

Barkman describes no less than 97 epiphytic associations and numerous subunits from Central Europe. The majority of these communities are characterised by several species and are thus floristically more independent than many woodland communities. This is even more remarkable because the wealth of epiphytes known from the tropics: orchids, bromeliads, araceae, ferns and other vascular plants, are entirely absent. Such cormophytes cannot survive our long cold winters mainly because, sitting on the tree bark, they loose more water on sunny days than can be replaced by their extremely restricted and often still frozen root systems.

Looked at ecologically the communities of thallophytes differ from those of the vascular plants mainly because of the following peculiarities which **predestine them for an epiphytic life**:

1 They do not have roots, but they may have rhizoids which serve to fix them to the tree bark.
2 Thus they take up water over the whole surface, and can take it from moist air.
3 They usually have a large capacity for water storage; in the case of mosses this can amount to between 650 and 1700% of the dry weight.
4 Their potential osmotic pressure is generally higher than that of vascular plants; in lichens it can be 1000 bar, in epiphytic mosses 20–90 bar (in terrestrial mosses in contrast only 5–50 bar, as in vascular plants).
5 Places which are too dry for vascular plants need not necessarily be too dry for cryptogams; these poikilohydrous plants are able to dry out into a dormant state from which they quickly recover when remoistened.
6 In their dried out condition they are also poikilothermic and can withstand large variations in temperature.
7 In their active state they achieve their biggest photosynthetic surplus at low temperatures (see fig. 369).
8 Their growth is very slow, especially when they are young and again when they reach a great age; in lichens the annual growth is usually less than 5 mm with a maximum of 14 mm achieved by *Bryoria* species (fig. 371a).
9 The atmosphere is an important source of nutrients for them, providing them with organic and inorganic dust, salts, NH_4, and NO_3 (see also sections C III 1 a and 4 a). Those lichens which contain cyanophyceae as their symbiotic partner are able to fix atmospheric nitrogen.
10 Their influence on the habitat is generally slight apart from the fact that mosses form a humus which is difficult to rot down.

2 A survey of the algal, lichen and moss communities

The epiphytic vegetation of Central Europe is divided into three large groups depending on whether algae, lichens or mosses predominate. In clean air the algae occupy mainly the highest part of the tree crown which at times can be very dry (see fig. 370). Lichens and mosses play a larger part lower down where, because of the shade, the air humidity is much higher. However most of the lichen species are easily killed by poisonous gases, and they are replaced by algae even on relatively moistened parts of the trunks. The more important orders and alliances are given here as examples, together with short notes on their ecological requirements:

I Epiphytic communities **rich in algae**, 3 ass., including: the green algal film (*Pleurococcetum vulgaris*). There must be scarcely a tree that does not carry some of this growth which is very resistant to air pollution and thrives well on tree bark in industrial areas and large towns.

II Epiphytic communities **rich in lichens**, 54 ass.:

1 Scab-lichen films (*Leprarietalia candelaris*) require high humidity and diffuse light. They are sensitive to wet conditions and prefer the montane climate.
2 Acid-loving tart-lichen coverings (*Lecanoretalia variae*, see fig. 76) pioneer communities on strongly acid smooth substrates, especially dead wood; in dry air, relatively immune to air pollution; e.g. the *Lecanoretum conizaeoidis*. *Lecanora varia* and *conizaeoides* are the most plentiful and widely distributed epiphytes in Central Europe today.
3 More or less acid-avoiding tart-lichen coverings (*Arthonietalia radiatae*); on trees with smooth bark it was formerly plentiful and widespread in Central Europe but is rapidly becoming rarer; with the alliances:
3.1 Script-lichen coverings (*Graphidion scriptae*), mainly on *Fagus*; extremely sensitive to pollution and to soluble nitrogen; therefore disappearing with the advance of industrialisation (see fig. 55).
3.2 Tart-lichen pioneer communities (*Lecanorion subfuscae*) mainly on *Fraxinus, Acer*, and other smooth-bark broadleaved trees, but less common on *Fagus* (see fig. 106); quick growing and less sensitive than 3.1).
4 Callus-lichen bark growths (*Physcietalia ascendentis*) on rough bark, with a wider amplitude regarding light and wind; needs nitrogen supply and supports dust deposition.
4.1 Disc-lichen bark growths (*Buellion canescentis*), mainly on *Ulmus, Tilia* and *Salix* but also on *Fagus* et al. Salt-loving and plentiful on the coast.
4.2 Yellow-lichen bark growths (*Xanthorion parietinae*), resemble 4.1 but less salt tolerant and less resistant to drying; mainly on the wetter side (facing S-NW) of free-standing trees; on Oaks, fruit-trees and other species with rough bark, especially in a humid climate (see fig. 8).
5 Rosette and beard-lichen growths (*Hypogymnetalia physodotubulosae*), strongly acidophilous, avoiding nitrogen; fairly to very sensitive to air pollution; light requiring (fig. 370).
5.1 Rosette-lichen bark growths (*Pseudevernion furfuraceae*); prefer relatively strong light and are resistant to drought, but less acidophile than the other two alliances. The *Pseudevernietum*, the most typical association, has been best studied as an environment indicator; it prefers the middle to upper part of the trunk and the lower part of the crown; in lowland and montane climate normally dominated by *Hypogymnia physodes*.
5.2 Beard-lichen hangings (*Usneion barbatae*, according to Wirth, 1980) on the branches of coniferous and broadleaved trees with acid bark; need foggy climate and are very sensitive to air pollution; therefore quickly disappearing except for high mountains (see fig. 371 a). Many associations and subunits correspond-

ing to the more or less oceanic climate and to the altitudinal belt.

III Epiphytic communities **rich in mosses** 39 ass.:

1 Acidophilous liverwort coverings (*Lophocoletalia heterophyllae*) on acid, damp substrates, especially rotten stumps and bases of trunks in Birch, Oak and conifer woodlands (*Quercion robori-petreue* and *Vaccinio-Piceion*, fig. 132), with 10 ass.

2 Acidophile moss carpets on tree boles and rocks (*Dicranetalia*), without character species (fig. 370).

2.1 Atlantic Claw-moss growth (*Isothecion myosuroides*), has its optimum in Ireland (there together with the Filmy Fern *Hymenophyllum peltatum*).

2.2 Central and northern European Claw-moss growth (*Dicrano-Hypnion filiformis*), in low lying areas on trunks of *Quercus, Fagus, Betula* and conifers: relatively resistant to occasional low air humidity, and for this reason more plentiful in continental areas (figs. 144 and 370).

3 Weakly acidophile moss bark coverings (*Neckeretalia pumilae*), growths of mosses on smooth tree bark, especially on *Fagus*; only moderately acid-tolerant, in atlantic and montane areas of Europe.

3.1 Fastidious pioneer moss growths (*Ulotion crispae*), mainly in *Fagetalia* communities as pioneers of the moss growths on *Fagus, Abies, Alnus incana, Corylus*

Fig. 370. Epiphytic growth on old Beech trees in a moder Beech wood (*Luzulo-Fagetum*) of the Belgian Ardennes, i.e. in a montane climate with little air pollution. After Duvigneaud and Kestemont (1977), somewhat modified.

The youngest twigs are very quickly occupied by green algae (*Chlorococcales*) and crustose lichens (e.g. *Lecanora*). Many foliose and fruticose lichens spread on the older branches, including *Parmeliopsis* which dominates the upper parts of the trunks. The lower parts of the trunks are festooned with mosses (*Hypnum cupressiforme*) while Cushion Mosses (*Dicranum*) and in some places ground-dwelling fruticose lichens (*Cladonia*) flourish on the boles. Similar growth is also found in the montane Woodrush-Beech woods of German mountains; however because of air pollution here the lichens have disappeared in many areas allowing the less-sensitive green algae to cover the trunks also.

In the case which has been investigated in the Ardennes it has been found that the total surface area of the branches and twigs making up the Beech crowns, on which epiphytes can settle, is almost five times the area of the ground on which the trees are growing (4.5 ha per ha). Even the surface of the trunks alone amounts to a third (3900 m^2/ha). If all the trunks, branches and twigs were placed end to end they would stretch for 1385 km per ha. Small wonder then that the biomass of the epiphytic cryptogams amounts to some 1.3 t/ha of dry matter.

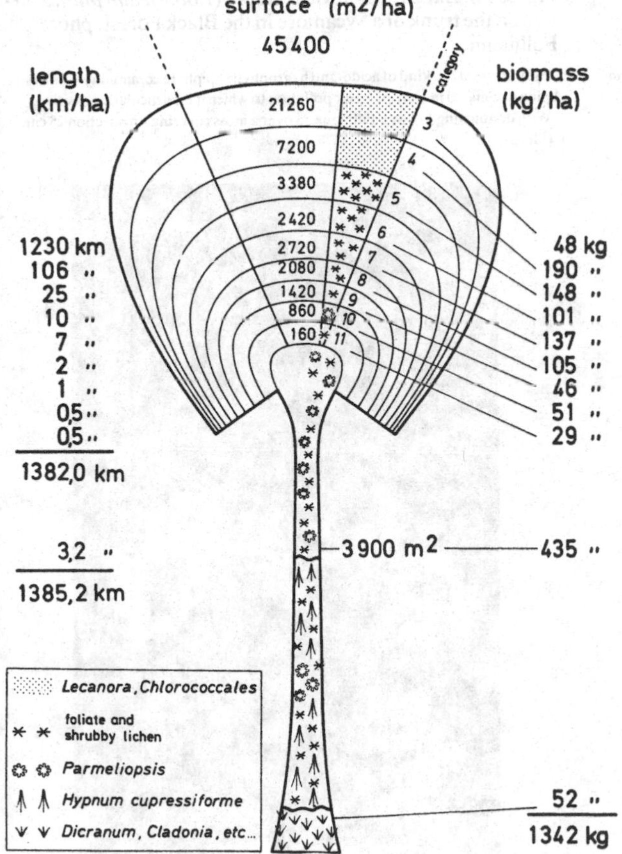

etc.; very sensitive to air pollution and heavier nitrogen supply. *Ulotetum crispae* is still relatively abundant in the mountains of Central and south-western Europe but is declining (fig. 61).

3.2 Counter-hair moss growth accompanying the Beech (*Antitrichion curtipendulae*), has a similar distribution to that of Beech woods (*Fagion sylvaticae*) in Europe; often found with 2.2 but is less acid-tolerant. The *Antitrichietum curtipendulae* reaches its optimum development on Beech bark (pH 5.5-7.1) in montane woodlands. It tolerates dry air better than the communities in the following alliance.

3.3 Lung-lichen moss carpet on tree trunks (*Lobarion pulmonariae*) prefers a relatively high air humidity and is best developed in the region of the oreal cloud level (see fig. 371b); needs more than 180 rainy days in the year. Plentiful on *Fagus, Alnus* and other broadleaved trees, but also on conifers, especially *Abies* (and on *Picea* in limestone areas). Very sensitive to poisons; became extinct in north and middle Germany because of inconsiderate over-collecting.

4 Neutrophilous moss growth on rough bark (*Leucodontetalia*); more strongly tolerant to pollution and nitrogen than other epiphytic moss communities. Generally found in the middle region of trunks on rough bark, often dovetailing in with lichen communities of the *Physcietalia ascendentis* (II 4).

4.1 Drought-tolerant, acid-avoiding Beard-moss growths (*Tortulion = Synchitrion laevipilae*). Of all the epiphytic moss communities these are the least sensitive to drought and therefore widely distributed in Central and southern Europe; however absent from trees with an acid bark (e.g. conifers and also *Fagus, Alnus, Quercus, Betula, Castanea* and *Sorbus aucuparia*) unless a base-rich dust has settled on them (see figs. 13 and 34).

4.2 Sub-continental False-tooth Moss growths (*Anomodontion europaeum*), typical for woodland communities of the *Carpinion* alliance with a corresponding distribution centred in the continental region of Central Europe, especially in Poland, as well as in Hungary, Estonia and Sweden (fig. 106).

4.3 Leske-moss growths (*Leskion polycarpae*), on stones and the bases of trunks in areas which are flooded by rivers but dry out in the summer (fig. 194, see also fig. 246).

Fig. 371 a. Beard Lichen community *Alectorietum sarmentosae* in the Stubai Alps (1500m). phot, Follmann.

An example of the community found in the needleleaved woods often shrouded in damp mists. It colonises acid substrates poor in nitrogen (with dominating *Bryoria subcana*, this area 80 cm wide).

Fig. 371 b. Lung Lichen-Moss carpet (*Lobarietum pulmonariae*) on the trunk of a Sycamore in the Black Forest. phot. Follmann.

An example of the kind of acido-and hygrophytic epiphyte community which is being decimated in Europe by air pollution to which it is particularly sensitive. (With dominating Lobaria pulmonaria over a moss covering, this section 25 cm wide.)

The algal and lichen alliances as well as the moss communities numbered III 1, 2.1 and 3.3 form, by and large, an ecological sequence from vegetation units which are relatively resistant to drought to those requiring extremely damp conditions. In this same sequence the sensitivity to SO_2, fluorine and probably some other toxins resulting from air pollution tends to increase. However each of the well over a hundred associations and other units has its own ecological niche. It can be stated without exaggeration then that the epiphytic plant combinations react more strongly to differences in the habitat than do the trees which serve as their substrate. Judging from examples provided by Barkman this is true as much for small local variations in the microclimate and other environmental factors as for large-scale geographical differences.

Amongst the bio-indicators, important for monitoring the environment, the natural epiphytic communities, which are being involuntarily altered by man's activities, occupy a key position. Their ecological investigation, with few exceptions (like that of H. Muhle 1977), is still in its infancy, and has thus become one of the most urgent tasks for the future.

The plant communities on old **thatched roofs** are related to some extent to the epiphytic vegetation. At first the straw of Rye, Wheat, or Reed has similar properties to those of the smooth-barked trees, but as it gets older it comes much more to resemble the humus layer on acid soil. This change in the substrate leads to a succession from algae, through lichen to moss communities. Frahm (1972) has devoted an instructive monograph on this subject with many examples from Schleswig-Holstein. Even on such an artificial habitat the sequence of cryptogam growth follows a regular pattern which is a function of the environmental conditions and time.

D Formations created and maintained largely by man's activities

I Arid and semi-arid grasslands in colline and montane climate

1 General review

(a) An ecological and phytosociological survey

During the past few thousand years Central Europe has gradually been denuded of most of its natural woodland cover. This process has had a great effect on the climate near to the soil surface which in turn has had a decisive influence on the living conditions of plants and animals. Instead of the solar radiation being converted at the level of the woodland canopy it is now reaching the lower growing plants and even the ground surface. Over wide areas this has greatly increased the range of temperature, humidity and wind speeds at this level (see fig. 373). Because of this, where the ground was previously fairly moist through the year it now dries out at certain times. It also becomes much warmer in summer and colder in winter when compared with the relatively cool moist conditions in the woodland interior. In other words the local climate now takes on many of the features of the general climate of continental steppes and mediterranean karst, or in some respects also of alpine meadows.

During the course of this change many plants and animals which need ample light and warmth and at the same time are adapted to survive periods of drought and cold could come in, or return, to Central Europe. Under the influence of man, these could build up new communities together with some hardy species of the original flora and fauna living in the few open places which were left amongst the almost continuous natural woodland. As we already know from the preceding sections such places are rocks and screes, dunes which had not yet become overgrown with trees, eroded river banks and temporary islands of sand or gravel in the larger rivers, or those areas of fens and bogs which dry out to some extent each season.

The 'xerothermic grasslands' of Central Europe are thus **relatively young**; in their present species composition they did not exist in the natural landscape. One indication of this may be seen in the fact that 'there is no subspecific or semispecific differentiation restricted to these formations'. This was the conclusion arrived at by Nagel (1975) for beetles (Coleoptera) in an exhaustive study of the calcareous dry grasslands in the region of the Saar and Mosel. The same probably applies also to other groups of animals as well as to the majority of phanerogams and cryptogams. Exceptions can be seen in certain critical species, such as the *Thymus*, which easily produce new minor taxa (see P. Schmidt 1974). The 'sandy dry grasslands', for instance on the former very extensive dunes in the Upper Rhine plain have also come into existence relatively recently. According to Philippi (1971) who studied them in the Schwetzinger Hardt, their character species appear to have come in between the New Stone Age and the late Middle Ages.

Where the moderating effect of a tree canopy has been removed a number of very different habitats can arise even within quite a small area. Differences in the depth of fine soil over the underlying rock, variations in the grain size and in the chemical properties of the soil, which under trees have scarcely any influence on the species composition of the community, give rise to very different combinations in grassland which is mown or grazed. One particular woodland community then is usually replaced by several anthropo-zoogenic communities, especially when in addition the soil has been changed by treading or erosion. In fact the number of associations, subassociations and variants which have been described from Central European grasslands is so large that, as in the cases of alpine swards or other types of vegetation so far dealt with, we are not able to discuss them all in detail. So we shall confine our attentions to the classes, orders and alliances and only exceptionally discuss lower units. This is advisable also because the sociological-systematic aspect is not yet sufficiently cleared up within all the alliances.

The steppe-like grasslands and heaths have interested Central European ecologists for many years, and a lot of very informative work has been carried out on these more or less extreme habitats and plant formations. Primarily those species which demonstrate a high degree of adaptation have been studied from an ecophysiological viewpoint. However there are still many questions of a causal nature not yet cleared up. In our discussions we shall be concerned more often with such ecological problems than with species lists, the more so as these may contain many names with which the reader is not familiar. To get an idea of the wealth of different vegetation types we will look first at the site conditions which mainly decide the particular species composition.

All the grassland types we are concerned with here show a certain poverty in nutrients resulting from many years of removal of material through grazing and mowing without any manuring. Thus they are used and treated just as the dwarf-shrub heaths (see section D

Fig. 372. Various forms of the suboceanic dry Bromegrass sward (*Xerobrometum*) on a south-west slope, covered with loess in places, near Limburg on the edge of the Kaiserstuhl. The scrub indicates a development towards a mixed Oak wood which would have dominated the slope under natural conditions. On the right is a woodland on the Rhine plain, which is flooded from time to time.

II), and often are also called heathlands or grass heaths. As long as such poor grasslands are free from ground water or flooding they are usually referred to as 'arid grasslands' or 'semi-arid grasslands'. The greatest contrasts in their species composition are brought about by the lime content of the soil, which is frequently tied up with the clay content. As in the case of the alpine natural grasslands then a class of **calcareous grasslands** (*Festuco-Brometea*) which are particularly rich in species, can be distinguished from a class of acid or **siliceous and sand grasslands** (*Sedo-Scleranthetea*). Such poor grasslands are almost never to be met with on 'average' or fertile soils, especially not on loamy brown earths, since these are generally used for arable cultivation or intensively in some other way (see tab. 88).

The subdivisions of the classes reflect more than anything climatic differences, especially the degree of continentality. This depends only to a certain degree on the distance from the sea; it is much more a result of local climate and soil conditions. Therefore 'subcontinental' and 'suboceanic' units may often be found quite close together (see section 2 a). Within the class *Sedo-Scleranthetea* the most significant is the **continental** order of the Stonecrop-Knawel sward (*Sedo-Scleranthetalia*). Its place is taken in the **oceanic** north-west by communities, poorer in species, of the order of the Sheep's Fescue-Biting Stonecrop sward (*Corynephoretalia*). We have already touched upon such sandy grass heaths in connection with the dune succession, especially the Grey Hairgrass communities, which occupy small wind damaged areas on grey dunes as well as on inland dunes. These have been dealt with in section C V 2.

Today the more or less dry calcareous grasslands (*Festuco-Brometea*) occupy far larger areas of Central Europe than the siliceous and sand grasslands. In this class too there is a sub-continental and a sub-oceanic order (see tab. 89). The Wallis Sheep's Fescue sward (*Festucetalia valesiacae*) leads on to the continental steppes, whereas the Upright Brome swards (*Brometalia*) are reminiscent of the grass heaths of west and southern Europe. The latter include numerous submediterranean elements in their broadest

Tab. 88. A synopsis of the habitats and systematics of the more or less dry sward communities using as an example the region east of the Harz, which has a relatively continental climate

Soil:	**acid**, poor in nutrients	intermediate	**rich in bases** (carbonate or silicate)
Class Order	*Sedo-Scleranthetea* or [a]) *Sedo-Scleranthetalia*		*Festuco-Brometea* *Festucetalia valesiacae*, or [b])
on **sand** or loose stones	**Sedo-Veronicion** (?) – *Galio-Agrostietum tenuis* race with *Eryngium campestre*		–
on **rock**	(separate alliance?)		**Seslerio-Festucion**
– very shallow soil		– *Poa badensis-Sedum acre* pioneer communities	
– shallow soil	– *Thymo-Festucetum cinereae*		– *Teucrio-Festucetum cinereae*
– stony	– *Cynancho-Festucetum ovinae*		– *Erysimo-Melicetum ciliatae*
on **gravel**, poor in fine soil			**Xerobromion**[b]) – *Teucrio-Melicetum ciliatae*
fine soil cover			**Astragalo-Stipion**
shallow over rock		– *Geranio-Stipetum capillatae*	
somewhat deeper			– *Teucrio-Stipetum capillatae*
good depth		– *Festuco valesiacae-Stipetum capillatae*	
average to deep soil	**Nardo-Galion** (Nardetalia)[1]) – *Filipendulo-Helictotrichetum*		**Cirsio-Brachypodion** – *Bupleuro-Brachypodietum*
deep soil		– *Festuco rupiculae-Brachypodietum*	
Footnotes:	[a])Class *Nardo-Callunetea*		[b])Order *Brometalia*

Source: From data in Mahn (1965).

sense, but have so many species in common with the *Festucetalia valesiacae* that the *Festuco-Brometea* can be counted as one of the best characterised classes in Central Europe.

In each of the two orders of the calcareous grass swards there is a group of communities which require somewhat warmer conditions, and at the same time are able to withstand drier periods than the others (tab. 88). In the more or less oceanic order *Brometalia* such 'real' **arid grasslands** (alliance *Xerobromion*) are very rare, whereas **slightly arid grasslands** (*Mesobromion*) are still frequent in most of the limestone areas. Arid grass heaths appear to be open above ground although the soil is completely filled with the root systems (fig. 372). On the other hand the slightly arid grass heaths form meadowlike closed stands which contain many relatively broad-leaved mesomorphic plants (fig. 383, see section 2b and 3b).

The alliance within the more continental order of the Wallis Sheep's-Fescue swards which corresponds to the *Xerobromion* is the *Festucion valesiacae* in that it also occupies this relatively extreme dry habitat. Many of the associations belonging to it resemble the true steppes of south-eastern Europe, and like them are dominated by species of the Sheep's-Fescue group with wiry leaves (*Festuca rupicola*, *F. valesiaca* etc). The continental parallel to the alliance *Mesobromion* is that of the Thistle swards (*Cirsio-Brachypodion*, see fig. 373 and table 88). Within the siliceous grass heath the alliances are more numerous and do not depend mainly on the degree of aridity (for details see section E III 1). Summing up schematically, we get the following hierarchy of the xerothermic grassland units in Central Europe:

classes	**orders**	**alliances**
Festuco-Brometea (**= calcareous**)	*Brometalia* (suboceanic	*Xerobromion* (arid, see No. 5.321 in E III 1)
		Mesobromion (slightly arid, 5.322)
	Festucetalia valesiacae (subcontinental)	*Festucion valesiacae* (arid, 5.311)
		Cirsio-Brachypodion (slightly arid, 5.312)
Sedo-Scleranthetea (**siliceous**)	*Corynephoretalia* (suboceanic)	*Koelerion glaucae* (rich in lime, 5.225)
		Koelerion albescentis (on coastal dunes, (5.223)
		Corynephorion (acid, 5.221)
		Sedo Scleranthetalia (subcontinental) (4 alliances, site conditions 5.211 5.214)

Amongst the factors leading to a particular combination of species within the xerothermic grasslands the influence of man and his domestic animals should not be overlooked (see section 5). Many of the communities appear in one facies which is produced by **frequent grazing** and another which has arisen by **occasional mowing**. In earlier times almost all grassland on sufficiently deep soils was ploughed up and cultivated from time to time. This type of husbandry is no longer practised but it has left its traces in the floristic make-up of the sward; e.g., according to Krause (1940), it has excluded the species which cannot disperse themselves quickly. The majority of the species taking part in the xerothermic grasslands are noted for the speed with which they are able to move across the country. In this respect they resemble the weeds of arable land which have also originated in the east and south (Hard 1964).

Any quick survey must contain simplifiations. The following sections will allow some of the details to be filled in and where necessary to be altered. Emphasis will be placed on the *Brometalia* not just because it has been investigated more than the other units but above all because it is characteristic of the largest part of Central Europe. In this it resembles the *Fagion* from which almost all the communities of the *Mesobromion* have arisen.

In our survey and in the following accounts we are keeping, by and large, to the basic classification given in Oberdorfer's flora (1979). Several suggestions have been made for more or less different groupings of the arid and slightly arid grass heath communities. In almost all cases this has resulted in an increase in the number not only of associations and alliances but also of higher units of classification. Since this development can be followed only with apprehension within the relatively narrow boundaries of Central Europe when looked at from a European or global viewpoint, we have once more consciously remained conservative. However the comprehensive survey on 'The structure and site conditions of the xerothermal grasslands in middle Germany' by Mahn (1965, 1966a) may be recommended. Here numerous 'ecologic-sociologic groups' have been used, ranging from species which are indicators of extremely dry and poor conditions to those which live in better ones or in semi-shade or on screes. Krausch (1968) was the first to give a systematic survey of the sandy grass heaths. Kolbeck (1975) has produced a monograph on the *Festucetalia valesiacae* communities with examples from the eastern mountains of Bohemia.

The wealth of different xerothermic grassland communities which still exist in southern Baden (the far south-west of F.R. Germany) was recently depicted by Witschel (1980) emphasising their spatial and successional relations to ligneous and dealpine plant communities. Depending on their micro-habitats these vegetation types form an extremely diverse mosaic including woodlands, scrub, herbaceous 'skirt communities' (see section D IV 2), small stands of therophytes or leaf succulents as well as rocks and screes with their more or less scarce thallophytic and cormophytic plant cover.

For nature conservancy the most problematic ecosystems are the *Xerobromion* communities. Their tiny areas left in Germany obviously become more and more invaded by taller herbs because they are no longer grazed, cut or burnt. Soon the remaining area will be smaller than the minimum necessary for maintaining their complete species combination. The same could be the fate of the arid grasslands in the hill region of southern Alsace, e.g. near Rouffach where remarkable remnants of the areas classically described by Issler (1937) are still grazed (Leuschner 1982). On the other hand the equally peculiar associations on dry calcareous gravels in the Rhine plain have already disappeared.

The once rather frequent *Mesobromion* communities in Central Europe are also regressing very quickly because their soil is relatively productive and more favourable for trees. If they cannot be managed by grazing or cutting on sufficiently large areas they will soon be a thing of the past (Witschel 1980, Kienzle 1979, Pfadenhauer and Erz 1980). In the county of Stuttgart for instance which includes part of the Swabian Jura these 'steppe heathlands', as Gradmann (1898, 1950) called them, around 1900 covered 7000 ha. Mattern, Wolf and Mauk (1980) found them shrivelled up to less than 2000 ha. Since shrubs and trees are reconquering most of these remaining areas quite rapidly their light-loving plants and animals will be unable to persist.

(b) Seasonal aspects of the xerothermic grasslands

Before we analyse the species combination of the individual communities mention should be made of the life forms and structural types they have in common (see section A I 3 a and C) and also of the sequence of changes affecting their appearance throughout the year.

In all grass heaths on more or less dry and poor sites slow-growing hemicryptophytes and chamaephytes are the dominating life forms among the higher plants. These are perennials which are generally

Fig. 373. The average monthly maximum and minimum temperatures of the soil surface distinguish the woodland interior, the edge of the wood, the dry sward and the rocky outcrop at all times of the year. After Baller (1974), modified.

In both the years 1966 (in which there were only moderated extremes of temperature) and 1968 (when there were big differences between the maxima and minima) the climate within the wood was evened out considerably. In contrast the bare soil, protected from the wind by an open dry sward, often warmed up to over 40–50°, and in one month (VI 1968) the average maximum reached 55°C. The minima in the early hours of the morning were nearly always below 10°C; the rocky areas which had scarcely any protection by plants cooled down most of all during the nights. With respect to the average temperatures the herbaceous (the 'skirt' community see figs. 430 – 433) community along the edge of the woodland is about midway between the open grassland and the woodland interior.

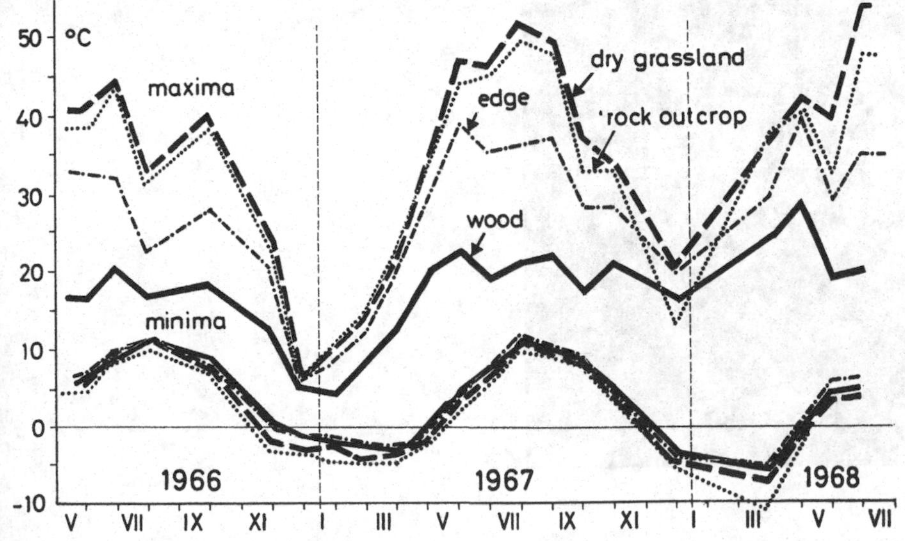

herbaceous with varying amounts of lignified tissue. Many of them are really scleromorphic, having small leaves with large numbers of stomata per unit area, a small ratio of stem to root, and a high proportion of conducting and strengthening tissue. Because of this quite a number retain their shape when the aerial parts die. In spite of this the arid grass heaths, like the natural steppeland in eastern Europe (see Walter 1975) presents a very different picture from one season to another–sometimes even from one week to the next.

In the early spring, while the ground is still moist from the winter but is becoming warmed by the increasing sunshine, the first therophytes burst into life and grow very quickly, soon to ripen their seed and die. The number of these tiny herbaceous plants is especially large in those places which dry out the most in summer since it is here that the turf is most open. The majority of sedges and grasses too begin to shoot long before the trees come into leaf; even in those later-flowering species the roots and shoots become active at this time. However the dry grass heath never appears a uniform bright green but always a mixture of yellows, browns, and greys in with the basic green. The many types of flowers however present a much gayer picture as they supersede each other right through into the summer in all shades of shining yellows, blues, violets, reds and whites each trying to surpass the others. As they change so do the almost incalculable hordes of bumble bees and other Hymenoptera, caterpillars, small butterflies, beetles, grasshoppers and spiders through the different seasons. There are even a large number of snails living in these steppe-like communities; during the heat of the day many of them climb up the highest stalks where it is cooler than on the ground and they do not dry out so quickly. Even in spring the surface temperature of the ground can exceed 50° C. on fine sunny days, thus producing very dry conditions.

Fig. 374. A subcontinental False Brome slightly arid sward near Lebus by the Oder. Appearance in spring with flowering *Adonis vernalis*; in the background the scrub suggests that the natural vegetation would be woodland. The river valley is flooded. Photo, Krausch.

In the height of summer flowering falls off and many leaves dry up. The nutrients, especially nitrogen, concentrate at the leaf bases which remain alive in the hemicryptophytes (P. Wagner 1972). Geophytes begin to accumulate phosphorus, nitrogen etc. in the subterranean organs as reserves for the recommencement of growth in the following year. Most of the plant-eating small animals retreat to sheltered corners or die off like the spring ephemerals. Even the crickets and other musicians, which have been broadcasting their active social life in such a meagre biotope ever since April, are now silent. Only the aromatic scent of the thyme and other herbs whose essential oils are released in the sun's heat remains clinging to hands and clothes. For most plants then even the late summer is not an absolute resting period, although it may force them into this if the year is exceptionally dry.

Here and there in the autumn flowering revives, but it dies completely in winter, which is a severe time for all the perennial members of the grassland communities. The snow cover remains thin and may be absent altogether for weeks or months at a time (fig. 125), so it offers no protection from the cold nor does it help to reduce transpiration. Only a few at least partially evergreen, deeprooted plants like *Teucrium chamaedrys* and *Artemisia campestris*, or small succulents such as the species of *Sempervivum* and *Sedum* are able to cope with the dry winters. It is worth noting that these are commoner in the relatively mild submediterranean climate of the south west than in the more continental eastern Europe where the danger of being frozen to death is so much greater.

Apart from the gaps in the sward, which are crowded with short-lived spring flowering plants there are, according to Grubb (1976), three mini-habitats in the calcareous slightly arid grassland where annuals and biennials normally come in, i.e. ant hills, the edges of footpaths and rabbit warrens. It is in the last of these that *Ajuga chamaepitys*, *Teucrium botrys* and other species whose seed can remain viable for a long time in the soil, have their best opportunity (see also section D VIII 1 c).

2 Supply of nutrients and bases in more or less dry poor-soil grasslands

(a) Nutrient supply and biomass production

It has already been stressed in the introduction that the one thing all the arid and semi-arid grassland communities have in common is their poverty in plant nutrients, which has been brought about by the removal of plant material. The shortage of nitrogen is at once apparent from the yellowish colour of the leaves. According to Gigon (1968) even in the relatively strong growing and dense sward of the slightly arid grassland no more than 20-30 kg N/ha/year are available, i.e. less than in many woodland communities and manured meadows where at least 50-100 kg N can be counted on. The other important nutrients, especially phosphorus are in similar short supply.

Without doubt the lack of nitrogen is a direct result of the regular mowing for hay and grazing which has been practised for centuries, but it is accentuated by the fact that the top soil frequently dries out. Because of this the activity of the mineralising bacteria is restricted in the same way that it is in the slope Sedge-Beech woods (see fig. 63). The nitrogen supply would be even less were it not for the leguminous plants present in the arid and slightly arid grass heaths, which are able to enrich the soil with their root nodule bacteria. It is this type of nitrogen fixation that has made possible the development of a primitive form of agriculture: since neolithic times a few years of cultivation have been alternating with many years of pasture. Blue-green bacteria living on the soil surface also contribute to nitrogen fixation, but in this case it is very little (about 2 kg/ha/year) since they have to keep to the illuminated soil surface and are restricted in their activity frequently and for long periods by drought (Vlassak, Paul and Harris 1973).

Nitrogen in the grass heath soils is offered mainly as nitrate, i.e. in a form which is easily assimilated by the majority of species. According to Gigon and Rorison (1972) there is an antagonism between the uptake of K^+ and that of NH_4^+ in *Scabiosa columbaria* and *Sesleria* and also in *Rumex acetosa*. On the other hand grassland plants normally growing on a strongly acid soil, e.g. *Avenella flexuosa* can make use of NH_4^+ just as well as if not better than NO_3^- within a wide

Fig. 375. The distribution of unmanured grassland communities depends on the depth of soil and the relief. These examples are on the limestones of the Swabian Jura, semi-schematic. After Ellenberg (1952). The deep soils (right) would be the most productive if they were dunged since they are better able to retain the rainwater. The potential natural vegetation would be slightly-moist limestone Beech wood (in place of the Brome Grass type of slightly arid sward), richer and typical brown-mull Beech wood (in place of the False Brome type or the degradation stage) and moder Beech wood (instead of the Matgrass sward).

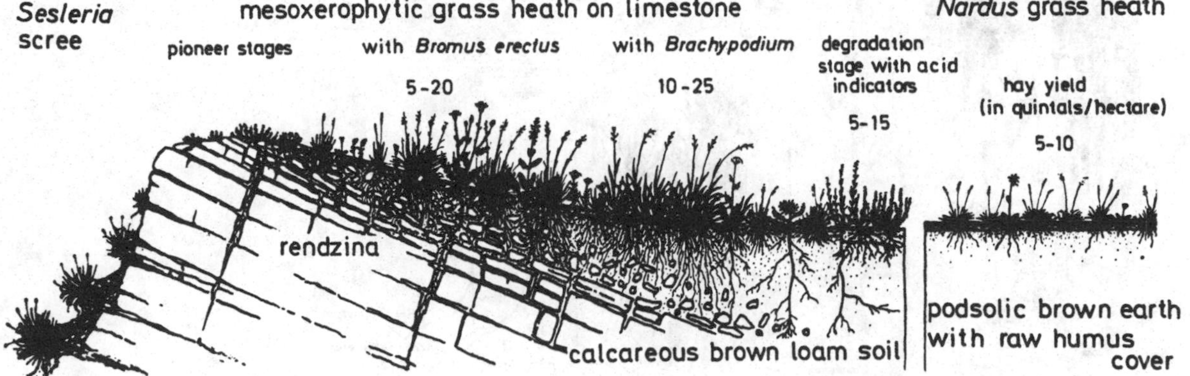

range of pH-values (4.2-7.2, see fig. 94).

Where the supply of nutrients is smaller than normal plants which can use them economically are indirectly favoured. Such plants have a marked internal cycling (Ellenberg 1977). For example the Upright Brome (*Bromus erectus*) stores a large part of its nitrogen and phosphorus in the bases of its leaves and in buds before the leaves are killed by drought or frost. These 'green remains' as P. Wagner (1972) calls the storage tissue are probably the reason why *Bromus* is readily eaten by sheep at all times of the year. The Torgrass (*Brachypodium pinnatum*) on the other hand stores its capital of nutrients mainly in its twitch-like rhizomes from which it can withdraw it in the spring for the rebuilding of its photosynthesising organs. Many of the representatives of the arid and slightly arid grass heath communities are of the '*Bromus* type' or the '*Brachypodium* type'.

The shortage of nutrients and the drought periods so restrict the production from these grassland types that they have always been considered poor yielding and today in many parts of Central Europe they are looked upon as wastelands. Klapp (1965) gives the yearly hay yield as between 1 and 3 t/ha, compared with at least 5 to 7 t/ha for good-feeding meadows. Gluch (1973) determined the net primary production of above ground material in unused semi-arid grassland in the nature conservation area 'Neutratal' near Jena and arrived at the following weights of dry matter per hectare (in tonnes):

Blue Moorgrass slope (*Epipactis-Seslerietum*)	0.28
Slightly arid grassland (*Onobrychi-Brometum*)	0.88
Brome-False Oatgrass meadow (*Arrhenatheretum salvietosum, Bromus erectus*-variant)	1.08

However the growth above ground does not give a true picture of the total production since with xeromorphic plants the underground parts are more strongly developed than the aerial ones. For example Pílát (1969) working near Prague found during the course of a year that the underground biomass varied between the following values (same units as above, also compare fig. 376):

Slightly arid grassland	(*Mesobrometum stipetosum*) 15.82–25.92
Foxtail-False Oatgrass meadow	(*Arrhenatheretum alopecuretosum*) 6.77–10.50

Thus the underground increment must be at least equal to the difference between the two extremes, i.e.:

Slightly arid grassland	10.10
Foxtail-False Oatgrass meadow	3.73

Fig. 376. The more or less xeromorphic plants (b) of the unmanured slightly-dry grassland have a very extensive root system whose biomass far exceeds that of the overground parts. Where the soils are richer in nutrients and not so dry the ratio of subterranean to aerial mass is less.

a. The root systems of Red Fescue, Oatgrass, Field Scabious and Ribwort Plantain in a False Oatgrass meadow (*Arrhenatheretum alchemilletosum*) of average fertility. After W. Schubert (1963), somewhat modified.
b. Underground rhizomes and roots of the Germander in a Blue Moorgrass sward over limestone. Only a proportion of the roots has been drawn. As a. Rhizomes and roots of Torgrass, also the root system of Red Fescue in a slightly-arid sward (*Origano-Brachypodietum*). After Kotanska (1970), modified.
c. Annual changes in the proportions of the underground and above-ground phytomass of a slightly-arid sward and a False Oatgrass meadow. After Pílát (1969), modified.
By carefully extracting all the roots it can be shown that the ratio of roots to shoots in a slightly-arid grassland is very high (5:1) even at the time of maximum shoot growth just before it is mown. After mowing and also at the end of the winter it exceeds 15:1, because a large part of the organs above ground has disappeared.

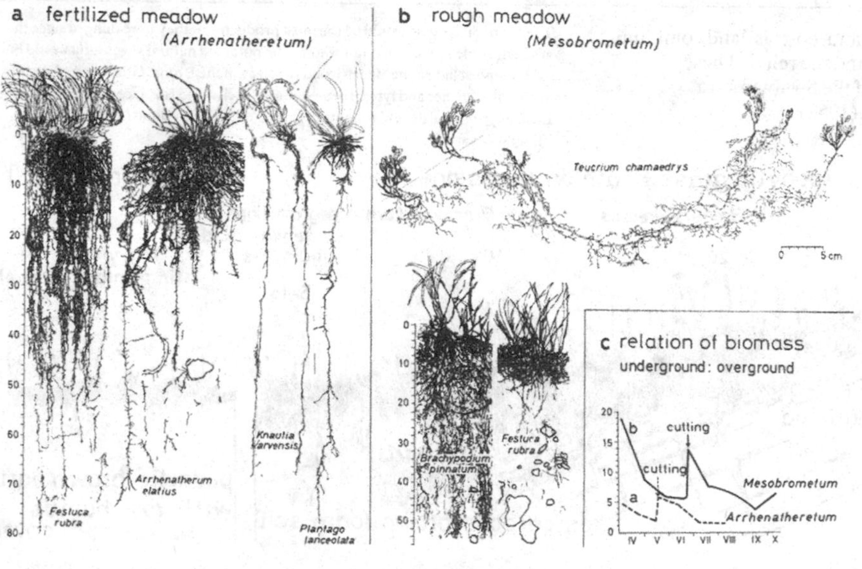

Wagner (see tab. 96) obtained figures of a similar order for *Bromus erectus*. In particular he discovered that the **finest roots**, which to a large extent are washed away with the soil in the normal method of cleaning, when carefully separated out amounted to more than half of the biomass. *Brachypodium pinnatum* and other species of the slightly arid grassland also gave figures approximating to these. Since the finest roots are usually short lived, the amount of material produced underground in such plant communities must be much larger than was at first supposed. Because research of this kind demands a large expenditure of work we are still at the beginning of some important findings. It is possible that the 'poor' grass heaths are scarcely any less productive than manured grassland or woodland on similar sites but that they take the larger part of their assimilates below ground where they are out of reach of both the grazing animal and the scythe.

(b) More or less calcicole arid and slightly arid grasslands

The grassland communities of dry ground in lowland areas react more strongly to the carbonate content of the substrate than they do at the alpine level (see section C VI 1 f). The calcareous xerothermic grassland (*Festuco-Brometea*) is separated from that of the siliceous and sand grasslands (*Sedo-Scleranthetea*) by an unusually large number of character species and intermediate forms between the two are relatively few in number. As was already indicated in the introduction the reason for this is that soils which are chemically intermediate between the two extremes, i.e. weakly calcareous and with a near neutral reaction, are generally deeper than either the limestone and dolomite rendzinas on the one hand or the rankers formed from siliceous rocks on the other. This deeper soil has a higher water-holding capacity and is hardly ever allowed to remain as a poor grassland but is generally used for arable cultivation, manured grassland or woodland. The removal of nutrients over the course of many centuries has accentuated the chemical difference between the calcareous and siliceous soils. Above all it has speeded up the removal of lime and the general impoverishment of sandy or sandstone soils which originally contained some lime.

The floristic difference between the grasslands on calcareous and siliceous soils can be explained by the physiological behaviour of the majority of their species (see Kinzel 1982). Experiments with pure cultures have produced the same results as they do with the lime and acid indicators in alpine grasslands and in woodlands: practically all vascular plants have a very wide pH tolerance when they are relieved of competition. Most plants from the acid-soil grass heaths thrive well if they are transplanted into a lime-containing soil with a pH value of over 7.5, e.g. *Festuca ovina tenuifolia* (= *capillata*), *Danthonia* (= *Sieglingia*) *decumbens* and *Rumex acetosella* (Bournerias 1959). The opposite is also true; many of the plants from a calcareous soil will grow on acid soil too. Of course some of the acid-soil plants soon become chlorotic on soils containing lime because they suffer from a shortage of iron. Such plants would be suppressed in competition with less-sensitive species. Grime (1965) for instance found that this was the case with *Lathyrus linifolius* which is confined to acid soils whereas *Lathyrus pratensis* which is widespread on less-acid soils shows no chlorophyll defect. Lime chlorosis is increased by dryness since the iron is changed into a form more difficult to assimilate. This is probably one of the main reasons why many of the character species of the *Sedo-Scleranthetea* never manage to penetrate the *Festuco-Brometea* (see also section C VI 1 f).

The grassland communities on dolomite (which is a limestone with a higher magnesium content) are so similar to those found on limestone under comparable conditions that they can at best be separated sociologically only at the level of association or subassociation. Since dolomite is relatively resistant to weathering the soils are usually less deep and have a lower water-holding capacity. Because of this the grass heaths on the dolomite rendzinas are more often of the arid than the semi-arid type provided the climate is not too humid. Like most of the limestone plants the dolomite plants too have a wide amplitude when grown in a pure culture, thriving both in soils with calcium carbonate as well as in quite acid soils. In fact Boukhris (1967) found that many of them even germinate better in soil which is deficient in Mg than on dolomite. On the other hand several species appear to be absent from the dolomite communities because they are not able to tolerate high concentrations of Mg. Cooper and Ethrington (1974) showed this to be the case for *Lophochloa* (= *Koeleria*) *cristata, Plantago media* and other limestone plants.

The influence of heavy metals such as zinc and lead is apparent in the species composition of some grasslands. Since there have been a considerable number of recent publications on this subject we shall deal with it as a special problem under a separate heading (D I 7).

Where the fine earth cover is more than 30–40 cm deep the sward is insulated from the underlying calcareous rock, and the soil is more acid but at the same time more water retentive. Thus one finds here all intermediate types leading to Matgrass swards or dwarf shrub heaths such as those which we shall deal with in section D II (see figs. 374 and 377).

'Limestone heaths' or 'chalk heaths', i.e. intimate mixtures of calcicoles and calcifuges growing on a more or less uniform weakly acid soil have been intensively studied in southern Britain by Grubb, Green and Merrifield (1969), Etherington (1981) and other authors. Similar *Mesobromion* grassland communities with *Calluna* and further heathland plants are rare in Central Europe; but they could or even can be found today, at least in the more oceanic parts, e.g. around Göttingen. In such habitats both groups of species 'have roots growing healthily in soil at pH 5–6 and both almost certainly regenerated from seed at that pH in the low turf maintained by ..... grazing' (Grubb et al. 1969). However 'strict calcicoles' are excluded by 'competition, combined with physical factors', as these authors suggest (see also fig. 330).

3 *Climatic variations in the calcareous poor grasslands*

(a) Effects of the degree of continentality

As we have already indicated in the introduction the different calcareous poor grasslands vary in their proportions of mainly continental or oceanic and submediterranean elements. Both the climatic and floristic differences are particularly noticeable on relatively warm sites. So in tab. 89 we restrict ourselves to colline habitats, but we shall select examples from very different parts of Central Europe so that it can be seen how the extremes are connected by intermediate types. In this respect too the sequence of xerothermic grasslands reminds us of the gradient in the xerothermic mixed broadleaved woodlands (see tab. 32); however in the grass heaths it is much more marked. marked.

The 'stony steppes' of Wallis (no.1) and slightly arid grassland near Göttingen (no.9) still have only a few class-character species and their companions in common. The

Fig. 377. A section through a slightly arid sward with *Brachypodium pinnatum* on deep loam over prophyry near Halle. After G. Mahn (1957).

The soil is short of lime but it has a relatively good water-holding capacity. From left to right: *Brachypodium p.* (sterile). *Filipendula hexapetala, B.p., Eurphorbia cyparissias, Festuca ovina sulcata, Salvia pratensis, B.p., Scabiosa canescens, B.p., Achillea millefolium, Centaurea scabiosa, Potentilla alba, B.p., Plantago lanceolata*, B.p.

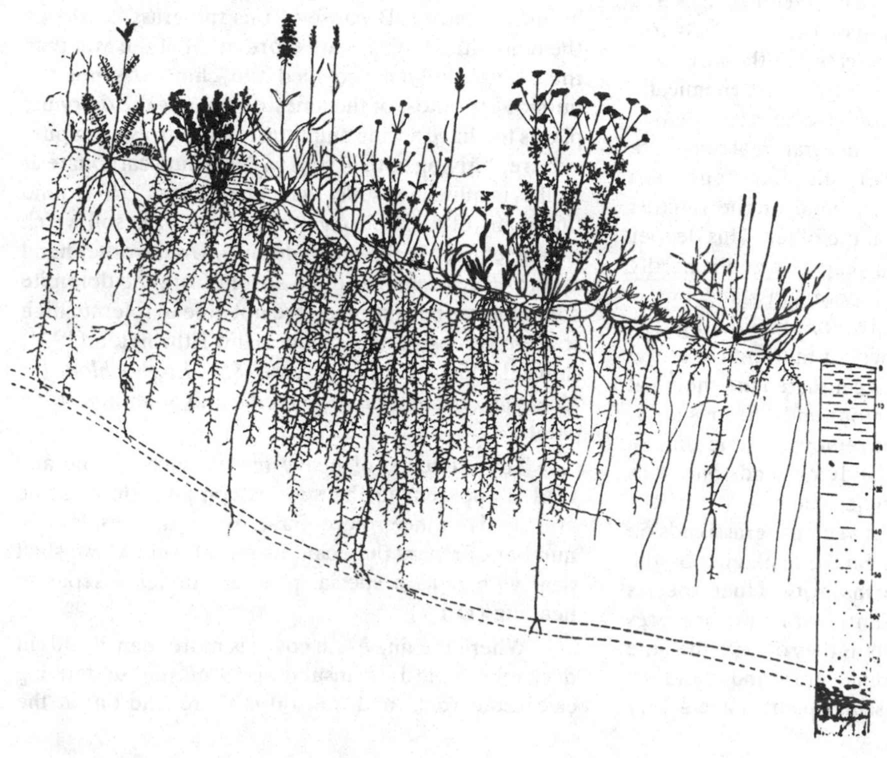

Tab. 89. More or less arid grasslands on basic rocks in different parts of Central Europe.

	Order:	Festucetalia valesiacae				Brometalia erecti				
	Alliance or suballiance:	St.-P.	Festucion valesiac.			Xero-brom.			Meso-brom.	
	Association no.:	1	2	3	4	5	6	7	8	9
C	*Carex liparicarpos*	5								
F	*Potentilla pusilla*	5								
P	*Poa alpina carniolica*	4								
P	*Koeleria vallesiana*	4								
F	*Pulsatilla montana*	4								
F	*Petrorhagia saxifrago*	4								
P	*Asperula aristata*	3								
F	*Scabiosa grammuntia*	3								
F	*Scorzonera austriaca*	3								
F	*Festuca valesiaca*	5	5							
	Agropyron intermedium	1	3							
C	*Veronica praecox*	1	3							
C	*Astragalus excapus*	·	3							
F	*Achillea setacea*	1	2							
F	*Centaurea stoebe*		5							
S	*Myosotis stricta*		5							
C	*Thymus glabrescens*		4							
S	*Achillea collina*		4							
F	*Pulsatilla prat. nigra*		3							
F	*Stipa penn.* and *joannis*	3	2	3						
	Syntrichia ruralis M	2	3	4						
S	*Poa bulbosa*	1	3	1						
F	*Oxytropis pilosa*	1	2	1						
S	*Sedum rupestre*	2	1	1						
K	*Melica transsilvanica*		3	1						
F	*Carex supina*		3	2						
F	*Veronica spicata*	2	1	1	1					
F	*Erysimum crepidifolium*		4	5	2					
S	*Festuca cinerea*		2	1	2					
	Sesleria varia				5					
	Buphthalm. salicifol(ium)				3					
F	*Erysimum odoratum*				2					
K	*Thymus praecox*		1	5	1	5				
F	*Festuca rupicola*		1	3		5				
F	*Thesium linophyllum*			2		3				
F	*Stipa capillata*	5	4	5		2	3			
K	*Allium sphaerocephal.*	4	2	3			3			
F	*Potentilla arenaria*		5	5	3	3	3			
K	*Galium glaucum*		4	5	3	1	1			
S	*Alyssum montanum*		1	5	1	1	2			
F	*Seseli hippomaratrum*		3	1			1			
K	*Stipa pulcherrima*		1		2	2	2			
S	*Artemisia campestris*	5	4	5	3	3	4	1		
K	*Stachys recta*	2	3	4	3	2	2	3		
K	*Koeleria macrantha*	1	5	3	2	2		3		
S	*Medicago minima*	2	2		1		2	3		
K	*Botriochloa ischaemum*	2	2			1	3	1		
K	*Silene otites*	4	3		1			2		
B	*Globularia punctata*	2			3		5	4		
K	*Lactuca perennis*	1		1	2	1		2		
S	*Melica ciliata*	1						3		
	Teucrium chamaedrys	3	3	5	4	4	5	4	1	

	Order:	Festucetalia valesiacae				Brometalia erecti				
	Alliance or suballiance:	St.-P.	Festucion valesiac.			Xero-brom.			Meso brom.	
	Association no.:	1	2	3	4	5	6	7	8	9
K	*Aster linosyris*	3	4	5	2	5	5		1	
K	*Carex humilis*	3	4	1	5	5	5	5	4	
K	*Phleum phleoides*	3	3	4	4		3	4	3	
K	*Asperula cynanchica*	1	2	5	5	4	4		4	
K	*Odontites lutea*	4	1		1			1	1	
K	*Peucedan. oroselinum*	4			2				3	
F	*Euphorbia seguierana*	4					1		1	
K	*Euphorbia cyparissias*	1	5	5	5	4	5	4	5	1
K	*Arenaria serp. leptoclad.*	2	4	5	2		3	2		1
S	*Acinos arvensis*	1	2	1	2			4		1
	Erophila verna	1	5	1	1			3		1
	Hierac. pilosella	1	1	3	3	3	5		4	4
B	*Centaurea scabiosa*	3	1		2	4	1	2	5	2
B	*Hippocrepis comosa*	2			5	3	3	1	5	1
B	*Helianthemum ovatum*	5		3	·	·	·	4	5	1
B	*Bromus erectus*	4		·	·	·	5	5	5	3
K	*Brachypod. pinnatum*	1	1	·	2	3	·	3	5	5
K	*Sanguisorba minor*	1	1	·	2	3	1	·	3	5
B	*Anthyllis vulnereria*	1			2	4	1	3	5	3
B	*Linum tenuifolium*	1	1			5	3	4		1
S	*Alyssum alyssoides Pium*	2	2				1	3		1
K	*Eryngium campestre*		5			3	2	3	1	
K	*Dianthus carthusianor.*		5	5	5	2	1	3	4	
K	*Pimpinella saxifraga*		1	·	2	4		2	4	5
S	*Sedum album*		2	4	2			3		
S	*Sedum acre*		2	5	2		1			1
	Medicago falcata		2		1	2	1		1	3
	Scabiosa canescens		2	1		1	1		1	
K	*Salvia pratensis*		1	·	3	4	5	3	4	1
B	*Avenochloa pratensis*		1		1	5		3	1	4
C	*Cerastium pumilum*		1	1			1	4		
	Galium verum		1		3	1			3	1
	Thymus pulegioides			5	4	1	2	4		4
K	*Festuca trachyphylla*			5	3	3	4	5	5	2
S	*Pulsatilla vulgaris*			1	5	5	1	4		
	Potentilla verna			1	5	3	1	4	1	
B	*Teurcrium montanum*				3	3	4	1	1	
B	*Helianthem. nummular.*				5	2	5		5	
B	*Fumana procumbens*				1		5	1		
	Anthericum ramosum				5	1			3	
B	*Scabiosa columbaria*				4	1	1	3	5	4
B	*Arabis hirsuta*				2		1		2	1
K	*Prunella grandiflora*					5		3	2	1
B	*Koeleria pyramidafa*					3	1		3	
M	*Ranunculus bulbosus*					1		2	5	3
M	*Cirsium acaule*					3		1		4
M	*Carlina vulgaris*					2		1	1	2
M	*Ononis spinosa*					2			2	2
K	*Carex caryophyllea*					·	·		5	4
M	*Medicago lupulina*					·		3	5	4

Tab. 89. (Continued).

		1	2	3	4	5	6	7	8	9
C	*Trinia glauca*					5				
C	*Onobrychis arenaria*					4				
D	*Cladonia furcata* F					4				2
D	*C. endivaefolia* F					4				
C	*Ononis natrix*		1					4		
S	*Petrorhagia prolifera*	2						4		
C	*Coronilla minima*							3		
C	*Micropus erectus*							3		
S	*Cerast. brachypetalum*							3		
S	*Trifolium scabrum*							3		
	Linum catharticum							4	5	4
	Trifolium pratense							4	1	3
D	*Briza media*							3	5	4
D	*Plantago media*							2	5	4
M	*Onobrychis viciaefolia*							1	5	1
M	*Anacamptis pyramidolis*								3	
D	*Potentilla heptaphylla*								5	
D	*Silene nutans*								5	
D	*Anthoxanth. odoratum*								5	
D	*Polygala amara*								4	

		1	2	3	4	5	6	7	8	9
D	*Leucanth. vulgare*								4	1
D	*Lotus corniculatus*								5	5
D	*Plantago lanceolata*								5	4
D	*Campan. rotundifolia*								5	3
D	*Dancus carota*								5	3
D	*Leontodon hispidus*								4	3
D	*Dactylis glomerata*								5	2
D	*Centaurea jacea*								4	2
D	*Carex flacca*								2	3
D	*Knautia arvensis*								2	3
D	*Poa angustifolia*									4
D	*Achillea millefolium*									3
D	*Agrimonia eupatoria*									3
M	*Gentianella ciliata*									2
	A few additional mosses:									
K	*Rhytidium rugosum*			4	5	3	2		5	
K	*Abietinella abietina*				4	2	1		5	1
K	*Camptothec. lutescens*						2		3	1
K	*Pleurochaete squarrosa*				2		4			

Note:

C = character species in the communities where they achieve a high degree of constancy,
D = differential species of a community or an alliance with few character species,
P = character species of the alliance *Stipo-Poion xerophilae*,
F = character species of the *Festucetalia* or *Festucion valesiacae*,
B = character species of the *Brometalia* or *Bromion*,
M = character species of the suballiance of the slightly arid swards (*Mesobromion*), or differential species against the suballiance of the more arid swards (*Xerobromion*),
K = Species characterising the class *Festuco-Brometea*,
S = character species of the class (or its subunits) of the open sand and rock swards (*Sedo-Scleranthetea*) which spread into K.

No. 1: Inner Alpine **dry rocky Feather Grass-Crested Hairgrass swards** (*Stipo-Koelerietum valesianum)* in the Lower **Wallis.** After Braun-Blanquet (1961).

Nos. 2–4: Examples of more or less continental **dry Wallis Fescue swards** *(Festucion valesiacae)*,

2: Dry rocky Wallis Fescue-Treacle Mustard sward (*Festuco valesacae-Erysimetum crepidifolii)* in the **Bohemian** mountains. After Preis (1939),

3: Dry rocky Treacle Mustard-Feathergrass sward (*Erysimo-Stipetum*) in the **Nahe** valley and in Rheinhesse. After Oberdorfer (1957),

4: Sheep-s Fescue-Blue Moorgrass slope (*Seslerio-Festucetum)* in the **Frankonian Jura.** After Gauckler (1938) in Oberdorfer (1957). Inclined towards the Brometalia.

Nos. 5–7: Examples of the more or less suboceanic **dry Brome swards** (*Bromion erecti*, suballiance *Xerobromion*):

5: Dry limestone Honeywort-Ground Sedge sward (*Trinio-Caricetum humilis*) near **Würzburg**. After Volk (1937) in Oberdorfer (1957); inclined towards the *Festucetalia*,

6: Dry limestone Brome sward ('*Xerobrometum rhenanum*') on the **Kaiserstuhl**. After von Rochow (1951) in Oberdorfer (1957),

7: Dry limestone Rest Harrow-Brome sward ('*Xerobrometum lugdunense*') in the **West Jura**. After Quantin (1935) in Braun-Blanquet and Moor (1938).

Nos. 8 and 9: Examples of suboceanic limestone **slightly arid swards** (*Bromion erecti*, suballiance Mesobromion):

8: Slightly arid Pyramidal Orchid-Brome sward ('*Mesobrometum collinum*') in the **Kraichgau**. After Oberdorfer (1957),

9: Slightly arid Gentian-Hairgrass sward (*Gentianello-Koeleriektum*) in the **Göttingen** area. After Bornkamm (1960).

Source: Based on tables by Braun-Blanquet, Preis, Oberdorfer, Quantin, Bornkamm and others. Some less-constant species have been omitted including the orchids which sometimes turn up in nos. 8 and 9, viz: *Aceras anthropophorum*, *Himantoglossum hircinum*, *Ophrys apifera*, *holosericea* and *sphegodes*, *Orchis militaris*, *morio*, *simia* and *ustulata*.

Tab. 90. Climatic data and indicator values for the examples of arid and semi-arid grassland given in Tab. 89.

Vegetation alliance community no. (see tab. 89) region, stations	altitude (m)	Precipitation (mm) year	Precipitation (mm) IV-IX	Order	Mean temp. (°C) year	Mean temp. (°C) July	Quotient[a]) July temp.: yearly precipitation	Average ind. values[b]) mT	mK	mF
Stipo-Poion carneolicae										
1. Wallis: Sitten	540	638	318		9.6	20.6	30.5	**6.5**	4.9	2.2
Siders	532	536	268		9.3	20.8	36			
Festucion valesiacae			<350				>30			
2. Böhmen: Leitmeritz[c])	177	502	318	**Festucetalia**	8.6	18.6	37	**6.5**	5.3	2.6
3. Rheinhessen: Mainz	94	512	290		10.0	19.2	37.5	6.2	4.7	2.5
Oberlahnstein	77	590	332		10.3	19.0	32.5			
4. Fränk. Jura: Nürnberg	320	585	356	**tending**	8.7	18.3	31.5	5.8	4.6	2.8
Amberg (Opf.)	525	677	399	**tow. B**	6.9	16.5	24.5			
Xerobromion				**tending**						
5. Mainfranken: Würzburg	179	560	318	**tow. F**	9.0	18.3	33.0	6.1	4.7	2.6
6. Kaiserstuhl: Oberrotweil	222	672	422		9.7 >9	18.7	28	6.3	4.3	2.7
7. Westjura: Genf	405	859	480		9.5	19.5	22.5	6.4	4.0	2.6
St Denis-Laval	398	747	443		10.0	18.5	27			
Mesobromion			<350				<30			
8. Kraichgau: Bretten	214	734	417	**Brometalia**	8.8	18.1	24.5	**5.5**	3.9	3.4
Pforzheim	258	728	431		8.6 <9	17.5	24			
9. Around Gö.: Göttingen	155	607	378		8.5	17.2	28.5	**5.5**	**3.8**	**3.6**
Herzberg a. Harz	242	802	436		7.6	17.3	20.5			

Note: [a] Quotient obtained by dividing the average July air temperature × 1000 by the total annual precipitation (see tab. 27) as a measure of the hygric continentality of the climate.
[b] Calculated from the indicator values of the species listed in tab. 89. (see sections B I 4 and E III).

[c] after Preis; now Litomerice.

Source: From data in Maurer et al. (1910), Reichsamt für Wetterdienst (1939) and Quantin (1935). Indicator values calculated according to Ellenberg (1974).

floristic connection of the Wallis example with Brome arid grassland of the south-west Jura, with its submediterranean flavour (no.7) is a little closer, but still surprisingly slight when one considers how close the two areas nos. 1 and 7 are. Much more closely related to the inner alpine arid grassland (no.1) is the 'steppe grass heath' of the Bohemian mountains near Leitmeritz (no.2), although it is much further away, and the rocky steppe at the edge of the Rhine valley to the west of Mainz. It is in these three communities that the Feather grasses (*Stipa capillata* and *S. pennata* with ssp. *joannis*) play a noteworthy part. Other species of the order *Festucetalia valesiacae* are also only found here in tab. 89 (see also fig. 378), whereas the species of *Brometalia* (B) are more frequent on the right side of the table.

One can understand these floristic similarities and differences if one characterises the particular area with climatic data and indicator values (tab. 90). In areas 1, 2 and 3 the climate is fairly 'continental'. Very little rain falls, at least during the growing period when the vegetation depends on it. The submediterranean reduction in rainfall during the summer is so noticeable that in Sion the average total from April to September falls to the same value as in Leitmeritz although the yearly total is much greater in Sion. The temperatures, especially the July average, are very high in all three areas. They are of the same order too in the areas of the Brome arid grass heaths (nos. 5-7). In south-east England where, according to Shimwell (1971) both *Xero-* and *Mesobromion*-communities occur, the July temperatures are lower.

The submediterranean and subatlantic arid grass heaths do not enjoy higher summer temperatures than the continental ones. They only receive relatively more rainfall and live under more favourable generally damper conditions. In order to express these two factors in one figure we can use the quotient which was explained in tab. 27. For all the areas with mainly *Brometalia* communities (nos. 6–9) it is less than 30, but for areas with *Festucetalia valesiacae* communities it is more than 30 (see tab. 90).

In the rocky steppe heath of the inner alpine valleys the conditions are even more continental than those of inner Bohemia, although they are further west and nearer the sea. The degree of thermal continentality is expressed in our table by the yearly variation in the average monthly air temperatures. This amounts to over 20° C for Wallis because sharp frosts often occur whereas in all the other areas, including Bohemia, the variation is less. But it is not for this reason alone that list no. 1 in tab. 89 contains so many unique species. The fact that the inner Alpine valleys are very isolated must also contribute to this. Each of these warm dry valley basins within the cool mountain world has its own floristic and phytosociological peculiarities which Braun-Blanquet has studied so thoroughly for a number of decades (1961). He has discovered that the climate in the eastern Alpine valleys is less continental than that in the valleys further west. Since this fact is expressed in the species composition of the arid grass heaths he distinguishes two parallel alliances, the eastern *Stipo-Poion xerophilae* and a western *St. -P carniolicae*, of which list no. 1 gives the example situated furthest to the north-east, i.e. towards the centre of Europe.

The sequence of submediterranean – subatlantic communities starts in our table with the arid grass heath in the neighbourhood of Würzburg (tab. 89, no. 5). On the basis of its species composition as well as the corresponding climatic data there is some doubt as to whether it would not be better included in the continental order of the *Festucetalia valesiacae*. This and the previous list illustrate very well how the two extremes are united by a series of gradually differing intermediates.

If one compares the examples from the *Xerobromion* (nos. 5–7) with those from the *Mesobromion* (nos. 8 and 9) it is obvious that the temperature data are lower for the less arid grasslands than they are for the true arid grass heaths, whereas there is very little difference in the amount of precipitation. The slightly arid grasslands are poorer in warmth-loving species of Mediterranean origin, which is consistent with their somewhat lower average temperature. On the other hand they are richer in mesophilic meadow plants such as *Dactylis glomerata*, *Lotus corniculatus*, *Plantago lanceolata* and *P. media*, *Daucus carota*, etc., which could also be a consequence of their better balanced water economy, since a lower temperature never leads to such a low relative humidity as for instance occurs on the slopes of the Kaiserstuhl or even of the Swiss and French Jura where the hot sun beats directly down.

A remarkable number of rare orchids thrive in the slightly arid grasslands of the *Mesobromion* on warm but not too dry calcareous soils of western and south-western Central Europe (list no. 8). They include for example *Orchis morio*, *militaris*, *ustulata* and *simia*, *Ophrys fuciflora*, *apifera* and *sphegodes*, *Herminium monorchis* and *Anacamptis pyramidalis*. However, their frequency is so low that they hardly appear at all in tab. 89, and apart from a few exceptions they are entirely absent from the *Mesobromion* swards of north-west Germany (no. 9), presumably because it is too cold for most of them here (Tüxen 1937). Other species too which hardly receive a second glance in the south are sought out as rarities in the northern part of Central Europe if they dare to venture up there at all.

As far as measurements from the normal meteorological stations can be used to define the local climate in the sunny grasslands the data given in tab. 90 are in general agreement with the floristic differences shown in tab. 89. Lacoste (1964) arrived at similar results for the French Alps where the separation of *Brometalia* and *Festucetalia valesiacae* communities corresponds to overall climatic differences in the same way as in Central Europe.

Fig. 378. A section through a 'Feathergrass steppe meadow' on deep gypsum soil on the Kyffhäuser. After Meusel (1939).

From left to right: *Scorzonera purpurea*, *Stipa pennata*, *Scabiosa canescens*, *Carex humilis*, *Festuca ovina cinerea*, *Fumana procumbens*, *Stipa P.*, *Aster lino synis*. *Stipa capillata*, *C. h.*, *Fest. o. c.* (*Festuca* and *Fumana* occupy open spaces caused by cattle treading. The steps correspond to previous cattle paths, see fig. 17.)

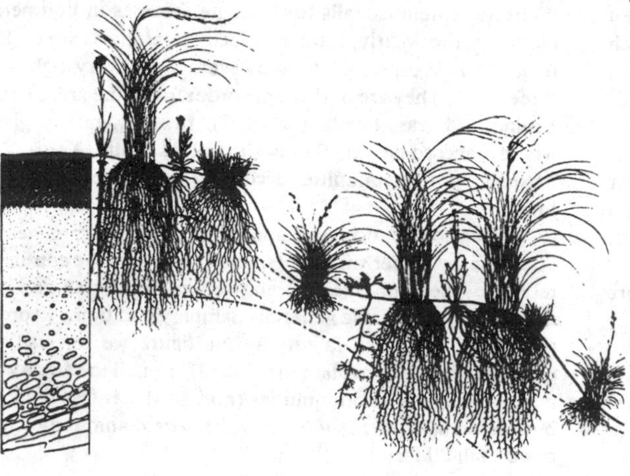

Tab. 91. Life forms, microclimate and soil conditions in the more or less arid swards on gently sloping ground in the French Jura.

Exposed rock communities (Anthylli- Teucrietum, Order Sedo-Scleranthetalia)
Dry grassland (Xerobrometum, Order Brometalia)
Semi-arid grassland (Mesobrometum, Order Brometalia)

Plant cover and climate (1932)	Anth. Teucr.	Xero-brom.	Meso-brom.
Life forms (%)			
chamaephytes	35.0	18.6	10.6
hemicryptophytes	**25.0**	**49.3**	**76.3**
geophytes	2.5	8.7	7.8
therophytes	37.5	21.9	5.3
parasites	–	1.5	–
Average temperatures (°C)			
spring (3.–23.4.)	12.8	9.7	7.5
summer (26.6–16.7.)	**23.4**	**20.3**	**15.3**
autumn (18.9–8.10.)	17.6	16.8	12.0
winter (11–31. 12.)	6.1	5.8	0.7
Average, 12 weeks	**15.0**	**13.1**	**8.9**
Daily fluctuations in temperature (°C)			
summer (26.6–16.7.)	21.5	13.0	9.2
Average, 12 weeks	**17.6**	**10.9**	**8.9**
Evaporation (cm^3 Piche, average daily totals)			
spring (3.–23.4.)	5.4	3.3	2.6
summer (26.6.–16.7.)	**8.3**	**6.5**	**4.6**
autumn (18.9.–8.10.)	6.5	4.3	2.4
winter (11.–31.12.)	4.4	3.2	1.5
12 week mean	**6.2**	**4.3**	**2.8**

All measurements *ca.* 15 cm above ground.
Source: From data in Quantin (1935).

Soil factors (1932)	Anth. Teucr.	Xero-brom.	Meso-brom.
Soil particle sizes (%)			
Coarse gravel	23.6	28.3	24.3
Fine gravel	30.1	32.6	32.5
Coarse sand	19.0	14.0	19.5
Fine sand	19.0	14.6	13.5
Silt and Clay	**8.3**	**10.5**	**10.3**
Water content (vol. %)			
Spring (25.4)	17.4	23.5	26.9
(16.5)	20.2	27.7	33.6
Summer (11.6.)	18.6	24.9	27.9
(19.7.)	10.9	20.2	23.3
(13.8.)	**9.3**	**13.2**	**14.9**
Autumn (17.9.)	10.6	10.9	16.0
Winter (21.12.)	24.8	28.4	32.3
max. (25.4.32–	**25.8**	**30.0**	**33.6**
min. 23.10.33)	**8.7**	**10.9**	**14.9**
Chemical factors			
pH-Value, max.	7.9	7.9	7.3
min.	**7.5**	**7.1**	**6.9**
Lime content (%), max.	51.5	67.1	28.2
min.	**12.7**	**12.0**	**4.4**
Humus content (%), max.	4.6	17.1	21.4
min.	**1.7**	**7.1**	**11.4**

All measurements in 0–10 cm depth.

(b) Differences in the Brome swards related to local and microclimates

The majority of the arid grass heaths occupy slopes or plateaux which are not quite flat, i.e. places where the local climate differs to some extent from the general climate for the district. Such local conditions can be ascertained only by studying the habitat itself. Measurements taken over just a few days or weeks in each of the four seasons give a good idea of the microclimate affecting a particular grassland stand. Quantin (1935) has demonstrated this in an exemplary manner (tab. 91). In limestone mountains such as the French or Swiss Jura true aris grassland *(Xerobromion)* can be found alongside slightly arid grasslands *(Mesobromion)* on a similar geological substrate and even on soils which are very similar in morphology and texture (tab. 91, right). The arid communities occupy gentle south or south-west slopes whereas the slightly arid ones are on slopes of similar gradient but facing north or north-east. Only at higher altitudes do the *Mesobrometa* take over on the sunny slopes.

Quantin compared typical *Xero-* and *Mesobrometa* of the lower south-west Jura both with each other and with additional communities in the area, especially with the stony stands on the driest and warmest sites (*Anthylli-Teucrietum*). We shall return to this special form of arid sward where the thinness of the top soil is a determining factor, in the next section. The columns of figures collected together in tab. 91 show how much the three neighbouring grass heath communities differ in their micro-climates and the habitat factors which depend on them. At all times of the year the air temperature (at a height of 15 cm) is lower in the *Mesobrometum* than in the *Xerobrometum* or even in the rock heath (tab. 91, 2). The daily, monthly and yearly fluctuations in temperature also increase in the same order as can already be seen from the few figures selected for tab. 91.

Such great temperature differences must have their effect on the water economy of plants and other organisms. Evaporation over the slightly arid grassland is in fact significantly less than over the *Xerobromion* communities (tab. 91, left). One can see clearly from the water content in the root run (at a depth of 0–10 cm) that the *Mesobromion* sward makes less demand on the available supplies than does the *Xerobrometum* (tab. 91, right). The figures for the volumes calculated by Quantin can be compared directly because the soil texture is much the same for all the sites under investigation.

The more frequent periods of extreme drought in the shallow soil of the rocky heath and in the *Xerobrometum* mean that the springtime therophytes can find more room here than in the *Mesobrometum*, and so in percentage terms their representation is increased (tab. 91). The chamaephytes too make up a larger proportion of the life form spectrum as they are not shaded so much as in the *Mesobrometum* by grasses which have generally a quicker growth and denser leaves than most of these partially evergreen plants. The hemicryptophytes, here mainly grasses and grass-like sedges make up more than three quarters of all the species in the slightly arid grasslands.

Summing up it may be stressed that the contrasts between the arid and slightly arid grass heaths which are determined over large areas by the general climate, are repeated on a smaller scale because of differences in the local climate. However this is true only for the south (to the middle) of Central Europe; on the northern edge, in southern Sweden, *Xerobromion* communities are absent, and *Mesobromion* ones occur only on the warmest and driest places, e.g. on the steep sunny slopes of the gravelly Oser ridges.

Many authors have already pointed out or verified by measurements how sensitively the species composition of the dry and poor grasslands responds to local climatic differences and even to very small variations within the same stand. We need only to be reminded of the classical work of G. Kraus (1911), as well as the publications of Heilig (1931/32, Müller-Stoll (1936), Volk (1937), Preis (1939), Dörr (1941), Zoller (1954), and Bornkamm (1958 1960) altogether involving a large number of different plant communities (see also fig. 379).

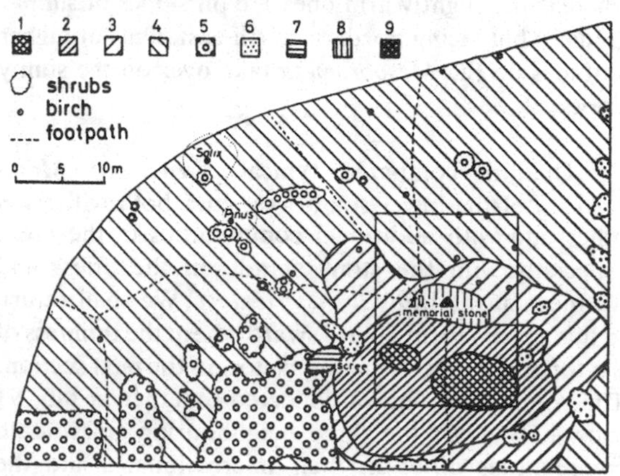

Fig. 379. A vegetation map of the slightly arid swards on the 'Kartoffelstein', a limestone mound with soil ranging from shallow to deep, to the east of Göttingen. After Bornkamm (1959), somewhat modified.

1 = *Gentiano-Koelerietum* typical subass. *Linum* variant, 2 = the same typical var. 3 = the same *Trisetum* var., 4 = G.K., Subass. of *Prunella vulgaris*, 5 = *Pruno-Carpinetum* (scrub), 6 = single pioneer shrubs, 7 = scree community, 8 = *Lolio-Cynosuretum luzuletosum*, 9 = *Caloplacetum murorum* (lichens covering the stone memorial).

1 and 2 on a gently southern slope, 4 more on shaded slope.

Fig. 380. Sunshine, air and soil temperatures, relative humidity and evaporation throughout the day in a *Stipa capillata* sward on a southern slope above the Oder valley. After Brzoska (1963), in Ellenberg (1963).

Broken lines = air temperatures at heights of 2.5, 30 and 200 cm; continuous lines = soil temperatures at depths of 1, 2, 3, 5, 10 and 15 cm, RH = relative humidity. Solar radiation as measured by a black-bulb thermometer. Below = evaporation measured by a Livingstone atmometer (12 July 1932).

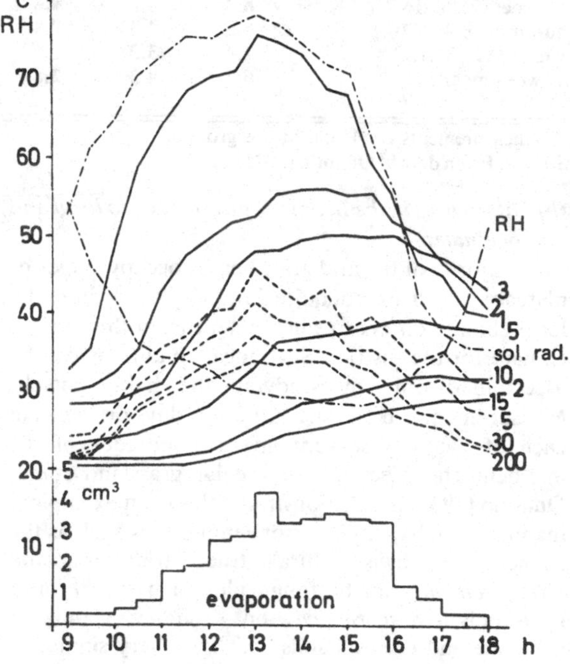

The title Kraus (1911) chose for his book was a signpost for the ecological field work: 'Soil and climate over very small areas'. He took as an example an arid grassland on 'Wellenkalk' (one of the thin layered fazies of the lower mussel beds) in the warm dry valley of the Main below Würzburg. Supported by measurements of temperature and soil factors he came to a conclusion, which is universally applicable 'to arid grassland on shallow stony ground with varying relief, and also to other habitats and plant formations of the open countryside: practically every single plant lives under unique conditions and the close proximity of different species with different requirements can be explained by a similar close proximity of micro-habitats differing in the water-holding capacity, the content of lime and nutrients, and the degree to which they are warmed by the sun etc. In many swards these micro-habitats form a characteristic mosaic following the different layers of rock or some other irregularities in the soil formation. Greig-Smith (1979) recently summarised further research: 'Although heterogeneity in the physical and chemical environment may determine patterns of all scales and intensities, a wide variety of other causes may produce patterns, though all these, except perhaps some cases of historically- and chance-determined patterns, are ultimately mediated by changes in the environment'.

Even the shade of a large plant can temper the scorching sun's rays to such an extent that organisms sensitive to heat and drought can find a suitable place to live. In such a way the mosaic of plants making up the sward creates numerous micro-environments where organisms with quite different physiological requirements can exist side by side. For example: on the southern slope of the Badberg (Kaiserstuhl) the surface temperature of the unshaded ground reached 50–60°C as early as 1st May 1950 while only 20–30 cm to the north in a dense clump of *Bromus erectus* it had not reached more than 15–17°C. At night the bare surface cooled down to below 5°C and was moistened with dew next morning whereas in the grass tussock the temperature did not drop below 10°C. Thus its range inside the clump was only one tenth of that in the open where only drought-resistant cryptogams were growing.

In addition to the horizontal variations are vertical gradients such as those shown in fig. 380. In an open Feather-grass sward on a marginal slope of the Oder valley during the course of a summer's day the temperature at the soil surface ranged from 30 to almost 70°C, but only 5 cm above and below this the highest temperature recorded was 40°C, while at a depth of 15 cm it did not even reach 30°C. The dry leaf remains with which the hemicryptophytes surround their younger leaves and buds protect these sensitive parts from being damaged by the scorching heat which could happen if they came into direct contact with the soil.

(c) Blue Moorgrass slopes as dealpine communities

On steep limestone slopes in more or less shaded sites, i.e. in edaphically and climatically unique habitats, Blue Moorgrass dominated communities are found in mountains down to quite a low altitude. These are reminiscent of the Blue Moorgrass-Sedge slopes in the alpine belt (see section C VI 2 a, figs. 333 and 375). As in these cases the Blue Moorgrass is in no way a pioneer but settles on stabilised slopes or on stepped rocks, and also on moist calcareous soils.

W. Schubert (1963) gives an elucidating survey of the communities dominated by *Sesleria albicans* in central Germany. At the colline level the following can be distinguished with reducing dryness and warmth:

Dwarf Rockrose-Blue Moorgrass slope	(*Fumano-Seslerietum*)
Juneberry-Blue Moorgrass heath	(*Amelanchier-Seslerietum*)
Dark-red Helleborine-Blue Moorgrass slope	(*Epipactis atrombens-Seslerietum*)
Grass-of-Parnassus-Blue Moorgrass slope	(*Parnassio-Seslerietum*).

The first named occupies sunny steep slopes with steps of shallow soil, the last one is found on shaded hillsides or cliffs where from time to time water will trickle through the soil and the air is always humid. *Sesleria albicans* is the dominating species in all of them, as it is in the Blue Moorgrass-Beech woods in the colline to montane belts and also on the montane Blue Moorgrass slopes in central Germany (*Helianthemo-Seslerietum*). On soils which are always moist, often with springs, this taxon is replaced by *Sesleria uliginosa*. Examples are the Blue Moorgrass-Blunt-flowered Rush spring swamp (*Juncetum subnodulosi seslerietosum*) and other calcareous Small-Sedge swards (see section C II 2 a).

From the plant geographers viewpoint the Blue Moorgrass slopes are worth noting because, like the Blue Moorgrass-Beech woods, they contain species which are centred in the Alps, e.g. *Carduus defloratus*, *Thesium alpinum*, and *Aster bellidiastrum*. One could see in these 'dealpine' species relicts of the Ice Age which for a time at least, brought a climate similar to that now found in the alpine belt. Outside the Alps, *Sesleria albican* itself may also be looked upon as a glacial relict. At any rate, according to W. Schubert, the Blue Moorgrass has a very restricted dispersal capability. It germinates badly and where vineyards, next to a Blue Moorgrass slope, have been abandoned they have not been taken over even though the site conditions are similar. Thus, where the Blue Moorgrass is found today it must have had a long period of time available for it to have come in and settled.

4 Effects of soil depth and texture

(a) Rock ledge and debris vegetation

In an area with a uniform general climate the composition of arid and slightly arid grass heathlands changes not only with the aspect, gradient and stability of the slope, but also with the soil depth. As we have already stressed in section 3 b this factor can change over a very short distance and brings about the formation of quite dissimilar communities with regard to their physiognomy and floral composition. One usually comes across five habitat types which are linked through intermediates (fig. 375):

- almost bare rock
- debris devoid of fine soil
- fine soil with a lot of stones
- deep loamy soil
- deep sandy soil

each in positions with different amounts of irradiation

These different habitats are occupied by separate communities often forming mosaics or more or less continuous series, except for the deep loamy soil which is normally ploughed.

Fig. 381. Acid-soil dry swards and other communities on a porphyry rock outcrop near Münster on Stein (Rotenfels). After Haffner (1968), modified.

1 Crustose and foliose lichen communities (*Parmelion saxatilis*), 2 *Asplenietum septentrionalis*, 3 *Coteneastro-Amelanchieretum*, 4 fragmentary communities of the order *Sedo-Scleranthetalia*, 5 *Erysimo-Stipetum*, 6 *Teucrio-Melicetum ciliatae*, 7 *Rumicetum scutati*.

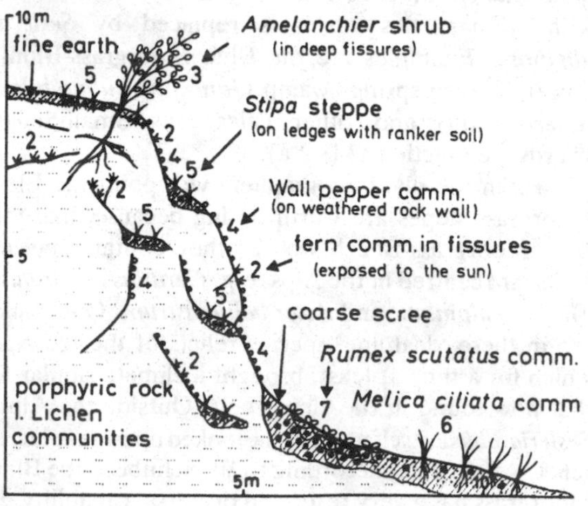

Only the rocks that are almost bare of soil cannot support woody vegetation. Most of these special sites are in the form of narrow edges, ledges or knolls, i.e. of limited area (fig. 381, see also fig. 382). Crustose and foliose lichens and cushion mosses cover the stones and account for the first humus formed on these raw substrates (protorendzina). In crevices, clefts and small depressions sufficient humus, dust and blown mineral soil can collect for vascular plants to establish a foothold. The big danger in such habitats is that the root system will dry out. Succulents and semi-succulents, e.g. *Sedum* and *Sempervivum* species have the best chance of survival. On the other hand spring-flowering annuals are able to produce seed here as we have already seen in section 1c. R. Schubert (1974) has described examples of such 'Wall Pepper pioneer communities' from the GDR.

The vegetation of rock ledges and debris (order *Sedo-Scleranthetalia*) has been thoroughly surveyed by Korneck (in Oberdorfer 1978) in southern Central Europe where it finds many favourable sites. He distinguished four alliances differing in their climatic and edaphic demands:

1. **Subalpine and alpine leaf succulent cushions** (*Sedo-Scleranthion*) thrive best on the tops of ridges of silicate rocks;
2. **Suboceanic silicate debris swards** in the lowlands (*Sedo-Veronicion dillenii*) grow on the disintegrated material accumulated on ledges of acid rocks;
3. **Suboceanic limestone-debris swards** (*Alysso-Sedion albi*), the corresponding associations on calcareous substrate, equally prefer a rather oceanic and warm climate;
4. **Subcontinental rock-ledge swards** (*Festucion pallentis*) on the other hand are best developed in the more continental parts of Central Europe on ledges of steep rocks on mountains as well as on lowland sites.

These inconspicuous low and rather open plant stands living on mostly very limited and isolated spots have often been overlooked, the more so since the best representatives are not within easy reach. But ecologists fond of floristic rarities will be rewarded for all efforts to get there. In southern Germany alone – including the Alps – Korneck described not less than 20 associations. The list of their character species is astonishingly long and comprises many otherwise rare species or subspecific taxa (see section E III 1, No. 2.51), the third alliance mentioned above being the richest. This exceptional position is due to the fact that the habitats of *Sedo-Scleranthetalia* communities are inhospitable for most other plants. The soils are poor in nutrients, at least in mineral nitrogen, and dry out often and rapidly. Neither trees and shrubs nor large herbs can get a footing here. Thus competition is not strong; even tiny plants and newly developed taxa have a chance to survive if they do reach these isolated places.

From such refuges only a few of the species are able to make inroads into the surrounding anthropo-zoogenic grasslands. Surely they are not the original sites of *Xero*- or *Mesobromion* communities or more continental arid and slightly arid grasslands.

Debris swards or other 'rock heaths' are often seen as initial stages in a succession leading first to a stony rendzina sward and later to grasslands and woody vegetation on deeper soils. However it is impossible to observe such a succession directly as it goes on very slowly, if at all. It can only be inferred from the proximity of different stages and their intermediates. As Wendelberger (1953) and others point out we are dealing in most cases with stable zones which are rarely if ever displaced by the next stage of the hypothetical succession series (see fig. 375). In fact where a replacement is recognised it is usually a step backwards through damage to the soil where grazing animals have increased the rate of erosion. One such loss of fine soil far exceeds the rate at which it can be replaced by weathering. What has been described here for calcareous rock heaths also applies to those on acid soils which are physically similar, such as the ones Moravec (1967, see fig. 383) has studied.

Fig. 382. The zonation of vegetation and soil on a sunny sandstone rock outcrop in south-western Bohemia. After Moravec (1967), somewhat modified.

Fine earth cannot collect along the dry edge of the rock because of continuous erosion. Soil development remains at the stage of a protoranker and plant succession at the Forked Spleenwort fissure community. Where more fine soil has accumulated there is a brown ranker carrying a Hair Moss-Knawel community. The dense Bent-grass sward indicates a soil of average depth, i.e., a more or less well developed sandy brown earth. Woody plants could obtain a footing here but were prevented from doing so by grazing sheep.

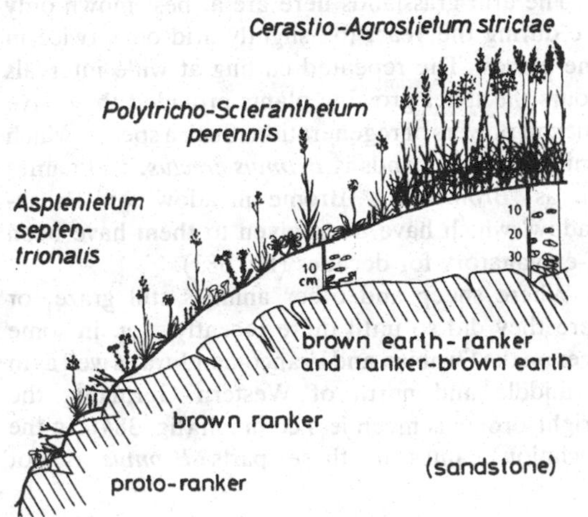

(b) Arid and slightly arid grasslands on stony fine soils

It is only where fine earth covers the rocks sufficiently deeply to hold enough rainwater that grass-rich heathlands can be formed (figs. 375, 382). Arid as well as slightly arid grass heaths (see tab. 89) need a more developed soil than the rock heaths treated in the preceding section. On calcareous rock the soil type corresponds to a mull-rendzina, and on siliceous rocks to a pararendzina although exceptionally it may be a ranker or a eutrophic brown earth. Sometimes one also finds limestone-brown loam or a fossil red loam from the Tertiary which has been 'topped' or laid bare by erosion.

The natural vegetation of such habitats in Central Europe is always a type of woodland, perhaps a Downy Oak scrub or a xerothermic mixed broadleaved woodland (section B III 4), or an open Pine wood (section B IV 6). Due to loss of fine earth by erosion after deafforestation the soil could become so shallow that trees would find it difficult to reestablish themselves even after man had stopped interfering. However there are mostly a few trees which manage to grow in the rock steppes today, even in such dry continental areas as the central Bohemia (Klika 1933) or middle Wallis (Braun-Blanquet 1961).

Even where there has been no heavy erosion the humus content of the top soil under arid or semi-arid grassland is generally less than under woodland on morphologically comparable soils (Quantin 1935) The A-horizon appears to be lighter in colour whether the background is yellowish, brownish or reddish. The grass heath rendzina then has nothing to do with the black earth (chernosem) of the continental loess steppelands where the humus content is considerably higher, but falls when the steppe is planted with trees. This humus is formed mainly from the grass roots which die every year, and are eaten by earthworms converting them into stable clay-humus complexes. In contrast even the 'rock steppe' in the Wallis which is dominated by Feather Grass (*Stipa capillata*), according to the author's observations, does not produce a humus-rich soil.

The difference in the humus content of woodland and grass heath soils over pervious, base-rich rocks in Central Europe also has a lot to do with the fact that earthworms find very different living conditions in the two ecosystems. These animals which are so important for the humus stabilisation can be active in shallow calcareous soils only when these are damp enough (see Zuck 1952), which is so for a much longer time in the woodland shade than under grassland in full sun.

(c) Sandy dry grasslands in different sites

The majority of the arid and semi-arid grasslands in Central Europe are found growing on more or less stony soils whose finer particles are rich in clay. However, poor grassland also grows on calcareous dunes and other sandy places which have been denuded of trees and turned over to grazing. These do not belong to the primary dune vegetation which has already been discussed in section C V. Even the grasslands of the blown sand areas south of Mainz, which are famous because of the large number of rare species growing there, are secondary. Of course the rainfall here is low, but not low enough to prevent tree growth. Since these sand dunes have ceased to be grazed and have become a conservation area the amount of scrub growing on them has increased noticeably, in a similar way to the Schwetzinger Hardt which lies in front of Heidelberg and has been reported on by Philippi (1971, see section 1a).

It is no accident that work on the plant sociology of the sandy dry grasslands originated on the one hand in the Upper Rhine plain and on the other in the lowlands of eastern and south-eastern Central Europe. Apart from the authors already listed Korneck (1974, 1975, 1978 in Oberdorfer), deserves a special mention for his monograph on the xerothermic vegetation of Rheinland-Pfalz and neighbouring areas. In the sandy plains of this region there are many places still free from trees but no longer used for farming and Korneck could study several communities demanding quite different conditions. These vegetation units can be arranged in the following way:

1 **Calcareous dune swards** (*Koelerion glaucae*, see section E III 1, No. 5.224) in the warm-dry climate of the Rhine plain are richest in species. Already Volk (1931) has investigated them fundamentally, especially the Blue-green Hairgrass sward (*Jurineo cyanoidis-Koelerietum glaucae*) on stabilised but still lime-rich sand. Winterhoff (1975) added a surprisingly large number of fungi to the species list.

2 **Acid-soil dune swards** and pioneer phases (*Corynephorion canescentis*, No. 5.221), in contrast to the foregoing, comprise only a few species. These do not need a warm climate and are distributed on open acid sands in many parts of Central Europe except for those in very winter-cold continental regions (see also section C V 1 and 3). Numbers 1 and 2 are connected through intermediate stages and have several species in common which characterise the order *Corynephoretalia* (according to Korneck).

3 **Short-lived Silver Hairgrass swards** (*Thero-Airion*, single alliance in the order *Thero-Airetalia*, No. 5.24) indicate sand, gravel or debris soils drying out early in summer.

All of these more or less sandy grasslands need full light and disappear where man and grazing animals no longer create and maintain open areas, or at least some patches of wind-blown sand. Since most inland dunes have been fixed and planted with trees the sites favourable for dune plants are shrinking quickly. Even the unforested nature reserves are losing their typical flora and fauna because most of them today are surrounded by forests. These reduce the wind strength, raising the air humidity and adding their litter to the nutrient cycles of the sand sward ecosystems, thus moderating the extremes of the former site conditions. Only the **clear felling** of wind-breaking forests and repeated rigorous **mechanical damage** to vegetation and soil can help to conserve some of these interesting and highly diverse community complexes.

5 Changes in the species composition brought about by farming

(a) Mowing and grazing of dry poor-soil grassland

In earlier centuries all the arid and slightly arid grassland was grazed. The feeding of cattle in stalls began in the southern part of Germany and in northern Switzerland about 100-150 years ago and some 50-70 years ago sheep grazing too gradually waned and lost its significance. Because of this certain grassland areas suited for mowing were grazed only after this or not interfered with at all. These were mainly the moist meadows (section D V 2) and the slightly arid grasslands. In the eastern Swiss Jura, the central area of Switzerland, on the Kaiserstuhl and in other parts of the Upper Rhine plain even the xerobrometa are no longer grazed since sheep keeping became unprofitable here (fig. 383).

The arid grasslands here are at best mown only once during the year, the slightly arid ones twice in some places. The repeated cutting at wide intervals favours the taller growing plants provided they have sufficient powers of regeneration. Such a species which dominates most stands is *Bromus erectus*, and names such as '*Brometum*', 'Brome meadow' or 'Brush-meadow' which have been given to them have been self-explanatory for decades (fig. 383).

Where sheep and other animals still graze, or where they did so until quite recently, e.g. in some places on the Swabian and Franconian Jura as well as in the middle and north of Western Germany, the Upright Brome is much less common (fig. 384). In the association names in these parts *Bromus* is not appropriate, e.g. *Gentianello-Koelerietum* or *Euphorbio-Brachypodietum* of the Leine district (Bornkamm 1960). When grazing stopped however *Bromus* began to dominate here too e.g. around Göttingen where it had been one of the rarest plants as recently as the thirties.

The main reason for the absence of *Bromus* is selection by the grazing animals (see section A II 2). This grass is palatable (section 2 a) and where it is repeatedly eaten down tightly it will soon be completely eliminated. According to Lohmeyer (1953) it began to return to the middle Weser region when grazing was discontinued. Today it is commonest here on stony ground where the grazing animals always found difficulty in treading. In the Swabian Jura where *Bromus erectus* maintained many refuges it returned to dominate the slightly arid grasslands which were no longer grazed after only a few years (K. Kuhn 1937).

Many other species which are plentiful in the mown semi-arid meadows are completely absent from the pastures, e.g. most of the tall orchids. These are not eaten by sheep or cattle, but as soon as the tender flowering shoots start to grow up they are killed by treading. Unless orchids can regenerate by means of their seeds they gradually die out. In north-west Germany the slightly arid grasslands which were richest in orchids, e.g. the meadows of the Ith near Koppenbrügge, were previously used for mowing in the same way that the famous orchid meadows on the hills surrounding the Upper Rhine plain were. However, mowing is actually only one of the factors favouring the development of the light-loving grassland orchids. The majority require a somewhat moister habitat than is offered by most of the dry grasslands, and for this reason they are character species of the *Mesobromion* (see section E III 2). Marls which are soaked by spring or standing water early in the year are preferred by species of *Ophrys*. These became progressively rarer in the slightly arid grassland on slopes of the Irchel (north of Zürich) as long as it was being grazed (Studer 1962), but has again become frequent since it is now cut from time to time (Klingler, unpublished). Also the *Colchico-Mesobrometum* described by Zoller (1954), which occupies deep lime-rich brown earths and often goes over into a moist-soil meadow, is noted amongst all the grassland associations of the Swiss Jura for the variety and constancy of its orchids (including *Orchis mascula*, *militaris*, *ustulata* and *pallens*, *Dactylorhiza maculata*, *Gymnadenia conopsea*, *Platanthera chlorantha* and *Listera ovata*). Near Jena, too, Reichhoff (1974) found most of the Orchids on not really dry sites. A further important condition, at least for sub-mediterranean genera such as *Ophrys*, is a sufficiently high temperature (or not too low a winter temperature?). For example in the Swiss Jura the *Teucrio-Mesobrometum*, especially the *Ophrys-Globularia punctata* subassociation does not go up as far as the high montane belt, and prefers sunny sites. In addition many orchids are helped by the occasional supply of manure, provided this is not so much as to favour the tall-growing grasses which could compete with them. The reasons for this behaviour and for the presence or absence of the Central European autotrophic orchids however are still not entirely explained.

In contrast to the mown grass heathlands those which are closely grazed by sheep contain prostrate and rosette forms as well as poisonous, unpalatable or prickly 'pasture weeds' (fig. 384). In the case of the

Fig. 383. A limestone slightly-arid meadow which is regularly mown for hay near Mellingen in the centre of Switzerland with a lot of *Bromus erectus* and *Salvia pratensis*.

Fig. 384. A calcareous slightly-arid pasture which is regularly grazed by sheep near Würmlingen in the Württemberg lowland in which *Brachypodium pinnatum* and *Carlina acaulis* are widespread.

Pasque Flower (*Pulsatilla vulgaris*) sheep grazing has led to the selection of a genetic dwarf variety, e.g. in the Nördlinger Ries (Gotthard 1965). Apart from the Juniper (*Juniperis communis*) which is hardly ever absent from these rough grazings, some of the most obvious plants are thistles (such as *Carlina acaulis*, *C. vulgaris* and *Cirsium acaule*), Gentians (e.g. *Gentiana verna*, *Gentianella ciliata* and *germanica*) and spurges (especially *Euphorbia cyparissias*). Another typical pasture weed is the Torgrass (or False Brome, *Brachypodium pinnatum*) which is only eaten by sheep when it is very young. To some extent it is the opposite of the highly palatable, tussock-forming Upright Brome (*Bromus erectus*) in that it can spread vegetatively by means of extensive rhizomes and is not compelled to ripen its seeds in order to reproduce as the Brome must do.

Since, during the course of time, the same piece of grassland may be used alternately for grazing and hay making there are, for example in the Swabian Jura (Kuhn 1937) not only 'pasture-*Mesobrometa*' rich in *Brachypodium* and 'meadow-*Mesobrometa*' rich in *Bromus* but also many intermediates the floristic separation of which is hardly possible.

All the examples discussed so far are communities belonging to the order *Brometalia*. This is not surprising since the continental grassheaths normally are not cut because of their low and irregular productivity. This is true for those in the inner Alps as well as for the alliance *Festucion valesiacae* in the lower parts of eastern Central Europe. So to a large extent they have a definite 'pasture character', especially where they are growing on shallow soils (see the two middle columns in tab. 89). When certain species become dominant in them these will be either Sheep's Fescue (varieties or subspecies of *Festuca ovina*, see tab. 89), whose dense clumps can withstand close grazing, or Feather Grasses (*Stipa* species, see fig. 378) which, like *Brachypodium* are only eaten when young. Representatives of the genera *Teucrium*, *Thymus*, *Artimesia*, *Helianthemum*, *Asperula*, *Allium*, *Anthericum* and *Sedum* only exceptionally show traces of having been grazed.

The Blue Moorgrass (*Sesleria albicans*) which has been mentioned so many times already can also be included amongst the pasture weeds. *Sesleria* can carry out its role as a stabiliser of steep slopes and screes largely by virtue of its being protected from grazing animals. A similar situation arises with *Molinia arundinacea* which is able to form a thick root mass in the steep marl slopes of the Jura mountains where the water supply fluctuates so much. In the Swabian Jura there is a *Mesobromion* community which Kuhn (1937) calls *Tetragonolobo-Mesobrometum* after the Lotus-like Dragon's-teeth *Tetragonolobus maritimus*. This corresponds to the *Tetragonolobo-Molinietum* which Zoller (1954) has described from the Swiss Jura. Because they have such little food value the Blue Moorgrass and Purple Moorgrass swards on steep slopes are seldom grazed.

The majority of the more or less arid grass heaths in Central Europe are indebted to grazing animals as well as to the scythe and fire; and this is the case not purely and simply for their existence, but also for many unique members of their communities. Since it is difficult to measure such anthropogenic habitat factors which can be determined only by field experiments over several years we unfortunately know far too little about them.

The spread and collapse of the Rabbit population (*Oryctolagus cuniculus* L.) was a unique natural experiment in the dry grass heaths of southern England and a few areas of Central Europe (see Thomas 1963, Myers and Poole 1963). F. Runge (1963) maintained a number of permanent quadrats in a Gentian-Torgrass sward in north west Germany in order to follow the changes in dominance of species. He found that Rabbits preferred to graze *Brachypodium pinnatum* so in this way they behave differently from domestic animals.

(b) The formation of steppe-like swards on fallow land

Before man started to influence the vegetation in Central Europe rock heaths and many Blue Moorgrass slopes were the last refuges of light-loving steppe plants which had been very widespread just after the ice receded and before the forests came in (see Frenzel 1968). Since the Late Stone Age they have again greatly extended their range. Secondary dry grasslands, i.e. ones which have arisen from former woodland, scrub or ploughed land, differ from the primary grasslands in the presence or absence of certain species (Krause 1940, who gives a list of further references). However as the stands get older these differences tend gradually to disappear. Each individual case must be carefully examined in order to obtain a clear history of the grassland area in question.

Many plants of the dry grasslands can spread surprisingly quickly into open new sites such as rock falls, roadsides, burnt areas or fallow land. In most cases grazing animals assist in this secondary succession. They speed up the dissemination of many endo- or epizoochorous grassland plants, e.g. *Trifolium repens* and *Plantago media* as well as those whose propagules are carried in the dirt on animals hoofs or coat. The seeds of other species e.g. many sedges are dispersed by ants (Krause 1940); however the majority of the partners in these grasslands, in particular the pioneers of the secondary communities, are brought in by the wind, especially the Feather Grasses and many of the Compositae. Quantin (1935) found that 51.2% of the species in the xerobrometa of the western Jura were anemochorous, 15.8% were zoochorous while the rest did not have any special dispersal mechanism. In the shallow-soil rock heaths (tab. 91), which from their nature always occur as isolated small islands, the wind-dispersed species can amount to as much as 68.3% of the total.

The occupation of new land starts with the formation of various **'patterns'** resulting from the concentric spread of those species which have arrived by chance on different spots, such as Kershaw (1963) observed and treated mathematically. Depending on their cause these patterns vary morphologically (produced by specific types of growth), ecologically (through small-scale differences in the habitats), and sociologically (by the mutual interactions of the plants which managed to arrive). Obviously heterogenous patterns are found chiefly in pioneer stages and rarely in a mature stand.

Abandoned vineyards and old arable fields are covered after very few years with large areas of *Stipa capillata*, *S. pennata*, *Festuca valesiaca* and other grasses. Such Feather Grass swards may appear very steppe-like with their tall blades and their waving beards shining brightly in the sun, but they are far from being natural steppes. For example the famous Feather Grass slopes on the Mont d'Orge below Sion in Wallis had previously been ploughed, as the name 'barley mountain' indicates. Even on the steppes of the southern Ukraine the dominance of *Stipa* species is an indication of arable farming one or two decades ago. In the Askania Nova reserve all grazed swards which are more than 25 years old are dominated by shorter grasses such as Sheep's Fescue. These relatively species-rich real steppes contain a number of slower-growing chamaephytes and virtually no more *Stipa* specimens (as was demonstrated during an excursion in 1975). H. Frey (1934) had already shown that *Stipa* clumps at first grow very strongly but are gradually smothered out by other plants which have not been able to establish themselves so quickly. During the course of the years *Sempervivum tectorum* for example can completely overgrow *Stipa capillata*. More and more species come in until a varied mosaic is formed in which no single species can be considered dominant any longer.

In their manner of dispersal, as well as in other respects, there are many differences between the species which make up the more or less arid grass heaths, but however they manage to arrive and settle in the new area they still have to survive the competition of many other species and the site conditions, which may often be severe and can change markedly with the weather from one year to another (see section 6a).

(c) Effects of burning and chemical control

Burning off the old dead grass before the start of the growing season used to be a common practice on many poor grasslands in Central Europe – as in the African Savanna. This factor plays an important part in the pattern of the sward-plant species. It served as a means by which the shepherds could combat the spread of trees and shrubs in the pasture. If carried out in good time and over only small areas, which are changed from year to year, the burning of grass and heather damages neither plants nor animals, as is often erroneously feared. It is true that here and there the odd individual plant or animal will be killed, but on the other hand burning helps to safeguard the habitat for the population as a whole. Species thriving on light and warmth have been, and still are, assured of open space at little cost, whereas without the fire a shade-giving litter would accumulate, and finally a woodland would reassert itself.

Chemical means of controlling 'weeds' are often thoughtlessly used; they spell disaster for most forms of life which can recover only slowly. Such means should never be used for 'maintaining' a poor grassland with its richness of species because they have a much worse effect on the flora and fauna as well as on the appearance of the landscape than the burnt areas which are quickly turned green again by the fresh growth.

Fanned by a gentle breeze the fire quickly passes over the turf scarcely heating the soil surface so that nothing below ground is killed. However uncontrolled burning can occasionally have as catastrophic effects as the chemical control methods. This was shown in a test quadrat set up by Mahn (1966b) in a dry grassland which had been damaged by fire (fig. 385). Regeneration began hesitatingly since all the plants and most of the seeds had been destroyed. However after 5 years practically all the perennial species present in the original grassland were represented once more. Amongst the exceptions was *Calluna vulgaris*. In a slightly arid grassland (*Onobrychi-Brometum*) on the Kaiserstuhl, Zimmermann (1976) noticed a more strongly selective effect of repeated burning off. Species with subterranean runners and rhizomes e.g. *Brachypodium pinnatum* are favoured in the competition. Shrubs and bushes also tend to come in. Thus a limestone grassland will not be kept as such, rich in its numbers of character species, merely by burning, but only by mowing or grazing regularly. However fire management is a suitable way of maintaining a heathland with dwarf shrubs (see section D II 3 c).

(d) Successions following the cessation of man's influence

The majority of other arid and slightly arid grass heaths in Central Europe have not been utilised in the former way for quite a long time now, and have become so-called 'social fallow land'. Since the removal of man's influence not all of them have reverted to woodland immediately as might have been expected theoretically. Rapid progress in this direction has only been made where, at the time grazing or mowing stopped, there were already nuclei of woody plants, either in the form of pasture weeds (such as Hawthorn or Blackthorn) or as groups of trees. On the contrary grass heathland without shrubs or trees, or with just a few junipers has remained for decades without any tree or shrub being able to establish a footing. For example the *Xerobrometum* on the upper part of the Rebberg on the Hohentwiel remained for more than 30 years just as it was in 1930 when Braun-Blanquet and co-workers recorded the flora there (Th. Müller 1966). There are also some slightly arid grasslands on deep slightly sloping soil near Göttingen which have remained free from trees for over 40 years.

The main barrier to the penetration of woody plants, in addition to the small amount of seeds they produce, may be the mass of dead leaves, which is pressed down by the snow from time to time and covers the ground like a thick felt. Woody plants require a more or less bare soil in order to germinate. Occasional burning then which removes the mass of dead plant material will improve their chances of occupation. Treading by cattle will cause so many bare patches that the process will be even easier. Thus a moderate influence of men and grazing animals favours the establishment of the same woody plants which heavy grazing and frequent burning cause to retreat. In order to keep the social fallow land free from trees and shrubs there is obviously no simpler way than **doing nothing at all for as long as possible**! This recommendation of course will not hold good over a period of many decades or centuries. Chance occurrences such as the digging of a hole by some animal or the accidental spread of a fire will allow a tree or shrub to come in here and there in whose shadow other woody plants can become established.

Those trees which can spread by subterranean runners such as Aspen, Privet, Blackthorn and other *Prunus* species including escapes of cultivated plums, as well as the Wig Tree (*Cotinus coggygria*) are particularly effective in this pioneer role. Such colonies of stems (polycorms) are the literal forerunners of woodland (see fig. 386) especially on relatively dry places. This is impressively seen in the mapping with which B. and S. Stephan (1971) have traced the vegetation development in the nature reserve at Stolzenburg.

Herbaceous plants which are able to spread out beneath the ground are also very successful colonisers of fallow land. This is particularly true of *Brachypodium pinnatum* in that this grass is able to tolerate semi-shade. In the orchid slightly arid grassland near Jena its polycorm has a homogenising effect, according to Reichoff (1974). Over an area of 2500 cm^2 the number of species fell from 16.8 to 11.7 (on average) while the living phytomass above ground rose from 55.8 to 76.4 g dry matter, but the weight of dead leaves remained about the same (67.3 to 76.9 g) How quickly being overshadowed by woody plants or even tall broadleaved grasses such as Torgrass (*Brachypodium*) leads to the reduction of light-loving herbaceous plants is well illustrated by Tamm (1972) in southern Sweden. For example *Primula veris* has a 'half-life' in the open of about 50 years; under light shadow this is reduced to 6.2 years and under an even greater light deficiency to 2.9 years.

Fig. 385. The succession following an intense fire in a sandy dry sward east of the Harz; mapped in a permanent 1 x 1 m quadrat in the spring and summer of 1962 and in the spring of 1963 and of 1964. After Mahn (1966), modified.

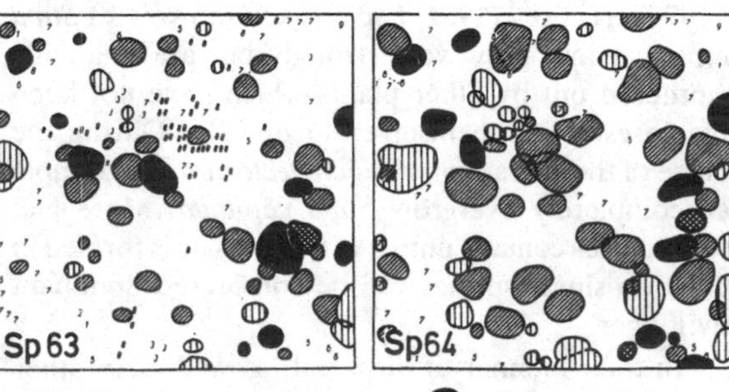

6 Investigations into the water factor in arid and slightly arid grasslands

(a) The behaviour of different species during dry periods

The species composition of the more or less dry grass heaths depends, as we have seen, on many factors which interact in various ways. By far the most important is the water supply. These heaths are quite rightly referred to as relatively arid or slightly arid.

However one should not misunderstand this description and imagine the habitats as being permanently short of water. If that were so the majority of higher plants would not be able to thrive there. Actually the soils of the dry grasslands are as damp as those of normal meadows throughout the greater part of the year, more particularly in spring and autumn but also in summer after prolonged heavy rain. However the periods without rain occurring during the warmer times of the year are very critical and in Central Europe these can last from a few days to many weeks. During these drought periods the water reserve in the shallow soils under the strong sun becomes exhausted very rapidly (fig. 63). Such times of shortage often occur in communities belonging to the units *Festucetalia valesiacae* and *Xerobromion* (see section I 1 a). One does not have to think only of exceptionally dry years such as 1911, 1919, 1933, 1947, 1952, 1959, 1970, 1973, 1975 and 1976 which were also catastrophic for other plant communities. How warm and dry the habitat of a subcontinental Feather Grass sward can become in a normal summer we have already seen in fig. 380. However the arid and slightly arid grasslands are capable of a relatively quick recovery even after an extremely dry period (fig. 387). Its members are very well adapted to the fluctuating water supply.

The vascular plants of the Central European dry grasslands take up the water they require almost entirely through their roots. Thus their ability to overcome drought periods depends in the first place on the organisation and efficiency of their subterranean parts. The reactions of the main and lateral roots of xerophytic dicots to the water tension of the surrounding medium as determined by Kausch (1955) appear to be very functional. While the lateral roots develop best when the soil is very moist the main root grows fastest when the suction pressure is 6-7 bar. These values are often reached in a dry soil and are near to the permanent wilting point when water uptake is at a minimum. According to Barth (1978) the zone of drying out gradually moves downwards, initiating an accelerated prolongation of the roots which reach right into the slightest cracks of the underlying rock. In a loess soil sheltered from any water supply Upright Brome did not wilt even when the upper 120 cm became totally dry (more than 1000 bar), because its roots went still deeper to get available water. The extent to which a sufficiently damp soil is filled with fine roots has already been shown in section 2a with the example of *Bromus erectus*.

Water from dew or mist falling on the ground surface and on the leaves cannot improve the water balance for the dry grassland plants so effectively as Arvidson (1951) observed on the Baltic island of Öland. It is just on the places most liable to drought, e.g. the sunny slopes of the Wallis rock-steppes that

Fig. 387. The potential osmotic pressure throughout the year in the leaves of dry sward plants on shallow limestone soil (1) and on deep loess soil (2) in the Kraichgau as an expression of their changing water supply. After Müller-Stoll (1935/36) from Walter (1962).
Hipp. = *Hippocrepis comosa*, Hel. = *Helianthemum nummularium*, Anther. = *Anthericum ramosum*.

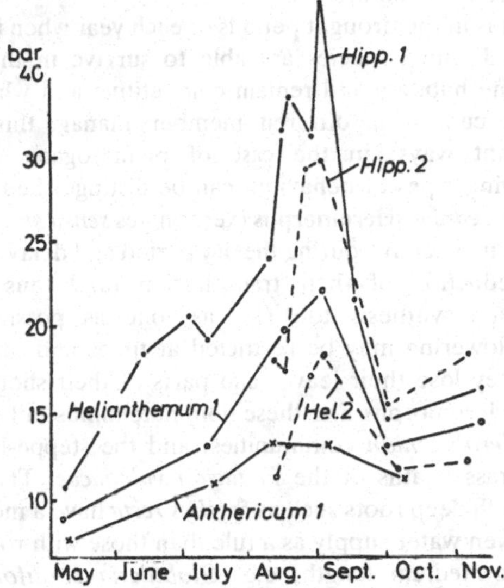

Fig. 386. The vegetative spread of shrubs often accelerates the succession into woodland, especially on previously grazed swards. Modified from Jakucs (1969) who has made an intensive study of this 'polycorm' succession in northern Hungary.

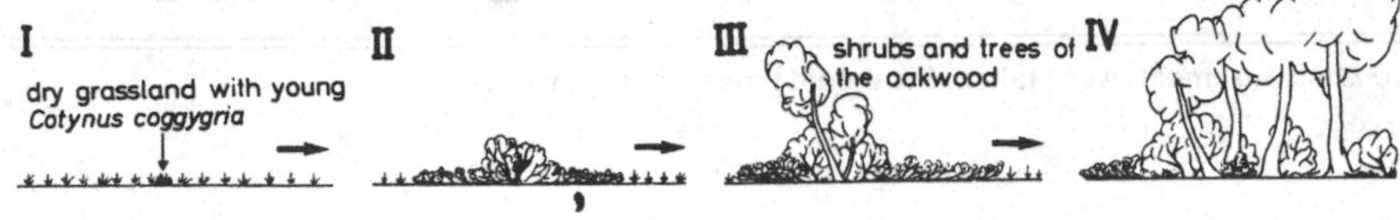

the dew point is hardly ever reached in summer. In any case the uptake of water by the xerophytes in the communities we have been looking at must be insignificant because nearly all have a small surface area covered by a thick cuticle which would make water exchange very difficult except through the numerous stomata.

It is in the drought periods of each year when it is decided which species are able to survive in these extreme habitats and remain competitive and which will succumb. The different members manage this in different ways; in the case of phanerogams the following types of behaviour can be distinguished:

1 **Perennial scleromorphs** (xerophytes *sensu strictu*) remain active during the dry period and delay the reduction of their transpiration (and thus of photosynthesis too) for as long as possible. Flowering may be restricted at times and often they lose their leaves and parts of their shoots. Scleromorphs like these dominate almost all the *Xerobromion* communities, and the steppe-like grass heaths of the *Festucion valesiacae*. Those with deep roots such as *Stachys recta* have a more even water supply as a rule than those with roots of medium length, e.g. *Globularia cordifolia*, Shallow-rooted perennial scleromorphs do not often occur, but one of them is *Hieracium pilosella* whose hair felt on the leaves may contribute to its water supply from dew. Since the scleromorphs use most of their assimilated material for building up their root systems and strengthening their cell walls they grow only slowly above ground; so where there is a somewhat better water supply they are suppressed by species from the following group.

2 **Perennial mesomorphs and weakly scleromorphic plants** have a better metabolism and grow more quickly than group 1 during a wet year. On the other hand in a dry year their aerial parts die back to some extent and they reduce their activity or go completely dormant, relying on rhizomes or some other subterranean organ or on seeds which are produced in favourable years. Many of the characteristic and differentiating species of the *Mesobromion* behave like this, as does *Brachypodium pinnatum*, which is plentiful in many slightly arid grasslands.

Bromus erectus comes somewhere between the two groups, and some other species also show intermediate behaviour taking on a more or less scleromorphic structure depending on the actual water supply; their reproduction varies in a similar way (tab. 92).

3 **Spring ephemerals** such as *Erophila verna* are also mesomorphic or at most only weakly scleromorphic but complete their development as a rule before the summer, thus avoiding the drought. In extraordinarily dry springs or in very arid micro-habitats they are stunted, but with a better water supply they shoot up luxuriantly and remain green longer into the warmer season. The few summer ephemerals (e.g. the semi-parasite *Odontites lutea*) behave in a similar manner to the group 2 species while the rare spring geophytes are like those in group 3.

4 **Succulents and half-succulents** because of the water they have stored in their tissues are able to endure a period of drought, but they grow even more slowly than the plants in group 1 so they only come into their own in places which frequently become so dry that other life forms are suffering badly, i.e. on stony rock-heaths (see section 3 a).

Cryptogams, as **poikilohydrous plants**, only assimilate during damp periods which are in the early spring and autumn as a rule. At other times of the year the primary production of mosses like *Hypnum purum* is very slight (Kilbertus 1970). The lichens mentioned in section I 1 a may also be active during the morning hours when they can get some water from the humid air cooled down during clear nights, as O.L Lange,

Tab. 92. The vitality of Bromus erectus in habitats of different degrees of moisture.

Degree of moisture	Community	Haulm length (cm)	**Panicle length**[a] **(mm)**	Spikelet length (mm)
very dry	rock comm.	15– **36**– 47	51– **62**– 72	10–**15**–21
dry	*Xerobrometum*	42– **68**– 81	83– **90**– 97	17–**21**–24
fairly dry	*Mesobrometum*	102–**113**–121	105–**125**–140	20–**23**–23

Note: [a] In each case more than 1000 measurements were taken. **Semi-bold** type = average values.

Source: From data by Quantin (1960).

Schulze and Koch (1970) measured in the Negev desert.

The species within each of the four groups of phanerogams can be distinguished by the form of their root system, the anatomy of their leaves, the relative area of their conducting tissue (according to Müller-Stoll 1936 this is relatively small in *Centaurea scabiosa*) or other features influencing their water economy. With regard to the latter, and along with Bornkamm (1958) we can subdivide the groups 1 and 2 in the following way:

I Species with a **high maximum transpiration rate** (and a correspondingly higher rate of CO_2 assimilation)
- a) with normally only slight daily variation in the saturation deficit (e.g. *Bromus erectus*)
- b) with a greater daily variation (e.g. *Brachypodium pinnatum*)

II Species with a **lower maximum transpiration rate**
- a) with a small daily variation in the deficit (e.g. *Anthyllis vulneraria*)
- b) with an average to large variation (e.g. *Lotus corniculatus*).

All the species named as examples root fairly deeply. With the shallower-rooting plants there are presumably unlikely to be any in group 1a and very few in group IIa.

It is true that this division is based solely on experiences acquired in slightly arid grassland in a relatively humid period (1953-1957). In principle however it may also be used for the phanerogams of the *Xerobromion* and other dry-sward alliances. Most species of the true arid grass heaths may belong to subgroups Ia and IIa since these are better adapted to repeated dry spells than are plants with a very variable water balance.

Fig. 388. The effect of a dry year (1947) and a wet one (1948) on a 1 x 1 m test area in a slightly-arid sward (*Mesobrometum salvietosum*). After Lüdi and Zoller (1949a), in Ellenberg (1963).
1 = tussock-forming grasses, mostly *Bromus erectus*, 2 = *Picris hieracioides*, 3 = *Salvia pratensis*, 4 = *Daucus carota*, 5 = *Centaurea jacea*, 6 = *Lotus corniculatus*, 7 = *Medicago sativa*. In both cases the mapping was carried out in August near Villnachern in the Aargau canton.

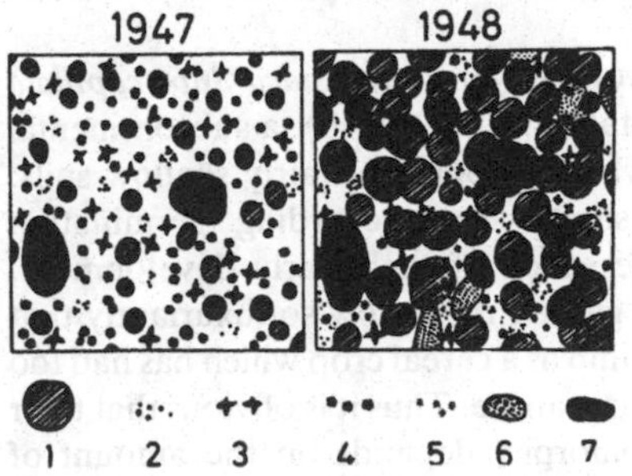

Although almost all the plants in group I have a deep rooting system the potential osmotic pressure in their cell sap, which has an optimal value of between 10 and 20 bar, can rise in late summer to as high as 30 or 40 bar (fig. 387). The highest osmotic potentials which have been measured in arid grassland plants of Central Europe are 102 bar for *Aster linosyris*, 81 for *Potentilla arenaria* and 80 for *Carex humilis*. Such values have rarely been exceeded even in desert plants. The majority of grasses, e.g. *Stipa pennata*, *S. capillata* and *Festuca valesiaca* hardly reduce their transpiration rate, according to Florineth (1974). Therefore their water potential increases rapidly and often overtakes their osmotic potential so that a 'negative turgor' must be supposed (Lösch and Franz 1974). This result though may be an artefact because the osmotic potential of living cells is higher than that determined in the sap pressed out of a killed tissue. During the dry summer of 1973 the water potential reached values of about 80 bar also in *Bromus erectus* growing in the slightly arid grassland near Göttingen, according to Barth (unpublished). The grasses suffer badly in such drought periods; however they recover (fig. 388).

The effectiveness with which a leaf reduces the rate of water loss under extreme conditions can be estimated approximately by measuring the speed with which it dries out when removed from the plant. Researches by Bornkamm (1958, see tab. 93) showed that species which are found mainly in the *Mesobromion* lose their water more quickly than the ones which are also found in the *Xerobromion*. For example it took *Avenochloa pratensis* eight times as long to reach the 'sublethal deficit' as *Brachypodium pinnatum*.

Although most of the plants living in more or less arid grasslands show some adaptations to temporary water shortages they should not be labelled as xerophilous. Certainly none of them is drought **loving** in the literal sense since they do not **require** a water shortage in order to thrive. Among other evidence this is shown by experiments carried out by Smetankova (1959) on *Carex humilis* (tab. 94), a species particularly resistant to drought. Plants which were watered did better than those which were not, and a light shade was favourable to their productivity since it reduced transpiration and made possible the maximum rate of photosynthesis in the circumstances. In the driest rocky habitat not only was total productivity much less but flowering was also reduced below that on deeper soils; the leaves had already begun to die at the beginning of July (tab. 94). It is worth noting that the large differences in the appearance and structure of *Carex humilis* in these trials were brought about by modifications of genetically homogeneous material, as

Tab. 93. The length of time taken for leaves of plants of a slightly-arid sward to dry out.

a Time required to lose half the water from the fresh leaves (hours)
b Time required to reach a sublethal deficit (i.e. irreversibly damaged to about 10% of the leaf area) (hours)

Herbs and leguminoses	a 50% loss	b sublethal deficit	Grasses	a 50% loss	b sublethal deficit
Scabiosa columbaria	11.3	29.5	*Avenochloa pratenis*	7,9	20,8
Anthyllis vulneraria	15.3	20.1	*Festuca valesiaca*	5.2	16.0
Knautia arvensis	8.1	18.9	*Bromus erectus*	4.9	8.5
Hieracium pilosella	4.9	13.6	*Brachypodium pinnatum*	1.6	2.4
Pimpinella saxifraga	5.9	11.4			
Lotus corniculatus	4.6	7.8			

Source: From a *Gentianello-Koelerietum* near Göttingen. After Bornkamm (1958). Average values of 16 tests carried out in a chamber at 20°C and a Piche evaporation rate of 0.09 cm^3 per hour.

Tab. 94. Variability in the growth of Carex humilis in relation to habitat moisture.

Plant community, slope and soil	Total length of shoots (cm)[a] watered	not wat.	Length of dried out leaf ends	Starting to die off	Flowering clumps (%)	Flowers per clump
Rock comm. similar to that in tab. 91 slope 18–20° WSW, very shallow soil (7.5cm)	–	–	32%	June	**70%**	6–10
Dry rocky sward similar to no. 2 in tab. 89; slope 36–44° SW, rendzina (13 cm fine soil)	333	198	24%	July	45%	**10–20**
Open woodland with *Quercus petraea* and *pubescens;* slope 24–31° NW (14 cm fine soil)	**353**	251	9%	Sept.	2%	2– 3

Note: [a] i.e. the sum of all the shoot lengths from the same number of watered and unwatered clumps.
The Ground Sedge grows best vegetatively in the woodland but produces more seed in the rocky ground community.

Source: From data by Smetankova (1959).

shown by transplantation and pot-culture, and was not because of any genetic differences. Other things being equal the Upright Brome also grows better where there is water always easily available. Numerous measurements by Quantin (1960) showed that progressively reducing the water supply decreased the haulm length significantly while the size and number of the reproductive parts were less affected.

Krause (1950) carried out extensive trials in which he sowed the seed from grassland plants of various communities under good conditions of moisture etc. in beds in the botanic garden, keeping the species separate. All the character species of the slightly arid grasslands, and many of their regular companions, succeeded in flowering in the first year after sowing. Their germination capacity, which Krause tested at the start of the trials, was high in almost all cases (tab. 95, 1a). Thus such plants are able to regenerate quickly from seed after being disturbed or following a catastrophic drought, once the soil becomes moist again.

The species of the *Festuca valesiaca-Stipa capillata* community (cf. tab. 89 and 95, 1 b), a subcontinental arid grassland which is found on fairly shallow soils, behaves in a similar way regarding germination capacity and the speed of their early development. However in the garden they grow so luxuriantly that they become as limp as a cereal crop which has had too much nitrogenous manure. Thus it is obvious that their degree of scleromorphy depends on the amount of

Tab. 95. Germination percentage and seedling development of dry-sward plants in pure culture on a good garden soil.

1. Species having **large numbers** of germinating seedlings which quickly grow up into a closed stand about 50 cm high:

a) forming dense enduring stands; mainly flowering in the first season after seeding: typical *Mesobrometum*

grasses:	others:
69 *Brachypodium pinnatum*	100 *Plantago media*
54 *Koeleria pyramidata*	90 *Pimpinella saxifraga*
37 *Briza media*	88 *Arabis hirsuta*
leguminous plants:	84 *Asperula cynanchica*
86 *Lotus corniculatus*	72 *Carlina vulgaris*
74 *Ononis spinosa*	64 *Sanguisorba minor*
54 *Anthyllis vulneraria*	41 *Agrimonia eupatoria* and many others

b) forming closed but lax stands which tend to gappiness through the death of many individuals; up to 75% flower in the first season after seeding: *Festuca valesiaca-Stipa capillata* community

grasses:	other:
75 *Stipa capillata*	83 *Seseli hippomarathrum*
68 *Andropogon ischaemum*	61 *Alyssum montanum*
leguminous plants:	57 *Verbascum phoeniceum*
84 *Astragalus excapus*	38 *Scabiosa canescens*
79 *A. danicus*	32 *Aster linosyris*
4 *Oxytropis pilosa*	2 *Adonis vernalis* and others

2. Species producing **small numbers** of individuals forming open stands from the outset[a]:

a) remaining short but spreading rapidly; flowering in the first season after seeding: *Festuca duvalii-Thymus serpyllum* community.

grasses:	herbs:
68 *Festuca valesiaca*	85 *Veronica spicata*
58 *Festuca duvalii*	71 *Sedum sexangulare*
9 *Poa badensis*	26 *Thymus serpyllum* and others

b) only slowly covering the ground; not flowering until the second season after seeding, or even later:

α) *Carex humilis-Pulsatilla vulgaris* community

grasses and grass-like:	others:
43 *Stipa joannis*	82 *Odontites lutea*
20 *Carex humilis*	65 *Globularia punctata*
leguminous plants	63 *Helianthemum canum*
72 *Hippocrepis comosa*	60 *Pulsatilla vulgaris* and many others

β) *Sesleria-Teucrium montanum* community and *Seslerio-Mesobrometum*

grass:	others:
52 *Sesleria albicans*	78 *Carlina acaulis*
leguminous plant:	33 *Teucrium montanum*
95 *Coronilla vaginalis*	6 *Anthericum lilago* and others

Note: [a] All the communities named under 2 occupy fairly shallow soils and impoverished habitats in nature.

Source: From data by Krause (1950). Figures in front of the plant names are the germination percentages found in the laboratory using the same samples of seed.

water and nutrients available, a fact which has been demonstrated experimentally with very different kinds of plants (see also fig. 282). During the course of time many of the originally established individuals die back, presumably because stronger ones have deprived them of light.

Almost all the species which play a part in plant communities on shallow calcareous soils such as rock heaths tend to make open stands right from the start (tabs. 95, 2). They also grow more luxuriantly in garden soil than in their natural habitat, but not to the extent that they become weak and lose their characteristic form. The slowest developers are representatives of the *Carex humilis-Pulsatilla vulgaris* community and of the Blue Moorgrass slopes. These species are thus least competitive on more fertile soils. Summarising the results of Krause it can be said that none of the dry grassland plants tested under garden conditions died because the environment was not dry enough for it.

In the final analysis then the presence or absence of distinct species in the 'dry' grassland communities in natural habitats of varying degrees of dampness depends mainly on **competition**. It is true that many species live in the arid or even in the semi-arid grass heaths unmolested by superior competitors, but nevertheless they are near the limit of their ability to live here. Expressed in another way their ecological optimum lies within the area of their physiological minimum. This could be seen very well in the dry years 1947, 1949, 1952 and (partly) 1973. It was not the cultivated meadows but the steppe-like dry grasslands which were most damaged in those catastrophic drought periods. *Bromus erectus*, one of their most plentiful species, and also other grasses were killed over large areas in the south-west of Central Europe. This was most evident in 1952 on the marly Keuper soils in Württemberg in which they form a shallow root system. Alongside the grasses which were dried up and bleached to the colour of straw, deep-rooting plants like *Salvia pratensis* and *Centaurea scabiosa* remained quite green. Those who expected this to lead to a shift in the balance of species however were disappointed during the following wet summer of 1953. *Bromus erectus*, *Koeleria pyramidata* and other plants which had been killed by drought in the previous year now regenerated surprisingly quickly from older seed. Where the soils were less rich in colloids many of the *Bromus* clumps had not been entirely killed, and here regeneration was even quicker (fig. 388).

Fig. 389. In pure culture plants which indicate dry and poor conditions, such as the Upright Brome, 'mesophilous' types like False Oatgrass, and those usually associated with damp or wet habitats, of which Meadow Foxtail and Marsh Poa are examples, all show maximum production at about the same optimum moisture, namely when the water table in sandy soil is about 25 – 35 cm below the surface. By competition in a mixed culture however these four species displace each other in the direction of their 'ecological behaviour' prevailing in the semi-natural meadow communities (see also figs. 39 and 57). After Ellenberg (1953), modified.

Dotted = physiological optimum area; lines = optimal area when in competition. The arrows indicate the direction of the displacement.

Sharifi (1978) arrived at substantially similar results in his lysimeter experiments with *Bromus*, *Arrhenatherum* and *Alopecurus* on loamy soils. Because of capillary water rising, however the optimum depth of the groundwater table was greater.

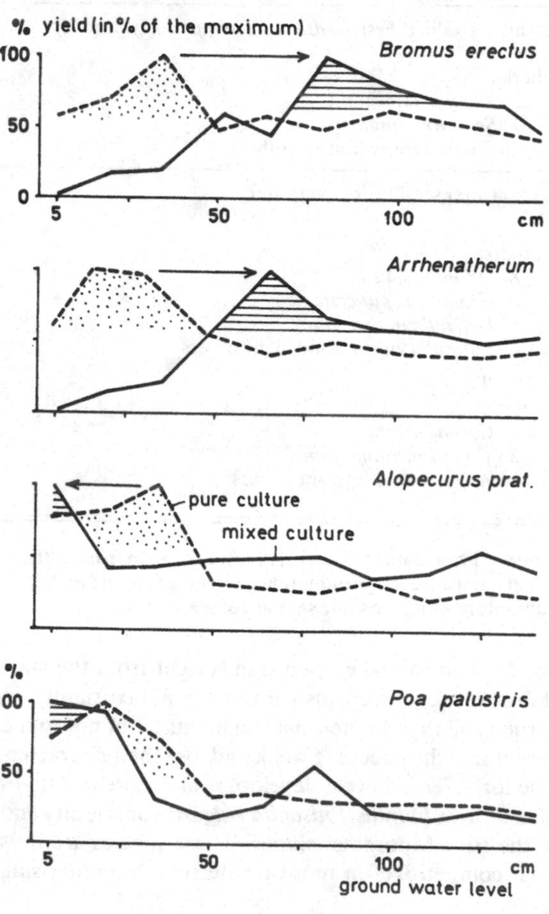

Through their observations in the areas of the Rhône, Rhine and Po Wilczek, Beauverd and Dutoit (1928) had already come to the conclusion that the Upright Brome was 'in the main a light loving and relatively warmth loving' plant. It would be wrong to see this grass as a 'pronounced xerophile' since in the drier areas it behaved 'more like a hygrophile'. Similarly in Wallis it is only dominant on shade slopes (Braun-Blanquet 1961), and in the submediterranean area it retires to the half-shade of an open *Buxus* scrub. Even on the much less arid Swabian Jura the Upright Brome always prefers those habitats which are not the driest (Kuhn 1937).

In spite of these and many confirmative observations or records in the literature, *Bromus erectus* and other plants of the semi-arid and arid grasslands are constantly being referred to even in recent publications as 'xerophilous', 'liking dry places', 'preferring dry sunny slopes' etc. It may not be superfluous then to look into a series of experiments which will help us to recognise more clearly the reasons for the occurrence or absence of such species in nature.

(b) The competitiveness of the Upright Brome under different conditions

Bromus erectus is one of the best-researched wild plants from the physiological and ecological points of view. Its behaviour in the field has already been described in the previous sections so now we need only to expand on some culture and regeneration experiments which have helped to determine more exactly its amplitude and competitiveness under different conditions and throw some light on the ecology of the arid and semi-arid grasslands in general. As a comparison with *Bromus* we will also get to know how other well-researched species behave.

When *Bromus* is grown as a pure culture without any competition its maximum production is always achieved when the soil is rather on the damp side. According to Ellenberg (1953b, cf. fig. 389) the optimal depth of the ground water in a sandy soil is 35 cm provided the water is not too poor in lime. In this respect *Bromus* behaves much like *Arrhenatherum* and *Dactylis glomerata*, i.e. definite mesophilous meadow grasses. It resembles these too in having a wide amplitude regarding water supply.

When growing in a waterlogged soil *Bromus erectus* forms large intercellular spaces in its root cortex (fig. 390) whereas in normally damp soils the roots take on a much more scleromorphic structure i.e. with thick walled parenchyma and very small or no air spaces. The Upright Brome could behave then like a swamp plant, and actually does this under certain conditions in nature. In the Upper Rhine plain and in the marshlands to the west of the Bodensee for example it grows in small-sedges and Moorgrass meadows on calcareous soils which from time to time get very wet (Lang 1973). In the steppes of southern Russia it keeps only to wet hollows, and generally speaking has become entirely helophytic (Gams, verbal communication)

Fig. 390. Cross-sections of *Bromus erectus* roots of the same age taken from soils where the amount of water had been kept constant but at different levels. From drawings by Rehder in Ellenberg 1963.

In the normal, relatively dry soil (upper sector), the cortical parenchyma has thick cell walls and no lacunae. In moist and very moist poorly aerated soils (two sectors on the right) all the roots form intercellular air spaces of varying size (dotted), and these are even larger when the soil is waterlogged (upper left). However small roots floating in the water have practically no intercellular spaces and very thin cell walls (lower left). Thus the adaptability of this dry-soil indicator to more or less wet conditions is surprisingly great (see also fig. 389).

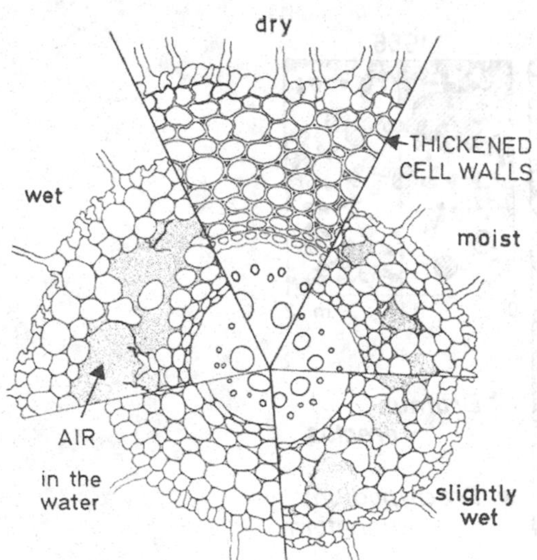

The reason for its varied behaviour may lie in the splitting up of the species into a number of physiological races. This may be true to some extent in the marsh- and steppelands, but in the investigations to which we refer the material used was a clone and therefore entirely uniform genetically. Moreover in pot cultures it was noticed that the same individual plant, growing in a damp soil, started to form lacunae in its roots at the point where these became submerged. Most of the older roots died following this drastic change in the micro-environment but younger roots adapted themselves to the shortage of oxygen by producing a root cortex with large intercellular air spaces. Still further evidence comes from those roots which grow straight down in well aerated soils and then reach a water table; at this point they begin to form the cortical spaces. There is no doubt then that the Upright Brome demonstrates a high degree of morphological, anatomical and physiological plasticity even without genetic differentiation. The same is true of other grasses which were tested experimentally, e.g. *Arrhenatherum elatius* and *Dactylis glomerata* (see also Armstrong 1981).

As we have already mentioned in section 6a *Bromus erectus* develops scleromorphic stalks and leaves on dry sites. This reduces its growth potential; but nevertheless it is on such sites that it is able to compete successfully with other tall growing meadow grasses. When sown together with *Arrhenatherum*, *Alopecurus pratensis* and *Poa palustris* it is almost entirely suppressed under the moist conditions which correspond with its physiological optimum and contributes noticeably to the sward only when the ground water level is more than 75 cm below the surface of a sandy soil (fig. 389). However it can also compete on soils which are kept permanently waterlogged (Ellenberg 1963). Relative to other species in a mixed culture it shows two optima with regard to the water factor, one close to its physiological minimum and the other close to its maximum. It behaves in a very similar fashion in nature provided the other habitat factors, e.g. soil pH and temperature, are not too unfavourable for it.

Although the Upright Brome can be suppressed by the Oatgrass and other tall grasses when there is a good supply of water, so in its turn, in their absence, it suppresses the Torgrass (*Brachypodium pinnatum*) whose leaves are usually shorter. This was shown to be the case by Bornkamm (1961a) who carried out some long-term observations in a slightly arid grassland (*Gentianello-Koelerietum* sub-assoc. of *Trisetum*, see fig. 391) on a deep calcareous soil, sloping 2° towards the SSW near Göttingen.

In the year 1953 Bornkamm took an area of 4 m ^{2}; from one half he removed all plants except *Bromus* and *Brachypodium* (the left side in each case in fig. 391) and from the other side all the plants including these two (right side). After only three years the two grasses had covered the whole area with a sward nearly as dense as the original, further evidence for the speed at which they can spread with their seeds and rhizomes. As soon as the growth had become fairly dense they began to compete with each other and with plants such as *Ononis spinosa* which had come in in the meantime. It appears from Bornkamm's investigations that in those years when the spring was moist and cool, especially in May, the proportion of the ground covered by *Bromus* increased at the cost of *Brachypodium*. This was particularly true in 1958 which was the wettest spring. On the other hand when the early part of the year was dry *Bromus* was weakened and *Brachypodium* was able to recover. This happened in 1954, 1956, 1957 and especially in 1959.

It appears then that the weather during the spring, when the leaves are still tender and the plants' water-conducting capacity is small, decides the relative dominance of the two species during the subsequent summer. The factor of primary significance may be the extent to which *Bromus* is able to form new shoots and occupy the ground with large clumps quickly. By spreading out from many centres it suppresses the *Brachypodium* shoots which grow up singly from creeping underground rhizomes.

It is worth noting that the relatively scleromorphic Upright Brome which is more resistant to drought overcomes the rather mesomorphic Torgrass more successfully just in wet years. If *Bromus* were really xerophilous then surely it must have been favoured by the dry springtimes and not the wet ones.

As we have already seen in section 5a the main reason for the increase in *Bromus erectus* has been a rapid falling off in the sheep population grazing these slightly arid swards. For hundreds of years these have been preferentially grazing and weakening *Bromus*, thus indirectly favouring the mostly disdained *Brachypodium*. Today the Brome Grass can spread unhindered wherever it gains a footing and conquer the semi-arid grasslands more or less completely.

Figure 391 illustrates the **ceaseless dynamics** which are going on in the poor dry grasslands and indeed in many other grassland communities. From one year to the next the balance of the species is displaced; their relative importance at any one time is not permanent. This dynamic state is due partly to the fact that the

Fig. 391. The spread of Upright Brome and of Torgrass on 2 x 2 m experimental areas in a limestone slightly-arid sward near Göttingen. In 1953 all plants except these two species were weeded out of the left half of the plot and the right half was entirely cleared of all vegetation. After Bornkamm (1961), modified. It took four years for the two species to encroach over the cleared area, but a change in dominance from year to year can be clearly seen already from 1955 onwards. Following a wet spring (1955, 1958) *Bromus* dominates and suppresses *Brachypodium*. In dry years (1957, 1959) the latter spreads at the expense of *Bromus*.

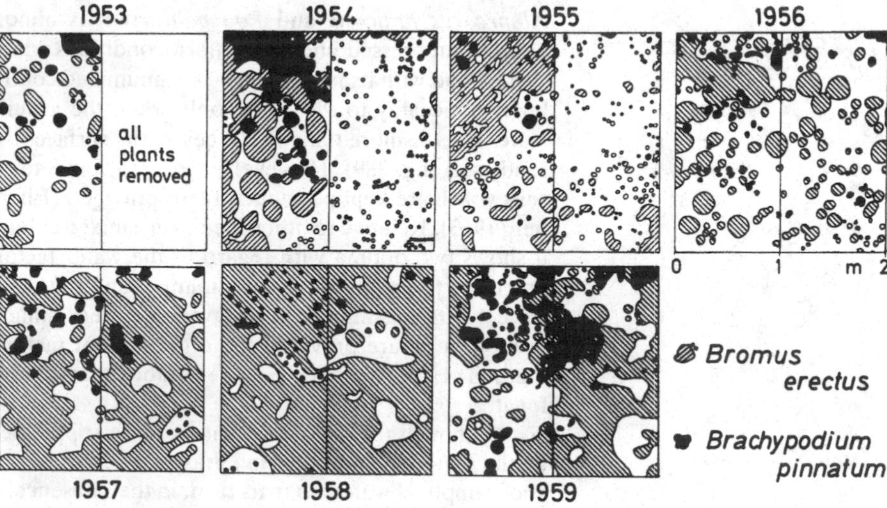

species under observation are short lived. Of 27 clumps of brome identified and marked in 1953 only 10 remained in 1959; the rest of those present were newcomers. In the case of the Tor Grass only 2 remained of the original 18 groups of shoots. The same is true of many other meadow plants, especially the tussock- or clump-forming grasses, most individuals of which die after a few years. They are replaced by younger plants which have germinated near the parent clump. Lieth and Ellenberg (1958) have observed this small-scale exchange of growing places many times during investigations on *Bromus erectus, Arrhenatherum* and other grasses at Hohenheim. On the other hand many dicotyledonous plants of the dry grasslands remain in the same place for many years, and individuals can reach a great age as for example Zoller and Stäger (1949) stated in the rock-steppes near Sion in Wallis.

However it would be premature to conclude that just because there is some exchange of places within the communities that there is a necessity for a 'natural crop rotation'. It is sufficient for us to look upon it as an expression of the plant's ability to regenerate and spread which helps some important members of the slightly arid and arid grasslands to overcome the catastrophic drought periods unavoidably occurring where these weak competitors are allowed to live. Many of them regenerate from the seed pool which they could build up in favourable years (Rabotnow 1969).

The behaviour of *Bromus erectus* however is still not completely explained by taking into account the water factor, the influence of grazing animals and its ability to regenerate. Of at least equal significance is its reaction to different levels of nitrogen supply. As with all grasses investigated so far, and indeed for the vast majority of higher plants, its most vigorous growth is attained with a nitrogen supply far higher than that which is available to it under natural conditions (over 1000 kg per ha, compared with less than 50, according to M. Vogel 1981). The difference however is smaller in the case of *Bromus* than it is with *Arrhenatherum, Dactylis* and other 'more demanding' tall grasses. Thus on the same calcareous soil with the same amount of moisture, one single heavier manuring may be sufficient to turn the balance in favour of the Oatgrass and its satellites (Ellenberg 1952a). Zoller (1954) too noticed that in the Swiss Jura *Bromus erectus* progressively lost ground to the other members of the Meadow Clary-Oatgrass meadow where manure had been applied, although *Bromus* itself, grown alone, is not damaged by additional nutrients; on the contrary its growth is considerably improved. It is only its ability to grow in poorer conditons which enables it to overcome its competitors in poorer habitats.

Calcareous grassland soils in our climate are poor in mineral nitrogen since ammonification and nitrification at root depth are often restricted during drought periods (see section 2a). It is true of course that the supply of nitrogen is even less in permanently saturated soil where the low oxygen content favours denitrification (tab. 56). Shortage of nitrogen is expressed not only by reduced productivity but also by an obvious yellow-green colour of the leaves, as is typical for arid grass heath and for small-sedges swamps too. It could well be this same factor, i.e. the shortage of nitrogen, which provides a refuge for the Brome in both very dry and very wet habitats where competition is reduced.

So it is **not dryness but poverty** which may be the main reason why these shallow calcareous soils as well as many calcareous small-sedges beds and Purple Moorgrass meadows enable *Bromus erectus* indirectly to become dominant. The arguments that have just been put forward apply only to grassland which has been mown, not grazed, and only under the influence of the **suboceanic climate of western Central Europe**. A continental climate brings catastrophic droughts more frequently and is less acceptable to *Bromus erectus*. Neither does the Upright Brome do well in a true mediterranean climate because the summers are too dry. Only where the ground is moist or the microclimate humid can it ripen its seed often enough to maintain itself in the grasslands here.

Bromus erectus manages very economically with the nutrients available to it. During the winter a considerable proportion of them is stored in the roots, to be mobilised again the following spring. This is especially so regarding nitrogen (P. Wagner verbal communication) and potash (see tab. 96). In common with many other plants of the poor and occasionally dry habitats its subterranean biomass is very large compared with that above ground (tab. 96 and fig. 376). It is also longer living, and an enrichment of nutrients takes place in it. For example the amount of potassium in the larger roots increases from 18.4 to 23.8 (in the top layer), or from 5.3 to 6.9 kg/ha (at a depth of 10-25 cm) while the total mass of the roots remains the same or even falls slightly. It seems that not only woodland trees but also grassland plants hold the easily leached potassium, which is so important for stomatal regulation, in the tightest possible cycle. This is well understood since Germann (1976) pointed out that the root activity in even a manured grassland soil filtered out more electrolyte than in a comparable woodland soil.

Tab. 96. The phytomass and P. K. and Ca content of the Upright Brome above and below ground in a slightly-arid sward near Göttingen.

Parts of *Bromus erectus*	**Phytomass** (dry weight) summer	winter	Amount of these elements in the phytomass **P** S	W	**K** S	W	**Ca** S	W
a Green parts	4,151	816	5.3	2.3	47.0	11.0	30.0	5.8
b Bases of stems	1,623	1,459	2.2	2.3	6.2	4.6	14.0	2.6
c Dead parts	1,292	4,650	1.1	2.5	1.9	3.9	10.0	17.9
d Litter	5,605	3,742	4.9	3.2	3.8	3.2	38.7	20.1
e a + b + c + d, **above ground**	**12,671**	10,667	**13.5**	10.2	**58.9**	22.7	**93.0**	55.5
f Roots 0–10 cm	7,033	6,884	7.0	5.3	18.4	23.8	64.8	64.3
g Roots 10–25 cm	1,937	1,968	1.8	1.5	5.3	6.9	21.3	18.0
h Finest roots 0–10 cm	4,477	4,451	4.6	3.4	13.5	12.3	50.4	46.4
i Finest roots 10–25 cm	2,464	3,582	2.7	2.7	8.1	12.1	29.2	40.8
k f + g + h + i, **below ground**	**15,911**	16,886	**16.1**	**12.9**	**45.5**	55.1	**165.7**	169.5
Proportion e : k	0.80	0.63	0.84	0.79	1.3	0.41	0.56	0.33
Total dry matter (e + k)	28,582	27,533	29.7	23.1	104.2	77.8	258.7	225.0
Difference summer – winter	1,029		6.6		26.4		33.7	

Source: After Wagner (1972) and Gonschorek (1971); in kg/ha.

Along with the habitat factors which we have considered up to now there are still other climatic and edaphic conditions important for the life of *Bromus erectus*, e.g. the temperature regime and the length of the growing period. The Upright Brome is absent from those places in Europe where these are insufficient; in particular it does not climb above the upper montane belt, and in the Scandinavian countries it is confined to the southern parts. As Steen (1957) showed other species of the Central European semi-arid grassland too will grow here only on decidedly southerly facing slopes or on shallow soils which are easily warmed up.

In addition it has long been known that the Upright Brome prefers calcareous soils, or at least ones without a strong acid reaction. All the experiments which have been discussed above were carried out on weakly acid to neutral soil. *Bromus erectus* and *Arrhenatherum elatius*, the two most important competitors in the relatively dry meadows of western Central Europe have practically the same pH amplitude. Both of them, and probably also many of their partners, when they are planted on soils poor in lime, can grow in pure cultures; but they become suppressed by species whose amplitude extends further to the acid side, especially representatives of the Matgrass swards and the dwarf-shrub heaths (see section II 3 and 4).

The investigations which have been carried out on *Bromus erectus* serve to endorse the significance of the part played by competition in the maintenance of certain species combinations, and show how careful we must be not to jump to conclusions about any species' demands just because it occurs in a habitat where there are these particular conditions. Experimental tests have shown that the **Upright Brome is a mesophilous grass requiring fairly warm conditions, and sufficient nitrogen supply, and avoiding very acid soils;** also that it is palatable and soon weakened by grazing. From an ecophysiological point of view it is not a xerophile, it does not avoid nitrogen and it does not have to grow on calcareous soil. It is only because of its competitors that it has become an indicator of dry, poor and lime-rich soils in Central Europe. Its behaviour has been treated so extensively because it serves as a good example of the importance of biotic factors in the causal analysis of plant communities.

Bradshaw, Lodge, Jowett and Chadwick (1960) showed that there is a similar wide physiological amplitude in the **Matgrass** (*Nardus stricta*, fig. 412) which is the direct opposite of the Upright Brome in being found on acidic poor pastures and being generally looked upon as an acidophile. However its growth is not restricted in soil with a lime content of even 80 ppm and an almost neutral reaction of pH 6.5, provided the pH value is kept constant and competition removed (see also fig. 330). However if productivity is used as a basis for comparison with the growth on acid soils, then other pasture grasses such as Common Bent-grass (*Agrostis tenuis*) or Crested Dog'stail (*Cynosurus cristatus*) are favoured by the addition of lime much more than is the Matgrass. Thus these grasses will become dominant under better management. Extensive grazing by sheep on the other hand favours the Matgrass as a pasture weed (see section A II 2c) to such an extent that it can even increase on soils which are only slightly acid and of higher fertility. As in the case of *Bromus erectus* and many other grassland plants so also with *Nardus stricta*: if one wishes to understand its distribution and its role in the species mosaic of plant communities then one must take into account the influence of competition as well as of other biotic factors. These are not easily measurable, but may be more important than many of the abiotic factors that we are accustomed to determine in order to characterise a site.

7 Grasslands on soils rich in heavy metals

(a) The nature and origin of heavy metal vegetation

Amongst the ecologically most interesting plant combinations are the isolated patches of so-called heavy metal vegetation. These sparse stony islands in the dense sea of our vegetation are generally associated with siliceous or calcareous poor grasslands, which is why we are dealing with them at this point. We need only look at them briefly, picking out the fundamental points and those applying particularly to Central Europe since there is an all round monograph by Ernst (1974) available which also contains an ample list of further references. The physiological background of the tolerance for distinct heavy metals was recently reviewed by Ernst (1982) in Kinzel's book.

Fig. 392. Almost all plant species growing on soil with a high zinc content take up more of the element (in terms of dry matter leaf) than when they are growing in normal soil which has a very low zinc content. After Denayer-de Smet (1970), modified.

Apart from the zinc-adapted Alpine Pennycress there is no difference in this respect in the behaviour of the heavy-metal plants (*Silene*, *Viola* and *Armeria*) and that of other species, including many trees. However the former are able to tolerate a high Zn concentration better over a longer period.

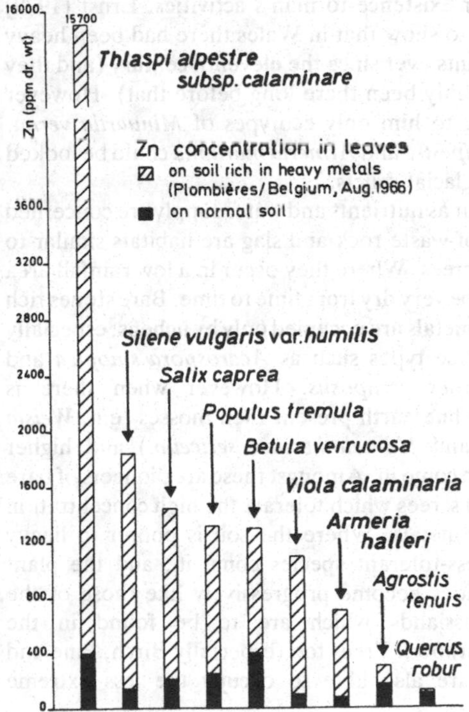

Heavy metals such as zinc and copper (i.e. those with a specific gravity of over 5) are indispensable as trace elements for phanerogams and for most cryptogams, but in higher concentrations they cause damage even to the plants which are best adapted to them. A prealpine Pennycress (*Thlaspi alpestre*) taken from a soil rich in zinc for example grew quite well in a nutrient solution with 100 mg Zn per litre but remained vegetative. At 250 mg it grew and flowered normally as it did at 500 mg with a slightly lower production. One thousand milligrams per litre was obviously nearing the limit at which it would succomb although this concentration could be met with in a heavy metal area. Plants not adapted to Zn can no longer live under such conditions, and even at the lower concentrations they were stunted. Thus while they normally supress the small light-loving Pennycress under less extreme conditions, their competition is removed on soils with a high Zn concentration. The same is true of other heavy metal indicators treated in the next section (see also fig. 392). Another example is the Bladder Campion which Baumeister and Ernst (1978) cultivated in a series of nutrient solutions with increasing zinc concentrations (see tab. 97). The zinc form of *Silene vulgaris* achieved its highest production at a concentration of 50 mg Zn per litre. However even at this level it yielded less than the normal form which needs virtually no zinc and was relatively more affected than the zinc form. When the zinc concentration was doubled the normal form was killed, but the zinc form was able to go on thriving quite well. If the two forms come into competition in nature the zinc concentration in the soil water will determine which strain becomes dominant even though, in the absence of competition, both can grow well without the addition of extra zinc.

Basically the heavy metal forms of plants behave in a similar way to the halophytes (which were discussed in section C IV 1). These require salt in only small quantities, but are better able to cope with higher concentrations than the glycophytes. Many of the plants living in heavy metal areas behave like *Juncus gerardii*, for example, in getting rid of the excess poison by concentrating it in their older leaves which are eventually shed.

Tab. 97. The influence of increasing concentrations of zinc on the normal and zinc-adapted froms of Silene vulgaris in culture solutions.

mg Zn per litre solution	Normal form (botanic garden, Nantes)			**Heavy-metal form (Silberberg near Osnabrück)**		
	shoot	root	total	shoot	root	total
0	4.1	1.2	5.3	1.8	0.4	2.2
10	7.0	1.7	8.7	1.8	0.7	2.5
50	3.2	0.9	4.1	2.6	1.0	**3.6**
100	dead	dead	dead	1.9	0.4	2.3

Source: After Baumeister (1967). Values in g of fresh weight for each plant.

Physiologically it is not quite correct to talk of 'heavy metal plants' in general terms since usually we are dealing with plants which, by selection, became tolerant of only one particular metal occurring in excess in its particular habitat. Nevertheless taxa resistant to large amounts of zinc frequently are also resistant to copper and other heavy metals. Incidentally the resistant plants do not restrict the absorption of these metals. Instead of this the Zn-tolerant plants form relatively large amounts of oxalate and malate; Cu tolerance is usually connected with the formation of phenolic compounds. In each case the maintainance of tolerance requires energy in quantities which reduce the biomass production and therefore the competitiveness of the plant in habitats where the tolerance is not needed. The physiological basis for tolerance appears to be inherited but is increased by environmental factors.

In contrast to the salt plants which are usually quite distinct species the heavy metal plants are minor taxa which, according to Ernst, cannot be distinguished morphologically from the normal type when grown under similar conditions. Apart from a few exceptions they can only be recognised because of their greater resistance to the heavy metals, i.e. ecophysiologically; so they are simply known as **ecotypes**. In a habitat rich in heavy metal they certainly show restricted growth, an obvious scleromorphic structure, a high content of anthocyanin, an unusually well developed root system (fig. 393) and quite often other anomalies such as distorted petals. However these features disappear in normal soil and are thus seen to be modifications.

Soils rich in heavy metals have been caused by the mining of ore and the dumping of waste and rock on the surface. This has been going on since the Middle Ages and even since the Bronze Age on a smaller scale. At first the tips were small and scattered but later they covered much larger areas. The earlier inefficient methods of extraction left a considerable amount of metal in the slag so that tips made of this material also contain large amounts of heavy metals. In the Harz valleys such slag heaps are to be found a long way from the source of the ore. This is because the wood for smelting was soon used up near the mines and had to be brought in from further afield, e.g. from the central and southern Harz valleys to Goslar in the north-east. On the return journey the mules would take back a load of ore to be smelted near to the source of timber. Natural outcrops of crystalline rock rich in heavy metal are rare in the lowlands of Central Europe since all this area has been covered by sedimentary deposits. Only in the mountains, especially the Alps, are there sites and communities of heavy metal plants which do not owe their existence to man's activities. Ernst (1979) was able to show that in Wales there had been heavy metal plants ever since the eleventh century (and they had probably been there long before that). However according to him only ecotypes of *Minuartia verna, Thlaspi alpestre* and *Armeria maritima* could be looked upon as glacial relicts.

As far as nutrients and water supply are concerned the tips of waste rock and slag are habitats similar to natural screes. Where they occur in a low rainfall area they can be very dry from time to time. Bare stones rich in heavy metals are occupied only by lichens, especially the crustose types such as *Acarospora sinopica* and *Diploschistes scruposus*. However when there is sufficient fine earth present then mosses (e.g. *Weisia viridula* and *Homalothecium sericeum*) and higher plants can come in. Amongst these are pioneers of bare earth and screes which tolerate the high concentration of heavy metals. Where the soil is poorer in heavy metals less-tolerant species come in, and the plant communities become progressively like those of the poor grasslands which are to be found in the neighbourhood. Trees too (especially Birch, Pine and Spruce) are also able to occupy the less extreme habitats and lead the succession to woodland provided the area is no longer being grazed. Whether the different stages from the bare rock to grassland and finally woodland really represent a succession in time or whether they are just a stable pattern of different habitats and communities must be decided for each case individually. Often it is nothing to do with succession in a narrower sense, even if the proximity of the different communities appears to indicate it (see also section C VI 8 a).

(b) Classification of the heavy metal communities
The vegetation of the true heavy metal habitats is very poor in species as it is in salty ground and other extreme sites. Between the extreme habitats and neighbouring grassland communities e.g. *Xerobrometum* and *Mesobrometum* there are intermediate stages which get progressively richer in the number of species but still contain some of the tolerant taxa. For this reason the heavy metal vegetation was previously placed in the order *Brometalia* or at least in the class *Festuco-Brometea*. However according to Braun-Blanquet and Tüxen (1943) they form their own class in Europe and western Siberia. This bears the name *Violetalia calaminariae* but as Ernst points out the Calamine Violet is not a good character species. The European heavy metal vegetation is much better characterised by ecotypes of the Spring Sandwort (*Minuartia verna*) and of the Bladder Campion (*Silene vulgaris*).

All the western and Central European phanerogam communities on soils rich in heavy metals also belong to the order *Violetalia calaminariae*, which can best be recognised by the absence of southern European species. Possibly distinct ecotypes of Sheep's Fescue and Common Bent (*Festuca ovina* and *Agrostis tenuis*) can be seen as character species for the order; Ernst distinguishes three alliances:

1 **Alpine** heavy metal vegetation (*Galio anisophylli-Minuartion vernae*) with ecotypes of *Galium anisophyllum. Poa alpina* and *Dianthus sylvestris;*
2 **Western European**, with units coming into the western part of Central Europe (*Thlaspion calaminaris*) with *Thlaspi alpestre*;
3 Units scattered throughout **Central Europe** (*Armerion halleri*) with ecotypes of *Armeria maritima*, but without *Thlaspi alpestre* (although its distribution extends as far as Poland).

The Calamine Violet sward of the Rhineland (*Violetum calaminariae rhenanicum*, in the 2nd alliance) and the heavy metal Thrift sward (*Armerietum halleri*, 3rd alliance) in the Harz and its northern foothills are amongst the best researched associations. In both communities subassociations can be distinguished corresponding to the soil dampness. In the latter for example there is *Cardaminopsis halleri* which also turns up on soils poor in heavy metals, e.g. in montane meadows, but in any case needs a good water supply. On the other hand *Plantago media* and *P. lanceolata* along with some plants of the semi-arid grasslands indicate drier conditions.

The species mentioned above are practically all the vascular plants which make up the heavy metal plant communities in Central Europe. Some of them may be absent from individual places just because they have not yet been able to reach them. *Thlaspi alpestre* var. *calaminare* for example has extended its range in recent years. *Viola calaminaria*, which usually has yellow flowers, is replaced in Westphalia by a blue subspecies, *westfalica*. It is obvious that accidents of isolation and dissemination have played and are still playing a part in the species combinations of the heavy metal communities.

(c) Effects of heavy metal emissions on the vegetation
In recent times air pollution has given plants tolerating

Fig. 393. The root systems of *Minuartia verna* ssp. *hercynica* when growing on a copper slag heap (right) and on a soil which is not unusually rich in heavy metals (left). After Schubert (1954). The heavy-metal form is much more strongly xeromorphic and its root system more extensive and thicker.

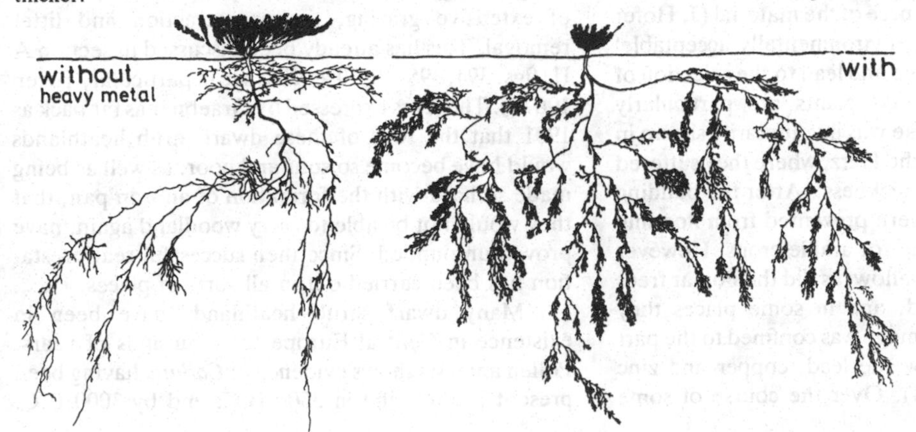

heavy metals further opportunities for extending their range. Along congested roads as well as in large towns the lead content of the ecosystems is increasing since the grasses and herbaceous plants as well as the soils accumulate lead exhausted from the motor vehicles. Already the appearance of *Cardaminopsis halleri* has been noticed in some places. This same meadow plant has also been found near smelting works, e.g. the lead works at Nordenham on the lower Weser where it became more and more frequent in cultivated pastures and meadows of the order *Arrhenatheretalia* (see section D V 1 a, VI 1 a). Most other plants however are suffering from emissions of heavy metals since these are accumulated in unrestricted amounts (Ernst, Mathys, Salaske and Janiesch 1974.). Three years after the start of operations in a zinc-smelting factory for example severe damage was being caused not only by zinc but also by Cu, Pb and Cd (Ernst 1973). Those species which are unable to produce resistant races, and these are the majority, suffer long-lasting damage. One of these is the Reed-mace (*Typha latifolia*) the subject of an intensive study by McNaughton (1968).

Some species which are relatively tolerant can be used as 'accumulators' to measure the amount of heavy metal emission. Annual Ryegrass (*Lolium multiflorum*) is particularly useful for this purpose when grown in standardised culture pots with soil free from heavy metals and then exposed for a few weeks (Schönbeck 1974). Zn, Cu, and Pb etc. can also be recovered from mosses such as *Polytrichum piliferum*, *P. juniperinum* or *Ceratodon purpureum*, the amount corresponding to the degree of air pollution. The accumulation in the gametophytes of *Atrichum undulatum* is 2-3 times more than in grasses, although in the sporophyte it is less (Lötschert, Wandtner and Hiller 1975).

When sewage sludge or waste compost is spread on garden or farm soil the Cu, Zn, Pb, Cd and Cr content can be raised very noticeably in quite a short time depending on the source of the material (J. Hofer and Jäggli 1975). Even 'environmentally acceptable' manures can cause damage and lead to the selection of resistant plants. Cultivated plants are particularly susceptible, as for example was involuntarily shown in the Oker valley north of the Harz, where they suffered from the so-called 'Oker sickness'. After the building of a dam the meadows were prevented from flooding and could be ploughed up for arable crops. However these crops soon turned yellow, as did the Poplar trees which had been planted, and in some places they actually died off. This damage was confined to the part of the river valley below the lead, copper and zinc oxide works (Horn 1974). Over the course of some decades before the dam construction the sediments brought by flooding had resulted in a sharp rise in the amount of these heavy metals in the soils.

II Dwarf-shrub heaths and commons on strongly acid soils

I A general account of heaths on the lowlands and mountains

(a) How the heaths arose and are destroyed

Treeless stands of dwarf shrubs on poor sandy soils in north-western Germany, poor calcareous grasslands in the mountains and hills of southern Germany and open pine woods on the sandy plains of eastern Central Europe are all known as heaths (German 'Heide'). In each case they are former 'commons' where the farmers were entitled to graze their domestic animals. Such common land consisted of open areas as well as of woodland remains, and much of it was woodland pasture through which the animals were allowed to wander (see section A II 2). The term 'heath' was originally much more to do with the law of the land than the type of landscape or vegetation, and had the same meaning as 'steppe', 'garigue' or 'macchia' have in other countries. A large part of the heathland in north-west Germany, Holland, Belgium and Jutland has been afforested with pines during this century and has taken on much of the appearance of the Pine heaths in eastern Germany. This development has lead to the disappointment of those visitors to the 'Lüneburg Heath' who might be expecting a vast sea of dwarf shrubs glowing purple and red in the August sunshine.

Natural dwarf-shrub heaths occur in only a few places in Central Europe, either on the coasts or on mires in the north-west or above the tree line in the higher mountains. The broad heathlands between northern Belgium and Denmark, including Lower Saxony which still covered up to 90% of the area about 100 years ago, were produced by the former practices of extensive grazing, timber extraction and litter removal. This has already been discussed in section A II (figs. 394, 395, 18, 19 and 21, see in particular, Tüxen 1967b). The fears expressed by Graebner as far back as 1901 that the soils of these dwarf-shrub heathlands would have become so sour and poor, as well as being made so hard with the formation of an iron pan, that they would not be able to carry woodland again, have proved unfounded. Since then successful reafforestation has been carried out in all sorts of places.

Many dwarf shrub heathlands have been in existence in Central Europe for thousands of years. Pollen analysis shows evidence of *Calluna* having been present in the Eifel in 3500 B.C. and by 3000 B.C.

Fig. 394. Fresh cut turves of heath with its litter from the *Genisto-Callunetum typicum* stacked by the side of the road from Wilsede to Niederhaverbeck in 1951, the last time turves were cut here (see fig. 19).

Fig. 395. An old sheep barn under the Birches in the Lüneburg Heath near Wintermoor. *Calluna* avoids the shade of the trees. It flowers well because it is being continually nibbled and rejuvenated by the animals.

Fig. 396. The retreat of heathland in eastern Netherlands (near Almelo, Twente) between 1843 and 1943. After Westhoff (1956), from Ellenberg (1963). Today almost all the remaining heaths have disappeared and much larger areas have been afforested.

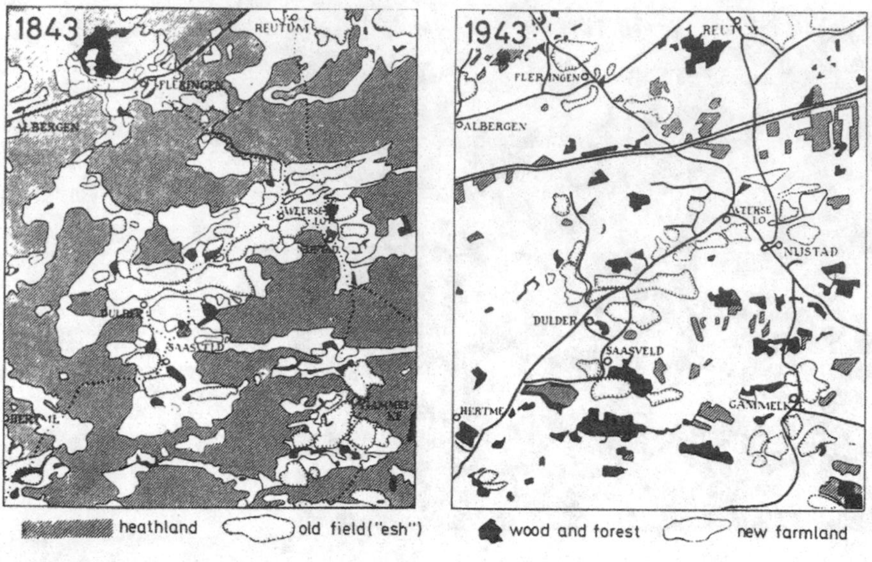

Fig. 397. Vegetation succession in the area of the *Calluna* heaths on pure sandy soils free from groundwater in north-west Germany. On the **left** where sheep are grazing, on the **right** after grazing has been stopped.

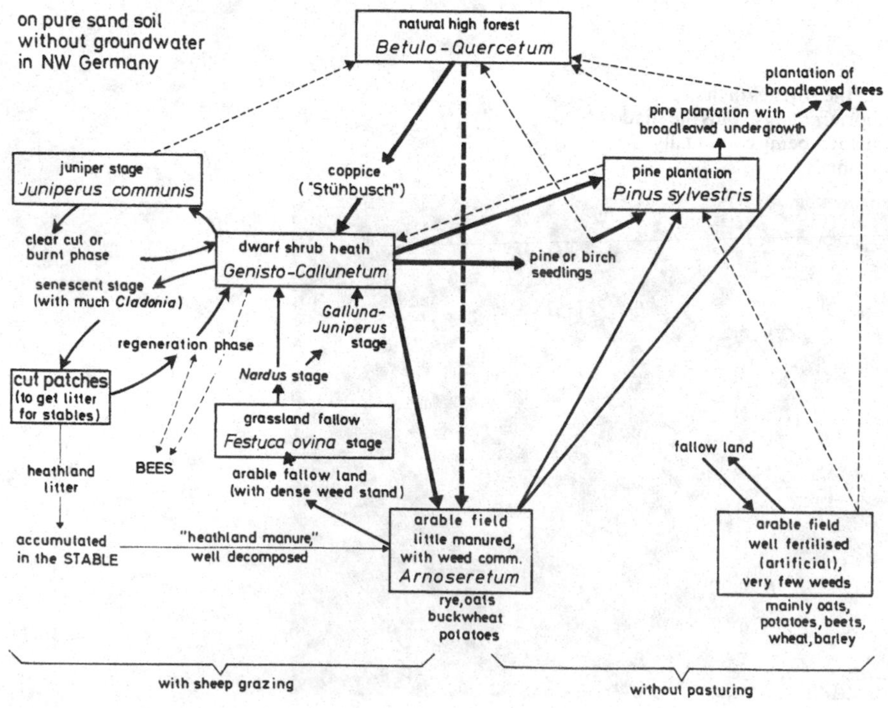

heaths were very widespread (Straka 1973). A parallel increase in the pollen of grain crops points to a rapid increase in human influences. Another maximum can be traced to about the fourth and fifth centuries A.D. Before and since this period it appears that much of the heath returned to woodland, especially during the thirty years war (1618–1648). In the lowlands close to the coast too the extent to which heaths were produced oscillated a number of times. It reached a maximum about 150 years ago, covering an area the size of which is difficult to imagine today; but the final retreat came about even more rapidly.

Figure 396 gives an example of how much the landscape has changed in the sandy plains along the North Sea because of the reduction in the moorland area since 1850. Even the few small remains of heathland which are still marked on the map of 1943 have since been planted with trees or cultivated. The main reason for this rapid and fundamental alteration has been the introduction of chemical fertilisers which have increased the yields of fodder and straw rendering superfluous the removal of litter from the heath for livestock bedding (see section A II 2c). Even more than this they have made it possible to bring the infertile heathlands themselves into cultivation. At the same time wool prices fell because of cheap imports from Australia, completely destroying the economy of the heathland flocks. A hundred years ago, for example, there were over 750,000,000 sheep which used to find a good living on the Lüneburg Heath; by 1900 there were barely 250,000,000 left and in 1950 less than 25,000. Thus the dwarf-shrub heathlands, which for many centuries had been the basis of a successful farming system on this poor sandy land and a necessary component of the cultivated landscape up to the nineteenth Century, became a wasteland within a few decades. At the last minute a small part of the remaining areas of this anthropo-zoogenous dwarf-shrub formation has been rescued from the plough and from the planting or natural regeneration of trees (see fig. 397) by the creation of heathland nature reserves.

There are only three site types in the lowland of Central Europe where the growth of woodland is restrained by natural factors allowing dwarf shrubs such as *Calluna vulgaris*, *Erica tetralix* and *Empetrum nigrum* to become more competitive:

1 Leached **dunes of blown sand** where wind prevents the growth of trees (see section C V 3),
2 Hummocks and marginal slopes of **raised bogs**, which have been discussed in section C III,
3 **Very acid swamp soils** where the growth of trees is prevented by oxygen-deficient ground water which frequently reaches the surface (section 2 a).

It is only on the last-named habitat that true heathland can develop since on raised bogs the stands of dwarf shrubs are interrupted by numerous wetter hollows, and on dunes the natural heathland is restricted to narrow belts by the encroachment of trees which, in spite of frequent storms, are able to develop quite near to the coast (see section C V 1 d). Unfortunately the majority of the natural dwarf-shrub heaths which existed between about 1880 and 1960 have either been brought into cultivation by draining and seeding pastures or have been afforested so that it is now difficult to determine their previous extent.

Dwarf-shrub heaths have never been so extensive in the hills and mountains as on the broad sandy plains, although here too there were communities rich in dwarf shrubs. Apart from a few small remains these have now disappeared, the majority having been planted up with Spruce in the same way as the Matgrass heaths (section 4 a) which played a greater part than *Calluna* heaths on the very acid soils at the montane level (see section 4 a).

Rhododendron heaths at the subalpine level are also generally the result of man's activities. We have met these already in connection with the natural heathland of high mountains (section C VI 4 c).

(b) On the classification of dwarf-shrub and Matgrass heaths

The classification of moorland communities has changed a number of times from the phytosociological aspect without altering the ecological grouping. Today most of the *Calluna* and *Nardus* heaths are put together in a class of man-made heaths (*Nardo-Callunetea*, see also tab. 98).

This has only a few character species, and even these also occur in poor meadows, mires and woodland, namely *Calluna vulgaris, Luzula campestris, Danthonia decumbens*, and *Cuscuta epithymum*, with *Carex pilulifera* and *Potentilla erecta* as possibilities. Many of the species also invade the slightly arid grasslands which have already been dealt with (sections D I 1 and 2), e.g. grasses such as *Agrostis tenuis* and *Briza media* and composites including *Hieracium pilosella, Hypochoeris radicata, Leontodon hispidus, L. saxatilis* and *Scorzonera humilis*. The classes *Nardo-Callunetea*, *Sedo-Scleranthetea* and *Festuco-Brometea* form a group of related communities with a similar history. They have many species in common and are often found growing next to one another or intermixed.

Within the Nardo-Callunetea the order of the Matgrass swards *(Nardetalia)* leads over into the more demanding grassland communities. Character species include the one from which it gets its name, the wiry leaved *Nardus stricta,* also *Arnica montana* and other acid-soil dwellers (see tab. 98). The Matgrass swards of the high mountains which have already been discussed in section C VI 3 c are put into the

Tab. 98. Calluna heaths on the coast and further inland in Schleswig-Holstein and Jutland.

Species groups	Inland heaths						Coastal heaths			Indicator val.		
Community no.:	1	2	3	4	5	6	7	8	9	**T**	**K**	**R**
Basic heath species:												
K *Calluna vulgaris*	*5*	*5*	*5*	*5*	*5*	*5*	*5*	*5*	*5*	×	3	1
Dicranum scoparium (M)	*3*	*3*	*3*	*3*	*3*	*2*	*5*	*5*	*4*	–	–	1
Pleurozium schreberi (M)	*5*	*4*	*4*	*5*	*4*	*5*	*3*	*3*	*3*	–	–	1
Hypnum cupressiforme (M)	*1*	*2*	*2*	*4*	*4*	*3*	*2*	*4*	*3*	–	–	×
S *Carex arenaria*	*2*	*2*	*5*	*5*	*3*	*2*	*5*	*5*	*5*	×	2	2
Campanula rotundifolia	*3*	*2*	*3*	*2*	*2*	*2*	*3*	*1*	*3*	×	×	×
S *Hypochoeris radicata*	*4*	*2*	*3*	*2*	*2*	*2*	*2*	*3*	*3*	5	3	4
Hieracium umbellatum	*1*	*2*	*3*	*1*	*4*	*2*	*3*	*4*	*3*	×	×	4
Agrostis tenuis	*5*	*3*	*3*	*2*	*1*	*1*	*2*	*2*	*4*	×	3	3
K *Luzula campestris*	*2*	*2*	*2*	*1*	*1*	*1*	*3*	*3*	*3*	×	3	3
Inland heath species:												
Avenella flexuosa	*5*	*5*	*5*	*5*	*4*	*5*	*3*	*1*	*2*	5	4	2
V_c *Genista pilosa*	*3*	*2*	*3*	*3*	*3*	*2*		*1*		5	1	4
V_c *G. anglica*	*5*	*3*	*3*	*4*	*4*	*2*		*2*	*2*	4	2	3
K *Carex pilulifera*	*2*	*3*	*3*	*1*	*2*	*2*	*1*			4	2	3
Molinia caerulea	*1*	*3*	*1*	*2*	*4*	*2*		*1*		×	3	×
N *Nardus stricta*	*1*	*1*	*2*	*2*	*3*	*1*	*1*	*1*	*1*	×	3	2
Quercus robur (young)	*4*	*3*	*3*	*2*	*1*	*2*				6	×	×
Agrostis canina	*3*	*2*	*2*	*1*	*2*	*1*		*1*		×	5	3
Sorbus aucuparia		*2*	*2*	*1*	*1*	*2*				×	×	4
V_s *Cytisus scoparius*	*1*	*2*	*2*	*1*	*1*			*1*	*1*	5	2	3
'Warm heath' species:												
S *Spergula morisionii*	*5*		*1*							5	4	×
S *Pulsatilla pratensis*	*3*									6	5	7
Carex ericetorum	*3*					*1*				5	7	×
S *Artemisia campestris*	*2*									6	5	5
Vicia cassubica	*2*									7	4	5
'South-eastern' species:												
Betula pendula	*4*	*2*		*1*	*1*	*1*	*1*			×	×	×
Rubus fruticosus	*2*	*2*	*1*	*1*			*1*			×	×	×
Hypericum perforatum	*3*	*1*	*1*		*1*					×	5	×
Pinus sylvestris	*3*	*2*	*2*	*1*		*1*				×	7	×
'North-western' species:												
N *Arnica montana*			*1*	*1*	*4*	*3*		*1*		4	4	3
Scorzonera humilis		*1*	*1*	*1*	*3*	*2*		*1*	*1*	6	5	5
Succisa pratensis		*1*	*1*	*1*	*3*	*1*				5	3	×
Northern woodland species:												
Populus tremula	*1*	*1*	*1*	*1*	*2*	*2*				5	5	×
Trientalis europaea			*1*	*1*		*4*				×	7	3
Vaccinium myrtillus	*1*	*1*	*1*	*1*	*1*	*3*				×	5	2
Solidago virgaurea	*2*	*1*	*1*	*1*	*2*	*3*				×	×	×
Jutlandic heath species:												
Vaccinium vitis-idaea		*1*				*4*	*1*			×	5	2
Arctostaphylos uva-ursi		*1*		*1*	*2*	*3*				3	5	×
Juniperus communis					*1*	*3*		*1*		×	×	×
Crowberry-heath species:												
V_e *Empetrum nigrum*		*1*	*2*	*5*	*5*	*5*		*5*	*5*	×	3	×
Salix repens			*2*	*2*	*2*	*2*	*1*	*2*	*3*	5	×	×
Coastal heath species:												
A *Ammophila arenaria*[a]	*1*		*3*	*2*	*1*		*2*	*4*	*5*	6	3	7
S *Jasione montana*	*3*	*1*	*3*	*1*	*1*		*1*	*3*	*3*	5	3	5
Lotus corniculatus	*1*	*1*	*1*	*1*	*1*	*1*	*1*	*3*	*2*	×	3	7
S *Festuca rubra arenaria*				*1*	*1*		*1*	*3*	*3*	×	5	7
Galium verum						*1*	*2*	*2*	*3*	5	×	7

[a] in part planted

Tab. 98. (continued)

Average indicator values:	mT	mK	mR
Inland heaths			
1. 'warm heaths' in s.e. Holstein	5.1	3.5	3.1
2. typical Holstein heaths	4.8	3.4	2.7
3. inland dune heaths in Holstein	5.0	3.1	3.0
4. inland dune heaths in Schleswig	4.9	2.9	2.8
5. *Arnica* heaths in Schleswig	4.7	3.2	2.9
6. typical Jutland heaths	4.7	3.9	2.6
Coastal heaths			
7. on the Baltic in Schleswig-Holstein	5.1	3.0	3.1
8. on the North Sea in Schleswig-Holstein	5.1	3.0	3.6
9. in southern Jutland	5.0	3.0	3.6

Note: The coastal heaths contain a relatively large number of species with a tendency to an oceanic distribution (mK 3.0) together with a greater requirement for warmth (mT 5.0–5.1) and a supply of bases (mR 3.1–3.6).

Of the inland heaths the ones which come nearest to the coastal ones with regard to temperature and soil reaction are the 'warm heaths' of south-eastern Holstein; however there are more species here with a relatively continental distribution (mK 3.5). This is also the case with the Jütland heaths although their mean temperature values, like those of other inland heaths are mostly below 5.0. The typical heaths in both Holstein and Jütland are characterised by species which can tolerate extremes of soil acidity with average reaction values of 2.7 and 2.6.
(Explanation of the method of calculation in tab. 12.)

These differences are rather small; thus it is obvious from the average indicator values that all the *Calluna* heaths have much in common not only floristically but also ecologically.

Some species, which are not constant in any of the examples have been omitted. Raabe's lists (1964) do not contain any Lichens, which surely occurred in his sample areas.

K = character species of the class of acid-soil heaths (*Nardo-Callunetea*)
N = order of Matgrass swards (*Nardetalia*)
V_c, V_e, V_s = order of Genista-Heather heaths (*Calluno-Ulicetalia*)
V_c = alliance of Genista-Heather heaths (*Calluno-Genistion*)
V_e = alliance of Crowberry heaths (*Empetrion boreale*)
V_s = alliance of Broom heaths (*Cytision scoparii*)
S = class of dry sandy swards (*Sedo-Scleranti*) which overlap with the heaths,
A = class of the Marram Grass dunes.

Source: From data by Raabe (1964).

alliance *Nardion* (or *Eu-Nardion*) which has numerous species indicating that its centre of distribution is at the subalpine level. At lower levels the Matgrass swards become poorer in species and can clearly be distinguished from the *Nardion*, but less easily characterised. So the name *Violion caninae* is given to it even though, the heath Dog Violet is not invariably present, while *Nardus* is as plentiful here as in the '*Nardion*'. In order to avoid misunderstanding it is recommended that the lowland Matgrass swards are given the name '*Violo-Nardion*'.

While the majority of the Matgrass swards are relatively rich in species, the acid-soil dwarf-shrub heaths (*Genisto-Callunetalia*) are amongst the least diverse of the phanerogam communities. Their distribution centre lies in Atlantic north-west Europe. They can be distinguished from the *Nardetalia* communities only by the absence of some of their character species. There are three alliances within the *Genisto-Callunetalia* which are of importance to us:

1 **Petty Whin-Heather heaths** (*Genisto-Callunion*) are the most widespread.
2 **Broom-heaths** (*Cytision scoparii*) in the hills and mountains of western Central Europe.
3 **Crowberry heaths** (*Empetrion boreale*) which radiate into Europe from their Scandinavian centre.

There are stands belonging to the latter which can be said to be natural, but all other communities of the class *Nardo-Callunetea* have come into existence only with the help of man. They also need the constant attention of man to maintain them.

On the other hand the order of **swamp heaths** (*Erico-Sphagnetalia*) contains mainly naturally tree-free dwarf-shrub communities. These together with the raised bogs (*Sphagnetalia*) form a class of the heather bogs and swamp-heaths (*Oxycocco-Sphagnetea*). Apart from the swamp-heaths which we shall deal with in section 2 a we have already thoroughly discussed these in section C III.

In order to complete the survey we just have to remember that the **Alpenrose heaths** are placed in the sub-alliance *Rhododendro-Vaccinion*, i.e. included with the acid-soil coniferous woods (*Vaccinio-Piceion*, class *Vaccinio Piceetea*).

Table 98 brings together examples of the floristic composition of the *Calluna* and *Empetrum* heaths. At one time each of these communities was distributed in Schleswig-Holstein in its own way (fig. 402). Soon there will be practically none left because it is impossible to maintain them without sheep. Outside Central Europe dwarf-shrub heathlands still play a more important part, as was recently shown by Gimingham, Chapman and Webb (1979).

2 Natural tree-deficient dwarf-shrub heaths in the north-west

(a) Cross-leaved Heath swamps near the sea

Dwarf-shrub heaths of undoubted natural origin are becoming rarer, the few remaining ones being found on waterlogged and very acid sandy soils in the coastal plain, often in connection with raised bogs which were once widespread in this area. The Cross-leaved Heath (*Erica tetralix*, fig. 398) dominates in these oligotrophic swamp-heaths and is accompanied by other atlantic or subatlantic species, e.g. the Heath Rush (*Juncus squarrosus*), Bog Asphodel (*Narthecium ossifragum*) and the Deergrass (*Scirpus cespitosus* ssp. *germanicus*). *Sphagnum* species not concerned in the formation of bogs are also characteristic especially *Sph. compactum* and *Sph. molle* (Dierssen 1972).

Fig. 398. Heather on peaty soil (*Ericetum tetralicis typicum*) in a dune valley on the island of Sylt. Photo, Lüdi.

Important conditions for the *Ericetum tetralicis* are cool summers and mild winters, a high stagnant water table and a low pH (4 or less). The extremely wet conditions under which this heath community is formed are inimical to the growth of trees. Menke (1963) was able to show that it began to spread at the end of the Postglacial Warm Period, and very often at the cost of the swamp woodland. According to him, 'there is no basis for thinking that their formation has been brought about by man'. This question was left open in the first edition, but has now been answered through a study of the vegetation history.

Menke's findings only apply to the typical Cross-leaved Heath community (*Ericetum tetralicis typicum*, fig. 399) and for its subassociations on even wetter soils, especially the Peatmoss *Erica* heath (*E.-t. sphagnetosum*) which is characterised by *Sphagnum papillosum, Andromeda polifolia* and a large amount of *Narthecium*. *Erica*-heaths rich in lichens on the other hand lead on to *Calluna*-heaths and like these have arisen from woodland. Since the surface dries out from time to time the growth of fruticose lichens is favoured, e.g. *Cladonia uncialis, C. gracilis* var. *chordalis* and *C. squamosa*.

The Cross-leaved Heath is avoided by cattle and for this reason was disliked by the herdsmen. However it was valued as a source of bedding material for the animals when they were confined to their stalls. This litter rich in mosses was regularly cut at intervals of 10-20 years (fig. 400). The bare swampy soil thus exposed was soon occupied by a **Beak Sedge** commun-

Fig. 399. Fluctuations in groundwater levels throughout the year under *Erica* heaths, *Calluna* heaths and acid sandy grassland (*Festuca* communities) at the southern edge of the Wümme valley. After Lache (1974), modified. From November to June the water in the *Eriophorum* variant of the *Erica* heath lies close to or above the ground level. Such a persistent high water table excludes trees. The habitat of the other *Erica* heaths (e.g. the *Pleurozium* variant) however could support woodland. Thus a part of the *Ericetum* must be seen as resulting from the activities of man and his animals.

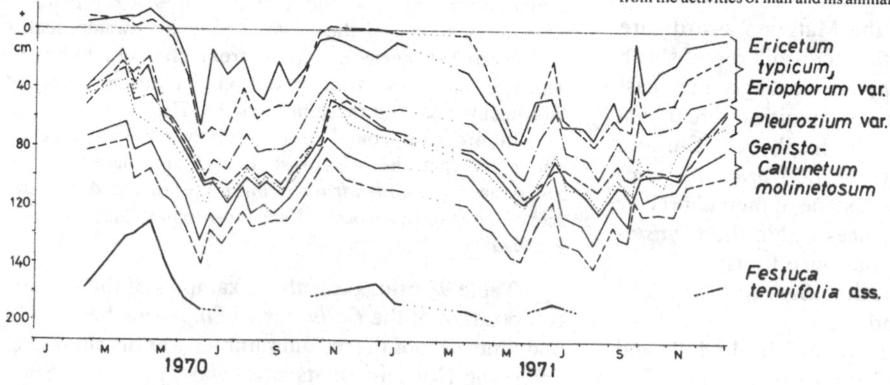

Tab. 99. The amount of available mineral nitrogen on inland dunes and on dwarf shrub heaths in Lower Saxony.

Plant communities and soils	Net mineralisation kg N: ha^{-1}, y^{-1}	Nitrification level
1. Poor grass on inland dunes, poor in lime		
a Silvergrass communities, ± gappy		
pioneer stage on blowing sand (*Corynephoretum typicum*)	14 – 21	IV
stable stage with Lichens (*Corynephoretum cladonietosum*)	13 – 19	I
b. Sheep's Fescue sward, relatively dense and closed *Festuca tenuifolia* sward on stable dry sand	12 – 19	III
2. Dwarf shrub heaths on podsols, influenced by human activity		
sand heath (*Genisto-Callunetum typicum*), relatively dry	5 – 17[a]	I
Crowberry heath (*Genisto-Callunetum empetretosum*), shaded	11 – 19	I
moist sand heath Z(*Genisto-Callunetum molinietosum*), groundwater near the surface, on gley-podsol	21 – 31	I
3. ± natural dwarf shrub heaths on acid peaty soils		
Moss-rich Erica heath (*Ericetum tetralicis, Pleurozium*-var.	21 – 30	III
Cottongrass-Erica heath (*Ericetum tetralicis, Eriophorum* var.) very wet, flooded at times, in depressions	29 –49[b]	I

Note: [a] Very impoverished by centuries of litter removal (turf cutting) and grazing.
[b] No litter ever removed. Occasional additional nutrients received in water from rain and melting snow.

Source: From data by Lache (1974) in Ellenberg (1977) (see tabs. 19 and 25).

ity (*Rhynchosporetum*) which in addition to *Rhynchospora alba* contained other light-loving representatives of the raised bog hollows, e.g. *Lycopodiella* (= *Lepidotis inundata*) and *Zygogonium ericetorum* (the latter is a reddish-green alga covering the soil with a tough skin which tears up when it dries out). Today the repeated succession from *Rhynchosporetum* to *Ericetum* is no longer to be seen although the numerous moss species and other cryptogams which dominated the different stages in the succession are still to be found here and there eking out a miserable existence.

The *Ericetum* humus from livestock stables was used as manure for the poor sandy arable land in the moorland areas. It must have made a significant contribution to the soil fertility because Lache (1974) found rather surprisingly that the soils of the *Erica* heaths produced more mineral nitrogen than those of the *Calluna* heaths (see tab. 99). Additional manuring with nitrogen for almost ten years had little effect on the species composition of a *Molinia-Erica* heath on the Hohes Venn (Sougnez 1965). There was just a change in the relative cover ratio. *molinia* and other grasses were encouraged while the peat mosses almost disappeared. On the other hand Loach (1966) found that there was a noticeable shortage of phosphorus in the soils of English *Erica* heaths similar to that found in raised bogs and oligotrophic waters. That this may be the deciding factor in the inability of trees to grow on such heath swamps cannot be excluded.

The occasional high groundwater table can be ruled out as the limiting factor right from the start as it is no lower in the fen Birch and Alder woods. There are probably factors at work connected with the stagnant

Fig. 400. Stages in the development of the *Erica* swamp heath after the cutting of turves which was the normal practice at one time. After Vanden Berghen (1952).

1 = *Ericetum* before being cut. 2 = *Rhynchosporetum* on the bared earth, 3 = transition stage, 4 = *Ericetum* which has resumed its typical form after about 10 years.

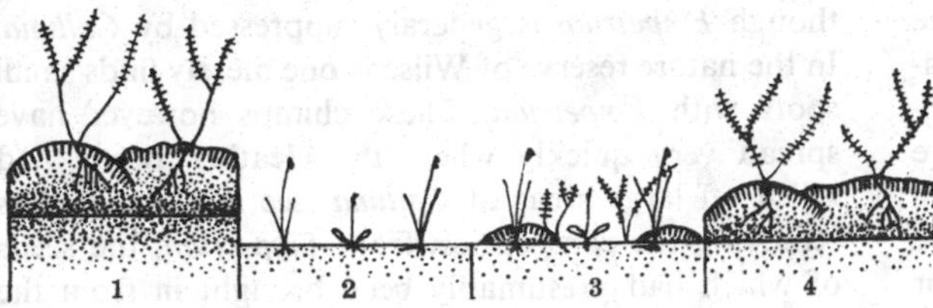

nature of the ground water that has already been mentioned as one of the important conditions. All wet humus soils are poor in oxygen (section B V 2 d) but this also cannot be held responsible for the absence of *Betula pubescens* or other swamp-dwelling trees from the *Ericetum*. However the high concentration of bivalent iron or of carbon dioxide in the soil water may be of significance. Webster (1962) drew attention to the last-named possibility from experiments with Purple Moorgrass (*Molinia caerulea*), found in abundance in Birch swamp woods, but thriving less well in wet *Erica* heaths. This grass grows much better when the soil water is flowing than when it is stagnant, other conditions being the same. The stagnant water contained 0.7 g CO_2 per litre while the flowing water had significantly less at 0.42 g/l. The oxygen content was nil in each case. Another poisonous substance present in higher concentration is hydrogen sulphide; this increases so much in the stagnant water of the *Ericetum* soil that a freshly dug hole will smell of bad eggs. However experimental proof is still needed to show whether higher concentrations of $Fe(OH)_2$, CO_2 and H_2S either separately or together are damaging to trees in the sites of *Ericetum*.

The productivities of *Erica* and *Calluna* heaths are similar if a general conclusion can be reached from the findings of Tyler, Gullstrand, Holmquist and Kjellstrand (1973) from southern Sweden. The yearly production here was 3.0 and 3.1 t/ha respectively and the total biomass 15.4 and 16.3 t/ha. They differed however in the proportion of the biomass above ground to that below, namely 9.2 : 6.2 and 6.3 : 10.0.

Most of the *Erica* heaths which were so extensive in the north-western sand plains area have been drained and are now used as grassland. They can still be studied quite well in the wet valleys of some coastal dune landscapes such as those in the Netherlands (Smidt 1966, Zonneveld 1965), in Lower Saxony (Tüxen 1967b, Lache 1974) and in Schleswig-Holstein (Raabe 1964). Even in southern Sweden Malmer (1965) still found *Ericetum* and *Rhynchosporetum* along with other Central European heathland communities which were unspoilt. The best examples though are to be met with on many of the islands in the North Sea where *Erica* swamp-heaths dovetail in with Crowberry dune-heaths, e.g. on Sylt. The *Erica* heaths soon fade out along the Baltic coast although the Cross-leaved Heath itself pushes as far as north-eastern Poland (Wojterski 1964b). On westerly mountains, especially on the Hohes Venn, the Ericeta are found right up to the montane belt (Schwickerath 1944, Sougnez 1965). Here too some of them have been protected from drainage, but are no longer cut for litter.

(b) Wind-hardy Crowberry heaths on the North Sea coast

Wind-hardy heath communities are to be found on old dunes and in valleys of the North and East Friesian islands as well as in some places on the continental coast, e.g. to the south of Cuxhaven. According to Tüxen (1956b) they belong to the alliance of the Crowberry heaths (*Empetrion boreale*) which is widespread in northern Europe. A subunit, the Creeping Willow-Crowberry heath (*Salici repentis-Empetretum*) which can withstand more drought and grows on the island dunes (fig. 301) is considered to be a natural dwarf-shrub heath since pioneer trees cannot easily find a footing here because of the high winds. However some species from the Birch-Oak wood do manage to establish themselves here and there, especially the Rowan (*Sorbus aucuparia*), and it is always possible that parts of these heaths also have been created and kept free from trees by man and his grazing animals.

The *Empetrum* heaths in northern Central Europe contain almost all the species present in the *Calluna* heaths, together with relicts from the dune grasslands from which they have usually developed. The Crowberry itself is always dominant and it differs from Heather in retaining its short growth even when it gets older. Because of this it is less liable to damage from the strong winds. It is also able to tolerate, and even to fix, a certain amount of blown sand into which it sends out adventitious roots. In this respect however it is not so efficient as the Creeping Willow which also thrives well in other communities in the coastal dunes. *Carex arenaria*, an efficient pioneer on leached blown sand, is also present in many Crowberry heaths.

Alongside the relatively dry Crowberry heath already mentioned there is also one in wetter habitats confined to the dune valleys. In contrast to the dry *Calluna* heaths further inland (section 3 a) the soils of Crowberry heaths are hardly podsolised at all and have no hard pan. The higher humidity near the coast encourages the growth of mosses, especially *Hypnum cupressiforme* var. *ericetorum* and *Pleurozium schreberi*. Here they are more numerous than the lichens which give the *Calluna* heaths their more impoverished appearance.

Crowberry heaths also occur further inland, especially on sand dunes. In a less oceanic climate though *Empetrum* is generaly suppressed by *Calluna*. In the nature reserve of Wilsede one merely finds small spots with *Empetrum*. These clumps however have spread very quickly where the Heather Beetle had killed off large areas of *Calluna* (see section 3c) thus removing the competition from *Empetrum*, the seeds of which had presumably been brought in from the coast by birds (endozoochorous!). In the areas where

Calluna is healthy, *Empetrum* grows, if at all, only on shady slopes or in the half shadow of a woodland's edge, i.e. in a relatively cool microclimate, and here it simply forms a subassociation of the heather community *(Genisto-Callunetum empetretosum)*.

By comparing numerous measurements Lache (1974, fig. 401) was able to demonstrate that the soil temperatures under a Crowberry heath were 1-3°C lower on average than those under the neighbouring *Calluna*-heath. Thus even over small areas the 'northern' character of this community is clearly expressed. Since *Empetrum* plants are generally more open and shorter than those of *Calluna* the temperature maxima in the vicinity of the leaves are about the same for both dwarf shrubs. In winter the Crowberry near the woodland's edge enjoys the protection of a snow cover for longer than the surrounding heather communities. Such narrowly localized lowland habitats are similar in many respects to the subalpine and alpine heaths.

Fig. 401. A comparison of the temperature extremes in adjoining experimental areas of Heather and Crowberry heaths on different days throughout the growing period in the north-west of Central Europe. After Lache (1974), somewhat modified.

Each block corresponds to a day on which measurements were taken and shows the maximum and minimum temperatures for each of the distances above and below ground. The part of the block with a line down the centre represents temperatures common to both heaths. The part left white refers to the *Calluna* heath only, that shown black to the *Empetrum* heath. This makes it clear that the ground under the *Empetrum* heath is generally cooler than that under the *Calluna* heath.

There is a less clear distinction between the two heaths regarding temperatures above ground. For 5 cm and even 150 cm above the ground the white sections frequently extend beyond the part with the central line, showing that in the *Calluna* heath the maximum is higher and the minimum lower than in the *Empetrum* heath. When the sun is lower in the sky in autumn and spring the *Empetrum* stands are often a little cooler because the experimental areas here are slightly inclined to the north and east while those of the *Calluna* are mostly level.

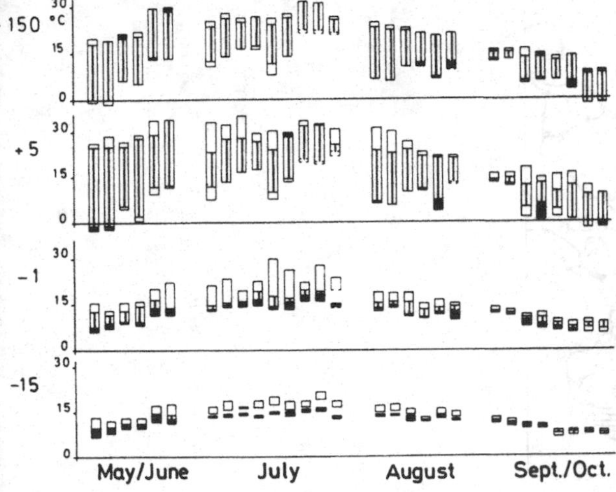

Nevertheless the Crowberry heaths of the Central European mountains have little floristic connection with those of the coastal dune heaths, especially since the dominant plant there is not *Empetrum nigrum* but the closely related *E. hermaphroditum* (see section C VI 4 b).

3 Lowland heaths brought about by farming

(a) Dry sandy Heather heaths and their soils

Wood cutting, burning and the browsing of animals are necessary but not sufficient conditions for the formation and maintenance of *Calluna*-heaths. Where the climate is not so oceanic, or the grouund is not so poor, the same influences lead to the establishment, not of dwarf-shrub heaths but of different grassland communities. Both the *Erica*- and *Calluna*-heaths form a raw humus layer which in former times was periodically removed for litter, leading to increasing impoverishment of the heathland soil. Both are found only in a damp climate which is relatively mild in winter, since both cannot survive long dry periods very well even though *Calluna* is in a position to reduce its transpiration losses.

In 1923 Stocker made careful measurements of the water loss from plants which were placed in pots but kept in their natural surroundings. These clearly proved that 'related to a root system of comparable size the mire and heath summer-green plants transpire on average to the same extent as plants in other habitats...., while *Erica, Empetrum* and even more so *Calluna* show a transpiration rate which is two or three times as great'. Nevertheless *Calluna* in particular can reduce its transpiration very considerably whenever the roots became dry. The differences in the microclimates on sunny and shady slopes of the Wilsede heathland which Lötschert (1962) measured had a marked effect on the water economy of *Calluna*. Its transpiration-regulating mechanism operates so efficiently that Breitsprecher (1935) often registered differences of up to 100% within a short space of time. It also reacts very efficiently to variations in the temperature and atmospheric saturation deficiency which can be very large during the course of a sunny day (Lache 1974). Lemée (1946) also came to similar conclusions working on heathland in Alsace (see tab. 100). In relation to the fresh weight of leafy shoots the transpiration rate in *Calluna* is somewhat less than that of some other plants since its stem is lignified and its leaves have very thick-walled cells. With a good water supply however the daily water loss does not differ significantly from that of other species.

Like most dwarf-shrub heaths anthropo-zoogenous *Calluna* heaths of the north-western plains are rather poor in

Tab. 100. Total transpiration from heath plants during fine summer days from July to September in Alsace.

Calluna vulgaris	sunny, dry soil	0.4 .... 4.1 g/g fresh weight
	sunny, damp soil	2.3...........4.5
	shaded	1.0 - 1.5
Genista pilosa	sunny, dry soil	1.1 .... 7.3
	shaded	1.8..........3.7
Teucrium scorodonia	sunny, dry soil	3.8 ...... 7.3
	shaded	1.8...........4.0
Pinus sylvestris	sunny, dry soil	1.7 .... 3.9
Oxalis acetosella	shaded	0.9—1.3

Source: After Lemée (1946).

their species diversity. In the **typical sand heath** (*Genisto-Callunetum typicum*, figs. 394, 395, 18 and 19) there are few phanerogams apart from *Calluna* and none of them can be considered as good character species. The name '*Calluno-Genistetum*' chosen by Tüxen (1937) leads to errors since the two atlantic *Genista* species *G. anglica* and *G. pilosa* are met with only exceptionally in the typical composition of this community and are in fact confined to the more fertile subassociations (see section b). In the typical sandy heath the only phanerogams which are frequently found accompanying the heather are a few small tussock-forming grasses such as *Festuca ovina, Avenella flexuosa* and *Danthonia decumbens*.

The humus layer under this small-leaved dwarf shrub is normally more or less thickly overgrown with cryptogams. Provided there is no heavy pollution with carbon dioxide the fruticose **lichen communities** (*Cladonion arbusculae*) form loose grey patches where there is more light. The much-branched grey forms such as *Cladina mitis, C. arbuscula* (=

Fig. 402. The potential distribution of different groups of heath communities in Schleswig Holstein. After Raabe (1964), somewhat modified (see also tab. 98).

The hatched areas indicate where the presence of the different heath types is possible, not where they are (or have been) really distributed. Even in Schleswig-Holstein nearly all the dwarf shrub heaths have been afforested or brought under cultivation, and today are in process of disappearing totally.

Acid-soil heath communities are absent from the coastal marshland in the west and from the young moraine landscape in the east (apart from some sandy areas). The heaths on the coastal dunes are rich in *Empetrum nigrum* and dune grasses (see tab. 98). The inland heaths in northern *Schleswig-Holstein* resemble those in Jutland and Scandinavia and also contain a lot of *Empetrum*. Further south they are more like those in the Lüneburg Heath with *Genista anglica* and *pilosa*. Because of the more continental climate in the south-east of Holstein the heaths on southern slopes favour warmth-loving and other more-demanding species.

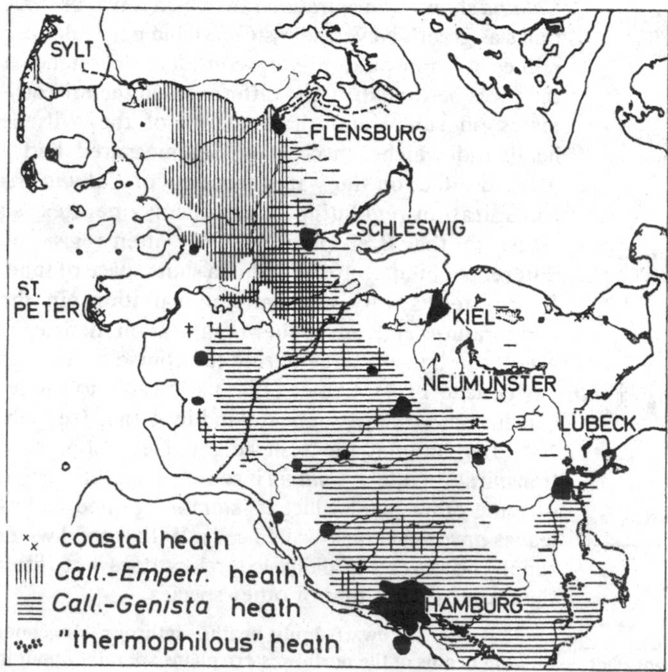

Cladonia sylvatica) and *C. portentosa* (= *Cladonia impexa*) attract attention. Lichens with cup-shaped apothecia are less common, e.g. *Cladonia chlorophaea*; the brown branched *Cornicularia aculeata* is usually found singly and then in small very dry micro-habitats. Mosses play only a small part on the relatively dry sands. The most frequent is *Pleurozium schreberi* which is also found in acid-soil woods. Apart from this mention need only be made of *Dicranum scoparium*, *Hypnum cupressiforme* var. *ericetorum*, *Ceratodon purpureum* and *Polytrichum juniperinum*. The liverwort *Ptilidium ciliare* var. *ericetorum* could serve as a character species, although it is relatively rare. Recently most of the lichens have disappeared, probably because of SO_2 immission.

Heinemann (1956) has compiled a list of the higher fungi found in the Belgian *Calluna* heaths some of which are characteristic of them. This list is extensive, but less so than that of the woodland communities from which the heaths originated. As in the case of the higher plants the variety of fungi is reduced when the woodlands are down-graded and the habitat conditions become more extreme. Pirk and Tüxen (1957) see a few of the higher fungi as possible character species of the *Genisto-Callunetum* or its superior units, e.g. the following which occur fairly regularly:

Polystictus perennis	*Cortinarius mucosus*
Clavaria argillacea	*Rhyzopogon virens*
Boletus variegatus	

There are also a few species characterising the *Quercion robori-petraeae*, i.e. the natural woodland from which the sand heaths are derived, e.g.:

Amanita muscaria	*Boletus scaber*
Lactarius rufus	*Paxillus involutus*
	et al.

As Pirk and Tüxen have stressed, the phytosociological pecularity of the *Calluna* heaths is most marked in those areas where they continue to be grazed. If the animals are withdrawn the communities degenerate and become poorer in character species.

The large variety of *Calluna* heath types (on soils with a low groundwater table) which could still be found a few decades ago is shown in tab.98 for Schleswig -Holstein. The florisitic variations corresponding to the climatic gradient on this peninsula express themselves, amongst other things, in the average values for temperature and continentality indicators (see also fig. 402).

The entry of floral elements from the west, north and east are not the only geographic nuances experienced by the *Genisto-Callunetum* over its relatively large area of distribution. Especially obvious is the north-western boundary of Juniper (*Juniperus communis*) in north-west Germany. This slender coniferous shrub has spread as a pasture weed both in the acid-soil *Calluna* heaths and in the calcareous slightly arid grasslands as well as in the Pine heaths on the more continental plains of Central Europe (see figs. 493, 27a and 184). In the nature reserves around Wilsede and

Fig. 403. Junipers on the Wilseder Berg have now grown to form impressive groups since they are no longer burned off by the shepherds. The light-coloured grasses (*Festuca ovina*, *Agrostis tenuis*) indicate the marginal parts of previous arable areas. In the left foreground is a Spruce which has been heavily browsed by sheep.

north-east of Gifhorn as well as in many heathland reserves in the Netherlands it forms dense stands which in places have grown so high as to resemble tree groups (fig.403). Yet it is completely absent from the Fischbeck heathland reserve near Harburg and other *Calluna* heaths only 20–30 km north of Wilsede. It is still not clear why this otherwise very widespread shrub is almost absent to the north of a line connecting Papenburg, Delmenhorst, Verden and Harburg, yet is again abundant in Holland, Denmark, Great Britain and Sweden. The grazing regime, on which the spread of Juniper probably depends (Gilbert 1980) did not differ on both sides of this line. The ripening and dissemination of seed could not give rise to such a sharp boundary since the Juniper berries are endozoochorous, being dispersed by birds, especially the Black Cock, doves and others. Since the Juniper is killed off whenever the heath is burned this human influence may have played a part. On the other hand there has been no known difference in the methods of moorland management on the two sides of the boundary line. Possibly pests not yet investigated should be held responsible; but the question remains open.

Fig. 404. A podsol profile under a *Calluna* heath free of groundwater, slightly-schematic. After Ellenberg (1937), somewhat modified.

A sequence of horizons under the thin surface layer of raw humus comprises a dark-grey humous bleached sand (Aeh), an ash-grey bleached sand deficient in humus (Ae), a dark coffee-brown humus-illuviated pan (Bh) and a shining rust-coloured iron-humus-pan (Bsh). Black bands of humus are also formed under the influence of the heath vegetation (BS^2). These merge into 'double bands' (Bht), so called because the upper side is darker, the lower a lighter brown. The thin light-ochre bands (Bt) had already been formed under woodland (*Betulo-Quercetum*), and would be present from a depth of 40–60 cm if the heathland podsol had not replaced them. The sand making up the matrix of the B-horizons is a whitish yellow colour. Another indication of previous woodland are the vertical plugs of Bh filled with bleached sand; these have been formed where strong tree tap-roots had loosened the ground before the woodland was degraded into heathland.

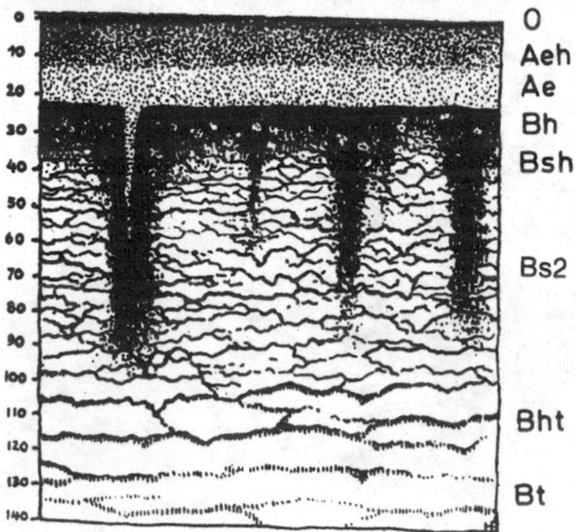

The soils of the typical *Calluna* heath are so characteristic of them and provide so much information about their history that we must consider them a little further here. Tüxen (1957) even spoke of the 'writing in the soil', and this is particularly easy to read in sandy soil.

Under the influence of the raw humus which is produced by the dwarf shrubs and cryptogams and as a consequence of the repeated removal of material by grazing and litter or turf cutting over the course of many centuries a **soil profile** has been formed which is very impressive (fig. 404). This heath podsol is conspicuous mainly because of the sharp differences in colour between the horizons. These start with a thin top layer of blackish-brown raw humus (0), followed by a layer rich in humus but with a good deal of sand (the 'humus-bleached sand horizon', A_eh), and a layer of ash-grey sand much poorer in humus (A_e) the colour of which gave rise to the Russian name 'podsol' (=ash soil). Under this sequence of eluvial horizons begins a still thicker series of illuvial ones in which leached material has been deposited. It can be subdivided into a number of distinct layers: the uppermost and relatively hardest, the coffee coloured 'humus hard pan' B_{sh}). Gradually the ground colour of the sand average depth of the heather roots, i.e. 20 to 40 cm below the soil surface. It can harden, but the formation of a stone-like pan is much less common than was at one time supposed (see section 1a). Here and there one comes across a 'stone pavement' just within the hard pan (although it may be above or below it, see fig. 406). This has nothing to do with the soil formation; it contains wind-polished stones and marks the boundary between the morains of an older diluvial period which are often rather sandy and liable to be driven away by the wind, and the wind-blown sand with which they were covered in early Post-glacial times. Further down the humus-rich hard pan becomes lighter in colour and softer, gradually merging into the rust-coloured 'iron pan' B_sh). Gradually the ground colour of the sand becomes a light ochre with irregular black bands or even seams of humus from 0.5 to 2.0 cm thick (B_{s2}). Below about 1 m the bands take on an ochre brown colour and are clearly harder than the yellowish white sand lying between them (B_t). Such bands are also found from 50-70 cm downwards under woodland which has not been converted into heathland, as in the typical Birch-Oak woods, but in these the A_e–B_{s2} horizons are absent. The latter are thus the result of the influence of the dwarf shrub heath, whereas the light brown bands are an indication of a previous woodland (Tüxen 1930, 1957, see figs. 404 and 133). The break up of the B-horizon into thin layers in sandy soils is due to air pockets holding up the drainage water which is carrying clay or humus colloids (B.Meyer verbal communication). It can be clearly seen in some 'double bands' (the upper part of which is rich in humus and the lower pale-ochre and free from humus) that the dark humus bands have been formed later than the bands of the woodland parabrown earth (B_{ht}).

Figure 405 shows the chemical changes accompanying the podsolisation of a *Calluna* heath soil beyond the influence of ground water. Under the extremely acid humus cover the pH value of the topsoil falls gradually. The bases and finally also the humus along with the phosphorus are deposited in the B-horizon and become, to a large extent, no longer available to the plants.

Heath podsols require at least a few centuries for their formation (Ellenberg 1969) but are often considerably older. For example Zotz (1930, quoted by Ellenberg 1963, see fig.406) found hard pans both above and below a burial place dating from the end of the Stone Age or the beginning of the Bronze Age (3–4000 year ago) near Nienburg on the Weser. Guillet (1968) using pollen analysis dated an iron-humus podsol in the lowland of Lothringia as having been formed under *Calluna* during the Bronze Age. According to Mückenhausen, Scharpenseel and Pietig (1968) the practice of cutting heathland litter for manuring the arable land started in north-west Germany around the tenth century simultaneously with the intensification of rye cultivation.

Fig. 405. The chemical properties of podsols of different ages in sand with a low silicate content under *Calluna* heath free of groundwater. After Kundler (1956), somewhat modified.

The older the soil the less it is saturated with bases (mb in % = mineral base index) but the pH value, humus and phosphoric acid content all increase, and in the final stage they have been largely translocated from the bleached sand and deposited in the lower layer Bsh.

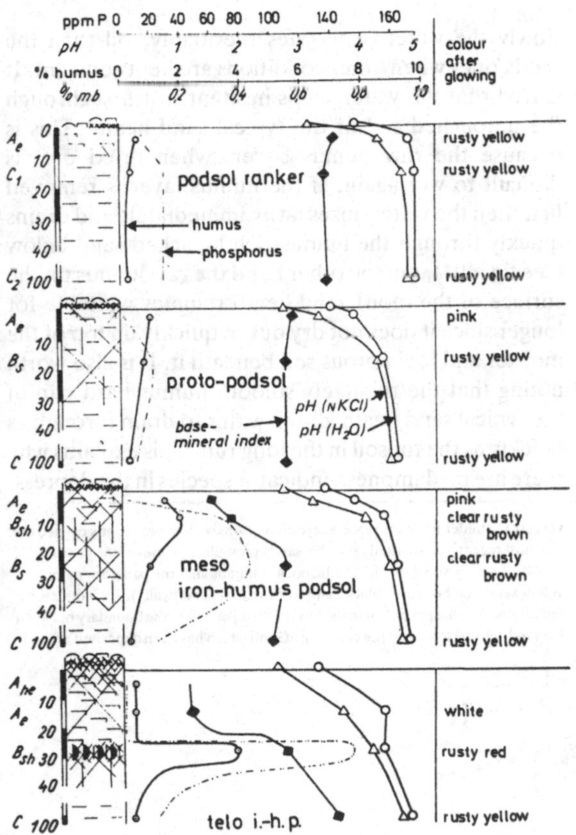

(b) Moist sandy heaths and loamy heaths

There is a continuous sequence of intermediates between the typical or 'dry' sand *Calluna* heath and the typical *Erica* swamp heath. Next to the lichen-rich subassociation of the latter (see section 2 a) stands the **moist sand heath** (*Genisto-Callunetum molinietosum*) which differs from the typical subunit in containing a number of dampness indicators. The most obvious is *Molinia caerulea* whose large tussocks shine out a pale yellow from the dark brown heathers during the cooler half of the year, and *Erica tetralix* whose larger bells begin to open in June whereas *Calluna* does not show its rose-red colour until August. Other representatives of the dwarf-shrub communities on peaty moorland, e.g. *Juncus squarrosus* and *Scirpus cespitosus* ssp. *germanicus* occasionally make it difficult to draw a distinction between the moist sandy *Calluna* heath and the relatively dry *Erica* heath.

All the moisture indicators are helped by the fact that their root run is waterlogged from time to time. In pure sandy soil the water level under the moist sandy *Calluna* heath varies from about 10 to 50 cm in the spring and 40 to 100 cm in late summer. Thus it does not approach the surface as closely as under the *Ericetum*, while the typical *Genisto- Callunetum* is only exceptionally if ever influenced by ground water (Lache 1974).

When one walks through the *Calluna* heathland park from Niederhaverbeck to Wilsede it is surprising to see specimens of *Erica* just on the tops of some hills where it cannot be particularly moist. The reason why *Erica* can thrive here lies in the greater water-holding capacity of the soil. As can be seen from tab.101 the soil in such **loam Heather heaths** (or briefly loam heaths) does not consist of coarse sand as it does in most of the *Calluna* heaths, but of fine sand, silt and clay. This kind of material forms oligotrophic parabrown earths but never a heath podsol. The upper layers may be podsolised to some extent but the lower horizons are seen to be changed in a gley-like manner

Tab. 101. The texture, humus content and pH of sand and loam heath soils (as %age weights).

		Coarse sand >0.2 mm	Fine sand 0.2–0.02	Silt 0.02–0.002	Clay >0.002	Organic content	pH value
Topsoil							
A_h	sand heath[a]	67	26	1.6	2.0	4.2	4.5
A	*clay heath*[b]	–	*68*	*25*	*4.7*	*2.3*	*4.7*
A_e	sand heath[a]	72	22	2.5	1.7	0.1	5.0
-	*clay heath*[b]	–	–	–	–	–	–
Subsoil							
B	*sand heath*[a]	57	38	1.0	**4.0**	0.5	5.0
(B)	*clay heath*[b]	–	*67*	*23*	*7.5*	*2.4*	*4.8*
C	sand heath[a]	73	24	2.5	0.7	0.1	6.0
C	*clay heath*[b]	–	*62*	*24*	*13.5*	*0.3*	*4.4*

Note: [a] Sand heath = '*Calluno-Genistetum typicum*, typical variant'
[b] *Loam heath* = '*Calluno-Danthonietum typicum*' (after Heinemann)
The coarse sand fraction dominates all the layers in the sand heath soils but is absent from the loam heaths which have the highest proportions of fine sand and clay. The amount of clay in the B-horizon of the sand heath is somewhat higher, and this has presumably been washed down from the bleached A_e-horizon.
The organic content of this A_e-horizon too has almost entirely disappeared. All the pH values in this table are rather too high as they were determined using dried soils; but they do show that the sand heath soils are not more acid than those of the *Loam* heath.

Source: Two typical examples from Belgium; from data in Heinemann (1956).

by the water suspended in them. Apart from the bleaching and the greater amount of acid humus in the topsoil similar profiles are found under Oak-Beech woods (section BII4) from which the majority of the loam heaths have probably come at one time.

The water-holding effect of loam-heath soil was demonstrated very well in a drainage investigation set up by Heinemann (1956). He put 10 cm diameter metal cylinders on the surface, poured 100 ml water into them and measured the time it took for the water to drain away. As soon as the water had disappeared into the soil he poured in another 100 ml and went on repeating this process until the time taken for this amount to run away became practically constant. Figure 407 gives the curves which were constructed from his figures and shows clearly how much more slowly the water penetrates the loamy soil than the sandy ones when other conditions are kept constant. It is true that the water seeps hesitantly at first through the untouched soil of the typical sand heath. This is because the raw humus cover, when dried out, is difficult to wet again. If the humus layer is removed first then the water oozes away immediately and drains quickly through the humus-rich bleached sand below (see fig. 404). On the other hand the raw humus on the surface of the moist sand heath remains wettable for longer since it does not dry out so quickly on top of the moister and less-porous soil beneath it. It is also worth noting that the relatively smooth humus hard pan of the typical sand heath allows water to drain through as quickly as the topsoil in the long run. This explains why there are no dampness-indicator species in the depress-

Fig. 406. Section through a burial mound near Gross-Varlingen (Nienburg-Weser district) constructed some 3000 to 4000 years ago. After Zotz (1930), from Ellenberg (1963).

During the construction of the burial chamber in the centre of the shallow mound the older B-horizon pan was disturbed but a new illuvial layer has since built up in the pale sand of the mound parallel to the present surface. This pan has become very strong underneath the place where cremations were subsequently carried out. 1 = most recent wind-drifted sand, 2 = sandy grey-black humus of the heath topsoil, 3 = greyish-white bleached sand, 4 = humus and iron pans, 5 = pale yellow sand, 6 = brownish-black bands (drawn in schematically), 7 = wooden inclusions. A stone pavement in the lower Bsh layer marks the boundary between the sandy glacial loam and the covering of sand which has been deposited above it.

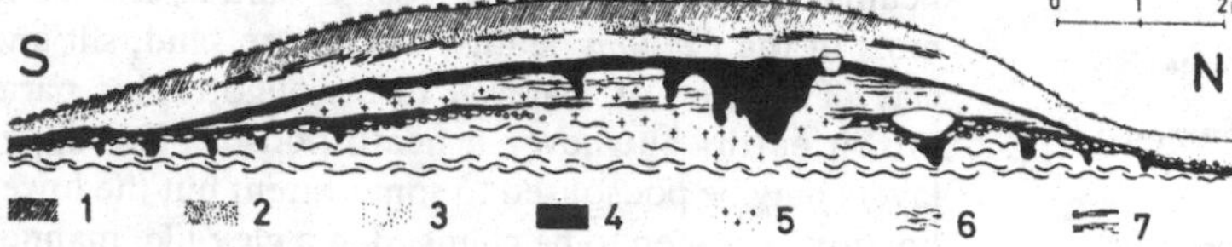

ions of the hilly *Calluna* heathland near Wilsede though the hard pan forms a continuous layer (see Lötschert 1962). On the other hand where the hard pan is really consolidated the seepage water is held as in a trough so that moist soil heaths are formed (Zonneveld 1965).

Loamy soils are generally more fertile than ones consisting mostly of coarse sand. For this reason plants demanding a higher level of fertility are relatively more numerous in the loamy heathlands, especially

Genista anglica	*Polygala serpyllifolia* and *vulgaris*
Potentilla erecta	*Orchis maculata*
Galium saxatile	*Platanthera bifolia*
Arnica montana	*Gentiana pneumonanthe*

Only the five first named occur on the hillocks in the heathland reserve near Wilsede, where only the subsoil is loamy. Where even the top layer contains some clay the stands are richer in species and also include orchids after which Tüxen named this subcommunity (the subassociation of *Orchis maculata*).

At one time loam heaths were much more common in north-west Germany than they are today since the loamy moraines and especially the loess-like cover of fine sand (the so-called 'Flott-sand') were partly dominated by heathland, e.g. in the region of Syke-Neubruchhausen. However since it was only the topsoil which had become impoverished by grazing and turf cutting, it has been quite easy to restore the productivity of this land by liming and fertilising. It is hardly credible that many of the fertile arable plains in the north of Lower Saxony were just as poor a heathland as those on pure sand only about 100 years ago.

Fig. 407. The permeability of three heath soils in Belgium on 6 September 1942. These are a typical sand heath, a moist sand heath and a loam heath (further explanations in the text). From data by Heinemann (1956) in Ellenberg (1963).

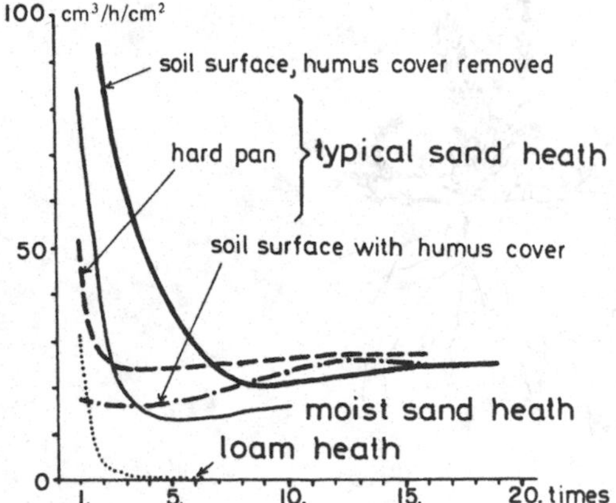

(c) The life cycle of Heather and measures for maintaining its communities

The *Calluna* heaths, once so widespread on the coastal plains adjoining the North Sea, can now only be studied in nature reserves, especially in the 'Naturschutzpark Lüneburger Heide' around Wilsede (Tüxen 1967b), in the Dutch National Park 'Veluwezoom' to the north-east of Arnhem (Westhoff 1958) and on the 'Randbøl Hede' in central Denmark (Böcher 1941). Even here though they are not managed as they were in former centuries. It is true that they are grazed by the frugal breeds of sheep such as the 'Heidschnucken' which are used to living on the dwarf shrubs. But these are no more cut for their turf which used to be skimmed off the surface regularly (see section AII2c and fig. 19). As a consequence the Heather does not regenerate so well and, in spite of the browsing animals, trees appear, especially Pine and Birch, which have to be removed from time to time (Beyer 1968). In the Netherlands as well as in Scotland repeated burning of small areas has proved to be effective if carefully controlled, and has largely replaced grazing (see fig. 408). In Denmark, too, *Calluna* heaths regenerate quickly after burning (K. Hansen 1964); even Matgrass is favoured by burning (King 1960).

Mechanical damage is not only tolerated by the heath shrubs but actually stimulates the regeneration of the whole stand. If their growth is left undisturbed the *Calluna* plants die of old age in any case at about 25-30 years. According to Gimmingham (1960) the oldest plant recorded lived to be 58. Even in Heather stands which have not been burnt off for over 100 years and not been grazed for decades a definite cycle can be seen (Barclay-Estrup 1970). After 2–3 decades the *Calluna* heath breaks down in patches; but it regenerates from seed during a pioneer and building phase until the greatest biomass is achieved in the mature phase, to be followed by another degeneration after a further 20–30 years. Thus the heathland community follows the same pattern as the original forest except that the latter takes several centuries to complete the cycle (see section B IV 2 d and fig. 409).

The primary effect of burning or turf and litter cutting on the heath is to leave the surface like a desert with no obvious sign of plant life (fig. 408, 48). However the following summer, given sufficient rainfall, innumerable tiny seedlings of *Calluna* or *Erica* shoot up in the bare soil. At the same time, or even somewhat earlier the rootstocks of some old Heather plants, which retained their viability, also start to send up new shoots as do some of the grass clumps so that the following year the whole area is green again. No other parts of the heath flower so evenly and abundantly as these which 2–5 years previously had been 'ruined' by cutting or burning. If Heather is allowed to get too old it takes on a very dishevelled appearance and finally dies back. Also it becomes more susceptible to the attacks of Heather Beetle (*Lochmaea)* which is rarely fatal to strongly growing *Calluna* plants. Its larvae feed on the tips and edges of young Heather leaves which then lose more water and dry out easily during a drought (Blanckwaardt 1968).

In the heathland reserves *Calluna* heath still covers hill and valley like a fur with occasional 'mangy' patches where Heather has become very old and has partly died. Here the dead twigs are (or were before the beginning of heavy SO_2 emissions) thickly covered by crustose and foliose lichens especially the grey *Hypogymnia physodes*, while the fruticose and cup species named in section 3 cover (or covered) the black humus in large numbers. However it would be quite wrong to blame these humble cryptogams for the death of Heather. They are only taking advantage of the additional light which they enjoy here for a time without competition from mosses. Practically all the lichens are (or were) also to be found in dense Heather which is growing vigorously, but here they are not so obvious. Pieces of the thallus which easily break off when dry are dispersed by the wind so that freshly cut or burnt heath areas are quickly reinvaded within a very few years. As a rule mosses take rather longer to become reestablished in spite of their dust-like spores being blown far and wide. Their protonema require a higher and constant humidity for their development which is only provided in the shade of larger Heather plants. The mosses develop best in heathland with a relatively moist atmosphere such as on north- or east-facing slopes, or where the soil is sometimes waterlogged. In the *Genisto-Callunetum*, *Leucobryum glaucum* can only be found in such places (see Lötschert and Horst 1962).

Fig. 408. Regrowth on a square metre of moist sand heath (*Genisto-Callunetum molinietosum*) following a superficial fire in 1944 and a more intense one in 1948. After Heinemann (1956), somewhat modified.

In 1948 the *Calluna* had to rely on seedlings for regeneration whereas in 1944 the plants had not been killed and were able to send out fresh shoots. Intense fires favour the Purple Moorgrass (*Molinia*) indirectly since the bases of its shoots where nutrients are stored withstand the fire and quickly shoot again.

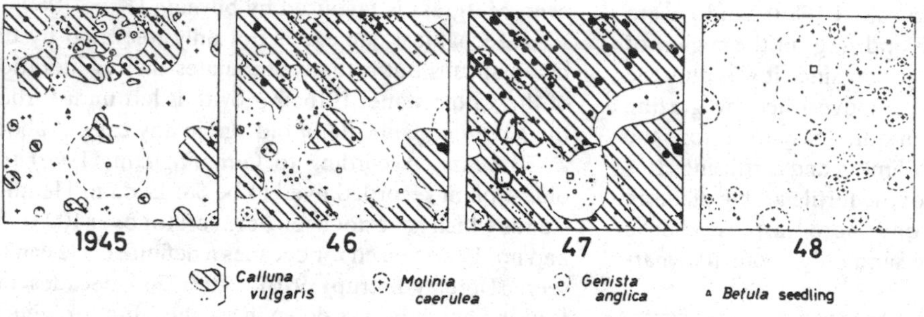

Fig. 409. The development of *Calluna vulgaris* on a porphyry brown earth near Halle. After G. Mahn (1957).

The roots of the seedlings and one-year-old plants penetrate only a few cm. Even in the following years the main root is still clearly recognisable. The older plants have been drawn on a reduced scale, especially the one which is flowering.

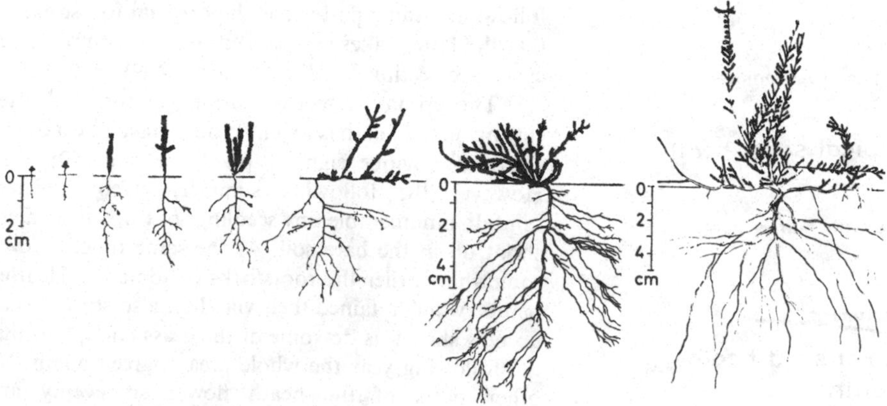

We must return once more to the problems presented by the **rejuvenation of Heather** for it is on this that the whole heathland ecosystem depends including the lichens, mosses, fungi and bacteria, the other higher plants, many small animals and above all man with his grazing animals and his pollinating bees.

A Heather stand which is old and gone into decay is of little value to the beekeeper whereas an actively growing one yields a rich harvest of honey two or three years after burning or cutting when the heather is in full bloom. Sheep browsing causes the Heather to keep on sending out new shoots which flower well. In addition the moorland sheep wandering to and fro over the heath tend to break up the spider webs that otherwise catch a large number of the bees. Stretches of heath without grazing sheep are often densely covered with such webs. For their part the bees ensure a heavy crop of seed for the Heather, one of the most important prerequisites for rejuvenation and maintainance of the heath community and its characteristic landscape. In these sandy areas of north-western Central Europe the keeping of bees and sheep and the arable cultivation relying on heather turfs have been so densely interwoven that the agricultural revolution since the beginning of our century has destroyed this old rural economy and with it the most important condition for the existence of *Calluna* heaths.

According to Beijerinck (1940) *Calluna* can produce up to 800 000 seeds per m^2 These seeds are so light that they may be dispersed by the wind over considerable distances. A storm wind can even tear off the whole capsule with its dried up perianth and in winter drive it across the snow surface. However, although the seeds germinate quickly and in large numbers hardly a single one will develop into a plant on the surface humus. During a long-term experiment Heinemann (1956) counted the number of *Calluna* seeds which had germinated on a 2500 cm^2 quadrat in Belgium and found that on 18.4.42 there were 175, on 13.7.42 a further 125, on 21.1.43 200, on 7.2.43 225 and on 7.5.44 a further 125, yet he did not discover a single larger young plant. Thus the raw humus layer does not restrict germination, but it certainly does restrict further development of the seedlings. Clear evidence of the advantage gained by Heather seedlings from the exposure of bare mineral soil can be seen where occasional traffic has broken through the continuous humus cover in withered *Calluna* stands; soon the wheel tracks are marked by rows of fresh green young plants.

The reason why Heather seedlings find it so much harder to grow on the raw humus than on the sand beneath obviously does not lie in a chemical difference between the two seedbeds. For firstly the uppermost sand is also rich in extremely acid humus, and secondly *Calluna* regenerates from seed very well on raw humus in wet years. When seeds are sown under laboratory conditions in culture pans with raw humus and with humus-rich mineral soil, and both are always kept well moistened then there is no significant difference in the establishment and growth of the young seedlings. Auer (1947) got similar results when he investigated the germination and subsequent growth of Larch seedlings on both Larch humus and mineral soil. Thus it is first and foremost the water content of the surface layer which determines the success following germination. Even in a suboceanic climate the raw humus dries out relatively quickly and often. Every one accustomed to wandering over the sandy heathlands knows that even just a few hours after a heavy rain it is possible to sit down on the humus-covered ground without getting damp provided there are no water-retaining mosses there. However underneath the humus layer the tightly packed sand is poorer in unfavourable colloids and therefore able to make the water more easily available to plants. The seed of *Calluna* is very small with little food reserve so that the first root grows slowly and is in danger of drying out in the raw humus (fig. 409). In the more constantly humid sand however it can develop indisturbed. This then appears to be the real reason why burning off the humus or mechanically removing it is favourable for the successful regeneration of *Calluna* heath. Burning also improves the germination of the heather seed directly as was demonstrated by Whittaker and Gimmingham (1962). They exposed seeds to temperature of 40–80°C for one minute, values which are rarely exceeded at the soil surface by a rapidly moving fire, and found that germination was improved compared with seed maintained at normal temperatures. Temperatures of 80-120°C for half a minute produced a similar result and it took a temperature of 200°C before the seeds were killed. *Erica tetralix* and *Empetrum* seeds react in a similar manner and are found to germinate better when exposed to raised temperatures (K. Hansen 1964). Matgrass (*Nardus stricta*) rejuvenates well after a fire but is quickly suppressed by *Calluna* if the two are germinating together (King 1960).

Gimingham (1960) found that older Heather plants were sensitive to drying out but according to Beijerinck (1940) they can withstand drought much better than they can a sharp spring frost. Very dry years can kill off the Heather over large areas as happened in north-west Germany in 1959. In such years germination is also completely suppressed.

Tab. 102. The loss of nutrients from dwarf shrub heaths when subjected to normal and to more intense fires.

Elements	Sand heath loss through smoke (%) during a. normal,	b. intensive burning	Peat heath c. total loss (in kg/ha) during normal burn	Causes
C	60.5	67.5	(much)	(smoke)
N	67.8	76.1	45	smoke
P	0.6	3.5	0.1	leaching
S	50.2	56.5	5	smoke
K	1.4	4.9	1	smoke and leaching
Ca	0.1	2.4	<0.1	leaching
Mg	0.4	2.1	<0.2	leaching

Note: a. and b. in %age of the amounts contained in the dry weight before the fire; a. after a normal fire (550–650°C), b. after a more intense fire (800–825°C) on a sand heath; c. in kg/ha after a normal fire on a peat heath.

Source: From data in Gimingham (1972) and in Muhle (1974, compare also Muhle and Röhrig 1979).

Tab. 103. Enrichment of the soil in a Calluna heath by sea birds.

	Nutrients (in mg/10g soil dry weight) N (as NH_3)	P (as P_2O_5)	Ca	pH-value
I Normal heath	0.43 – 0.88	2.27 – 3.03	139.8 – 246.1	3.9 – 5.5
II Overmanured heath	13.94 – 15.04	6.15 – 7.66	255.2 – 355.7	3.8 – 5.9

Note: I Normal heath growing on the Fähr island near Hiddensee

II Heath killed by Gulls' nesting sites. After Gessner (1932); two analyses in each case.

Fig. 410. A cross-section through a *Calluna vulgaris* community in the dry region near Halle. After G. Mahn (1957).
From left to right: *Festuca ovina, Polygala vulgaris, Polytrichum formosum, Calluna vulgaris. Dianthus carthusianorum, Genista pilosa, Danthonia decumbens, Helianthemum nummularium,* f.o., *Hieracium umbellatum, Luzula campestris.*

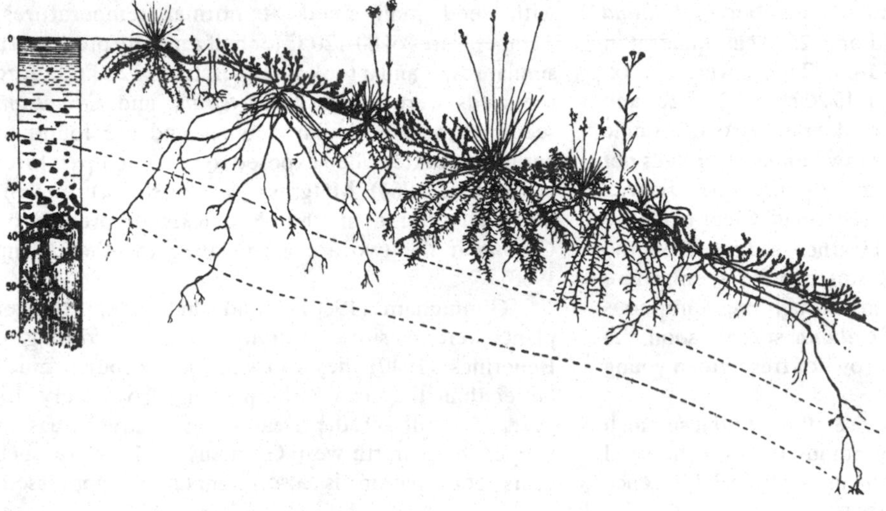

Nevertheless there are semi-arid grasslands even in the dry area east of the Harz which have become impoverished and are now dominated by *Calluna* (fig.410). These are related floristically to the mountain heaths rich in Matgrass which we shall be considering in the next section.

In addition to the water supply the amount of nutrients available plays an important part in the development of *Calluna vulgaris*. In nature it is largely dependent on mycorrhiza fungi (see Burgeff 1961) the importance of which lies in their ability to make nutrients available for the Heather. Heathland soils are always very poor in mineral nitrogen (Lache 1974). Kriebitzsch (1978) includes them with a few other very acid soils as ones where only ammonium and not nitrate is available to plants. Indirectly the lack of nitrogen is a deciding habitat quality for the members of the *Calluna* heath community since it keeps out possible competitors requiring more fertile conditions. According to Miller (1979) the amount of nutrients (N,K.P,Ca, Mg etc.) in whole stands of *Calluna* increases with age until it levels out at 15–20 years. Just at this age they were cut for making litter and manure in former times (see section A II 2 c). The cutting of turfs was undoubtedly an important factor in the impoverishment of the soil, but so too was burning to a certain extent According to O. Muhle (1974) and Groves (1981) most of the nutrients are removed in the smoke, not by rain washing out the bare soil (see tab. 102). Critical experiements carried out by Allen (1964) proved that the latter accounts for much less loss than was previously supposed.

An excess of nitrogen, at least in the form of urea, can be dangerous for older Heather plants. Gessner (1932) noticed this on Fähr island (a small low island lying to the east of Hiddensee near Rügen) where each year 700–800 pairs of Black-headed Gulls and 100–120 pairs of Common Gulls use to nest in tall, vigorous Heather stands. A continuous rain of bird drippings over many weeks completely kills off the Heather over an area of about 100–150 m by 50 m. The following year the birds seek out another area of tall Heather which in its turn is killed, and so on. The effect on the heathland is similar to that of turf cutting in that a few years later the dwarf shrubs regenerate and after some 10–20 years the heather again provides a good shelter for the nesting birds so that the cycle can start all over again. Bird droppings are particularly rich in N, P and Ca, and the heathland soil is considerably enriched for a time (tab.103). The increase in calcium however is so slight that it has scarcely any effect on the soil pH and it is unlikely that this is the reason for the Heather dying back. Black-headed Gulls also nest in the Zwillbrocker Venn (Westphalia) in heathland but in this case in a *Molinia-Erica* community on damp soil derived from a drained raised bog. In thirty years their numbers have grown to 10 000 or 12 000 which has resulted in a guano- trophication, completely killing off the heath vegetation and partly replacing it with hummocks of Soft Rush (*Juncus effusus*, see Burrichter 1968).

One question which has not yet been finally answered is whether *Calluna* really requires acid soil for its germination and development. According to Lötschert and Horst (1962) the pH of the soils in *Calluna* heath communities lies around 4 but varies considerably during the course of the year. These soils are only weakly buffered and rain washing sulphur dioxide out of the air sometimes increases the acidity. *Calluna* can obviously tolerate this extreme acidity. Nevertheless Jones(1911, see also Rayner 1913) was able to cultivate it in pure chalk, i.e. in a neutral medium, but one poor in nutrients. Rooted cuttings of *Calluna* cultivated by Marrs and Bannister (1978) grew best on more or less acid soil, but could live also on calcareous ones. Even in nature *Calluna* occurs, along with other 'lime avoiding' plants on limestone, and here we are not dealing with specially adapted ecotypes (Grime 1963). Older records by Eberle that *Calluna* and Bracken (*Pteridium aquilinum*), which is also held to be acidophilous, occur in soils with pH values ranging from 3.5 to 8.5 on relatively dry slopes above the lower reaches of the Trave river north of Lübeck, have been confirmed by Raabe (1960). Thus an acid soil is not necessary for the growth of *Calluna* and is certainly of only secondary importance. Böcher (1941) made some observations on an arable field which had been marled and later abandoned. He noticed that Heather quickly came in and grew particularly well even though the soil was just about neutral at first (pH 6.8) and there was no raw humus. In itself then Heather is nothing like so 'lime-hating' and 'raw humus-loving' as is usually accepted. However on soils which remain rich in nutrients it is soon suppressed by more-demanding plants able to overgrow it and to shade it out. Grime (1963) considers this to be the main reason why Heather is confined to poor acid soils.

A quick invasion of fallow land by *Calluna* is only possible where the sheep or deer do not browse too much. Otherwise it is very slow to get established in competition with grasses and herbs which are not affected so much by grazing and therefore the abandoned fields in sand heath areas are normally covered with a **Sheep's Fescue-Thyme sward** (*Festuca ovina – Thymus angustifolius* – association). In the

Randbøål Hede which has been investigated by Böcher there are a number of such grassed down arable stretches which have been fallow since 1870. Between the 'Wilseder Berg' and the village of Wilsede almost all stages in this reoccupation process can be followed because parcels of arable land have been purchased by the Nature Reserve Association successively between 1910 and 1957 'in order to return them to the heath' (fig. 403). The venture did not entirely come up to expectations because it was just these areas which the moorland sheep found most attractive, and they tended to concentrate on them, hampering the establishment of Heather. On the other hand it was relatively easy to restore the heath community from seed on freshly rotavated poor sandy ground if sheep were excluded (Tromp 1968). This method also suggests itself where the heath surface has been damaged through troop manoevres or otherwise.

In order to maintain and encourage *Calluna* heath the young plants must always be allowed sufficient light. According to the laboratory experiments carried out by Grace and Woolhouse (1970) **light is the most important factor** in the productivity of *Calluna*. Then follow the current and the previous temperatures and only after these come the age of the shoots and the presence of flowers, i.e. internal factors which set in motion the *Calluna* cycle already discussed. Other factors act only indirectly by discriminating for or against any competitors via-a-viz the Heather. Miles (1974) showed by seeding experiments that many more species can germinate and develop in heaths than are usually found there. Thus the lack of floral diversity in *Calluna* heaths results from competition, and if the dwarf shrubs are weakened in any way then the species composition changes rapidly. This can be seen particularly well in montane heaths which form a varying mosaic in both space and time with acidic grassland communities.

4 Heaths and Matgrass swards in the mountains

(a) Shrubby mountain heaths and Matgrass swards

Dwarf shrub heaths have never dominated the landscape in the mountains of Central Europe to such an extent as they once did in the sandy lowland plains or as they still do in parts of north-west Europe, e.g. in the Scottish Highlands. There are nevertheless even today, heath communities comparable with the lowland *Calluna* heaths in many of the Central European mountains. Here the dwarf shrub stands generally dovetail in with Matgrass swards (see section 1 b). They have rarely if ever been cut for turfs, but were often mown or cut with a sickle to provide bedding litter for the animals. Their maintainance and gradual extension was due to grazing by sheep, cattle and goats while the herdsmen continually strove to increase the area of heathland for grazing by means of axe and fire.

The larger stretches of *Calluna* heath were only to be found on sandstones, slates or other parent material similar to the old diluvial quartz sand in being deficient in bases and nutrients from the outset. For example there used to be large areas in the Harz foothills and in the Eifel (see section a) which are now mostly planted

Fig. 411. A montane *Calluna* heath on a deep acid loam cover ('firestone loam') in the Swabian Jura near Rötenbach. In the background are Birch and Rowan trees alongside the road, and behind that a Spruce plantation.

Fig. 412. Details of a Matgrass heath in the uplands of Bohemia-Moravia. Photo, Iltis-Schutz.

The clump of *Nardus stricta* on the right has been pulled up to show the short tillers with the hard leaves sticking out of them at an angle. *Vaccinium vitis-idaea* in the foreground. One fifth natural size.

up with Spruce. *Calluna* and *Nardus* heaths can also be formed on limestone mountains where the rocks are covered with acid loam. The remains of such heaths are still maintained quite well on the higher parts of the Swabian Jura, e.g. near Böhmenkirch (fig.411). One could imagine one had been transported to the Lüneburg Heath so similar is the landscape until one notices that, instead of Pines, Spruce have been planted around the heathland reserve.

In the mountains, as in the plain, the requirements of *Calluna* heath are met by both the poor acid soil and a cool humid climate. So the montane site favours not only the formation of raised bogs but also of *Calluna*-rich dwarf-shrub heathland right away from the coastal region which must be looked upon as the centre of their distribution.

If grazing is more intensive the mountain heath is changed from its typical form into a subcommunity with more Matgrass (fig.412). Such intermediate stands are more common in the Central European highlands than those with more of the dwarf shrubs. On the plains, where *Calluna* heaths rich in Matgrass are also to be found, the opposite is the case. Here it is only where sheep are browsing and treading very heavily, along footpaths or at the resting places, that Heather gives way to Matgrass, which not only withstands the treading better but is not eaten down so readily by the animals. Results obtained by Kruijne (1965) showed that the success of *Calluna* or *Nardus* does not depend on differences in the soil quality. Regarding water supply too *Nardus* has a similar amplitude to *Calluna*; in response to this factor the anatomy of the leaves and roots of *Nardus* can vary considerably (Smarda et al. 1963). On most of the acid-soil mountains pure grass pastures were more frequent than those with varying amounts of dwarf shrubs. Up to 200 years ago Matgrass swards of the alliance *Violo-Nardion* (see section 1b) covered practically all the higher parts of the Harz, from which the trees had been removed to supply the ore mines. They also occupied large areas in the Solling and Sauerland, on the Hohes Venn and other relatively low mountains away from the Alps. The higher mountains further to the south carried grassland the species combination of which was more like that in the Alps (*Eu-Nardion* alliance, see section C IV 3 c). These remarks are necessarily based on restricted observations since practically all the Mat-grass heaths have either been planted up with conifers or changed into fertilised grassland. In more recent times the practice of mountain grazing on unmanured pastures has been revived to some extent in order to maintain open spaces for skiing and other leisure-time activities.

In the high mountains *Calluna* is found only as far up as the lower alpine belt and rarely occurs even in the subalpine although the soil and moisture conditions may still be quite suitable. It does not go so high in the cloudy and rainy outer chain as it does in the central Alps with their decidedly continental climate. All these observations point to the need *Calluna* has for a relatively warm and fairly long growing period to reach its full development. According to Grace and Woolhouse (1973) its buds first break in the spring when the average of the minimum and maximum temperature reaches 7.2°C and the leaves stop growing when it falls beneath 7.4°C. Söyrinki (1954) also noticed that Heather does not ripen seed if there is insufficient warmth. Thus stands rich in *Calluna* can form only on slopes which are clear of snow early in the spring; places where the snow persists are taken over by *Vaccinium* or *Rhododendron* species (fig. 335).

This tendency is already being indicated on the Wilseder Berg where *Vaccinium myrtillus* dominates on steep cool north-facing slopes. The further north the heathland lies in Europe the more its microclimate favours *Vaccinium myrtillus* as well as *Vaccinium uliginosum* and *Empetrum nigrum* (see also fig.401). Damman (1957) points out that in southern Sweden *Calluna* stands free from *Vaccinium* are very rare. Thus dwarf shrub heaths dominated by *Calluna* are limited in their total distribution by a temperate oceanic to weakly continental climate between the planar and montane belts.

The same is true of the **Broom heath** (*Cytision scoparii*) which occurs in the Schwarzwald and other western mountains, and has been described, for example by Oberdorfer (1957). Further to the north-west it is occasionally found on the plains, but little is known about it ecologically. Its total distribution corresponds approximately to that of the Birch-Oak woods (see section B III 5), even regarding their southern boundary. According to Hofer (1967) the *Cytisus scoparius – Calluna* heaths of the Tessin valleys, in common with almost all other acid-humus communities there, are rich in *Molinia arundinacea*. The Beardgrass-Heather heath (*Gryllo-Callunetum*) also belongs to the alliance of the Broom heaths (Antonietti 1970).

(b) Alpenrose heaths in the subalpine belt

On extensive acid pastures in the upper montane and subalpine belts the *Calluna* heaths are replaced by communities rich in *Rhododendron ferrugineum* or *Vaccinium uliginosum*. These, in spite of many parallels with the lowland dwarf-shrub heaths, also show a number of important differences (see section C VI 1 and D II 1 b).

The Alpenrose heaths (*Rhododendro-Vaccinietum*) have their natural distribution centre in the uppermost woodland region of the Alps, although they do also occur above the treeline (see section C VI 4 c and E.Pignatti 1970b). Between the closed woodland proper and isolated trees they form the undergrowth between groups of Larch, Arolla Pine or Spruce which are struggling for their existence. If these trees were removed it would have hardly any effect on the species composition of the dwarf-shrub layer. This, together with its dependence on sufficient snow cover, has already been dealt with in section B IV 4.

The Alpenrose species are worthless as animal fodder, in contrast to those of *Calluna* and *Vaccinium*. They are avoided even by sheep and goats and will spread at the cost of the more edible plants unless they are deliberately checked. Therefore the herdsmen used to chop them down from time to time for fuel which was becoming increasingly scarce within easy reach of their huts.

The repeated cutting of old branches of the Alpenroses improved their regeneration in the same way that cutting of turfs improved the *Calluna* heaths of the north-western plains. Young trees also did not escape the axe, so regeneration of the woodland was also prevented. Like *Calluna* and *Erica* the *Rhododendron* species are able to send out fresh shoots from the rootstock, and also similarly, removal of the surface humus benefitted the germination of its seed. In spite of thorough searching Söyrinki (1954) could find neither seedling nor young plant in undisturbed stands of Alpenrose, whereas they appeared in numbers wherever the humus layer had been damaged by erosion, fire or in other ways.

Heavy cutting followed by grazing damages *Rhododendron* heaths in the same way that it does *Calluna* heaths, even to the point where they give way to a community rich in grasses. Very often this is a Matgrass sward (alliance *Eu-Nardion*). The Matgrass heaths here though are much richer in species than those on the lower mountains or plains, and they merge into the natural Crooked Sedge swards of the alpine belt (section C VI 3).

Since the *Rhododendron* and *Vaccinium* species cannot withstand treading so well as *Calluna* or *Erica tetralix* they may finally be driven into corners near boulders, between stones and other places where the animals find it difficult to go. Thus the practice of grazing animals which in previous centuries had enabled the Alpenrose heaths to extend far beyond their natural area of distribution has recently reduced the beautiful stands of Alpenrose in those places where the grazing had been intensified.

The Rust-red Alpenrose community which can tolerate very acid soils is replaced on calcareous rocks by the Hairy Alpenrose heath. This occurs mainly under the influence of extensive wood cutting and grazing, but also naturally. Like *Rhododendron ferrugineum* the Hairy Alpenrose (*R. hirsutum*) is found near the tree line and in avalanche tracks as natural permanent communities free from trees. It also has a similar relationship with open stands of conifers (see section B IV 4 a). Where the two species meet, their hybrid *Rhododendron intermedium* is common. The pH requirement of this plant is also intermediate between that of the two parents as Knoll (1929/30, see Ellenberg 1963) and other authors have established.

III Forestry plantations and woodland clearings

1 'Planted-forest communities' compared with natural woodland communities

(a) The classification of planted-forest communities

More or less artificial coniferous forests cover much greater areas in Central Europe than do the near natural broadleaved woodlands. Forest 'monocultures' dominate the landscape in the northern diluvial lowlands, on the Harz and most of the other mountains, on the Upper Rhine plain, in the Nürnberg area, in the whole of the northern pre-Alps and in the south-west of the Vienna basin. However these are only pure stands as regards the trees. Their undergrowth of bushes and dwarf shrubs, herbaceous plants,

Fig. 413. A *Calluna*-Pine plantation on dune sands of the Upper Rhine plain south-west of Heidelberg. The broadleaved woody undergrowth is an indication that the natural vegetation here would have been a mixed Oak wood. Broom and Heather (*Cytisus scoparius* and *Calluna*) are plentiful in lighter places. To the right the clear-felled area has been ploughed over; in the background a young plantation of *Pinus sylvestris*. This tree does not regenerate naturally here.

mosses and fungi can be very rich in species, and the animal life too is not always less diverse than the more naturally managed stands (see section e and fig. 413).

The species composition of such conifer plantations has not yet become stable. It contains relicts of the heaths and grasslands or of the weed flora of old fields which have been planted up with trees. They are mixed with plants whose seeds have been brought from neighbouring communities by wind or animals. Some of the natural components of coniferous woodlands have managed to come in very quickly even from quite a distance, especially mosses, club-mosses and ferns. *Diphasium complanatum* for example had already begun to spread through a Pine stand in Finland only 15 years after it had been planted on former arable land (Oinonen 1967). Northern or mountain vascular plants may behave in a similar way, e.g. *Linnaea borealis* which has now spread into some of the Pine forests in the Netherlands and north-western Germany after its fruits had been brought over by birds.

Where conifers have been planted in a zone where the potential natural vegetation is broadleaved they rarely regenerate spontaneously. It is much more common to see broadleaved trees and shrubs growing happily under the coniferous canopy, especially where the ground is relatively moist and fertile. This undergrowth indicates the direction the succession would take if there was no interference from the forester. However it would take more than one generation of trees before a sufficient number of character species had become established to enable one to assign the vegetation to a particular woodland community.

In spite of their unstable composition and lack of character species these artificial forests can be recorded and classified by the plant sociologist. Køie (1938) has already distinguished a number of 'sociations' (i.e. types of ground vegetation) in Danish coniferous plantations and shown that these were not just chance collection but clearly reflected the soil condition such as pH. Hauff, Schlenker and Krauss (1950, see also the 'Oberschwäbische Fichtenreviere' for 1964) have described similar '**ground vegetation types**' and used them to map the Spruce plantations of the Swabian pre-Alps. The designation 'type' reminds one of the 'woodland types' established by Cajander on the basis of the dominant ground plants for the purposes of forest organisation in the natural coniferous woodlands of Finland.

At about the same time and quite independently Tüxen (1950c, see also Ellenberg 1969) coined the term 'Forstgesellschaft' (ground flora community in a forest plantation) which may be translated as '**planted-forest community**' (or, when combined with the name of a doubtless planted tree briefly: 'forest community'). This always refers to the undergrowth of afforested pure stands of Pine, Spruce and Fir or other trees 'foreign to the potential natural community', i.e. of species that at most would play a very small part in the natural woodland of the area. In this sense for example a plantation of Oaks on a Matgrass heath in the montane Beech area of Solling would be as much a 'planted-forest community' as a stand of Spruce planted under similar conditions. In contrast to heathland, poor grassland, meadows and other replacement communities the planted-forest communities have no character species but are recognised by the combination of several differential species, whether dominating or not. Nevertheless they indicate the peculiar properties of the habitat scarcely less accurately than the more natural woodland communities. This was confirmed among others in the D.R. Germany by Scamoni (1967a) and Hofmann (1968), in the Netherlands by Zonneveld (1966) and in western Germany by Ellenberg (1969, see fig.414). Most planted-forest communities have their parallels in certain grassland or arable-weed communities (Rodi 1968), but also in woodland communities towards which they tend to develop (see tab. 104).

Since many phytosociologists today, at least in northern Central Europe work on the lines introduced by Tüxen, and a large number of forest communities been described and listed by them we shall use their classification in the following sections. Nevertheless it must be expressly pointed out that the demarcation of planted-forest communities and semi-natural woodland communities in the southern and eastern parts of Central Europe causes some difficulties because too little is known for certain about the ratio of the different tree species in the natural woodlands. The use of this method is only really without problems in the case of Central European conifer species, but planted in an area where broadleaved woodland surely is the climax, or when exotic species such as Douglas Fir, Sitka Spruce, Red Oak or Robinia are involved. In Switzerland where clear-felling is prohibited and afforestations are mostly carried out with indigenous tree species one does not need the notion 'Forstgesellschaft' (see Ellenberg 1967b).

(b) Planted Pine forests on sandy and calcareous soils

How many-sided the species combinations even in the most monotonous coniferous monoculture may be, and how clearly they relate to the newly formed living conditions, has been well demonstrated in the Pine plantations of north-western Germany by Meisel-Jahn

(1955b, see tab. 104, also figs. 414 and 415) According to Rheinheimer (1957) the moss species are fair indicators of the particular site qualities. Most communities belong to the large group of the **'moss-Pine planted-forests'** where

Hypnum cupressiforme *Dicranum scoparium*
Pleurozium schreberi

i.e. acid-loving species widely distributed in heathland and woodland play a part. The two last-named begin to disappear recently from those forests which are exposed to heavy air pollution, e.g. north-east of the Ruhr industrial region (Haustein 1982).

Within the moss-Pine forests one can distinguish several subgroups. The *Cladonia*-Pine planted-forest occurs on the poorest sandy soils; floristically and physiognomically it is reminiscent of the lichen-Pine woods of the east (see section B IV 6 c). The *Danthonia*-Pine planted-forest corresponds to the loamy *Calluna* heaths with:

Danthonia decumbens *Agrostis tenuis*
Nardus stricta *Potentilla erecta*
Galium harcynicum *Trientalis europaea*
Dryopteris carthusiana *Lycopodium clavatum*

Between the *Cladonia* and the *Danthonia* forests comes the *Dicranum*-Pine forest (*sensu strictu*) which lacks these, relatively demanding differential species but also the majority of the lichens. The *Empetrum*-Pine forest is found in dune valleys near the coast (fig.415). As in the case of the Birch-Oak woods (section B III 5 a) the damper habitats can be recognised by the presence of

Molinia caerulea *Betula pubescens*
Erica tetralix

For example there is a *Molinia-Dicranum* and a *Molinia-Danthonia*-Pine forest. On even wetter acid humus soils the *Sphagnum* species of the Birch swamp wood (section B V 2 c) also develop in the Pine planted forests.

All the 'richer' habitats in the Pine plantations can be recognised not only by the absence of the very acid-tolerant mosses but by the presence of *Oxalis acetosella* and other more or less nitrogen-loving plants. However since Pines are rarely planted in such places we shall not go into their floristic differentiation here. In heathland areas the species-rich forests can often be traced back to former arable fields (see fig. 414). Where Pine and Spruce are mixed together here the latter quickly gets the upper hand.

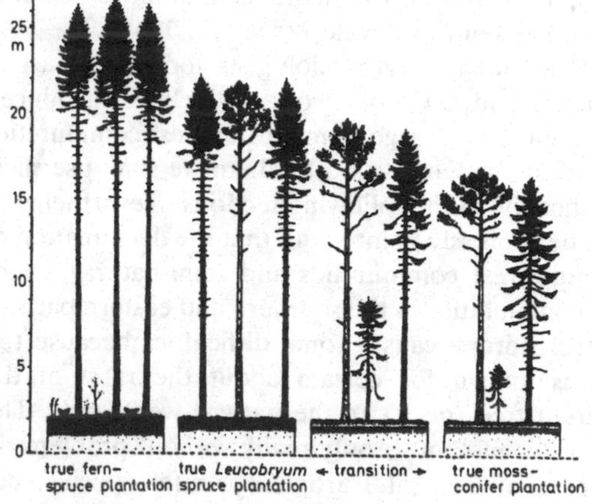

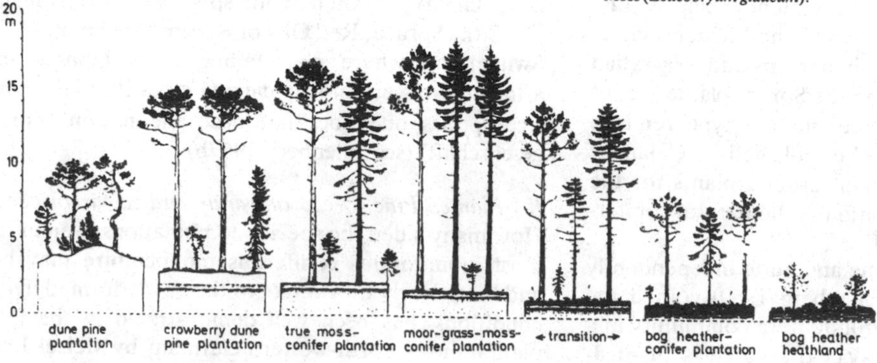

Fig. 414. Plantations of needle-leaved trees develop very differently depending on the type of soil. These examples from the Knyphausen Forest (northern Oldenburg) were planted at the same time and in the same way. After Ellenberg (1969).

On 'normal' **heathland** (below) with a podsolised soil (*Genisto-Callunetum typicum*) the mixed Spruce-Pine plantation remained a mixed stand reaching a height of 17 – 20 m after about 70 years ('typical conifer forest'). On damper soils with *Genisto-Callunetum molinietosum*) Spruce thrives better but still tolerates the Pine ('Moorgrass-moss-conifer forest'). On the other hand the much wetter *Erica* heaths restricted the growth of trees (Bell Heather – conifer forest') or even prevented it almost entirely (Bell Heather community). On dunes and heaths covered with blown sand (*Genisto-Callunetum empetretosum*) only *Pinus sylvestris* can establish itself, but does not grow so well as on normal heathland soils ('dune Pine forest').
Land which has been **cultivated** (above) for varying lengths of time and covered with a more or less deep layer of humous soil through manuring with Heath litter produces a stronger growth of trees when used for afforestation. On the best soils *Picea abies* completely dominates the Shield Fern-Spruce forest'. In the Knyphausen Forest a good indicator of somewhat more fertile soils is the White Moss (*Leucobryum glaucum*).

Tab. 104. Pine forest communities and their origins in the north-west German plain.

Vegetation present at the time of planting[a]	Pine forest plantation	Nat. veg[b]
I. **Silvergrass community** *(Corynephoretum)*	*Caladonia-Pine forest*	
II. **Heather heath** *(Calluno-Gernistetum)*[c]		
1. Subass. of *Cladonia*		
a typical variant	pure *Dicranum*-Pf	
b *Molinia* variant	*Empetrum-Dicranum*-Pf	III 1a
	Molinia-Danthonia-Pf	III 1b
2. Subass. of *Danthonia*		
a Typical variant	pure *Danthonia*-Pf	III 2a
b *Molinia*-variant	*Molinia-Danthonia*-Pf	III 2b
III. Acid-soil mixed Oak woods		*Hypnum*-Pine forest plantations
1. Birch-Oak wood (*Betulo-Quercetum*)		
a typical subass.	pure *Drypoteris*-Pf	
b *Molinia*-subass.	*Molinia-Dryopteris*-Pf	
2. Oak-Beech wood (*Fago-Quercetum*)		
a typical subass.	pure *Rubus*-Pf	
b *Molinia*-subass.	*Molinia-Rubus*-Pf	
IV. Birch swamp wood (*Betuletum pubescentis*)	*Sphagnum*-Pf	
V. Moist-soil Oak-Hornbeam wood (*Stellario-Carpinetum* subass.)	*Oxalis*-Pf	
VI. Alder swamp wood (*Alnetum glutinosae*)	*Lysimachia*-Pf	

[a] In the case of the communities III to VI we may also be dealing with a mosaic of heathland and poor grassland which has been impoverished by grazing.
[b] The potential natural vegetation when this differs from the first column

[c] 1a corresponds to Tüxen's (1937) *C.-G. typicum*, 1b to his *C.-G. molinietosum*, 2 to his subass. of *Orchis maculata*.
Source: From data in Meisel-Jahn (1955b).

Hofmann (1969) has produced a very similar sociological and ecological classification to the one proposed by Meisel-Jahn (1955b) for the north-east German lowlands. He distinguishes 15 Pine forest comunities, some of which are close to the natural Pine woods (Hofmann 1964a). On more fertile soils species requiring more nitrogen such as the Blackberry (*Rubus* spp.) or the Wood Small-reed (*Calamagrostis epigeios*) often form quite a dense undergrowth.

The planted Pine forests on the Upper Rhine plain, like those in the north-east of Central Europe give a much more 'natural' impression than the ones in the north-west. But appearances are deceptive because *Pinus sylvestris* was only introduced into the Upper Rhine plain towards the end of the sixteenth century, and before that had played only a minor role in the northern part. According to Philippi (1970) the natural woodland on lime-deficient sands would have been Oak-Beech wood (*Fago-Quercetum*), and on the ones richer in lime, Beech wood (*Carici-Fagetum* and others) would have dominated. In their place there are Moss-Pine planted-forests, *Teucrium scorodonia* – Pine forests (somewhat richer in bases) or planted 'calcareous sand Pine forests' which remind one of the Polish *Peucedano-Pinetum*.

In an area with a uniform climate and history the planted-forest communities can be used to evaluate the productive potential of the different habitats; fig. 416 shows an example. Incidentally in spite of their low absolute production it is just the poorest Pine forests, e.g. those with *Calluna* and *Dicranum* on sandy heathland soils, where economically one is forced to keep to a system of pure coniferous forestry because all the other trees, including Oak, Birch and similar natural species yield less here than the undemanding Pine (see Kremser 1974). For this reason alone there will be Pine plantations in the northern part of Central Europe for a long time yet (fig.417).

Fig. 415. A schematic vegetation profile through Pine plantations on dunes in north-western Germany. After Meisel-Jahn, 1955).

1 = typical *Dicranum*-Pine forest, *Avenella flexuosa* fazies, 2 = typical *Cladonia*-Pine forest (very gaunt), 3 = *Empetrum*-Pine forest (on the shaded slope), 4 = *Empetrum*-Pine forest, *Vaccinium* form (in the dune valleys).

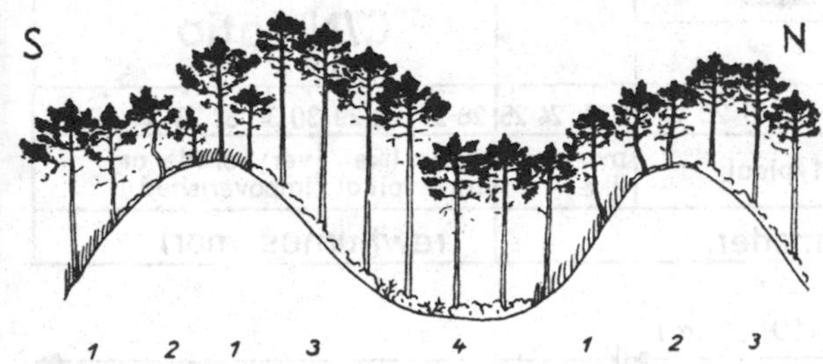

On fertile soils it is generally not difficult to draw conclusions about the natural woodland communities from a study of the planted-forest communities which have replaced them. Where the soils are poorer it is necessary to take into account the forest history or resort to pollen analysis. Adjoining communities (so-called 'contact communities') will supply additional information, but often one has to try to find parallels between the planted-forest communities and the natural woodlands in a roundabout way using soil profiles. Hofmann (1957) found in the case of planted Pine forests on calcareous soils in South Thuringia that, in spite of employing all such aids, it was only possible to determine the association, a fairly all embracing unit, to which the planted-forest community belonged and not any particular subunit which would identify it more exactly.

The mapping of planted-forest communities is made all the more difficult because the ground flora composition depends not only on the site and the planted tree species, but also on the **age of the forest** and the number of generations of the introduced tree species. In the majority of forest plantations the species combination is developing, quickly at first and later more slowly, into a balanced community which, on the basis of actual experience, cannot be exactly predicted.

According to Meisel-Jahn (1955b) in the Pine plantations of north-western Germany the 'stages' in their development correspond approximately to the number of tree generations. After a heath has been planted up for example the *Dicranum*-Pine forest develops first into a *Ptilidium* stage in which the cryptogams of the *Calluna* heath are still to be found:

Ptilidium ciliare	*Cladina portentosa*
Dicranum spurium	*Cladonia chlorophaea*
Polytrichum juniperinum	*Cladina tenuis*

Fig. 416. Natural Pine woods and Pine plantation communities in the German Democratic Republic arranged according to the amount of nitrogen available and the type of surface humus along with the yield levels of the trees. After G. Hofmann (1968), modified.

The crosses indicate the mean values for the C/N ratio determinations and the length of the horizontal lines the range for a 5% probability. Roman figures indicate the yield class from I = very good to V = very poor (with one exception the subdivisions 1 – 5 are not shown).

Pinus sylvestris is naturally dominant only on very poor dry soils where it grows slowly and yields badly. In better habitats it would normally be crowded out by Oak, Beech or other broadleaved trees. In forestry plantations where it takes the place of the natural mixed broadleaved woods the Pine does much better than in the Lichen-, Heather- or Bilberry- Pine woods. Exceptions to this rule are the plantations on calcareous and relatively dry soils, especially the Torgrass-Pine forest which replaces limestone Beech wood on shallow rendzina.

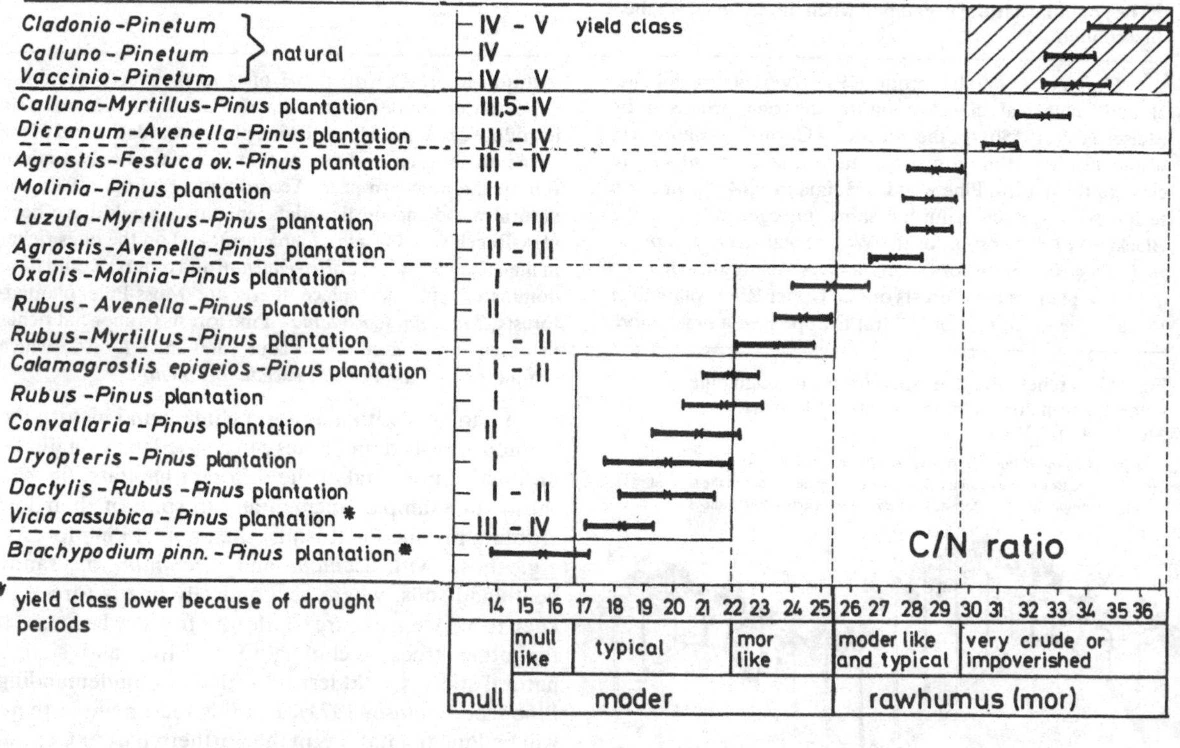

If the Pine forestry is continued the second generation of trees sees the invasion of scattered berried shrubs whose seeds have been brought in by birds. As Christiansen (1968) emphasised, such zoochorous plants play a major part in the newly arrived phanerogams. At the latest during the third generation of trees the pure *Dicaranum* Pine forest changes into a fairly stable *Vaccinium* stage in which

Vaccinium myrtillus M *Hylocomium splendens*
V. vitis-idaea

determine the mosaic of the ground cover. Planted Pine forests with other starting points, (e.g. *Cladonia*- or *Viola*-stages) often end up with this stage too. The Pine forests in north-eastern Germany and in the Upper Rhine area seem to require a somewhat longer time to form relatively stable communities resembling those of the natural woodland. In any case this is what Philippi (1970) found after a thorough investigation in the Schwetzinger Hardt.

If it is not heathland which has been afforested but a **broadleaved woodland** which has been **converted into a coniferous planted-forest** (see fig. 419) then the development of the ground flora on the same site is quite different. Even where the original woods have been severely degraded by the extraction of timber and litter and by grazing they still contain real woodland plants absent from the open *Calluna*-heaths.

Because of these, the former Birch-Oak wood habitat does not become a *Dicranum* Pine forest, but a **Dryopteris-Pine forest** with the following differential species:

S *Dryopteris carthusiana* *Frangula alnus*
S *Trientalis europaea* *Epilobium angustifolium*
S *Galium harcynicum* M *Scleropodium purum*

Those species marked 'S' are not found in the afforested heathland until at least three generations of trees have been

Fig. 417. The planned distribution of the main tree species in the state forests in Lower Saxony in different sites. After Kremser (1974), somewhat modified.

Broadleaved trees will be taken more into account, or at least they should no longer be replaced by conifers except in the poorest places where only needle-leaved trees give a useful yield. The Douglas Fir (*Pseudotsuga menziesii*), which is planned to make up 25% of total plantings, is worth special mention. It is recommended mainly for level sites to replace Scots Pine which is very prone to storm and fire damage. Where the climate is sufficiently humid Douglas Fir has produced astonishingly good yields up to now with few disease problems. Its influence on the undergrowth is similar to that of the Silver Fir, i.e. more like that of the broadleaved trees than of the Pine or Spruce.

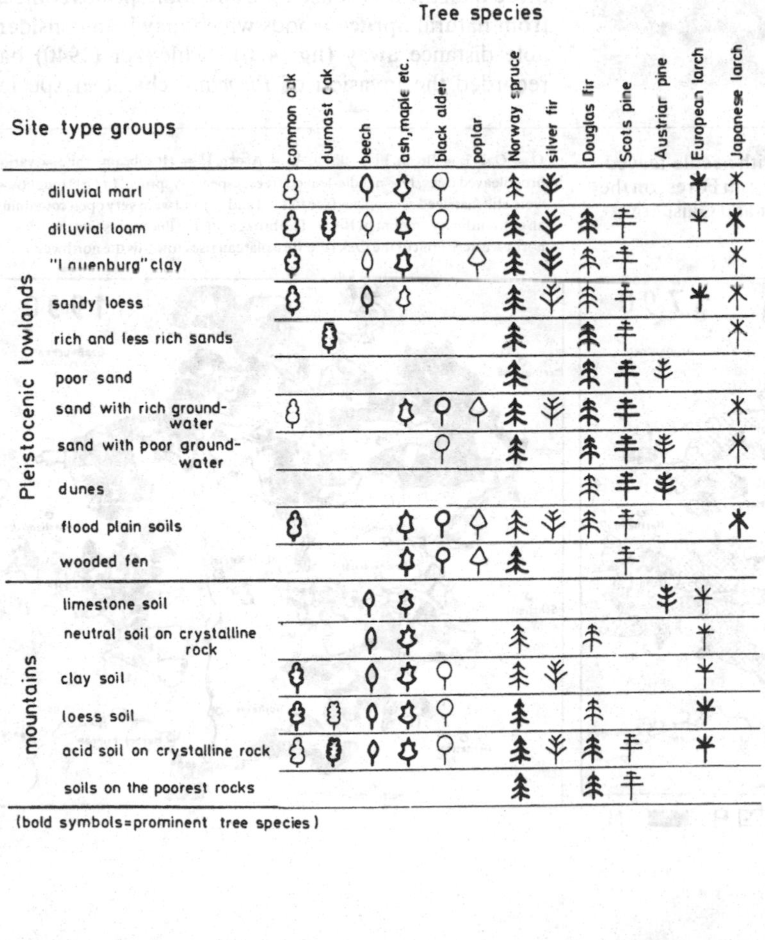

Tree species: common oak, durmast oak, beech, ash, maple etc., black alder, poplar, Norway spruce, silver fir, Douglas fir, Scots pine, Austrian pine, European larch, Japanese larch

Site type groups

Pleistocenic lowlands: diluvial marl; diluvial loam; "Lauenburg" clay; sandy loess; rich and less rich sands; poor sand; sand with rich ground-water; sand with poor ground-water; dunes; flood plain soils; wooded fen

mountains: limestone soil; neutral soil on crystalline rock; clay soil; loess soil; acid soil on crystalline rock; soils on the poorest rocks

(bold symbols=prominent tree species)

growing there, and then only in the relatively fertile habitats of *Danthonia*-Pine forests.

The planting of Pine forests on slightly arid and arid grassland, or on land which previously carried a limestone Beech or Oak wood naturally has a quite different outcome. This is shown by records kept in the Thuringia area by Hofmann (1957, 1969), to whose classification we can now make brief reference.

For example in place of a slope Sedge-Beech wood there developed a *Veronica chamaedrys*-Pine forest or a *Helianthemum*-Pine forest. In both of these *Brachypodium pinnatum* is dominant as it is the only member of the *Mesobromion* communities which can tolerate heavier shade (see fig. 418). These and other 'Torgrass-Pine forests' are distributed widely in Pine and Larch plantations on calcareous soils of central and southern Germany (see fig. 416).

Fig. 418. Heavy browsing of young Beech by Deer in an area of needle-leaved forest on limestone to the west of Stuttgart. The grass sward consists mainly of *Brachypodium pinnatum*.

(c) Forestry plantations of Spruce and other conifers

On most of the Central European mountains and in the higher parts of the pre-Alps preference is given to the Norway Spruce, which does not grow well on pure sand nor in regions of low rainfall. Like the Scots Pine it yields much more in the majority of planted forests than it does in the natural woodlands where it is dominant. In Central Europe the *Piceeta* are confined to colder and more continental climates (see section B IV 3). When Spruce is planted in warmer, less-continental areas its growth is generally quicker and more readily exploited than that of Beech and other broadleaved trees which form the natural woodland on such sites. For this reason Spruce is not only preferred for planting on Matgrass swards and mountain heaths, but is often put in after the clear felling of these woodlands (see tabs. 105 and 106, and fig. 419).

In both cases Spruce makes its presence felt much more than the less dense Pine and soon species come in from natural Spruce woods which may be a considerable distance away (fig. 420). Schlenker (1940) has recorded the invasion of *Piceetum* character species

Fig. 419. The increase in the area planted with needle-leaved trees between 1790 and 1930 in the Heidenheim Forest on the eastern Swabian Jura. After Koch, Schairer and Gaisberg (1939) from Ellenberg (1963).

O = Oak, B = Beech, Bi = Birch, As = Aspen, H = Hornbeam, white = various broadleaved trees, N = needle-leaved trees, especially spruce. In 1790 the stands were still managed as coppices (see figs. 22 and 23) and were very open containing an extraordinary amount of Birch. The broken double line represents the north-western limit of the Oak (the Jura plateau rises towards the northwest).

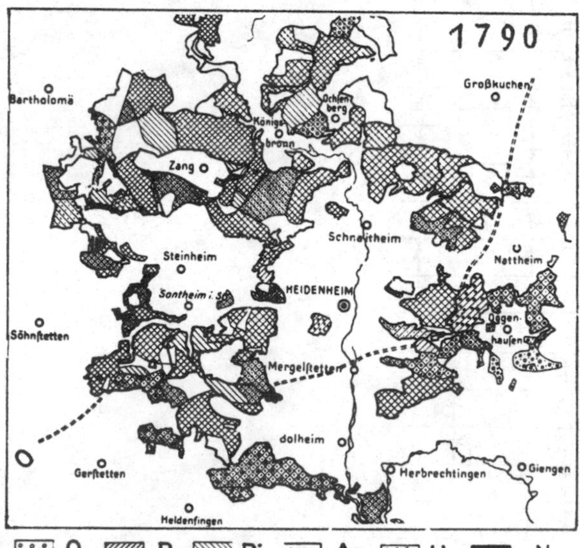

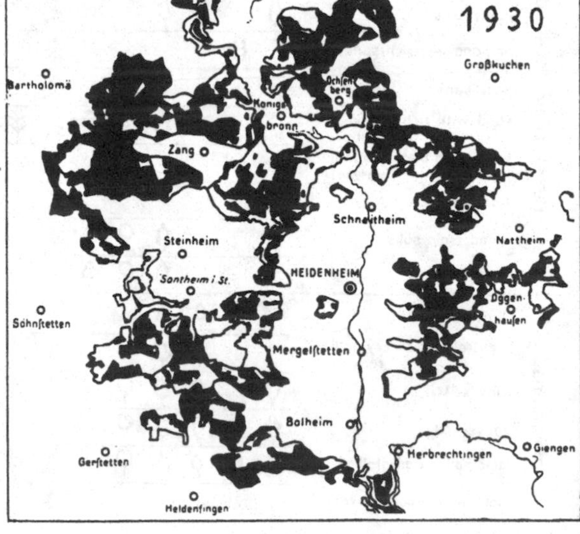

O B Bi As H N

Tab. 105. The annual production of timber from different species of needle- and broadleaved trees.

Tree species	Height (m)	Average diameter (cm)	Proportion of bark (%)	Total timber yield (fm/ha) with bark	Total timber yield (fm/ha) without bark	Density (kg/m^3)	Annual timber yield (kg/ha)
N **Spruce** 100 y	35.5	42.1	10	1570	1413	390	**5510**[a]
Fir "	31.8	40.9	10	1506	1356	370	5020
Pine "	30.5	37.3	13	828	721	420	3030
L **Beech** "	32.0	33.7	7	929	864	570	**4920**
Oak "	26.7	34.9	15	706	600	570	3420
Ash 80 y	28.0	31.0	10	504	454	600	2720
Alder "	27.7	36.0	10	719	647	430	2780
Birch "	26.0	32.0	10	389	350	510	2230

Note: [a] This calculation does not take account of the greater probability of storm damage to which the needle-leaved trees are prone. Therefore the advantage of conifer plantations, looked upon financially, is smaller than it may appear by considering merely the yield.

Source: After Trendelenburg and Mayer-Wegelin (1955).

into planted Spruce forests in the Stuttgart region. Hauff (1965) reports how plantations of Spruce on former sheep pastures in the Swabian Jura, i.e. in the area of the natural *Fagus* woods have developed surprisingly quickly into communities which resemble the natural montane *Picea* woodlands on higher mountains. Thus *Orthilia secunda*, *Pyrola minor*, *P. chlorantha*, *Moneses uniflora*, *Corallorhiza trifida*, *Listera cordata* and *Goodyera repens* as well as *Rhytidiadelphus loreus*, *Bazzania trilobata* and other mosses have come in. All of these species which were previously very rare or absent from the area. Schlüter (1964) compiled similar records for the Thüringia Forest and for Schleswig-Holstein showing that boreal and boreo-alpine species turned up in many Spruce plantations, e.g. *Huperzia selago*, *Diphasium alpinum*, *Listera cordata* and the fungi *Russula paludosa*, *R. rhodopoda* and *Boletus elegans*. Even quite rare plants can spread quickly and invade suitable habitats when conditions are favourable. At the same time their

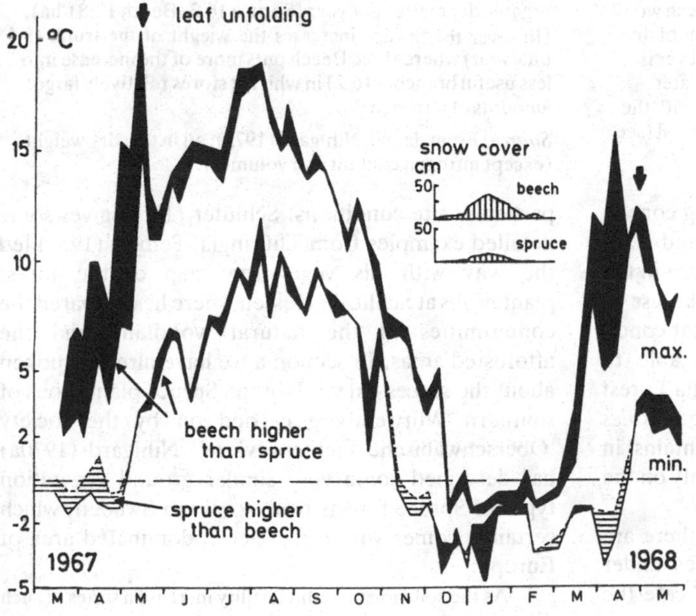

Fig. 420. The ground temperatures in a Beech wood are somewhat higher as a rule than those in a Spruce forest in the same area. This is true both for the maxima and minima. After Nihlgård (1969), somewhat modified. It is only before the Beech leaves unfold in spring that the minimum temperature is lower here than in the Spruce plantation where the evergreen canopy reduces the radiation at night. The difference between the two is least during the months from August to October.

Tab. 106. Reserves and production of phytomass together with chemical data in a Beech wood compared with that in a Spruce plantation in southern Sweden in 1968.

			Reserves[a]		Production[b]	
		Tree species Age:	Beech 100	Spruce 55	Beech 100	Spruce 55
Phytomass						
above ground:	trunk wood	t/ha	212	240	4.6	8.6
	trunk bark	”	9	22	0.2	0.8
	branches with bark	”	103	46	**6.2**	0.9
	twigs with leaves	”			3.9	3.3
	summer leaf fall	”			0.2	0.1
	tree layer total	”	324	308	15.1	13.7
	herb layer (July '67)	”	0.2	<0.1	0.3	<0.1
below ground:	tree roots (calculated)	”	49	58		
	additional fine roots (estimated)	”	0.5	0.8		
	tree layer roots total	”	50	59		
	herb layer roots	”	1.5	<0.1		
total living phytomass		”	375	367	17.8	16.3
Volume of tree layer, above ground						
	trunk + branches + bark	m^3/ha	553	**802**	18.7	26.3
Chemical composition (July '67)						
Organic matter:	phytomass	t/ha	375	367	7.6	6.8
	litter	”	2.7	9.1	3.0	2.7
	soil	”	207	207	–	–
Water content:	phytomass	”	284	**510**	–	–
	litter	”	6	19	–	–
	soil	”	**1920**	1400	–	–
Nitrogen content:	phytomass	kg/ha	1121	860	**204**	67
	litter	”	90	250	69	58
	soil	”	7800	6900	–	–

Note: [a] During the course of its 55 years the Spruce plantation had produced just about as much biomass as the Beech wood in 100 years in the same site, when measured in terms of dry matter (367 against 375 t/ha DM). The volume however is considerably greater (802 against 553 m^3/ha). The water content of the phytomass is correspondingly higher with the Spruce and so is the amount of water taken from the soil (see also tab. 23).

[b] Both stands produce approximately the same amount of organic dry matter per year (Spruce 16.3; Beech 17.8 t/ha). However the Spruce increases the wieght of the trunk (8.6 t/ha/year) whereas the Beech puts more of the increase into less useful branches (6.2) in which it stores relatively large amounts of nitrogen.

Source: From data in Nihlgård (1972); all in t/ha dry weight (except nitrogen content and volumes).

presence emphasises the difficulty of drawing conclusions about the natural broadleaved woodland communities from a study of the ground vegetation in the Spruce forests. For this it is essential to make use of historical evidence and to consider the habitat conditions. For example Schlüter (1966a) was able to demonstrate that a large part of the Thüringia Forest was previously dominated by Beech, and that *Abies alba* also played a part on the slate mountains in Thuringia. In both areas *Picea* was present only on the edges of oligotrophic mires.

As we found with the Pine plantations there are numerous different groundflora communities under the Spruce stands which can be used to indicate the particular site conditions. Schlüter (1965) gives some detailed examples from Thüringia. Seibert (1954) led the way with his vegetation map of the forest plantations at Schlitz in Hessen where he compared the communities of the natural woodland and the afforested areas. In section a we have already spoken about the successful work in the Spruce plantations of southern Württemberg carried on by the society 'Oberschwäbische Fichtenreviere'. Nihlgård (1970a) has described some very similar ground vegetation types of Spruce forests from southern Sweden, which certainly comes within the Beech-dominated area of Europe.

As the soil moisture and fertility increase a series of such

types can be recognised which Nihlgard named after the dominating species:
Avenella flexuosa
mosses
Avenella- Oxalis acetosella
Sambucus
Rubus-Oxalis
Sambucus-Oxalis
Dryopteris carthusiana- Oxalis
However the communities defined in this way are only recognisable when sufficient light is getting through to the forest floor, i.e. in somewhat older stands.

Most forest plantations are dominated by either Pine or Spruce. For this reason little information is available from stands of other conifers with the exception of the **Larch groves** near the villages in the central Alps. The **Silver Fir planted forests** near Lütetsburg in East Friesland which Tüxen and Ellenberg (1947) recorded are mostly past their maturity and are breaking down. A study of the **Douglas Fir forests** would be worth while because, after the catastrophic storms and fires in the northern part of Lower Saxony which destroyed large areas of Pine, *Pseudotsuga menziesii* should be preferred and the proposed mixtures of tree species (see fig. 417) will enable studies to be made of the vegetation development associated with each (Kremser and Otto 1973). Plantations of *Picea sitchensis* such as those described by Hill and Jones (1978) from South Wales, are relatively rare in Central Europe except for some former heathlands near the North Sea coast.

(d) Planted broadleaved forests, Poplar and Robinia stands

Up to now phytosociological studies in plantations of broadleaved trees appear to have been neither necessary nor sufficiently interesting. An exception lies in the plantations established on wet ground which have been investigated by Buchwald (1951) in the Syke Forest near Bremen. Stands of Alder, Ash, Birch and Oak which have been planted on meadow land or drained mires develop basically in a similar manner to the Pine forests in poor grassland or heathland. However they adjust themselves more quickly to the broadleaved woodland community appropriate to the particular habitat.

Poplar plantations are somewhat of an exception in growing very quickly but remaining relatively open. Thus species of the former grassland community are able to remain much longer, provided they tolerate a small amount of shade. Such plantations are quite frequent in different parts of Central Europe, and merit a closer study than they have had especially on successions of the ground vegetation in different sites. Generally the hybrid Canadian Poplar is the one planted, even in river valleys where Black and White Poplars occur naturally (Wendelberger-Zelinka 1952 see also section B V 1 d).

From several points of view the planting of **Robinia** (*Robinia pseudacacia*) on hillsides and embankments is a special case. This member of the Leguminosae from North America enriches the soil, not only because of its root nodules, but also through the dead leaves falling on the ground (Hofmann 1961). Nitrogen fixation is so active that many ruderal plants are able to thrive under the open canopy (see section D VIII). On bombed sites after the war it came in spontaneously as a pioneer tree e.g. in Berlin. Kohler and Sukopp (1964a) distinguish between a Common Bent type (with *Agrostis tenuis*) on base-deficient dry soils and a Greater Celandine type on damper soils richer in humus (with *Chelidonium majus, Galium aparine, Urtica dioica* and *Geranium robertianum*). On warm dry habitats *Robinia* can spread without the help of man and in some cases it has crowded out the native trees in southern and eastern parts of Central Europe (Kohler 1968). For this reason Jurko (1963) does not see *Robinia* stands as planted-forest communities but as natural associations. Among others he describes a *Balloto-Robinietum* on wind-blown sand, a *Solidagino-Robinietum* on river valley soils and a *Chelidonio-Robinietum* corresponding to the Greater Celandine type just mentioned. With the *Robinia* planations or woodlands we are witnessing the establishment of a new group of plant communities, the sort of thing which may have happened in earlier times in the case of limestone grasslands, Oatgrass meadows or many of the plant communities on cultivated land (see fig. 499).

(e) Effects of converting broadleaved woodlands into coniferous monocultures

Apart from Switzerland where a type of semi-natural forestry has been practised for many generations the monoculture of coniferous trees is taking up an ever increasing area of land and is thus becoming more and more a problem for the care of the countryside. There are at least three sides to the problem – an aesthetic, an economic and an ecological. Here we must confine ourselves to the last one with a few brief comments. Do these artificially created pure stands have a harmful effect on the biological balance of the ecosystem, especially on soil and climate? This question is not easy to answer nor would any answer be universally applicable.

As we have pointed out already in section a, although planted-forest communities certainly include pure stands of particular tree species, they are biocenoses in which many other organisms take part. We have already discovered in sections B and C that even in nature there are many communities with a low species diversity, for example those in water, mire, marshland and dune habitats, as well as in many woodlands growing under extreme conditions. These often form 'natural monocultures' in which a single species is absolutely dominant whether it be *Alnus glutinosa* in the fen woodland, *Pinus sylvestris* in some very different habitats 'on the edge' of natural woodland sites (see fig. 39), or *Picea abies* in the high montane and *Pinus cembra* in the subalpine belts of the central Alps. The question then is just whether pure stands which have been planted in areas where the tree population is naturally more varied have had a particularly deleterious effect, especially where conifers now dominate in habitats previously ruled by broadleaved species.

Conifers are more in danger of being damaged by **storms**, which usually occur in the autumn or early spring, than the broadleaved trees which have bare crowns at these times. This was clearly seen following the catastrophic storms in Lower Saxony on 13th November 1972 (see e.g. Kremser 1974). The greater **danger from fire** in conifer forests is also undisputed especially when the dead wood is not removed and the needle litter dries out during a rainless period. Forests, mostly Pine stands, destroyed by fire totalled 8213 ha in Lower Saxony in August 1975. Average figures covering a number of years show that 69% of all forest fires occurred in Pine stands up to 40 years old, a further 18% in older Pine forests, compared with only 5% in broadleaved woodlands ('Unser Wald', 1975, 27). In the remaining Pine forests of F.R. Germany an average of 5 ha per 10 000 are burnt down each year in such places as the sandy plains of Schleswig-Holstein, the Nürnberg basin and the Oberpfalz hills. As Walter (1967) points out, Pine is adapted to such fires, and in northern and eastern Europe its successful spread under natural conditions is helped by burning. Woodland fires occurring from time to time not only open up areas for pioneer shrubs and trees, but actually assist in the germination of their seeds. Pure coniferous forestry is thus just an extension of the state of affairs which must be looked upon as normal in the boreal and continental areas of Europe and North America.

Both in a natural woodland and in a planted forest a pure stand is much more susceptible to attacks from **pests and diseases**, whether insect, fungus or other organism, than a mixed tree population. A coniferous forest in a climate where broadleaved trees normally dominate is put even further at risk by the milder winters and damper summers than the evergreen needle trees would encounter in their natural area of distribution. According to Hesmer and Schroeder (1963) for example the killing of young Pine trees by the fungus *Lophodermium pinastri* was a significant factor in the retreat of *Pinus sylvestris* from north-western Germany during the Post-glacial warm period when it was only able to hold its own against the invasion of broadleaved trees along the edges of the raised bogs and other poor habitats where the competition was weaker. A disease such as this would always mean a higher risk when *Pinus sylvestris* is grown as a pure stand. The introduced Weymouth Pine (*P. strobus*) is subject to a rust fungus and is not worth planting extensively in most parts of Central Europe. The same sort of trouble could affect other aliens, perhaps even the Douglas Fir, whereas native trees, like *Pinus sylvestris* and *Picea abies* which have never been entirely absent from the area, are much less likely to be affected (see fig. 21).

Taking the risks of storm, fire and pest damage into account, as long as these are compensated for by an improved forestry technique and a higher yield, there still remains the question of the effect these pure coniferous forests have on the soil which previously carried broadleaved woodland. Up to 10 or 20 years ago it was feared that there would be a loss of fertility through increased acidification and podsolisation, i.e. a process which meant a long-term loss in value. These fears appeared to have been well grounded when it was noticed that on the Lüneburg Heathland for example the Pines were occupying extremely podsolised soils while neighbouring Beech and Oaks were growing on parabrown earth which, although acid was hardly podsolised at all. Later however it was recognised that this was a false comparison because most of the Pines had been planted on former heathland that was already podsolised before the afforestation (see section D II 3 c), whereas the remains of deciduous woodlands for various reasons had suffered less from the effects of extensive grazing during the previous centuries. Quite often the boundary between Pine forest and Beech or Oak wood could be traced back to a dividing line between two occupations (Ellenberg 1969, see fig. 414). In addition the soils in the broadleaved woodland are generally somewhat richer in clay and so less liable to podsolisation than in the adjacent sandy area where the natural woodland could have been more rapidly degraded to heathland. Where soils have been formed through glacial action it is not uncommon to find quite marked differences in the clay and silt content over short distances.

In general where a critical approach to the comparison has been possible **no significant impover-**

ishment of the soil could be established up to now after the replacement of broadleaved woodland by pure conifer stands. According to investigations carried out by Genssler (1959) in the Harz, Solling and other mountains the pH value of the surface humus was certainly lower under Spruce than under Beech. Nevertheless even after 250 years there were no indications of podsolisation under spruce, nor was there any reduction in their yield during this time. The same thing was seen by Holmsgaard (1968) and Holstener-Jørgensen (1968) in the Danish young moraine region where no difference in height or mass could be detected between the Spruce of the first and those of the second generation. The humus layer which had been formed during the first decades later remained constant both in thickness and in its characteristics. Genssler (1959) also came to a conclusion which may seem surprising at first: The differences between the soils under neighbouring coniferous forest and broadleaved woodland were least marked where the soils were originally the poorest. His explanation for this was that on such soils Beech and even Oaks also tend to form a surface layer of raw humus. Even with richer brown earths Fiedler (1967, see also Nihlgård 1971) could never say with certainty that the coniferous forests had influenced the soil in a detrimental way. Certainly a moder layer was accumulated on these soils whereas it was relatively thinner or absent from broadleaved woods, but after some decades there was a balance between the addition and removal of organic matter. The biological activity is increased in such a moder layer. For example Rawald and Niemann (1967) found that the amount of respiration in it was just about twice that in a comparable soil of a deciduous wood, and also the rates of dehydrogenase and polyphenoloxidase activity reached higher values. Often the humus layer under a Spruce plantation produced more mineral nitrogen than the top soil of the adjacent Oak-Beech stand, as Schönhaar (1955, see also Froment and Remacle 1975) had been able to show. The enrichment of the humus in phosphorus (Kriebitzsch 1976) and trace elements (Delecour 1968) could possibly play a part here. Where water drips from the crown the supply of nutrients to the undergrowth is especially good, as can be seen for example in the productivity of the mosses (fig. 421).

Since Spruce is more shallow rooting on average than Beech and many other broadleaved trees, in some waterlogged habitats this may lead to a consolidation of the sub-soil. In order to make a careful study of this possibility Miehlich (1970) chose what was originally a homogeneous pseudogley in the northern part of the Iller-Lech plain. Here he compared an old Oak-Beech woodland with a 120-year-old Spruce plantation now in its second generation. Although the nature of the soil would make it very susceptible to progressive consolidation he found no indication of such a process having taken place. However he could confirm the chemical findings of the authors quoted above. The planting of conifers appears to have just as little effect on the water balance of the soil as it does on its structure, at least in certain regions. According to Noirfalise (1967) there is usually no indication of an unfavourable influence in humid climates of the kind that can certainly be traced in climates with drought periods. On the other hand however Nihlgård (1970b) found in southern Sweden that the ground dried out more under Spruce than under Beech because the Spruce transpired longer during the year and also intercepted more of the rainfall in the canopy. Benecke (1976) got similar results in the montane belt of Solling where the rainfall is higher than 1000 mm/year on average. Here the **amount of ground water formed under a Spruce forest is significantly less than**

Fig. 421. The amount of production in the moss layer of a Spruce forest increases with the amount of light (left); but where the light is constant at 50% or more relative illumination the moss production again falls off as the distance from the edge of the canopy increases. After Tamm (1953), somewhat modified.

The reason for the latter falling off lies in the smaller quantity of nutrients reaching the mosses in water dropping from the canopy. (The different symbols represent measurements made at different times.)

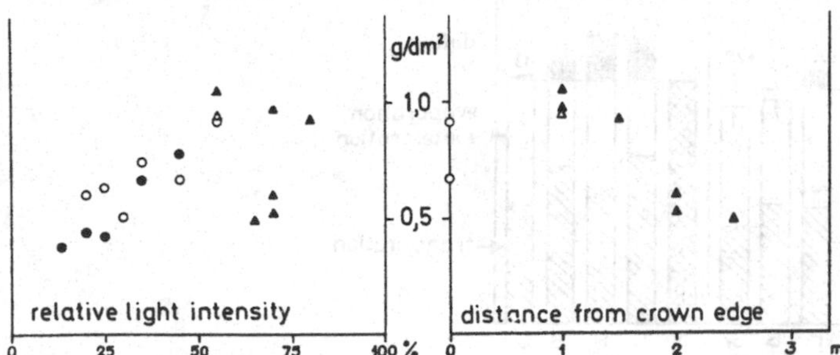

Tab. 107. The reserves and increases in the dry matter and of the surface area of a Spruce plantation in the Ebersberg Forest near Munich.

	Total	Trunks	Growing branches	Needles	Dead branches
Above ground dry matter reserves (t/ha)	322	268	28.3	15.9	10.0
Above ground annual dry matter increase (t/ha)	15.5	5.9	3.3	6.3	–
Surface index (m^2 per m^2 ground area)	26.2	1.3	2.8	21.6	0.5

Source: From data in von Droste zu Hülshoff (1969).

that from a semi-natural Woodrush-Beech wood (see tabs. 23 and 106). Brechtel (1975) found that Beech deals with its water supply very efficiently (fig. 422). In this respect it is better than Oak, whereas there is no significant difference between Oak and Pine, at least in the area of Frankfurt a.M. (see also H.M. Keller 1968 and Kiese 1975.)

Evergreen conifers, especially the shade trees, usually have a larger leaf surface area than the broadleaved trees under similar conditions. This is mainly due to the fact that the needles remain functional for a number of years (up to 12!) especially in the upper part of the canopy. Detailed measurements by Droste zu Hülshoff (1969) show that the total surface area above ground of a Spruce plantation is 26.2 times the area of the ground on which it is growing (of this the needles alone account for an area 21.6 times ground surface, see tab. 107). The highest values obtained for a Beech stand on similar brown earth are 14 to 17 times. This means that the filtering action of Spruce in removing pollution from the air is significantly greater than that of Beech, especially in winter. According to Ulrich and Mayer (1973) a Beech stand in Solling removed 6–17 kg of sulphur per ha per annum from the air, but from a neighbouring stand of Spruce the amount removed was 31–50 kg/ha/annum. Since this is in the form of sulphurous or sulphuric acid the increased amount has a noticeable effect on the soil acidity. Knabe (1975) showed, using the Ruhr area as an example, that not only do the **conifers remove more SO_2 from the air** but also that **they are more heavily damaged by this**. Where there are heavy emissions from urban and industrial concentrations the planting of conifers is not to be recommended, however desirable their filtering effect may be. Conifers have practically disappeared from the central area of the Ruhr and from other highly industrialised areas in Central Europe. The resistance of different tree

Fig. 422. The water balance of Oak, Pine and Beech stands and of an area of grassland (Gr) for four years with different amounts of precipitation (N). Average values from young, middle-aged and old stands in the surroundings of Frankfort on the Main. After Brechtel (1975), somewhat modified.
In the wet 'water economy year' (1 May 1969 to 30 April 1970) the total evaporation for all the stands was less than the total precipitation. Thus the ground was able to store water (34 to 147 mm). In the following years which were drier more water was used up than was received resulting in substantial deficits at times.
The woodlands always lost more water than the repeatedly mown grassland; of the tree stands the more open Oak and Pine lost more than the Beech. In the grassland the 'productive evaporation' (i.e. transpiration) was always less than the 'unproductive evaporation' (the water lost through direct evaporation from the soil or that intercepted by the plants). In the cases of Oak and Pine the two amounts were about the same but with Beech the interception and evaporation were relatively low.
In other words the ecosystem of the Beech wood was the most economical in the use of water in all years (see also tab. 23).

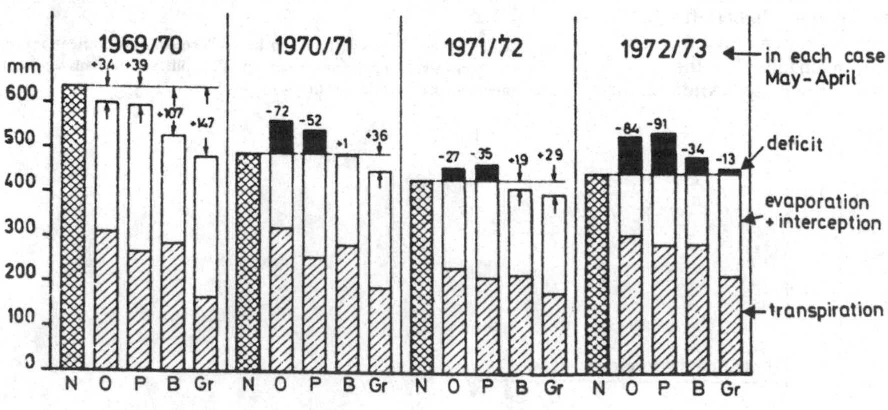

species to SO_2 and HF emissions has been studied by Dässler, Rauft and Rehn 1972.

(f) *Effects of fertilising on the ground flora of forest plantations*

It is true that planted coniferous forests have not impaired the fertility of the soils (see section e). Nevertheless it is also true that many of the soils on which such forests grow are very poor in bases and other nutrients, but this is because of their impoverishment through grazing and/or removal of litter and turf in the preceding heathland. In order to increase their productivity they have been, and still are, fertilised in many areas. This should not be considered an unnatural practice since those soils have been exploited by farmers for many centuries and the foresters are only returning what their agricultural forebears had removed (see fig. 20).

The productivity of Pine and Spruce forests can be considerably increased on very acid raw humus soils with a liberal dressing of fertiliser containing sulphate of ammonia. The results are not so good if lime is also given because this increases the breakdown of the raw humus and some of the nitrogen is lost through NH_3 release and NO_3 washout (Duchaufour and Turpin 1960). Zöttl (1959, 1964) too insists that an increase in the nitrogen supply is of primary importance. However Kriebitzsch (1978) points out that phosphorus and potassium are key requirements for nitrogen mineralisation and that these nutrients also must be available in sufficient amounts.

Even just a dressing of lime can lead to at least a temporary improvement in the amount of nitrogen available. In a *Dicranum*-Spruce forest of the Thuringia Forest, according to Schlüter (1966b) liming always encouraged species which are definitely nitrogen indicators such as:

Urtica dioica — *Cirsium vulgare*
Rubus idaeus — *Moehringia trinervia*
Senecio nemorensis subsp. *fuchsii*

and also some other 'demanding' plants such as:

Epilobium montanum — *Oxalis acetosella*
Mycelis muralis — *Athyrium filix-femina*
Luzula pilosa

The balance of species already began to change even after the first year of liming, and after ten years there were clear floristic differences. The application of nitrates or of ammoniacal liquor also brings about noticeable changes. Even a single application of fertiliser has a persistent effect because it significantly increases the nitrogen content of the needles and this in turn enriches the litter when the needles fall (Zöttl and Kemmel 1963). At the same time it improves the living conditions for animals such as myriapods, isopods and worms inhabiting the litter (Traitteur-Ronde 1961). The harvesting of the timber removes very little by way of nutrients, and the manuring does not have to be repeated so often as in the case of a meadow or arable field (see also section B II 4 e).

2 The vegetation of woodland clearings and burnt areas

(a) *Development and decline of the vegetation in forest clearings*

The management of a forest monoculture also involves the raising of young plants to replace the trees which are harvested. In most cases the forester does not rely on a natural regeneration of the preferred tree species or race but replants the whole area from nursery stock. This method involves the clear-felling of large tree stands and provides an opportunity for a group of quick-growing, nitrogen-requiring tall herbs to build up their luxurious but shortlived stands (fig. 423).

Such **woodland clearing communities** also occur in natural woodland and their character species belong to the original flora of Central Europe (see section A I 2). In the ancient woodlands however they only occupied small areas where senile trees had died or fallen (sections B II 2 a and B IV 2 d). Even when there was a catastrophic storm, such as occurred from time to time in Central Europe, there always remained in the undergrowth a few young trees, shrubs and herbs which were immediately able to make use of the extra amount of light available. In addition the fallen trees provided some shade enabling the true woodland plants to remain competitive. Similar habitat conditions are present for a time in the small clearings resulting from the removal of groups of trees (section A II 3 b). This has been the usual method for a long time in Switzerland where forestry practice does not involve large scale clear felling.

In contrast when forestry plantations are **clear-felled**, all trees are removed and the process of trimming and hauling out the trunks destroys so much of the undergrowth that little remains. All at once the whole area is bathed in sunshine which encourages the growth of light- and warmth-requiring species whose seeds may reach the forest floor every year but are unable to grow there because of the dense forest canopy. They thrive all the better because more nitrogen is mineralised from the disturbed humus layer (see fig. 424). If the cleared area is still surrounded by stands of tall trees, which is often the case, then its local climate takes on the nature of that in a hollow, protected from the wind. At night the relatively heavy cooled air collects here pushing the temperature on the ground below that in the surrounding woodland as well as in the wind-swept open country. In contrast the days can become much hotter in this sun trap than in the

Fig. 423. A clear-felled area with Willow Herb in a Spruce forest in the Bohemian-Moravian upland. *Epilobium angustifolium*, *Calamagrostis epigejos* and *Juncus conglomeratus* (indicating waterlogging) can be recognised. Photo Iltis.

Fig. 424. The production of inorganic nitrogen in the humus and in the mineral topsoil is not always greater after clear-felling than in neighbouring woodland on similar soils. From sample measurements by Clausnitzer (unpublished).
This certainly appears to be true for limestone Beech woods (the three columns on the extreme left) and for moder Beech woods (next three columns), where in any case it was not possible to obtain figures for a freshly clear-felled area.
In Spruce forests on base-rich soils the mineralisation of nitrogen is greatly increased directly after the clear-felling, especially in the moder layer which had been formed from dead needles during the life of the plantation (left column in the third group of three). A corresponding effect however could not be detected on very acid soils (four columns on the right).
The dynamics of inorganic nitrogen formation after clear-felling cannot definitely be clarified merely by a comparison of adjacent areas. An investigation lasting several years in the same experimental area is really required (see however Glavač and Koenies 1978).

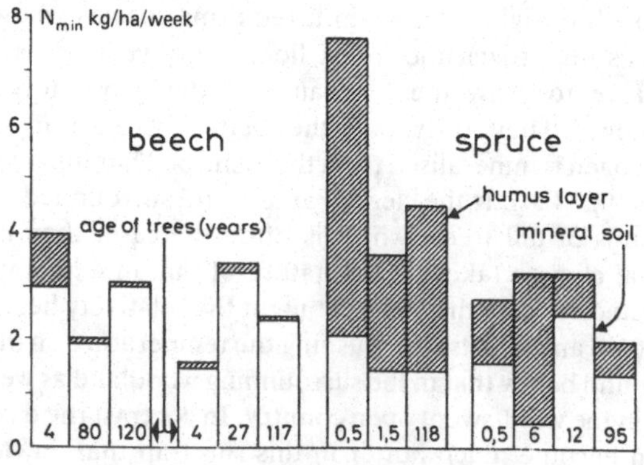

neighbouring habitats (Bjor 1972, see fig. 425 and compare Lützke 1961). In small clearings, leaving not more than a hole in the forest (Schlüter 1966c), these extreme conditions do not occur as the surrounding trees cut down the radiation in both directions.

The ground in a clear-felled area is often somewhat damper than that in the neighbouring forest especially in the relatively humid climate of Central Europe. This is because the trees are no longer drawing out the water while herbaceous plants have not yet become sufficiently numerous to exhaust the available supplies (fig. 426). All these favourable conditions mean that a particularly luxuriant growth of mesomorphic to hygromorphic herbaceous plants can develop here, and this often contains a number of dampness indicators (fig. 423). The vegetation of clear-felled areas is often very reminiscent of the tall-herb communities found in the subalpine belt and contains some common species, e.g. *Epilobium angustifolium* (see section C VI 7 a).

This favourable state of affairs does not last long however, since the quickly mobilised supply of nutrients is used up within two or three years (fig. 424). Perennial herbaceous plants, most of which had already germinated in the first year, gradually crowd out the shorter-lived ones. Shrubs and young trees also put in an appearance including early pioneers such as Elder, Birch and Willow. As the main crop of trees increases in size the species composition takes on more and more the character of a woodland until only the stunted remains of the light-loving plants are left. The forester considerably shortens the length of time needed for this succession by planting saplings of the new generation of conifers soon after clear-fellings. As these grow slowly in the first years they must be protected from the competition of tall herbs and grasses by mowing the latter down at intervals or killing them with a herbicidal spray. Shortage of labour means that the last measure is resorted to more and more as a general practice (see Günther 1965). According to Passarge (1970) a rational programme of weed control is possible only if one first understands the natural succession which follows clear-felling. The succession proceeds at its slowest pace in habitats which are deficient in nutrients and rather dry.

One is continually surprised how quickly the herbaceous plants of woodland clearings become established even over the large area of a clear-felling. According to Karpov (1960) this is because many seeds lie on the ground even in a dense continuous woodland waiting for favourable conditions before they can develop successfully. Hill and Stevens (1981) agree with him; in 30 – 40 year old plantations they counted 1000 – 5000 viable seeds per m^2 on brown earth and 500 – 2500 on peaty gleys. Karpov recorded 1300 – 5000 viable seeds per m^2 in a Bilberry-Spruce wood in western Russia. The majority of these had not come from members of the woodland community itself but were from plants of the clearings e.g. *Epilobium angustifolium*, *Deschampsia cespitosa*, and *Rubus idaeus*. Each year they are brought in afresh by the wind, birds or rodents, but after they have germinated they are normally smothered out by the root competition of trees. As Thompson and Grime (1979) pointed out, *Epilobium angustifolium* is not capable of developing a persistent seed bank because its seeds are no longer viable after one year. R.H. Wagner (1965) demonstrated how heavy the 'rain' of seeds is in a radiation-damaged Oak wood in North America by using sticky traps. He calculated that the fruits of

Fig. 425. Isotherms obtained by exact temperature measurements on a clear summer day in southern Norway show how extremely 'continental' the local climate can be in a clear-felled area of a Spruce wood. Covering the surface with Spruce branches (right) moderates the extremes of temperature producing a less continental and more typically woodland microclimate much more favourable for the growth of the young trees. After Bjor (1972), modified. The vertical scale is logarithmic covering a range of measurements between 2 m above the soil surface to 7 cm below it. The areas where the temperature is below 0°C are shaded with horizontal lines and those below -3°C are black. Areas over 40°C have vertical lines and those over 50°C are again black. During the course of a single day the temperature just above the ground may be below freezing shortly before sunrise and rise to desert heat by midday. (This kind of extreme variation can be experienced only in large clear felled areas on level ground protected from the wind. In small clearings, e.g. where just a few trees have been removed the lateral shading reduces the amount of radiation both inwards and outwards.)

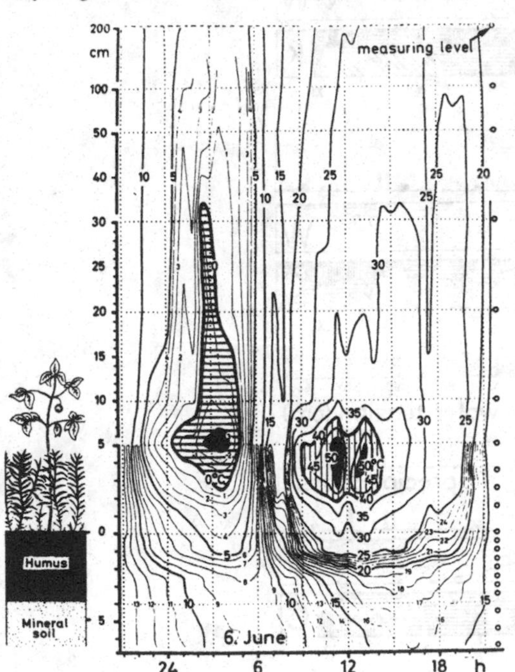

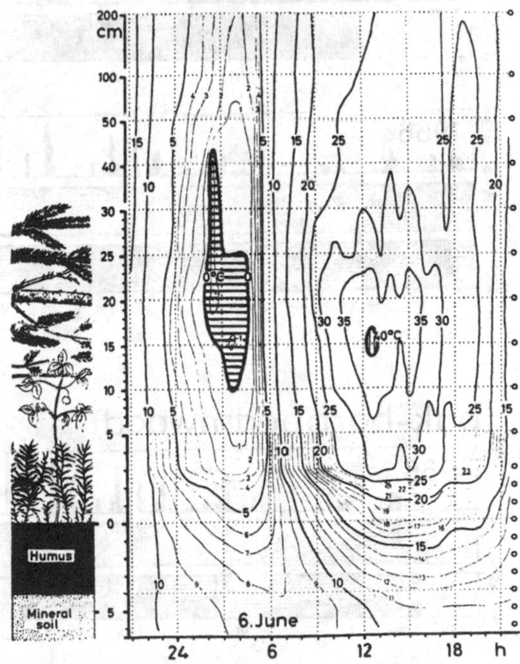

Conyza canadensis arrived at the rate of 91.6 per m^2 along with numerous fruits and seeds of grasses, composites and other species which did not grow in the wood. Even species completely alien to the habitat and ones from a considerable distance were represented in quite large numbers, e.g. *Phragmites australis* with 21.7 and *Typha latifolia* with 8.4 propagules per m^2.

The most important prerequisite for the establishment of woodland-clearing plants is undoubtedly sufficient light, a subordinate rôle being played by soil and climatic factors. The Wood Small-reed (*Calamagrostis epigeios*), which has been thoroughly investigated by Höpfner (1966) serves as a good example. The competitive strength of this pioneer grass is independent of soil properties such as grain size, organic and nitrogen content or microbial activity. The acidity can have any value it is possible to find in Central Europe, from pH 1.85 to 8.5 (measured in KCl,1M). The soil moisture can vary considerably too, provided there is sufficient water present in spring, which is always so in Central Europe. Only persistent flooding or heavy mowing (or other damage) is not tolerated by the Wood Small-reed but it rapidly disappears when it becomes shaded.

To a certain extent the vegetation of clear-felled areas delays the growth of the planted trees, but only where it is very thick. Even quite short grasses like the Wavy Hairgrass (*Avenella flexuosa*) are able to inhibit tree seedlings. According to Jarvis (1964) root secretions play a part; at least they have been shown to be effective against Birch and Lupin. The humus under *Avenella* actually contains fewer mycorrhizal rootlets of Birch and Oak. Probably it is the root competition from grasses and low shrubs which is responsible for the death of tree seedlings, and this may not necessarily always involve the presence of some

Fig. 426. Broadleaved wood land (W) at the colline level always draws more water from the ground than a neighbouring cleared area (K), especially during the summer following a dry winter (1966). In this respect a Beech wood is more effective than an Oak-Hornbeam wood. From a coloured chart by Brülhardt (1966), modified (see also fig. 422).

The area tested lies in the northern part of the Zürich canton. The Beech wood was growing on deep-leached loess soil and the Oak-Hornbeam wood on very sandy soil over gravel. The soil water was measured by tensiometers at different depths and expressed as suction pressure in cm of a water column. The capacity of the tensiometers used was limited to about 800 cm, i.e. less than one bar. In any case then the tree roots were only under water stress for short times. However it is possible that in the top 10 cm of soil, where most of the roots are living, but where tensiometers could not be used, the suction did rise above 1 bar more often and for longer periods.

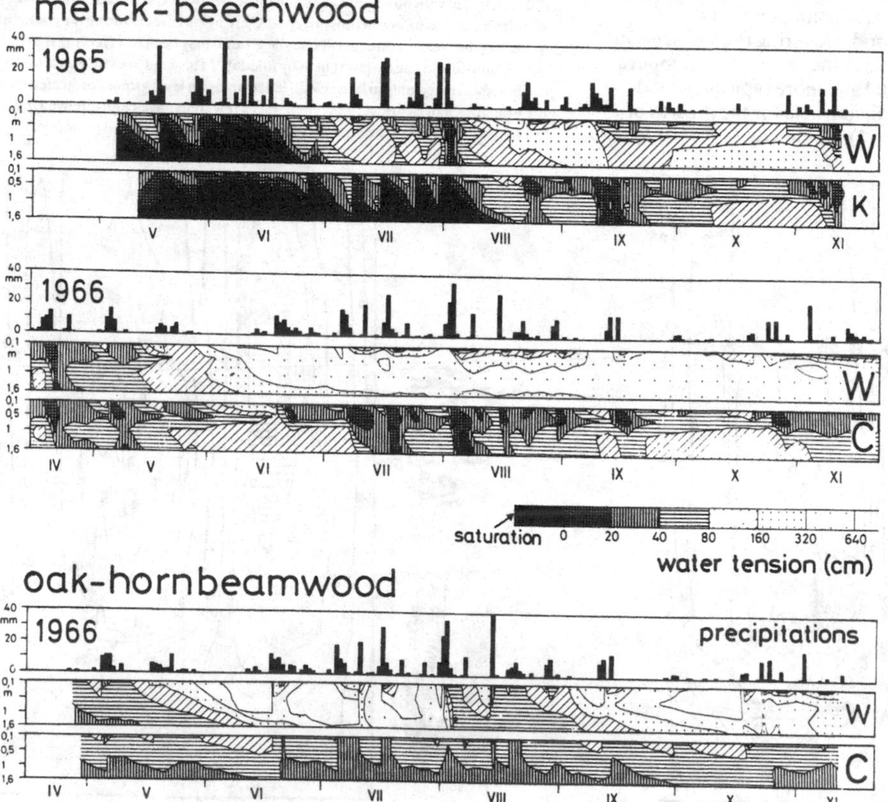

Tab. 108. The influence of competition on the growth of young Beech under an opened up old stand of Beech.

Beech saplings[a]	Leaf area[a] cm²/m²	Leaves[a] DM g/m²	Phytomass[a] DM g/m²	Increase[b] DM g/m²
without competition	17,405	150	363	208
competition from older trees	15,447	168	362	211
competition from older trees and weeds[c]	11,750	171	281	169
competition from weeds[c]	10,464	159	232	139

Note: [a] after two years.
[b] in 1963, above and below ground.
[c] Meaning weeds and grasses growing in the clearings. In the shade of the old Beeches or in small clearings these do not grow so strongly that they could hinder the Beech regeneration from seed; nevertheless they do affect the growth of the young trees to a certain extent.
Source: After Burschel and Schmaltz (1965).

allelopathic influence. Research carried out by Leibundgut (1964) has shown that dense stands of *Nardus stricta* or *Vaccinium myrtillus*, such as become established in moderately grazed woodland clearings, inhibit the growth of *Pinus sylvestris* and *Larix* from seed, even where there is no covering of raw humus.

In smaller and more-shaded clearings grasses and other herbs rarely become so vigorous that they interfere with the regeneration of the trees (see tab. 108). There is also less activation of the ground microflora, evidence of which is only traceable during the first two years (Faille 1975). Typical herbaceous communities such as the ones described in the next section are not found in small clearings, but only where clear felling has taken place over large areas or where there has been an extensive fire.

A quickly moving fire, especially one doing a lot of damage to the canopy and killing the trees and shrubs, but not destroying the surface humus, has similar consequences for the woodland soil and the sequence of vegetation as those following a clear-felling. Since a fire immediately mineralises the organically bound nutrients, the concentration of these is very high just after the burning. Some of them however are washed away by rain and are lost to the herbaceous vegetation. A **slow** moving fire destroys the humus layer, and kills most organisms living in it so that afterwards these species have to come in from the surrounding area. In addition if the ground is sloping this can lead to increased erosion. Wherever the mineral soil is laid bare either by the fire or by erosion the tree seeds germinate more quickly and in greater numbers than on the remains of the humus cover (see also section II 3 c). On the other hand the seeds which have survived the fire grow out from the humus. Yli-Vakkuri (1961) made both these observations in Finland, and they were confirmed many times following the huge fires which swept through the Pine stands on the Lüneburg Heathland in August 1975. Here such protein-rich plants as Willow Herb (*Epilobium angustifolium*) and Raspberry (*Rubus idaeus*) were only able to grow when they were fenced off from the deer which returned seeking food in the area.

Grabherr (1936, see also Ellenberg 1963) produced a classical exposition of the different effects of fire on the soils and on the successions of vegetation started by it, using as an example the semi-natural coniferous woods of the Karwendel. Here the subject can be studied particularly well because the drying föhn winds spread the fires. The first plants to appear both in the montane belt and lower down are the fire-moss communities, followed by Willow Herb and Blackberry stages. Repeated fires may result in the formation and maintainance of a Purple Moorgrass sward on marl soils, and this could become a dry grassland if it were grazed. In the absence of heavy grazing or further fires broadleaved shrubs come in, and after them trees take over once more so that the woodland community tends to approach the natural climax vegetation.

(b) Herb and shrub communities of woodland clearings

The succession following forest cutting or burning can take different courses depending on the soil and climatic conditions and other circumstances. The species composition of the successive communities clearly mirrors the environmental conditions and enables us also to draw certain conclusions concerning the potential natural vegetation.

Tüxen (1950b) put the short-lived herbaceous communities into the class and order of **Willow Herb communities** (*Epilobietea angustifolii*, *Atropetalia*, see section E III 1 and fig. 423). The best examples of these are found on acid soils with raw humus (*Epilobion angustifolii*). In a relatively warm climate Wood Groundsel (*Senecio sylvaticus*) is prominent, but on higher suboceanic mountain sites Foxglove (*Digitalis purpurea*) plays a larger part. On mull soils the clear-felling floras are richer in species (*Atropion*). Burdock communities (*Arctietum nemorosi*) dominate on brown earths, whereas on soils rich in lime, especially on rendzinas, it is the Deadly Nightshade scrub (*Atropetum belladonnae*, fig. 427) which comes into its own.

The herbaceous communities are always followed by a **shrub community** but there is no sharp division between this and the herb stand which precedes it or the woodland community into which it develops. For this reason it took a long time before woodland clearing scrubs were described as distinct associations, and the separation of them is fairly arbitrary. On soils with a raw humus layer the Willow Herb community is replaced by bushes in which Brambles play a large part (*Lonicero-Rubion sylvatici*). On mull soils on the other hand a temporary dominance is achieved by Elder and Goat Willow (*Sambuco-Salicion capreae*) or other more demanding species. Descriptions of individual communities can be found particularly in works by Tüxen (1950a), Oberdorfer (1957) and Passarge (1957b). The bush communities of woodland clearings have many members in common with those of woodland edges which will be discussed in the next section.

Fig. 427. Deadly Nightshade (*Atropa belladonna*) amongst the vegetation of a cleared area of a limestone Fir-Beech wood in the Swiss Jura. Young Ash form a pioneer woodland; in the foreground is *Stachys sylvatica*.

IV Woodland edges, bushes, hedges and their herbaceous margins

1 Communities rich in shrubs lying below the subalpine belt

(a) The origin and character of broadleaved shrub communities

Large areas of Central Europe have been deforested and completely cleared and there are only diminishing remains of the parkland that were grazed in previous centuries (see section A II 2 a). However in many parts of the country we still find groups of trees and shrubs between the fields or along the banks of waterways. In Schleswig-Holstein and some other regions moreover there is a network of hedges or rows of trees and bushes acting as windbreaks. Today people who care about the countryside are very concerned that such woody stands with their attendant herbaceous communities should be preserved and extended in as many forms as possible, adding character and diversity to the landscape. Happily the shrubs and trees used for these purposes are for the most part native and suited to the habitats (see Ehlers 1960, also figs. 428 and 16).

Our flora already contains a wealth of species which can form part of these relatively small areas of woody plants. Depending on the climate, soil and method of management they form their own individual communities which can be classified in a way similar to that used for woodlands. There are however several points of difference between these broadleaved bush communities, or fragments of woodland rich in shrubs, and both the semi-natural woodlands and the forestry plantations:

1 The isolated small woody stands consist almost entirely of **light-requiring woody species** which would at best be dwarfed in the shade of a natural high forest. Under natural conditions these would be competitive only where river banks have broken away, on sharp ridges, on steep slopes with a shallow soil or on rocky screes.
2 Since they are often low-growing and on their outside edges carry their leaves right down to the ground, and also come into leaf early in the year, they **only allow a little light to penetrate to the ground**. Requiring a lot of light themselves they are, nevertheless, intolerant of other light-loving species. Because of this young trees often perish in their shade and herbaceous plants too do not thrive.
3 Thus they consist almost entirely of **woody plants**, a state of affairs reminiscent of humid tropical woodland. Vegetation relevés which have been strictly confined to the interior of such stands

consist mainly of shrubs or semi-shrubs. Amongst these are climbers and lianas which, although unable to grow up to the top of taller trees, can easily cover the bushes and reach more light, e.g. *Clematis vitalba* and *Lonicera periclymenum*.

4 Because of their limited height and the light requirement of most of the members there is **hardly any stratification** in the community. Instead they tend to be arranged in a mosaic. Only if individual trees are preserved whether planted or spontaneous, e.g. Oaks, Elms or Ash does the community take on the form of a 'coppice-with-standards'. Such stands lead directly on to the woodland communities.

5 Because of the coppice management in the past when they were cut down every 5–15 years the development of these communities has followed a **rhythmical pattern**. All the shrubs regenerate very rapidly by growth from the base so that the light phase lasts for only 1–2 years. It is only during this time that annuals are found which have come in from neighbouring communities. There may also be species coming from felled forest areas.

6 Only rarely does the soil under these bush communities resemble that of the surrounding land or possess the profile of an old woodland soil. Generally it is **poorer and drier** because almost all the old hedges were planted on piles of stones cleared from the fields or on artificial earth banks. Sometimes however they **can even be richer**, if fertile silt soil is blown along and trapped in them, or if manured arable soil is washed off the fields and caught up in the hedge bottom. This is particularly liable to happen in places where there is a thick layer of loess, e.g. in the eastern foothills of the Harz or in the Kaiserstuhl (south-western Germany).

These floristic, physiognomic and ecological peculiarities justify separating the broadleaved bush communities completely from the woodlands and putting them in an order of their own (*Prunetalia spinosae*). The broadleaved shrub communities of the flood plains and swamps have already been separated off systematically. *Prunetalia spinosae* have many species in common with the mixed deciduous woodlands (class *Querco-Fagetea*) so it does not appear to be advisable to separate them from these (see section E III 1, No. etc.). In contrast to the woodlands however they owe their existence much more to cultural than natural influences and are generally found in contact with farmed areas or managed forests from which they get some of their species.

As Groenman-van Waateringe (1975) pointed out there were hedgerows even in Neolithic times acting as natural fences to keep the animals out of the cultivated fields. At the same time they could provide many kinds of additional food such as nuts, stone fruits and berries. In Switzerland and south-western Germany there are some quite old hedges which have developed spontaneously along the edges of terraces or on banks of stones picked up from the adjacent cultivated fields (Steiner-Haremaker and Steiner 1961). Many hedgerows however are of a much more recent date than is normally assumed. The network of hedges-on-banks in Schleswig-Holstein for instance was constructed during the seventeenth to nineteenth centuries. These have been thoroughly studied by Weber (1967, see

Fig. 428. Hedgerows on banks in southern Holstein. Photo, Schmidt.

The Hazel in the foreground has recently been cut, that in the centre has made some regrowth. Since firewood is no longer needed many such hedges are allowed to grow on, and eventually rows of Oak (background right) or other trees may result.

fig. 429) while similar hedges in Westphalia have been dealt with by Wittig (1976). Originally none of these hedges had been planted as a windbreak. They served mainly as a kind of stabilised fence between individual fields after a general consolidation of farmland. Moreover they provided the farmers in these widely deforested areas with firewood, which was cut regularly before turning the grazing land into arable land for some years.

(b) 'Mantle' and 'fringe' communities at woodland edges

A clear division of hedges, groups of bushes and shrubby edges of broadleaved woods on the one hand from the real woodland communities on the other is made more difficult by the fact that in former times the woodlands were opened by grazing and coppicing. Owing to this the light-requiring shrubs had established themselves well inside the woods, and in many places they are still surviving there. They reached their greatest distribution about 200 years ago just before the start of modern forestry methods. Moreover the phytosociologist who aims at separating the shrub and woodland communities along the forest borders has to cope with a steep ecological gradient. The shrubs form only a narrow outer cover (fig. 430), and woodland plants are found growing right to the margin because of the supply of seeds from the interior. It is only since Tüxen (1952) proposed a comprehensive system of classification for the shrub communities of central and western Europe and collected together all the relevant material distributed throughout the literature that more careful attention has been paid to the special character of the woodland edges.

Not all the woodland communities in Central Europe have a surrounding **'mantle'** of shrubs showing characteristics different from those of the wood itself. Many shade trees, especially Spruce and Fir, but also Beech, are inclined to put out low branches along the edge of their stand right down to the ground making it impossible for light-loving shrubs to grow. The richest edge communities are found around mixed Oak woods, especially the xerothermic ones (see section B III 4).

This is also the best place to see the herbaceous woodland margin along the front of the shrubs (fig. 431). Such a 'strip' of herbs is distinguished from the adjoining poor pasture or dunged meadowland by the taller growth and larger leaves of its members, in part because they are protected against mechanical damage. So many species enjoy their ecological optimum along the woodland edges that Müller (1962) proposed they should be seen as character species of separate associations or higher units. These were called 'Saum-Gesellschaften' (herbaceous woodland edge or **'fringe'** communities). Müller's idea has been opposed, especially by the Hungarians (Jakucs 1972) because in those areas where the woods are relatively open and rich in light-loving herbs there can be no sharp distinction drawn between the main part of the wood and the communities along its edge. A clear separation does nevertheless appear to be justified in the cooler, moister parts of Central Europe where the natural woods tend to be very dark. This view is supported by the thoroughgoing ecological investigations of the 'habitat gradient at woodland edges' carried out by Dierschke (1974b, see figs. 430, 432 and 433). We shall keep mainly to his scheme in section 2.

Fig. 429. The zonation of the vegetation in a banked hedge running east-west in eastern Schleswig-Holstein, slightly-schematic. After Weber (1967), modified.

Inside the hedge itself, consisting mainly of Hazel and Blackthorn bushes, there are woodland plants, especially early developing geophytes. Along the edges Blackberry shoots hook themselves up on the shrubs. Rhizome geophytes such as Couch Grass and Creeping Thistle flourish on the sunny side of the bank while the shaded side, with its much damper atmosphere, favours woodland plants which require a fair amount of light such as wood Poa, Wood Stitchwort and Male Fern. At the bottom of the bank on this side grassland plants such as the Roughstalked Meadowgrass and Creeping Buttercup find a rather moist soil from which they can invade the adjoining arable or grass field.
According to measurements made by Weber on 13.8.1965 near Segeberg when the air temperature in the open was about 20°C ground temperatures of over 25°C (up to a maximum 58°C) were recorded on the south side of the bank while on the north side the highest surface temperature was just over 15°C. There was a corresponding difference, but in the opposite direction, in the humidity of the air.

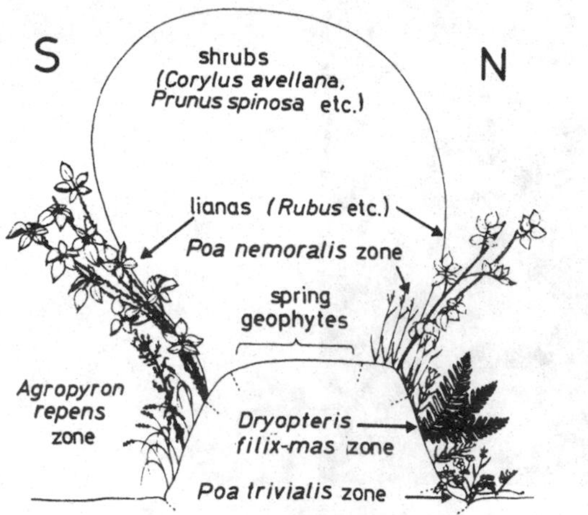

(c) Shrub communities in Central Europe

The amount of light available and the mechanical effects of cutting and burning are such deciding factors in the life of most of the hedge and bush communities that the influences of soil and climate are of minor importance. Very many woody plants therefore are to be met with throughout almost the whole of Central Europe (the character species for the order and class are given in tab. 109). However the four alliances within the order *Prunetalia* do have a tendency to be

Fig. 430. The gradient of light intensity and plant species composition along a SW – NE transect from grassland (*Arrhenatheretum*) through a 'fringe' of herbaceous plants (*Trifolio-Melampyretum*), a 'mantle' of scrub (*Carpino-Prunetum*) and on into the wood itself (*Melico-Fagetum*) on similar deep loess soil. After Dierschke (1974), modified.

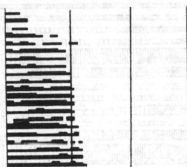

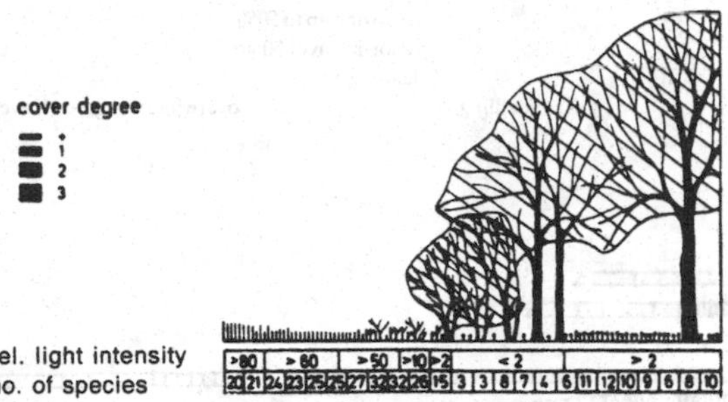

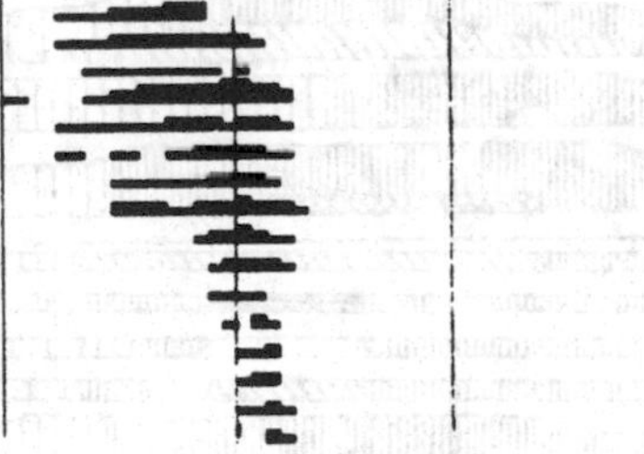

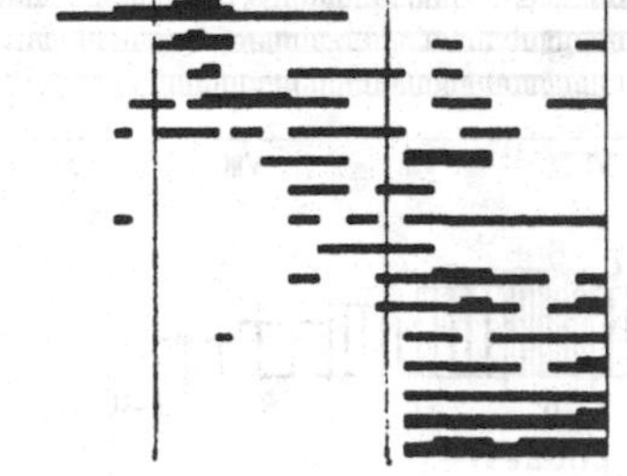

distributed according to the climate:

1 **Subatlantic** shrub communities on acid, mostly sandy soils dominated by Blackberry (*Rubion subatlanticum*),
2 **True Central European** shrub communities on more or less loamy soils *Prunion spinosae*),
3 **Warmth-loving** shrub communities on sunny and stony slopes, especially in the south (*Berberidion*),
4 **Continental** shrub communities on loamy soils (*Prunion fruticosae*) which are better developed in south-eastern than in Central Europe.

The units with a more westerly distribution are rich in species of *Rubus* yet it would be misleading to translate *Rubion subatlanticum* as 'bramble hedges' since there are often taller bushes in it which overshadow the Blackberry to a large extent. It is only on poorer sandy soils that the physiognomy of the hedge is dominated by the **Blackberry** in a way similar to that seen in England, Ireland, northern France, Belgium and Holland in grazed areas, where hedges of *Rubion subatlanticum* are a feature of the landscape and have a number of associations (see Weber 1967). Natural bush communities of *Rubion subatlanticum* are as good as absent from the Central European area.

In the middle and south of Central Europe and also on the fertile loamy soils of eastern Schleswig-Holstein the alliance *Prunion spinosae* is represented by a community rich in species, that has been named by Tüxen on systematic grounds after the **Blackthorn and Hornbeam** even though they may be less obvious (tab. 109). A number of nitrogen-loving plants distinguish this association from others which do not grow so luxuriantly. If the hedges are not cut down very often Hornbeams and other woodland trees become prominent, and in fact the longer the interval between cuts the more like a woodland the hedge becomes, especially if it is growing on deep soil rather than on a bank. Thus many hedges in the Aachen area (Ruhrberg-Monschau, see Schwickerath 1953) from an altitude of 500 m upwards have Beech as the dominant tree, whereas those at lower altitudes can be looked

Fig. 431. A fringe of warmth-loving herbaceous plants with *Dictamnus albus* along the edge of a Downy Oak coppice in the south-western part of the Kaiserstuhl.

Fig. 432. A phenological diagram of a herbaceous fringe community (*Trifolio-Melampyretum veronicetosum*) along the man-made edge of a wood near Göttingen in 1970. Two sample areas were investigated. After Dierschke (1974). The figures 1 – 10 indicate the stages in the vegetative development of each species. (See the list and the vertical hatching.) The cover degree is indicated by the height of the strip diagram. The period of flowering is marked with thicker lines, and the colours of the flowers by different signs.

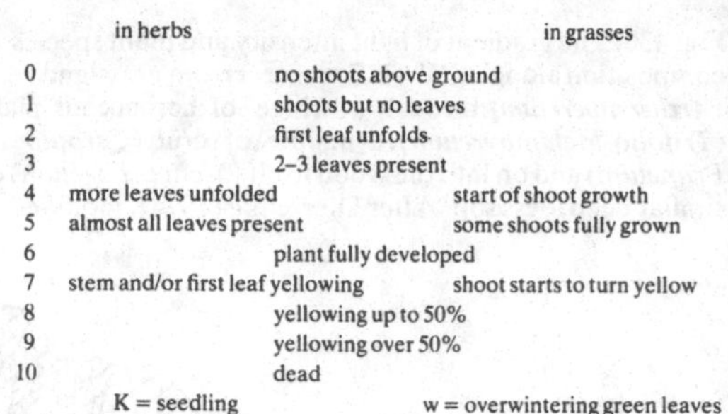

	in herbs		in grasses
0		no shoots above ground	
1		shoots but no leaves	
2		first leaf unfolds	
3		2–3 leaves present	
4	more leaves unfolded		start of shoot growth
5	almost all leaves present		some shoots fully grown
6		plant fully developed	
7	stem and/or first leaf yellowing		shoot starts to turn yellow
8		yellowing up to 50%	
9		yellowing over 50%	
10		dead	
	K = seedling		w = overwintering green leaves

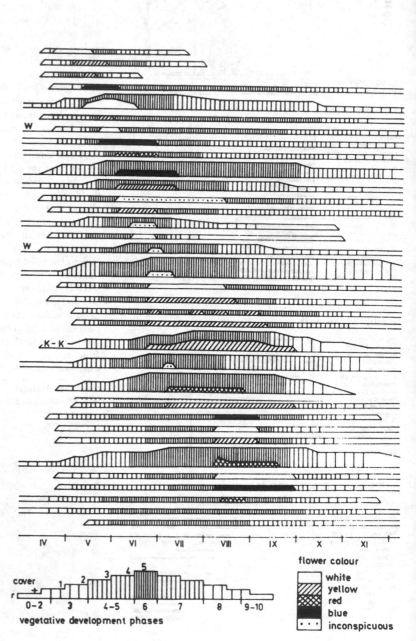

Anemone nemorosa
Ranunculus auricomus
Ranunculus ficaria
Viola reichenbachiana
Stellaria holostea
Taraxacum officinale
Fragaria vesca
Ajuga reptans
Rumex acetosa
Veronica chamaedrys
Ranunculus polyanthemos
Plantago lanceolata
Ranunculus bulbosus
Poa angustifolia
Phyteuma spicatum
Dactylis glomerata
Festuca rubra
Plantago media
Lathyrus pratensis
Leontodon hispidus
Lotus corniculatus
Melampyrum nemorosum
Brachypodium pinnatum
Trifolium medium
Hypericum perforatum
Knautia arvensis
Pimpinella major
Agrimonia eupatoria
Betonica officinalis
Pimpinella saxifraga
Scabiosa columbaria
Cirsium acaule
Achillea millefolium
Viola hirta

W
W
K - K

IV V VI VII VIII IX X XI

cover
r + 1 2 3 4 5
0–2 3 4–5 6 7 8 9–10
vegetative development phases

flower colour
white
yellow
red
blue
inconspicuous

Tab. 109. A synopsis of the species making up the scrub and hedge communities in Central Europe.

Species occurring in many hedges and scrub:

Prunetalia spinosae Character species of the order:	*Querco-Fagetea* Char. spec. of the class:	Widespread accompanying species:
Clematis vitalba	*Acer campestre*	*Ilex aquifolium* (subatlantic)
Cornus sanguinea	*Carpinus betulus*	*Juniperus communis*
Crataegus monogyna	*Cornus mas* (in the South)	*Quercus petraea*
Crataegus laevigata	*Corylus avellana*	*Quercus robur*
Euonymus europaeus	*Fagus sylvatica*	*Salix caprea*
N *Humulus lupulus*	*Fraxinus excelsior*	N *Sambucus nigra*
Prunus spinosa	*Hedera helix*	*Sorbus aucuparia*
Rhamnus cathartica	*Lonicera xylosteum*	*Brachypodium pinnatum* (on limestone)
Rosa canina	*Quercus pubescens* (in the South)	*Origanum vulgare*
Rosa corymbifera	*Ulmus minor* (in the South and East)	*Viola riviniana*
Rosa pimpinellifolia	N *Aegopodium podagraria*	etc.
Rosa rubiginosa	N *Geranium robertianum*	
Rosa div. spec.	N *Geum urbanum*	
Rubus canescens	*Melica nutans*	
Rubus div. spec. etc.	N *Mercurialis perennis* (on limestone)	

Subatlantic (-atlantic)	Submediterran.-Central European		South-east and east European
Rubion subatlanticum Char. spec. of the alliance: *Malus sylvestris* *Rubus affinis* *Rubus vulgaris* N *Rubus* div. spec. ? *Viburnum opulus* etc.	**Prunion spinosae** Important association *Prunus spinosa-Carpinus betulus*-Ass., Differential species: N *Chaerophyllum temulentum* *Dactylis glomerata* N *Galium aparine* N *Glechoma hederacea* *Poa nemoralis* N *Urtica dioca* etc.	**Berberidion vulgaris** Char. spec. of the alliance: *Amelanchier ovalis* *Berberis vulgaris* *Cotoneaster tomentosus* *Prunus mahaleb* *Rhamnus alpinus* etc. Differential species: *Ligustrum vulgare* *Viburuum lantana* etc.	**Prunion fruticosae** Char. spec. of the alliance: (still not clarified) *Prunus fruticosa* etc. *Anemone sylvestris* etc.
Remark: Tüxen (1952) recognises a 5th alliance of the coastal dunes, *Salicion repentis*, which however has few species in common with the other four (compare section CV 1d).		Important associations: *Prunus spinosa-Ligustrum*-Ass. (merging into the *Prumus spinosa-Carpinus*-Ass.) *Cotoneaster integerrima-Amelanchier ovalis*-Ass. (natural steppe-heath scrub) etc.	

Note: N = species requiring a high nitrogen supply. (See also section E III 1, No. 8.41.)

Source: Mainly from data in Tüxen (1952).

upon as fragments of the Oak-Hornbeam woodland. According to Wollert (1970) the boundary hedges in Mecklenburg do not grow on banks either and they are like woodland, especially as individual trees are generally spared when the shrubs are cut, and some have become quite old (e.g. *Milio-Euonymo-Coryletum*).

Between the subatlantic hedges and the *Berberidion* shrub communities, which prefer a warm habitat but are able to tolerate drier periods, there comes the **Blackthorn-Privet scrub** (*Prunus spinosa-Ligustrum* ass.). This is the most frequent scrub community in central and western Germany as well as in the central area of Switzerland. It reaches its best development on calcareous soils either as a field boundary or at woodland edges. Within the *Berberidion* alliance (tab. 109) there are a few shrub communities which remain free from trees naturally. The **Rock Pear scrub** (*Cotoneaster-Amelanchier* ass.) for example develops on very shallow soils between *Xerobromion* – rock heaths and open xerothermic Oak woods and is part of the steppe-heath mosaic. However in Central Europe most of the shrub communities owe their existence to man's activity, especially hedges on field boundaries and thorn thickets which come in when the pastures are not tended. The alliance gets its name from the Barberry (*Berberis vulgaris*) which is plentiful in these patches of pasture scrub. It is characteristic of the dry valleys in the inner Alps where it was defined by Braun-Blanquet (1961, consult for previous literature).

The hedges and woodland margins are already beginning to take on a more continental look in the dry area to the east of the Harz. Field Elm (*Ulmus minor*) plays a conspicuous part here as it does in many submediterranean and south-eastern landscapes. Here one also occasionally meets the **Dwarf Cherry** (*Prunus fruticosa*) which is otherwise quite rare. Hedge communities in the alliance *Prunion fruticosae* are distributed right from Bulgaria through to Finland. Its western outposts lie in the dry parts of Rhein-Hessen, which also contain many other continental plant communities.

Bush and woodland edge communities in the flood plains are interspersed with the natural woodlands and are included in the riverside willow communities (*Salicetea, Salicetalia, Salicion purpurea*, see section B V 1 d).

According to Moor (1958) there is a **Bird Cherry-Hazel bush** (*Pado-Coryletum*) corresponding to the Ash-Elm wood (*Fraxino-Ulmetum*), a **Willow-Guelder Rose bush** (*Salici-Viburnetum*) corresponding to the Grey Alder wood (*Equise-*

Fig. 433. Daily movements of the air temperature, saturation deficit, evaporation and wind on the sunny slope of the Kohnstein, a limestone terrace in the Werra valley, during clear weather in summer (9 July). After Dierschke (1974), somewhat modified.

It is very warm and dry throughout the day on the stony slope of the Blue Moorgrass sward (*Teucrio-Seslerietum*). Compared with this the microclimate in the herbaceous fringe (*Geranio-Peucedanetum*) is already more equable. Near the edge of the adjoining stunted Sedge-Beech wood (*Carici-Fagetum*) the conditions are almost as evened out as in the Tall Melick-Beech wood (*Melico-Fagetum*) growing on the loam-covered limestone plateau. The Blue Moorgrass slope is naturally free from trees and shrubs.

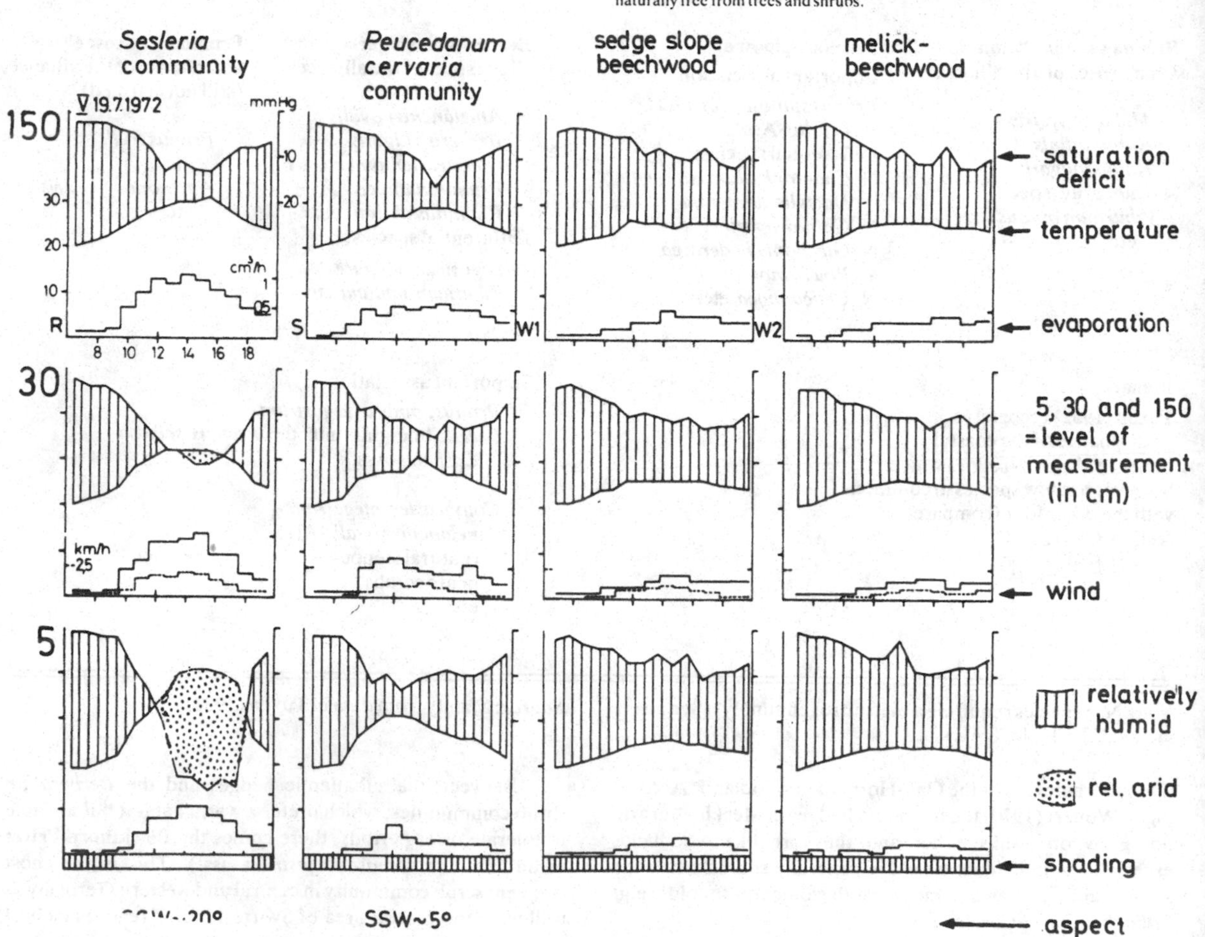

to-Alnetum) and an **Almond Willow-Osier scrub** (*Salicetum triandro-viminalis*) corresponding to the white Willow wood (*Salicetum albo-fragilis*, section C V 1 d). The last named can be considered as the natural community occupying the edge of flood-plain woodlands.

On the other hand the mire willow scrubs which live on very wet peaty soils have a strong floristic connection with the class *Alnetea glutinosae*; according to Oberdorfer (1979) they form a separate order (*Salicetalia auritae*) and alliance, the **Alder Buckthorn-Eared Willow** shrub communities (*Frangulo-Salicion auritae*).

As can be seen particularly from the last comments the hedges and shrub communities are good indicators of the potential natural woodland communities in the place of which they have developed (see e.g. Faliński, Hornkievicz-Sudnik and Fabiszewski 1963, Jurko 1964). A word of caution however since the habitat may be considerably changed artificially, and shade trees like Beech cannot exercise their full competitive strength in a narrow hedge, to say nothing of coniferous trees most of which are very susceptible to being cut back.

This outline may suffice for now, and we must also forgo adding to the lively discussion on the value of hedges in improving the climate or treating practical questions as for instance the choice of woody species for new planting. The reader should refer to the publications by Troll (1951 a, b), Tischler (1951), Ellenberg (1954a) Steubing (1960), Th. Müller (1964), Mestel (1965), Ehlers (1960) and Wittig (1976).

2 Herbaceous edging to woodland and shrub communities

(a) Warmth-requiring, drought-resistant fringe communities

Herbaceous plants growing along the edges of shrub or tree stands require the protection afforded by the woody plants against too much exposure to the sun or from grazing animals or mowing. At the same time they avoid the deep shadow of the bushes and trees which means that they are restricted to a narrow strip, often less than 1 m wide (see fig. 430). Thus they border the shrub 'mantle' like a 'fringe' (German: Saum). Nevertheless a large number of species meet here and form many different communities. Depending mainly on the aspect of the site they can be grouped into two quite opposite floristic and ecological classes:

1 **Warmth-requiring, drought-resistant** fringe communities (class *Trifolio-Geranietea*, with the order *Trifolio-Origanetalia*) on the sunny side where the nutrient supply is rather limited and reminds one of the slightly arid grasslands.
2 **Nitrate and higher humidity** requiring fringe communities (belonging to the class *Artemisietea vulgaris*, especially to the order *Galio-Calystegietalia*). They require a very uniform temperate microclimate and a fertile soil.

According to Dierschke (1974) who discusses all previous literature the distribution centre for the first named class lies in Central Europe but it can also be traced to southern England, central Scandinavia, western Russia, Austria, northern Italy and the greater part of France. Important species by which the units can be recognised and separated are *Vincetoxicum hirundinaria*, *Tanacetum corymbosum*, *Bupleurum falcatum*, *Peucedanum cervaria*, *Anthericum ramosum* and other more or less tall, weakly scleromorphic perennials (see fig. 431 and section E III 1, No. 6.1). Most of these were previously considered as character species of the xerothermic mixed Oak woods (*Quercetalia pubescentis*, see section B III 4 b); at least they can serve as its differential species to distinguish the xerotolerant woodlands from the mesophilous mixed broadleaved ones (*Fagetalia*).

Now that the grazing of woodlands and the unregulated cutting of timber and firewood have stopped and the wood canopy has become more closed, causing the light-loving species to suffer or to disappear, it is possible to separate the warmth-loving **herbaceous fringe communities** from the mixed Oak woods. It is only on rocky places such as the Isteiner Klotz in the southern Rhine plain, or the Kohnstein in the Werra valley that intermediate stages are found between Stone Heath patches, fragments of arid grassland, herbaceous and bushy woodland edges, stunted open woodland and dense high forest. In such places the pattern of plant formations is drawn by the natural habitats according to the depth of soil and the availability of water during dry periods. On the other hand the majority of the woodland edges in Central Europe have without doubt been designed by man and placed on uniformly deep soil. The habitats here are less extreme with regard to water supply and nutrients, but the amount of sunshine, and with it the species combination, shows a more abrupt gradient than in the natural mosaic which has just been described, or in areas with a relatively low rainfall, e.g. the Hungarian lowland.

The natural decidedly xerothermic fringe communities form the basis of an alliance which Dierschke (1974) has recently designated *Vincetoxico-Geranion sanguinei* **(Swallow-Wort- Bloody Cranesbill fringe).** But even within this alliance there are many communities owing their existence to man. On very sunny shallow calcareous soils can be found the Bloody Cranesbill – Stagwort fringe (*Geranio-Peucedanetum cervariae*) which is scattered throughout the distribution area of *Trifolio-Geranietea*. Its unusual habitat accounts for a few really faithful character species, especially *Libanotis sibirica, Thesium bavarum* and *Coronilla coronata*. The remaining communities, e.g. the strongly thermophilous *Geranio-Dictamnetum* play only a small part (fig. 431).

On more or less deep soils and exclusively along the edge of man-made woodland there are communities of the mesophilous alliance *Trifolion medii* (**Zigzag Clover**). It is only weakly characterised by *Trifolium medium* and *Agrimonia eupatoria* but it does lodge a number of meadow plants which distinguish it from the *Vincetoxico-Geranion*, especially *Dactylis glomerata*, *Achillea millefolium*, *Knautia arvensis*, *Lathyrus pratensis, Veronica chamaedrys,* etc. Dierschke could find no character species for the commonest association, the **Zigzag Clover-Agrimony** (*Trifolio-Agrimonietum*). The less common communities are recognised by character species, e.g. *Trifolio-Melampyretum nemorosi* (in the north-east of Central Europe as far as the Leine-Werra mountains, see fig. 432, *Vicio cassubicae-Agrimonietum* (in the east), *Vicietum sylvaticae-dumetorum* (more in the west) and *Teucrio-Centauretum nemoralis* (in the south-west).

In view of the small number of really exclusive species it must be questioned whether raising the status of the fringe communities to that of separate alliances or even classes is really justified. Apart from the fact that they could only be included in other classes under pressure there are several reasons for giving prominence to them, particularly ecological ones. In the first place their unique position is a result of the special microclimate brought about by the radiation which is quite different from that within the woodland or on open grassland. During the course of the day the plants along the edge of woody stands may at one minute be fully exposed to the intense sunlight and a few minutes later be shaded by a bush or tree branch, only to be exposed again a little later. Thus it is subject to abrupt and frequent changes in humidity and temperature (see fig. 433). The microclimate varies as it were from that of the open country to that of the woodland instead of being intermediate between the two as may be inferred from a consideration of daily, monthly or yearly averages. With regard to soil factors Dierschke could find no significant peculiarities about the sunny herbaceous fringe communities. The nitrogen supply for example is similar to that in the neighbouring wood or that of the poor grassland, if adjoining, on the other side.

(b) Fringe communities requiring nitrate and a high humidity

In contrast to the sunny fringe communities those on the shaded side of the woody stand enjoy a more even microclimate as well as a soil which rarely dries out and provides ample nitrogen. Both factors mean that the communities of the order *Galio-Calystegietalia* are dominated by large-leaved, more or less hygromorphic herbs, whereas *Trifolio- Origanetalia* communities are composed mainly of tough-leaved or at best mesomorphic species. Umbelliferous representatives of the former were studied by Janiesch (1973b) from the ecophysiological viewpoint.

Amongst the commonest character species for the order are *Galium aparine*, *Glechoma hederacea*, and *Geum urbanum* which are also to be found in flood plain woodland. Plants of waste places with a similar structure such as *Urtica dioica*, *Galeopsis tetrahit* and *Artemisia vulgaris* are also typical for the class *Artemisietea*. In addition nearly always accompanying species whose home is the open woodlands of river valleys play a part, but these only reach their full development in manured meadows (section D V), especially *Dactylis glomerata*, *Poa trivialis*, *Anthriscus sylvestris*, *Ranunculus repens* and *Vicia sepium*.

The natural habitat of such herbaceous communities then is also to be sought in flood plains (section B V). The alliance of the **Ground Elder** (*Aegopodion*) also includes for example the Butterbur community (*Petasitetum hybridi*, see fig. 204) which is found on the banks of small rivers or streams, but is absent in the north of Central Europe. Sissingh (1973) put this striking largeleaved association along with other semi-shade herbaceous communities into a montane suballiance named after the Red Campion as *Sileno dioicae-Aegopodion*. Some high-montane and sub-alpine herbaceous fringes also belong here, such as the *Chaerophyllum hirsutum* community along streams in semi-shade, and the *Rumici alpini-Aegopodietum* described from the Tatra.

The **Curled Thistle-Golden Chervil** (*Carduo crispi-Chaerophylletum aurei*) may have occurred naturally here and there in the flood valleys of the planar and colline levels, but most of the herbaceous fringe communities of Central Europe requiring high nitrogen supply and humidity have originated in the semi-shade of managed woodlands or hedges. Sissingh also put this community in a suballiance with White Dead-nettle (*Lamio albi-Aegopodion*), which is mainly distributed from the planar to the submontane belts. By far the most frequent fringe community in Central Europe, the **Stinging Nettle-Ground Elder** (Urtico-Aegopodietum) belongs here. As the 'central association' (in the sense of Dierschke) this has none of its own character species, just like the *Trifolio-Agrimonietum* mentioned in section a. It corresponds most closely to the picture of the shaded fringe communities given in the introduction. Because of its very wide distribution several subassociations can be distinguished (Tüxen 1967c).

Communities dominated by the tall Chervil species require rather more warmth and light than the *Urtico-Aegopodietum*, and they fade out towards the north of Central Europe. The most important are *Chaerophylletum aurei* (subcontinental, submediterranean) and *Chaerophylle-*

tum aromatici (strongly continental). The *Urtico-Cruciane-tum* of the Leine-Werra mountains added by Dierschke (1974a) also makes greater demands in this direction.

Although all the fringe communities mentioned so far consist mainly of perennial and biennial species or are at least dominated by them, there are many associations among the nitrate- and humidity-loving communities where short-lived species play the main part. These wayside communities are found only in the lowlands, and Sissingh (1973) put them together in a separate alliance which Dierschke (1974a) named after the Nipplewort and Herb Robert (*Lapsano-Geranion robertiani*; but it may be preferable to join them with the *Geo-Alliarion*, see section E III 1, No. 3.522).

These two species, along with perhaps *Alliaria petiolata*, are the only character species. The latter often forms **Garlic Mustard-Rough Chervil fringes** (*Alliario-Chaerophylletum temulenti*) along strongly eutrophic woodland edges, and these are the chief habitat of the Greater Celandine (*Chelidonium majus*).

This rather thermophilous community is replaced by the **Broadleaved Willowherb-Herb Robert fringe** in somewhat coller and damper habitats (*Epilobio montani-Geranietum robertiani*). Some other communities of the *Lapsano-Geranion* (or *Geo-Alliarion*) require more light and these lead on to the herbaceous vegetation of woodland clearings (section D III 2), e.g. *Torilidetum japonicae* and *Cephalarietum pilosae*. On damp nitrogen-rich soils such as the habitats of the Alder-Ash woods (section B V 1 f) the Hemp Agrimony forms a community of tall perennials (*Eupatorietum cannabini*).

V Hay and litter meadows

1 *A general survey of farmed grasslands*

(a) *Grazing and mowing as decisive factors in the habitat*

Without the scythe and the grazing animal there would be neither meadow nor intensive pasture in Central Europe where the climate favours woodland growth (see section A II 6). It is only the regularly repeated interference by man, either directly or indirectly, which prevents the trees and shrubs from reestablishing themselves, sooner or later, on the land which has been wrested from them.

As in the arid and slightly arid grasslands it is the light-loving easily regenerating hemicryptophytes which dominate cultivated grassland. Their growth here is so strong that the therophytes can scarcely find a plate to develop, especially as the hemicryptophytes often remain green throughout the winter or begin to produce fresh shoots early in the spring. Geophytes and chamaephytes too are seldom to be found in the majority of meadow communities.

Of all the life forms it is only those species which can adapt to the rhythm of the husbandry methods that are able to grow and spread. However the advantage they get from the mowing and grazing is in any case relative. In themselves all plants and the majority of individuals are more or less badly damaged. However the characteristic meadow and pasture plants withstand the damage better than other species living in the open country. The mechanical interference removes the competitors which would otherwise take away their light. Different life forms and species are favoured by different methods of husbandry (fig. 435).

Mowing is a sudden cutting off and only the leaves and stems close to the ground escape. Afterwards all the plants have an equal start and the ones growing most quickly then will win the competition. In meadows cut only once or twice in the season it is the tall grasses and other herbaceous perennials which become dominant. The more often the stand is mown the more numerous lower growing herbs and grasses become. Regularly cut lawns and playing fields consist entirely of short grasses, Creeping White Clover and a few rosette plants.

The so-called **litter meadows** (fig. 434), which today are confined to a few places in the Alpine foothills, are cut once only for getting straw, and this is done in the autumn after nearly all plants have become strawy and most of them have already ripened their seeds. In fact they have had time to transfer most of their assimilated material down into storage organs on or below the soil surface, e.g. buds, rhizomes or roots. There is little advantage in being able to spread vegetatively, and quite often it is the tussock-forming plants like *Molinia caerulea*, *Bromus erectus* or *Carex davalliana* which become dominant. These live for only a few years and must be renewed from seed. The Purple Moorgrass which is at home in woodlands on more or less damp soils has become the litter grass par excellence. Although it is late to start into growth and is the last of all the grasses to flower it can collect large reserves of nitrogen, phosphorus and other nutrients in the swollen internodes at the base of its apparently node-free stem ready for the following season. It soon loses its competitiveness if it is cut too soon or if grazing animals damage its storage organs lying close above the ground. The harvesting of dead parts removes such a small quantity of essential nutrients that these areas will continue to provide straw year after year without

Fig. 434. A litter meadow with ricks in eastern Estonia shortly after being mown in October. The shrubs in the foreground and the woodland behind are occupying drier places where less grass grows. Left to nature the whole area would become thickly wooded.

Fig. 435. A schematic synopsis of the ways in which management of grassland affects the plant communities.

The **litter meadow** is cut so late that the growth has turned to straw and consists of little more than carbohydrates. As a result the yield can be maintained without any addition of plant nutrients whereas in order to keep up the production from **hay meadows** cut once or twice a year for feed they must be regularly manured. The number of plant species per unit area – including the 'weeds' – is greatest under a less-intensive system where the field is cut for hay only once in the season and then grazed.

The extremes regarding number of species, abundance of 'pasture weeds', shortage of nutrients and poor yield are found in **rough grazing** or common land which was formerly the usual means of husbandry. In contrast the fields used for rotational grazing, which may also be mown occasionally, give the highest output and have the least variety of species with few weeds. However this method also requires more work and a higher level of manuring. Pastures which are set **stocked** are being continuously manured by the animals which remain in them throughout the season. Weeds are common under this method, especially in those places where the animals will not graze for some time because of their dung. With cattle these patches are small but spread over the whole area; horses instead tend to dung in one area which becomes covered with weeds and coarse grass.

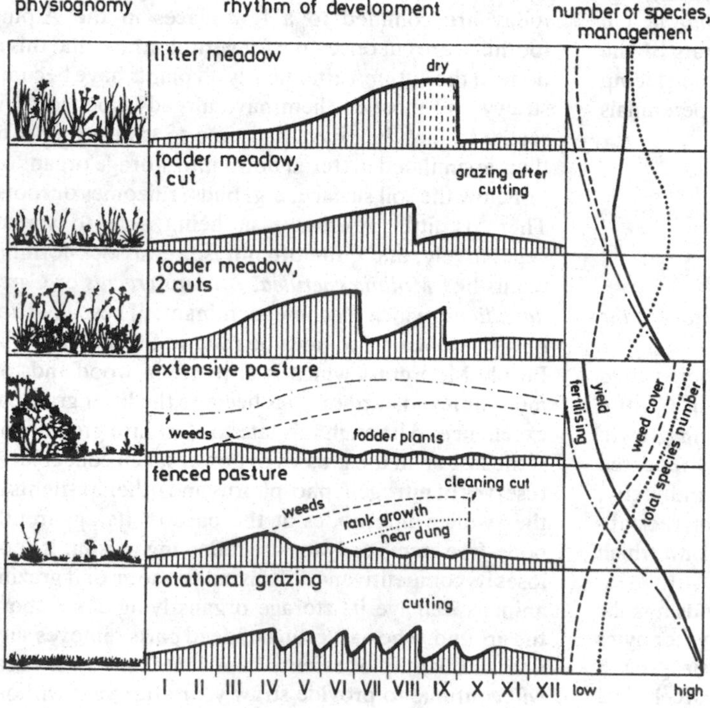

manure, provided there is sufficient water available. The litter which is removed consists mainly of cellulose, other carbohydrates and silica; it contains hardly any protein or other nutrient-rich compounds.

If a meadow is cut early in the summer when it is still green in order to obtain a protein-rich fodder then the field is soon impoverished by the removal of bioelements, and small slow-growing plants requiring little in the way of food are able to spread. **Fodder meadows** therefore must be manured regularly, even if they are cut only once a year, and this is even more necessary if two or three crops are taken each year. Under such circumstances species are encouraged which benefit most from the addition of nutrients, especially nitrogen (fig. 437), and grow quicker than the plants of the litter meadows (fig. 435). A characteristic representative is *Arrhenatherum elatius*, one of the most valuable hay grasses. Like *Molinia* the False Oatgrass is a tall growing tussock former and relies on occasional regeneration from seed, but it flowers much earlier, ripens its seeds even more rapidly and lays up fewer reserves, even if the straw is allowed to die naturally on the plant.

Repeated mowing results in an alternation of tall growth with short growth phases (fig. 435). The life rhythm of grasses and herbs has to fit in with the given sequence of shade and light phases. Spring flowering plants like *Primula acaulis*, *Leucojum vernum*, *Crocus albiflorus* (fig. 436) or *Narcissus* species (fig. 450) only develop during the first period of short growth. They adorn the meadows especially where a deep layer of snow has pressed the grass down to the ground in winter and on into the spring, but is then rapidly melted away by the warm föhn wind. Grasses and other hemicryptophytes require a little while to recover whereas *Crocus* and most other geophytes profit by this interval to shoot up unhindered. The deciding factor then is the ability to make this rapid growth at the critical time which, according to Duhme and Kaule (1970) also favours the sub-spontaneous spread of the Daffodil (*Narcissus pseudo-narcissus*). In the lowlands with their cold winters but small amounts of snow bulb geophytes cannot enjoy this competitive advantage and so are unable to produce the spring display which makes our mountain meadows so attractive. Dandelions, Daisies and low-growing leguminous plants flower in each period when the grasses are cut short, and the more often this is done the more freely they flower. The Autumn Crocus (*Colchicum autumnale*) does not produce its pale flowers until the last period of short growth; much later, in the early summer, it sends up its fruiting stalk with tulip-like leaves amongst the grasses. The Eyebright (*Euphrasia rostkoviana*) is a well known example of so-called seasonal dimorphism. Its longer unbranched form grows up with the grasses and manages to flower and fruit in the first tall growth. Its branched form however, also genetically determined, does the same in the later short growth. Tall grasses and other perennials in the hay meadows become dominating in the tall growth phases, some of them making fresh growth each time, others tending to concentrate their growth into one of the phases. For example Cow

Fig. 436. A Yellow Oatgrass meadow (*Trisetetum*) in the Upper Engadine shortly after the snow has melted in spring. *Crocus albiflorus* is starting to flower. This geophyte is so common here that hundreds of corms are thrown out of their burrows by the field voles.

Fig. 437. A False Oatgrass meadow (*Arrhenatheretum typicum*) near Gross-Bieberau in Hesse shortly before the first cut. *Anthriscus sylvestris* is in flower, *Cerastium fontanum holosteoides* (left) and *Trifolium pratense* can be seen in the undergrowth.

Parsley (*Anthriscus sylvestris*) generally flowers before the first cut (fig. 437) while *Heracleum*, *Cirsium oleraceum* and other large-leaved herbs only flower after the first, but before the second-cut where the meadow is mown twice (figs. 453 and 454).

Within the sequence: litter meadow, once-cut hay meadow, twice-cut hay meadow and finally grassland which is cut several times, the number of species first rises and then falls (see fig. 435). Because of the earlier cutting many low-growing plants find conditions more favourable in the hay meadow than in the late-mown litter meadow. Repeated cutting progressively reduces the number of taller plants. In addition species sorting their nutrients in seeds on which they rely for reproduction also tend to become less frequent as the years go by.

While the scythe or mowing machine cuts everything impartially the grazing animal can exercise selection. In the case of **extensive rough grazing** as described in section A II 2 this finally leads to 'pasture weeds' getting the upper hand. Also where **set stocking** is practised, i.e. a constant number of animals have use of the whole area for the whole season, certain plants are eaten less than others and thereby favoured indirectly. With the modern method of **rotational grazing** however all the plants are eaten or trampled down at the same time, and then have an equal chance of recovery while the area is being rested. The more intensive the grazing and manuring the greater the tendency for the 'carpet plants' such as *Trifolium repens*, *Agrostis stolonifera* or *Poa trivialis* to cover the ground. Only species which remain fairly short and are capable of rapid growth and regeneration can withstand the heavy and frequent mechanical impacts. For this they require frequent applications of good dressings of manure or chemical fertilisers. In general the more the grassland is cropped the more it will need compensating for nutrient losses.

(b) Survey of the meadow communities in Central Europe

Alongside the mechanical factors like mowing, grazing and treading there are the climatic and edaphic ones, especially the temperature and the length of the growing period together with the supply of water and nutrients. Each of these has a distinct effect on the composition of the grassland communities.

First of all we will look at the true meadows, i.e. grassland which is only or mainly cut and in which neither selection by grazing animals nor the effect of their treading plays any part. In order to survey at a glance the ecological position of the various plant communities and to relate them to the woodland communities whose sites they have taken we shall use the same ecogram as we did in section B I (figs. 40, 53, 438 and 439). In doing so we shall look first at the submontane belt, where the grasslands occupy large areas, as we did when we discussed the woodlands.

Basically we must distinguish between grassland which is not manured and that in which the nutrients removed are replaced either by natural or artificial manures (fig. 439). The wettest meadows lie in sites where lakes and other still waters have become silted up. The communities which are naturally without trees have already been discussed in sections C I and II. In the geobotanical sense the only **natural grasslands** are the beds of reeds and sedges together with some intermediate mires. The farmer however calls all the wet or damp meadows 'natural' grassland, even though they would be capable of supporting trees. What he means is that they cannot be converted into ploughed land without being drained (fig. 439). Arable cultivation is not possible at the other extreme of soil water conditions either i.e. on very stony dry ground or on pure sand liable to blowing. Therefore most of the arid and slightly arid grasslands too are considered to be 'natural' grassland by the farmers.

Habitats which would naturally be occupied by swamp woodlands, those on the highest parts of fens and on slopes saturated with spring water become small-sedges swamps under extensive cutting. These too have already been discussed as they are very close to the natural swamp communities (see sections C II 2 and C VI 6 a). Where mixed deciduous woodland has been replaced on damp mineral soils or on drained peat there may be litter meadows of Moorgrass (alliance *Molinion*), or, where manure has been applied a moist-soil meadow community of the alliance *Calthion*. *Molinion* and *Calthion* are so similar floristically that they are put together in the order *Molinietalia*. Where the 'wet' and 'damp' meadows receive no manure they produce only a very hard fodder, poor in protein, and should only be cut for litter. They are, or were up to only a few decades ago, in fact almost exclusively used as litter meadows. For the same reason the slightly arid swards (*Mesobromion*) were also often used as a source of bedding for animals.

There is practically no unmanured grassland today on average soils, especially those ranging from fairly damp to fairly dry. These are mostly used for arable farming or for forestry since they are very suitable for practically all field crops or trees. Where they are used as grassland they are tended and manured as intensively as possible. Consequently we are no longer in a position today to give an account of

Fig. 438. An ecogram of the alliances of unmanured meadow communities in the submontane belt of Central Europe (compare the ecogram of the corresponding woodland communities in fig. 53). Today there are practically no extensively managed grassland communities left on soils which have an average amount of water and are not extremely acid since it pays better to use these as arable fields or high-yield grassland.

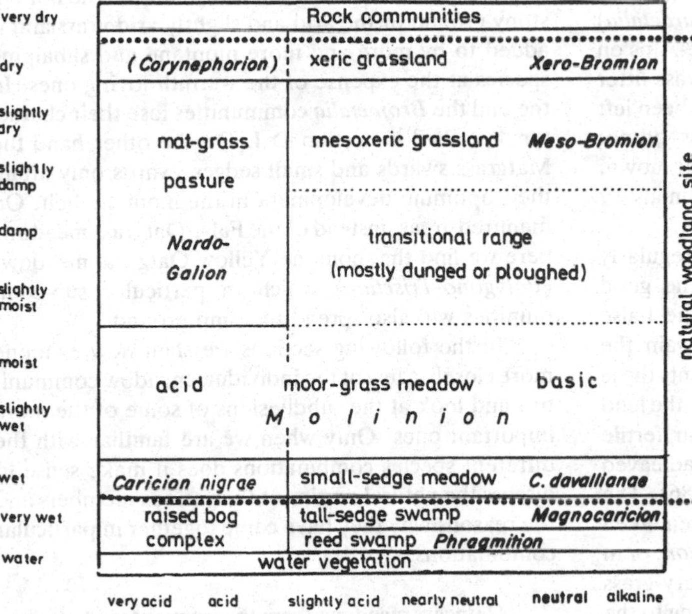

Fig. 439. When fertility is increased through manuring or when flood water brings in plant nutrients other meadow plant alliances are formed in place of the ones shown in fig. 438. Dunged (or fertilised) meadows are absent from very dry habitats since the extra nutrients have little effect under these conditions, and from very acid soils because the very act of introducing plant nutrients also increases the base content of the soil.

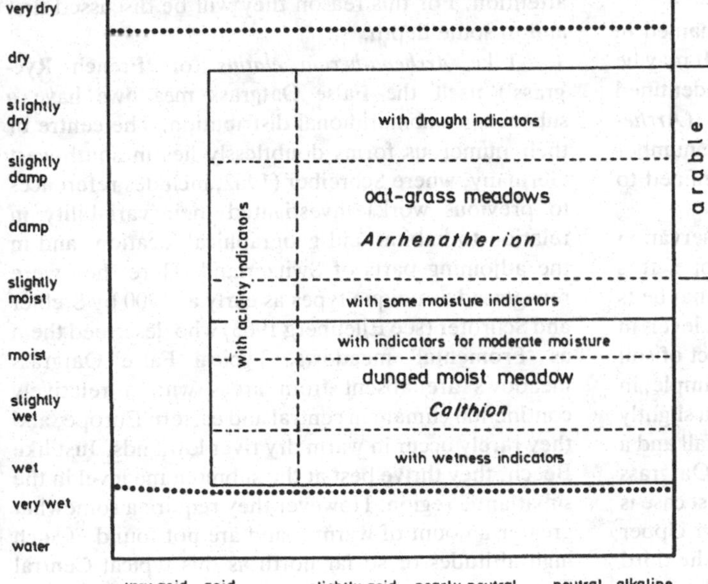

the species composition of an unmanured once-cut meadow on a soil of average water regime. Very probably they would include species from both the damp litter meadows (*Molinion*) and the slightly arid grasslands (*Mesobromion*), together with perhaps some from the acid-soil Matgrass swards (*Nardetalia*) such as we still see here and there in the pre-Alps on drier knolls in the vicinity of Purple Moorgrass litter meadows. Recently some hay meadows have been left lying untended, but these 'social fallow fields' do not develop into meadows since they are no longer mown. Instead they become thickets of tall perennials or shrubs (see section X 2 a).

If a poor meadow is manured and cut regularly and has an average amount of moisture the good fodder plants, such as some species from the False Oatgrass meadows (*Arrhenatherion*), soon gain the upper hand. Without artificially adding nutrients these and similar communities can occur only where the land is occasionally flooded by rivers bringing in fertile sediments, especially on the level of the broadleaved woodland (see section B V 1 g and h, fig. 186). The sites of Willow flood-plain woodlands have been taken over by 'manured' moist-soil meadows (*Calthion*, or in the east *Deschampsion cespitosae*). Reed Canary-grass swamps (*Phalaridetum arundinaceae*) belonging to the *Phragmition* alliance (see section B V 1 b) occupy still wetter sites near some lowland rivers, e.g. large areas by the Havel. In depressions on all flood-plain levels we may find associations or at least fragments of the alliance *Agrostion stoloniferae* (see section VI 2 b) which tolerate frequent inundations by rain or ground-water.

The character species of the alliances named in figs. 438 and 439 are listed in section E III 1. It may be seen that many of the meadow units can be identified very clearly. The False Oatgrass meadows (*Arrhenatherion*) in particular contain an impressive number of species which are more or less closely confined to these mainly anthropogenic communities.

Our ecograms can be applied without reservation only to the submontane belt in Central Europe. It is possible to use them also to the colline and planar belts as long as one bears in mind that the dampness levels in figs. 438 and 439 relate to the combined effect of soil and climatic factors. A deep loess soil, for example, in a warm climate with a low rainfall, may carry a slightly arid grassland community; with a higher rainfall and a cooler climate the same soil may support an Oatgrass meadow or even a moist-soil meadow. The first case is realised e.g. in the Kaiserstuhl hills (southern Upper Rhine plain), the second one is normal, and the third one typical for some rain-rich areas north of the Alps.

Montane and subalpine climates restrict the development of many herbaceous species just as they do that of the woodlands. Thus other communities appear in place of the ones found at the submontane and colline levels. With increasing altitude the flora of stony ground or the arid and slightly arid grassland is added to by more and more montane and subalpine species at the expense of the warmth-loving ones. In the end the *Brometalia* communities lose their character described in section D I. On the other hand the Matgrass swards and small sedges swards only attain their optimum development in the montane belt. On manured areas, instead of the False Oatgrass meadows here we find the montane Yellow Oatgrass meadows (*Polygono-Trisetion*) which in particular sub-communities will also spread to damp ground.

In the following sections we shall now examine more closely a few of the individual meadow communities and look at the subdivisions of some of the more important ones. Only when we are familiar with the different species combinations does it make sense to discuss the natural origins of the various members and the reasons why they have come together in particular combinations.

2 Oatgrass meadows from the submontane belt down to the lowlands

(a) Local variations in the False Oatgrass meadows

Although the False Oatgrass meadows are of rather recent origin in Central Europe they are by far the most plentiful in south-west Germany and in the lower parts of the Alps, and have received the most attention. For this reason they will be discussed first and in some depth.

Like *Arrhenatherum elatius* (or 'French Rye-grass') itself the False Oatgrass meadows have a suboceanic submeridional distribution. The centre of their numerous forms doubtlessly lies in south-west Germany, where Schreiber (1962, includes references to previous work) investigated their variability in relation to habitat and geographical location, and in the adjoining parts of Switzerland. Here they were recognised as unique types as early as 1900 by Stebler and Schröter (see Ellenberg 1963) who described them as 'Fromental' meadows. Typical False Oatgrass meadows are absent from areas with a relatively continental climate in central and eastern Europe, and they rarely occur in warm dry river lowlands. Just like Beech, they thrive best at the submontane level in the subatlantic region. However they require a somewhat greater amount of warmth and are not found at such high altitudes or so far north as this typical Central European tree.

The False Oatgrass meadows are richest in species and best characterised when they are cut twice in the summer and receive a dressing of mainly farmyard manure. In earlier days this was the general custom, mainly in the southern parts of Central Europe. Cutting more often and using a heavier dressing of manure, generally artificial, along with occasional grazing certainly makes them more productive, but this occurs at the expense of the species diversity until finally they lose their character altogether. Since the old methods of farming are being rapidly replaced by more intensive ones, or the grass land is used no more at all there are fewer and fewer typical False Oatgrass meadows. As in the case of the *Calluna*-heaths, which were once so widespread in northwestern Germany, so it is with the south German and northern Swiss *Arrhenatherum* meadows; if some examples are to be handed down to posterity they will have to be protected and managed according to traditional methods. At the present time there are in fact still enough well maintained False Oatgrass meadows which we can use as a basis for studying and classifying them.

Where the loamy soil is deep and moist enough and where there is sufficient manure applied the False Oatgrass meadow remains partly green throughout a normal winter and begins to grow immediately the snow has melted. While the grasses are still short the Lady's Smock (*Cardamine pratensis*) with its countless flowering heads throws a pale violet haze over the deep green. From about the middle of April or early May the False Oatgrass meadows are spotted with yellow as Dandelions (*Taraxacum officinale*) open up their big heads towards the sun. Later on Meadow Buttercups (*Ranunculus acris*) begin to shine out from the grasses which are by now shooting up, and finally the Goat'sbeard (*Tragopogon pratensis*) with its grass-like leaves mixes its large golden-yellow flowers in with the tall grass panicles. When the Oxeye Daisy (*Leucanthemum vulgare*) spreads its white ray florets towards the end of May or in early June, the Cow Parsley (*Anthriscus sylvestris*) hangs its umbels like bright clouds over the abundant variety of leaves and the Hedge Bedstraw (*Galium mollugo*) throws out its white blossoms. Soon the Cocksfoot (*Dactylis glomerata*) begins to shed its pollen and the False Oatgrass (*Arrhenatherum elatius*) to spread out its panicles which were narrow and nodding before. Finally the monotonous dull green of the grasses takes over, relieved only by the soft silvers, golds and purples of the awns.

Suddenly one sunny morning in June sees the whole magnificent display being sacrificed to the mower's blades. Only a pale stubble remains of the tall grasses with hardly any leaf; the smaller grasses, leguminous plants and lowly herbs are recovering for a time from the shade which had been increasing as the weeks went by and which will build up again as the tall grasses and herbs renew their growth. Before the second cut Rough Hawk's-beard (*Crepis biennis*) and Hogweed (*Heracleum sphondylium*), also in some places the Great Burnet Saxifrage (*Pimpinella major*) and Wild Parsnip (*Pastinaca sativa*) have reached their full development. Mainly in the south they are joined by Spreading Bellflower and Meadow Cranesbill (*Campanula patula* and *Geranium pratense*) with their clear blue and violet flowers. However before leaves and stems turn to straw they are cut down again in mid-summer, and it is only on the warmest sites that the tall growing plants are able to flower for a third time.

All the faithful character species of the False Oatgrass meadows in the wider sense (*Arrhenatherion* alliance) belong to the middle and upper layers. They have just been mentioned in the phenological description (see also tab. 110). Amongst the members less closely bound up with the False Oatgrass meadows are various tall and middle-sized grasses, papilionaceous plants and other herbs. Creeping White Clover (*Trifolium repens*) and grasses flowering nearer the ground (such as *Lolium perenne*) are suppressed since they require more light. Only shade-tolerant species, like *Bellis perennis*, *Ajuga reptans* and *Lysimachia nummularia* can live through the time when the meadow has grown tall and wait for the light after haymaking. The low grasses have a lot of leaves and are able to make use of restricted amounts of light just as the plants on the woodland floor do. (see figs. 50 and 117). Intensities of only 1–5% of the full light above the stand are sufficient for them, at least in order to survive (fig. 440). Because of this the lower meadow layers contribute significantly to the total yield, especially in late spring and during the time after cutting when the stand is still short.

The amount and nature of the manuring affect the type of False Oatgrass meadow as does the number of cuts per year (fig. 441). If not manured enough the tall grasses and herbs do not thrive so well as is shown in the figures. Medium-sized grasses such as *Trisetum flavescens* or *Holcus lanatus* and more easily satisfied shorter grasses like *Poa pratensis*, *Festuca rubra* or even *Agrostis tenuis* are then able to spread because there is more light. In the poorest of the False Oatgrass meadows these modest plants become dominant along with *Plantago lanceolata* and other low-growing herbs. If one could not find here and there an odd starved

Tab. 110. False Oatgrass and Marsh Ragwort meadows in south-west and north-west Germany.

Region:	A. Danube				B. L. Rhine			
Serial no.:	1	2	3	4	1	2	3	4
Arrhenatheretum sens. lat.:								
Arrhenatherum elatius	5	5	5	5	5	5	1	
Trisetum flavescens	5	5	5	5	5	4	1	
Galium mollugo	5	5	5	5	3	2	2	
Heracleum sphondylium	4	4	5	5	4	5	2	
Leucanthemum vulgare	5	5	5	4	3	3	3	1
Dactylis glomerata	4	5	5	4	5	5	2	
Bellis perennis	2	5	1	2	2	3	3	
Veronica chamaedrys	5	4	5	2	3	2	1	
Achillea millefolium	5	4	3	2	3	2		
Vicia sepium	3	3	3	2	2	3		
Lolium perenne	1	4	1	1	3	3	1	
Avenochloa pubescens	4	4	4	4	2	1		
Lolium multiflorum	4	4	3	2		1		
Knautia arvensis	5	4	5	5				
Anthriscus sylvestris		3	2	5	2	3		
Pimpinella major		1	1	4	2	3	1	
Tragopogon pratensis	3	2	1		1	1		
Campanula patula	2		1					
Crepis biennis	.	.	.	.	2	3	1	
Pastinaca sativa		.	.	.	2	2		
Geranium pratense			1	1				
Dryness indicators:								
Salvia pratensis	5							
Thymus pulegioides	5		1					
Scabiosa columbaria	5		2					
Bromus erectus	4	1						
Koeleria pyramidata	4							
Lotus corniculatus	4	2						
Festuca ovina coll.	4		2					
Luzula campestris	4	2	2		1			
Ranunculus bulbosus	5				5			
Briza media	3				2		1	1
Silene vulgaris	4	4						
Medicago lupulina	5	5	1					
Plantago media	4	4	1	1	1			
Campanula rotundifolia	5	4	5		1	1		
Moisture indicators:								
Glechoma hederacea	1	3	3	4	1	3	2	1
Ajuga reptans		4	4	4	1	1	3	2
Silene dioica		3	5	4				
Festuca pratensis	1	1	1	1	4	5	5	5
Deschampsia cespitosa			5	5	1	1	2	1
Alopecurus pratensis		1	3	4	1	5	3	3
Poa trivialis		2	2	5	2	5	5	5
Cardamine pratensis			2	4		2	5	5
Ranunculus repens			3	5		5	5	4
Rumex crispus			1	2		2	1	1
Lysimachia nummular.				2		1	3	3

Region:	A. Danube				B. L. Rhine			
Serial no.:	1	2	3	4	1	2	3	4
Moist and wet indicators:								
M *Lychnis flos-cuculi*			1	2		1	5	5
M *Filipendula ulmaria*			3	5		1	4	5
M *Angelica sylvestris*			1	5			1	3
C *Cirsium oleraceum*			5	5				
M *Geum rivale*			5	5				
M *Sanguisorba officinalis*			5	4				
C *Polygonum bistorta*			3	5				
Alchemilla vulgaris coll.			4	2				
Phalaris arundinacea			1	4				
M *Lythrum salicaria*				4				
Carex gracilis				4		1	3	5
C *Myosotis palustris*				2			3	3
C *Bromus racemosus*						1	4	4
C *Senecio aquaticus*							2	4
C *Lotus uliginosus*							2	4
Juncus articulatus							2	4
C *Caltha palustris*							1	3
M *Equisetum palustre*						1	3	3
Carex disticha				1			2	3
Widespread meadow plants:								
Festuca rubra ssp. *rubra*	4	5	4	5	5	5	5	5
Rumex acetosa	4	4	5	5	4	5	5	5
Ranunculus acris	4	4	5	5	5	5	5	5
Holcus lanatus	4	4	4	5	5	5	5	5
Cerastium fontan. hol.	5	4	3	4	4	4	4	5
Poa pratensis	5	5	5	4	2	4	3	2
Trifolium pratense	5	5	5	3	3	2	2	3
Centaurea jacea	5	4	3	2	4	3	2	2
Lathyrus pratensis	4	3	5	4	4	2	3	4
Vicia cracca	5	1	1	1	2	3	3	4
Prunella vulgaris	4		2	1	1	1	1	1
Phleum pratense		2		2	1	1	1	1
Daucus carota	3				2	1		
Cynosurus cristatus				1	2	1	3	4
Trifolium dubium	3				3	2	3	
Bromus hordeaceus			1		2	3	3	
Other plants:								
Plantago lanceolata	5	5	5	5	5	5	5	4
Taraxacum officinale	4	5	5	5	4	5	.	3
Trifolium repens	3	5	3	2	3	3	4	4
Anthoxanthum odorat.	3	2	2		2	2	5	5
Agropyron repens		2	2	2		1		
Agrostis tenuis			3		3	1		

Note: Nos. A 1–4 and B 1 and 2 are False Oatgrass meadows belonging to the alliance *Arrhenatherion* and the order *Arrhenatheretalia*. Their character species are included with those of the *Arrhenatheretum*. The moist-soil meadows, nos. B 3 and 4, certainly contain many plants indicative of damp or wet conditions; but only a few of these are characteristic of the alliance *Calthion* (C) and the order *Molinietalia* (M). Some of these avoid very acid soils and so are absent from the Marsh Ragwort meadows (see section D V 4 b).
Groundwater does not reach the plant roots in units A1 and B1 and only occasionally does so in A2 and B2. In A3 and B3 the topsoil remains frozen from groundwater, but in A4 and B4 it is often completely saturated.
A **Danube valley** near Herbertingen (after Eskuche 1955)
No. 1: Salvia-False Oatgrass meadow
No. 2: Typical False Oatgrass meadow, described by Eskuche as 'subass, of *Alopercurus pratensis*, var. of *Deschampsia cespitosa*, subvar. of *Silene cucubalus*.'
No. 3: Cabbage Thistle-False Oatgrass meadow
No. 4: Sedge-False Oatgrass meadow
B **Lower Rhine** district near Moers (after Meisel 1966)
No. 1: Quaking Grass-False Oatgrass meadow
No. 2: Meadow Foxtail-False Oatgrass meadow
No. 3: Yellow Trefoil-Marsh Ragwort meadow '*Bromus racemosus-Senecio aquaticus* Ass., subass of *Trifolium dubium*' (corresponds in soil dampness to the Cabbage Thistle-False Oatgrass meadow of the Danube valley)
No. 4: Pure Marsh Ragwort meadow '*Bromus racemosus-Senecio aquaticus* ass, typical subass.' (in the dampness of its soil this corresponds to the Sedge-False Oatgrass meadow in the Danube valley).

In addition the following are present in Danube no. 1: *Carex caryophyllea (3), Linum catharticum (3), Euphrasia rostkoviana (3)*. Species which do not reach a constancy of 2 in any unit have generally been omitted.

Source: From data in Eskuche (1955) and Meisel (1960).

specimen of the False Oatgrass or other character species from the *Arrhenatherion* one would doubt how to classify these run-down meadows. However the poorest facies which are dominated by *Festuca rubra* or *Poa pratensis* are connected through all possible intermediates with the False Oatgrass facies (fig. 441).

Treating the meadows with slurry (i.e. cow dung diluted with water) which is unbalanced in plant nutrients results in a dominance of umbelliferous plants, *Anthriscus* before the first cut and *Heracleum* before the second. Their vigorous growth quickly smothers the grasses with the deep shade produced by their more or less horizontal leaves. Such umbelliferous facies (figs. 437 and 441) yield less fodder than most of the grass facies because the stems of Cow Parsley and Hogweed become woody and their leaves dry out and break up. In addition they are poorer in protein than one might imagine because they store a lot of the nitrogen they take up in the cell sap (Janiesch 1973). These tall umbellifers however can only achieve full dominance if they are allowed to ripen their seed, i.e. when the fields are regularly mown rather late.

Fig. 440. The above-ground phytomass of a well fertilised False Oatgrass meadow is mostly made up of the bottom grasses. This is well demonstrated if the grasses are cut off in 10 cm deep layers (the lowest one starting 6 cm above the ground). After Weniger in Koblet (1966), modified.

The two lowest layers comprise almost half the phytomass and each of them has a leaf-surface index greater than 1, i.e. the area of the leaves (one side only) is greater than that of the area of ground on which the plants are growing. A proportion of the leaves in any case has already become yellow since the strength of light getting to them is insufficient for a positive CO_2 balance.
The amount of dry matter harvested in the first cut amounts to 575 g/m^2, i.e. about 5.8 t/ha. At least twice this amount would be taken in three cuts. For comparison the production of above-ground phytomass in a Beech wood on the same ground would be about 12 – 16 t/ha of dry matter throughout the growing season, i.e. nearly the same as in a well manured False Oatgrass meadow (see also fig. 36). However in the wood the herbaceous plants near the ground contribute only a small part and the woodland needs no fertilising to maintain its productivity.

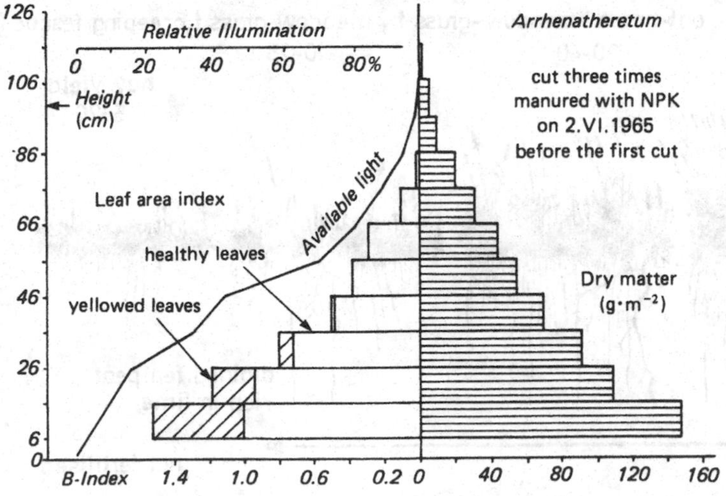

Anthriscus and *Heracleum* are both biennial. In the first year they just form a cluster of leaves building up the swollen storage root; in the second a flowering stem is sent up from this root, and the whole plant dies after the seed has ripened. Where cutting is repeated more often these species cannot spread even with high fertility because their seeds are never allowed to mature.

The better manured the meadows are the more tall-growing species become dominating and the deeper do the roots penetrate (see fig. 441, 446 and e.g. Kotanska 1970). According to Ellenberg (1952a) the consequence of this is rather surprising in that they can survive the crisis of a drought better than the less fertile

Fig. 441. **Above**: *Arrhenatheretum*, *Polygono-Cirsietum* and *Caricetum gracilis*. The sequence of hay meadow communities as the soil water content increases in well manured soils (in a drained fen valley to the west of Brunswick).
The 'slightly-moist' False Oatgrass meadow of northern Germany is known as the subass. *Alopecurus pratensis* (Foxtail-False Oatgrass meadow); it corresponds to the typical False Oatgrass meadow of southern Germany (see fig. 443).

Below: Different fazies of the Meadow Foxtail-False Oatgrass meadow (one of the communities shown above) brought about by different levels of manuring. After Ellenberg (1952), modified.

The figures indicate the yield of hay in 'doppelzentner' (i.e. 0.1 tonne) per ha). The range of yields produced by different manurial levels is greater than that produced by differences in the water supply.
In the upper series the communities are shown just before the second cut; in the lower series before the first cut.

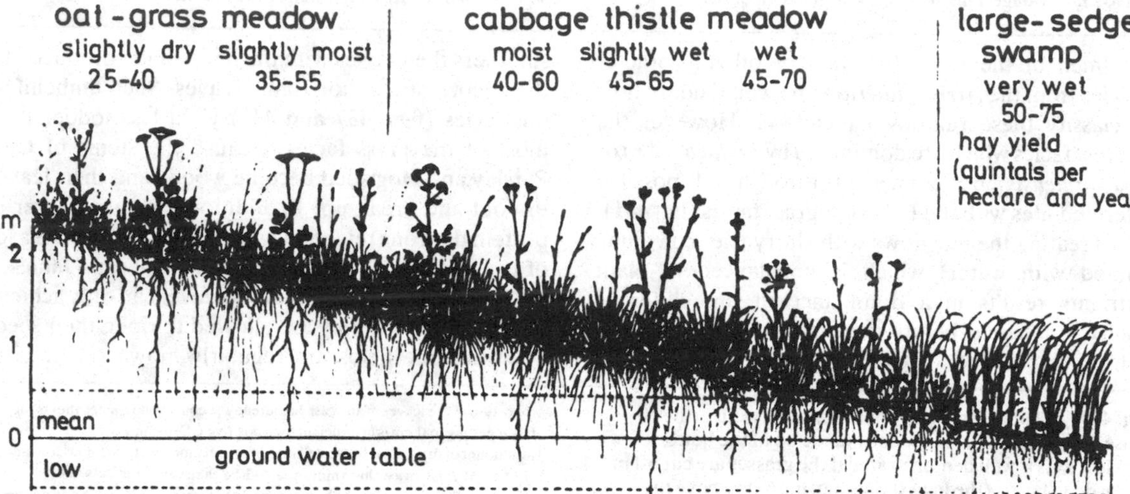

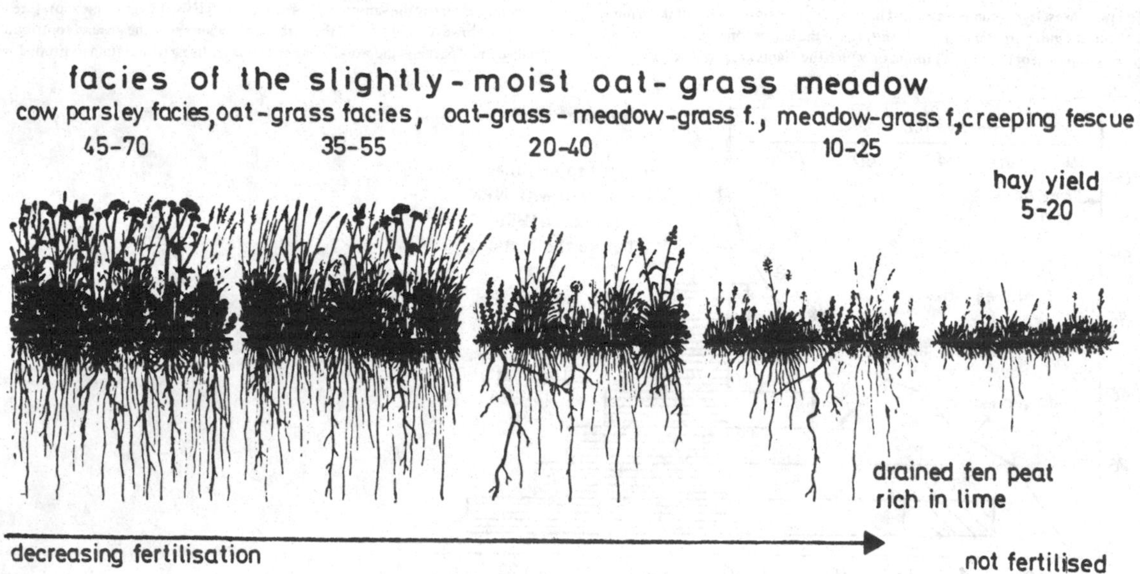

meadows in which shorter grasses predominate (see fig. 441). The uneven depth of rooting may in part have something to do with the large and deeply burrowing earthworms which respond positively to manuring. With their vertical burrows they take the microbe-rich dung down into the ground, at the same time making it easier for the roots to penetrate the subsoil (Graff 1971). In general the high soil microbe activity is one of the characteristics of the False Oatgrass meadows (Hülsenberg 1966). However a full description of all the different life forms is no more possible with these meadow communities than it was with the woodlands and other groups which have been dealt with so far. Such detail would burst the bounds of this book.

The picture of the False Oatgrass meadow which has emerged so far only applies fully to the south-western part of Central Europe and even there only to the typical subassociation of the *Arrhenatheretum* i.e. on soils with a medium water regime (see tab. 110, A 2). On damper soils some species tolerating a lower oxygen supply come to be mixed in with the normally characteristic species combination of the False Oatgrass meadows, e.g.:

Angelica sylvestris	*Filipendula ulmaria*
Cirsium oleraceum	*Geum rivale*
Deschampsia cespitosa	*Sanguisorba officinalis*

Under similar management these **'Cabbage Thistle-False Oatgrass meadows'** are more productive than the **typical** subassociation. Since they usually contain many good fodder plants and are found in sites unsuited to arable farming they are particularly valuable as farmed grassland.

According to the amount of manure given the damp-soil *Arrhenatheretum* is also met with in a number of different forms. A Foxtail facies corresponds to the False Oatgrass facies in the typical *Arrhenatheretum*. In it the False Oatgrass and the Meadow Fescue also play a part. Where manuring is less stands dominated by *Holcus lanatus* and *Festuca rubra* are formed. Umbellifer facies are seldom met with in the damp-soil *Arrhenatheretum*, and they indicate neglect; *Angelica sylvestris*, which also thrives in the unmanured litter meadows, is the one generally found here (fig. 454).

In warmer districts with a low rainfall, such as the Upper Rhine plain, on soils with a high water table, False Oatgrass meadows are often found containing large sedges and other definite swamp plants in addition to the dampness indicators already named. They include:

Caltha palustris	*Lythrum salicaria*
Carex acutiformis	*Phalaris arundinacea*
Carex gracilis	

These **'Sedge-False Oatgrass meadows'** (fig. 441, tab. 110, A4) produce large quantities of poor quality hay. If well managed, Meadow Foxtail and Reed Canary-grass are encouraged and either could assume complete dominance. Ground which is even wetter is taken over by the Cabbage Thistle meadow or other types of moist-soil meadow (section 4).

Going from the False Oatgrass meadows in the direction of drier habitats one sees more and more species which are generally associated with the slightly arid grasslands. The first dry-soil indicator to be met with in south-west Germany is usually the Bulbous Buttercup closely followed by the Meadow Clary (*Salvia*). Well established **'Salvia-False Oatgrass meadows'** (tab. 110, A 1) have a number of differential species, amongst them:

Bromus erectus	*Salvia pratensis*
Ranunculus bulbosus	*Scabiosa columbaria*

This subassociation of *Arrhenatheretum* is especially rich in character species. It contains a wealth of herbaceous plants with eye-catching flowers, and its mixture of colourful species coupled with its constantly changing appearance make it one of the most attractive plant communities in Central Europe. Unfortunately because of its low yield this meadow community too is fast disappearing.

(b) Geographic variability of the False Oatgrass meadows and their subunits

It is in the region which includes south-western Germany and the adjoining central area of Switzerland that the False Oatgrass meadows are best developed. Here they are found in many different forms from the Salvia-False Oatgrass meadows, through the typical subassociation and its variants (according to the degree of dampness) to the Cabbage Thistle-False Oatgrass, and Sedge-False Oatgrass meadows. It is only in this region that all these subassociations are found equally well developed, especially in the foothills of the Swabian Alb and of the Schwarzwald. Provided the soil contains enough lime, they can often be found close together in a relatively small area. The subassociations and variants can be easily distinguished here, many of them can even be further subdivided (see Ellenberg 1952a, Eskuche 1955, Schreiber 1962). In particular there is no difficulty in finding the typical subassociation, i.e. an *Arrhenatheretum* without either dampness or dryness indicator species, as this is spread over large areas (see fig. 442 No.1 a) while it is lacking in more northern and eastern parts of Central Europe.

In the Upper Rhine plain and other low-rainfall districts the typical False Oatgrass meadow is confined to a narrow strip between the drier and damper subcommunities (Oberdorfer 1952, 1957). Almost all the habitats which are free from groundwater as well as the deep soils with a level surface carry the Salvia-False Oatgrass meadow or similar 'dry' *Arrhenathereta*. On the other hand where the topsoil is sandy but the ground influenced by a high water table the dampness- or wetness- indicator species are able to grow alongside many *Mesobromion* species. On heavy marls where the water supply is not dependent on groundwater a variant of the typical subassociation is found which is indicative of fluctuating dampness and contains a few scattered indicators like *Lychnis flos-cuculi* and *Ajuga reptans*.

In those parts of the upper Rhine plain and of the Tauber region with the lowest rain fall and highest temperature (No. 2 in fig. 442) the manured mown grasslands are dominated by Salvia-False Oatgrass meadows. Here the dryness indicators are competitive even on the deepest loess loams with their good water-retaining properties, and all other subassociations of the False Oatgrass meadows are absent or found only on very small areas.

Fig. 442. The various regions of the False Oatgrass meadows in Southwest Germany where the six subunits of the *Arrhenatheretum* brought about by different soil moisture levels are to be found to varying extents (see the legend top left). After Schreiber (1962), modified.

The habitat of the typical variant of the Salvia-False Oatgrass meadow is relatively dry, that of the typical variant of the typical False Oatgrass meadow slightly moist (or 'fresh'). The variants with fluctuating soil moisture are found growing on soils subject to temporary waterlogging, whether due to clay or for some other reason. The Cabbage Thistle-False Oatgrass meadow indicates moist ground while the Sedge-False Oatgrass meadow shows that the ground is relatively wet (see fig. 441).

Each subunit is allocated to one of the six outer segments of the circle, the amount of shading indicating its abundance in any district; (white = rare or absent; hatched = intermediate; black = plentiful). The inner circle indicates the number of montane species present (hatched = with, black = without such species).

By taking the different outer sectors and the inner circle the diversity regions are characterised in the following way: 1a = All six subunits present and well grown with no montane species (average altitude and not too dry a climate), 1b = Typical variant of the typical False Oatgrass meadow not so well grown (drier climate); 2 = Few hay meadows present apart from the Salvia-False Oatgrass meadow (relatively dry climate, free draining soil); 3 = The first four subunits present along with montane species. Damp-soil meadows absent (Swabian Jura with free draining soil); 4a = Salvia-False Oatgrass meadows absent because of humid climate (Black Forest and Odenwald), 4b = The same but predominating pasturing has completely suppressed the False Oatgrass; 5 = Too high altitude (i.e. too low temperatures) for False Oatgrass meadows (parts of the Black Forest).

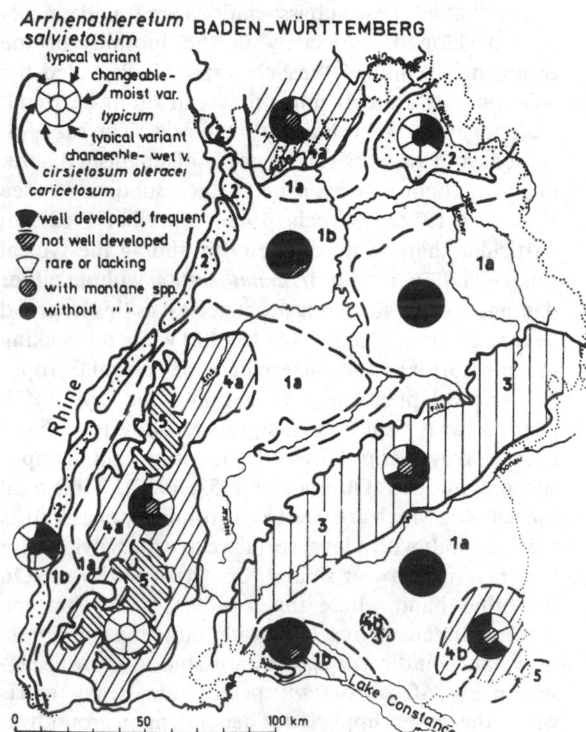

The typical as well as the Salvia subassociation is commonly met with on the Swabian Jura where the climate is somewhat damper and cooler (No. 3 in fig. 442). In the higher parts of this plateau the *Arrhenatheretum* begins to take on a montane look (see section 3) but this does not change the way in which it reacts to soil moisture. Cabbage Thistle and Sedge-False Oatgrass meadows are virtually absent from the Jura plateau because of the free-draining nature of the limestone rock.

On the other hand Salvia-False Oatgrass meadows and other dry types of *Arrhenatheretum* could not develop in the coolest parts of south-western Germany which also have the highest rainfall. Occasionally they may be found on steep calcareous sunny slopes but everywhere else in the regions No. 4 a one finds either the typical or the damp subassociations. There are no Sedge-False Oatgrass meadows here either, because the group of their character species does not appear sufficiently competitive to succeed on constantly wet ground. As in north-western Central Europe, which we shall be looking at shortly, the *Arrhenatheretum* is restricted to moist or fairly damp sites, but due to the higher altitude in the Black Forest and the Odenwald it contains some montane species. On the highest levels of the Black Forest the False Oatgrass meadows are absent altogether.

Not only in south-western Germany but also in the other parts of Central Europe the species composition of the *Arrenatheretum* and its subdivisions brought about by differences in soil-water regime and altitude are further affected by the degree of continentality of the general climate. The False Oatgrass meadows become progressively poorer in species as they get nearer to the north-western edge of their natural distribution area. Whereas 4 subassociations and numerous variants can be distinguished in the warmer sites of south-western Central Europe (fig.443, bottom) only 2 can be separated from each other in north-west Germany and the adjoining parts of Holland, the 'Quaking Grass-False Oatgrass meadow' and the 'Foxtail-False Oatgrass meadow' (fig.443, top). What is more, their differential species are so few that if they had occurred in south-western Germany they would have only been given the status of variant. For the relatively dry **Quaking Grass-False Oatgrass meadow** Tüxen (1937) named:

Avenochloa pubescens	*Pimpinella saxifraga*
Briza media	*Plantago media*
Luzula campestris	

These species indicate that the ground is poor rather than that it is dry. The *Briza media* subassociation of the north-western *Arrhenatheretum* thus corresponds to a poorly manured form of the typical False Oatgrass meadow of south-western Germany, and especially to a moderately dry variant of it. The commonest of the *Mesobromion* species present here is *Ranunculus bulbosus* (compare B1 with A1 in tab. 110) but even this does not occur in all the Quaking Grass False Oatgrass meadows. The **Foxtail-False Oatgrass meadow** of northern Germany can be compared with a fairly damp variant of the typical False Oatgrass meadow (compare B2 with A2). Tüxen has named only the following differential species in it:

Alopecurus pratensis	*Glechoma hederacea*

According to Meisel (1960) other weak dampness indicators found in the Lower Rhine region can also be used (see tab. 10,B2).

There are no longer any False Oatgrass meadows in north-western Central Europe in those sites which are more strongly influenced by groundwater (fig.443). In the Donau-Ried (south-west Germany) however all the character species of the *Arrhenatheretum* are found along with many dampness or wetness indicators (e.g. in columns A3 and A4 in tab. 110); in the Lower Rhine area these two species groups are mutually exclusive (B3 and B4). On the moist soil which would be occupied by a Cabbage Thistle-False Oatgrass meadow in the south-west one finds a Cabbage-Thistle meadow or some other 'damp' meadow community, e.g. a Marsh Ragwort meadow (*Senecio aquaticus* association) in the north-west (B3). These contain only scattered individuals of the characteristic False Oatgrass meadow species, e.g. *Trifolium dubium*. An even better parallel to the Cabbage Thistle-False Oatgrass meadow is one described from a similar habitat in north-west Germany by Tüxen (1937) as a 'dry' subassociation of the *Cirsium oleraceum* – *Angelica sylvestris* association (see section 4 a and fig.441).

In the north-west *Arrhenatherion* species do not venture on to wet soils, such as those where large sedges, like *Carex gracilis* thrive. Here there are no counterparts of the Sedge-False Oatgrass meadows (A4), but instead we find pure wet-soil communities like the typical Marsh Ragwort meadow (B4).

What is the reason for the amplitude of the lowland False Oatgrass meadows in respect of soil wetness being so much narrower in the northern part of Central Europe than in the south? Ellenberg (1954b) thinks it probable that the drying out of the top soil may have something to do with it. In a relatively warm and continental climate there must be greater demands on the water supply in the top-soil with its dense mat of roots. As a result, even when the water table is high, the upper soil layer is not permanently saturated and devoid of air, and the grasses and herbs whose roots require more soil oxygen are able to spread. In the more even cool and damp oceanic climate on the other hand only swamp plants with an internal aeration system are able to persist under such conditions.

Fig. 443. The classification of False Oatgrass meadows and moist soil meadows according to the groundwater level in free-draining soils in north-west and south-west Germany. 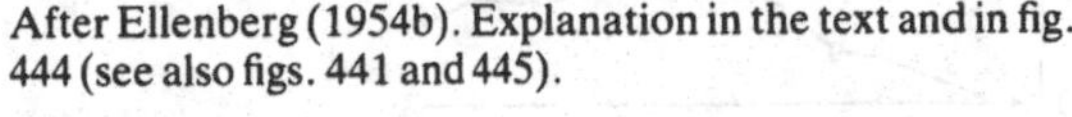
After Ellenberg (1954b). Explanation in the text and in fig. 444 (see also figs. 441 and 445).

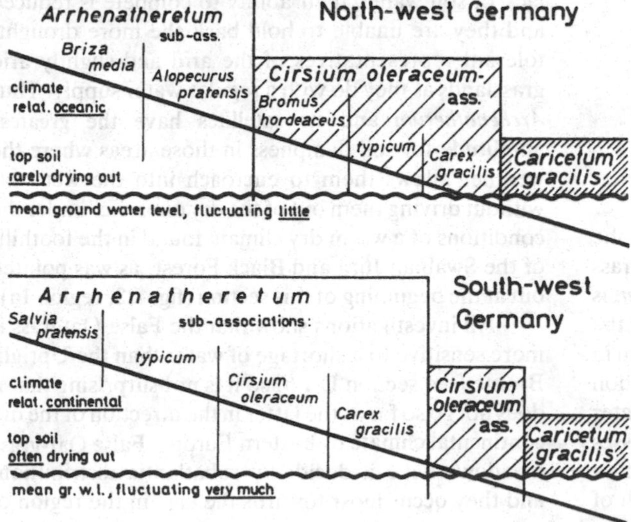

Fig. 444. The production of some of the meadow grasses (expressed as dry matter) in pure cultures in a sandy soil where the water table was kept constant but at different heights in the dry summer of 1952 and (from a fresh sowing) in the wetter summer of 1953 (cf. fig. 445). After Ellenberg (1954b).

During the dry year all species gave their highest production when the water level was relatively high since the topsoil soon dried out and by the end of the growing period contained very little water (see fig. 445, below). This type of situation occurs frequently in south-west Germany allowing the *Arrhenatheretum* species to occupy soils in which the groundwater is near the surface (fig. 443, lower diagram).

On the other hand frequent saturation of the soil such as happened in the experimental container in 1953 (fig. 445, above) discouraged the deep-rooting grasses such as False Oatgrass and Cocksfoot which could not tolerate the badly aerated soil, but favoured the moisture-loving species such as Foxtail which normally have air passages in the root cortex (aerenchyma, see fig. 390).
In the comparatively humid north-west Germany this is so often the case that the *Arrhenatheretum* communities are restricted to soils with a low water table (fig. 443, upper diagram).

The Upright Brome, with its relatively shallow root system, reacted surprisingly weakly to the height of the water table and in this experiment too did not show itself to be xerophilous (see also figs. 389 and 390).

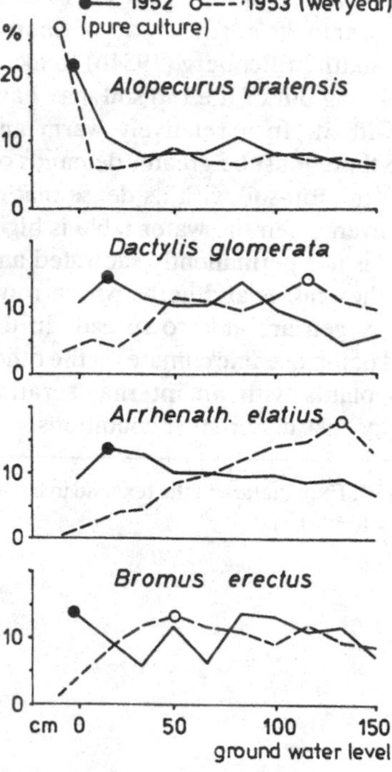

Fig. 445. The water content of the sandy soil used in the experiment described in fig. 444 in a dry year and in a wet one (expressed as a percentage of the fresh weight). After Ellenberg (1954b).

Ordinate = groundwater level, abscissa = distance from the lower edge of the concrete experimental tank.

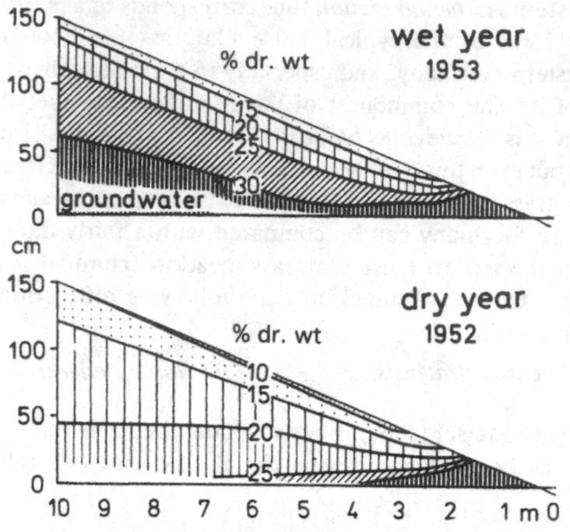

The extent to which the amount of water in the top-soil and with it the well being of the False Oatgrass and other character species of the *Arrhenatheretum* is dependent on the weather conditions is shown in the results of research into ground water at Hohenheim in two successive years (see figs. 444 and 445, also section D I 6 b). In the dry year 1952 the groundwater represented a significant source of water for the grasses on sandy soil, and in pure culture they achieved their best growth when the groundwater was at a depth of 10–20 cm. A parallel experiment in 1953, a wet year, showed that *Arrhenatheretum* and *Dactylis* grew best in the dry part of the research tank because there was no difficulty in the water supply but the roots were well above the water table and so were able to get more air than in a position nearer to the ground water (compare figs. 444 and 445). Almost all meadow grasses grow best when the water content of the soil is about 85% of its total capacity (Kanter 1933, see Ellenberg 1963).

Thus the competitive strength of *Arrhenatherum, Dactylis* and other meadow plants growing on soil with a high water table will always be greater in a dry climate. At the same time if there is a frequent lack of soil water, their ability to compete is reduced and they are unable to hold back the more drought-tolerant representatives of the arid and slightly arid grasslands as they do with a regular water supply. Thus *Arrhenatherum* and its satellites have the greatest amplitude for soil dampness in those areas where the weather allows them to encroach into the wet soils without driving them out of the dry ones, i.e. under the conditions of a warm dry climate found in the foothills of the Swabian Jura and Black Forest, as was pointed out at the beginning of this section (fig.422, region 1a).

All investigations show that the False Oatgrass is more sensitive to a shortage of water than the Upright Brome (see section D I 6) so it is not surprising that it does not go so far as the latter in the direction of the dry continental climate of Eastern Europe. False Oat-grass meadows have a decidedly subatlantic distribution, and they occur most towards the east in the region of the outer Alps where the climate is of a relatively oceanic nature (see figs.4 and 42). They are quite well developed with a rich variety of character plants in Steiermark (Eggler 1958) and in the Karawanken (Aichinger 1933) for example, but are absent from the valleys of the inner Alps (see Braun-Blanquet 1961). To the north of the Alps they die out completely in the middle of western Poland.

A relatively dry, a typical and a damp subassociation of the *Arrhenatheretum* can readily be distinguished in the Unstrut valley near Strausfurt. Hundt (1957) named them the 'Upright Brome-False Oatgrass meadow' (since *Bromus erectus* is strongly represented), the 'Pure False Oatgrass meadow' (fig.446) and the'Cabbage Thistle-False Oatgrass meadow'. He also recorded a 'Bedstraw-False Oatgrass meadow' whose habitat has a very changeable amount of dampness and corresponds to a variant of the typical False Oatgrass meadow in southern Germany already mentioned (see fig. 442).

By the time we get as far eastwards as the middle reaches of the Elbe and the lower Mulde the False Oatgrass meadows have become rarer and poorer in species. Here Hundt (1954) only separates a 'dry' subcommunity from a 'moist' one. The first occupies rather sandy soils and is distinguished by the following differential species:

Festuca ovina	*Armeria maritima* ssp. *elongata*
Pimpinella saxifraga	*Cerastium arvense*
Ranunculus bulbosus	*Equisetum arvense*
Thymus serpyllum	*Hieracium pilosella*
	Sedum acre, etc.

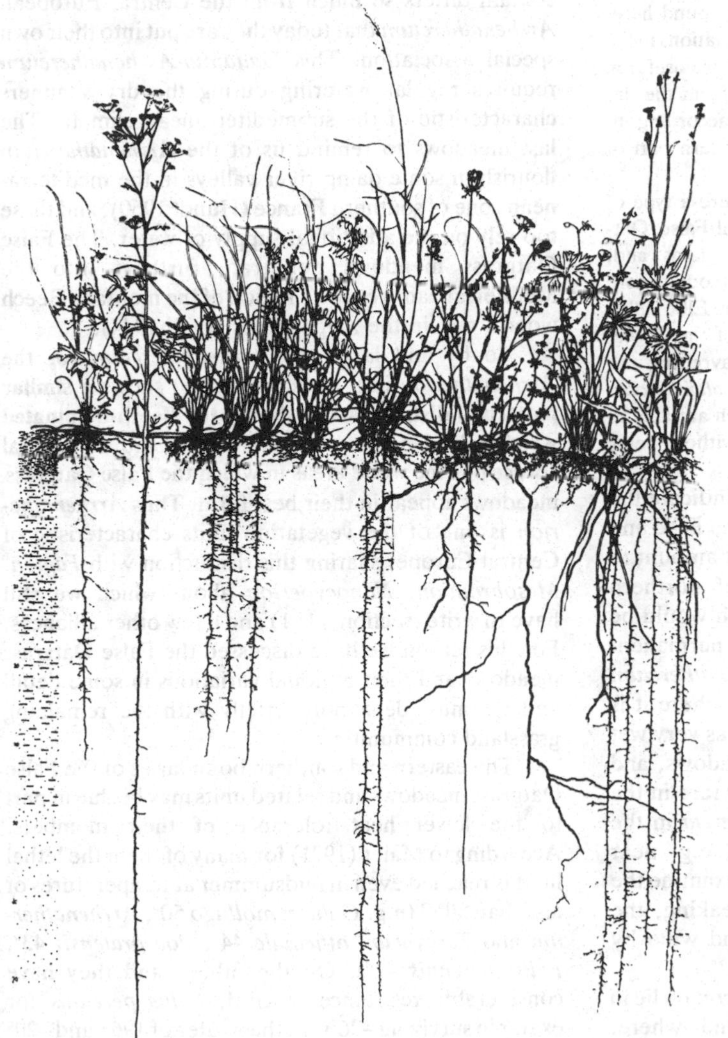

Fig. 446. Section through a typical Oatgrass meadow (*Arrhenatheretum*, slightly-moist subass.) in the Saale valley above Wörmlitz. After Hundt (1958).

From left to right: *Arrhenatherum elatius*, *Pastinaca sativa*, *Poa pratensis*, *A.e.*, *Vicia sepium*, *A.e.* flowering *P.p.*, *Daucus carota*, *Galium mollugo*, *Geranium pratense*, *Crepis biennis*, Total depth of the soil profile 75 cm.

Those listed in the left column are also found in the Salvia-False Oatgrass meadows of south-western Central Europe, while the Thrift, the Field Mouse-ear and the others in the right hand column must be looked upon as indicating a rather poor sandy habitat. *Briza media* is only found in a variant of this 'Thrift-False Oatgrass meadow' which grows on acid soil and can be recognised from the presence of *Hypochoeris radicata*, *Crepis capillaris*, *Rumex acetosella* and other acid indicators. The moist, or rather variably moist, False Oatgrass meadows of the Elbe and Mulde region also contain a number of poverty indicators which are nevertheless helped by damp conditions, e.g.

Silaum silaus	*Ranunculus auricomus*
Selinum carvifolia	*Veronica serpyllifolia*
Lychnis flos-cuculi	*Leontodon saxatilis*
Deschampsia cespitosa	*Agrostis stolonifera*
Glechoma hederacea	

The species in the left-hand column are also often found in the corresponding False Oatgrass meadows of the Upper Rhine region whereas those on the right are seldom found here. *Alopecurus pratensis* is plentiful in both subassociations in the Elbe valley; so in this eastern region it is no more useful in separating the damp False Oatgrass meadow from the dry one. Real Foxtail-False Oatgrass meadows according to Tüxen are therefore confined to the north-western part of Central Europe.

Only in the north of East Germany is there a type of meadow resembling the north-western Foxtail-False Oatgrass subassociation restricted to the flood-plains and, according to Freitag and Körtge (1958) it is flooded more often (see section D VI 3). In the lower part of the Elbe valley near Hamburg flooding damages the topsoil of the *Arrhenatheretum* habitat quite extensively, thus favouring the spread of species such as *Agropyron repens*, *Calamagrostis epigeios* and other rhizomatous geophytes which are able to grow out into the bare patches of sandy soil without delay (Meyer 1957).

These examples may be sufficient to indicate the variability of False Oatgrass meadows in the Elbe and Havel region. In area however they do not amount to much today in these and other parts of northern Central Europe, as most of the land which would be suitable for them has been taken for permanent pasture or for arable cropping. Pure *Arrhenatheretum* meadows only occur now in those places where it is inconvenient or inadvisable to graze, such as very wet sites with sedge fen or other swampy meadows, and steep slopes which naturally are extremely rare in the lowlands. If one wants to find *Arrhenathereta* in the coastal area of north-western Germany, e.g. near Hamburg, or in Holland, then one should examine the old river dykes. Otherwise, generally speaking, the permanent grassland is dominated far and wide by Perennial Ryegrass (see section VI 1 a).

The eastern outposts of the *Arrhenatheretum* lie in northern Romania and in North Poland where, according to Passarge (1963b) they still correspond rather well to the False Oatgrass meadows lying further west. In the Soviet Union, the Eastern Baltic countries, but also in Slovakia, other meadow types replace them, e.g. those dominated by the Tufted Hairgrass (*Deschampsia cespitosa*). In the Balkans the best False Oatgrass meadows are found typically in Croatia, especially in the vicinity of Zagreb where the climate and the general character of the vegetation are most like those in Central Europe (see Horvat, Glavač and Ellenberg 1974). In the East and the North the *Arrhenatherion* alliance barely reaches the boundary of Central Europe, but in the West and South it goes far beyond it. False Oat-grass meadows had already been described by J. Braun in 1915 from the southern Cevennes, i.e. from a mountain district near the Mediterranean. However the collection of species they contain differs so much from the Central European *Arrhenatheretum* that today they are put into their own special association. This *Gaudinio-Arrhenatheretum* requires regular watering during the dry summers characteristic of the submediterranean climate. The last meadows to remind us of the *Arrhenatheretum* flourish in some damp river valleys in the mediterranean zone of southern France (Hundt 1960), and these too rely on the additional supply of water. The False Oatgrass meadows penetrate furthest into the Mediterranean area in he region of the montane Beech woods, e.g. in the Appennines (Lüdi 1944).

So by and large the *Arrhenatheretum* (or the *Arrhenatherion* in a narrower sense) shows a **similar pattern of distribution to that of the Beech-dominated woodlands**. On the other hand there are no natural Spruce or Pine wood areas in which the False Oatgrass meadows appear in their best form. Thus *Arrhenatherion* is one of the vegetation units characteristic of Central Europe, sharing this distinction with *Fagion*, *Mesobromion*, *Nanocyperion* about which we still have to write (section VII 1) and a few other alliances. For this reason we have discussed the False Oatgrass meadows and their regional variations in some detail and we must deal more briefly with the remaining grassland communities.

The eastern and southern boundaries of the False Oatgrass meadows and related units may be due in part to the lower heat tolerance of their members. According to Maier (1971) for many of them the lethal limit is reached even in midsummer at temperatures of less than 50°C (e.g. *Galium mollugo* 50°, *Arrhenatherum* and *Taraxacum officinale* 44°, *Poa pratensis* 43°, *Bellis perennis* 42°). On the other hand they have considerable resistance to cold, *Bellis perennis* for example surviving –26°C in the winter of 1961 and –20°

in 1962. The sensitivity to heat may also indirectly become a handicap in that grasses such as *Arrhenatherum* have stomata which react very strongly to relative air humidity and close completely during a dry period, stopping the photosynthesis but also depriving the plant of the cooling effect of transpiration (Stocker 1967). Their photosynthetic apparatus is much **more adapted to low temperatures** which prevail in Central Europe during the spring and autumn and also during the early morning in summer. The compensation point is exceeded even at 0°C, i.e. as the hoar frost is melting from the leaves. Species which are present in larger numbers in the mountains than they are in the lowlands, such as the Red Fescue (*Festuca rubra*) which was thoroughly investigated by Ruetz (1973), reach their optimum net photosynthesis at temperatures below 10°C while many others have an optimum up to 25° (Kalckstein 1974).

In this behaviour the grasses of the temperate zone (which are all C 3 plants) differ fundamentally from those of the tropics whose optimum lies around 35°C according to Kalckstein. Since these C 4 plants deal with the carbon dioxide more economically they never reach the point at which light becomes limiting under normal conditions. The European meadow grasses however on a bright day (with 100 klux or more) are not able to use all the light because they reach their saturation point at about 30–40 klux. The maximal photosynthetic potential of C 3 grasses (such as *Festuca rubra*, *Lolium multiflorum* or *Bromus erectus*) is of the order of 90 mg CO_2 per gramme of dry weight per hour whereas that of C 4 grases (e.g. *Echinochloa crus-galli* and *Panicum miliaceum*) is 130–150. According to Cartledge and Connor (1973) the effectiveness of the energy-fixing process in temperate grassland is inversely proportional to the intensity of the light, whereas in tropical grass stands it remains nearly constant as the intensity of sunlight increases. In this respect too our meadows are very well adapted to the weather in Central Europe which is often cloudy. This is particularly true of mountain meadows which we shall turn to next.

3 False Oatgrass and Yellow Oatgrass meadows in the mountains

a Changes in the Oatgrass meadows with increasing altitude

The great variety of Oatgrass meadows in Central and South Europe is added to by the changes which take place in them with increasing altitude. The shortening of the growing period, the reduction in the summer temperature, the increase in precipitation resulting in more soil leaching, and also the reduction in the farming intensity all operate unfavourably on the competitiveness of False Oatgrass and its satellites (fig. 447). Instead the less-demanding grasses, which are suppressed in the lowland by the tall-growing species, are allowed to come into their own, especially *Trisetum flavescens*, *Holcus lanatus* and others of medium height along with grasses having their leaves near the ground like *Festuca rubra*, *Agrostis tenuis* etc.

In the more northerly mountains in Central Europe such as Vogelsberg or Rhön odd individual specimens of *Arrhenatherum* may be found up to 600 m above sea level but the association named after it has already been replaced by the 'Yellow Oatgrass meadow' at an altitude of 350–400 m (Speidel 1972). Up to about 500 m, that is where there are still Oaks present in the Beech woods, the **Meadowgrass-Yellow Oatgrass meadow** (*Poa-Trisetetum*) dominates in manured grassland. It is replaced by the **Cranesbill-Yellow Oatgrass meadow** at the level of pure Beech woods (*Geranio-Trisetetum*, see tab. 111). In some places in the Alps on the other hand the False Oatgrass can still be dominant as high as 1200 m especially on base-rich soils, some individuals getting up to 1500 m. Since other species characteristic of the lowland False Oatgrass meadows can also occasionally reach the montane belt. Baeumer (1956) and Oberdorfer (1957) speak here of 'mountain False Oatgrass meadows' (*Arrhenatheretum montanum*, see figs. 447 and 448). Intermediate situations are found in other mountains, for example the Polish West Carpathians. Here there are true *Arrhenathereta* up to about 500–600 m but above these there are only poor mountain pastures with an admixture of eastern species, a special association which Kornaś (1967) designated *Gladiolo-Agrostietum* (fig. 456).

Just as in woodlands, the productivity of meadows falls off with increasing altitude; in spite of the most intensive husbandry the reduction amounts to some 6% for every 100 m height difference (Spatz 1970 in the Allgäu between 955 and 1555 m). This is mainly a function of the length of the growing season. During that season however, according to Speidel and Weiss (1971, 1974) and Ruetz (1973), the conditions are relatively favourable. These authors found in the case of a *Trisetetum* with a fair amount of Red Fescue that the time of maximum daily net-production was progressively earlier as more manure was applied. The roots in well manured stands lived for less than 13 months and 54% of them for less than 1 month. The ones which lived longer were those produced in the second half of the year, and they ensured not only that the plant could survive the winter, but that it could start growing immediately the snow had gone from the leaves which had remained green throughout the

Tab. 111. Climatic data relevant to slightly-moist meadows at different altitudes.

Community	Altitude (m)	Precipitation (mm)	Mean air temperature (January)	(year)
Trisetetum:				
Rumici-T. (high montane)	900 –1500	1240–1720	−6.7 to −3.4	2.7–5.7
Geranio-T. (montane)	600 –1400	1100–1700	−3.8 " −2.0	5.2–6.3
Poo-T.	200[a]– 600	600–1050	−3.0 " −0.4	6.3–7.0
Arrhenatheretum:				
Alchemillo-A.	100[a]– 800	575– 900	−4.0 " +1.7	5.5–8.7
Dauco-A. (planar-colline)	20 – 150	475– 650	−0.8 " +0.3	8.6–9.4

Note: [a] The lowest value in northern Poland; otherwise 400 (*Poo-T.*) and 350 (*Alch.-A.*).

Source: Literature data taken from northern Poland to Belgium and from Czechslovakia to Switzerland; simplified after Passarge (1969a).

winter. In sites where the plants are short of nutrients many grass roots remain alive for up to 3 years, in fact those of *Nardus stricta* as long as 9 years (Ellenberg, unpublished). A starved meadow is also more economical with the nutrients it has been able to assimilate, leaving fewer dead remains for the soil organisms. This results in an accelerated reduction of the natural soil fertility. Such connections can best be studied in mountain meadows but in principle they also hold good for those in the lowlands.

(b) Montane and subalpine Yellow Oatgrass meadows

Typical Yellow Oatgrass meadows like the ones Marschall (1947) described from Switzerland are first encountered in the upper montane belt (figs. 449 and 450). Most authors ascribe these 'real' Yellow Oatgrass communities to their own alliance, the *Polygono-Trisetion* within the order *Arrhenatheretalia*. Alongside character species such as:

Alchemilla vulgaris coll.	*Narcissus poëticus* ssp. *radiiflorus*
Centaurea pseudophrygia	*Pimpinella major* var. *rubra*
Crepis mollis	*Thlaspi alpestre*
Crocus albiflorus	*Viola tricolor* ssp. *subalpina*

and others they can be recognised by a number of montane and subalpine species distinguishing them from the *Arrhenatherion*, e.g.:

Primula elatior (plentiful)	*Muscari botryoides*
Silene dioica (plentiful)	*Phyteuma halleri*
Astrantia major	*Phyteuma orbiculare*
Campanula scheuchzeri	*Poa chaixii*
Geranium sylvaticum	*Rumex arifolius* et al.

Fig. 447. The extent to which the presence of meadow plants depends on altitude in the Harz and in the Thuringia Forest (in a percentage of all the stands examined at each height). From data in Hundt (1966).
The following species behave in a similar way to those:

on the left	in the centre	on the right
Anthriscus sylvestris	*Achillea millefolium*	*Agrostis tennuis*[a])
Avenochloa pubescens	*Alchemilla vulgaris*	*Arnica montana*[a])
Bellis perennis	*Alopecurus pratensis*	*Avenella flexuosa*[a])
Centaurea jacea	*Anthoxanthum odoratum*	*Carex pilulifera*[a])
Cirsium oleraceum	*Briza media*	*Chaerophyllum hirsutum*
Colchicum autumnale	*Campanula rotundifolia*	*Cirsium heterophyllum*
Cynosurus cristatus	*Centaurea pseudophrygia*	*Crepis mollis*
Festuca pratensis	*Cirsium palustre*	*Galium harcynicum*[a])
Geranium pratense	*Deschampsia cespitosa*	*Geranium sylvaticum*
Holcus lanatus	*Festuca rubra*	*Nardus stricta*[a])
Lotus corniculatus	*Hypericum maculatum*	*Pedicularis sylvatica*[a])
Lysimachia nummularia	*Lathyrus montanus*	*Phyteuma spicatum*
Poa pratensis	*Leontodon autummale*	*Potentilla erecta*[a])
P. trivialis	*Luzula campestris*	
Ranunculus auricomus	*Lychnis flos-cuculi*	
Tragopogon pratensis	*Myosotis palustris*	
Trifolium dubium	*Ranunculus acris*	
T. repens etc.	*Trollius europaeus* etc.	

[a]) These species occur at greater altitudes mainly because here the soils are poorer and more acid.

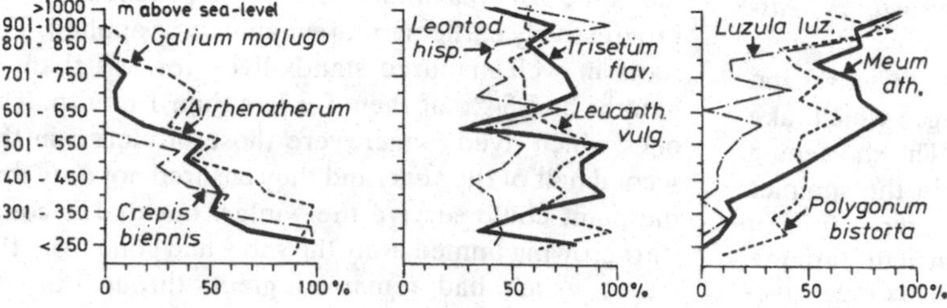

The Yellow Oatgrass itself (*Trisetum flavescens*) however cannot be used either as a character species nor as a differential one because it occurs too frequently in the *Arrhenatheretum* at lower levels (fig. 447, see also Baeumer 1956). There is still controversy over the synsystematic role of the Common Bistort (*Polygonum bistorta*, see fig. 455). Its overall distribution indicates that it is a boreal and montane species, but in northern Germany and in many other parts of Central Europe it also occurs in the lowland and turns up so regularly in manured damp meadows that many authors consider it to be a character species of the order *Molinietalia* (section 4 a). Hundt (1972) even records it from the island of Rügen along with the Globeflower (*Trollius europaeus*) which it resembles in its behaviour. Where *Polygonum bistorta* does occur in the lowland there are often local climatic 'montane' conditions. For example Scheel (1962) found that to the north of Berlin the soils which carried *Polygonum* were 1–2°C cooler than comparable soils without it. In the mountains it is not confined to damp habitats as it is in the lowland, but even here it avoids the drier sites of the Yellow Oatgrass meadows and is only very occasionally found in them.

The Yellow Oatgrass meadows also dovetail in with other grassland communities especially as the relief in the mountains changes so much over quite small distances. If it gets a little more manure than normal it can pass over into a tall-herb community (see section V1 7). In a similar way to the Hogweed-False Oatgrass meadow the large-leaved herbaceous plants suppress the young growth of grasses and leguminous by depriving them of sufficient light (fig. 441). In mountains with strongly acid soils there are all possible transitions to the Matgrass sward (section II 4 a) e.g. in the Black Forest. The more extensive the form of husbandry practised and the less frequent the applications of manure the more of the undemanding members of the *Nardeta* come to take the place of the taller and less easily satisfied species of the *Triseteta*.

Fig. 448. Ecological grassland regions of the Harz. After Hundt (1964). The main divisions are determined by altitude, subdivisions by soil type and other factors.

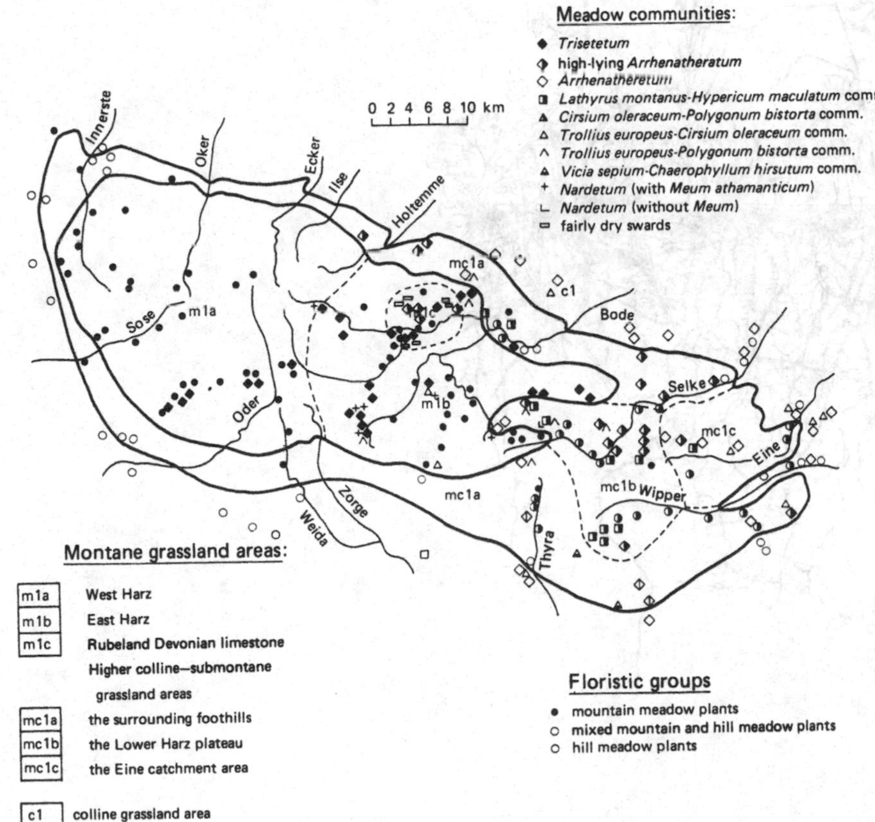

Intermediate species combinations are often dominated by the Common Bent (*Agrostis tenuis*) or the Tussock Red Fescue (*Festuca nigrescens*), Such **Bent-Red Fescue meadows** cover larger areas than the pure Yellow Oatgrass and False-oatgrass meadows together in many mountainous districts (figs. 451 and 452). In high calcareous mountains there is often a mingling of the Yellow Oatgrass meadows and the Blue Moorgrass slopes or related communities (see section C VI). Because the grasslands here are used mainly for grazing intermediates between these and the Rough Hawkbit meadows are even more common (see section D VI 1 d).

Compared with the Alpine Yellow Oatgrass meadows (e.g. *Astrantio-Trisetetum*, see Oberdorfer 1957) those of all the lower mountains appear to be impoverished. This is brought out particularly well in the literature survey by Hundt (1964 and 1966) who has worked on the meadows of the Thuringia Forest and the Harz. (see figs. 448 and 449).

4 Manured moist-soil meadows and related communities

(a) Cabbage Thistle meadows and other hay meadows on moist base-rich soils

Both the False Oatgrass and Yellow Oatgrass

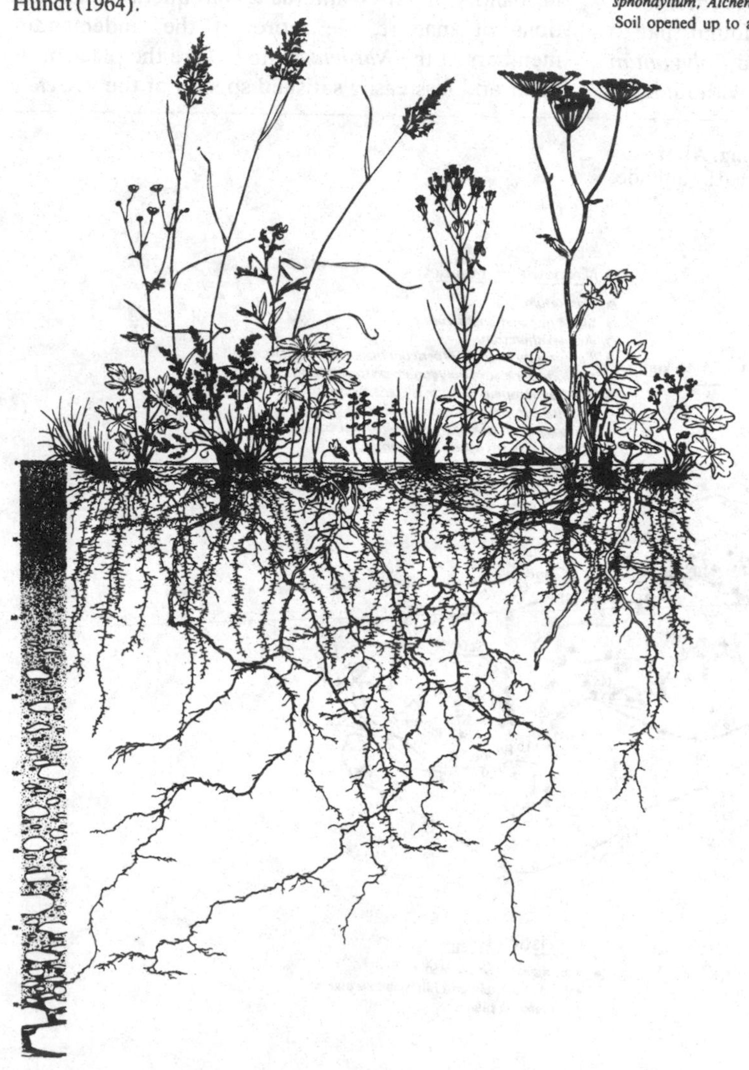

Fig. 449. Section through a Yellow Oat meadow (*Trisetetum*) in the Lower Harz, which was developed from a Matgrass sward by fertilising and mowing twice every year. After Hundt (1964).

From left to right: *Nardus stricta, Ranunculus acris, Meum anthamanticum, Trisetum flavescens* (3 flowering stalks), *Lathyrus linifolius, Trollius europaeus, Hypericum maculatum, Festuca nigrescens, Hypericum perforatum, Heracleum sphondylium, Alchemilla vulgaris.*
Soil opened up to a depth of 75 cm.

Fig. 450. *Narcissus radiiflorus* in a slightly moist Yellow Oatgrass meadow in spring below the Moléson in the Alpine foothills of western Switzerland. Photo. Perrochet.

meadows are connected to the 'moist' meadows on equally fertile but damper soils by many intermediate types. Therefore it will be best if we take a look at the 'manured moist-soil meadows' first. They assume many forms and their survey is not easy but they are all included in the alliance *Calthion*. This of course is not altogether a happy choice of name since the Marsh Marigold is not always present nor is it confined to these communities (see e.g. section C I 2 a). However there is no other species which can be better used to characterise these types of meadow. Those marked with C in tab. 110 can be looked upon as possible candidates. At best the *Calthion* can be distinguished in a negative way by the absence of *Arrhenatherion* or *Trisetion* species on the one hand and of *Molinion* species on the other (see section 6). Many of the communities placed in *Calthion* alliance have affinities with small-sedges or large-sedges fens with which they often form a mosaic of habitats. This is true also for English *Calthion* fen meadows, for which Wheeler (1980) found it 'difficult to produce a fully satisfactory classification'.

Along with a nucleus of generally distributed meadow plants (the species by which the class *Molinio-Arrhenatheretea* is recognised,see tab. 110) a lot of helophytes flourish in the damp meadows. These include a number by which *Molinietalia* is recognised; they are also given in tab. 110. Different ones are found depending on the base or lime content of the soil, on the climatic conditions and on the water supply. The method of farming also has a large influence on the species composition.

Cabbage Thistle meadows (*Angelico-Cirsietum oleracei*, see Meisel 1969b, figs. 443 and 453) are widespread in the south-west of Central Europe. They prefer relatively base-rich wet mineral soils or drained peat soils with similar properties. If they are cut twice a year and given an occasional dressing of manure they are fairly rich in grasses especially *Alopecurus pratensis*, *Festuca pratensis*, *Holcus lanatus* and *Poa trivialis*.

In spring the dominating flowers in a Cabbage Thistle meadow are first Lady's Smock and Marsh Marigold (see fig. 240). Later on it is covered in carmine, yellow, pink and brownish-red, as the flowers of *Lychnis flos-cuculi*, *Ranunculus acris* and *Polygonum bistorta* or the anthocyanin-rich shoots of *Rumex acetosa* and *Holcus lanatus*, colour over the hesitant early growth of the grasses. There are practically no white or blue flowers, nor later when the stand is dominated by the grasses. After the first cut however the Yellowish green Cabbage Thistle (*Cirsium oleraceum*) with its soft leaves grows up and begins to flower, and in places white is now brought in by *Angelica sylvestris* (fig. 454). The second cut is often as productive as the first because these meadows are slower in getting into their full growth than the Oatgrass meadows, the more so as there is still a good supply of water in the soil after the first cut.

Cabbage Thistle meadows are similar to the Oatgrass meadows in the amount of nutrients they need and contain, especially of nitrogen (Williams 1968). However part of the nitrate that has been formed in the soil is lost through denitrification. According to Schaefer (1964) this is due to the conditions produced when water begins to fill up the large pores in the soil. Meadows such as the Sharp Sedge swamp (*Caricetum gracilis*, section C I 1 g) which are even wetter are for this reason always relatively badly supplied with nitrogen (Kovács 1964, León 1968). The wettest subassociation of the *Angelico-Cirsietum,* the Sedge- Cabbage Thistle meadow (*A.-C. caricetosum*) tends in this direction. At the other extreme the Cock'sfoot- Cabbage Thistle meadow (*A.-C. dactyletosum*) connects up to the *Arrhenatheretum*, and gives the highest yields of all subunits.

With increasing altitude the Cabbage Thistle meadows are replaced gradually by the **Globe Flower-Brook Thistle meadow** (*Trollio-Cirsietum rivularis*, see also fig. 455) described by Oberdorfer (1957). This too can be subdivided into a number of subunits and comes into contact with some very different grassland communities.

In the east and south-east of Central Europe the

Tufted Hairgrass (*Deschampsia cespitosa*) plays such an obviously important part in the moist-soil hay meadows that its name is used for the *Calthion* communities which are found in those areas, e.g. *Stellario-Deschampsietum* near Warsaw (Traczyk 1968) and *Deschampsietum* or *Cirsietum rivularis deschampsietosum* (Špániková 1971, see also fig. 456). To some extent at least the Tufted Hairgrass must thank the livestock for its position of dominance. These animals are allowed to graze after the meadows have been cut but they avoid this hard grass. On the other hand climatic factors must play a part for, according to Winterhoff (1971) *Deschampsia cespitosa* has already reached the south-western limit of its distribution in the Swabian Jura, at least as far as *Calthion* meadows are concerned. Tracyck (1968)

Fig. 451. Bentgrass-Red Fescue meadows alternating with Yellow Oatgrass meadows (near the farmstead, right centre), Matgrass swards and Heaths on the Feldberg (Black Forest, compare fig. 452). In the foreground *Gentiana lutea* as a pasture weed.

Fig. 452. The same Feldberg landscape as shown in fig. 451 covered with winter snow; a view from the Feldberg-Mittelbuck at about 1500 m looking towards Hinterwald-Kopf and Kandel. All mountain meadows are well protected from the cold by a permanent covering of snow in winter and like the alpine pastures remain green for the most part.

discovered that the amount of dead plant material in a stand of damp Tufted Hairgrass meadow is unusually high when compared with that in other meadow types. Not only the dominating grass but sedges like *Carex nigra* and *C. panicea* contribute to this material. Such meadows respond to an increased amount of manure and more frequent mowing by turning into Cabbage Thistle meadows, e.g. in central and southern Poland.

(b) Hay meadows on base-deficient moist soils

Manured meadows occur in base-deficient habitats both in the silicate mountains and on the diluvial plains in such variety that we can only mention a few examples here. The majority have come from Purple Moorgrass litter meadows (section 5) during the course of the past century. Others have resulted from the draining of small-sedges or large-sedges fens. It is possible that there exist meadow-like communities of herbaceous plants and acid-tolerant grasses which have come about without human influence, but such natural communities on acid wet soil must be very rare and limited to such places as the sources of springs (see section C II 2).

Hay meadows on base-deficient soils are generally very poor in species. The chief type in the northern lowland is the **Marsh Ragwort meadow** (*Bromus racemosus-Senecio aquaticus* association see tab. 110, B 4). According to Tüxen and Preising (1951) *Senecio*

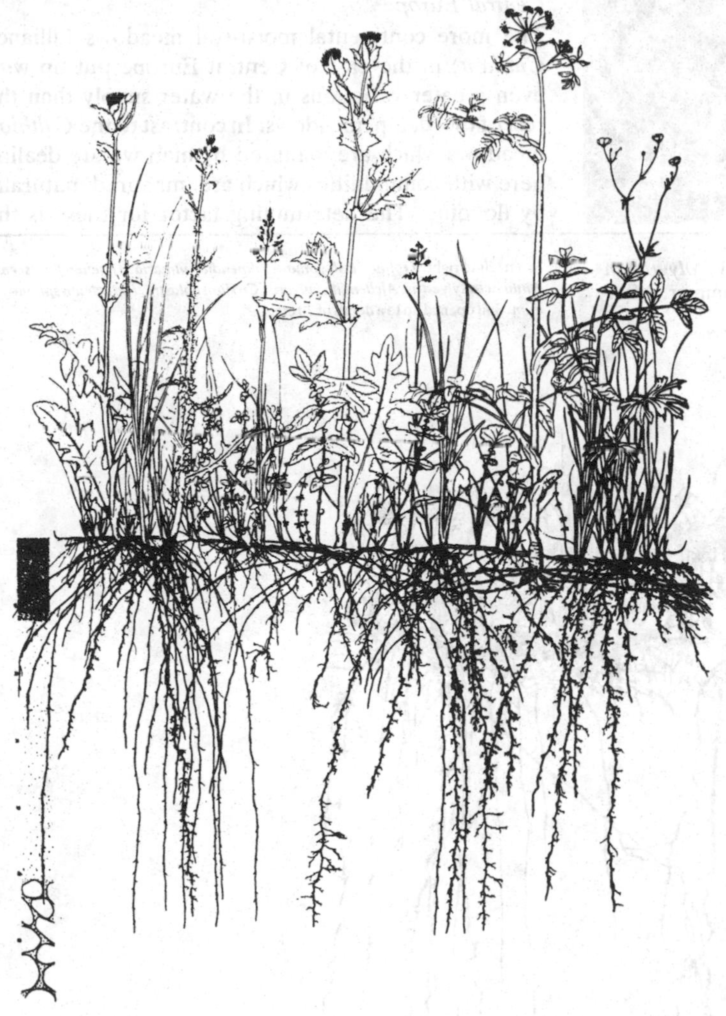

Fig. 453. A section through a wet Cabbage Thistle meadow (*Cirsium oleraceum – Polygonum bistorta* ass., *Carex acutiformis* subass.) on sand over rubble to the south of the Fläming. After Hundt (1958). Left to right: *Cirsium oleraceum*, *Carex acutiformis*, *Cirsium palustre*, *Lotus uliginosus*, *Holcus lanatus*, *Galium plaustre*, *Filipendula ulmaria*, *C.o.*, *H.l.*, *C.a.*, *Geum rivale*, *Angelica sylvestris*. *Ranunculus acris*. *Poa trivialis*. Soil opened up to a depth of 70 cm.

Fig. 454. Flowering time in a moderately well manured Cabbage Thistle meadow just before the second cut (near Amelinghausen, Lüneburg Heath). *Angelica sylvestris* dominates the picture; in the foreground (left) are leaves of *Filipendula ulmaria*.

aquaticus can be considered as the only faithful character species. As Meisel points out (1969b) the Smooth Brome (*Bromus racemosus*) which was unfortunately included in the name of the association, is absent from the really wet meadows. In many respects the Marsh Ragwort meadows are similar to the Tufted Hairgrass meadows and like these they occupy a place between the Cabbage Thistle and the Purple Moorgrass communities. But there is no difficulty in subdividing them into a relatively dry, a typical and a wet subassociation by means of differential species. As can also be seen in tab. 110 these units growing on soils of similar texture are distinguished by differences in the groundwater level. The fluctuations in soil moisture during the course of the year along with the variations in the weather conditions are quite considerable for all the subassociations.

(c) Moist-soil meadows of the river valleys in eastern Central Europe

The more continental moist-soil meadows (alliance *Cnidion*) in the east of Central Europe put up with even greater variations in the water supply than the rest of the 'damp' meadows. In contrast to the *Calthion* meadows which are manured by man we are dealing here with communities which are 'manured' naturally by flooding. The determining factor for these is the

Fig. 455. A section through a *Trollius europaeus-Polygonum bistorta* meadow in the south-eastern Harz in summer. After Hundt (1964). From left to right: *Trollius*, *Polygonum*, *Filipendula ulmaria*, *Trisetum flavescens*, *Anthriscus sylvestris*, *Alchemilla vulgaris*, *Cirsium palustre*, *Heracleum sphondylium*. Soil opened up to a depth of 70 cm.

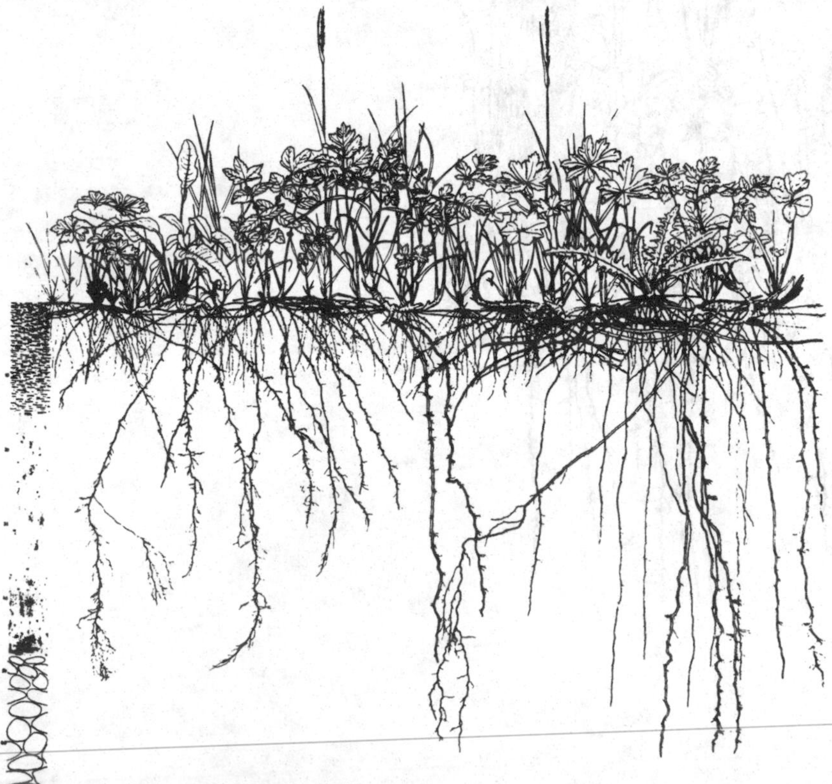

amount of alluvial material brought down by the rivers when they are in spate. Several times in the year the large rivers supply them with both water and nutrients in this way. According to Balátová Tuláčková (1969) there are different associations to be found in eastern Germany, southern Moravia, Austria and north-eastern Croatia. As characteristic for the alliance she named species whose distribution centres are in a continental climate:

Cnidium dubium	*Juncus atratus*
Allium angulosum	*Leucojum aestivum*
Carex praecox var. *suzae*	*Lythrum virgatum*
Gratiola officinalis	*Oenanthe silaifolia* et al.

There is no doubt that with regard to the supply of nutrients the *Cnidion* lies somewhere between the *Calthion* which is dependent on additional nutrients and the *Molinion* which loses its character species if it is fertilised regularly. *Cnidion* associations show most similarity with the Meadowsweet stream bank communities (*Filipendulion*, section 5 c), including the water relationships of these tall-herb stands.

(d) Rush and Clubrush spring swamp meadows

Sites which are permanently saturated with spring water stand out from the surrounding damp meadow because of the dark green colour of their vegetation. In such places rushes are dominant, *Juncus subnodulosus* where lime is present and the very similar *Juncus acutiflorus* (= *sylvaticus*) where it is not. The **Blunt-flowered Rush meadow** and the **Sharp-flowered Rush meadow** were previously classified with the small-sedges fens and were considered as natural grassland in which trees were unable to grow. However, as Oberdorfer (1957) rightly pointed out they are generally cut and are floristically close to the *Molinietalia*. They rapidly degenerate if regular mowing stops because the Common Reed (*Phragmites*) soon smothers out the light-requiring rushes. Then in its turn the reed is overgrown by Birches, Willows and other pioneers of the Alder-Ash wood or the wet Birch-Oak wood which would be the natural communities in these habitats.

The *Juncetum subnodulosi* is still quite plentiful on marl soils in the Alpine foothills. While the character species of the *Molinietalia* are stunted and sparsely represented in it there are always more species from the calcareous small-sedges swards (*Caricion davallianae*), and it often leads to a calcareous Black Bog-rush fen (see section C II 2 a). The Sharp-flowered Rush swards of the acid-spring swamps (*Juncetum acutiflori* and related communities) form their own alliance within the *Molinietalia*, according to Meisel (1969b) and older authors. Today Oberdorfer and his co-workers (1977) have returned it to the alliance of the Common Sedge and Cottongrass communities (*Caricion nigrae*) along with other man-

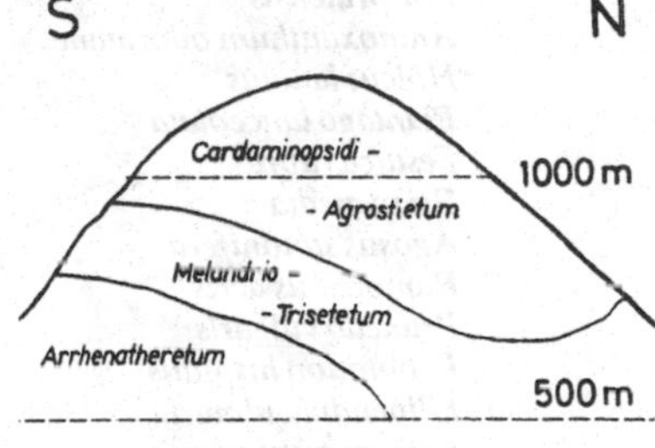

Fig. 456a. The distribution of meadow communities according to altitude and aspect in the mountains of Bohemia. After Moravec (1965), somewhat modified (see also Kornaś 1967).

The False Oatgrass, Yellow Oatgrass and Common Bent meadows are distributed in a similar fashion in the mountains of western Central Europe, but here the lower limit of the montane associations is usually higher

Fig. 456b. The arrangement of meadow communities in a Bohemian river valley, on the one hand under extensive grazing and on the other by regular manuring and mowing following draining. Schematic. After Moravec (1965), modified.

Almost all these communities are or were to be found in south-western Central Europe except that here a Cabbage Thistle meadow (*Polygono-Cirsietum*) or a moist-soil False Oatgrass meadow would dominate in place of the Tufted Hairgrass meadow (*Sanguisorbo-Deschampsietum*).

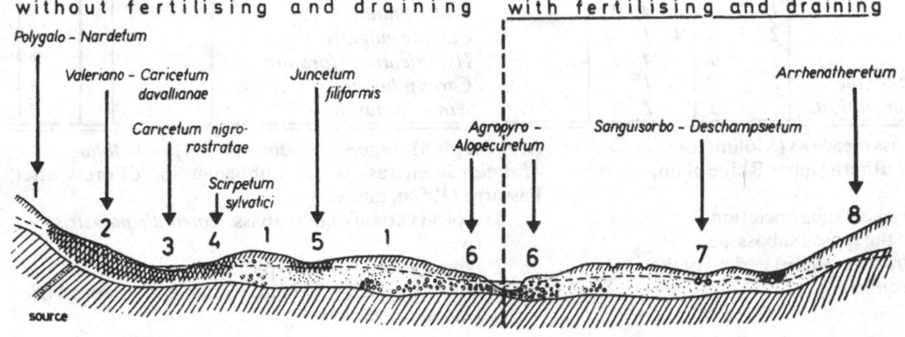

Tab. 112. Calcareous and acid Moorgrass meadows on soils ranging from wet to dry.

	Region: A = Upper Rhine, B = Spreewald	A			B		
	No.	1	2	3	1	2	3
O	*Silaum silaus*	5	5	5			
O	*Sanguisorba officinalis*	5	5	4			
	Centaurea jacea	5	4	5			
V	*Selinum carvifolia*	4	5	4			
V	*Betonica officinalis*	4	4	4			
A	*Galium boreale*	4	4	4			
A	*Cirsium tuberosum*	4	3	5			
V	*Serratula tinctoria*	4	3	4			
	Galium verum	4	2	4			
	Lathyrus pratensis	4	3	3			
	Lotus corniculatus	3	3	5			
A	*Carex tomentosa*	2	3	4			
A	*Inula salicina*	2	4	2			
A	*Tetragonolobus maritimus*	2	1	5			
	Dactylis glomerata	2	1	4			
	Ononis spinosa	2	1	4			
A	*Dianthus superbus*	3	1	1			
A	*Filipendula hexapetala*		1	3			
O	*Juncus subnodulosus*	2	3	2			
O	*Equisetum palustre*	2	2	2			
	Carex flacca	1		3			
	Allium angulosum	2	1	1			
	In all Moorgrass meadows						
V	*Molinia caerulea*	**5**	**5**	**5**	**5**	**5**	5
V	*Succisa pratensis*	5	4	3	5	5	5
	Deschampsia cespitosa	5	3	3	4	5	4
B?	*Potentilla erecta*	4	3	4	3	5	5
	Danthonia decumbens	2	3	3	3	4	5
	Poa pratensis	1	3	2	4	5	4
	Anthoxanthum odoratum	1	2	2	1	4	1
	Holcus lanatus	3	3	3		5	5
	Plantago lanceolata	2	2	4		4	5
	Festuca rubra	3	4		1	4	2
	Briza media	4	2	5		2	3
	Agostis stolonifera	5	2	4	3	2	
	Ranunculus acris	3		5	2	5	3
	Prunella vulgaris	3		3	1	4	2
	Leontodon hispidus	2		3	3	4	1
O	*Filipendula ulmaria*	4	1		3	5	2
	Linium catharticum	1	2	5	4	2	
	Vicia cracca	2	1	3		1	2
O?	*Lysimachia vulgaris*	5	3	3	1	1	
O?	*Symphytum officimale*	3	3		3		1
	Achillea millefolium	2		3			3
O	*Achillea ptarmica*	1	1		2	4	5
O	*Thalictrum flavum*	2	1		4	1	
	Valeriana dioica	1		1		4	1
V	*Gentiana pneumonanthe*	2		1		1	
	Cerastium fontanum holost.			1		2	2
B?	*Viola persicifolia*				5	5	4
O	*Galium uliginosum*				4	5	4
O	*Luzual multiflora*	1			2	5	5
B	*Ophioglossum vulgatum*				1	4	3
	Rumex acetosa				1	5	5
B	*Salix repens*				3	2	1
	Cardamine pratensis				4	2	1
O	*Cirsium palustre*				3	2	2
	Potentilla anserina				2	3	2
	Taraxacum officinale				3	2	1
B	*Inula britannica*				1	1	2
	Leontodon saxatilis				1	1	2
O	*Lotus uliginosus*				1	4	
	Moisture indicators:[a]						
	Mentha aquatica	3			5		
	Caltha palustris	2			4		
	Lythrum salicaria	3		1	1	1	
	Phragmites australis	4	4	3	1		
	Carex panicea	4		1	**5**	4	5
	Carex acutiformis	1		1	4		1
	Hydrocotyle vulgaris				**3**	3	1
	Galium palustre				5		
	Potentilla palustris				4		
	Ranunculus flammula				4		
	Phalaris arundinacea				4		
	Iris pseudacorus				4		
	Ranunculus repens				5	3	
	Indicators of dryness and a poor soil:[b]						
	Bromus erectus			**5**			
	Koeleria pyramidata			5			
	Brachypodium pinnatum			5			
	Plantago media			3			
	Trifolium montanum			3			
	Pimpinella saxifraga			3			
	Festuca ovina coll.			4		2	2
	Nardus stricta					4	**5**
	Polygala vulgaris					3	4
	Carex pallescens					2	5
	Hypochoeris radicata					3	3
	Campanula patula					2	3
	Agrostis tenuis						5
	Dianthus deltoides						5
	Viola canina						5
	Calluna vulgaris						5
	Hypericum perforatum						5
	Carex pilulifera						3
	Armeria maritima						3

A: **Calcareous Purple Moorgrass meadows** (Molinietum medioeuropaeum) of the southern Upper Rhine plain, after Philippi (1960, tab.1).
No. 1: a wet variant of the typical subassociation,
2: the typical variant of the typical subass.,
3: a moderately dry (*Brachypodium*) variant of the subassociation of *Bromus erectus*.

B: **Acid Purple Moorgrass meadows** (*Viola persicifolia-Molinia caerulea* ass.) of the Lübbenau-Spree Forest; after Passarge (1956c, tab.XII).
1: typical variant of the subass. *Potentilla palustris* (wet),
2: typical variant of the typical subass.,
3: moderately dry (*Calluna*) variant of the subass. of *Dianthus deltoides*.

made wet meadows. In contrast to the *Juncetum subnodulosi* the *Juncetum acutiflorae* has a mainly oceanic to suboceanic distribution.

The position of the **Wood Clubrush spring meadow** (*Scirpetum sylvatici*) is not very clear either. Its appearance reminds one of the large-sedges fens but in species composition it is much nearer the Marsh Marigold meadow (*Calthion*). Wherever stands of *Scirpus sylvaticus* occur, they indicate the presence of lime-deficient, but relatively nutrient-rich water (Yerly 1970, Amani 1981, see also fig. 456 b).

5 Purple Moorgrass litter meadows and related herb communities

(a) Purple Moorgrass meadows in different sites

Each year the litter meadows in Central Europe have become fewer and now they are only to be found in the plains north of the Alps and in valleys of the outer Alps themselves. Even here they are rapidly becoming less common since the modern tendency to house cattle without litter has gained ground. This is especially true of the Purple Moorgrass meadows (*Molinion*, see tab. 112) which occur on the same sort of soil as the hay meadows and, as we have seen in section 1 a, can be easily converted into these. Those classical authors in grassland sociology Stebler and Schröter devoted extensive works to the litter meadows and even W. Koch (1926) was able to study large areas in the Swiss Alpine foothills such as on the Linth flood plain, but in the forseeable future they will be confined to small remains in a few reserves. Even here they will have to be managed in the traditional way by being mown every year or at least every second year; this cut must not be made before the end of September, and the straw will have to be removed from the meadow. Otherwise the *Molinion* meadows will soon lose their particular floristic composition.

Long after the manured wet-soil meadows have turned lush green and become gay with flowers the stubble on the *Molinia* meadows is still a pale straw yellow. In the mountains it is often Anemones (*A. nemorosa*) or Oxlips (*Primula elatior*) i.e. really woodland plants, which take advantage of this delayed reaction to the coming of spring. When many of the hay meadows have already been mown *Molinia* starts to send up its shoots with their blue buds, and the accompanying plants also start to flower. By the late summer there is a glorious symphony of colour with the dark blue bells of the Gentians, the yellowish white umbels of *Silaum silaus*, the carmine heads of *Serratula tinctoria* and many other gay flowers. When finally the haulm and leaves of the Purple Moorgrass turn golden yellow and copper, the litter meadow stands out shining from the 'evergreen' hay meadows to compete with the autumn colours of the mixed broadleaved woodlands.

It is only the **calcareous Purple Moorgrass meadows** (*Cirsio tuberosi-Molinietum*), i.e. the 'real' Central European *Molinieta* on base-rich damp soils which are so rich in species and so colourful (tab. 112). Compared with these the acid-soil Purple Moorgrass meadows (*Junco-Molinietum*), which can be looked upon as the substitute communities of Birch swamp woods and moist-soil Birch-Oak woods (sections B V 2 c and B III 5 a), appear to be impoverished floristically (figs. 457 and 458).

Both the **calcareous and the lime-deficient Purple Moorgrass meadows** can be subdivided into a number of subunits reflecting the degree of soil moisture. The typical ones (e.g. A2 and B2 in tab. 112) have no particular differential species. To some extent the same swamp plants appear on both lime-rich and lime-deficient substrates in the relatively wet subcommunities (A 1, B 1). As one would expect the drier

Tab. 112. (Continued)

Note: [a] In addition in no. B1 with a constancy of 3: *Juncus effusus*, *Carex vesicaria*, *Lycopus europaeus* and *Stellaria palustris*.
[b]In no. A3 with a constancy of *3: **Phyteuma tenerum***.
Species which do not achieve a constancy of 2 in any of the units have for the most part been omitted.

Meaning of the letters in front of the species name:
A = Character species of the unit A
B = Character species of the unit B.
V = character species of the alliance *Molinion* (Moorgrass meadows)
O = character species of the order of *Molinietalia* (moist-soil meadows).

subunits differ most (A3 and B3). The 'dry calcareous Purple Moorgrass meadow' has many species in common with the calcareous slightly arid grassland (section D I 2 b) and is particularly rich in species. On the other hand the *Junco-Molinietum* on acid soils shows affinities with the Matgrass meadows or even the *Calluna* heaths (section D II 4 and 3), this tendency being stronger the lower the water table is below the surface. On waterlogged soil the *Junco-Molinietum* may pass over to an *Erica tetralix* mire just as has been described from England (Rutter 1955).

The plants listed in tab. 112 are extreme examples of calcareous and acid-soil Purple Moorgrass meadows but both types can intermingle. In addition many of other *Molinion* communities can be distinguished. They form an examplary pattern of geographical differentiation of plant communities as Philippi (1960) has shown in the Upper Rhine plain. Kovács (1962) gives a review of the literature as seen from the Hungarian side. Wagner (1950b) has already pointed out that one cannot use the same method of classification for Purple Moorgrass meadows throughout the whole of Central Europe as W. Koch (1926) has done in his classical work which received a good deal of attention. This is shown particularly well in monographs covering small areas, e.g. those produced by Korneck (1962/63), Rodi (1963), Fritsch (1962) and many other authors who worked before these interesting vegetation types lost their economic importance.

Fig. 458. A Soft Rush-Purple Moorgrass meadow (*Junco-Molinietum*) on a wet acid soil in Bohemia with *Succisa pratensis*, *Juncus effusus*, *Mentha arvensis* and *Ranunculus repens* (in front). Photo. Iltis-Schultz.

Fig. 457. A section through a calcareous Purple Moorgrass meadow in the centre of Switzerland. After M. Mayer (1939).
Left to right: *Serratula tinctoria*, *Carex panicea*, *Molinia caerulea* (*M.*), *Selinum carvifolia*, *Potentilla erecta*, *Gentiana pneumonanthe*, *M.*, *Thalictrum flavum*, *M.*, *Sanguisorba officinalis*, *Succisa pratensis*, *Allium angulosum*, *Epipactis palustris*, *Iris sibirica*, *M.*
The depth of soil shown here is about 10 cm but only the beginnings of the roots are depicted. They go down several decimetres and form a dense network.

Special *Molinion* associations occur in the valleys of the large rivers. For example in wet basins in the Upper Rhine plain there is a *Cnidio-Violetum* which has character species such as *Viola persicifolia* (= *stagnina*), *Cnidium dubium*, *Lathyrus palustris* and *Gratiola officinalis* (Philippi 1960). This connects up with the *Cnidion* meadow which was discussed briefly in section 4 c.

(b) How to maintain litter meadows

One of the most pressing problems for nature conservation has become that of preserving at least some of the large unmanured litter meadow complexes. The easiest and most obvious method would be to go on managing them as they have been farmed up to now, but there is a problem in that with the modernisation of cow stalls there is much less demand for litter for bedding. The simple mowing and removal of the straw has thus become too expensive. However if mowing is discontinued or if the cut herbage is allowed to lie on the ground then the species composition will be upset. The rarer species would be particularly susceptible while *Filipendula* and other tall herbs would spread or the grasses and sedges would form thick layers of dead leaves. Up to a point the mowing can be replaced by burning off and this certainly maintains the dominance of the fire-tolerant Purple Moorgrass. As Miles (1971) reports this is eaten by Deer when it shoots again after a fire. Thus a *Molinietum* can be subjected to a certain amount of grazing by Deer. However after many years of research carried out in Switzerland it has been discovered that the only way of being sure that the species combination is maintained is to mow the Purple Moorgrass meadow in the autumn at least every second year by machine and remove the cut material to a place where it can be burned.

(c) Meadowsweet bank strips and similar communities

Many members of the Purple Moorgrass meadows, Meadowsweet (especially large-leaved tall herbs such as *Filipendula ulmaria*), are in some respect similar to the nitrogen-loving species of woodland clearings and grow strikingly luxuriantly when they are well supplied with mineral nitrogen (Grabherr 1942b, see fig. 459). Thus they are found equally plentifully if not more so in manured wet-soil meadows (*Calthion*).

The sides of small streams or ditches in the meadowland are especially rich in nutrients which accumulate from the mud and weeds regularly cleaned out of the water courses. Provided these **stream-bank tall-herb communities** (*Filipendulo-Geranietum*) are not mown too often they are mostly dominated by *Filipendula ulmaria*. Their species make-up has affinities with the *Molinion* but recently they have been put into an alliance of their own, the *Filipendulion* (see Meisel 1969b). Physiognomically they are more reminiscent of the tall-herb communities in the high mountains (section C VI 7 a) with their strong growth and their flowers. M. Mayer (1939) working in Switzerland has devoted an ecological and phytosociological monograph to them which still merits attention. Bálatová-Tuláčková (1979) distinguishes several *Filipendulion* associations corresponding to the altitudinal belts.

Fig. 459. A section through a Meadowsweet river-bank community in central Switzerland. After M. Mayer (1939).

Left to right: *Carex gracilis*, *Phalaris arundinacea*, *Carex acutiformis*, *Geranium palustre*, *Equisetum palustre*, *Filipendula ulmaria* with *Calystegia sepium*, *Caltha palustris*, *Galium mollugo*, *Colchicum autumnale*, *C.g.* Soil opened up to a depth of 15 cm, only a part of the roots are shown.

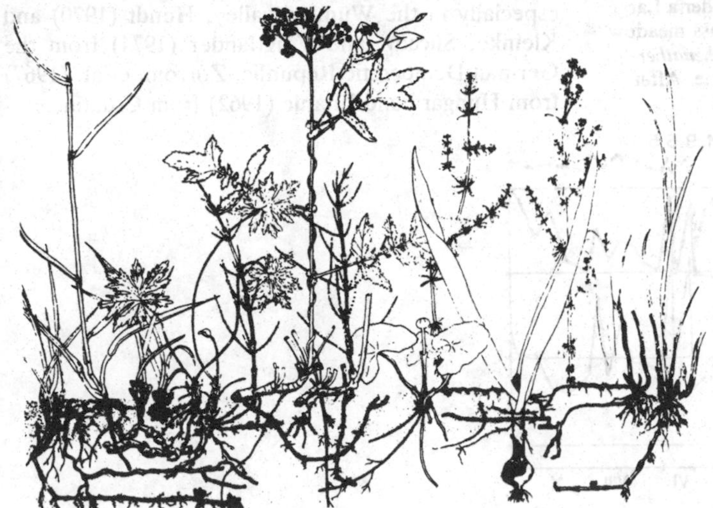

The methods of meadow management in previous centuries have been largely responsible for the present-day distribution of the *Filipendulo-Geranietum*, especially in the lowlands. In spite of this one could imagine that there were Meadowsweet stream bank communities here and there in Central Europe even before the intrusion of man, perhaps along small streams in the area of the Alder-Ash woods. In the southern part of Fennoscandia too one finds similar communities which are presumably natural, at least in part. In the east of Central Europe the Meadowsweet-Cranesbill community is replaced by one richer in species where besides the Meadowsweet itself important parts are played by the **Long-leaved Speedwell** (*Veronica longifolia*) and the **Spear-leaved Skullcap** (*Scutellaria hastifolia*). This *Veronico-Scutellarietum* is very much associated with flood plains where it occurs not only on river banks but more especially at the edges of streams and ditches. From here it extends into meadows of the *Cnidion* alliance which also has an eastern distribution, just as the *Filipendulo-Geranietum* spreads into the *Molinion* meadows further west.

In many areas extensive **Meadowsweet stands** can be found where meadows have been abandoned or neglected (see the previous section). These never show the characteristic *Filipendulion* combination of species but they represent intermediate stages in the succession from the wet-soil meadow towards a natural woodland, although it may be a long time before the first Willows, Alders or other pioneers become established in the shade-producing tall herbaceous flora. The creation of an open space by fire, rooting of Wild Boar or other disturbance will finally allow the woody plants to gain a footing.

Fig. 460. Fluctuations of the groundwater level under a Large Sedge bed (*Caricetum gracilis*), a Purple Moorgrass meadow (*Molinietum*) and a False Oatgrass meadow (*Arrhenatheretum*) near Cracow in a dry year and in a wet one. After Zarzycki (1956), modified, from Ellenberg (1963).

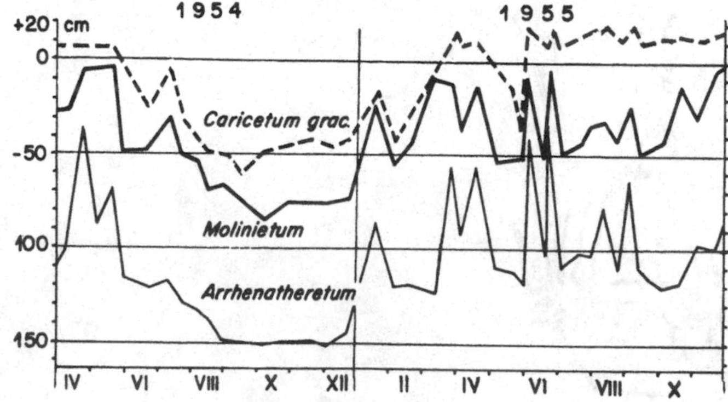

6 *Effects of habitat and competition on the species composition*

(a) Results of pot and field experiments with meadow plants

Since meadow plants are relatively easy to cultivate and since a large number of manuring experiments have been carried out on meadowland the causal analysis of these communities is more advanced than with any other plant formations. Therefore it may be interesting to summarise the knowledge gained from these experiments concerning the reasons why certain combinations of species have come about. This final survey is concentrated on the orders *Arrhenatheretalia* and *Molinietalia*, especially on the *Arrhenatheretum* and its various subassociations and variants which have already been described in section 2 a, and also on the *Angelico-Cirsietum*.

On soils with an average base content the following communities form a sequence according to increasing dampness of the habitat (figs. 374, 441 and 460): *Xerobrometum*, *Mesobrometum*, dry subassociations of the *Arrhenatheretum* with and without *Bromus erectus*, typical *Arrhenatheretum*, damp subcommunities of the same, moist-soil meadows rich in *Cirsium oleraceum* in its moderately moist, typical and wet subassociations, and finally *Caricetum gracilis* and *Phragmitetum*. Such an ecological series can be seen in many river valleys often within quite a short distance, e.g. repeatedly in the valley of the Itz, which is divided up by several mill dams, and has been mapped by Vollrath (1963). Individual species have different ranges within this series but no single one extends over the whole range of available water supply. This has been confirmed by various authors from many parts of Central Europe, e.g. Ellenberg (1952a, b, 1974, 1979) from the German Federal Republic, Baeumer (1962) especially in the Wümme valley, Hundt (1970) and Kleinke, Succow and Voigtländer (1974) from the German Democratic Republic, Zólyomi et al. (1967) from Hungary and Ilijanic (1962) from Croatia.

As we have already argued in section D I 6 b without experimental evidence it would be wrong to conclude from this series that species such as *Bromus erectus*, *Avenochloa pratensis* and *Phleum phleoides* are xerophilous or that they would have their optimal physiological growth in dry habitats. Nor, without further information, would there be any justification in assuming that *Alopecurus pratensis*, *Poa palustris* and other dampness or wetness indicators could not thrive on drier soils. Their amplitude in grassland communities appears to depend on the presence of stronger competitors which can sometimes even push them right outside the range of their physiological optimum.

During a period of drought, which occurs in Central Europe from time to time, the tall grasses suffer more than other grassland plants in the Oatgrass meadows, just as they do in the *Brometalia* swards. This was clearly shown by Lüdi and Zoller (1949, see fig. 387) from their observations using long-term quadrats. After the extremely dry year of 1947 both the *Arrhenatheretum* and the *Mesobrometum* stands recovered surprisingly quickly because, although the majority of the individual plants were damaged they remained alive. In years which were dry but not so severe there was a sufficient reserve of water in the soil for the members of the Oatgrass community which are nearly all deep rooting (see fig. 461, also 376 and 446). This is probably even more true of Yellow Oatgrass meadows in the montane belt where real droughts are not likely. Nevertheless the mountain meadows may also sometimes find difficulty in getting sufficient water; but this is due in some degree to the fact that the roots are less active during the summer months (fig. 462).

Fig. 462. Root activity in a Yellow Oatgrass meadow in the Solling falls off in the summer both in manured and unmanured areas. After Speidel and Weiss (1974), somewhat modified.

As a measure of root activity the number of living root tips were counted against a slightly overhanging glass wall, 50 x 80 cm in size, in an otherwise light-proof pit.

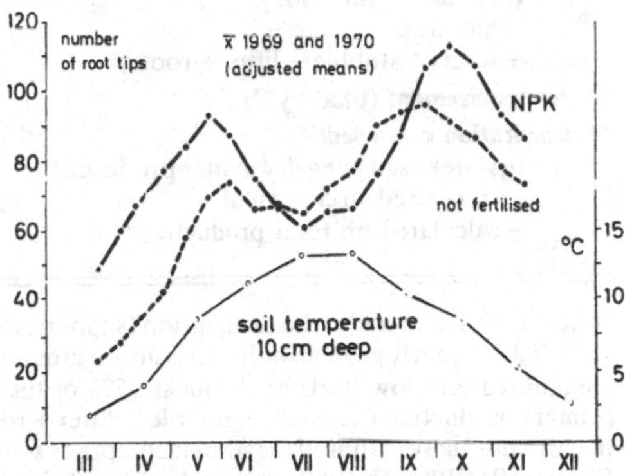

Other factors being equal the supply of nitrate in a neutral or weakly acid soil is greatest when there is an average amount of soil moisture, i.e. sufficient water but also sufficient oxygen available. Both drier and

Fig. 461. Even in dry years such as 1964 the soil of the False Oatgrass meadow (1 m depth of flood-plain loam over clay) near Kelbra in the Helme valley holds sufficient water available for the plants (see fig. 376). After Hundt (1970), modified.

Upper diagram: the average precipitation for five-day periods and the 'available water reserve' calculated from the water present in the top 50 cm, taking into account the permanent wilting percent (PWP, between 5.4 and 16.6%).
Lower diagram: the lines represent equal amounts of available water (volume percent) down to a depth of 120 cm between April 1963 and June 1965.
During the growing period even heavy storm rain (as in August 1964) is not able to replenish the water reserves to any marked degree. This happens only during the winter, especially when the snow melts releasing all the small amounts of precipitation which have accumulated during the previous months. (The same applies for woodland soils also; see fig. 426.)

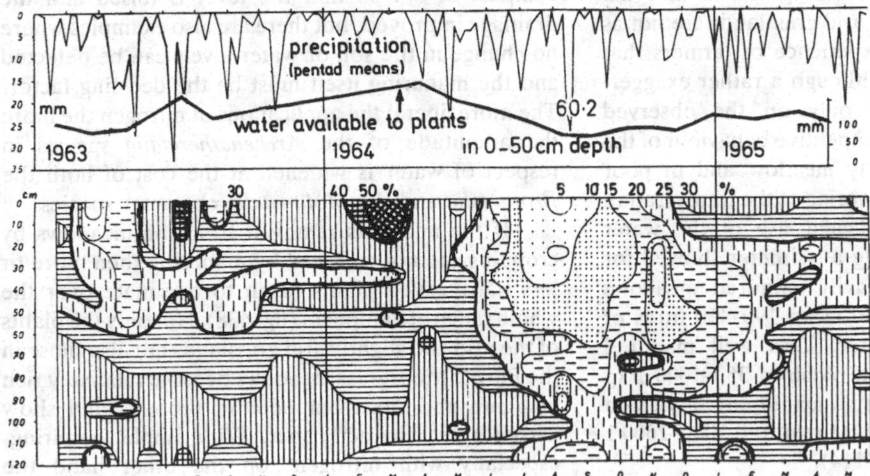

Tab. 113. The net primary production and the transpiration coefficient of meadows in lysimeters receiving different amounts of fertiliser.

Fertilising (* standard fertilising)	0	PK	NPK*	PK + 520 kg N/ha
Net primary production				
(dry matter int (mm·y^{-1}				
– hay crop	3.6	7.2	8.0	**16.2**
– total (+ stubble + litter + roots)[a]	14.4	18.0	16.0	24.3
Water requirement (t·ha^{-1}·y^{-1})	451	452	497	469
'Transpiration coefficient'[b]				
(g water used per g dry matter produced)				
– calculated on crop yield	1253	630	621	**290**
– calculated on total production[a]	313	251	311	193

Note: [a] Calculated under the assumption (supported in tab. 114) that the yearly growth of the parts above ground in an unmanured meadow (0) is at the most 25% of the total net primary production (i.e. yield + stubble + litter + roots) but that in a normally manured meadow it comprises some 40 (PK) or 50 (NPK), and with a heavy dressing of nitrogen (PK + 520 kg N/ha) as much as 66%. If these assumptions are correct then the transpiration coefficients are of the same order as those in woodlands (see tab. 24).

[b] There are no significant differences between the levels of manuring in the amount of water used by the meadow stands (in transpiration and evaporation). The proportion lost through direct evaporation from the soil under a meadow is mostly so small as to be scarcely measurable. If it were possible to calculate the transpiration coefficient by making allowance for evaporation all the transpiration coefficients would be slightly smaller (1 mm corresponds to 10 t water per ha).

Source: After Klapp (1971), modified and extended.

wetter habitats are poorer in nitrate and so they favour plants which are less demanding against species such as *Arrhenatherum* which require more nutrients. If large amounts of manure are given to a *Mesobrometum*, or even a *Xerobrometum*, a *Salvia-Arrhenatheretum* can be produced. Aichinger (1933) had already reported that this was the case and Ellenberg (1952a, 1963) included an example from the Swabian Jura where this change had taken place within a few years. All the Salvia-False Oatgrass meadows have been formed in this way from calcareous slightly arid swards by manuring during the last centuries or even decades.

The effect of extra manuring shows that the habitats of arid and slightly arid grasslands are not as dry as they appear. The experience of farmers that 'nitrogen replaces water', although a rather exaggerated claim, is based not only on the observed differences in the relative competitive behaviour of the same species in a good hay meadow and in poor grassland, but also on the lowering of the transpiration coefficient, i.e. a more efficient use of the water available. Experiments using a lysimeter showed for example that the most heavily manured meadows yielded about 10 t/ha of hay using 190–450 litres of water per kg of dry matter. Less well manured meadows giving about 5 t/ha required 350–1000 1/kg DM and meadows receiving no manure at all yielded barely 1 t/ha but each kg of DM required 1000–2600 1 of water (see tabs. 113 and 114).

Looking now towards the wet end of our series of communities we can add to the above saying another which at first looks equally paradoxical: 'nitrogen replaces oxygen'. Good manuring favours the species of the *Arrhenatheretum* in wet habitats too where under less-intensive methods swamp plants would get the upper hand. In parcels of land on homogenous moist soils it has been noticed repeatedly that those meadows which are more heavily manured carry a plant community indicating a relatively dry, better aerated soil; but these apparently fail as indicators here. In such cases the soil may have had a dressing of compost or silt so that the level is raised and the drainage improved, but there are also examples where no change in the soil or water levels can be detected and the manuring itself must be the deciding factor. The more liberal the applications of nitrogen the more the amplitude of the *Arrhenatheretalia* species in respect of water is widened at the cost of both the *Brometalia* and the *Molinietalia* character species.

If the nutrients removed from the meadows by cropping are not replaced by manuring then more or less hard-leaved plants begin to dominate over the whole range of dampness (fig. 438). Not only the plants of the arid and slightly arid grasslands but also those in the Purple Moorgrass meadows become relatively rich in lignin and silica but poor in proteins and show xeromorphic characteristics. With ample manuring, especially with nitrogen, on the other hand the

Tab. 114. The net primary production of Red Fescue-Yellow Oatgrass meadows subject to different levels of fertilising in the Solling after five years of dressings with amounts of nutrients which experience has shown to be effective.

Fertilising		0	PK	NPK
Net primary production (10^5 kcal ha^{-1}·y^{-1})				
Green crop (used):		98	204	**306**
Litter and stubble (above ground, unused)		194	220	198
Roots, apart from the finest[a]		**140**	134	110
Total[a]		432	558	614
Green crop as a fraction of total	%	22.7	36.6	49.8
Efficiency of the net primary production				
as a % age of the total *radiation*[b] during the year:	%	0.54	0.62	**0.82**
the same during the growing period:	%	0.77	0.88	1.14

Note: [a] The finest roots which cannot be recovered by normal washing make up a large percentage of the total production. This is significantly larger when the level of fertility is low than it is with better manuring. Thus the sum of the net primary production above and below ground at all manurial levels is probably about the same (estimated at 700–800 kcal/ha per annum, see fig. 36).
[b] **As a % age of the photosynthetically effective radiation** about twice as much.
Source: From data in Runge (1973).

mesomorphic, protein rich fodder plants of the *Arrhenatheretum*, which are the ones which respond most clearly to the supply of nutrients (see fig. 439), tend to spread themselves. Consequently the Oatgrass meadows can be considered as mostly mesophilous in contrast to the more or less arid grasslands and the damp or wet meadows. The factor above all which determines their distribution though is their love of nitrogen, or, in more general terms, their higher nutrient requirement. If this is satisfied they are able to cover the wide range of dampness from the *Mesobromion* right to the borders of the *Magnocaricion*.

Usually in fact a few of the species of the community which have been replaced by increased fertility still persist in the *Arrhenatheretum*, and these can be used as differential species for the dry, damp and wet subassociations. However even their amplitude can be restricted by generous applications of manure, so if one wishes to use the meadow communities as **indicators of dampness** then one must **also take account of the fertility level**. Provided there is sufficient water then the meadow plants will arrange themselves according to the nutrients available in the soil. The amount of nitrogen in particular gives many plants a competitive advantage when compared with others (fig. 463). In pure stands however even the 'poverty indicators' are helped by extra nitrogen (fig. 464) whereas in mixed cultures tall-growing species such as Cocksfoot gain more ground than for example Red Fescue or Dock. Rather surprisingly the Dandelion (*Taraxacum*) proved to be relatively unresponsive to the amount of nitrogen in an otherwise fertile meadow, but it responded strongly to applications of potassium as the right hand side of fig. 464 shows clearly. Unfortunately we still know far too little about the influence of K, P and other bioelements on the species composition and how this compares with that of nitrogen. Nor do we have enough experimental evidence to form a judgement on the effect of temperature, water and other habitat factors.

The experience of farmers that leguminous plants are made more competitive against grasses by the application of phosphorus and potassium has been confirmed by hundreds of manurial trials on grassland (Klapp 1965). In the final analysis this is really a nitrogen effect and depends on the ability of clovers and other legumes to fix gaseous nitrogen with the help of the bacteria in their root nodules. They are the only plants of the meadow communities capable of doing this; when their roots die and decay the proteins are mineralised and the nitrogen is made available to the grasses and other plants in the community.

Starting with the behaviour of the Upright Brome and other grasses we arrived in section D 16 at correlations which contributed to the causal understanding of the the *Arrhenatheretalia* and *Molinietalia* meadows. There we learnt to recognise the False Oatgrass as one of the most important competitors of *Bromus erectus*. In all the situations set up by Bornkamm, Salinger and Strehlow (1975) it was superior to the Upright Brome after only the first 6 weeks. When conditions for growth are good *Arrhenatherum* grows higher and produces a thicker shear of leaves than most of the other grasses of *Molinio-*

Arrhenatheretea meadows. When the soil is too dry or too wet, or short in mineral nitrogen however, or even with a slightly unfavourable combination of these factors the productivity of Oatgrass falls off particularly rapidly. Its ability to compete then varies considerably depending on the conditions and cannot be assessed as a specific value, as e.g. Caputa (1948) tried to do.

The causal analysis of the species composition of grassland communities is made all the more difficult because at first all the species are encouraged by manuring, including the so-called 'poverty indicators' (Künzli 1967, etc.). Even Matgrass and Sheep's Fescue (*Nardus stricta* and *Festuca ovina*) in a pure culture increase their production as the quantity of nitrogenous manure is increased, and investigations by M. Vogel (1981) showed that they could take concentrations of up to 3600 kg/ha of N per annum, i.e. many times the normal dressing for a meadow (50–100 kg/ha). Just like the species of the manured meadows they absorb large quantities of NO_3 and store it in their cell sap; their chlorophyll content is increased, the structure becomes less xeromorphic and other changes result which were noted by Kurkin (1975) in cultivated grasses. Extra phosphate is also valued by all meadow plants and this too reduces the degree of xeromorphism they show, although to a lesser extent than with nitrogen (Steubing and Alberdi 1973). Davies and Snaydon (1974) found that the dry matter production of the Sweet Vernal-grass (*Anthoxanthum odoratum*),

Fig. 463. Amplitudes of meadow and river bank plants arranged according to the gradient of available mineral nitrogen in the topsoil taken from sample plots in the Elbe valley above Hamburg. After F. H. Meyer (1957), modified and with indifferent species omitted.

To the right of the plant name is the nitrogen indicator value according to Ellenberg's nine point scale, to the left according to a five point scale derived from the area under investigation (U) by Meyer. Along the top above the line are the average nitrogen values calculated according to Meyer's scale from the complete list of species present in each of the relevés of the sample plots no. 1–59. The sequence of relevés was determined from the level of nitrogen mineralisation of the soils measured under constant and favourable conditions in the laboratory. The height of the black blocks indicates the cover degree of the species in question.

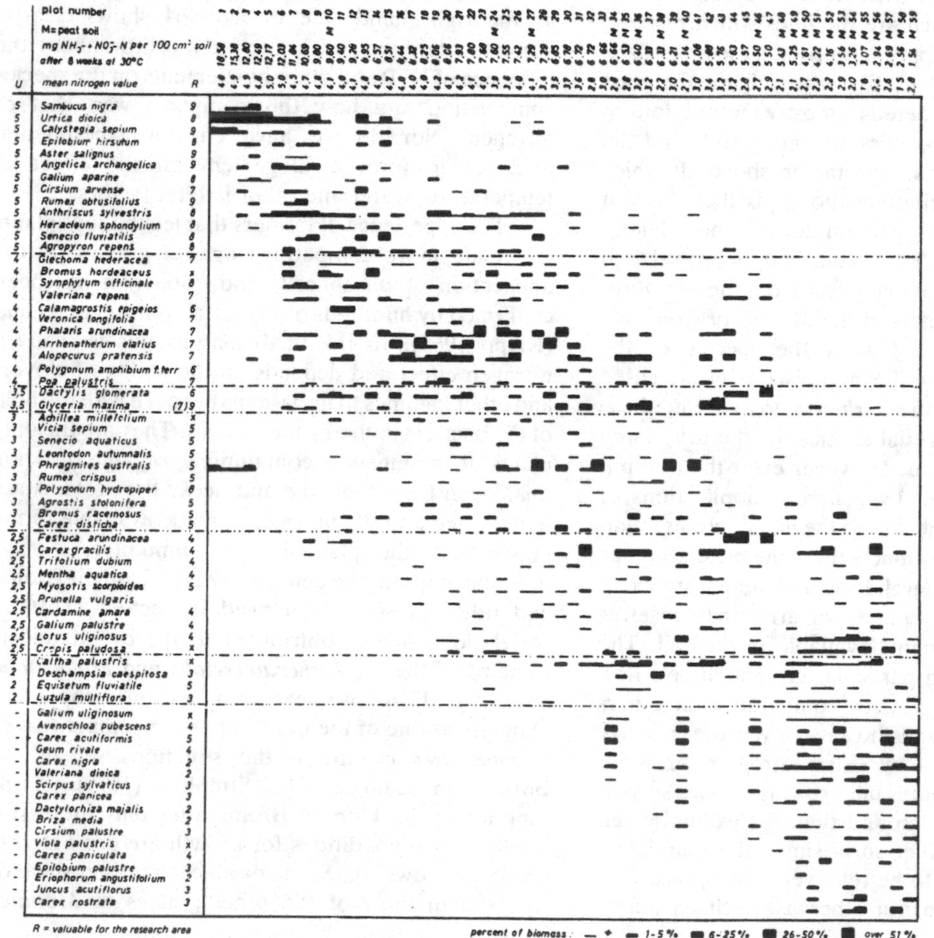

Fig. 464. The effect of additional nitrogen (left) and potassium (right) on the yields of pure and mixed stands of meadow plants, which otherwise are receiving normal amounts of nutrients (near Zürich). After Hofer (1970), somewhat modified; (yield given in g of dry matter per m^2).

The **pure stands** are shown by diagonal hatching. In each of the three years all species responded favourably to the additional nitrogen and to the potash. The Dandelion (*Taraxacum*) and the Meadow Buttercup (*Ranunculus acris* ssp. *friesianus*) showed a better response to K than to N. The Sorrel (*Rumex acetosa*) was helped by both equally but not to any great extent. The grasses (*Festuca rubra* and *Dactylis glomerata*) gave a substantially better yield with both the additional N and K.

When **in competition with one of the other species** the growth varied according to the competitiveness of each. In most cases *Dactylis* was successful because of its relatively rapid growth. Only the Dandelion and the Buttercup were able to compete successfully with it when additional K was given. The low-growing Red Fescue was only helped with K against the Sorrel and with N against this and the Dandelion.

Sorrel is actually very rare in well manured meadows whereas it spreads in neglected ones along with the equally poorly competitive Red Fescue (see also fig. 441).

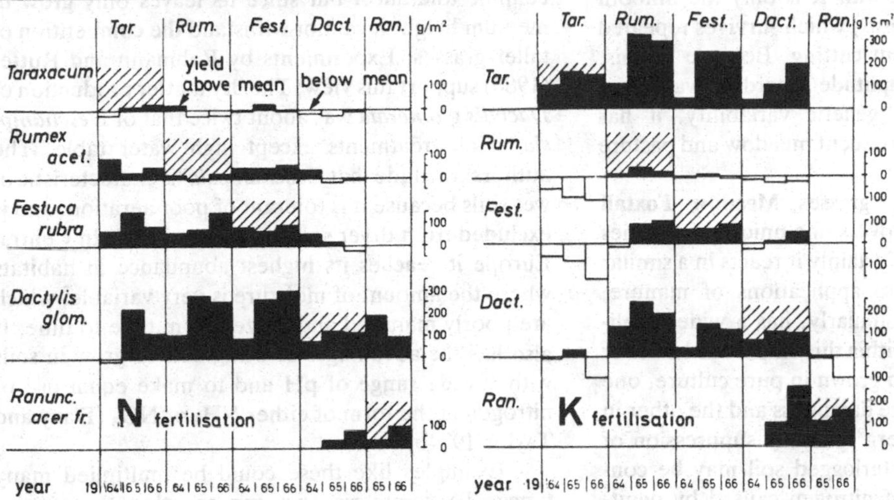

which is considered a species with modest demands, was in close correlation with the amount of soluble phosphate supplied. The same was found in the Sorrel (*Rumex acetosa*) by Gupta and Rorison (1975) although in meadows this herb, as well as *Anthoxanthum*, is able to come through against the competition only when fertility is poor. What goes for N and P is also true of the third main nutrient element, potassium. According to Wild, Skarlon, Clement and Snaydon (1974) and H. Hofer (1970) additional dressings of K stimulated the growth of grasses as well as that of legumes and other herbs in culture experiments. Under competition however their behaviour can vary considerably. For example low-growing *Ranunculus acris* is able to get the better of Cock'sfoot, which would otherwise smother it, because its roots become active earlier (Osman 1971).

The majority of species present in fertilised meadows are fairly indifferent to the soil pH. With less manure however *Arrhenatherum*, *Bromus erectus* and numerous other grasses as well as legumes and herbs thrive better on a neutral or calcareous soil than on an acid one. To some extent this too may be acting through the supply of nitrogen since the mineralisation is quicker as a rule in less acid soils, other conditions being equal.

False Oatgrass and Upright Brome also resemble each other in being sensitive to frequent as well as to

Fig. 465. The influence of mowing and grazing on the relative amounts of the important grasses in identically seeded grassland plots on dry, moderately manured sandy soil. From data in Könekamp and König (1929) and from Schwarz (1933) in Ellenberg (1963).

On the **left** three years with several cuts per year and then grazed for three years with sheep. On the **right** the management was reversed. *Dactylis glomerata* and *Festuca pratensis* suffer under grazing but increase when the field is mown. The behaviour of *Lolium perenne* and *Phleum pratense* is the opposite to this. *Poa pratensis* increases under both systems. Ordinates = percentage proportion of the green weight. The 'other species' comprise also pasture weeds.

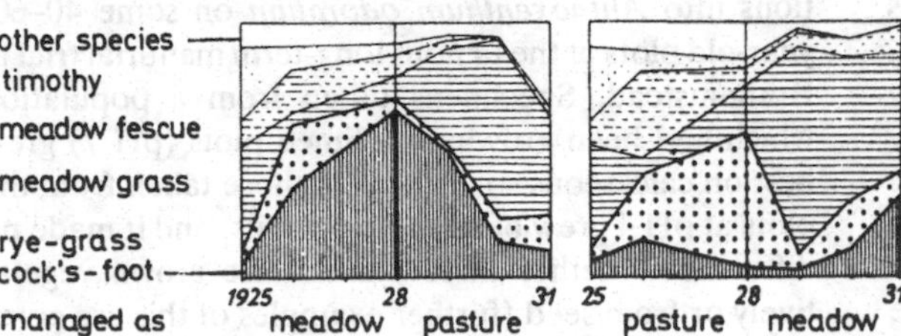

very early cutting, and to grazing. Under such conditions they never produce seed and finally die out completely, whereas species with strong vegetative reproductive capacities such as *Agrostis tenuis* on poor pastures or *Lolium perenne* and or *Poa trivialis* on fertile ones do better than the clump-forming grasses. Researches by Könekamp and others (see fig. 465) have shown that the relative abundance of the two groups can be altered at will. It is only the Smooth Meadowgrass (*Poa pratensis*) which survives repeated as well as very infrequent cutting. Because of this, coupled with its wide amplitude regarding water and nutrient supply and its genetic variability, it has become one of the most frequent meadow and pasture grasses.

Amongst the taller grasses **Meadow Foxtail** (*Alopecurus pratensis*) shows some unique properties in a number of respects. Certainly it reacts in a similar way to *Arrhenatherum* to applications of manure, especially nitrogen and, similarly, has a wide amplitude for soil dampness. Within this amplitude however it shows two optima when grown in pure culture, one where the soil is of average dampness and the other in waterlogged soil (Ellenberg 1963). A suppression of yield on permanently waterlogged soil may be connected with a shortage of nitrogen caused by denitrification (see tab. 56) as we have already explained in section D 1 6). Nevertheless fertilising favours the Meadow Foxtail so effectively on moist and wet soils that it becomes dominating. According to Sharifi (1978) This is due mainly to the supply of nitrogen.

Alopecurus pratensis is a true helophyte in that it has elongated air spaces in the root cortex which help in the respiration of the roots in waterlogged soil. This aerenchyma is present even when the plant is growing in slightly damp, well aerated soils. Freshly severed pieces of root about 10 cm in length offer less resistance to air under pressure than a similar piece of *Arrhenatherum* root grown under wet conditions even though this shows large lacunae (like those represented in fig. 390 for *Bromus erectus*). In competition with *Arrhenatherum* and other grasses in waterlogged soils *Alopecurus* comes out the undisputed winner, but on rather dry soils it is suppressed, although it does not disappear entirely.

The **Swamp Meadowgrass** (*P. palustris*) behaves in a similar way to *Alopecurus pratensis* in pure culture. In a mixture however, being a short grass, it is strongly suppressed by the tall grasses and is banished to very wet sites. The competition of just a few other grasses then is sufficient to restrict the physiological amplitude of a species under test so that its behaviour in a model community is approximately the same as that in nature (see fig. 389).

The Tufted Hairgrass (*Deschampsia cespitosa*), whose role in more or less wet meadows of eastern Europe has already been described in section 4a, is thus perhaps not so frequent in those sites because it makes greater demands on the water supply. On the contrary, according to Davy and Taylor (1974a) it can survive a dry time without damage and shows a wide edaphic tolerance, but since its leaves only grow to medium height it cannot withstand the competition of taller grasses. Experiments by Rahmann and Rutter (1980) support this view: The dry matter production of *Dactylis glomerata* was about twice that of *Deschampsia* in all treatments except high water table. The authors conclude that *Deschampsia* 'is characteristic of wet soils because it is tolerant of poor aeration, and is excluded from dryer soils by competition'. In Central Europe it reaches its highest abundance in habitats where the amount of moisture is very variable, which are poorly manured and grazed from time to time. It also has the advantages of being able to grow in soils with a wide range of pH and to make equal use of nitrogen in the form of either NH_4 or NO_3 (Davy and Taylor 1974b).

Examples like these could be multiplied many times. To summarise we can say that the species composition of our meadowland communities by no means reflects simply the 'demands on the habitat' of the individual species but rather their competitiveness under particular circumstances. **The competitive strength of any one species is not a constant amount** which can be just entered into any calculation (as, for example Grime 1973 has tried to do). It varies a good deal according to the other species involved in the competition and according to the environmental conditions which vary both in time and place (Ellenberg 1954b, Hofer 1970). Recently also Austin and Austin (1980) stated that 'ecological optima may be very sensitive to the particular species composition'.

Any assessment is made the more difficult by the fact that many species have several or even many physiological races with different patterns of behaviour in competition. The Sweet Vernal Grass is a good example of a species showing such genetically divergent races. Snaydon (1970) carried out investigations into *Anthoxanthum odoratum* on some 40–60-year-old plots of the famous long-term manurial trial at Rothamstead. Specimens taken from a population which had been growing on limed plots (pH 7) grew best on calcareous soils whereas those taken from the plots at pH 4 grew better on acid soils, and it made no difference whether these plants were grown vegetatively or from seed (further examples of this are given

Tab. 115. The influence of acidity and fertilising on the growth of Moorgrasses.

	Molinia	*caerulea*		*arundinacea*		
pH-value (measured in KCl)	3.7	5.3	7.0	3.7	5.3	7.0
Average plant yield (g dry matter)						
unmanured	0.68	**0.77**	0.69	**4.52**	3.34	2.38
with a complete manure[a]	**4.14**	3.44	2.04	5.50	**5.94**	2.07
Increase in yield as a result of manuring (%)	609	447	296	121	178	− 87

[a] Each experimental (Mitscherlich) container received 0.6 g N, 0.6 g K_2O and 0.8 g P_2O_5 as well as a heavy dressing of $CaCO_3$
Source: After Grabherr (1942).

in section D IV 1 b). The present day state of knowledge concerning the importance of competition supports Besson (1972, page 371) who states, after a piece of comprehensive research in the frequency of mowing with *Dactylis glomerata*, *Anthriscus sylvestris*, *Ranunculus acris*, *Taraxacum officinalis* and *Trifolium pratense* that 'the ways in which species perform in a mixed culture cannot be predicted from their behaviour in pure cultures'.

This is also applicable when one considers allelopathic influences which are often thought to be of great significance in the competition between plant species (see e.g. Whittaker and Feeny 1971 and Grodzinsky 1973). Although many species of grass and clover produce and set free phytotoxic substances under laboratory conditions, according to Pork (1975) these are not effective in meadow soils in humid climates because they are soon broken down or modified by microorganisms. **Light, water, nutrients and other primary habitat factors, together with the competition for them, are the main conditions which determine the species composition** of a plant community. Allelopathy plays a secondary role at most. These findings obtained from the study of meadow communities may also apply to the majority of the other phanerogam communities in Central Europe.

(b) On the ecology of Purple Moorgrass meadows

The False idea that the habitats of the *Molinion* meadows are characterised by a very varying water supply, while the Cabbage Thistle meadows and other fertilised moist-soil meadows have a more even water economy stubbornly persists. This assumption cannot be justified on two counts. On the one hand many of the soils occupied by *Calthion* communities have in fact a variable water regime (see e.g. fig. 470). The water content and the groundwater level can both vary as much as those in soils of the Purple Moorgrass meadows. On the other hand, as Eskuche (1962) has already shown, there are *Molinion* communities on soils the moisture of which varies surprisingly little under Central European conditions, (see fig. 466). Of course a continually waterlogged soil restricts the development of *Molinia* and other members of the moist-soil meadows and even that of swamp plants such as *Eriophorum vaginatum* (Gore and Urquahrt 1966). The previous findings have been confirmed by Klötzli (1969b) who studied the groundwater conditions in the litter and hay meadows of north-eastern Switzerland over a wide area and supported by many measurements.

In southern Germany, because of different management on adjacent meadow areas *Molinion* and *Calthion* communities can be studied side by side in an arbitrary pattern. So the deciding cause of their differences must be looked for **in the way they have been manured and cut** and not in their water regime. Remnants of the old methods of cutting litter for cow-sheds from the meadows can still be found. The most likely sites are on marl or clay slopes or around some of the lakes north of the Alps where there is the possibility of flooding, i.e. in places where it does not

Fig. 466. Not every *Molinietum* has a fluctuating water supply. After Eskuche (1962), modified.

The soil moisture (in % by volume) and the water table fluctuated surprisingly little during the years 1956–58 in the soil of the Soft Rush-Purple Moorgrass meadow (*Junco-Molinietum*, *Carex nigra* variant) in the Erft lowland (see fig. 460).

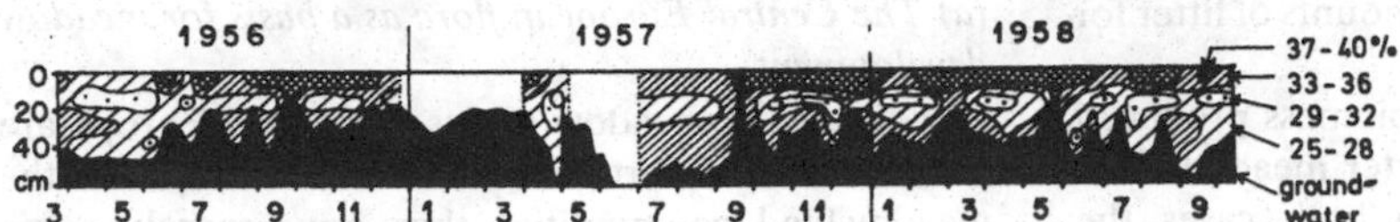

pay to practise an intensive form of husbandry with applications of manure. These sites do in fact have a very variable water regime, i.e. in spring they are exceptionally wet but in the late summer they are often exhausted to the point of permanent wilting. From such special cases however no conclusions can be drawn about the behaviour of all the *Molinion* meadows.

Other ecological misconceptions which are widespread in the literature have already been shown by Grabherr (1942b) to be wrong. Using pot cultures with different pH values he found that the typical species of the Purple Moorgrass aggregate (*Molinia caerulea*, see tab. 115) is quite indifferent to the soil reaction and that it is not divided up into races with different pH optima. The same is true of the stronger, taller species *M. arundinacea* (= *litoralis*) which has its natural centre of distribution in somewhat warmer climate and on marl with a high pH, but is also found on soils poorer in lime.

Although *Molinia* only dominates in those meadows where no manure has been applied for years nevertheless both species in pot culture react positively to a full application of nitrogen, phosphorus and potassium (tab. 115). In a meadow however, where other grasses get the same amount of manure, because they start growth in the spring much earlier than the late, slow-growing *Molinia* they can easily suppress it. This process is often speeded up because the manured meadows are usually cut more than once and relatively early so that they do not become strawy, which seals the fate of the Purple Moorgrass and all other true litter meadow plants.

As León (1968) has shown from numerous different subcommunities of *Molinietum* in Switzerland the soils on which they grow are extremely poor in nitrogen. Neither nitrate nor ammonium is produced in them in any great quantity apparently because the dead leaves and root remains are themselves poor in nitrogenous compounds (see tab. 116). In this respect they resemble the *Calluna* heaths in north-western Germany (see section D II 3), from which nutrients (especially N and P) in the form of litter have also been regularly removed over the past centuries. A third very informative parallel can be drawn with those tropical and subtropical savannahs which have been formed and maintained by annual burning (De Rham 1970). The fire gets rid of the dead plants above the ground surface, leaving behind only small amounts of litter for mineralisation.

In spite of this the amount of biomass produced above ground every year in both litter meadows and savannah is surprisingly large. In both cases the constantly high productivity is made possible by the nutrients which are taken down to the subterranean parts for storage before the leaves and stems turn yellow. These nutrient reserves can be remobilised for the new growth the following year. In the case of *Molinia* Chwastek (1963) has demonstrated such economy concerning nitrogen as well as phosphorus and other elements. The nutrients are stored mainly in the compressed internodes at the bottom of the long joint-free haulm of *Molinia* (Ellenberg unpublished). This 'internal' circulation enable the Moorgrass as well as the savannah grasses to produce a large amount of carbohydrate using the same capital resources of nutrients over and over again.

Such **internal nutrient economy** represents an extreme case of what is the normal behaviour of every perennial vascular plant. A certain amount of storage takes place in all of them before the leaves and other parts die off. But in the leaf litter of a broadleaved woodland for example there is still quite a large amount of protein remaining and through the ammonification of this an 'external' nitrogen cycle can operate. This is characteristic for most of the woodlands and for many other land ecosystems, (see section B II 4 d). According to De Rham every small group of trees in the savannah starts this cycle operating again and initiates a return of nitrogen to the soil by litter decomposition. The same sort of thing takes place if a Purple Moorgrass meadow becomes neglected so that dead material collects on the ground as a resource available to soil animals, fungi and bacteria. According to León there is a considerable amount of nitrogen mineralisation in the topsoil after a former Purple Moorgrass meadow has gone out of use. Along with the restoration of the external nitrogen cycle there is a change in the balance of the species. The Moorgrass itself disappears increasingly quickly while Meadowsweet (*Filipendula ulmaria)* and other large perennials or Lesser Pond-sedge (*Carex acutiformis*) and other grasslike plants get the upper hand. So without any further action on man's part the Purple Moorgrass meadow turns into a community requiring a higher level of nutrients. It often resembles the Meadowsweet river bank association which we have already discussed in section 5c.

7 The origin of grassland plants and of meadow communities

(a) The Central European flora as a basis for meadow development

Almost all the meadow plants in Central Europe are native and take part in the natural landscape. In the meadowland communities they have merely come

Tab. 116. The amount of mineral nitrogen available in litter meadows in northern Switzerland and in some West African savannahs.

Plant communities and soils	Net mineralisation ($kgN \cdot ha^{-1} \cdot y^{-1}$)	Nitrification level
1. Moorgrass litter meadows and related communities		
a Acid-soil Moorgrass meadows		
Rush-Moorgrass meadows (*Junco-Molinieteum*), relatively dry	0	
the same on fairly damp soil	1	IV
the same on wet soil	8	IV
b Calcareous Moorgrass meadows		
species-rich Moorgrass meadows (*Molinietum*), relatively dry	0	
the same on fairly damp soil	2–14[a]	IV
the same on wet soil	5–40[a]	IV
Bogrush-Moorgrass meadow (*Molinietum schoenetosum*) very wet	0	
2. Tropical savannah on the Ivory Coast (for comparison)		
a Savannah tall grasses, often burnt		
Brachiara brachylopha ass., *Loudetia*-subass., treeless	0–2	IV
the same with occasional trees	0–5	IV
b Groups of trees on the savannah	30[b]	IV

Note: [a] The higher values refer to stands which have not been mown for some time.
[b] The nitrogen cycle is restored by the trees with the decomposition of the leaf litter.

Source: From data by León (1968) and De Rahm (1970) in Ellenberg (1977); see tabs. 99 and 19.
The method is also reliable for wet soils (Gerlach 1978)

Fig. 467. The Cabbage Thistle (*Cirsium oleraceum*) in a seminatural woodland habitat and in a managed meadow.

a. In the Alder-Ash wood (*Pruno-Fraxinetum*) of the Olsberg forest near Basel; *Impatiens noli-tangere* in the foreground.

b. In a Cocksfoot-Cabbage Thistle meadow (*Polygono-Cirsietum, Dactylis glomerata* subass.) in the Wäggi valley. The white umbels are *Heracleum sphondylium*; the flowering spike (in front left) is *Polygonum bistorta*. The Cabbage Thistle here is less hygromorphic than in the semishade of the woodland.

a

b

together in new combinations under human influence, and through this they have achieved prominence in the Central European countryside.

This is particularly true for the members of the *Molinietum* and other **litter meadows**. The dominating grass species in most of the Purple Moorgrass meadows, *Molinia caerulea*, grows in open woodland which is on the wet side, but where the moisture content can fluctuate considerably, and also on the hummocks of some raised bogs. G. Weise (1960a and b) has shown experimentally that the type of *Molinia* growing in Purple Moorgrass meadows on acid soils (*Junco-Molinietum*) in the neighbourhood of Dresden is essentially similar to that which appears there in the moist-soil Birch-Oak wood. Plants from both communities showed the same sort of variability in their development, in the amount of green matter they produced, in morphological features such as the form of the inflorescence, in their reaction to additional nitrogen and in the way productivity falls with increasing altitude, as recorded by measurements of gas exchange. The meadow Purple Moorgrass differs slightly physiologically from the woodland Purple Moorgrass in that it develops more quickly and its net photosynthetic production is somewhat greater on acid moorland soil whereas the gas exchange of the woodland form is greater on calcareous gravel. The differences are in any case so slight that they could well have been produced by selection from the same population.

Like *Molinia* other species found in the Purple Moorgrass meadows have originated in natural woodland, e.g. *Serratula tinctoria* and *Galium boreale* which are common in the mixed Cinquefoil-Oak wood of eastern Central Europe (see section B III 3).

The dampness indicators which are widespread in the manured meadows are found growing under natural conditions, some in flood-plain woods and some in other woodland communities of damp or wet habitats, e.g. *Cirsium oleraceum* (see figs. 467a and b), *Angelica sylvestris*, *Lysimachia vulgaris*, *Crepis paludosa*, *Filipendula ulmaria* and *Deschampsia cespitosa*. Several species are also found in reedbeds and sedge fens, especially *Phalaris arundinacea*, *Carex acutiformis* and *Caltha palustris*.

Members of the False Oatgrass meadows are also to be found in river valleys. These include Cow Parsley, Hogweed, Bedstraw and Cocksfoot. However the rest of the species characteristic of the *Arrhenatherion* have hardly any suitable natural habitats in Central Europe. *Arrhenatherum* itself is at home on rocky screes rich in mosses and with a good nitrogen supply e.g. in South Germany (Sebald 1980) and on similar thinly occupied slopes in western and southern parts of West Europe. It was first cultivated in France and was relatively late in being sown in Central European meadows as 'French Ryegrass'. Once here though it spread rapidly and now occupies all suitable habitats. *Tragopogon*, *Crepis biennis* and *Campanula patula* may have come to us from the steppeland meadows of the southern part of Eastern Europe. The different origins for the species are fresh evidence that the Oatgrass meadows are dependent on man to a greater extent than are the Moorgrass communities and other litter meadows which are cut less often and not manured.

The character species of the class *Molinio-Arrhenatheretea* widely distributed in Central Europe are known also to thrive individually in different natural formations, e.g. *Poa trivialis* grows in swamp and flood-plain woods, *Poa pratensis* on rock heaths, *Anthoxanthum odoratum* in certain Birch-Oak woods, and heaths, *Lathyrus pratensis*, *Viccia cracca* and *Trifolium pratense* in skirt communities along the edges of woods, etc.

It is quite likely that Beaver, Red and Roe Deer and other large herbivores created clearings in the natural woodland or kept them open so that grassland plants would find conditions much more favourable there than in a dense wood. Beaver meadows like those which can still be studied in North America are however always wet and frequently flooded. Natural clearings in drier woodland would have been rather small, and sufficiently sunny for grassland plants for only a few years,. More permanent open places within the natural woodland cover could have existed at most around the edge of sunny rock outcrops. Large openings caused by wind damage may also be considered as possible habitats, but it is unlikely that communities resembling meadows would have grown up here because it takes only a few years for trees and shrubs to overshadow the ground again. Much more likely is that small and large woodland clearings would first have been settled by nitrophilous herb communities made up of true Central European floral elements which are only exceptionally found in meadowland (see sections A I 2 and D III 2).

Before man began to make effective impacts on the landscape the only areas which had been naturally free from woodland for a long time were raised bogs and many areas of fenland or intermediate mires, the alpine and higher belts above the tree line and those parts affected by the salt-laden North Sea. In all these habitats however the living conditions are too extreme for meadow plants of the *Arrhenatheretalia* to survive.

Pollen analysis of peat accumulations shows that

in Post-glacial times not only grain crops but also wild grasses became more plentiful after man had opened up larger clearings in the woodland and maintained them for at least some decades. *Plantago* and *Rumex* pollen in particular can be looked upon as indicators of human settlements. However macroscopic remains of fodder grasses were apparently rare right up the the Bronze Age, showing that the making of meadow hay for winter foddering of housed animals played no great part in the farming economy in those days (see the comments on the use of leafy shoots from woody plants as winter fodder in section A II 2).

According to the summary compiled by Lüdi (1955) and others the macroscopic remains of more than 50 species of meadow plants have been identified from Bronze Age remains in the Swiss midlands, i.e. from the colline to submontane belts in south-western Central Europe. These lists show that not only the grasses but all the character species of the False Oatgrass meadows were still absent more than 2000 years ago whereas those of the damp and dry meadows were well represented. If what we now know as meadow communities existed at that time they must have corresponded to the ones shown in fig.438 as 'unmanured'. It is probable though that the open grassland was maintained more by grazing than mowing, and rather than occupying large areas it was confined to a parklike countryside resulting from the extensive grazing of woodlands (see section A II 2). According to Willerding (1977, 1979) there are also no remains of typical *Arrhenatherion* meadow plants from the Iron Age and early Middle Ages. Even in the late Middle Ages only a small number of records have been verified.

The slightly arid grassland too was used only as pasture and not mown with the scythe or sickle. Meadows dominated by *Bromus erectus* then may be amongst the latest-formed communities in Central Europe although the species of which they are made up have come from natural rock heaths and from the pastureland communities of the order *Brometalia* which, it must be conceded, are much older though also anthropogenous. Some plants of the dry grassland probably managed to save themselves during the late Ice Age and the period immediately afterwards by moving from the steppeland into the regularly used grassland.

In contrast to the weeds of arable land (section D IX1 a) or waste places (section D VIII 1 a) not one plant species brought in by shipping from overseas has been able to establish itself either in the meadows or in the woods of Central Europe. The communities in these areas appear to be so permanent and stable that newcomers cannot find a place in them.

On the other hand it must be said that there are almost no plants in the temperate zone of the southern hemisphere which could go to make up a meadow or a pasture rich in grasses. As Klapp (1965) pointed out European and North American plants have had to be introduced into countries like Chile in order to produce meadows. In some cases such plants have rapidly become naturalised (Oberdorfer 1960). Ellenberg (unpublished) has studied many Oatgrass meadows which have arisen subspontaneously in the neighbourhood of San Carlos de Bariloche (in the Argentinian National Park Nahuel Huapí) especially at the *Nothofagus dombeyi* level. These already contain almost all the species characteristic of Central European Oatgrass meadows but, apart from the rosaceous genus *Achaena* there are practically no representatives from the native flora. The European species could only have been introduced during the past few decades, probably along with the seeds of *Arrhenatherum*, *Dactylis* and *Poa pratensis* and other grasses which were widely sown. They thrive very well in their new environment the climate of which is rather like that of the region betweeen south-western Central Europe and the submediterranean zone.

Compared with most other parts of the Earth then the **flora of Central Europe appears to have been predestined for the formation of meadow communities**. This is even more surprising because, as already pointed out, meadow-like communities in nature must have been extremely rare. The same is true for North America and especially for the temperate parts of Asia where 'real' meadow plants are also native. What is it that distinguishes those regions having their own meadow flora from those where this has not been able to develop? To answer this question we must solve the problem of how soils, in a sufficiently warm climate, which are not too dry and have a good nutrient status could be kept treeless under natural conditions, that means without the help of man.

In the author's opinion the only places which offer habitats of this kind are the avalanche tracks in the high mountains which cut through the wooded slopes in the montane and submontane belts and are swept free of trees by the snow masses crashing down each year (fig. 316 and 317). The avalanches originate in an extensive area above the timber line where a lot of snow can accumulate. If the topography allows this area to warm up quickly in the spring the snow becomes unstable and suddenly rushes down to the valley following the same course every time. These conditions exist in the higher parts of the Alps and in similar sites in North America, but above all in the high mountain complexes in Asia which were in existence long before the formation of similarly high mountains in Europe during the later Tertiary Period. In the mountains of the tropics and subtropics, as well as those of the extremely oceanic southern Andes there are no marked seasonal changes which could favour the formation of avalanches. The boreal and polar mountains have such a short summer that meadow plants are not able to grow in the avalanche tracks. The low mountains in the temperate zone do not have avalanche tracks, or only narrow ones, as Jeník (1958) has described from the Karknosze. These favour meadow plants just as little. Thus the **high mountains** of the temperate zone, principally those in Asia and Europe but to a lesser extent also in North America, must be considered as the probable centres of evolution of the mesophilous meadow flora.

(b) The time required for a new meadow community to become established

Grassland communities can be formed very quickly on new land or where the living conditions are changed, provided the members are present in the neighbourhood or some of them are sown there. The previous sections have already given us several examples of such communities. Going on from these we shall now give further consideration to the question of how many years are required before a particular meadow community achieves its normal characteristic species composition.

The observations of numerous authors indicate that where arable land in south-west Germany or adjacent parts of Switzerland have been sown with *Dactylis glomerata*, Lucerne, Red Clover or a grassland and Clover mixture, it has often not taken more than 3 to 4 years for it to become enriched with species to the extent that it resembled a Salvia-Oatgrass meadow or a typical Oatgrass meadow. However a necessary proviso for this is that the grass shall be cut two or three times each year and that it shall receive a dressing of manure from time to time. The seeds of many species are brought in by the wind or by animals from neighbouring older meadows or from the sides of footpaths and roadways if these are cut regularly and so have a similar flora. Or they could be brought into the field directly with the farmyard manure. It takes a lot of effort to keep a plot of a particular grass entirely clean and pure because not only do garden and farm weeds soon appear in the seedbed but many different grassland plants also. If these are allowed to develop unhindered, whether they have come from outside or from seeds already present, then, according to the observations of Lieth and Ellenberg (1958) they enter into competition with each other and with the sown grass species so that in just one or two years they have formed a combination suited to the particular habitat. This selection is speeded up even further by frequent cutting.

According to relevés made in the meadow communities by Ellenberg (1952b) and taken before and after a large-scale lowering of the groundwater table in a peaty valley near Braunschweig, seven years were sufficient for Sedge fen to become Cabbage Thistle meadow or for the latter to turn into Foxtail-Oatgrass meadow. Where the water level had been lowered by 60 cm on average in a Narrow-leaved Sedge meadow in 1939, by 1946 there were only diminishing relicts of the swamp plants even though during the intervening years it had only been subject to the normal management, i.e. without any soil cultivation or reseeding (see tab. 117). Similar results have been obtained from many other grassland communities (Walther 1977, Ellenberg 1952b, Meisel 1960, and others). The sandy-peaty valley of the Ems for example has changed from a moist or wet meadow countryside, dominated by acid-soil *Calthion* meadows, into one of arable fields and grazing cattle (fig. 468). It is not necessary to sow new seed in order to bring about even this degree of change in the species composition of grassland communities. The lowering of the groundwater level together with the grazing alter the balance of competition to such a degree that practically all the species originally present give way to new ones some of which grow from seed brought in by the grazing animals.

As a rule the more fertile the soil the quicker the species combination changes. Good dressings of farmyard or artificial manure so favour the tall-growing, strongly competitive species that they are able to suppress all the others unless the latter too are adapted to the particular habitat, i.e. the group factors enable them to live alongside or beneath the dominant species. Immediate changes like this are possible only if the new species coming in are already present in the vicinity as seed or as stunted individuals at the time the conditions are altered. The formation of new meadows happening so quickly on land which has been arable for a time (as was mentioned in the introduction) is also partly to be explained by seeds of the meadow plants reaching the neighbouring arable land each season and thus being already present when the farmer decides to use the field for grassland. Where there is no meadow with the corresponding species combination in the area then it takes longer for a complete meadow community to build up on former arable land. Nevertheless, even in these cases many plant species of farmed meadows arrive very soon as they are surprisingly capable of dispersing themselves.

The ability to produce large quantities of seed together with the relatively short life span of even the most 'permanent' of the meadow plants favours both

Tab. 117. The drainage of a peat fen in the river valley to the west of Brunswick has brought about the change from a Sharp Sedge bed to a cabbage Thistle meadow during the years 1939 to 1946 (compared to section E III 1, no. 1.413 and 5.415).

Year 19	39	46	F	N
Eliminated species:				
C *Glyceria fluitans*	2		9	5
C *Equisetum fluviatile*	1		10	6
C *Ranunculus flammula*	1		9	2
Juncus articulatus	+		8	2
Epilobium palustre	+		9	3
K *Prunella vulgaris*	1		×	×
K *Ranunculus acris*	+		×	×
Reduced species:				
C *Carex gracilis*	3	1	9	4
C *Calliergon cuspidatum* (Moss)	3	+	9	4
Ranunculus repens	3	2	7	×
C *Galium palustre*	2	+	9	4
C *Glyceria maxima*	1	+	10	7
K *Anthoxanthum odoratum*	1	+	×	×
Mentha aquatica	1	+	9	4
Unchanged species:				
K *Holcus lanatus*	2	2	6	4
K *Festuca rubra* ssp. *rubra*	1	1	×	×
Festuca arundinacea	1	1	7	4
C *Phalaris arundinacea*	+	+	8	7
K *Bellis perennis*	+	+	×	5
M *Juncus effusus*	+	+	7	3

Year 19	39	46	F	N
Favoured species:				
K *Alopecurus pratensis*	1	4	6	7
Lysimachia nummularia	+	3	6	×
K *Rumex acetosa*	+	2	×	5
M *Angelica sylvestris*	+	1	8	×
Caltha palustris	+	1	8	×
M *Lychnis flow-cuculi*	+	1	6	×
K *Cardamine pratensis*	+	1	7	×
M *Galium uliginosum*	+	1	8	×
Additional species:				
Agrostis stolonifera		3	6	5
M *Cirsium oleraceum*		2	7	5
Glechoma hederacea		2	6	7
K *Poa pratensis*		2	5	×
M *Filipendula ulmaria*		1	8	4
Deschampsia cespitosa		1	7	3
M *Festuca pratensis*		1	6	6
K *Cerastium fontanum holost.*		1	5	5
M *Selinum carvifolia*		1	7	2
Stellaria palustris		1	8	2
K *Bromus hordeaceus*		+	×	×
Cirsium vulgare		+	5	8
Total numbers	28	33		

Note: [a]C = local character and differential species of the Narrow-leaved Sedge bed (*Caricetum gracilis*);
M = character species of the order of moist-soil meadows (*Molinietalia*) and local character species of the Cabbage Thistle meadow (*Cirsium oleraceum - Angelica sylvestris* – ass.);
K = character species of the class of cultivated meadows (*Molinio-Arrhenatheretea*).

[b] i.e. species with abundance 4 are included four times in the calculation of the average values, those with abundance 3 three times and so on; 1 and + both once.

Source: After Ellenberg (1952 and 1974), modified.

Changes in the average indicator values (see section B I 4) year 19	mF 39	mF 46	mN 39	mN 46
— calculated according to presence:	8.0	7.1	4.5	4.8
— calc. acc. to abundance of species[b]:	8.0	6.8	4.4	5.1

The average moisture value has decreased because of the disappearance of some moisture indicators. The average nitrogen value has increased because some more-demanding species have come in. Both changes are much more apparent when the abundance of the species is taken into account and not just their presence.

the formation of new communities and the changes from one type to another should the habitat conditions alter. Long-lived grassland plants which reproduce mainly vegetatively are also able to react quickly to changing conditions. This is apparent especially in pastures and on heavily trodden swards, and we would now like to turn our attention to these.

VI Manured pastures, trodden swards and those subject to flooding

1 *Intensively managed pastures*

(a) *Ryegrass-White Clover pastures and grassland alternately grazed and cut*

Whereas the manured grassland in the planar to

submontane belts of southern Central Europe had almost exclusively been used for mowing up to about 1960, in the northern and especially in the north-western parts it has been the custom for many centuries to graze this type of grassland constantly. The monotonous green of the intensively grazed pastures is only relieved by the swamp meadows and by some old river dykes whose steep banks with their Oatgrass meadows are like a greeting from the south.

The main reason for this difference in the grassland management from south to north (figs. 437, 469 and 472) lies in the increasingly oceanic climate which tempers the summer drought and makes it possible to graze the fields intensively throughout the relatively long vegetation period. In western Central Europe where there is more cloud grazing is normal in the lowlands but is also carried up into the hills and mountains. In the east and south on the other hand grazing is confined to the higher parts of the mountains and to the outer Alps with their higher rainfall.

Fig. 469. A permanent pasture of *Lolio-Cynosuretum typicum* near Osterholz-Scharmbeck. In the foreground can be seen a number of dung patches, in the centre Hawthorn hedges. This kind of picture was and still is rarely seen in the southern parts of Central Europe.

Grazing is also favoured near the coast because the winters here are milder and shorter, thus extending the grazing season. According to Klapp (1965) about 200–220 days per year are available for grazing in the

Fig. 468. The effect of lowering the water table on the grassland communities of the Ems flood plain south of Haren (the position in 1956–57 and in 1974). After Meisel and Hübschmann (1975), modified.

1. Open standing water and waterplant communities. 2. Vegetation bordering the water (*Glycerietum maximae, Caricetum gracilis, Pedticulari-Juncetum filiformis*). 3. Wet meadows and pastures (*Rumici-Alopecuretum, Senecioni-Brometum*, subassociation of *Carex nigra* and *Potentilla palustris; Lolio-Cynosuretum lotetosum; Junco-Molinietum*, various subassociations). 4. Moist-soil meadows and pastures (*Senecioni-Brometum*) remaining subassociations). 5. Slightly-moist and typical Ryegrass meadows (*Lolio-Cynosuretum lotetosum*), typ. variant; *Luzulo-Cynosum typicum*. var. with *Cardamine pratensis*). 6. Unfertile pastures and dry sandy swards (*Luz. Cyn. typ.* remaining variants; *Diantho-Amerietum* in various forms; *Cauno Genistetum*. 7. Arable land with various weed communities 8. Scrub (*Salicetum*) and plantations. 9. Houses, farms etc. 10. Areas disturbed by embankments 11. Course of the river and unmapped areas. 12. Position of the profile shown in section above. The previous flood plain is now for the most part cultivated and the remaining grassland is occupied by communities almost all of which grow in relatively dry habitats.

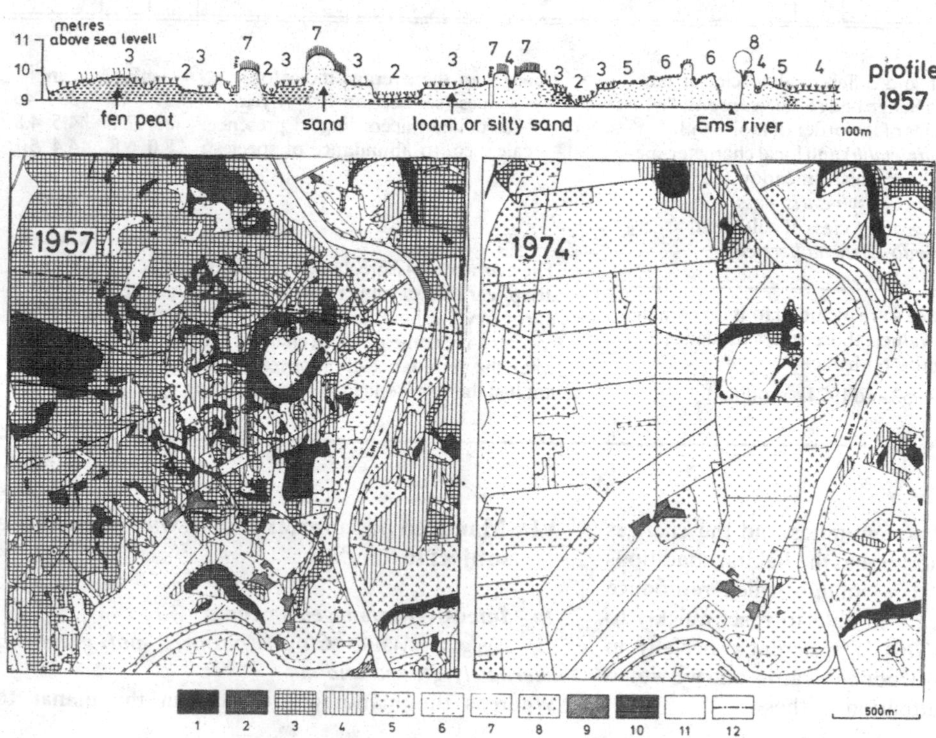

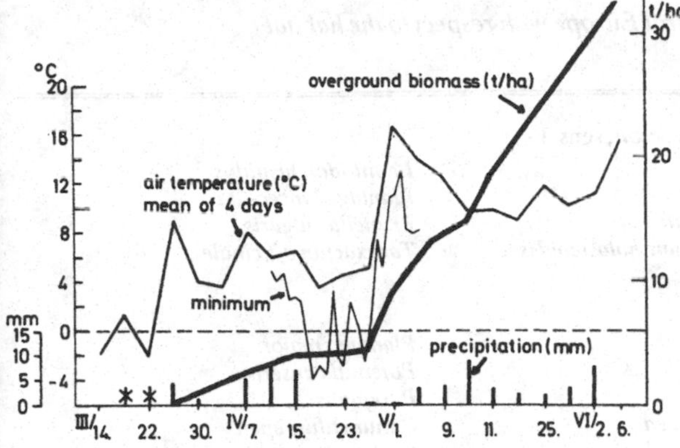

Fig. 470. The 'yield progress' in an intensively managed Ryegrass-White Clover pasture in relation to the temperature and rainfall (columns below). After Zillmann and Kreil (1957), in Ellenberg (1963).

The 'yield progress' was determined by weighing the amount of greenstuff cut from separate plots on the dates shown. The plots were not grazed during the trial year; 1955 in Jühnsdorf (Zossen district).

warmest parts of western Germany. A drop in temperature such as is frequently experienced in the more strongly continental climate considerably delays the time of putting the cattle out to grass (fig. 470). Only a modern system of rotational grazing and moving with longer resting periods, coupled with artificial irrigation makes possible a high yield from pasture in the low rainfall areas and eliminates the consequences of a drought.. These management changes are progressively reducting the differences in the character of the landscape between the north-west and the southern and eastern parts of Central Europe which were so obvious at one time.

The generally decreasing profitability of cattle rearing has brought with it further rapid changes during the past few years. Because of this previously intensively grazed pastures have fallen to disuse, especially on the marshes and mires where the maintainance of the drainage system has proved to be too costly. Instead management of the remaining grassland has been intensified, and it is so heavily manured and stocked that these two factors have become the dominating influence on the species composition. Irrespective of whether the soil is sandy, loamy or peaty the **Perennial Ryegrass-White Clover pasture** (*Lolio-Cynosuretum*, tab. 118) has become dominant. Klapp (1950) especially has described many thoroughly investigated examples of this independence regarding soil type in the planar and submontane belts. With increasing altitude the soil and climatic factors show a much clearer influence on the species composition even of the intensively used pastures (see tab. 119).

Only plants which are able to withstand trampling and to regenerate strongly can survive in such swards. According to Lieth (1954, see fig.471) the soil in pastures is not as porous as that in meadows because of the continual treading. In this respect they are only exceeded by the soils along footpaths or in sportsgrounds. In gateways, round drinking troughs and milking places too the pasture swards are replaced by communities of trodden plants (see fig.472 and tab. 120). Moreover many of the differential species named in tab.118 indicate that the soil of Ryegrass-White Clover pastures is rich in nitrogen, e.g. *Agropyron repens* and *Poa annua*. They originate from natural or seminatural flooded swards which we shall soon be discussing.

Although the farming methods have a levelling out effect on the species composition this does express many of the edaphic and microclimatic peculiarities of the habitat. In Holland where the Perennial Ryegrass-White Clover pasture reaches its greatest expression, as well as in north-western Germany, a number of corresponding subassociations, variants and other units can be distinguished (tab. 118, fig.473). The most plentiful are the typical (a) and the damp (e) subcommunities since the habitats of the drier ones (b and c) are largely arable land today. The Saltmarsh Rush-White Clover pasture (g) is confined to a narrow transition strip leading to the salt marsh communities of the North Sea and Baltic coasts (see sections C IV 1 e and 2 a).

If manuring is discontinued the turf of the cattle pastures becomes broken up quite quickly and poverty indicators such as the Mouse-ear Hawkweed (*Hieracium pilosella*) and the Field Woodrush (*Luzula campestris*) come in, whether the soil is relatively wet or dry. Most of the differential species named in tab. 118 for the montane pastures (c) are also such indicators of deficient nutrition. At one time these must have been distributed throughout all the lowland pastures, but for the most part they have now disappeared from these following the increased use of artificial manure.

Dörrie (1958) made repeated relevés on long-term test areas showing how quickly the heavier dressings of manure, together with an increased stocking rate, affected a typical Ryegrass-White Clover pasture. After only two years several species had disappeared, among them even such obstinate weeds as *Cirsium arvense*, which had generally served as a differential species for the intensive pastures. On the other hand there were only a few new arrivals, e.g. *Stellaria media* indicating high phosphate and nitrogen, and *Polygonum aviculare*, a plant which can

Tab. 118. A classification of White Clover pastures in Central Europe with respect to the habitat (see section E III 1, no. 5.423).

a Species present in practically **all** White Clover pastures (*Cynosurion*, sens. lat.)		
G: *Cynosurus cristatus*	L: *Trifolium repens*	*Leontodon hispidus*
Festuca rubra	*T. pratense*	*Plantago lanceolata*
Phleum pratense	H: *Achillea millefolium*	*Prunella vulgaris*
Poa pratensis	*Cerastium fontanum holosteoides*	*Taraxacum officinale*
b *Present in Ryegrass-White Clover pastures* (*Lolio-Cynosuretum*), **well manured** and mainly in the lowlands:		
G: *Lolium perenne*	H: *Bellis perennis*	*Plantago major*
Poa annua	*Cirsium arvense*	*Potentilla anserina*
P. trivialis	*C. vulgare*	*P. reptans*
Agropyron repens	*Glechoma hederacea*	*Ranunculus repens*
Agrostis stolonifera	*Lysimachia nummularia*	*Sagina procumbens*
Hordeum secalinum (coastal)	*Odontites rubra*	*Veronica serpyllifolia*
c In Fescue-White Clover pastures (*Festuco-Cynosuretum*), **badly manured** and mainly in the **uplands**:		
G: *Agrostis tenuis*	H: *Alchemilla vulgaris*	*Plantago media*
Festuca nigrescens	*Campanula rotundifolia*	*Polygala vulgaris*
Anthoxanthum odoratum	*Euphrasia rostkowiana*	*Potentilla erecta*
Briza media	*Hieracium pilosella*	*Ranunculus bulbosus*
Danthonia decumbens	*Hypochoeris radicata*	*Stellaria graminea*
Luzula campestris	*Leucanthemum vulgare*	*Thymus serpyllum*
L: *Lotus coriculatus*	*Pimpinella saxifraga*	etc.
d In pastures rich in Matgrass (*F. -C. nardetosum*) neglected and very **acid**; mostly in upland sites:		
G: *Nardus stricta*	H: *Antennaria dioica*	*Calluna vulgaris*
Avenella flexuosa	*Hypericum maculatum*	*Vaccinium myrtillus*
Carex pilulifera	*Veronica officinalis*	etc.
e Moisture indicators:	**f Sand** indicators (coastal):	**g Salt** indicators (coastal):
(.. *lotetosum uliginosi*)	(.. *armerietosum*)	(.. *juncetosum gerardii*)
G: *Carex leporina*	G: *Festuca rubra arenaria*	G: *Juncus gerardii*
L: *Lotus uliginosus*	K: *Armeria maritima*	*Festuca rubra litoralis*
Juncus effusus	*Galium verum*	L: *Trifolium fragiferum*
H: *Cirsium palustre*	*Viola canina*	H: *Glaux maritima*
Lynchis flos-cuculi	etc.	etc.

Note: G = grasses and grass-like plants, L = leguminous plants, H = herbaceous plants and dwarf shrubs.
Source: Passarge (1969b), Jurko (1974) and others.

Fig. 471. The relationship between soil porosity and the management of permanent grassland, (in brackets: number of measurements taken in the top soil to a depth of 10 cm). After Lieth (1954) modified.

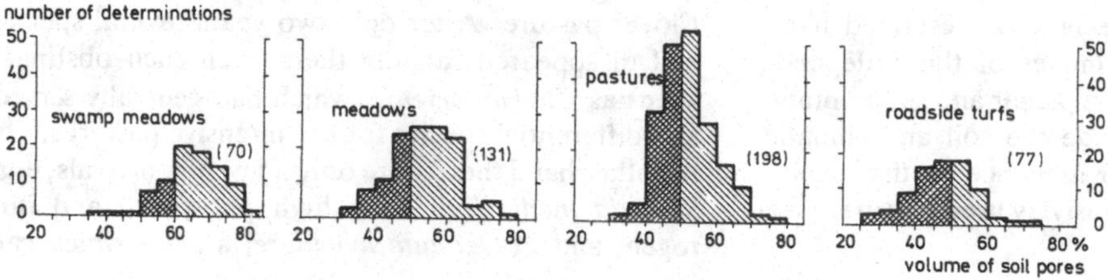

Fig. 472. An intensively used and heavily stocked Ryegrass-White Clover pasture with *Lolium perenne*, *Poa annua* and *Plantago major*. About one sixth natural size. Photo. Iltis-Olbert.

withstand treading. From the estimated biomass proportions it could be seen that *Lolium perenne*, *Phleum pratense* and other grasses had spread at the expense of the leguminous plants to such an extent that after some years they were almost completely dominant. A continuation of the heavier manuring and carefully timed rotational grazing and mowing would probably result in nearly all other species being eliminated eventually.

The more intensive the management the poorer in species does the pasture become. Ryegrass, Timothy and a few other grasses, e.g. *Poa trivialis* and *Dactylis glomerata*, are able to form dense leafy growth during the resting periods, as Ruthsatz (1970) showed near Göttingen. While low-growing pasture plants are disappearing the sward is invaded by tall meadow plants such as *Anthriscus sylvestris*, *Heracleum sphondylium*, *Crepis biennis* and *Arrhenatherum*. From the viewpoint of the plant sociologist therefore rotational grazing and mowing lead to the formation of a community which in any case is poor in species and cannot be included in either the *Lolio-Cynosuretum* or the *Arrhenatheretum* without stretching a point. Vollrath (1970) came to a similar conclusion after analysing the vegetation of paddocks where rotational grazing was carried out in Bavaria (see tab. 120). Intensively managed grassland on raised bogs also causes difficulties in classification, according to Schwaar (1973). Looked at ecologically however the species composition reacts in each case to every nuance of the environmental conditions since the high level of nutrition speeds up the changes in relative competitiveness.

(b) Experimental investigations into the Ryegrass-White Clover pastures

Even such floristically 'boring' communities as intensively farmed pastures can serve as attractive objects for experimental research into their phytosociology and their ecosystems. The Ryegrass-White Clover pasture with its single layer immediately presents itself

Tab. 119. Yields and stocking rates of pasture communities at various altitudes in the Bavarian outer Alps.

Altitude and plant communites	Yield (t/ha)	Stocking capacity (%)	Species number A[a]	Species number P[b]
1400–1700 m				
Poor calcareous sward (*Carlino-Caricetum sempervirentis*)	0.9	5	**77**	**30**
Grazed poor calcareous sward (*Crepido-Caricetum*, *Thymus*-form)	1.3	16	56	13
Subalpine Matgrass sward (*Nardetum alpinum*)				
Ranunculus montanus-form	1.7	11	48	11
Gentian form (with *Gentiana punctata*)	2.2	6	37	12
Poor Crested Dogstail pasture (*Crepido-Cynosuretum*)	1.8	24	44	9
1000–1200 m				
Calcareous spring-watered sward (*Caricetum davallianae*)	3.1	10	57	7
Montane Matgrass sward (*Nardetum*)	3.4	12	55	6
Lady's Mantle-Crested Dogstail pasture (*Alchemillo-Cynosuretum*)	4.1	71	43	3
Ryegrass-Crested Dogstail pasture (*Lolio-Cynosuretum*)	**6.2**	**100**	32	0

Note: [a] As a relative measure of the importance of the different communities for nature conservation the average number of species in the individual stands is given (A) along with the total number of protected species (P).

Source: From data in Spatz (1974); yield expressed in dry matter cut (t/ha); possible stocking rate for cattle as a percentage of the best pasture calculated on the basis of the yield and quality of the fodder. The Gentian form of the Matgrass sward contains a particularly large number of unpalatable or poisonous plants.

Tab. 120. The influence of treading by cattle on the species composition of a well manured Ryegrass pasture.

	Serial No.	1	2	3	4	F	N
	Distance from the yard gate (up to m)	6	18	63	360		
A	*Echinochloa crus-galli*	4				5	8
B	*Glyceria plicata*	3				10	8
A	*Atriplex patula*	3				5	7
T	*Juncus tenuis*	2				6	5
A	*Chenopodium glaucum*	3	1			6	9
T	*Matricaria discoidea*	5	4			5	8
T	*Polygonum aviculare*	**5**	4		1	×	×
T	*Poa annua*	**5**	5	4	2	6	8
T	*Plantago major*	5	5	5	5	5	6
	Agrostis stolonifera	5	4	4	5	6	5
	Poa pratensis	5	**5**	**5**	5	5	6
	Trifolium prepens	5	5	5	**5**	×	7
A	*Rumex obtusifolius*	4	5	4	4	6	9
A	*Capsella bursa-pastoris*	4	5	5	5	×	7
	Lolium perenne	4	5	**5**	**5**	5	7
	Taraxacum officinale	3	5	5	5	5	7
A	*Stellaria media*	3	4	5	5	4	8
	Ranunculus repens	3	1	3	5	7	×
	Dactylis glomerata	2	4	5	5	5	6
	Festuca pratensis	2	2	5	4	6	6

	Serial No.	1	2	3	4	F	N
	Distance from the yard gate (upto m)	6	18	63	360		
	Potentilla anserina		4		2	6	7
	Agropyron repens		3	4	5	5	8
	Poa trivialis		2	4	5	7	7
	Phleum pratense		1	2	2	5	6
	Leontodon autumnale		1		2	5	5
	Bellis perennis		1	4	5	×	5
	Deschampsia cespitosa			4	2	7	3
	Achillea millefolium			2	3	4	5
	Plantago lanceolata			2	3	×	×
	Trifolium pratense				5	×	×
	Bromus hordeaceus				4	×	3
	Alopecurus pratensis				3	6	7
	Ranunculus acris				3	×	×
	Glechoma hederacea				3	6	7
	Cerastium fontan. holost.				2	5	5
	Heracleum sphondylium				2	5	8
	Carum carvi				2	5	6
M	*Crepis biennis*				2	5	5
	Average plant cover (%)	**46**	91	98	97		
	Average number of species in each	15	16	17	28		

Note: From data by Vollrath (1970) on the pastures of the Veitshof (Upper Bavaria) which are rotationally grazed. The plots examined are shown in the order of increasing distance from the paddock gate, and the figures are each based on a number of counts.

A = arable weed or plant of waste ground (*Chenopodieta, Artemisietea*) B = stream Reed bed plant (*Glycerio-Sparganion*), T = plant of trampled swards (*Plantaginetalia*) M = fertile meadow plant (*Arrhenatherion*). Some species which occur very infrequently have been omitted.

No. 1: **Near the gateway**, most heavily trodden, more than 50% bare earth; waterlogged in places (*Glyceria*!);

2: **Heavily trampled** and grazed, sward short and in some places with gaps. The massive Broad-leaved Dock (*Rumex obtusifolius*), a pasture weed requiring high nitrogen levels, is locally abundant;

3: **'Normal' pasture**, dense sward dominated by short grasses; like 1 and 2 relatively poor in species;

4: **Least grazed** by cattle and thus more like a meadow, less heavily manured with dung (mN relatively low) and noticeably richer in species.

[b] mF = average moisture value (the soil dampness is by and large independent of the amount of grazing);

mN = average nitrogen value (the greater the distance from the gate the more numerous are the easily satisfied species, nevertheless those requiring high nitrogen levels are to be found in all areas).

ΣT = sum of the constancy numbers of the 'track' plants;

ΣA = sum of the constancy numbers of arable weeds and ruderal plants (which at the same time are indicators of mechanical disturbance).

for the analysis of the **small-scale pattern and physiological differentiation** of species with wide amplitudes. *Trifolium repens* especially is very well suited to this since it spreads vegetatively, and genetically identical clones tend to form local populations adapted to peculiarities of the habitat. For instance small-scale variations in the potash and phosphate contents of loamy soil clearly have an effect on the distribution of *Trifolium repens* (Snaydon 1962a, Snaydon and Bradshaw 1962, see also Turkington and Harper 1979). Clones from a soil rich in phosphate do not grow so well in a soil deficient in P as do clones from a soil poor in phosphate. The proportion of the ground covered by White Clover has a particularly marked correlation with the lime content of the soil. Snaydon (1962b) explained this by the smaller growth of the limestone soil races when they are grown on acid soil or even when they are just grown mixed together with the acid-soil races. In contrast the latter thrive just as well on calcareous soil as on acid soil. The 'edaphic ecodemes' within a species react in the same sort of way as different species which are found chiefly on calcareous or on acid soils as the case may be (see section C VI 2). In any case the differences in growth reported by Snaydon are quite small (most of them less than 10% in 2 months). Nevertheless in the long run they may affect the competition. Burdon (1980), having studied 50 clones of *Trifolium repens* in permanent grassland of North Wales, emphasises that 'the nature of the local environment is very complex and ever-changing. Genotypes which are well adapted to one particular position at one time may, as a result of temporal and spatial interaction with a variety of different neighbours, be less well adapted at a subsequent time or in a different position in the community'. Discussing reciprocal transplants of *Plantago lanceolata* between field sites Antonovics and Primack (1982) even conclude: 'In most cases environmental differences were much more important than the genetic ones'.

Fig. 473. The soil water content under different Ryegrass-White Clover pastures near Stolzenau on the Weser. From a coloured diagram by A. von Müller (1956) from Ellenberg (1963).

1 = *Lolio-Cynosuretum plantaginetosum mediae*, *Avenochloa pubescens* variant (relatively dry habitat), 2 = *L.-C. typicum*, *Dactylis glomerata* variant, 3 = *L.-C. lototetosum*, typical variant.

The amount of water present above the permanent wilting point (PWP) is based on lysimeter measurements and expressed as a %age by volume. The growth was cut three times, not grazed.

In their habitats the Hoary Plantain-White Clover pasture corresponds to the Quaking Grass-Oatgrass meadow, the slightly-moist Ryegrass-White Clover pasture to the Foxtail-Oatgrass meadow (see section V 2 b) and the moist Ryegrass-White Clover pasture to the Marsh Ragwort meadow (section V 4 b).

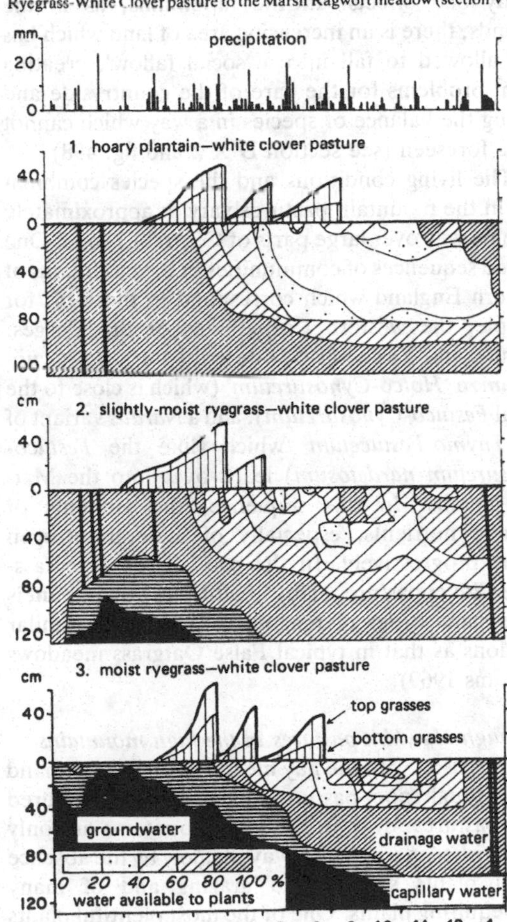

K. Schäfer (1972) found that with Perennial Ryegrass the optimum temperature for photosynthesis varies with the amount of light. In a relatively cool climate the plant requires less light and so to some extent is adapted to montane conditions. In this respect a tetraploid cultivar is more productive than the diploids. Of all the grasses *Lolium perenne* is particularly responsive to applications of nitrogenous and phosphatic manure. Bradshaw, Lodge, Jowett and Chadwick (1960) and Bradshaw, Chadwick, Jowett and Snaydon (1964) carried out culture experiments which showed that increase in the growth rate is stepwise lower in the sequence: *Lolium perenne*, *Agrostis stolonifera*, *A. tenuis*, *Cynosurus cristatus*, *Festuca ovina* and *Nardus stricta*. These grasses appear in approximately the same sequence in pastures of decreasing fertility (see tab. 118), although M. Vogel (1981) found in an investigation into the effects of increasing the amounts of nitrogen that even *Nardus* was able to stand relatively high nitrogen concentrations (up to 3600 kg/ha/year). So here again it is competition which works as the deciding factor in the distribution of species.

Plewczyńska-Kurras (1974) measured the biomass production in a *Lolio-Cynosuretum* grazed by sheep on the northern edge of the Carpathians. During the year 1971 the amount of dry matter reached a maximum of 163g/m^2 (1630 kg/ha) above ground and 805 g/m^2 below ground. In a grazed sward new shoots are constantly being produced so the amount produced during the season would be several times that of the standing crop. Andrzejewska (1974), studying pastures from which the grazing animals had been temporarily excluded, harvested 471 g/m^2 during the growing period from the manured plots but only 143 g/m^2 from the unmanured. The production beneath the surface could well have been much greater, but it is difficult to measure. Earthworms enable some conclusions to be drawn about this since they live principally on dead roots. According to Czerwiński, Jakubczyk and Nowak (1974) 3534 g/m^2 were produced in the heavily manured plots during the season against 781 g/m^2 in the unmanured ones. The humus content of the soil brought about by the activity of earthworms and of the bacteria in their casts increased by 145 g/m^2 during the year in the manured plots compared with only about 28 g/m^2 in the ones without manure.

The sheep ate some 80–90% of the phytomass made during the growing season, i.e. about 400 g/m^2 from the manured plots and about 120 g/m^2 from the unmanured. Their dung was taken into the soil by 16 species of beetle (Scarabaeidae investigated by Breymeyer 1974) to the extent of 216 g/m^2 dry weight. This was used as food by earthworms to some extent and, along with the worm casts was invaded by micro-organisms. The process of decomposition, including the mineralisation of the nitrogen on which the plants so much depend, is therefore concentrated just here (Kojak 1974). Even these few measurements and references serve to indicate how important the activity of the heterotrophic organisms is for an ecosystem, even for a fertilised pasture.

According to Whittaker and Feeny (1971) chemical interactions also play a part in the dynamic balance between the members of an ecosystem. Without these allomones the living roots and other parts of the plants would be eaten in much larger quantities and by many more animals than is actually the case. Specific secondary plant substances are very often the reason why certain plants are avoided by some or all herbivores and thus have an advantage in competition. Pasture weeds do not necessarily have to possess spines like Thistles or other obvious physical protection against being eaten. Particularly in mountain pastures with their wealth of species (see section c and d) there are numerous examples of such 'allochemical effects', the ecological consequences of which have for the most part not been investigated.

(c) Tussock Red Fescue-White Clover pastures in the mountains

As the growing season becomes shorter the manuring and tending of the pastures becomes less profitable. This means that the intensity of farming used for both the pastures and meadows falls off by and large with increasing altitude (and also at higher geographical latitudes). In both directions the Ryegrass-White Clover pastures become less frequent and finally poor pastures reminiscent of the Common Bent-Red Fescue meadows become completely dominant where cattle are still grazed (see fig. 451). They can be described as Tussock Red Fescue-White Clover pastures (*Festuco-Cynosuretum*) since *Festuca nigrescens* and other low-growing undemanding grasses predominate (tab. 118). Alternative names sometimes given to them are *Luzulo-Cynosuretum* (after *L. campestris*, Meisel 1966b) and *Anthoxantho-Agrostietum* (Jurko 1974). In any case it is difficult to fix a line which separates them from the more heavily manured Ryegrass-White Clover pastures (*Lolio-Cynosuretum*) as none of these pasture communities possesses good character species and there is a continuous sequence of intermediate types. Because of this tab. 118 is limited to a grouping based on habitat, and a number of subcommunities are given of which the ones rich in Matgrass is frequent in the mountains. Higher up the White Clover pastures are replaced by the Rough Hawkbit pastures (section d) and like these contain more and more montane and subalpine species, or they become more like the calcareous slightly arid grasslands (tab. 119). The productivity of these mountain pastures is inversely proportional to the variety of species they contain, especially the number of protected plants.

Up to a few years ago the danger that many people feared was reduction in the diversity of these alpine pastures because of increasingly rapid intensification. More recently there has been a complete reversal of this trend and now the tendency is for grazing animals to be withdrawn from the high mountains, and for the former pastures to be given over to recreation with equally bad effects. In the mountains, as in the lowlands, there is an increasing area of land which has been allowed to fall into a 'social fallow', creating urgent problems for the care of the countryside and altering the balance of species in a way which cannot yet be foreseen (see section D X 2 and fig. 498).

The living conditions and the species combinations in the mountain pastures vary in approximately the same way over large parts of Central Europe. One can find sequences of communities in the mountains of northern England which correspond to those of, for example, the western Carpathians or the Vosges. Williams and Yarley (1967) described a *Lolio-Cynosuretum*, a '*Holco-Cynosuretum*' (which is close to the typical *Festuco-Cynosuretum*), and a *Nardus* variant of the '*Thymo-Festucetum*' which (like the *Festuco-Cynosuretum nardetosum*) leads over into the Matgrass heaths. In this sequence, the amounts of available nutrients, especially of mineral nitrogen decline progressively. In the case of the Ryegrass-White Clover pastures at lower altitudes this amount is of the same order of magnitude and shows similar variations as that in typical False Oatgrass meadows (Williams 1969).

(d) Rough Hawkbit pastures in the high mountains

In comparison with the hay meadows on the one hand and the rough, poor pastures on the other the manured cattle pastures, even in the high mountains are only distinguished in a negative way, namely by the absence of tall plants sensitive to grazing and of many light-requiring plants. One of the most plentiful plants in manured alpine pastures is the Rough Hawkbit

Fig. 474. *Leontodon hispidus* (right), *Crepis aurea* (centre) and *Trifolium badium* (left) in a mown 'Milkweed pasture' (Rough Hawkbit community) in the Alpine garden Schynige Plate. Photo. Lüdi.

Leontodon hispidus which the local people call 'Milkweed' and for this reason Schröter and other older geobotanists used the name 'milkweed pastures' (see fig. 474). These are the most valuable grassland communities for Alpine dairy farming in the subalpine and the lower alpine belt, especially on level deep soils (see fig. 475 and compare section C VI 2 d).

In spite of their shortage of character species we agree with Marschall (1947) that these Rough Hawkbit pastures represent a separate association. Like the Tussock Red Fescue-White Clover pastures they dovetail in with the Matgrass swards and other Alpine grassland communities. There are a number of differential species in the latter which enable a fairly sharp line to be drawn between the Rough Hawkbit pastures and the Tussock Red Fescue pasture.

Fig. 475. Hawkbit pastures immediately above the tree line in the Lechtal Alps. On the less well manured slopes they merge into Violet Fescue swards. In the foreground centre are traces of a resting-place community.

2 Swards influenced by treading or by occasional flooding

(a) The vegetation of footpaths and other trodden areas

Not the least reason why intensive pastures are so poor in species is the fact that many plants cannot withstand the mechanical damage resulting from the treading by the grazing animals. Bothmer (1953) investigated the resistance to treading of different species by measuring how far away from the bare area in the gateway each species first appeared in the sward of an intensively grazed pasture. Repeatedly he found that it was the same few species which ventured towards the bare heavily-trodden zone while others were always found at a distance. Haessler (1954) measured the resistance to a mechanical stress similar to treading and reached equal conclusions. His results have been discussed so thoroughly by Walter (1962, pages 557 and 558) that we can restrict ourselves here to a few remarks.

The treading factor operates most strongly on those well used footpaths and poorly managed football pitches which have a loamy soil. In such habitats throughout the whole of Central and West Europe there is one of the most uniform plant communities in the world, the Ryegrass – Greater Plantain carpet (*Lolio-Plantaginetum majoris*). In it one always finds a combination of the following species (see fig. 476):

Lolium perenne	*Poa annua*
Matricaria discoidea	*Polygonum aviculare*
Plantago major	

Coronopus procumbens and in warmer districts *Cynodon dactylon* are less constant. All these species can be seen as faithful character species of the trodden plants alliance (*Polygonion avicularis*) even though they are occasionally found in some other communities (Berset 1969, Faliński 1961, 1963, Oberdorfer 1971). However some species which are frequently found with them have their main distribution in grassland or in the weed communities of arable land, e.g.:

Agrostis tenuis — *Taraxacum officinale*
Capsella bursa-pastoris — *Trifolium repens*
Leontodon autumnalis

The 10 plants just mentioned include annuals, biennials and perennials, and their ability to withstand this high degree of treading rests with their small size, the fact that they branch close to the ground, the toughness and elasticity of their tissues, rapid regeneration and other properties of their vegetative organs. Where there is very heavy treading the plants do not produce seed but generally the vegetation in the neighbourhood supplies sufficient seed. These are usually epizoochorous or tiny and easily dispersed in the mud adhering to people's shoes, animals' hooves, or to wheels (tab.121). White Clover and some other species are dispersed by seed-eating birds, such as the Rook (Krach 1959) and by grazing animals. According to Clifford (1959) the wheels of motor vehicles are playing an increasing part in the distribution of seeds.

However seeds are not always able to develop just where they happen to arrive; the conditions for germination may be even more important than the conditions met with by the grown plant (Sagar and Harper 1961). For example *Plantago major* and *Polygonum aviculare* have seeds which require really wet conditions for germination, (Ellenberg and Snoy 1957) and so the rain water collecting on consolidated loam creates a favourable site for this. Grasses like *Poa annua* also germinate best where a moist condition is maintained. *Trifolium repens* requires a surface moist enough for its runners to root into. On pure sandy soils where the surface dries out rapidly the Ryegrass-Plantain community can never be typical. Long dry periods are the main reason why the *Lolio-Planta-*

Fig. 476. The normal zonation of trodden plant communities alongside a track in manured grassland on loamy soil. After Oberdorfer (1971), somewhat modified.

Therophytes such as Pineappleweed and Knotgrass which only partially cover the ground can withstand the most treading. The White Clover which spreads by runners joins them in the transition zone towards the White Clover-Broadleaved Plantain track sward into which there soon come Perennial Ryegrass and other species.

The phenological development of these communities, which are widespread in Central Europe has first been described quite recently by F. Runge (1981). They are by no means 'evergreen' as had been supposed up to now, but disappear nearly totally in winter, even in the rather oceanic climate around Münster in Westphalia.

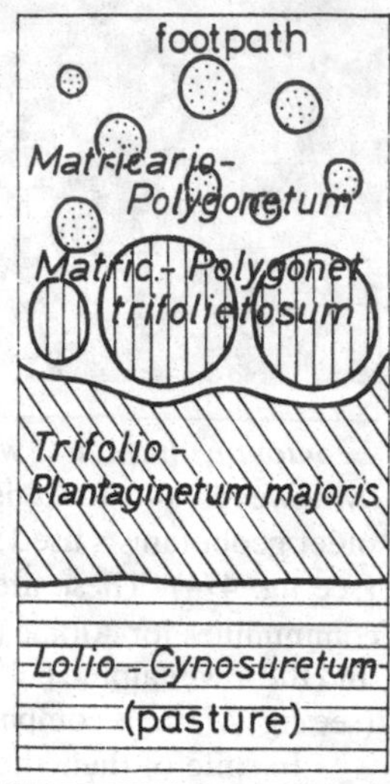

Tab. 121. Seed germination in heavily trodden soil near a drinking trough.

From about 3,000g of fresh earth taken up in December, kept moist and frequently turned, seeds germinated in the following numbers:

Plants of trodden ground, sens. lat.[a]	**Meadow plants,** sens. lat.	**Arable weeds,** sens. lat.
37 *Poa annua*	10 *Agrostis tenuis*	5 *Stellaria media*
31 *Plantago major*	10 *Poa trivialis*	4 *Juncus bufonius*
15 *Trifolium repens*	6 *Juncus articulatus*	1 *Sonchus arvensis*
13 *Matricaria discoidea*	3 *Poa pratensis*	*Sonchus oleraceus*
10 *Sagina procumbens*	2 *Cerastium fontanum holosteoides*	1 *Urtica urens*
3 *Capsella bursa-pastoris*	1 *Festuca rubra*	
2 *Polygonum aviculare*	1 *Stellaria graminea*	
2 *Veronica serpyllifolia*	1 *Achillea millefolium*	
1 *Taraxacum officinale*		

Note: [a] *Lolium perenne* is absent here nor did it appear in similar investigations or at other times of the year.

Source: From data in Boeker (1959).

ginetum rarely appears in the Mediterranean area whereas the humid coastal climate of the north-western part of Central Europe is ideal to the trodden plants. In drier regions like the northern part of the Upper Rhine plain the sap-green carpet of the Ryegrass-Plantain community is replaced by the parched Hardgrass-Knotgrass carpet *(Sclerochloo-Polygonetum avicularis*, see Korneck 1969) which is only mentioned here as it is rather rare.

It can be seen from fig. 477 that treading influences many different properties of the habitat, some of which could be decisive for the extent to which a particular species can develop. A favourable side effect of mechanical damage is that more light is available and many competitors are removed. In addition there may be another important factor present although this is not necessarily connected with the treading as such, but is generally operating in this kind of habitat. That is the good supply of nutrients, especially nitrogen, which the nearness to human habitation ensures. Without this the damaged plants would not regenerate so strongly nor the seedlings develop so quickly.

It must seem very surprising that Boeker (1959) was hardly ever able to get the seed of *Lolium perenne*, that most plentiful of trodden plants, to germinate in the soil taken from footpaths, around drinking troughs or places where animals gathered to rest. The other species of the trodden carpet grew from seed in large numbers (tab. 121). Apparently ryegrass does not produce any seed in these heavily trodden places and relies on being brought in from adjacent sites. On the other hand Boeker found that there were viable seeds of many meadow plants and arable weeds in the consolidated ground. These are not normally able to develop in such an extreme habitat either because the radicles will not penetratre the hard earth or they are destroyed soon after germination.

Besides the widespread **Perennial Ryegrass-Plantain carpet** there are a few other communities resistant to treading which occur much less frequently; only a few examples will be named here. Oberdorfer (1957) and others have described a **Slender-Rush trodden community** from shady woodland paths in which the neophyte *Juncus tenuis* has soon made itself at home.

Mosses often develop in the trodden plant communities between paving stones or on cinder and gravel paths where they are relatively safe from mechanical damage but the soil is very poor. The commonest is the small Cushion Moss *Bryum argenteum* that can stand being dried out occasionally. The **Pearlwort-Cushion Moss cranny community** (*Sagino-Bryetum*) is often well developed but only in an oceanic climate, especially in Holland and the western part of North Germany.

All the trodden plant communities in Central Europe are so similar floristically and ecologically that they can be put together in the same alliance (*Polygonion avicularis*). However because of the extreme effect of an unusual factor this is so isolated

Fig. 477. Treading by animals and people has partly a direct effect and partly an indirect one on the plant cover. The 'track plants' are damaged less than other species and so hold an advantage in the invasion and competition on trodden places. After Lieth (1954), somewhat modified.

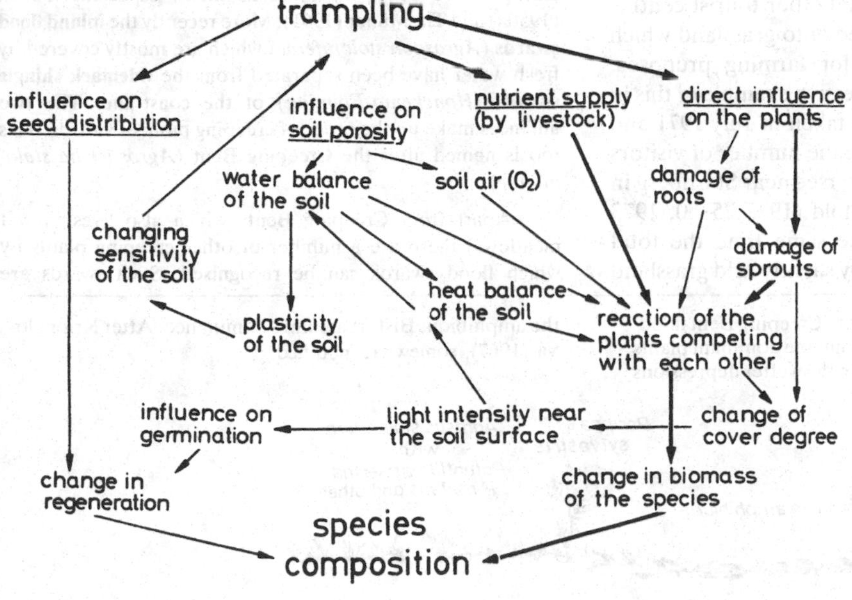

Fig. 478. 'Flood swards' on the bank of the Wäggital reservoir (Switzerland) in summer. In the foreground are Couch Grass (*Agropyron repens*), Curled Dock (*Rumex crispus*) and the Greater Yellow Rattle (*Rhinanthus alectorolophus*, pale flowers).

that it must also serve as a separate order and class (*Plantaginetalia* and *Plantaginetea majoris*).

If the vegetation development were to become completely natural most trodden plant communities would disappear from the landscape of Central Europe just as would most of the grasslands, wasteland communities and arable weeds. The mechanical effect of man and his animals preserves them from the shade of the woodland which would be the vegetation climax.

Because of the rapidly increasing recreational traffic in the region of lakes and other tourist centres trampling pressure has intruded onto grassland which was previously used purely for farming pruposes. Seibert (1974) gives us an impressive example of this in the form of vegetation relevés taken in July 1971 and again in 1973. During this time the number of visitors to the bathing beach on the Ostersee near Starnberg in Upper Bavaria increased ten-fold (1972 25–50, 1973 250–300 per day) while at the same time the total number of species in the nearby slightly arid grassland fell from 59 to 30 and the number of character species from 27 to 7. In their places the number of species typically associated with farmed grassland rose from 3 to 7, and 4 representatives of trodden plant communities found a footing. This succession is also a good example of how quickly the plants of trodden ground are transported. F. Runge (1970) noticed the first specimens of *Poa annua* and *Lolium perenne* in May, only six weeks after the completion of a road shoulder. By August the *Lolio-Plantaginetum* was fully formed. In addition representatives of the arable weed and wasteland communities together with some from the sandy dry grassland had arrived, but because of heavy treading they had been unable to establish themselves. Vegetation dynamics are nowhere more easily observed than alongside roads or waterways which tend both to destroy the plant cover and to enrich it with other species.

(b) Creeping communities of occasionally flooded sites

On the banks of rivers and lakes, and also along the sea coast are found vegetation types growing under natural conditions, encouraged by flooding, wave action or standing water, which are reminiscent of trodden plant carpets. Such 'flood swards' in Central Europe are quite often walked over by ducks and geese, hoofed animals or men and this may make them even more like the *Plantaginetalia* communities than would otherwise be the case (see figs. 478 and 479).

Flood swards near the coast are particularly well developed in Northern Europe, and it is from here that they were first described (as *Agropyro-Rumicion crispi* by Nord hagen 1940). Since that time a number of authors in Central Europe have been interested in them, especially Tüxen (1950a) and Th. Müller (1961). More recently the **inland flood swards** (*Agrostion stoloniferae*) which are mostly covered by fresh water have been separated from the **tidemark shingle swards** (*Honckenio-Elymion*) of the coast-line. The two alliances make up the order of creeping pioneer swards; This too is named after the Creeping Bent (*Agrostietalia stoloniferae*).

Apart from Creeping Bent (which also lives in wet meadows) there are a number of other creeping plants by which flood swards can be recognised. Such swards are

Fig. 479. In the Creeping Yellowcress-Creeping Bent flood swards of Czechoslovakian flood plains the dominant plants are usually grasses or Cinquefoils. In the wetter depressions the amphibious Bistort assumes dominance. After Krippelová (1967), somewhat modified.

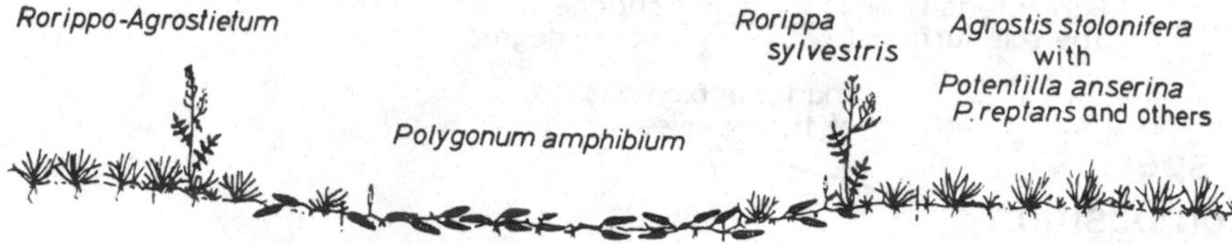

frequently disturbed, and they comprise only a small number of species. These however have runners growing out along the surface, quickly covering the bare earth and sending down roots whenever it is damp enough; they include:

Mentha longifolia	*Potentilla reptans*
Mentha pulegium	*Ranunculus repens*
Potentilla anserina	*Scirpus radicans* (rare)

Other species are successful in such habitats by producing rhizomes below the surface, e.g.:

Agropyron repens ssp. *maritimum*

A. junceum	*Honkenya peploides*
Carex hirta	*Rorippa sylvestris*

Agropyron repens subspecies *repens* is a plant characterising somewhat ruderal slightly arid grasslands (*Agropyretalia intermedii-repentis*, see section E III 1, No. 3.6).

Many of these plants certainly appear in other communities, e.g. on trodden loam and on rubble, in arable fields or in pastures, but their main distribution is in the natural flooded swards. The same is true of a few tussock-forming hemicryptophytes and the Curled Dock:

Festuca arundinacea	*Rumex crispus*
Juncus inflexus	

Some of the species named above are characteristic for particular associations, e.g. *Rorippa sylvestris* (for the *Rorippo-Agrostietum*, fig. 479, on gravel beds in the rivers), *Festuca arundinacea* (for the *Dactyli-Festucetum* on somewhat higher ground though still subject to flooding), *Mentha longifolia* (for the *Junco-Menthetum longifoliae* on marls saturated with water from time to time) and *Mentha pulegium* (for the warmth-loving *Potentillo-Menthetum*).

The communities along the sea shore that correspond ecologically to the flood swards inland are difficult to distinguish from the more strongly nitrophilous tidemark communities *Salsolion*, see section C V 1 b) and from the pre-dunes (*Agropyro-Honkenion*). For this reason they have already been considered in the section mentioned. In the neighbourhood of salty places inland, which have been described in section C IV 3 the flood sward dovetails in with more or less halophilous swards like that shown for example in fig. 480.

In some places fragments of flood swards are interspersed with the grassland communities of our loamy river valleys. In wet depressions where rainwater tends to collect and where the water remains for a longer time after the land has been flooded the true meadow plants are handicapped through shortage of air. Frequently all the tall-growing plants die off, and the places they leave empty are first taken over by members of the creeping sward community. The proportion of the ground they occupy varies from one year to another and depends on the weather and the length of time the ground is under water. Tüxen aptly compared this to and fro movement of *Agrostion stoloniferae* swards with an accordion. The **Marsh Foxtail flood sward** (*Rumici-Alopecuretum geniculati*) is particularly involved in this irregular backward and forward succession which is characteristic of grassland in lowland river valleys. The grass which gives its name to this community not only has a useful ability for vegetative reproduction but it produces large numbers of seeds which can be carried by the flooding water and

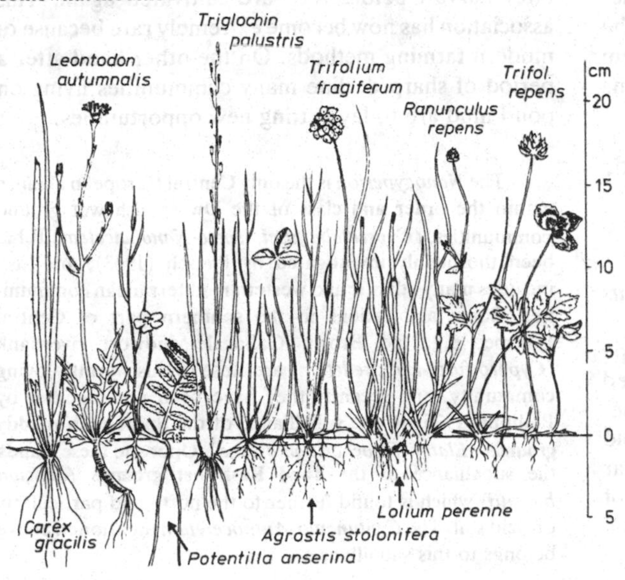

Fig. 480. Cross-section through a slightly brackish flood sward (*Triglochino-Agrostietum stoloniferae*) along the Schwielow lake in the Potsdam Havelland. After Konczak (1968), somewhat modified.

Apart from the Autumn Hawkbit and the Marsh Arrowgrass all the species spread quickly vegetatively, the Clovers, Silverweed and Creeping Buttercup by overground runners.

Fig. 481. Dwarf plants growing on the mud of a pond still containing some water in the Bohemian-Moravian upland. *Gnaphalium uliginosum*, *Juncus bufonius* and, in the water, *Callitriche palustris*. Photo. Iltis-Schulz.

in this way succeed in reaching all suitable habitats. The probable reason why other communities of the *Agrostietalia stoloniferae* are also able to use quickly even very isolated small areas is their ability to disperse themselves by means of water-transported fruits or parts of shoots.

Agrostis stolonifera has an exceptionally large number of different forms and is found not only in flood swards but also in farmed pastures, weed communities and others. Aston and Bradshaw (1966) have demonstrated that this variation is genetically determined because it persisted even when the plants were cultivated under identical conditions. It includes such morphological and physiological variations as the length of the stolons, the height of the flowering stem and the tolerance to drought and to salt concentration.

VII Man-influenced vegetation of muddy ground, lakesides and banks

1 *Short-lived Dwarf Rush communities on ground with fluctuating water content*

(a) Distribution and dispersal of the dwarf plants on pond mud

Extensive areas in the lowlands of Central Europe today are occupied by therophytes leading an unsettled life as weeds reluctantly tolerated by man. Nevertheless these species, whether they come from the old European flora or are newcomers, form regular groupings which can be put into a number of sociological units and arranged into a hierarchical system.

The short-lived communities growing on mud, especially the dwarf rushes (*Nanocyperion*) are rather insignificant representatives of this group. The sites where they flourish are covered with water from time to time although they may not be waterlogged for months, or even for several years. We would like to discuss these first because although weed communities they can also occur naturally, because they are amongst the best identified plant communities and because to a considerable extent they are characteristic of Central Europe (see fig. 481).

According to Moor (1936) the area in which the *Nanocyperion* alliance occurs embraces the whole of Central Europe and extends a little around the boundary. The map of its distribution resembles that of Durmast Oak or Beech in that its centre of frequency lies more to the west and south. However within these wide geographical boundaries *Nanocyperion* communities occur only sporadically and erratically on relatively small areas. As light-requiring, low-growing and therefore uncompetitive stands they are confined to bare ground which is wet during the time they are germinating, e.g. on the mud of drained ponds, around the edges of pools, on bare banks of slowly flowing streams where the water level fluctuates a good deal, in shallow ditches which dry out in summer, on humus-rich puddles in sand or gravel pits as well as on footpaths through marshes or woods, Deer tracks and wallows liable to flooding from time to time. The **Chaffweed-Hornmoss community** (*Centuncula-Anthoceretum punctati*, see tab. 122) has specialised in the consolidated surface of wet cornfields forming a carpet in the stubble if sufficient time elapses after harvest before they are cultivated again. This association has now become extremely rare because of modern farming methods. On the other hand after a period of sharp decline many cummunities living on pond mud are today getting new opportunities.

The *Nanocyperion* is the only Central European alliance within the order and class of the Dwarf-rush wet ground communities, *(Cyperetalia fusci, Isoeto-Nanojuncetea)*. It has been thoroughly investigated by Pietsch (1963), and also includes many atlantic and western mediterranean communities. Some associations in the southern part of Central Europe, e.g. the Brown Galingale-Mudwort riverbank (*Cypero fusci-Limoselletum*) are close to this warmth-loving community, and recently they have been put together by Philippi (1975) in the suballiance of the **Spike-rush muddy grounds** *(Elatini-Eleocharition ovatae)*. Opposite these stands the suballiance of the **Toad Rush wet grounds** *(Juncion bufonii)* which is found further to the north and particularly on acid soil. The *Centunculo-Anthoceretum* mentioned above belongs to this suballiance.

Tab. 122. The character species of dwarf plant communities in temporarily wet habitats.[b]

Class and order character species of all **fluctuating wetness dwarf plant communities** (*Isoëto-Nanojuncetea*[a], *Cyperetalia fusci*):

	Centaurium pulchellum		*Gnaphalium uliginosum* (?)		*Lythrum hyssopifolia*
	Cyperus fuscus		*Gypsophila muralis*		*Schoenoplectus supinus*
	Elatine hexandra		*Juncus bufonius* (?)		
	E. hydropiper		*J. tenageia*	M	*Riccia* spp.
	Gnaphalium luteo-ablum	a	*Limosella aquatica*		and other mosses

Character species of the **Central and Western European** alliance *(Nanocyperion)*:

d	*Blackstonia perfoliata*		*Illecebrum verticillatum*		*Sagina apetala*
b	*Carex bohemica*	f	*Isolepis setacea*		*S. nodosa*
h	*Centunculus minimus*	c	*Lindernia procumbens*		*Scirpus radicans*
g	*Cicendia filiformis*		*Marsilea quadrifolia*		*Veronica acinifolia*
e	*Cyperus flavescens*		*Montia chondrosperma*		
	Elatine alsinastrum		*Peplis portula*	A	*Botrydium granulatum*
	Eleocharis ovata		*Plantago major* ssp. *intermedia*	M	*Fossombronia* spp.
	Hypericum humifusum		*Radiola linoides*	M	*Riccia glauca*

Note: [a] The name '*Isoeto*'-*Nanojuncetea* refers to the mediterranean alliance *Isoetion* described by Braun-Blanquet (1935, see Moor 1936). Its character species are some *Isoetes* which are absent from Central Europe (*I. duriaei, setaceum* and *adspersum*).

[b] The letters indicate the character species for the following communities (named as examples):

a Brown Galingale-Mudwort riverbank ground (*Cypero-Limoselletum*)
b Spike Rush-Bohemian Sedge mud community (*Eleochari-Caricetum bohemicae*).
c Spike Rush-Boxweed mud community (*Eleochari-Lindernietum*)
d Lesser Centaury-Yellowwort wet ground (*Centaurio-Blackstonietum*)
e Yellow Galingale sandy trodden ground (*Cyperetum flavescentis*)
f Trodden Bristle Clubrush spring ground (*Isolepidetum*)
g Yellow Centaury wet ground (*Cicendietum filiformis*)
h Chaffweed-Hornmoss wet ground (*Centunculo-Anthoceretum*).

Source: After Moor (1936), Oberdorfer (1957, 1970) and others.

Although the places where the *Nanocyperion* communities grow are not related to each other and the habitats vary from one year to the next depending on weather and management, the communities themselves are surprisingly homogeneous. Where the same conditions prevail one always finds the same group of species, even though elsewhere in the district they may be extremely rare. In most parts of Central Europe they are looked upon as floristic gems. The reader will only exceptionally if ever have come face to face with most of the alliance and order character species named in tab. 122.

Moor (1936), who devoted a first monograph to these modest but valuable communities, believed that the uniformity of their species composition points to a previous wider distribution. In fact because of increasing cultivation and the 'drying out' of the Central European countryside the sites available to them have become fewer. They had their heyday in the Middle Ages when the use of small artificial lakes was at its height and the level valleys were filled with hundreds and thousands of fish ponds to meet the demand on fast days. Even these series of ponds did not form a complete network over the whole of Central Europe which might have made the exchange of plant species easier.

The homogeneous species composition and the almost spontaneous appearance of these tiny communities in every place which offers the right conditions depends much more on their potentially universal presence. 90–95% of their species produce huge numbers of seed, and these are so small that they are blown like fine dust by the wind. Many of the seeds are also transported by water, but the most effective method and one which has a certain element of directional guidance is the dispersal by limicolous birds spending most of their time around the water and on the mud. As Kerner von Marilaun reported as long ago as 1887 mud containing numerous seeds sticks to the feet of wading birds and in this way can be carried to the next feeding ground which may be a considerable distance away. They are thus taken to the very sites where they are most likely to germinate and survive. Kerner observed the germination of many plants from the mud collected from the beaks and feet of such birds; and of these plants more than half were character species of the *Nanocyperion* alliance, e.g. *Centaurium pulchellum, Centunculus minimus Limosella aquatica, Lindernia procumbens, Elatine hydropiper, Isolepis setacea, Juncus bufonius, Cyperus fuscus* and *C. flavescens*.

It is probable that a lot of seeds are also carried on the hooves of Deer or domestic animals and on wheels along with the mud from the tracks. Deer walks and earth roads are frequent habitats for *Nanocyperion* communities. Dispersal in mud also explains why most of the communities of these small plants are found on soils rich in colloids although they are not confined to them. For example in the dry year 1959 Burrichter (1960) found them to be well developed on pure sand or loess along the dams in some valleys of Westphalia. The substrate type does not appear to be important provided it offers sufficient moisture for germination and early growth (see fig.482). The seeds of many species are able to retain their viability while lying dormant at the bottom of ponds, sometimes even for many years, only to germinate in large numbers when the pond has been drained. This can be seen in those places where the ponds are no longer managed regularly as they used to be, and where the Dwarf Rush communities do not have the opportunity to ripen their seed every year.

The adaptation which literally decides the existence of the Dwarf Rushes, cyperaceous and caryophyllaceous plants, Gentians, and other small herbs in their insecure life is their ability to use wind, water and animals for dispersal. It would be wrong to consider them as solely anemochorous just because they have tiny seed, since it is only exceptionally that the wind plays a part in disseminating such low-growing plants over long distances. Here we have an example of how dubious the usual method of grouping plants according to the means of seed dispersal is when this is based just on the morphology of the seeds or fruits. It would be much better to test the **effectiveness** of the dispersal mechanism by careful observations in the field and by experiment.

(b) Particular dwarf-plant communities of Central Europe

The Spike Rush-Bohemian Sedge mud community (*Eleochari-Caricetum bohemicae*) can serve as the type association in southern Central Europe of the dwarf plant communities living where there is a varying water supply. Klika (1935) described it as the 'community of exposed pond mud' since it passes its fleeting existence on the muddy bottoms of drained ponds or along the lakesides from which the water has drawn back in summer. The Spiked Rush (*Eleocharis ovata*) is often dominant, and quite often it is accompanied by the Mudwort (*Limosella aquatica*). Alongside these will be most of the character species for the alliance, order and class to be found in tab. 122. Other, less-common association character species are:

Elatine alsinastrum f. *terrestre*	*Lindernia pyxidaria*
Elatine hexandra f. *terrestre*	*Marsilea quadrifolia*
Elatine hydropiper f. *terrestre*	*Schoenoplectus supinus*

The Spiked Rush community appears as soon as the ground is free from standing water. According to Moor (1936) almost all the species require light for germination, and their further growth too will only go on normally in full light. The role of temperature has so far not been determined. Presumably most of the species require warm conditions since they originate from the mediterranean region.

These dwarf plants never exceed 10 cm in height

Fig. 482. A vegetation transect on the southern bank of the Möhne reservoir with dwarf plant communities and nitrophilous Goosefoot communities in the unusually dry summer of 1959. The water content and the water-holding capacity of the soil, as well as the length of time it is submerged, all fall off from left to right. After Burrichter (1960), somewhat modified. *Eleocharitetum ovatoe* (Pond Bottom community) and *Stellario-Isolepidetum setacei* (Bristle Scirpus community in the area of frequent wave action) belong to the *Nanocyperion*. The *Polygono-Chenopodietum rubri* is a *Bidention* community. Its *Corrigiola* fazies, now dry, indicates drift lines formed during the years when the water level was normal. *Limosella aquatica* predominates in the *Eleocharitetum*.

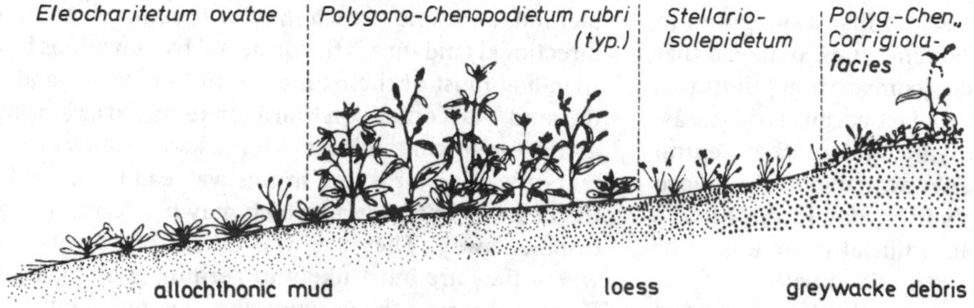

even under favourable conditions and more usually they are less than 5 cm. They are thus microphytes in the true sense of the word. However even species of normally taller-growing communities remain tiny if they occasionally occur in a dwarf plant community, e.g. species from the Trifid Bur-marigold community which is often found closely associated with it (compare section 2). This could be due to the shortage of nitrogen following the long time when the ground had been waterlogged and so deprived of oxygen, conditions favouring denitrification (see section D I 6 b). The pond bottoms and muddy lakesides usually consist of finer soil particles which have been subject to weathering and soil formation over a long period of time. They are thus short of lime and their reaction is weakly to moderately acid (pH 5.4 – 6.8, but rarely less than 5.0).

Stands that became established early in the year are often overshadowed during the later summer by nitrogen-loving Trifid Bur-marigold communities (*Bidention tripartitae*, section 2a), the members of which grow much taller and denser whenever they find suitable conditions. This very obvious sequence has often been considered a succession (e.g. by Burrichter 1960). However, as Moor has already suggested, it is not a case of a further development which is benefitted by the presence of the *Eleochari-Caricetum*, but the emergence of two communities, independent of one another, in the same place. The Trifid Bur-marigold community does not require any preparation of the soil but can turn up as the first occupier wherever there is sufficient nitrogen and moisture. Burrichter himself noticed that it developed at the same time as the Spiked Rush community but occupied ground which was somewhat higher (fig. 482). Whenever members of a Trifid Bur-marigold community are found growing amongst those of a dwarf plant community, or have come in after the latter, it is because the habitat itself has changed in that the soil has dried out thus allowing more air to penetrate, speeding up the process of nitrification. The members of the Trifid Bur-marigold communities also take rather longer to reach their full development than those of the *Nanocyperion* and because of this they get their chance somewhat later.

The deciding factor in the formation of dwarf plant or Trifid Bur-marigold communities, which both consist of therophytes, is the presence of waterlogged bare soil allowing for only a very sparse plant growth. Where flooding no longer destroys all plant growth each year, then hemicryptophytes soon appear. At about the same time seeds of Alder and other trees germinate leading to the development of a swamp or flood plain woodland. Here we really are dealing with a succession which denies the dwarf plants any further possibility of returning.

A few similarly well characterised units of the remaining *Nanocyperion* communities only penetrate the edge of the Central European area, e.g. the atlantic-subatlantic **Yellow Centaury community** (*Cicendietum*) which has been thoroughly described by Diemont, Sissingh and Westhoff (1940), and the atlantic-mediterranean **Common Centaury -Yellow-wort community** (*Centaurio-Blackstonietum*) which Oberdorfer (1957) has met with in the Upper Rhine plain. The *Cicendietum* grows on fairly poor sandy, peaty or even loamy ground in the area of the Birch-Oak woods. Like the latter it fades away towards the east.

The **Yellow Galingale community** (*Cyperetum flavescentis*) should also be mentioned. It is found on wet fenland paths but has only been observed in the south-west, e.g. in the Swiss lowlands. It is the only *Nanocyperion* community of Central Europe which grows on neutral to alkaline soils (pH 6.6-7.8).

The most commonly encountered in western Central Europe is the **Bristle Clubrush-Bog Stitchwort community** (*Stellario uliginosae-Isolepidetum setacei*). It lives on roadways or paths which are only occasionally used in shady *Fagetalia* woods, although in the oceanic climate further north-west it can grow without shade. For example Diemont, Sissingh and Westhoff found it near cattle drinking places and on paths in pastureland, while Burrichter (1960) saw it along some storage dams in the mountains of Westphalia. So it is not the shade which confines this community to woodland paths in less oceanic parts of Central Europe, but the protection from drying out. The Bog Stitchwort community is generally very poor in species and often does not contain any good character species. Nevertheless it sould be considered as a separate association since it differs from the other *Nanocyperion* communities both ecologically and floristically. Its natural habitat could well be on deer tracks but it can develop fully on roadways used for the extraction of timber if these are waterlogged from time to time. When these tracks are no longer in use there appear, depending on the amount of light available, Water Pepper (*Polygonum hydropiper*) and other representatives of the Trifid Bur-marigold community, or plants of woodland clearings, followed quite soon by shrubs and trees. Some species from the neighbouring woodland or grassland communities are also to be found in the Bristle Clubrush-Bog Stitchwort community, but these should be looked upon as relicts, left behind from a previous community, or as pioneers; they have nothing really to do with the therophyte community.

Tab. 123. A synopsis of the summer annual nitrophilous pond and river bank communities and their character species.

Class and order character species of the **intermittently flooded nitrophilous annual communities** (Bidentetea, compare section E III 1, no. 3.2)

Alopecurus aequalis (tending more to 1)
Bidens radiata
B. tripartita
Polygonum mite

Character species of the alliances:

	1. **Bur Marigold communities** (ruderal) (Bidention tripartitate)		2. **Goosefoot communities** (near natural) (Chenopodion rubri)
	Bidens cernua		*A triplex hastata* (?) (only in mealy-leaf form!)
	B. connata		*Bidens frondosa*
	Catabrosa aquatica (?)		*Chenopodium glaucum* (?)
	Leersia orizoides (?)		*Ch. rubrum* (?)
B	*Polygonum hydropiper* (üppig)	P	*Polygonum lapathifolium* ssp. *danubiale*
B	*P. minus*	X	*Xanthium albinum*
R	*Ranunculus sceleratus*		
	Rorippa amphibia (?)		
R	*Rumex maritimus*		

Some of these are at the same time characteristic of several associations:

B	Water Pepper-Bur Marigold community (*Polygono-Bidentetum*) on pond mud, recently removed ditch sediment and similar places	P	Knotgrass-Goosefoot riverside comm. (*Polygono danubiale-Chepodietum rubri*) in western Central Europe
R	Celery-leaved Crowfoot comm. (*Rumici-Ranunculetum scelerati*) on similar sites, but even richer in salts and nitrogen, by cattle drinking places	X	Cocklebur-Goosefoot riverside comm. (*Xanthio albini-Chenopodietum*), distributed more towards the east.

Source: After Poli and J. Tüxen (1960) and Oberdorfer (1970).

2 Nitrophilous communities on the edges of standing or running waters

(a) Semi-ruderal Trifid Bur-marigold communities

Where ponds and ditches contain nutrient-rich effluent e.g. in and below villages which lack modern sewerage, or near frequently used cattle drinking places, a luxuriant growth of annuals flourishes around the edges from which the water has receded in summer and where there is little treading. These communities constitute the Trifid Bur-marigold alliance (*Bidention tripartitae*, see fig.483 and tab. 123, left-hand column).

As we saw in the previous section these strongly nitrophilous mud dwellers often grow in close contact with the dwarf plant communities occupying poorer habitats with a fluctuating amount of water, and in some cases they may replace these later in the summer. Their normal appearance however is the exact opposite of that of the inconspicuous *Nanocyperion* carpet: the dominant plants are more than knee-high sapgreen polygonums and composites growing so luxuriantly that we can hardly believe that they will only live for a single short summer. However if we were to wander through them in late autumn, after the first frosts we would see them merely as dried skeletons or as easily decomposed straw which has collapsed onto the cracked dry mud. At this time the boot-jack shaped fruits of the Trifid Bur-marigold cling fast to stockings and clothes.

Countrymen and animals help a good deal in dispersing these semi-ruderal weeds (i.e. ones which are assisted by man but also appear under natural conditions) near and far. Wet habitats which at the same time are rich in nutrients seldom occur in nature, and where they do they only occupy small areas, such as wallows and watering places of deer within flood plains or swamp woods, or in the vicinity of salt sources which are visited by many animals. Whether the habitats are natural or artificial they must offer the

Trifid Bur-marigold community not only water and nutrients in plenty but also good light and bare ground on which the seeds can get a good start. Otherwise there will very soon be competition for space from perennial plants which germinate at the same time or even later. In this respect they behave like the dwarf plant communities and the likewise short-lived ruderal communities of drier habitats which we shall discuss in section VIII.

In its narrow sense the *Bidention* alliance contains several associations of which two are commonly met with in Central Europe, the **Water Pepper-Trifid-Bur-marigold community** (*Polygono-Bidentetum*, fig. 483) and the **Celery-leaved Buttercup community** (*Rumici-Ranunculetum scelerati*). Both are often found incomplete because the habitats are unstable and sometimes the conditions are not exactly right; and both occupy similar habitats, especially the nutrient-rich mud of ponds and pools which dry out in the summer, the spoil from ditches carrying a fair amount of effluent, or equally fertile and wet waste land. They have also been seen in the vicinity of seabird nesting places (Niemi 1967). However the *Rumici-Ranunculetum scelerati* appears to require a higher level of nutrients than the *Polygono-Bidentetum*, which in turn is more demanding than the *Nanocyperion* communities (see section I b). The Celery-leaved Buttercup community for example grows well on sludge beds of sewage disposal farms excessively rich in nutrients, which most species find intolerable. *Ranunculus sceleratus* and the less common *Rumex maritimus* can also tolerate brackish water (see section C VI 2). However there are no results available at present of any exact investigations into the chemical factors and other habitat conditions of these communities.

In the Göttingen area Grosse-Brauckmann (1953b) reported that around 1950 the Water Pepper-Trifid-Bur-marigold community only grew well in the backward villages. Today it has disappeared from practically all the settlements in this area now that the drainage systems have been improved and extended. Increasing urbanisation is also driving them out of their former habitats throughout the whole of Central Europe.

(b) More or less natural Goosefoot riverside communities

While the habitat of the Trifid Bur-marigold communities in the hamlets and villages is shrinking the level of nutrition enjoyed by the very similar riverside communities is steadily improving. So much sewage from settlements and effluent from heavily manured farms is finding its way into the majority of streams and rivers in Central Europe today that the ground exposed when the water level falls in summer is becoming increasingly rich in plant nutrients.

Poli and J. Tüxen (1960) have grouped together the short-lived nitrophilous riverside communities into the alliance of the Goosefoot riverside communities (*Chenopodion fluviatile*). They are dominated by the more demanding species of Goosefoot together with special forms of Orache (*Atriplex hastata*). There are also species of Bur-marigold and Knotgrass which the *Bidention* and the *Chenopodion* have in common (see tab. 123).

In the western part of Central Europe, e.g. by the Weser, one comes across the **Knotgrass-Red Goosefoot riverside community** (*Polygono danubiale-Chenopodietum rubri*), see fig. 482) which has been thoroughly investigated by Lohmeyer (1950), along with a few other less common ones. By the time we get to the Elbe and rivers further east this gives way to the **Cocklebur-Goosefoot riverside community** (*Xanthio albini-Chenopodietum*). The rivers which are fed mainly from the high mountains and have their highest water levels in summer offer different conditions. For this reason the *Polygono-Chenopodietum* described by Moor (1958) is not identical with that in north-western Germany. This annual community which germinates on the gravel banks of rapidly flowing mountain streams has to rely on the arable weed and rubble communities in the flooded and catchment areas for its fresh supplies of seed. It thus takes on a much more anthropogenic character.

Fig. 483. A nitrophilous Water Pepper-Bur Marigold bank beside a drainage ditch to the northwest of Bratislava. About one tenth natural size. Photo. Iltis-Schulz.
Bidens tripartita (left), *Polygonum hydropiper* (in front and behind), *Rumex conglomeratus* (right) and *Ranunculus repens* can be seen.

Lohmeyer (1950) and R. Tüxen (1050a) maintained that the Goosefoot riverside communities growing in the lowlands far away from the mountains are part of the natural vegetation. However Poli and J. Tüxen (1960) argue against this, pointing out that these communities are well developed only where the waterways pass through arable land and pastures so that they receive a good deal of manure as well as seeds from the arable weeds. On the other hand along the Hunte river, whose course runs mostly through unmanured meadows, there are practically no Goosefoot riverside communities in habitats which would be quite suitable physically. As was emphasised in the introduction, sewage getting into the waterways also plays an important part. Most of the character species and accompanying plants in the *Chenopodion fluviatile* require an abnormally high level of nutrition if they are to grow quickly, reach the large size typical of them and complete their development all within a short summer season. Another argument against the mainly natural origin of these communities is that the Central European rivers have only carried large quantities of nutrient-rich clay and organic mud since the extensive settlement by human populations had taken place (see section B VI b); and it is only when the river banks are covered with this mud that they offer a suitable substrate for the development of Goosefoot communities.

Although at first sight the *Bidentetalia* communities alongside running waters do look very natural they must then be looked upon as 'ruderally influenced'. The question as to whether the *Chenopodion fluviatile* alliance is one of the places from which our ruderal and arable weed communities have drawn their components or whether the opposite is the case can never be reliably answered today (Krause 1956). Probably both directions have been taken by different species. *Polygonum lapathifolium* ssp. *danubiale* is an old resident in our flora and confined to the river edges. *Chenopodium glaucum* and *rubrum* as well as *Atriplex hastata* may also have been growing there for a long time along the line where the flotsam is deposited, i.e. in locally very fertile and open habitats. *Chenopodium polyspermum* is an archeophyte and was brought to Central Europe by man, probably in prehistoric times. *Bidens frondosa* like most of the Goosefoot species is a neophyte which has made use of the 'migration routes' along river banks (see section B V 1 j). *Xanthium albinum*, a very demanding subtropical species, only arrived here in the nineteenth century (see section VIII 1 a). On the other hand *Xanthium strumarium* counts as an archaeophyte in the south-east of Central Europe (Krippelová 1974). The original limited number of the species therefore in the Goosefoot communities (and similarly in the Bur-marigold communities along still waters) has gradually been added to with the help of man. Since man has also increased the fertility of the habitats and, by building groynes and narrowing the rivers, has increased the water level fluctuations, he has created more sites which these semi-ruderal communities can occupy.

VIII Ruderal communities of drier sites

1 Summer and winter annual ruderal communities

(a) Origin and development of the ruderal flora

Ruderal communities on building rubble, rubbish tips, roadsides and similar relatively dry habitats are even more heavily dependent on man than the nitrophilous weed vegetation on more or less wet soils. The term 'ruderal' was first used by botanists of the last century and comes from the plural form of the latin word 'rudus' meaning rubble, ruins or heaps of mortar. The opposite word was 'segetal' communities meaning the weed stands of arable land and gardens. Krause (1958) proposed using the word ruderal to describe all weed communities, even those of flooded ground (see section D VI 2) and those on woodland clearings (D III 2). We shall restrict outselves to the original narrower sense of the word.

The ruderal communities in this particular sense can be divided into two large groups, the short-lived (*Sisymbrion*, tab. 124) and the more or less perennial (*Onopordetalia*, tab. 124). Both include numerous vegetation units in which the climatic and soil conditions are reflected as clearly as they are in woodland, heaths, arid grasslands and other communities on soils with a low water table.

Before the ruderal flora was investigated more closely many people were of the opinion (Rübel 1926, quoted by Ellenberg 1963, is an example) that they were the result of accidental distribution and hardly merited study by phytosociologists or ecologists. We now know that their species composition responds as sensitively to differences in the habitats as does that of natural communities, thanks to the work of Tüxen (1937, 1950a and b), Lohmeyer (1950), Oberdorfer (1957), Grosse-Brauckmann (1953a and b), Düll and Werner (1956), Krause (1956, 1958), Ubrizsy (1956), Oberdorfer et al. (1967), Tillich (1969b) and many others.

Not even the new arrivals some of which almost every year succeed in gaining a foothold in Central Europe, are allowed to spread indiscriminately. They can fit into the framework of the ruderal communities only after being selected by the factors operating at the various habitats and by the competition of the flora

Tab. 124. Short- and long-lived ruderal communities on soils poor to rich in nitrogen in relatively dry sites of Central Europe, together with their character species.

Weakly nitrophilous	**Fairly** nitrophilous	**Strongly** nitrophilous
A Short-lived pioneer communities		
– (The first settlers always find sufficient nutrients on building and rubbish tips and the like as well as on roadsides or road and railway embankments)	**1 Goosefoot pioneer community** Chenopodietum ruderale (raw soils in towns) *Amaranthus albus* *A. hybridus* *Chenopodium strictum* *Ch. opulifolium* *Nicandra physaloides* *Xanthium strumarium*	**2 Goosefoot-Mallow comm.** Urtico-Malvetum (raw soils in villages) *Chenopodium murale* *Ch. vulvaria* *Malva neglecta* *Urtica urens* (?) (often leads to 14)
B More permanent annual communities		
3 Prickly lettuce comm. Conyzo-Lactucetum (frequent successor to 1) *Diplotaxis tenuifolia* *Lactuca serriola* (opt.) *Lepidium densiflorum*	5 Rocket-Orache comm. Sisymbrio-Atriplicetum (continental, otherwise as 3) *Atriplex acuminata* *A. oblongifolia* etc.	7 Madwort niche comm. Sisymbrio-Asperuginetum (at the entrance to limestone caves) *Asperugo procumbens* *Sisymbrium austriacum*
4 Wall Barley comm. Bromo-Hordeetum (building plots, sandy unploughed land) *Bromus sterilis* *Cnicus benedictus* *Hordeum murinum* *Lepidium graminifolium*	6 Submediterranean Rocket comm. Descurainietum (warmth loving) *Descurainia sophia* (?) *Sisymbrium loeselii* *S. altissimum* etc.	8 Lappula niche comm. Lappulo-Asperuginetum (as 7, subalpine) local: *Asperugo procumbens* *Lappula deflexa* (7 and 8 are very rare)
C Perennial hemicryptophyte communities		
9 Woolly Thistle comm. Cirsietum eriophori (calcareous roadsides) *Cirsium eriophorum*	**10 Scotch Thistle comm.** Onopordetum (in villages, ± warm-continental) *Anchusa officinalis* *Carduus acanthoides* *Echinops sphaerostachya* *Hyoscyamus niger* *Onopordum acanthium* *Verbascum densiflorum*	**15 Alpine Dock resting place comm.** Rumicetum alpini (cattle resting places, subalpine) *Rumex alpinus* (opt.) *Senecio alpinus*
11 Grey Cress comm. Bertroetum incanae (roadsides, embankments etc.) *Berteroa incana*		16 Subalpine Good King Henry comm. Chenopodietum subalpinum local: *Chenopodium bonus-henricus*
12 Viper's Bugloss comm. Echio-Melilotetum (limestone gravel etc.) *Echium vulgare* (opt.) *Melilotus albus* *M. officinalis* *Oenothera biennis (?)*	**13 Tansy-Mugwort comm.** Tanaceto-Artemisietum (embankments, roadsides etc.) *Artemisia vulgaris* (opt.) *Linaria vulgaris* (?) *Tanacetum vulgare*	**14 Good King Henry comm.** Balloto-Chenopodietum (sides of village streets) *Ballota nigra* (opt.) *Chenopodium bonus-henricus*

Note: [a] The arrangement of the classes has not yet been satisfactorily clarified; for this reason their character species have not been shown here. Each of the associations No. 1 – 16 contains, in addition to its character species, some of the species mentioned on the next page, as well as several 'companions' which are not confined to any particular type of community, e.g. *Poa* species, *Dactylis glomerata*, *Urtica dioica* or *Chenopodium album*.

Source: From data supplied by Lohmeyer, Tüxen, Oberdorfer, F. Runge and others. The names of the more common communities are in heavier type. A summary of the alliances (and orders) with their character species is to be found on the next page.

already present. Many of them perish while others may become character species for small or even large units. Among others H. Scholz (1960) has described the steps of these historical changes in the flora, using the example of the built-up areas of Berlin:

1 **Native** ruderal plants belonged to the Central European flora before it felt any influence of man. The majority of these are species belonging to the perennial or winter annual communties or the Bur-marigold community. Examples, arranged according to their families, include: *Urtica dioica*, *Rumex obtusifolius*, *Chenopodium glaucum* and *album*, *Stellaria media*, *Silene alba*, *Chelidonium majus*, *cynoglossum officinale*, *Galeopsis tetrahit*, *Verbascum thapsus*, *Tanacetum vulgare*, *Artemisia vulgaris*, *Carduus nutans*, *Cirsium arvense* and *Lapsana communis*.

2 **Archaeophytic** ruderal plants came in with the help of man but in ancient times. Members of the more or less permanent ruderal communities form the bulk of this group, but they are mostly therophytes, e.g. *Bromus sterilis* and *tectorum*, *Hordeum murinum*, *Urtica urens*, *Bilderdykia convolvulus*, *Chenopodium polyspermum*, *Echium vulgare*, *Sisymbrium officinale*, *Descurainia sophia*, *Melilotus albus* and *officinalis*, *Malva neglecta*, *Verbena officinalis*, *Ballota nigra*, *Onopordum acanthium* and *Arctium lappa*.

3 **Neophytic** ruderal plants have only found their way here in recent times, helped by the increased amount of traffic. Most of these are species which we meet in the short-lived ruderal communities (tab. 124). From herbarium specimens and references in the literature H. Scholz has subdivided this group into three periods which clearly bring out the continuing nature of this immigration:

a) 1500–1787: *Atriplex hortensis*, *Armoracia rusticana*, *Oenothera biennis*, *Leonurus cardiaca*, *Lactuca serriola* and many others.

b) 1787–1884: *Atriplex tatarica*, *Chenopodium opulifolium*, *Sisymbrium altissimum*, *irio* and *loeselii*, *Reseda lutea*, *Matricaria discoidea* and many others.

c) 1884–1959: *Atriplex nitens*, *Chenopodium botrys*, *Amaranthus albus* and *blitoides*, *Lepidium densiflorum*, *Solidago canadensis*, *Carduus acanthoides* etc.

Tab. 124 (Continued)

Alliances and orders with their character species[a)] (see also section E III 1, no. 3.31, 3.32 and 3.51)		
A and B, 1–8 **Rocket communities** sens, lat.: Sisymbrion, *Sisymbrietalia* (Class *Chenopodietea*)		
Anthemis austriaca	*Cardaria draba*	*Lepidium virginicum*
A. cotula	*Conyza canadensis*	*Malva sylvestris*
Barbarea verna	*Crepis tectorum* (?)	*Plantago indica*
Bromus arvensis	*Kochia laniflora*	*Sisymbrium irio*
B. tectorum (?)	*Lappual squarrosa*	etc.
C 9 and 10 **Scotch Thistle communities** sens. lat.: Onopordion, *Onopordetalia* (Class *Artemisietea*)		
Artemisia absinthium	*Cynoglossum officinale*	*Reseda luteola*
Asperula arvensis	*Marrubium vulgare*	*Stachys germanica*
Carduus nutans	*Potentilla intermedia*	*Verbascum blattarioides*
C 11 and 12 **Melilot communities:** Dauco-Melilotion, *Onopordetalia*		
Avena nuda	*Daucus carota* (?)	*Picris hieracioides*
Cichorium intybus	*Pastinaca sativa* (?)	*Rumex thyrsiflorus*
C 13 and 14 **Burdock communities:** Arction, *Artemisietalia* (Class *Artemisietea*)		
Arctium lappa	*Conium maculatum*	*Malva alcea*
A. minus	*Cruciata glabra*	*Parietaria officinalis*
A. tomentosa	*Dipsacus fullonum*	*Silene alba*
Armoracia rusticana	*Geranium pyrenaicum*	*Solidago canadensis*
Carduus crispus	*Lamium album*	*S. gigantea*
Chelidonium majus (?)	*Leonurus cardiaca*	etc.
C 15 and 16 **Alpine Rock communities** sens. lat.: Rumicion alpini, *Artemisietalia*		
Cerinthe glabra	*Cirsium spinosissimum*	

Even these few examples show that the Polygonaceae, Chenopodiaceae, Brassicaceae, Lamiaceae and Asteraceae are represented in practically all the groups. The Malvaceae, Onagraceae and Schrophulariaceae also play an eye-catching role as ruderal plants, whereas many other families are widespread in Central Europe but occur here very sparsely if at all. For example Scholz does not name any Cyperaceae (except *Carex hirta* which however is absent from true ruderal communities), Juncaceae, Liliaceae, Orchidaceae, Ranunculaceae or Gentianaceae since these families contain very few or no quick-growing nitrophiles. As expected the nutrition specialists which normally grow in impoverished habitats, e.g. Ericaceae and Pyrolaceae, or slow-growing succulents such as the Crassulaceae (apart from *Sedum telephium*) are completely absent.

These immigrant plants have a variety of origins but most of the archaeophytes and neophytes originate from lands with a warmer and drier climate than Central Europe. On the other hand many of the native ruderal plants have a distribution area extending far to the north (tab. 124). Along the Scandinavian coasts for example there are long mounds of seaweed which Nordhagen (1940) sees as the natural home of many ruderal plants and arable weeds. The natural vegetation of Central Europe offers almost no site where the plant cover would be open enough for ruderal plants to find a footing. As Fenner (1978) demonstrated by sowing experiments in a *Festuca rubra* turf the germination of ruderal plant species is greatly reduced when the turf is 8 cm tall, but is not appreciably reduced in a very short turf (1 cm tall). The establishment of *Verbascum thapsus* seedlings in an old field, according to K. L. Gross (1980) was limited to the most open plots; only in plots devoid of other plants did individuals survive to flower and produce seeds. Thus the main habitat factor for the development of ruderal plants is a loosely covered or bare soil.

(b) Short-lived communities on rubble and refuse dumps in towns

The short-lived ruderal plants had never found such extensive areas of fresh house rubble, their favourite habitat, as they did during the years 1943–45. These ruined sites began to green over surprisingly quickly. Although they had previously been completely free from plants the first seedlings of higher plants appeared on the heaps of bricks and plaster and amongst the charred timber within a very few months.

Most of these were composites and other wind-dispersed plants from the alliance *Sisymbrion*, or they came from woodland clearings e.g. *Epilobium angustifolium* which grew particularly luxuriantly on wood ashes. Rubble lacking organic matter offered an unfavourable substrate but nevertheless it had sufficient nutrient at first to allow some pioneer plants to make rapid growth. From the outset the conditions in these habitats varied according to the proportion of fine earth in the rubble and the extent to which it allowed water to drain through it, to the slope and aspect, and to the general climate. This meant that the spontaneous growth over the rubble in the different destroyed towns did not follow the same course.

In Berlin the succession of plant communities had a decidedly continental character, as Düll and Werner (1956) were able to show. Open types of vegetation remained dominant for years. Mesophilous meadow plants and mosses were very hesitant to come in, and only then where the local climate created a moister habitat. Garden escapes requiring a high fertility did not spread on to the rubble to any great extent, nor did the arable or garden weeds that had already been rare in the city centre. The succession to woodland on the rubble could make scarcely any progress as the work of clearance cut short the spontaneous spread and development of slower-growing species. On untouched places drought-resistant shrubs and trees were most successful, e.g. *Buddleia davidii* and *Robinia pseudacacia* (Kohler and Sukopp 1964a, b, Sukopp 19 72, Kunick 1974).

In the town of Münster in Westphalia on the other hand, which is much nearer the sea, there were more shrubs and trees coming in right from the start, according to Engel (1949). The first impression was created by a number of moisture-requiring ruderal species which were quite absent from Berlin. These were soon replaced by representatives of manured grassland (*Arrhenatheretalia*). Mosses covered even places in the full sun. Amongst woody plants demonstrating their natural superiority, woodland pioneers and garden escapes were well represented, especially the bird-dispersed species with pulpy fruits as well as many wind-dispersed ones such as *Betula pubescens*, *B. pendula*, *Populus tremula* and *Alnus glutinosa*.

Braunschweig, Dresden and Stuttgart lay somewhere between these two extremes as far as their rubble vegetation was concerned. Düll and Werner's comparison (1956, includes many references) brings out the differences. In each case however the vegetative occupation began with short-lived ruderal communities which we should like to talk about next.

According to Oberdorfer (1957) the first community to settle on dry refuse dumps in the towns of the

Upper Rhine area is the **Goosefoot pioneer community** *(Chenopodietum ruderale)* to which the warmth-loving *Amaranthus* and *Chenopodium* species belong (see tab. 124). Its summer annuals are soon replaced by hibernating annuals, i.e. ones which germinate in the autumn and remain green throughout the winter, such as *Lactuca serriola* and other character plants of the **Prickly Lettuce community** (*Erigerono-Lactucetum*). In Stuttgart these are joined by the **Rocket community** (*Sisymbrietum sophiae*). Düll and Werner have described a similar Rocket population as the commonest pioneer community of dry rubble heaps. Probably the Goosefoot species cannot come in here because of the lack of nutrients. *Atriplex nitens* plays a prominent part in Halle/Saale; Oberdorfer (1957) names this Orache as characteristic of a **Rocket-Orache community** (*Sisymbrio-Atriplicetum*, see tab. 124) with a decidedly eastern distribution.

(c) Short-lived communities in villages and in front of limestone caves

All the ruderal communities mentioned so far contain numerous neophytes (tab. 124 and sect. a.), a few archaeophytes and practically no species native to Central Europe. Apparently the plants settling on open stony rubble from houses and streets are a relatively young element in our flora and vegetation just as are most of the towns. On the other hand the corresponding village habitats are invaded mainly by archaeophytes.

The members of the **Dwarf Mallow wayside** *(Urtico-Malvetum neglectae,* tab. 124) however are not only older inhabitants but are more strongly nitrophilous than most of the ruderal plants in towns. They are bound to be absent from the larger towns for this reason alone. The rubbish dumps, yard corners, wall bottoms and similar places in villages contain more dung and other material rich in plant nutrients than those in towns where cattle are no longer free to wander about. Also the rubble from old village buildings is much richer in fine particles since in earlier times they were plastered with clay rather than mortar and many were built without stone at all. Their remains produced a fertile loamy soil able to hold a lot of water.

Probably even older than the Mallow-Stinging Nettle community of our villages are the peculiar **Madwort stands** at the entrances to limestone caves or overhangs of the southern Jura mountains and the Alps. Oberdorfer (1957) distinguishes between one in the colline-montane belts (*Sisymbrio-Asperuginetum*) and a subalpine *Lapulo-Asperuginetum*.

The Madwort communities like the Mallow-Stinging nettle, Goosefoot and Rocket communities are dependent on open ground which from time to time has been almost if not entirely cleared of plants. Where the vegetation is not repeatedly being destroyed perennial species gradually take up more and more ground until they are finally obliterated by more permanent ruderal communities, grassland or shrub and tree stands.

The **Wall Barley sward** (*Bromo-Hordeetum*, tab. 124) occupies a place in between the short-lived and the more permanent ruderal communities. It is found both in villages and towns but always on level or slightly sloping sandy ground, e.g. on arable land intended for building, along roads and paths which are little used, or on recently levelled out rubble. This is one of the few ruderal communities rich in grass; in the late summer its straw colour stands out from the other weeds most of which are still of a sappy green colour. In relatively cool and wet areas *Hordeum murinum* shows a predisposition for urban habitats; according to Davison (1977) this is probably due to the occurrence of warmer, drier microhabitats. Drooping Brome, Barren Brome and Wall Barley in particular can form quite a dense sward here, and they may persist for a number of years in their relatively poor habitat even though they must renew themselves each year from new seed. Finally however, the *Bromo-Hordeetum* changes, like the other *Sisymbrion* communities, into a grassland, a trodden sward or a ruderal community dominated by biennials and perennials.

2 Perennial ruderal communities

(a) Warmth-loving Thistle and Viper's Bugloss communities

The more or less perennial ruderal communities occur in many more forms than the short-lived ones. They can be divided into three alliances according to their climatic requirements (see tab. 124): the Cotton Thistle communities (*Onopordion acanthii)* which are warmth loving and relatively less sensitive to drought; the more mesophilous burdock communities (*Arction);* and the ruderal and cattle resting place communities of the high mountains which in part are related to the tall-herb communities *(Chenopodion subalpinum)*. The three groups have so many species in common that they can be put together into the order *Onopordetalia,* (tab. 124, C 9–12).

The *Onopordion* communities lie next to the *Sisymbrion* which we have just discussed. They too occupy free-draining sites, where activities of man protect them for a long time from the entry of woodland plants. Tillich (1969b) found them near Potsdam even on waterlogged refuse. Frequently the *Onopordion* communities develop from pioneer stages belonging to the *Sisymbrion*.

According to Düll and Werner the **Viper's Bugloss community** (*Echio-Melilotetum*, tab. 124, fig. 484) has taken over large areas from the Rocket community. Up to a short time ago it was the most common *Onopordion* community in Central Europe, occupying the ballast on railway embankments and storage yards provided these contained lime and had not been treated with chemical weedkiller. They can also be found in limestone quarries whenever more open drier waste is present and not just a wet marly floor. Presumably their natural habitat is in the lime-rich gravel of Alpine rivers when its surface is unusually high above the normal water level. They can also obtain a temporary footing on the screes below limestone cliffs in the colline and submontane belts before these are eventually occupied by grassland or shrubs. Both of these natural habitats however are very rare. In addition it must be borne in mind that many of the members of this community are not native so it can hardly be considered a part of our natural vegetation.

The Viper's Bugloss community is one of the finest of our ruderal communities both in colour and form. The intense violet-blue of the Viper's Bugloss itself shines through the tender white and yellow racemes of the tall Melilots, pushing up through it in the summer. Also loosely grouped may be the greenish-gold flowered Wild Mignonette, the pale yellow Evening Primrose scented at night, the nodding carmine heads of the Musk Thistle and other plants with colourful flowers or bizarre forms. Their mostly small leaves, their pale green colour and the unusually large presence of legumes leads one to presume that there is not much nitrogen available, and Kronisch (1975) confirmed this in comparison with other ruderal vegetation types.

The **Cotton Thistle community** grows more luxuriantly (*Onopordetum*, fig. 485, tab. 124). This is the type association of the *Onopordion* alliance. It reaches its best development on the deep fertile black earths (chernozem) of eastern and south-eastern Europe and in similar habitats in the middle and eastern parts of Central Europe. This community is very characteristic of the courtyards and village greens of the continental loess soils. Its appearance is dominated by thistles lower or even taller and stronger than *Onopordon* itself – *Carduus* and *Cirsium* species or the rare *Echinops*. Once they have occupied an area they hold on stubbornly, but it is surprisingly difficult for them to establish themselves on new sites. Krause (1950)

Fig. 484. Viper's Bugloss community along the dry edge of a pathway to the north of Bratislava. *Echium vulgare* (in front), *Pimpinella saxifraga* (right) etc. About one sixth natural size. Photo. Iltis-Schulz.

These and other ruderal plant associations are disappearing faster and faster today because of chemical weedkillers. According to Brandes (1981) their continuance is already threatened as are the *Onopordetum* (Fig. 485) and *Sisymbrion* communities.

carried out experiments on the seed and found that with most of the members germination is poor, or following germination the seedlings often died. Krause considered that a fungal disease which he had observed on the young plants had a lot to do with this. It is possible however that the danger from this would be less in the drier warmer habitats of the *Onopordetum* than in the temperate climate of the botanic garden where the investigations were carried out.

The **Woolly Thistle community** (*Cirsietum eriophori*, tab. 124), is found on limestone sheep grazings but has received little attention. It probably also belongs to the *Onopordion*. With its ornamental deeply dissected leaves and its long-haired pappus bracts the Woolly Thistle looks more like an ornamental plant than a weed. Like almost all the *Onopordion* representatives and by far the greater number of ruderal plants it is avoided by grazing animals. This is one of the main reasons why the villages in which cattle and sheep are still driven are

Fig. 485. A Cotton Thistle community by a garden wall in Gurdau (Moravia). *Onopordon acanthium* (right), *Leonurus cardiaca* (left), *Arctium minus* (broad leaves) etc., behind them *Sambucus nigra*. Photo. Iltis.

Fig. 486. The shoot and root systems of some of the ruderal plants to be found growing alongside village streets. After Grosse-Brauckmann (1953), modified.

Left (facing south): *Verbena officinalis*, *Chenopodium bonus-henncus*, *Artemisia vulgaris*, *Ballota nigra*.
Right (in shade): *Rumex obtusifolius*, *Urtica dioica* and *Lamium album*.

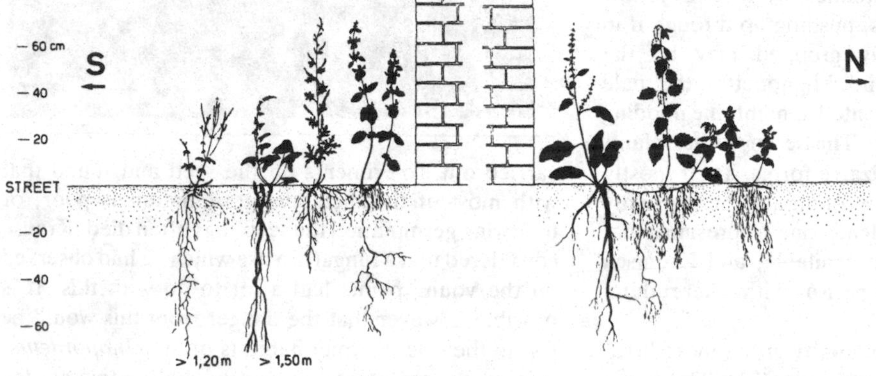

rich in *Onopordion* communities.

(b) Mugwort scrubs and other Burdock communities
In a moister cooler climate or where the soil is less free-draining Burdock communities (*Arction*, tab. 124, fig. 485) replace those of the Cotton Thistle. Only the Lesser Burdock (*Arctium minus*) will also thrive in the warm dry places where its large-leaved relatives (*Arctium lappa* and *tomentosum*) are rarely seen. A similar mesophilous impression is also given by the soft-leaved Greater Celandine, the shade-tolerant Nipplewort as well as by Hemlock, Horse-radish, Stinging Nettles, Dead Nettles and other members of the *Arction* communities (see fig. 486).

Investigations carried out by Grosse-Brauckmann (1953a) have shown however that in their numbers of stomata per unit leaf area and in their transpiration pattern these species are more like xerophytes than mesophytes. This is particularly true of *Artemisia vulgaris*, *Ballota nigra*, *Chenopodium bonus-henricus*, *Lamium album*, *Urtica dioica* and *Verbena officinalis*, but also of *Malva neglecta* and *Urtica urens* which belong to the *Sisymbrion*. Grosse-Brauckmann described the habitat of these ruderal plants as a ‘typical

Central European xerophyte habitat' insofar as the relative humidity could fall to a very low level because of the heat reflected from the house walls. When they are compared with the real dry-site vegetation of Central Europe however (e.g. *Xerobromion*, *Mesobromion*, *Festucion valesiacae* and *Onopordion* communities) it would be better just to refer to them as 'sun plants'. Even the driest of the sites tested by Grosse-Brauckmann on village streets near Göttingen corresponded at most to the *Prunella* subassociation of *Koelerio-Gentianetum* (see section D 1 2 b), i.e. to a relatively mesophilous community of *Mesobromion* (and not *Xerobromion*).

Nevertheless the fact remains that ruderal plants, even though they have the appearance of mesophytes, are able to live in relatively dry habitats and have a type of transpiration similar to that of xerophytes. In complete contradiction to this, of course, is the development of a large leaf area which in its surface area/weight ratio is more like that of shade plants. Grosse-Brauckmann rightly proposed that this tender leaf form was due directly to the large amount of nitrogen available. The ruderal plants then are an interesting contrast to the plants of raised bogs which show xeromorphic structures because of the shortage of nitrogen even though they live in wet ground (see section C III 4 a).

The most widespread community of the *Arction* alliance is that of the **Mugwort scrub** *(Tanaceto-Artemisietum)* which can exist for years on rubbish tips and builders' rubble as well as roadside embankments. It generally follows the *Sisymbrion* communities, especially that of the Wall Barley (tab. 124). Where village streets and wall bottoms are more fertile it is replaced by the **Dwarf Mallow wayside** (*Urtico-Malvetum neglectae*) and finally by the **Good King Henry community** (*Balloto-Chenopodietum*). The latter is mainly distributed in the western and northern parts of Central Europe. In areas with a more continental climate its place is taken by the **Motherwort-Downy Burdock community** (*Leonuro-Arctietum tomentosi*). *Leonurus cardiaca*, *Conium maculatum* and *Arctium tomentosum* serve as character species for this community. The most noticeable thing about it is the smell of absinthe given off by the grey leaves of the True Wormwood (*Artemisia absinthium*), a garden escape now finding a home here. This continental community has developed very well in warm-dry parts of Wallis (see Braun-Blanquet 1961).

Most ruderal habitats are rich not only in available nitrogen but also in phosphorus, potash and other nutrients which are always found where excrement, household refuse and other rubbish collect. Whether these other elements are required in such large quantities for the welfare of ruderal plants is not yet clear. For Stinging Nettle (*Urtica dioica*) P appears to be more important than N (Pigott and Taylor 1964). Investigations by Mayser (1954, communicated by Walter 1963) showed that the form in which nitrogen was supplied was immaterial. For the same N concentration ruderal plants given nitrate grew just as well as those which had ammonium (see fig. 492).

(c) Subalpine and alpine ruderal communities

The ruderal community of Good King Henry mentioned in the previous section is found in all belts right up to the highest settlements in the Alps. However as the altitude increases the companions of *Chenopodium bonus-henricus* change so that one must speak of different associations (Oberdorfer 1957). The subalpine community (*Chenopodietum subalpinum*) is nearer to that of the Mugwort communities lower down. Like the *Rumicetum alpini* therefore it is assigned to a subalpine-alpine alliance *Chenopodion subalpinum* (see tab. 124).

These subalpine communities appear to be more strongly mesophilous than the Mugwort communities. However no comparative studies have been made of their water or nutrient economies.

IX Weed communities of arable land, gardens and vineyards

1 General remarks about the weeds of cultivated soils

(a) Origin and growth forms of field and garden weeds

Today large areas of Central Europe are under cultivation as arable fields or gardens. As we saw in section A II 5 some of this land has been worked since prehistoric times. Equally old are the communities of 'weeds' i.e. wild plants which the farmer has had to contend with amongst his crops and many of which are still troublesome in spite of modern methods of combating them. Then as now the planted crops have never been able to occupy the ground completely, smothering out all the other species. Seeds of weeds still accompanying the cultivated plants have been discovered in archeological remains of Neolithic and Bronze Age settlements (tab. 125).

Many weeds are native to the European flora even though in the woodland-dominated natural landscape they only played an insignificant part. We have already met them in the short-lived herbaceous communities of dried up river beds, ponds and tide lines, on coastal sandbanks as well as in woodland clearings brought

Tab. 125. Arable weed species known to have existed in prehistoric times.

a In winter corn	b In hoed crops and gardens	c Other weeds (in a and b)
1. Previously cultivated species:		
S *Avena fatua*[a]	*Aethusa cynapium*	*Agropyron repens*
S *A. nuda*	*Chenopodium album* (→ 3)	*Fallopia convolvulus* J
Matricaria chamomilla J[a)]	*Panicum miliaceum*	
Valerianella locusta	*Setaria italica* J	
	S. viridis	
2. Only occuring in **weed** communities:		
S *Agrostemma githago* J	*Euphorbia helioscopia*	*Anagallis arvensis*
S *Bromus secalinus* J	*Fumaria officinalis* J	*Atriplex patula*
S *Centaurea cyanus* J	*Geranium pusillum*	*Myosotis arvensis*
S *Lolium temulentum* (in Flax fields)	*Lamium purpureum*	*Raphanus raphanistrum*
Galium spurium J	*Senecio vulgaris*	*Sinapis arvensis*
Vaccaria hispanica	*Sonchus asper*	*Thlaspi arvense* J
Valerianella dentata J	*S. oleraceus*	*Viola tricolor* ssp. *tricolor*
V. rimosa	*Stachys arvensis*	
	Vicia hirsuta	
	V. tetrasperma	
3. Also occuring in **natural** plant communities:		
(none)	*Chenopodium polyspermum* J	*Arenaria serpyllifolia*
	Polygonum lapathifolium	*Cirsium arvense* J
	P. persicaria J	*Galeopsis tetrahit* J
	Stellaria media J	*Galium aparine*
4. Moisture indicators in cereals and hoed crops, which also occur in natural plant communities:		*Lapsana communis* J
Mentha arvensis		*Polygonum aviculare* J
Polygonum hydropiper		*Rumex acetosella*
Ranunculus repens		*R. conglomeratus*
Stachys palustris		*R. crispus*
		Taraxacum officinale

Note: [a] Have become more widespread again today. All other cereal weeds are becoming very rare or have died out, especially the 'seed weeds' (S) which are involuntarily sown together with the corn.

Source: From data by Rytz (1949) and Lüdi (1955); arranged according to their present day sociological behaviour.
J = present in the New Stone Age; the rest in the Bronze Age.

about by fire or storm. They also turn up where bare earth is exposed by the collapse of steep river banks, or where the waves of the sea or large lakes have undermined the shoreline, where rockfalls and screes contain a proportion of fine soil or where waterlogged ground on steep mountain slopes has become unstable and formed a mud stream.

According to Krause (1956) before ever man came on the scene burrows of wild animals or Deer tracks could have enabled those plants which would have found it impossible to grow in woods or swampy meadows and other perennial dense plant formations, to survive and spread.

However there was no site to offer in the natural countryside of Central Europe resembling those of our arable land. So as in the cases of meadows and manured pastures we are dealing here with newly created formations. In fact these are even further removed from nature in that they contain many plants which would certainly be absent from our flora without the help of man (see Weinert 1973).

Some weeds of cereal crops, like the wild forms of our cereals and the methods of cultivation, came originally from the steppes, mountain steppes and semi-deserts of the Near East. Others have their home in the Mediterranean area and would have very little chance of surviving in Europe if the farmers did not create the necessary open spaces for them. Following the modern increase in world traffic an increasing number of species from overseas have become established, especially from North and Central America. The immigration of elements from foreign floras is still

Fig. 487. The weeds in a Rye stubble on acid loamy sand in the Bohemian-Moravian upland with *Aphanes arvensis* (at the very front) *Rumex acetosella* (rosette), *Anthemis arvensis* (right and far left), *Sagina procumbens* (left) and *Cerastium holosteoides* (centre). The photograph covers a width of about 40 cm. Photo. Iltis-Schulz.

going on. For example a composite, brought in about a hundred years ago, *Galinsoga ciliata*, has only begun to spread in the western part of Central Europe since 1950 but today is well established in many areas. Such newcomers gain a footing on arable land and in gardens because they always find bare soil and are protected from the competition of most of our perennial native plants. In order to survive over the long term as an **arable or garden weed**, i.e. as a plant which is undesirable from man's point of view, it must possess the following **attributes**:

1. If it is a short-lived therophyte it must develop rapidly from germination to fruiting.
2. If it is not able to germinate or grow at any time then its rhythm of development must fit in with that of the crop and the cultivations it receives, while at the same time being adapted to the climatic rhythm of the district.
3. It must be in a position to tolerate shade for a time or, like the Bindweed, Vetch or Pea, to escape from it by climbing (figs. 488–490).
4. Whether it is an annual or perennial it must be able to regenerate easily following mechanical damage or when clods of earth have been moved to cover it.
5. Today, following the increased use of herbicides (chemical weed killers) during the past 30 years, it must have some way of circumventing these by later-germinating individuals or by the ability to regrow.

Fig. 488. Light intensities throughout the course of a day near the ground under stands of various cultivated plants. After Tranquillini (1960), somewhat modified.

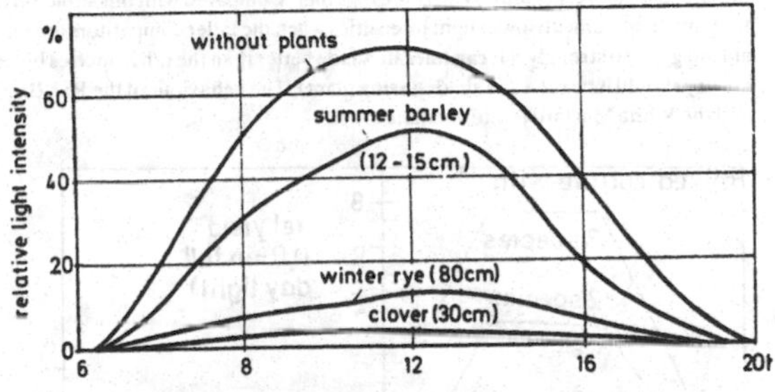

Fig. 489. The reduction in light intensity going vertically downwards through a standing crop of Winter Rye (with mainly narrow leaves not held horizontally) and of hemp (with a layer of broad, more or less horizontal leaves). After Rademacher, in Tranquillini (1960).

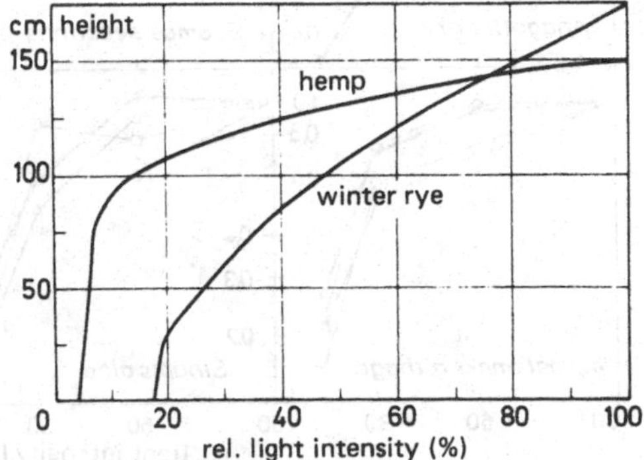

Ellenberg (1950) has divided all the weeds, therophytes as well as hemicryptophytes and geophytes, into four groups depending on the time of year when the leaves are assimilating:

f **Springtime plants** – all their growth, leaves and flowers, takes place in spring, and by midsummer they are dried up (e.g. the therophyte *Veronica triphyllos,* as well as *Ornithogalum umbellatum, Tulipa silvestris* and other rare geophytes of mediterranean origin);

v **Green from early spring** – also germinating or shooting in the early spring but remaining green throughout the year and carrying on their assimilation into the autumn. Their leaves are killed by a sharp frost (e.g. *Chenopodium album* and *Agropyron repens);*

Fig. 490. Effect of different levels of illumination on the dry matter production of four arable weeds grown in pure culture and also in competition with one, two or three other species. After Bornkamm (1961), modified.

In **pure culture** all the species thrive best in full daylight and worst when the light is reduced to the greatest extent.

In **competition** with other species the tall-growing Corn Cockle (*Agrostemma githago*) keeps on top and maintains about the same rate of production as in the pure culture. At the other extreme is the Scarlet Pimpernel (*Anagallis arvensis*). It is such a low-growing plant that it is strongly suppressed in full light by each of the other plants, and especially by all three together. Compared with this situation it fares much better with lower light intensities when the taller competitors are not able to grow so strongly. (It **can tolerate** shade better than the other species but in no way should it be seen as a shade-**loving** plant). The behaviour of the Rye Brome and the White Mustard is intermediate.

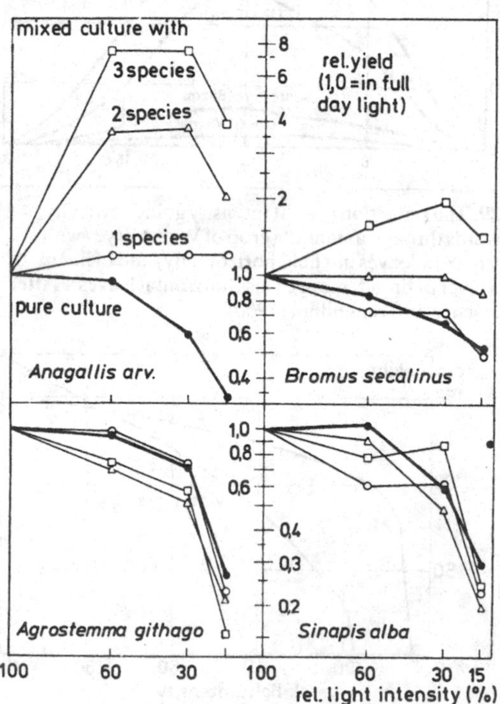

s **Summergreen** i.e. therophytes which do not germinate until the late spring or early summer and overwinter as seed (e.g. species of *Setaria* and *Panicum*, *Solanum nigrum*). This group also includes geophytes starting out late into growth (e.g. *Convolvulus arvensis*)

w **Overwintering** i.e. therophytes withstanding the winter in their summer dress (e.g. *Stellaria media*, *Capsella bursa-pastoris*, *Senecio vulgaris*); also many perennials (e.g. *Ranunculus repens*).

The strength with which each life form (therophyte, geophyte and hemicryptophyte and their various subgroups) is represented in the weed population depends on the intensity of cultivation and the rotation of crops. Hemicryptophytes are the most successful in long-term crops such as Clover or Lucerne whereas in hoed crops therophytes and geophytes do best.

Since the therophytes are often destroyed by harrowing, hoeing and harvesting or by herbicides before their seeds are ripe it is an advantage to them if their seeds do not all germinate at the same time, but a proportion retain their viability over as long a time as possible. For this reason species with an irregular germination dominate the arable weeds. In the case of a few tall weeds (e.g. Corncockle, *Agrostemma githago*) man has been carrying on an unconscious selection over the course of some thousands of years, establishing in them properties which have worked against him. According to Thellung (1925) and others such **'features of cultivated plants'** include:

1 No delay of germination, i.e. all seeds develop promptly after sowing (e.g. *Agrostemma githago* and *Bromus secalinus*),
2 Loss of viability after less than a year even when kept dry (this is the case with the species named in 1),
3 Failure of natural means of dehiscence or seed dispersal (e.g. *Camelina alyssum*, a weed of Flax, whose fruits will only open under pressure, *Agrostemma* where the seeds do not leave the capsule easily, and *Avena nuda* where the grain is not shed spontaneously),
4 Large seeds (or other propagules of similar size) which are difficult to separate from those of the cultivated plant (e.g. fruits of *Camelina alyssum*, one-seeded fruit joints of *Raphanus raphanistrum*).

Agrostemma githago, *Bromus secalinus* and other species whose seeds lose their power of germination after just a few months in the ground rely on being sown with the crop each year. In the past few decades machines for cleaning seed have become more and

Tab. 126. Plant growth and seed reserves in a three-year-old social fallow on acid ranker brown earth in the Dill district at 380 m above sea level.

	Species found mainly in	A	S
	Weed communities:		
d	*Centaurea cyanus*	2	
d	*Bromus secalinus*	+	
	Sonchus asper	+	
	Apera spica-venti	3	2
	Tripleurospermum inodorum	2	64
	Myosotis arvensis	+	105
	Viola tricolor	+	104
	Chenopodium album	+	35
	Vicia hirsuta	+	3
	Galeopsis tetrahit	+	2
	Raphanus raphanistrum	+	2
	Scleranthus annuus	+	2
d	*Veronica persica*		66
d	*Arabidopsis thaliana*		49
d	*Gnaphalium uliginosum*		40
	Capsella bursa-pastoris		6
	Spergula arvensis		5
	Thlaspi arvense		1
	Lamium purpureum		1
	Various combinations:		
v	*Agropyron repens*	1	
	Galium aparine	1	
v	*Linaria vulgaris*	+	
	Verbascum nigrum	+	
	Plantago major	+	
	Poa annua	+	12
v	*Cirsium arvense*	1	8
	Rumex obtusifolius	+	2
	Polygonum aviculare	+	1
	Fallopia convolvulus	+	1
	Rumex crispus		11
v	*Rumex acetosella*		2
	Veronica hederifolia		2
	Woodland clearings:		
	Epilobium montanum		37
	Epilobium tetragonum ssp. *lamyi*		28
	Gnaphalium sylvaticum		4

	Species found mainly in	A	S
	Grassland:		
	Dactylis glomerata	+	
	Phleum pratense	+	
v	*Alopecurus pratensis*	+	
	Anthoxanthum odoratum	+	
v	*Leucanthemum vulgare*	+	
	Ranunculus acris	+	
	Alchemilla vulgaris	+	
	Cirsium palustre	+	
	Trifolium pratense	+	
	Trifolium hybridum	+	
v	*Trifolium repens*	+	5
v	*Agrostis tenuis*	3	590
v	*Poa trivialis*	3	370
	Holcus lanatus	2	104
v	*Festuca rubra*	+	4
v	*Galium mollugo*	+	2
v	*Poa pratensis*		25
	Hypericum perforatum		6
	Hypericum maculatum		1
	Veronica serpyllifolia		1

Note:

A – plants actually growing on a 100 m² plot after a previous arable field had lain fallow for three years. The figures indicate the cover degrees according to Braun-Blanquet.

S = number of seeds which have germinated from soil samples taken to a depth of 10 cm on a total area of 800 cm² (1 = 12.5 per m²).

v = species with an efficient method of vegetative reproduction.

d = quickly decreasing 'seed weeds' and other annuals.

There are still no **woody** plants!

Source: From data in Borstel (1974).

more efficient, and these cereal weeds, previously universally common have now practically disappeared from Central Europe. Species like *Centaurea cyanus* behave in a similar way but their seed tends to retain its viability a bit longer. During the last few decades these 'seeded weeds' too have fallen victim to the improved methods of seed cleaning; the Blue Cornflower which at one time could never be got rid of has today become a botanical rarity, especially as it is easily killed by herbicides.

We should be conscious that weed communities have altered considerably over the past centuries. As already pointed out in section A II 4, in prehistoric times right through to the Middle Ages with the three-field system, the weed communities must have resembled grassland. Every year a portion of the land was not cultivated but grazed. Also with the primitive ploughs, especially the hook plough, the ground could be broken up but not turned over. Therefore short-lived weeds did not have enough open space to develop abundantly. It is true however that even by the tenth century there were some 'cereal weeds' (e.g. *Agrostemma*, *Anthemis arvensis*, *Centaurea cyanus* and *Neslia paniculata*) as well as 'weeds of intercultivated crops' such as *Aethusa cynapium*, *Euphorbia helioscopia* and *Solanum nigrum* (Willerding 1973). Weed communities resembling those of the Middle Ages may only be met with today in the Black Forest, in parts of Schleswig-Holstein, and in other regions where a grass and clover ley is included in the rotation. Here some meadow plants are still living as arable weeds; the majority of them however cannot stand repeated use of the complete turning over of the topsoil. Since effective ploughs have been introduced the proportion of short-lived species in arable weed communities has gradually increased, especially those like *Thlaspi arvense* and *Capsella bursa-pastoris* which are able to germinate at any time of the year (Salzmann 1939).

Renewed changes in the species composition have taken place as a result of the use of chemical weed killers and modern harvesting machines. Since the corn must stand longer in the field to become ripe enough for the combine harvester this allows many species of weeds more time in which to ripen their seeds. The Wild Oat, *Avena fatua*, for example had almost been exterminated during the decades before 1960, but actually is making a come-back as a troublesome weed, and has even extended its range into areas where it was previously unknown. In general however things have been made much more difficult for nearly all of the weeds. Their communities have become poorer both in number of individuals and of species, some of which now are rarities (see section 2f). The minimum area covered by the majority of weed communities has increased from about 25–50 m^2 to more like 200–500 m^2. In spite of this spreading uniformity they can still be used today as indicators of site conditions.

(b) Weed communities as regular species combinations

In contrast to woods, mires, grasslands and other natural or semi-natural plant formations, segetal as well as ruderal communities have been considered for a long time as being chance meetings of species which it would not pay to investigate more closely in a scientific way. However since the work by Volkart (1933), Buchli (1936), Salzmann (1939), Tüxen (1937, 1950a), Ellenberg (1950), Sissingh (1950) and many others there is no longer any doubt that arable weed communities too are determined in a regular manner by the environment and that, in spite of their short existence and the many ways in which they are constantly being fought, they have a long-term structure. They can be grouped into vegetation types just as well as can the meadows, pastures, forest plantations or other species combinations which owe their existence to man's activities, and they are scarcely any more difficult to characterise than many of the natural communities.

When the soil is broken up or a seed bed has been freshly prepared for a new crop it gives the appearance that the arable land has no life and that the weed community would have to come in again afresh. In truth however it is always present even though for the time being it is in the form of seeds or of subterranean organs able to regenerate. In this respect too there is no fundamental difference between arable weed and many natural plant formations, e.g. herbaceous vegetation of a dried up river bed or semi-desert dominated by herbs and therophytic grasses. Even the ground under many broadleaved woodlands appears to be dead in winter. Nevertheless the plants are there, ready to brighten it up again in the spring. It is true though that from the latent plant material in the topsoil of an arable field different species combinations will arise from one year to the next depending on what crop is being grown and which ground preparations and weed control methods are put into operation.

The stock of seed and other propagules from which the weed community is recruited is often surprisingly large according to records made by Korsmo (1930), Wehsarg (1954), Borstel (1974) and others. Usually several hundreds have been counted from each m^2 and in individual cases this number has risen to over 5000 (see tab. 126). Most of these seeds retain their viability for several years as the investigations of Kozma (1922), Korsmo (1939), Salzmann (1939) and others have shown. Many have survived as much as several decades during which the former arable field had been used as grassland or tree plantation or received a correspondingly long period of chemical treatment, and then have surprisingly reappeared. It is only when they are deeply buried – according to Kozma more than 50 cm deep – that the seeds of these persistent species die more quickly because of lack of oxygen.

When a former woodland, a drained mire or any other area which has not previously been under cultivation is ploughed up the new land is at first almost entirely devoid of arable weeds. If there are no weed communities in the vicinity it takes a long time until a rich weed community appropriate to the new habitat is formed. The first to arrive are species of the genera

Cirsium, *Sonchus*, *Senecio* or *Taraxacum* which have wind-dispersed seeds. Others whose seeds are capable of withstanding the putrifying bacteria arrive with dressings of farmyard manure, e.g. *Chenopodium album*, *Stellaria media* and many leguminous plants. Yet others may be brought in on the feet of wild or domestic animals or clinging to the wheels of farm machinery or sown with the seed if this has not been very efficiently cleaned. However according to J. Tüxen (1958) the species composition still indicates after a number of decades that this is newly broken land.

Even if the arrival of the seeds is 'by chance' in the first instance the habitat soon begins to make a selection. Of the total of some 300 arable weed species in Central Europe only about 15–40 find themselves growing together in a particular site to form a community. This selection is more than anything a result of competition. Soon after they come up the seedlings are crowded together; it is rare that the weedkiller or cultivation leaves a single plant to grow alone, and so competition between neighbouring plants still goes on. The keenest competition however for the weeds is from the crop which is being assisted by the farmer (see figs. 488 and 489). This plays a significant part in deciding which of the weeds will succeed.

Just as in the case of broadleaved woodlands, the amount of light under shooting cereals or a rapidly growing root crop falls off as the summer progresses (see fig. 488). Also like the trees the crop deprives the plants living underneath it of water and nutrients. The quicker the crop forms a closed stand and the taller it grows the more the weeds are adversely affected. Spring Barley for example suppresses the weeds less than Rye, which in its turn is not so effective as Red Clover, Hemp or other dicots (figs. 488 and 489).

With regard to light intensity, soil acidity and other factors many of the arable weeds are forced by competition into habitats offering less than their optimum requirements (see fig. 490). For example all the species tested experimentally by Bornkamm (1961b) grow best in full light. In a mixed culture however the taller plants such as Corncockle and White Mustard suppress the low-growing ones so that with the combined competition of the crop and taller weeds the latter have to put up with relatively dark places. They do not need the shade but tolerate it to a certain extent.

Several inorganic factors operate either directly or through the competition of other plants on the species composition of the weed community. The chief of these are the dampness of the soil and its lime content, although the availability of nitrogen and other chemicals show up more or less strongly. Climatic peculiarities, especially the temperature, also have their effect. This is true even today, i.e. after decades of chemical weed killing (see section f).

Since many of the species are short-lived the arable weed communities react more quickly than other plant communities in Central Europe to the changes in the habitat factors which take place from one year to the next. A wet spring for example will favour plants germinating best under moist conditions, whereas a dry summer helps the deep-rooting perennial weeds since the annuals are not able to ripen so much seed as during a normal season (F. Koch 1955, see tab. 127). A heavy manuring will enable the more demanding plants to spread whereas under poor fertility they would be stunted and lose ground (Zoldan 1981). Above all though it is the crop rotation and the different sequence of cultivation and weed control methods which have the greatest effect. One can thus speak of different 'phenological aspects' arising from the permanently present weed community part of which is latent in the soil in the form of seeds and subterranean perennating organs (Rademacher 1948). In spite of intermediate forms these can be clearly differentiated and so they deserve to be described in detail.

2 Arable weed communities and their habitats

(a) Weed communities under winter and summer crops

The classification of arable and garden weed communities has been changed a number of times over the past 50 years. However most phytosociologists have always been agreed on one point: they divide them into two main groups, separating the tilled or 'root' crops from the cereals or 'straw' crops sown in the autumn. Maize and summer cereals such as oats, with respect to their weed vegetation, incline to the first group. This division still represents the highest rank of the systematic hierarchy in many publications though more and more authors discard it. Oberdorfer (1979) distinguishes two classes with two and three orders respectively:

1 **Weeds of winter cereals** *(Secalietea)*
  on acid soil *(Aperetalia)*
  on calcareous soil *(Secalietalia)*
2 **Weeds of summer crops and ruderal communities** *(Chenopodietea)*
  Summer crop communities *(Polygono-Chenopodietalia)*
  Short-lived ruderal communities *(Sisymbrietalia)*
  Perennial ruderal communities *(Onopordetalia)*[1]

([1]*Note. This applies to a temperate climate. The tilled weed community of warm-dry climates (Eragrostietalia)* is barely represented even in the warm Upper Rhine plain.)

Tab. 127. Effects of the unusually dry weather conditions in 1952 on the arable weeds in long-term manurial trials.

Badly affected (1952 < 70% of 1951)		**Moderately** affected (1952 70–90% of 1951)		**Hardly** affected at all (1952 > 90% of 1951)
Anthemis arvensis	P	*Agropyron repens*		*Anagallis arvensis*
Aphanes arvensis		*Capsella bursa-pastoris*	B	*Arenaria serpyllifolia*
Arabidopsis thaliana	P	*Equisetum arvense*		*Chenopodium album*
Conyza canadensis		*Galeopsis tetrahit*	P	*Cirsium arvense*
Euphorbia exigua		*Galinsoga parviflora*	P	*Convolvulus arvensis*
Geranium dissectum		*Galium aparine*		*Fallopia convolvulus*
Gnaphalium uliginosum		*Polygonum persicaria*	P	*Mentha arvenis*
Lamium amplexicaule		*Senecio vulgaris*		*Papaver rhoeas*
Lamium purpureum	P	*Sonchus arvensis*	B	*Plantago major*
Matricaria chamomilla		*Stellaria media*	B	*Taraxacum officinale*
Myosotis arvensis		*Thlaspi arvense*		*Viola arvensis*
Poa annua		*Veronica persica*		
Raphanus raphanistrum				
Sinapis arvensis				
Sonchus oleraceus				
Veronica hederifolia				
Vicia angustifolia				
Vicia hirsuta				

Note:
P = perennial species
B = biennial or perennials (all the rest are short-lived summer or winter annuals).

Source: After F. Koch, modified. Herbicide had not been applied in 1951/52.

Thus he brings the ruderal communities of relatively dry ground, which have already been dealt with in section VIII, into a closer relationship with the weed community of, for example, a turnip field than with that of winter wheat which is occupying the same field the following year. The close connection with the ruderal communities lies more than anything in the fact that numerous nitrogen indicators also do well in a root crop or in a garden both of which are as a rule manured more heavily than a cereal crop. Amongst the character species of the class *Chenopodietea* are:

Aethusa cynapium	*Chenopodium album*
Amaranthus albus	and other *Ch.* species
and other *A.* species	*Capsella bursa-pastoris*
Echinochloa crus-galli	*Tripleurospermum inodorum*
Portulaca oleracea	*Senecio vulgaris*
Senecio vernalis	*Sonchus oleraceus*
Solanum nigrum	*Stellaria media*
Urtica urens	etc.

Other nutrient-demanding species can serve as characteristic of the **summer crop weed communities** (*Polygono-Chenopodietalia*):

Chenopodium polyspermum	*Anagallis arvensis*
Euphorbia peplus	*Euphorbia helioscopia*
Galinsoga ciliata	*Galinsoga parviflora*
Mercurialis annua	*Geranium dissectum*
Oxalis corniculata	*Lamium amplexicaule*
Setaria pumila	*L. purpureum*
S. verticillata	*Sonchus arvensis* and *asper*
S. viridis	*Spergula arvensis*
Veronica agrestis etc.	*Veronica perisca* etc.

The species listed in the right-hand column in each case are by no means confined to the summer crops but are found almost as frequently in the winter cereals, if perhaps with decreased vitality. The same sort of thing in reverse applies to the species named by Oberdorfer (1979) as characteristic of **winter cereal weed communities** *(Secalietea)*:

Alopecurus myosuroides	*Anthemis arvensis*
Buglossoides arvensis	*Papaver rhoeas*
Galium spurium	*Sherardia arvensis*
G. tricornutum	*Sinapis arvensis*
Ranunculus arvensis	*Valerianella locusta*
etc.	*V. rimosa*
	Vicia sativa angustifolia
	Viola arvensis etc.

Here the right-hand column is even longer than that on the left, and here too the species named in it are not only found in winter cereals. The calcareous cereal weed communities on the other hand are well characterised – or rather, it would be more accurate to say, they were, since most of their character species have become vary rare following the intensification of farming methods. Their names are given in section 2 c. The same is true of the character species of the acid-soil cereal weed communities which can also be referred to in section 2 c. Alongside the species listed as characteristic of particular vegetation units there are many others which seem to be equally at home in ruderal and in arable weed communities, e.g. *Agropyron repens, Galium* ***aparine, Bilderdykia convolvulus, Poa annua, Tussilago farfara*** and *Vicia hirsuta.*

As Rademacher (1948), Ellenberg (1950, 1963) and many authors quoted by Schubert and Mahn (1968) pointed out, and as Oberdorfer had already conceded in 1957, there are very many cross-connections between the weed communities of winter and summer crops. The logical conclusions should be drawn from this fact, and they **should be united systematically** or at least be put closer together, especially since the more intensive husbandry is making the habitats more and more alike (see section 2 e). The first step in this direction has been taken by Géhu, Richard and Tüxen (1973) who have put all arable weed communities back into a single class – the *Stellarietea mediae*.

Since further changes in the classification are expected, and since the changes in the weed communities under the pressure of modern farming methods are still going on, we shall refrain from describing any individual units. We shall confine ourselves to a few larger groups and concentrate on plant-habitat interrelations. To begin with though we should point out that numerous recent works have appeared dealing with arable weeds from all parts of Central Europe but especially from the eastern side. As examples we can mention Brun-Hool (1963), Burrichter (1963), Hilbig (1962, 1966, 1967a, b), Hilbig and Mahn (1979), Hofmeister (1970), Holzner (1970, 1971), Kutschera (1966), Mahn and Schubert (1962), Meisel (1966a, 1967, 1969a), Meisel and Hübschmann (1976), Nezadal (1972, 1975), Passarge and Jurko (1975), Rodi (1966), Schubert and Köhler (1964), Schubert and Mahn (1968), Tillich (1969a), Wedeck (1972) and Wiedenroth and Mörchen (1964).

(b) Causes of the floristic alternations between winter and summer crop weeds

The appearance of either so-called 'root crop weeds' or 'winter cereal weeds' in one and the same field is doubtless connected with the fact that, depending on which crop is being grown, first one and then the other group of weeds finds conditions more favourable. The question is, what factors are operating these changes in the 'phenological aspect' of the weed community?

Apparently the nature of the crop as such plays only a minor role. Spring wheat, oats and maize for example are without doubt cereals but they harbour 'typical' weed communities which are much closer to those of the root crops or gardens. Where, exceptionally, a field for some reason is not carrying a crop one may occasionally see a well characterised 'summer' or 'winter crop' type of weed community. This was also shown in research at Göttingen into plant successions which we shall return to in section X 1 b.

More important than the crop as such are the supply of nutrients and the methods of cultivating the soil. Yet there does not seem to be a direct interrelation between the mechanical methods used and the species combination of the weed stand. The 'tilled crop weeds' are no more resistant to repeated hoeing than are the 'cereal weeds' in general terms. In fact according to an (unpublished) trial carried out on the farm 'Monrepos' to the north of Stuttgart, comprising once to 16 times hoeing the opposite is the case: The 'root crop' and garden weeds disappeared quicker than the 'winter cereal' and indifferent ones did. Much more effective than the **number** of mechanical impacts was the '**timing**' of the last (or only) hoeing or other cleaning operation. If this is in the late autumn or very early spring then a 'winter cereal weed community' results, whether or not there is any cereal actually present. If the hoeing is carried out in May or June or even later, or if all the weeds are attacked with chemical herbicides before this time, then the same field will carry a 'tilled crop or garden weed community', provided that for the previous decades the field had been cropped under a rotation which included both root crops and cereals and that it was situated in the planar, colline or submontane belts (see also section X 1 b).

Most tilled-crop weeds do not germinate during the cooler months. Salzmann (1939, see also Ellenberg 1963) has given a number of examples showing this. For species investigated by Lauer (1953) using a series of thermostats the optimum temperature for germination was over 20°C. and for some the minimum was over 15°C (fig.491). On the other hand when she tested the 'winter cereal weeds' they proved to be capable of germinating at very low temperatures, and their optimum percentage germination was below 10°C. If these 'cold-germinators', along with some indifferent species whose germination is not so much affected by temperature, are already in possession of the ground by the spring when the soil temperature first becomes high enough for the 'warmth-germinators' then the latter will scarcely be able to grow because of shortage of light. It is only after the corn harvest that they will have enough light to grow up. On the other hand if the last cultivations on the field are not carried out until the soil has already warmed up then the warmth-germinators will get a good start and since, according to Lauer, most of them grow quickly they will get the upper hand. Lauer's results have been confirmed and added to by W. Koch (1970).

It appears then that the **germinating temperatures play a significant part in determining the current composition of the arable weed community**. They also explain why the tilled-crop weeds play an altogether larger part in those areas with warm summers than in the relatively cool districts. In the oceanic climate of Ireland for example there are none of the species which Lauer showed to be warmth-germinators except for their occasional appearance on roadsides or in ruderal communities as temporary casuals. On the other hand, according to Passarge (1959b) in the relatively continental Havelland with its warm spring and summer *Setaria* species and other 'root crop weeds' appear regularly in the fields of winter wheat and rye, and not only in summer crops. Brun-Hool (1963) found the same thing in the Upper Rhine plain where the temperatures are already rising early in the spring. Depending on the environmental conditions then the tilled-crop weed communities are often found flourishing in cereals, but the opposite is much rarer.

A natural experiment which was evaluated by Seibert (1965) proved very informative in this connection. This was an unusually persistent high water in June in the Danube area. Immediately afterwards meadow plants and representatives of communities from moist and wet habitats appeared on the arable fields. In addition species of the tilled-crop and garden weeds, i.e. mainly warmth-germinators, were much more plentiful in the winter cereal fields so that the two weed communities could hardly be distinguished.

Not all the species which can be considered as characteristic of the order *Polygono-Chenopodietalia* are ones which need warmth for germination. In fact two of the most frequent are not, *Chenopodium album* and *Stellaria media* (see fig. 491). Such species with a wider amplitude for the germination temperatures can be found regularly in cereal crops although they do not thrive so well there as in the 'root' crops. This is because they are plants which demand a higher level of fertility, and as a rule they find this in the spring sown crops which are generally manured more heavily than winter corn. Mayser (1954, see Walter 1963, and figs.492 and 493) showed experimentally that ruderal plants and tilled-crop weeds (e.g. *Amaranthus retroflexus*, *A. lividus*, *Solanum nigrum*, *Setaria verticillata*, *S. glauca* and *Chenopodium polyspermum*) both in their germination and in their growth benefitted from extra nitrogen applications more than did winter cereal weeds or indifferent ones (e.g. *Ranunculus arvensis*, *Bupleurum rotundifolium*, *Sherardia arvensis* and *Sonchus arvensis*). These last-named are therefore scarcely able to compete successfully with the strong-growing chenopodiaceous or similar large-leaved plants amongst the garden and summer crop weeds.

While those members of the *Polygono-Chenopodietalia* requiring a high temperature for germination do not climb up into the mountains, the nitrogen

Fig. 491. Germination rates at various constantly maintained temperature levels in some weeds which are found mainly in summer crops and gardens and in others which are more common in winter cereals. From data by Lauer in Ellenberg (1963), modified.

The curves show the percentages of seeds germinating at each of the constant temperatures shown along the abscissa. 'Warmth germinators' prefer a temperature of 20 to 30°C (narrow lines) or even over 30°C (black). 'Cold germinators' are favoured by temperatures below 20°C (broad lines) and in some cases even below 7°C (dotted).

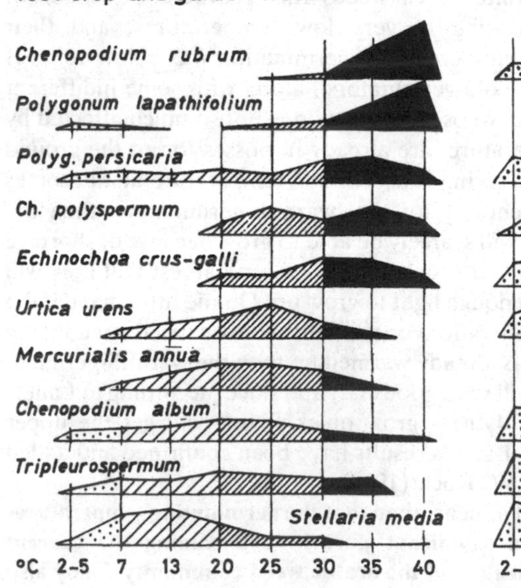

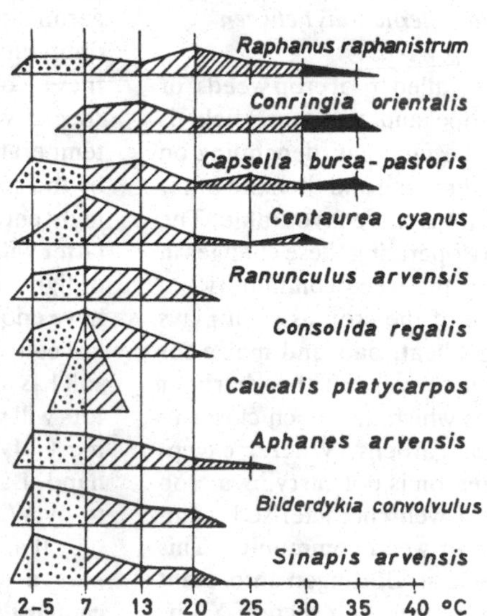

indicators which are indifferent to temperature are particularly well represented in arable fields at the montane level. Where winter corn is no longer grown because the snow is too deep or persists too long, or for some other reason, then many typical winter cereal weeds are absent. On the other hand communities of the *Polygono-Chenopodietalia*, becoming more and more depleted of species, are found right to the upper limit of arable cultivation. In contrast the *Secalietalia* communities only go up to about 1200 m above sea-level. Today arable farming in the valleys of the high mountains is quite unprofitable and is visibly disappearing where it has not already died out.

(c) Weed communities and soil acidity

Just as in Beech woods, Oak-Hornbeam woods, poor

Fig. 492. The effect of increased applications of nitrate and of ammonium on pure stands of weed and ruderal plants in pot cultures. From data by Mayser in Walter (1963), modified.

The pot cultures were carried out with a loam and sand mixture the water being carefully maintained at 60% saturation. A basic addition of P and K was included. Some pots had no nitrogen added, others had high levels of nitrate or of ammonium. The block diagram gives the increase in yield as a multiple of the control (no N) pot measured as the weight of dry matter produced above ground. Apart from *Urtica urens* and *Sisymbrium* both forms of nitrogen had about the same effect on the yield. Ruderal plants responded most (*Lappa*, *Amaranthus* and *Sisymbrium*). Some weeds of hoed crops and gardens also responded strongly (*Urtica urens*, *Echinochloa*, *Setaria*) but not all (*Chenopodium polyspermum*). The least response to a high nitrogen supply was seen in *Silene nutans* and *S. vulgaris*, two species which are found in poor grassland. Under natural conditions *Capsella bursa-pastoris* has a wide amplitude.

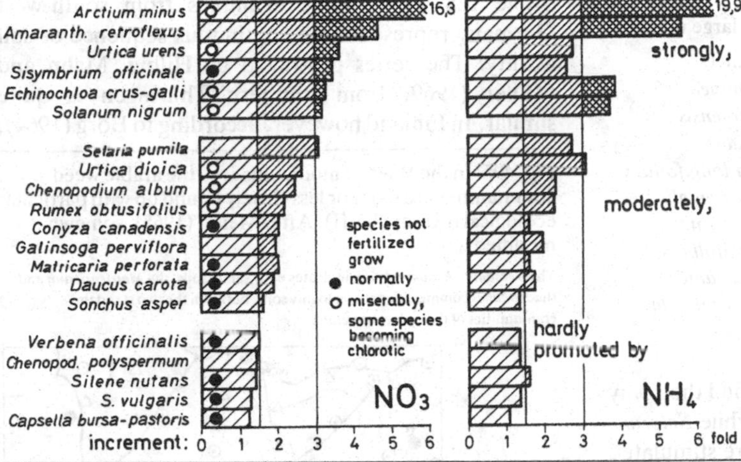

Fig. 493. Effects of increased applications of ammonium and of nitrate on a mixed stand of arable weeds. From data by Mayser in Walter (1963), modified.

The same number of individuals of the participating species was planted out on each of the trial plots of 1 m² of garden soil. Some plots had no manure, others were given increasing amounts of nitrogen in the form of either nitrate or ammonium. The proportion of the total dry weight contributed by each species is expressed as a percentage. The dominant species both on the unmanured plots and those with additional nitrogen was the White Goosefoot or Fat-hen (*Chenopodium album*), and this was also the species which responded most to the nitrogen supply. The Many-seeded Goosefoot managed to hold its own while the Black Nightshade and the Cockspur gained some ground with the extra nitrogen. The other species were increasingly suppressed although without competition they each benefit from higher nitrogen levels. From the results shown in fig. 492 it also appears that in pure culture *Chenopodium album* actually responds less strongly to nitrogen than do *Echinochloa* and *Solanum nigrum*.

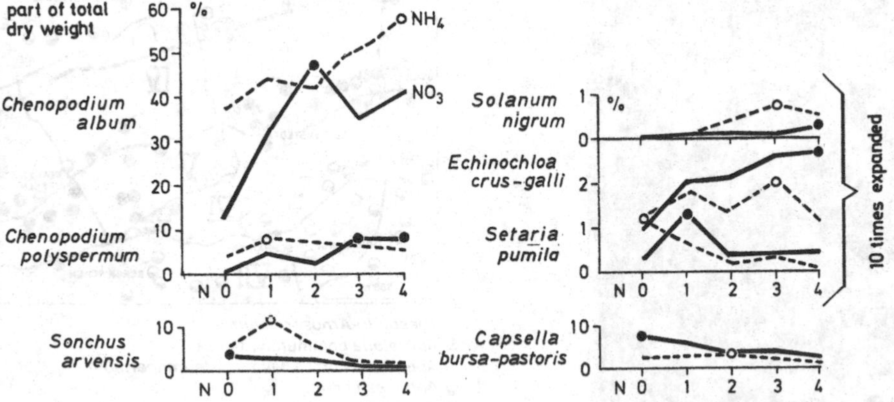

grasslands and many other communities in Central Europe so it is with arable weed communities that the species composition reflects the lime content or the acidity of the soil. This is especially true of the winter cereal weed communities which most authors today divide into two groups – the **acid-soil Silky-bent communities** and the **lime-loving Bur-parsley communities**. Oberdorfer (1979) considers the first group to merit a separate order although it has few character species and these are not very constant, namely:

Apera spica-venti	*Aphanes arvensis*
Scleranthus annuus	

However many other acid-indicators can be used as differential species. In addition the Silky-bent communities can be recognised easily in a negative way by the absence of the character species of the order *Secalietalia* (or of the alliance *Caucalion*) of which there are quite a large number:

plentiful species:	almost extinct:
Anagallis foemina	*Adonis flammea*
Consolida regalis	*Asperula arvensis*
Euphorbia exigua	*Bifora radians*
	Bupleurum longifolium
infrequent:	*Conringia orientalis*
Adonis aestivalis	*Legousia hybrida*
Caucalis platycarpos	*Turgenia latifolia*
Lathyrus tuberosus	*Vaccaria hispanica*
Legousia speculum-veneris	*Vaccaria pyramidata*
Melampyrum arvense	
Scandix pecten-veneris	

Many of these 'lime-indicators' also find their way into acid fields e.g. *Euphorbia exigua*, while *Scleranthus annuus* and other 'acid-indicators' are stimulated rather than suppressed when a field is limed. Their

Fig. 494. A weed community in a sandy Rye field in the uplands of Bohemia-Moravia with *Trifolium arvense* and *repens*, *Spergula arvensis* (left), *Scleranthus annuus* (in front left) and *Ranunculus repens* (right). About one third natural size. Photo. Iltis-Schulz.

physiological amplitude therefore is greater than one would expect from their normal ecological behaviour. It is only after a year-long competition that the difference between the species becomes apparent. It could be that factors other than the pH value of the soil are the deciding ones. Within the Silky-bent communities two groups can be distinguished:

Lamb's Succory communities (alliance *Arnoseridion*) of acid sandy soils, extremely poor in nutrients and in a more or less oceanic climate. These communities take the place of the Birch-Oak woods (see figs. 494 and 495).

Parsley-piert communities (alliance *Aphanion*) need soils somewhat richer in colloids and nutrients which may be more or less acid.

Table 128 shows some examples from south-west Germany representing a continuous sequence of soil acidity. The series protrayed by Hilbig, Mahn and Müller (1969, from southern Thüringen is quite similar. In Finland however, according to Borg (1964),

Fig. 495. In the Wiehen mountains area the arable weed communities are more or less closely bound up with particular ecological regions (I-VII). After Meisel (1965), somewhat modified.

The *Teesdalio-Arnoseridetum* indicates poor sandy soils, the *Matricarietum* and the *Aphanes* communities are on loamy soils; between the two stand the communities of the order *Aperetalia*.

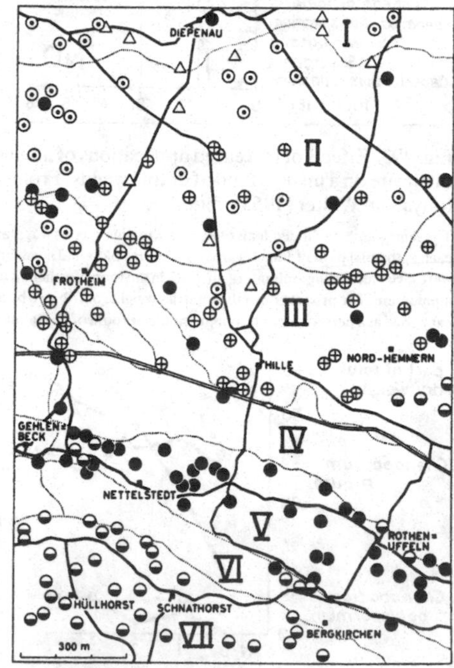

△ *Teesdalio-Arnoseridetum*
⊙ *Aperetalia* community, typical
⊕ *Aperetalia* comm. with *Tripleurospermum*
◒ *Aphanes* comm.
● *Chamomilletum*

Tab. 128. The acidity amplitudes of some arable weeds in south-west Germany.

Serial no.	1	2	3	4	5	6	7	8	9	10	11	12	13	14	15	16	17	18	19	20	21	22	23	24	25	Evaluation 1950	1974
Average **pH** value units	**4,**				**5,**						**6,**								**7,**							**1950**	**1974**
for 1948 tenths	5	6	6	8	1	2	3	5	7	8	1	3	4	4	5	5	7	9	0	2	2	2	2	3	3	R1-5	R1-9
Pronounced acid indicators:																											
Rumex acetosella	1	1																								1	2
Scleranthus annuus	1		2	+	1		2																			1	2
Spergula arvensis	2		+		1		2																			1	2
Acid indicators sens. lat.:																											
Aphanes arvensis			2	+					+					+												2	4
Raphanus raphanistrum	2	2				2		2		1	3		+		2											2	4
Apera spica-venti	1	2	2	2	2	+	2	+	1		+	3	+			+	+	1								3	4
Indifferent:																											
Matricaria chamomilla						2				2	+		2		2	2	1									3	5
Poa annua	1	1	+	+	3	2	2	1	1	2			1	1	+		1		+							3	×[a]
± Lime prefering:																											
Sinapis arvensis						1		2		1	2		1	2	2	1	2	1	2	1	3	1	1	2	1	4	8[b]
Fumaria officinalis										1			1	1	2					1	1	+				4	6
Papaver rhoeas										+		+	+	+	3	+	2		3	+	1	2	+	+		4	7
Sonchus oleraceus												+			1		+	+		1	+	+		1	1	4	8[b]
Lime indicators																											
Consolida regalis																			1	3	3	1	+		1	5	8
Galium tricornatum																					2	+	2	1		5	8
Caucalis platycarpos																			1			3	3	1		5	9
Anagallis foemina																					+		1	1	1	5	9

Note: [a] Rated as indifferent since it is also found on calcareous soils when they are in good heart.

[b] Based on the present results should be graded R7 but in general found mostly on base-rich soils.

Source: After Ellenberg (1950), somewhat modified. The pH values are the averages from many determinations during 1948 from each of the long term experimental areas. The figures in columns 1–25 indicate the highest cover degrees reached during the year.

some of the same species behave in a rather different way. There are weed communities on sandy acid arable fields from Portugal to Estonia which show up the oceanic-continental gradient of the climate as clearly as the woodland communities (Malato-Belitz, J. Tüxen and R. Tüxen 1960).

(d) Effects of soil dampness and limited aeration on the species composition

Since a profitable arable field can be neither very wet nor short of water the majority of the weed communities thrive under average moisture conditions suitable for the majority of species. From a systematic point of view any deviations from these conditions show up only as subassociations, variants and subvariants of the associations whose make up is primarily determined by other factors. In spite of this the weed communities can be used just as well as grassland or woodland communities to map the countryside according to the amount of water available to the plants (Meisel and Wattendorf 1962).

Where the soil is waterlogged and badly aerated for part of the season some plant species which would otherwise play only a very small part in arable land or gardens become competitive. Three groups can be distinguished, according to their behaviour and their type of growth. Their representatives as differential species of the subunits are often found growing together but also on occasion they are found separately:

1 Deep-rooted perennial species indicating that the **sub-soil is sometimes wet** (see fig. 496):

Equisetum arvense	*Polygonum amphibium* var. *terrestre*
Tussilago farfara	

2a Perennial species rooting mainly in the top soil and indicating some **waterlogging** (see fig. 479):

Agrostis stolonifera	*Potentilla anserina*
Equisetum sylvaticum	*Ranunculus repens*
Mentha arvensis	*Rorippa sylvestris*
Poa trivialis	*Stachys palustris*

2b Species behaving rather like those in the previous list but **not so** closely confined to waterlogged soils:

Aphanes arvensis	*Poa annua*
Alopecurus myosuroides	*Sonchus arvensis*
Apera spica-venti	*Sonchus asper*
Matricaria chamomilla	*Ranunculus arvensis*

3 Shallow-rooted, short-lived species which germinate under wet conditions and will only develop when there is **sufficient water in the topsoil**, but do not necessarily indicate waterlogging:

Gnaphalium uliginosum	*Polygonum hydropiper*
Juncus bufonius	*Sagina apetala*
Plantago major ssp. *intermedia*	*Sagina procumbens*

Ellenberg and Snoy (1957) have tested experimentally the physiological behaviour of many of the species named as well as that of several 'dryness indicators' at the time of germination, during young growth and during the later developmental stages. They got the surprising result that the water requirements of many of these plants changed throughout the course of their life and could change completely round (see fig.497).

Fig. 496. The Common Horsetail has deep air-filled rhizomes which enable it to grow in waterlogged soils, e.g. over an impervious layer or in a broken drainpipe. From these it continues to send shoots up to the surface. Many other weeds and cultivated plants (e.g. Red Clover) also have deep root systems. After Kutschera (1966), somewhat modified.

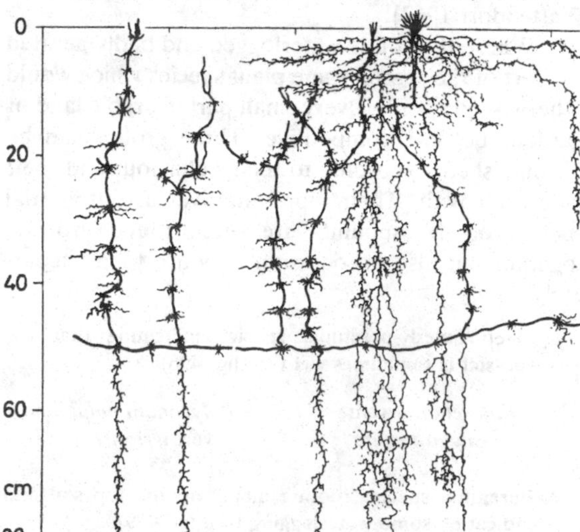

Gnaphalium uliginosum for example, the most plentiful representative of group 3 is very definitely a plant germinating in the wet, best of all when the soil is saturated right up to the surface. However as the plant becomes older it suffers more and more from the shortage of air in the soil. If these conditions persist it forms a very shallow root system and easily falls a victim to fungi which prevent it from fruiting. *Juncus bufonius* and *Polygonum hydropiper* behave in a similar manner. Actually these species are never found on continuously waterlogged soils and they thrive best on deep soils that are well aerated but are always moist. It would be wrong then merely to look upon them as indicating a 'waterlogged' soil as happens repeatedly when based just on people's opinions. It would be more correct to speak of them as **indicators of 'topsoil moisture'** especially since, in a sufficiently wet year, they take over soils which are certainly not waterlogged.

On the other hand the perennial species in groups 1 and 2a, especially *Ranunculus repens* and *Equisetum arvense*, are insensitive to the lack of air associated with permanent waterlogging. Their root cortex and rhizomes contain large air spaces like those of swamp plants. This aerenchyma makes it possible for them to push their roots deep into the waterlogged soil, in contrast to the shallow root system which the species of group 3 and many other weeds as well as most cultivated plants must adopt in these circumstances. Because of this the waterlogged soil indicators are relatively favoured in wet soils and rightly deserve the name.

On the other hand, as Ellenberg and Snoy demonstrated *Ranunculus repens* (and to a lesser extent *Agrostis stolonifera*) are able to survive a drought lasting several weeks. Since the majority of plants on a water-saturated soil have shallow rooting systems their water supply is exhausted more rapidly in a dry period than those in a normally aerated and physiologically deeper soil. The ability to withstand extreme fluctuations in the water supply and then to spread vegetatively after recovering from a dry period make species like *Ranunculus repens* very competitive. They play such a large part in **temporarily waterlogged sites** that they have already been discussed in section VI 2 b.

Waterlogging is a typical property of pseudogleys and similar soil types. However even in parabrown earths ploughing during wet weather can lead to a relatively impervious layer, a 'plough pan', being formed. The waterlogging effect of the pan and the effectiveness of measures taken to break it up are usually clearly seen in the increase or decrease of waterlogging indicators (groups 2a and b, Ellenberg 1950). This may thus be taken as experimental evidence that the species mentioned above are deriving some advantage from the waterlogging or from the more severe fluctuations in the available moisture.

The majority of arable weeds are not able to germinate in wet soil because it is so deficient in oxygen. For example Müllverstedt (1963a) found that *Avena fatua*, *Chenopodium album*, *Galium aparine*, *Sonchus asper* and *Stellaria media* germinated only when there was a high partial pressure of oxygen. However many plants growing well on soils which are inclined to be waterlogged will germinate under a low partial pressure of oxygen, especially:

Alopecurus myosuroides	*Poa annua*
Apera spica-venti	*Rumex obtusifolius*
Matricaria chamomilla	*Veronica persica*

All seeds die in the complete absence of oxygen. The reason why so many weeds suddenly appear after ground has been ploughed or harrowed is because of the better oxygen supply. According to Müllverstedt (1963b) light or the depth of the seed in the soil play no part in this phenomenon.

We still do not know very much about the behaviour of dryness indicators on arable fields. In any case they are not dryness-'loving' plants in the literal sense (xerophilous), but plants which are better able than others to survive a drought period. The Longleaf (*Falcaria vulgaris*), in trials described by Ellenberg and Snoy showed that it was obviously a plant requiring moisture, and not only for germination (see fig.497). In spite of this it is found almost exclusively on fairly unproductive dry calcareous or loess soils, and this is mainly due to the fact that it needs a lot of light and it also has the ability to survive drought periods because its very long main root grows quickly down into the subsoil where there is always water available.

(e) Effects of nitrogen supply and other fertilising measures

While the dampness of the soil and its degree of acidity along with the methods of cultivation are clearly indicated by the species present in weed communities this is only true in exceptional cases regarding the main plant nutrients. Generally these factors have a similar effect on the vitality of all the members of the community. For instance when there is a good supply of nitrogen all weed species grow more luxuriantly than when this element is in short supply provided they can still get enough light. In garden and field soils where vegetables are grown and which are always heavily manured the nutrients may reach such a concentration that they offer very good conditions over a period of years. In this case the more demanding species become dominant, without however completely eliminating the more modest ones.

Fig. 497. Germination, young growth and subsequent vegetative growth are influenced by soil moisture in different ways, even in one and the same species. This is true both for moisture indicators (indicator values F7 and 8) and for dryness indicators (F4 and 3) and also for species which are found most commonly on average soils (F5 or 6). From data by Snoy (see Ellenberg and Snoy 1957).

Pot experiments were carried out in three series. The dry soil (d) was kept at a suction pressure of about 10 bar; in the wet soil the water table was kept almost at the surface (m), and the slightly-moist soil pots were watered every day, but not excessively (w). The yield from the latter served as a base for comparison (= 100%).

Gnaphalium uliginosum and *Sagina procumbens* germinated best under wet conditions (the germination rate in w was many times that in m). However even in the seedling stage the growth was inhibited by too much water (average dry matter production per pot in m, not more than 50% of that in w). Later they suffered so much from excess water that they died. *Polygonum hydropiper* and above all *Plantago intermedia*, *Poa annua* and *Alopecurus myosuroides* behaved in quite the opposite manner. *Chaenarrhinum minus* and *Falcaria vulgaris* which are mainly to be found on relatively dry free-draining soils only germinate under conditions of average dampness. *Falcaria* can tolerate more water at a later stage but it does not make use of this capability under field conditions.

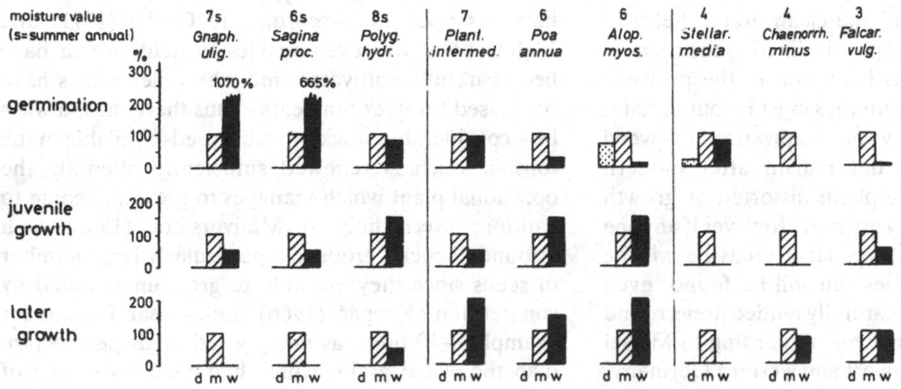

Tab. 129. Changes in constancy of winter cereal weeds between 1937 and 1975 on sandy soils of the north-west German plain.

	R	N		R	N
Marked reduction 1937–75			Increase 1937–75		
Achillea millefolium	×	5	*Cirsium arvense*	×	7
Aphanes microcarpa	4	4	*Galeopsis tetrahit*	×	7
Anthoxanthum puëlii	2	3	*Galium aparine*	6	8
Arnoseris minima	3	4	*Myosotis arvensis*	×	6
Teesdalea nudicaulis	1	1	*Poa annua*	×	8
etc.			*Polygonum aviculare*	4	8
			P. hydropiper	4	8
Reduction 1937–75			*Stellaria media*	7	8
Centaurea cyanus	×	×	etc.		
Rumex acetosella	2	2			
Scleranthus annuus	2	4	Marked increase		
etc.			*Polygonum persicaria*	×	7
			Tripleurospermum inodorum	6	6
			Vicia hirsuta	×	×

Note: R = reaction value, N = nitrogen value (see section E III 2 and B I 4).
Species indicating acid soils or a shortage of nitrogen have given way to the more demanding or the indifferent.
Source: After Meisel and Hbschmann (1976).

Ellenberg (1950) has divided the weeds into 6 and later into 10 (1974, 1979) groups according to their ecological behaviour in respect of nitrogen supply, based on many fertilising trials and comparative observations. It is not easy to confirm this grouping experimentally since it is very difficult to determine the amount of nitrogen available in arable soil. But more recently Zoldan has shown that most of Ellenberg's ratings correspond quite well to the degree of nitrogen supply in a series of field plots studied around Göttingen (see Zoldan 1981). As for phosphate and other nutrients there is no clear evidence that they could influence the species composition of weed communities. The ecological research into these communities has made little progress since they are being increasingly affected by herbicides.

(f) Effects of modern weed control on the species present

Whoever contemplates all the clean arable fields in large areas of Central Europe is inclined to the opinion that everything which has been said in the previous sections about weed communities must be out of date, and that it could hardly be a question of weed 'communities' since all that remain after modern control methods would be plants distorted by growth substances and a few hardy grasses. However if anyone takes the time to search over large areas he will be surprised how many species can still be found, even when he excludes the less carefully tended areas round the field edges from his relevés. According to Meisel (1967), referring to north-west and western Germany, 'it is always possible, in spite of intensive weed control measures, to recognise not only small-scale differences in the soil conditions but also variations in the climate, even in stands which have become floristically impoverished'. Tillich (1969a) even found that near Potsdam the fields of spring wheat which had been treated with herbicides were no poorer in species, because although some species had disappeared others had come in to replace them. Research subjects such as 'The arable vegetation of Lower Austria as an indicator of the environment' (Holzner 1971), or 'The significance of arable weed communities for plant-geographical classification in Thuringia' (Hilbig 1966, Schubert, Hilbig and Mahn 1973) are still topical.

It is always surprising that in field trials involving herbicides the control plots, which receive no treatment, show a luxuriant growth of weeds after just a few months, containing many species which seemed to have disappeared. According to Grube (1971) and Mahn (1973) this even applies to fields which have been carefully cultivated and where herbicides have been used for over ten years. Thus there must always be a considerable stock of viable seeds available in the topsoil which is renewed sufficiently often by the occasional plant which manages to grow and come to fruition. According to Müllverstedt (1966) such 'ground keepers' produce a particularly large number of seeds since they are able to grow unrestricted by competition. Kropáč (1966) found near Prague for example 8–10 times as many weed seeds per hectare than the wheat grains which had been sown, and of these 93–98.5% were those of annuals. Thus the effect of herbicide on the species populations hardly lasts beyond the first year (Mahn 1969). Even more unexpected is that after ten years of continuous use of herbicides the number of soil organisms had been reduced by no more than 5% (Rademacher 1967).

However time does not pass without some changes being noticed in the weed communities. Apart from the very obvious reduction in numbers of individuals there has been some loss of species since the middle of this century. The weeds mentioned in section 1 a whose seeds used to be so troublesome as contaminants of seed corn, especially *Agrostemma githago* and *Centaurea cyanus*, have died out over large areas. The brightly coloured character species are now almost entirely absent from calcareous soils (see section 2 b), probably because they require a lot of light and cannot tolerate the dense cereal crop grown nowadays with the use of heavier dressings of manure. The same is true for some character species in the Lamb's Succory community on sandy soils which were formerly so extremely acid and poor in nutrients. These have been limed and artificially manured to such an extent that they are now capable of carrying even a crop of sugarbeet. Correspondingly *Aperetalia* communities have formed on them which are without character species of any association. Therefore Meisel (1969a) has put them altogether into one broad association, the *Anthoxantho-Arnoseridetum* (see also tab. 129). Within large areas of the German Democratic Republic however, the weed communities still reflect quite well their site conditions (Hilbig and Schubert 1980).

There are some species which, in spite of all opposition, have managed to survive in arable fields and gardens. On the one hand these are perennial indicators of the soil water conditions (see section d) and on the other hand annual nitrogen-loving 'universal weeds' such as *Thlaspi arvense*, *Capsella bursa-pastoris*, *Euphorbia helioscopia*, *Stellaria media* and *Chenopodium album*. According to Williams (1964) the last-named is able to hold its ground under relatively unfavourable conditions and ripen numerous seeds even when it remains very tiny. In general we can state that the **difference between the weed communities of winter and summer crops is becoming progressively smaller**. This is due mainly to the increased amounts of fertiliser being given to modern cereal crops while at the same time the number of individual weeds and therefore the competition between them is reduced. The classification which we discussed at the beginning of section 2a appears increasingly to be less justified.

Although the majority of the weed species have suffered quantitative losses and a few have disappeared there are some which have benefited from modern methods of farming. We have already indicated how the use of the combine harvester has led to a renewed spread of the Wild Oat (section 1 b). The Loose Silky-bent (*Apera spica-venti*) too is able to distribute its seeds better because of the later harvesting by combining (Petzold 1959). It is very difficult to control grasses in cereal crops because the growth hormones used as herbicides primarily damage dicots. The Couch Grass (*Agropyron repens*) is one which has been quick to take advantage of this; Rademacher (1962, 1967) names other species. Perennial herbs with a deep root system have also increased in some places as they are often able to shoot again after the parts above the soil have been damaged.

Thus changes are to be recorded in almost all the communities, and this makes it necessary to review the present classification. The changes are not yet finished, nor indeed have they only recently begun. For practical as well as scientific reasons it is advisable to set up as many **permanent test areas** as possible under different climatic conditions and on different soils, where the respective fates of the weed communities can be traced with more precision and carefully documented. Some authors (e.g. Nezadal 1980) even recommend at least a partial protection of weed communities. How to realise this has been well demonstrated by Schlenker and Schill (1979) near Münsingen on the Swabian Jura.

(g) Weed communities of gardens and vineyards

The weed communities of permanent special types of cultivation, especially of vineyards, resemble those of gardens, since they are frequently being manured and hoed. However in the past there were several species growing in the warm dry habitats of the grapevine slopes which were not usually found in gardens. A special splash of colour was provided by some spring flowering geophytes of Mediterranean origin, e.g.:

Allium vineale	*Muscari racemosum*
Gagea villosa	*Ornithogalum umbellatum*
Muscari comosum	*Tulipa sylvestris*

Today these early blooming aliens are missing from most Central European vineyards, having fallen victim to modern weed control methods. Along with *Geranium rotundifolium* a few of them are still to be found on the loess soils of the Kaiserstuhl in the *Geranio-Allietum* which has been studied by Wilmanns (1975). Since 1942 this community has been getting less abundant and poorer in species. Also, according to Hilbig (1967a), in the vineyards of East

Germany it pays to remove the weeds. However in most places only Chickweed (*Stellaria media*) and a few other quick-growing perpetual flowering remnants of the formerly very rich vineyard weed communities have survived because they are able to ripen seed in the short intervals between cultivations. Nowadays they are tolerated in a growing number of vineyards because they shade the ground and maintain the tilth as well as giving some protection against erosion on the steep slopes. At the same time their roots are so shallow that they do not compete seriously with the vines. In some regions however their role is taken over by a carpet of sown grass (e.g. *Lolium multiflorum*).

There is always the question as to whether keeping a soil completely free from weeds is not harmful in the long run because the animal, fungal and bacterial life in the soil is progressively impoverished. However pleasant it might be for the farmer if all weeds could be exterminated, and however much this may be possible through our chemical and technical knowledge, this should not be the aim before sufficient research has been carried out to try and assess its value. It may be that there will come a time when our vineyards, gardens and even the arable fields will carry virtual monocultures. At the present time however nature still asserts her right to produce living more or less diverse biocenoses in every type of habitat.

X Vegetation development on abandoned land

1 Successions on arable fallow land

(a) 'Social fallows' as problems for landscape management

While large fields, favourably sited, have been farmed more and more intensively during the past few decades, in the vicinity of large towns and in places where crop yields are becoming marginally profitable more and more arable fields are going out of cultivation. Such 'social wasteland' (see Bühring 1970, Stählin and Schäfer 1972) – brought about partly by rationalisation and partly by the drift to the cities – includes also meadow and pastureland, especially in inaccessible valleys and in the inclement mountains, but also in mires and marshlands where it is too expensive to maintain the drainage system. In many parts of Central Europe the amount of cultivated land which has gone out of production is so great that it is presenting a major problem for those responsible for the care of the countryside. In some of the districts of north Hesse, on the lower Main, along the Mosel, in the Pfalz and in the Saarland for example it comprises more than 30% of the agricultural land, especially where smallholdings lie near to industrial agglomerations. Meisel and Melzer (1972) have mapped the actual state of these changes in West Germany.

Should these idle fields just be left to go wild naturally? Are they going to be a danger to the soil quality or to neighbouring fields which are still under cultivation? How long will it take them to turn into woodland, the natural climax? In recreation areas can they be kept free from woodland in order to maintain the stimulating variety of the landscape, including both wooded and open areas? Up to now, the experiences of vegetation scientists and ecologists are not wide enough to answer all these and other questions connected with social wasteland. In any case it has become clear that the ideas about successions on fallow land which have been prevalent in the literature are much too theoretical and schematic. Succession takes a different course depending on the environmental factors and the plant cover at the time the area is abandoned. In almost all cases woodland takes longer to become established than one would have expected in the Central European climate which is so favourable to the growth of trees. Meanwhile some of the experiments, long-term observations and comparisons between areas which have lain idle for different lengths of time have provided useful information.

(b) Succession phases on abandoned ploughed land

As an example we can look at a many-sided investigation set up in 1966/67 in the new Botanic Garden in the University of Göttingen on deep silt loess with a neutral reaction (see tab. 130). Where a previous cornfield was just allowed to lie fallow after being ploughed over, numerous therophytes appeared from seed which had remained viable in spite of several decades of weed control measures (see section IX 2 f). At the same time many perennial plants germinated, but they were not much in evidence during the first two years. Only a few deeprooted arable weeds, especially the Creeping Thistle (*Cirsium arvense*) and the Coltsfoot (*Tussilago farfara*) immediately started to spread out unhampered. As far as the initial stages of the succession are concerned it makes a difference when the last cultivations were carried out. If this was in May, as was the case with some of the Göttingen plots, then the weed community springing from the stock of seed in the topsoil resembles a summer crop community (section IX 2 b). However if the land had been last touched in late summer then the species composition was more like a winter corn weed community, although the difference between this and the first is not very clear cut. The difference can be maintained and enlarged up to a certain point by annual ploughing or cultivating on the same dates.

Tab. 130. Fallow land succession on arable ground which was heat-sterilised in 1968 and left untouched up to 1977.

Serial No.:	1	2	3	4	5	6	7	8	9	10				
Year 19	68	69	70	71	72	73	74	75	76	77				
Shrub layer, cover %	–	–	–	–	–	2	6	12	17	**28**			Ecological and sociological particulars	
Herb layer ,,	1	5	43	58	80	83	87	**89**	86	81				
Moss layer ,,	+	8	35	**78**	53	9	1	1	3	15				
Number of species (excl. mosses)	24	36	40	43	51	56	**60**	57	53	57	R	N	community	dispersal
Persicaria group														
T *Polygonum persicaria*	+	×									×	7	Ah	WZ
T *Bilderdykia convolvulus*	+	×	×								×	×	Ac	Za
T *Anagallis arvensis*		×		+	+						×	6	Ah	W
Canadian Fleabane group														
T *Chaenorrhinum minus*	+	×	1								8	4	Ah	W
T *Stellaria media*	×	×	1		+						7	8	Ah	W
T *Senecio vulgaris*	×	×	1	1							×	8	Ah	W
T *Sonchus asper*	×	×	2	1		+					7	7	Ah	W
T *Capsella bursa-pastoris*	+	1	1	×	+						×	7	Ah	WZ
T *Papaver rhoeas*	+	+	1	×	+						7	6	Ac	W
T *Atriplex patula*		×	1	+							7	7	Ah	W
T *Chamomilla suaveolens*		×	1		+						7	8	Rt	Zf
T *Conyza canadensis*		+	15	2	2						×	4	Ra	W
Prickly Lettuce group														
T *Sonchus oleraceus*	+	×	1	1							8	8	Ah	W
T *Viola tricolor arvensis*	×	×	1	1	×	+	+				×	×	Ac	S
H *Epilobium tetragonum*	×	×	1	7	7	×	1	1	+	×	5	5	A+R	W
G *Equisetum arvense*	+		+	×	×	×	+	+	×	×	×	5	A+R	W
T *Arenaria serpyllifolia*		×	1	1	×		+				×	×	A+R	W
T *Poa annua*		×	1	1	1	1	×				×	8	Rt	W
H *Lactuca serriola*			1	3	1						×	4	Rl	W
T *Matricaria perforata*			+	1	1	×	×		+		6	6	Ah	W
Willow Herb group														
G *Cirsium arvense*	+	×	1	1	4	1	1	1	×	×	×	7	A+R	W
H *Myosotis arvensis*	+	+	×	+	2	+	×	+		+	×	6	Ac	Zf
T *Galium aparine*		×	+		2	+	×	+			6	8	Rp	Zf
H *Epilobium parviflorum*		×	1	5	7	2	1	+			8	5	P	W
H *E. adenocaulon*		+	2	4	7	5	1	×	+	×	–	–	A+R	W
H *E. angustifolium*		×	3	5	7	5	3	3	4	2	3	8	C	W
H *E. hirsutum*			×	1	1	×		+		+	8	8	M	W
H *Cirsium vulgare*			1	1	2	×	1	1	1	+	×	8	Rp	W
C *Cerastium font. holosteoid.*					1	×	+	+	+	×	×	5	M	W
Dandelion group														
H *Taraxacum officinale*	+	×	2	7	11	19	18	13	10	7	×	7	M	W
H *Platago major*		+	+	+	+	×	+		+	+	×	6	Rt	Zf
Coltsfoot group														
G *Tussilago farfara*	+	×	1	3	6	10	12	10	11	7	8	6	A+R	W
H *Poa trivialis*	+		×	3	5	10	13	10	4	4	×	7	M	W
Oxtongue group														
H *Picris hieracioides*		+	+	3	4	12	15	21	21	16	8	4	Rl	W
H *Senecio jacobaea*						+	×	×	+	×	7	5	M	W
H *Torilis japonica*								×			8	8	C	Zf
Carrot group														
H *Daucus carota*				+	+	+	1	1	2	1	×	4	Rl	Zf
H *Phleum pratense*					×	×	×	1	1	1	×	6	M	WZ
H *Solidago gigantea*						+	×	×	×		×	6	Rp	W
H *Arrhenatherum elatius*						+	×	×	1	1	7	7	M	W
Golden Rod group														
H *Solidago canadensis*		×	3	3	7	11	14	20	22	24	×	6	Rp	W
H *Crepis biennis*			+	+	+	1	1	×	×	1	6	5	M	W

Tabl. 130. (continued)

		Serial No.:	1	2	3	4	5	6	7	8	9	10	R	N	comm.	disp.
H	*Geum urbanum*				+	+	+	×	+	×	×	1	×	7	R + Wl	Tk
H	*Crepis capillaris*					+	×	1	1	×	+	1	5	3	M	W
H	*Fragraria vesca*					+	×	×	×	1	2	2	×	6	C	Tv
H	*Dactylis glomerata*					+	+	+	×	2	2	4	×	6	M + W	W
H	*Epilobium montanum*						+	×	+	×	+	×	6	6	W	WT
T	*Trifolium dubium*						+	+	×	×	×	×	5	4	M	W
H	*Agrostis stolonifera*						+	+	+	+	+	×	×	5	Rt	W
H	*Poa palustris*							×	+	+	×	×	8	7	P	W
H	*P. pratensis*							+	×	+	1	1	×	6	M	W
H	*Festuca rubra*							+	×	+	×	1	×	×	M	W
H	*Prunella vulgaris*							+	+	+	×	1	4	×	M	Tm
H	*Deschampsia cespitosa*							+	+	+	+	×	×	3	M + W	W
H	*Luzula luzuloides*							+	+	+	+	×	3	4	W	Ta
H	*Hieracium sylvaticum*								+	+		×	5	4	W	W
T	*Trifolium campestre*											×	×	3	Grass	WT
Woody plants																
P	*Fraxinus excelsior*	S							×	×	1	1				
		H		×	×	×	1	1	2	2	3	2	7	7	Wl	W
P	*Betula pendula*	S						1	2	4	6	9				
		H		×	×	1	1	1	×	×	+	+	×	×	Wl	W
N	*Salix caprea*	S						1	4	6	10	15				
		H		×	×	1	1	1	×	×	+	×	7	7	C	W
P	*Acer platanoides*	S										+				
		H			+	+	+	+	+	+	+	×	×	×	Wl	W
L	*Clematis vitalba*	S							×	×		1				
		H				+	×	×	1	×	×	1	7	7	Sc	Ze
N	*Cornus sanguinea*	S									+	1				
		H					+	+	×	×	1	×	8	×	Sc	Ze
N	*Rosa* cf. *canina*	S								+	1	1				
		H					+	+	+		+	×	×	×	Sc	Ze
P	*Sorbus aucuparia*	S										+				
		H					+	+	+	+	+	×	4	×	Sc	Ze
N	*Crataegus monogyna*	S									+	1				
		H							+	+		+	8	3	Sc	Ze
Number of other species			7	5	2	4	5	8	11	11	11	13	(each of these less than 1%)			

Note: [a] The figures in columns 1–10 indicate the rounded off average values for the amount of cover measured during 4–6 inspections of the vegetation each year. These were evaluated as percentages of the total experimental area : + = less than 0.1%, × = 0.1 – **0.4%. S = in the shrub layer (0.5–5 m high),** H = in the herbaceous layer.

The ecological evaluation (according to Ellenberg 1974) refers to the soil reaction (R) and the available nitrogen (N, see section B I 4); the sociological evaluation refers to the plant formation to which the species normally belongs, roughly divided into:

- Ah = **arable weeds, mainly in hoed crops**
- Ac = " " " " **cereals**
- Ra = ruderal plants in short-lived (annual) communities
- Rl = " " in longer surviving communities
- Rp = " " in permanent communities rich in perennials
- Rt = " " in trodden communities
- A + R = in arable and ruderal communities
- C = woodland clearings and clear felled area plants
- P = **Reedbed plants (*Phragmitetea*)**
- M = meadow plants
- **M + Wl = in meadows and woods**
- D = in dry sandy swards
- Sc = shrubs in scrub or woodland mantle communities
- Wl = woodland plants

The last column contains information about the means of dispersal (according to W. Schmidt 1976).

- S = self dispersal (autochorous)
- Za = ant dispersed (myrmecocorous)
- Z = carried by animals (zoochorous)
- Zm = by man (anthropochorous)
- Zf = burr fruits (epizoochorous)
- Ze = eaten and eliminated viable (endozoochorous)
- WZ = wind and animal dispersal
- W = wind carried (anemochorous)

The letters in front of the plant names indicate the life forms **(see section A I 3 a) T = therophyte, H =** hemicryptophyte, G = geophyte C = herbaceous chamaephyte, N = nanophanerophyte, P = phanerophyte, L = liana.

This parcel of former arable land had been thoroughly cleared of all plants, seeds and other propagules by heat sterilisation. The first plants to appear were mainly short lived herbs (T) and from the fourth year longer living ones (H, G). Not until 9 or 10 years had passed did the woody plants gain a larger proportion of the increasingly dense plant cover although many of them had already germinated as early as the second year. Those trees which germinated in the early years were able to develop well since there was little competition with quickly growing herbaceous plants which still covered less than 50%. Mosses (not shown in the table) soon covered the bare soil in the beginning, but later disappeared through shortage of light. From the ninth year onwards woodland mosses formed a ground layer.

(continued foot of next page)

By the third year tall hemicryptophytes, especially ruderal and grassland plants (*Artemisietalia* and *Molinio-Arrhenatheretea*) began to dominate the plots which had not been disturbed. The short-lived species retreated into a subordinate role. From the fourth year onwards the greatest biomass was that of the Golden Rod (*Solidago canadensis*) and it has remained so until today, i.e. more than ten years after the start of the investigation. No meadow-like stage, rich in grasses has developed, and most of the species present after 4-5 years had not been there in the arable field.

Amongst the newcomers there were a few woody plants and these had been present right from the first year. Ash in particular germinated because its seeds were received repeatedly from a nearby avenue. Pioneers such as Goat Willow and Weeping Birch were also included in the first relevés. Many trees which germinated later did not survive the sapling stage as they were overshadowed by the quicker-growing herbaceous plants. Isolated Willows, Birch and Ash have succeeded in reaching a height over 2 or even 5 m. More trees will probably come in quite quickly in their shadow. This can be deduced from the succession on one of the plots which was overshadowed by a solid wooden fence right from the start. In this plot woody plants gained the upper hand even in the first year and maintained it. The higher humidity in the shade soon enabled ferns and other woodland plants to germinate and persist. However several more years must pass before the plot in the full sun becomes covered with a pioneer woodland since the tall herbaceous perennials make it very difficult for tree seedlings to come through. The successful trees for the most part are those first-year settlers against which the short arable weeds could do little harm.

Woody plants had a better chance from the outset on those plots which had been sterilised before the start of the experiment (see tab. 130). This was carried out either by heating to about 100°C or chemically. The topsoil was heat sterilised to a depth of 30 cm, and the deeper roots or rhizomes were killed chemically. Thus the ground was devoid of any active life, and did not contain any viable seed. However it had not become a raw soil but was still like arable soil with plenty of humus and nutrients. Tree seedlings soon appeared alongside herbaceous immigrants which in the first two years were isolated and covered only 1 to 5% of the ground. Thus the young trees had to suffer much less competition than on the plots where the soil had not been sterilised. Certainly after three years there had been a succession of therophytes, hemicryptophytes and geophytes very similar to that which has been described above, i.e. on 'normal' fallow land. However the woody plants were able to capitalise on their good start and form groups which today (1986) cover more than 70% of the surface area and have reached a height of more than 10 m. On the 'normal' waste land they will take much longer competely to dominate this.

W. Schmidt (1976, 1982) ascertained that 226 species had taken part in the succession. These comprise some 63% of the vascular plants growing within a radius of 1 km around the experimental area, and 25% of the total flora of the Göttingen area. At least 160 species must have come into the experimental plots from outside since they had not been present in the original arable weed community. The mobility of these plant species is thus shown to be surprisingly great even when one considers that there are a sufficient number of seed producers in the neighbourhood. Within the 1 km radius one can find almost all the formations which are present on limestone and loess soils around Göttingen namely Beech woods and Oak-Hornbeam woods, Ash and Alder stands along the streams, communities of woodland clearings, mantles and skirts, scrub, semi-arid swards, managed pastures and meadows, ruderal communities, arable land and gardens as well as abandoned fields which have lain unused for varying lengths of time and have been partly taken over by Golden Rod. About 42% of the vascular plants which have come into the research area were essentially wind dispersed and about 41% were brought in by animals. It is worth noting that some representatives of the remainder (without recognisable method of dispersal) appeared right from the start in the plots which had been sterilised so they could not have been lying dormant in the soil. However since W. Schmidt had seen both Roe Deer and Hares on the plots which at that time had not been fenced the assumption is that the seeds came in on their

(Tab. 130 cont.)

During the succession lime was leached from the topsoil and the pH value sank from 7.8 in 1969 to 7.3 in 1974. Also the very intensive nitrogen mineralisation released by the heat sterilisation fell away during the years. In 1970 it amounted to 67 kg N/ha in 30 weeks but by 1974 had fallen to only 32 (confirmed samples by W. Schmidt). These **changes in the soil** brought about a corresponding reduction in the more demanding species and an increase in the more modest ones, a trend which is clearly seen in the average reaction and nitrogen values (these figures would be scarcely affected by including the 'other species' in the calculation).

Running number	1	2	3	4	5	6	7	8	9	10
Average R-value	7.1	6.9	6.8	6.7	6.3	6.2	6.1	6.1	6.1	6.1
Average N-value	6.5	6.6	6.5	6.2	6.3	6.0	5.8	5.7	5.7	5.6

Source: From data by W. Schmidt (1976 and unpublished) in the New Botanic Garden at Göttingen.

feet as in the case of sheep and wading birds mentioned in sections D I 5 and D VII 1.

Basically similar results to those recorded by W. Schmidt in detailed observation of successions were obtained by F. Runge (1968a), Hard (1972), Borstel (1974), Surber, Amiet and Kobert (1975) and others from their observations on abandoned arable land of different ages but with comparable habitats. Hard summarised his observations on the advance of trees and woodland on to young social fallows in the following way: 'If large areas are suddenly given up then the wind-dispersed woody plants which can germinate on bare soil and require a lot of light, provided they are on hand, will come in during very few years (2-4) and over a moderate distance (100 m), but only on land that has been ploughed and is free of herbaceous plants from the start. Any further succession towards woodland after that goes on extremely slowly'. On bare quarry heaps woody pioneers also turn up very quickly provided the soil suits them. It is much more difficult for them after a grass sward has become established.

If one wishes to avoid the establishment of woody stands for as long as possible one must make sure that there is **a dense competitive plant cover present on the fields** which have fallen out of use, or that such is established as quickly as possible. Abandoned grassland is much better in this respect than a bare fallow, as we shall see in the next section.

2 Successions on abandoned grassland

(a) Development of the vegetation on unused meadows

Meadows where mowing is discontinued are so dense that they remain inimical to trees for a long time (Borstel 1974, Surber, Amiet and Kobert 1975, and others). Flattened down to the ground by the winter snows the dead grass leaves form a felted mat which allows very little light through. The grasses themselves are able to penetrate this by producing adventitious shoots. Many of the other herbs too can reproduce vegetatively, and either push through or grow over the straw covering. Shrubs and trees however meet this obstacle in their most vulnerable stage of development, during the germination and the slow early growth of the seedling. On less-active soils in particular then where the straw covering persists for a long time the chances of woody plants establishing themselves is very slight.

This can be seen very well in previously mown slightly arid swards as well as on moist-soil litter meadows. They remain free from trees and shrubs unless there happen to be some in the near vicinity producing root suckers (such as *Prunus spinosa* and *Populus tremula*). Purple Moorgrass litter meadows develop into Meadowsweet communities whose further development has been dealt with in section D V 5 b. In Cabbage Thistle meadows too and in other previously manured damp meadows the Meadowsweet will become dominant although sometimes thistles or rushes will get the upper hand. In many cases however a Sedge stage will be formed in which *Carex nigra*, *elata*, *paniculata* or other species appear, depending on the amount of moisture in the soil. With their hard straw these make it particularly difficult for trees and shrubs to gain a footing. A successful coloniser of wet neglected meadows is the Eared Sallow (*Salix aurita*) which is met with occasionally. Once it has gained a foothold it can spread out in concentric circles with its branches bowing down to the ground. However the strongest conqueror of wet meadow fallows proves on all sides to be the Reed *Phragmites australis* once it is not kept down by annual mowing. It can spread far and wide with its aerated rhizomes, and its pointed young shoots can penetrate the mat of dead leaves wherever this is not too thick.

The tallest grasses, sedges or perennial herbs happening to be dominant at the time can form a stage in the neglected meadow that may persist for decades. In a long-term stage of this kind some species which were previously present in the meadow community may still persist. However the members which were not so tall and required more light and therefore were indirectly encouraged by the regular mowing now disappear. Finally the community becomes poorer in species as a rule even though a few new ones may have come in (see also K. Schäfer 1975).

It is also a general rule that the proportion of species requiring wet or damp conditions increases in comparison with that in the properly managed meadow. One reason for this is that the mat of dead leaves allows the rain to penetrate through to the soil but keeps away much of the effects of wind and sun so that the soil surface stays much wetter, especially where there is clay present and it is already inclined to be waterlogged (Gisi and Oertli 1981). Another reason can be looked for in the leaves of the uncut meadow not being so dense. Because of this transpiration is less and it goes on for a shorter period so that the soil water is used up more slowly than with a farmed grassland or even a woodland. From the standpoint of the landscape management this water conservation should be welcomed in many cases because our countryside has been impoverished for many decades by drainage schemes. The majority of rare plants however do not return owing to the shortage of light under the tall stands. This as well as further changes in the habitat conditions also affect the other components of the ecosystem, especially the animals (Campino 1978).

If one wishes to prevent the accumulation of dead leaves in neglected meadows it appears that a controlled burning would be the simplest method (Riess 1976), but one must bear in mind that this will favour rather than hold back occupation by trees since they will find more bare ground after the burning and also experience less competition than before. In the present state of our knowledge the following statement still applies: the best thing to do if you want to keep abandoned grassland as open country is to **leave it alone**.

Litter meadows on mires that have been drained but not fertilised will be conquered more rapidly by shrubs and trees than other meadows when they are no longer cut. If these woody invaders remain untouched they soon form a secondary woodland, e.g. in the northern foreland of the Alps where this kind of succession was studied by Briemle (1980).

(b) Successions on previous pastures

The advice given at the end of the previous section applies equally well to former cattle pastures. If White Clover pastures are no longer eaten down the thistles and rushes present, along with Stinging Nettle, spread rapidly. Docks too may often take over (*Rumex crispus* and *obtusifolia*). The remaining grasses form a surface mat of dead leaves similar to that in abandoned meadows. Couch Grass (*Agropyron repens*) may spread temporarily to cover large areas and suppresses almost every other species. As the years go by however the course of the succession comes to resemble more and more that of the abandoned meadows on moist or damp soil.

In high mountains stopping the grazing on steep slopes may have dangerous consequences since snow can slide easily over the mat of dead leaves and cause avalanches or lead to soil erosion (fig. 498). Here too the number of species is reduced in most cases as the light-requiring plants are suppressed.

On dry soils, i.e. in semi-arid swards, there are often woody plants present even before the pastures are given up, namely shrubs like Hawthorn, Sloe, Roses, Juniper and other pasture weeds. Starting with these the succession may lead to a scrubland within 3–4 decades if the habitat conditions are not too unfavourable. The shrub stage will soon be infiltrated with Ash, Beech and other shade trees. In bringing this about the individual species employ quite different strategies as Lloyd (1975) observed in southern England. *Crataegus*, for example propagates itself entirely by seed and has finely branched roots able to compete with grass roots. *Cornus sanguinea* on the other hand has a deeply penetrating main root and extensive lateral roots forming suckers by which it can extend its range step by step. Proceeding from the 'fringe' communities

Fig. 498. Recent damage caused in a subalpine sward which is no longer grazed in the Bavarian outer Alps at 1500 m. The stalks and leaves from the previous year's growth of *Carex sempervirens* form a slippery path, like a thatched roof, down which the winter snow slides pulling out young Spruce trees and large stones which erode the soil. Photo. Spatz.

around the shrub stages some herb species may rapidly conquer the former pastures. According to Schwabe-Braun (1980) this is true even for the Matgrass swards on very acid soils in the Black Forest. In the Swiss Jura and in the Napf mountain area Kienzle (1979) distinguished numerous succession series according to the different site conditions. Most of the unfertilised swards are rather quickly settled by scrub and woodland stages.

In order to maintain neglected pasture as open grassland without resorting to mowing, sheep are used here and there to keep it grazed down. From what we know at present we have to question whether this is the correct action to take. Grazing animals are very selective when they are allowed a choice and they rarely touch groups of stinging nettles, rushes, tough grasses, thistles and many other plants which are not to their taste (see section A II 2 b). In other words they tend to avoid the very plants which are most likely to spread on the disused pastures and in no way do they hinder the occupation of the ground by thorn bushes and other woody pasture weeds. If these are not controlled as the shepherd formerly did by cutting and burning then the waste land will become a scrub more quickly than if it were left alone. On steep slopes this process will be speeded up even further by the animals damaging the ground surface and providing bare places for the seeds of woody plants to germinate. Without the direct help of man this type of grazed wasteland will not be kept free from shrubs and trees. It is true that in earlier times herds of cattle and sheep contributed significantly to the opening up of woodland in Central Europe and keeping it at bay. However even at that time man cooperated in this by cutting out firewood and timber, ringing the trees and burning down groups of woody plants.

Amongst the trees which have been able to gain a footing on untended embankments, burnt areas and similar wasteland the False Acacia (*Robinia pseudacacia*, see fig. 499 and section D III 1 d) played a special role to the extent that it has become more and more naturalised in those sites where the climatic conditions suit it in the south and east of Central Europe. Other neophytes too, like the Golden Rod (*Solidago canadensis*) from America, have also seized the chance offered them by the lack of attention on many previously cultivated fields. However if the succession is allowed to take its natural course they must finally give way to the native trees of Central Europe along with their related communities which we came to know

Fig. 499. The False Acacia (*Robinia pseudacacia*), coming from North America, is one of the few exotic trees which have become established in parts of Central and south-east Europe. After the war it began to spread spontaneously over the bomb-damaged sites in the towns from the specimens which had been planted earlier, but only in the relatively continental regions with their warm summers. After Kohler and Sukopp (1964), somewhat modified.

'Subsarmatic' climatic district according to Werth, B = Bohemian inland climate, P = edge of the pannonic climate. Stuttgart lies in a warm valley basin.

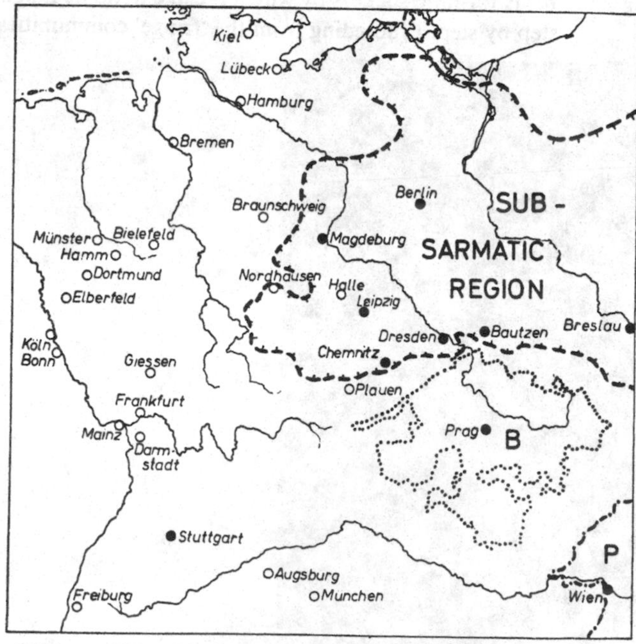

in some detail earlier in this book.

Of course the return of the zonal woodland associations in most fallow areas would take at least several centuries. After 150 years the succession merely reaches a pioneer stage dominated by short-lived tree species, e.g. in the insubrian part of Switzerland (Alther and Stählin 1977). On the Swabian Jura where Beech is the undoubted climax tree, as a rule the semi-arid swards with some *Juniperus* groups, when abandoned, are first settled by Scots pine and Norway Spruce neither of which would play any part in the natural vegetation of those sites (Götz 1979). Most of the former arable fields in Central Europe however which had been left unused in the Middle Ages are today dominated by Beech or other woodland types looking fairly natural. Examples may be found in the 'Sachsenwald' south-east of Hamburg, in the Ebergötzen Forest near Göttingen, on the 'Zürichberg' in the eastern part of Zurich and in many other forest districts. The climax communities of these sites then are able to regenerate relatively quickly and rather completely whereas those in the humid tropics (e.g. on the Mexican peninsula Yucatan) even after 500 years of an undisturbed succession are still poorer in species than the climax stage in comparable sites. This contrast is due mainly to the fact that the potential natural vegetation of Central Europe is made up of only a few tree species. They may complete their numbers much more easily than the hundreds of species taking part in tropical zonal vegetation concerned.

Postscript to the third edition

Concern for the future of the countryside and uncertainty about further changes in its flora and vegetation compel me to add my personal postscript. Others too, having dipped into this book, will share my concern, even though they may not be able to look back over fifty years of intensive study of the vegetation.

When I started my investigations under the inspiring tuition of Reinhold Tüxen and Josias Braun-Blanquet, extensive areas of heathland and brightly flowering meadows abounded. Raised bogs and swamps were still wild lonely places, even though some had already been drained and brought into cultivation. Clean rivers and clear lakes invited one to investigate their living communities. Curious associations of lichens in woodlands and fens, on heathland and pasture, presented the opportunity to examine the relationship between the habitat on the smallest scale.

Much of this can still be found today, but it has to be searched for and in that search we are aware how its very existence is threatened. Where changes in plant communities have been recorded they can no doubt be explained in ecological terms. In most cases the result has shown an impoverishment. Even the highest mountains are not entirely protected from this trend though they have been able to preserve a wealth of variety to a greater extent.

Should we then resign ourselves to this state of affairs? In my opinion we have no right to do so, for none of us can escape responsibility for what has occurred. Action is called for.

It is high time we took steps to save what remains. Above all we must establish reserves which are large enough to enable the living communities to survive in their rich variety. These must be managed in the appropriate manner since many communities have been created with the help of man. It is also urgent that pollution of the air and waters be considerably reduced since pollution does not respect the boundaries of our reserves. In fact we are ourselves directly affected by it.

We must also put into practice now the lessons from our experience as ecologists. We should set up permanent experimental areas in all parts of Central Europe where exact measurements can be taken in order to document both present-day and future changes in the vegetation. Wherever possible we should cooperate with animal ecologists in making repeated accurate observations in these areas. It also appears to be necessary to look afresh at some of the plant communities which were described before about 1950 and, if need be, reclassify them. Only in this way can we decide how far the changes have gone, and only then can we continue to use concepts corresponding to the real vegetation.

However we must not forget that behind such mainly descriptive operations there are many questions of causality remaining to be answered. Ecophysiological and experimental verification must increasingly become the main objectives of our studies, especially with regard to the problems which a heavily industrialised environment inflicts upon us. Thus for botanists and ecologists there is no diminution in the work to be undertaken and the problems to be solved. In the future too every such effort will bring its reward.

E

Tabular summary and index

This bibliography consists mainly of publications quoted or otherwise used. Some works which have appeared since 1978 could not be referred to in the text. An asterisk (*) in front of the authors' names indicates a comprehensive work with an extensive reference list. A plus sign (+) after the reference indicates an informative local account, not mentioned in the text, or similar supplementary literature.

New works on vegetation and ecology are referred to in *Progress in Botany* (previously *Fortschritte der Botanik:* Springer, Berlin). New and supplementary bibliographies can be looked for in *Excerpta botanica,* section B. sociologica (Gustav Fischer, Stuttgart). The modified vowels ä, ö, ü and all accented letters are included in the alphabetical order along with the unmodified ones. The term 'ebenda' refers only to the previous publication containing the work of the same author(s).

I Bibliography

Aartolahti, T., 1965: Oberflächenformen von Hochmooren in Südwest-Häme und Nord-Satakunta. Fennia 93, 1, 268 S.

Abel, W.O., 1956: Die Austrocknungsresistenz der Laubmoose. Sitz. ber. Österr. Akad. Wiss., Math.-nat. Kl., Abt. I 165: 619–707.

Adams, D.A., 1963: Factors influencing vascular plant zonation in North Carolina salt marshes. Ecology 44: 445–456.

*Adriani, M.J., 1958: Halophyten. Handb. Pflanzenphysiol. 4: 709–736.

Aichinger, E., 1933: Vegetationskunde der Karawanken. Pflanzensoziol. (Jena) 2: 329 S.

– 1951: Lehrwanderungen in das Bergsturzgebiet der Schütt am Südfuß der Villacher Alpe. Angew. Pflanzensoziol. (Wien) 4: 67–118. +

– 1957: Die Zwergstrauchheiden als Vegetationsentwicklungstypen. Ebenda 12: 124 S., 13: 84 S., u. 14: 171 S.

–, Gams, H., Wagner, H., Wendelberger, G., u.a., 1956: Exkursionsführer für die XI. Internationale Pflanzengeographische Exkursion durch die Ostalpen 1956. Ebenda 16: 151 S. +

Aletsee, L., 1967: Begriffliche und floristische Grundlagen zu einer pflanzengeographischen Analyse der europäischen Regenwassermoorstandorte. Beitr. Biol. Pflanzen 43: 117–160.

Allen, S.E., 1964: Chemical aspects of heather burning. J. Appl. Ecol. 1: 347–367.

Altehage, C., Rossmann, B., 1940: Vegetationskundliche Untersuchungen der Halophytenflora binnenländischer Salzstellen im Trockengebiet Mitteldeutschlands. Beih. Botan. Cbl. 60, Abt. B: 135–180.

Alther, E.W., Stählin, A., 1977: Entwicklung von Böden und Pflanzenbeständen auf Brachland und ihre Dynamik während 150 Jahren. Z. „Das wirtschaftseigene Futter" 23: 144–167.

Ambros, W., Kneitz, G., 1961: Die Regenwürmer und ihre waldhygienische Bedeutung. Waldhygiene (Würzburg) 4: 34–53. +

Ambrož, Z., Balátová-Tuláčková, E., 1968: Zur Kenntnis der biologischen Aktivität und des Humus-Anteils in Böden der Magnocaricetalia- und Molinietalia-Gesellschaften im Gebiet der SW-Slowakei (Záhorie). Preslia (Praha) 40: 80–93.

Anderson, F., 1970: Ecological studies in a Scanian woodland and meadow area, Southern Sweden. I. Vegetational and environmental structure. Opera Botan. (Lund) 27: 190 S. II. Plant biomass, primary production and turnover of organic matter. Bot. Not. 123: 8–51. +

Anderson, J.M., 1973: The breakdown and decomposition of sweet chestnut (Castanea sativa Mill.) and beech (Fagus silvatica L.) leaf litter in two deciduos woodland soils. Oecologia (Berl.) 12: 251–274.

Anderson, R.C., Loucks, O.L., 1973: Aspects of the biology of Trientalis borealis Raf. Ecol. 54: 798–808.

Andrzejewska, L., 1974: Analysis of a sheep pasture ecosystem in the Pienine mountains (the Carpathians) V. Herbivores and their effect on plant production. Ekol. Pol. 22: 527–534.

Ant, H., Diekjobst, H., 1967: Zum räumlichen und zeitlichen Gefüge der Vegetation trockengefallener Talsperrenböden. Arch. Hydrobiol. 62: 439–452. +

Antonietti, A., 1968: Le associazioni forestali del 'orizzonte submontano del Cantone Ticino su substrati pedogenetici ricchi di carbonati. Mem. Ist. Svizz. Ric. Forest. 44: 85–226.

Apinis, A., 1940: Untersuchungen über die Ökologie der Trapa L. Acta Horti Bot. Univ. Latviens. 13: 7–145.

Arbeitsgemeinschaft „Oberschwäbische Fichtenreviere" (Hrsg.), 1964: Standort, Wald und Waldwirtschaft in Oberschwaben. Stuttgart: 323 S.

Armstrong, W., 1975: The wetland condition and the internal aeration of plants. XII. Internat. Botan. Congr. (Leningrad) Abstr. II, 347.

–, Boatman, D.J., 1967: Some field observations relating the growth of bog plants to conditions of soil aeration. J. Ecol. 55: 101–110.

Arvidsson, J., 1951: Austrocknungs- und Dürreresistenzverhältnisse einiger Repräsentanten öländischer Pflanzenvereine nebst Bemerkungen über Wasserabsorption durch oberirdische Organe. Oikos, Suppl. 1: 181 S.

Arzani, Gh., 1974: Ökophysiologische Untersuchungen über die SO_2-, HCl- und HF-Empfindlichkeit verschiedener Flechtenarten. Diss. Univ. Gießen: 136 S.

Aston, J.L., Bradshaw, A.D., 1966: Evolution in closely adjacent plant populations. II. Agrostis stolonifera in maritime habitats. Heredity 21: 649–664.

Atkinson, D., 1973: Observations on the phosphorus nutrition of two sand dune communities at Ross Linhs. J. Ecol. 61: 117–133.

Auer, C., 1947: Untersuchungen über die natürliche Verjüngung der Lärche im Arven-Lärchenwald des Oberengadins. Mitt. Schweiz. Anst. Forstl. Versuchsw. 25: 3–140.

Aulitzki, H., 1961a: Die Bodentemperaturen in der Kampfzone oberhalb der Waldgrenze und im subalpinen Zirben-Lärchenwald. Mitt. Forstl. Bundes-Versuchsanst. Mariabrunn 59: 153–208.

–, 1961b: Die Bodentemperaturverhältnisse an einer zentralalpinen Hanglage beiderseits der Waldgrenze. I. Die Bodentemperatur oberhalb der zentralalpinen Waldgrenze. Arch. Meteorol., Geophys. u. Bioklimatol. B 10: 445–532.

–, 1962: Welche bioklimatischen Hinweise stehen der Hochlagenaufforstung heute zur Verfügung? Wetter u. Leben 14: 95–117.

–, 1963a: Bioklima und Hochlagenaufforstung in der subalpinen Stufe der Inneralpen. Schweiz. Z. Forstwes. 114: 1–25.

–, 1963b: Grundlagen und Anwendung des vorläufigen Wind-Schnee-Ökogrammes. Mitt. Forstl. Bundes-Versuchsanst. Mariabrunn 60: 765–834.

–, 1965: Waldbau auf bioklimatischer Grundlage in der subalpinen Stufe der Innenalpen. Cbl. Gesamte Forstwes. 82: 217–245.

Aust, E., 1937: Die Verbreitung, Zusammensetzung und Nutzung der schlesischen Waldungen. Z. Wirtschaftsgeogr. Deut. Ost. 13. 150 S.

Averdieck, F.-R., 1971: Zur postglazialen Geschichte der Eibe (Taxus baccata L.) in Nordwestdeutschland. Flora 160: 28–42.

Bach, R., 1950: Die Standorte jurassischer Buchenwaldgesellschaften mit besonderer Berücksichtigung der Böden (Humuskarbonatböder. und Rendzinen). Ber. Schweiz. Botan. Ges. 60: 51–152.

–, Kuoch, R., Iberg, R., 1954: Wälder der Schweizer Alpen im Verbreitungsgebiet der Weißtanne II: Entscheidende Standortsfaktoren und Böden. Mitt. Schweiz. Anst. Forstl. Versuchsw. 30: 261–314.

Backhuys, W., 1968: Der Elevations-Effekt bei einigen Alpenpflanzen der Schweiz. Blumea 16: 273–320.

Backmund, F., 1941: Der Wandel des Waldes im Alpenvorland. Eine forstgeschichtliche Untersuchung. Schr.R. Akad. Deut. Forstw.. 4: 126 S.

Baer, J.-G., 1962: Un demi-siècle d'activité scientifique dans le Parc national. Act. Soc. Helvét. Sci. Nat., Scuol: 50–62. +

Baeumer, K., 1956: Verbreitung und Vergesellschaftung des Glatthafers (Arrhenatherum elatius) und des Goldhafers (Trisetum flavescens) im nördlichen Rheinland. Decheniana Beih. 3: 1–77.

–, 1962: Die Wasserstufen-Karte der Wümme-Niederung. Abh. Naturw. Ver. Bremen 36: 118–168.

Baig, M.N., Tranquillini, W., 1976: Studies on upper timberline: morphology and anatomy of Norway spruce *(Picea abies)* and stone pine *(Pinus cembra)* needles from various habitat conditions. Canad. J. Botany 54: 1622–1632.

Balátová-Tuláčková, E., 1957: Wiesengesellschaften mit Bezug auf die Bodenfeuchtigkeit. Eine Studie aus den Wiesen der Umgebung von Brünn. Sborn. Českosl. Akad. Zeměd. Věd, Rostl. Výr. 30: 529-557.

–, 1963: Zur Systematik der europäischen Phragmitetea. Preslia 35: 118–122.

–, 1969: Beitrag zur Kenntnis der tschechoslowakischen Cnidion venosi-Wiesen. Vegetatio 17: 200–207.

–, 1972: Flachmoorwiesen im mittleren und unteren Opava-Tal (Schlesien). Vegetace ČSSR. A 4: 201 S. +

–, 1976: Rieder- und Sumpfwiesen der Ordnung Magnocaricetalia in der Záhorie-Tiefebene und dem nördlich angrenzenden Gebiete. Vegetácia ČSSR B 3: 258 S.

–, 1978: Die Naß- und Feuchtwiesen Nord-West-Böhmens mit besonderer Berücksichtigung der Magnocaricetalia-Gesellschaften. Rozpr. Československ. Akad. Ved. R. Mat. Prír. 88, 3, 113 S.

–, 1979: Synökologische Verhältnisse der Filipendula ulmaria-Gesellschaften NW-Böhmens. Folia Geobot. Phytotax. (Praha) 14: 225–258.

–, Zelena, V., Tesarova, M., 1977: Synökologische Charakteristik einiger wichtiger Wiesentypen des Naturschutzgebietes Zd'árské Vrchy. Rozpr. Českosl. Akad. Ved. R. Mat. Prír. 87, 5, 115 S.

Baller, A., 1974: Ökologische Untersuchungen im xerothermen Vegetationsmosaik des NSG »Hohe Lehden« bei Jena. Mitt. Sekt. Geobot. u. Phytotax. Biol. Ges. DDR 1974: 93–108.

Baradziej, E., 1974: Net primary production of two marsh communities near Ispina in the Niepotomice Forest (Southern Poland). Ekol. Pol. 22: 145–172.

Barclay-Estrup, P., 1970: The description and interpretation of cyclical processes in a heath community II. Changes in biomass and shoot production during the Calluna cycle. J. Ecol. 58: 243–249.

*Barkman, J. J., 1958: Phytosociology and ecology of cryptogamic epiphytes. Van Gorcum u. Comp., Assen (Niederlande): 628 S.

*–, 1962: Bibliographia phytosociologica cryptogamica. Pars I. Epiphyta. Excerpta Botan., Sect. B, 4: 59–86.

–, 1968: Das systematische Problem der Mikrogesellschaften innerhalb der Biozönosen. In: R. Tüxen(Hrsg.) Pflanzensoziologische Systematik. Ber. über d. internat. Symp. in Stolzenau/Weser 1964. Den Haag, Verl. Dr. W. Junk, 1968: 21–48.

Bartsch, J. u. M., 1940: Vegetationskunde des Schwarzwaldes. Pflanzensoziol. (Jena) 4: 229 S.

*– u. –, 1952: Der Schluchtwald und der Bach-Eschenwald. Angew. Pflanzensoziol. (Wien) 8: 109 S.

Bauch, E., 1970: Die Buchenwälder im Elm und ihre Standorte. Diss. T. U. Braunschweig 1970: 167 S.+

Bauer, E., 1973: Zur ökologisch-physiologischen Indikation von Immissionsschäden im Stadtgebiet von Eßlingen. Diss. Hohenheim: 106 S.

*Baumeister, W., Ernst, W., 1978: Mineralstoffe und Pflanzenwachstum. 3. Aufl. Verlag Gustav Fischer, Stuttgart, 416 S.

*Baumgartner, A., 1967: Entwicklungslinien der forstlichen Meteorologie. Forstwiss. Cbl. 86: 156–175.

*–, 1969: Forstliche Meteorologie, Klimatologie und Hydrologie. In: K. Mantel: Stand und Ergebnisse der forstlichen Forschung 1965–1968. Freiburg i.Br., Forschungsrat f. Ernährung, Landw. u. Forsten 1969: 93–125.+

Baumgartner, A., Klemmer, L., Raschke, E., Waldmann, G., 1967: Waldbrände in Bayern 1950–1959. Allg. Forstz. 1967: Nr. 13.

Becher, R., 1964: Experimentelle Untersuchungen über die Bedeutung der Licht-Intensität und der Wurzel-Konkurrenz für Lebensmöglichkeiten von Sauerklee (Oxalis acetosella) in Waldbeständen. Ber. Oberhess. Ges. Natur- u. Heilkunde Gießen, N.F., Naturw. Abt. 33: 145–148.

Beeftink, W.G., 1965: De zoutvegetatie van ZW-Nederland beschouwd in europees verband. Hydrobiol. Inst. Afd. Delta-Onderz., Yerseke, Nederl. 30: 167 S.

–, 1968: Die Systematik der europäischen Salzpflanzengesellschaften. In: Tüxen, R. (Hrsg.), Pflanzensoziologische Systematik. Den Haag, Verl. Dr. W. Junk: 239–272.

–, Daane, M.C., de Munck, W., 1971: Tin jaar botanisch-oecologische verkenningen langs het Veerse Meer. Nat. en Landschap 25: 50–65. +

Beekman, A.A., 1932: Nederland als Polderland. 3. Aufl., Zutphen: 510 S.

Behre, K., 1956: Die Algenbesiedlung einiger Seen um Bremen und Bremerhaven. Veröff. Inst. Meeresforsch. Bremerhaven 4: 221–383.

–, 1966: Zur Algensoziologie des Süßwassers (unter besonderer Berücksichtigung der Litoralalgen). Arch. Hydrobiol. 62: 125–164.

Behre, K.E., 1970: Die Entwicklungsgeschichte der natürlichen Vegetation im Gebiet der unteren Ems und ihre Abhängigkeit von den Bewegungen des Meeresspiegels. Habil. Schr. Math.-Nat. Fakult. Univ. Göttingen, Hildesheim: 47 S.

–, 1979: Zur Rekonstruktion ehemaliger Pflanzengesellschaften an der Nordseeküste. Ber. Internat. Symposien Internat. Vereinig. Vegetationskunde (Rinteln 1978): 181–214.

–, 1980: Zur mittelalterlichen Plaggenwirtschaft in Nordwestdeutschland und angrenzenden Gebieten nach botanischen Untersuchungen. Abh. Akad. Wiss. Göttingen, Phil.-Hist. Kl., Dritte F. 116: 30–44.

*Beijerinck, W., 1940: Calluna. A monograph on the Scotch heather. Verh. Kon. Nederl. Akad. Wetensch., Afd. Natuurk., 2. Sect. 38: 4, 180 S.

*Beldie, A., 1967: Flora si vegetatia muntilor Bucegi. Ed. Acad. R.S. Romania, Bukarest: 578 S.

Bell, J.N.B., Tallis, J.H., 1974: The response of Empetrum nigrum L. to different mire water regimes, with special reference to Wybunbury moss, Cheshire and Featherbed moss, Derbyshire. J. Ecol. 62: 75–93.

Bellot Rodríguez, F., 1964: Sobre Phragmitetea en Galicia. Ann. Inst. Bot. A. J. Cavanilles (Madrid) 22: 61–80.

Benecke, P., 1976: Der Wasserhaushalt von Buchen- und Fichtenbeständen. Vortragsreferat, Ges. f. Ökologie, Göttingen, 20.-24.9.76.

Berger-Landefeldt, U., Sukopp, H., 1965: Zur Synökologie der Sandtrockenrasen, insbesondere der Silbergrasflur. Verh. Botan. Ver. Prov. Brandenburg 102: 41–98.

Berset, J., 1969: Pâturages, prairies et marais montagnards et subalpins des Préalpes Fribourgeoises. Fribourg/Schweiz: Ed. Univ., ca. 1969; 55 S.

Bertsch, K., 1925: Das Brunnenholzried. Veröff. Staatl. Stelle f. Naturschutz b. Württemb. Landesamt f. Denkmalspflege 2: 67–172.

–, 1959: Moosflora von Südwestdeutschland. 2. Aufl. Stuttgart: 234 S.

Besson, J.-M., 1972: Nature et manifestations des relations sociales entre quelques espèces végétales herbacées. Ber. Schweiz. Botan. Ges. 81: 319–397.

Beug, H.-J., 1967: Probleme der Vegetationsgeschichte in Südeuropa. Ber. Deut. Botan. Ges. 80: 682–689.

–, 1977. Waldgrenzen und Waldbestand in Europa während des Eiszeitalters. Göttinger Rektoratsreden 61: 23 S.

Beyer, H., 1968: Versuche zur Erhaltung von Heideflächen durch Heidschnucken im Naturschutzgebiet „Heiliges Meer". Natur u. Heimat 28: 145–148.

Bittmann, E., 1953: Das Schilf (Phragmites communis Trin.) und seine Verwendung im Wasserbau. Angew. Pflanzensoziol. (Stolzenau/Weser) 7: 44 S.

Bjor, K., 1972: Micro-temperature profiles in the vegetation and soil surface layers on uncovered and twig covered plots. Medd. Norske Skogforsøksvesen 30: 203–218.

Blankwaardt, H.F.H., 1968: De heidekever. Tijdschr. Kon. Nederl. Heidemaatschappij 1968: 30–35.

Blaser, P., 1973: Die Bodenbildung auf Silikatgestein im südlichen Tessin. Schweiz. Anst. Forstl. Versuchswes., Mitt. 49: 253–340.

Bliss, L.C., 1962: Coloric and lipid content in alpine tundra plants. Ecology 43: 753–757.

–, 1966: Plant productivity in alpine microenvironments on Mt. Washington, New Hampshire. Ecol. Monogr. 36: 125–155.

Blöchliger, G., 1931: Mikrobiologische Untersuchungen an verwitternden Schrattenkalkfelsen. Diss. ETH Zürich: 102 S.

Blume, H.-P., Friedrich, F., Neumann, F., Schwiebert, H., 1975: Dynamik eines Düne-Moor-Biotops in ihrer Bedeutung für die Biozönose. Verh. Ges. Ökol., Erlangen 1974: 89–101.

Boatman, D.J., 1962: The growth of Schoenus nigricans on blanket bog peats. I. The response to pH and the level of Potassium and Magnesium. J. Ecol. 50: 823–832.

Böcher, T.W., 1941: Die Vegetation der Randböler Heide mit besonderer Berücksichtigung des Naturschutzgebietes. Kong. Danske Videns. Selsk., Biol. Skrift, 1: 3, 234 S.

Bodenmann, A., Eiberle, K., 1967: Über die Auswirkungen des Verbisses der Gemse im Aletschwald. Schweiz. Z. Forstwes. 118.

*Bodeux, A., 1955: Alnetum glutinosae. Mitt. Florist.-Soziol. Arb. gem. N. F. 5: 24 S.

Boeker, P., 1959: Samenauflauf aus Mist und Erde von Triebwegen und Ruheplätzen. Z. Acker- u. Pflanzenbau 108: 77–92.

Bogenrieder, A., Werner, H., 1979: Experimentelle Untersuchungen an zwei Charakterarten der Eisseggenflur des Feldberges (Carex frigida All. und Soldanella alpina L.). Beitr. Naturk. Forsch. Südwestdeut. 38: 61–69

Bogner, W., 1966: Experimentelle Prüfung von Waldbodenpflanzen auf ihre Ansprüche an die Form der Stickstoffernährung. Diss. Hohenheim: 131 S., und: Mitt. Ver. Forstl. Standortskunde u. Forstpflanzenzüchtung 18: 3.

Böhm, H., 1966: Die geländeklimatische Bedeutung des Bergschattens und der Exposition für das Gefüge der Natur- und Kulturlandschaft. Erdkunde 20: 81–93.

Bohn, U., 1981: Vegetationskarte der Bundesrepublik Deutschland 1:200000 – Potentielle natürliche Vegetation – Blatt CC 5518 Fulda. Schriftenr. Vegetationskunde (Bonn-Bad Godesberg) 15, 330 S.

Bohus, G., Babos, M., 1967: Mycocoenological investigation of acidiphilous deciduous forests in Hungary. Botan. Jb. 87: 304–360.

Bolleter, R., 1920: Vegetationsstudien aus dem Weißwassertal. Jahrb. St. Gallisch. Naturw. Ges. 57, Beilage: 141 S.

Borg, P., 1964: Über die Beziehungen der Ackerunkräuter zu einigen bodenökologischen Faktoren in der Landgemeinde Helsinki. Ann. Botan. Fenn. 1: 146–160.

Borhidi, A., 1970: Ökologie, Wettbewerb und Zönologie des Schilfrohrs (Phragmites communis L.) und die Systematik der Brackröhrichte. Acta Botan. Acad. Sci. Hung. 16: 1–12.

Bornkamm, R., 1958: Standortsbedingungen und Wasserhaushalt von Trespen-Halbtrockenrasen (Mesobromion) im oberen Leinegebiet. Flora 146: 23–67.

–, 1960: Die Trespen-Halbtrockenrasen im oberen Leinegebiet. Mitt. Florist.-Soziol. Arb. gem. N. F. 8: 181–208.

–, 1961 a: Zur Konkurrenzkraft von Bromus erectus. Ein sechsjähriger Dauerversuch. Botan. Jb. 80: 466–479.

–, 1961 b: Zur Lichtkonkurrenz von Ackerunkräutern. Flora 151: 126–143.

–, 1974: Die Unkrautvegetation im Bereich der Stadt Köln I. Die Pflanzengesellschaften. II. Der Zeigerwert der Arten. Decheniana 126, 267–306 u. 307–332.

–, Eber, W., 1967: Die Pflanzengesellschaften der Keuperhügel bei Friedland (Kr. Göttingen). Schriftenreihe f. Vegetationskunde (Bonn) 2: 135–160.+

–, Salinger, S., Strehlow, H., 1975: Substanzproduktion und Inhaltsstoffe zweier Gräser in Rein- und Mischkultur. Flora 164: 437–448.

Borstel, U.-O. von, 1974: Untersuchungen zur Vegetationsentwicklung auf ökologisch verschiedenen Grünland- und Ackerbrachen hessischer Mittelgebirge (Westerwald, Rhön, Vogelsberg). Diss. Univ. Gießen: 159 S.

Borza, A., 1963: Pflanzengesellschaften der rumänischen Karpaten. Biológia (Bratislava) 18: 856–864.

Bothmer, H.J., 1953: Der Einfluß der Bewirtschaftung auf die Ausbildung der Pflanzengesellschaften niederrheinischer Dauerweiden. Z. Acker- u. Pflanzenbau 96: 4, 457–476.

Boukhris, M., 1967: Sur l'écologie et la nutrition des végétaux croissant sur dolomie dans le sud de la France. Centre Nation. Rech. Sci., Centre d'Études Phytosociol. et Ecol., Doc. 40: 104 S.

Bournérias, M., 1959: Le peuplement végétal des espaces nus. Bull. Soc. Botan. France Mém. 106: 300 S.

Bracker, H.H., 1960: Zweijährige Gefäßversuche über den Einfluß unterschiedlicher Wasserversorgung auf das Wachstum und die Ertragsleistung von 14 Kulturgräsern. Der Kulturtechn. 47: 60–72.

Bradshaw, A.D., 1976: Pollution and Evolution. In: Mansfield, T.A. (ed.): Effects of air pollutants on plants. Cambridge University Press 1976: 135–159.

–, Chadwick, M.J., Jowett, D., Lodge, R.W., Snaydon, R.W., 1960: Experimental investigations into the mineral nutrition of several grass species. III. Phosphate level. J. Ecol. 48: 631–637.

–, Chadwick, M.J., Jowett, D., Snaydon, R.W., 1964: Experimental investigations into the mineral nutrition of several grass species. IV. Nitrogen level. J. Ecol. 52: 665–676.

–, Lodge, R.W., Jowett, D., Chadwick, M.J., 1960: Experimental investigations into the mineral nutrition of several grass species. Part. II. Calcium and pH. J. Ecol. 48: 143–150.

Brandes, D., 1981: Gefährdete Ruderalgesellschaften in Niedersachsen und Möglichkeiten zu ihrer Erhaltung. Göttinger Florist. Rundbr. 14: 90–98.

Braun, J., 1913: Die Vegetationsverhältnisse der Schneestufe in den Rätisch-Lepontischen Alpen. Ein Bild des Pflanzenlebens an seinen äußersten Grenzen. Neue Denkschr. Schweiz. Naturf. Ges. 48: 347 S.

–, 1915: Les Cévennes méridionales (Massif de l'Aigoual). Arch. Sci. Phys. et Nat. Genève, 4. Sér. 39/40: 207 S.

Braun-Blanquet, J., 1932: Zur Kenntnis nordschweizerischer Waldgesellschaften. Beih. Botan. Cbl. 49: 7–42.

–, 1948/50: Übersicht der Pflanzengesellschaften Rätiens. Vegetatio 1: 29–41, 129–146, 285–316; 2: 20–37, 214–237, 341–360.

–, 1957: Ein Jahrhundert Florenwandel am Piz Linard. Bull. Jard. Botan., Bruxelles, Vol. Jubil. W. Robyns, 221–232.

–, 1958: Über die obersten Grenzen pflanzlichen Lebens im Gipfelbereich des schweizerischen Nationalparks. Ergebn. Wiss. Unters. Schweiz. Nationalpark N. F. 6: 119–142.

*–, 1961: Die inneralpine Trockenvegetation. Geobotanica selecta 1: 273 S.

*–, 1964: Pflanzensoziologie. Wien 1928, 2. Aufl. Wien 1951: 631 S.; 3. Aufl. 1964: 865 S.

–, 1969: Die Pflanzengesellschaften der rätischen Alpen im Rahmen ihrer Gesamtverbreitung. I. Teil. Chur: Bischofberg u. Co., 1969: 100 S.

–, Jenny, H., 1926: Vegetationsentwicklung und Bodenbildung in der alpinen Stufe der Zentralalpen (Klimaxgebiet des Caricion curvulae). Denkschr. Schweiz. Naturf. Ges. 63: 183–349.

–, de Leeuw, W.C., 1936: Vegetationsskizze von Ameland. Nederl. Kruidk. Arch. 46: 359–393.

–, Pallmann, H., Bach, R., 1954: Pflanzensoziologische und bodenkundliche Untersuchungen im schweizerischen Nationalpark und seinen Nachbargebieten. II. Vegetation und Böden der Wald- und Zwergstrauchgesellschaften (Vaccinio-Piceetalia). Ergebn. Wiss. Unters. Schweiz. Nationalpark N. F. 4: 200 S.

–, Roussine, N., Nègre, R., 1951: Les groupements végétaux de la France méditerranéenne. Centre Nation. Rech. Sci., Serv. Carte des Groupements Végétaux: 297 S.

– u. G., Trepp, W., Bach, R., Richard, F., 1964: Pflanzensoziologische und bodenkundliche Beobachtungen im Samnaun. Jahresber. Naturf. Ges. Graubündens 90: 3–50.

–, Tüxen, R., 1943: Übersicht der höheren Vegetationseinheiten Mitteleuropas (unter Ausschluß der Hochgebirge) Comm. SIGMA 84, 11 S.

Braun, W., 1968: Die Kalkflachmoore und ihre wichtigsten Kontaktgesellschaften im Bayerischen Alpenvorland. Diss. Botan. 1: 134 S.

*Bray, J.R., Gorham, E., 1964: Litter production in forests of the world. Adv. Ecol. Res. 2, 101–157.

Brechtel, H.M., 1969: Gravimetrische Schneemessungen mit der Schneesonde „Vogelsberg". Die Wasserwirtschaft 59: 323–327.

–, 1975: Niederschlagsmessungen und Auswertung im Rahmen gebietshydrologischer Untersuchungen. Deut. Verb. Wasserwirtschaft (Bad Ems), 45 S.

–, Balázs, Á., 1975: Auf- und Abbau der Schneedecke im westlichen Vogelsberg in Abhängigkeit von Höhenlage, Exposition und Vegetation. Beitr. Hydrol. (Freiburg i.Br.) 3, 35–107.

Brehm, K., 1968: Die Bedeutung des Kationenaustausches für den Kationengehalt lebender Sphagnen. Planta (Berlin) 79: 324–345.

–, 1971: Ein Sphagnum-Bult als Beispiel einer natürlichen Ionenaustauschersäule. Beitr. Biol. Pflanzen 47: 287–312.

Breitsprecher, G., 1935: Vergleichende Transpirationsmessungen an Pflanzen der Hiddenseer Dünenheide. Mitt. Naturw. Ver. Neuvorpommern u. Rügen 62: 5–91.

Breymeyer, A., 1974: Analysis of a sheep pasture ecosystem in the Pieniny mountains (the Carpathians) XI. The role of coprophagous beatles (Coleoptera, Scarabaeidae) in the utilization of sheep dung. Ekol. Pol. 22: 617–634.

Briemle, G., 1980: Untersuchungen zur Verbuschung und Sekundärbewaldung von Moorbrachen im südwestdeutschen Alpenvorland. Diss. Botan. 57, 286 S.

Brockmann, C., 1935: Diatomeen und Schlick im Jadegebiet. Abh. Senckenberg. Naturf. Ges. 430: 1–64.

Brockmann-Jerosch, H., 1907: Die Flora des Puschlav (Bezirk Bernina, Kanton Graubünden). Diss. Univ. Zürich: 236 S.

–, 1913: Der Einfluß des Klimacharakters auf die Verbreitung der Pflanzen und Pflanzengesellschaften. Botan. Jb. 49, Beibl. 109: 19–43.

–, 1936: Futterlaubbäume und Speiselaubbäume. Ber. Schweiz. Botan. Ges. 46, Festbd. Rübel: 594–613.

Brouillard, Ch., 1911: Le traitement du bois en France. 3. Aufl. Paris u. Nancy: 685 S.

Brown, R.T., Mikola, P., 1975: Influence of reindeer lichens on growth of mycorrhizae and seedling growth of trees. Abstr. XIVth Internat. Botan. Congr. Leningrad 1975, 1: 139.

Brülhardt, A., 1969: Jahreszeitliche Veränderungen der Feuchtigkeit im Boden. Mitt. Schweiz. Anst. Forstl. Versuchswes. 45: 127–232.

Brun-Hool, J., 1963: Ackerunkrautgesellschaften der Nordwestschweiz. Beitr. Geobot. Landesaufn. Schweiz 43.

Brzoska, W., 1973: Stoffproduktion und Energiehaushalt von Nivalpflanzen. In: Ellenberg (Hrsg.): Ökosystemforschung. Springer-Verlag Heidelberg, Berlin, New York: 225–234.

Bublinec, E., 1974: Bodenpodsolierung unter Kiefernbeständen. Pedologica (Bratislava) 8: 119 S.

Buchli, M., 1936: Ökologie der Ackerunkräuter in der Nordostschweiz. Beitr. Geobot. Landesaufn. Schweiz 19, 354 S.

Buchwald, K., 1951: Wald- und Forstgesellschaften der Revierförsterei Diensthoop, Forstamt Syke b. Bremen. Angew. Pflanzensoziol. Stolzenau/Weser) 1: 72 S.

Bücking, W., 1970: Nitrifikation als Standortfaktor von Waldgesellschaften. Diss. Freiburg i.Br., 1970: 84 S.

–, 1972: Zur Stickstoffversorgung von südwestdeutschen Waldgesellschaften. Flora 161: 384–400.

–, 1975: Nährstoffgehalte in Gewässern aus standörtlich verschiedenen Waldgebieten Baden-Württembergs. Mitt. Ver. Forstl. Standortskunde u. Forstpflanzenzücht. 24: 47–67.

BURCKHARDT, H., BURGSDORF, H. L., 1962: Floristische und pflanzensoziologische Betrachtung des Naturschutzgebietes „Schwarzes Wasser" bei Wesel. Gewässer u. Abwässer 1962, 36–98.

*BURGEFF, H., 1961: Mikrobiologie des Hochmoores mit besonderer Berücksichtigung der Erikazeen-Pilz-Symbiose, Stuttgart: 197 S.

*BURGER, K., u. Mitarb., 1974: Auswertung von Untersuchungen und Forschungsergebnissen zur Belastung der Landschaft und ihres Naturhaushaltes. Schr. R. Landschaftspflege u. Naturschutz (Bonn-Bad Godesberg) 10: 119 S.

BÜRING, H., 1970: Sozialbrache auf Äckern und Wiesen in pflanzensoziologischer und ökologischer Sicht. Diss. Univ. Gießen: 81 S.

BURNAND, J., 1976: Quercus pubescens-Wälder und ihre ökologischen Grenzen im Wallis (Zentralalpen). Veröff. Geobot. Inst. ETH, Stiftg. Rübel, Zürich 59, 158 S.

BURRICHTER, E., 1960: Die Therophyten-Vegetation an nordrhein-westfälischen Talsperren im Trockenjahr 1959. Ber. Deut. Botan. Ges. 73: 24–37.

–, 1963: Das Linarietum spuriae Krusem. et Vlieger 1939 in der Westfälischen Bucht. Mitt. Florist.-Soziol. Arb. gem. N. F. 10: 109–115.

–, 1968: Überblick über die Vegetation des Zwillbrocker Venns. Mitt. Flor.-Soz. Arb. gem. N. F. 13: 275–279.

–, 1969: Das Zwillbrocker Venn, Westmünsterland, in moor- und vegetationskundlicher Sicht. Abh. Landesmus. Naturk. Münster in Westf. 31, 1: 60 S.

–, 1973: Die potentielle natürliche Vegetation in der Westfälischen Bucht. Erläuterungen zur Übersichtskarte 1 : 200 000. Landeskundl. Karten u. H. Geogr. Kommiss. Westf., R. Siedlung u. Landsch. Westf. 8: 58 S.

–, 1976: Vegetationsräumliche und siedlungsgeschichtliche Beziehungen in der Westfälischen Bucht. Abh. Landesmus. Naturk. Münster i. W. 38: 3–14.

BURSCHEL, P., 1966 a: Untersuchungen in Buchen-Mastjahren. Forstwiss. Cbl. 85: 193–256.

–, 1966 b: Untersuchungen über die Düngung von Buchen- und Eichen-Verjüngungen, Teil 2. Allg. Forst- u. Jagdztg. 137: 221–236.

–, HUSS, J., KALBHENN, R., 1964: Die natürliche Verjüngung der Buche. Schr. R. Forstl. Fak. Univ. Göttingen 34: 186 S.

–, SCHMALTZ, J., 1965: Untersuchungen über die Bedeutung von Unkraut- und Altholzkonkurrenz für junge Buchen. Forstwiss. Cbl. 84: 201–264.

CALDWELL, M. M., 1970: The wind regime at the surface of the vegetation layer above timberline in the Central Alps. Cbl. Ges. Forstwes. 87: 65–74.

CAMPELL, E., TREPP, W., 1968: Vegetationskarte des schweizerischen Nationalparks. W. TREPP: Beschreibung der Pflanzengesellschaften. Ergebn. Wiss. Unters. Schweiz. Nationalpark 11: 19–42.

CAMPINO, I., 1978: Einfluß der Nutzungsintensität auf Kompartimente von Grünlandökosystemen. Diss. Univ. Gießen, 249 S.

CAPUTA, J., 1948: Untersuchungen über die Entwicklung einiger Gräser und Kleearten in Reinsaat und Mischung. Diss. ETH Zürich: 127 S.

CARBIENER, R., 1963: Les sols du massif du Hohneck. Aspects physiques, biologiques et humains. Edit. Ass. Philomat. Alsace et Lorraine, Straßburg 1963: 103–154.

–, 1964: La détermination de la limite naturelle de la forêt par des critères pédologiques et géomorphologiques dans les hautes Vosges et dans le Massif Central. Compt. Rend. Acad. Sc. Paris 258: 4136–4138.

–, 1969: Subalpine primäre Hochgrasprärien im herzynischen Gebirgsraum Europas, mit besonderer Berücksichtigung der Vogesen und des Massif Central. Mitt. Florist.-Soziol. Arb. gem. N. F. 14: 322–345.

–, 1970 a: Un exemple de type forestier exceptionel pour l'Europe occidentale: La forêt du lit majeur du Rhin au niveau du fossé Rhénan (Fraxino-Ulmetum Oberd. 53). Vegetatio 20: 97–148.

–, 1970 b: Frostmusterboden, Solifluktion, Pflanzengesellschafts-Mosaik und -Struktur, erläutert am Beispiel der Hochvogesen. In: Gesellschaftsmorphologie (Strukturforschung). Den Haag: Verl. Dr. W. Junk, 1970: 187–217.

–, 1975: Die linksrheinischen Naturräume und Waldungen der Schutzgebiete von Rhinau und Daubensand (Frankreich), eine pflanzensoziologische und landschaftsökologische Studie. In: Das Taubergießengebiet. Die Natur- und Landschaftsschutzgebiete Baden-Württembergs 7: 438–535.

–, OURISSON, N., BERNARD, A., 1975: Erfahrungen über die Beziehungen zwischen Großpilzen und Pflanzengesellschaften in der Rheinebene und den Vogesen. Beitr. Naturk. Forsch. Südw. Deut. 34: 37–56.

CARTELLIERI, E., 1940: Über Transpiration und Kohlensäureassimilation an einem hochalpinen Standort. Sitz. ber. Akad. Wiss. Wien, Math.-Nat. Kl., I 149: 95–154.

CARTLEDGE, O., COMOR, D. J., 1973: Photosynthetic efficiency of tropical and temperate grass canopies. Phytosynthetica 7: 109–113.

CASPARIE, W. A., 1969: Bult- und Schlenkenbildung in Hochmoortorf. Vegetatio 19: 146–180.

CASPERS, H., 1975: Pollution in coastal waters. DFG Research Rep., Boppard (Harald Boldt Verl.), 142 S.

CATE, C. L. TEN, 1972: Wan god mast gift ... Bilder aus der Geschichte der Schweinezucht im Walde. Wageningen: Centre for Agricultural Publishing and Documentation 1972: 300 S.

CELIŃSKI, F., FILIPEK, M., 1958: The flora and plant communities in the forest-steppe reserve in Bielinek on the Oder. Badan. Fizjogr. Pol. Zach. 4: 198 S.

–, KRASKA, M., 1969: L'influence des vents dominants et des expositions des pentes sur la formation des habitats forestiers de la grande forêt de hêtre près de Szczecin. Ebenda 22: 53–67.

CERNUSCA, A. (Hrsg.), 1977: Alpine Grasheide Hohe Tauern. Ergebnisse der Ökosystemstudie 1976. Univ. Verlag Wagner, Innsbruck: 175 S.

ČEŘOVSKÝ, J., 1960: Über die Felsenpflanzen (Petrophyten). Ochrana Přírody (Prag) 15: 97–114.

CHALON, M.-P., DEVILLEZ, F., DUMONT, J. M., 1977: Recherches sur les variations de la teneur en eau chez Leucobryum glaucum (Hedw.) Schimp. In: DUVIGNEAUD u. KESTEMONT 1977: 69–72.

CHAPMAN, V. J., 1975: The salinity problem in general, its importance, and distribution with special reference to natural halophytes. Ecol. Studies 15: 7–24.

–, 1979: Some interrelationship between soil and root respiration in lowland Calluna heathland in southern England. J. Ecol. 67: 1–20.

CHRISTIANSEN, W., 1955: Salicornietum. Mitt. Florist.-Soziol. Arb. gem. N. F. 5: 64–65.

–, 1960: Vegetationsstudien auf Helgoland. Schr. Naturw. Ver. Schlesw.-Holst. 31: 3–24.

CHWASTEK, M., 1963: The influence of nutritional soil resources, especially phosphorus content, on the dominance of Molinia coerulea (L.) Moench in the meadow sward. Poznán Soc. Friends of Sci., Sect. Agric. and Sylvicult. Sci. 14: 277–356.

CLATWORTHY, J. N., HARPER, J. L., 1962: The comparative biology of closely related species living in the same area. V. Inter- and intraspecific interference within cultures of Lemna ssp. and Salvinia natans. J. Exper. Botan. 13, No. 38: 307–324.

CLAUSNITZER, I., 1976: Stickstoffmineralisation in verschiedenen Entwicklungsphasen von Buchenwäldern und Fichtenforsten. Diplomarb. Göttingen: 58 S.

CLIFFORD, H. T., 1959: Seed dispersal by motor vehicles. J. Ecol. 47: 311–315.

CLYMO, R. S., 1964: The origin of acidity in Sphagnum bogs. The Bryol. 67: 427–431.

–, 1973: The growth of Sphagnum: Somme effects of environment. J. Ecol. 61: 849–869.

–, REDDAWAY, E. J. F., 1974: Growth rate of Sphagnum rubellum Wils. on pennine blanket bog. J. Ecol. 62: 191–196.

COLTERMAN, H. L., 1975: Physiological Limnology. An approach to the physiology of lake ecosystems. Elsevier Sci. Publ. Comp. Amsterdam u. New York: 490 S.

CONVAY, V. M., 1937: Studies in the autecology of Cladium mariscus R. Br. III. The aeration of the subterranean parts of the plant. New Phytol. 36: 64–96 (1937), siehe auch 35: 177–204, 359–380 (1936) und 36: 312–328.

COOPER, A., ETHERINGTON, J. R., 1974: The vegetation of carboniferous limestone soils in South Wales. I. Dolomitization, soil magnesium status and plant growth. J. Ecol. 62: 179–190.

COULON, M. DE: 1959: Résineux et feuillus. La Forêt 13: 2–10.

COULT, D. A., 1964: Observations on gas movement in the rhizome of Menyanthes trifoliata L., with comments on the role of the endodermis. J. Exper. Botan. 15: 205–218.

CSAPODY, I., 1964: Die Waldgesellschaften des Soproner Berglandes. Acta Bot. Acad. Sci. Hung. 10: 43–85.+

CSÜRÖS, S., 1963: Kurze allgemeine Kennzeichnung der Pflanzendecke Siebenbürgens. Acta Botan. Horti Bucurest. 1961/62, Fasc. II: 825–854 (1963).+

CURL, H., HARDY, J. T., ELLERMEIER, R., 1972: Spectral absorption of solar radiation in alpine snowfields. Ecology 53: 1189–1194.

CZERWIŃSKI, Z., JAKUBCZYK, H., NOWAK, E., 1974: Analysis of a sheep pasture ecosystem in the Pieniny mountains (the Carpathians) XII. The effect of earthworms on the pasture soil. Ekol. Pol. 22: 635–650.

DAFIS, S. A., 1962: Struktur- und Zuwachsanalysen von natürlichen Föhrenwäldern. Beitr. Geobot. Landesaufn. Schweiz 41: 86 S.

DAHL, E., 1951: On the relation between summer temperature and the distribution of alpine vascular plants in the lowlands of Fennoscandia. Oikos 3: 22–52.

DAMMAN, A. W. H., 1957: The South-Swedish Calluna-heath and its relation to the Calluneto-Genistetum. Botan. Not. (Lund) 110: 363–398.

DÄSSLER, H.-G., RANFT, H., REHN, K.-H., 1972: Zur Widerstandsfähigkeit von Gehölzen gegenüber Fluorverbindungen und Schwefeldioxid. Flora 161: 289–302.

DAVIES, M. S., SNAYDON, R. W., 1974: Physiological differences among populations of Anthoxanthum odoratum L. collected from the park grass experiment, Rothamsted. III. Response to phosphate. J. Appl. Ecol. 11: 699–707.

DAVY, A. J., TAYLOR, K., 1974 a: Water characteristics of contrasting soils in the Chiltern Hills and their significance for Deschampsia caespitosa (L.) Beauv. J. Ecol. 62: 367–378.

–, –, 1974 b: Seasonal patterns of nitrogen availability in contrasting soils in the Chiltern Hills. J. Ecol. 62: 793–807.

DELECOUR, F., 1968: Distribution des oligo-éléments cuivre, zinc et molybdène dans les sols forestiers de l'Ardenne belge. Pédologie (Gand) 18: 43–62 u. 156–175.

*DELZENNE-VAN HALUWYN, C., 1976: Bibliographia societarum lichenorum. Suppl. Bibliogr. Phytosoc. Syntax. (Vaduz) 1, 177 S.

DENAEYER-DE SMET, S., 1966: Bilan annuel des apports d'éléments minéraux par les eaux de précipitation sous couvert forestier dans la forêt mélangée caducifoliée de Blaimont (Virelles-Chimay). Bull. Soc. Roy. Botan. Belg. 99: 345–375.

–, DUVIGNEAUD, P., 1972: Comparaison du cycle des polyéléments biogènes dans une hêtraie (Fagetum) et une pessiere (Piceetum) établies sur même roche-mère, à Mirwart (Ardenne Luxembourgeoise). Bull. Soc. Roy. Botan. Belg. 105: 197–205.

DENAEYER, S., LEJOLY, J., DUVIGNEAUD, P., 1968: Note sur la spécifité biogéochimique des halophytes du littoral belge. Bull. Soc. Roy. Botan. Belg. 101: 293–301.

DENGLER, A., 1904/1912: Untersuchungen über die natürlichen und künstlichen Verbreitungsgebiete einiger forstlich und pflanzengeographisch wichtiger Holzarten in Nord- und Mitteldeutschland. I. Die Horizontalverbreitung der Kiefer (Pinus silvestris L.). Mitt. Forstl. Versuchsw. Preußens 1904. II. Die Horizontalverbreitung der Fichte (Picea excelsa Lk). III. Die Horizontalverbreitung der Weißtanne (Abies pectinata DC). Ebenda 1912.

–, 1930: Waldbau auf ökologischer Grundlage. Berlin: 560 S.

DENISIUK, Z., 1963: Vegetation of deciduous forests in the regions of Leśna Podlaska. Poznań Soc. Friends of Sci., Dep. Math. Nat. Sci., Sect. Biol. 17, 2: 132 S.+

DE RHAM, P., 1970: L'azote dans quelques forêts, savanes et terrains de culture d'Afrique tropicale humide (Côte-d'Ivoire). Veröff. Geobot. Inst. ETH, Stiftg. Rübel, Zürich 45: 124 S.

DE SLOOVER, J., LE BLANC, F., 1968: Mapping of atmosphere pollution on the basis of lichen sensivity. Proc. Sympos. Recent Advances Trop. Ecol. 1968: 42–56.

DETHIOUX, M., 1969: La hêtraie à mélique et aspérule des districts mosan et ardennais. Bull. Rech. Agron. Gembloux, N. S. 4: 471–481.

DIELS, L., 1914: Die Algen-Vegetation der Südtiroler Dolomitriffe. Ein Beitrag zur Ökologie der Lithophyten. Ber. Deut. Botan. Ges. 32: 507–531.

–, 1918: Das Verhältnis von Rhythmik und Verbreitung bei den Perennen des europäischen Sommerwaldes. Ebenda 36: 337–351.

DIEMONT, W. H., 1938: Zur Soziologie und Synökologie der Buchen- und Buchenmischwälder der nordwestdeutschen Mittelgebirge. Mitt. Florist.-Soziol. Arb. gem. Niedersachsen 4: 5–182.

–, SISSINGH, G., WESTHOFF, V., 1940: Het dwergbiezen-verbond (Nanocyperion flavescentis) in Nederland. Nederl. Kruidk. Arch. 50: 215–284.

DIEREN, J. W. VAN, 1934: Organogene Dünenbildung. Den Haag: 304 S.

*DIERSCHKE, H., 1967: Excerpta Botanica B. Sociologica, eine weltumfassende Bibliographie der Vegetationskunde. Mitt. Florist.-Soziol. Arb. gem. 11/12: 251–254.

–, 1969: Vegetationskundliche Beobachtungen im Fimbertal (Silvretta-Unterengadin). In: Bericht über die Alpenexkursion des Systematisch-Geobotanischen Institutes in das Fimbertal, 17.-31. 7. 1969. Göttingen 1969.+

–, 1971: Stand und Aufgaben der Pflanzensoziologischen Systematik in Europa. Vegetatio 22: 255–264.

*–, 1974 a: Saumgesellschaften im Vegetations- und Standortsgefälle an Waldrändern. Scripta Geobot. (Göttingen) 6: 246 S.

–, 1974 b: Zur Syntaxonomie der Klasse Trifolio-Geranietea. Mitt. Florist.-Soziol. Arb. gem. N. F. 17: 27–38.

–, 1974 c: Zur Abgrenzung von Einheiten der heutigen potentiell natürlichen Vegetation in waldarmen Gebieten Nordwestdeutschlands. In: TÜXEN, R. (Hrsg.): Tatsachen und Probleme der Grenzen in der Vegetation. Verlag J. Cramer, Lehre: 305–325.

–, 1979: Die Pflanzengesellschaften des Holtumer Moores und seiner Randgebiete (Nordwest-Deutschland). Mitt. Florist.-Soziol. Arb. gem. N. F. 21: 111–143.+

–, TÜXEN, R., 1975: Die Vegetation des Langholter- und Rhauder Meeres und seiner Randgebiete. Mitt. Florist.-Soziol. Arb. gem. N. F. 18: 157–202.+

DIERSSEN, K., 1972: Sphagnum molle Sull., übersehene Kennart des Ericetum tetralicis. Ber. Naturhist. Ges. Hannover 116: 143–150.

–, 1973: Die Vegetation des Gildehauser Venns (Kreis Grafschaft Bentheim). Beih. Ber. Naturhist. Ges. Hannover 8: 120 S.

*–, 1975: Littorelletea uniflorae Br.-Bl. et Tx. 1943. Prodromus der europäischen Pflanzengesellschaften. Vaduz (J. Cramer): 149 S.

DIETERICH, H., MÜLLER, S., SCHLENKER, G., 1970: Urwald von morgen. Bannwaldgebiete der Landesforstverwaltung Baden-Württemberg. Stuttgart: Verl. Eugen Ulmer, 174 S.+

DIETL, W., 1972: Die Vegetationskartierung als Grundlage für die Planung einer umfassenden Alpverbesserung im Raume von Glaubenbüelen (Obwalden). In: Alpwirtschaft u. Landschaftspflege im Gebiet Glaubenbüelen/OW, herausgegeben vom Oberforstamt Obwalden, 6060 Sarnen (Schweiz): 115 S.+

DIETRICH, H., 1958: Untersuchungen zur Morphologie und Genese grundwasserbeeinflußter Sandböden im Gebiet des nordostdeutschen Diluviums. Arch. Forstwes. 7: 577–640.

DISTER, E., 1980: Geobotanische Untersuchungen in der Hess. Rheinaue als Grundlage für die Naturschutzarbeit. Diss. Univ. Göttingen, 170 S.

DOBAT, K., 1966: Die Kryptogamenvegetation der Höhlen und Halbhöhlen im Bereich der Schwäbischen Alb. Abh. Karst- u. Höhlenkunde, Reihe E, 3: 153 S.

DOING KRAFT, H., WESTHOFF, V., 1959: Die Stellung der Buche im west- und mitteleuropäischen Wald. Jb. Niederl. Dendrol. Ver. 21: 226–254.

DÖRR, M., 1941: Temperaturmessungen an Pflanzen des Frauensteins bei Mödling. Beih. Botan. Cbl. 60, 679–728.

DÖRRIE, A., 1958: Das Leistungsvermögen einer Marschweide bei intensiver Bewirtschaftung. Landwirtsch.-Angew. Wiss. 88: 78 S.

DOXTADER, K. G., ALEXANDER, M., 1966: Nitrification by heterogrophic soil microorganisms. Soil Sci. Soc. Am. Proc. 30: 351–355.

DREYLING, G., 1973: Spezifische und infraspezifische Mannigfaltigkeit der Gattung Puccinellia Parlatore (Poaceae) von der deutschen Nordseeküste. Diss. Hamburg: 150 S.

DROSTE ZU HÜLSHOFF, B. VON, 1969: Struktur und Biomasse eines Fichtenbestandes auf Grund einer Dimensionsanalyse an oberirdischen Baumorganen. Diss. Univ. München: 209 S.

DUCHAUFOUR, PH., BALANDREAN, J., QUELEN, D., 1971: Minéralisation de l'azote dans deux types de sols bruns acides vosgiens. Bull. École Nation. Sup. Agron. Nancy 13: 3–6.

–, TURPIN, P., 1960: Essais de fertilisation sur humus brût et contrôlés par l'analyse foliaire sur pin sylvestre et épicéa. Ann. École Nation. Eaux et Forêts et Stat. Rech. et Expér. 17: 209–233.

DUHME, F., KAULE, G., 1970: Zur Verbreitung der gelben Narzisse (Narcissus pseudonarcissus L.) auf Primär- und Sekundärstandorten in Mittel- und Nordwesteuropa. Ber. Deut. Botan. Ges. 83: 647–659.

DÜLL, R., o. J. (1973?): Neuere Untersuchungen über Moose als abgestufte ökologische Indikatoren für die SO_2-Immissionen im Industriegebiet zwischen Rhein und Ruhr bei Duisburg. VDI-Kommission Reinhaltung der Luft, o. J.

–, 1980: Die Moose (Bryophyta) des Rheinlandes (Nordrhein-Westfalen, Bundesrepublik Deutschland). Decheniana Beih. 24, 365 S.

–, WERNER, H., 1956: Pflanzensoziologische Studien im Stadtgebiet von Berlin. Wiss. Z. Univ. Berlin 5: 321–331.

DU RIETZ, E. G., 1925: Zur Kenntnis der flechtenreichen Zwergstrauchheiden im kontinentalen Südnorwegen. Svenske Växtsoc. Sällsk. Handl. 4: 3–80.

–, 1954: Die Mineralbodenwasserzeigergrenze als Grundlage einer natürlichen Zweigliederung der nord- und mitteleuropäischen Moore. Vegetation 5/6: 571–585.

DUVIGNEAUD, J., 1967: Flore et végétation halophiles de la Lorraine orientale (Dép. Moselle, France). Mém. Soc. Roy. Botan. Belg. 3: 122 S.

–, MULLENDERS, W., 1961: La végétation forestière des Cotes lorraines: La forêt du Mont-Dieu (Département des Ardennes, France). Bull. Soc. Roy. Botan. Belg. 94: 91–130.+

DUVIGNEAUD, P., 1946: La variabilité des associations végétales. Bull. Soc. Roy. Botan. Belg. 78: 107–134.

–, DENAEYER-DE SMET, S., 1962: Distributions de certains éléments minéraux (K, Ca et N) dans les tapis végétaux naturels. Bull. Soc. Franç. Physiol. Végét. 8: 1–8.

–, –, 1967: Biomass, productivity and mineral cycling in deciduous mixed forests in Belgium. In: Symposium on Productivity and Mineral Cycling in Natural Ecosystems. Univ. of Maine (USA) 1967: 167–186.

–, –, 1970 a: Biomasse, productivité et phytogéochimie de la végétation riveraine d'un ruisseau Ardennais. Bull. Soc. Roy. Botan. Belg. 103: 355–396.

–, –, 1970 b: Phytochimie des groupes ecosociologiques forestiers de Haute-Belgique. I. Essai de classification phytochimique des espèces herbacées. Oec. Plant. 5: 1–32.

–, –, 1971: Cycle des éléments biogènes dans les écosystèmes forestiers d'Europe (principalement forêts caducifoliées). Ecol. and Conserv. 4: 527–542.

–, KESTEMONT, P. (Éd.), 1977: Productivité biologique en Belgique. SCOPE, Trav. Sect. Belge Progr. Biol. Internat. (Paris-Gembloux): 617 S.

DYKYJOVÁ, D., 1971: Ecomorphoses and ecotypes of Phragmites communis Trin. Preslia (Praha) 43: 120–138.

–, HRADECKÁ, D., 1976: Production ecology of Phragmites communis 1. Relations of two ecotypes to the microclimate and nutrient conditions of habitat. Folia Geobot. Phytotax. Praha 11: 23–61.

–, ONDOK, J. P. 1973: Biometry and the productive stand structure of coenoses of Sparganium erectum L. Preslia, Praha 45: 19–30.

–, ONDOK, P. J., HRADECKÁ, D., 1972: Growth rate and development of the root/shoot ratio in reedswamp macrophytes grown in winter hydroponic cultures. Folia Geobot. Phytotax., Praha 7: 259–268.

–, VÉBER, K., PŘIBÁŇ, K., 1971: Productivity and root/shoot ratio of reedswamp species growing in outdoor hydroponic cultures. Folia Geobot.

Phytotax. (Praha) 6: 233–254.

Džatko, M., 1972: Synökologische Charakteristik der Waldgesellschaften im nördlichen Teil des Donauflachlandes. Biol. Práce 18: 4, 95 S. +

–, 1974: Ecological aspects of the differences between chierniczas and chernozems. Ved. Práce Výskum. Úst. Pôdozn. (Bratislava) 1974: 31–39.+

Eber, W., 1972: Über das Lichtklima von Wäldern bei Göttingen und seinen Einfluß auf die Bodenvegetation. Scripta Geobot. 3: 150 S.

Eggelsmann, R., 1967: Oberflächengefälle und Abflußregime der Hochmoore. Wasser u. Boden 19: 247–252.

Eggers, Th., 1969: Über die Vegetation im Gotteskoog (Nordfriesland) nach der Melioration. Mitt. Arb. gem. Floristik Schlesw.-Holst. u. Hamburg 17: 95 S.

Eggler, J., 1955: Beitrag zur Serpentinvegetation in der Gulsen bei Kraubath in Obersteiermark. Mitt. Naturw. Ver. Steiermark 85: 27–72.

–, 1958: Wiesen und Wälder des Saßtales in der Steiermark. Ebenda 88: 23–50.

Ehlers, M., 1960: Baum und Strauch in der Gestaltung der deutschen Landschaft. Berlin u. Hamburg: 279 S.

Ehrendorfer, F. (Hrsg.), 1973: Liste der Gefäßpflanzen Mitteleuropas. 2. Aufl. Gustav Fischer-Verlag, Stuttgart: 318 S.

Ehrhardt, F., 1961: Untersuchungen über den Einfluß des Klimas auf die Stickstoff-Nachlieferung von Waldhumus in verschiedenen Höhenlagen der Tiroler Alpen. Forstwiss. Cbl. 80: 193–215.

Ehwald, E., 1957: Über den Nährstoffkreislauf des Waldes. Sitz. ber. Deut. Akad. Landwirtsch. wiss. Berlin 6: 56 S.

Eichrodt, R., 1969: Über die Bedeutung von Moderholz für die natürliche Verjüngung im subalpinen Fichtenwald. Diss. E.T.H. Zürich: 122 S.

Ellenberg, H., 1937: Über die bäuerliche Wohn- und Siedlungsweise in NW-Deutschland in ihrer Beziehung zur Landschaft, insbesondere zur Pflanzendecke. Mitt. Florist. Soziol. Arb. gem. Niedersachsen 3: 204–235.

–, 1939: Über Zusammensetzung, Standort und Stoffproduktion bodenfeuchter Eichen- und Buchen-Mischwaldgesellschaften Nordwestdeutschlands. Ebenda 5: 3–135.

–, 1950: Unkrautgemeinschaften als Zeiger für Klima und Boden. Landwirtschaftliche Pflanzensoziologie I, Verlag Eugen Ulmer, Stuttgart: 141 S.

–, 1952 a: Wiesen und Weiden und ihre standörtliche Bewertung. Ebenda II, Stuttgart, 143 S.

–, 1952 b: Auswirkungen der Grundwassersenkung auf die Wiesengesellschaften am Seitenkanal westlich Braunschweig. Angew. Pflanzensoziol. (Stolzenau/Weser) 6: 46 S.

–, 1953 a: Führt die alpine Vegetations- und Bodenentwicklung auch auf reinen Karbonatgesteinen zum Krummseggenrasen (Caricetum curvulae)? Ber. Deut. Botan. Ges. 66: 241–246.

–, 1953 b: Physiologisches und ökologisches Verhalten derselben Pflanzenarten. Ebenda 65: 351–362.

–, 1954 a: Naturgemäße Anbauplanung, Melioration und Landespflege. Landwirtschaftliche Pflanzensoziologie III, Stuttgart: 109 S.

–, 1954 b: Über einige Fortschritte der kausalen Vegetationskunde. Vegetatio 5/6: 199–211.

–, 1956: Aufgaben und Methoden der Vegetationskunde. Eugen Ulmer Verl., Stuttgart: 156 S.

*–, 1958: Bodenreaktion (einschließlich Kalkfrage). Handb. Pflanzenphysiol. 4: 638–708.

–, 1959: Typen tropischer Urwälder in Peru. Schweiz. Z. Forstwes. 110: 169–187.

*–, 1963: Vegetation Mitteleuropas mit den Alpen (1. Auflage dieses Buches). Verlag Eugen Ulmer, Stuttgart: 943 S.

–, 1964: Stickstoff als Standortsfaktor. Ber. Deut. Botan. Ges. 77: 82–92.

–, 1966: Leben und Kampf an den Baumgrenzen der Erde. Naturw. Rundschau 4: 133–139.

–, 1967 a: Internationales Biologisches Programm. Beiträge der Bundesrepublik Deutschland. Bad Godesberg (DFG): 28 S.

–, (Hrsg.), 1967 b: Vegetations- und bodenkundliche Methoden der forstlichen Standortskartierung. Veröff. Geobot. Inst. ETH, Stiftung Rübel, Zürich 39: 296 S.

–, 1968: Wege der Geobotanik zum Verständnis der Pflanzendecke. Naturwissenschaften 55: 462–470.

–, 1969: Wald- und Feldbau im Knyphauser Wald, einer Heideaufforstung in Ostfriesland. Ber. Naturhist. Ges. Hannover 112: 17–90.

–, (Ed.), 1971: Integrated experimental ecology. Ecol. Stud. 2: 214 S.

–, 1973 a: Folgen der Belastung von Ökosystemen. DFG-Mitteilungen 73, 2: 11–17.

–, (Hrsg.), 1973 b: Ökosystemforschung. Springer-Verlag, Heidelberg, Berlin, New York: 280 S.

–, 1974: Zeigerwerte der Gefäßpflanzen Mitteleuropas. Scripta Geobot. (Göttingen) 9: 97 S., 2. Aufl. 1979, 122 S.

–, 1976: Zur Rolle der Pflanzen in natürlichen und bewirtschafteten Ökosystemen. Bayer. Landw. Jb. 53: 51–59.

–, 1977: Stickstoff als Standortsfaktor, insbesondere für mitteleuropäische Pflanzengesellschaften. Oecol. Plant. 12: 1–22.

–, Haeupler, H., Hamann, U., 1968: Arbeitsanleitung für die Kartierung der Flora Mitteleuropas (Ausgabe für die Bundesrepublik Deutschland). Mitt. Florist.-Soziol. Arb.gem. N.F. 13: 284–297.

–, Klötzli, F., 1967: Vegetation und Bewirtschaftung des Vogelreservates Neeracher Riet. Ber. Geobot. Inst. ETH, Stiftg. Rübel, Zürich 37: 88–103.

–, –, 1972: Waldgesellschaften und Waldstandorte der Schweiz. Mitt. Schweiz. Anst. Forstl. Versuchswes. 48: 388–930.

–, Mueller-Dombois, D., 1967 a: A key to Raunkiaer plant life forms with revised subdivisions. Ber. Geobot. Inst. ETH, Stiftg. Rübel, Zürich 37:

–, –, 1967 b: Tentative physiognomic-ecological classification of plant formations of the earth. Ebenda 37: 21–55.

–, Rehder, H., 1962: Natürliche Waldgesellschaften der aufzuforstenden Kastanienflächen im Tessin. Schweiz. Z. Forstwes. 113: 128–142.

–, Snoy, M.-L., 1957: Physiologisches und ökologisches Verhalten von Ackerunkräutern gegenüber der Bodenfeuchtigkeit. Mitt. Staatsinst. Allg. Botan. Hamburg 11: 47–87.

Elster, H.-J., 1963: Die Stoffwechseldynamik der Binnengewässer. Verh. Deut. Zool. Ges. (München) 1963: 335–387.

–, 1966: Über die limnologischen Grundlagen der biologischen Gewässer-Beurteilung in Mitteleuropa. Verh. Internat. Verein. Limnol. 16: 759–785.

–, 1974: Das Ökosystem Bodensee in Vergangenheit, Gegenwart und Zukunft. Schr. Ver. Geschichte Bodensees u. seiner Umgebung 92: 233–250.

–, 1977: Der Bodensee – Bedrohung und Sanierungsmöglichkeiten eines Ökosystems. Naturwissenschaften 64: 207–215.

Emanuelsson, A., Eriksson, E., Egnér, H., 1954: Composition of atmosphere precipitation in Sweden. Tellus 6: 261–267.

Engel, H., 1949: Trümmerpflanzen von Münster. Natur u. Heimat 9: 2.

Ennik, G.C., 1976: De invloed van stikstofbemesting en oogstfrequentie op de bevorteling van gras. De Buffer (Wageningen) 22, 0–4 u. 5–18.

Ernst, W., 1965: Ökologisch-soziologische Untersuchungen der Schwermetall-Pflanzengesellschaften Mitteleuropas unter Einschluß der Alpen. Abh. Landesmus. Naturkunde Münster i. Westf. 27: 3–54.

–, 1969: Beitrag zur Kenntnis der Ökologie europäischer Spülsaumgesellschaften. I. Mitt.: Sand- und Kiesstrände. Mitt. Florist.-Soziol. Arb. gem. N. F. 14: 86–94.

–, 1973: Zink- und Cadmium-Immissionen auf Böden und Pflanzen in der Umgebung einer Zinkhütte. Ber. Deut. Botan. Ges. 85: 295–300.

*–, 1974: Schwermetallvegetation der Erde. Gustav Fischer Verlag, Stuttgart: 194 S.

–, 1979: Population biology of Allium ursinum in northern Germany. J. Ecol. 67: 347–362.

–, Mathis, W., Salaske, J., Janiesch, P., 1974: Aspekte von Schwermetallbelastungen in Westfalen. Abh. Landesmus. Naturk. Münster in Westf. 36: 3–30.

Eskuche, U., 1955: Vergleichende Standortsuntersuchungen an Wiesen im Donauried bei Herbertingen. Jber. Ver. Vaterl. Naturk. Württemb. 109: 33–135.

–, 1962: Herkunft, Bewegung und Verbleib des Wassers in den Böden verschiedener Pflanzengesellschaften des Erfttales. Arb. Bundesanst. Vegetationskartierung Stolzenau/Weser 1962: 72 S.

Etter, H., 1943: Pflanzensoziologische und bodenkundliche Studien an schweizerischen Laubwäldern. Mitt. Schweiz. Anst. Forstl. Versuchsw. 23: 5–132.

–, 1947: Über die Waldvegetation am Südostrand des schweizerischen Mittellandes. Ebenda 25: 141–210.

–, 1949: Über die Ertragsfähigkeit verschiedener Standortstypen. Ebenda 26: 91–152.

Eurola, S., 1962: Über die regionale Einteilung der südfinnischen Moore. Ann. Botan. Soc. „Vanamo" 33: 243 S.

–, 1968: Über die Ökologie der nordfinnischen Moorvegetation im Herbst, Winter und Frühling. Ann. Botan. Fenn. 5: 83–97.+

–, Ruuhijärvi, R., 1961: Über die regionale Einteilung der finnischen Moore. Arch. Soc. „Vanamo" 16 Suppl.: 49–63.

Evers, F.H., 1964: Die Bedeutung der Stickstoff-Form für Wachstum und Ernährung der Pflanzen, insbesondere der Waldbäume. Mitt. Ver. Forstl. Standortskunde u. Forstpflanzenzüchtg. 14: 19–37.

–, Schöpfer, W., Mikloss, J., 1968: Die Zusammenhänge zwischen Stickstoff, Phosphor- und Kalium-Mengen (in kg/ha) und den C/N-, C/P- und C/K-Verhältnissen der Oberböden von Waldstandorten. Mitt. Ver. Forstl. Standortskunde u. Forstpflanzenzüchtung 18: 59–71.

Fabijanowski, J., 1950: Untersuchungen über Zusammenhänge zwischen Exposition, Relief, Mikroklima und Vegetation in der Fallätsche bei Zürich. Beitr. Geobot. Landesaufn. Schweiz 29: 104 S.

Fabiszewski, J., 1967: Associations de lichens arboricoles dans les forêts des Sudètes orientales. Vegetatio 15: 137–165.

–, 1968: Les lichens du Massif Śnieżnik et des Montagnes Białskie dans les Sudètes orientales. Monogr. Botan. (Warszawa) 26: 115 S.

–, 1975: Stagnant stages of peat bogs in Central Europe and Canada. XII. Internat. Botan. Congr. (Leningrad) Abstr. 1: 143.

Faille, A., 1975: Recherches sur les écosystèmes des réserves biologiques de la forêt de Fontainebleau V. Evolution à court terme des humus à la suite de l'ouverture de clairières. Oecol. Plant. 10: 43–62.

Faliński, J.B., 1961: Végétation des chemins forestiers du Parc National de Białowieza (Pologne). Acta Soc. Bot. Polon. 30: 163–185.

–, 1963: Groupements piétinés des parties occidentales du terrain bas Grande Pologne-Cuiavie. Ebenda 32: 81–99.

–, 1968: Park Narodowy uv Puszczy Białowieskiej. Państw. Wyd. R. i Leśne, Warszawa: 350 S.

–, 1972: Végétation potentielle naturelle du pays des lacs de Masury (Partie Centrale). Phytocoenosis (Warszawa-Białowieża) 1: 79–94.

–, 1975: Anthropogenic changes of the vegetation of Poland. Ebenda 4: 97–116.

–, 1977: Research on vegetation and plant population dynamics conducted by Białowieża Geobotanical Station (1952–1977). Ebenda 6 (1 and 2).

–, Hrynkievicz-Sudnik, J., Fabiszewski, J., 1963: Broussailles champêtres (ordre Prunetalia) de la plaine de Kutno comme indicateur de la végétation potentielle de cette région. Acta Soc. Botan. Polon. 32: 693–714.

Fekate, G., 1974: Relative light intensity and distribution of herb layer species in oakwoods. Studia Bot. Hung. 9: 87–96.

*Feth, J.H., 1966: Nitrogen compounds in natural water – a review. Water Resources Research, Washington 2: 41.

Fetzmann, E.L., 1956: Beiträge zur Algensoziologie. Sitz. ber. Österr. Akad. Wiss., Math.-Nat. Kl., Abt. I, 165: 709–783.

–, 1961: Ein Beitrag zur Algenvegetation des Filzmooses bei Tarsdorf (Oberösterreich). Österr. Botan. Z. 108: 217–227.

Fiedler, H.-J., 1967: Zur Systematik der Braunerden im Mittelgebirgsbereich. Wiss. Z. T. U. Dresden 16: 1591–1602.

–, Nebe, W., 1969: Über Ernährung und Höhenwachstum von Fichtenbeständen in den oberen Berg- und Kammlagen des Osterzgebirges. Arch. Forstwes. 18: 747–756.

Filzer, P., 1951: Die natürlichen Grundlagen des Pflanzenertrages in Mitteleuropa. Gustav Fischer Verlag, Stuttgart: 198 S.

Firbas, F., 1931: Untersuchungen über den Wasserhaushalt der Hochmoorpflanzen. Jb. Wiss. Botan. 74: 457–696.

*–, 1949 u. 1952: Spät- und nacheiszeitliche Waldgeschichte von Mitteleuropa nördlich der Alpen, 1. Bd.: Allgemeine Waldgeschichte. Gustav Fischer, Jena: 480 S.; 2. Bd.: Waldgeschichte der einzelnen Landschaften. Jena: 256 S.

–, Willerding, U., 1965: Zur jüngeren Vegetationsgeschichte des Leinetals. Veröff. Max Planck-Inst. Geschichte 11, 2: 78–82.

Fischer, W., 1962: Die Quellflurgesellschaften der Ruppiner Schweiz (Nord-Brandenburg). Limnologica (Berlin) 1: 255–262.

Fischlin, A., Baltensweiler, W., 1979: Systems analysis of the larch bud moth system. Part 1: the larch-larch bud moth relationship. Mitt. Schweiz. Entomol. Ges. 52: 273–289.

Flach, E., 1963: Grundzüge einer spezifischen Bewölkungsklimatologie. Arch. Meteorol., Geophys. u. Bioklimatol., Ser. B, 12: 357–403.

–, 1967: Zur klimatologischen Charakteristik des Hochgebirges. Arch. Physikal. Therapie 19: 277–290.

Flohn, H., 1954: Witterung und Klima in Mitteleuropa. 2. Aufl., Forsch. Deut. Landeskunde 78: 214 S.

Florineth, F., 1974: Wasserhaushalt von Stipa pennata ssp. eriocaulis, Stipa capillata und Festuca vallesiaca im Steppengebiet des oberen Vinschgaus. Oecol. Plant 9: 295–314.

Follmann, G., Nakagava, M., 1968: Keimhemmung von Angiospermensamen durch Flechtenstoffe. Naturwiss. (Berlin) 50: 696–697.

Fonda, R.W., Bliss, L.C., 1969: Forest vegetation of the montane and subalpine zones, Olympic Mountains, Washington. Ecol. Monogr. 39: 271–301.

Förster, M., 1968 a: Über xerotherme Eichenmischwälder des deutschen Mittelgebirgsraumes. Diss. Hann.-Münden/Göttingen: 424 S.

–, 1968 b: Neufund von Quercus pubescens Willd. in Hessen. Hess. Florist. Rundbr. 17: 43–44.

–, 1975: Kennarten der Staudensäume oder der xerothermen Eichenwälder? Mitt. Florist.-Soziol. Arb. gem. N. F. 18: 258–264.

*Förstner, U., Müller, G., 1974: Schwermetalle in Flüssen und Seen als Ausdruck der Umweltverschmutzung. Springer-Verlag, Berlin-Heidelberg, New York: 225 S.

Fossati, A., 1980: Keimverhalten und frühe Entwicklungsphasen einiger Alpenpflanzen. Veröff. Geobot. Inst. ETH, Stiftg. Rübel, Zürich, 73, 193 S.

Frahm, J.-P., 1972: Die Vegetation auf Rethdächern. Diss. Univ. Kiel: 212 S.

*Franz, H., 1979: Ökologie der Hochgebirge. Verlag Eugen Ulmer, Stuttgart, 495 S. +

Frehner, H.K., 1963: Waldgesellschaften im westlichen Aargauer Mittelland. Beitr. Geobot. Landesaufn. Schweiz 44: 96 S.

Freitag, H., Körtge, U., 1958: Die Pflanzengesellschaften des Zarth bei Treuenbrietzen. Wiss. Z. Pädagog. Hochsch. Potsdam, Mat.-Nat. 4: 29–53.+

Frenzel, B., 1964: Über die offene Vegetation der letzten Eiszeit am Ostrande der Alpen. Verh. Zool.-Botan. Ges. Wien 103/104: 110–137.

*–, 1968: Grundzüge der pleistozänen Vegetationsgeschichte Nord-Eurasiens. Franz Steiner-Verlag, Wiesbaden: 326 S.

Frey, E., 1933: Die Flechtengesellschaften der Alpen. Ber. Geobot. Forsch. Inst. Rübel, Zürich 1932: 36–51.

–, 1947: Älteste Gipfelbewohner. Flechten als Pioniere der alpinen Vegetation. Die Alpen 23: 345–354.

–, 1958: Die anthropogenen Einflüsse auf die Flechtenflora in verschiedenen Gebieten der Schweiz. Veröff. Geobot. Inst. Rübel, Zürich 33: 91–107.

–, 1969: Alpin-nivale Flechten der Tauernketten. Verh. Zool.-Botan. Ges. Wien 108/9: 75–98.

Frey, H., 1934: Die Walliser Felsensteppe. Diss. Univ. Zürich: 218 S.

Frey-Wyssling, A., 1941: Die Guttation als allgemeine Erscheinung. Ber. Schweiz. Botan. Ges. 51: 321–325.

Friedel, H., 1938 a: Die Pflanzenbesiedlung im Vorfeld des Hintereisferners. Z. Gletscherkunde 26: 215–239.

–, 1938 b: Boden- und Vegetations-Entwicklung im Vorfelde des Rhonegletschers. Ber. Geobot. Forsch. inst. Rübel, Zürich 1937: 65–76.

–, 1956: Die alpine Vegetation des obersten Mölltales (Hohe Tauern). Erläuterung zur Vegetationskarte der Umgebung der Pasterze (Großglockner). Wiss. Alpenvereinshefte 16: 153 S.

–, 1967: Verlauf der alpinen Waldgrenze im Rahmen anliegender Gebirgsgelände. Mitt. Forstl. Bundesvers.anst. Wien 75: 81–172 (u. 5–55).

Frischknecht-Tobler, U., Traub, F., Hohl, H.R., 1979: Ökologische Beziehungen zwischen zellulären Schleimpilzen und mikrobieller Aktivität eines Waldbodens im Jahresverlauf. Vierteljahrschr. Naturforsch. Ges. Zürich 124: 77–155.

Fritsch, H., 1962: Die Pfeifengraswiesen und andere Grünlandgesellschaften des Teufelsbruches bei Henningsdorf. Wiss. Z. Pädagog. Hochsch. Potsdam 7: 151–166.

Fritz, A., 1970: Die pleistozäne Pflanzenwelt Kärntens. (Mit einem Beitrag zur pleistozänen Verbreitungsgeschichte der Rotbuche, Fagus sylvatica L., in Europa). Naturwiss. Beitr. Heimatkunde Kärntens 29 (Sdh. Carinthia II, Klagenfurt).

Fröde, E.T., 1950: Die Pflanzengesellschaften der Insel Hiddensee. Diss. Univ. Greifswald (unveröff., zit. nach Voderberg 1955).+

Froment, A., 1970: Étude expérimentale de la minéralisation de l'azote organique dans les différents types d'humus. Bull. Soc. Roy. Belg. 103: Contrib. 19: 311–319.

–, Remacle, J., 1975: Évolution de l'azote minéral et de la microflore dans le sol d'une pessière (Piceetum) à Mirwart. Bull. Soc. Roy. Botan. Belg. 108: 53–64.

Fukarek, F., 1961: Die Vegetation des Darß und ihre Geschichte. Pflanzensoziol. (Jena) 12: 321 S.

Fukarek, P., 1964: Die Tannen und Tannenmischwälder der Balkanhalbinsel. Schweiz. Z. Forstwes. 115: 518–533.

Füllekrug, E., 1971: Über den Jahresgang der Bodenfeuchtigkeit in verschiedenen Buchenwaldgesellschaften der Umgebung Bad Gandersheims. Diss. Botan. 13: 136 S.

Funk, S., 1927: Die Waldsteppenlandschaften, ihr Wesen und ihre Verbreitung. Veröff. Geogr. Inst. Univ. Königsberg 8: 65 S.

Funke, W., 1972: Energieumsatz von Tierpopulationen in Land-Ökosystemen. Verh. Deut. Zool. Ges., 65: 95–104.

Furrer, E., 1923: Kleine Pflanzengeographie der Schweiz. Zürich, 331 S. (2. Aufl. 1942, 127 S.)

–, 1955: Probleme um den Rückgang der Arve (Pinus cembra) in den Schweizer Alpen. Mitt. Schweiz. Anst. Forstl. Versuchsw. 31: 669–705.

–, 1958: Die Edelkastanie in der Innerschweiz. Mitt. Schweiz. Anst. Forstl. Versuchsw. 34: 89–182.

–, 1966: Kümmerfichtenbestände und Kaltluftströme in den Alpen der Ost- und Innerschweiz. Schweiz. Z. Forstwes. 1966: 720–733.

Gadow, A. von, 1975: Ökologische Untersuchungen an Ahorn-Eschenwäldern. Diss. Univ. Göttingen: 76 S.

Galoux, A., 1966: La variabilité génécologique du Hêtre commun (Fagus silvatica L.) en Belgique. Trav. Stat. Rech. Eaux et Forêts Belg., Sér. A, 11: 121 S.

Gams, H., 1927: Von den Follatères zur Dent de Morcles. Beitr. Geobot. Landesaufn. Schweiz 15: 760 S.

–, 1930: Über Reliktföhrenwälder und das Dolomitphänomen. Veröff. Geobot. Forsch.-inst. Rübel, Zürich 6: 32–80.

–, 1962: Das Gurgler Rotmoos und seine Stellung innerhalb der Gebirgsmoore. Veröff. Geobot. Inst. ETH, Stiftung Rübel, Zürich 37: 74–82.

Gauckler, K., 1938: Steppenheide und Steppenheidewald der Fränkischen Alb in pflanzensoziologischer, ökologischer und geographischer Betrachtung. Ber. Bayer. Botan. Ges. 23: 3–134.

–, 1954: Serpentinvegetation in Nordbayern. Ebenda 30: 19–26.

Gäumann, E., 1935: Der Stoffhaushalt der Buche (Fagus sylvatica L.) im Laufe eines Jahres. Ber. Schweiz. Botan. Ges. 44: 157–334.

Géhu, J.M., Chestem, A., 1965: La minéralisation expérimentale de l'azote organique de deux systèmes pédologiques littoraux naturels (dunes et prés salés). Ann. Inst. Pasteur 109, Suppl.: 136–152.

–, Richard, J.-L., Tüxen, R., 1973: Compte-rendu de l'excursion de l'Association Internationale de Phytosociologie dans le Jura en Juin 1967. Documents Phytosociologiques. Fasc. 2: 1–44 (Lille).

Geiger, R., 1950: Die meteorologischen Voraussetzungen der Sturmgefährdung. Forstwiss. Cbl. 69: 71–81.

*–, 1969: Das Klima der bodennahen Luftschicht. Die Wissenschaft 78, 4. Aufl. Braunschweig: 646 S.

Geister, H., 1975: Stickstoffzufuhr durch die Luft an Orten verschiedener geographischer Lage in Europa. Hausarbeit f.d. Lehramt an Realschulen, Göttingen: 54 S. (unveröff.)

Gensac, P., 1967: Les forêts d'épicéa de Tarentaise. Rev. Génér. Botan. 74: 425–528.

–, 1968 a: Les groupements forestiers de l'étage collinéen en Tarentaise moyenne et supérieure. Ann. Centre d'Enseignem. Sup., Sect. Sci. 6: 103–122.

–, 1968 b: La végétation des entonnoirs du gypse: cas de la Haute Tarentaise. Bull. Soc. Botan. France 115: 91–99. +

–, 1970: Les pessières de Tarentaise comparées aux autres pessières alpestres. Veröff. Geobot. Inst. ETH, Stiftg. Rübel, Zürich 43: 104–139.

Genssler, H., 1959: Veränderungen von Boden und Vegetation nach generationsweisem Fichtenanbau. Diss. Hann. Münden: 191 S. u. Tab.-Anhang.

Gerlach, A., 1973: Methodische Untersuchungen zur Bestimmung der Stickstoffnettomineralisation. Scripta Geobot. (Göttingen) 5: 115 S.

–, 1978: Zur Bestimmung der Stickstoff-Nettomineralisation in mehr oder minder nassen Böden. Oecol. Plant 13: 163–174.

Germann, P., 1976: Wasserhaushalt und Elektrolytverlagerung in einem Wald und einem mit Wiese bestockten Boden in ebener Lage. Mitt. Schweiz. Anst. Forstl. Versuchsw. 52: 163–309.

Gessner, F., 1932: Die Entstehung und Vernichtung von Pflanzengesellschaften an Vogelnistplätzen. Beih. Botan. Cbl. 49, Erg.-Bd.: 113–128.

–, 1933: Nährstoffgehalt und Planktonproduktion in Hochmoorblänken. Arch. Hydrobiol. 25: 394–406.

–, 1939: Die Phosphorarmut der Gewässer und ihre Beziehung zum Kalkgehalt. Int. Rev. Ges. Hydrobiol. u. Hydrograph. 38: 203–211.

–, 1951: Untersuchungen über den Wasserhaushalt der Nymphaeaceen. Biol. General, 19: 247–280.

*–, 1956: Die Binnengewässer. Handb. Pflanzenphysiol. 4: 179–232.

*–, 1957: Meer und Strand, 2. Aufl. Deut. Verlag der Wissenschaften, Berlin: 426 S.

Geyger, E., 1964: Methodische Untersuchungen zur Erfassung der assimilierenden Gesamtoberflächen von Wiesen. Ber. Geobot. Inst. ETH, Stiftg. Rübel, Zürich 35: 41–112.

Giacomini, V., Pirola, A., Wikus, E., 1964: I pascoli di altitudine dello Spluga (con carta della vegetazione all'1 : 12500). Delpinon (Napoli) N. S. 4: 233–317. +

Gies, T., Lötschert, W., 1973: Untersuchungen über den Kationengehalt im Hochmoor. II. Jahreszeitliche Veränderungen und Einfluß der Sphagnen-Vegetation. Flora 162: 244–268.

Gigon, A., 1968: Stickstoff- und Wasserversorgung von Trespen-Halbtrockenrasen (Mesobromion) im Jura bei Basel. Ber. Geobot. Inst. ETH, Stiftg. Rübel, Zürich 38: 28–85.

–, 1971: Vergleich alpiner Rasen auf Silikat- und auf Karbonatboden. Veröff. Geobot. Inst. ETH, Stiftung Rübel 48: 164 S.

–, Rorison, I. H., 1972: The response of some ecologically distinct plant species to nitrate- and to ammonium-nitrogen. J. Ecol. 60: 93–102.

Gillham, M. E., 1970: Seed dispersal by birds. In: F. Perrings (Ed.), The flora of a changing Britain. Hampton, Middlesex: E. W. Classey, Ltd.: 90–98.

Gillner, V., 1960: Vegetations- und Standortsuntersuchungen in den Strandwiesen der schwedischen Westküste. Acta Phytogeogr. Suecica 43: 198 S.

Gilomen, H., 1938: Carex curvula All. ssp. nov. rosae Gilom. (Kalk-Krummsegge). Ber. Geobot. Forsch. Inst. Rübel, Zürich 1937: 77–104.

Gimingham, C. H., 1960: Biological flora of the British Isles. Calluna Salisb. A monotypic genus. J. Ecol. 48: 455–483.

–, 1972: Ecology of Heathlands. Chapman and Hall, London, 266 S.

Gisi, U., Oertli, J. J., 1981: Ökologische Entwicklung in Brachland verglichen mit Kulturwiesen IV. Veränderungen im Mikroklima. Oecol. Plant. 16: 233–249.

Glahn, H. von, Tüxen, J., 1963: Salzpflanzen-Gesellschaften und ihre Böden im Lüneburger Kalkbruch vor dem Bardowicker Tore. Naturw. Ver. Fürstentum Lüneburg 28: 1–32.

Glässer, E., 1969: Zur Frage der anthropogen bedingten Vegetation, vor allem in Mitteleuropa. Die Erde 100: 37–45.

Glavač, V., Bohn, U., 1971: Quantitative vegetationskundliche Untersuchungen zur Höhengliederung der Buchenwälder im Vogelsberg. Schriftenr. Vegetationskunde (Bonn) 5: 135–186.

–, Krause, A., 1969: Über bodensaure Wald- und Gebüschgesellschaften trockenwarmer Standorte im Mittelrheingebiet. Schriftenr. Vegetationskunde (Bonn) 4: 85–102.

–, –, Wolff-Straub, R., 1971: Über die Verteilung der Hainsimse (Luzula luzuloides) im Stammabflußbereich der Buche im Siebengebirge bei Bonn. Schriftenr. Vegetationskunde (Bonn) 5: 187–192.

Gluch, W., 1973: Die oberirdische Netto-Primärproduktion in drei Halbtrockenrasengesellschaften des Naturschutzgebietes „Leutratal" bei Jena. Arch. Naturschutz u. Landschaftsforsch. 13: 21–42.

Glück, H., 1934: Wasserpflanzen. Handwörterb. Naturw., 2. Aufl., 10: 575–590.

Gökçeoğlu, M., 1975: Untersuchungen über Produktion und Nährstoffumsatz in Rasengesellschaften von Carex sempervirens und Carex ferruginea. Diss. T.U. München: 108 S.

Goldsmith, F. B., 1973: The vegetation of exposed sea cliffs at South Stack, Anglesey. II. Experimental studies. J. Ecol. 61: 819–829.

Golubić, S., 1967: Algenvegetation der Felsen. Eine ökologische Algenstudie im dinarischen Karst. Die Binnengewäss. (Stuttgart) 23: 183 S.

Gonschorek, L., 1971: Untersuchungen über den P-, K- und Ca-Umsatz in einem Halbtrockenrasen. Staatsexamensarbeit Göttingen Nr. 6595: 22 S.

Goodman, P. J., 1960: Investigations into „dieback" in Spartina townsendii agg. II. The morphological structure and composition of the Lymington sward. J. Ecol. 48: 711–724.

Gore, A. J. P., Urquhart, C., 1966: The effects of waterlogging on the growth of Molinia caerulea and Eriphorum vaginatum. J. Ecol. 54: 617–633.

Gorham, E., 1974: The relationship between standing crop in sedge meadows and summer temperature. J. Ecol. 62: 487–491.

Görs, S., 1960: Das Pfrunger Ried. Die Pflanzengesellschaften eines oberschwäbischen Moorgebietes. Veröff. Württ. Landesst. Naturschutz u. Landschaftspflege Baden-Württemb. 27/28: 5–45. +

–, 1963: Beiträge zur Kenntnis basiphiler Flachmoorgesellschaften (Tofieldietalia Preisg. apud Oberd. 49). I. Teil. Das Davallseggen-Quellmoor (Caricetum davallianae W. Koch 28). Ebenda 31: 7–30.

–, 1964: Beiträge zur Kenntnis basiphiler Flachmoorgesellschaften. 2. Teil: Das Mehlprimel-Kopfbinsenmoor. Veröff. Landesstelle Naturschutz Baden-Württ. 32: 7–42.

*–, 1967: Der Wandel der Vegetation im Naturschutzgebiet Schwenninger Moos unter dem Einfluß des Menschen in zwei Jahrhunderten. In: „Das Schwenninger Moos". Die Natur- u. Landschaftsschutzgebiete Baden-Württ. 5: 190–284. +

–, 1975: Das Cladietum marisci All. 1922 in Süddeutschland. Beitr. Naturk. Forsch. Südw.-Deut. 34: 103–123.

–, Müller, Th., 1969: Beitrag zur Kenntnis der nitrophilen Saumgesellschaften Südwestdeutschlands. Mitt. Flor.-Soz. Arb. gem. N. F. 14: 153–168.

Gorshina, T. K., 1975: Ecological studies of herbaceous cover in foreststeppe oakwood. XII. Internat. Botan. Congr. (Leningrad) Abstr. 1: 147.

Gotthard, W., 1965: Die Küchenschelle (Pulsatilla vulgaris Mill.) im Ries. Botan. Jb. 84: 1–50.

Göttsche, D., 1972: Verteilung von Feinwurzeln und Mykorrhizen im Bodenprofil eines Buchen- und Fichtenbestandes im Solling. Diss. Univ. Hamburg: 102 S.

Götz, V., 1979: Pflege von Wacholderheiden auf der Münsinger Alb. Mitt. Ver. Forstl. Standortskunde u. Forstpflanzenzüchtg. 27: 49–54.

Grabherr, W., 1934: Der Einfluß des Feuers auf die Wälder Tirols in Vergangenheit und Gegenwart. Cbl. Ges. Forstwes. 60: 260–273 u. 289–302.

–, 1936: Die Dynamik der Brandflächenvegetation auf Kalk- und Dolomitböden des Karwendels. Beih. Botan. Cbl. 55 B: 1–94.

–, 1942 a: Bodenkundlich-nährstoffökologische und pflanzensoziologische Beiträge zur Frage der Waldbodendüngung. Mitt. Forstwirtsch. u. Forstwiss. 1942: 248–278.

–, 1942 b: Über die Nährstoffökologie und das Formbildungsvermögen der Gräsergattung Molinia (Schrank) in Abhängigkeit von Nährstoffgehalt und Reaktion des Bodens. Ebenda 1942: 172–196.

*Grable, A. R., 1966: Soil aeration and plant growth. Advanc. Agron. 18: 57–106.

Grace, J., Woolhouse, H. W., 1970: A physiological and mathematical study of the growth and productivity of a Calluna-Sphagnum community. I. Net photosynthesis of Calluna vulgaris L. Hull. J. Appl. Ecol. 7: 363–378.

–, –, 1973: A physiological and mathematical study of the growth and productivity of a Calluna-Sphagnum community. II. Distribution of photosynthesis in Calluna vulgaris L. Hull. Ebenda 10, 77–91.

Gradmann, R., 1932: Unsere Flußtäler im Urzustand. Z. Ges. f. Erdk. Berlin 1932: 1–17.

–, 1950: Das Pflanzenleben der Schwäbischen Alb. 1. Aufl. 1898, 4. Aufl. Stuttgart 1950, 2 Bde.: 407 u. 449 S. (Schwäb. Albverein).

Graebner, P., 1901: Die Heide Norddeutschlands und die sich anschließenden Formationen in biologischer Betrachtung. Die Vegetation der Erde 5: 320 S.

Graff, O., 1971: Beeinflussen Regenwurmröhren die Pflanzenernährung? Landbauforschung Völkenrode 21: 103–108.

Grime, J. P., 1963: Factors determining the occurence of calcifuge species on shallow soils over calcareous substrata. J. Ecol. 51: 375–390.

–, 1965: The ecological significance of lime-chlorosis. An experiment with two species of Lathyrus. New Phytol. 64: 477–487.

–, 1973: Competitive exclusion in herbaceous vegetation. Nature (London) 242, No. 5396: 344–347.

Grimme, K., 1975: Wasser- und Nährstoffversorgung von Hangbuchenwäldern auf Kalk in der weiteren Umgebung von Göttingen. Diss. Univ. Göttingen, 190 S. (Scripta Geobot. Göttingen 12, 1977: 58 S.)

Grisebach, A., 1846: Über die Bildung des Torfs in den Emsmooren aus deren unveränderter Pflanzendecke. Göttinger Studien 1845: 118 S.

Grodzińska, K., 1971: Acidification of tree bark as a measure of air pollution in Southern Poland. Bull. Acad. Polon. Sci., Sér. Biol. Cl. II, 19: 189–195.

–, Pancer-Kotejowa, E., 1965: Forest communities in the Bukowica Range (Low Beskids, Polish Western Carpathians). Fragm. Florist. Geobot. 11, 4: 563–599. +

Grodzinsky, A. M., 1973: Fundamentals of chemical interactions of plants. Kiev: 206 S. (ukr.).

Groenman-van Waateringe, W., 1975: Prunetalia scrub: early neolothic field enclosures in Europe. XII. Internat. Botan. Congr. (Leningrad) Abstr. 1: 113.

Gronwald, W., 1953: Welche Erkenntnisse zur Frage der vermuteten neuzeitlichen Nordseeküstensenkung hat die Wiederholung des Deutschen Nordseeküsten-Nivellements gebracht? Die Küste (Heide i. Holst.) 1953: 66–82.

Gross, H., 1935: Der Döhlauer Wald in Ostpreußen. Eine bestandesgeschichtliche Untersuchung. Beih. Botan. Cbl. 53 B: 405–431.

*Grosse-Brauckmann, G., 1953 a: Untersuchungen über die Ökologie, besonders den Wasserhaushalt von Ruderalgesellschaften. Vegetatio 4: 245–283.

–, 1953 b: Über die Verbreitung ruderaler Dorfpflanzen innerhalb eines kleinen Gebietes. Mitt. Florist.-Soziol. Arb. gem. N. F. 4: 5–10.

–, 1962: Moorstratigraphische Untersuchungen im Niederwesergebiet (Übergangs-Moorbildungen am Geestrand und ihre Torfe). Veröff. Geobot. Inst. ETH, Stiftg. Rübel, Zürich 37: 100–116.

–, 1965: Vom Hochmoor und seiner Pflanzenwelt. Materia Medica Nordmark, 4. Sonderh. 1965: 26 S.

–, 1968: Einige Ergebnisse einer vegetationskundlichen Auswertung botanischer Torfuntersuchungen, besonders im Hinblick auf Sukzessionsfragen. Acta Botan. Neerl. 17: 59–69.

–, Puffe, D., 1967: Über Zersetzungsprozesse und Stoffbilanz im wachsenden Moor. 8th Internat. Congr. Soil. Sci. Bukarest, Romania 1964, 5: 635–649.

Grosser, K. H., 1956: Die Vegetationsverhältnisse an den Arealvorposten der Fichte im Lausitzer Flachland. Arch. Forstwes. 5: 285–294.

–, 1964: Die Wälder am Jagdschloß Weißwasser (OL). Abh. u. Ber. Naturk. mus. Görlitz 39, Nr. 2: 102 S. +

–, 1966: Altteicher Moor und Große Jeseritzen. Brandenburg. Naturschutzgebiete (Potsdam) 1: 31 S. +

Grossmann, H., 1927: Die Waldweide in der Schweiz. Diss. ETH Zürich: 123 S.

–, 1934: Der Einfluß der alten Glashütten auf den schweizerischen Wald. Ber. Geobot. Forsch. Inst. Rübel, Zürich 1933: 15–32.

Grubb, P. J., 1976: A theoretical background to the conservation of ecologically distinct groups of annuals and biennials in the chalk grassland ecosystem. Biol. Conserv. 10: 53–76.

Grube, H.-J., 1975: Die Makrophytenvegetation der Fließgewässer in Süd-Niedersachsen und ihre Beziehungen zur Gewässerverschmutzung. Arch. Hydrobiol. Suppl. 45: 376–456.

Gruber, P., 1973: Zusammenhänge zwischen Klimaunterschieden, Bodenchemismus und Bodenwassergehalt auf Lockersedimentböden des Wiener Raumes. Mitt. Österr. Bodenk. Ges. 17: 123 S.

Grulois, J., 1968 a: Recherches sur l'écosystème forêt. Contrib. No. 20. Réflexion, interception et transmission du rayonnement de courtes longueurs d'onde: Varitation au cours d'une année. Bull. Soc. Roy. Botan. Belg. 102: 13–26.

–, 1968 b: Contrib. no. 21. Flux thermiques et évaporation au cours d'une journée serreine. Ebenda 102, 27–41.

Grünig, A., 1980: Unsere Seeufer in Vergangenheit und Gegenwart. Jber. Verb. Schutze d. Landschaftsbildes am Zürichsee 53: 15–34 (s. auch 35–69!).

Guderian, R., 1970: Untersuchungen über quantitative Beziehungen zwischen dem Schwefelgehalt von Pflanzen und dem Schwefeldioxidgehalt der Luft. 1. Teil. Z. Pflanzenkrankh. Pflanzenschutz 77: 200–399. +

–, 1977: Air Pollution. Ecol. Studies 22, 127 S.

Guillet, B., 1968: Essai de détermination de l'âge de deux podsols Vosgiens par la palynologie. Oecol. Plant. 3: 101–119.

Günther, G., 1965: Die chemische Unkrautbekämpfung in der Forstwirtschaft bei Bestandesbegründung und Bestandespflege. Diss. Landw. Hochsch. Hohenheim: 104 S.

Gupta, P. L., Rorison, I. H., 1975: Seasonal differences in the availability of nutrients down a podzolic profile. J. Ecol. 63: 521–534.

Guyan, W. U., 1955: das jungsteinzeitliche Moordorf von Thayngen-Weier. Monogr. Ur- und Frühgesch. Schweiz 11: 223–272.

Haber, W., 1965: Zusammenhänge zwischen Bakterienbesatz des Bodens und Vegetation. In: Biosoziologie. Verl. Dr. W. Junk, Den Haag: 1965: 284–289.

–, 1966: Über die ursprüngliche Vegetation auf den höchsten Erhebungen des Sauerlandes. Naturkunde in Westfalen 66: 11–17.

Hadač, E., 1962: Übersicht der höheren Vegetationseinheiten des Tatragebirges. Vegetatio 11: 46–54. +

–, Váňa, J., 1967: Plant communities of mires in the western part of the Krkonoše mountains. Folia Geobot. Phytotax. (Praha) 2: 213–254. +

Haefner, H., 1963: Vegetation und Wirtschaft der oberen subalpinen und alpinen Stufe im Luftbild, dargestellt am Beispiel des Dischmatales und weiteren Teilen der Landschaft Davos, Schweiz. Landeskundl. Luftbildauswertung im mitteleurop. Raum (Bad Godesberg) 6: 117 S.

Haeupler, A., 1974: Statistische Auswertung von Punktrasterkarten der Gefäßpflanzen Süd-Niedersachsens. Scripta Geobotan. (Göttingen) 8: 141 S.

Haessler, K., 1954: Zur Ökologie der Trittpflanzen. Diss. Landw. Hochsch. Stuttgart (unveröff., s. Walter 1962, Bd. III, 1).

Haffner, W., 1968: Die Vegetationskarte als Ansatzpunkt landschaftsökologischer Untersuchungen. Erdkunde 22: 215–225.

–, 1969: Das Pflanzenkleid des Naheberglandes und des südlichen Hunsrück in ökologisch-geographischer Sicht. Decheniana Beih. 15, 145 S.

Hafsten, U., 1965: The norwegian Cladium mariscus communities and their post-glacial history. Acta Univ. Bergen, Ser. Math.-Nat. 1965, No. 4: 55 S.

Hagedorn, J., 1969: Beiträge zur Quartärmorphologie griechischer Hochgebirge. Göttinger Geogr. Abh. 50: 135 S.

Hagel, H., 1966: Gesteinsmoosgesellschaften im westlichen Wienerwald. Verh. Zool.-Botan. Ges. Wien 105/6: 137–167.

Hampel, R. (Hrsg.), 1963: Ökologische Untersuchungen in der subalpinen Stufe zum Zwecke der Hochlagenaufforstung. Teil II. Mitt. Forstl. Bundes-Versuchsanstalt Mariabrunn 60: 433–887.

Hansen, B., 1966: The raised bog Draved Kongsmose. Botan. Tidsskr. 62: 146–185.

Hansen, K., 1964: Studies on the regeneration of heath vegetation after burning-off. Botan. Tidsskr. 60: 1–41.

Hard, G., 1964: Kalktriften zwischen Westrich und Metzer Land. Ann. Univ. Sarav. (Heidelberg), R. Philos, Fak. 2: 176 S.

–, 1972: Wald gegen Driesch. Das Vorrücken des Waldes auf Flächen junger „Sozialbrache". Ber. Deut. Landeskunde 46: 49–80.

Harper, J. L., 1961: Approaches to the study of plant competition. Symposia Soc. Exper. Biol. (Cambridge) 15: 1–39.

Harrison, A. F., 1971: The inhibitory effect of oak leaf litter tannins on the growth of fungi, in relation to litter decomposition. Soil Biol. Biochem. 3: 167–172.

Hartl, H., 1963: Die Vegetation des Eisenhutes im Kärntner Nockgebiet. Carinthia II (Klagenfurt) 73: 294–336. +

–, 1967: Die Soziologie der Urwälder Scatlé und Derborence. Schweiz. Z. Forstwes. 1967: 737–743.

–, 1978: Vegetationskarte der Großfragant (Hohe Tauern). Carinthia II, 168: 339–367.

Hartmann, F. K., 1933: Zur soziologisch-ökologischen Charakteristik der Waldbestände Norddeutschlands. Forstl. Wochenschr. Silva 21: 161–168, 241–147 u. 249–318.

–, 1941: Über den waldbaulichen Wert des Grundwassers. I. Mitt. Forstwirtsch. u. Forstwiss. 1930: 385–457; II. Ebenda 1941: 91–218.

–, van Eimern, Jahn, G., 1959: Untersuchungen reliefbedingter kleinklimatischer Fragen in Geländequerschnitten der hochmontanen und montanen Stufe des Mittel- und Südwestharzes. Ber. Deut. Wetterdienst 7, 50: 39 S.

*–, Jahn, G., 1967: Waldgesellschaften des mitteleuropäischen Gebirgsraumes nördlich der Alpen. Gustav Fischer Verlag, Stuttgart: 635 S. u. Tabellenteil.

Haslam, S. M., 1975: Plant populations of wetlands and streams. XII. Internat. Botan. Congr. (Leningrad) Abstr. 1: 148.

Hauff, R., 1937: Die Buchenwälder auf den kalkarmen Lehmböden der Ostalb und die nacheiszeitliche Waldentwicklung auf diesen Böden. Jb. Ver. Vaterl. Naturk. Württemb. 1937: 51–97.

–, 1964: Erläuterungen zur vegetationskundlichen Karte 1 : 25 000 Blatt 8123 Weingarten. Stuttgart (Landesvermess. amt Baden-Württ.): 47 S. (m. farb. Karte). +

–, 1965: Die Bodenvegetation älterer Fichtenbestände auf aufgeforsteten Schafweiden der Mittleren Alb. Mitt. Ver. Forstl. Standortskunde u. Forstpflanzenzücht. 15: 39–43.

–, Schlenker, G., Krauss, G. A., 1950: Zur Standortsgliederung im nördlichen Oberschwaben. Allg. Forst- u. Jagdztg. 122: 27 S.

Hausrath, H., 1907: Der deutsche Wald. Aus Natur und Geisteswelt 153: 130 S.

Havas, P. J., 1965: Pflanzenökologische Untersuchungen im Winter. I. Zur Bedeutung der Schneedecke für das Überwintern von Heidel- und Preisselbeere. Aquilo (Oulu, Finnland), Ser. Botan. 4: 1–36.

–, 1971: The water economy of the bilberry (Vaccinium myrtillus) under winter conditions. Rep. Kevo Subarctic Res. Stat. 8: 41–52.

Havill, D. C., Lee, J. A., Stewart, G. R., 1974: Nitrate utilization by species from acidic and calcareous soils. New Phytol. 73: 1221–1231.

Havinga, A. J., 1972: A palynological investigation in the Pannonian climate region of Lower Austria. Rev. Palaeobot. Palynol. 14: 319–352.

Havranek, W., 1972: Über die Bedeutung der Bodentemperatur für die Photosynthese und Transpiration junger Forstpflanzen und für die Stoffproduktion an der Waldgrenze. Angew. Botan. 46: 101–116.

Heer, O., 1884: Über die nivale Flora der Schweiz.

Denkschr. Schweiz. Ges. f. d. Gesamten Naturwiss. 29: 4–144.

Heerdt, P. F. van, Mörzer Bruijns, M. F., 1960: A biocenological investigation in the yellow dune region of Terschelling. Tijdschr. Entomol. 103: 225–275.

Hegg, O., 1965: Untersuchungen zur Pflanzensoziologie und Ökologie im Naturschutzgebiet Hohgant (Berner Voralpen). Beitr. Geobot. Landesaufn. Schweiz 46: 188 S.

*Hegi, G., 1908 ff.: Flora von Mitteleuropa. 7 Bde., München, teilweise mit Neuauflagen bis heute.

Heilig, H., 1930/31: Untersuchungen über Klima, Boden und Pflanzenleben des Zentralkaiserstuhls. Z. Botan. 24: 225–279.

Heinemann, P., 1956: Les landes à Calluna du district picardo-brabançon de Belgique. Vegetatio 7: 99–147.

Heinrich, W., Hilbig, W., Niemann, E., 1972: Zur Verbreitung, Ökologie und Soziologie der Roten Pestwurz, Petasites hybridus (L.) Gaertn., Meyer et Scherb. Wiss. Z. Univ. Jena, Math.-Nat. R. 21: 1099–1124.

Hejný, S., 1960: Ökologische Charakteristik der Wasser- und Sumpfpflanzen in den slowakischen Tiefebenen (Donau- und Theißgebiet). Bratislava: 487 S.

–, 1962: Über die Bedeutung der Schwankungen des Wasserspiegels für die Charakteristik der Makrophytengesellschaften in mitteleuropäischen Gewässern. Preslia (Praha) 34, 359–367.

Heller, H., 1963: Struktur und Dynamik von Auenwäldern. Beitr. Geobot. Landesaufn. Schweiz 42: 75 S.

–, 1969: Lebensbedingungen und Abfolge der Flußauenvegetation in der Schweiz. Mitt. Schweiz. Anst. Forstl. Versuchswes. 43: 123 S.

Herlitzius, R., 1975: Streuabbau in Laubwäldern. Untersuchungen in Kalk- und Sauerhumusbuchenwäldern. Dipl. arb. Math.-Nat. Fak. Univ. Göttingen, 59 S.

*Hermes, K., 1955: Die Lage der oberen Waldgrenze in den Gebirgen der Erde und ihr Abstand zur Schneegrenze. Kölner Geogr. Arb. 5: 277 S.

–, 1964: Der Verlauf der Schneegrenze. Geogr. Taschenb. 1964/65: 38–71.

Herz, K., 1962: Zustand und Leistungsfähigkeit der Agrarflächen Mittelsachsens im 18./19. Jahrhundert. Wiss. Veröff. Deut. Inst. Länderkunde N. F. 19/20: 233–242.

Herzog, T., 1943: Moosgesellschaften des höheren Schwarzwaldes. Flora, N. F. 36: 263–308.

Hesmer, H., 1932: Waldentwicklung im nordwestdeutschen Flachland. Z. Forst- u. Jagdwes. 64: 577–607.

–, 1936: Die Bewaldung Deutschlands. Dargestellt an Hand von Karten der einzelnen Holz- und Betriebsarten. VI. Die Buche. Forstl. Wochenschr. Silva 24: 169–176; VII. Die Eiche. Ebenda 24: 409–424.

*–, Schroeder, F.-G., 1963: Waldzusammensetzung und Waldbehandlung im Niedersächsischen Tiefland westlich der Weser und in der Münsterschen Bucht bis zum Ende des 18. Jahrhunderts. Decheniana (Bonn), Beih. 11: 304 S.

Hess, E., 1942: Études sur la répartition du mélèze en Suisse. Z. Schweiz. Forstwes., Beih. 20: 1–80.

Hesselman, H., 1910: Über den Sauerstoffgehalt des Bodenwassers und dessen Einwirkung auf die Versumpfung des Bodens und das Wachstum des Waldes. Medd. Stat. Skogsförs. Anst. 7: 91–130.

–, 1917: On the effect of our regeneration measures on the formation of salpetre in the ground and its importance in the regeneration of coniferous forests. Ebenda 13/14: 925–1076.

Heuberger, H., 1968: Die Alpengletscher im Spät- und Postglazial. Eiszeitalter u. Gegenwart 19: 270–275.

Heynert, H., 1964: Das Pflanzenleben des hohen Westerzgebirges. Th. Steinkopff, Dresden u. Leipzig: 141 S.

Hilbig, W., 1962: Vegetationskundliche Untersuchungen in der mitteldeutschen Ackerlandschaft. VII. Die Pflanzengesellschaften der Umgebung von Dehlitz (Saale), Kr. Weißenfels. Wiss. Z. Univ. Halle, Math.-Nat. 11: 817–866.

–, 1966: Die Bedeutung der Ackerunkrautgesellschaften für die pflanzengeographische Gliederung Thüringens. Feddes Repert. 73: 108–140.

–, 1967 a: Die Unkrautbestände der mitteldeutschen Weinberge. Hercynia N. F. 4: 325–338.

–, 1967 b: Die Ackerunkrautgesellschaften Thüringens. Feddes Repert. 76: 83–191.

–, 1971 a: Übersicht über die Pflanzengesellschaften des südlichen Teiles der DDR. I. Die Wasserpflanzengesellschaften. II. Die Röhrichtgesellschaften. Hercynia N. F. 8: 4–33.

–, 1971 b: Kalkschuttgesellschaften in Thüringen. Hercynia N. F. 8: 85–95.

–, 1972: Beitrag zur Kenntnis einiger wenig beachteter Pflanzengesellschaften Mitteldeutschlands. Wiss. Z. Univ. Halle 21: 83–98.

*–, 1980: Übersicht über die Pflanzengesellschaften des südlichen Teiles der DDR. Hercynia N. F. (Leipzig) 17: 375–435.

–, Mahn, E. G., 1974: Zur Verbreitung von Ackerunkräutern im südlichen Teil der DDR. Wiss. Z. Univ. Halle-Wittenberg, Math.-Nat. R. 23: 5–57.

–, –, Müller, G., 1969: Zur Verbreitung von Ackerunkräutern im südlichen Teil der DDR. Wiss. Z. Univ. Halle 18: 211–270.

–, Schubert, R., 1980(?): Grünlandvegetation – Segetalvegetation. Atlas DDR Blatt 14. VEB Hermann Haack, Gotha-Leipzig.

Hild, J., Rehnelt, K., 1971: Öko-soziologische Untersuchungen an einigen niederrheinischen Meeren. Ber. Deut. Botan. Ges. 84: 19–39.

Hillgarter, F., 1971: Waldbauliche und ertragskundliche Untersuchungen im subalpinen Fichtenwald Scatlé/Brigels. Diss. ETH Zürich: 80 S.

Hofer, H., 1970: Über die Zusammenhänge zwischen der Düngung und der Konkurrenzfähigkeit ausgewählter Naturwiesenpflanzen. Diss. E. T. H. Zürich: 70 S.

–, Jäggli, F., 1975: Probleme bei der umweltgerechten Anwendung von Düngemitteln. Mitt. Schweiz. Landwirtsch. 23: 89–111.

Hofer, H. R., 1967: Die wärmeliebenden Felsheiden Insubriens. Botan. Jb. 87: 176–251.

Höfler, K., 1951: Zur Kälteresistenz einiger Hochmooralgen. Verh. Zool.-Botan. Ges. Wien 92: 234–242.

Hofmann, G., 1957: Zur Soziologie einiger Kiefernforsten im Bereich der Kalk-Trockenlaubwälder Südthüringens. Arch. Forstwes. 6: 233–249.

–, 1958 a: Die eibenreichen Waldgesellschaften Mitteldeutschlands. Ebenda 7: 502–558.

–, 1958 b: Vegetationskundliche Untersuchungen an wärmeliebenden Gebüschen des Meininger Muschelkalkgebietes. Ebenda 7: 369–387.

–, 1959: Die Wälder des Meininger Muschelkalkgebietes. Feddes Repert. Beih. 138: 56–140.

–, 1961: Die Stickstoffbindung der Robinie (Robinia pseudoacacia L.). Arch. Forstwes. 10: 627–631.

–, 1962: Synökologische Untersuchungen im Waldschutzgebiet Gellmersdorfer Forst/Oder. Arch. Naturschutz u. Landschaftsforsch. 2: 3–139.

–, 1963: Der Hainbuchen-Buchenwald in den Muschelkalkgebieten Thüringens. Arch. Forstwes. 12: 706–716.

–, 1964 a: Kiefernforstgesellschaften und natürliche Kiefernwälder im östlichen Brandenburg. I. Kiefernforstgesellschaften. II. Natürliche Kiefernwälder und -gehölze. Arch. Forstwes. 13: 641–664 u. 717–732.

–, 1964 b: Die Höhenstufengliederung des nordöstlichen Rhöngebirges. Archiv Naturschutz 4: 191–206.

–, 1965: Die Vegetation im Waldschutzgebiet „Hainich" (Westthüringen). Landschaftspflege u. Naturschutz Thüring. 2: 1–12.

–, 1968: Über die Beziehungen zwischen Vegetationszeit, Humusform, C/N-Verhältnis und pH-Wert des Oberbodens in Kiefernbeständen des nordostdeutschen Tieflandes. Arch. Forstwes. 17: 845–855.

–, 1969: Zur pflanzensoziologischen Gliederung der Kiefernforsten des nordostdeutschen Tieflandes. Feddes Repert. 80: 401–412.

Hofmann, W., 1966: Laubwaldgesellschaften der Fränkischen Platte. Würzburg (Selbstverlag Naturw. Ver. Würzburg): 194 S.

–, 1968: Vitalität der Rotbuche und Klima in Mainfranken. Feddes Repert. 78: 135–137.

Hofmeister, H., 1970: Pflanzengesellschaften der Weserniederung oberhalb Bremens. Diss. Botan. 10: 116 S.

Hohenester, A., 1960: Grasheiden und Föhrenwälder auf Diluvial- und Dolomitsanden im nördlichen Bayern. Ber. Bayer. Botan. Ges. 33: 30–85.

–, 1967: Silbergrasfluren in Bayern. Mitt. Florist.-soziol. Arb.gem. N. F. 11/12: 11–21.

–, 1978: Die potentielle natürliche Vegetation im östlichen Mittelfranken (Region 7). Erläuterungen zur Vegetationskarte 1:200000. Mitt. Fränk. Geogr. Ges. 23/24: 57 S.

Hollmann, H., 1972: Verbreitung und Soziologie der Schachblume Fritillaria meleagris L. Abh. u. Verh. Naturw. Ver. Hamburg N. F. 15 Suppl.: 82 S.

Holmsgaard, E., 1968: Ertragskundliche Untersuchungen in Fichtenbeständen erster und zweiter Generation im dänischen Jungmoränengebiet. Tagungsber. Deut. Akad. Landwirtschaftswiss. Berlin 84: 25–35.

Holstener-Jørgensen, H., 1968: Bodenkundliche Untersuchungen in Fichtenbeständen erster und zweiter Generation im dänischen Jungmoränengebiet. Tagungsber. Deut. Akad. Landwirtschaftswiss. Berlin 84: 37–45.

–, 1970: Fertilizing experiments in six Norway spruce plantations in Jutland. Det Forstl. Forsøgsv. Danmark 32: 297–311.

–, 1971: A nitrogen-dose experiment on single tree plots of 68–75 year-old beech in the Rude Skov. Ebenda 32: 369–378.

Holtmeier, F.-K., 1967: Die Waldgrenze im Oberengadin in ihrer physiognomischen und ökologischen Differenzierung. Diss. Univ. Bonn: 163 S.

–, 1968: Entgegnung zu: „Über Schneeschliff in den Alpen" von Hans Turner. Wetter u. Leben 20: 201–205.

–, 1971: Der Einfluß der orographischen Situation auf die Windverhältnisse im Spiegel der Vegetation. Erdkunde 25: 178–195.

Holub, J., Hejný, S., Moravec, J., Neuhäusl, R., 1967: Übersicht der höheren Vegetationseinheiten der Tschechoslowakei. Rozpr. Československ. Akad. Věd, Rada Mat. Přírodn. Věd 77: 75 S.

Holzner, W., 1970: Die Ackerunkrautvegetation des nördlichen Burgenlandes. Wiss. Arb. Burgenland 44: 196–243.

–, 1971: Niederösterreichs Ackervegetation als Umweltzeiger. Die Bodenkultur (Wien) 22: 397–414.

–, Hübl, E., 1977: Zur Vegetation der Kalkalpengipfel des Westlichen Niederösterreich. Jahrb. Ver. Schutze Bergwelt 42: 247–270.+

Hope-Simpson, J. F., Jefferies, R. L., 1966: Observations relating to vigour and debility in marram grass (Ammophila arenaria [L.] Link). J. Ecol. 54: 271–274.

Höpfner, B., 1966: Ökophysiologische Untersuchungen in einem extrem sauren Calamagrostidetum epigeios. Diss. Univ. Gießen: 243 S.

Hoppe, I. M., 1972: Untersuchungen über den Stickstoff- und Phosphorgehalt in Strandwällen verschiedenen Alters bei Heiligenhafen/Ostsee. Dipl.-Arb. Math.-Nat. Fak. Göttingen: 30 S. (unveröff.)

Horak, E., 1963: Pilzökologische Untersuchungen in der subalpinen Stufe (Piceetum subalpinum und Rhodoreto-Vaccinietum) der Rätischen Alpen (Dischmatal, Graubünden). Mitt. Schweiz. Anst. Forstl. Versuchsw. 39, 1: 112 S.

Horak, J., 1971: Westliche Tatra-Geobiozönosen der oberen Wald- und Krummholzgrenze. Acta Sc. Nat. Acad. Sc. Bohemosl. Brno NS 5: 47 S.

Horikawa, Y., Suzuki, H., Yokogawa, H., Matsumura, T., 1959: Moorland vegetation in the Yawata Highland, in the northwestern part of Hiroshima Prefecture, SW Japan. Sci. Res. Sandankyo Gorge and Yawata Highland (Hiroshima, Jap.) 1959: 121–152.

Horn, H.-H. von, 1974: Die Oker- oder Steinfeldkrankheit. Diss. Univ. Göttingen: 144 S.

Hornsberger, H., 1905: Streu und Stickstoff. Z. Forst- u. Jagdwes. 37: 71–82.

Horvát, A. C., 1972: Die Vegetation des Mecsek-Gebirges und seiner Umgebung. Budapest (Verl. Ungar. Akad. Wiss.): 360 S.+

*Horvat, I., Glavač, V., Ellenberg, H., 1974: Vegetation Südosteuropas. Stuttgart, Gustav Fischer-Verlag: 752 S.

Hübl, E., 1959: Die Wälder des Leithagebirges. Verh. Zool.-Bot. Ges. Wien 98/99: 96–167.+

Hübschmann, A. von, 1967: Über die Moosgesellschaften und das Vorkommen der Moose in den übrigen Pflanzengesellschaften des Moseltales. Schriftenr. Vegetationskunde (Bonn) 2: 63–121.

*–, Tüxen, R., 1964: Bibliographia phytosociologica cryptogamica, Pars III: Musci (sine Epiphyta). Excerpta Botan., Sect. B 6: 179–207.

Hueck, K., 1928: Die Vegetations- und Oberflächengestaltung der Oberharzer Hochmoore. Beitr. Naturdenkmalpflege 12: 153–214.

–, 1932: Erläuterung zur vegetationskundlichen Karte der Lebanehrung (Ostpommern). Beitr. Naturdenkmalpflege 15: 99–133.+

–, 1934: Erläuterung zur vegetationskundlichen Karte des Memeldeltas (südlicher Teil). Ebenda 16: 161–224.

–, 1939: Botanische Wanderungen im Riesengebirge. Pflanzensoziol. (Jena) 3: 116 S.+

–, 1942: Die Pflanzenwelt des Naturschutzgebietes Krumme Lanke bei Rahnsdorf. Arb. Berliner Provinzstelle f. Naturschutz 3: 3–87.+

Hügin, G., 1962: Wesen und Wandlung der Landschaft am Oberrhein. Beitr. Landespflege 1, Festschr. Prof. Wiepking: 186–250.

Hülsenberg, C., 1966: Bodenökologische Untersuchungen in einem Arrhenatheretum elatioris. Diss. Univ. Gießen: 163 S.

Hundt, R., 1954: Grünlandgesellschaften an der unteren Mulde und mittleren Elbe. Wiss. Z. Univ. Halle, Math.-Nat. 3: 883–928.

–, 1957: Pflanzensoziologie im Dienste der Grünlandwirtschaft. Die Deut. Landwirtsch. 1957, Nr. 6, 7 u. 11.

–, 1958: Beiträge zur Wiesenvegetation Mitteleuropas. I. Die Auewiesen an der Elbe, Saale und Mulde. Nova Acta Leopoldina N. F. 20, 135: 206 S.

–, 1960: Einige Beobachtungen über die Höhenstufen-Differenzierung der Mähwiesen in der mediterranen Quercus-ilex-Stufe von Montpellier. Wiss. Z. Univ. Halle, Math.-Nat. 9: 251–258.

–, 1963: Die Entwicklung der Grünlandwirtschaft und der Naturschutz. Ebenda 3: 37–58.

*–, 1964: Die Bergwiesen des Harzes, Thüringer Waldes und Erzgebirges. Pflanzensoziol. (Jena) 14: 264 S.

–, 1966: Ökologisch-geobotanische Untersuchungen an Pflanzen der mitteleuropäischen Wiesenvegetation. Botan. Studien 16: 176 S.

–, 1970: Untersuchungen zum Wasserfaktor im Arrhenatheretum elatioris. Arch. Naturschutz u. Landschaftsforsch. 10: 241–267.

–, 1972: Die Trollius europaeus-Polygonum bistorta-Ges. am Ossen auf der Insel Rügen. Natur u. Naturschutz Mecklenburg 10: 25–33.

–, 1974: Les relations phytogéographiques entre les associations de pelouses du sud de l'Angleterre et celles de l'Europe Centrale. Documents Phytosoc. (Lille) 7–8: 65–93.

–, 1975: Zur anthropogenen Verbreitung und Vergesellschaftung von Geranium pratense L. Vegetatio 31: 23–32.

Hürlimann, H., 1951: Zur Lebensgeschichte des Schilfs an den Ufern der Schweizer Seen. Beitr. Geobot. Landesaufn. Schweiz 30: 232 S.

Ilijanić, L., 1962: Beitrag zur Kenntnis der Ökologie einiger Niederungswiesentypen Kroatiens. Acta Botan. Croat. 20/21: 95–167.

Ingestad, T., 1973: Mineral nutrient requirements of Vaccinium vitis-idaea and V. myrtillus. Physiol. Plant. 29: 239–246.

Issler, E., 1937: Les associations végétales des Vosges méridionales et de la plaine rhénane avoisinante. Les tourbières. Bull. Soc. Hist. Nat. Colmar 43: 5–53.

–, 1942: Vegetationskunde der Vogesen. Pflanzensoziol. (Jena) 5: 192 S.

Iversen, J., 1929: Studien über die pH-Verhältnisse dänischer Gewässer und ihren Einfluß auf die Hydrophyten-Vegetation. Botan. Tidskr. 40: 277–333.

–, 1936: Biologische Pflanzentypen als Hilfsmittel in der Vegetationsforschung. Kopenhagen: 224 S.

–, 1953: The zonation of the salt marsh vegetation of Skallingen in 1931–34 and in 1952. Geogr. Tidskr. 52: 113–118.

–, 1958: Pollenanalytischer Nachweis des Reliktcharakters eines jütischen Lindenmischwaldes. Veröff. Geobot. Inst. Rübel, Zürich 33: 137–144.

*Ives, J. D., Barry, R. G. (Ed.)., 1974: Arctic and alpine environments. London (Methuen and Co. Ltd.): 999 S.

Izdebska, M., Szynal, T., 1961: Geobotanical investigations in the forest reserve of Obrocz in Central Roztocze. Ann. Univ. Lublin-Polonia 26: 351–386.+

*Jaag, O., 1945: Untersuchungen über die Vegetation und Biologie der Algen des nackten Gesteins in den Alpen, im Jura und im schweizerischen Mittelland. Beitr. Kryptogamenflora d. Schweiz 9, 3: 560 S.

Jäger, E., 1968: Die pflanzengeographische Ozeanitätsgliederung der Holarktis und die Ozeanitätsbindung der Pflanzenareale. Feddes Repert. 79: 157–335.

Jäger, K.-D., 1967: Eine pliozäne Vegetationsgemeinschaft und ihre Fortentwicklung bis zur Gegenwart. Abh. Zentr. Geol. Inst. Berlin 10: 99–112.

–, 1970: Beiträge zur Bodensystematik unter besonderer Berücksichtigung reliktischer und rezenter Merkmale. Tag. Ber. Deut. Akad. Landwirtschaftswiss. Berlin 102: 109–122.

Jahn, G., 1975: Die Waldgesellschaften im nordwestdeutschen Pleistozän. Vortr. u. Tagungen Arb.gem. Forstl. Vegetationskunde 5: 58–69.+

–, 1979: Werden und Vergehen von Buchenwald-Gesellschaften. Ber. Internat. Symposien Internat. Vereinig. Vegetationskunde (Rinteln 1978): 339–362.

Jahn, S., 1952: Die Wald- und Forstgesellschaften des Hils-Berglandes (Forstamtsbezirk Wenzen). Angew. Pflanzensoziol. (Stolzenau/Weser) 5: 77 S.

Jahns, W., 1962: Zur Kenntnis der Pflanzengesellschaften des Großen und Weißen Moores bei Kirchwalsede (Krs. Rotenburg/Hann.). Mitt. Florist.-Soziol. Arb. gem., N. F. 9: 88–94.

–, 1969: Torfmoos-Gesellschaften der Esterweger Dose. Schriftenr. Vegetationskunde (Bonn) 4: 49–74.

Jakucs, P., 1961 a: Die Flaumeichenwälder in der Tschechoslowakei. Veröff. Geobot. Inst. ETH, Stiftg. Rübel, Zürich 36: 91–118.

*–, 1961 b: Die phytozönologischen Verhältnisse der Flaumeichen-Buschwälder Südostmitteleuropas. Monographie der Flaumeichen-Buschwälder I. Akad. Kiadó, Budapest: 314 S.

–, 1968: Comparative and statistical investigations on some microclimatic elements of the biospaces of forests, shrub stands, woodland margins and open swards. Acta Botan. Acad. Sci. Hung. 14: 281–314.

–, 1969: Die Sproßkolonien und ihre Bedeutung in der dynamischen Vegetationsentwicklung (Polycormonsukzession). Acta Botan. Croat. (Zagreb) 28: 161–170.

–, 1970: Bemerkungen zur Saum-Mantel-Frage. Vegetatio 21: 29–47.

–, 1972: Dynamische Verbindung der Wälder und Rasen. (Quantitative und qualitative Untersuchungen über die synökologischen, phytozönologischen und strukturellen Verhältnisse der Waldsäume). Akad. Kiadó, Budapest: 228 S.

–, Jurko, A., 1967: Querco petraeae-Carpinetum waldsteinietosum, eine neue Subassoziation aus dem slowakischen und ungarischen Karstgebiet. Biológia (Bratislava) 22: 321–335

–, Kovács, M., Précsényi, I., 1970: Complex investigations on some soil characteristics of the bio-units sward–woodland margin–shrub–forest. Acta Botan. Acad. Sci. Hung. 16: 111–116.

Janiesch, P., 1973: Beitrag zur Physiologie der Nitrophyten: Nitratspeicherung und Nitratassimilation bei Anthriscus sylvestris. Flora 162: 479–491.

–, 1973 b: Ökophysiologische Untersuchungen an Umbelliferen nitrophiler Säume. Oecol. Plant. 8: 335–352.

–, 1981: Ökophysiologische Untersuchungen an Carex-Arten aus Erlenbruchwäldern. Habilitationsschr. Fachber. Biol. Univ. Münster i. W., 123 S. (im Druck).

*Jankuhn, H., 1969: Vor- und Frühgeschichte vom Neolithikum bis zur Völkerwanderungszeit. Verlag Eugen Ulmer, Stuttgart: 300 S.

Jannson, S. L., 1958: Tracer studies on nitrogen transformations in soil with special attention to mineralisation-immobilisation relationships. Kgl. Lantbrukshögskol. Ann. 24: 101–361.

Járai-Komlódi, M., 1960: Beiträge zur Kenntnis der Vegetation des Moorgebietes Hansag. Ann. Univ. Sci. Budapest, Sect. Biol. 3: 229–234.+

Jarvis, P. G., 1964: Interference by Deschampsia flexuosa (L.) Trin. Oikos 15: 56–78.

Jeník, J., 1955: Die Sukzessionen der Pflanzen auf den Flußalluvionen des Flusses Bela in der Hohen Tatra. Acta Univ. Carol. Pragae 4, 58 S.

–, 1958: Geobotanische Untersuchungen einer Lawinenbahn im Tale Blaugrund im Riesengebirge. Acta Univ. Carol. Pragae, Biol. 5: 47–91.

–, Lokvenc, T., 1962: Die alpine Waldgrenze im Krkonoše Gebirge. Rozpr. Československ. Akad. Věd, Řada Mat. Přírodn. Věd 72: 65 S.

Jenny-Lips, H., 1930: Vegetationsbedingungen und Pflanzengesellschaften auf Felsschutt. Beih.

Botan. Cbl. 46: 119–296.
JENSEN, U., 1961: Die Vegetation des Sonnenberger Moores im Oberharz und ihre ökologischen Bedingungen. Naturschutz u. Landschaftspflege in Niedersachsen 1: 85 S.
JESCHKE, L., 1961: Die Vegetation des Naturschutzgebietes „Mümmelken-Moor" auf der Insel Usedom. Arch. Naturschutz und Landschaftsforsch. 1: 54–84.+
–, 1962: Vegetationskundliche Beobachtungen in Listland (Insel Sylt). Beitr. Naturkundemus. Stralsund 1: 67–84.
–, 1963: Die Wasser- und Sumpfvegetation im Naturschutzgebiet „Ostufer der Müritz". Limnologica (Berlin) 1: 475–545.+
–, 1964: Die Vegetation der Stubnitz (Naturschutzgebiet Jasmund auf der Insel Rügen). Natur u. Naturschutz Mecklenb. 2: 154 S.+
JESSEN, O., 1937: Heckenlandschaften im nordwestlichen Europa. Mitt. Geogr. Ges. Hamburg 45, 7–58.
JOCHIMSEN, M., 1963: Vegetationsentwicklung im hochalpinen Neuland. Ber. Naturw.-Mediz. Ver-Innsbruck 53: 109–123.
–, 1970: Die Vegetationsentwicklung auf Moränenböden in Abhängigkeit von einigen Umweltfaktoren. Veröff. Univ. Innsbruck 46: 5–20.
JONES, K., 1974: Nitrogen fixation in a salt marsh. J. Ecol. 62: 553–565.
JURASZEK, H., 1928: Pflanzensoziologische Studien über die Dünen bei Warschau. Bull. Acad. Polon. Sci. et Lettr., Cl. Sci. Mat. Nat., Sér. B 1927: 565–610.
JÜRGING, P., 1975: Epiphytische Flechten als Bioindikatoren der Luftverunreinigung. Bibliotheca Lichenol. 4: 164 S.
JURKO, A., 1958: Bodenökologische Verhältnisse und Waldgesellschaften der Donautiefebene. Slov. Akad. Vied: 264 S.
–, 1963: Die Veränderung der ursprünglichen Waldphytozönosen durch die Introduktion der Robinie. Českosl. Ochrana Príody 1: 56–75.
–, 1964: Feldheckengesellschaften und Uferweidengebüsche des Westkarpatengebietes. Biol. Práce (Bratislava) 10, H. 6: 100 S.
–, 1974: Prodromus der Cynosurion-Gesellschaften in den Westkarpaten. Folia Geobot. Phytotax. Praha 9: 1–44.
–, DUDA, M., 1970: Research project Báb (IBP). Progress report 1. Botan. Inst. Slovak Acad. Sci., Bratislava 1970: 240 S.
–, PECIAR, V., 1963: Pflanzengesellschaften an schattigen Felsen in den Westkarpaten. Vegetatio 11: 199–209.

KADLUS, Z., 1967: Gipfelphänomen im Gebirge Orlické hory (Adlergebirge). Opera Corcontica 4: 55–77.
KALB, K., 1970: Flechtengesellschaften der vorderen Ötztaler Alpen. Diss. Botan. 9: 118 S.
KALCKSTEIN, B., 1974: Gaswechsel, Produktivität und Herbizidempfindlichkeit bei verschiedenen tropischen, subtropischen und europäischen Gramineen. Diss. Univ. Wien (mscr.), 119 S.
KALELA, A., 1937: Zur Synthese der experimentellen Untersuchungen über Klimarassen der Holzarten. Comm. Inst. Forest. Fenn. 26: 445 S.
KAMBACH, H. H., WILMANNS, O., 1969: Moose als Strukturelemente von Quellfluren und Flachmooren am Feldberg im Schwarzwald. Veröff. Landesstelle Naturschutz u. Landschaftspflege Baden-Württemb. 37: 62–80.
KÄMMER, F., 1974: Klima und Vegetation auf Tenerife, besonders im Hinblick auf den Nebelniederschlag. Scripta Geobot. (Göttingen) 7, 78 S.
KAPPEN, L., 1975: The lichen symbiosis, a special adaptation to the ecological conditions of extreme environments. XII. Internat. Botan. Congr. (Leningrad) Abstr. 1. 151.
–, LANGE, O. L., 1970: Kälteresistenz von Flechten aus verschiedenen Klimagebieten. Deut. Botan. Ges. N. F. 4: 61–65.
–, ULLRICH, W. R., 1970: Verteilung von Chlorid und Zuckern in Blattzellen halophiler Pflanzen bei verschieden hoher Frostresistenz. Ber. Deut. Botan. Ges. 83: 265–275.
KÁRPÁTI, I. u. V., 1971 a: Die Hochwassertoleranz der ungarischen Donauauen-Vegetation. Schriftenr. Raumforsch. u. Raumplanung (Klagenfurt) 11: 146–148.
–, 1971 b: Methodological problems of the research on the production of the primary phytobiomass of lake Balaton. Hydrobiologia 12: 155–158.
–, V., 1963: Die zönologischen und ökologischen Verhältnisse der Wasservegetation des Donau-Überschwemmungsraums in Ungarn. Acta Bot. Acad. Sci. Hung. 9: 323–385.+
–, V. u. I., 1961: Winter dormancy of hungarian trees and shrubs. I. Trees and shrubs of natural groves. Acta Biol. Acad. Sci. Hung. 11: 359–385.
KARPOV, V. G., 1960: On the quantity and species composition of the viable seeds in the soil of spruce forests of Piceetum myrtillosum type. Transact. Moscov Soc. Naturalists 3: 131–140.
–, 1975: Competition for nutrients in plant communities. XII. Internat. Botan. Congr. (Leningrad) Abstr. 1, 151.
KÄSTNER, M., 1941: Über einige Waldsumpfgesellschaften, ihre Herauslösung aus den Waldgesellschaften und ihre Neueinordnung. Beih. Botan. Cbl. 61 B: 137–207.
KAULE, G., 1969: Vegetationskundliche und landschaftsökologische Untersuchungen zwischen Inn und Chiemsee. (Manuskriptdruck). Freising-Weihenstephan: Inst. f. Landschaftspflege der TH München, 1969: 153 S. +
–, 1973: Typen und floristische Gliederung der voralpinen und alpinen Hochmoore Süddeutschlands. Ber. Geobot. Inst. ETH, Stiftg. Rübel, Zürich 51: 127–143.
–, 1974: Die Übergangs- und Hochmoore Süddeutschlands und der Vogesen. Diss. Botan. 27, 345 S.
–, PFADENHAUER, J., (1973): Vegetation und Ökologie eines Hochmoorrandbereichs im Naturschutzgebiet Eggstätt-Hemhofer Seenplatte. Ber. Bayer. Botan. Ges. 44: 201–210.
–, SCHALLER, J., SCHOBER, H.-M., 1979: Auswertung der Kartierung schutzwürdiger Biotope in Bayern. Allgemeiner Teil – Außeralpine Naturräume. Verlag R. Oldenbourg, München-Wien, 94 S.
KAUSCH, W., 1955: Saugkraft und Wassernachleitung im Boden als physiologische Faktoren. Planta (Berlin) 45: 217–263.
KAUTNER, A., 1933: Beiträge zur Kenntnis des Wurzelwachstums der Gräser. Ber. Schweiz. Botan. Ges. 42: 37–108.
KAŻMIERCZAKOWA, R., 1971: Ecology and production of Potentillo albae-Quercetum and Tilio-Carpinetum ground flora in two forest reserves on the Matopolska upland. Zakł. Ochrony Przyr. Polsk. Akad. Nauk 5: 104 S.
KELLER, B., 1925: Halophyten- und Xerophyten-Studien. J. Ecol. 13, 224–261.
KELLER, Hs. M., 1968: Der heutige Stand der Forschung über den Einfluß des Waldes auf den Wasserhaushalt. Schweiz. Z. Forstwes. 119: 364–379.
KELLER, M., 1972: Kleinräumige Verbreitung von Pflanzenarten im Luzerner Seetal im Vergleich zu Gesamtverbreitung und Umwelt. Mitt. Naturforsch. Ges. Luzern 23: 189 S.
KELLER, TH., 1968: Der Einfluß der Luftverunreinigungen auf den Wald im Lichte der neuesten Literatur. Schweiz. Z. Forstwes. 119: 353–363.
–, 1970: Über die Assimilation einer jungen Arve im Winterhalbjahr. Bündnerwald (Chur) 23: 49–54.
–, 1971: Der Einfluß der Stickstoffernährung auf den Gaswechsel der Fichte. Allg. Forst- u. Jagdztg. 142: 89–93.
KELLER, W., 1974: Der Lindenmischwald des Schaffhauser Randens. Ber. Schweiz. Botan. Ges. 84: 105–120.
KEPCZYŃSKI, K., 1965: Die Pflanzenwelt des Diluvialplateaus von Dobrzyń. Univ. Toruń: 821 S.+
*KERFOOT, O., 1968: Mist precipitation and vegetation. Forestry Abstr. 29: 8–20.
KERN, K. G., 1970: Ertragskundlich-ökologische Untersuchungen an Pappeln im Überschwemmungsbereich des Rheins. Allg. Forst- u. Jagdztg. 141: 83–86.
KERNER VON MARILAUN, A., 1863: Das Pflanzenleben der Donauländer. Innsbruck, 350 S.
–, 1887: Pflanzenleben. Bd. 1: Gestalt und Leben der Pflanze. Leipzig, 744 S.
KERSHAW, K. A., 1963: Pattern in vegetation and its causality. Ecology 44: 377–388.
KICKUTH, R., 1970: Ökochemische Leistungen höherer Pflanzen. Naturwissenschaften 57: 55–61.
KIENZLE, U., 1979: Sukzessionen in brachliegenden Magerwiesen des Jura und Napfgebietes. Diss. Univ. Basel, 104 S.
KIESE, O., 1972: Bestandsmeteorologische Untersuchungen zur Bestimmung des Wärmehaushalts eines Buchenwaldes. Ber. Inst. Meteor. Klimatol. TU Hannover 6: 132 S.
–, 1975: Zur Frage der Verdunstungsleistung von Wäldern. Ebenda 10: 125–137.
KILBERTUS, G., 1970: La production primaire et la décomposition des mousses (Eubrya) comparés à celles d'autres végétaux. Bull. Acad. et Soc. Lorraines Sci. 9: 136–145.
KING, J., 1960: Observation on the seedling establishment and growth of Nardus stricta in burned Callunetum. J. Ecol. 48: 667–677.
KIRSCHBAUM, U., 1972: Kartierung des natürlichen Flechtenvorkommens. Regionale Planungsgemeinschaft Untermain, Frankfurt a. M., 4. Arbeitsber.: 76–80.
KISSER, J., 1966: Forstliche Rauchschäden aus der Sicht des Biologen. Mitt. Forstl. Bundesversuchsanst. Mariabrunn 73: 7–46.
KLAPP, E., 1950: Dauerweiden West- und Süddeutschlands. Z. Acker- u. Pflanzenbau 91: 265–305.
–, 1965: Grünlandvegetation und Standort. Verl. Paul Parey, Berlin-Hamburg: 384 S.
–, 1971: Wiesen und Weiden. 4. Aufl. Verl. Paul Parey, Berlin-Hamburg: (3. Aufl. 1956), 519 S.
*KLAUSING, O., LOHMEYER, W., WALTHER, K., 1963: Bibliographie zum Thema Produktionspotentiale von Pflanzengesellschaften. Excerpta Bot., Sect. B, 5: 161–202.
*KLEBELSBERG, R. V., 1948 u. 1949: Handbuch der Gletscherkunde und Glazialgeologie. Bd. I. Allgemeiner Teil: 403 S. Wien 1948. Bd. II. Historisch-regionaler Teil: 625 S. Wien 1949.
KLEINKE, J., SUCCOW, M., VOIGTLÄNDER, U., 1974: Der Wasserstufenzeigerwert von Grünlandpflanzen im nördlichen Teil der DDR. Arch. Naturschutz und Landschaftsforschung 14: 139–146.
KLEMENT, O., 1953: Die Vegetation der Nordseeinsel Wangerooge. Veröff. Inst. Meeresforsch. Bremerhaven 2: 279–379.
–, 1955: Prodromus der mitteleuropäischen Flechtengesellschaften. Feddes Repert. Beih. 135: 5–194.
–, 1971: Über Flechten der Eilenriede. Beih. Ber. Naturhist. Ges. Hannover 7: 139–142.
KLIKA, J., 1933: Studien über die xerotherme Vegetation Mitteleuropas. II. Xerotherme Gesellschaften in Böhmen. Beih. Botan. Cbl. 50 B: 707–773.
–, 1935: Die Pflanzengesellschaften des entblößten Teichbodens in Mitteleuropa. Ebenda 53 B: 286–310.
–, u. Mitarb., 1943: Die Durchforschung des Naturschutzgebietes Velká hora bei Karlštein. Sborn. Čes. Akad. Techn. 16: 497–520, 560–610 u. 644–658.+
KLIX, W., KRAUSCH, H.-D., 1958: Das natürliche Vorkommen der Rotbuche in der Niederlausitz. Wiss. Z. Pädagog. Hochsch. Potsdam, Math. Nat. 4: 5–27.
KLOSE, H., 1963: Zur Limnologie von Lemna-Gewässern. Wiss. Z. Univers. Leipzig 12, Math.-Nat. R. 1: 233–259.
KLOSS, K., 1965: Schoenetum, Juncetum subnodulosi und Betula pubescens-Gesellschaften der kalkreichen Moorniederungen Nordost-Mecklenburgs. Feddes Repert. Beih. 142: 65–117.
KLÖTZLI, F., 1965: Qualität und Quantität der Rehäsung in Wald- und Grünland-Gesellschaften des nördlichen Schweizer Mittellandes. Diss. E. T. H. Zürich 1965: 186 S.
–, 1968 a: Über die soziologische und ökologische Abgrenzung schweizerischer Carpinion- von den Fagion-Wäldern. Feddes Repert. 78: 15–37.
*–, 1968 b: Wald und Umwelt. Schweiz. Z. Forstwes. 1968: 264–334.
*–, 1969 a: Zur Ökologie schweizerischer Bruchwälder unter besonderer Berücksichtigung des Waldreservates Moos bei Birmensdorf und des Katzensees. Ber. Geobot. Inst. ETH, Stiftg. Rübel, Zürich 39: 56–123.
–, 1969 b: Die Grundwasserbeziehungen der Streu- und Moorwiesen im nördlichen Schweizer Mittelland. Beitr. Geobot. Landesaufn. Schweiz 52: 1–296.
–, 1971: Biogenous influence on aquatic macrophytes, especially Phragmites communis. Hidrobiologia (Bucureşti) 12: 107–111.
–, 1973: Waldfreie Naßstandorte der Schweiz. Ber. Geobot. Inst. ETH, Stiftung Rübel, Zürich 51: 15–39.
–, 1975 a: Zum Standort vom Edellaubwäldern im Bereich des südlichen borealen Nadelwaldes. Mitt. Eidg. Anst. Forstl. Versuchsw. 51: 49–64.
–, 1975 b: Ökologische Besonderheiten Pinus-reicher Waldgesellschaften. Schweiz. Z. Forstwes. 126: 672–710.
–, MEYER, M., ZÜST, S., 1973: Exkursionsführer. Ber. Geobot. Inst. ETH, Stiftung Rübel, Zürich 51: 40–95.
–, ZÜST, S., 1973: a) Nitrogen regime in reed-beds. Polsk. Arch. Hydrobiol. 20: 131–136. b) Conservation of reed-beds in Switzerland. Ebenda 20: 229–235.
KNABE, W., 1975: Luftverunreinigungen und Forstwirtschaft. Forstarchiv 46: 59–62.
KNAPP, H. D., 1979/80: Geobotanische Studien an Waldgrenzstandorten des herzynischen Florengebietes. Teil 1 u. 2. Flora 168: 276–319 u. 468–510. Teil 3. Ebenda 169: 177–215.
KNAPP, R., 1942: Zur Systematik der Wälder, Zwergstrauchheiden und Trockenrasen des eurosibirischen Vegetationskreises. Arb. Zentralst. Veget.kartierung d. Reiches, Beil. z. 12. Rundbrief (als Manuskr. gedr.).
–, 1958 a: Untersuchungen über den Einfluß verschiedener Baumarten auf die unter ihnen wachsenden Pflanzen. Ber. Deut. Botan. Ges. 66: 167–178.
–, 1958 b: Pflanzengesellschaften des Vogelsberges unter besonderer Berücksichtigung des „Naturschutzparkes Hoher Vogelsberg". Schriftenr. Naturschutzstelle Darmstadt 4: 161–220. +
–, 1960: Die Bedeutung der Dauer der Schneebedeckung für die Vegetation in subalpinen Lagen. Ber. Bayer. Botan. Ges. 33: 89–93.
–, 1967: Die Vegetation des Landes Hessen. Gießen u. Göttingen (Syst.-Geobot. Inst.): 148 S. +
–, STOFFERS, A. L., 1962: Über die Vegetation von Gewässern und Ufern im mittleren Hessen und Untersuchungen über den Einfluß von Pflanzen auf Sauerstoffgehalt, Wasserstoff-Ionenkonzentration und die Lebensmöglichkeit anderer Gewächse. Ber. Oberhess. Ges. Natur- u. Heilkunde Gießen, Naturw. Abt. 32: 90–141.
KOBLET, H., 1972: Über die Entwicklung und die Stoffproduktion von Wiesenpflanzen in Abhängigkeit von der Artenkombination und von Umweltfaktoren. Angew. Botan. 46: 59–74.
KOCH, F., 1955: Die Auswirkungen der anormalen Trockenheit des Sommers 1952 auf die Ackerunkrautgemeinschaften deutscher Dauerdüngungsversuche. Z. Pflanzenbau u. Pflanzenschutz 1955: 32–40.
KOCH, H. G., 1958: Der Holzzuwachs der Waldbäume in verschiedenen Höhenlagen Thüringens in Abhängigkeit von Niederschlag und Temperatur. Arch. Forstwes. 7: 27–49.
KOCH, WALO, 1926: Die Vegetationseinheiten der Linthebene, unter Berücksichtigung der Verhältnisse in der Nordostschweiz. Jb. Naturw. Ges. St. Gallen 61: 144 S.
KOCH, W., 1970: Temperaturansprüche von Unkräutern bei der Keimung. Saatgutwirtschaft (Hohenheim) 22: 85–86.
–, KÖCHER, H., 1968: Zur Bedeutung des Nährstoffaktors bei der Konkurrenz zwischen Kulturpflanzen und Unkräutern. Z. Pflanzenkrankheiten (Pflanzenpathologie) u. Pflanzenschutz 1968: 79–87.
KOHL, F. (Red.), 1971: Kartieranleitung, Anleitung und Richtlinien zur Herstellung der Bodenkarte 1:25000. Arb.-Gem. Bodenkunde (Hannover): 169 S.
KOHLER, A., 1968: Zum ökologischen und soziologischen Verhalten der Robinie (Robinia pseudoacacia) in Deutschland. In: R. TÜXEN (Hrsg.) Pflanzensoziologie und Landschaftsökologie. Den Haag 1968: 402–411.
–, 1976: Makrophytische Wasserpflanzen als Bioindikatoren für Belastungen von Fließwasser-Ökosystemen. Verh. Ges. Ökol. Wien 1975: 255–276.
–, BRINKMEIER, R., VOLLRATH, H., 1974: Verbreitung und Indikatorwert der submersen Makrophyten in den Fließgewässern der Friedberger Au. Ber. Bayer. Botan. Ges. 45: 5–36.
–, SUKOPP, H., 1964 a: Über die soziologische Struktur einiger Robinienbestände im Stadtgebiet von Berlin. Sitz. ber. Ges. Naturforsch. Freunde Berlin N. F. 4: 74–88.
–, –, 1964 b: Über die Gehölzentwicklung auf Berliner Trümmerstandorten. Ber. Deut. Botan. Ges. 76: 389–406.
–, ZELTNER, G., BUSSE, M., 1972: Wasserpflanzen und Bakterien als Verschmutzungsanzeiger von Fließgewässern. Umschau 72: 158–159.
KØIE, M., 1938: The soil vegetation of the Danish conifer plantations and its ecology. Kong. Danske Vidensk. Selsk. Skr., Naturw. Math. Afd., 9. R. 7, 2: 85 S.
KOJAK, A., 1974: Analysis of a sheep pasture ecosystem in the Pieniny mountains (the Carpathians). XVII. Analysis of the transfer of carbon. Ecol. Pol. 22: 711–732.
KOLBEK, J., 1975: Die Festucetalia valesiacae-Gesellschaften im Ostteil des Gebirges Ceské středohoří (Böhmisches Mittelgebirge). 1. Die Pflanzengesellschaften. Folia Geobot. Phytotax. (Praha) 10: 1–57.
KOLUMBE, E., 1931: Spartina townsendii-Anpflanzungen im schleswig-holsteinischen Wattenmeer. Wiss. Meeresunters., Abt. Kiel 21: 67–71.
KONCZAK, P., 1968: Die Wasser- und Sumpfpflanzengesellschaften der Havelseen bei Potsdam. Limnologica (Berlin) 6: 147–201.
KÖNIG, D., 1972: Diatom investigations at the West coast of Schleswig-Holstein. Beih. Nova Hedwigia 39: 127–137.
KOPECKÝ, K., 1960: Phytocoenologische Studie der Kalk-Flachmoorwiesen in Nordostböhmen. Rozpr. Československ. Akad. Věd, Mat. Přírodn. Věd, 70, 4: 64 S.
–, 1966: Ökologische Hauptunterschiede zwischen Röhrichtgesellschaften fließender und stehender Binnengewässer Mitteleuropas. Folia Geobot. Phytotaxon. Bohemoslov. 1: 193–242.
–, 1967 a: Die flußbegleitende Neophytengesellschaft Impatienti-Solidaginetum in Mittelmähren. Preslia (Praha) 39: 151–166.
–, 1967 b: Mitteleuropäische Flußröhrichtgesellschaften des Phalaridion arundinaceae-Verbandes. Limnologica (Berlin) 5: 39–79.
–, 1969: Klassifikationsvorschlag der Vegetationsstandorte an den Ufern der tschechoslowakischen Wasserläufe unter hydrologischen Gesichtspunkten. Arch. Hydrobiol. 66: 326–347.
KOPPE, F., 1969: Moosvegetation und Moosflora der Insel Borkum. Natur u. Heimat 29: 41–84.
KÖRBER-GROHNE, U., 1967: Geobotanische Untersuchungen auf Feddersen Wierde. Steiner-Verlag, Wiesbaden: 357 S.
KORNAŚ, J., 1955: Caractéristique géobotanique des Gorces (Karpates Occidentales Polonaises). Monogr. Botan. 3: 216 S. +
–, 1957: Les associations végétales du Jura Cracovien. III. Acta Soc. Botan. Polon. +
–, 1967: Montane hay-meadow Gladiolo-Agrostietum in the Polish Western Carpathians. Contrib. Botan. (Cluj) 1967: 167–176.
–, 1968: Der Linden-Eichen-Hainbuchenwald (Tilio-Carpinetum) in den polnischen Karpaten. Feddes Repert. 77: 143–153.
–, PANCER, E., BRZYKI, B., 1960: Studies on sea-bottom vegetation in the bay of Gdańsk off Rewa. Fragm. Flor. Geobot. 6: 1–92.
KORNECK, D., 1962/63: Die Pfeifengraswiesen und ihre wichtigsten Kontaktgesellschaften in der nördlichen Oberrheinebene und im Schweinfurter Trockengebiet. II. Die Molinieten feuchter Standorte. Beitr. Naturk. Forsch. Südw.-Deut. 21: 83–86 u. 22: 19–44.
–, 1969: Das Sclerochloo-Polygonetum avicularis, eine seltene Trittgesellschaft in Trockengebieten Mitteleuropas. Mitt. Florist.-Soziol. Arb. gem. N. F. 14: 193–210.
–, 1974: Xerothermvegetation in Rheinland-Pfalz und Nachbargebieten. Schr. Reihe Vegetationskunde 7: 196 S.
–, 1975: Beitrag zur Kenntnis mitteleuropäischer Felsgras-Gesellschaften (Sedo-Scleranthetea). Mitt. Florist.-Soziol. Arb. gem. N. F. 18: 45–102.
KORSMO, E., 1930: Unkräuter im Ackerbau der Neuzeit. Berlin: 580 S.
KÖSTLER, J. N., BRÜCKNER, E., BIBELRIETHER, H., 1968: Die Wurzeln der Waldbäume. Verlag Paul Parey, Hamburg-Berlin: 284 S.
–, Mayer, H., 1970: Waldgrenzen im Berchtesgadener Land. Jahrb. Ver. Schutze Alpenpflanzen u. -Tiere 35: 1–35.
KOTAŃSKA, M., 1970: Morphology and biomass of the underground organs of plants in grassland communities of the Ojców National Park. Zakł. Ochrony Przyr. Polsk. Akad. Nauk 4: 167 S.
KÖTTER, F., 1961: Die Pflanzengesellschaften im Tidegebiet der Unterelbe. Arch. Hydrobiol. Suppl. 26: 106–185.
*KOVÁCS, M., 1962: Die Moorwiesen Ungarns. Die Vegetation ungarischer Landschaften 3, Budapest: 214 S.
–, 1964: Ökologische Untersuchungen von

Sumpf- und Mähwiesen in der Umgebung von Galgamácsa. Acta Agron. Acad. Sci. Hung. 13: 61–91.

–, 1968 a: Die Acerion pseudoplatani-Wälder (Mercuriali Tilietum und Phyllitido-Aceretum) des Matra-Gebirges. Ebenda 14: 331–350.

–, 1968 b: Die Vegetation im Überschwemmungsgebiet des Ipoly (Eipel)-Flusses. II. Die okologischen Verhältnisse der Pflanzengesellschaften. Ebenda 14: 77–112.

–, 1975: Beziehung zwischen Vegetation und Boden. Die Bodenverhältnisse der Waldgesellschaften des Mátragebirges. Die Vegetation Ungarischer Landschaften 6: 365 S.

Kozma, D., 1922: Über das Verhalten der Unkrautsamen im Ackerboden. Kisérl. Közlem. 25: 1–79.

Krach, K.E., 1959: Untersuchungen über die Ausscheidung unverdauter Klee-, Gras- und Unkrautsamen durch Vögel und die Beeinflussung ihrer Keimwerte durch die Magen- und Darmpassage. Z. Acker- u. Pflanzenbau 107: 405–434.

Kral, F., 1979: Spät- und postglaziale Waldgeschichte der Alpen auf Grund der bisherigen Pollenanalysen. Österr. Agrarverlag, Wien, 175 S.

–, Mayer, H., Zukrigl, K., 1975: Die geographischen Rassen der Waldgesellschaften in vegetationskundlicher, waldgeschichtlicher und waldbaulicher Sicht. Beitr. Naturk. Forsch. Südw.-Deut. 34: 167–185.

Krauklis, A.A., 1975: Self-regulation, stability and productivity of boreal forest ecosystems. Vortrag XII. Internat. Botan. Congr. (Leningrad) Abstr. 1, 154.

Kraus, G., 1911: Boden und Klima auf kleinstem Raum. Gustav Fischer Verlag, Jena, 184 S.

Krausch, H.D., 1960: Die Pflanzenwelt des Spreewaldes. A. Ziemsen, Wittenberg: 124 S.

–, 1964–1970: Die Pflanzengesellschaften des Stechlinsee-Gebietes. Limnologica (Berlin) 2: 423–482, 5: 331–366, 6: 321–380, 7: 397–454.+

–, 1965 a: Zur Gliederung des Scirpo-Phragmitetum medioeuropaeum W. Koch 1926. Limnologica (Berlin) 3: 17–22.

–, 1965 b: Vegetationskundliche Beobachtungen im Donaudelta. Ebenda 3: 271–313.

–, 1965 c: Zur Gliederung des Scirpo-Phragmitetum medioeuropaeum W. Koch 1926. Limnolo- 3: 17–22.

–, 1968: Die Sandtrockenrasen (Sedo-Scleranthetea) in Brandenburg. Mitt. Florist.-Soziol. Arb. gem. N.F. 13: 71–200.

–, 1969: Geobotanische Exkursionen in die Niederlausitz, das Odertal, zum Plagefenn bei Chorin und in andere brandenburgische Landschaften. Potsdam: Fernstudium der Lehrer, Biologie, 1969: 142 S.+

Krause, W., 1940: Untersuchungen über die Ausbreitungsfähigkeit der niedrigen Segge (Carex humilis Leyss.) in Mitteldeutschland. Planta (Berlin) 31: 91–168.

–, 1950: Über Keimung und Jugendwachstum im Hinblick auf die Entwicklung der Pflanzendekke. Ebenda 38: 132–156.

–, 1956: Über die Herkunft der Unkräuter. Natur u. Volk 86: 109–119.

–, 1957: Pflanzengesellschaften als Anzeiger der Standortsbedingungen. Umschau (Frankfurt) 1957: 78–81.

*–, 1958: Ruderalpflanzen. Handb. Pflanzenphysiol. 4: 737–754.

–, 1963: Eine Grünland-Vegetationskarte der südbadischen Rheinebene und ihre landschaftsökologische Aussage. Untersuchung über den Wasserhaushalt der Ebene und seine Empfindlichkeit gegen Eingriffe. Arb. Rhein. Landeskunde (Bonn) 20: 77 S.

–, 1969: Zur Characeenvegetation der Oberrheinebene. Arch. Hydrobiol; Suppl. 35: 202–253.

Kreeb, K., 1965: Die ökologische Bedeutung der Bodenversalzung. Angew. Botan. 39: 1–15.

Kremser, W., 1974: Säen und Pflanzen. Das langfristige regionale Waldbauprogramm der Niedersächsischen Landesforstverwaltung und ihre Walderneuerungspläne nach der Sturmkatastrophe vom 13. November 1972. Neues Arch. Niedersachs. 23: 256–286.

–, Otto, H.-J., 1973: Grundlagen für die langfristige, regionale waldbauliche Planung in den niedersächsischen Forsten. Aus dem Walde; Mitt. Niedersächs. Landesforstverw. 20: 496 S.

Kriebitzsch, W.-U., 1978: Stickstoffnachlieferung in sauren Waldböden Nordwestdeutschlands. Scripta Geobot. (Göttingen) 14, 66 S.

Krieger, H., 1937: Die flechtenreichen Pflanzengesellschaften der Mark Brandenburg. Beih. Botan. Cbl. 57, B: 1–76.

Krippelová, T., 1967: Vegetation des Žitny Ostrov (Schüttinsel). Biol. Práce 13: 108 S.

–, (ed.), 1974: Synanthropic flora and vegetation. Acta Inst. Bot. Acad. Sci. Slov. Ser. A 1: 1–306.

Krisch, H., 1968: Die Grünland- und Salzpflanzengesellschaften der Werraaue bei Bad Salzungen. Teil II: Die salzbeeinflußten Pflanzengesellschaften. Hercynia 5: 49–95.

Krolikowska, J., 1975: Water budget of helophytes. XII. Internat. Botan. Congr. (Leningrad) Abstr. 1: 154.

Kronfuss, H., 1972: Kleinklimatische Vergleichsmessungen an zwei subalpinen Standorten. Mitt. Forstl. Bundes-Versuchsanst. Wien 96: 159–176.

Kronisch, R., 1975: Zur Stickstoff-Versorgung von Ruderalpflanzen-Gesellschaften in Göttingen. Schriftl. Hausarbeit f.d. Lehramt an Gymnasien, Göttingen (unveröff.): 75 S.

Kropáč, Z., 1966: Estimation of weed seeds in arable soil. Pedobiologia 6: 105–128.

Kruijne, A.A., 1965: Nardus stricta L. as a grassland species in the Netherlands. Netherl. J. Agric. Sci. 13, 2: 171–177.

Krzymowski, R., 1939: Geschichte der deutschen Landwirtschaft. Stuttgart: 309 S.

Kubiena, W.L., 1948: Entwicklungslehre des Bodens. Wien: 215 S.

–, 1953: Bestimmungsbuch und Systematik der Böden Europas. Stuttgart: 392 S.

*Küchler, A.W. (Hrsg.), 1966: International bibliography of vegetation maps. 2. Europe. Univ. Kansas Publ., Libr. Ser. 26: 584 S.

Kuhn, K., 1937: Die Pflanzengesellschaften der Schwäbischen Alb. Öhringen: 340 S.

Kuhn, N., 1967: Natürliche Waldgesellschaften und Waldstandorte der Umgebung von Zürich. Veröff. Geobot. Inst. ETH, Stiftg. Rübel, Zürich 40: 84 S.+

Kujala, V., 1926: Untersuchungen über den Einfluß von Waldbränden auf die Waldvegetation in Nord-Finnland. Comm. Inst. Quaest. Forest. Finland 10: 41 S.

Kulczynski, S., 1949: Peat bogs of Polesie. Mem. Acad. Sci. Cracovie, Ser. B 1949: 356 S.

Kümmel, K., 1950: Das mittlere Ahrtal. Eine pflanzengeographisch-vegetationskundliche Studie. Pflanzensoziol. (Jena) 7: 192 S.+

Kundler, P., 1956: Beurteilung forstlich genutzter Sandböden im nordostdeutschen Tiefland. Arch. Forstwes. 5: 585–672.

Kunick, W., 1974: Veränderungen von Flora und Vegetation einer Großstadt, dargestellt am Beispiel von Berlin (West). Diss. T.U. Berlin: 472 S.

Künne, H., 1969: Laubwaldgesellschaften der Frankenalb. Diss. Botan. 2: 177 S.+

Künstle, E., Mitscherlich, G., 1975: Photosynthese, Transpiration und Atmung in einem Mischbestand im Schwarzwald. 1. Teil: Photosynthese, 2. Teil: Transpiration. Allg. Forst- u. Jagdztg. 146: 45–63 u. 88–100.

Landesanstalt für Umweltschutz Baden-Württemberg (Hrsg.), 1978: Der Rußheimer Altrhein, eine nordbadische Auenlandschaft. Natur- u. Landschaftsschutzgebiete Bad.-Württ. 10, 622 S. +

Kunze, M., 1972: Emittentenbezogene Flechtenkartierung aufgrund von Frequenzuntersuchungen. Oecologia (Berlin) 9: 123–133.

–, 1974: Mathematischer Zusammenhang zwischen der Frequenz epiphytischer Flechten und der Fluor-Immissionsrate am Beispiel der Aluminiumhütte Rheinfelden. Beih. Veröff. Landesstelle Naturschutz u. Landschaftspflege Baden-Württ. 5: 5–13.

Künzli, W., 1967: Über die Wirkung von Hof- und Handelsdüngern auf Pflanzenbestand, Ertrag und Futterqualität der Fromentalwiese. Schweiz. Landw. Forsch. 6: 34–130.

Kuoch, R., 1954: Wälder der Schweizer Alpen im Verbreitungsgebiet der Weißtanne. Mitt. Schweiz. Anst. Forstl. Versuchswes. 30: 133–260.

–, 1970: Die Vegetation auf Stillberg (Dischmatal, Kt. Graubünden). Ebenda 46: 329–342.+

–, Amiet, R., 1970: Die Verjüngung im Bereich der oberen Waldgrenze der Alpen mit Berücksichtigung von Vegetation und Ablegerbildung. Ebenda. 46: 159–328.

Kurkin, K.A., 1975: Approach to studies of nitrate regime in grassland biogeocenoses. XII. Internat. Botan. Congr. (Leningrad) Abstr. 1: 155.

Kurth, A., Weidmann, A., Thommen, F., 1960: Beitrag zur Kenntnis der Waldverhältnisse im schweizerischen Nationalpark. Mitt. Schweiz. Anst. Forstl. Versuchswes. 36: 221–378.

Kutschera, L., 1966: Ackergesellschaften Kärntens als Grundlage standortsgemaßer Acker- und Grünlandwirtschaft. Verlag BVA Gumpenstein, Irdning (Österreich): 194 S.

Laatsch, W., Grottenthaler, W., 1973: Stabilität und Sanierung der Hänge in der Alpenregion des Landkreises Miesbach. (Zit. nach Spatz 1974).

Labrone, L., Lascombes, G., 1971: Minéralisation de l'azote organique dans les sols alpins du Pic du Midi de Bigorre. Oecol. Plant. 6: 247–270.

Lache, D.-W., 1974: Wasser- und Stickstoff-Versorgung sowie Mikroklima von Heide- und Binnendünen-Gesellschaften NW-Deutschlands. Diss. Univ. Göttingen: 64 S. (Scripta Geobot., Göttingen 11: 96 S., 1976).

Lacoste, A., 1964: Premières observations sur les associations subalpines des Alpes matitimes: étude phytosociologique des pelouses sèches basophiles. Bull. Soc. Botan. France 111: 61–69.

*Landesstelle für Naturschutz und Landespflege Baden-Württemberg, 1966: Der Spitzberg bei Tübingen. Ludwigsburg: 1140 S.

Lang, G., 1967 a: Über die Geschichte von Pflanzengesellschaften auf Grund quartärbotanischer Untersuchungen. In: R. Tüxen. Pflanzensoziologie und Palynologie. Ber. üb. d. Internat. Sympos. in Stolzenau/Weser 1962: 24–37.

*–, 1967 b: Die Ufervegetation des westlichen Bodensees. Arch. Hydrobiol. Suppl. 32: 437–574.

–, 1968: Vegetationsänderungen am Bodenseeufer in den letzten hundert Jahren. Schr. Ver. Gesch. Bodensees u. seiner Umgebung 86: 295–319.

*–, 1973: Die Vegetation des westlichen Bodenseegebiets. Pflanzensoziologie (Stuttgart) 17: 451 S.+

–, Oberdorfer, E., 1960: Vegetationskundliche Karte des oberen Wutachgebietes (Ostschwarzwald-Baar). Karlsruhe.+

Lange, E., 1971: Botanische Beiträge zur mitteleuropäischen Siedlungsgeschichte. Ergebnisse zur Wirtschaft und Kulturlandschaft in frühgeschichtlicher Zeit. Deut. Akad. Wiss. Berlin, Schr. Ur- u. Frühgesch. 27: 142 S.

–, Schlüter, H., 1972: Zur Entwicklung eines montanen Quellmoores im Thüringer Wald und des Vegetationsmosaiks seiner Umgebung. Flora 161: 562–585.

Lange, O.L., 1966: Der CO_2-Gaswechsel von Flechten nach Erwärmung im feuchten Zustand. Ber. Deut. Botan. Ges. 78. 441–454

–, 1969: CO_2-Gaswechsel von Moosen nach Wasserdampfaufnahme aus dem Luftraum. Planta (Berlin) 89: 90–94.

–, Kanzow, H., 1965: Wachstumshemmung an höheren Pflanzen durch abgetötete Blätter und Zwiebeln von Allium ursinum. Flora, Abt. B, 156: 94–101.

–, Schulze, E.-D., Koch, W., 1970: Experimentell-ökologische Untersuchungen an Flechten der Negev-Wüste. II. CO_2-Gaswechsel und wasserhaushalt von Ramalina maciformis (Del.) Bory am natürlichen Standort während der sommerlichen Trockenperiode. Flora 159: 38–62.

Larcher, W., 1953: Frostschäden und Frostschutz bei Pflanzen. „Pyramide" (Innsbruck) 1953, 10: 4 S.

–, 1957: Frosttrocknis an der Waldgrenze und in der alpinen Zwergstrauchheide auf dem Patscherkofel bei Innsbruck. Veröff. Ferdinandeum Innsbruck 37: 49–81.

–, 1963 a: Zur spätwinterlichen Erschwerung der Wasserbilanz von Holzpflanzen an der Waldgrenze. Ber. Naturw.-Mediz. Ver. Innsbruck 53: 125–137.

–, 1963 b: Zur Frage des Zusammenhanges zwischen Austrocknungsresistenz und Frosthärte bei Immergrünen. Protoplasma 57: 569–587.

–, 1980 a: Ökologie der Pflanzen auf physiologischer Grundlage. 3. Aufl. Verlag Eugen Ulmer, Stuttgart, 399 S. (2. Aufl. 1976)

–, 1980 b: Klimastreß im Gebirge – Adaptationstraining und Selektionsfilter für Pflanzen. Rhein.-Westf. Akad. Wiss. Vortr. N. 291: 49–88.

–, Mair, B., 1968: Das Kälteresistenzverhalten von Quercus pubescens, Ostrya carpinifolia und Fraxinus ornus auf drei thermisch unterschiedlichen Standorten. Oecol. Plant. 3: 225–270.

–, Schmidt, L., Tschager, A., 1973: Starke Fettspeicherung und hoher Kaloriengehalt bei Loiseleuria procumbens (L.) Desv. Oecol. Plant 8: 377–383.

Lauer, E., 1953: Über die Keimtemperaturen von Ackerunkräutern und deren Einfluß auf die Zusammensetzung von Unkrautgesellschaften. Flora 140: 551–595.

Le Blanc, F., 1969: Epiphytes and air pollution. Air Pollution. Proc. First Eur. Congr. Influence of Air Pollution on Plants and Animals, Wageningen 1968 (Pudoc): 211–221.

Lee, J.A., Stewart, G.R., Harvill, D.C., 1975: Nitrate utilization and succession. XII. Internat. Botan. Congr. (Leningrad) Abstr. 1: 156.

Leibundgut, H., 1953: Beobachtungen über den Streuabbau einiger Baumarten im Lehrwald der ETH. Schweiz. Z. Forstwes. 104: 1–14.

–, 1959: Über Zweck und Methodik der Struktur- und Zuwachsanalyse von Urwäldern. Schweiz. Z. Forstwes. 110: 111–124.

–, 1962: Waldbauprobleme in der Kastanienstufe Insubriens. Mitt. Schweiz. Anst. Forstl. Versuchsw. 113: 164–188.

–, 1964: Einfluß von Borstgras und Heidelbeere auf die Ansamung von Föhre und Lärche. Schweiz. Z. Forstwes. 115: 331–336.

Leiningen, W., 1907: Die Waldvegetation praealpiner bayerischer Moore, insbesondere der südlichen Chiemseemoore. Naturw. Z. Land- u. Forstwirtsch. 5: 1–79.

Leippert, S., 1978: Ökologische Untersuchungen zu den Ursachen des Schilfrückgangs im Dümmer See. Manuskr. Inst. Vegetationskde TU Hannover, 113 S.

Lemée, M.G., 1946: Recherches sur l'économie de l'eau chez les sousarbrisseaux xéromorphes de landes. Ann. Sci. Nat., Botan. 11, VII: 53–64.

–, 1966: Sur l'intéret écologique des réserves biologiques de la Forêt de Fontainebleau. Bull. Soc. Botan. France 113: 305–323.

–, 1968: Investigations sur la minéralisation de l'azote et sqn évolution annuelle dans les humus forestiers in situ. Oecol. Plant. 2: 285–324.

–, 1974: Recherches sur les écosystèmes des réserves biologiques de la forêt de Fontainebleau. IV. Entrées d'éléments minéraus par les précipitations et transport au sol par le pluviolessivage. Oecol. Plant. 9: 187–200.

–, 1975: Recherches sur les écosystèmes des réserves biologiques de la Forêt de Fontainebleau. III. Influence du peuplement graminéen sur les caractères et l'activité biologique du mull acide. Rev. Écol. Biol. Sol 12: 157–167.

–, Bichaut, N., 1973: Recherches sur les écosystèmes des réserves biologiques de la forêt de Fontainebleau. II. Décomposition de la litière de feuilles des arbres et libération des bioéléments. Oecol. Plant 8: 153–174.

Lenz, O., 1967: Action de la neige et du gel sur les arbres de montagne, en particulier sur leur forme et l'anatomie de la tige. Mitt. Schweiz. Anst. Forstl. Versuchswes. 43: 293–316.

León, R., 1968: Balance d'eau et d'azote dans les prairies à litière des alentours de Zurich. Veröff. Geobot. Inst. ETH, Stiftg. Rübel, Zürich 41: 2–68.

Leuthold, Ch., 1980: Die ökologische und pflanzensoziologische Stellung der Eibe (Taxus baccata) in der Schweiz. Veröff. Geobot. Inst. ETH, Stiftg. Rübel, Zürich 67, 217 S.

Lienenbecker, H., 1971: Die Pflanzengesellschaften im Raum Bielefeld-Halle. Ber. Naturw. Ver. Bielefeld 20: 67–170.+

Lieth, H., 1954: Die Porenvolumina der Grünlandböden und ihre Beziehungen zur Bewirtschaftung und zum Pflanzenbestand. Z. Acker- u. Pflanzenbau 98: 453–460.

–, (Hrsg.), 1962: Die Stoffproduktion der Pflanzendecke. Gustav Fischer Verlag, Stuttgart: 156 S. +

–, Ellenberg, H., 1958: Konkurrenz und Zuwanderung von Wiesenpflanzen. Z. Acker- u. Pflanzenbau 106: 205–223.

Lillieroth, S., 1950: Über Folgen kulturbedingter Wassersenkungen für Makrophyten- und Planktongemeinschaften in seichten Seen des südschwedischen Oligotrophiegebietes. Acta Limnol. 3: 288 S.

Lindner, A., 1974: Die Makrophytenvegetation im Ökosystem des Brackwassers. Mitt. Sekt. Geobot. Phytotax. Biol. Ges. DDR 1974: 31–35.

Lindquist, B., 1931: Den skandinaviska bokskogens biologi. Svenska Skogsvardsför. Tidskr. 3: 117–532.

Linhard, H., 1964: Die natürliche Vegetation im Mündungsgebiet der Isar und ihre Standortsverhältnisse. Ber. Naturw. Ver. Landshut 24: 3–70.

–, Stückl, E., 1972: Xerotherme Vegetationseinheiten an Südhängen des Regen- und Donautales im kristallinen Bereich. Denkschr. Regensburg. Botan. Ges. 30: 245–279.

Linke, O., 1939: Die Biota des Jadebusenwattes. Helgoländ. Wiss. Meeresunters. 1: 201–348.

Lippert, W., 1966: Die Pflanzengesellschaften des Naturschutzgebietes Berchtesgaden. Ber. Bayer. Botan. Ges. 39: 67–122 u. Anhang 1–70. +

Lloyd, P.S., 1975: An experimental study of scrub development. XII. Internat. Bot. Congr. (Leningrad) Abstr. 1: 156.

Lohammar, G., 1938: Wasserchemie und höhere Vegetation schwedischer Seen. Symb. Botan. Upsal. 3, 1: 252 S.

Lohmeyer, W., 1950: Das Polygoneto brittingeri-Chenopodietum rubri und das Xanthieto riparii-Chenopodietum rubri, zwei flußbegleitende Bidention-Gesellschaften. Mitt. Florist.-Soziol. Arb. gem. N.F. 2: 12–19.

–, 1951: Die Pflanzengesellschaften der Eilenriede bei Hannover. Angew. Pflanzensoziol. (Stolzenau/Weser) 3: 72 S.

–, 1953: Beitrag zur Kenntnis der Pflanzengesellschaften in der Umgebung von Höxter a.d. Weser. Ebenda 4: 59–76.

–, 1957: Der Hainmieren-Schwarzerlenwald (Stellario-Alnetum glutinosae [Kästner 1938]). Angew. Pflanzensoziol. (Stolzenau/Weser) 6/7: 247–257.

–, 1962: Zur Gliederung der Zwiebelzahnwurz (Cardamine bulbifera)-Buchenwälder im nördl. Rheinischen Schiefergebirge. Mitt. Florist.-Soziol. Arbeitsgem., N.F. 9: 187–193.

–, 1967: Über den Stieleichen-Hainbuchenwald des Kern-Münsterlandes und einige seiner Gehölz-Kontaktgesellschaften. Schriftenr. Vegetationskunde (Bad Godesberg) 2: 161–180.

–, 1970 a: Über einige Vorkommen naturnaher Restbestände des Stellario-Carpinetum und des Stellario-Alnetum glutinosae im westlichen Randgebiet des Bergischen Landes. Schriftenreihe Vegetationskunde (Bonn) 5: 67–74.

–, 1970 b: Über das Polygono-Chenopodietum in Westdeutschland unter besonderer Berücksichtigung seiner Vorkommen am Rhein und im Mündungsgebiet der Ahr. Ebenda 5: 7–28.

–, Bohn, U., 1973: Wildsträucher-Sproßkolonien (Polycormone) und ihre Bedeutung für die Vegetationsentwicklung auf brachgefallenem Grünland. Natur u. Landschaft 48: 75–79.

–, Krause, A., 1974: Über den Gehölzbewuchs an kleinen Fließgewässern Nordwestdeutschlands und seine Bedeutung für den Uferschutz. Natur u. Landschaft 49: 323–330.

–, Rabeler, W., 1965: Aufbau und Gliederung der mesophilen Laubmischwälder im mittleren und oberen Wesergebiet und ihre Tiergesellschaften. In: Tüxen, R. (Hrsg.): Biosoziologie. Dr. W. Junk, Den Haag: 238–257.

Lösch, R., Franz, N., 1974: Tagesverlauf von Wasserpotential und Wasserbilanz bei Pflanzen verschiedener Standorte des fränkischen Wellenkalks. Flora 163: 466–479.

Lötschert, W., 1952: Vegetation und pH-Faktor auf kleinstem Raum in Kiefern- und Buchenwäldern auf Kalksand, Löß und Granit. Biol. Zbl. 71: 327–348.

–, 1962: Beiträge zur Ökologie der subatlantischen Zwergstrauchheide NW-Deutschlands. I. Vegetation und Bodenfaktoren. II. Mikroklima und Transpiration. Beitr. Biol. Pflanzen 37: 331–410.

–, 1963: Keimzahlgehalt, CO_2-Gehalt der Bodenluft und CO_2-Abgabe des Bodens in verschiedenen Ausbildungsformen des baltischen Perlgras-Buchenwaldes. Mitt. Florist.-Soziol. Arb. gem. N.F. 10: 188–200.

–, 1964: Vegetation, Trophiegrad und pflanzengeographische Stellung des Salemer Moores. Beitr. Biol. Pflanzen 40: 65–111.

–, 1968: Krähenbeerheiden und Dünenbildung durch die Krähenbeere (Empetrum nigrun L.). Natur u. Museum 98: 425–429.

–, Gies, T., 1973: Untersuchungen über den Kationengehalt im Hochmoor. I. Abstufungen in den Vegetationskomplexen. Flora 162:

215–243.

–, Horst, K., 1962: Zur Frage jahreszeitlicher pH-Schwankungen. II. Untersuchungen an Heide- und Waldstandorten. Flora 152: 689–701.

–, Köhm, H.-J., 1973: pH-Wert und S-Gehalt der Baumborke in Immissionsgebieten. Oecol. Plant. 8: 199–209.

–, Ullrich, C., 1961: Zur Frage jahreszeitlicher pH-Schwankungen an natürlichen Standorten. Flora 150: 657–674.

–, Wandtner, R., Hiller, H., 1975: Schwermetallanreicherung bei Bodenmoosen in Immissionsgebieten. Ber. Deut. Botan. Ges. 88: 419–431.

Loach, K., 1966: Relations between soil nutrients and vegetation in wetheaths. 1. Soil nutrient content and moisture conditions. J. Ecol. 54: 597–608.

Loub, W., Url, W., Kiermayer, O., Diskus, A., Hilmbauer, K., 1954: Die Algenzonierung in Mooren des österreichischen Alpengebietes. Sitz. ber. Österr. Akad. Wiss., Math.-Nat. Kl., Abt. 1, 163: 447–494.

Lüdi, W., 1921: Die Pflanzengesellschaften des Lauterbrunnentales und ihre Sukzession. Beitr. Geobot. Landesaufn. Schweiz 9: 350 S.

–, 1936: Experimentelle Untersuchungen an alpiner Vegetation. Ber. Schweiz. Botan. Ges. 46: 632–681.

–, 1944: Die Gliederung der Vegetation auf der Apenninhalbinsel, insbesondere der montanen und alpinen Höhenstufen. In: Rikli, M., Das Pflanzenkleid der Mittelmeerländer, Bern: 573–596.

–, 1945: Besiedlung und Vegetationsentwicklung auf den jungen Seitenmoränen des großen Aletschgletschers. Ber. Geobot. Forsch. Inst. Rübel, Zürich 1944: 35–112.

–, 1948: Die Pflanzengesellschaften der Schinigeplatte bei Interlaken und ihre Beziehungen zur Umwelt. Veröff. Geobot. Inst. Rübel, Zürich 23: 400 S.

–, 1955: Beitrag zur Kenntnis der Vegetationsverhältnisse im Schweizer Alpenvorland während der Bronzezeit. In: W. U. Guyan u.a., Das Pfahlbauproblem. Monogr. Ur- u. Frühgesch. Schweiz 11: 91–109.

–, 1958 a: Beobachtungen über die Besiedlung von Gletschervorfeldern in den Schweizeralpen. Flora 146: 386–407.

–, 1958 b: Bericht über den 11. Kurs in Alpenbotanik. Ber. Geobot. Forsch. Inst. Rübel, Zürich 1957: 15–32.

–, (Hrsg.), 1961: Die Pflanzenwelt der Tschechoslowakei. Ergebnisse der 12. internationalen pflanzengeographischen Exkursion (IPE) durch die Tschechoslowakei 1958. Veröff. Geobot. Inst. ETH, Stiftg. Rübel, Zürich 36: 170 S. +

–, Zoller, H., 1949: Einige Beobachtungen über die Dürreschäden des Sommers 1947 in der Nordschweiz. Ber. Geobot. Forsch. Inst. Rübel, Zürich 1948: 69–85.

Lumiala, O. V., 1945: Über die Standortsfaktoren bei den Wasser- und Moorpflanzen sowie deren Untersuchung. Ann. Acad. Sci. Fenn. Ser. A IV Biol. 6: 47 S.

Lundegårdh, H., 1954: Klima und Boden in ihrer Wirkung auf das Pflanzenleben. 4. Aufl. Verlag Gustav Fischer, Jena, 598 S.

Luther, H., 1950: Beobachtungen über die fruktifikative Vermehrung von Phragmites communis Trin. Acta Botan. Fenn. 46: 3–18.

Lutz, J. L., 1956: Spirkenmoore in Bayern. Ber. Bayer. Botan. Ges. 31: 58–69.

–, u. Mitarb., 1957: Zur Charakterisierung von Biocoenose und Biotop des Übergangs-Moorwaldes. Forstwiss. Cbl. 76: 257–275.

Lützke, R., 1961: Das Temperaturklima von Waldbeständen und -lichtungen im Vergleich zur offenen Feldflur. Arch. Forstwes. 10: 17–83.

Lux, H., 1964: Die biologischen Grundlagen der Strandhaferpflanzung und Silbergrasansaat im Dünenbau. Angew. Pflanzensoziol. (Stolzenau/Weser) 20: 6–53.

Maas, F. M., 1959: Bronnen, bronbeken en bronbossen van Nederland, in het bijzonder die van de Veluwezoom. Meded. Landbouwhogesch. Wageningen 59: 1–166.

Macko, S., 1960: The National Park in the Karkonosze mountains and its vegetation. Ann. Silesiae 1: 331–376. +

Mägdefrau, K., Wutz, A., 1951: Die Wasserkapazität der Moos- und Flechtendecke des Waldes. Forstwiss. Cbl. 70: 103–117.

Mager, F., 1961: Der Wald in Altpreußen als Wirtschaftsraum Ostmitteleuropas in Vergangenheit und Gegenwart. Bd. I: 391 S., Bd. II: 328 S. Verlag Böhlau, Köln.

Magic, D., 1968: Waldgesellschaften der Eichen-Hainbuchen- und Buchenwälder mit Festuca drymeja Mert. et Koch im Slowakischen Erzgebirge. Biol. Práce (Bratislava) 14: 71–107.

Mahn, E.-G., 1965: Vegetationsaufbau und Standortsverhältnisse der kontinental beeinflußten Xerothermrasengesellschaften Mitteldeutschlands. Abh. Sächs. Akad. Wiss. Leipzig, Math. Nat. Kl. 49, 1: 138 S.

–, 1966 a: Die ökologisch-soziologischen Artengruppen der Xerothermrasen Mitteldeutschlands. Botan. Jb. 85: 1–44.

–, 1966 b: Beobachtungen über die Vegetations- und Bodenentwicklung eines durch Brand gestörten Silikattrockenrasenstandortes. Arch. Naturschutz u. Landschaftsforsch. 6: 61–90.

–, 1969: Untersuchungen zur Bestandsdynamik einiger charakteristischer Segeltalgesellschaften unter Berücksichtigung des Einsatzes von Herbiziden. Ebenda 9: 3–42.

–, 1973: Zum Einfluß von Herbiziden auf Agroökosysteme. In: Schubert, Hilbig u. Mahn: Probleme der Agrogeobotanik. Wiss. Beitr. Univ. Halle-Wittenberg 1973, 11 (P2): 131–138.

–, Schubert, R., 1962: Vegetationskundliche Untersuchungen in der mitteldeutschen Ackerlandschaft. VI. Die Pflanzengesellschaften nördlich von Walzleben (Magdeburger Börde). Wiss. Z. Univ. Halle, Math.-Nat. R. 11: 765–816.

Mahn, G., 1957: Über die Vegetations- und Standortsverhältnisse einiger Porphyrkuppen bei Halle. Wiss. Z. Univ. Halle, Math.-Nat. R. 6: 177–208.

Maier, R., 1971: Einfluß von Photoperiode und Einstrahlungsstärke auf die Temperaturresistenz einiger Samenpflanzen. Österr. Botan. Z. 119: 306–322.

–, 1973: Produktions- und Pigmentanalysen an Utricularia vulgaris L. In: H. Ellenberg (Hrsg.), Ökosystemforschung. Springer-Verlag, Berlin-Heidelberg-New York: 87–102.

Mair, N., 1967: Zuwachs- und Ertragsleistung subalpiner Wälder. Mitt. Forstl. Bundesversuchsanst. Wien 75: 385–424.

Maļato-Beliz, J., Tüxen, J., Tüxen, R., 1960: Zur Systematik der Unkrautgesellschaften der west- und mitteleuropäischen Wintergetreide-Felder. Mitt. Florist.-Soziol. Arb. gem. N. F. 8: 145–147.

Málek, J., 1961: Zur Frage des ursprünglichen Fichtenareals in den böhmischen Ländern. Sborn. Československ. Akad. Zeměd. Věd, Lesn. 34: 35–54.

–, 1979: Zur Frage des Weideeinflusses in der Urzeit auf die Waldzusammensetzung im Böhmerwaldvorgebirge. Preslia (Praha) 51: 255–270.

Malmer, N., 1960: Some ecologic studies on lakes and brooks in the South Swedisch Uplands. Botan. Not. (Lund) 113: 87–116.

–, 1961: Ecologic studies on the water chemistry of lakes in South Sweden. Ebenda 114: 121–144.

–, 1962 a u. b: Studies on mire vegetation in the archaean area of southwestern Gotaland (South Sweden). I. Vegetation and habitat conditions on the Akhult mire. Opera Botan. Soc. Botan. Lund 7, 1: 322 S. II. Distribution and seasonal variation in elementary constituents on some mire sites. Ebenda 7, 2: 67 S.

–, 1965: The south-western dwarf shrub heaths. Acta Phytogeogr. Suecica 50: 123–130.

–, 1968: Über die Gliederung der Oxycocco-Sphagnetea und Scheuchzerio-Caricetea fuscae. In: Pflanzensoziologische Systematik. Dr. W. Junk, Den Haag: 293–305.

–, 1974: On the effects on water, soil and vegetation of an increasing atmospheric supply of sulphur. National Swed. Environm. Protection Board (Stockholm): 98 S.

–, 1980: Supply and transport of mineral nutrients in a subarctic mire. Ecol. Bull. (Stockholm) 30: 63–95.

Manil, G., 1963: Niveaux d'écosystèmes et hierarchie de facteurs écologiques. Un exemple d'analyse dans les hêtraies ardennaises de Belgique. Bull. Cl. Sci. Acad. Roy. Belg. 49: 603–623.

–, Delecour, F., Fouget, G., el Attar, A., 1963: L'humus, facteur de station dans les hêtraies acidophiles de Belgique. Bull. Inst. Agron. et Stat. Rech. Gembloux 31: 1–114.

Marcet, E., 1971: Versuche zur Dürreresistenz inneralpiner „Trockentannen" (Albies alba Mill.). Schweiz. Z. Forstwes. 122: 117–134.

Margl, H., 1971: Die Ökologie der Donauauen und ihre naturnahen Waldgesellschaften. In: Naturgeschichte Wiens, Verlag Jugend und Volk, Wien-München: Bd. 2: S. 1–42.

Markgraf, F., 1931: Aus den südosteuropäischen Urwäldern. I. Die Wälder Albaniens. Z. Forst- u. Jagdwes. 63: 1–32.

–, 1932: Der deutsche Buchenwald. Veröff. Geobot. Inst. Rübel, Zürich 8: 15–62.

Markgraf-Dannenberg, I., 1979: Festuca-Probleme in ökologisch-soziologischem Zusammenhang. Bundesvers. Anst. Alpenländische Landw. Gumpenstein 1979: 373–386.

Marschall, F., 1947: Die Goldhaferwiese (Trisetetum flavescentis) der Schweiz. Beitr. Geobot. Landesaufn. Schweiz 26: 168 S.

Martin, W. E., 1959: The vegetation of Island Beach State Park, New Jersey. Ecol. Monogr. 29: 1–46.

Mason, C. F., Bryant, R. J., 1975: Production, nutrient content and decomposition of Phragmites communis Trin. and Typha latifolia L. J. Ecol. 63: 71–95.

Mathey, A., 1900: Pâturage en forêt. Besançon: 172 S.

•Mathey, W., 1964: Observations écologiques dans la tourbière du Cachot. Bull. Soc. Neuchâtel. Sci. Nat. 87, 3ᵉ Sér.: 103–135.

Mattern, H., Wolf, R., Mauk, J., 1980: Heiden im Regierungsbezirk Stuttgart – Zwischenbilanz im Jahre 1980. Veröff. Naturschutz Landschaftspflege Bad.-Württ. 51/52: 153–165.

Matuszkiewicz, A., 1958: Zur Systematik der Fagion-Gesellschaften in Polen. Acta Soc. Botan. Polon. 27: 675–725.

– u. W., 1954: Die Verbreitung der Waldassoziationen des Nationalparks von Białowieża. Polska Akad. Nauk, Kom. Ekol. 2: 33–60.

Matuszkiewicz, W., 1962: Zur Systematik der natürlichen Kiefernwälder des mittel- und osteuropäischen Flachlandes. Mitt. Florist.-Soziol. Arb. gem. N. F. 9: 145–186.

–, 1963: Zur systematischen Auffassung der oligotrophen Bruchwaldgesellschaften im Osten der Pommerschen Seenplatte. Ebenda 10: 149–155.

– u. A.: 1956 a: Pflanzensoziologische Untersuchungen im Forstrevier „Ruda" bei Puławy (Polen). Acta Soc. Botan. Polon. 25: 331–400.

– u. A.: 1956 b: Zur Systematik der Quercetalia pubescentis-Gesellschaften in Polen. Ebenda 25: 27–72.

– u. A.: 1973: Pflanzensoziologische Übersicht der Waldgesellschaften in Polen. Teil 1. Die Buchenwälder. Phytocoenosis (Białowieża) 2: 143–202.

–, Borowik, M., 1957: Zur Systematik der Auenwälder in Polen. Acta Soc. Botan. Polon. 26: 719–756.

–, u. Mitarb., 1963: Internationale pflanzensoziologische Exkursion durch NO-Polen. Mater. Zakł. Fitosocjol. Stosow. (Warszawa-Białowieża) 2: 90 S. +

–, A., Traczyk, H. u. T., 1958: Zur Systematik der Bruchwaldgesellschaften (Alnetalia glutinosae) in Polen. Acta Soc. Botan. Polon. 27: 21–44.

Maurer, W., 1966: Flora und Vegetation des Serpentingebietes bei Kirchdorf in Steiermark. Mitt. Abt. Zool. u. Botan. Landesmus. Graz 25: 15–76. +

Mayer, H., 1957: An der Kontaktzone des Lärchen- und Fichtenwaldes in einem Urwaldrest der Berchtesgadener Kalkalpen. Jb. Ver. z. Schutze Alpenpflanzen u.-Tiere 22.

–, 1959: Waldgesellschaften der Berchtesgadener Kalkalpen. Mitt. Staatsforstverw. Bayerns 30: 163–215. +

–, 1960: Bodenvegetation und Naturverjüngung von Tanne und Fichte in einem Allgäuer Plenterbestand. Ber. Beobot. Inst. ETH, Stiftg. Rübel, Zürich 31: 19–42.

–, 1964: Bergsturzbesiedlungen in den Alpen. Mitt. Staatsforstverw. Bayerns 34: 191–203. +

–, 1966: Analyse eines urwaldnahen, subalpinen Lärchen-Fichtenwaldes (Piceetum subalpinum) im Lungau. Cbl. Ges. Forstwes. 83: 129–151.

–, 1969: Aufbau und waldbauliche Beurteilung des Naturwaldreservates Freyensteiner Donauwald. Ebenda 86: 3–59.

–, 1970: Zum Reliktvorkommen von Alnus viridis und Rhododendron ferrugineum in Tieflagen der Ostalpen. Mitt. Ostalpin-Dinar. Sekt., Internat. Ver. Vegetationskunde (Wien) 10: 59–63.

–, 1971: Das Buchen-Naturwaldreservat Dobra Kampleiten im niederösterreichischen Waldviertel. Schweiz. Z. Forstwes. 122: 45–66.

•–, 1974: Wälder des Ostalpenraumes. Gustav Fischer Verlag, Stuttgart: 344 S.

•–, 1976: Gebirgswaldbau-Schutzwaldpflege. Gustav Fischer-Verlag, Stuttgart: 436 S.

–, Schlesinger, B., Thiele, K., 1967: Dynamik der Waldentstehung und Waldzerstörung auf den Dolomit-Schuttflächen im Wimbachgries (Berchtesgadener Kalkalpen). Jahrb. Ver. Schutze Alpenpflanzen u.-Tiere 32: 1–29.

Mayer, M., 1939: Ökologisch-pflanzensoziologische Studien über die Filipendula ulmaria-Geranium palustre-Assoziation. Beitr. Geobot. Landesaufn. Schweiz 23: 64 S.

Mayer, R., 1971: Bioelement-Transport im Niederschlagswasser und in der Bodenlösung eines Wald-Ökosystems. Göttinger Bodenkdl. Ber. 19: 1–119.

McVean, D. N., 1959: Ecology of Alnus glutinosa (L.) Gaertn. VII. Establishment of alder by direct seeding of shallow blanket bog. J. Ecol. 47: 615–618.

McNaughton, S. J., 1968: Autotoxic feedback in relation to germination and seedling growth in Typha latifolia. Ecology 49: 367–369.

Medwecka-Kornaś, A., 1952: Les associations forestières du Jura Cracovien. Ochr. Przyrody 20: 133–236. +

–, 1960: Poland's steppe vegetation and its conservation. State Council f. Conserv. of Nature, Publ. (Krakow) 6: 30 S.

–, 1962: Jak powstała mapa roślinności Ojcowskiego Parku Narodowego i co z niej mozna odczytać. Chrońmy Przyr. Ojczysta N. S. 28, 4: 3–12. +

–, (Ed.), 1967: Ecosystem studies in a beech forest and meadow in the Ojców National Park. Studia Naturae, Ser. A (Kraków): 213 S.

–, Kornaś, J., 1964: Plant communities in the Jaszcze and Jamne valleys. Zakł. Ochron. Przyrody Polsk. Akad. Nauk, Studia Nat. Ser. A, 2: 49–91. +

Meier, H., Braun-Blanquet, J., 1934: Classe des Asplenietales rupestres, groupements rupicoles. Prodromus d. Pflanzengesellschaften 2: 47 S.

Meisel, K., 1960: Die Auswirkung der Grundwasserabsenkung auf die Pflanzengesellschaften im Gebiete um Moers (Niederrhein). Arb. Bundesanst. Vegetationskartierung (Stolzenau/Weser) 1960: 105 S.

–, 1966 a: Ergebnisse von Daueruntersuchungen in nordwestdeutschen Ackerunkrautgesellschaften. In: R. Tüxen (Hrsg.): Anthropogene Vegetation. Dr. W. Junk, Den Haag: 1966: 86–96.

–, 1966 b: Zur Systematik und Verbreitung der Festuco-Cynosureten. Ebenda: 202–211.

–, 1967: Über die Artenverbindungen des Aphanion arvensis J. et R. Tx. 1960 im west- und nordwestdeutschen Flachland. Schriftenr. Vegetationskunde (Bad Godesberg) 2: 123–133.

–, 1969 a: Verbreitung und Gliederung der Winterfrucht-Unkrautbestände auf Sandböden des nordwestdeutschen Flachlandes. Ebenda 4: 7–22.

–, 1969 b: Zur Gliederung und Ökologie der Wiesen im nordwestdeutschen Flachland. Ebenda 4: 23–48.

–, Hübschmann, A. von, 1975: Zum Rückgang von Naß- und Feuchtbiotopen im Emstal. Natur u. Landschaft 50: 33–38.

–, –, 1976: Veränderungen der Acker- und Grünlandvegetation im nordwestdeutschen Flachland in jüngerer Zeit. Schriftenr. Vegetationskunde (Bonn) 10, 109–124.

–, Melzer, W., 1972: Nicht mehr landwirtschaftlich genutzte Fläche (Sozialbrache) in v. H. der landwirtschaftlichen Nutzfläche (LN) in der BRD (Karte 1 : 1 000 000). Bundesanstalt f. Vegetationskunde, Naturschutz- u. Landschaftspflege, Bonn.

–, Wattendorff, J., 1962: Über eine von der Wirtschaftsart unabhängige Wasserstufenkarte. Mitt. Florist.-Soziol. Arb. gem. N. F. 9: 230–238.

Meisel, S., 1969: Ackernutzung und Unkrautgesellschaften der Naturräume in der Umgebung des Wiehengebirges. Vegetatio 18: 246–256.

Meisel-Jahn, S., 1955 a: Die pflanzensoziologische Stellung der Hauberge des Siegerlandes. Mitt. Florist.-Soziol. Arb. gem. N. F. 5: 145–150.

–, 1955 b: Die Kiefern-Forstgesellschaften des norddeutschen Flachlandes. Angew. Pflanzensoziol. (Stolzenau/Weser) 11: 128 S.

Menke, B., 1963: Beiträge zur Geschichte der Erica-Heiden Nordwestdeutschlands. Flora 153: 521–548.

Merkel, J., 1979: Die Vegetation des Meßtischblattes 6434, Hersbruck. Diss. Botan. 51: 174 S. +

Merxmüller, H., Poelt, J., 1954: Beiträge zur Florengeschichte der Alpen. Ber. Bayer. Botan. Ges. 30: 91–101.

–, Zollitsch, B., 1967: Über die Sonderstellung der Vegetation auf Kalkschieferschutt. Aquilo (Oulu, Finnland) Ser. Botan. 6: 228–240.

Mestel, E., 1965: Windschutz im schleswig-holsteinischen Küstengebiet. Die Holzzucht 19: 15–19.

Meusel, H., 1935: Die Waldtypen des Grabfeldes und ihre Stellung innerhalb der Wälder zwischen Main und Werra. Beih. Botan. Cbl. 53: 175–251.

–, 1939: Die Vegetationsverhältnisse der Gipsberge am Kyffhäuser und im südlichen Harzvorland. Hercynia 2: 372 S.

–, 1940: Die Grasheiden Mitteleuropas. Versuch einer vergleichend-pflanzengeographischen Gliederung. Botan. Arch. 41: 357–519.

–, 1942: Der Buchenwald als Vegetationstyp. Ebenda 43: 305–321.

–, 1952 a: Vegetationskundliche Studien über mitteleuropäische Waldgesellschaften 3. Über einige Waldgesellschaften der Insel Rügen. Ber. Deut. Botan. Ges. 64: 223–241.

–, 1952 b: Über die Elyneten der Allgäuer Alpen. Ber. Bayer. Botan. Ges. 29: 47–55.

–, 1952 c: Die Eichen-Mischwälder des mitteldeutschen Trockengebietes. Wiss. Z. Univ. Halle, Math.-Nat. 1: 49–72.

–, 1954: Vegetationskundliche Studien über mitteleuropäische Waldgesellschaften. 4. Die Laubwaldgesellschaften des Harzgebietes. Angew. Pflanzensoziol. (Wien), Festschr. Aichinger 1: 437–472.

–, 1955: Die Laubwaldgesellschaften des Harzgebietes. Wiss. Z. Univ. Halle, Math.-Nat. 4: 901–908.

Meyer, F. H., 1957: Über Wasser- und Stickstoffhaushalt der Röhrichte und Wiesen im Elballuvium bei Hamburg. Mitt. Staatsinst. Allg. Botan. Hamburg 11: 137–203.

–, 1959: Untersuchungen über die Aktivität der Mikroorganismen in Mull, Moder und Rohhumus. Arch. Mikrobiol. 33: 149–169.

–, 1961: Die Entwicklung von Buchenjungpflanzen in unterschiedlichem Bodenmilieu. Ber. Deut. Botan. Ges. 74: 292–299.

–, 1974: Physiology of mycorrhiza. Ann. Rev. Plant Physiol. 25: 567–586.

– u. Mitarb., 1978: Bäume in der Stadt. Verlag Eugen Ulmer, Stuttgart, 327 S.

Mezera, A., 1956/1958: Mitteleuropäische Tieflandauen und die Bewirtschaftung von Auenwäldern. Teil I, Prag 1956: 301 S.; Teil II 1958: 364 S.

Michael, G., 1966: Untersuchungen über die winterliche Dürreresistenz einiger immergrüner Gehölze im Hinblick auf eine Frosttrocknisgefahr. Flora, Abt. B, 156: 350–372.

Michaelis, G. u. P., 1934: Ökologische Studien an der alpinen Baumgrenze. III. Über die winterlichen Temperaturen der pflanzlichen Organe, insbesondere der Fichte. Beih. Botan. Cbl. 52 B: 333–377.

Michaelis, P., 1932: Ökologische Studien an der alpinen Baumgrenze. Das Klima und die Temperaturverhältnisse der Vegetationsorgane im Hochwinter. Ber. Deut. Botan. Ges. 50: 31–42.

–, 1934 a: Ökologische Studien an der alpinen Baumgrenze. IV. Kenntnis des winterlichen Wasserhaushaltes. Jb. Wiss. Botan. 80: 169–247.

–, 1934 b: Ökologische Studien an der alpinen Baumgrenze. V. Osmotischer Wert und Wassergehalt während des Winters in den verschiedenen Höhenlagen. Ebenda 80: 337–362.

Michalko, J., Dzatko, M., 1965: Phytocoenologische und oekologische Charakteristik der Pflanzengesellschaften des Waldkomplexes Dubník bei der Gemeinde Sered. Biol. Práce (Bratislava) 11: 47–113. +

•Miehlich, G., 1970: Veränderung eines Lößlehm-Pseudogleys durch Fichtenreinanbau. Diss. Univ. Hamburg 1970: 231 S.

Mikyška, R., 1956: Eine phytosoziologische Studie der Terrassenwälder in den unteren Flußgebieten der Orlice und Loučna. Sborn. Československ. Akad. Zeměd. Věd. Lesn. 29: 313–370. +

–, 1963: Die Wälder der ostböhmischen Tiefebene. Rozpr. Československ. Akad. Věd, Řada Mat. Přírodn. Věd. 73, Heft 15: 91 S.

–, 1964: Beitrag zur Phytosoziologie der Reliktkiefernwälder des Böhmerwaldes. Časopis Národn. Muz. Přírodn. 133: 185–195.

–, 1967: Vegetations Rekonstruktionen der Wälder im Gebiete Zálabi (Elbegebiet) der Ostböhmischen Tiefebene. Preslia (Praha) 39. 312–318 u. 403–420. +

–, 1968: Wälder am Rande der ostböhmischen Tiefebene. Eine pflanzensoziologische Studie. Rozpr. Československ. Akad. Věd, Řada Mat. Přírodn. Věd. 78: 3–122. +

–, 1972: Die Wälder der böhmischen mittleren Sudeten und ihrer Vorberge. Rozpr. Československ. Akad. Věd, Řada Mat. Přírodn. 82: 162 S.

Miles, J., 1971: Burning Molinia-dominant vegetation for grazing by red deer. J. Brit. Grassland Soc. 26: 247–250.

–, 1974: Experimental establishment of new species from seed in Callunetum in North-east Scotland. J. Ecol. 62: 527–551.

Milthorpe, F. L., 1975: Competition for water in plant communities. XII. Internat. Botan. Congr. (Leningrad) Abstr. 1: 159.

Mitscherlich, G., 1955: Untersuchungen über das Wachstum der Kiefer in Baden. 2. Teil: Die Streunutzungs- und Düngungsversuche. Allg. Forst- u. Jagdztg. 126: 193–204.

–, 1975: Wald, Wachstum und Umwelt, 3. Bd. Boden, Luft und Produktion. J. D. Sauerländers Verlag, Frankfurt a. M.: 352 S.

Miyawaki, A., Ohba, T., 1965: Studien über Strand-Salzwiesengesellschaften auf Ost-Hokkaido (Japan). Sci. Rep. Yokohama Nation. Univ., Sect. II, 12: 25 S.

Möller, H., 1970: Soziologisch-ökologische Untersuchungen in Erlenwäldern Holsteins. Mitt. Arb. gem. Floristik Schlesw.-Holst. u. Hamburg 19: 109 S.

Mondino, G. P., 1963: Boschi planiziari a Pinus silvestris ed Alnus incana nelle alluvioni del torrente Bardonecchia (Piemonte). Allionia (Torino) 9: 43–64.

Montfort, G., Brandrup, W., 1927: Physiologische und pflanzengeographische Seesalzwirkungen. II. Ökologische Studien über die Keimung und erste Entwicklung bei Halophyten. Jb. Wiss. Botan. 66: 902–946.
III. Die Salzwachstumsreaktion der Wurzeln. Ebenda 67: 105–173.

Moor, M., 1936: Zur Soziologie der Isoëtetalia. Beitr. Geobot. Landesaufn. Schweiz 20: 148 S.

–, 1952: Die Fagion-Gesellschaften des Schweizer Jura. Ebenda 31: 201 S.

–, 1958: Pflanzengesellschaften schweizerischer Flußauen. Mitt. Schweiz. Anst. Forstl. Versuchswes. 34: 221–360.

–, 1968: Der Linden-Buchenwald. Vegetatio 16: 159–191.

–, 1970: Adenostylo-Fagetum, Höhenvikariant des Linden-Buchenwaldes. Bauhinia (Basel) 4: 161–185.

–, 1972: Versuch einer soziologisch-systematischen Gliederung des Carici-Fagetum. Vegetatio 24: 31–69.

–, 1973: Das Corydalido-Aceretum, ein Beitrag zur Systematik der Ahornwälder. Ber. Schweiz. Botan. Ges. 83: 106–132.

–, 1975 a: Die soziologisch systematische Gliederung des Hirschzungen-Ahornwaldes. Beitr. Naturk. Forsch. Südw.-Deut. 34: 215–223.

–, 1975 b: Der Ulmen-Ahornwald (Ulmo-Aceretum Issler 1926). Ber. Schweiz. Botan. Ges. 85: 187–203.

–, 1975 c: Ahornwälder im Jura und in den Alpen. Phytocoenologia 2: 244–260.

–, 1976: Gedanken zur Systematik mitteleuropäischer Laubwälder. Schweiz. Z. Forstwes. 127: 327–340. +

Moravcová-Husová, M., 1964: Beitrag zur Ökologie der Laubwälder im südlichen Teil des mittelböhmischen Granit-Hügellandes. Preslia 36: 55–63. +

Moravec, J., 1965: Wiesen im mittleren Teil des Böhmerwaldes. Vegetace (Praha) A 1: 498–508.

–, 1966: Zur Syntaxonomie der Carex davalliana-Gesellschaften. Folia Geobot. Phytotaxon. Bohemoslov. 1: 3–25.

–, 1967: Zu den azidophilen Trockenrasengesellschaften Südwestböhmens und Bemerkungen zur Syntaxonomie der Klasse Sedo-Scleranthetea. Ebenda 2: 137–178.

–, Rybníčková, E., 1964: Die Carex davalliana-Bestände im Böhmerwaldvorgebirge, ihre Zusammensetzung, Ökologie und Historie. Preslia (Praha) 36: 376–391.

Moser, M., 1967: Die ektotrophe Ernährungsweise an der Waldgrenze. Mitt. Forstl. Bundesversuchsanst. Wien 75: 357–373.

Moser, W., 1967: Einblicke in das Leben von Nivalpflanzen. Jahrb. Ver. Schutze Alpenpflanzen u.-Tiere 32: 11 S.

–, 1970: Ökophysiologische Untersuchungen an Nivalpflanzen. Mitt. Ostalp.-Dinar. Ges. Vegetationskunde 11: 121–134.

–, 1973: Licht, Temperatur und Photosynthese an der Station „Hoher Nebelkogel" (3184 m). In: Ellenberg (Hrsg.): Ökosystemforschung. Springer-Verlag Heidelberg, Berlin, New York: 203–224.

Moss, R., Miller, G. R., 1976: Production, dieback and grazing of heather (Calluna vulgaris) in relation to numbers of red gouse (Lagopus l. scoticus) and mountain hares (Lepus timidus) in north-east Scotland. J. Appl. Ecol. 13: 369–377.

Mráz, K., 1958: Beitrag zur Kenntnis der Stellung des Potentillo-Quercetum. Arch. Forstwes. 7: 703–728.

–, Šika, A., 1965: Böden und Vegetation der Auewaldstandorte. Feddes Repertorium, Beih. 142: 5–64.

Mückenhausen, E., 1970: Fortschritte in der Systematik der Böden der Bundesrepublik Deutschland. Mitt. Deut. Bodenkundl. Ges. 10: 246–279.

–, 1977: Entstehung, Eigenschaften und Systematik der Böden der Bundesrepublik Deutschland. DLG-Verlag, Frankfurt a. M., 3. Aufl.: 300 S., 60 Farbtaf.

–, Scharpenseel, H. W., Pietig, F., 1968: Zum Alter des Plaggeneschs. Eiszeitalter u. Gegenwart 19: 190–196.

*Mueller-Dombois, D., Ellenberg, H., 1974: Aims and methods of vegetation ecology. John Wiley and Sons, New York: 547 S.

Muhle, H., 1977: Ein Epiphytenkataster niedersächsischer Natuwaldreservate. Mitt. Florist.-Soziol. Arb.gem. NF 19/20: 47–62.

*Muhle, O., 1974: Zur Ökologie und Erhaltung von Heidegesellschaften. Allg. Forst- u. Jagdztg. 145: 232–339.

–, Röhrig, E., 1979: Untersuchungen über die Wirkungen von Brand, Mahd und Beweidung auf die Entwicklung von Heide-Gesellschaften. Schriftenr. Forstl. Fak. Univ. Göttingen 61: 72 S.

Müller, A. von, 1956: Über die Bodenwasser-Bewegung unter einigen Grünland-Gesellschaften des mittleren Wesertales und seiner Randgebiete. Angew. Pflanzensoziol. (Stolzenau/Weser) 12: 85 S.

Müller, F., 1980: Glaciers and their fluctuations. Nature and Resources 16: 5–15.

Müller, G., 1964: Die Bedeutung der Ackerunkrautgesellschaften für die pflanzengeographische Gliederung West- und Mittelsachsens. Hercynia 1: 82–313.

Müller, K., 1965: Zur Flora und Vegetation der Hochmoore des nordwestdeutschen Flachlandes. Schr. Naturw. Ver. Schlesw.-Holst. 36: 30–77.

–, 1968: Ökologisch-vegetationskundliche Untersuchungen in ostfriesischen Hochmooren. Ber. Deut. Botan. Ges. 81: 221–237.

–, 1973: Ökologische und vegetationsgeschichtliche Untersuchungen an Niedermoorpflanzen-Standorten des ombrotrophen Moores unter besonderer Berücksichtigung seiner Kolke und Seen in NW-Deutschland. Beitr. Biol. Pflanzen 49: 147–235.

*Müller, K. (Hrsg.), 1948: Der Feldberg im Schwarzwald. Naturwissenschaftliche, landwirtschaftliche, forstwirtschaftliche, geschichtliche und siedlungsgeschichtliche Studien. Freiburg i. Br.: 586 S. +

Müller, K. M., 1929: Aufbau, Wuchs und Verjüngung der südosteuropäischen Urwälder. Hannover: 323 S.

Müller, P., 1955: Verbreitungsbiologie der Blütenpflanzen. Veröff. Geobot. Inst. Rübel, Zürich 30: 152 S.

Müller, Th., 1961: Einige für Süddeutschland neue Pflanzengesellschaften. Beitr. Naturk. Forsch. Südw.-Deut. 20: 15–21.

–, 1962: Die Saumgesellschaften der Klasse Trifolio-Geranietea sanguineae. Mitt. Florist.-soziol. Arb. gem. N. F. 9: 95–140.

–, 1964: Ergebnisse von Windschutzversuchen in Baden-Württemberg, Veröff. Landesstelle Naturschutz Baden-Württemb. 32: 71–126.

–, 1966: Vegetationskundliche Beobachtungen im Naturschutzgebiet Hohentwiel. Veröff. Landesstelle Naturschutz u. Landschaftspflege Baden-Württemb. 34: 14–62.

–, 1967: Die geographische Gliederung des Galio-Carpinetum und des Stellario-Carpinetum in Südwestdeutschland. Beitr. Naturk. Forsch. Südw.-Deut. 26: 47–65.

–, 1968: Die südwestdeutschen Carpinion-Gesellschaften. Feddes Repert. 77: 113–116.

–, 1975: Natürliche Fichtengesellschaften der Schwäbischen Alb. Beitr. Naturk. Forsch. Südw.-Deut. 34: 233–249.

–, Görs, S., 1958: Zur Kenntnis einiger Auenwaldgesellschaften im württembergischen Oberland. Beitr. Naturk. Forsch. Südw.-Deut. 17: 88–165.

–, –, 1960: Pflanzengesellschaften stehender Gewässer in Baden-Württemberg. Ebenda 19: 60–100.

–, –, 1969: Halbruderale Trocken- und Halbtrockenrasen. Vegetatio 18: 203–215.

–, Oberdorfer, E., 1974: Die potentielle natürliche Vegetation von Baden-Württemberg. Beih. Veröff. Landesstelle Naturschutz u. Landschaftspflege Baden-Württemb. 6: 46 S.

Müller-Stoll, W. R., 1936: Ökologische Untersuchungen an Xerothermpflanzen des Kraichgaus. Z. Botan. 29: 161–153.

–, 1947: Der Einfluß der Ernährung auf die Xeromorphie der Hochmoorpflanzen. Planta (Berlin) 35: 225–251.

–, Gruhl, K., 1959: Das Moosfenn bei Potsdam, Vegetationsgeographie eines märkischen Naturschutzgebietes. Wiss. Z. Pädagog. Hochsch. Potsdam, Math.-Nat. 4: 151–180.

–, Krausch, H.-D., 1968: Der azidophile Kiefern-Traubeneichenwald und seine Kontaktgesellschaften in Mittel-Brandenburg. Mitt. Florist.-Soziol. Arb. gem. N. F. 13: 101–121.

Müller-Suur, A., 1972: Vegetations- und Standortsuntersuchungen im Rantum-Becken auf Sylt. Diss. Univ. Göttingen 1972: 116 S.

Müllverstedt, R., 1963 a: Untersuchungen über die Keimung von Unkrautsamen in abhängigkeit vom Sauerstoffpartialdruck. Weed Res. 3: 154–163.

–, 1963 b: Untersuchungen über die Ursachen des vermehrten Auflaufens von Unkräutern nach mechanischen Unkrautbekämpfungs-Maßnahmen (Nachauflauf). Weed Res. 3: 298–303.

–, 1966: Vergleich der Auswirkungen mechanischer und chemischer Unkrautbekämpfung auf Massenwachstum und Samenproduktion der verbliebenen Unkräuter. Z. Pflanzenkrankheiten u. Pflanzenschutz 73: 598–603.

Myczkowski, S., 1973: Ecology of spruce (Picea abies Karst.) near the timber line in the forest association Piceetum tatricum in the Polish Tatra National Park. In: Internat. Sympos. Biol. Woody Plants, Bratislava: 511–526.

Myers, K., Poole, W. E., 1963: A study of the biology of the wild rabbit, Oryctolagus cuniculus (L), in confined populations. IV. The effects of rabbit grazing on sown pastures. J. Ecol. 51: 435–451.

*Nagel, P., 1975: Studien zur Ökologie und Chorologie der Coleopteren (Insecta) xerothermer Standorte des Saar-Mosel-Raumes mit besonderer Berücksichtigung der die Bodenoberfläche besiedelnden Arten. Diss. Univ. Saarbrücken: 225 S.

Nägeli, W., 1969: Waldgrenze und Kampfzone in den Alpen. Hespa Mitt. (Luzern) 19, 1: 44 S.

Nanson, A., 1962: Quelques éléments concernant le bilan d'assimilation photosynthétique en hêtraie ardennaise. Bull. Inst. Agron. et Stat. Rech. Gembloux 30: 320–331.

Nash, T. H., Nash, E. H., 1974: Sensitivity of mosses to sulfur dioxide. Oecologia 17: 257–263.

Naumann, E., 1927: Ziel und Hauptprobleme der regionalen Limnologie. Botan. Not. (Lund) 1927: 81–103.

Nees, J. C., Dugdale, R. C., Goering, J. J., Dugdale, V. A., 1963: Use of nitrogen-15 for measuremtent of rates in the nitrogen cycle. In: Radioecology (Ed. V. Schultz and A. W. Klement), New York and Washington D.C.: 481–484.

Neuhäusl, R., 1972: Subkontinentale Hochmoore und ihre Vegetation. Stud. Čs. Akad. Věd, Praha 13: 1–121.

–, 1975: Hochmoore am Teich Velké Dářko. Vegetace ČSSR A 9, Prag: 267 S.

–, Neuhäuslová-Novotná, Z., 1964: Vegetationsverhältnisse am Südrand des Schemnitzer Gebirges. Biol. Práce (Bratislava) 10, 4: 5–77. +

–, –, 1967: Syntaxonomische Revision der azidophilen Eichen- und Eichenmischwälder im westlichen Teile der Tschechoslowakei. Folia Geobot. Phytotaxon. (Praha) 2: 1–42.

–, –, 1968 a: Mesophile Waldgesellschaften in Südmähren. Rozpr. Českosl. Akad. Věd, Řada Mat. Přírodn. Věd 70, 11: 83 S.

–, –, 1968 b: Übersicht der Carpinion-Gesellschaften der Tschechoslowakei. Feddes Repert. 78: 39–56.

–, –, 1969: Die Laubwaldgesellschaften des östlichen Teiles der Elbeebene, Tschechoslowakei. Folia Geobot. Phytotax. (Praha) 4: 261–301.

Neuhäuslová-Novotná, Z., Neuhäusl, R., 1971: Beitrag zur Kenntnis der Carpinion-Gesellschaften im subkontinentalen Teil Europas. Preslia (Praha) 43: 154–167.

Neuwohner, W., 1938: Der tägliche Verlauf von Assimilation und Atmung bei einigen Halophyten. Planta (Berlin) 28: 644–677.

Nezadal, W., 1972: Getreideunkrautgesellschaften des Fränkischen Stufenlandes in der Umgebung Erlangens. Denkschr. Regensburg. Botan. Ges. 30: 21–71.

–, 1975: Ackerunkrautgesellschaften Nordostbayerns. Ebenda 34: 17–149.

–, 1980: Naturschutz für Unkräuter? Zur Gefährdung der Ackerunkräuter in Bayern. Schriftenr. Naturschutz u. Landschaftspflege 12: 17–27.

Niemann, E., 1964: Beiträge zur Vegetations- und Standortsgeographie in einem Gebirgsquerschnitt über den mittleren Thüringer Wald. Arch. Naturschutz u. Landschaftsforsch. 4: 3–50. +

–, 1965: Submontane und montane flußbegleitende Glanzgras-Röhrichte in Thüringen und ihre Beziehungen zu den hydrologischen Verhältnissen. Limnologica (Berlin) 3: 399–438.

Nietsch, H., 1939: Wald und Siedlung im vorgeschichtlichen Mitteleuropa. Mannus-Bücherei 64: 254. S.

Nihlgård, B., 1969: The microclimate in a beech and a spruce forest – a comparative study from Kongalund, Scania, Sweden. Botan. Not. (Lund) 122: 333–352.

–, 1970 a: Vegetation types of planted spruce forest in Scania, Southern Sweden. Ebenda 123: 310–337.

–, 1970 b: Precipitation, its chemical composition and effect on soil water in a beech and a spruce forest in South Sweden. Oikos 21: 208–217.

–, 1971: Pedological influence of spruce planted on former beech forest soils in Scania, South Sweden. Oikos 22: 302–314.

–, 1972: Plant biomass, primary production and distribution of chemical elements in a beech and a planted spruce forest in South Sweden. Oikos 23: 69–81.

Niklfeld, H., 1964: Zur xerothermen Vegetation im Osten Niederösterreichs. Verh. Zool.-Botan. Ges. Wien 103/104: 152–181.

–, 1973: Natürliche Vegetation. Atlas der Donauländer. Franz Deuticke, Wien, 4. Lief., Karte 127.

Noirfalise, A., 1952: La frênaie à Carex. Verh. Kön. Belg. Inst. Naturw. 122.

–, 1956: La hêtraie ardennaise. Bull. Inst. Agron. et Stat. Rech. Gembloux 24: 208–240.

–, 1967: Conséquences écologiques de la monoculture des conifères dans la zone des feuillues de l'Europe tempérée. Publ. UICN N. S. 9: 61–71.

–, 1968: Le Carpinion dans l'Quest de l'Europe. Feddes Repert. 79: 69–85.

–, 1969: La chênaie mélangée à jacinthe du domaine atlantique de'Europe (Endymio-Carpinetum). Vegetatio 17: 131–150.

–, Sougnez, N., 1956: Les chênaies de l'Ardenne verviétoise. Pédologie (Gand) 6, 119–143.

Nordhagen, R., 1940: Studien über die maritime Vegetation Norwegens I. Bergens Mus. Aarb. 7, Naturw. R 2: 123 S.

Nosova, L. I., 1975: Seed productivity of the high montain Pamirs plants. XII. Internat. Botan. Congr. (Leningrad) Abstr. 1: 160.

Oberdorfer, E., 1950: Beitrag zur Vegetationskunde des Allgäu. Beitr. Naturk. Forsch. Südw.-Deut. 9: 29–98. +

–, 1952: Die Wiesen des Oberrheingebietes. Ebenda 11: 75–88.

–, 1956: Die Vergesellschaftung der Eissegge (Carex frigida All.) in alpinen Rieselfluren des Schwarzwaldes, der Alpen und der Pyrenäen. Veröff. Landesstelle Naturschutz u. Landschaftspflege Baden-Württemb. 24: 452–465.

–, 1957: Süddeutsche Pflanzengesellschaften. Pflanzensoziol. (Jena) 10: 564 S.

–, 1959: Borstgras- und Krummseggenrasen in den Alpen. Beitr. Naturk. Forsch. Südw.-Deut. 18: 117–143.

–, 1964: Der insubrische Vegetationskomplex, seine Struktur und Abgrenzung gegen die submediterrane Vegetation in Oberitalien und in der Südschweiz. Beitr. Naturk. Forsch. Südw.-Deut. 23: 141–187.

–, 1968: Studien in den Wäldern des Carpinion-Verbandes im Apennin an der Südwestgrenze des Vorkommens von Carpinus betulus. Feddes Repert. 77: 65–74.

–, 1971: Zur Syntaxonomie der Trittpflanzen-Gesellschaften. Beitr. Naturk. Forsch. Südw.-Deut. 30: 95–111.

–, 1979: Pflanzensoziologische Exkursionsflora, 4. Aufl. Verlag Eugen Ulmer, Stuttgart: 997 S.

–, (Hrsg.), 1977–1982: Süddeutsche Pflanzengesellschaften. Gustav Fischer Verlag, Stuttgart. Teil I 1977, 311 S., Teil II 1978, 355 S., Teil III u. IV folgen.

–, Lang, G., 1957: Vegetationskundliche Karte des Schwarzwaldes bei Freiburg i. Br. Beilage zu: Ber. Naturf. Ges. Freiburg i. Br. 47. +

–, u. Mitarb., 1967: Systematische Übersicht der westdeutschen Phanerogamen- und Gefäßkryptogamen-Gesellschaften. Schriftenreihe Vegetationskunde (Bad Godesberg) 2: 7–62.

–, Müller, Th., 1974: Vegetation. In: Das Land Baden-Württemberg, Bd. I. Verlag W. Kohlhammer, Stuttgart, S. 74–93.

Odum, E. P., 1980: Grundlagen der Ökologie. 2 Bde. Verlag Georg Thieme, Stuttgart, 836 S.

Oechel, W. C., Sveinbjörnsson, B., 1975: Primary production processes in arktic mosses at Barrow, Alaska. XII. Internat. Botan. Congr. (Leningrad) Abstr. 1: 160.

Oelkers, K. H., 1970: Die Böden des Leinetales, ihre Eigenschaften, Verbreitung, Entstehung und Gliederung, ein Beispiel für die Talböden im Mittelgebirge und dessen Vorland. Beih. Geol. Jahrb., Bodenkundl. Beitr. 99: 71–152.

Oettli, M., 1904: Beiträge zur Ökologie der Felsflora. Jb. St. Gall. Naturw. Ges. 1903: 1–171.

Ogasahara, K., 1964: Snow survey of Mt. Tateyama and Mt. Tsurugi of the Japanese North Alps. An essay on the relation of the topographic distribution of snowfall with vegetation. Synthetic Sci. Res. Org. Toyama Univ., Japan 1964: 1–32.

Ohba, T., 1974: Vergleichende Studien über die alpine Vegetation Japans. 1. Carici rupestris-Kobresietea bellardii. Phytocoenologia 1, 339–401.

Oinonen, E., 1967: Sporal regeneration of ground pine (Lycopodium complanatum L.) in southern Finland in the light of the dimensions and the age of its clones. Acta Forest. Fenn. 83: 85 S.

Önal, M., 1971: Der Einfluß steigender Natriumchlorid-Konzentrationen auf den Transpirationskoeffizient einiger Halophyten. Rev. Fac. Sci. Univ. Istanbul, Sér. B 36: 1–8.

Ondok, J. P., 1970: The horizontal structure of reed stands (Phragmites communis) and its relation to productivity. Preslia (Praha) 42: 256–261.

Onno, M., 1969: Laubstreunutzungs-Versuche in Waldgesellschaften des Wienerwaldes. In: R. Tüxen (Hrsg.): Experimentelle Pflanzensoziologie. Dr. W. Junk, Den Haag: 206–212.

Osman, A. Z., 1971: Seasonal pattern of root activity of Dactylis glomerata L. and Ranunculus frieseanus Jor. grown in mono- and mixed culture, measured by using P 32. Diss. E. T. H. Zürich 1971: 89 S.

Osvald, H., 1925: Die Hochmoortypen Europas. Veröff. Geobot. Inst. Rübel, Zürich 3: 707–723.

Ottar, S., 1972: Saure Niederschläge in Nordeuropa. Naturw. Rundschau 25 (11): 449.

Overbeck, F., 1961: Die Zeitstellung des „Grenzhorizontes" norddeutscher Hochmoore und ihre Bedeutung für die Vorgeschichte. Ber. V. Internat. Kongr. Vor- u. Frühgesch., Hamburg 1958. Berlin: 631–635.

*–, 1975: Botanisch-geologische Moorkunde unter besonderer Berücksichtigung Nordwestdeutschlands usw. Karl Wachholtz Verlag,

Neumünster: 719 S.

–, Happach, H., 1957: Über das Wachstum und den Wasserhaushalt einiger Hochmoorsphagnen. Flora 144: 335–402.

Overbeck, J., 1965: Die Meeresalgen und ihre Gesellschaften an den Küsten der Insel Hiddensee (Ostsee). Botanica Marina 8: 218–233.

–, 1970: Distribution pattern of phytoplankton and bacteria, microbial decomposition of organic matter and bacterial production in eutrophic, stratified lake. In: Productivity Problems of Freshwaters. Warszawa-Kraków 1972: 229–237.

–, 1972: Zur Struktur und Funktion des aquatischen Ökosystems. Ber. Deut. Botan. Ges. 85: 553–577.

Ovington, J.D., 1963: Flower and seed production. Oikos 14: 148–153.

Påhlsson, L., 1966: Vegetation and microclimate along a belt transect from the esker Knivsas. Botan. Not. (Lund) 119: 401–418.

Pałczyński, A., 1980: Die natürlichen Gegebenheiten der Moore des Biebrza-Tales und die Probleme ihres Schutzes. Telma 10: 205–226. +

Pallmann, H., Haffter, P., 1933: Pflanzensoziologische und bodenkundliche Untersuchungen im Oberengadin mit besonderer Berücksichtigung der Zwergstrauchgesellschaften der Ordnung Rhodoreto-Vaccinietalia. Ber. Schweiz. Botan. Ges. 42: 357–466.

Pancer-Kotejowa, E., 1965: Forest communities of the Gubałowka Elevation (West Carpathian Mts.). Fragm. Flor. Geobot. 11: 239–304. +

Paschinger, V., 1954: Zur Statik und Dynamik der Höhengrenzen in den Ostalpen. Angew. Pflanzensoziol. (Wien), Festschr. Aichinger 2: 785–801.

Passarge, H., 1953 a: Waldgesellschaften des mitteldeutschen Trockengebietes. Arch. Forstwes. 2: 1–58, 182–208, 340–383 u. 532–551.

–, 1953 b: Schädlingsbefall und Standort. Ebenda 2: 245–254.

–, 1956 a: Vegetationskundliche Untersuchungen in Wäldern und Gehölzen der Elbaue. Ebenda 5: 339–358.

–, 1956 b: Die Pflanzengesellschaften der Wiesenlandschaft des Lübbenauer Spreewaldes. Feddes Rep. Beih. 135: 194–231.

–, 1957 a: Vegetationskundliche Untersuchungen in der Wiesenlandschaft des nördlichen Havellandes. Feddes Repert. Beih. 137: 5–55.

–, 1957 b: Über Kahlschlaggesellschaften im baltischen Buchenwald von Dargun (Ost-Mecklenburg). Phyton 7: 142–151.

–, 1957 c: Waldgesellschaften des nördlichen Havellandes. Wiss. Abh. Deut. Akad. Landwirtsch. wiss. Berlin 26: 139 S.

–, 1958: Vergleichende Betrachtungen über das soziologische Verhalten einiger Waldpflanzen. Arch. Forstw. 7: 302–315.

–, 1959 a: Vegetationskundliche Untersuchungen in den Wäldern der Jungmoränenlandschaft um Dargun/Ostmecklenburg. Arch. Forstwes. 8: 1–74.

–, 1959 b: Gliederung der Polygono-Chenopodion-Gesellschaften im nordostdeutschen Flachland. Phyton 8: 10–34.

–, 1960: Pflanzengesellschaften der Elbauwiesen unterhalb Magdeburg zwischen Schartau und Schönhausen. Abh. u. Ber. Naturk. u. Vorgesch. Magdeburg 11: 19–33.

–, 1962: Waldgesellschaften des Eichenwaldgebietes von SW-Mecklenburg und der Altmark. Arch. Forstwes. 11: 199–241.

–, 1963 a: Zur soziologischen Gliederung von Kiefernwäldern im nordöstlichen Mitteleuropa. Ebenda 12: 1159–1176.

–, 1963 b: Beobachtungen über Pflanzengesellschaften landwirtschaftlicher Nutzflächen im nördlichen Polen. Feddes Repert. 140: 27–69.

–, 1964 a: Pflanzengesellschaften des nordostdeutschen Flachlandes I. Pflanzensoziol. (Jena) 13: 324 S.

–, 1964 b: Beobachtungen zur soziologischen Gliederung masurischer Hainbuchenwälder. Arch. Forstwes. 13: 667–689.

–, 1965: Beobachtungen über die soziologische Gliederung baltischer Buchenwälder in S-Schweden. Arch. Forstwes. 14: 1133–1148.

–, 1968: Neue Vorschläge zur Systematik nordmitteleuropäischer Waldgesellschaften. Feddes Repert. 77: 75–103.

–, 1969 a: Zur soziologischen Gliederung mitteleuropäischer Frischwiesen. Ebenda 80: 357–372.

–, 1969 b: Zur soziologischen Gliederung mitteleuropäischer Weißklee-Weiden. Ebenda 80: 413–435.

–, 1970: Zur Kenntnis der Vegetationsabfolge nach Kahlschlag, eine Voraussetzung für die rationelle Unkrautbekämpfung. Arch. Forstwes. 19: 269–276.

–, 1971: Zur soziologischen Gliederung mitteleuropäischer Fichtenwälder. Feddes Repert. 81: 577–604.

–, 1975: Über Wiesensaumgesellschaften. Ebenda 86: 599–617.

–, 1978: Über Erlengesellschaften im Unterharz. Hercynia N.F. 15: 399–419.

–, Hofmann, G., 1964: Soziologische Artengruppen mitteleuropäischer Wälder. Arch. Forstwes. 13: 913–937.

–, –, 1968: Zur soziologischen Gliederung nordmitteleuropäischer Hainbuchenwälder. Feddes Repert. 78: 1–13.

–, Jurko, A., 1975: Über Ackerunkrautgesellschaften im nordslowakischen Bergland. Folia Geobot. Phytotax. (Praha) 10: 225–264.

Paul, H., Lutz, J., 1941: Zur soziologischen Charakterisierung von Zwischenmooren. Ber. Bayer. Botan. Ges. 25: 1–28.

Paul, K.H., 1944 u. 1953: Morphologie und Vegetation der Kurischen Nehrung I. Gestaltung der Bodenformen in ihrer Abhängigkeit von der Pflanzendecke. Nova Acta Leopoldina N.F. 13: 215–378 (1944). II. Entwicklung der Pflanzendecke von der Besiedlung des Flugsandes bis zum Wald. Ebenda 16: 261–378.

Paul, Ph., Richard, Y., 1968: Études expérimentales sur le déterminisme de la composition floristique des pelouses xérophiles. Oecol. Plant. 3: 29–48.

Pawłowski, B., 1931: Altitudes maxima de plusieurs plantes vasculaires dans les monts Tatra, Spraw. Kom. Fizjogr. Polsk. Akad. Umiej. 65: 153–158.

–, 1935: Über die Klimaxassoziation in der alpinen Stufe der Tatra. Bull. Acad. Polon. Sci. Lettr., Cl. Sci. Math.-nat., Sér. B 1935: 115–146.

–, 1959: Szata róslinna gór polskich. In: W. Szafer u. Mitarb.: Szata róslinna polski II: 189–253. Warszawa.

–, 1969: Der Endemismus in der Flora der Alpen, der Karpaten und der balkanischen Gebirge im Verhältnis zu den Pflanzengesellschaften. Mitt. Ostalp.-Dinar. Pflanzensoz. Arb. gem. (Camerino) 9: 167–178.

–, Pawłowska, S., Zarzycki, K., 1960: Les associations végétales des prairies fauchables de la partie septentrionale des Tatras et de la Région subtatrique. Fragm. Florist. Geobot. (Kraków) 6: 95–222. +

–, Sokołowski, M., Wallisch, K., 1929: Die Pflanzenassoziationen des Tatra-Gebirges. VII. Die Pflanzenassoziationen und die Flora des Morskie Oko-Tales. Bull. Acad. polon. Sci. et Lettr., Cl. Sci. Math.-Nat., Sér. B, 1928: 205–272.

Pearson, M.C., Rogers, J.A., 1962: Hippophaë rhamnoides. J. Ecol. 50: 501–513.

Pegtel, D.M., 1976: On the ecology of two varieties of Sonchus arvensis L. Proefschr. Rijksuniv. Groningen: 148 S.

Pemadasa, M.A., Lowell, P.H., 1974: The mineral nutrition of some dune annuals. J. Ecol. 62: 403–416, 647–657, 869–880.

Petermann, R., 1970: Montane Buchenwälder im westbayerischen Alpenvorland zwischen Iller und Ammersee. Diss. Botan. 8: 227 S.

–, Seibert, P., 1979: Die Pflanzengesellschaften des Nationalparks Bayerischer Wald. Nationalpark Bayer. Wald (Grafenau) 4, 142 S.

Petzold, K., 1959: Wirkung des Mähdruschverfahrens auf die Verunkrautung. Z. Acker- u. Pflanzenbau 109: 49–78.

Pfadenhauer, J., 1969: Edellaubholzreiche Wälder im Jungmoränengebiet des bayerischen Alpenvorlandes und in den bayerischen Alpen. Diss. Botan. 3: 212 S. +

–, 1971: Vergleichend ökologische Untersuchungen an Plateau-Tannenwäldern im westlichen Aargauer Mittelland. Veröff. Geobot. Inst. ETH, Stiftung Rübel, Zürich 47: 74 S.

–, 1973: Versuch einer vergleichend-ökologischen Analyse der Buchen-Tannen-Wälder des Schweizer Jura (Weissenstein und Chasseral). Ebenda 50: 60 S.

–, 1975: Beziehungen zwischen Standortseinheiten, Klima, Stickstoff-Ernährung und potentieller Wuchsleistung der Fichte im bayerischen Flyschgebiet, dargestellt am Beispiel des Teisenbergs. Habilitationsschr. TU München: 239 S.

–, Erz, F., 1980: Standort und Gesellschaftsanbindung von Ophrys apifera und Ophrys holosericea im Naturschutzgebiet „Neuffener Heide". Veröff. Naturschutz Landschaftspflege 51/52: 411–424.

–, Kaule, G., 1972: Vegetation und Ökologie eines Waldquellenkomplexes im bayerischen Inn-Chiemsee-Vorland. Ber. Geobot. Inst. ETH, Stiftung Rübel, Zürich 41: 74–87.

Philippi, G., 1960: Zur Gliederung der Pfeifengraswiesen im südlichen und mittleren Oberrheingebiet. Beitr. Naturk. Forsch. Südw.-Deut. 19: 138–187.

–, 1963 a: Zur Kenntnis der Moosgesellschaften saurer Erdraine des Weserberglandes, des Harzes und der Rhön. Mitt. Florist.-Soziol. Arb. gem. N.F. 10: 92–108.

–, 1963 b: Zur Gliederung der Flachmoorgesellschaften des Südschwarzwaldes und der Hochvogesen. Beitr. Naturk. Forsch. Südw.-Deut. 22: 113–135.

–, 1965 a: Moosgesellschaften des morschen Holzes und des Rohhumus im Schwarzwald, in der Rhön, im Weserbergland und im Harz. Nova Hedwigia (Weinheim) 9: 185–232.

–, 1965 b: Die Moosgesellschaften der Wutachschlucht. Mitt. Bad. Landesver. Naturkunde u. Naturschutz, N.F. 8: 625–668.

–, 1966: Sporenkeimung und Protonemawachstum von Moosen verschiedener Standorte in Abhängigkeit vom pH-Wert. Flora, Abt. B 156: 319–349.

–, 1969 a: Laichkraut- und Wasserlinsengesellschaften des Oberrheingebietes zwischen Straßburg und Mannheim. Veröff. Landesstelle Naturschutz u. Landschaftspflege Baden-Württemb. 37: 102–172.

–, 1969 b: Zur Verbreitung und Soziologie einiger Arten von Zwergbinsen- und Strandlingsgesellschaften im badischen Oberrheingebiet. Mitt. Bad. Landesver. Naturkde. u. Naturschutz N.F. 10: 139–172.

–, 1970: Die Kiefernwälder der Schwetzinger Hardt (nordbadische Oberrheinebene). Veröff. Landesstelle Naturschutz u. Landschaftspflege Baden-Württemb. 38: 46–92.

–, 1971: Sandfluren, Steppenrasen und Saumgesellschaften der Schwetzinger Hardt (nordbadische Rheinebene). Ebenda 39: 67–130.

–, 1975: Quellflurgesellschaften der Allgäuer Alpen. Beitr. Naturk. Forsch. Südw.-Deut. 34: 259–287.

Phillipson, J., Putman, R.J., Steel, J., Woodell, S.R.J., 1975: Litter input, litter decomposition and the evolution of carbon dioxide in a beech woodland – Wytham Woods, Oxford. Oecologia 20: 203–217.

Pietsch, W., 1965: Die Erstbesiedlungsvegetation eines Tagebau-Sees. Synökologische Untersuchungen im Lausitzer Braunkohlen-Revier. Limnologica (Berlin) 3: 177–222.

–, 1963: Vegetationskundliche Studien über die Zwergbinsen- und Strandlingsgesellschaften in der Nieder- und Oberlausitz. Abh. u. Ber. Naturkundemus. Görlitz 38, Nr. 2: 80 S.

–, 1974: Ökologische Untersuchung und Bewertung von Fließgewässern mit Hilfe höherer Wasserpflanzen – ein Beitrag zur Belastung aquatischer Ökosysteme. Mitt. Sekt. Geobot. u. Phytotax. Biol. Ges. DDR 1974: 13–29.

–, 1976: Vegetationsentwicklung und wasserchemische Faktoren in Moorgewässern verschiedener Naturschutzgebiete der DDR. Arch. Naturschutz u. Landschaftsforsch. Berlin 16: 1–43.

–, Müller-Stoll, W.R., 1968: Die Zwergbinsen-Gesellschaft der nackten Teichböden im östlichen Mitteleuropa, Eleocharito-Caricetum bohemicae. Mitt. Florist.-Soziol. Arb. gem. N.F. 13: 14–47.

Pignatti, E., 1970 a: Über die subnivale Vegetationsstufe in Osttirol. Mitt. Ostalp.-Dinar. Ges. f. Vegetationskunde 11: 167–174.

–, 1970 b: Le brughiere subalpine a Rhododendron ferrugineum nel versante meridionale delle Alpi Orientali. Atti Istit. Veneto Sci., Lett. Arti 128: 195–212.

Pignatti, S., 1970: Die Fichtenwälder Norditaliens. Mitt. Ostalpin-Dinar. Arbeitsgem. 6 (Wien) 1970.

Pignatti-Wikus, E., 1959: Pflanzensoziologische Studien im Dachsteingebiet. Boll. Soc. Adriat. Sci. Nat. Trieste 50: 89–168.

Pigott, C.D., 1975: Natural regeneration of Tilia cordata in relation to forest-structure in the forest of Białowieża, Poland. Philos. Transact. Roy. Soc. London, B. Biol. Sci. 270: 151–179.

–, Taylor, K., 1964: The distribution of some woodland herbs in relation to the supply of nitrogen and phosphorus in the soil. J. Ecol. 52 (Suppl.): 175–185.

Pikula, J., 1963: Die Disseminationsweisen in den Pflanzengesellschaften der Belaer Tatra. Sborník Prác Tatransk. Národn. Parku 6: 27–42.

Pílát, A., 1969: Underground dry weight in the grassland communities of Arrhenatheretum elatioris alopecuretosum Vicherek 1960. Folia Geobot. Phytotax. (Praha) 4: 225–234.

Pineau, M., 1968: Observations phénologiques et morphologiques sur le comportement de quelques essences forestieres soumises à l'action d'engrais N, P et K. Diss. ETH Zürich: 136 S.

Piotrowska, H., 1955: Les associations forestières de l'Ile de Wolin. Poznánsk. Towar. Przyj. Nauk, Wydz. Math.-Przyr., Práce Kom. Biol. 16, 5: 168 S. +

Pirk, W., Tüxen, R., 1957: Höhere Pilze in nw-deutschen Calluna-Heiden (Calluneto-Genistetum typicum). Mitt. Florist.-Soziol. Arb. gem. N.F. 6/7: 127–129.

Pisek, A., 1950: Frosthärte und Zusammensetzung des Zellsaftes bei Rhododrendron ferrugineum, Pinus cembra und Picea excelsa. Protoplasma 39: 129–146.

–, 1963: An den Grenzen des Pflanzenlebens im Hochgebirge. Jb. Ver. Schutze Alpenpflanzen u. -Tiere 28: 112–129.

–, Berger, E., 1938: Kutikuläre Transpiration und Trockenresistenz isolierter Blätter und Sprosse. Planta (Berlin) 28: 124–155.

–, Larcher, W., Moser, W., Pack. I., 1969: Kardinale Temperaturbereiche der Photosynthese usw. III. Temperaturabhängigkeit und optimaler Temperaturbereich der Netto-Photosynthese. Flora B 158: 608–630.

Pitschmann, H., Reisigl, H., Schiechtl, H., 1959: Bilderflora der Südalpen. Vom Gardasee zum Comersee. Stuttgart: 278 S.

Plewczýnska-Kurraś, U., 1974: Analysis of a sheep pasture ecosystem in the Pieniny montains (the Carpathians) IV. Biomass of the upper and underground parts and of organic detritus. Ekol. Pol. 22: 517–526.

Poelt, J., 1963: Flechtenflora und Eiszeit in Europa. Phyton 10: 206–215.

*Pohl, D., 1975: Bibliographie der niedersächsischen Naturschutzgebiete. Naturschutz u. Landschaftspflege Niedersachs. 4: 290 S.+

Poldini, L., 1969: Le pinete di pino austriaco nelle Alpi Carniche. Boll. Soc. Adriat. Sci., Trieste 57: 3–65.

Poli, E., Tüxen, J., 1960: Über Bidentetalia-Gesellschaften Europas. Mitt. Florist.-Soziol. Arb. gem. N.F. 8: 136–144.

Ponnamperuma, F.N., 1972: The chemistry of submerged soils. Advanc. Agron. 24: 29–96.

Pop, E., 1964: Über die Herkunft der ombrogenen Moore und ihrer Flora. Ber. Geobot. Inst. ETH, Stiftg. Rübel, Zürich 35: 113–118.

–, u. Mitarb., 1964: Effects of atmospheric precipitations on the pollen and spores concentration from the aeroplankton. Rev. Roum. Biol., Sér. Bot. 9: 329–334.

Pork, K., 1975: Allelopathie relations between species in meadow plant communities. In: Laasimer, L. (Ed.): Some aspects of botanical research in the Estonian SSR. Tartu: 137–157.

Portmann, A., Antonietti, A., Klötzli, F., u.a., 1964: Le Bolle di Magadino. Quaderu Ticinesi (Locarno) 7: 39 S.+

Post, L. von, 1925: Einige Aufgaben der regionalen Moorforschung. Sver. Geol. Onders. Ser. C 337, Arsb. 19, 4, 41 S.

Praag, H., Weissen, F., 1973: Elements of a functional definition of oligotroph humus based on the nitrogen nutrition of forest stands. J. Appl. Ecol. 10: 569–583.

–, –, Brigode, N., Dufour, J., 1973: Évaluation de la quantité d'azote mineralisée par an dans un sol de hêtraie ardennaise. Bull. Soc. Roy. Botan. Belg. 106: 137–146.

Preis, K., 1939: Die Festuca vallesiaca-Erysimum crepidifolium-Assoziation auf Basalt. Glimmerschiefer und Granitgrus. Beih. Bot. Cbl. 59 B: 478–530.

Preising, E., 1943: Die Waldgesellschaften des Warthe- und Weichsellandes. Arb. Zentralstelle f. Veget. Kart. d. Reiches (Manuskr. Druck) 1943, 142 S.

–, 1950: Nordwestdeutsche Borstgras-Gesellschaften. Mitt. Florist.-Soziol. Arb. gem. N.F. 2: 33–41.

–, 1953: Süddeutsche Borstgras- und Zwergstrauch-Heiden (Nardo-Callunetea). Ebenda 4: 112–123.

Prott, R., 1979: Die Wasser- und Sumpfvegetation eutropher Gewässer in der Westfälischen Bucht – pflanzensoziologische und hydrochemische Untersuchungen. Abh. Landesmus. Naturkunde Münster Westf. 42, 156 S.

Putzer, J., 1967: Pflanzengesellschaften im Raum von Brixen unter besonderer Berücksichtigung der Trockenvegetation. Diss. Univ. Innsbruck.

Quantin, A., 1935: L'évolution de la végétation à l'étage de la chênaie dans le Jura méridional. Thèse Paris, Comm. SIGMA (Montpellier) 37: 382 S.

–, 1960: in: G. Viennot-Bourgin. Rapports du sol et de la végétation. Paris: 110–113.

Rabe, E.-W., 1950: Über die Vegetationsverhältnisse der Insel Fehmarn. Mitt. Arb. gem. Floristik Schlesw.-Holst. 1: 106 S.

–, 1954: Sukzessionsstudien am Sandkatener Moor. Arch. Hydrobiol. 49: 349–375.

–, 1960: Über die Vegetationstypen am Dummersdorfer Ufer, dem linken Ufer der Untertrave. Ber. Ver. „Natur u. Heimat" u. Naturhist. Mus. Lübeck 2: 5–78.

–, 1964: Die Heidetypen Schleswig-Holsteins. „Die Heimat" (Neumünster) 71: 169–175.

–, 1965: Salzwiesen in der Treene-Niederung bei Sollbrück. Jb. Schleswigsche Geest 13: 1–10.

–, 1972: Über den Stand der Vegetationskartierung in Schleswig-Holstein 1971. Schr. Naturw. Ver. Schlesw.-Holst. 42: 70–85.

Rabeler, W., 1962: Die Tiergesellschaften von Laubwäldern (Querco-Fagetea) im oberen und mittleren Wesergebiet. Mitt. Florist.-Soziol. Arb. gem. N.F. 9: 200–229.

Rabotnov. T.A., 1969: Plant regeneration from seed in meadows of the USSR. Herbage Abstracts 39, No. 4: 269–277.

Rädel, J., 1962: Die Reste naturnaher Waldgesellschaften im Landschaftsschutzgebiet Kriebstein/Sa. Ber. Arb. gem. Sächs. Botan., N.F. 149–186.+

Rademacher, B., 1948: Gedanken über Begriff und Wesen des „Unkrauts". Z. Pflanzenkrankh. (Pflanzenpathol.) u. Pflanzenschutz 55: 1–10.

–, 1962: Grasartige Unkräuter und ihre Bekämpfung. Arb. D.L.G. 86: 5–21.

–, 1967: Beobachtungen in Dauerversuchen mit Unkräutern und Herbiziden. Mitt. Biol. Bundesanst. Land- u. Forstwirtsch. 1967: 177–185.

Ramaut, J.-L., Corvisier, M., 1975: Effects inhibiteurs des extraits de Cladonia impexa Harm., C. gracilis (L.) Willd. et Cornicularia muricata (Ach.) Ach. sur la germination des graines de Pinus sylvestris L. Oecol. Plant. 10, 295–299.

Ranwell, D.S., 1972: Ecology of Salt Marshes and Sand Dunes. Chapman and Hall, London: 258 S.

–, 1974: The salt marsh to tidal woodland transition. Hydrobiol. Bull. (Amsterdam) 8: 139–151.

Rauh, W., 1939: Über polsterförmigen Wuchs. Nova Acta Leopoldina N.F. 7, 49: 267–508.

Rawald, W., Niemann, E., 1967: Über bodenmikrobiologische und vegetationskundliche Untersuchungen im Naturschutzgebiet „Prinzenschneise" bei Weimar. Arch. Naturschutz u. Landschaftsforsch. 7: 191–246.

Rayner, M.C., 1913: The ecology of Calluna vulgaris. New Phytol. 12: 59–77.

Redinger, K., 1934: Studien zur Oekologie der Moorschlenken. Beih. Bot. Cbl. 52 B: 231–309.

Rehder, H., 1960: Saugkraftmessungen an Stachys silvatica im frischen und welken Zustand. Ber. Deut. Botan. Ges. 73, 73–82.

–, 1962: Der Girstel – ein natürlicher Föhrenwaldkomplex am Albis bei Zürich. Ber. Geobot. Inst. ETH, Stiftg. Rübel, Zürich 33: 17–64.

–, 1965: Die Klimatypen der Alpenkarte im Klimadiagramm-Weltatlas (Walter u. Lieth) und ihre Beziehungen zur Vegetation. Flora, Abt. B 156: 78–93.

–, 1970: Zur Ökologie, insbesondere Stickstoffversorgung subalpiner und alpiner Pflanzengesellschaften im Naturschutzgebiet Schachen (Wettersteingebirge). Diss. Botan. 6: 90 S.

–, 1975: Phytomasse- und Nährstoffverhältnisse einer alpinen Rasengesellschaft (Caricetum firmae). Verh. Ges. Ökol., Wien 1975: 93–99.

–, 1976: Nutrient turnover studies in Alpine ecosystems. II. Phytomass and nutrient relations in the Caricetum firmae. Oecologia (Berlin) 23: 49–62.

REHFUESS, K.E., 1968: Beziehungen zwischen dem Ernährungszustand und der Wuchsleistung südwestdeutscher Tannenbestände. Forstwiss. Cbl. 87: 36–58.

REICHELT, G., 1966: Anthropogene Veränderungen der Pflanzendecke und ihre Folgen an Beispielen aus Mitteleuropa. Der Math.-naturwiss. Unterricht (Frankfurt/M.) 19: 61–71.

REICHHOFF, L., 1974: Untersuchungen über den Aufbau und die Dynamik des Orchideen-Halbtrockenrasens im Naturschutzgebiet „Leutratal" bei Jena/Thüringen. Mitt. Sekt. Geobot. Phytotax. Biol. Ges. DDR 1974: 115–125.

REINHOLD, F., 1956: Das natürliche Waldbild der Baar und der angrenzenden Landschaften. Schr. Ver. Gesch. u. Naturgesch. Baar usw., Donaueschingen 24: 224–268.

REINKE, J., 1903: Die Entwicklung der Dünen an der Westküste von Schleswig. Sitzber. Kön. Preuß. Akad. Wiss. Berlin 1903, 1. Hälfte.

–, 1912: Studien über die Dünen an unserer Ostseeküste III. Wiss. Meeresunters. Abt. Kiel, N.F. 14.

REISIGL, H., PITSCHMANN, H., 1958: Obere Grenzen von Flora und Vegetation in der Nivalstufe der zentralen Ötztaler Alpen (Tirol). Vegetatio 8: 93–128.

–, –, 1959: Zur Abgrenzung der Nivalstufe. Phyton 8: 219–224.

–, SCHIECHTL, H.M., STERN, H., 1970/71: Karte der aktuellen Vegetation von Tirol 1:100000. 1. Teil: Blatt 6, Innsbruck-Stubaier Alpen. Documents Carte Végét. Alpes 8, Saint Martin-d'Héres/Grenoble 1970. 2. Teil: Blatt 7, Zillertaler Alpen. Ebenda 9: 1971. +

REJMENT-CROCHOWSKA, I.: Concentration of heavy metals in mosses. XII. Internat. Botan. Congr. (Leningrad) Abstr. 1: 86.

REMACLE, J., 1975: Microbial N-transformation in forest soils. Comité Nation. Belge P.B.I., Sect. PT-PF, Projet Mirwart, Contrib. 28: 12 pp.

–, FROMENT, A., 1972: Teneurs en azote minéral et numérations microbiologiques dans la chênaie calcicole de Virelles (Belgique). Oecol. Plant. 7: 69–78.

RICHARD, J.L., 1961: Les forets acidophiles du Jura. Beitr. Geobot. Landesaufn. Schweiz 38, 164 S.

–, 1968 a: Quelques groupements végétaux à la limite supérieure de la forêt dans les hautes chaînes du Jura. Vegetatio 16: 205–219.

–, 1968 b: Les groupements végetaux de la réserve d'Aletsch. Beitr. Geobot. Landesaufn. Schweiz 51: 30 S.

–, 1972: La végétation des crêtes rocheuses du Jura. Ber. Schweiz. Botan. Ges. 82: 68–112.

–, 1973: Dynamique de la végétation au bord du grand glacier d'Aletsch (Alpes suisses). Ebenda 83: 159–174.

–, 1975: Les groupements végétaux du clos du Doubs (Jura suisse). Beitr. Geobot. Landesaufn. Schweiz 57, 71 S.+

RICHARD, L., 1969: Une interprétation éco-physiologique de la répartition de l'aune vert (Alnus viridis Chaix). Docum. Carte Végét. Alpes 7 (Saint Martin d'Héres, Grenoble).

RICHTER, W., 1965: Die natürliche Begrünung der erzgebirgischen Bergwerkshalden. Hercynia (Leipzig) 3: 114–146.

RIED, A., 1960: Stoffwechsel und Verbreitungsgrenzen von Flechten. I. Flechtenzonierungen an Bachufern und ihre Beziehungen zur jährlichen Überflutungsdauer und zum Mikroklima. Flora 148: 616–638.

RIEDE, U., 1973: Die Bestimmung der maximalen Biomasse und deren Gehalt an Gesamtstickstoff verschiedener Pflanzengesellschaften des Graswarders vor Heiligenhafen/Ostsee. Dipl. Arb. Math.-Nat. Fak. Göttingen 1973 (unveröff.): 44 S.

RIKLI, M., 1909: Die Arve in der Schweiz. Neue Denkschr. Schweiz. Naturf. Ges. 44, 455 S.

RIVAS MARTINEZ, S., 1964: Esquema de la vegetación potenciál y su correspondencia con los suelos en la España peninsular. An. Inst. Botan. A. J. Cavanillas (Madrid) 22: 343–405.

ROCHOW, M. VON, 1951: Die Pflanzengesellschaften des Kaiserstuhls. Pflanzensoziol. (Jena) 8: 140 S.

RODI, D., 1960: Die Vegetations- und Standortsgliederung im Einzugsgebiet der Lein (Kreis Schwäbisch Gmünd). Veröff. Württ. Landesst. Naturschutz und Landschaftspflege 27/28: 76–167.+

–, 1963: Die Streuwiesen- und Verlandungsgesellschaften des Welzheimer Waldes Veröff. Landesstelle f. Naturschutz u. Landschaftspflege Baden-Württemb. 31: 31–67.

–, 1966: Ackerunkrautgesellschaften und Böden des westlichen Tertiär-Hügellandes mit besonderer Berücksichtigung des Kreises Schrobenhausen. Denkschr. Regensburger Botan. Ges. 26: 161–198.

–, 1968: Die Pflanzendecke. In: Erläuterungen zur Bodenkarte von Bayern 1:25000, Blatt Nr. 7433 Schrobenhausen 1968: 36–56.

–, 1974: Trockenrasengesellschaften des nordwestlichen Tertiärhügellandes. Ber. Bayer. Botan. Ges. 45: 151–172.

–, 1975: Die Vegetation des nordwestlichen Tertiärhügellandes (Oberbayern). Schriftenr. Vegetationskunde (Bonn-Bad Godesberg) 8: 21–78. +

*RODIN, L.E., BAZILEVIČ, N.I., 1966: The biological productivity of the main vegetation types in the northern hemisphere of the old world. Forestry Abstr. 27: 357–372.

*RÖHRIG, E., 1964: Über die gegenseitige Beeinflussung der höheren Pflanzen. Forstarch. 35: 25–39.

ROISIN, P., 1961: Reconnaissances phytosociologiques dans les hêtraies atlantiques. Bull. Inst. Agron. et Stat. Rech. Gembloux 29: 356–385.

ROLL, H., 1938. Die Pflanzengesellschaften ostholsteinischer Fließgewässer. Arch. Hydrobiol. 34: 159–305.

–, 1939: Isoëtes, Lobelia und Litorella in kalkarmem und kalkreichem Wasser. Beih. Botan. Cbl. 59: 345–358.

ROMELL, L.G., 1967: Die Reutbetriebe und ihr Geheimnis. Studium Generale 20: 362–369.

ROMPEL, J., 1928: Beobachtungen über die bis zum Aufblühen alpiner Arten verstreichende Aperzeit. Österr. Botan. Z. 77: 178–194.

RÖNICKE, G., KLOCKOW, D., 1974: Der Grundpegel der Schwefelkonzentration der Luft in der Bundesrepublik Deutschland 1967–1972. DFG-Kommission Erforsch. Luftverunreinigung, Mitt. 12: 38 S.

ROSSKOPF, G., 1971: Pflanzengesellschaften der Talmoore an der Schwarzen und Weißen Laber im Oberpfälzer Jura. Denkschr. Regensburg. Botan. Ges. 28, 115 S. +

ROTHE, E., 1963: Beiträge zur Kenntnis der Färbungen und Farbstoffe bei Torfmooren (Sphagnum). Beitr. Biol. Pflanzen 38: 331–381.

ROTHMALER, W., MEUSEL, H., SCHUBERT, R., 1972: Exkursionsflora für die Gebiete der DDR und der BRD, Gefäßpflanzen. Berlin: Volk u. Wissen VEV 1972: 612 S.

ROTTENBURG, W., KOEPPNER, T., 1972: Die Wirkung der Faktoren Licht und Wasser auf den Spaltöffnungszustand bei Koniferen. Ber. Deut. Botan. Ges. 85: 353–362.

ROUSSEAU, L.Z., 1960: De l'influence du type d'humus sur le développement des plantules de sapin dans les Vosges. Ann. École Nation. Eaux et Forêts et Stat. Rech. et Expér. Nancy 17: 15–118.

ROUX, C., 1978: Complément á l'étude écologique et phytosociologique des peuplements lichéniques saxicoles-calcicoles du SE de la France. Bull. Mus. Hist. Nat. Marseille 38: 65–186.

RÜBEL, E., 1912: Pflanzengeographische Monographie des Berninagebietes. Leipzig: 615 S.

–, 1922: Curvuletum. Mitt. Geobot. Inst. Rübel, Zürich 1922: 1–15.

RUBNER, K., 1926: Die forstlichen Verhältnisse Rumäniens in pflanzengeographischer Betrachtung. Forstwiss. Cbl. 48: 145–258.

RÜDENAUER, B., RÜDENAUER, K., SEYBOLD, S., 1974: Über die Ausbreitung von Helianthus- und Solidago-Arten in Württemberg. Jb. Ges. Naturkde. Württemb. 129: 65–77.

RUDOLPH, H., 1963: Die Kultur von Hochmoor-Sphagnen unter definierten Bedingungen. Beitr. Biol. Pflanzen 39: 153–177.

–, BREHM, K., 1966: Kationenaufnahme durch Ionenaustausch? Neue Gesichtspunkte zur Frage der Ernährungsphysiologie der Sphagnen. Ber. Deut. Botan. Ges. 78: 484–491.

RUETZ, W.F., 1973: The seasonal pattern of CO_2 exchange of Festuca rubra L. in a montane meadow community in Northern Germany. Oecologia (Berlin) 13: 247–269.

RÜHL, A., 1959: Über das soziologische Verhalten der schlanken Segge (Carex strigosa Huds.). Decheniana 111: 27–31.

–, 1960: Über die Waldvegetation der Kalkgebiete nordwestdeutscher Mittelgebirge. Decheniana 111, Beih. 8: 1–50.

–, 1964: Vegetationskundliche Untersuchungen über die Bachauenwälder des Nordwestdeutschen Berglandes. Decheniana 116: 29–44.

–, 1973: Waldvegetationsgeographie des Weser-Leineberglandes. Veroff. Nieders. Inst. Landeskunde u. Landesentwickl. Göttingen, R. A. 101: 95 S.

RUNGE, F., 1950: Vergleichend pflanzensoziologische und bodenkundliche Untersuchungen von bodensauren Laubwäldern im Sauerland. Abh. Landesmus. Naturk. Prov. Westfalen 13: 3–48.

–, 1963: Die Artmächtigkeitsschwankungen in einem nordwestdeutschen Enzian-Zwenkenrasen. Vegetatio 11: 237–240.

–, 1968 a: Vegetationsänderungen nach Auflassung eines Ackers. Natur u. Heimat (Münster i. Westf.) 28: 111–125.

–, 1968 b: Schwankungen der Vegetation sauerländischer Talsperren. Arch. Hydrobiol. 65: 223–239.

–, 1969 a: Die Verlandungsvegetation in den Gewässern des Naturschutzgebietes „Heiliges Meer". Naturkunde in Westf. 5: 89–95.

–, 1969 b: Vegetationsschwankungen in einem Melico-Fagetum. Vegetatio 17: 151–156.

–, 1969 c u. 1980: Die Pflanzengesellschaften Mitteleuropas. Eine kleine Übersicht. (1. Aufl. 1961: Die Pflanzengesellschaften Westfalens. 3. Aufl. 1969: D. Pfl. Deutschlands ...), 6./7. Aufl. 1980: 278 S. Aschendorff, Münster i. W.

–, 1970: Die pflanzliche Besiedlung eines Straßenbanketts. Natur u. Heimat 30: 54–56.

RUNGE, M., 1965: Untersuchungen über die Mineralstickstoff-Nachlieferung an nordwestdeutschen Waldstandorten. Flora 155: 353–386.

–, 1970: Untersuchungen zur Bestimmung der Mineralstickstoff-Nachlieferung am Standort. Flora, Abt. B 159: 233–257.

–, 1973 a: Der biologische Energieumsatz in Land-Ökosystemen unter Einfluß des Menschen. In: ELLENBERG 1973 b: 123–141.

–, 1973 b: Energieumsätze in den Biozönosen terrestrischer Ökosysteme. Scripta Geobot. 4 (Göttingen): 78 S.

–, 1974: Die Stickstoff-Mineralisation im Boden eines Sauerhumus-Buchenwaldes I. Mineralstickstoff-Gehalt und Netto-Mineralisation. II. Die Nitratproduktion. Oecol. Plant. 9: 201–218 u. 219–230.

–, 1981: Die Bedeutung des Aluminiums für die Ausbildung der natürlichen und naturnahen Vegetation. Mitt. Ergänzungsstudium Ökol. Umweltsicherung (Kassel) 7: 16–38.

RUTHSATZ, B., 1970: Die Grünlandgesellschaften um Göttingen. Scripta Geobot. (Göttingen) 2: 31 S.

RUTTNER, F., 1962: Grundriß der Limnologie. Hydrobiologie des Süßwassers. 3. Aufl. Berlin: 332 S.

RUUHIJÄRVI, R., 1963: Zur Entwicklungsgeschichte der nordfinnischen Hochmoore. Ann. Botan. Soc. „Vanamo" 34, No. 2: 40 S.

RUŽIČKA, M., 1961: Flechten-Kiefernwald auf den Flugsanden der Tiefebene Záhorská Nížina (Cladonio-Pinetum zahoricum). Biológia (Bratislava) 16: 881–894.

–, 1964: Geobotanische Verhältnisse der Wälder im Sandgebiete der Tiefebene Záhorská Nížina (Südwestslowakei). Biol. Práce (Bratislava) 10, 1: 119 S.+

RYBNIČEK, K., 1970: Rhynchospora alba (L.) Vahl, its distribution, communities and habitat conditions in Czechoslovakia, Part 2. Folia Geobot. Phytotax. (Praha) 5: 221–263.

–, 1964: Die Braunmoosgesellschaften der Böhmisch-mährischen Höhe (Tschechoslowakei) und die Problematik ihrer Klassifikation. Preslia (Praha) 36: 403–415.

–, 1974: Die Vegetationsverhältnisse der Moore im südlichen Teil der Böhmisch-Mährischen Höhe. Veget. ČSSR Ser. A 6: 243 S.

RYTZ, W., 1949: Die Pflanzenwelt. In O. TSCHUMI: Urgeschichte der Schweiz (Frauenfeld) Bd. 1: 15–119.

SAGAR, G.R., HARPER, J.L., 1961: Controlled interference with natural populations of Plantago lanceolata, P. major and P. media. Weed Res. 1: 163–176.

SALISBURY, E.J., 1922: The soils of Blakeney Point: A study of soil reaction and succession in relation to the plant covering. Ann. Botan. 36: 391–431.

SALISBURY, F.B., 1975: The active growth of plants under snow. XII. Invernat. Botan. Congr. (Leningrad) Abstr. 1: 166.

SALZMANN, R., 1939: Die Anthropochoren der schweizerischen Kleegraswirtschaft; die Abhängigkeit ihrer Verbreitung von der Wasserstoffionen-Konzentration und der Dispersität des Bodens mit Beiträgen zu ihrer Keimungsbiologie. Diss. ETH Zürich: 82 S.

*SAUBERER, F., HÄRTEL, O., 1959: Pflanze und Strahlung. Leipzig: 268 S.

SCAMONI, A., 1954: Waldvegetation des Unterspreewaldes. Arch. Forstwes. 3: 122–161 u. 230–260.

–, 1957: Vegetationsstudien im Wildschutzgebiet „Fauler Ort" und in den angrenzenden Waldungen. Feddes Repert. Beih. 137: 55–109. +

–, 1960: Waldgesellschaften und Waldstandorte, dargestellt am Gebiet des Diluviums der Deutschen Demokratischen Republik. Berlin: 326 S. +

–, (Hrsg.), 1964: Vegetationskarte der Deutschen Demokratischen Republik (1:500000) mit Erläuterungen. Berlin: Akademie-Verlag, 106 S. +

–, 1965: Vegetationskundliche und standortskundliche Untersuchungen in mecklenburgischen Waldschutzgebieten. Natur Meckl. (Stralsund-Greifswald) 3: 15–142.

–, 1967 a: Vegetation–Standort. Methodenvergleich in der Oberförsterei Chorin bei Eberswalde. Arch. Naturschutz u. Landschaftsforsch. 6: 167–206.

–, 1967 b. Der Waldmeister-Buchenwald (Asperulo-Fagetum). Botan. Jb. 86: 494–521.

–, PASSARGE, H., 1959: Gedanken zu einer natürlichen Ordnung der Waldgesellschaften. Arch. Forstwes. 8: 386–426.

–, GROSSER, K.H., GÜRTLER, CH., HOFMANN, G., HURTTIG, H., PASSARGE, H., SIEFKE, A., WEBER, H., 1963: Natur, Entwicklung und Wirtschaft einer jungpleistozänen Landschaft, dargestellt am Gebiet des Meßtischblattes Thurow (Kreis Neustrelitz). Wiss. Abh. Deut. Akad. Landwirtschaftswiss. Berlin 56: 340 S., 2 farb. Karten. +

SCHAEFER, R., 1964: Influence du régime thermique d'incubation, en particulier d'un gel répété, sur la réduction dissimilative des nitrates dans un mull et un hydromull calciques. Ann. Inst. Pasteur 107: 534–549, u. 1964, Suppl. au No. de Sept.: 282–292.

SCHÄFER, H., WITTMANN, O. (Hrsg.), 1966: Der Isteiner Klotz. Zur Naturgeschichte einer Landschaft am Oberrhein. Verlag Rombach, Freiburg i. Br.: 445 S.+

SCHÄFER, K., 1972: Temperaturkurven der Nettoassimilation und Dunkelatmung einiger Sorten von Lolium perenne L. Landwirtsch. Forsch. 25: 191–202.

, 1975: Über die Entwicklung der Pflanzenbestände von ehemaligem Grünland auf grundwassernahen und grundwasserfernen Standorten. In: SCHMIDT, W. (Hrsg.): Sukzessionsforschung. J. Cramer, Vaduz: 527–533.

SCHARFETTER, R., 1938: Das Pflanzenleben der Ostalpen. Wien: 419 S.

SCHAUER, TH., 1979: Die Vegetation des Spitzingsees. Jb. Verein Schutz d. Bergwelt (München) 44: 137–154.

SCHEEL, H., 1962: Moor- und Grünlandgesellschaften im oberen Briesetal nördlich von Berlin. Wiss. Z. Pädag. Hochsch. Potsdam 7: 201–230.+

SCHEFFER, F., ULRICH, B., 1960: Lehrbuch der Agrikulturchemie und Bodenkunde III. Humus und Humusdüngung. Bd. 1. Morphologie, Biologie, Chemie und Dynamik des Humus. 2. Aufl., Stuttgart: 266 S.

SCHIECHTL, H.M., 1958: Grundlagen der Grünverbauung. Mitt. Forstl. Versuchsanst. Mariabrunn 55: 273 S.

–, 1970: Die Ermittlung der potentiellen Zirben-Waldfläche im Ötztal. Mitt. Ostalpin.-Dinar. Ges. Vegetationskunde 11 (Innsbruck): 1970.

SCHLENKER, G., 1940: Erläuterungen zum pflanzensoziologischen Kartenblatt Bietigheim. Tübingen.+

–, SCHILL, G., 1979: Das Feldflora-Reservat auf dem Beutenlay bei Münsingen. Mitt. Ver. Forstl. Standortskunde u. Forstpflanzenzüchtg. 27: 55–59.

SCHLÜTER, H., 1955: Das Naturschutzgebiet Strausberg. Feddes Repert. Beih. 135: 260–350.

–, 1959: Waldgesellschaften und Wuchsbezirksgliederung im Grenzbereich der Eichen-Buchen-zur Buchenstufe am Nordwestabfall des Thüringer Waldes. Arch. Forstwes. 8: 427–493.

–, 1964: Zur Waldentwicklung im Thüringer Gebirge, hergeleitet aus Pollendiagrammen, Archivquellen und Vegetationsuntersuchungen. Arch. Forstwes. 13: 283–305.

–, 1965: Vegetationskundliche Untersuchungen an Fichtenforsten im Mittleren Thüringer Wald. Die Kulturpflanze (Berlin) 13: 55–99.

–, 1966 a: Abgrenzung der natürlichen Fichtenwälder gegen anthropogene Fichtenforste und die Ausweitung des Fichtenwaldareals in Zusammenhang mit dem Tannenrückgang im Thüringer Wald. In: R. TÜXEN: Anthropogene Vegetation. Dr. W. Junk, Den Haag: 263–274.

–, 1966 b: Untersuchungen über die Auswirkung von Bestandskalkungen auf die Bodenvegetation in Fichtenforsten. Die Kulturpflanze (Berlin) 14: 47–60.

–, 1966 c: Licht- und Temperaturmessungen an den Vegetationszonen einer Lichtung („Lochhieb") im Fichtenforst. Flora, Abt. B 156: 133–154.

–, 1967: Buntlaubhölzer in kollinen Waldgesellschaften Mittelthüringens. Die Kulturpflanze (Berlin) 15: 115–138.

–, 1968: Zur systematischen und räumlichen Gliederung des Carpinion in Mittelthüringen. Feddes Repert. 77: 117–141.

–, 1970: Vegetationskundlich-synökologische Untersuchungen zum Wasserhaushalt eines hochmontanen Quellgebietes. Wiss. Veröff. Geogr. Inst. Deut. Akad. Wiss. N.F. 27/28: 23–144.

SCHMEIDL, H., 1962: Kleinklimatische Vergleiche in Moorgebieten. Wetter u. Leben (Wien) 14: 77–82.

–, 1964: Bodentemperaturen in Hochmoorböden. Bayer. Landw. Jb. 41: 115–122.

–, SCHUCH, M., WANKE, R., 1970: Wasserhaushalt und Klima einer kultivierten und unberührten Hochmoorfläche am Alpenrand. Schriftenr. Kurator. f. Kulturbauwesen 19: 174 S.

SCHMEISKY, H., 1974: Vegetationskundliche und ökologische Untersuchungen in Strandrasen des Graswarders vor Heiligenhafen/Ostsee. Diss. Univ. Göttingen: 103 S.

SCHMID, E., 1936: Die Reliktföhrenwälder der Alpen. Beitr. Geobot. Landesaufn. Schweiz 21: 190 S.

–, 1961: Erläuterungen zur Vegetationskarte der Schweiz. Beitr. Geobot. Landesaufn. Schweiz 39: 52 S.

SCHMID, H., ZEIDLER, H., 1953: Beobachtungen und Gedanken zum Rückgang der Tanne. Forstwiss. Cbl. 72: 101–110.

SCHMID, J., 1955: Der Bodenfrost als morphologischer Faktor. Heidelberg: 144 S.

SCHMIDT, K.W., 1957: Studien über das Verhalten von 14 kalkmeidenden Pflanzen der Bauernwälder auf dem Kalkwildboden eines Berghanges. Botan. Jb. 77: 158–192.

SCHMIDT, L., 1974: Stoffproduktion und Energiehaushalt von alpinen Zwergstrauchgesellschaften. Diss. Univ. Innsbruck, 153 S.

SCHMIDT, P., 1974: Das soziologische Verhalten der mitteleuropäischen Thymus-Arten als Beispiel für die Bedeutung „kritischer Sippen" in der Geobotanik. Mitt. Sekt. Geobot. Phytotax. Biol. Ges. DDR 1974: 49–59.

SCHMIDT, W., 1970: Untersuchungen über die Phosphorversorgung niedersächsischer Buchenwaldgesellschaften. Scripta Geobot. (Göttingen) 1: 120 S.

*–, 1976: Ungestörte und gelenkte Sukzession auf Brachäckern. Habil. Schr. Göttingen: 276 S., sowie 1980: Scripta Geobot. 15: 199 S.

SCHMITHÜSEN, J., 1934: Der Niederwald des linksrheinischen Schiefergebirges. Beitr. Landeskunde Rheinlande 4: 106 S.

–, 1948: Wirkungen des trockenen Sommers 1947 als Forschungsaufgabe. Ber. Deut. Landeskunde 5: 37–52.

–, 1950: Die Dürreempfindlichkeit der mitteleuropäischen Wirtschaftslandschaft in Vergangenheit und Gegenwart. Deutscher Geographentag München 1948, 27: 129–145.

*–, 1959: Allgemeine Vegetationsgeographie. Berlin, 261 S., 3. Aufl. 1968.

–, 1976: Atlas zur Biogeographie. Bibliographisches Institut, Mannheim/Zürich: 80 S.

SCHMITT, R., 1936: Die waldbauliche und bodenkundliche Bedeutung der Bodenflora des Buchenwaldes im Hoch-Spessart. Würzburg: 74 S.

SCHMUCKER, TH., 1942: Die Baumarten der nördlich-gemäßigten Zone und ihre Verbreitung. Berlin, 250 S.

–, DRUDE, G., 1934: Verbreitungsgesetze bei Pflanzen, besonders Allium ursinum. Beih. Botan. Cbl. 52 A: 240–565.

SCHNETTER, M.-L., 1965: Frostresistenzuntersuchungen an Bellis perennis, Plantago media und Helleborus niger im Jahreslauf. Biol. Zbl. 84: 469–487.

SCHNOCK, G., 1967 a: Recherches sur l'écosystème forêt. Contrib. No. 11 Bull. Inst. Roy. Sci. Nat. Belg. 43, No. 35: 15 S.

–, 1967 b: Contrib. No. 12. Thermisme comparé de l'habitat, dans la forêt et la prairie permanente (1). Ebenda 43, 36: 17 S.

–, 1972: Interception des précipitations par les colonies de Mercurialis perennis L. Bull. Soc. Roy. Botan. Belg. 105: 151–156.

SCHOLZ, H., 1960: Die Veränderungen in der Ruderalflora Berlins. Willdenowia (Berlin-Dahlem) 2: 379–397.

SCHOLZ, J., 1980: Mineralstickstoff-Nachlieferung in sauren Buchenwäldern Niedersachsens. Staatsexamensarbeit (Geobotanik) Göttingen, 94 S.

SCHÖNBECK, H., 1974: Nachweis schwermetallhaltiger Immissionen durch ausgewählte pflanzliche Indikatoren. VDI-Ber. 203: 75–87.

SCHÖNENBERGER, W., 1978: Ökologie der natürlichen Verjüngung von Fichte und Bergföhre in Lawinenzügen der nördlichen Voralpen. Mitt. Schweiz. Anst. Forstl. Versuchsw. 54: 217–320.

SCHÖNFELDER, P., 1967: Das Galeopsietum angustifoliae Büker 1942 – eine Kalkschuttpioniergesellschaft Nordbayerns. Mitt. Florist.-Soziol. Arb. gem. N.F. 11/12: 5–10.

–, 1968: Adalpin – dealpin, ein historisch-chorologisches Begriffspaar. Ebenda 13, 5–9.

–, 1970: Die Blaugras-Horstseggenhalde und ihre arealgeographische Gliederung in den Ostalpen. Jahrb. Ver. Schutze Alpenpflanzen u. -Tiere 35: 10 S.

–, 1972: Systematisch-arealkundliche Gesichtspunkte bei der Erfassung historisch-geographischer Kausalitäten der Vegetation, erläutert am Beispiel des Seslerio-Caricetum sempervirentis in den Ostalpen. In: TÜXEN, R. (Hrsg.): Grundfragen und Methoden der Pflanzensoziologie. Dr. W. Junk, Den Haag: 279–290.

–, 1976: Vegetationsverhältnisse auf Gips im südwestlichen Harzvorland. Naturschutz u. Landschaftspflege Niedersachsen, 149 S.

SCHÖNHAR, S., 1952: Untersuchungen über die Korrelation zwischen der floristischen Zusammensetzung der Bodenvegetation und der Bodenazidität sowie anderen chemischen Bodenfaktoren. Mitt. Ver. Forst. Standortskartierung 2: 1–23.

–, 1953: Die ökologischen Artengruppen. Ebenda 3: 26–28.

–, 1954: Die Bodenvegetation als Standortsweiser. Allg. Forst- u. Jagdztg. 125: 259–265.

SCHÖNNAMSGRUBER, H., 1959: Mineralstoffuntersuchungen an Waldgesellschaften Baden-Württembergs. Ber. Deut. Botan. Ges. 72: 220–229.

SCHRATZ, E., 1934–36: Beiträge zur Biologie der Halophyten I–III. Jb. Wiss. Bot. 80: 113–189.

SCHREIBER, K. F., 1962: Über die standortsbedingte und geographische Variabilität der Glatthaferwiesen in Südwest-Deutschland. Ber. Geobot. Inst. ETH, Stiftg. Rübel, Zürich 33: 65–128.

SCHREITLING, K.-T., 1959: Beiträge zur Erklärung der Salzvegetation in den nordfriesischen Kögen. Mitt. Arb. gem. Floristik Schlesw.-Holst. u. Hamburg 8: 98 S.

SCHRETZENMAYR, M., 1957: Die Wald- und Forstgesellschaften im westthüringischen Buntsandsteinbezirk. Arch. Forstwes. 6: 481–573.

SCHROEDER, F.-G., 1973: Westerhof, ein natürliches Fichtenvorkommen westlich des Harzes. Mitt. Deut. Dendrol. Ges. 66: 9–38.

–, 1974: Waldvegetation und Gehölzflora in den Südappalachen (USA). Ebenda 67: 128–163.

SCHRÖTER, C., 1926: Das Pflanzenleben der Alpen. 1. Aufl. 1912, 2. Aufl. Verl. Albert Raustein, Zürich: 1288 S.

SCHROTT, R., 1974: Verlandungsgesellschaften der Weiher um Eschenbach und Tirschenreuth und Vergleich der Verlandungszonen. Hoppea, Denkschr. Regensburg. Botan. Ges. 33: 247–310.

SCHUBERT, R., 1954: Die Schwermetallpflanzengesellschaften des östlichen Harzvorlandes. Wiss. Z. Univ. Halle, Math.-Nat. 3: 52–70.

–, 1960: Die Zwergstrauchreichen azidiphilen Pflanzengesellschaften Mitteldeutschlands. Pflanzensoziol. (Jena) 11: 235 S.

–, 1969: Die Pflanzengesellschaften der Elster-Luppe-Aue und ihre voraussichtliche Strukturänderung bei Grundwasserabsenkung. Wiss. Z. Univ. Halle 18: 125–162.+

–, 1972: Übersicht über die Pflanzengesellschaften des südlichen Teiles der DDR. III. Wälder, Teil 3. Hercynia N.F. 9: 197–228.+

–, 1974: Übersicht über die Pflanzengesellschaften des südlichen Teiles der DDR. IX. Mauerpfefferreiche Pionierfluren. Hercynia N.F. 11: 201–214.

–, HILBIG, W., MAHN, E.-G. (Hrsg.), 1973: Probleme der Agrogeobotanik. Wiss. Beitr. Univ. Halle-Wittenberg 1973/11 (P 2): 213 S.

–, KLEMENT, O., 1961: Die Flechtenvegetation des Brocken-Blockmeeres. Arch. Naturschutz u. Landschaftsforsch. 1: 18–38.

–, KÖHLER, H., 1964: Vegetationskundliche Untersuchungen in der mitteldeutschen Ackerlandschaft. Die Pflanzengesellschaften im Einzugsgebiet der Luhne im Bereich des oberen Unstruttales. Wiss. Z. Univ. Halle 13, Sonderb. Botan.: 3–51.

–, MAHN, E.-G., 1968: Übersicht über die Ackerunkrautgesellschaften Mitteldeutschlands. Feddes Repert. 80: 133–304.

–, u. Mitarb., 1961: Botanische Exkursionen im Ostharz und im nördlichen Thüringen. Akademischer Verlag, Halle (Saale): 109 S.+

SCHUBERT, W., 1963: Die Sesleria varia-reichen Pflanzengesellschaften in Mitteldeutschland. Feddes Repert. Beih. 140: 71–199.

SCHÜTTE, K., 1939: Sinkendes Land an der Nordsee? Ferd. Rau, Öhringen: 144 S.

SCHULZE, E.-D., 1970: Der CO_2-Gaswechsel der Buche (Fagus silvatica L.) in Abhängigkeit von den Klimafaktoren im Freiland. Flora 159: 177–232.

–, LANGE, O.L., 1968: CO_2-Gaswechsel der Flechte Hypogymnia physodes bei tiefen Temperaturen im Freiland. Flora B 158: 180–184.

SCHWAAR, J., 1973: Hochmoorgrünland, seine pflanzensoziologische und ökologische Zuordnung. Z. Kulturtechnik u. Flurbereinigung 14: 197–203.

SCHWABE, A., 1975: Dauerquadrat-Beobachtungen in den Salzwiesen der Nordseeinsel Frischen. Mitt. Florist.-Soziol. Arb. gem. N.F. 18: 111–128.

SCHWABE-BRAUN, A., 1980: Eine pflanzensoziologische Modelluntersuchung als Grundlage für Naturschutz und Planung. Urbs et Regio (Kassel) 18, 212 S.

SCHWARZ, W., 1968: Der Einfluß der Temperatur und Tageslänge auf die Frosthärte der Zirbe. Deut. Akad. Landwirtschaftswiss. Berlin, Tagungsber. 100: 55–63.

–, 1970: Der Einfluß der Photoperiode auf das Austreiben, die Frosthärte und die Hitzeresistenz von Zirben und Alpenrosen. Flora 159: 258–285.

SCHWEINGRUBER, F., 1972: Die subalpinen Zwergstrauchgesellschaften im Einzugsgebiet der Aare (schweizerische nordwestliche Randalpen). Mitt. Schweiz. Anst. Forstl. Versuchswes. 48, 2: 200–504.

–, 1974: Föhrenwälder im Berner Oberland und am Vierwaldstättersee. Ber. Schweiz. Botan. Ges. 83: 175–204.

SCHWENKE, H., 1964: Vegetation und Vegetationsbedingungen in der Kieler Ostsee (Kieler Bucht). Kieler Meeresforsch. 20: 157–168.

–, 1969: Meeresbotanische Untersuchungen in der westlichen Ostsee als Beitrag zu einer marinen Vegetationskunde. Internat. Rev. Ges. Hydrobiol. 54: 35–94.

*SCHWERDTFEGER, F., 1975: Ökologie der Tiere, Bd. III. Synökologie. Verlag Paul Paray, Hamburg-Berlin: 451 S.

SCHWICKERATH, M., 1942: Bedeutung und Gliederung des Differentialartenbegriffs in der Pflanzensoziologie. Beih. Botan. Cbl. 61 B: 351–383.

–, 1944: Das Hohe Venn und seine Randgebiete. Pflanzensoziol. (Jena) 6: 278 S.

–, 1951: Letzte Hartauenwälder der Erfttrockenmulde. Naturschutz u. Landschaftspflege Nordrhein-Westf. 1951: 1–32.

–, 1953: Die Studienfahrt zum Kermeter und Schülerbeiträge zu dem Thema „Westdeutsche Hecken und Heckenlandschaften". In: Wasser u. Boden in der Landschaftspflege. Ratingen: 60–103.

–, 1954: Die Landschaft und ihre Wandlung, auf geobotanischer und geographischer Grundlage entwickelt und erläutert im Bereich des Meßtischblattes Stolberg. Aachen: 118 S.+

–, 1958: Die wärmeliebenden Eichenwälder des Rheinstromgebietes und ihre Beziehungen zu den verwandten Wäldern Österreichs. Schr. Ver. Verbreitung Naturw. Kenntnisse Wien 98: 85–112.

–, GALLHOFF, E., RADKE, G. J., 1969: Die florengeographische und vegetationskundliche Gliederung des Naturschutzgebietes „Wollerscheider Venn", Kreis Monschau. Schr.reihe Landesstelle Naturschutz u. Landschaftspflege Nordrhein-Westfalen 6: 39–68.+

SCHWOERBEL, J., 1980: Methoden der Hydrobiologie. 2. Aufl., UTB 979, Gustav Fischer Verlag Stuttgart.

SEAWARD, M.R.D. (Ed.), 1977: Lichen ecology. Academic Press, London, 550 S.

SEBALD, O., 1956: Über Wachstum und Mineralstoffgehalt von Waldpflanzen in Wasser- und Sandkulturen bei abgestufter Azidität. Mitt. Württemb. Forstl. Vers. Anst. 13: 3–83.

–, 1964: Zur Ökologie und Soziologie der Simsenlilie (Tofieldia calyculata L. Wahl) im Muschelkalkgebiet des oberen Neckarlandes. Jh. Ver. Vaterl. Naturk. Württemb. 118/119: 287–292.

–, 1975: Zur Kenntnis der Quellfluren und Waldsümpfe des Schwäbisch-Fränkischen Waldes. Beitr. Naturk. Forsch. Südw.-Deut. 34: 295–327.

–, 1980: Über einige interessante Ausbildungen der Vegetation auf moosreichen Felsschutthalden im oberen Donautal (Schwäbische Alb). Veröff. Naturschutz Landschaftspflege Bad.-Württ. 51/52: 451–477.

SEIBERT, P., 1954: Die Wald- und Forstgesellschaften im Graf Görtzischen Forstbezirk Schlitz. Angew. Pflanzensoziol. (Stolzenau/Weser) 9: 63 S.

–, 1958: Die Pflanzengesellschaften im Naturschutzgebiet „Pupplinger Au". Landschaftspflege u. Vegetationskunde (München) 1: 79 S.

–, 1965: Die Auswirkung des Donau-Hochwassers 1965 auf Ackerunkrautgesellschaften. Mitt. Florist.-Soziol. Arb. gem. N.F. 14: 121–135.

–, 1966: Der Einfluß der Niederwaldwirtschaft auf die Vegetation. In: R. TÜXEN (Hrsg.) Anthropogene Vegetation. Dr. W. Junk, Den Haag: 336–346.

–, 1968: Vegetation und Landschaft in Bayern. Erläuterungen zur Übersichtskarte der natürlichen Vegetationsgebiete von Bayern. Erdkunde 22: 294–313.+

–, 1969: Über das Aceri-Fraxinetum als vikariierende Gesellschaft des Galio-Carpinetum am Rande der bayerischen Alpen. Vegetatio 17: 165–175.

–, 1974: Die Belastung der Pflanzendecke durch den Erholungsverkehr. Forstwiss. Cbl. 93: 35–43.

–, 1975: Veränderung der Auenvegetation nach Anhebung des Grundwasserspiegels in den Donauauen bei Offingen. Beitr. Naturk. Forsch. Südw.-Deut. 34: 329–343.

–, HAGEN, J., 1974: Zur Auswahl von Waldreservaten in Bayern. Forstw. Cbl. 93: 274–284.+

SEIDEL, K., 1966: Reinigung von Gewässern durch höhere Pflanzen. Naturwissenschaften 53: 289–297.

SEIFERT, J., 1962: The effects of winter on the number of bacteria and nitrification power of soils. II. Acta Univ. Carol. Biol. Suppl. 1962: 41–49.

SENDTNER, O., 1860: Die Vegetations-Verhältnisse des Bayerischen Waldes … Literarisch-artistische Anstalt, München, 505 S.

ŠERCELJ, A., 1970: Das Refugialproblem und die spätglaziale Vegetationsentwicklung im Vorfeld des Südostalpenraumes. Mitt. Ostalp.-Dinar. Pflanzensoz. Arbeitsgem. 10: 76–78.

SHARIFI, M.R., 1978: Ökologisches und physiologisches Verhalten von Alopecurus pratensis, Arrhenatherum elatius und Bromus erectus bei unterschiedlicher Wasser- und Stickstoff-Versorgung. Diss. Univ. Göttingen: 109 S.

SHIMWELL, D.W., 1971: Festuco-Brometea. Br.-Bl. et R. Tx. 1943 in the British Isles: The phytogeography and phytosociology of limestone grasslands. Vegetatio 23: 1–60.

SIEGHARDT, H., 1973: Strahlungsnutzung von Phragmites communis. In: H. ELLENBERG (Hrsg.): Ökosystemforschung. Springer-Verl. Berlin-Heidelberg-New York: 79–86.

SIEGRIST, R., 1913: Die Auenwälder der Aare mit besonderer Berücksichtigung ihres genetischen Zusammenhanges mit anderen flußbegleitenden Pflanzengesellschaften. Jb. Aargauisch. Naturf. Ges. 1913: 182 S.

SIMONIS, W., 1948: CO_2-Assimilation und Xeromorphie von Hochmoorpflanzen in Abhängigkeit vom Wasser- und Stickstoffhaushalt des Bodens. Biol. Zbl. 67: 77–83.

SISSINGH, G., 1950: Onkruid associaties in Nederland. Versl. Landbouwk. Ond. 56, 15, 224 S.

–, 1970: Dänische Buchenwälder. Vegetatio 21: 245–254.

SJÖGREN, E., 1964: Epilithische und epigäische Moosvegetation in Laubwäldern der Insel Öland. Acta Phytogeogr. Suecica 48: 184 S.

SJÖRS, H., 1950: On the relation between vegetation and electrolytes in North Svedish mire waters. Oikos 2: 241–258.

–, 1954: Meadows in Grangärde Finnmark, SW Dalarna, Sweden. Acta Phytogeogr. Suecica 34: 135 S.

SKYKE, E., 1968: Lichens and air pollution. A study of cryptogamic epiphytes and environment in the Stockholm region. Acta Phytogeogr. Suecica 52: 123 S.

SLAVÍK, B., LHOTSKÁ, M., 1967: Chorologie und Verbreitungsbiologie von Echinocystis lobata (Michx) Torr. et Gray mit besonderer Berücksichtigung ihres Vorkommens in der Tschechoslowakei. Folia Geobot. Phytotax. (Praha) 2: 255–282.

SLAVÍKOVÁ, J., 1958: Einfluß der Buche (Fagus silvatica L.) als Edifikator auf die Entwicklung der Krautschicht in Buchenphytozönosen. Preslia 30: 19–42.

SLOBODDA, S., 1979: Die Moosvegetation ausgewählter Pflanzengesellschaften des NSG „Peenewiesen bei Gützkow" unter Berücksichtigung der ökologischen Bedingungen eines Flußtalmoor-Standortes. Feddes Repert. 90: 481–518.

ŠMARDA, J., u. Mitarb., 1963: Sekundäre Pflanzengesellschaften im Schutzgebiet der Hohen Tatra. Biblioth. Samml. Studien Tatra-Nationalpark 4: 219 S.

SMETÁNKOVÁ, M., 1959: Dry matter production and growth in length of overground parts of Carex humilis Leyss. Biol. Plant. (Praha) 1: 235–247.

SMIDT, J.T. DE, 1966: The inland-heath communities of the Netherlands. Wentia 15: 142–162.

SNAYDON, R.W., 1962 a: Micro-distribution of Trifolium repens L. and its relation to soil factors. J. Ecol. 50: 133–143.

–, 1962 b: The growth and competitive ability of contrasting natural populations of Trifolium repens L. on calcareous and acid soils. J. Ecol. 50: 439–447.

–, 1970: Rapid population differentiation in a mosaic environment. I. The response of Anthoxanthum odoratum populations to soils. Evolution 24: 257–269.

–, BRADSHAW, A.D., 1962: Differences between natural populations of Trifolium repens L. in response to mineral nutrients. J. Exper. Botan. 13: 422–434.

SØCHTING, U., JOHNSON, I., 1974: Changes in the distribution of epiphytic lichens in the Copenhagen area from 1936 to 1972. Botan. Tidskr. 69: 60–63.

SOKOŁOWSKI, A.W., 1962: L'épaisseur de la couverture de neige et la profondeur de la congélation du sol dans les associations forestières du parc national de Białowieża. Ochrony Przyrody 28: 111–135.

–, 1963: Waldgesellschaften des südöstlichen Teiles des Masowien-Tieflandes. Monogr. Botan. 16: 176 S.+

–, 1965: Forest associations of the Laska Forestry District in Bory Tucholskie (Tuchola Forest). Fragm. Florist. Geobot. 11: 97–119.+

–, 1966 a: Vegetation in the forest reserve „Debowo" on the territory of the forest district Sadłowo, Olsztyn Voivodship. Prace Inst. Badaw. Leśn. (Warszawa) 303: 2–44.

–, 1966 b: Phytosociological characteristic of coniferous woods of Dicrano-Pinion alliance of Białowieża forest. Ebenda 305: 72–105.

–, 1966 c: Phytosociological character of spruce woods of Białowieża forest. Ebenda 304: 46–69.

–, 1970: Phytosociological characteristics of Cladonia-pine woods in Poland and their position in systematics. Ebenda 368: 3–13.

SOMMER, W.-H., 1970: Das „cladonietosum-Problem" in Silikattrockenrasen. Herzogia 2: 116–122.

–, 1971: Wald- und Ersatzgesellschaften im östlichen Niedersachsen. Diss. Botan. 12: 101 S.

SOÓ, R., 1930: Vergleichende Vegetationsstudien – Zentralalpen-Karpathen-Ungarn – nebst kritischen Bemerkungen zur Flora der Westkarpathen. Veröff. Geobot. Inst. Rübel, Zürich 6: 237–322.

–, 1960: Übersicht der Waldgesellschaften und Waldtypen Ungarns. Különlenyomat Az Erdö (Budapest) 1960: 321–340

–, 1962, 1963: Systematische Übersicht der pannonischen Pflanzengesellschaften V. u. VI. Die Gebirgswälder. Acta Bot. Acad. Sci. Hung. 8: 335–366, u. 9, 123–150.

*–, 1964 a: Synopsis systematico-geobotanica florae vegetationisque Hungariae I. Akadémiai Kiadó, Budapest: 589 S.

*–, 1964 b: Die regionalen Fagion-Verbände und Gesellschaften Südosteuropas. Studia Biol. Hung. (Budapest) 1: 104 S.

*–, 1974: Die Pflanzengesellschaften der mitteleuropäischen Buchenwälder in Ungarn. Acta Botan. Acad. Sci. Hung. 20: 335–377.

SOUGNEZ, N., 1965: Réactions floristiques d'une lande humide aux fumures minérales. Oecol. Plant. 1: 219–234.

SÖYRINKI, N., 1954: Vermehrungsökologische Studien in der Pflanzenwelt der Bayerischen Alpen I. Ann. Botan. Soc. Vanamo 27, 1: 232 S.

ŠPÁNIKOVÁ, A., 1971: Phytozönologische Studie der Wiesen des südwestlichen Teils des Talkessels Košická Kotlina. Biol. Práce (Bratislava) 17, 2: 103 S.

SPARLING, J.H., 1967: The occurence of Schoenus nigricans L. in blanket bogs. I. Environmental conditions affecting the growth of S. nigricans in blanket bog. II. Experiments on the growth of S. nigricans under controlled conditions. J. Ecol. 55: 1–14 u. 15–32.

SPATZ, G., 1970: Pflanzengesellschaften, Leistungen und Leistungspotential von Allgäuer Alpweiden in Abhängigkeit von Standort und Bewirtschaftung. Diss. T.H. München 1970: 155 S.

–, 1974: Die wirtschaftliche und ökologische Bedeutung der Almweiden. Antrittsvorlesung Weihenstephan (unveröff.): 13 S.

–, 1975: Die Almen in ihrer Bedeutung als bewirtschaftete Ökosysteme. „Das wirtschaftseigene Futter", DLG-Verlag, 21: 264–273.

–, PLETL, L., MANGSTL, A., 1979: Programm OEK-SYN zur ökologischen und synsystematischen Auswertung von Pflanzenbestandsaufnahmen. Scripta Geobot., Göttingen 9, 2. Aufl.: 29–36.

SPEIDEL, B., 1972: Das Wirtschaftsgrünland der Rhön. Vegetation, Ökologie und landwirtschaftlicher Wert (mit einer Vegetationskarte). Ber. Naturwiss. Ges. Bayreuth 14: 201–240. +

–, WEISS, A., 1971: Zur ober- und unterirdischen Stoffproduktion einer Goldhaferwiese bei verschiedener Düngung. Angew. Botan. 46: 75–93.

–, –, 1974: Untersuchungen zur Wurzelaktivität unter einer Goldhaferwiese. Ebenda 48: 137–154.

STÄHLIN, A., BOMMER, D., 1958: Grünlandwirtschaftliche Untersuchungen an binnendeutschen Salzstandorten. Z. Acker- u. Pflanzenbau 106: 321–336.+

–, STÄHLIN, L., SCHÄFER, K., 1972: Über den Einfluß des Alters der Sozialbrache auf Pflanzenbestand, Boden und Landschaft. Z. Acker- u. Pflanzenbau 136: 177–199.

STAMM, E., 1938: Die Eichen-Hainbuchenwälder der Nordschweiz. Beitr. Geobot. Landesaufn. Schweiz. 22: 163 S.

STEEN, E., 1957: The influence of exposure and slope upon vegetation and soil in a natural pasture. Stat. Jordbruksförs. Medd. 86: 54 S.

STEFFEN, H., 1931: Vegetationskunde von Ostpreußen. Pflanzensoziol. (Jena) 1: 406 S.

–, 1936: Ostpreußens Eichenwälder. Beih. Botan. Cbl. 55 B: 182–250.

STEINER-HAREMAKER, I., STEINER, D., 1961: Zur Verbreitung und geographischen Bedeutung der Grünhecken in der Schweiz. Geographica Helvet. 1961: 61–76.

STEINFÜHRER, A., 1945: Die Salzgesellschaften der Schleiufer und ihre Beziehungen zum Salzgehalt des Bodens. Diss. Univ. Kiel.

STEINHARDT, U., 1973: Input of chemical elements from the atmosphere. A tabular review of literature. Göttinger Bodenkundl. Ber. 29: 93–132.

STEPHAN, B. u. S., 1971: Die Vegetationsentwicklung im Naturschutzgebiet Stolzenburg und ihre Bedeutung für die Schutzmaßnahmen. Decheniana 123: 281–305.

STERN, R., 1965: Anlage und Ergebnisse von Versuchspflanzungen in der subalpinen Entwaldungszone Nordtirols. Mitt. Forstl. Bundes-

Versuchsanst. Mariabrunn 66: 215–239.

Steubing, I., 1949: Beiträge zur Ökologie der Wurzelsysteme von Pflanzen des flachen Sandstrandes. Z. Naturforsch. 4 b: 114–123.

–, 1960: Wurzeluntersuchungen an Feldschutzhecken. Z. Acker u. Pflanzenbau 110: 332–341.

–, Alberdi, M., 1973: The influence of phosphorus deficiency on the sclerophylly. Oecol. Plant. 8: 211–218.

–, Dapper, H., 1964: Der Kreislauf des Chlorids im Meso-Ökosystem einer binnenländischen Salzwiese. Ber. Deut. Botan. Ges. 77: 71–74.

–, Westhoff, V., 1966: Kationenaustauschkapazität der Wurzeln und Nährstoffpotential des Bodens in psammophilen und halophilen Pflanzengesellschaften der niederländischen Meeresküste. Vegetatio 13: 293–301.

*Stewart, W.D.P., 1970: Algal fixation of atmospheric nitrogen. Plant and Soil 32: 555–588.

Stöcker, G., 1962: Vorarbeit zu einer Vegetationsmonographie des Naturschutzgebietes Bodetal. Wiss. Z. Univ. Halle, Math. Nat. 11: 897–936. +

–, 1965 a: Eine neue Zwergstrauch-Gesellschaft aus dem Naturschutzgebiet „Oberharz". Arch. Naturschutz 5: 111–115.

–, 1965 b: Vegetationskomplexe auf Felsstandorten, ihre Auflösung und Systematisierung der Komponenten. Feddes Repert. 142: 222–236.

–, 1967: Der Karpatenbirken-Fichtenwald des Hochharzes. Eine vegetationskundlich-ökologische Studie. Pflanzensoziol. (Jena) 15: 123 S.

–, 1968: Konzentration löslichen NH_4-N in organischen Horizonten naturnaher Berg-Fichtenwälder. Flora B 158: 41–59.

–, 1979: Pflanzenverfügbarer Stickstoff in naturnahen Berg-Fichtenwäldern. Dynamik und Regulation. Abstracta Botan. (Budapest) 6, Suppl. 1: 53–88.

–, 1980: Beiträge zur ökologischen Charakterisierung naturnaher Berg-Fichtenwälder. Arch. Naturschutz Landschaftsforsch. 20: 65–90.

–, Bergmann, A., 1975: Ergebnisse eines Modellversuchs zur quantitativen Erfassung von Umweltänderungen I. Versuchsobjekt, Methodik, univariate Analyse. Flora 164: 145–167.

Stocker, O., 1923: Die Transpiration und Wasserökologie nordwestdeutscher Heide- und Moorpflanzen am Standort. Z. Botan. 15: 1–41.

–, 1925: Beiträge zum Halophytenproblem II. Standort und Transpiration der Nordsee-Halophyten. Z. Botan. 17: 1–24.

–, 1928: Das Halophytenproblem. Ergebn. Biol. 3: 265–353.

–, 1960: Einige Bemerkungen über die Salzstandorte östlich des Neusiedler Sees. Verh. Zool.-botan. Ges. Wien 100: 106–111.

–, 1967: Der Wasser- und Photosynthese-Haushalt mitteleuropäischer Gräser, ein Beitrag zum allgemeinen Konstitutionsproblem des Grastypus. Flora, Abt. B 157: 56–96.

–, 1970: Transpiration und Wasserhaushalt in verschiedenen Klimazonen. IV. Untersuchungen an Sandpflanzen der Ostseeküste. Flora 159: 367–409.

Stoffler, H.-D., 1975: Zur Kenntnis der Tannen-Mischwälder auf Tonböden zwischen Wutach und Eyach (Pyrolo-Abietetum Oberd. 1957). Beitr. Naturk. Forsch. Südw.-Deut. 34: 357–370.

Straka, H., 1963: Über die Veränderungen der Vegetation im nördlichen Teil der Insel Sylt in den letzten Jahrzehnten. Schr. Naturw. Ver. Schlesw.-Holst. 34: 19–43.

–, 1973: La végétation des landes d'Europe occidentale (Nardo-Callunetea). In: Géhu, J.-M. (Ed.). Coloque de l'Amicale Phytosociologique, 1–3 Octobre 1973, Lille: 243–245.

Straškraba, M., 1963: Share of the littoral region in the productivity of two fishponds in Southern Bohemia. Rozpr. Československ. Akad. Věd, Řada Mat. Přír. Věd 73, No. 13: 64 S.

Strautz, W., 1959: Früheisenzeitliche Siedlungsspuren in einem älteren Auelehm des Wesertales bei Wellie (Kreis Nienburg). „Die Kunde" (Hildesheim), N.F. 10: 69–86.

–, 1962: Auelehmbildung und -gliederung im Weser- und Leinetal mit vergleichenden Zeitbestimmungen aus dem Flußgebiet der Elbe. Beitr. Landschaftspflege (Stuttgart) 1: 273–314.

Strebel, O., 1960: Mineralstoffernährung und Wuchsleistung von Fichtenbeständen (Picea abies) in Bayern. Forstw. Cbl. 79: 17–42.

Streitz, H., 1967: Bestockungswandel in Laubwaldgesellschaften des Rhein-Main-Tieflandes und der Hessischen Rheinebene. Diss. Hann. Münden 1967: 304 S.

–, 1968: Verbreitung, Standortansprüche und soziologisches Verhalten der Wimpersegge (Carex pilosa Scop.) in Oberhessen. Hess. Florist. Briefe 17: 11–18.

Studer, P., 1962: Grünlandgesellschaften bei Dättlikon am Irchel und ihr Schutz. Manuskr. Geobot. Inst. ETH Zürich (unveröff.).

Succow, M., 1967: Pflanzengesellschaften der Zieseniederung (Ostmecklenburg). Natur u. Naturschutz Meckl. 5: 79–108.

Sukopp, H., 1959: Vergleichende Untersuchungen der Vegetation Berliner Moore unter besonderer Berücksichtigung der anthropogenen Veränderungen. Botan. Jb. 79: 36–126. +

–, 1962: Das Naturschutzgebiet Teufelsbruch in Berlin-Spandau. Sitzber. Ges. Naturforsch. Freunde Berlin, N.F. 2: 38–49. +

–, 1969: Der Einfluß des Menschen auf die Vegetation. Vegetatio 17: 360–371.

–, 1972: Wandel von Flora und Vegetation in Mitteleuropa unter dem Einfluß des Menschen. Ber. Landwirtsch. 50: 112–139.

–, Kunick, W., 1969: Veränderungen des Röhrichtbestandes der Berliner Havel 1962–1967. Berliner Naturschutzblätter 13: 303–313 u. 332–344.

–, Kunick, W., Runge, M., Zacharias, F., 1973: Ökologische Charakteristik von Großstädten. Dargestellt am Beispiel Berlins. Verh. Ges. Ökol. Saarbrücken 1973: 383–403.

–, Markstein, B., Trepl, L., 1975: Röhrichte unter intensivem Großstadteinfluß. Beitr. Naturk. Forsch. Südw.-Deut. 34: 371–385.

–, Trautmann, W., 1976: Veränderungen der Flora und Fauna in der Bundesrepublik Deutschland. Schr. R. Vegetationskde. 10: 409 S.

Summerfield, R.J., Riley, J.O., 1974: Growth of Narthecium ossifragum in relation to the dissolved oxygen concentration of the rooting substrate. Plant and Soil 41: 701–705.

Surber, E., Amiet, R., Kobert, H., ca. 1975: Das Brachenproblem in der Schweiz. Eidg. Anst. Forstl. Versuchswes., Ber. 112.

Sutter, R., 1962: Das Caricion austroalpinae. Ein neuer insubrisch-südalpiner Seslerietalia-Verband. Mitt. Ostalpin-Dinar. Pflanzensoz. Arb. gem. 2: 18–22.

–, 1969: Ein Beitrag zur Kenntnis der soziologischen Bindung südsüdostalpiner Reliktendemismen. Acta Botan. Croat. 28: 349–366.

Szabó, I.M., 1974: Microbial communities in a forest-rendzina ecosystem. Akadémiai Kiadó, Budapest, 415 S.

Szafer, W., 1924: Zur soziologischen Auffassung der Schneetälchenassoziationen. Veröff. Geobot. Inst. Rübel, Zürich 1: 300–310.

*–, (Edit.), 1966: The vegetation of Poland. Warszawa (PWN) 1966: 738 S.

*–, u. Mitarb., 1962: Tatrzánski Park Narodowy. Polska Akad. Nauk, Wydawn. Popularn. 21, Kraków 1962: 675 S. +

*–, Zarzycki, K. (Hrsg.) 1972: Szata roślinna Polski. 2 Bde. Warszawa, Państwowe Wydawnictwo Naukowe: 615 u. 347 S.

Tamm, C.O., 1953: Growth, yield and nutrition in carpets of a forest moss (Hylocomium splendens). Medd. Stat. Skogsf. Inst. 43, 1: 140 S.

–, 1965: Some experiences from forest fertilization trials in Sweden. Silva Fenn. 117: 24 S.

–, 1972: Survival and flowering of perennial herbs III. The behaviour of Primula veris on permanent plots. Oikos 23: 159–166.

–, 1974: Experiments to analize the behaviour of young spruce forest at different nutrient levels. 1st Internat. Congr. Ecology, The Hague, Sept. 1974: 7 S.

–, 1975: The behaviour of young spruce and pine forest kept at different nitrogen regimes. XII. Internat. Botan. Congr. (Leningrad) Abstr. 1: 170.

–, Holmen, H., 1967: Some remarks on soil organic matter turn-over in Swedish podzol profiles. Medd. Norske Skogsforsøkswes. 85: 69–88.

Tanghe, M., 1970: La végétation forestière de la vallée de la Semois ardennaise, 2me et 3me partie. Bull. Inst. Roy. Sci. Nat. Belg. 46/16: 60 S., u. 46/30: 76 S.

–, 1971: Étude d'un transect topo-lithologique de la région d'Éprave-Rochefort (vallée de la Lomme) pour la délimitation des groupes écologiques forestiers de la Calestienne. Bull. Soc. Roy Belg. 104: 333–371.

Tansley, A.G., 1939: The British Islands and their vegetation. Cambridge Univ Press: 930 S.

Thellung, A., 1925: Kulturpflanzen-Eigenschaften bei Unkräutern. Veröff. Geobot. Inst. Rübel, Zürich 3: 745–762.

Thiele, A., 1974: Luftverunreinigungen und Stadtklima im Großraum München, insbesondere in ihrer Auswirkung auf epixyle Testflechten. Bonner Geogr. Abh. 49: 175 S.

–, K., 1978: Vegetationskundliche und pflanzenökologische Untersuchungen im Wimbachgries. Aus den Naturschutzgebieten Bayerns (München) 1, 73 S.

Thienemann, A., 1922: Hydrobiologische Untersuchungen an Quellen. Arch. Hydrobiol. 14, 1.

*–, 1925: Die Binnengewässer Mitteleuropas. Die Binnengewässer 1: 225 S.

–, 1950: Verbreitungsgeschichte der Süßwassertierwelt Europas. Die Binnengewässer XVIII: 142–165.

–, 1956: Leben und Umwelt. Vom Gesamthaushalt der Natur. Rowohlts Deutsche Enzyklopädie. Hamburg: 153 S.

Thill, A., 1964: La flore et la végétation du Parc National de Lesse et Lomme. Publ. A.S.B.L. „Ardenne et Gaume", Monogr. 5: 51 S. +

Thomas, A.S., 1963: Further changes in vegetation since the advent of myxomatosis. J. Ecol. 51: 151–183.

Thomaser, J., 1967: Die Vegetation des Peitlerkofels in Südtirol. Veröff. Mus. Ferdinandeum (Innsbruck) 47: 67–119.

Thorn, K., 1958: Die dealpinen Felsheiden der Frankenalb. Sitz. ber. Phys.-Med. Soz. Erlangen 78: 128–199.

Till, O., 1956: Über die Frosthärte von Pflanzen sommergrüner Laubwälder. Flora 143: 499–542.

Tillich, H.-J., 1969 a: Die Ackerunkrautgesellschaften in der Umgebung von Potsdam. Wiss. Z. Pädag. Hochsch. Potsdam 13: 273–320.

–, 1969 b: Über einige interessante Onopordion-Gesellschaften in der Umgebung von Potsdam. Ebenda 13: 321–329.

Timm, R., 1930: Die Frostschäden des Winters 1928/29 in Nordwestdeutschland. Mitt. Florist.-Soziol. Arb. gem. Niedersachsen 2: 116–145.

Tischler, W., 1951: Die Hecke als Lebensraum für Pflanzen und Tiere, unter besonderer Berücksichtigung ihrer Schädlinge. Erdkunde 5.

–, 1980: Biologie der Kulturlandschaft. Gustav Fischer Verlag, Stuttgart: 253 S.

Tolonen, K., 1966: Stratigraphic and rhizopod analysis on an old raised bog, Varrassuo, in Hollola, South Finland. Ann. Botan. Fenn. 3: 147–166.

Traczyk, H. u. T., 1965: Phytosociological characteristics of the research areas of the Institute of Ecology, Polish Academy of Sciences, at Dziekanów Leśny (Kampinos Forest near Warsaw). Fragm. Florist. Geobot. 11: 547–562. +

Traczyk, T., 1968: Studies on the primary production in meadow communities. Ekol. Polska, Ser. A 16: 59–100.

Traitteur-Ronde, G., 1961: Bodenzoologische Untersuchungen von Stickstoff-Formen-Vergleichsversuchen in Baden-Württemberg. Allg. Forst- u. Jagdztg. 132: 303–311.

Tranquillini, W., 1955: Die Bedeutung des Lichtes und der Temperatur für die Kohlensäureassimilation von Pinus cembra-Jungwuchs am hochalpinen Standort. Planta (Berlin) 46: 154–178.

–, 1957: Standortsklima, Wasserbilanz und CO_2-Gaswechsel junger Zirben (Pinus cembra L.) an der alpinen Waldgrenze. Planta (Berlin) 49: 612–661.

–, 1958: Die Frosthärte der Zirbe unter besonderer Berücksichtigung autochtoner und aus Forstgärten stammender Jungpflanzen. Forstwiss. Cbl. 77: 65–128.

–, 1960: Das Lichtklima wichtiger Pflanzengesellschaften. Handb. Pflanzenphys. 5, 2: 304–338.

–, 1968: Dürreresistenz und Anpflanzungserfolg von Junglärchen verschiedenen Entwicklungszustands. In: Deut. Akad. Landwirtschaftswiss. Berlin, Tagungsber. 100: 123–129.

–, 1970: Einfluß des Windes auf den Gaswechsel der Pflanzen. Umschau 1970: 860–861.

–, 1974: Der Einfluß von Seehöhe und Länge der Vegetationszeit auf das cuticuläre Transpirationsvermögen von Fichtensämlingen im Winter. Ber. Deut. Botan. Ges. 87: 175–184.

–, 1979: Physiological ecology of the alpine timberline. Ecol. Studies 31, 137 S.

–, Machl-Ebner, I., 1971: Über den Einfluß von Wärme auf das Photosynthesevermögen der Zirbe (Pinus cembra L.) und der Alpenrose (Rhododendron ferrugineum L.) im Winter. Rep. Kevo Subarctic Res. Stat. 8: 158–166.

– u. Mitarb., 1980: Über das Höhenwachstum von Fichtenklonen in verschiedener Seehöhe. Mitt. Forstl. Bundes-Versuchsanst. Wien 129: 7–25.

Trautmann, W., 1966: Erläuterungen zur Karte der potentiellen natürlichen Vegetation der Bundesrepublik Deutschland, Blatt 85 Minden, Schriftenr. Vegetationskunde (Hiltrup i.W.) 1: 138 S. +

–, 1976: Stand der Auswahl und Einrichtung von Naturwaldreservaten in der Bundesrepublik Deutschland. Natur u. Landschaft 51, 67 bis 72. +

–, 1980: die Bedeutung der Naturwaldreservate für Schutzgebietssysteme. Natur u. Landschaft 55: 132–134 (u. anschließende Berichte).

–, Lohmeyer, W., 1960: Gehölzgesellschaften in der Fluß-Aue der mittleren Ems. Mitt. Florist.-Soziol. Arb. gem., N.F. 8: 227–247.

Tregubov, S.S., 1941: Les forêts vierges montagnardes des Alpes Dinariques. Comm. SIGMA (Montpellier) 78: 116 S.

Trepp, W., 1947: Der Lindenmischwald (Tilieto-Asperuletum taurinae). Beitr. Geobot. Landesaufn. Schweiz 27: 128 S.

–, 1950: Ein Beitrag zur Bonitierungsmethode von Alpweiden. Schweiz. Landwirtsch. Monatsh. 28: 366–371.

Treskin, P.P., Abrazho, V.I., Karpov, V.G., 1975: An attempt of causal analysis of spruce forests ecosystems. XII. Internat. Botan. Congr. (Leningrad) Abstr. 1: 171.

Troels-Smith, J., 1955: Pollenanalytische Untersuchungen zu einigen schweizerischen Pfahlbauproblemen. Monogr. Ur- u. Frühgesch. Schweiz 11: 40–58.

Troll, C., 1951 a: Die Problematik der Heckenlandschaft. Ihr geographisches Wesen und ihre Bedeutung für die Landeskultur. Erdkunde 5: 105–110.

–, 1951 b: Heckenlandschaften im maritimen Grünlandgürtel und im Gäuland Mitteleuropas. Ebenda 5: 152–157.

Tromp, P.H.M., 1968: Aanleg, regeneratie en onderhoud van heidevelden en -tuinen. Tijdschr. Kon. Nederl. Heidemaatschappij 1968: 5–23.

Tschermak, L., 1935 a: Die natürliche Verbreitung der Lärche in den Ostalpen. Mitt. Forstl. Versuchswes. Österr. 43: 361 S.

–, 1935 b: Die wichtigsten natürlichen Waldformen der Ostalpen und des heutigen Österreich. Forstl. Wochenschr. Silva 23: 293–298 u. 401–407.

Turner, H., 1958: Über das Licht- und Strahlungsklima einer Hanglage der Ötztaler Alpen bei Obergurgl und seine Auswirkung auf das Mikroklima und auf die Vegetation. Arch. Meteorol. Geophys. u. Bioklimatol., Ser. B 8: 273–325.

–, 1968: Über den „Schneeschliff" in den Alpen. „Wetter und Leben" 20: 192–200.

–, 1970: Grundzüge der Hochgebirgsklimatologie. In: „Die Welt der Alpen"; Umschau Verlag, Frankfurt a.M.: 170–182.

–, Rochat, P., Streule, A., 1975: Thermische Charakteristik von Hauptstandortstypen im Bereich der oberen Waldgrenze (Stillberg, Dischmatal bei Davos). Mitt. Eidg. Anst. Forstl. Versuchsw. 51: 95–119.

–, Tranquillini, W., 1961: Die Strahlungsverhältnisse und ihr Einfluß auf die Photosynthese der Pflanzen. Mitt. Forstl. Bundes-Versuchsanst. Mariabrunn 59: 59–103.

Tüxen, J., 1958: Stufen, Standorte und Entwicklung von Hackfrucht- und Garten-Unkrautgesellschaften und deren Bedeutung für Ur- und Siedlungsgeschichte. Angew. Pflanzensoziol. (Stolzenau/Weser) 16: 164 S.

Tüxen, R., 1928: Über die Vegetation der nordwestdeutschen Binnendünen. Jber. Geogr. Ges. Hannover 1928: 71–93.

–, 1930: Über einige nordwestdeutsche Waldassoziationen von regionaler Verbreitung. Ebenda 1929: 3–64.

–, 1933: Klimaxprobleme des nw-europäischen Festlandes. Nederl. Kruidk. Arch. 43: 293–309.

–, 1937: Die Pflanzengesellschaften Nordwestdeutschlands. Mitt. Florist.-Soziol. Arb. gem. Niedersachsen 3: 170 S.

*–, 1950 a: Grundriß einer Systematik der nitrophilen Unkrautgesellschaften in der Eurosibirischen Region Europas. Mitt. Florist.-Soziol. Arb. gem., N.F. 2: 94–175.

–, 1950 b: Wanderwege der Flora in Stromtälern. Ebenda 2: 52–53.

–, 1950 c: Neue Methoden der Wald- und Forstkartierung. Ebenda 2: 217–219.

–, 1952: Hecken und Gebüsche. Mitt. Geogr. Ges. Hamburg 50: 85–117.

–, 1954: Über die räumliche, durch Relief und Gestein bedingte Ordnung der natürlichen Waldgesellschaften am nördlichen Rande des Harzes. Vegetatio 5/6: 454–477.

–, 1955: Das System der nordwestdeutschen Pflanzengesellschaften. Mitt. Florist.-Soziol. Arb. gem., N.F. 5: 155–176.

–, 1956 a: Die heutige potentielle natürliche Vegetation als Gegenstand der Vegetationskartierung. Angew. Pflanzensoziol. (Stolzenau/Weser) 13: 5–42.

–, 1956 b: Die Pflanzengesellschaften Nordwestdeutschlands. Bremen (Gartenbauamt): 119 S.

–, 1956 c: Vegetationskarte der Ostfriesischen Inseln: Baltrum. Bundesanst. f. Vegetationskartierung, Stolzenau (Weser).

–, 1957: Die Schrift des Bodens. Angew. Pflanzensoziol. (Stolzenau/Weser) 14: 41 S.

–, 1958: Pflanzengesellschaften oligotropher Heidetümpel Nordwestdeutschlands. Veröff. Geobot. Inst. Rübel, Zürich 33: 207–231.

–, 1960: Zur Systematik der west- und mitteleuropäischen Buchenwälder. Bull. Inst. Agron. et Stat. Rech. Gembloux, hors Ser. 1: 45–58.

*–, 1964: Bibliographia phytosociologica cryptogamica, Pars IV: Lichenes (sine Epiphyta). Excerpta Botan., Sect. B. 6: 208–244.

–, 1967 a: Die potentielle natürliche Vegetation der Dorumer Geest. In: Körber-Grohne, U., 1967: 331–334.

–, 1967 b: Die Lüneburger Heide, Rothenburger Schriften (Rotenburg/Wümme) 26: 3–32.

–, 1967 c: Ausdauernde nitrophile Saumgesellschaften Mitteleuropas. Contrib. Botan. (Cluj) 1967: 431–453.

–, 1974 a: Die Haselünner Kuhweide. Die Pflanzengesellschaften einer mittelalterlichen Gemeindeweide. Mitt. Florist. Soziol. Arb. gem. N.F. 17: 69–102.

*–, 1974b, 1979: Die Pflanzengesellschaften Nordwestdeutschlands. 2., völlig neu bearb. Aufl., Lief. 1. Lehre (Verl. J. Cramer): 207 S., Lief. 2. Ebenda 1979, 212 S.

–, (Hrsg.), 1975: Prodromus der europäischen Pflanzengesellschaften. Lief. 2: Littorelletea uniflorae. J. Cramer, Vaduz, 149 S.

–, Böckelmann, W., 1957: Scharhörn. Die Vegetation einer jungen ostfriesischen Vogelinsel. Mitt. Florist.-Soziol. Arb. gem. N.F. 6/7: 183–204.

–, Diemont, H., 1937: Klimaxgruppe und Klimaxschwarm. Jber. Naturhist. Ges. Hannover 88/89: 73–87.

–, Hülbusch, K.-H., 1971: Bolboschoenetea maritimi. Fragm. Florist. Geobot. 17: 391–407.

–, Knapp, R., 1979: Pflanzensoziologische und vegetationskundliche Länder-Bibliographien 1959–1979. Bearbeitungsstand und Perspektiven. Excerpta Botan. Sect. B 19: 1–10.

–, u. Mitarb., 1957: Die Pflanzengesellschaften des Außendeichslandes von Neuwerk. Mitt. Florist.-Soziol. Arb. gem., N.F. 6/7: 205–234. +

–, Oberdorfer, E., 1958: Eurosibirische Phanerogamen-Gesellschaften Spaniens, mit Ausblicken auf die alpine und die Mediterranregion dieses Landes. Veröff. Geobot. Inst. Rübel, Zürich 32: 328 S.

–, Ohba, T., 1975: Zur Kenntnis von Bach- und Quell-Erlenwäldern (Stellario nemori-Alnetum glutinosae und Ribo sylvestris-Alnetum glutinosae). Beitr. Naturk. Forsch. Südw.-Deut. 34: 387–401.

–, Preising, E., 1951: Erfahrungsgrundlagen für die pflanzensoziologische Kartierung des westdeutschen Grünlandes. Angew. Pflanzensoziol. (Stolzenau/Weser) 4: 28 S.

–, Westhoff, V., 1963: Saginetea maritimae, eine Gesellschaftsgruppe im wechselhalinen Grenzbereich der europäischen Meeresküsten. Mitt. Florist.-Soziol. Arb. gem., N.F. 10: 116–129.

Tyler, G., Gullstrand, Ch., Holmquist, K.-A., Kjellstrand, A.-M., 1973: Primary production and distribution of organic matter and metal elements in two heath ecosystems. J. Ecol. 61: 251–268.

Ubrizsy, G., 1956: Die ruderalen Unkrautgesellschaften Ungarns. II. Studien über Ökologie und Sukzession. Acta Agron. (Budapest) 5: 393–418.

*ULBRICHT, H., BÜTTNER, R., FUNKE, H., GUTTE, P., HEMPEL, W., MÜLLER, G., SCHRETZENMAYR, M., WEISE, G., 1965: Die Pflanzenwelt Sachsens. Ber. Arb.gem. Sächs. Bot. N.F. 5/6, H. 2: 472 S. +

ULLMANN, J., 1977: Die Vegetation des südlichen Maindreiecks. Denkschr. Regensburg. Botan. Ges. 36: 5–190.

ULMER, W., 1937: Über den Jahresgang der Frosthärte einiger immergrüner Arten der alpinen Stufe, sowie der Zirbe und Fichte. Jb. Wiss. Botan. 84: 553–592.

ULRICH, B., MAYER, R., 1973: Systemanalyse des Bioelement-Haushaltes von Wald-Ökosystemen. In: ELLENBERG, H. (Hrsg.), Ökosystemforschung: 165–174.

–, –, KHANNA, P.K., 1979: Deposition von Luftverunreinigungen und ihre Auswirkungen in Waldökosystemen im Solling. Schr. Forstl. Fak. Univ. Göttingen 58, 291 S.

UNGAR, I.A., 1962: Influence of salinity on seed germination in succulent halophytes. Ecology 43: 763–764.

UNGER, F., 1836: Über den Einfluß des Bodens auf die Verteilung der Gewächse, nachgewiesen in der Vegetation des nordöstlichen Tirol's. Rohrmann u. Schweigerd, Wien: 367 S.

VAARAMA, A., 1941: Die Winterschäden im botanischen Garten der Universität Helsinki im Frostwinter 1939–40. Ann. Botan. Soc. Vanamo 16, 4: 48 S.

VALK, Van der, A.G., 1974: Mineral cycling in coastal foredune plant communities in Cape Hattras National Seashore. Ecology 55: 1349–1358.

VANDEN BERGHEN, C., 1951: Landes tourbeuses et tourbières à sphaignes de Belgique (Ericeto-Sphagnetalia Schwickerath 1940). Bull. Soc. Roy. Botan. Belg. 84: 157–226.

–, 1952: Contribution à l'étude des bas-marais de Belgique (Caricetalia fuscae W. Koch 1926). Bull. Jard. Botan. de l'État, Bruxelles 22: 1–64.

–, 1953: Contribution à l'étude des groupements végétaux notés dans la vallée de l'Ourthe en amont de Laroche-en-Ardenne. Bull. Soc. Roy. Botan. Belg. 85: 195–277.

VICHEREK, J., 1962: Typen von Phytozönosen der alluvialen Aue des unteren Thaya-Gebietes mit besonderer Berücksichtigung der Wiesenpflanzengesellschaften. Folia Přírodn. Fakult. Univ. Brno 3, 113 S. +

–, 1964: Phytozönologische Charakteristik der subhalophytischen Wiesenpflanzengesellschaften des pannonischen Gebietes der ČSSR. Publ. Fac. Sci. Univ. Brno 463: 233–248.

VIERECK, L.A., 1966: Plant succession and soil development on gravel outwash of the Muldrow Glacier, Alaska. Ecol. Monogr. 36: 181–199.

VILMOS, M., 1965: Einige ökologische Beziehungen der Periodizität des Samenertrags bei Eiche und Buche. Erdészeti Kutatások (Budapest) 1965: 99–121.

VINŠ, B., 1964: Einfluß von waldbaulichen Eingriffen auf die Tannenverjüngung. Práce Vyzkum. Ústavu Lesn. ČSSR 28: 223–279.

VISCHER, W., 1946: Naturschutz in der Schweiz. Schweiz. Naturschutzbücherei 3: 380 S. +

VLASSAK, K., PAUL, E.A., HARRIS, R.E., 1973: Assessment of biological nitrogen fixation in grassland and associated sites. Plant and Soil 38: 637–650.

VODERBERG, K., 1955: Die Vegetation der neugeschaffenen Insel Bock. Feddes Repert. Beih. 135: 232–260.

–, FRÖDE, E., 1958: Die Vegetationsentwicklung auf der Insel Bock. Ebenda 138: 214–229.

–, –, 1967: Abschließende Betrachtung der Vegetationsentwicklung auf der Insel Bock in den Jahren 1946–1966. Feddes Repert. 74: 171–176.

VOGLER, P., 1904: Die Eibe (Taxus baccata L.) in der Schweiz. Jb. St. Gall. Naturw. Ges. 1903: 56 S.

VOLGER, C., 1958: Waldwirtschaft vergangener Zeiten im Reinhardswald. Allg. Forstztg. 13: 781–784.

VOLK, O.H., 1931: Beiträge zur Ökologie der Sandvegetation der oberrheinischen Tiefebene. Z. Botan. 24: 81–185.

–, 1937: Über einige Trockenrasengesellschaften des Würzburger Wellenkalkgebietes. Beih. Botan. Cbl. 57 B: 577–598.

VOLKART, A., 1933: Untersuchungen über den Ackerbau und die Ackerunkräuter im Gebirge. Landw. Jb. Schweiz 47: 77–138.

VOLLRATH, H., 1963: Die Morphologie der Itzaue. Ausdruck hydro- und sedimentologischen Geschehens. Mitt. Fränk. Geogr. Ges. 10: 297–309.

–, 1970: Unterschiede im Pflanzenbestand innerhalb der Koppeln von Umtriebsweiden. Bayer. Landw. Jb. 47: 160–173.

–, 1974: Flora und Vegetation des Helmberges nördlich von Straubing. Denkschr. Regensburg. Botan. Ges. 33, 98 S.

WACHTER, H., 1964: Über die Beziehungen zwischen Witterung und Buchenmastjahren. Forstarch. 35: 69–78.

WAGNER, H., 1950 a: Die Vegetationsverhältnisse der Donauniederung des Marchlandes. Bundesvers. Inst. Kulturtechn. u. Techn. Bodenk. Petzenkirchen, Nieder-Österr., Mitt. 5: 32 S.

–, 1950 b: Das Molinietum coeruleae (Pfeifengraswiese) im Wiener Becken. Vegetatio 2: 128–165.

–, 1958: Regionale Einheiten der Waldgesellschaften in Niederösterreich 1:500000. Atlas von Niederösterreich. Wien. 2. Aufl. 1972.

–, 1965: Die Pflanzendecke der Komperdellalm in Tirol. Docum. Carte Végét. Alpes (Grenoble) 3: 7–59.

–, 1966: Ost- und Westalpen, ein pflanzengeographischer Vergleich. Angew. Pflanzensoz. (Wien) 18/19: 265–278.

–, 1970: Zur Abgrenzung der subalpinen gegen die alpine Stufe. Mitt. Ostalp.-Dinar. Ges. Vegetationskunde 11: 225–234.

WAGNER, P., 1972: Untersuchungen über Biomasse und Stickstoffhaushalt eines Halbtrockenrasens. Dipl.-Arb. Math.-Nat. Fak. Göttingen (unveröff.): 54 S.

WAGNER, R.H., 1965: The annual seed rain of adventive herbs in a radiation damaged forest. Ecology 46: 517–520.

WAHLENBERG, G., 1813: De vegetacione et climate in Helvetia septentrionali. Zürich.

WALAS, J., 1938: Wanderungen der Gebirgspflanzen längs der Tatra-Flüsse. Bull. Acad. Polon. Cl. Sci. Math.-Nat. Sér. B 1938: 59–80.

WALKER, D., 1961: Peat stratigraphy and bog regeneration. Proceed. Linn. Soc. London 172 (zit. nach HANSEN, B., 1966).

WALLENTINUS, H.-G., 1973: Above-ground primary production of a Juncetum gerardii on a Baltic sea-shore meadow. Oikos 24: 200–219.

WALTER, H., 1954/1962: Einführung in die Phytologie Bd. I: Grundlagen des Pflanzenlebens. 4. Aufl. Stuttgart 1962: 494 S. – Bd. III: Grundlagen der Pflanzenverbreitung, Teil 1: Standortslehre. 2. Aufl. Verlag Eugen Ulmer, Stuttgart 1962: 525 S.

–, 1963: Über die Stickstoffansprüche (die Nitrophilie) der Ruderalpflanzen. Mitt. Florist.-Soziol. Arb. gem. N.F. 10: 56–69.

–, 1967: Das Feuer als natürlicher klimatischer Faktor. Aquilo (Oulu, Finnland), Ser. Botan. 6: 113–119.

–, 1968: Die Vegetation der Erde in ökophysiologischer Betrachtung. Bd. II: Die gemäßigten und arktischen Zonen. VEB Gustav Fischer Verlag Jena: 1001 S. +

–, 1971: Vegetationszonen und Klima. Verlag Eugen Ulmer, Stuttgart: 244 S., 3. Aufl. 1977.

–, 1974: Die Vegetation Osteuropas, Nord- und Zentralasiens. Gustav Fischer Verlag, Stuttgart: 452 S.

–, 1975: Besonderheiten des Stoffkreislaufs einiger terrestrischer Ökosysteme. Flora 164: 169–183.

–, 1979: Vegetationszonen und Klimazonen. Verlag Eugen Ulmer, Stuttgart, 342 S.

–, 1981: Über Höchstwerte der Produktion von natürlichen Pflanzenbeständen in N.O. Asien. Vegetatio 44: 37–41.

– u. E., 1953: Das Gesetz der relativen Standortskonstanz, das Wesen der Pflanzengesellschaften. Ber. Deut. Botan. Ges. 66: 227–235.

–, LIETH, H., 1960 usw.: Klimadiagramm-Weltatlas. 1. Lieferg., 11 Karten. VEV Gustav Fischer, Jena und spätere Lieferungen.

–, STRAKA, H., 1970: Arealkunde. Floristisch-historische Geobotanik. 2. Aufl. Verlag Eugen Ulmer, Stuttgart: 478 S.

WALTHER, K., 1977: Die Vegetation des Elbetales. Die Flußniederung von Elbe und Seege bei Gartow (Kr. Lüchow-Dannenberg). Abh. Verh. Naturw. Ver. Hamburg N.F. 20 (Suppl.): 123 S.

WANDTNER, R., 1981: Indikatoreigenschaften der Vegetation von Hochmooren der Bundesrepublik Deutschland für Schwermetallimmissionen. Diss. Univ. Frankfurt a. M. (im Druck).

WARMING, E., 1906: Dansk plantevaekst. 1. Strandvegetation. Kopenhagen u. Oslo: 325 S. +

–, 1907: Dansk plantevaekst. 2. Klitterne. Kopenhagen: 376 S.

WASSÉN, G., 1965: Lost and living lakes in the upper Ume Valley. In: The plant cover of Sweden. Acta Phytogeogr. Suecica 50: 233–239.

WATT, A.S.: 1932: On the ecology of British beechwoods with special reference to their regeneration. J. Ecol. 11: 1–48.

–, 1934: The vegetation of the Chiltern Hills, with special reference to the beechwoods and their seral relationships. 22: 230–270.

–, Fraser, G.K., 1933: Tree roots and the field layer. J. Ecol. 21: 404–414.

WATTENDORFF, J., 1964: Über Hartholz-Auenwälder im nordwestlichen Münsterland (Kreis Steinfurt/Westfalen). Abh. Landesmus. Naturk. Münster Westf. 26: 2–33.

WEBER, D.F., GAINEY, P.L., 1962: Relative sensitivity of nitrifying organisms to hydrogen ions in soils and in solutions. Soil Sci. 94: 138–145.

WEBER, H.E., 1967: Über die Vegetation der Knicks in Schleswig-Holstein. Mitt. Arb. gem. Floristik Schlesw.-Holst. u. Hamburg 15: 196 S.

–, 1976: Die Vegetation der Hase von der Quelle bis Quakenbrück. Osnabrücker Naturwiss. Mitt. 4: 131–190.

WEBER-OLDECOP, D.W., 1969: Wasserpflanzengesellschaften im östlichen Niedersachsen. Diss. T.H. Hannover 1969: 171 S.

–, 1977: Fließwassertypologie in Niedersachsen auf floristisch-soziologischer Grundlage. Göttinger Florist. Rundbr. 10: 73–80.

WEBSTER, J.R. (1), 1962: The composition of wet-heath vegetation in relation to aeration of the ground-water and soil. I. Field studies of ground-water and soil aeration in several communities. Ecol 50: 619–637.

WEDECK, H., 1972: Unkrautgesellschaften der Hackfruchtkulturen in Osthessen. Philippia (Kassel) 1, 4: 194–212.

WEHSARG, O., 1954: Ackerkräuter. Akademie-Verlag, Berlin: 294 S.

WEIHE, K. VON, DREYLING, G., 1970: Kulturverfahren zur Bestimmung der Salz- und Überflutungsverträglichkeit von Puccinellia spp. (Gramineae). Helgoländer Wiss. Meeresunters. 20: 157–171.

–, REESE, G., 1968: Deschampsia wibeliana (Sonder) Parlatore. Beiträge zur Monographie einer Art des Tidegebietes. Botan. Jb. 88: 1–48.

WEINERT, E., 1973: Herkunft und Areal einiger mitteleuropäischer Segetalpflanzen. Arch. Naturschutz und Landschaftsforschung 13: 123–139.

WEINITSCHKE, H., 1963: Beiträge zur Beschreibung der Waldvegetation im nordthüringer Muschelkalk. Hercynia 2: 1–58. +

WEISE, G., 1960 a: Experimentelle Untersuchungen zur Kenntnis des Verhaltens von Molinia coerulea Moench in Reinkultur und in Vergesellschaftung. Biol. Zbl. 79: 285 bis 311.

–, 1960 b: Experimentelle Beiträge zur Frage der Ökotypenbildung von Molinia coerulea Moench. Ebenda 79: 427–454.

WEISSENBÖCK, G., 1969: Einfluß des Bodensalzgehaltes auf Morphologie und Ionenspeicherung von Halophyten. Flora Abt. B 158: 369–389.

WEISSER, P., 1970: Die Vegetationsverhältnisse des Neusiedlersees. Wiss. Arb. Burgenland 45: 83 S.

WENDELBERGER, G., 1950: Zur Soziologie der kontinentalen Halophytenvegetation Mitteleuropas unter besonderer Berücksichtigung der Salzpflanzengesellschaften am Neusiedler See. Österr. Akad. Wiss., Math.-Nat. Kl., Denkschr. 108, 5: 180 S.

–, 1953 a: Die Trockenrasen im Naturschutzgebiet auf der Perchtoldsdorfer Heide bei Wien. Angew. Pflanzensoziol. (Wien) 9: 51 S.

–, 1953 b: Über einige hochalpine Pioniergesellschaften aus der Glockner- und Muntanitzgruppe in den Hohen Tauern. Verh. Zool.-Botan. Ges. Wien 93: 100–109.

–, 1954: Steppen, Trockenrasen und Wälder des pannonischen Raumes. Angew. Pflanzensoziol. (Wien), Festschr. Aichinger 1: 573–634.

–, 1962: Die Pflanzengesellschaften des Dachstein-Plateaus (einschließlich des Grimming-Stockes). Mitt. Naturw. Ver. Steiermark 92: 120–178. +

–, 1963 a: Die Relikt-Schwarzföhrenwälder des Alpenostrandes. Vegetatio 11: 265–287.

–, 1963 b: Standorte und Pflanzengesellschaften am Beispiel der Rätischen Gebirge. Ebenda 11: 235–236. +

–, 1964: Sand- und Alkalisteppen im Marchfeld. Jb. Landeskunde Niederösterreich 36: 942–964.

–, 1971: Die Pflanzengesellschaften des Rax-Plateaus. Mitt. Naturw. Ver. Steiermark 100: 197–239.

WENDELBERGER-ZELINKA, E., 1952: Die Vegetation der Donauauen bei Wallsee. Schr. Oberösterr. Landesbaudirektion 11: 196 S.

WESTHOFF, V., 1952: Gezelschappen met houtige gewassen in de duinen en langs de binnenduinrand. Jb. Nederl. Dendrol. Ver.: 9–49.

–, 1958: De plantengroei van het Nationale Park Veluwezoom. Kon. Nederl. Natuurhist. Ver., Wetensch. Med. 26: 1–40.

WESTLAKE, D.F., 1966: The biomass and productivity of Glyceria maxima I. Seasonal changes in biomass. J. Ecol. 54: 745–753.

–, 1975: Storage organs growth and productivity of some aquatic macrophytes. XII. Internat. Botan. Congr. (Leningrad) Abstr. 1: 174.

WETSCHAAR, R., 1968: Soil organic nitrogen mineralization as affected by low soil water potentials. Plant and Soil 29: 9–17.

WETTER, E., 1918: Ökologie der Felsflora kalkarmer Gesteine. Diss. ETH Zürich: 176 S.

WHITE, E.J., TURNER, F., 1970: A method of estimating income of nutrients in a catch of airborne particles by a woodland canopy. J. Appl. Ecol. 7: 441–461.

WHITTACKER, R.H., FEENY, P.P., 1971: Allochemics: Chemical interactions between species. Science 171: 757–770.

WHITTAKER, E., GINNINGHAM, C.H., 1962: The effects of fire on regeneration of Calluna vulgaris (L) Hull. from seed. J. Ecol. 50: 815–822.

WIEDENROTH, E.-M., MÖRCHEN, G., 1964: Wurzeluntersuchungen im Aphano-Matricarietum Tx. 37 im Parthegebiet (Bezirk Leipzig). Wiss. Z. Humboldt-Univ. Berlin, Math.-Nat. R. 13: 645–651.

WIEGLEB, G., 1976: Untersuchungen über den Zusammenhang zwischen Chemismus und Makrophytenvegetation stehender Gewässer in Niedersachsen. Diss. Univ. Göttingen: 113 S.

–, 1977: Vorläufige Übersicht über die Pflanzengesellschaften der niedersächsichen Fließgewässer. Gutachten Nieders. Landesverwaltungsamt, Hannover (unveröff.): 41 S.

WIELGOLASKI, P.E., 1975: Productivity of tundra and forest tundra. XII. Internat. Botan. Congr. (Leningrad) Abstr. 1: 174.

WIEMANN, P., DOMKE, W., 1967: Pflanzengesellschaften der Ostfriesischen Insel Spiekeroog. 1. Teil Dünen. Mitt. Staatsinst. Allg. Botan. Hamburg 12: 191–353.

WIESNER, J., 1907: Der Lichtgenuß der Pflanzen. Verl. Wilh. Engelmann, Leipzig: 322 S.

WIKUS, E., 1960: Die Vegetation der Lienzer Dolomiten (Osttirol). Arch. Botan. e Biogeogr. Ital. 34–37 (1965–60): 189 S.

WILCZEK, E., BEAUVERD, G., DUTOIT, D., 1928: Le compartement écologique de Bromus erectus. Vierteljahrsschr. Naturf. Ges. Zürich 73: 468–508.

WILD, A., SKARLOU, V., CLEMENT, C.R., SNAYDON, R.W., 1974: Comparison of potassium uptake by four plant species grown in sand and in flowing solution culture. J. Appl. Ecol. 11: 801–812.

WILLERDING, U., 1968: Beiträge zur Geschichte der Eibe (Taxus baccata L.). Plesse-Archiv (Göttingen) 3: 96–155.

–, 1973: Frühmittelalterliche Pflanzenreste aus Braunschweig. Nachr. Niedersachs. Urgesch. 42: 358–359.

–, 1977: Das Dorf der Eisenzeit und des frühen Mittelalters. Abh. Akad. Wiss. Göttingen, Phil.-Hist. Kl., Dritte Folge 101: 357–405.

–, 1979: Paläo-ethnobotanische Untersuchungen über die Entwicklung von Pflanzengesellschaften. Ber. Internat. Symposien Int. Vereinig. Vegetationskunde (Rinteln 1978): 61–109.

–, 1980 a: Zum Ackerbau der Bandkeramiker. Materialh. Ur- u. Frühgesch. Nieders. 16: 421–456.

–, 1980 b: Untersuchungen zur eisenzeitlichen und frühmittelalterlichen Flur in Mitteleuropa und ihrer Nutzung. Abh. Akad. Wiss. Göttingen, Phil-Hist. Kl., Dritte F. 116: 126–196.

WILLIAMS, J.T., 1964: A study of the competitive ability of Chenopodium album. I. Interference between Kale and C. album grown in pure stands and in mixtures. Weed Res. 4: 285–295.

–, 1968: The nitrogen relations and other ecological investigations on wet fertilized meadows. Veröff. Geobot. Inst. ETH, Stiftg. Rübel, Zürich 41: 69–193.

–, 1969: Mineral nitrogen in British grassland soils. I. Seasonal patterns in simple models. Oecol. Plant. 4: 307–370.

–, YARLEY, Y.W., 1967: Phytosociological studies of some British grasslands. I. Upland pastures in northern England. Vegetatio 15: 169–189.

WILLIAMS, W.T., BARBER, D.A., 1961: The functional significance of aerenchyma in plants. Symposia Soc. Exper. Biol. (Cambridge) 15: 132–144.

WILLIS, A.J., 1963: Braunton Burrows: The effects on the vegetation of the addition of mineral nutrients to the dune soils. J. Ecol. 51: 353–374.

WILMANNS, O., 1956: Pflanzengesellschaften und Standorte des Naturschutzgebietes „Greuthau" und seiner Umgebung (Reutlinger Alb). Veröff. Landesstelle Naturschutz u. Landschaftspflege Baden-Württemb. 1956: 317–451. +

–, 1958: Zur standörtlichen Parallelisierung von Epiphyten- und Waldgesellschaften. Beitr. Naturk. Forsch. Südw.-Deut. 17: 11–19.

–, 1966: Anthropogener Wandel der Kryptogamen-Vegetation in Südwestdeutschland. Ber. Geobot. Inst. ETH, Stiftung Rübel, Zürich 37: 74–87.

–, 1971: Verwandte Züge in der Pflanzen- und Tierwelt von Alpen und Südschwarzwald. Jb. Ver. Schutze Alpenpflanzen u.-Tiere 36: 36–50.

–, 1973: Ökologische Pflanzensoziologie. UTB 269, Heidelberg: 288 S.

–, 1975: Wandlungen des Geranio-Allietum in den Kaiserstühler Weinbergen? Pflanzensoziologische Tabellen als Dokumente. Beitr. Naturk. Forsch. Südw.-Deut. 34: 429–443.

–, BAUMERT, J., 1965: Zur Besiedlung der Freiburger Trümmerflächen – eine Bilanz nach zwanzig Jahren. Ber. Naturf. Ges. Freiburg i. Br. 55: 399–411.

–, RUPP, S., 1966: Welche Faktoren bestimmen die Verbreitung alpiner Felsspaltenpflanzen auf der Schwäbischen Alb? Veröff. Landesstelle Naturschutz u. Landschaftspflege Baden-Württemb. 34: 62–86.

WILSON, A.T., 1959: Surface of the ocean as a source of air-borne nitrogenous material and other plant nutrients. Nature 184: 99–101.

WILZEK, 1934: Die Pflanzengesellschaften des mittelschlesischen Odertales. Beitr. Biol. Pflanzen 23: 1–96. +

WINKLER, E., 1963: Beiträge zur Klimatologie hochalpiner Lagen der Zentralalpen. Ber. Naturw.-Mediz. Ver. Innsbruck 53: 209–223.

WINKLER, O., 1933: Forstgeschichtlich bedingte Wandlungen in den Gebirgswäldern des St. Galler Oberlandes. Schweiz. Z. Forstwes. 84: 109–120.

–, 1943: Waldbrände, ihre Ursachen, Verhütung und Bekämpfung. In: „Mein Einsatz – Deine Sicherheit", Zürich: 5 S.

WINTELER, R., 1927: Studien über Soziologie und Verbreitung der Wälder, Sträucher und Zwergsträucher des Sernftales. Vierteljahrsschrift Naturf. Ges. Zürich 72: 1–185.

WINTERHOFF, W., 1963: Vegetationskundliche Untersuchungen im Göttinger Wald. Nachr. Akad. Wiss. Göttingen II. Math.-Phys. Kl. 1962: 21–79.

–, 1965: Die Vegetation der Muschelkalkfelshänge im hessischen Werrabergland. Veröff. Württemb. Landesstelle Naturschutz u. Landschaftspflege 33: 146–197.

–, 1971: Zur Verbreitung und Soziologie von Carex cespitosa L. auf der Schwäbischen Alb. Jb. Ges. Naturk. Württemb. 126: 270–279.

–, 1975: Die Pilzvegetation der Dünenrasen bei Sandhausen (nördlich Oberrheinebene). Beitr. Naturk. Forsch. Südw.-Deut. 34: 445–462.

–, HÖLLERMANN, P., 1968: Morphologie, Flora und Vegetation des Bergsturzes am Schickeberg (Nordhessen). Nachr. Akad. Wiss. Göttingen, II. Math.-Phys. Kl. 1968, 7: 110–170. +

WIRTH, V., 1972: Silikatflechten-Gemeinschaften im außeralpinen Zentraleuropa. Diss. Botan. 17: 326 S.

–, 1975: Die Vegetation des Naturschutzgebietes Utzenfluh (Südschwarzwald), besonders in lichenologischer Sicht. Beitr. Naturk. Forsch. Südw.-Deut. 34: 463–476. +

–, 1980: Flechtenflora. Ökologische Kennzeichnung und Bestimmung der Flechten Süddeutschlands und angrenzender Gebiete. UTB 1062 Verlag Eugen Ulmer, Stuttgart, 552 S.

WITTICH, W., 1951: Der Einfluß der Streunutzung auf den Boden (Untersuchungen an diluvialen Sandböden). Forstwiss. Cbl. 70: 65–92.

–, 1954: Die Melioration streugenutzter Böden. Ebenda. 73: 211–232.

–, 1963: Bedeutung einer leistungsfähigen Regenwurmfauna unter Nadelwald für Streuzersetzung, Humusbildung und allgemeine Bodendynamik. Schr. reihe Forstl. Fak. Univ. Göttingen 30: 3–60.

WITTIG, R., 1976: Die Gebüsch- und Saumgesellschaften der Wallhecken in der Westfälischen Bucht. Abh. Landesmus. Naturkunde Münster i.W. 38: 78 S.

–, 1980: Die geschützten Moore und oligotrophen Gewässer der Westfälischen Bucht. Schriftenr. Landesanst. Ökol., Landschaftsentw. u. Forstplanung 5: 226 S.

WŁODEK, J., STRZEMIENSKI, K., RALSKI, E., 1931: Untersuchungen über die Böden der Mischassoziationen im Gebiete der Czerwone Wierchy und Bielskie Zstry (Tatra-Gebirge). Bull. Acad. Polon. Sci. et Lettr. Sér. B: Sci. Nat. 1: 1–5.

WLOTZKA, F., 1961: Untersuchungen zur Geochemie des Stickstoffs. Geochimica et Cosmochimica Acta 24: 106–154.

WOHLENBERG, E., 1931: Die Grüne Insel in der Eidermündung. Eine entwicklungsphysiologische Untersuchung. Arch. Deut. Seewarte 50: 3–34.

–, 1937: Die Wattenmeer-Lebensgemeinschaften im Königshafen von Sylt. Helgoländer Wiss. Meeresunters. 1: 1–92

–, 1963: Der Deichbruch des Ülvesbüller Kooges in der Februar-Sturmflut 1962. Versalzung – Übersandung – Rekultivierung. „Die Küste" (Heide i.H.) 11: 52–89.

–, 1965: Deichbau und Deichpflege auf biologischer Grundlage. Ebenda 13: 73–103.

WOJTERSKI, T., 1963: Wet pine-woods in the West Kashubian coastal region. Badan. Fizjograf. Polska Zachodnia 12: 139–191.

–, 1964 a: Pine forests on sand dunes at the polish Baltic coast. Poznań Soc. Friends of Sci., Dep. Math. Nat. Sci., Sect. Biol. 28, No. 2: 217 S.

–, 1964 b: Schemata of the zonal vegetation system on the southern coast of Baltic Sea. Badan. Fizjograf. Polska Zachodnia 14: 87–105.

–, LESZCZYNSKA, M., PIASZYK, M., 1973: Potential nutural vegetation of the lakeland of Lubusz. Ebenda 26, Ser. B.: 107–142.+

WOŁEK, J., 1974: A preliminary investigation on interactions (competition, allelopathy) between some species of Lemna, Spirodela, and Wolffia. Ber. Geobot. Inst. ETH, Stiftg. Rübel, Zürich 42: 140–162.

WOLLERT, H., 1967: Die Pflanzengesellschaften der Oser Mittelmecklenburgs unter besonderer Berücksichtigung der Trockenrasengesellschaften. Wiss. Z. Univ. Rostock, Math.-Nat. R. 16: 43–95.

–, 1970: Zur soziologischen Gliederung und Stellung der Grenzhecken Mittelmecklenburgs und deren Säume. Naturschutzarbeit in Mecklenburg 13: 92–100.

WOLTER, M., DIERSCHKE, H., 1975: Laubwald-Gesellschaften der nördlichen Wesermünder Geest. Mitt. Florist.-Soziolog. Arb. gem. N.F. 18: 203–217. +

WRABER, M., 1952: Sur l'importance, pour la sylviculture et l'économie forestière, des surfaces réservées à l'étude de la forêt vierge. Ponatis Biol. Vestn. 1: 38–66.

–, 1963: Die Waldgesellschaft der Fichte und der Waldhainsimse in den slowenischen Ostalpen (Luzulo silvaticae-Piceetum Wraber 1953). Acad. Sci. Art. Sloven., Diss. 7: 75–176.

YERLY, M., 1970: Écologie comparée des prairies marécageuses dans les préalpes de la Suisse occidentale. Veröff. Geobot. Inst. ETH, Stiftg. Rübel, Zürich 44: 119 S.

YLI-VAKKURI, P., 1961: Emergence and initial development of tree seedlings on burnt-over forest land. Acta Forest. Fenn. 74: 5–51.

ZACHARIAE, G., 1965: Spuren tierischer Tätigkeit im Boden des Buchenwaldes. Forstwiss. Forsch. 20: 68 S.

–, 1967: Die Streuzersetzung im Köhlgartengebiet. In: GRAFF, O. and SATCHELL, J.E. (Ed.), Progress in soil biology. Vieweg u. Sohn, Braunschweig: 490–506.

ZAHLHEIMER, W.A., 1979: Vegetationsstudien in den Donauauen zwischen Regensburg und Straubing als Grundlage für den Naturschutz. Denkschr. Regensburg. Botan. Ges. 38: 398 S.

ZARZYCKI, K., 1963: The forests of the western Bieszczady Mts. (Polish eastern Carpathians). Acta Agr. et Silv., Ser. Leśna 3: 3–132. +

–, 1964: Biological and ecological studies in Carpathian beechwood. Bull. Acad. Polon. Sci., Cl. II, 12: 15–21.

–, 1968: Experimental investigation of competition between forest herbs. Acta Soc. Botan. Polon. 37: 393–411.

ZEIDLER, H., 1953: Waldgesellschaften des Frankenwaldes. Mitt. Florist.-Soziol. Arb. gem. N.F. 4: 88–109. +

–, 1970: Edaphisch und anthropogen bedingtes Vegetationsmosaik in Wäldern. In: R. TÜXEN (Hrsg.) Gesellschaftsmorphologie (Strukturforschung). Dr. W. Junk, Den Haag: 322–333.

–, STRAUB, R., 1967: Waldgesellschaften mit Kiefer in der heutigen potentiellen natürlichen Vegetation des mittleren Maingebietes. Mitt. Florist.-Soziol. Arb. gem. N.F. 11/12: 88–126.

ZELLER, W., ZUBER, E., KLÖTZLI, F., 1968: Das Schutzgebiet Mettmenhaslisee, Niederhasli. Vierteljahresschr. Naturf. Ges. Zürich 113: 373–405. +

ZELNIKER, J.L., 1968: Dürreresistenz von Baumarten unter Steppenbedingungen. In: Deutsche Akad. Landwirtschaftswiss. Berlin, Tagungsber. 100. (Vorträge u. Diskussionen des ersten Baumphysiologen-Symposiums): 131–140.

ZIELONKOWSKI, W., 1973 a: Wildgrasfluren der Umgebung Regensburgs. Denkschr. Regensburg. Botan. Ges. 31: 181 S.

–, 1973 b: Vegetationskundliche Untersuchungen im Rotwandgebiet zum Problemkreis Erhaltung der Almen. Schriftenr. Naturschutz u. Landschaftspflege (München) 5: 28 S.

ZIMMEK, G.-E., 1975: Die Mineralstickstoff-Versorgung einiger Salzrasen-Gesellschaften des Graswarders vor Heiligenhafen/Ostsee. Diss. Univ. Göttingen: 66 S.

*ZIMMERMANN, W. (Hrsg.), 1961: Der Federsee. Die Natur- und Landschaftsschutzgebiete Baden-Württembergs 2. Stuttgart: 411 S. +

ZIMMERMANN, R., 1976: Feuer als Pflegemaßnahme in Halbtrockenrasen des Kaiserstuhls. Vortragsreferat, Ges. f. Ökologie, Göttingen, 20.–24.9.76.

ZINGG, T., 1952: Gletscherbewegungen in den letzten 50 Jahren in Graubünden. Wasser- u. Energiewirtsch. (Zürich) 1952: H. 5–7.

–, 1954: Die Bestimmung der klimatischen Schneegrenze auf klimatologischer Grundlage. Angew. Pflanzensoziol. (Wien), Festschr. Aichinger 2: 848–854.

ZIOBROWSKI, S., 1933: Über den Einfluß des harten Winters 1928/29 auf die Holzgewächse im Rabaflußtale. Acta Soc. Botan. Polon. 10: 49–111.

ZLATNÍK, A., 1928: Études écologiques et sociologiques sur la Sesleria coerulea et le Seslerion calcariae en Tchécoslovaquie. Trav. Soc. Roy. Sci. Bohême, Cl. Sci., N.S. 8: 116 S.

–, 1935: Entwicklung und Zusammensetzung der Naturwälder in Podkarpatská Rus und ihre Beziehung zum Standort. Rec. Trav. Inst. Rech. Agron. Républ. Tchécosl. 127: 205 S.

–, 1958: Waldtypengruppen der Slowakei. Schr. Wiss. Labor. Biogeozönol. u. Typol. d. Waldes, Forstw. Fak. Landw. Hochsch. Brno 4: 195 S.

–, 1961: Großgliederung der slowakischen Wälder in waldtypologischer und pflanzensoziologischer Auffassung. Veröff. Geobot. Inst. ETH, Stiftg. Rübel, Zürich 36: 52–90. +

ZOBRIST, L., 1935: Pflanzensoziologische und bodenkundliche Untersuchung des Schoenetum nigricantis. Beitr. Geobot. Landesaufnahme Schweiz 18: 144 S.

ZOLLER, H., 1951: Das Pflanzenkleid der Mergelsteilhänge im Weißensteingebiet. Ber. Geobot. Forsch.inst. Rübel, Zürich 1950: 67–95.

–, 1954: Die Typen der Bromus erectus-Wiesen des Schweizer Jura. Beitr. Geobot. Landesaufn. Schweiz 33: 309 S.

–, 1958: Die Vegetation und Flora des Schaffhauser Randens. Mitt. Naturf. Ges. Schaffhausen 26: 1–36. +

–, 1960: Pollenanalytische Untersuchungen zur Vegetationsgeschichte der Schweiz. Denkschr. Schweiz. Naturf. Ges. 83, Abh. 2: 45–157.

–, 1961: Die kulturbedingte Entwicklung der insubrischen Kastanienregion seit den Anfängen des Ackerbaues im Neolithikum. Ber. Geobot. Inst. ETH, Stiftg. Rübel, Zürich 32: 263–279.

–, 1964: Flora des schweizerischen Nationalparks und seiner Umgebung. Ergebn. Wiss. Unters. Schweiz. Nationalpark 9, 51: 408 S. +

–, 1974: Flora und Vegetation der Innalluvionen zwischen Scuol und Martina (Unterengadin). Ebenda 12: 209 S. +

–, KLEIBER, H., 1971: Vegetationsgeschichtliche Untersuchungen in der montanen und subalpinen Stufe der Tessintäler. Verhandl. Naturf. Ges. Basel 81: 90–154.

–, STÄGER, R., 1949: Beitrag zur Altersbestimmung von Pflanzen aus der Walliser Felsensteppe. Ber. Geobot. Forsch. Inst. Rübel, Zürich 1948: 61–68.

ZOLLITSCH, B., 1966: Soziologische und ökologische Untersuchungen auf Kalkschiefern in hochalpinen Gebieten. Teil I. Die Steinschuttgesellschaften der Alpen. Ber. Bayer. Botan. Ges. 40: 38 S.

ZÓLYOMI, B., 1953: Die Entwicklungsgeschichte der Vegetation Ungarns seit dem letzten Interglazial. Acta Biol. Acad. Sci. Hung. 4: 367–413.

–, u. Mitarb., 1967: Einreihung von 1400 Arten der ungarischen Flora in ökologische Gruppen nach TWR-Zahlen. Fragmenta Botan. Mus. Hist.-Nat. Hung. 4: 101–142.

ZONNEVELD, I.S., 1960: De Brabantse Biesbosch. A study of soil and vegetation of a freshwater tidal delta. Meded. Stichting Bodenkartiering, Bodenk. Stud. 4: 210 S. (mit 2 Bänden als Anlage).

–, 1965: Studies van landshap, bodem en vegetatie in het westelijke deel van de Kalmthoutse Heide. Boor en Spade 14: 216–238.

–, 1966. Zusammenhänge Forstgesellschaft – Boden – Hydrologie und Baumwuchs in einigen niederländischen Pinus-Forsten auf Flugsand und auf Podsolen. In: TÜXEN, R., (Hrsg.), Anthropogene Vegetation. Dr. W. Junk, Den Haag: 312–335.

ZÖTTL, H., 1951 a: Die Vegetationsentwicklung auf Felsschutt in der alpinen und subalpinen Stufe des Wettersteingebirges. Jb. Ver. Schutze Alpenpflanzen u.-tiere 16: 10–74.

–, 1951 b: Experimentelle Untersuchungen über die Ausbreitungsfähigkeit alpiner Pflanzen. Phyton 3: 121–125.

–, 1952 a: Beitrag zur Ökologie alpiner Kalkschuttstandorte. Phyton 4: 160–175.

–, 1952 b: Zur Verbreitung des Schneeheide-Kiefernwaldes im bayerischen Alpenvorland. Ber. Bayer. Botan. Ges. 29: 92–95.

–, 1953: Untersuchungen über das Mikroklima subalpiner Pflanzengesellschaften. Ber. Geobot. Forsch. Inst. Rübel, Zürich 1952: 79–103.

–, 1959: Voraussetzungen für eine wirkungsvolle Verbesserung der Stickstoffversorgung von Nadelholzbeständen. Z. Pflanzenernähr., Düng., Bodenkunde 48: 116–122.

–, 1960: Dynamik der Stickstoffmineralisation im organischen Waldbodenmaterial. I–III. Plant and Soil 13: 166–223.

–, 1964: Wirksamkeit der Forstdüngung in Süddeutschland. 8th Internat. Congr. Soil Sci. Bukarest, Romania. Forest soils 2: 1009–1017.

–, 1966: Kalkböden der Alpen. Jb. Ver. Schutze Alpenpflanzen u.-Tiere 31: 5 S.

–, KEMEL, R., 1963: Ernährungszustand und Wachstum von Fichten-Altbeständen nach Ammoniakgas- und Stickstoffsalzdüngung. Forstwiss. Cbl. 82: 76–100.

ZUBER, E., 1968: Pflanzensoziologische und ökologische Untersuchungen an Strukturrasen (besonders Girlandenrasen) im Schweizerischen Nationalpark. Ergebn. Wiss. Unters. Schweiz. Nationalparks 60: 79–157.

ZUCK, W., 1952: Untersuchungen über das Vorkommen und die Biotope einheimischer Lumbriciden. Jahresh. Ver. Vaterl. Naturk. Württemb. 107: 95–132.

ZUKRIGL, K., 1973: Montane und subalpine Waldgesellschaften unter mitteleuropäischem, pannonischem und illyrischem Einfluß. Mitt. Forstl. Bundesversuchsanst. Wien 101: 386 S.

–, ECKHARDT, G., NATHER, J., 1963: Standortskundliche und waldbauliche Untersuchungen in Urwaldresten der niederösterreichischen Kalkalpen. Mitt. Forstl. Bundesversuchsanst. Mariabrunn 62: 244 S.

ZWÖLFER, H., 1974: Das Goldrutenproblem: Möglichkeiten für ein biologisches Unkrautbekämpfungsprojekt in Europa. Memorandum Ludwigsburg (unveröff.), 13 S.

The following references have been additionally cited in the text of the English version. Some important recent publications are included without citation.

Albrecht, J., 1969: Soziologische und ökologische Untersuchungen alpiner Rasengesellschaften, insbesondere an Standorten auf Kalksilikat-Gestein. Diss. Botan. (Lehre) *5*: 91 pp.

Alexander, V., 1975: Nitrogen fixation by blue-green algae in polar and subpolar regions. In: STEWART, W.D.P. (ed.): Nitrogen fixation by free–living micro–organisms. Cambridge University Press, London: 175–188.

Amani, M.R., 1980: Vegetationskundliche und ökologische Untersuchungen im Grünland de Bachtler um Suderburg. Diss. Univ. Göttingen: 116 pp.

Antonietti, A., 1970: Su un'associazione di brughiera del piede meridionali delle Alpi. Ber. Geobot. Inst. ETH, Stiftg. Rübel, Zürich *40*: 9–27.

Antonovics, J., Primack, R.B., 1982: Experimental ecological genetics in Plantago. VI. The demography of seedling transplants of *P. lanceolata*. J. Ecol. *70*: 55–75

Armstrong, W., 1978: Root aeration in the wetland condition. In: HOOK, D.D., CRAWFORD, R.M.M. (eds.): Plant life in anaerobic environments. Ann. Arbor Science Publ. Inc., Ann. Arbor, Michigan, p. 269–298.

–, 1981: The water relations of heathlands: general physiological effects of waterlogging. In: SPECHT, R.C. (ed.): Heathlands and related shrublands. Analytical studies. Elsevier Sci. Publ. Comp., Amsterdam etc.: 11–121.

Aulitzky, H., Turner, H., Mayer, H., 1982: Bioklimatische Grundlagen einer standortsgemäßen Bewirtschaftung des subalpinen Lrchen–Arvenwaldes. Mitt. Eidgen. Anstalt Forstl. Versuchswes. *58*: 325–580.

Austin, M.P., Austin, B.O., 1980: Behaviour of experimental plant communities along a nutrient gradient. J. Ecol. *68*: 891–918.

Barnes, R.S.K., 1980: Coastal lagoons. The natural history of a neglected habitat. Cambridge University Press, 106 pp.

Barth, H., 1978: Untersuchungen zum Wasserhaushalt von einigen Halbtrockenrasen–Pflanzen unter kontrollierten Feuchtebedingungen. Diss. Univ. Göttingen: 109 pp.

Bates, J.W., 1978: The influences of metal availability on the bryophyte and macrolichen vegetation of four rock types on Skye and Rhum. J. Ecol. *66*: 457–482.

Becker, B., 1982: Dendrochronologie und Paläoökologie subfossiler Baumstmme aus Flußablagerungen. Ein Beitrag zur nacheiszeitlichen Auenentwicklung im südlichen Mittleleuropa. Mitt. Komm. Quartrforsch. sterr. Akad. Wiss. *5*: 120 pp.

Bibliographia Phytosoliologica Syntaxonomica, 1971–1982: Lieferung 1–35, Verlag Cramer, Lehre, ab 1976 Vaduz.

Boatman, D.J., Goode, D.A., Hulme, P.D., 1981: The Silver Flowe. III. Pattern development on Long Loch B and Craigeazle mires. J. Ecol. *69*: 897–918.

–, Tomlaison, R.W., 1977: The Silver Flowe. II. Features of the vegetation and stratigraphy of Brishie boy, and their bearing on pool formation. J. Ecol. *65*: 531–545.

Böcker, R., Kowarik, L., Bornkamm, R., 1984: Untersuchungen zur Anwendung der Zeigerwerte nach ELLENBERG. Verh. Ges. kol. *11*: 35–56.

Bradshaw, R.H.W., 1981: Quantitative reconstruction of local woodland vegetation using pollen analysis from a small basin in Norfolk, England. J. Ecol. *69*: 641–955.

Brandes, D., 1982: berblick über die Literaturinformation der Pflanzensoziologie. Phytocoenologia *10*: 375–381.

Braun, W., 1968: Die Kalkflachmoore und ihre wichtigsten Kontaktgesellschaften im bayerishcen Alpenvorland. Diss. Botan. (Lehre) *1*: 134 pp.

Bücking, W., 1981: Kulturversuche an azidophytischen Waldbodenpflanzen mit variierter Stickstoff–Menge und Stickstoff–Form. Mitt. Ver. Forstl. Standortskde. u. Forstpflanzenzücht. *29*: 42–57.

–, 1983: Kulturversuche an azidophytischen Waldbodenpflanzen mit variierter Stickstoff–Menge und Stickstoff–Form. II. Nhrelementgehalte kultivierter Pflanzen im Vergleich zu Freilandpflanzen. Mitt. Ver. Forstl. Standortskunde u. Forstpflanzenzücht. *30*: 40–53.

Burdon, J.J., 1980: Intra–specific diversity in a natural population of *Trifolium repens*. J. Ecol. *68*: 717–735.

Burgeff, H., 1956: Mikrobiologie des Hochmoores, mit besonderer Berücksichtigung der Ericaceen–Pilsymbiose. Ber. Deut. Bot. Ges. 69: 257–262. (See also: Book, title identical, Stuttgart 1961).

Chapin III, F.S., Cleve, K. van, Chapin, M.C., 1979: Soil temperature and nutrient cycling in the tussock growth form of *Eriophorum vaginatum*. J. Ecol. *67*: 169–189.

Chapman, V.J. (ed.), 1977: Wet coastal ecosystems. Elsevier Scientific Publ. Comp., 428 pp.

Coenen, H., 1981: Flora und Vegetation der Heidegewsser und–moore auf den Maasterrassen im deutsch–niederlndischen Grenzgebiet. Arb. Rhein. Landeskunde *48*: 217 pp.

Coulson, J.C., Butterfield, J., 1978: An investigation of the biotic factors determining the rates of plant decomposition on blanket bog. J. Ecol. *66*: 631–650.

Davison, A.W., 1977: The ecology of *Hordeum murinum* L. III. Some effects of adverse climate. J. Ecol. *65*: 523–530.

Den Hartog, C., Segal, S., 1964: A new classification of water plant communities. Acta Botan. Neerl. *13*.

Denayer–de Smet, S., 1970: Considerations sur l'accumulation du zinc par les plantes poussant sur sols calaminaires. Bull. Inst. Roy. Sci. Nat. Belg. *46*, 11, 13 pp.

Dierschke, H. 1981a: Zur syntaxonomischen Bewertung schwach gekennzeichneter Pflanzengesellschaften. In: DIERSCHKE, H. (Red.): Syntaxonomie. J. Cramer, Vaduz: 109–122.

–, 1981b: Syntaxonomische Gliederung der Bergwiesen Mitteleuropas (Polygono – Trisetion). In: DIERSCHKE, H. (Red.): Syntaxonomie. J. Cramer, Vaduz: 311–341.

Dierssen, K., 1982: Die wichtigsten Pflanzengesellschaften der Moore NW–Europas. Königstein (Koeltz): 382 pp.

Dietrich, H., 1958: Untersuchungen zur Morphologie und Genese grundwasserbeeinflußter Sandböden im Gebiet des nordost–deutschen Diluviums. Arch. Forstwes. *7*. 577–640.

Durwen, K.–J., 1981: Zur Nutzung von Zeigerwerten und artspezifischen Merkmalen der Gefßpflanzen Mitteleuropas für Zwecke der Landschaftsökologie und –planung mit Hilfe der EDV – Voraussetzungen, Instrumentarium, Methoden und Möglichkeiten. Diss. Math.–Nat. Fak. Univ. Münster, 194+42 pp.

El Ayouty, E.Y.M., 1966: Systematik und Stickstoffbindung einiger Blaualgen in Lehmböden aus einem humiden und einem semi–ariden Gebiet. Diss. Gießen: 143 pp.

Elster, H.–J., 1982: Neuere Untersuchungen über die Eutrophierung und Sanierung des Bodensees. gwf – wasser + abwasser *123*: 277–287.

Ernst, W.H.O., 1982: Schwermetallpflanzen. In: KINZEL, H. (ed.), 1982: Pflanzenökologie und Mineralstoffwechsel. Eugen Ulmer Verlag, Stuttgart: 472–506.

Etherington, J.R., 1981: Limestone heaths in south–west Britain: their soils and the maintenance of their calcicole–calcifuge mixtures. J. Ecol. *69*: 277–294.

Eurola, S., 1965: Beobachtungen über die Flora und Vegetation am südlichen Ufersaum des Saimaa–Sees in Südostfinnland. Aquilo (Oulu), Ser. Botan. *2*: 1–56.

Fenner, M., 1978: A comparison of the abilities of colonizers and closed–turf species to establish from seed in artificial swards. J. Ecol. *66*: 583–963.

Franz, H., 1960: Feldbodenkunde als Grundlage der Standortsbeurteilung und Bodenwirtschaft. Wien u. München: 583 pp.

Gilbert, O.L., 1980: Juniper in upper Teesdale. J. Ecol. *68*: 1013–1024.

Gimingham, C.H., Chapman, S.B., Webb, N.R., 1979: European heathlands. In: SPECHT, R.L. (ed.): Heathlands and related shrublands. Descriptive Studies. Elsevier Sci. Publ. Comp., Amsterdam etc.: 365–413.

Glavac, V., Koenies, H., 1978: Vergleiche der N–Nettomineralisation in einem Sauerhumus-Buchenwald (Luzulo–Fagetum) und einem benachbarten Fichtenforst am gleichen Standort vor und nach dem Kahlschlag. Oecol. Plant. *13*: 219–226.

Goldsmith, F.B., 1978: Interaction (competition) studies as a step towards the synthesis of sea cliff vegetation. J. Ecol. *66*: 921–931.

Greig–Smith, P., 1979: Pattern in vegetation. J. Ecol. *67*: 755–779.

Grime, J.P., 1979: Plant strategies and vegetation processes. John Wiley and Sons, Chichester, New York etc.: 222 pp.

Gross, K.L., 1980: Colonization by *Verbascum thapsus* (Mullein) of an old–field in Michigan: experiments on the effects of vegetation. J. Ecol. *68*: 919–927.

Groves, R.H., 1981: Nutrient cycling in heathlands. In: SPECHT, R.L. (ed.): Heathlands and related shrublands. Analytical studies. Elsevier Sci. Publ. Comp., Amsterdam etc.: 151–163.

Grubb, P.J., Green, H.E., Merrifield, R.C., 1969: The ecology of chalk–heath: its relevance to the calcicole–calcifuge and soil acidification problems. J. Ecol. *57*: 175–213.

Grube, H.–J., 1971: Auswirkungen von Herbiziden auf Getreide–Unkrautgesellschaften in Feldversuchen des Pflanzenschutzamtes Northeim. Staatsexamensarbeit, Göttingen: 52 pp.

Hackett, C., 1962: Stimulative effects of aluminium on plant growth. Nature *195*: 471–472.

Harper, J.L., 1977: Population biology of plants. Acad. Press, London, 892 pp.

Haustein, B., 1982: Luftverschmutzung und Moosvorkommen in Kiefernforst–Gesellschaften des niederrheinischen Tieflands. Dipl. Arb. Syst.–Geobot. Inst. Göttingen: 77 pp.

Heal, O.W., Perkins, D.F. (eds.), 1978: Production ecology of British moors and montane grasslands. Ecol. Studies *27*: 426 pp.

Heinrichfreise, A., 1981: Aluminium–Toleranz von *Luzula albida* und *Milium effusum*, Pflanzen saurer und basischer Laubwaldböden. Oecol. Plant. N.S. *2*: 87–100.

Hilbig, W., 1982: Pflanzengeographische Landschaftsgliederung auf der Grundlage der Ackervegetation. Arch. Naturschutz u. Landschaftsforsch. (Berlin) *22*: 131–144.

Hill, M.O., Jones, E.W., 1978: Vegetation changes resulting from afforestation of rough grazings in Caeo Forest, South Wales. J. Ecol. *66*: 433–456.

–, Stevens, P.A., 1981: The density of viable seed in soils of forest plantations in upland Britain. J. Ecol. *69*: 693–709.

Holtmeier, F.–K., 1981: What does the term 'Krummholz' really mean? Observations with special reference to the Alps and the Colorado Front Range. Mountain Res. and Developm. *1*: 253–260.

Holzner, W., 1981: Ackerunkruter. Bestimmung, Verbreitung, Biologie und kologie. Leopold Stocker, Graz u. Stuttgart: 178 pp.

–, Numata, N. (eds.), 1981: Biology and ecology of weeds. Dr. W. Junk Publ., The Hague.

Hook, D.D., Scholtens, J.R., 1978: Adaptations and flood tolerance of tree species. In: HOOK, D.D., CRAWFORD, R.M.M. (eds.): Plant life in anaerobic environments. Ann. Arbor Science Publ. Inc., Ann. Arbor, Michigan, pp. 299–332.

Horvat, A.O., 1978: Potentillo–Quercetum (sensu latissimo). Wlder. Janus Pannonius Mūz. Evkönyve *22*: 23–70.

Huse, S., 1965: Strukturformer hos urskogbestand i Ovre Pasvik Meld. Norges landbr. Uogsk. *44*: 31.

Ivimey–Cook, R.B., PROCTOR, M.C.F., 1966: The plant communities of the Burren, Co. Clare. Proc. Roy. Irish Acad. *64*, Sect. B, No 15: 211–301.

Jacobs, R.P.W.N., 1983: Biomass potential of Eelgrass (*Zostera marina* L.). See: CRC Critical Rev. *2*.

Jefferies, R.L., Davy, A.J., Rudmik, T., 1981: Population biology of the salt marsh annual Salicornia europaea agg. J. Ecol. *69*: 17–31.

Jenik, J., 1961: Alpinska vegetace Krkonos, Kralickeho Snezniku a Hrubeho Jeseniku. Teorie anemo–orografickych systemu. Naklad. Ceskoslov. Akad. Ved, Praha: 409 pp.

Jugowiz, R.A., 1908: Wald und Weide in den Alpen I. Wien: 98 pp.

Kinzel, H., 1982: Pflanzenökologie und Mineralstoffwechsel. Verlag Eugen Ulmer, Stuttgart: 534 pp.

Knorre, D. von, 1974: kosystembindung von Asseln und Mollusken. Mitt. Sekt. Geobot. Phytotax. Biol. Ges. DDR *1974*: 137–149.

Knuchel, H., 1914: Spektrophotometrische Untersuchungen im Walde. Mitt. Schweiz. Centralanst. Forstl. Versuchsw. *11*: 1–94.

Koblet, H., 1966: ber Fragen der Bestandesbildung im Futterbau. Z. Acker– u. Pflanzenbau *124*: 165–178.

Koenies, H., 1982: ber die Eigenart der Mikrostandorte im Fußbereich der Altbuchen unter besonderer Berücksichtigung der Schwermetallgehalte in der organischen Auflage und im Oberboden. Diss. Gesamthochsch. Kassel, 288 pp.

Körner, C., 1982: CO2 exchange in the alpine sedge *Carex curvula* as influenced by canopy structure, light and temperature. Oecologia (Berl.) *53*: 98–104.

–, Jussel, V., Schiffer, K., 1978: Transpiration, Diffusionswiderstand und Wasserpotential in verschiedenen Schichten eines Grünerlenbestandes. Veröff. sterr. Maß–Hochgebirgsprogr. Hohe Tauern (Innsbruck) *2*: 81–98.

–, Mayr, R., 1980: Stomatal behaviour in alpine plant communities between 500 and 2600 metres above sea level. In: GRACE J., FORD, E.D., JARVIS, P.G. (eds.): Plants and their atmospheric environment. Blackwell Sci. Publ., Oxford etc.: 205–218.

–, Moraes, J.A.P.V. de, 1979: Water potential and diffusion resistance in alpine cushion plants on clear summerdays. Oecol. Plant. *14*: 109–120.

–, Wieser G., Guggenberger, H., 1980: Der Wasserhaushalt eines alpinen Rasens in den Zentralalpen. Veröff. sterr. Maß–Hochgebirgsprogr. Hohe Tauern (Innsbruck) *3*: 243–264.

Künstle, E., Mitscherlich, G., Hdrich, F., 1979: Gaswechseluntersuchungen in Kiefernbestnden im Trockengebiet der Oberrheinebene. Allg. Forst– u. Jagdztg. *150*: 205–227.

Kyriakopoulos, E., Richter, H., 1977: A comparison of methods for the determination of water status in *Quercus ilex* L. Pflanzenphys. *82*: 14–17.

Lang, G., 1981: Die submersen Makrophyten des Bodensees – 1978 im Vergleich mit 1967. Internat. Gewsserschutzkommission Bodensee, Ber. *26*: 64 pp.

Lee, J.A., 1977: The vegetation of British inland salt marshes. J. Ecol. *65*: 673–698.

Leibundgut, H., 1983: Die waldbauliche Behandlung wichtiger Waldgesellschaften der Schweiz. Mitt. Eidgen. Anst. Forstl. Versuchsw. *59*: 3–78.

Leuschner, C., 1982: kologische Untersuchungen in den Halb– und Volltrockenrasen des Oberelsaß. Dipl.–Arb. Syst.–Geobot. Inst. Göttingen, 176 pp.

Livett, E.A., Lee, J.A., Tallis, J.H., 1979: Lead, zinc, and copper analyses of British blanket peats. J. Ecol. *67*: 865–891.

Lohmeyer, W., 1970: ber das Polygono–Chenopodietum in Westdeutschland unter besonderer Berücksichtigung seiner Vorkommen am Rhein und im Mündungsgebiet der Ahr. Schr. Vegetationskunde (Bonn) *5*: 7–28.

Lüdi, W., 1934: Beitrag zur Kenntnis der Beziehungen zwischen Vegetation und Boden im östlichen Aarmassiv. Ber. Geobot. Forsch. Inst. Rübel, Zürich *1933*: 41–54.

Luther, H., 1983: On life forms, and above–ground and underground biomass of aquatic macrophytes. Acta Botan. Fenn. *123*: 1–23.

Mkirinta, U., 1978: Ein neues ökomorphologisches Lebensformen–System der aquatischen Makrophyten. Phytocoenol. *4*: 446–470.

Mars, R.H., Bannister, P., 1978: Response of several members of the ericaceae to soils of contrasting pH and base status. J. Ecol. *66*: 829–834.

Matthes, H., 1978: Der Tannenhher im Engadin. Studien zu seiner kologie und Funktion im Arvenwald. Münstersche Geogr. Arb. *2*: 87 pp.

McRoy, C.P., Helfferich, C. (eds.), 1977: Seagrass ecosystems. Marine Science (New York) *4*: 314 pp.

Meusel, H., 1951: ber Pflanzengemeinschaften: Probleme der Vegetationskunde, behandelt an einigen Pflanzenvereinen der Heimat. Urania *14*: 95–106.

Miller, G.R., 1979: Quantity and quality of the annual production of shoots and flowers by *Calluna vulgaris* in North–east Scotland. J. Ecol. *67*: 109–129.

Mitscherlich, G., Künstle, E., 1970: Untersuchungen über die Bodentemperatur in einigen Nadel– und Laubholzbestnden in der Nhe von Freiburg/Br. Allg. Forst– u. Jagdztg. *141*: 129–134.

Mittnacht, A., 1980: Segetalflora der Gemarkung Mehrstetten 1975–78 im Vergleich zu 1948/49. Diss. Univ. Hohenheim: 95 pp.

Moll, W., 1959: Bodentypen im Kreis Freiburg i.Br. Ber. Naturf. Ges. Freiburg i. Br. *49*: 5–58.

Morrison, M.E.S., 1959: The ecology of a raised bog in Co. Tyrone, northern Ireland. Proc. Roy. Irish Acad. *60*, Sect. B No. 9: 291–308.

Müller, S., 1981: Oberbodenstörungen nache Weide und Streunutzung, Mitt. Ver. Forstl. Standortskde.u. Forstpflanzenzücht. *29*: 3–6.

Neuwinger, I., 1967: Zum Nhrstoffhaushalt in Vegetationseinheiten der subalpinen Entwaldungszone. Mitt. Forstl. Bundesversuchsanst. Wien *75*: 269–303.

Niemi, A., 1967: Hemerophilous plants on gull skerries in the archipelago SW of Helsingfors. Mem. Soc. Fauna et Flora Fenn. *43*: 8–16.

Oberdorfer, E., 1960: Pflanzensoziologische Studien in Chile. Ein Vergleich mit Europa. Flora et Vegetatio Mundi *2*: 208 pp.

Ott, F., 1961: ber den Einfluß von Flechtensuren auf die Keimung verschiedener Baumarten. Schweiz. Z. Forstwes. *1961*: 303–304.

Otte, A., 1983: nderungen in Ackerwildkraut–Gesellschaften als Folge sich wandelnder Feldbaumethoden in den letzten 3 Jahrzehnten – dargestellt an Beispielen aus dem Raum Ingolstadt. Diss. TU München, 159 pp. +Anhang.

Overrein, L.N., Seip, H.M., Tollan, A., 1980: Acid precipitation – effects on forest and fish. Final Report of the SNSF–Projekt 1972–1980 (Oslo), 175 pp.

Packham, J.R., Willis, A.J., 1977: The effects of shading on *Oxalis acetosella*. J. Ecol. *65*: 619–642.

Peace, W.J.H., Grubb, P.J., 1982: Interaction of light and mineral nutrient supply in the growth of *Impatiens parviflora*. New Phytol. *90*: 127–150.

Persson, S., 1981: Ecological indicator values as an aid in the interpretation of ordination diagrams. J. Ecol. *69*: 71–84.

Pietsch, W., 1973: Beitrag zur Gliederung der europischen Zwergbinsengesellschaften (Isoto–Nanojuncetea Br.–Bl. et Tx. 1943). Vegetatio *28*: 401–438.

–, 1980: Zeigerwerte der Wasserpflanzen Mitteleuropas. Feddes Repert. *91*: 106–213.

–, 1982: Makrophytische Indikatoren für die ökochemische Beschaffenheit der Gewsser. In: BREITIG, G., TMPLING, W. von (Hrsg.): Ausgewhlte Methoden der Wasseruntersuchung. Bd. II: Biologische, mikrobiologische und toxikologische Methoden. VEB Gustav Fischer Verlag Jena, p. 67–89.

Pijl, I. van der, 1969: Principles of dispersal in higher plants. Springer–Verlag, Berlin–Heidelberg–New York: 154 pp.

Pisek, A., Winkler, E., 1958: Assimilationsvermögen und Respiration der Fichte (*Picea excelsa* LINK.) in verschiedener Höhenlage und der Zirbe (*Pinus cembra* L.) an der alpinen Waldgrenze. Planta (Berlin) *51*: 518–543.

Pott, R., 1983: Geschichte de Hude– und Schneitel–wirtschaft in Nordwestdeutschland und ihre Auswirkungen auf die Vegetation. Oldenburger Jb. *83*: 357–376.

–, Burrichter, E., 1983: Der Bentheimer Wald. Geschichte, Physiognomie und Vegetation eines ehemaligen Hude– und Schneitelwaldes. Forstwiss. Cbl. *102*: 350–361.

Rahman, M.S., Rutter, A.J., 1980: A comparison of the ecology of *Deschampsia cespitosa* and *Dactylis glomerata* in relation to the water factor. II. Controlled experiments in glass–house conditions. J. Ecol. *68*: 479–491.

Rheineimer, G., 1957: ber die Standorte der Moosvegetation in Nadelholzforsten bei Hamburg.

Mitt. Staatsinst. Allg. Botan. *11*: 89–136.

Rogister, J.E., 1980: Ekologische en floristische vergelijking van de belangrijkste bosplantengezelschappen in Belgisch Lorreinen en in het G.H. Luxemburg. Trav. Stat. Rech. Eaux et Forts Groenendaal–Hocilaart, Belg. Ser. A *23*: 40 pp.

– , 1981: Rangschikking van de belangrijkste boskruidsoorten volgens humuskwaliteit en bodemvochtigheid. Trav. Stat. Rech. Eaux et Forts Groenendaal–Hocilaart, Belg., Ser A *25*: 22 pp.

Roller, M., 1965: Kleinklimatische Unterrsuchungen in einem alpinen Urwaldbestand. Geofis. Meteorol. (Genova) 11: 20 pp.

Rozema, J., Blom, B., 1977: Effects of salinity and inundation on the growth of *Agrostis stolonifera* and *Juncus gerardii*. J. Ecol. 65: 213–222.

Runge, F., 1981: Zur Phnologie des Weidelgras–Breitwegerichrasens. Natur u. Heimat (Münster i.W.) *41*: 28–32.

Rutter, A.J., 1955: The composition of wet–heath vegetation in relation to the water table. J. Ecol. 43: 507–543.

Samek, V., Javurek, M., 1964: Svetlostni stadia bucin a smrkobucin ve vztahu k prirozene obnove drevin. (Zerfallsphasen und Naturverjüngung in den Buchenmischwldern). Lesn. Casopis, Rocn. *10*: 173–194.

Schiefer, J., 1981: Bracheversuche in Baden–Württemberg. Beih. Veröff. Naturschutz u. Landschaftspflege Baden–Württ. *22*: 328 pp.

Schinner, F., 1978: Die subalpine Waldgrenze und Bedeutung der Mykotrophie im Gasteiner Tal. Veröff. sterr. Maß–Hochgebirgsprogr. Hohe Tauern (Innsbruck) *2*: 311–314.

Schlüter, H., Baller, A., 1982: Vegetationskundliche und ökologische Untersuchungen im Naturschutzgebiet 'Hohe Lehde' bei Jena. Landschaftspflege u. Natursch. i. Thür. *19*: 1–10.

Schmidt, W., 1982: Ungestörte und gelenkte Sukzession auf Brachckern. Scripta Geobot. 15: 199 pp.

– , 1983: Experimentelle Syndynamik – Neuere Wege zu einer exakteren Sukzessionsforschung, dargestellt am Biespiel der Gehölzentwicklung auf Ackerbrachen. Ber. Deut. Botan. Ges. *96*: 511–533.

Schoof–Van Pelt, M.M., 1973: Littorelletea. A study of the vegetation of some amphiphytic communities of western Europe. Thesis Univ. Nijmegen: 216 pp.

Seaward, M.R.D., 1977: Lichen Ecology. Academic Press, London, New York, San Francisco: 550 pp.

Sharifi, M.R., 1983: The effects of water and nitrogen supply on the competition between three perennial meadow grasses. Acta Oecol., Oecol. Plant. *4*(18): 71–82.

Sissingh, G., 1973: ber die Abgrenzung des Geo–Alliarion gegen das Aegopodion podagrariae. Mitt. Florist.–soziol. Arb. gem. NF. *15/16*: 60–65.

Sloboda, S., 1982: Pflanzengesellschaften als Kriterien zur ökologischen Kennzeichnung der Standortsmosaike. Arch. Naturschutz u. Landschaftsforsch. (Berlin) *22*: 79–101.

Small, E., 1972: Photosynthetic rates in relation to nitrogen recycling as an adaptation to nutrient deficiency in peat bog plants. Can. J. Botany *50*: 2227–2233.

Smith, A.J.E., 1978: The Moss–Flora of Britain and Ireland. Cambridge.

Smith, C.J., 1980: Ecology of the English chalk. Academic Press, London, 573 pp.

Sorg, J.–P., 1980: Vegetation et rejeunissement naturel dans la pessiere subalpine de Vels (GR). Mitt. Eidg. Anst. Forstl. Versuchswes. *56*: 1–94.

Sparling, J.M., 1967: The occurrence of *Schoenus nigricans* L. in blanket bogs. I. Environmental conditions affecting the growth of *S. nigricans* in blanket bog. J. Ecol. *55*: 1–13.

Steenis, C.G.G.J. van, 1961: An attempt towards an explanation of the effect of mountain mass elevation. Proc. Acad. Sci. Amsterdam 64: 435–442.

Sukopp, H., Markstein, B., 1981: Vernderungen von Röhrict–bestnden und –pflanzen als Indikatoren von Gewssernutzungen, dargestellt am Beispiel der Flavel in Berlin (West.) Limnologica (Berlin) *13*: 459–471.

Sydes, C., Grime, J.P., 1981: Effects of tree leaf litter on herbaceous vegetation in deciduous woodland. II. An experimental investigation. J. Ecol. *69*: 249–262.

Sykora, K.V., 1983: The Lolio–Potentillion anserinae Tüxen 1947 in the northern part of the Atlantic domain. Proefschrift Kathol. Univ. Nijmegen, Publication VI + VII 1983: 118 pp.

Tensley, A.G., 1968: Britain's Green Mantle. Past, Present and Future. 2nd edn revised by M.C.F. PROCTOR. London: George Allen and Unwin Ltd., 327 pp.

Thompson, K., Grime, J.P., 1979: Seasonal variation in the seed banks of herbaceous species in ten contrasting habitats. J. Ecol. 67: 893–921.

Tranquillini,. W., 1967: ber die physiologischen Ursachen der Wald- und Baumgrenze. Mitt. Forstl. Bundesversuchsanst. Wien *75*: 457–487.

– , 1982: Frost–drought and its ecological significance. Encyclopedia of Plant Physiology, N.S. *128*: 379–400.

Tschager, A., Hilscher, H., Franz, S., Kull, U., Larcher, W., 1982: Jahreszeitliche Dynamik der Fettspeicherung von *Loiseleuria procumbens* und anderen Ericaceen der alpinen Zwergstrauchheide. Oecol. Plant. *17*: 119–134.

Turkington, R., Cahn, M.A., Vardy, A., Harper, J.L., 1979: The growth, distribution and neighbour relationships of *Trifolium repens* in a permanent pasture. III. The establishment of *Trifolium repens* in natural and perturbed sites. J. Ecol. *67*: 231–243.

– , Harper, J.L., 1979: The growth, distribution and neighbour relations of Trifolium repens in a permanent pasture. I. Ordination, pattern and contact. J. Ecol. *67*: 201–218. – IV. Fine scale biotic differentiation: 245–254.

Tüxen, R., Ellenberg, H., 1947: Erluterungen zur pflanzensozio–logischen Karte der Fürstlich knyphausenschen Forsten in Lütetsburg bei Norden (Ostfriesland). Manuskr. Stolzenau/Weser.

– , Wilmanns, O., Schwabe–Braun, A., 1981: Querceto–Fagetea. Bibliogr. Phytosociol. Syntax (Vaduz) *35*: 1118 pp.

Umezu, Y., 1964: ber die Salzwasserpflanzengesellschaften in der Nhe von Yukuhasi, Nordksy, Japan. Jap. J. Ecol. *14*: 153–160.

Vogel, M., 1981: kologische Untersuchungen in einem Phragmites–Bestand. Diss. Univ. Marburg/Lahn: 97 pp.

Wagenitz, G., Meyer, G., 1981: Die Unkrautflora der Kalkcker bei Göttingen und im Meißnervorland und ihre Vernderungen. Tuexenia, Mitt. Florist.–soziol. Arb.gem. N.S. *1*: 7–23.

Walter, J.–M.N., 1981: Architectural profiles of flood–forests in Alsace. Manuscr. Lab. Ecol. Veget. Univ. Strasbourg, 27 pp.

Wardle, P. 1974: Alpine timberlines. In: IVES, J.D., BARRY R.G. (eds.): Arctic and alpine environments. Methuen, London: 371–402.

Warnke, E., 1980: Spring areas: ecology, vegetation, and comments on similarity coefficients applied to plant communities. Holarctic Ecol. *3*: 233–333.

Waterhouse, D.F., 1974: The biological control of dung. Sci. Amer. *230*: 101–109.

Weihe, K. von, 1980: Konkurrenzvorgnge bei der Aussüßung von Soden des Puccinellion maritimae. Jber. Inst. Angew. Botan., Hamburg *95/96*: 232–250.

Westhoff, V., Sykora, K.V., 1979: A study of the influence of desalination on the Juncetum gerardii. Acta Botan. Neerle. *28*: 505–512.

Wheeler, B.D., 1980: Plant communities of rich–fen systems in England and Wales. I. Introduction. Tall sedge and reed communities. J. Ecol. *68*: 365–395.

– , 1980: Plant communities of rich–fen systems in England and Wales. II. Communities of calcareous mires. J. Ecol. *68*: 405–420.

– , 1980: Plant communities of rich–fen systems in England and Wales. III. Fen meadow, fen grassland and fen woodland communities. J. Ecol. *68* 761–788.

– , Giller, K.E., 1982: Species richnmess of herbaceous fen vegetation in Broadland, Norfolk in relation to the quantity of above–ground plant material. J. Ecol. *70*: 179–200.

Wierzchowska, U., 1981: Ecological amplitude and the regional variation of soil conditions in oak–hornbeam forests, Tilio–Carpinetum Tracz. 1962, in Poland. Ekol. Pol. *29*: 469–498.

Wilmanns, O., Brun–Hool, J., 1982: Irish Mantel and Saum vegetation. J. Life Sci. (Dublin) *3*: 165–174.

Zimmermann, A., 1982: Erica–reiche Silikat–Föhrenwlder in den östlichen Zentralalpen (III): überregionaler Vergleich. Phyton (Austria) *22*: 283–316.

Zoldan, J.–W., 1981: Zur kologie, insbesondere zur Stickstoff–Versorgung von Ackerunkraut–Gesellschaften in Süd–Niedersachsen und Nordhessen. Diss. Univ. Göttingen, 96 pp.

II Remarks on the changes in the system of plant sociology

It has already been mentioned in the foreword to this new edition that section E II of the second edition 'Notes on the vegetation descriptions of interesting regions' has now been omitted, mainly to provide more space for an extension of the bibliography and of the plant community system, including the character species (E III 1). It has been necessary to make several changes in the latter, compared with the second edition, in order to bring it into line as far as possible with the new edition of *South German Plant Communities* by Oberdorfer and his co-workers (1977, 1978, 1979). However I cannot go along with all the changes of names or status and would like briefly to give the reasons for this and for some important rearrangements of the groups. I have done this here so that the format of the book can be adhered to. The changes have also been taken into consideration in sections B, C and D.

In the class *Lemnetea* (1.1) neither the alliances *Lemnion gibbae* (1.111) and *Lemnion trisulcae* (1.112) nor the alliance *Hydrocharition* (1.113) from the survey by Th. Müller (in Oberdorfer 1977) have been adhered to since none of the character species named, according to Tüxen (1974b) in the second edition, are true for any of them. As stated in older literature all Duckweed and Frogbit communities must be included in a single alliance *Lemnion minoris* whose character species in Central Europe are also those of the order and class of the floating plant communities. On the other hand, because of the special floristic characteristics, the alliance of Peatmoss–Bladderwort bog pools (1.114 in the second edition) is raised to the rank of order.

Conversely the brackish-water Clubmoss communities (*Bolboschoenetea*, 2.7 in the second edition) are downgraded to class rank. Since, like many Reed beds (1.4 *Phragmitetea*), they are often found in only slightly brackish water, they are best included with these. Here they form the alliance 1.414 *Bolboschoenion* (= *Scirpion maritimi* of Oberdorfer). Other such alliances, poor in species, have recently been raised to the rank of class or order by Oberdorfer and his co-workers, but I prefer not to take this line.

In my opinion it is unfortunate that several units in the class of Matgrass swards and grass heaths (5.1 *Nardo–Callunetea*) have been renamed. For this reason I have placed them in parentheses in the survey in section E III 1. To designate Matgrass swards of low ground *'Violion caninae'* instead of *'Violo–Nardion'*, when they are dominated by *Nardus* just as much as those in the mountains (5.111 *Eu–Nardion*), I consider a mistake. On the other hand I welcome the new alliance of the 'Matgrass moist sward with Heath Rush' (*Juncion squarrosi*, 5.113). Since the Western European *Erica cinerea* and *Ulex nanus* heaths are included in a separate order (*Erico–Ulicetalia*) the Central European Calluna heaths must likewise form their own order (previously they were included with the former as *Calluno–Ulicetalia*, 5.12). The name *'Vaccinio–Genistetalia'* proposed by Schubert (1960) however disregards the dominating Heather. I suggest that the term *'Genisto–Callunetalia'* (5.12) corresponds more closely to the previous alliance *Genisto–Callunion* (5.121 which should now be called *Genistion*).

The thorough investigation of the 'Wall Pepper tracks, sandy swards, stony ground and rocky outcrops' (5.2 *Sedo–Scleranthetea*) by Korneck (in Oberdorfer 1978) led to the dissolution of the previous order *Festuco–Sedetalia* (5.23). The Crested Hairgrass sandy steppes of the Upper Rhine plain (*Koelerion glaucae*, previously 5.231) are better included in the Silvergrass communities (*Corynephoretalia*, 5.22) as 5.224. On the other hand the coastal Sea Pink sward (*Armerion elongatae*, previously 5.232, now 5.323) is better placed in the class of poor soil swards (*Festuco–Brometea*, 5.3). Korneck puts the Hairgrass communities, consisting mainly of short-lived species, in a separate order (*Thero–Airetalia*, 5.24). I am glad to say there have been few amendments in the remaining grassland communities.

In the group 'woody perennials and scrub' (6) only a few insignificant changes have taken place. In the class *Betulo–Adenostyletea* two new alliances have been added to the order *Adenostyletea* (6.312 *Salicion arbusculae* and 6.313 *Calamagrostion arundinaceae*).

I do not feel any justification for including the extreme acid-soil Birch–Oak woods (8.311 *Quercus robori–petraeae* and 8.31 *Quercetalia r–p*) in the 'rich-soil deciduous woods and shrubland' (8.4 *Querco–Fagetea*) which is already an extremely all-embracing class. It is true that there are intermediates between the mull and moder Beech woods (8.431.1 and 8.431.2) and also between the latter and the Birch–Oak woods. The extremes however no longer have anything in common and, apart from *Anemone nemorosa* and *Convallaria majalis* there are none of the long list of class-character species of 8.4 to be found spreading into the Birch–Oak woods. Furthermore the acid-soil Oak woods play such an independent rôle in the more strongly oceanic western Europe that one should not deny them a special place in their Central European marginal area. In the order *Fagetalia* the shade-slope and ravine woods, together with some Lime or Elm mixed woods, have recently been given the status of their own alliance (8.434 *Tilio–Acerion*) which they undoubtedly deserve.

A systematic listing and ecological evaluation such as those on pages 645–676 can never be final and entirely satisfactory. Oberdorfer and his colleagues concentrated on the German Federal Republic and its neighbours, particularly in the south. The author's own experience has been obtained mainly in the north-west and south-west of Germany, in Switzerland and western Austria, as well as in parts of Poland, Czechoslovakia, the German Democratic Republic and the Netherlands. Consequently the species are arranged mainly according to their behaviour in the western parts of Central Europe. Information relating to aberrant systematic and/or ecological behaviour - preferably with details of the locality - would be gratefully received (under the code name Species of Central Europe' and addressed to H. Ellenberg, Untere Karspüle 2, D. 3400, Göttingen).

III Summary of the vegetation units and species

1 The system of plant communities with character species

This summary has been brought up-to-date showing the position in the autumn of 1981. In many places there have been changes from that in the second edition. The order of the eight main groups and the classes within the groups corresponds to the sociological progression. Names of the authors have been omitted; they can be found in Oberdorfer (1979) and F. Runge (1980).

The suffix **-etea** denotes a class, **-etalia** an order, **-ion** an alliance, UV a suballiance. The character species in each case are printed in italics.

1 FRESHWATER AND MIRE VEGETATION

1.1 **Lemnetea,** free-floating still water communities *Lemna minor.*

1.11 **Lemnetalia**, Lemnion minoris, floating communities of waters more or less rich in nutrients. The following species characterise at the same time association, alliance and order (p. 296, 297).

(1.111 & 2) Duckweed and floating liverwort comm.: *Azolla filiculoides, Lemna gibba, L. trisulca, Riccia rhenana* (=*fluitans*, M), *Ricciocarpus natans* (M), *Salvinia natans, Spirodela polyrrhiza, Wolffia arrhiza.*

(1.113) Frogbit and Water Soldier comm. etc.: *Aldrovanda vesiculosa, Hydrocharis morsus-ranae, Stratiotes aloides. Utricularia australis, U. vulgaris.*

1.12 (1.114) **Ultriculariêtalia,** Sphagno-Utricularion (Utricularion intermedio-minoris), Peatmoss-Bladderwort bog pools (put in a separate class by Oberdorfer and co-workers, 1977) *Sparganium minimum Utricularia intermedia, U. Minor, U. ochroleuca*

Potamogetonetea, Potamogetonetalia[1]**), rooted waterplant comm:** *Ceratophyllum demersum, C. submersum, Elodea canadensis, Myriophyllun spicatum, Potamogeton angustifolius, P. crispus, P. lucens, P. pectinatus, P. perfoliatus, P. pusillus* (p. 293–295).

1.211 Potamogetonion[1]) Pondweed, Naiad and Horned Pondweed submerged communities: —mainly in Pondweed comm: *Potamogeton acutifolius, P. alpinus, P. compressus, P. gramineus, P. filiformis, P. nitens, P. obtusifolius, P. praelongus, P. salicifolius;* —mainly in Naiad comm: *Najas flexilis, N. intermedia, N. marina, N. minor;*
—mainly in Horned Pondweed comm: *Zannichellia palustris* ssp. *palustris.*

1.212 Nymphaeion, rooted floating-leaf cover: —mainly in Waterlily comm: *Myriophyllum verticillatum, Nuphar lutea, N. pumila, Nymphaea alba, N. candida, Potamogeton natans, Ranunculus peltatus;*
—mainly in Crowfoot-Water Violet comm: *Callitriche palustris, Hottonia palustris, Potamogeton friesii, Ranunculus aquatilis.*

1.212 Ranunculion fluitantis, Water Crowfoot running-water comm: *Butomus umbellatus f. valisneriifolius,* submerged forms of *Callitriche cophocarpa, hamulata, obtusangula and stagnalis. Groenlandia densa, Potomogeton colaratus, P. helveticus, P. natans* var. *prolixus, P. nodosus, Ranunculus fluitans, R. penicillatus, R. trichophyllus, Sagittaria sagittifolia* var. *vallisnoriifolia, Schoenoplectus lacustris* var. *fluitans, Sparganium emersum* ssp. *fluitans* (p. 307–311).

1.3 **Littorelletea, Littorelletalia,** Shore Hairgrass shallow-water swards and related communities: *Deschampsia setacea, Elatine hexandra, Juncus bulbosus, Littorella uniflora, Potamogeton oblongus, P. polygonifolius, Ranunculus ololeucos, Veronica scutellata* (p. 291).

1.311 Deschampsion littoralis, Shore Hairgrass-Bodensee shore sward: *Deschampsia littoralis* (=*rhenana*), *Myosotis rehsteineri, Ranunculus reptans.*

1.313 Isoëtion, Quillwort clear-water sward: *Isoëtes echinospora, I. lacustris, Lobelia dortmanna, Sparganium angustifolium, Subularia aquatica.*

1.314 Eliocharition acicularis, Spike Rush shallow-water swards: *Eleocharis acicularis, Luronium natans, Myriophyllum alternifolium.*

1.315 Hydrocotylo-Baldellion, Atlantic shore shallow-water sward: *Apium inundatum, Baldellia ranunculoides, Hypericum elodes, Pilularia globulifera.*

1.4 **Phragmitetea, Phragmitetalia,** Reed and Tall Sedge swamps: *Acorus calamus, Alisma plantago-aquatica, Eleocharis palustris, Equisetum fluviatile, Iris pseudacorus, Phragmites australis, Rumex hydrolapathum, Sagittaria sagittifolia, Schoenoplectus mucronatus* (p 297–301, 305–307).

[1]) The correct names are difficult to pronounce. Shortened forms are commonly used, namely Potametea, Potametalia and Potamion.

1.411 Phragmition, reeds in slow-moving water
—mainly in Bulrush and Reed Mace beds: *Schoenoplectus lacustris, Typha angustifolia, T. latifolia;*
—mainly in Reedgrass beds: *Glyceria maxima;*
—mainly in Bur-reed beds: *Sparganium erectum;*
—mainly in Saw Sedge reed beds: *Cladium mariscus;*
—mainly in Small Reed beds: *Carex pseudocyperus, Cicuta virosa;*
—mainly in Flowering Rush and Water Dropwort beds: *Butomus umbellatus, Oenanthe aquatica, Rorippa amphibia, Sparganium emersum;*
—in many reed bed communities of the alliance: *Hippuris vulgaris, Ranunculus lingua, Sium latifolium.*

1.412 Sparganio-Glycerion fluitantis, stream reed beds: *Apium nodiflorum, Berula erecta, Epilobium roseum, Glyceria fluitans, G. plicata, Nasturtium officinale, Scrophularia umbrosa, Veronica anagallis-aquatica, V. beccabunga* (p. 309–312).

1.413 Magnocaricion, Tall Sedge swamps: *Carex cespitosa, C. disticha, C. riparia, Cyperus longus, Galium palustre, Oenanthe fistulosa, Peucedanum palustre, Poa palustris, Rorippa anceps, Scutellaria galericulata* (p. 301–305).

1.413.1 UV Caricion elatae (Caricion rostratae) Tall Sedge mires:
—mainly in Tall Tussock Sedge mires: *Carex appropinquata, C. elata, C. paniculata, Lysimachia thyrsiflora, Senecio paludosus;*
—mainly in sward-forming Tall Sedge mires: *Carex rostrata;*

1.413.2 UV Caricion gracilis, Tall Sedge beds poor in peat: *Carex gracilis, C. vesicaria, C. vulpina* (p. 595).

1.413.3 UV (?) Phalaridion, stream reed beds, *Phalaris arundinacea, Rumex aquaticus* (see also p. 249, 250).

1.414 Bolboschoenion maritimi (?) brackish-water reed beds and related communities: *Bolboschoenus* (= *Scirpus*) *maritimus, Schoenoplectus americanus, S. carinatus, S. kalmussii, S. tabernaemontanus, S. triqueter* (p. 306, 307, 363, 366).

1.5 **Montio-Cardaminetea,** Montio-Cardaminetalia, **spring communities:** *Bryum schleicheri* (M), *Cardamine amara, Epilobium alsinifolium, Saxifraga stellaris* (pp. 313, 430, 431).

1.511 Montio-Cardaminion, soft-water spring communities: *Ranunculus hederaceus, Stellaria alsine.*

1.511.1 UV Montion, unshaded soft-water spring communities: *Bryum weigelii* (M), *Diobelon squarrosum* (M), *Epilobium nutans, Montia fontana, Philonotis fontana* (M), *Ph. sericea* (M), *Plectocolea obovata* (M), *Scapania paludicola* (M), *S. paludosa* (M).

1.511.2 UV Cardaminion, shaded soft-water spring comm.: *Cardamine amara, C. flexuosa, Chrysosplenium oppositifolium.*

1.512 Cratoneurion, tufa spring communities: *Arabis soyeri, Cochlearia pyrenaica, Cratoneuron commutatum* (M), *C. filicinum* (M), *Philonotis calcarea* (M), *Saxifraga aizoides* (weak).

1.6 **Scheuchzerio-Caricetea nigrae (fuscae),** Small Sedge intermediate mire and swamp swards: *Calliergon stramineum* (M), *Campylium stellatum* (M), *Carex dioica, C. nigra, C. panicea, Dactylorhiza traunsteineri, Drepanocladus revolvens* (M), *Eriophorum angustifolium, Fissidens adianthoides* (M), *Juncus triglumis, Menyanthes trifoliata, Parnassia palustris, Pedicularis palustris, P. sceptrum-carolinum, Potentilla palustris, Sphagnum subsecundum* (M), *Tomenthypnum nitens* (M), *Trichophorum alpinum, Triglochin palustre* (p. 318–322, 334–336, 428–429).

1.61 **Scheuchzerietalia,** intermediate mires and hollows: *Agrostis canina, Calliergon trifarium* (M), *Carex canescens, C. echinata, C. oederi, Drosera anglica, Meesia triquetra* (M), *Scheuchzeria palustris, Sphagnum contortum* (M).

1.611 Rhynchosporion albae, Beak Sedge hollows and related comm.: *Carex limosa, Drosera intermedia, Lycopodiella inundata, Rhynchospora alba, R. fusca* (p. 334–336).

1.612 Caricion nigrae, acid Small Sedge meadows: *Cardamine matthiolii, Carex paupercula, C. pulicaris, Eriophorum scheuchzeri* (p. 321, 322, 428).

1.613 Caricion lasiocarpae, intermediate mire Sedge swards: *Bryum neodamense* (M), *Calamagrostis stricta, Calliergon giganteum* (M), *Carex chordorrhiza, C. diandra, C. heleonastes, C. lasiocarpa, Cinclidium stygium* (M), *Eriophorum gracile, Sphagnum obtusum* (M), *Stellaria crassifolia* (p. 321–323).

1.62 **Tofieldietalia,** calcareous Small Sedge beds and related communities: *Bartsia alpina, Carex flava, Eleocharis quinqueflora, Juncus alpino-articulatus, Pinguicula vulgaris, Riccardia pinguis* (M), *Selaginella selaginoides.*

1.621 Caricion davallianae, calcareous fens and Small Sedge meadows: *Carex hostiana, Dactylorhiza incarnata, Epipactis palustris, Eriophorum latifolium, Liparis loeselii, Pinguicula leptoceras, Primula farinosa, Swertia perennis, Taraxacum palustre, Tofieldia calyculata* (p. 318–321, 429, 430).
—mainly in Bog Rush calcareous fens: *Gentiana utriculosa, Orchis palustris, Schoenus ferrugineus, S. intermedius, S. nigricans, Spiranthes aestivalis;*
—mainly in Davall's Sedge spring flush swards: *Carex davalliana.*

1.622 Caricion maritimae, alpine Rush-leaved Sedge calcareous swamp swards: *Carex bicolor, C. frigida, C. maritima, C. microglochin, Equisetum variegatum, Juncus arcticus, Tofieldia pusilla, Trichophorum pumilum, Typha minima* (p. 429, 430).

1.7 **Oxycocco-Sphagnetea,** raised bogs and wet heaths: *Aulacomnium palustre* (M), *Drosera rotundifolia, Kalmia angustifolia, Sphagnum tenellum* (M), *Trichophorum cespitosum, Vaccinium oxycoccus* (p. 339ff).

1.71 **Sphagnetalia magellanici,** Sphagnion magellanici, raised bog hummocks and related communities: *Andromeda polifolia, Betula nana, Calypogeia sphagnicola* (M), *Carex pauciflora, Cephaloziella connivens* (M), *Eriophorum vaginatum, Mylia anomala* (M), *Pohlia nutans* var. *sphagnetorum* (M),

Polytrichum strictum (M), *Rubus chamaemorus*, *Sphagnum angustifolium* (M), *S. fuscum* (M), *S. magellanicum* (M), *S. papillosum* (M), *S. rubellum* (M), *Talaranea setacea* (M).

1.72 **Sphagno-Ericetalia,** Ericion tetralicis, wet heaths: *Erica tetralix*, *Juncus balticus*, *Narthecium ossifragum*, *Sphagnum compactum* (M), *S. molle* (M), *Trichophorum germanicum* (p. 338, 510–513).

1.8 **Charetea fragilis,** Stonewort submerged swards (all Algae of which one species is usually completely dominant): *Chara fragilis*, *Nitella batrachosperma* (p. 291).

1.81 **Nitelletalia flexilis** (with the alliances Nitellion flexilis and Nitellion syncarpotenuissimae), more or less acid-loving communities: *Chara braunei*, *Nitella flexilis*, *N. mucronata*, *N. opaca*, *N. syncarpa*, *N. tenuissima*.

1.82 **Charatalia hispidae** (with the alliance Charion asperae and the suballiance Charion vulgaris), lime-loving communities: *Chara aspera*, *Ch. contraria*, *Ch. hispida*, *Ch. intermedia*, *Ch. strigosa*, *Ch. tomentosa*, *Nitellopsis obtusa*, *Tolypella glomerata*.

2 SALTWATER AND SEA COAST VEGETATION

2.1 **Zosteretea, Zosteretalia,** Zosterion marinae, Eelgrass sea-bed swards: *Zostera marina*, *Z. noltii* (p. 352, 353).

2.2 **Ruppietea, Ruppietalia**, Ruppion maritimae, Tasselweed comm.: *Eleocharis parvula*, *Ranunculus baudotii*, *Ruppia maritima* (p. 354).

2.3 **Spartinetea, Spartinetalia,** Spartinion, Cordgrass pioneer comm.: *Spartina townsendii* (p. 353, 354).

2.4 **Thero-Salicornietea, Thero-Salicornietalia,** short-lived annual mud flat comm.: *Salicornia europaea* (p. 349–358).

2.411 Salicornion dolichostachyae, Samphire mud flat comm.: *Salicornia dolichostachya* ssp *decumbens*.

2.412 Suaedion maritimae (Salicornion ramosissimae), Seablite comm.: *Bassia hirsuta*, *Salicornia ramosissima*, *Suaeda maritima*.

2.5 **Saginetea, Saginetalia,** Saginion maritimae, Sea Pearlwort comm.: *Sagina maritima* (p. 361).

2.6 **Asteretea, Asteretalia tripolii,** saltmarsh swards: *Aster tripolium*, *Carex secalina*, *Cochlearia officinalis*, *Glaux maritima*, *Juncus maritimus*, *Plantago maritima*, *Spergularia media*, *Triglochin maritimum* (p. 354–360, 364, 365).

2.611 Puccinellion (=Puccinellio-Spergularion marinae), Sea Poa swards: *Cochlearia angelica*, *Halimione pedunculata*, *Puccinellia distans*, *P. maritima*, *Spergularia marina*.

2.612 Armerion maritimae, Thrift swards and related comm.: *Armeria maritima*, *Artemesia maritima*, *Blysmus rufus*, *Carex extensa*, *Halimione portulacoides*, *Juncus gerardii*, *Limonium vulgare*, *Lotus tenuis*, *Odontites litoralis*, *Parapholis strigosa*, *Plantago coronopus*.

2.7 (previously Bolboschoenetea etc., brackish water reedbeds, see 1.414).

2.8 **Cakiletea, Cakiletalia,** tide-line comm.: *Tripleurospermum maritimum* (p. 371, 372).

2.811 Salsolion, Saltwort tideline comm.: *Atriplex calotheca*, *A. glabriuscula*, *Cakile maritima*, *Salsola kali*, *Sonchus arvensis* ssp. *uliginosus* (also p. 607).

2.812 Atriplicion littoralis, Orache tideline comm.: *Atriplex littoralis*, *Crambe maritima*.

2.9 **Ammophiletea,** Marram Grass dune comm.: *Ammophila baltica* (p. 368–374).

2.91 **Ammophiletalia,** Lyme Grass-Marram dunes: *Elymus arenarius*, *Oenothera ammophila*.

2.911 Ammophilion, Marram white dunes: *Ammophila arenaria*, *Calystegia soldanella*, *Eryngium maritimum*, *Lathyrus maritimus*.

2.912 Agropyro-Honkenion, Couch Grass pre-dunes: *Agropyron junceum*, *Honkenia peploides* (also p. 607).

3 HERBACEOUS VEGETATION OF FREQUENTLY DISTURBED SITES

3.1 **Isoëto-Nanojuncetea, Cyperetalia fusci,** dwarf plant communities, inundated in winter but drying out: *Centaurium pulchellum*, *Cyperus fuscus*, *Elatine triandra*, *Gnaphalium uliginosum* (weak), *G. luteoalbum*, *Juncus bufonius*, *J. capitatus*, *J. sphaerocarpus*, *J. tenageia*, *Lythrum hyssopifolia*, *Riccia* spp. (M), *Schoenoplectus supinus* (p. 608–611, 634).

3.111 Nanocyperion, Central and West European dwarf plant comm.: *Blackstonia acuminata*, *Carex bohemica*, *Cyperus flavescens*, *Hypericum humifusum*, *Illecebrum verticillatum*, *Isolepis setacea*, *Marsilea quadrifolia*, *Montia chondrosperma*, *Peplis portula*, *Riccia glauca* (M), *Scirpus radicans*, *Veronica acinifolia*.

3.111.1UV Juncion bufonii, Toad Rush-rich dwarf plant comm.: *Antheceros levis* (M), *A. punctatus* (M), *Centunculus minimus*, *Cicendia filiformis*, *Fossombronia wondraczekii* (M), *Gypsophila muralis*, *Radiola linoides*, *Sagina apetala*, *Spergularia segetalis*.

3.111.2 UV Elatini-Eceocharition ovatae, Spike Rush-rich and related dwarf plant comm.: *Botrydium granulatum* (A), *Elatine alsinastrum*, *E. hydropiper*, *Eleocharis ovata*, *Limosella aquatica*, *Lindernia procumbens*, *Ludwigia palustris*, *Physcomitrella patens* (M), *Riccia cavernosa* (M).

3.2 **Bidentetea, Bidentitalia tripartitae,** Bur Marigold mud banks: *Alopecurus aequalis*, *Bidens radiata*, *B. tripartita*, *Polygonum mite*, *Rorippa palustris* (p. 612–614).

3.211 Bidention tripartitae, Bur Marigold communities in the narrow sense: *Bidens cernua*, *B. connata*, *Catabrosa aquatica*, *Leersia oryzoides*, *Polygonum hydropiper* (weak), *P. minus*, *Ranunculus sceleratus*, *Rumex maritimus*, *R. palustris*, *Senecio congestus*.

3.212 Chenopodium rubri, Orache comm.: *Atriplex hastata*, *Bidens frondosa*, *Chenopodium glaucum*, *Ch. rubrum*, *Polygonum lapathifolium* ssp. *danubiale*, *Xanthium albinum*.

3.3 **Chenopodietea,** communities of waste ground and related arable and garden weed comm.: *Aethusa cynapium*, *Amaranthus albus*, *A. blitoides*, *A. hybridus*, *A. retroflexus*, *Atriplex hortensis*, *A. tarica*, *Capsella bursa-pastoris*, *Chenopodium album*, *Ch. bonus-henricus*, *Ch. botrys*, *Ch. ficifolium*, *Ch. foliosum*, *Ch. murale*, *Ch. vulvaria*, *Corrigiola litoralis*, *Datura stramonium*, *Diplotaxis muralis*, *Echinochloa crus-galli*, *Eragrostis minor*, *Geranium rotundifolium*, *Linaria vulgaris*,

Panicum capillare, Phleum paniculatum, Portulaca oleracea, Senecio vulgaris, Solanum luteum, S. nigrum, Sonchus oleraceus, Stellaria media, Tripleurospermum inodorum, Urtica urens (p. 616–631).

3.31 **Sisymbrietalia,** Sisymbrion, short-lived ruderal comm.: *Anthemis austriaca, A. cotula, Asperugo procumbens, Atriplex acuminata, A. oblongifolia, Barbarea verna, Brassica oleracea, Bromus arvensis, B. sterilis, B. tectorum, Cardaria draba, Chenopodium hybridum, Ch. opulifolium, Ch. strictum, Cnicus benedictus, Conyza canadensis, Crepis tectorum, Diplotaxis tenuifolia, Hordeum murinum, Kochia laniflora, Lappula deflexa, Lepidium densiflorum, L. graminifolium, L. virginicum, Malva neglecta, Nicandra physalodes, Plantago indica, Sisymbrium altissimum, S. austriacum, S. irio, S. loeselii, S. officinale, Torilis nodosa, Xanthium strumarium* (p. 616–618).

3.32 **Onopordetalia**[1]), persistent ruderal comm.: *Asperula arvensis, Cirsium eriophorum, C. vulgare, Echium vulgare, Malva sylvestris, Potentilla intermedia, Reseda lutea, Verbascum blattaria* (p. 615, 616, 618, 619).

3.321 Onopordion, Scotch Thistle comm.: *Anchusa officinalis, Artemisia absinthium, Carduus acanthoides, C. nutans, Cynoglossum officinale, Echinops sphaerocephalus, Hyoscyamus niger, Lappula squarrosa, Onopordum acanthium, Reseda luteola* (weak), *Verbascum densiflorum.*

3.322 Dauco-Melilotion, Melilot comm.: *Avena nuda, Berteroa incana* (weak), *Centaurea diffusa, Cichorium intybus, Daucus carota, Isatis tinctoria* (weak), *Marrubium vulgare, Melilotus alba, M. officinalis, Oenothera biennis, O. parviflora, Pastinaca sativa, Picris hieracioides, Rumex thyrsiflorus, Tanacetum vulgare* (weak).

3.33 **Polygono-Chenopodietalia,** rich soil arable and garden weed comm.: *Amaranthus lividus, Anagallis arvensis, Chenopodium polyspermum, Euphorbia helioscopia, E. peplus, Gagea villosa, Galinsoga ciliata, G. parviflora, Geranium dissectum, Lamium amplexicaule, L. purpureum, Mercurialis annua, Misopates orontium, Oxalis corniculata, Polygonum persecaria, Setaria glauca, S. verticillata, S. viridis, Sonchus asper, Spergula arvensis, Veronica agrestis, V. persica* (p. 621–646).

3.331 Fumario-Euphorbion, demanding arable and garden weeds on base-rich soils: *Allium vineale, Atriplex patula, Calendula arvensis, Erucastrum gallicum, Fumaria officinalis, Muscari racemosum, Thlaspi arvense, Tulipa sylvestris, Veronica opaca.*

3.332 Spergula-Oxalidion, on acid soils: *Anchusa arvensis, Chrysanthemum segetum, Digitaria ischaemum, D. sanguinalis, Oxalis fontana, Stachys arvensis.*

3.4 **Secalietea,** weed communities of cereal fields (increasingly mixed with 3.33): *Alopecurus myosuroides, Anthemis arvensis, Buglossoides arvensis, Euphorbia falcata, Galium spurium, G. tricornutum, Lathyrus hirsutus, Lolium temulentum, Odontites verna, Papaver rhoeas, Ranunculus arvensis, Rapistrum rugosum, Sherardia arvensis, Sinapis arvensis, Valerianella locusta, V. rimosa, Vicia angustifolia, V. tenuissima, Viola arvensis* (p. 621–640).

3.41 **Secalietalia,** Caucalion, cereal weeds on calcareous soils: *Adonis aestivalis, A. flammea, Ajuga chamaepytis, Anagallis foemina, Bifora radians, Bupleurum rotundifolium, Caucalis platycarpos, Conringia orientalis, Consolida regalis, Euphorbia exigua, Fumaria vaillantii, Kicksia elatine, K. spuria, Lathyrus aphaca, L. tuberosus, Legousia hybrida, L. speculum-veneris, Melampyrum arvense, Neslia paniculata, Nigella arvensis, Stachys annua, Thymelaea passerina, Torilis arvensis, Turgenia latifolia, Vaccaria hispanica* (p. 632).

3.42 **Aperetalia,** cereal weeds on lime-deficient soils: *Apera spica-venti, Bromus secalinus, Camelina microcarpa, C. pilosa, Raphanus raphanistrum, Scleranthus annuus, Vicia tetrasperma* (p. 632, 633).

3.421 Aphanion (Aperion) Parsley Piert comm.: *Aphanus arvensis, Malva moschata, Matricaria chamomilla, Papaver argemone, P. dubium, Veronica triphyllos, Vicia villosa.*

3.422 Arnoserion (Arnoseridion), Lamb's Succory comm. *Anthoxanthum puelii, Aphanes microcarpa, Arnoseris minima.*

3.43 **Lolio-Linetalia,** Lolio-Linion, weeds of flax fields (disappeared?): *Lolium remotum, Silene linicola* and others.

3.5 **Artemisietea**[1]), persistent nitrophilous ruderal comm.: *Aster laevis, A. lanceolatus, A. novae-anglicae, Cruciata glabra, Dipsacus fullonum, Solidago gigantea* (p. 615).

3.51 **Artimisietalia,** Mugwort and Burdock comm.: *Artemisia vulgaris, Rumex obtusifolius, Solidago canadensis, Urtica dioica* (weak) (p. 620, 621).

3.511 Arction, Burdock Comm.: *Arctium lappa, A. minus, A. tomentosum, Armoracia rusticana, Ballota nigra, Chelidonium majus, Conium maculatum, Fallopia dumetorum, Geranium pyrenaicum, Lamium album, Lavatera thuringiaca, Leonurus cardiaca, Malva alcea, Silene alba.*

3.512 Rumicion alpini, Monk's Rhubarb comm.: *Cerinthe glabra, Cirsium spinosissimum, Rumex alpinus, Senecio alpinus* (p. 432, 434, 459, 621).

3.52 **Calystegio-Alliarietalia** (Convolvuletalia): hedgerow veil comm. and semi-shade comm.: *Aster tradescantii, Calystegia* (=*Convolvulus*) *sepium, Cruciata laevipes, Galium aparine.*

3.521 Calystegion (=Senecion fluviatilis), veil and river bank comm.: *Angelica archangelica* ssp. *litoralis, Aster salignus, Carduus personata, Cucubalus baccifer, Cuscuta europaea, C. gronovii, C. lupuliformis, Echinocystis lobata, Helianthus tuberosus, Impatiens glandulifera, Melilotus altissima, Myosoton aquaticum, Reynoutria japonica, Rubus caesius, Saponaria officinalis, Senecio fluviatilis, Sisymbrium strictissimum* (p. 371).

3.522 Geo-Alliarion (Alliarion) Hedge Garlic and related comm.: *Alliaria petiolata* (weak), *Anthriscus cerefolium, Bryonia alba, B. dioica, Cardamine hirsuta, Chaerophyllum temulum, Euphorbia stricta, Galeopsis pubescens, Geum urbanum, Impatiens parviflora, Lapsana communis, Mycelis muralis, Parietaria officinalis, Polygonum dumetorum, Stellaria media* ssp. *neglecta, Torilis japonica, Viola alba, V. odorata* (p. 553).

3.523 Aegopodion, semi-shade fringe communities

[1]) Onopordetalia and Artemisietalia have recently been combined by Oberdorfer (1979) under 3.51.

adjacent to woody plants (see also 6.1): *Aegopodium podagraria, Barbaraea stricta, Carduus crispus* (weak), *Chaerophyllum aromaticum, Ch. aureum, Ch. bulbosum, Dipsacus pilosus, Epilobium hirsutum, E. parviflorum, Eupatorium cannabinum, Lamium maculatum, Petasites hybridus, Sambucus ebulus, Solidago graminifolia* (p. 264, 552).

3.6 **Agropyretea, Agropyretalia intermedii-repentis.** Convolvulo-Agropyrion, Couch Grass pioneer communities of dry habitats: *Agropyron intermedium, A. repens* ssp. *repens* var. *glaucum, Anthemis tinctoria, Bromus inermis, Cerastium arvense, Chondrilla juncea, Convolvulus arvensis, Falcaria vulgaris, Poa angustifolia, P. compressa* (weak).[2])

3.7 **Plantaginetea,** swards of pathways and flooded areas: *Plantago major, Poa annua, Potentilla anserina, Verbena officinalis* (p. 603–607).

3.71 **Plantaginetalia,** Polygonion avicularis, pathway swards: *Blysmus compressus, Coronopus squamatus, Juncus tenuis, Lepidium ruderale, Malva pusilla, Matricaria discoidea, Sagina procumbens, Schlerochloa dura.*

3.72 **Agrostietalia stoloniferae,** Agropyro-Rumicion (Agrostion stoloniferae), pioneer swards of flooded and damp places: *Agropyron pungens, Agrostis stolonifera, Alopecurus geniculatus, Apium repens, Cardamine parviflora, Carex hirta, C. hordeistichos, Cerastium dubium, Deschampsia media, Inula britannica, Juncus compressus, J. inflexus, Mentha longifolia, M. pulegium, Myosurus minimus, Plantago major* ssp. *intermedia, Potentilla reptans, Pulicaria dysenterica, Rorippa sylvestris, Rumex crispus* (weak), *R. stenophyllus, Sagina nodosa, Spergularia rubra, Trifolium fragiferum, T. hybridum, T. resupinatum* (p. 245–250, 606–608).

4 STONY SITES AND ALPINE GRASSLANDS

4.1 **Parietarietea, Parietarietalia judaicae,** Centrantho-Parietarion: warmth-loving wall comm.: *Cymbalaria muralis, Parietaria judaica,* as well as *Antirrhinum majus, Cheiranthus cheiri* and other garden escapes (p. 451).

4.2 **Asplenietea trichomanis,** communities of wall clefts and rock crannies: *Asplenium septentrionale, A. trichomanes, Ceterach officinarum, Draba dubia, Hieracium amplexicaule, Polypodium vulgare* (weak), *Saxifraga decipiens, Sedum dasyphyllum, Valeriana tripteris, Veronica fruticans* (p. 446–451).

4.21 **Potentilletalia caulescentis,** limestone rock cleft and wall comm.: *Adrosace lactea, Artemisia mutellina, Asplenium fontanum, A. ruta-muraria, Campanula cochleariifolia, Daphne alpina, Festuca alpina, F. pumila, F. stenantha, Poa glauca, Potentilla clusiana, Rhamnus pumilus, Saxifraga cotyledon, Silene saxifraga, Valeriana saxatilis* (p. 448–451).

4.211 Potentillion caulescentis, sunny limestone rocks and wall comm.: *Androsace helvetica, Arabis pumila, Biscutella laevigata, Cardaminopsis petraea, Carex mucronata, Draba tomentosa, Hieracium bupleurioides, H. humile, Kernera saxatilis, Minuartia rupestris, Potentilla caulescens* (weak), *Primula auricula, Saxifraga paniculata, Veronica lutea; Draba aizoides, D.ladina (?).*

4.212 Cystopteridion, shaded limestone rocks and wall comm.: *Asplenium viride, Aster bellidiastrum, Carex brachystachys, Cystopteris fragilis, C. regia, Moehringia muscosa, Phyllitis scolopendrium.*

4.22 **Androsacetalia vandellii,** clefts in silicate and serpentine rocks: *Androsace vandellii, Asplenium alternifolium, Epilobium collinum, Eritrichium nanum, Erysimum rhaeticum, Phytuma scheuchzeri, Primula glutinosa* (p. 448, 451).

4.221 Androsacion vandellii, sunny silicate rock cleft comm.: *Asplenium adiantum-nigrum, Erigeron gaudinii, Minuartia cherlerioides, Woodsia ilvensis.*

4.222 Anarrhinion bellidifolii (Asarinion procumbentis), shaded silicate rock cleft comm. (atlantic): *Asplenium billotii.*

4.223 Asplenion serpentini, serpentine rock cleft comm.: *Asplenium adulterrimum, A. cuneifolium, A. serpentini.*

4.3 **Violetea calaminariae, Violetalia calaminariae,** heavy metal mine spoil heap comm.: ecotypes of *Agrostis tenuis, Festuca ovina, Minuartia verna* and *Silene vulgaris* (p. 501–504).

4.311 Thlaspion calaminaris, mainly Western European heavy metal comm.: *Thlaspi alpestre, Viola calaminaria.*

4.312 Amerion halleri, Central European heavy metal comm.: *Armeria halleri* (incl. ssp. *bottendorfensis*).

4.313 Galio anisophyllo-Minuartion vernae, alpine heavy metal comm.: ecotypes of *Dianthus sylvestris, Galium anisophyllum* and *Poa alpina.*

4.4 **Thlaspietea rotundifolii,** stone scree and rubble comm.: *Aethionema saxatile, Arabis alpina, Campanula cespitosa, Chaenarrhinum minus, Galeopsis ladanum, Gymnocarpium robertianum, Gypsophila repens, Hieraceum glaucum, H. staticifolium, Linaria alpina, Rumex scutatus, Salix serpyllifolia* (p. 436–440).

4.41 **Thlaspietalia rotundifolii,** calcareous scree communities: *Achillea atrata, Doronicum grandiflorum, Moehringia ciliata, Poa cenisia, P. minor, Ranunculus hybridus, R. montanus.*

4.411 Thlaspion rotundifolii, Pennycress calcareous scree comm.: *Cerastium latifolium, Crepis terglouensis, Galium helveticum, Hutchinsia alpina, Leontodon montanus, Leucanthemum atratum, Papaver alpinum* ssp. *sendtneri, Ranunculus parnassifolius, Saxifraga aphylla, Thlaspi rotundifolium, Valeriana supina, Viola calcarata, V. cenisia* (p. 436–440).

4.412 Petasition paradoxi, Butterbur communities on damper calcareous screes: *Adenostyles glabra, Athamantha cretensis, Dryopteris villarii, Leontodon hispidus* ssp. *hyoseroides, Petasites paradoxus, Polystichum lonchitis, Trisetum distichophyllum, Valeriana montana* (p. 438).

4.42 **Drabetalia hoppeanae,** Drabion hoppeanae, Whitlow Grass calcareous slate screes: *Achillea nana, Artemisia genipi, Campanula cenisia, Crepis rhaetica, Dorycnium glaciale, Draba fladnizensis, D. hoppeana, Gentiana orbicularis, Pedicularis aspleniifolia, Saxifraga biflora, Sesleria ovata, Trisetum spicatum* (p. 439).

4.43 **Androsacetalia alpinae,** Androsacion alpinae, silicate scree comm.: *Androsace alpina, A. leucophylla, Cardamine resedifolia, Cerastium pedunculatum, C. uniflorum, Cryptogramma crispa,*

[2]) This class is poorly characterised and often mixed with other communities, especially with 3.31. Because of this it is not dealt with in the text.

Geum reptans, Oxyria digyna, Poa laxa, Ranunculus glacialis, Saxifraga bryoides, S. seguieri (p. 438, 439, 444, 451).

4.44 **Epilobietalia fleischeri,** Epilobion fleischeri, flowing gravel and damp scree communities of high mountains: *Calamagrostis pseudophragmites, Chondrilla chondrilloides, Epilobium dodonei, E. fleischeri, Erigeron acris* ssp. *angulosus, Erucastrum nasturtiifolium, Hieracium piloselloides, Scrophularia canina, Trifolium pallescens* (p. 256, 257).

4.45 **Achnatheretalia** (Stipetalia calamagrostis), Achnatherion, summer-warm Rough Grass calcareous screes: *Achnatherum calamagrostis, Calamintha nepetoides, Galeopsis angustifolia* (p. 440).

4.46 **Galeopsietalia segetum,** Galeopsion segetum, summer-warm Hemp Nettle silicate screes: *Anarrhinum bellidifolium, Epilobium lanceolatum, Galeopsis segetum, Senecio viscosus* (p. 440).

4.5 **Salicetea herbaceae,** snow patch communities: *Alchemilla fissa, Arenaria biflora, Cardamine alpina, Cerastium cerastoides, Epilobium anagallidifolium, Plantago atrata, Sedum alpestre, Soldanella pusilla, Taraxacum alpinum, Veronica alpina* (p. 424–428).

4.51 **Salicetalia herbaceae,** Salicion herbaceae, acid snow patch comm.: *Alchemilla pentaphyllea, Carex foetida, Gnaphalium supinum, Kiaeria starkei* (M), *Luzula alpino-pilosa, L. desvauxii, Orthotrichum incurvum* (M), *Pohlia drummondii* (M), *Polytrichum sexangulare* (M), *Ranunculus pygmaeus, Salix herbacea, Sibbaldia procumbens, Tanacetum alpinum.*

4.52 **Arabidetalia caeruleae,** Arabidion caeruleae, calcareous snow patch comm.: *Arabis caerulea, Carex parviflora, Gentiana bavarica, Gnaphalium hoppeanum, Potentilla brauneana, Ranunculus alpestris, Rumex nivalis, Salix reticulata, S. retusa, Saxifraga androsacea* (p. 427, 442).

4.6 **Caricetea curvulae** (L = Juncetea trifidi), **Caricitalia,** Caricion curvulae, alpine acid-soil swards: *Achillea erba-rotta, Agrostis rupestris, Androsace carnea, A. obtusifolia, Armeria alpina*[1]), *Avenochloa versicolor, Bupleurum stellatum, Carex curvula, Euphrasia minima, Festuca halleri, Hieracium glanduliferum, Juncus jacquinii, J. trifidus, Koeleria hirsuta, Laserpitium halleri, Luzula lutea, L. spicata, Lychnis alpina, Minuartia recurva, M. sedoides, Oreochloa disticha, Pedicularis kerneri, Phyteuma globulariifolium, Ph. hemisphaericum, Poa violacea, Potentilla frigida, vernalis, Ranunculus pyrenaicus, Senecio incanus, Silene excapa, Trifolium alpinum, Veronica bellidioides* (p. 408–411, 416–418, 458).

4.7 **Elyno-Seslerietea,** alpine calcareous soil swards: *Aster alpinus, Astragalus alpinus, A. australis, Carex rupestris, C. sempervirens, Gentiana nivalis, Oxytropis montana, Veronica aphylla* (p. 412).

4.71 **Seslerietalia,** alpine-subalpine Blue Sesleria and Rusty Sedge swards: *Acinos alpinus, Alchemilla hoppeana, Anemone narcissiflora, Anthyllis alpestris, Arabis ciliata, Astragalus frigidus, Carduus defloratus, Erigeron polymorphus, Euphrasia salisburgensis, Gentiana verna, Globularia cordifolia, Helianthemum grandiflorum, Hieracium bifidum, Leucanthemum maximum, Nigritella nigra, Pedicularis verticillata, Phyteuma orbiculare, Polygala alpestris, Potentilla crantzii, Pulsatilla alpina, Saxifraga moschata, Scabiosa lucida, Senecio doronicum, Sesleria varia* (weak), *Thesium alpinum, Thymus polytrichus* (p. 408–416, 458).

4.711 Seslerion, Blue Sesleria swards and related comm.: *Bupleurum ranunculoides, Carex firma, Chamorchis alpina, Crepis kerneri, Gentiana clusii, Globularia nudicaulis, Hedysarum hedysaroides, Helianthemum alpestre, Horminum pyrenaicum, Leontopodium alpinum, Pedicularis rostrato-capitata, Ranunculus thora, Saxifraga caesia, Sedum atratum, Silene acaulis.*

4.712 Caricion ferrugineae, Rusty Sedge swards: *Aquilegia alpina, Carex capillaris, C. ferruginea, Crepis pontana, Heracleum austriacum, Lathyrus laevigatus, Pedicularis foliosa, Phleum hirsutum, Traunsteinera globosa.*

4.72 Elynetalia[2]) *Elynion, Elyna wind-exposed swards: Antennaria carpatica, Carex atrata, Cerastium alpinum, Draba siliquosa, Elyna myosuroides, Erigeron uniflorus, Gentianella tenella, Ligusticum mutellinoides, Minuartia gerardii, Saussurea alpina* (p. 414–416, 458).

5 HEATHS AND GRASSLANDS DETERMINED BY HUMAN AND ANIMAL ACTIVITY

5.1 **Nardo-Callunetea,** Matgrass pastures and dwarf shrub heaths: *Calluna vulgaris, Carex pilulifera, Cuscuta epithymum, Danthonia decumbems, Luzula campestris, Potentilla erecta* (p. 504–524).

5.11 **Nardetalia,** poor Matgrass swards: *Ajuga pyramidalis, Antennaria dioica, Arnica montana, Botrychium lunaria, Carex pallescens, Coeloglossum viride, Euphrasia stricta* var. *subalpina, Festuca nigrescens, Hypericum maculatum, Meum athamanticum, Narcissus pseudo-narcissus, Nardus stricta, Ptilidium ciliare* (M), *Thesium pyrenaicum.*

5.111 Eu-Nardion, mountain Matgrass swards: *Alchemilla alpina, Campanula alpina, C. barbata, Crepis conyzifolia, Diphasium alpinum, Festuca supina, Gentiana acaulis, G. pannonica, G. punctata, G. purpurea, Geum montanum, Hieracium alpinum, H. hoppeanum, Hypochoeris uniflora, Leontodon helveticus, Phyteuma betonicifolium, Plantago alpina, Potentilla aurea, Pulsatilla alba, P. apiifolia* (p. 418, 419, 599).

5.112 Violo-Nardion, Matgrass swards at lower levels: *Centaurea nigra, Chamaespartium sagittale, Dianthus deltoides, Festuca tenuifolia, Galium pumilum* (weak), *Gentianella campestris, Hieracium lactucella, Jasione laevis, Polygala vulgaris, Viola canina* (p. 524, 525, 599).

5.113 Juncion squarrosi, Heath Rush-Matgrass swards: *Juncus squarrosus, Pedicularis sylvatica, Polygala serpyllifolia.*

5.12 **Genisto-Callunetalia,** subatlantic and atlantic acid-soil dwarf shrub heaths (p. 504–524).

5.121 Genisto-Callunion, Needle Whin-Heather heaths: *Genista anglica, G. germanica, G. pilosa, Lycopodium clavatum.*

5.122 Empetrion boreale, northern Crowberry heaths: *Empetrum nigrum* (p. 377).

5.2 **Sedo-Scleranthetea,** open swards of sandy and rocky ground: *Androsace septentionalis, Arabidopsis thaliana,*

[1]) According to Lippert (verbal communication) found on limestone, but Oberdorfer (1979) puts it in 4.6.

[2]) Recently Oberdorfer and his collaborators have raised this to a separate class (Carici rupestris-Kobresietea see Ohba 1974) with a world-wide distribution.

Brachythecium albicans (M), *Ceratodon purpureus* (M), *Cladonia alcicornis* (L), *Draba nemorosa*, *Helichrysum arenarium Herniaria glabra*, *Hieracium echioides*, *Holosteum umbellatum*, *Jasione montana*, *Medicago minima*, *Minuartia viscosa*, *Myosotis ramosissima*, *M. stricta*, *Peltigera rufescens* (L), *Petrorhagia prolifera*, *Poa bulbosa* (weak), *Polytrichum canescens* (M), *Rumex tenuifolius*, *Scleranthus perennis*, *Sedum anglicum* ssp. *anglicum*, *S. acre*, *S. forsteranum*, *S. rubens*, *S. sartorianum* ssp. *hillebrandtii*, *S. sexangulare*, *Syntrichia ruralis* (M), *Taraxacum laevigatum*, *Trifolium arvense*, *T. campestre*, *Valerianella carinata*, *V. dentata*, *Veronica verna*, *Vicia lathyroides* (p. 467, 485, 486).

5.21 **Sedo-Scleranthetalia,** weathered rock and outcrop comm.: *Allium montanum*, *A. strictum*, *Arenaria leptoclados*, *Cerastium glutinosum*, *C. pumilum*, *Erophila praecox*, *Jovibarba sobolifera*, *Poa concinna*, *Sedum album*, *S. ochroleucum* ssp. *montanum*, *S.o.* ssp. *ochroleucum*, *Sempervivum tectorum*, *Teucrium botrys*, *Thymus praecox* ssp. *praecox*, *Festuca cinerea*, *Sedum reflexum* (?).

5.211 Sedo-Scleranthion, Stonecrop and Houseleek communities of high mountains: *Arenaria serpyllifolia* var. *viscida*, *Cerastium arvense* ssp. *strictum*, *Festuca rubra* var. *microphylla*, *Paronychia kapela* ssp. *serpyllifolia*, *Plantago serpentina*, *Poa molinerii*, *Sedum annuum*, *Sempervivum arachnoideum*, *S. montanum*, *Silene rupestris*.

5.212 Alysso-Sedion albi, warmth-loving Stonecrop calcareous weathered rock comm.: *Alyssum alyssoides*, *Arabis auriculata*, *Cerastium brachypetalum*, *Clypeola jonthlaspi*, *Hornungia petraea*, *Linaria simplex*, *Micropus erectus*, *Minuartia fastigiata*, *M. glomerata*, *M. hybrida*, *Saxifraga tridactylites*, *Thlaspi perfoliatum*, *Veronica praecox*, *Trifolium scabrum*, *Trisetaria cavanillesii*.

5.213 Festucion pallentis, Pale Fescue rocky outcrop comm.: *Allium flavum*, *Alyssum montanum* ssp. *montanum*, *A. transsylvanicum*, *Artemisia campestris* var. *lednicensis*, *Aurinaria saxatilis*, *Dianthus gratianopolitanus*, *D. lumnitzeri*, *Erysimum crepidifolium*, *Euphorbia seguierana* ssp. *minor*, *Festuca pallens* var. *pallens*, *F.p.* var. *pannonica*, *F.p.* var. *styriaca*, *Medicago prostrata*, *Melica ciliata*, *Minuartia setacea*, *Seseli austriacum*, *S. osseum*, *Thlaspi goesingense*.

5.214 Sedo-Veronicion dillenii, warmth-loving silicate weathered rock comm.: *Androsace elongata*, *Cruciata pedemontana*, *Gagea bohemica*, *G. saxatilis*, *Poa bilbosa* ssp. *bulbosa*, *Scleranthus verticillatus*, *Spergula pentandra*, *Veronica dillenii*.

5.22 **Corynephoretalia,** Silvergrass-rich open swards on sand: *Carex arenaria*, *C. ligerica*, *Corynephorus canescens*, *Thymus serpyllum*, *Viola tricolor* ssp. *curtisii* (p. 486).

5.221 Corynephorion canescentis, Silvergrass dune comm.: *Conicularia stuppea* (L), *Spergula morisonii* (p. 375, 381–386).

5.223 Koelerion albescentis, coastal dune Hairgrass comm.: *Festuca rubra* ssp. *arenaria*, *Jasione montana* var. *litoralis*, *Koeleria albescens* (=*arenaria*), Lotus corniculatus var. *crassifolius* (p. 375).

5.224 Sileno conicae-Cerastion semidecandri, short-lived warmth-loving Mouse Ear dune comm.: *Cerastium diffusum* (=*tetrandrum*), *C. semidecandrum*, *C. subtetrandrum*, *Erodium lebelii*, *Phleum arenarium*, *Silene conica*.

5.225 Koelerion glaucae, warmth- and lime-loving Blue green Hairgrass sandy swards: *Alyssum montanum* ssp. *gmelinii*, *Astragalus arenarius*, *Festuca psammophila*, *Gypsophila fastigiata* ssp. *arenaria*, *Jurinea cyanoides*, *Koeleria glauca*, *Onosma arenarium*, *Silene chlorantha*, *Stipa sabulosa*, *Tragopogon floccosus*.

5.24 **Thero-Airetalia,** Thero-Airion, short-lived Hairgrass comm.: *Aira caryophyllea*, *A. praecox*, *Filago arvensis*, *F. gallica*, *F. lutescens*, *F. minima*, *F. pyramidata*, *F. vulgaris*, *Hypochoeris glabra*, *Moenchia erecta*, *Myosotis discolor*, *Nardurus lachenalii*, *Ornithopus perpusillus* (weak), *Scleranthus polycarpos*, *Teesdalia nudicaulis*, *Trifolium striatum*, *Tuberaria guttata*, *Vulpia bromoides*, *V. myuros* (p. 377).

5.3 **Festuco-Brometea,** more or less arid poor calcareous swards: *Abietinella abietina* (M), *Allium sphaerocephalum*, *Asperula aristata*, *A. cynanchica*, *Aster linosyris*, *Avenochloa pratensis*, *Botriochloa ischaemum*, *Campanula glomerata*, *Carex caryophyllea*, *C. humilis*, *Eryngium campestre*, *Euphorbia cyparissias*, *Filipendula vulgaris*, *Galium glaucum*, *Lophochloa cristata*, *Odontites lutea*, *Onobrychis arenaria*, *Orobanche caryophyllacea*, *Phleum phleoides*, *Pimpinella saxifraga*, *Polygala comosa*, *Prunella grandiflora*, *Rhytidium rugosum* (M), *Salvia pratensis*, *Sanguisorba minor*, *Stipa pulcherrima*, *Teucrium chamaedrys*, *Thesium linophyllon*, *Thymus praecox*, *Trifolium montanum* *Veronica spicata* (p. 467–501).

5.31 **Festucetalia valesiaceae,** continental more or less arid swards: *Adonis vernalis*, *Astragalus onobrychis*, *Centaurea stoebe*, *Oxytropis pilosa*, *Petrorhagia saxifraga*, *Potentilla arenaria*, *P. pusilla*, *Scabiosa canescens*, *Scorzonera austriaca*, *Silene otites*, *Stipa capillata*, *St. joannis*, *St. tirsa*, *Verbascum phoeniceum*, *Veronica prostrata*.

5.311 Festucion valesiacae, continental arid swards: *Achillea nobilis*, *A. setacea*, *Astragalus excapus*, *Carex supina*, *Erysimum odoratum* (weak), *Festuca rupicola*, *F. trachyphylla*, *F. valesiaca*, *Poa badensis*, *Seseli hippomarathrum*.

5.312 Cirsio-Brachypodion, continental fairly dry swards: *Astragalus danicus*, *Inula spiraeifolia*, *Scabiosa ochroleuca* (weak), *Scorzonera purpurea*, *Seseli annuum*.

5.32 **Brometalia erecti,** suboceanic more or less arid swards: *Arabis hirsuta*, *Bromus erectus*, *Centaurea scabiosa*, *Dianthus carthusianorum*, *Gentiana cruciata*, *Helianthemum nummularium*, *H. ovatum*, *Hippocrepis comosa*, *Koeleria pyramidata*, *Linum tenuifolium*, *L. viscosum*, *Ononis natrix*, *Potentilla tabernaemontani*, *Pulsatilla vulgaris*, *Scabiosa columbaria*, *Teucrium montanum*, *Thymus froelichianus* (p. 479–484).

5.321 Xerobromion, suboceanic arid swards: *Fumana procumbens*, *Globularia punctata*, *Helianthemum apenninum*, *Koeleria vallesiana*, *Minuartia verna*, *Orobanche teucrii*, *Trinia glauca* (p. 479–482).

5.322 Mesobromion, suboceanic fairly dry swards: *Aceras anthropophorum*, *Anacamptis pyramidalis*,

Carlina acaulis, C. vulgaris, Cirsium acaule, Erigeron acris, Euphorbia verrucosa, Gentianella ciliata, G. germanica, Herminium monorchis, Himanthoglossum hircinum, Onobrychis viciaefolia, Ononis repens, O. spinosa, Ophrys apifera, O. holosericea, O. insectifera, O. sphecodes, Orchis militaris, O. morio, O. simia, O. ustulata, Phyteuma tenerum, Primula veris, Prunella laciniata, Ranunculus bulbosus, Spiranthes spiralis (p. 477–483, 486–488, 491–501).

5.323 Koelerio-Phleion phleoides, acid-soil fairly dry swards (previously 5.232 Armerion elongatae): *Armeria elongata, Dactylorhiza sambucina, Koeleria macrantha* (=*gracilis*), and others (still not satisfactorily clarified regarding the coastal communities).

5.4 **Molinio-Arrhenatheretea,** cultivated meadow and pasture comm.: *Agrostis gigantea, Alopecurus pratensis, Cerastium (fontanum agg.) holosteoides, Colchicum autumnale, Dactylis glomerata, Euphrasia rostkoviana, Festuca pratensis, F. rubra* ssp. *rubra. Holcus lanatus, Lathyrus pratensis, Leontodon hispidus* ssp. *hispidus, Poa pratensis, P. trivialis, Prunella vulgaris, Ranunculus acris, Rhinanthus minor, Rumex acetosa, Trifolium dubium, Vicia cracca* (p. 553–608).

5.41 **Molinietalia,** moist meadow and stream bank comm.: *Achillea ptarmica, Angelica sylvestris, Chaerophyllum hirsutum, Cirsium palustre, Dactylorhiza maculata, D. majalis, Equisetum palustre, Filipendula ulmaria, Juncus conglomeratus, J. effusus, Lathyrus palustris, Lotus uliginosus, Lychnis flos-cuculi* (weak), *Lysimachia vulgaris, Platanthera chlorantha, Sanguisorba officinalis* var. *officinalis, Selinum carvifolium, Silaum silaus, Thalictrum flavum, Trifolium spadiceum, Trollius europaeus, Valeriana dioica* (weak) (p. 558, 578–582).

5.411 Molinion, Purple Moorgrass litter meadows: *Allium anguiosum, A. suaveolens, Betonica officinalis, Carex tomentosa, Cirsium dissectum, C. tuberosum, Dianthus superbus, Galium boreale, Gentiana asclepiadea, G. pneumonanthe, Gladiolus palustris, Inula salicina, Iris sibirica, Laserpitium prutenicum, Lathyrus pannonicus* ssp. *pannonicus, Serratula tinctoria, Succisa pratensis* (weak), *Tetragonolobus maritimus* (p. 578–581, 589–592).

5.412 Filipendulion ulmariae, Meadowsweet stream bank and similar comm.: *Euphorbia lucida, Geranium palustre, Hypericum destangsii, H. tetrapterum, Lythrum salicaria, Scutellaria hastifolia, Sonchus palustris, Stachys palustris, Telekia speciosa, Valeriana procurrens, Veronica longifolia* (p. 581, 582).

5.413 Cnidion, subcontinental Cnidium meadows: *Cnidium dubium, Gratiola officinalis, Juncus atratus, Oenanthe silaifolia, Viola elatior, V. persicifolia* (p. 576, 577).

5.414 Juncion acutiflori, subatlantic Sharp-flowered Rush meadows: *Anagallis tenella, Carum verticillatum, Juncus acutiflorus* (weak; can also dominate in 5.415 or 5.411), *Oenanthe peucedanifolia, Scutellaria minor, Wahlenbergia hederacea* (p. 577, 579).

5.415 Calthion, manured moist meadows: *Angelica palustris, Bromus racemosus, Caltha palustris, Cirsium canum, C. oleraceum, C. rivulare, Fritillaria meleagris, Juncus subnodulosus* (weak), *Scirpus sylvaticus* (weak), *Senecio aquaticus* (p. 573–576, 594, 595).

5.42 **Arrhenatheretalia,** manured slightly moist meadows and pastures: *Achillea millefolium, Anthriscus sylvestris, Bellis perennis, Carum carvi, Crepis capillaris, Gaudiana fragilis, Heracleum sphondylium, Leucanthemum vulgare, Pimpinella major, Ornithogalum umbellatum, Rhinanthus alectorolophus, Saxifraga granulata, Stellaria graminea, Tragopogon pratensis, T. oreintalis, Trifolium patens, Trisetum flavescens* (p. 557–572, 582–589).

5.421 Arrhenatherion, Oatgrass meadows: *Arrhenatherum elatius, Campanula patula, Crepis biennis, Galium mollugo, Geranium pratense, Knautia arvensis* (p. 560–571).

5.422 Polygono-Trisetion, Yellow Oat mountain meadows: *Alchemilla vulgaris, Campanula rhomboidalis, Cardaminopsis halleri, Centaurea pseudophrygia, Crepis mollis, Crocus albiflorus, Narcissus radiiflorus, Phyteuma nigrum, Ph. ovatum, Pimpinella major* ssp. *rubra, Viola tricolor* ssp. *subalpina, V.t.* ssp. *tricolor* (p. 569–572).

5.423 Cynosurion, Ryegrass-Crested Dogstail meadows: *Cynosurus cristatus, Hordeum secalinum, Leontodon autumnalis* (weak), *Lolium perenne, Odontites rubra, Phleum bertolonii, Ph. pratense, Trifolium repens, Veronica filiformis, V. serpyllifolia* (p. 596–603).

5.424 Poion alpinae, subalpine Hawkbit pastures: *Agrostis alpina, Cerastium fontanum, Crepis aurea, Phleum alpinum, Poa alpina, Trifolium badium, T. thalii* (p. 602, 603).

6 WOODLAND-RELATED HERBACEOUS PERENNIAL AND SHRUB COMM.:

6.1 **Trifolio-Geranietea, Origanetalia,** sunny fringe communities at woodland edges (see also 3.52): *Astragalus cicer, A. glycyphyllos, Clinopodium vulgare, Coronilla varia, Inula conyza, Lathyrus heterophyllus, L. sylvestris, Medicago falcata, Origanum vulgare, Silene armeria, S. nutans, Valeriana wallrothii, Verbascum lychnites, Vicia pisiformis, Viola hirta* (p. 177, 178, 547–552).

6.111 Trifolion medii, Zigzag Clover fringe comm.: *Agrimonia eupatoria, A. procera, Lathyrus latifolius, Trifolium medium, Vicia cassubica, V. dumetorum, V. orobus, V. sylvatica.*

6.112 Geranion sanguinei, drought-tolerating Bloddy Cransebill fringe comm.: *Anemone sylvestris, Anthericum ramosum, Aster amellus, Bupleurum falcatum, Campanula bononiensis, C. rapunculoides, Centaurea nigra* ssp. *nemoralis, Clematis recta, Coronilla coronata, Crepis praemorsa, Dictamnus albus, Fragaria viridus, Galium lucidum, Geranium sanguineum, Inula hirta, Lathyrus pannonicus* ssp. *collinus, Melampyrum cristatum, Peucedanum alsaticum, P. cervaria, P. oreoselinum, Polygonatum odoratum, Potentilla rupestris, Rosa pimpinellifolia, Seseli libanotis, Thalictrum minus, Thesium bavarum, Trifolium alpestre, T. rubens, Veronica teucrium, Vicia tenuifolia, Viola collina* (p. 551–553).

6.2 **Epilobietea angustifolii, Atropetalia,** woodland clearing comm.: *Calamagrostis epigeios, Centaurium erythraea, Epilobium angustifolium, Fragaria vesca, Myosotis sylvatica, Rubus fruticosus* coll., *Senecio fuchsii, Verbascum thapsus* (p. 539–544).

6.211 Epilobion angustifolii, Willow Herb clearing comm.: *Corydalis claviculata, Digitalis grandiflora, D. lutea, D. purpurea, Gnaphalium sylvaticum, Senecio sylvaticus, Verbascum nigrum.*

6.212 Atropion, Deadly Nightshade clearing comm.: *Arctium nemorosum, Atropa belladonna, Bromus ramosus, Hypericum hirsutum, Stachys alpina.*

6.213 Sambuco-Salicion capreae, woodland clearing shrub comm.: *Rubus affinis, R. fuscus, R. gratus, R. koehleri, R. rudis, R. schleicheri, R. sulcatus, Salix caprea, Sambucus nigra, S. racemosa, Sorbus aucuparia* ssp. *glabrata.*

6.3 **Betulo-Adenostyletea, Adenostyletalia,** tall perennial herb and shrub comm.: *Adenostyles alliariae, Athyrium distentifolium, Cicerbita alpina, C. plumieri, Crepis pyrenaica, Epilobium alpestre, Geranium sylvaticum, Milium effusum* var. *violaceum, Ranunculus platanifolius, Ribes petraeum, Rosa pendulina, Rumex alpestris, Salix foetida, Saxifraga rotundifolia, Streptopus amplexifolius, Tozzia alpina, Viola biflora* (p. 431–435).

6.311 Adenostylion alliariae, subalpine tall herb and shrub comm.: *Achillea macrophylla, Alnus viridis, Chaerophyllum hirsutum* ssp. *villarsii, Corthusa matthioli, Doronicum austriacum, Heracleum sphondylium* ssp. *elegans, Hieracium jurassicum, Peucedanum ostruthium, Poa hybrida, Salix appendiculata* (p. 431–435).

6.312 Salicion arbusculae (waldsteinianae), subalpine Bushy Willow scrub comm.: *Salix arbuscula, S. glabra, S. hastata* (p. 435).

6.313 Calamagrostion arundinaceae, subalpine Small Reed comm.: *Allium victorialis, Betonica officinalis* var. *alpestris, Centaurea montana, Gnaphalium norvegicum, Hieracium prenanthoides* (p. 435).

7 NEEDLE-LEAVED WOODLAND AND RELATED COMM.

7.1 **Erico-Pinetea, Erico-Pinetalia,** Erico-Pinion, calcareous pine woods: *Aquilegia atrata, Calamagrostis varia, Carduus crassifolius, Carex ericetorum, Chamaecytisus ratisbonensis, Cirsium erisithales, Coronilla vaginalis, Crepis alpestris, Daphne cneorum, D. striata, Dorycnium germanicum, Epipactis atrorubens, Erica herbacea, Festuca amethystina, Gymnadenia odoratissima, Helianthemum canum, Pinus nigra, Polygala chamaebuxus, Rhamnus saxatilis, Rhododendron hirsutum, Rhodothamnus chamaecistus, Saponaria ocymoides, Thesium rostratum, Thlaspi montanum* (weak) (p. 227–231, 236–238, 270–272).

7.2 **Vaccinio-Piceetea, Vaccinio-Piceetalia,** acid-soil needle-leaved woods and related comm.: *Arctostaphylos uva-ursi, Corallorhiza trifida, Diphasium tristachyum, Epipogium aphyllum, Homogyne alpina, Huperzia selago, Juniperus communis* ssp. *alpina, Linnaea borealis, Listera cordata, Lonicera caerulea, Lycopodium annotinum, Melampyrum sylvaticum, Moneses uniflora, Monotropa hypopytis* ssp. *hypopytis, Orthilia secunda, Pyrola media, P. minor, P. rotundifolia, Trientalis europaea, Vaccinium uliginosum, V. vitis-idaea* (p. 191–197).

7.211 Dicrano-Pinion, acid-soil pine woods: *Chimaphila umbellata, Diphasium complanatum, Goodyera repens, Pyrola chlorantha, Viola rupestris, Viscum laxum* (p. 183, 185, 186, 191, 192, 231–235, 238–243).

7.212 Vaccinio-Piceion, Spruce woods and related communities: *Arctostaphylos alpina, Blechnum spicant, Calamagrostis villosa, Larix decidua, Luzula luzulina, L. sieberi, Picea abies, Pinus mugo, Soldanella montana* (p. 208ff).

7.212.1 UV Piceion, Spruce woods (p. 207–218).

7.212.2 UV Ledo-Pinion, swamp and bog Pine woods: *Chamaedaphne calyculata, Ledum palustre* (p. 276, 279, 340).

7.212.3 UV Betulion pubescentis, Birch and Birch-Pine swamp woods: *Betula pubescens* (p. 278–283).

7.212.4 UV Rhododendro-Vaccinion, Arolla Pine and mountain dwarf shrub heaths: *Clematis alpina, Empetrum hermaphroditum, Loiseleuria procumbens, Pinus cembra, Rhododendron ferrugineum, R. intermedium, Vaccinium ulginosum* ssp. *pubescens* (p. 218–226, 409, 419–424, 525, 526).

7.212.5 UV Vaccinio-Abietion, acid-soil Silver Fir woods, according to Oberdorfer (1979) (p. 195, 207).

8 BROADLEAVED WOODLAND AND RELATED COMMUNITIES

8.1 **Silicetea purpureae, Salicetalia purpureae,** willow communities on flood plains *Salix purpurea* (p 245–250, 257–261).

8.111 Salicion eleagni, Mountain Willow scrub: *Myricaria germanica, Salix daphnoides, S. eleagnos, S. nigricans* (p. 257, 258).

8.112 Salicion albae, Willow communities of lowland flood plains: *Populus nigra, Salix alba, S. fragilis, S. rubens, S. triandra, S. viminalis* (p. 257–261).

8.2 **Alnetea glutinosae,** Alder swamp woods and mire Willow scrub: *Calamagrostis cunescens, Dryopteris cristata, Galium elongatum, Salix cinerea, S. pentandra* (p. 274–278, 279–283).

8.21 **Alnetalia,** Alnion glutinosae, Alder swamp woods: *Alnus glutinosa, Carex elongata, C. laevigata, Osmunda regalis, Ribes nigrum, Thelypteris palustris.*

8.22 **Salicetalia auritae,** Frangulo-Salicion auritae (Salicion cinereae), mire Willow scrub: *Frangula alnus, Myrica gale, Salix aurita, S. repens* ssp. *rosmarinifolia, S.r.* ssp. *repens.*

8.3 **Quercetea robori-petraeae, Quercetalia,** Quercion robori-petraeae, acid-soil mixed Oak woods: *Hieracium glaucinum, H. lachenalii, H. laevigatum, H. sabaudum, H. umbellatum, Holcus mollis, Hypericum pulchrum, Lathyrus linifolius, Lonicera periclymenum* (p. 180–191, 397).

8.4 **Querco-Fagetea,** broadleaved woods and scrub on more fertile soils: *Acer campestre, Amelanchier ovalis, Anemone nemorosa* (weak), *Aquilegia vulgaris, Aremonia agrimonoides, Brachypodium sylvaticum, Buglossoides* *purpurocaerulea, Buxus sempervirens, Campanula trachelium, Carex digitata, C. montana, Cephalanthera longifolia, C. rubra, Convallaria majalis* (weak), *Corylus avellana, Crataegus laevigata, Cypripedium calceolus, Daphne laureola, Fragaria moschata, Galanthus nivalis, Hedera helix, Hepatica nobilis* (weak), *Hypericum montanum, Lathraea squamaria, Lonicera xylosteum, Malus sylvestris, Melica nutans, M. picta, Moehringia trinervia, Ribes alpinum, Scilla bifolia, Ulmus minor, Viburnum opulus, Vinca minor, Viola mirabilis* (p. 69, 70).

8.41 **Prunetalia,** woodland mantle scrub and hedges: *Clematis vitalba, Cornus sanguinea, Crataegus monogyna, Euonymus europaea, Mespilus germanica, Prunus spinosa, Rhamnus catharticus, Rosa agrestis, R. canina, R. corymbifera, R. nitidula, R. obtusifolia, R. rubiginosa, R. stylosa, R. tomentosa, Rubus bifrons, R. canescens* (p. 544–551).

8.411 Prunion spinosae (=Rubo-Prunion), Blackthorn scrub: *Rubus adspersus, R. mucronulatus, R. procerus, R. radula, R. sprengelii, R. villicaulis, R. vulgaris.*

8.412 Berberidion, Barberry scrub: *Berberis vulgaris, Coronilla emerus, Cotinus coggygria, Cotoneaster integerrima, C. tomentosa, Hippophaë rhamnoides* ssp. *carpatica, Ligustrum vulgare, Prunus mahaleb, Rosa abietina, R. caesia, R. coriffolia, R. elliptica, R. glauca, R. micrantha, R. sherardii, R. villosa, R. vosagiana, Staphylea pinnata, Tamus communis, Viburnum lantana.*

8.413 Prunion fruticosae, Steppe Cherry scrub: *Prunus fruticosa.*

8.414 Salicion repentis (arenariae), seashore scrub: *Hippophaë rhamnoides* ssp. *rhamnoides, Salix repens* ssp. *argentea* (=*S. arenaria*) and others (p. 378).

8.415 UV Cytision scoparii (=Sarothamnion), Broom scrub: connection not clear: *Cytisus scoparius* and others (p. 526, 527).

8.42 **Quercetalia pubescenti-petraeae,** xerothermic mixed Oak woods: *Acer monspessulanum, Arabis pauciflora, A. turrita, Campanula persicifolia, Colutea arborescens, Cornus mas, Laburnum anagyroides, Lactuca quercina, Lathyrus niger, Limodorum abortivum, Melica ovata, Melittis melissophyllum, Mercurialis ovata, Orchis purpurea, Potentilla micrantha, Quercus cerris, Q. pubescens, Sorbus domestica, S. torminalis, Tanacetum corymbosum* (p. 172–181).

8.421 Quercion pubescenti-petraeae, Central European xerothermic mixed Oak woods: *Acer opalus, Helleborus foetidus, Potentilla alba, Pulmonaria angustifolia.*

8.43 **Fagetalia,** noble broadleaved woods and related comm.: *Acer platanoides, A. pseudoplatanus, Aconitum variegatum, Actaea spicata* (weak), *Adoxa moschatellina, Allium ursinum, Anemone ranunculoides, Aposeris foetida, Arum maculatum, Asarum europaeum, Asperula taurina, Bromus benekeni, Bupleurum longifolium, Cardamine impatiens* (weak), *Carex brizoides, C. pendula, C. pilosa, C. remota, C. sylvatica, Circaea alpina, C. lutetiana, Corydalis cava, C. intermedia, C. solida, Daphne mezereum, Dentaria bulbifera, Dryopteris filix-mas, D. pseudomas, Epipactis helleborine, E. microphylla, E. purpurata, Euphorbia amygdaloides, Festuca gigantea, Fraxinus excelsior, Gagea spathacea, Galium odoratum* (weak), *Impatiens nolitangere, Lamiastrum galeobdolon, Lathyrus vernus, Leucojum vernum, Lilium martagon* (weak), *Lonicera alpigena, Melica uniflora, Mercurialis perennis, Milium effusum, Neottia nidus-avis, Paris quadrifolia* (weak), *Phyteuma spicatum, Polygonatum multiflorum, Primula vulgaris, Prunus avium, Pulmonaria obscura, P. officinalis, Ranunculus lanuginosus, Salvia glutinosa, Sanicula europaea, Scrophularia nodosa* (weak), *Stachys sylvatica, Symphytum tuberosum, Viola reichenbachiana* (p. 72ff).

8.431 Fagion sylvaticae, Beech woods: *Cardamine trifolia, Cyclamen purpurascens, Dentaria ennaphyllos, D. heptaphylla, D. pentaphyllos, Euonymus latifolia, Euphorbia dulcis, Fagus sylvatica, Festuca altissima, Helleborus niger, Hordelymus europaeus, Omphalodes verna, Orchis pallens, Petasites albus* (weak), *Rubus hirtus, R. tereticaulis, Taxus baccata, Veronica urticifolia* (p. 73, 73, 139, 197).

8.431.1 UV Luzulo-Fagion, Woodrush-Beech woods: *Calamagrostis arundinacea, Luzula luzuloides* (p. 110–139).

8.431.2 UV Galio odorati-Fagion (Eu-Fagion) Sweet Woodruff-Beech woods (p. 76–86, 93–97, 101–110, 121–139).

8.431.3 UV Cephalanthero-Fagion, slope Sedge-Beech woods: *Cephalanthera damasonium* (p. 86–93).

8.431.5 UV Galio rotundifolii-Abietion, mixed Silver Fir woods: *Abies alba, Galium rotundifolium* (p. 197–207).

8.432 Carpinion betuli, mixed Hornbeam woods: *Carex fritschii, C. umbrosa, Carpinus betulus, Crategus lindmanii, Dactylis polygama, Festuca heterophylla, Galium schultesii, Melampyrum nemorosum, Ornithogalum pyrenaicum, Potentilla sterilis, Pulmonaria montana, Rosa arvensis, Scilla non-scripta, Stellaria holostea, Tilia cordata* (weak) (p. 147–172).

8.433 Alno-Ulmion, Alder and noble broadleaved woods of flood plains: *Agropyron caninum, Allium scorodoprasum, Alnus incana, Anthriscus nitida, Carex otrubae, C. strigosa, Chrysosplenium alternifolium, Circaea intermedia, Equisetum hyemale, E. pratense, E. telmateia, Gagea lutea, Hesperis matronalis, Matteucia struthiopteris, Omphalodes scorpioides, Polemonium caeruleum, Populus alba, Prunus padus, Ribes rubrum, R. spicatum, Rumex sanguineus, Stellaria nemorum, Thalictrum aquilegiifolium, Ulmus laevis, Veronica montana, Vitis vinifera* ssp. *sylvestris* (p. 245, 246, 249–256, 262–270).

8.434 Tilio-Acerion (formerly UV Aceri-Fagion), mixed Maple woods and Maple-Beech woods: *Aconitum paniculatum, Aruncus dioicus, Campanula latifolia, Lunaria rediviva, Polystichum aculeatum, P. braunii, Ranunculus serpens, Tilia platyphyllos, Ulmus glabra* (p. 95–100, 139–146).

2 *List of species referred to, their indicator values and life forms*

(a) Explanation

As far as possible the following information is given for each species:

Numbers of the tables in which the species occurs.

pp. = pages where the species occurs in the text and in figures.

decimal figures in brackets = community in which the species is characteristic; (see section E III 1, page 665).

In addition the ecological indicator values are given for most of the vascular plants. These are –light, temperature, continentality, –dampness, soil reaction, nitrogen, –salt (see also pp. 63–68).

Life forms: P. phanerophyte, N nanophanerophyte, Z woody chamaephyte, C herbaceous chamaephyte, H hemicryptophyte, G geophyte, T therophyte, A hydrophyte, li liane, p parasite, hp semiparasite, E epiphyte (further details on p. 11).

Explanation of the ecological indicator values

x = indifferent behaviour, i.e. a wide amplitude or different behaviour in the different parts of Europe.

y = behaviour not clear; I have not yet been able to speculate on this.

L = **Light value** (first number in the first triad) (occurrence in relation to the relative light intensity = R.L.)
For all species measurement of light intensity is taken where they are growing at the time when the deciduous plants are in full leaf (mid-June to mid-October)

1 Plants in deep shade, may be less than 1%, seldom more than 30% R.L.
2 between 1 & 3
3 Shade plants, mostly less than 5% R.L., but also in lighter places
4 between 3 & 5
5 Plants of half shade, rarely in full light but generally more than 10% R.L.
6 between 5 & 7
7 Plants generally in well lit places but also occur in partial shade
8 Light-loving plants, rarely found where there is less than 40% R.L.
9 Plants of full light, found only in full sun; rarely in less than 50% R.L.

T = **Temperature value** (second number in the first triad) (occurrence in the temperature gradients from the Mediterranean to the Arctic and from lowland to alpine levels)

1 Cold-indicator plants, found only in high mountains or in the boreal-arctic regions
2 between 2 & 3 (many alpines)
3 Indicators of cool conditions, mainly in high montane to subalpine or temperate-boreal sites
4 between 3 & 5 (especially mountain species)
5 Indicators of fairly warm conditions from lowland to high mountain sites, but especially in submontane to temperate regions.
6 between 5 & 7
7 Warmth indicators, only on lowland sites in the northern part of Central Europe.
8 between 7 & 9, mainly submediterranean
9 Indicate extreme warm conditions, spreading only into the warmest sites in Central Europe from the Mediterranean

K = **Continentality value** (third number of first triad) (occurrence in the gradient from the Atlantic coast to the inner parts of Eurasia, especially with regard to temperature ranges)

1 Extreme oceanic, only in a few outposts of Central Europe
2 Oceanic, mainly in the west including western Central Europe
3 between 2 & 4 (i.e. in most parts of Central Europe)
4 Suboceanic, mainly in Central Europe but spreading towards the East
5 Intermediate, weakly suboceanic to weakly subcontinental
6 Subcontinental, mainly in the east of Central Europe and the adjoining parts of Eastern Europe
7 between 6 & 8
8 Continental, spreading into Central Europe from the east only into particular sites
9 Extreme continental (virtually absent from Central Europe)

F = **Water value** (first number in second triad) (occurrence in the gradient from dry shallow-soil rocky slopes to marshy ground; also from shallow to deep water)

1 Indicators of extreme dryness, often restricted to places which dry out completely
2 between 1 & 3
3 Dry site indicators, more often found on dry ground than moist places, not found on damp soil

4 between 3 & 5
5 Moist-site indicators, mainly on soils of average dampness, absent from both wet ground and places which may dry out
6 between 5 & 7
7 Damp-site indicators, mainly on constantly damp, but not wet, soils
8 between 7 & 9
9 Wet-site indicators; often in water-saturated, badly aerated soils
10 Indicators of sites occasionally flooded but free from surface water for long periods
11 Plants rooting under water but at least for a time exposed above or floating on the surface
12 Submerged plants, permanently or almost constantly under water

R = **Reaction value** (second number of second triad) (occurrence in the gradient of soil acidity and lime content)
1 Indicators of extreme acidity, never found on weakly acid or alkaline soils
2 between 1 & 3
3 Acid indicators, mainly on acid soils but can also be found where there is a neutral reaction
4 between 3 & 5
5 Indicators of fairly acid soils, only occasionally found in more acid, or in neutral to slightly alkaline, situations
6 between 5 & 7
7 Indicators of weakly acid to weakly basic conditions. Never found on very acid soils
8 between 7 & 9, i.e. mostly seen on limestone or chalk
9 Basic reaction and lime indicators, always found on calcareous soils

N = **Nitrogen value** (third number of second triad) (occurrence in the gradient of available nitrogen during the growing period)
1 Indicators of sites poor in available nitrogen
2 between 1 & 3
3 More often found on nitrogen-deficient soils than on richer ones
4 between 3 & 5
5 Indicators of sites with average nitrogen availability, seldom found on either poorer or richer soils
6 between 5 & 7
7 More often found in places rich in available nitrogen than in poor or average situations
8 between 7 & 9
9 In extremely rich situations (cattle resting places; indicators of pollution)

S = **Salt indicator value** (in Roman numeral after the second triad)
No number given = glycophyte
I Occasionally in salt-containing soils
II Generally in saline soils
III Always in saline soils (true halophytes)

B = **Heavy metals tolerance** (letter after second triad)
b = **Resistant ecotypes**

(b) Species index

German species names have been retained in order to provide a link with the Central European literature.

- *erectus* Huds. = Aufrechte T. Tab. 89, 92, 96, 110, 112; p. 13, 271, 474f., 486ff., *487*, 491ff., *493*, 496f., *496*, *497f.*, *498ff.*, 553, 563, *566*, 567, 582, 585f. (5.32) – 852-383-H
- *hordeaceus* L. = Weiche T. Tab. 110, 117, 118, 120; p. *547*, *586* – 763-× ×3-T
- *inermis* Leyser = Unbegrannte T. (3.6) – 8×7-485-H, G
- *racemosus* L. = Trauben-T. Tab. 14; p. *586* (5.415) – 662-855-T
- *ramosus* Huds. = Wald-T. Tab. 10, 110 (6.212) – 652-686-H
- *secalinus* L. = Roggen-T. Tab. 125, 126; p. 624, *624* (3.42) – 6×3-× × ×-T
- *sterilis* L. = Taube T. Tab. 50, 124; p. 616 (3.31)-774-4×5-T
- *tectorum* L. = Dach-T. Tab. 124; p. 616 (3.31) – 867-384-T

Bryonia alba L. = Weiße Zaunrübe (3.522)
- *dioica* Jacq. = Zweihäusige Z. (3.522)

Bryoria subcana (Nyl. ex. Stizenb.) Brodo et Hawksw. = Bartflechte p. 464, *466*

Bryum argenteum Hedw. = Birnmoos p. 605
- *erythrocarpum* Schw. Tab. 10
- *kunzei* (Hornsch.) Schpr. p. 460
- *neodamense* C. Müll. (1.612)
- *schleicheri* Schw. Tab. 80; p. *430*, 430 (1.5)
- *ventricosum* Dicks. Tab. 80; p. *430*
- *weigelii* Spreng. (1.511.1)

Buellia canescens (Dickson) De Not. = Scheibenflechte Tab. 86

Buglossoides arvensis (L.) I. M. Johnst. = Acker-Steinsame p. 628 (3.4) – 5×5-×75-T
- *purpurocaerulea* (L.) I. M. Johnst. = Purpurblauer S. Tab. 10, 32 (8.4) – 574-484-C, H

Buphthalmum salicifolium L. = Gemeines Ochsenauge Tab. 42, 89 – 854-×93-H

Bupleurum falcatum L. = Sichelblättriges Hasenohr Tab. 32; S. 721 (6.112) – 666-393-H
- *longifolium* L. = Langblättriges H. p. 632 (8.43) – 556-495-H
- *ranunculoides* L. = Hahnenfuß-H. Tab. 75 (4.711) – 915-593-H
- *rotundifolium* L. = Acker-H. p. 630 (3.41) – 874-394-T
- *stellatum* L. = Sterndolden-H. Tab. 75 (4.6) – 835-433-H

Butomus umbellatus L. = Schwanenblume Tab. 49, 51; p. *250* (1.411) – 6 × 5-10 × 8-A
- – var. *vallisneriifolius* p. 310 (1.213) – 6 × 5-1177-A

Buxus sempervirens L. = Buchsbaum Tab. 32; S. 95, 248 (8.4) – 582-484-N

Cakile maritima Scop. = Meersenf p. 371, *372* (2.811) – 9×× -6×8 II-T

Calamagrostis arundinacea (L.) Roth. = Rohr-Reitgras Tab. 17, 32, 33; p. 112, 115, 156, 435, 452 (8.431.1) – 654-545-H
- *canescens* (Web.) Roth = Lanzettliches R. Tab. 10, 46; p. 64, 277 (8.2) – 645-955-H
- *epigejos* (L.) Roth = Sand-R. Tab. 20; S p. 374, *376*, 529, *540*, 542, 568, *586* – 757-× ×6-G, H
- *pseudophragmites* (Haller fil.) Koeler = Ufer-R. – H(4.44) – 778-89?-H, G
- *stricta* (Timm) Koeler = Moor-R (1.613) – 94×-9×2-H
- *varia* (Schrad.) Host. = Berg-R. Tab. 10, 13, 35, 42; p. *72*, 237, *271* 583-H
- *villosa* (Chaix) Gmel. = Wolliges R. Tab. 35, 37, 41, 42; p. 212, *212*, 227, *419* (7.212) – 644-722-H, G

Calamintha nepetoides Jord. = Bergminze Tab. 82 (4.45) – 873-393-H

Calendula arvensis L. = Acker-Ringelblume (3.331) – 783-385-T

Calliergon giganteum (Schimp.) Kindb. = Schönmoos (1.613)
- *sarmentosum* (Wahlenb.) Kindb. Tab. 79
- *stramineum* (Dicks.) Kindb. Tab. 57, 79 (1.6)
- *trifarium* (Web. et Mohr) Kindb. (1.61)

Callitriche cophocarpa Sendtn. = Stumpf-kantiger Wasserstern (1.213)
- *hamulata* Kütz. ex Koch = Haken-W. Tab. 49; p. *290* (1.213)
- *obtusangula* (pal.-agg) Le Gall = Nußfrüchtiger W. Tab. 53 (1.213)
- *palustris* L. = Sumpf-W. p. 428, *608* (1.212) –7× ×-11×6-A, T
- *platycarpa* Kütz. = Flachfrüchtiger W. Tab. 49
- *stagnalis* Scop. = Teich-W. (1.213)

Calluna vulgaris (L.) Hull. = Besenheide Tab. 10, 17, 33, 57, 58, 71, 84, 98, 100, 112, 118; p. 12, 117, 124, 128, *129*, 182, *240*, 279, *330*, *334*, *335*, 337, *338*, 340, 377, 489, 504, 507ff, 514, 519ff, *520*, *522* (5.1) – 8×3-×11-Z

Caloplaca cirrochroa (Ach.) = Schönflechte Tab. 86
- *decipiens* (Arn.) Blomb, et Forss. Tab. 86
- *ochracea* (Schaerer) Flagey Tab 86

Caltha palustris L. = Sumpf-Dotterblume Tab. 10, 45, 46, 80, 110, 112, 117; p. *169*, *307*, *308*, 430, 563, *581*, *586*, 592 (5.415) – 7× ×-8× ×-H

Calypogeia fissa (L.) Raddi = Bartkelch-moos Tab. 14
- *muelleriana* (Schiffn.) K. Müll. Tab. 58; p. *120*
- *neesiana* (Massal. et Carestia) K. Müll. Tab. 58; p. *120*
- *sphagnicola* (Arn. et Perss.) Warnst. et Loeske Tab. 58 (1.71)
- *suecica* Arn. et Perss. p. *120*

Calystegia sepium (L.) R. Br. = Zaunwinde Tab. 46; p. *250*, *306*, *581*, *586* (3.52) – 865-679-G, Hli
- *soldanella* L. = Strandwinde (2.911)

Camelina microcarpa Andrz. ex DC. = Kleinfrüchtiger Leindotter (3.42) – 767-383-H, T
- *pilosa* (D.C.) Zing. = Behaarter L. (3.42)

Campanula alpina Jacq. = Alpen-Glocken-blumes p. 460 (5.111) – 724-542-H
- *barbata* L. = Bärtige G. p. 417 (5.111) – 724-512-H
- *bononiensis* L. = Bologneser G. (6.112) – 766-382-H
- *cespitosa* Scop. = Rasen-G. Tab. 35, 42, 82 (4.4) – 925-583-H
- *cenisia* L. = Mont-Cenis-G. Tab. 72 (4.42) – 817-583-C
- *cochleariifolia* Lam. = Kleine G. Tab. 35, 42, 82; p. 273 (4.21) – 7×4-7×?-H
- *glomerata* L. = Knäuel-G. (5.3) – 7×7-47×-H
- *latifolia* L. = Breitblättrige G. Tab. 2; p. *58* (8.434) – 455-688-H
- *patula* L. = Wiesen-G. Tab. 2; 110, 112; p. 559, 592 (5.421) – 854-574-H

Datura stramonium L. = Weißer Stechapfel (3.3)–87 × - 4 × 8 I-T
Daucas carota L. = Möhre Tab. 89, 110, 124, 130; p. 480, *493, 567, 631* (3.322) – 865-4 × ×4-H
Dentaria bulbifera L. = Zwiebel-Zahnwurz Tab. 2, 14, 26; p. 150 (8.43) – 354-576-G
– *enneaphyllos* L. = Weiße Z. (8.431) – 444-577-G
– *heptaphylla* Vill. = Fieder-Z. Tab. 11 (8.431)–352-586-G
– *pentaphyllos* L. = *digitata* Lam. = Finger-Z. Tab. 11; p. *78* (8.431) – 352-576-G
Dermatocarpon lachneum (Ach.) A.L. Smith = Nabelflechte S. 618
– *miniatum* (L.) Mann Tab. 86
Deschampsia cespitosa (L.) P.B. = Rasen-Schmiele Tab. 10, 13, 14, 20, 26, 37, 45, 46, 80, 83, 110, 112, 117, 120, 130; p. *58*, 152, 306, 430, 541, 563, 568f, *576*, 574, *586*, 588, 592 – 6 × ×-7 × 3-H
– *litoralis* (Gaudin) Reuter = Ufer-S. p. *288* (1.311)–864-1072-H, A
– *media* (Gouan) R. et Sch. = Binsen-S. (3.72)–872-872-H
– *setacea* (Huds.) Hackel = Borsten-S. (1.3)–851-92l-H
– *wibeliana* (Sonder.) Parl. = Wibels S. p. 306 – 862-884 II-H
Descurainia sophia (L.) Webb. et Prantl = Besenrauke Tab. 124; p. 616 – 867-4 × 6-T
Dianthus armeria L. = Rauhe Nelke Tab. 2–663-533-T, H
– *carthusianorum* L. = Karthäuser-N. Tab. 89; p. *522* (5.32)–854-372-C
– *deltoides* L. = Heide-N. Tab. 112 (5.112)–8 × 4-432-C, H
– *glacialis* Haenke = Gletscher-N. p. 414
– *gratianopolitanus* Vill. = Pfingst-Nelke (5.213)
– *lumnitzeri* (Wiesb. = Feder-N. (5.213)
– *superbus* L. = Pracht-N. Tab. 112 (5.411)–7 × 7-882-H
– *sylvestris* Wulf. = Stein-N. p. 503 (s.auch 4.313)
Dicranella heteromalla (Hedw.) Schpr. = Kleingabelzahnmoos Tab. 10, 14, 17, 20; p. 112, *120*
Dicranum undulatum Brid. = Gabelzahnmoos Tab. 10, 33, 58
– *bonjeani* De Not. Tab. 58
– *mühlenbeckii* Br. Tab. 41
– *scoparium* (L.) Hedw. Tab 10, 14, 17, 20, 33, 37, 41, 42, 58, 71, 84, 98; p. 93, 94, 117, 378, 452, 515, 528
– *spurium* Hedw, Tab. 10; p. 530
– *undulatum* Brid. Tab. 10, 33, 58;
Dictamnus albus L. = Diptam Tab. 10; p. 64, *72, 547* (6.112) – 784-282-H
Digitalis grandiflora Mill. = Großblütiger Fingerhut (6.211) – 754-555-H
– *lutea* L. = gelber F. (6.211) – 762-575-H
– *purpurea* L. = Roter F. Tab. 20; p. 132, 544 (6.211)–752-536-H
Digitaria ischaemum (Schreber) Mühlenb. = Fadenhirse (3.332) – 764-523-T
– *sanguinalis* (L.) Scop. = Bluthirse (3.332)–773-354-T
Dimelaena oreina (Ach.) Norm. (Halbkru-stenflechte) Tab. 86
Diphasium alpinum (L.) Rothm. = Alphen Flachbärlapp Tab. 41; p. 533 (5-111) – 823-522-C
– *complanatum* (L.) Rothm. = Gemeiner F. p. 527 (7.211)–647-412-C
– *tristachyum* (Pursh.) Rothm. Dreiähriger F. (7.2)–875-51-C
Diphyscium foliosum (Hedw.) Mohr = Blasenmoos Tab. 14
Diplotaxis muralis (L.) D.C. = Mauer-Doppelsame (3.3)–883-485-T
– *tenuifolia* (Jusl.) D.C. = Schmalblättriger D. Tab. 124 (3.31) – 873-3 × 4-C, H
Dipsacus fullonum L. = Weber-Karde Tab. 2, 124 (3.5)–963-685-H
– *pilosus* L. = Behaarte K. Tab. 2 (3.523) – 755-776-H
Ditrichum pallidum (Hedw.) Hampe = Doppelhaarmoos Tab. 10
Doronicum austriacum Jacq. = Österreichische Gemwurz Tab. 81 (6.311) – 534-677-H
– *clussi* (All.) Tausch = Zottige G. p. 460 (4.43)–814-68?-H
– *glaciale* (Wulf.) Nyman = Eis-G. (4.42)
– *grandiflorum* Lam. = Großblütige G. Tab. 82 (4.41)–824-693-H
Dorycnium germanicum (Gremli) Rikli = Seidiger Backenklee p. *271* (7.1) – 764-291-Z
Draba aizoides L. = Immergrünes Felsenblümchen Tab. 85; p. 452 (4.21) – 8 × 4-39?-C
– *dubia* Suter = Kälteliebendes F. (4.2) – 815-3 × 2-C
– *fladnizensis* Wulf. = Fladnizer F. Tab 82, 87; p. 415, 460 (4.42) – 917-55?-C, H
– *hoppeana* Rchb. = Hoppe's F. Tab. 82 (4.42) – 81?-472-C
– *ladina* Br.-Bl. Ladinisches F. (4.21) – 827-493-C
– *muralis* L. = Mauer-Hungerblümchen (4.21?) – 762-482-T, H
– *nemorosa* L. = Hain-H. (5.2)
– *siliquosa* MB. = Kärntner F. Tab 75 (4.72) – 824-562-C
– *tomentosa* Clairv. = Filziges Felsenblümchen (4.211)–914-292-C
Drepanocladus aduncus (Hedw.) Warnst = Sichelmoos p. *290*
– *exannulatus* (Gümb.) Warnst. Tab 79; p. 428
– *revolvens* (C. Muell.) Warnst. (1.61); var. *intermedius* = *D. intermedius* (Lindw.) Warnst. Tab. 79
Drosera anglica Hudson = Langblättriger Sonnentau Tab. 58 (1.61) – 743-932-H
– *intermedia* Hayne = Mittlerer S. Tab. 58; p. 335 (1.611) – 952-922-H
– *rotundifolia* L. = Rundblättriger S. Tab 57, 58; p. *335*, 335, 337 (1.7) – 843-911-H
Dryas octopetala L. = Silberwurz Tab. 42, 75; p. *271*, 398, *402*, 413, 419, 422, 450 – 9 × 7-484-Z
Dryopteris carthusiana (Vill.) H.P. Fuchs = Dornfarn Tab. 10, 13, 14, 17, 20, 26, 33, 35, 37, 41, 46; p. 112, 150, *201*, 279, 528, 531 – 5 × 3- × 43-H
– *cristata* (L.) A. Gray = Kammfarn p. 278 (8.2) – 445-95 × -H
– *dilatata* (Hoffm.) A. Gray = Breitblättriger D. Tab. 10, 35, 37; p. *58*, 106, *107* 4 × 3-6 × 7-H
– *filix-mas* (L.) Schott = Gemeiner Wurmfarn Tab. 6, 10, 11, 13, 14, 26, 35, 37; p. *58*, 104, 106, *201; 214* (8.43)–3 × 3-556-H
– *pseudo-mas* (Woll.) Holub et Pouzar = Spreuschuppiger W. (8.43) – 352-656-H
– *villarii* (Bell.) Woyn. = Starrer W. Tab. 82 (4.412)–922-59?-H

Echinochloa crus-galli (L.) PB. = Hühnerhirse Tab. 120;

76 × -4 × 7-T
Sherardia arvensis L. = Ackerröte p. 628, 630 (3.4)–663-585-T
Sibbaldia procumbens L. = Alpen-Gelbling Tab. 78; p. 425, 460 (4.51) – 713-72?-H
Silaum silaus (L.) Sch. et Th. = Wiesensilge Tab. 112; p. 568, 579 (5.41) – 765-772-H
Silene acaulis (L.) Jacq. = Stengelloses Leimkraut Tab. 83 p. 413, 458, 460 (4.711) – 913-481-C
– *alba* (Mill.) E.H.L. Krause = Weiße Lichtnelke Tab. 124; p. 616 (3.511) – 8 × × -4 × 7-H
– *armeria* L. = Nelken-L. (6.1) – 774-452-T
– *chlorantha* (Willd.) Ehrh. = Grünliches L. (5.225)
– *conica* L. = Kegel-L. (5.224) – 974-252-T
– *dioica* (L.) Clairv. = Rote L. Tab. 2, 10, 45, 110; p. *58*, 141, 570 – × × 4-678-H
– *excapa* (All.) Vierh. = Polsternelke Tab. 75 (4.6)
– *linicola* C. Gmel. = Flachs-L. (3.43)
– *noctiflora* L. = Acker-Leimkraut (3.41) – 754-385-T
– *nutans* L. = Nickendes L. Tab. 10, 32, 89; p. *631* (6.1) -755-373-H
– *otites* (L.) Wib. = Ohrlöffel-L. Tab. 68, 89 (5.3) – 877-272-H
– *rupestris* L. = Felsen-L. (5.211) – 922-231-H
– *saxifraga* L. = Stein-L. Tab. 85 (4.21) – 866-483-H
– *vulgaris* (Moench) Garcke = Taubenkropf Tab. 97, 110; p. *501*, 501, 503, *631* (see 4.3) – 8 × × -472-H, C, ssp. *humilis* (Schub.) Rothm. B.
Sinapis alba L. = Weißer Senf. p. *624*
– *arvensis* L. = Acker-S. Tab. 125, 127, 128; p. 628, *630* (3.4) – 753- × 86-T
Sisymbrium altissimum L. = Ungarische Rauke Tab. 124 p. 616 (3.31) – 867-3 × × -T, H
– *austriacum* Jacq. = Österreichische R. Tab. 124 (3.31) -754-487-H, T
– *irio* L. = Schlaffe R. Tab. 124 p. 616 (3.31) – 883-375-T
– *loeselii* L. = Loesel's R. Tab. 124 p. 616 (3.31) – 777-375-H, T
– *officinale* (L.) Scop. = Wege-R. p. 616, *631* (3.31)–865-4 × 7-T, H
– *strictissimum* L. = Steife R. (3.521) – 774-687-H
Sium latifolium L. = Hoher Merk Tab. 49, 51 (1.411)–7 × 4-1078-A, H
Solanum dulcamara L. = Bittersüß Tab. 10, 45, 46; p. 262, 277 – 75 × -8 × 8-Nli
– *luteum* Mill. = Gelbfrüchtiger Nachtschatten (3.3)
– *nigrum* L. = Schwarzer N. p. 624, 626, 628, 630, *631f.*, (3.3) – 763-578-T
Soldanella alpina L. = Alpen-Troddelblume Tab. 75; p. 425, *427* – 724-78 × -H
– *montana* Willd. = Berg-T. (7.212) – 544-622-H
– *pusilla* Baumg. = Kleine T. Tab. 72, 78; p. 425 (4.5)–714-722-H
Solidago canadensis L. = Kanadische Goldrute Tab. 124, 130; p. 274, 616, 641, 644 (3.51) – 875- × × 6-H, G
– *gigantea* Ait. = Riesen-G. Tab. 124, 130; p. 274 (3.5)–875-6 × 6-H, G
– *graminifolia* (L.) Ell. = Grasblättrige G. p. 274 (3.523) – 8 × × -777-H, G
– *virgaurea* L. = Gemeine G. Tab. 10, 11, 12, 13, 14, 26, 32, 33, 35, 37, 41, 98; p. 112, 186 – 5 × × -5 × 5-H
Solenostoma crenulatum (Smith) Mittem. p. *120*
Sonchus arvensis L. = Acker-Gänsedistel Tab. 68, 121, 127; p. *373*, 628, 630, *631*, 634 – 75 × -57 × I-G, H
– – ssp. *uliginosus* (M. Bieb.) Nym. (2.811)
– *asper* (L.) Hill. = Rauhe G. Tab. 125, 126, 130; p. 628, *631*, 634, 635 (3.33) – 75 × -677-T
– *oleraceus* L. = Kohl-G. Tab. 121, 125, 127, 128, 130; p. 628 (3.3) – 75 × -488-T, H
– *palustris* L. = Sumpf-G, (5.412) – 766-877 I-H
Sorbus aria (L.) Cr. = Mehlbeere Tab. 9, 11, 12, 26, 32, 35, 42; p. *60*, 87 – 652-473-P, N
– *aucuparia* L. = Vogelbeere Tab. 9, 13, 14, 17, 20, 26, 32, 33, 35, 37, 84, 98, 109, 130; p. *60*, 61, 262, 278, 512–6 × × - × × × -P, N
– – ssp. *glabrata* (Wimm. et Grab.) Hayek (6.213)
– *chamae-mespilus* (L.) Cr. = Zwergmispel Tab. 42 – 734-483-N
– *domestica* L. = Speierling Tab. 9, 32; p. 150 (8.42) – 474-383-P
– *torminalis* (L.) Cr. = Elsbeere Tab. 26, 9, 12, 32; p. *60*, 150 (8.42) – 464-474-P, N
Sparganium angustifolium Michx. = Schmalblättriger Igelkolben Tab. 50; p. 428 (1.313) – 843-1131-A
– *emersum* Rehm. = Einfacher I. Tab. 51, 53 (1.411) – 753-11 × 5-A
– *erectum* L. = Astiger I. Tab. 51, 53; p. 300, *303*, *308*, 310 (1.411) – 765-10 × 5-A
– – ssp. *fluitans* (Gren. et Godr.) Arcang. (1.213)
– *minimum* Wallr. = Zwerg-I. (1.114) – 755-1153-A
Spartina townsendii Grov. = Schlickgras Tab. 62; p. 353, *358* (2.3) – 852-977 III-H, G
Spergula arvensis L. = Feld-Spark Tab. 125, 128; p. 628, *632* (3.33) – 6 × 3-526-T
– *morisonii Bareau* = Frühlings-S. Tab. 71, 98; p. 384 (5.221) – 954-2 × 2-T
– *pentandra* L. = Fünfmänniger S. (5.214)
Spergularia marina (L.) Griseb. = S. *salina* Presl.
– *Meerstrands-Schuppenmiere Tab. 63, 65 (2.611) – 7 × 8-6 × ? III-H*
– *media* (L.) C. Presl. = S. *marginata* (D.C.) Kittel = Flügelsamige S. Tab. 62, 64 p. 359 (2.6) – 7 × 8-7 × 5 II-H, C
– *rubra* (L.) J. et C. Presl = Rote S. (3.72) – 7 × × -634-T, H
– *segetalis* G. Don = *Delia s.* (L.) Dum. = Getreidemiere (3.111.1)
Sphagnum acutifolium Ehrh. = Bleichmoos, Torrfmoos Tab. 10, 35, 58, 59
– *angustifolium* Jens (1.71)
– *apiculatum* Lindb.f. Tab. 57, 58, 59; S. 450
– *balticum* (Russ.) C. Jens. Tab. 58, 59
– *compactum* D.C. Tab. 58; p. 326, 340, 510 (1.72)
– *contortum* Schultz (1.61)
– *cuspidatum* Hoffm. Tab. 10, 58, 59; p. 326, *331*, 334, *334*, 335f., *335f.*, 337
– *cymbifolium* Ehrh. Tab. 10
– *dusenii* (C. Jens.) Russ. et Warnst. p. 334
– *fuscum* Klinggr. Tab. 10, 58, 59; p. 326f., *330f.*, *333*, *334*, 337ff. (1.71?)
– *girgensohnii* Fuss. p. 326
– *imbricatum* Hornsch. Tab. 58, 59; p. *339*
– *magellanicum* Brid. Tab. 57, 58, 59; p. 326f., *333*, *334*, 335, *335*, 337f., *339*, 342, *343*, *344* (1.71)
– *molle* Sull. Tab. 58; p. 510 (1.72)
– *obtusum* Warnst. (1.613)
– *palustre* L. = *S. cymbifolium* Ehrh. p. 326

IV Subject index

The English plant names refer to communities. Species are indexed separately section E, III 2b, p. 676 with figures, other page numbers to the text and tables. Pages marked • indicate that the subject is dealt with in a more thoroughgoing manner.